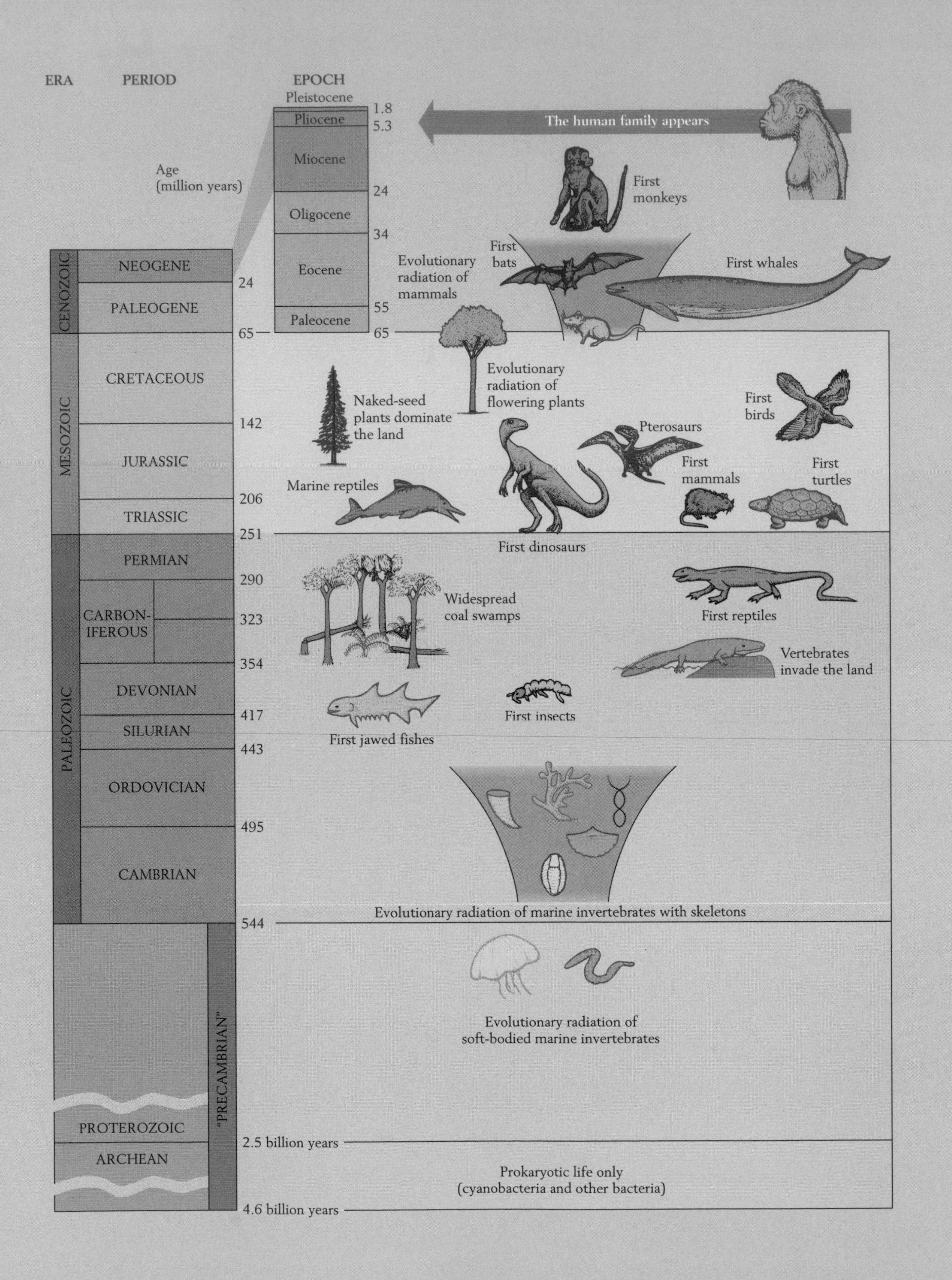

ERA
PERIOD
EPOCH
Pleistocene
Pliocene
Miocene
Oligocene
Eocene
Paleocene
1.8
5.3
24
34
55
65
The human family appears
Age
(million years)
First
monkeys
First
bats
Evolutionary
radiation of
mammals
First whales
CENOZOIC
NEOGENE
PALEOGENE
24
65
MESOZOIC
CRETACEOUS
JURASSIC
TRIASSIC
142
206
251
Evolutionary
radiation of
flowering plants
Naked-seed
plants dominate
the land
First
birds
Pterosaurs
First
mammals
First
turtles
Marine reptiles
First dinosaurs
PALEOZOIC
PERMIAN
CARBON-
IFEROUS
DEVONIAN
SILURIAN
ORDOVICIAN
CAMBRIAN
290
323
354
417
443
495
544
Widespread
coal swamps
First reptiles
Vertebrates
invade the land
First insects
First jawed fishes
Evolutionary radiation of marine invertebrates with skeletons
"PRECAMBRIAN"
PROTEROZOIC
ARCHEAN
Evolutionary radiation of
soft-bodied marine invertebrates
2.5 billion years
Prokaryotic life only
(cyanobacteria and other bacteria)
4.6 billion years

MIDDLE SILURIAN

Euramerica
Gondwanaland

Laurentia and Baltica are sutured to form Euramerica

MIDDLE MIOCENE

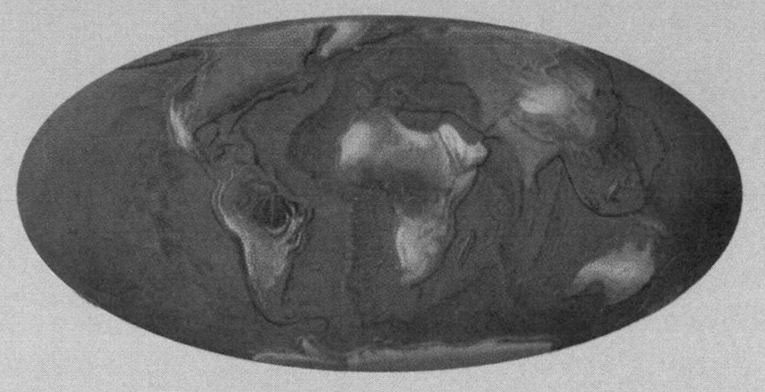

Continents are dispersing

MIDDLE ORDOVICIAN

LATE CRETACEOUS

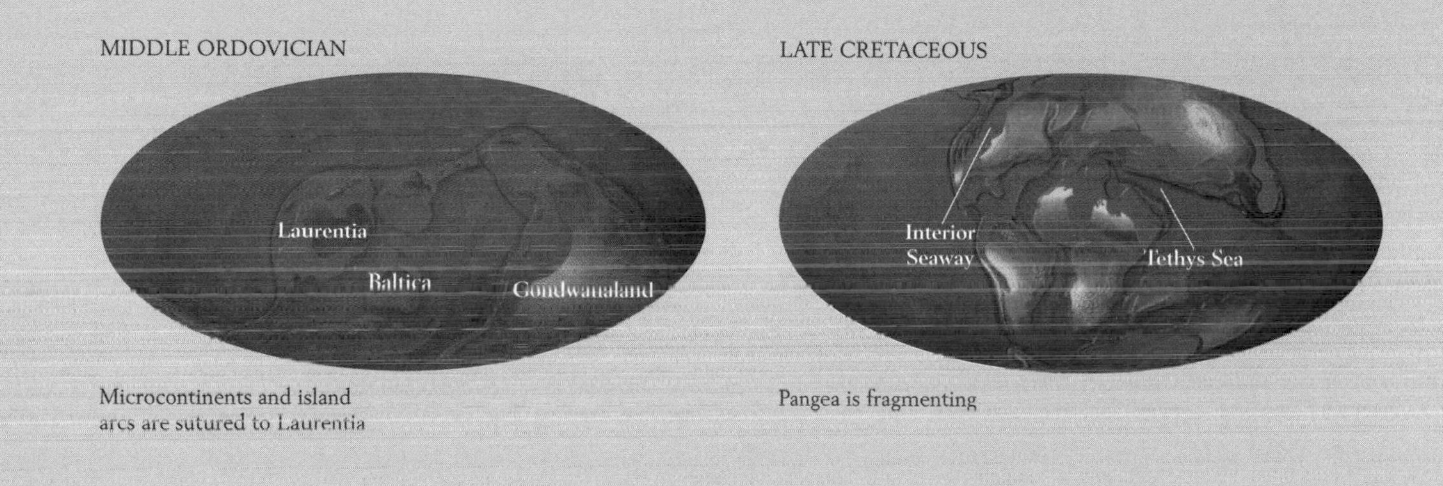

Microcontinents and island arcs are sutured to Laurentia

Pangea is fragmenting

LATE CAMBRIAN

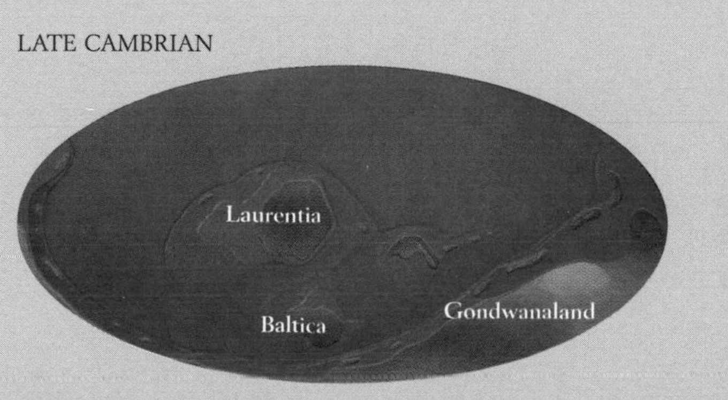

The Proterozoic supercontinent has fragmented

LATE PERMIAN

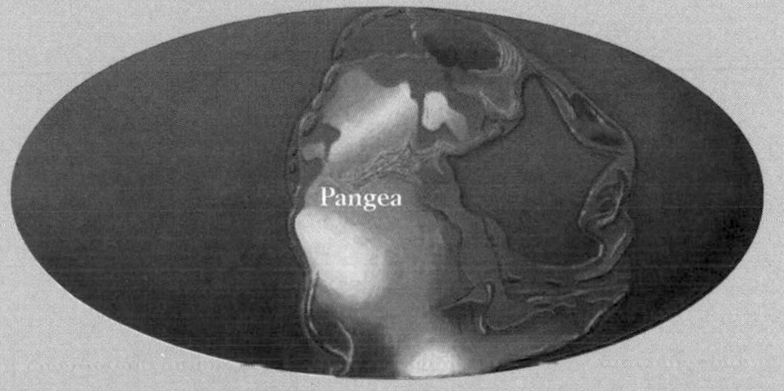

Gondwanaland is sutured to Euramerica to form Pangea

Earth System History

Earth System History

STEVEN M. STANLEY
Johns Hopkins University

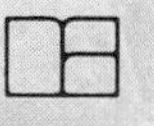 W. H. Freeman and Company • *New York*

To Nell and Sveta

About the Cover: Early Earth was a glowing, liquid planet. After it cooled, the ocean, atmosphere, and continental crust formed, and as continents coalesced and fragmented, life appeared. Throughout its subsequent evolution, life has interacted with Earth's dynamic crust and enveloping fluids. The cover for this first edition of *Earth System History* depicts these geologic and biological events, showing four stages of Earth's development as a planet.

Sponsoring Editors: Holly Hodder, Melissa Levine Wallerstein
Development Editors: Kay Ueno, Diana Gill
Project Editor: Georgia Lee Hadler
Cover Illustrator: Tomo Nagashima
Text Designer: Blake Logan
Illustration Coordinator: Susan Wein
Illustrations: Roberto Osti; Fine Line Studios
Photo Researcher: Kathy Bendo
Production Coordinators: Maura Studley, Susan Wein
Composition: Sheridan Sellers and Bonnie Stuart, W. H. Freeman Electronic Publishing Center; Progressive Information Technologies
Manufacturing: RR Donnelley & Sons Company
Supplements Coordinators: Patrick Shriner, Robert Christie
Marketing Manager: John Britch

Library of Congress Cataloging-in-Publication Data

Stanley, Steven M.
Earth system history / Steven M. Stanley.
p. cm.
Includes bibliographical references and index.
ISBN 0-7167-2882-6
1. Historical geology. 2. Physical geology. I. Title.
QE28.3.S735 1998
551.7—dc21
18–21861
CIP

Printed in the United States of America
First Printing 1998

Contents in Brief

Contents

Preface

The curriculum in the Earth sciences has been evolving rapidly in the last few years, after many decades of relative stability. Appropriately, the new approach to teaching reflects a shift in research toward integrative study of the entire Earth system and its many subsystems. This Earth system approach demands a rethinking of the field traditionally called historical geology: the only way to reconstruct and interpret Earth's history effectively is by treating the lithosphere, biosphere, hydrosphere, and atmosphere as parts of a single system. *Earth System History* reforges its successful predecessor, *Exploring Earth and Life Through Time*, to reflect this powerful new way of studying Earth's history. The traditional topics of historical geology still remain at the core of *Earth System History* but are enriched by placement in the broader framework of the planet as a whole.

Now more than ever, Earth's history deserves a central place in a balanced undergraduate curriculum. The rock record of Earth's deep history is what sets geology apart from all other sciences, and in an era of heightened concern about the environment it is imprudent to ignore this unique archive and to focus strictly on present-day geologic processes. The full geologic record of life and environments is more broadly relevant: it documents the myriad unique natural experiments that show how the Earth system functions. We will never duplicate these experiments in the laboratory or in the field. We must study them in rocks. Understanding changes of the past will help us—and our students—confront future environmental changes.

The history of life and the history of the physico-chemical environment are inextricably intertwined. This is not a unidirectional relationship: environmental changes affect life, but biologic changes in turn alter the environment. For example, my colleague Lawrie Hardie and I have found that simple changes in ocean chemistry, governed by global spreading rates along mid-ocean ridges, have determined what kinds of organisms have been the dominant reef formers and sediment producers during particular intervals of the Phanerozoic Eon. Thus, "For The Record 10-1: Seawater Chemistry and Chalk" presents strong evidence that the massive chalk deposits that gave the Cretaceous System its name accumulated because of the presence of "calcite seas," which resulted from unusually high rates of seafloor spreading. As another example, Chapters 10 and 14 describe how the advent of forests in Devonian time must have reduced carbon dioxide levels in the

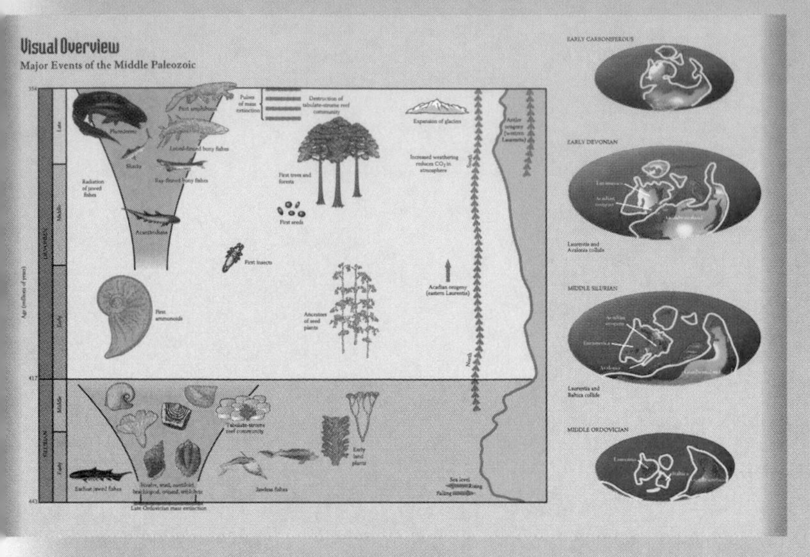

atmosphere and set the stage for the late Paleozoic ice age.

A survey of instructors revealed a strong consensus that courses covering Earth's history should explain how studies of stable isotopes shed light on events of the geologic past. Accordingly, Chapter 10 introduces the use of stable isotopes to study the history of key chemical cycles and other aspects of the Earth system including the greenhouse effect and the effects of plate tectonic activity on seawater. This new chapter is central to the Earth system approach, introducing concepts and techniques that later emerge in discussions of particular events in Earth's history. The Earth system approach culminates in the new final chapter on the Holocene, which ushers students into the modern ice age and raises issues about future global change.

How This Book Is Organized

This edition has been substantially revised to provide teachers and students with a comprehensive guide to Earth system history. It has been updated in accordance with the latest scientific knowledge.

Chapter 1 introduces the Earth system approach, emphasizing the rock cycle and the hydrological cycle as well as plate tectonic cycling of materials. It also contrasts the gradual and catastrophic processes that shape Earth.

New to this book, Chapter 2 introduces important groups of rocks and rock-forming minerals and explains how they relate to one another within the rock cycle. This chapter includes information on how chemical bonds and structures affect the properties of minerals. It can serve as a review for students who have completed a course in physical geology or as a primer for those with no background in physical geology.

Also new to this edition, Chapter 3 reviews the biology and classification of the six kingdoms of organisms and introduces methods of phylogenetic reconstruction. This chapter assumes no background in biology on the student's part, but provides a context for subsequent discussions of the history of life.

Chapter 4 includes an expanded discussion of the influence of topography on climate and a new treatment of thermal regimes of shallow seas.

Chapter 5 includes a new discussion of the use of fossils to delineate sedimentary environments.

Chapter 6 now introduces the field of stratigraphy by showing how it arose as a branch of geology. This chapter also integrates the definition of formal stratigraphic units with discussions of stratigraphic concepts.

Chapter 7 invokes the origin of whales to exemplify the origins of higher taxa. It also contains a new introduction to major mass extinctions.

Chapter 8 integrates an introduction to the basic kinds of faults with a discussion of plate tectonics, and Chapter 9 integrates folding with mountain building in a similar way.

Chapter 10 is an entirely new chapter that uses an approach never before seen in an introductory textbook. This chapter examines the oxygen and carbon cycles and describes the factors that govern levels of atmospheric oxygen and carbon dioxide. To comprehend Earth system history it is necessary to understand the impact of major chemical cycles on Earth, and Chapter 10 uses simple diagrams and real-world examples to clearly explain the effects of chemical cycles in the past and their possible consequences for the future.

Chapter 11 presents the widely favored explanation of the origin of the moon as the result of an asteroid impact and the likely origin of life in the heated zone along a mid-ocean ridge.

Chapter 12 presents new evidence about major events near the end of the Proterozoic Eon, including the assembly of a supercontinent, profound glaciation, the possible buildup of atmospheric oxygen, and the explosive diversification of multicellular organisms.

In response to instructors' comments, I have rewritten Chapters 13, 14, and 15 to place mountain-building episodes in the Appalachian region in the context of Earth's Paleozoic history. These chapters also provide up-to-date reviews of the great Paleozoic mass extinctions and show how the origin of forests and coal swamps probably led to late Paleozoic glaciation by reducing greenhouse warming.

Chapters 16 and 17 are highlighted by new discoveries that change our view of dinosaurs and Mesozoic land plants. Chapter 16 also focuses on new evidence that the arrival of a meteorite brought the Mesozoic Era to an end.

Chapters 18 and 19 present new evidence of climatic, biotic, and oceanographic changes during the Cenozoic, new chronologies of mountain building in the American West, and recent discoveries that place human origins in a new light.

Chapter 20, an entirely new review of the Holocene, describes Earth's complicated emergence from the most recent glacial maximum. This chapter emphasizes changes in climate and sea level in recent history, and their consequences for the environment. The present world emerges in this chapter: having gained a broad knowledge of Earth's past, the student becomes oriented toward its future.

A New Vision, a New Book

Several features highlight the new approach of *Earth System History*.

Visual Overviews

Visual Overviews provide students with vivid illustrations of the key themes and concepts of Chapters 2 through 20. Expanded from the Major Events diagrams in the previous edition, *Visual Overviews* concisely portray major geologic events and cycles.

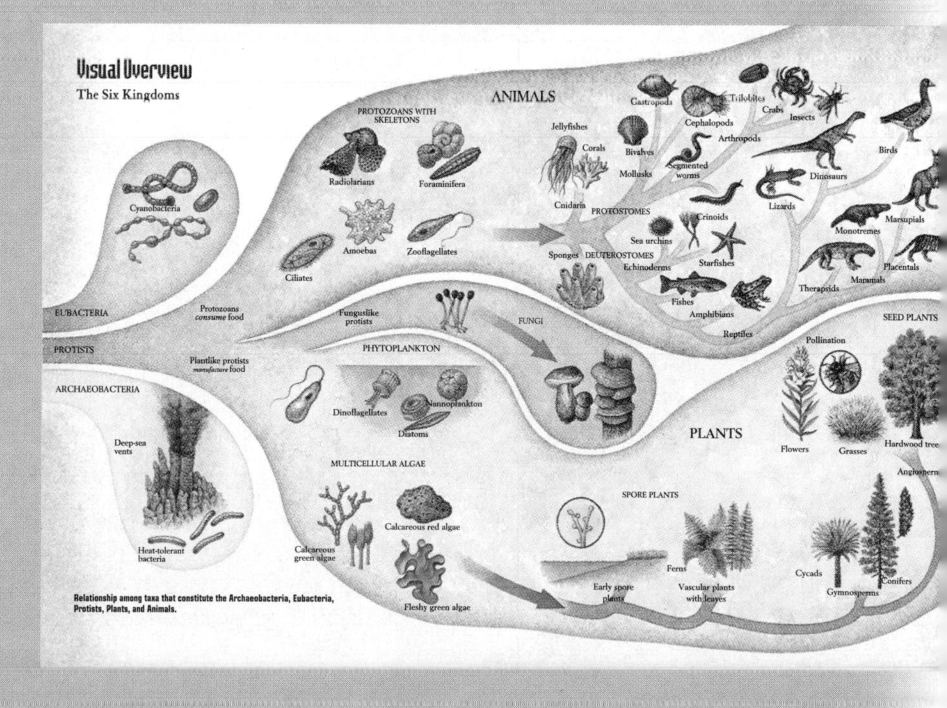

Relationship among taxa that constitute the Archaeobacteria, Eubacteria, Protists, Plants, and Animals.

Furthermore, a *Review Question* at the end of each chapter challenges the student to use the chapter's *Visual Overview* to synthesize key points. Each special *Review Question* is identified by a System Icon: ●.

Review Questions

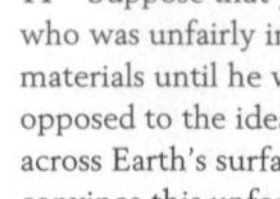

11 Suppose that you were to encounter a well-trained geologist who was unfairly imprisoned in 1955 and was deprived of reading materials until he was released last week. He entered prison firmly opposed to the idea that continents have moved large distances across Earth's surface. Given an hour of time, how would you convince this unfortunate geologist that continents have actually moved thousands of kilometers? Use the Visual Overview on p. 208 as a guide to develop your argument.

For the Record

Chapters 2 through 20 feature *For the Record*, an in-depth look at an environmental, evolutionary, or resource-oriented example in Earth system history. Ranging from the extinction of the dinosaurs to drilling for oil off the shore of New Jersey, *For the Record* boxes offer real-world examples of Earth system history in action.

Seawater Chemistry and Chalk

Chalk is a soft, white, fine-grained limestone that we use to write on blackboards, but it is also the formal name for a large Upper Cretaceous body in western Europe. Along the coast of France the Chalk forms cliffs that Allied troops scaled during the invasion of Normandy in 1944, and it rises up as the famous White Cliffs of Dover on the British side of the English Channel. The Cretaceous contains a vastly larger volume of chalk than any other geologic system. As it turns out, this un-usual abundance probably reflects the chemical composition of Cretaceous seas.

The most massive deposits of Cretaceous chalk are those of western Europe, which are typically about 200 meters (~660 feet) thick, but substantial bodies of chalk also accumulated in an epicontinental sea in North America. Between Alabama and Kansas, Upper Cretaceous chalks have yielded spectacularly well preserved fossils, including skeletons of huge marine reptiles that might be described as sea monsters. Fossils are exceptionally well preserved in chalk for the same reason that the chalk is soft: it is formed of minute grains of calcite. The small size of these grains makes

Illustration Program

A full-color illustration program enhances *Earth System History*'s new approach. Line illustrations have been clarified and recreated in full-color, and many black-and-white photographs have been replaced by striking new color photographs.

New three-dimensional block diagrams illustrate processes, cycles, and events in Earth system history.

For effective presentation of key subjects, line art and photographs have been paired frequently: line diagrams illustrate important patterns and accompanying photographs provide actual examples.

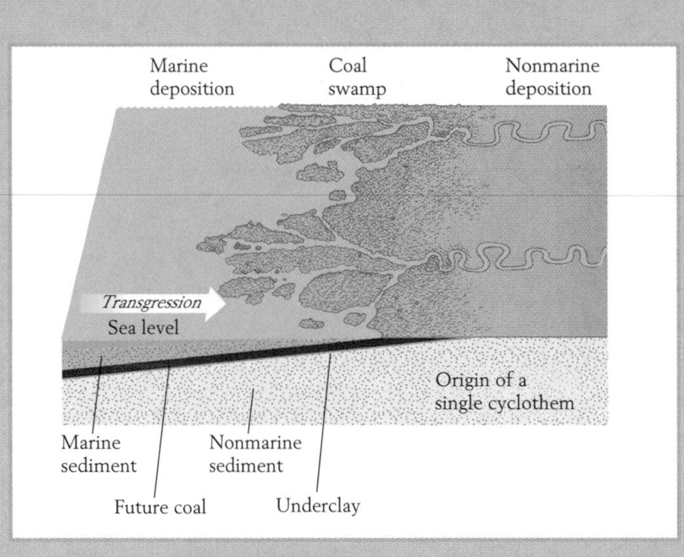

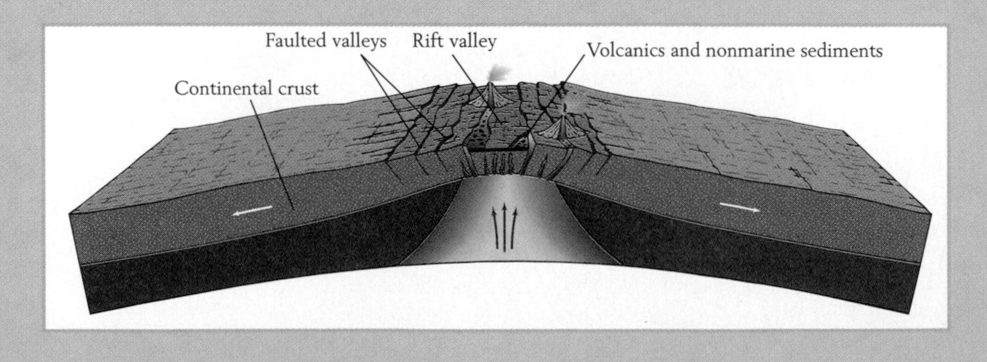

Chapter Summaries, Review Questions, and Additional Readings

End-of-chapter material aids students in identifying and retaining key themes of each chapter. *Chapter Summaries* reiterate integral concepts and facts, *Review Questions* help students understand and remember important ideas, and *Additional Readings* offer interested students resources for learning more about Earth system history.

Appendix and Glossary

The *Appendix* covers the major stratigraphic stages of Earth's history in one compact, comprehensive source. The *Glossary* includes concise definitions of important scientific and geologic terms that may be unfamiliar to the student. Scientific and geologic terms are printed in **boldface** when they first appear in the text and then are gathered with their definitions in the comprehensive *Glossary.*

Supplements

Earth System History provides a complete historical geology program for instructors and students.

Student CD-ROM

- Selected exercise modules and activities on key areas such as the evolution of the continents and sedimentary environments.
- Student study features include multiple-choice quizzes with built-in feedback and mnemonic exercises for learning the intervals of the geologic time scale.
- Original geological animations convey challenging but essential concepts of atmospheric chemistry, mountain building, stratigraphy, sedimentology, and paleoclimatology.

Instructor's Resource Manual with Test Questions

- Recommendations, written by Steven Stanley, explain how to emphasize the Earth system approach in covering the material in each chapter.
- Many of the test questions for each chapter, in a variety of formats, contain geologic diagrams, and all test questions provide page references.
- Original transparency masters are provided.

Slide Set with Lecture Notes

- A selection of 60 top-quality historical geology slides picked specifically to accompany the text.
- An accompanying booklet of lecture notes puts each slide in the context of the course material. The slide set was curated by renowned geology photographer Peter Kresan of the University of Arizona.

Overhead Transparency Set

- 50 full-color overheads, including key diagrams from the text as well as original illustrations.

Computerized Test Banks

- Amenable to editing and the addition of questions.
- Available for both Windows and Macintosh operating systems.

Freeman Geology Web Site

http://www.whfreeman.com/geology/

- Provides links to hundreds of Web sites related to geology.
- Includes expansion modules with additional information on historical and physical geology.
- Presents a complete list of geology newsgroups.

Acknowledgments

A large group of colleagues offered advice and reviewed the text as this edition was created. I thank them for their hard work and valuable suggestions, but accept full responsibility for any errors that remain. These colleagues are

Beth Nichols Boyd, *Yavapai College*
Fred Clark, *University of Alberta*
William C. Cornell, *University of Texas at El Paso*
John W. Creasy, *Bates College*
R. A. Davis, *College of Mount St. Joseph*
Steven R. Dent, *Northern Kentucky University*
Stanley C. Finney, *California State University, Long Beach*
John P. Grotzinger, *Massachussetts Institute of Technology*
Bryce M. Hand, *Syracuse University*
Leo J. Hickey, *Yale University*
Calvin James
Markes E. Johnson, *Williams College*
James O. Jones, *University of Texas at San Antonio*
Andrew H. Knoll, *Harvard University*
Karl J. Koenig, *Texas A&M University*
Michael T. May, *Western Kentucky University*
Cathryn R. Newton, *Syracuse University*
Geoffrey Norris, *University of Toronto*
Mark E. Patzkowsky, *Pennsylvania State University*
Gregory J. Retallack, *University of Oregon*
Charles A. Ross, *Western Washington University*
Frederic J. Schwab, *Washington and Lee University*
Peter Sheehan, *Milwaukee Public Museum*
John F. Taylor, *Indiana University of Pennsylvania*
David K. Watkins, *University of Nebraska, Lincoln*
Neil A. Wells, *Kent State University*

I am also grateful to colleagues who contributed their efforts to the previous edition: Barbara Brande, University of Montevallo; William Cornell, University of Texas, El Paso; Louis Dellwig, University of Kansas; Jay M. Gregg, University of Missouri, Rolla; Bryce Hand, Syracuse University; James Jones, University of Texas, San Antonio; Karl Koenig, Texas A&M University; David Liddell, Utah State University; Robert Merrill, California State University, Fresno; Anne Noland, University of Louisville; Lisa Pratt, University of Indiana, Bloomington; Randall Spencer; Old Dominion University.

I owe special thanks to many people at W. H. Freeman and Company who aided me in launching this book. Holly Hodder and her successor, Melissa Levine Wallerstein, provided invaluable guidance along the way, and Moira Lerner, Kay Ueno, and Diana Gill were a great help at various stages of the molding process. As always, Mary Shuford and Philip McCaffrey managed things beautifully, and Georgia Lee Hadler was her usual pillar of strength and reliability while assembling the book. Barbara Salazar copyedited deftly once again, offering improvements without offence. Chris Scotese kindly made available his splendid reconstructions of ancient worlds. Blake Logan created a splendid design, and I feel privileged to have had the brilliant watercolorist Roberto Osti render the magnificent Visual Overviews in the first part of the book. Kathy Bendo gathered many excellent photographs under great time pressure. Finally, Patrick Shriner and Robert Christie opened the door to wonderful new supplemental resources.

CHAPTER 1

Earth as a System

Few people recognize, as they travel down a highway or hike along a mountain trail, that the rocks they see around them have rich and varied histories. Unless they are geologists, they probably have not been trained to identify a particular cliff as rock formed on a tidal flat that once fringed a primordial sea, to read in a hillside's ancient rocks the history of a primitive forest buried by a fiery volcanic eruption, or to decipher clues in lowland rocks telling of a lofty mountain chain that once stood where the land is now flat. Geologists can do these things because they have at their service a wide variety of information gathered during the two centuries that the modern science of geology has existed. The goal of this book is to introduce enough of these geologic facts and principles to give you an understanding of the general history of our planet and its life. The chapters that follow describe how the physical world assumed its present form and where the inhabitants of the modern world came from. They also reveal the procedures through which geologists have assembled this information. Students of Earth history inevitably discover that the perspective this knowledge provides changes their perception of themselves and of the land and life around them.

Knowledge of Earth's history can also be of great practical value. Geologists have learned to locate petroleum reservoirs, for example, by ascertaining where the porous rocks of these reservoirs tend to form in relation to other bodies of rock. Geologists have also helped discover deposits of coal and ore and other natural resources buried within Earth.

A Hawaiian volcano from which fiery lava flows and black cinders of volcanic rock shoot into the air.

Exploring the Earth System

The rocks of Earth's outer regions amount to a vast archive that we can read and interpret in order to unravel the planet's long history. By studying Earth's history, we learn how our planet functions as a complex system. This is not a purely academic exercise. An understanding of the system will help us to address problems caused by changes that are now taking place in the world, or that will be occurring soon.

The Components of the System

The Earth system has both physicochemical and biological components. We can reconstruct many aspects of the planet's physical history, including the growth and destruction of mountains, the breakup and collision of continents, the flooding and reemergence of land areas, and the warming and cooling of climates. We can also trace the evolution of life from an early world inhabited largely by bacteria through the origins of plants and animals in ancient seas to the invasion of the land, the rise and fall of dinosaurs, and ultimately the ascendance of humans. We cannot understand either the physical history of Earth or the history of life in isolation, however, because the two have been tightly intertwined: the physical environment has influenced life, and life, in turn, has influenced the physical environment. For example, as we will see in Chapter 4, climatic patterns control distributions of plants on the land. At the same time, plant life affects climates. Forests warm regional climates by trapping heat, for instance, and plants also affect global climates by altering the chemistry of the atmosphere. The geologic record reveals that the histories of land plants and climates have shifted in concert for hundreds of millions of years. Many other factors, including continental movements and the rising and falling of seas, have influenced climates as well. Clearly, we can understand the Earth system only by studying how its many components have interacted over hundreds of millions of years. The present state of this system is a momentary condition that is the product of a long and complex history.

Armed with knowledge of Earth system history, we can more effectively address problems caused by changes that are now taking place in the world. Consider the shifting of coastlines as sea level rises or falls. The geologic record of the past few thousand years documents a general rise in sea level as huge glaciers have melted and released water into the ocean. The geologic record near the edge of the sea reveals how coastal marshes have shifted as sea level has changed. These marshes are very important to humankind; they cleanse marginal marine waters and sustain forms of animal life that are valuable to us. Study of the geologic history of coastal marshes will help us to predict their fate as sea level continues to change in the decades and centuries to come.

The Fragility of the System

The geologic record of the history of life also provides a unique perspective on the numerous extinctions of animals and plants that are now resulting from human activities. Humans are causing extinction by destroying forests and other habitats, but our collective behavior also affects life profoundly in less direct ways. Very soon human activities will cause average temperatures at Earth's surface to rise in many areas of the world. The geologic record of ancient life reveals how climatic change has affected life in the past—how some species have survived by migrating to favorable environments, for example, and how others that failed to migrate successfully have died out. To the surprise of many biologists, geologic evidence has revealed that many of the natural assemblages of species that populate the world today are not ancient associations of interdependent species. Instead, they are associations that have developed very recently (on a geologic scale of time) as climates have changed in ways that have shifted the distribution of species.

As we come to understand the speed and power of natural environmental change and the temporary nature of assemblages of species, we begin to appreciate the fragility of the world we live in. More generally, having studied the past, we can make more intelligent choices as we contemplate the future of our changing planet.

Before we launch into our detailed examination of the history of Earth and its life, however, an introduction to some of the basic facts and unifying concepts of geology is in order. The first ten chapters lay this groundwork, and the chapters that follow trace out Earth system history.

The Principle of Uniformitarianism

Fundamental to the modern science of geology is the principle of **uniformitarianism**—the belief that there are inviolable laws of nature that have not changed in the course of time. Of course, uniformitarianism applies not only to geology but to all scientific disciplines—physicists, for example, invoke the principle of uniformitarianism when they assume that the results of an experiment will be applicable to events that take place a day, a year, or a century after the experiment is conducted—but geologists hold this principle in particularly great esteem because, as we shall see, it was the widespread adoption of uniformitarianism during the first half of the nineteenth century that signaled the beginning of the modern science of geology.

Actualism: The Present as the Key to the Past

The principle of uniformitarianism governs geologists' interpretations of even the most ancient rocks on Earth. It is in the present, however, that many geologic processes are discovered and analyzed. The application of these analyses to ancient rocks, in accordance with the principle of uniformitarianism, is sometimes called **actualism.** When we see ripples on the surface of an ancient rock composed of hardened sand (sandstone), for example, we assume that they formed in the same way that similar ripples develop today—under the influence of certain kinds of water movement or wind (Figure 1-1). Similarly, when we encounter ancient rocks that closely resemble those forming today from volcanic eruptions of molten rock in Hawaii, we assume that the ancient rocks are also of volcanic origin. We cannot observe rocks

A

B

Figure 1-1 **Ripples in sediments and sedimentary rocks.** *A.* Sand along a beach, exposed to the air at low tide. *B.* A large block of sandstone showing similar ripples.

twisting into contorted configurations like those seen in mountains, but we can witness the breaking, bending, and uplift of rocks during earthquakes, and we can easily imagine that the immense forces that produce these effects can strongly contort rocks deep within Earth and elevate them into mountains.

Actualism is commonly expressed by the phrase "The present is the key to the past." This idea is only partly true, however. Although it is universally agreed that natural laws have not varied in the course of geologic time, not all kinds of events that occurred in the geologic past have been duplicated within the time span of human history. Many researchers believe, for example, that the impact of very large meteorites may explain certain past events, such as the extinction of the dinosaurs 65 million years ago.

In Chapter 17 we will review evidence that the dinosaurs' reign on Earth ended when a massive meteorite—one perhaps 10 kilometers (6 miles) in diameter—plunged through the atmosphere and ocean and penetrated the seafloor along the coast of Mexico. Such an impact should have produced a huge wave that would have crashed over coastlines thousands of kilometers from the impact site. Nonetheless, because we have never observed the arrival of such a large meteorite, we do not know exactly what else would have happened. Some scientists have suggested that dust injected into the upper atmosphere might have blocked the sun's rays from Earth's surface for many days. Others have asserted that when this dust later settled, the total kinetic energy (energy of motion) of the dust particles would have been converted to heat, warming Earth's surface dramatically. It is easy to imagine that these and other consequences of a huge meteorite impact would have wiped out many species around the world. Even so, because humans have never observed such an event, we must rely on theoretical considerations to surmise what actually happened. As a result, the details remain uncertain. In other words, in this case, actualism does not apply.

Similarly, geologists have found that certain types of rocks cannot be observed in the process of forming today. In such cases, geologists usually assume that

1. The rocks in question formed under conditions that no longer exist;
2. The conditions responsible for the formation of these rocks still exist, but at such great depths beneath Earth's surface that we cannot observe them; or
3. The conditions exist today but produce the rocks only over a long interval of geologic time.

Many iron ore deposits more than 2 billion years old, for example, are of types that cannot be found in the process of forming today. It is believed that when the iron ore formed, chemical conditions on Earth differed from those of the present world and, furthermore, that the rocks underwent slow alteration after they were formed. The existence of these iron ore deposits does not necessarily negate the principle of uniformitarianism inasmuch as there is no evidence that natural laws were broken. It does, however, present geologists with a problem they cannot solve by applying the principle of actualism, because a human lifetime is only a small fraction of the time needed to study the rocks' development.

In an attempt to address some of these problems, geologists have learned to form certain kinds of rocks in the laboratory by simulating the conditions that prevail at great depths within Earth. They expose simple chemical components to temperatures and pressures many times greater than those at Earth's surface. Such experiments indicate the range of conditions under which a particular type of rock could have formed in nature. In conducting these experiments, geologists are, in a sense, expanding the domain of actualism by using as a model not only what is happening in nature today, but also what happens under artificial conditions and may have happened under natural conditions long ago.

The End of Catastrophism

Until the early nineteenth century, many natural scientists subscribed to the concept of **catastrophism,** which proposes that floods caused by supernatural forces formed most of the rocks visible at Earth's surface. Late in the eighteenth century, Abraham Gottlob Werner, an influential German professor of mineralogy, claimed that most rocks had been formed as a result of the precipitation of minerals from a vast sea that periodically flooded and retreated from the surface of Earth. These ideas were largely speculative.

Near the end of the eighteenth century, not long after Werner published his ideas, James Hutton, a Scottish farmer, established the foundations of uniformitarianism in his writings on the origins of rocks in Scotland. Hutton believed the rocks had formed as a result of the same processes that were currently operating at or near the surface of Earth—processes such as volcanic activity and the accumulation of grains of sand and clay under the influence of gravity.

Central to Hutton's view of Earth's history was vast geologic time. For the processes that were constantly shaping and reshaping the planet he envisioned "no vestige of a beginning, no prospect of an end." Everyday processes, he claimed, had created and destroyed large bodies of rock, elevated and leveled mountains, and left remnants of their workings in an immense geologic record.

After extensive debate, Hutton's principle of uniformitarianism came to dominate the science of geology, gaining almost total acceptance after Charles Lyell, an Englishman, popularized it in the 1830s in a three-volume book titled *Principles of Geology*. Lyell was a more effective writer than Hutton, and the world was more receptive to the concept of uniformitarianism when he wrote than in Hutton's day. Eleven editions of Lyell's book appeared between 1830 and 1872. Lyell understood that volcanoes, floods, and earthquakes transform Earth. He claimed that such violent events are restricted to particular regions, however, and that their effects have been evenly distributed through geologic time. He argued that these and more subtle agents of change, such as the wearing away of old rocks and the accumulation of sand and mud to form new ones, operate incessantly. In the eyes of Hutton and Lyell, Earth resembled an enormous machine that was always churning but that retained its basic features.

Uniformitarianism, with its implication that Earth is very old, remains the central principle of modern geology. Nonetheless, as we shall see, Lyell's rejection of global catastrophes was unjustified, and he and Hutton also were mistaken in believing that the Earth system had remained unchanged in all of its basic features. Before addressing these points, we must learn more about the materials and processes of geology.

The Nature and Origin of Rocks

Rocks consist of interlocking or bonded grains of matter that are typically composed of single minerals. A **mineral** is a naturally occurring inorganic solid element or compound with a particular chemical composition or range of compositions and a characteristic internal structure. Quartz, which forms most grains of sand, is probably the most familiar and widely recognized mineral; the materials we call limestone, clay, and asbestos consist of other minerals. Most rocks are formed of two or more minerals.

Rocky surfaces that stand exposed and are readily accessible for study are generally designated as *outcrops* or *exposures.* Scientists also have access to rocks that are not visible in outcrops. Well drilling and mining, for example, allow geologists to sample rocks that lie buried beneath Earth's surface.

Basic Kinds of Rocks

On the basis of modes of origin, many of which can be seen operating today, early uniformitarian geologists, led by Hutton and Lyell, came to recognize three basic types of rocks: igneous, sedimentary, and metamorphic.

Igneous rocks, which form by the cooling of molten material to the point at which it hardens, or freezes (much as ice forms when water freezes), are composed of bonded grains, each consisting of a particular mineral (Figure 1-2). The igneous rock most familiar to nongeologists is granite. Molten material, or **magma,** that turns into igneous rock comes from great depths within Earth, where temperatures are very high. This material may reach Earth's surface through cracks and fissures in the crust and then cool to form **extrusive,** or **volcanic,** igneous rock, or it may cool and harden within Earth to form **intrusive** igneous rock (Figure 1-3).

Even intrusive rocks that form deep within Earth can eventually be exposed at the surface if they are uplifted by earth movements and overlying rocks are stripped away. **Weathering** is a collective term for the chemical and physical processes that break down rocks of any kind at Earth's surface (Figure 1-4). Water carries some products of weathering away in solution. Solid products are removed by **erosion,** the

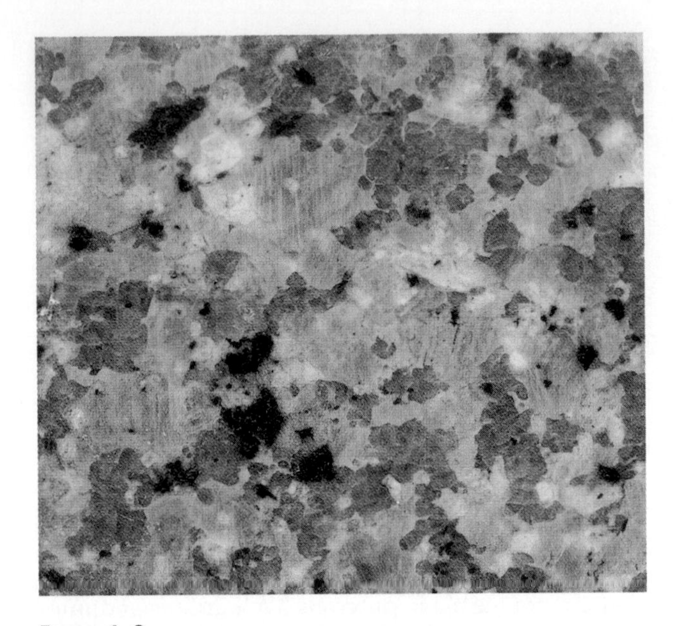

Figure 1-2 **Interlocking grains in granite.** The pink and white grains are two kinds of feldspar, the gray grains are quartz, and the black grains are mafic minerals.

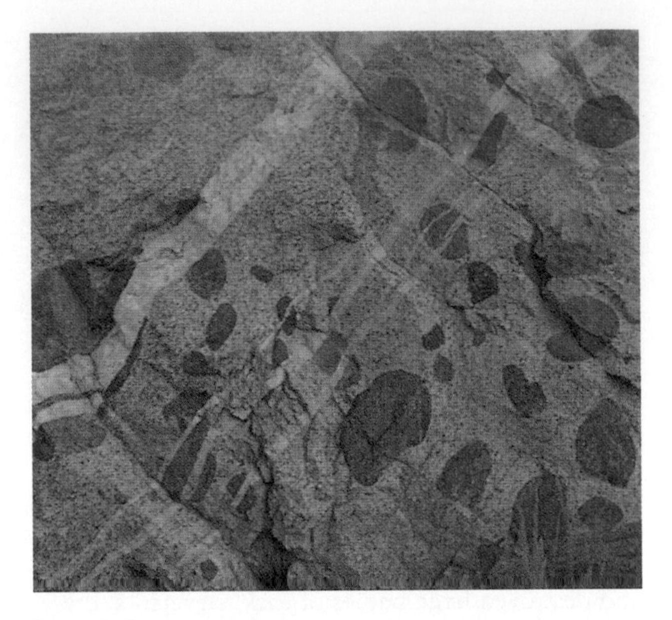

Figure 1-3 **Intrusive igneous rock.** The dark bodies are pieces of surrounding rock that magma incorporated before it solidified into igneous rock. The light-colored diagonal bands on the left are veins that formed when a second body of magma intruded the main body of igneous rock.

process that loosens pieces of rock and moves them downhill. After erosion sets these pieces of rock in motion, moving water, ice, or wind transports them to a site where they accumulate as sediments.

Sediment is material deposited on Earth's surface by water, ice, or air. **Sedimentary rocks** form by the accumulation of grains in any of a variety of settings, ranging from the bottoms of lakes and rivers to sandy beaches and the floor of the deep sea. After grains have accumulated as loose sediment in some such setting, they can become bonded together to form solid sedimentary rock by either of two processes: the grains may become mutually attached by compression of the sediment after burial; or, they may be glued together by precipitation of mineral cement from watery solutions that flow through the sediment. There are three principal kinds of rock-forming sediments:

1 Most sedimentary rocks are formed of the kind of sediment described above—debris generated by weathering of preexisting rocks. The most common grains produced in this way are particles of sand and clay. Sand particles are grains that weathering releases from preexisting rocks, generally without chemical alteration. Sand grains are globular, and they do not stick together well when compacted. Loose sand therefore becomes solid **sandstone** only when cement precipitates between adjacent grains, locking them together. Tiny clay particles form by the chemical breakdown of certain minerals: they are chemical products of weathering. Clay is a flaky material that compacts to form the soft rock known as **shale.**

2 Other sedimentary rocks consist of grains of skeletal debris. Many **limestones** are formed of such material, including bits of broken seashells. Cementation turns accumulations of this limey debris into solid rock.

3 Still other grains that form rocks are precipitated chemically from water. Salt deposits that we mine for a variety of purposes form in this way when bodies of water evaporite in dry climates.

Sediments usually accumulate in discrete episodes, each of which forms a tabular layer known as a **stratum** (plural, **strata**) or **bed.** A breaking wave

Figure 1-4 Granite weathering in a mountainous region of Morocco.

can create a stratum, for example, and so can the spreading waters of a flooding river. Even after lithification, a stratum tends to remain distinct from the one above it and the one below it because the grains of adjacent strata usually differ in size or composition. Because of their differences, the contacting surfaces of the strata usually adhere to each other only weakly, and sedimentary rocks often break along these surfaces. As a result, sedimentary rocks exposed at Earth's surface often can be seen to have a steplike configuration when they are viewed from the side (Figure 1-5). **Stratification** and **bedding** are the synonymous words used to describe the arrangement of sedimentary rocks in discrete layers.

Metamorphic rocks form by the alteration, or **metamorphism,** of rocks within Earth under condi-

Figure 1-5 **Horizontal bedding in the Kaibab Formation, bordering the Grand Canyon.** The top of this formation forms the Kaibab Plateau, which marks the horizon.

Figure 1-6 Metamorphic rock. The rock shown here is a coarsely crystalline kind known as gneiss.

melted to produce magma or weathered to produce sediment.

Classification of Bodies of Rock

Geologists also classify rocks into units called **formations.** Each formation consists of a body of rocks of a particular type that formed in a particular way—for example, a body of granite, of sandstone, or of alternating layers of sandstone and shale. A formation is formally named, usually for a geographic feature such as a town or river where it is well exposed.

The Kaibab Limestone is a typical formation. It forms the rim of a large portion of the Grand Canyon, and its upper surface forms much of the surface of the Kaibab Plateau, which borders the canyon and gives the formation its name (see Figure 1-5). The Kaibab Limestone is composed of fragments of shells and other skeletal debris from organisms. These and other distinctive features of the formation, including its color and the characteristic thickness of the beds within it, permit geologists to recognize the Kaibab wherever it occurs. Other limestones that occur

tions of high temperature and pressure. By definition, metamorphism alters rocks without turning them to liquid. If the temperature becomes high enough to melt a rock and the molten rock later cools to form a new solid rock, this new rock is by definition igneous rather than metamorphic. Metamorphism produces minerals and textures that differ from those of the original rock and that are characteristically arrayed in parallel wavy layers (Figure 1-6). The two groups of rocks that form at high temperatures—igneous and metamorphic rocks—are commonly referred to as **crystalline rocks.**

Figure 1-7 summarizes the various possible relationships among igneous rocks, metamorphic rocks, and sedimentary rocks that are composed of debris from other rocks. Any body of rock can be transformed into another body of rock belonging to the same group (metamorphic, igneous, or sedimentary) or to either of the other two groups. In other words, any kind of rock can be metamorphosed or it can be

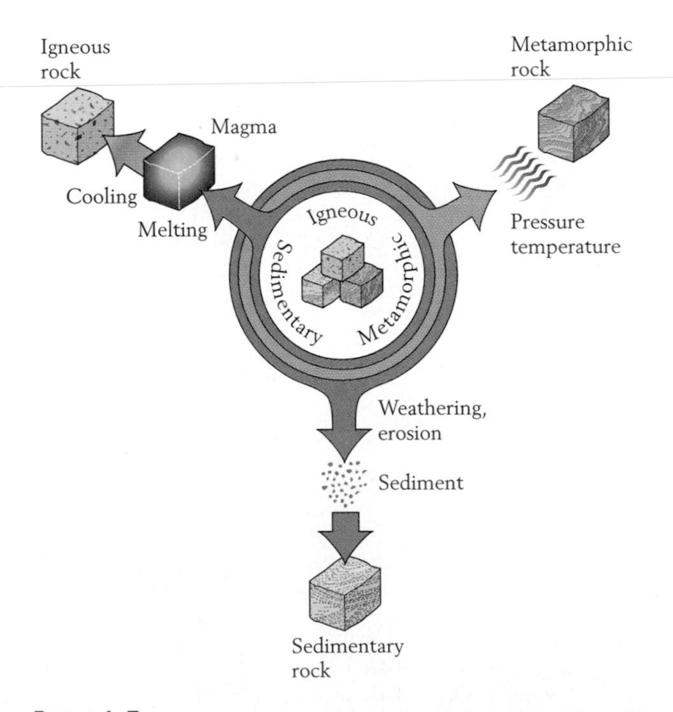

Figure 1-7 Relationships in the rock family. Any of the three basic kinds of rock—igneous, sedimentary, or metamorphic—can be transformed into another rock of the same kind or either of the other two kinds.

below the Kaibab in the Grand Canyon region display different features.

Smaller rock units called **members** are recognized within some formations. Similarly, some formations are united to form larger units termed **groups,** and some groups, in turn, are combined into **supergroups.**

Steno's Three Laws for Sedimentary Rocks

Forming on Earth's surface, sedimentary rocks provide most of our information about the history of life and environments on Earth. It is therefore important that we understand their distribution and their age relationships. In the seventeenth century, Nicolaus Steno, a Danish physician who lived in Florence, Italy, formulated three axioms for interpreting stratified rocks.

Steno's first principle, the **principle of superposition,** states that in an undisturbed sequence of strata, the oldest strata lie at the bottom and successively higher strata are progressively younger (Figure 1-8*A*). In other words, in an uninterrupted sequence of strata, each bed is younger than the one below and older than the one above. This is a simple consequence of the law of gravity, of course, as is Steno's second principle, the principle of original horizontality.

The **principle of original horizontality** states that all strata are horizontal when they form. As it turns out, this principle requires some modification. We now recognize that some sediments, such as those of a sand dune, accumulate on sloping surfaces, forming strata that lie parallel to the surface on which they were deposited. Sediments seldom accumulate at an angle greater than 45° to the horizontal, however, because they slide down slopes that are steeper than that. Therefore, a reasonable restatement of Steno's second principle would be that almost all strata are initially more nearly horizontal than vertical. Thus we can conclude that any strongly sloping stratum was tilted by external forces after it formed (Figure 1-8*B*).

Steno invoked his third principle, the **principle of original lateral continuity,** to explain the occurrence on opposite sides of a valley (or some other intervening feature of the landscape) of similar rocks that seem once to have been connected. Steno was, in effect, pointing out that strata originally are unbroken flat expanses, thinning laterally to a thickness of zero or abutting against the walls of the natural basin in which they formed. The original continuity of a stratum can be broken by erosion, as when a river cuts downward to form a valley (Figure 1-8C).

The Rock Cycle

After rocks form, they are subject to many kinds of change. Central to the uniformitarian view of Earth is the **rock cycle**, the endless pathway along which rocks of various kinds change into rocks of other kinds. Two simple principles are useful for recognizing steps of the rock cycle. The **principle of intrusive relationships** states that intrusive igneous rock is always younger than the rock that it invades. The

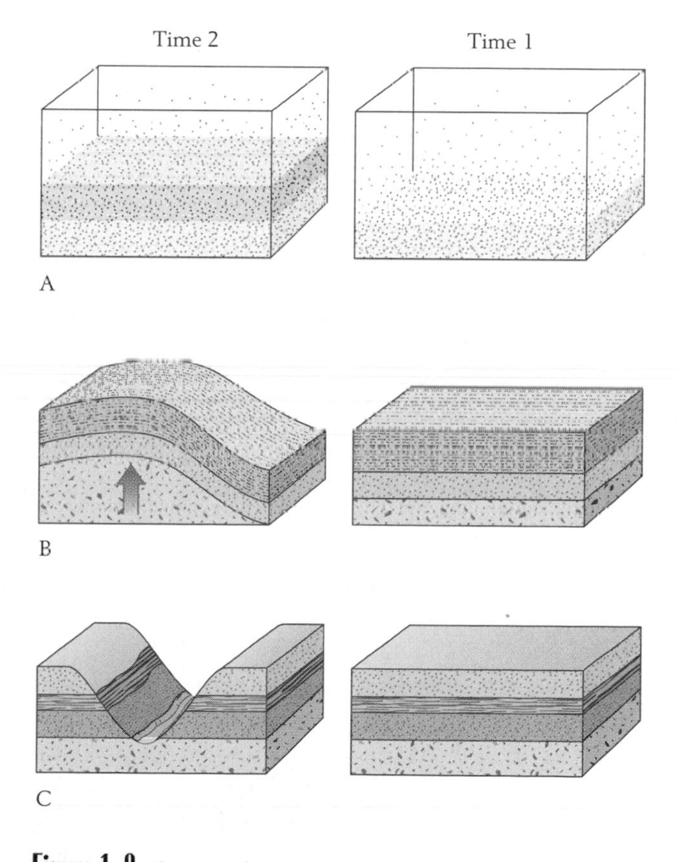

Figure 1-8 **Steno's three principles.** *A*. The principle of superposition: at time 2, sediment builds up on top of other sediment that was deposited earlier, at time 1. *B*. The principle of original horizontality: strata that were horizontal at time 1, shortly after being deposited, have been uplifted and tilted by time 2. C. The principle of original continuity: by time 2, strata that were continuous at time 1 have been divided into two bodies of strata by a river that has cut through them.

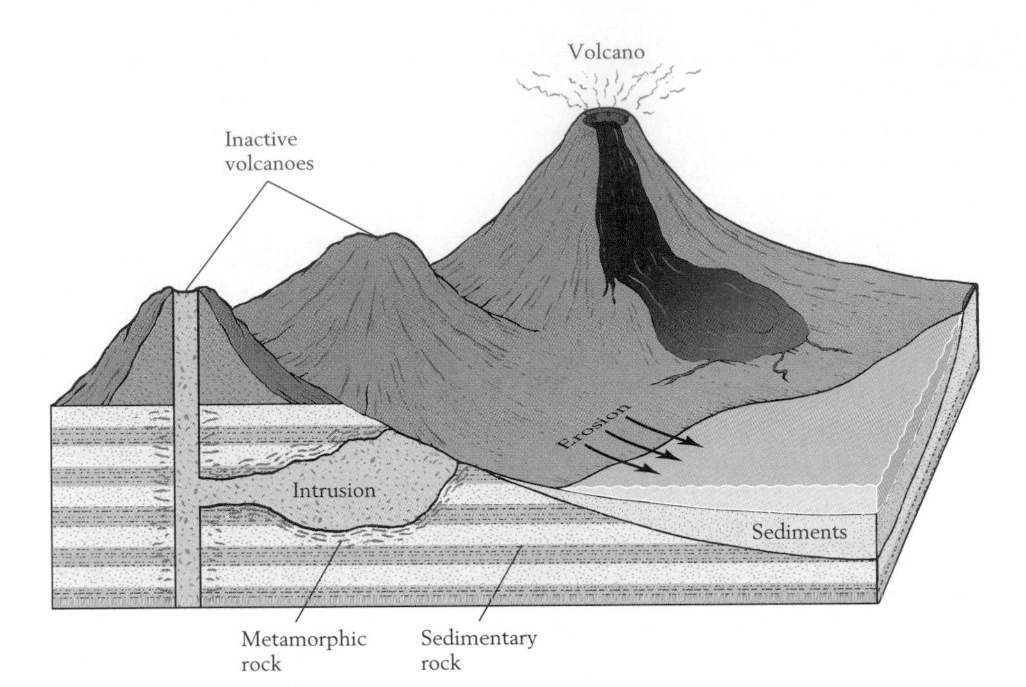

Figure 1-9 **The rock cycle.** On the left is an igneous intrusion formed by magma that melted sedimentary rock and incorporated it. Some of the magma containing the melted sedimentary rock was extruded from volcanoes. Those volcanoes are now inactive and are eroding, along with exposures of the intrusive igneous rock and of metamorphic rock formed during the intrusive activity. The resulting sediment is accumulating in water nearby. Thus sediments have become igneous and metamorphic rocks, and those rocks have yielded younger sediment, completing the cycle. The volcano on the right is still active.

principle of components states that when fragments of one body of rock are found within a second body of rock, the second body is always younger than the first; the second may be a body of sedimentary rock in which the pieces have come from another body of rock, or it may be a body of igneous rock that contains distinctive pieces of older rock that magma engulfed before it cooled (see Figure 1-3).

The rock cycle is actually a complex of many kinds of cycles in which components of any body of rock—whether igneous, sedimentary, or metamorphic—can become part of another body of rock of the same kind or either of the other two kinds. In other words, as partly illustrated by Figure 1-7, any rock may be (1) melted to form magma that later cools to form igneous rock, (2) incorporated in magma without melting, (3) weathered to form debris that becomes part of sedimentary rock, or (4) turned into metamorphic rock by exposure to high temperatures and pressures. Figure 1-9 illustrates the rock cycle with a hypothetical example that includes igneous, sedimentary, and metamorphic rocks.

Movements of Earth play a key role in the rock cycle. When mountains rise up, for example, weathering and erosion wear them down to expose rocks that formed deep within the planet. Over vast stretches of time, these destructive processes level mountains, and streams and rivers carry the resulting sediments to far away depositional settings.

Partly because movements of earth from time to time elevate low areas in which sediments accumulate, rocks are not deposited continuously anywhere; one or more breaks interrupt any local record of deposition. An **unconformity** is a surface between a group of sedimentary strata and the rocks beneath them; it is an ancient land surface that represents an interval of time during which erosion occurred rather than deposition. Because rock deformation is episodic, only rocks of a certain age will have been affected by a particular deformational episode. The surface between the deformed and undeformed rocks represents one of several kinds of unconformity. When a group of rocks has been tilted and eroded and younger rocks have been deposited on

Figure 1-10 **An angular unconformity in the Grand Canyon.** Horizontal Cambrian strata rest on tilted strata of Precambrian age.

top of them, the eroded surface is termed an **angular unconformity** (Figures 1-10 and 1-11*A*). Other unconformities are less dramatic. Sometimes the beds below an eroded surface are undisturbed, and only the irregular surface between groups of beds reveals a past episode of erosion. This kind of unconformity is called a **disconformity** (Figure 1-11*B*). An unconformity in which bedded rocks rest on an eroded surface of crystalline rocks (Figure 1-11C) is sometimes called a **nonconformity.**

Global Dating of the Rock Record

Thus far, we have discussed only age relations between bodies of rock that are in close proximity to one another. Geologists use additional techniques to piece together the history of Earth on a broader scale, showing, for example, that a body of rock located on one continent is older or younger than a body of rock located on another continent.

Fossils, Markers, and Relative Ages of Rocks

Remnants of ancient life that we call **fossils** are useful for comparing the ages of bodies of sedimentary rock throughout the world. The term fossil is usually restricted to tangible remains or signs of ancient organisms that died thousands or millions of years ago. Because few fossils can survive at the high

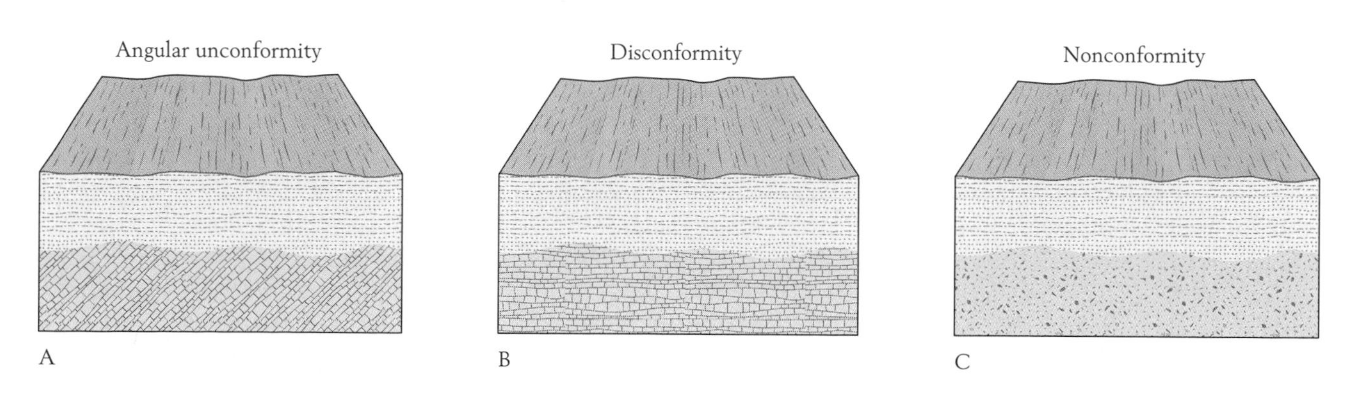

Figure 1-11 **An angular unconformity, a disconformity, and a nonconformity.** An angular unconformity *(A)* separates tilted beds below from flat-lying beds above. A disconformity *(B)* separates flat-lying beds below from other flat-lying beds above, but the upper beds rest on an erosion surface that developed after the lower beds were deposited. A nonconformity *(C)* separates flat-lying beds from crystalline (igneous or metamorphic) rocks.

temperatures at which igneous and metamorphic rocks form, almost all fossils are found in sediments and sedimentary rocks. Fossils range in size from cells of tiny bacteria to massive dinosaur bones. They include such things as shells of invertebrate animals and teeth and bones of vertebrate animals—and also leaves of plants and impressions of soft-bodied animals.

Fossils provide one valuable means of establishing the relative ages of rocks that lie far apart. William "Strata" Smith, a British surveyor, noted late in the eighteenth century that fossils are not randomly distributed in rocks. When Smith studied large areas of England and Wales, he found that fossils in sedimentary rocks occurred in a particular vertical order ("vertical" in terms of the succession of one layer above another). To the surprise of less experienced observers, Smith could predict the vertical ordering of fossils in areas he had never visited. We now recognize that this ordering, known as **fossil succession,** reflects the sequence of organic evolution—the natural appearance and disappearance of species through time. Figure 1-12 illustrates the fossil succession of bivalve mollusks, a group of animals that includes clams, scallops, and mussels. Each of the geologic eras during which bivalve mollusks lived was populated by its own unique kinds of these animals. Although Smith, like nearly all of his contemporaries, had no knowledge of evolution, he was able to use his knowledge of fossil succession to determine where isolated outcrops of sedimentary rocks fitted into the general sequence of strata in England and Wales.

Figure 1-12 **Characteristic bivalve mollusks from the Paleozoic (bottom), Mesozoic (middle), and Cenozoic (top) eras.**

Since the time of William Smith, geologists have extended the use of fossils to establish the relative ages of rocks on a global scale. They have also discovered other kinds of markers showing that strata in many parts of the world were deposited simultaneously. One such marker is a high concentration of the element iridium at a stratigraphic level just above the uppermost dinosaur fossils. Iridium is much more common in meteorites than in nearly all rocks at Earth's surface. The unusual iridium-rich layer occurs throughout the world. As Chapter 17 will explain, this occurrence is what first suggested that a meteorite struck Earth and killed off the dinosaurs. The impact sent up a huge cloud of dust, spiked with telltale iridium, that spread around the globe. In more practical terms, the layer that is rich in iridium, an element very rare on Earth, allows geologists to locate, in sedimentary deposits throughout the world, the precise stratigraphic level that marks the end of the Age of Dinosaurs.

Radioactivity and the Actual Ages of Rocks

Although the principles outlined in the preceding section allow us to establish the relative ages of many bodies of rock, they do not permit us to determine the actual ages of rocks measured in thousands or millions of years. As we will see in Chapter 6, some sedimentary beds are produced annually, like the rings in a tree trunk. Unless the latest of a continuous sequence of annual beds is currently forming, however, it is impossible to count backward to determine precisely how many years ago an older bed formed. In other words, if a sequence of this type formed long ago, we cannot tell the actual ages of its beds. Fortunately, "geologic clocks" in the form of chemical components that undergo radioactive decay provide us with good estimates of the actual ages of ancient rocks. Naturally occurring radioactive materials decay into other materials at known rates. By measuring the amount of radioactivity in a radioactive material that has been decaying since it became part of a rock, we can estimate the age of the rock.

The Geologic Time Scale

During the nineteenth century, long before the discovery of radioactivity, it became apparent that very old sedimentary rocks contain no identifiable fossils. Beginning with these rocks and examining progressively younger rocks in any region, early geologists discovered that fossils became abundant at a certain level. This level became the boundary at which all of geologic time was divided into two major intervals (Figure 1-13). The oldest rocks with conspicuous fossils were designated as Cambrian in age, and still older rocks became known as Precambrian rocks. Today the Precambrian designation is still used informally, but the Precambrian interval is formally divided into the Archean Eon and the Proterozoic Eon, with the boundary between these two eons placed at 2.5 billion years ago. Subsequent geologic time, from Cambrian on, constitutes the Phanerozoic Eon, meaning the "interval of well-displayed life." An eon is the largest formal unit of geologic time.

Phanerozoic time is divided into three primary intervals, or **eras,** which the history of life on Earth serves to define. The earliest is the "interval of old life," or the Paleozoic Era. This era is followed by the "interval of middle life," or the Mesozoic Era, which is commonly called the Age of Dinosaurs, and by the "interval of modern life," or the Cenozoic Era, which is informally designated as the Age of Mammals. Figure 1-13 depicts these eras and the intervals within them, known as the geologic **periods.** Periods are divided into **epochs.** Figure 1-13 lists epochs for the Cenozoic Era.

Figure 1-13 also indicates when each period began and ended, as determined by radioactive materials in rocks whose ages approximate period

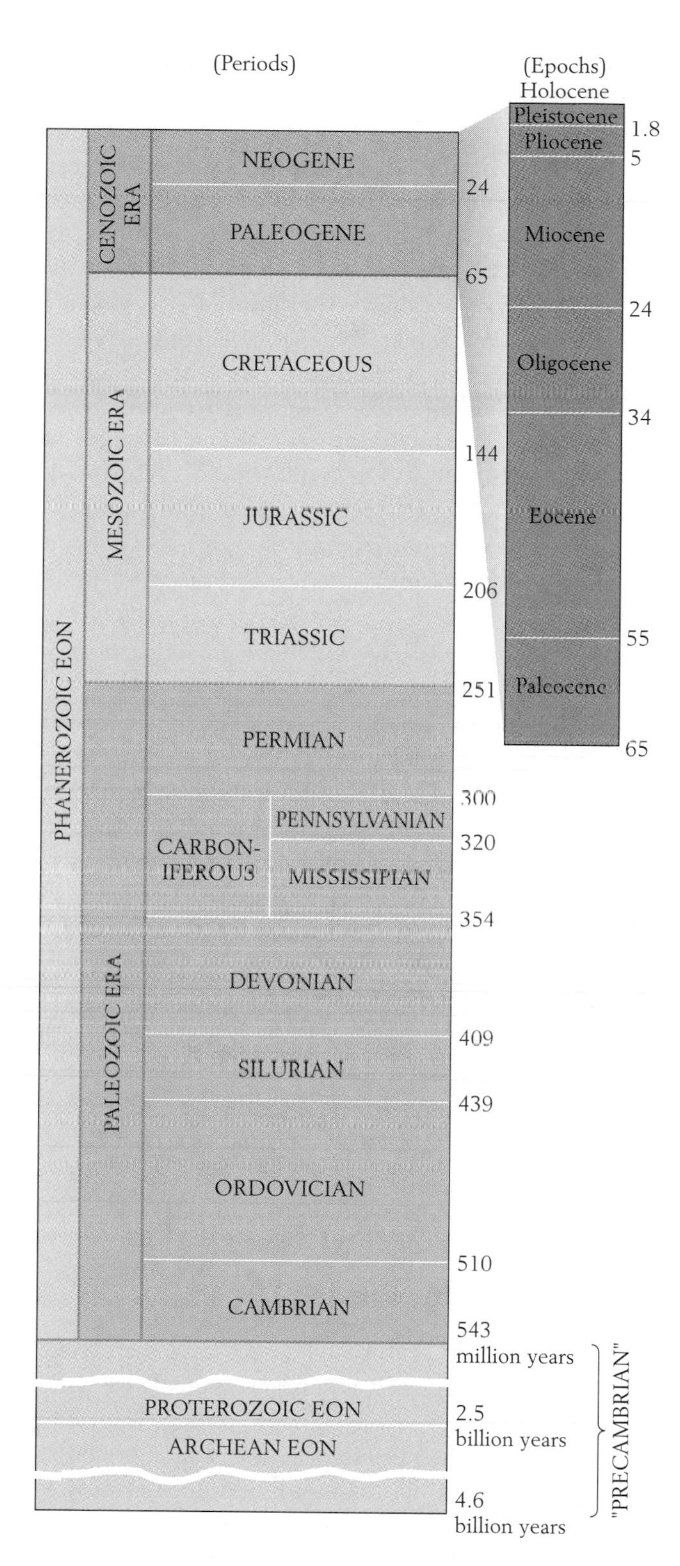

Figure 1-13 **The geologic time scale.** The numbers on the right represent the ages of the boundaries between periods and epochs in millions of years. The Holocene Epoch (the past 10,000 years or so) is also known as the Recent.

boundaries. Note that the Phanerozoic interval began about 543 million years ago. A human lifetime is so short in comparison that geologic time seems too vast for us to comprehend; experience does not permit us to extrapolate from the time scale we are familiar with, measured in seconds, minutes, hours, days, and years, to a scale suitable for geologic time. Geologists therefore use a separate scale when they think about geologic time—one in which the units are millions of years. If the Phanerozoic interval of time were compressed into a year, we would find animals with backbones crawling up onto the land for the first time in mid-April, dinosaurs inheriting Earth in early July but then suddenly dying out in late October, and humans appearing on Earth within two hours of midnight on New Year's Eve.

Dating of rocks by means of radioactive materials reveals that some rocks on Earth are more than 3.8 billion years old. Many major geologic events span millions of years, but on the scale of geologic time they are only brief episodes. We now know, for example, that the Himalayas, the tallest mountains on Earth, formed within the past 15 million years or so, but this period of time represents less than one-third of 1 percent of Earth's history. Destructive processes have also yielded enormous changes within a tiny fraction of Earth's lifetime. Mountains that were the precursors of the Rockies in western North America were leveled just a few million years after they formed, and much of the Grand Canyon of Arizona was cut by erosion within just the past 2 million or 3 million years. We will examine these events in greater detail in later chapters.

Development of the Time Scale

In the nineteenth century, when the geologic periods were first distinguished as unique intervals of time, geologists did not know even approximately how long ago each period had begun or ended. Each period was defined simply as the undetermined interval of time represented by a body of rock called a geologic system. The Cambrian Period, for example, was simply an interval of time that corresponded to those rocks that were designated as the Cambrian System.

Although the order in which the geologic systems were designated was haphazard, the total body of rock assigned to each system was not chosen arbitrarily. Two criteria were most important in these decisions. One was the occurrence of unique groups of fossils. Most systems contain many fossils that differ considerably from the fossils found below and above them. Major extinctions have caused the most striking contrasts between systems. A system representing an interval that followed a great extinction lacks many fossil groups that are well represented in the preceding system, and the younger system contains many new taxonomic groups that evolved to replace those that died out.

Another feature that led early workers to recognize some bodies of rock as systems was the nature of the rocks themselves. Most of the distinctive lithological features of systems have some relation to the history of life. The Cretaceous System, for example, was designated to include the thickest deposits of chalk in the world. Chalk is soft, fine-grained limestone. The abundance of chalk in the Cretaceous System reflects the fact that during the Cretaceous Period there was a great proliferation of the kinds of organisms that produce the particles of calcium carbonate that form chalk: small, single-celled organisms that still float in the sea today, but in reduced abundance. No early scientist had the means to study the entire sequence of rocks on Earth, from the most ancient to the most modern; a single person could study only promising rock sequences that were accessible. Thus the Cretaceous System was formally designated in 1822, whereas the much older Cambrian and Silurian systems did not gain formal recognition until 1835. System after system was added up and down the sequence until, finally, all the Phanerozoic rocks of Europe were included.

It seems remarkable today that all of the geologic systems of the Phanerozoic Eon were first designated during a brief interval of the nineteenth century in one small region of the world: Great Britain and nearby areas of western Europe. It was a lucky circumstance that modern geology came into being in this particular region. Few other geographic areas of comparable size display such large volumes of sediment and rock representing all of the Phanerozoic systems.

Imaging the Earth Below

Most of our knowledge about the structure of Earth's interior derives from the study of *seismic waves*, large vibrations that travel through Earth as a consequence of natural earthquakes or artificial disturbances, such as those produced by nuclear explosions.

An earthquake always begins at a *focus,* a place within Earth where rocks move against other rocks along a fault and produce seismic waves. A **fault** is a surface along which rocks have broken and moved. Earthquake foci lie within Earth's mantle and crust, far from its center—but the waves emitted from foci often pass great distances through Earth to emerge at the surface, where they can be detected with instruments called seismographs. Geophysicists can then evaluate movements deep within Earth by recording the times at which the waves of an earthquake arrive at many locations.

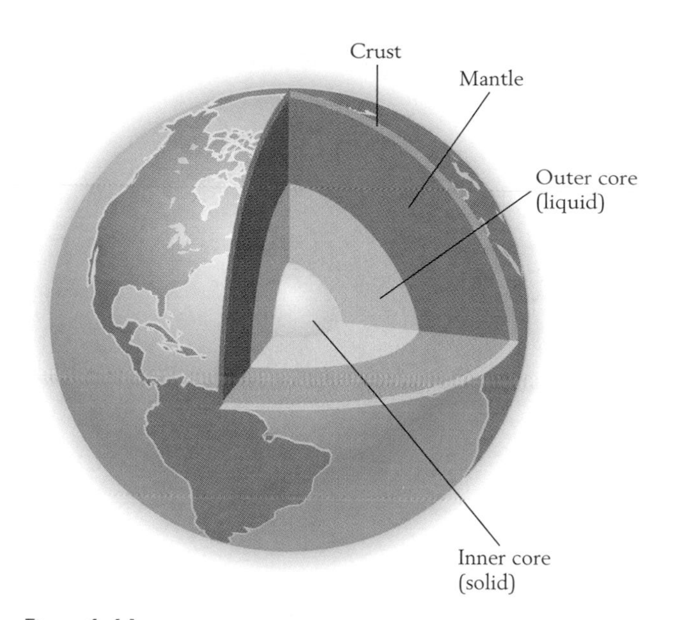

Figure 1-14 Zonation of Earth's interior. The crust, which includes Earth's continents, rests on the mantle. The mantle in turn rests on the core. The outer core is liquid, but the inner core is solid. *(After W. J. Kaufmann, Planets and Moons, W. H. Freeman and Company, New York, 1979.)*

The Density Gradient

The denser the material, the more rapidly seismic waves travel through it. Study of the rates at which waves travel in various directions from a focus reveals that the materials that form the central part of Earth are much more dense than those near the surface. The density gradient from the surface to the center of Earth is not gradual, however; instead, the planet is divided into several discrete concentric layers (Figure 1-14). At Earth's center is the **core,** whose solid, spherical inner portion and liquid outer portion are thought to consist primarily of iron. Forming a thick envelope around the outer core is the **mantle,** a complex body of less dense rocky material. Finally, capping the mantle is the **crust,** which consists of still less dense rocky material. As we shall see in Chapter 11, the density gradient from the core to the crust developed early in Earth's history, when molten materials of low density rose to float on materials of higher density. The passage of seismic waves from the rocks of the crust to the denser rocks of the mantle is signaled by an abrupt increase in velocity known as the **Mohorovičić discontinuity,** or **Moho** for short (Figure 1-15). Because continental crust is much

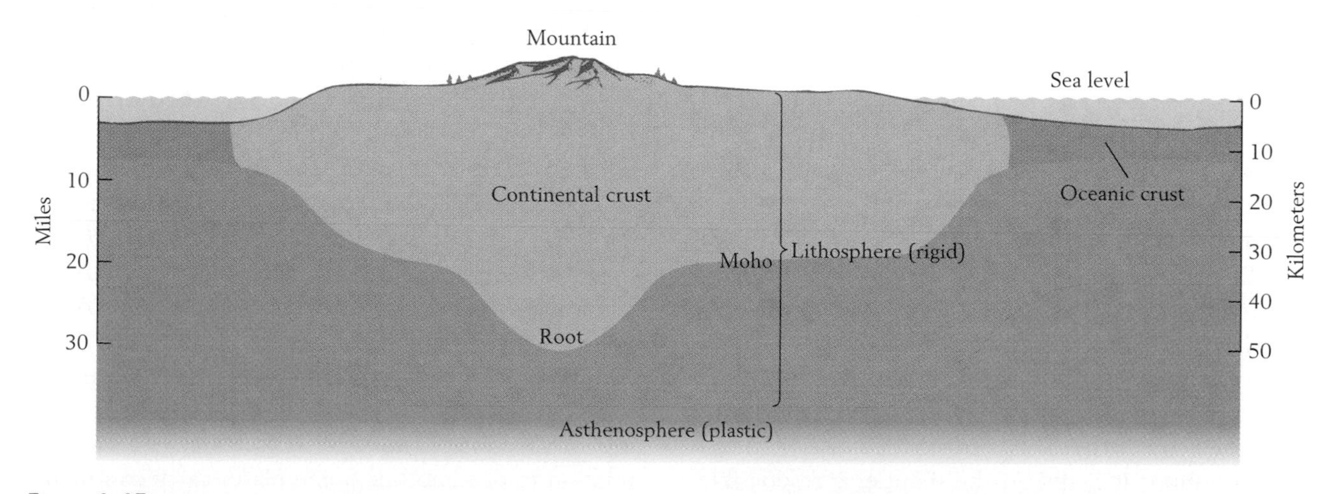

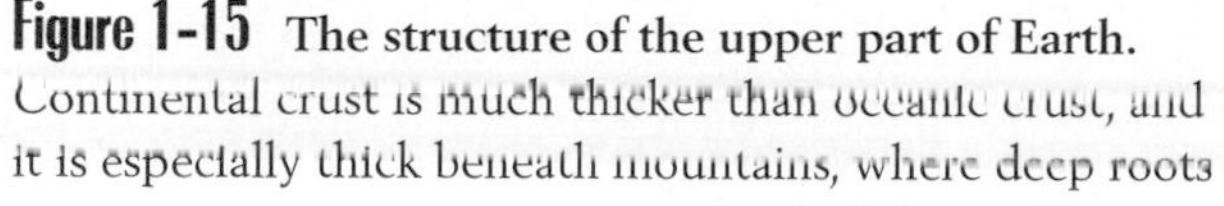

Figure 1-15 The structure of the upper part of Earth. Continental crust is much thicker than oceanic crust, and it is especially thick beneath mountains, where deep roots extend downward. The Moho separates the crust from the mantle. The crust and the upper mantle together form the rigid lithosphere.

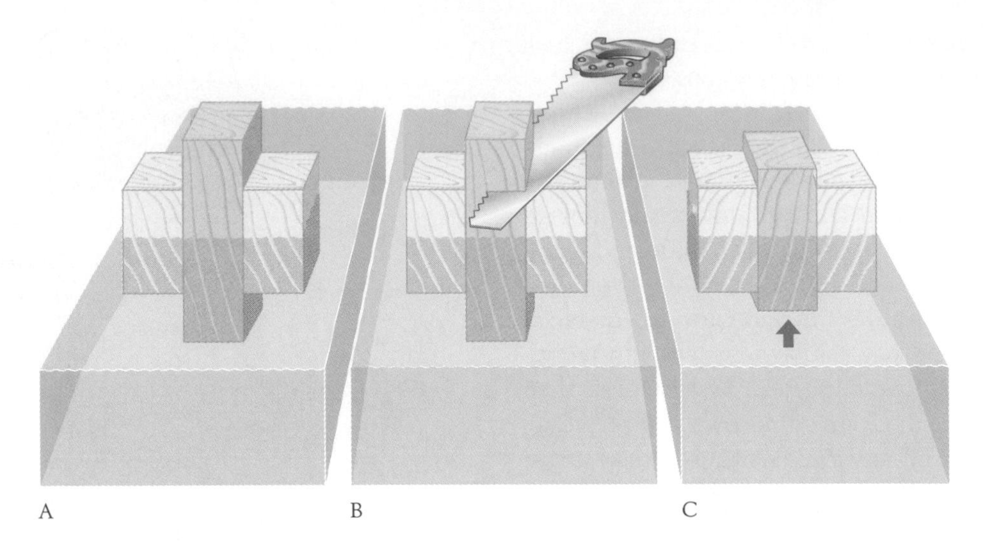

Figure 1-16 **The principle of isostasy.** *A*. Three blocks of wood float next to one another. Because the wood is half as dense as the water, half of each block lies above the surface of the water and half lies below. The weight of a block of wood is equivalent to the weight of the volume of water that it displaces. When the top of the center block is cut off *(B)*, the weight of this block no longer balances the weight of the displaced water. As a result, the block bobs upward until it is balanced once again, lying half above and half below the water *(C)*. The central block is like the part of a continent where a large mountain is present—the low-density elevated crust must be balanced by a root.

thicker than the crust beneath the oceans, the Moho dips downward beneath the continents.

Rocks that form oceanic crust are the type known as **mafic**—a label whose first three letters indicate that these dark rocks are rich in magnesium (Mg) and iron (Fe). Mafic rocks are much less common in continental crust than are the lighter-colored, less dense rocks known as **felsic**—an adjective derived from the first three letters of *feldspar,* the name of the most common mineral of continental crust. In comparison with mafic rocks, felsic rocks are rich in silicon and aluminum and poor in the heavier element iron. Rocks of the mantle are even richer in iron than the oceanic crust—hence their great density—and they are known as **ultramafic** rocks.

Continental crust not only stands above oceanic crust but also extends farther down into the mantle (see Figure 1-15). The continental crust extends even farther down beneath a mountain range than it does elsewhere. Isostatic adjustment, the upward or downward movement that keeps the crust in gravitational equilibrium as it floats on the mantle, is responsible for this phenomenon (Figure 1-16). In effect, the root beneath a mountain acts to balance the mountain; this balance is known as **isostasy**. An iceberg illustrates the same principle. Because it is only slightly less dense than the water in which it floats, only a small fraction of its mass stands above the surface.

Lithosphere and Asthenosphere

Although the crust and the upper mantle differ in density, they are firmly attached to each other, forming a rigid layer known as the **lithosphere.** Below the lithosphere is the **asthenosphere,** which is also known as the "low-velocity zone" of the mantle because seismic waves slow down as they pass through it. This property tells us that the asthenosphere is composed of partially molten rock—slushlike material consisting of solid particles with liquid occupying the spaces in between. Although the asthenosphere represents no more than 6 percent of the thickness of the mantle, the mobility of this layer allows the overlying lithosphere to move. The lithosphere does not move as a unit, however; instead it is divided into sectors called **plates,** which move in relation to one another. Some plates carry continents with them as they move, while others carry only oceanic crust.

Plate Tectonics

The movement of lithospheric plates is known as **plate tectonics.** Throughout the first half of the twentieth century, most geologists believed that Earth's continents occupied fixed positions above the mantle. During the 1960s, however, geologists established beyond a reasonable doubt that plates—some carrying continents—migrate around the globe. This discovery ushered in one of the greatest scientific revolutions of the twentieth century. Plate tectonics accounts for many kinds of geological phenomena that previously defied explanation, including the elevation of great mountain ranges.

Plate Movements

Figure 1-17 illustrates the distribution of lithospheric plates in the modern world. An average plate moves about as rapidly as your fingernails grow: about 5 centimeters (2 inches) per year. When we recognize that plates have moved tens of thousands of kilometers at this pace, we can appreciate the immensity of geologic time. One of the most important of the forces that drive plate movements is convection, which operates on a large scale within the asthenosphere. **Convection** is a process in which material that is heated deep within the asthenosphere rises to displace cooler, less dense material nearer the surface. The same kind of process causes rotational motion in a pot of water when it is heated (Figure 1-18).

Plate boundaries are dynamic regions of the lithosphere. In fact, most earthquakes originate when rocks move along faults at or close to these junctures. Three types of relative movement can occur between two lithospheric plates that are in contact: the plates can be sliding past each other, or they may be moving toward or away from each other (see Figure 1-17).

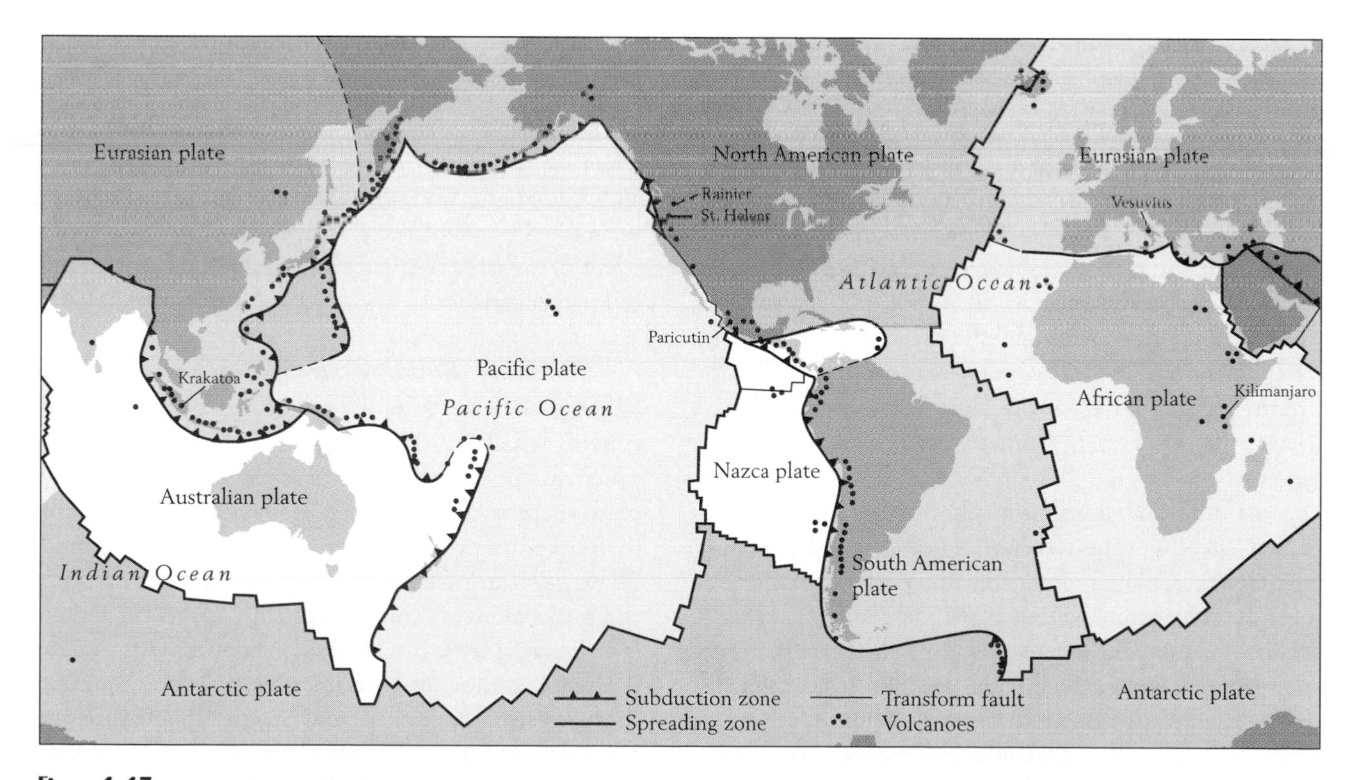

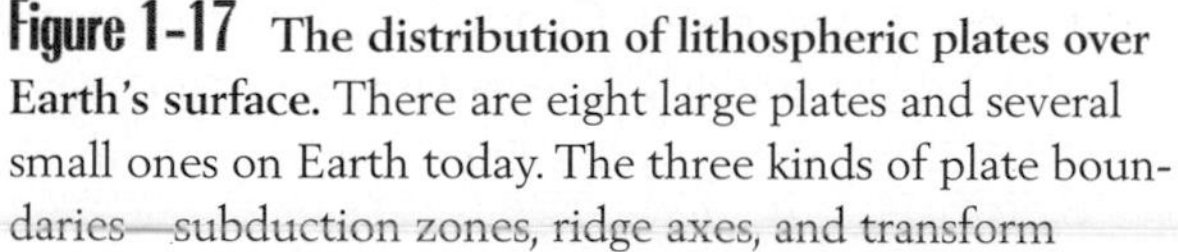

Figure 1-17 The distribution of lithospheric plates over Earth's surface. There are eight large plates and several small ones on Earth today. The three kinds of plate boundaries—subduction zones, ridge axes, and transform faults—are shown here (they will be discussed later in the chapter). Most volcanoes lie near subduction zones or ridges. *(After F. Press and R. Siever, Earth, W. H. Freeman and Company, New York, 1986.)*

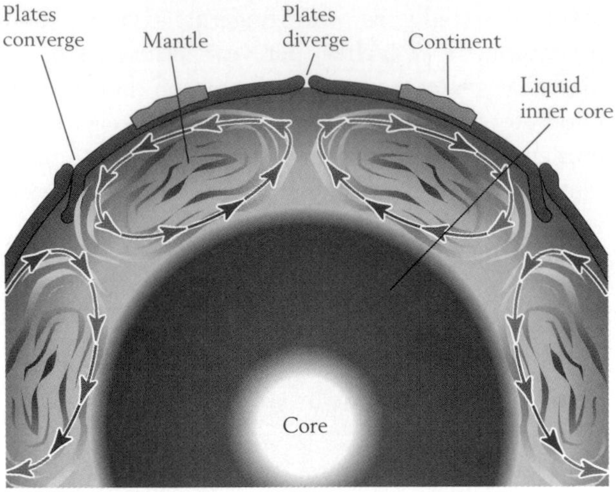

Figure 1-18 **Convection.** When a pot of water is heated from beneath, the water at the bottom warms and expands. Because of its reduced density, this water rises, and cooler, denser water from above replaces it. The result is the circular motion known as convection. Earth's mantle also experiences convection because it is heated from below, while the upper mantle loses heat to the crust.

Plates move apart along **spreading zones.** These are zones along which new lithosphere forms as mafic magma of relatively low density rises up from the ultramafic asthenosphere and cools. Heat that rises from the asthenosphere along a spreading zone swells the newly formed lithosphere to create a **mid-ocean ridge,** such as the Mid-Atlantic Ridge (Figure 1-19). The lithosphere on each side of the ridge moves laterally away from the axis. Thus two plates that are growing along a ridge are also moving apart.

At other locations, lithospheric plates move back down into the asthenosphere. They descend at deep-sea trenches, which floor the deepest zones of the oceans. The regions where plates descend are termed ***subduction zones.*** Because Earth is neither shrinking nor growing, on a global scale the total rate at which plate margins are descending into subduction zones must balance the rate at which they are growing at spreading zones.

Plates slide past each other along surfaces called ***transform faults.*** Figure 1-19 shows that these faults offset mid-ocean ridges throughout their length.

Volcanoes along Plate Boundaries Most volcanoes occur along those boundaries where plates are moving apart or together. Figure 1-19 shows the reason for this pattern. Large fractures that parallel the axis of a spreading zone, such as the Mid-Atlantic Ridge, act as conduits through which magma rises from the mantle and is added to the lithosphere. These fractures result from divergence at the spreading zone. A spreading zone stands above the seafloor as a ridge because it swells as heat flows upward along it as magma ascends. Some of the magma cools to form shallow intrusive rock and some makes its way through the lithosphere to form a zone of new seafloor along the spreading zone. In most places, the lava emerges along cracks as broad flows, but in some places it emerges through pipe-like conduits to build tall, cone-shaped volcanoes. As newly formed lithosphere moves away from a ridge, it cools and subsides, becoming part of the broad, nearly planar expanse that floors most of the deep sea.

Volcanoes are also abundant along subduction zones. There they result from partial melting of the lithosphere that is moving down into the asthenosphere. Temperature increases with depth in the planet, and a segment of descending lithosphere partly melts when it reaches a critical temperature. The resulting magma is less dense than the surrounding asthenosphere and rises. Much of this magma reaches the seafloor and erupts to form volcanoes, some of which eventually rise above the sea surface to form islands.

Mountain Building along Subduction Zones When a plate is subducted beneath continental lithosphere, which is thicker than oceanic lithosphere, much of the magma that rises from the subducted plate margin cools within the overlying lithosphere to form intrusive rock. Some of it, however, makes its way to the surface; thus volcanoes form many of the peaks along a mountain chain such as the Andes (see Figure 1-19). It is not only igneous activity that produces a mountain chain when a continent encounters a subduction zone. Under great pressure, the continental margin crumples into folds that elevate the land. As we will see in Chapter 9, the continental crust also breaks under the pressure here, and large slices of it slide along faults. These slices pile up on top or each other, thickening the crust.

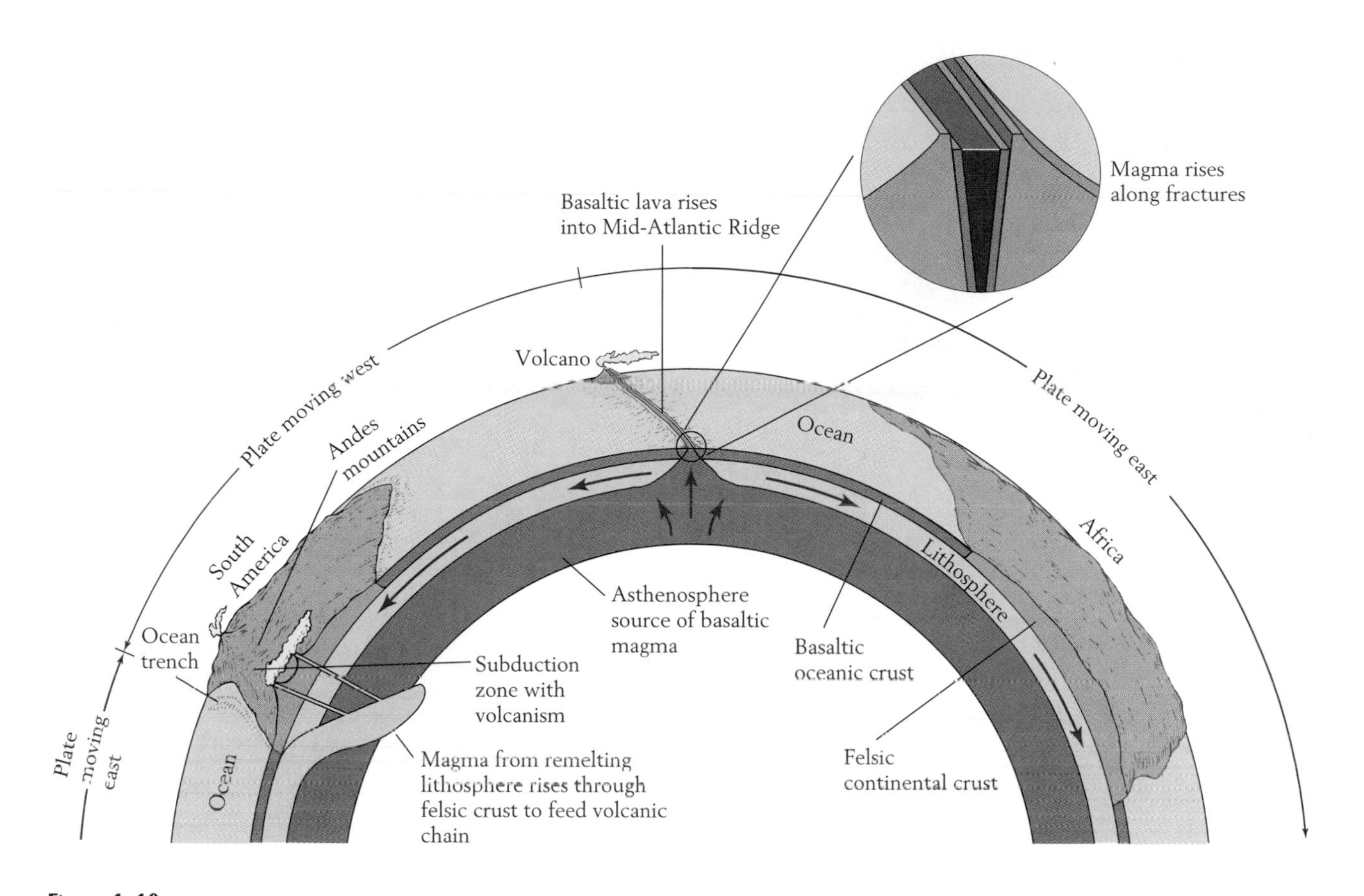

Figure 1-19 Cross section of the lithosphere and the asthenosphere in the vicinity of the South Atlantic Ocean. Note that the lithospheric plate that includes South America is moving westward from the Mid-Atlantic Ridge. At the same time, the lithospheric plate beneath the eastern Pacific Ocean is moving eastward; when it meets South America along a deep-sea trench, this oceanic plate moves downward into the asthenosphere, and the collision produces the Andes mountains. *(After F. Press and R. Siever, Earth, W. H. Freeman and Company, New York, 1986.)*

The Engine of Plate Tectonics

Heat from deep within the asthenosphere drives plate tectonic processes. It causes convection within the asthenosphere (see Figure 1-18) and supplies the magma that rises along spreading zones. For centuries, miners have known that Earth's interior is hotter than its surface, because the temperature increases as one descends through a mine shaft. Of course, hot springs and geysers, such as those of Yellowstone National Park, also give evidence of great heat deep within the planet. Late in the nineteenth century, it seemed that this was primordial heat—heat that remained after the planet cooled from a molten state early in its history. Soon after the Frenchman Henri Becquerel discovered radioactivity in 1896, however, geologists came to a new understanding of the Earth's heat budget. They learned that radioactive material is so abundant within Earth that its decay supplies an enormous amount of the planet's heat. Thus, while Earth has been gradually losing its original heat, it has been generating a large amount of new heat.

Plate Tectonics and the Rock Cycle

Plate tectonic processes exemplify the rock cycle on a grand scale (Figure 1-20). As Figure 1-20*A* illustrates, most oceanic lithosphere that forms where magma rises from the asthenosphere to form new seafloor ultimately descends into the asthenosphere again. This happens when that seafloor, having moved

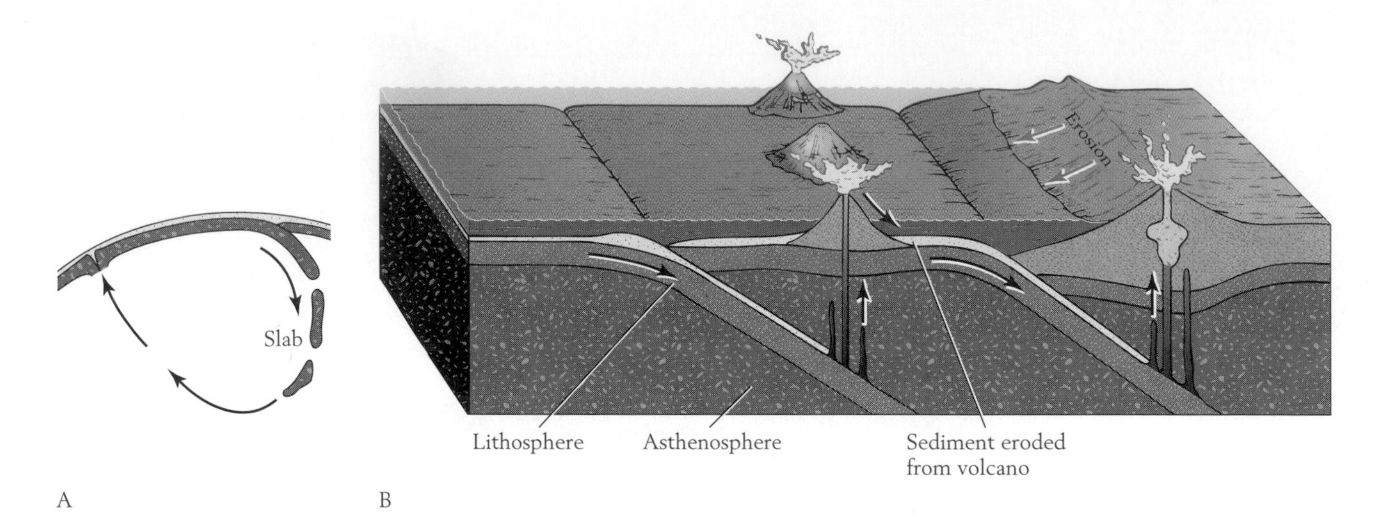

Figure 1-20 **Plate tectonics and the rock cycle.** *A.* A subducted slab of lithosphere sinks, melts, and is incorporated into the asthenosphere. Convection (see Figure 1-18) carries the combined material upward, and some of the subducted material emerges along a spreading zone as new lithosphere, completing the cycle. *B.* A shallower cycle operates where sediments are subducted and melted to contribute to magma that is extruded through volcanoes of an island arc (left). Sediment eroded from these volcanoes accumulates on the seafloor and is subducted and melted, completing a cycle. The cycling continues as melted material emerges through volcanoes along a mountain chain that borders a continent (right); erosion of the volcanic rock produces another generation of sediment.

away from the spreading zone, encounters a subduction zone. In time, the area of lithosphere that is subducted, termed a **slab,** breaks away from the plate to which it originally belonged. The isolated slab then sinks further and melts to become part of the asthenosphere. It may seem surprising that the slab sinks, because it consists partly of mafic material, whereas the asthenosphere is ultramafic, but the slab is denser than the ultramafic material because it is cold. If, perhaps hundreds of millions of years later, convection moves the melted material from the slab upward along a spreading zone, it will again serve as a source of new lithosphere and thus complete a large cycle of movement.

Figure 1-20*B* shows how plate tectonics also causes cycling of materials at shallower depths. Lithosphere that has formed along a spreading zone is eventually subducted, along with sediment. Some of the subducted material melts and returns to the surface through volcanism. Erosion of the resulting volcanoes yields new sediment, some of which is subducted and melts to produce a new generation of magma that rises to form new volcanic rock. This rock, in turn, erodes and sheds sediment, some of which is subducted. In this manner, the cycling continues.

The Water Cycle

Water that erodes rock flows over the land after falling from the atmosphere as rain or snow. In fact, water also moves over and through the earth in a complex cycle of its own. The water cycle is simpler that the rock cycle, however, in the sense that the cycled material is a single chemical compound. Water is so abundant on Earth that it is too easily taken for granted. Actually, its abundance is quite remarkable. Our planet is large so that its gravitational attraction holds water vapor in its **atmosphere,** the envelope of gases that surrounds Earth. At the same time, Earth's range of surface temperatures makes it the only planet of the solar system to have abundant liquid H_2O on its surface. Here this special liquid plays many crucial roles. Water is essential for the existence of life as we know it, since it is a major component of the cells that are the basic units of organisms. It also serves as an external medium for all those forms of life that we refer to as "aquatic." A high heat capacity is another important feature of water; in other words, water holds a large amount of heat for a given temperature. For this reason, water currents in the ocean spread heat much more effectively over Earth's surface than atmospheric winds can do.

The Reservoirs

The **water cycle** is the endless flow of H_2O between reservoirs at or near Earth's surface. Most of these reservoirs, including oceans, lakes, and rivers, contain liquid water. In addition, however, the atmosphere serves as a reservoir for water vapor, and **glaciers**—large, slowly flowing masses of ice—amount to solid reservoirs of H_2O.

Figure 1-21 illustrates the major features of the water cycle. Whereas heat within Earth drives plate movements and igneous processes, heat from the sun drives the water cycle. Solar heat evaporates water, sending water vapor into the atmosphere. It also causes the oceans and atmosphere to circulate, transporting H_2O over vast distances.

The oceans cover more than two-thirds of Earth's surface and contain more than 97 percent of the H_2O within the water cycle. It may seem surprising that glaciers contain most of the remaining H_2O: more than 2 percent. Glaciers are confined to high latitudes in lowland areas, but they occur throughout the world in mountainous regions, where the air is cool. Most of the water locked up in glaciers today occupies two ice caps, the one that covers most of Greenland and the one that covers most of Antarctica. If all of the glaciers now present on Earth were to melt, sea level would rise about 100 meters throughout the world.

Lakes, though more abundant than glaciers, contain about one-tenth of 1 percent of the H_2O of the water cycle. The atmosphere contains even less, only about one-hundredth of 1 percent, and rivers and streams account for only one-thousandth of 1 percent.

Much more water occupies pores and cracks within Earth than runs over its surface. Soil contains about 50 times as much water as do rivers and streams, and the upper 4 kilometers of the lithosphere, below the soil, contain more than 100 times as much water as the soil itself. Wells supply humans with what is known as *groundwater* from the vast lithospheric reservoir. Land plants also take up groundwater through their roots, and although they use some of this water to produce sugars for their use, they release most of it from their leaves. In this process, termed **transpiration,** plants convey moisture from subsurface reservoirs to the atmosphere, augmenting evaporation from bodies of water and moist soil. Although land plants contain only a minute fraction of the water in the water cycle, they transpire about as much water to the atmosphere

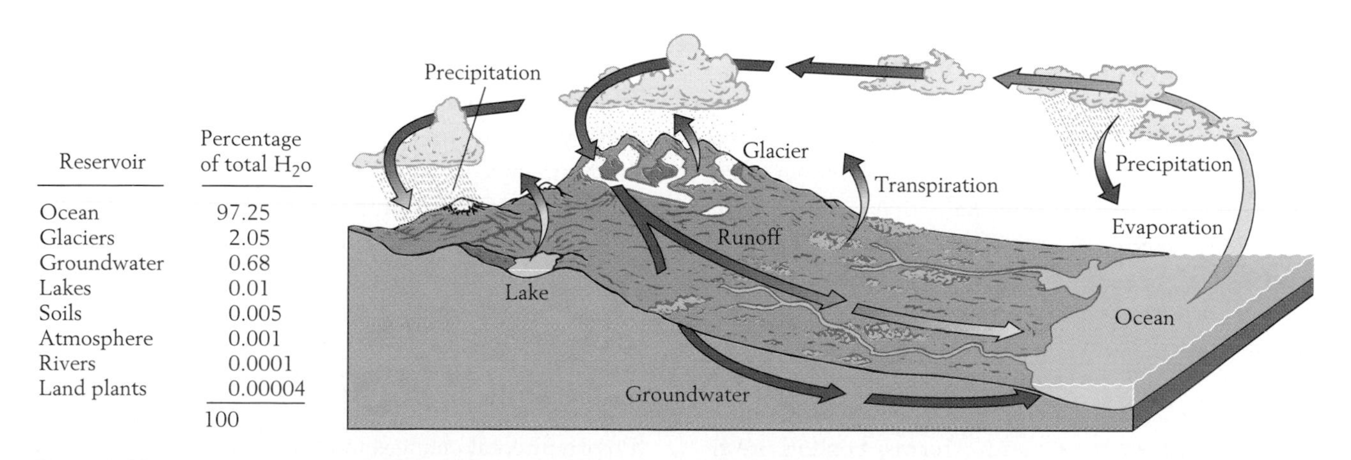

Reservoir	Percentage of total H_2O
Ocean	97.25
Glaciers	2.05
Groundwater	0.68
Lakes	0.01
Soils	0.005
Atmosphere	0.001
Rivers	0.0001
Land plants	0.00004
	100

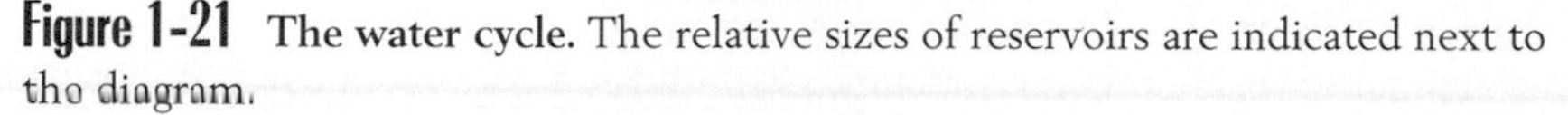

Figure 1-21 **The water cycle.** The relative sizes of reservoirs are indicated next to the diagram.

during any period of time as is discharged by all the world's rivers.

The sizes and locations of various reservoirs of the water cycle are constantly changing, and during many intervals of geologic time these reservoirs have differed markedly from their present configuration. At times, glaciers have expanded where none existed before, moist land has turned into desert, or the ocean has spread over dry land—and such changes have later been reversed or followed by equally profound changes of some other kind. The water cycle is constantly reshaping Earth, in concert with the rock cycle and plate tectonics. The study of Earth history focuses on these changes.

Interactions with the Rock Cycle

The water cycle and rock cycle are actually interconnected. In fact, water is fundamental to the rock cycle. Earth's surface is in a dynamic state not only because of plate movements and igneous activity, but also because of weathering, erosion, and deposition. In these latter processes H_2O plays a much larger role than air. Watery solutions weather rocks chemically and transport the dissolved products. Water and ice also erode rocks and sediments, and they transport the solid material released by weathering and erosion. Water contained within oceanic lithosphere and deep-sea sediments is even subducted into the asthenosphere—and then returned to the lithosphere and atmosphere by rising magma.

Water as pure as the distilled water produced in a laboratory does not exist in nature. All natural waters contain chemical constituents other than H_2O, which find their way into the hydrological cycle from such processes as weathering at Earth's surface and chemical leaching of subsurface rocks and sediments by groundwater. When natural waters evaporate, chemical components dissolved in them form what are termed salts. Salts that accumulate in layers constitute sedimentary deposits known as **evaporites** (see p. 6). Sodium chloride, which we use as table salt, is the most abundant of the evaporites. When mined from evaporite deposits, it is called rock salt. We refer to natural water as **freshwater** if it contains less than 0.5 percent salt by weight. Most lakes, like most rivers and streams, contain freshwater. In a dry climate, however, a lake from which no water flows may become quite salty. This process occurs over a long period of time as streams continually replace the water that evaporates from the lake's surface: although the inflowing waters contain only low concentrations of salt, dense concentrations eventually build up in the lake. Consider the Great Salt Lake of Utah. The salt concentration within it varies, but averages about five times that of the ocean. The ocean itself might be viewed as a giant salty lake that is billions of years old.

In Chapter 10 we will examine how chemical compounds necessary for life cycle through the lithosphere, the oceans, the atmosphere, and living things. We will also learn how changes in the abundance of these compounds profoundly affect environments and life.

Directional Change in Earth's History

Although Charles Lyell made an enormous contribution to science when he argued for uniformitarianism, his view of Earth history was flawed. Lyell viewed Earth simplistically, as a huge machine that underwent no net directional changes in the course of geologic time. In his view, the planet's changes were entirely cyclical: they were embodied in the rock cycle and the water cycle. Geologists recognize that, although these cycles are always operating, Earth has also experienced many long-term directional changes. Lyell also believed, erroneously, that life had not changed in major ways in the course of geologic time. He understood from the fossil record that species had disappeared and were somehow replaced by others, but he envisioned no progression from simple, primitive forms of life early in Earth's history to the present array of species, which includes many complex forms.

We will first examine what was wrong with Lyell's view of the history of life. Then we will review some of the physical aspects of our planet that have changed markedly in the course of geologic time.

The Evolution of Life

Environmental changes in the course of Earth history have had enormous impact on life. In fact, the word *environment* means "that which surrounds some form of life." Early in the nineteenth century, the prevailing view of literate people in Europe and North America was that, although many species of animals and plants had become extinct in the past, no species had ever given rise to another species before disappearing. According to this view, although environments on earth had undergone significant changes—at least cyclical changes—populations of plants and animals had not. Charles Darwin overturned this belief in 1859 by introducing the concept of organic evolution. In fact, he showed that life evolves even in the absence of changes in the physical environment.

Darwin offered a scientific explanation for fossil succession. Particular kinds of plants and animals, he concluded, gave rise to other kinds by a process that other scientists later termed **evolution.** Darwin reasoned that, although many forms of life have become extinct, many of them have first given rise to other forms that have outlived them. The preservation of sequences of parent and descendant species in the stratigraphic record, then, has produced the pattern of fossil succession that geologists use to date rocks.

Darwin's conception of organic evolution resulted from observations that he made as an unpaid naturalist aboard the *Beagle,* a ship that sailed around the world, making many landings along the way. Darwin embarked on the voyage in 1831, at the age of 23. He had been well tutored in geology by Adam Sedgwick, an expert on the early Paleozoic rocks and fossils of England.

The principle of uniformitarianism played a key role in Darwin's thinking. Just before he set sail on the *Beagle,* one of his teachers, J. S. Henslow, urged him to read *Principles of Geology,* Charles Lyell's new book on uniformitarianism. Henslow also cautioned Darwin not to believe Lyell's ideas, but Darwin was convinced by Lyell's arguments. On the voyage of the *Beagle,* Darwin observed processes operating in his day that clearly could have produced major features of Earth's crust. In Chile, for example, he witnessed violent volcanic eruptions in the Andes mountains, and during the period of eruption he experienced an earthquake and soon found that a strip of seafloor along the coast had abruptly risen to a position above sea level. It seemed obvious to Darwin that such processes, operating over millions of years, could produce a tall mountain range such as the Andes by elevating it a little at a time and occasionally piling volcanic rocks here and there on its surface. Lyell's uniformitarian picture of Earth's history provided Darwin with the vast stretches of geologic time required for the evolution of life.

Although Lyell's concept of uniformitarianism influenced Darwin profoundly, Darwin's concept of biological evolution differed in a significant way from Lyell's concept of Earth history. Lyell recognized that the upper part of the planet is locked in an endless rock cycle. Life also changes, Darwin concluded, but not in a cyclical fashion. Instead, it is constantly moving in new directions. These changes have introduced altogether new kinds of organisms during successive intervals of Earth history.

Physical and Chemical Trends

Not only living things but also physical and chemical features of Earth have undergone major directional changes in the course of geologic time. One of the best examples is the cooling of Earth since early in its history. Two factors have caused this decline in average temperature. First, Earth has been continuously losing the heat that was generated when it came into being. Second, because Earth's radioactive furnace has been weakening, it has been generating less and less heat to replace heat that escapes from the planet. As the radioactive materials that fire this furnace decay, fewer and fewer remain to decay at a future time; in effect, the fuel is being used up. As a result of Earth's declining temperature, plate tectonic processes have weakened in the course of geologic time. Fewer spreading and subduction zones exist today than existed early in Earth's history.

The concentration of oxygen in Earth's atmosphere has also undergone a long-term directional change. As later chapters will spell out more fully, little oxygen was present in the planet's early atmosphere. The concentration of oxygen increased substantially only after the evolution of bacteria that, like green plants, manufactured their own food by the process of photosynthesis. Oxygen is a product of photosynthesis, and in time it built up in the atmosphere to a level that allowed animals to exist. In other words, the evolution of life altered the environment of the entire planet, and the new environmental conditions led in turn to further evolutionary changes. In the interplay between life and environments we see the intimate intertwining of the physicochemical and biological histories of Earth.

The History of Ecosystems

To study the history of interactions between the biological and physicochemical aspects of our planet, it is useful to identify ecosystems. An **ecosystem** is an environment together with the group of organisms that live within it. Ecosystems come in all sizes. Earth and all the forms of life that inhabit it constitute an ecosystem, but so does a tiny droplet of water inhabited by only a few microscopic organisms. Obviously, then, large ecosystems can be divided into many smaller ecosystems, and the size of the ecosystem that is treated in a particular study depends on the type of research that is being conducted.

Modern-day ecosystems are the products of billions of years of Earth history. Even during the past few hundred million years, continents have broken apart and collided, and mountains have risen and succumbed to erosion. The ocean has repeatedly flooded vast areas of continents and receded again, and massive glaciers have spread across broad regions and melted away. The deep sea, now near freezing, has at times been much warmer. Likewise, climates on the land have warmed and cooled. For example, as we will learn in Chapter 18, 40 million years ago warm temperatures extended to Earth's poles, and palm trees grew in the northern United States; then these northern regions cooled, and they have never become so warm again. Not only has environmental change caused animals to migrate, but it has also profoundly influenced evolution and, in the course of geologic time, caused many forms of life to disappear.

As was noted at the beginning of this chapter, by reconstructing the patterns of change undergone by ancient ecosystems, geologists learn lessons that will benefit civilization as it confronts future environmental changes.

Episodic Change in Earth's History

Recall that when Charles Lyell described Earth as if it were a huge machine within which materials cycled, he portrayed the cycles as moving continuously and very slowly (p. 5). Geologists now recognize that sudden, episodic events have actually played large roles in Earth's physical and biological history.

Abrupt events have occurred on many scales. Sudden events of erosion or deposition can last only a few seconds, for example, and be confined to a few square centimeters. In contrast, the meteorite that apparently swept away the dinosaurs and many other forms of life seems to have produced this catastrophe by altering climates throughout the world within the space of a few days. On a geologic scale of time, even events that have spanned a few hundred thousand years appear abrupt.

Gaps in the Stratigraphic Record

The stratigraphic record is a great archive of past events, yet it is anything but continuous. Gaps within it reveal that sediments accumulate in pulses. We have already seen how erosion can produce unconformities in the rock record. The gap between the rocks below an unconformity and the rocks above it may represent many millions of years that are not recorded by strata. Upon close inspection, most sedimentary rocks reveal a history of discontinuous deposition on a much smaller scale (Figure 1-22). As

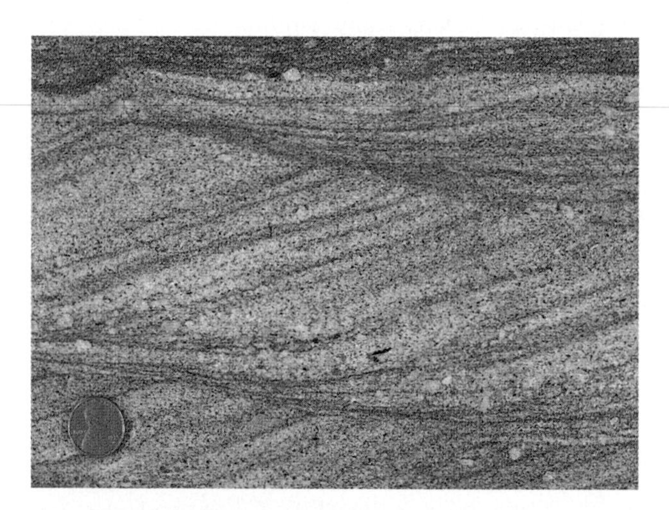

Figure 1-22 **Episodic deposition.** This portion of a sandstone displays four episodes of deposition. The penny rests on beds that represent the first episode; these beds were laid down along a surface that sloped to the left. Scouring of these beds produced the surface, just above the penny, that slopes to the right. Then additional beds accumulated along another surface that sloped to the left. After a second interruption of deposition, a third series of beds accumulated; these are nearly horizontal. After another interval of nondeposition, the dark sediment at the top of the photograph accumulated.

we have seen, strata in rocks represent discrete depositional events, and a surface that separates two such layers represents the interval between those events. Such gaps generally include not only intervals during which no sediments were deposited, but also intervals during which strata were built up but then eroded away. It is easy to picture how such a sequence of sedimentary layers can develop along a sandy beach, where the waters of broken waves wash up. Here one broken wave may deposit a new layer consisting of sand that the wave scoured from the seafloor farther offshore. Another broken wave may arrive later with greater force and scour away layers that were deposited earlier. A surface between two strata may represent a gap in deposition of only a few seconds, or if it was produced by severe scouring, it may represent hundreds or even thousands of years. So substantial are gaps within most large bodies of rock that the total time required for deposition of the preserved sedimentary layers was quite brief in comparison with the total time required for accumulation of the entire body of rock.

Figure 1-23 **A sequence of Late Cretaceous turbidite beds separated by thin shale beds in Alaska's Arctic National Wildlife Refuge.** The turbidite beds stand out because they consist of coarser sediment than the intervening shales. Each turbidite bed was deposited rapidly by a current that carried sediment from shallow to deep water.

Catastrophic Deposition

Some episodic deposition can be described as catastrophic. Hurricanes and other large storms produce strong waves and currents that erode sediment and then deposit it elsewhere. An event of this type can deposit a meter or more of sediment during a few hours or even a few minutes. Episodic deposition occurs even in deep water. Along the submarine margins of continents, for example, an event that amounts to a submarine landslide will occasionally send a slurry of sediment-laden water flowing down to the broad floor of the deep sea. There the flow spreads out, slows down, and deposits a layer of sediment that may be several centimeters thick. These so-called *turbidite* beds, about which we will learn more in Chapter 5, are often stacked one on top of another to great thicknesses (Figure 1-23).

Episodic Events in the History of Life

From time to time, individual species disappear from Earth, or undergo **extinction.** At many times in the history of the planet, extinctions have been clustered: many species have died out during brief intervals of geologic time. The largest of these events, known as **mass extinctions,** have amounted to global catastrophes in which a large percentage of the species on Earth have disappeared. The event that swept away the dinosaurs was only one of several mass extinctions that have occurred during the past half-billion years.

Major episodes of evolution have also occurred during relatively brief intervals of geologic time. Many of the basic kinds of multicellular animals that exist today evolved within just a few million years at the beginning of the Phanerozoic Eon. About half a billion years later, most of the basic groups of mammals that occupy the world today evolved within an interval of about 15 million years. Among them were the group that includes whales, the group that includes bats, and the group that includes humans.

As Chapters 12 through 19 will demonstrate, major pulses of evolutionary diversification and extinction have repeatedly transformed the fabric of life on Earth. One major task of earth scientists is to reconstruct these events and relate them to environmental changes that are recorded within ancient strata.

Chapter Summary

1 The principle of uniformitarianism, which is fundamental to natural science, asserts that the laws of nature do not vary in the course of time.

2 Actualism is the uniformitarian principle according to which events and processes of the geologic past are interpreted in light of events and processes observed in the modern world or recreated in the laboratory.

3 A mineral is an inorganic element or compound that is characterized not only by its chemical composition but also by its internal structure. Rocks are aggregates of mineral grains.

4 Igneous rocks form by the cooling of magma; sedimentary rocks are layered rocks that accumulate under the influence of gravity; metamorphic rocks form by the alteration of preexisting rocks at high temperatures and pressures.

5 The rock cycle is the endless circle in which rocks of various kinds change into rocks of other kinds.

6 The relative ages of rocks that come into contact with one another can often be determined by the principles of superposition, original horizontality, original lateral continuity, intrusive relationships, and components.

7 Fossils are remains or tangible evidence of ancient life found within rocks.

8 Changes in life on Earth during the course of geologic time are documented in the fossil record. The succession of fossils reveals the relative ages of rocks in different regions.

9 The decay of naturally occurring radioactive materials reveals the actual ages (in years) of some rocks.

10 The scale that is employed to divide the rock record into units representing discrete intervals of geologic time was developed in Europe during the nineteenth century.

11 Earthquakes and artificial explosions create seismic waves that reveal much about the structure of Earth's interior.

12 Earth's interior is divided into concentric layers. A central core of high density is surrounded by a less dense mantle, which is blanketed by a still less dense crust.

13 The parts of Earth's crust that form continents are thicker and less dense than the parts that lie beneath the oceans.

14 The crust and upper mantle constitute the rigid lithosphere, which is divided into discrete plates that move laterally over a partially molten zone of the mantle.

15 The lithosphere is divided into plates that move in relation to one another over the underlying asthenosphere. Lithosphere that underlies the ocean forms along narrow ridges and descends into Earth along deep-sea trenches. Heat from radioactive decay drives this movement.

16 Rocks break under stress along plate boundaries and also fold and contort.

17 Water moves through the upper part of Earth, the atmosphere, and bodies of water at Earth's surface in a complex cycle that is linked to the rock cycle.

18 Earth history entails not simply cyclical changes but also directional changes, including the evolution and extinction of life. Some of these changes are abrupt, even catastrophic.

Review Questions

1 Give general examples of the use of actualism to interpret ancient rocks.

2 In which of the three basic kinds of rocks do nearly all fossils occur? Why?

3 What is metamorphism?

4 Where is lithosphere formed along plate boundaries, and where does it disappear into the asthenosphere?

5 What is isostasy and how does it explain why mountains have roots?

6 What kinds of features distinguish one system of geologic time from another?

7 Name three kinds of unconformities. How does each type form?

8 Describe three kinds of relationships between two bodies of rock that indicate which of the two bodies is the younger one.

9 How do sedimentary rocks relate to igneous and metamorphic rocks in the rock cycle?

10 What drives the water cycle and how does this cycle relate to geologic, biological, and atmospheric processes near Earth's surface?

11 In what ways was Lyell's view of Earth's history flawed?

Additional Reading

Berry, W. B. N., *Growth of a Prehistoric Time Scale,* Blackwell Scientific Publications, Palo Alto, Calif., 1987.

Decker, R., and B. Decker, *Volcanoes,* 3rd ed., W. H. Freeman and Company, New York, 1998.

Erikson, J., *Plate Tectonics: Unraveling the Mysteries of the Earth,* Facts on File, New York, 1992.

Hallam, A., *Great Geological Controversies,* Oxford University Press, Oxford, 1989.

Press, F., and R. Siever, *Understanding Earth,* 2nd ed., W. H. Freeman and Company, New York, 1998.

van Andel, T. H., *New Views on an Old Planet,* Cambridge University Press, Cambridge, 1994.

2

Rock-Forming Minerals and Rocks

Rocks, the building blocks of Earth, provide evidence of conditions and events of the geologic past. The composition and configuration of rocks tell stories of mountain-building episodes, changing positions of land and sea, climatic changes, and many other aspects of Earth's history. In this chapter our emphasis is on materials that play prominent roles in the rock cycle. After we have investigated those minerals, we shall examine major groups of rocks.

The Structure of Minerals

Recall from Chapter 1 that a mineral is a naturally occurring inorganic solid element or compound (p. 5). In order to understand the properties of minerals, we need to examine the nature of atoms, the fundamental units of elements. Then we can see how atoms combine to form minerals and how minerals combine to form rocks.

Atoms, Elements, and Isotopes

At the center of an atom is a *nucleus,* in which nearly all of the atom's mass resides. Particles called *protons* and *neutrons* form the nucleus. The two particles have the same mass, which is designated *one atomic mass unit*. Protons are electrically charged,

Evaporite deposits of Oligocene age from Germany. The white and colorless crystals are halite.

Visual Overview

Rocks and Their Origins

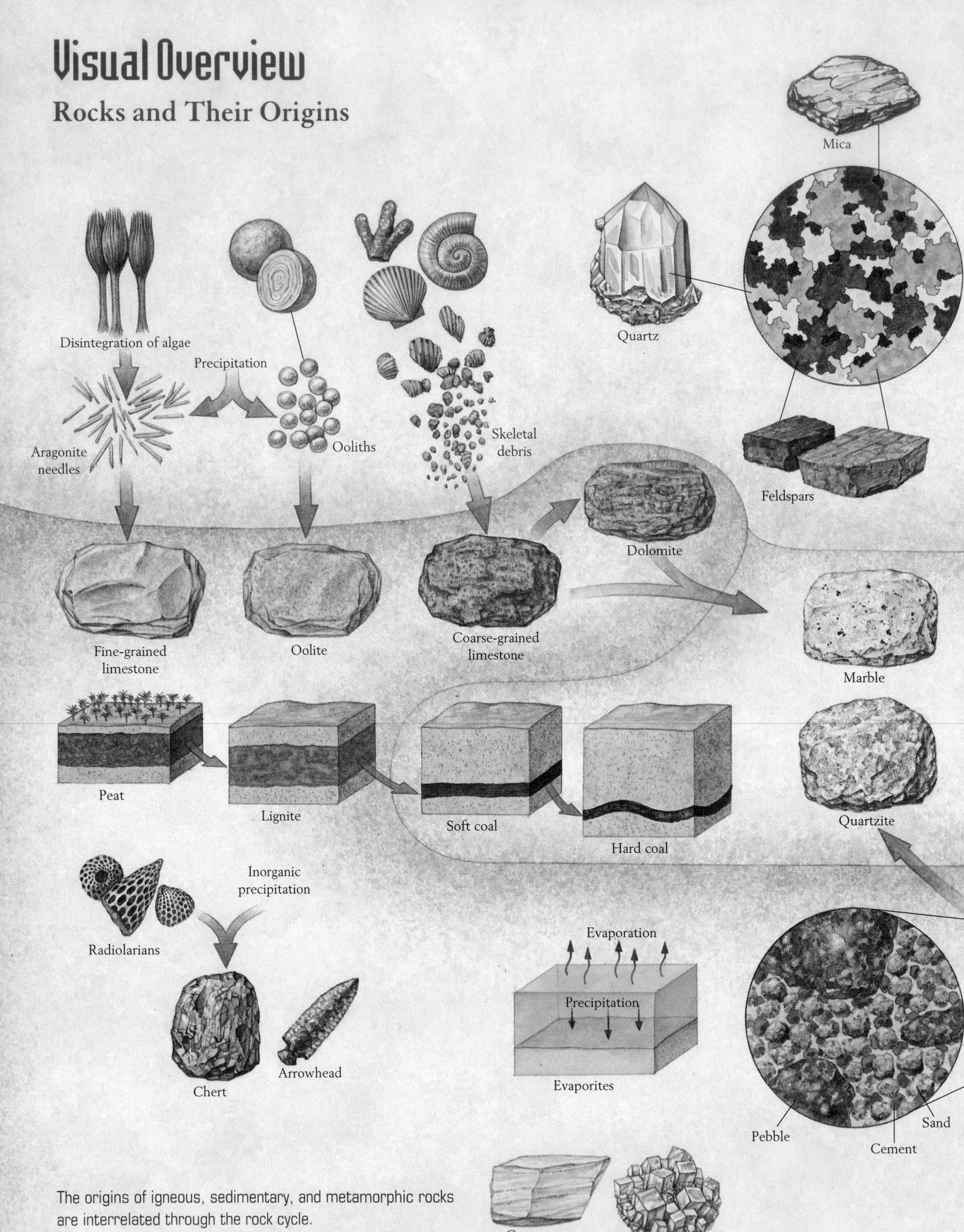

The origins of igneous, sedimentary, and metamorphic rocks are interrelated through the rock cycle.

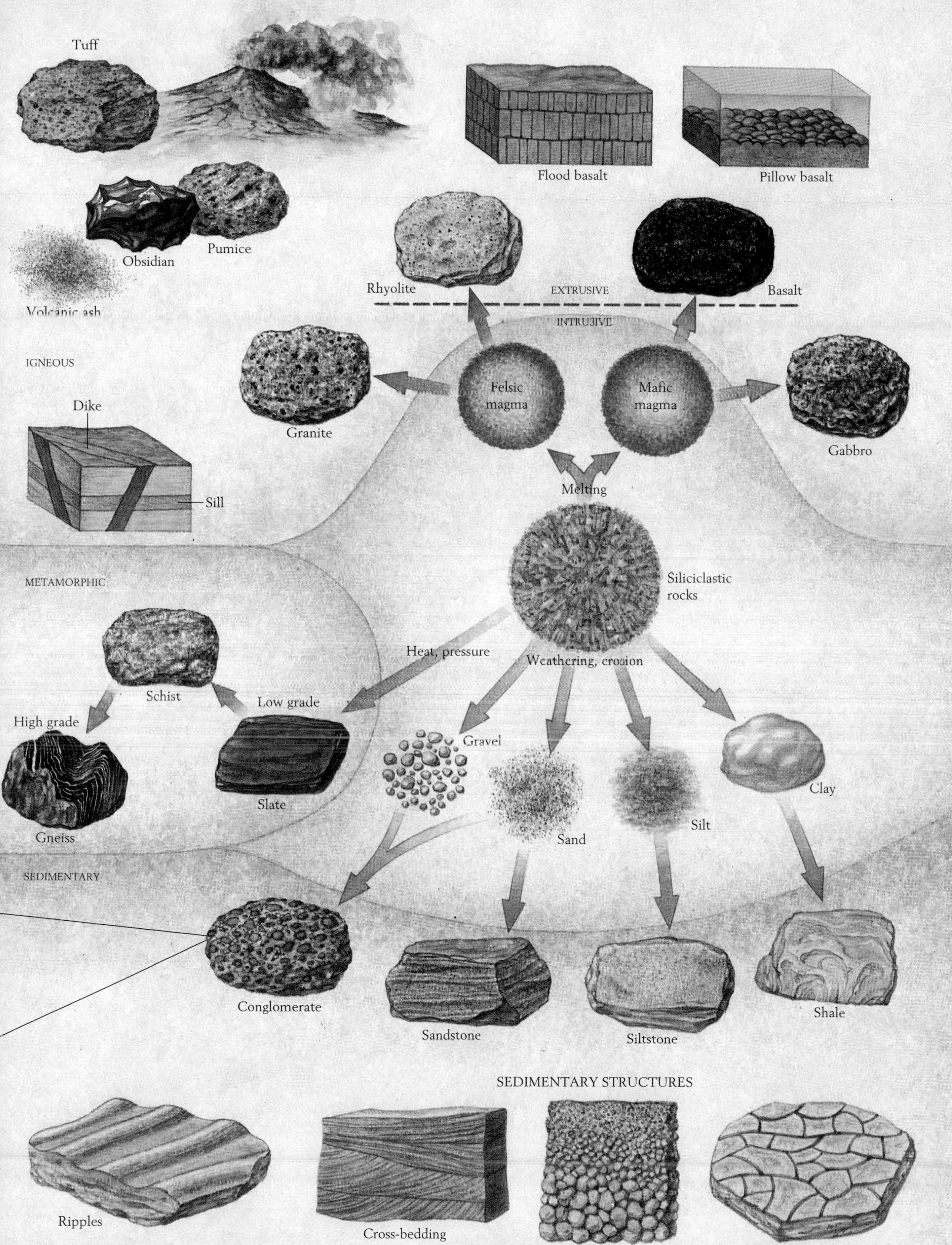
Tuff
Flood basalt
Pillow basalt
Pumice
Obsidian
Volcanic ash
Rhyolite
EXTRUSIVE
Basalt
INTRUSIVE
IGNEOUS
Felsic magma
Mafic magma
Granite
Gabbro
Dike
Sill
Melting
METAMORPHIC
Siliciclastic rocks
Heat, pressure
Weathering, erosion
Schist
Low grade
High grade
Gravel
Clay
Slate
Gneiss
Sand
Silt
SEDIMENTARY
Conglomerate
Sandstone
Siltstone
Shale
SEDIMENTARY STRUCTURES
Ripples
Cross-bedding
Graded bed
Mudcracks

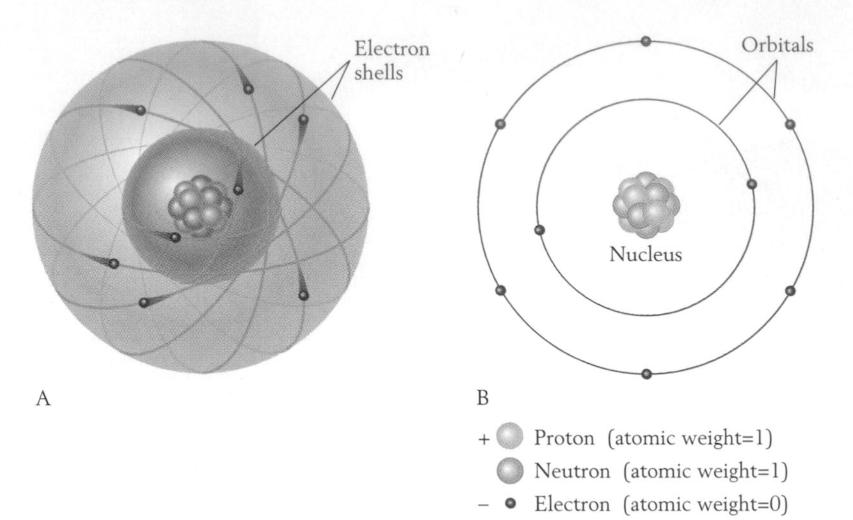

Figure 2-1 **Models of a carbon atom.** *A*. The central nucleus is surrounded by spherical zones called shells, in which the electrons move. *B*. Depiction of electrons as occupying two-dimensional circular orbitals rather than shells; protons and neutrons are clustered to form the nucleus.

each having a positive charge of +1. Neutrons, in contrast, have no charge; as their name suggests, they are electronically neutral. In orbit around the nucleus are electrons, which are negatively charged particles, each having a charge of –1. Electrons move about within spherical zones called *shells* (Figure 2-1*A*). To depict the positions of electrons graphically, however, scientists use simplified diagrams that show the electrons in circular *orbitals* around the nucleus (Figure 2-1*B*).

Every chemical element is characterized by atoms that have a particular number of protons. This is the element's unique atomic number. Hydrogen, with only one proton, has an atomic number of 1; for oxygen the number is 8, for silicon it is 14, and for iron it is 26 (Figure 2-2). As the number of electrons increases, the number of shells also increases, by steps. The innermost shell holds a maximum of two electrons, whereas the second and third shells hold a maximum of eight electrons each. Thus 8 of the 26 electrons of iron have to occupy a fourth shell.

Unlike protons, the neutrons of a given element may vary in number. Thus carbon, which has six protons, may have six, seven, or eight neutrons (Figure 2-3). The atomic masses of the protons and neutrons in a carbon atom may therefore add up to 12, 13, or 14. These are termed the *atomic weights* of the different kinds of carbon atoms. Each kind of atom, with its unique atomic weight, is called an **isotope** of its element. These carbon isotopes are designated carbon 12, carbon 13, and carbon 14.

Isotopes of certain elements have special importance in geology. Some are radioactive; that is, they decay to form other isotopes, informally termed daughter isotopes. Recall that the amount of decay that such isotopes have undergone while present in rocks allows geologists to date those rocks (p. 13). The rapid rate at which carbon 14 decays, for example, enables geologists to date carbon-bearing materials, such as wood, that are only a few hundred or a few thousand years old. Most other naturally occurring radioactive isotopes decay too slowly to produce measurable amounts of so-called daughter

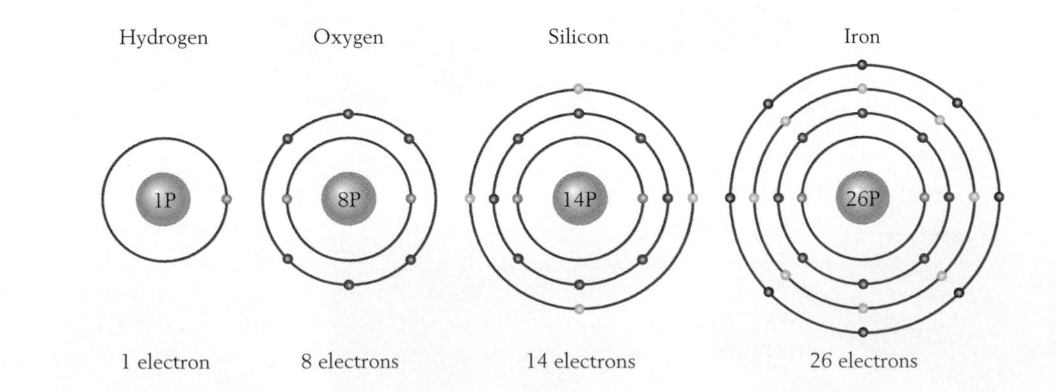

Figure 2-2 **Models of common elements, each with a different number of orbitals.**

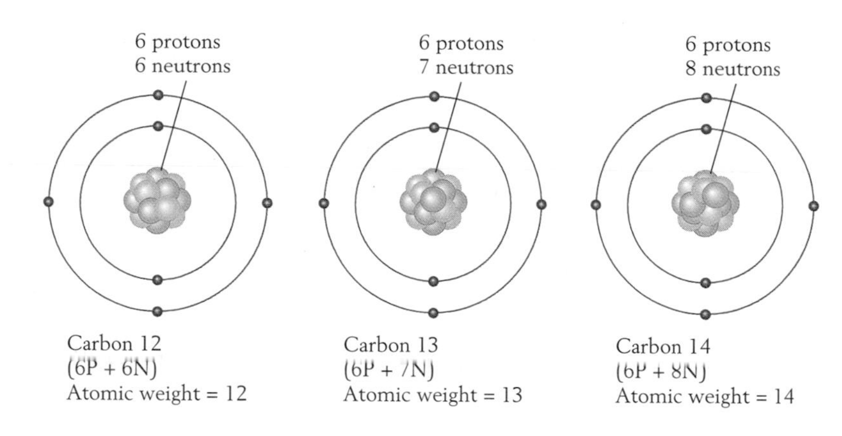

Figure 2-3 Models of three isotopes of carbon. Carbon 12 and carbon 13 are stable isotopes, whereas carbon 14 is radioactive.

isotopes during such short intervals. Some of these isotopes are widely used to date rocks hundreds of millions of years old. The total number of radioactive atoms in Earth is so enormous that their decay releases enough heat to warm the planet significantly (p. 19).

Carbon 12 and carbon 13, the nonradioactive or **stable isotopes** of carbon, are also useful to geologists. As we will see in Chapter 10, the relative abundances of these isotopes within organic matter and fossils in sediments shed light on aspects of Earth's history, including changes in the composition of Earth's atmosphere.

Chemical Reactions

In an undisturbed atom the number of electrons equals the number of protons, so that the negative and positive charges cancel out: the atom has no charge. This situation can change when an atom takes part in a chemical reaction. A *chemical reaction* occurs when two or more atoms interact to form a structure called a *molecule*. An atom of sodium and an atom of chlorine, for example, react chemically to form a molecule of sodium chloride (NaCl). This is the salt with which we flavor food and melt ice on roads. Sodium chloride occurs in nature as the mineral **halite,** informally called rock salt. Having been precipitated from ancient seas or salty lakes, halite forms large bodies of sedimentary rock that we can mine for our use (see p. 28).

Just as an atom is the basic unit of a chemical element, a molecule is the basic unit of a *chemical compound*. Most compounds have very different properties from the elements of which they are formed. Sodium, for example, is a metal, and chlorine is a gas; neither bears any resemblance to sodium chloride, which is a clear solid compound.

Chemical Bonds Chemical reactions take place through interactions between electrons of the reacting atoms. In these reactions atoms form molecules by attachments between them known as *chemical bonds.*

An *ionic bond* forms in the chemical reaction that produces a molecule of sodium chloride. In this kind of bond, one atom loses an electron to another atom. Elements such as sodium, which have one or two electrons in their outer shells, tend to lose electrons to other kinds of atoms. Sodium, for example, tends to lose its lone outer electron to chlorine. Chlorine, with an atomic number of 17, has seven electrons in its outermost (third) shell, just one short of the full complement for this shell. If a sodium atom and a chorine atom approach each other, a chemical reaction can occur in which sodium transfers an electron to chlorine (Figure 2-4). The result is a stable molecule of sodium chloride (NaCl), in which the outer shells of both the sodium and the chlorine are filled with electrons.

The electron transfer that forms an ionic bond has a substantial effect on the charges of the atoms that are bonded. Each develops a charge imbalance. In the formation of a sodium chloride molecule, for example, the sodium loses an electron, so that it is left with only ten electrons. With 11 protons remaining in the nucleus, sodium loses its electrical neutrality, ending up with a net charge of +1. In complementary fashion, chlorine develops a charge of

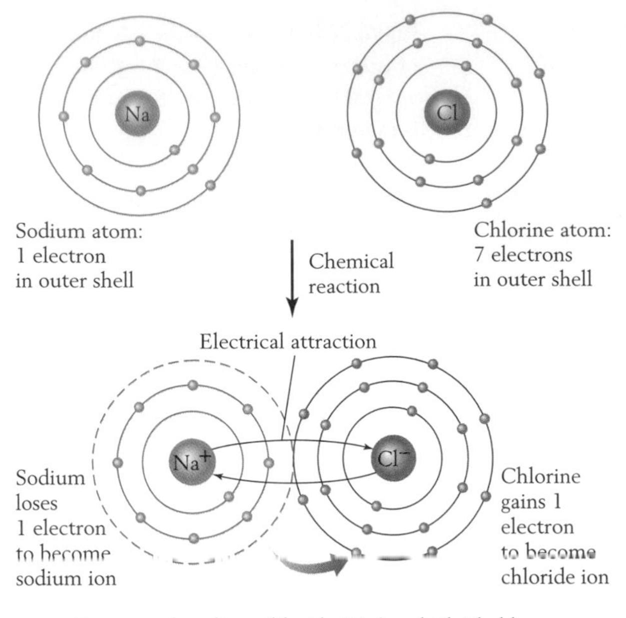

Figure 2-4 A chemical reaction forming sodium chloride. Sodium chloride (NaCl) forms by the transfer of the only electron in the outer orbital of sodium (Na) to the outer orbital of chlorine (Cl). The sodium ion (Na^+) and chloride ion (Cl^-) thus formed are held together by an electrical attraction resulting from their opposite charges.

–1 by gaining an electron. When an atom becomes charged, as sodium and chorine do when they form a molecule, it becomes an *ion*. It is the attraction between the positively charged sodium ion and the negatively charged chlorine ion that holds the molecule together.

Ions retain their identity when the compounds they form dissolve in water. When sodium chloride dissolves in seawater, for example, Na^+ and Cl^- ions are released to form a salt solution. In fact, the accumulation of these and other, less abundant ions is what makes seawater salty. When seawater evaporates in a dry climate, Na^+ and Cl^- can become too highly concentrated to remain in solution. Then Na^+ and Cl^- ions combine and salt precipitates. This is how halite deposits form in nature.

Sodium chloride displays the simplest kind of ionic bond, in that a single electron is transferred from sodium to chlorine. Many other kinds of molecules form by the transfer of more than one electron; in fact, many form by the union of three or more ions. Calcium chloride exemplifies both of these complex features; it consists of a Ca^{2+} ion that is attached to two Cl^{1-} ions.

Ions often group together to form *complex ions.* Examples are carbonate (CO_3^{2+}) and sulfate (SO_4^{2+}). Both of these complex ions are abundant in seawater, from which they can precipitate by combining with positive ions to form minerals.

Covalent Bonds Atoms can also form *covalent bonds,* in which electrons are shared rather than exchanged. The mineral diamond, which is cut to pro-

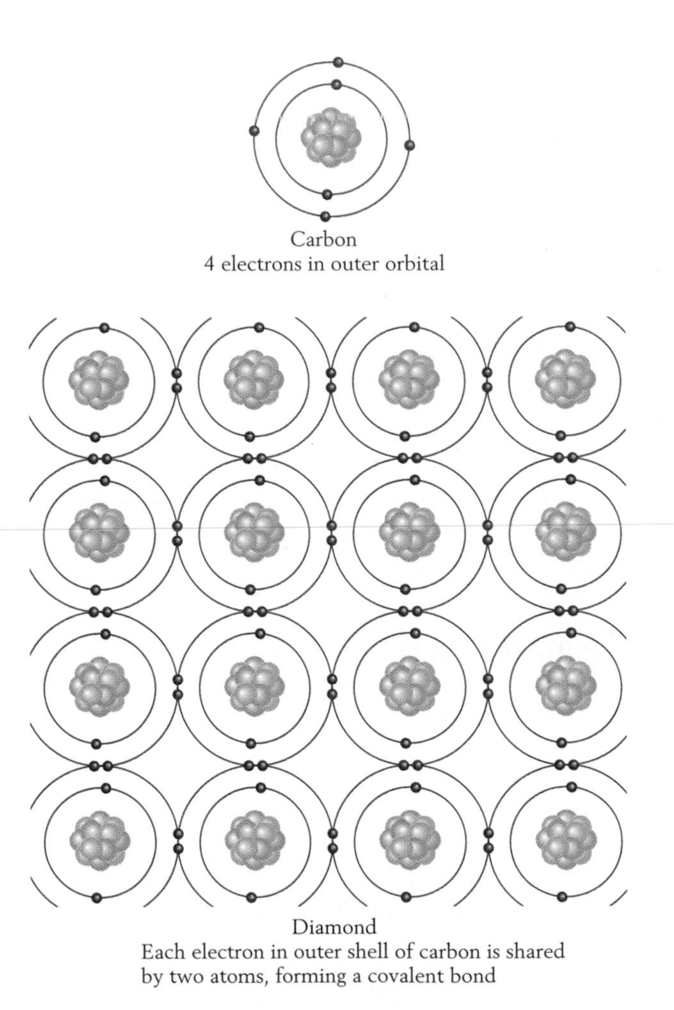

Figure 2-5 Covalent bonding in diamond. An isolated carbon atom (above) has four electrons in its outer shell. All atoms in diamond (below) share the electrons of their outer orbit equally. Each of these electrons is shared by two atoms, and each atom ends up with a total of eight electrons in its outer orbit instead of four. Because all of the electrons of the outer shell are shared, the atoms are strongly bonded, and diamond is very hard.

duce gems, consists entirely of carbon atoms united by covalent bonds. An isolated carbon atom, with an atomic number of 6, has four electrons in its outermost (second) shell. When carbon forms diamond, each of the electrons in its outer shell is shared by two atoms (Figure 2-5). Thus the number of electrons in the outer shell of each atom is doubled, and now it has the full complement of eight electrons there. The result is a stable molecular structure in which every carbon atom shares two electrons with each of four neighboring atoms.

Crystal Lattices

An ion or uncharged atom behaves like a sphere, with a diameter that varies from element to element. For ions to form a stable compound, not only must their ionic charges be in balance, but the ions must have relative diameters that allow them to fit together. In the mineral halite, for example, the small sodium ions fit comfortably between larger chloride ions (Figure 2- 6*A*).

The three-dimensional molecular structure of a mineral is known as a *crystal lattice.* The configuration of a particular mineral's crystal lattice reflects the relative sizes and numbers of the various kinds of ions that make up the mineral. The external form of the mineral, in turn, reflects the configuration of the mineral's crystal lattice. The crystal lattice of halite, for example, consists of equal numbers of Na^+ and Cl^- ions packed closely together in a regular pattern. It is easy to see how this pattern is reflected in the cubic shape that is the most common one for crystals of halite (Figure 2-6*B*).

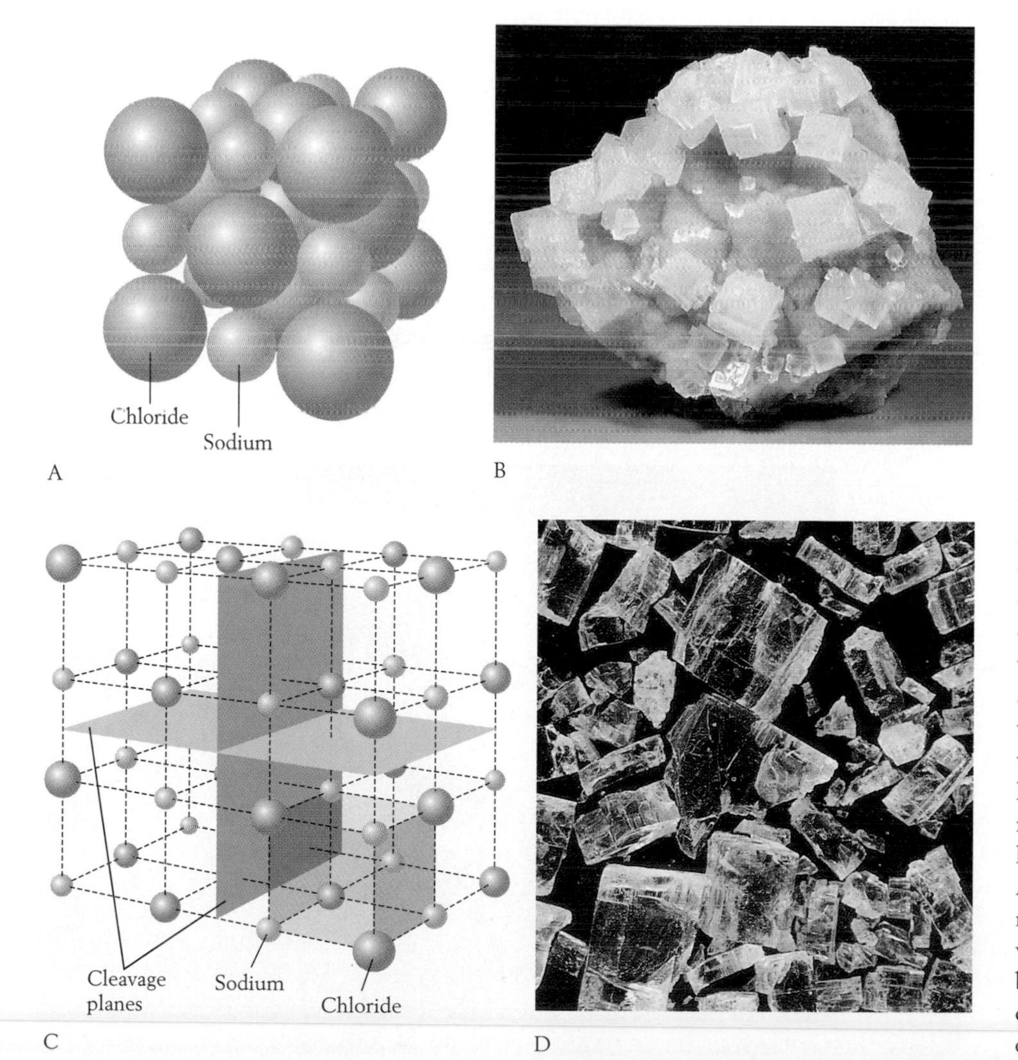

Figure 2-6 Sodium chloride (halite). *A.* Sodium and chloride ions are packed together in equal numbers, forming a cubic structure. *B.* Crystals of halite, whose cubic shape reflects the crystal structure of sodium chloride. C. Shown more widely separated, the sodium and chloride ions can be seen to form a crystal lattice in which ions form parallel layers. These layers are separated by planes of weakness, known as cleavage planes. *D.* Table salt consists of tiny rectangular solids formed when halite is crushed and breaks along cleavage planes oriented at right angles to one another.

Figure 2-7 Calcite and aragonite, two forms of calcium carbonate ($CaCO_3$). *A.* Calcite crystals. *B.* Aragonite crystals in the form of very small needles; needles like these form carbonate mud on shallow tropical seafloors.

Two or more minerals can have identical chemical compositions but altogether different crystal lattices. **Calcite** and **aragonite,** for example, are common minerals of this kind that consist of calcium carbonate ($CaCO_3$). Both of these minerals precipitate from watery solutions in nature, and a variety of organisms secrete one or both in the form of a skeleton. Calcite forms blocky or tooth-shaped crystals (Figure 2-7*A*). Aragonite precipitates directly from shallow tropical seas as tiny needle-shaped crystals that accumulate as what is termed **carbonate mud** (Figure 2-7*B*). This mud and the aragonitic and calcitic skeletons of marine animals are major components of sediment that hardens to form limestone. Limestones more than a few million years old, however, consist almost entirely of calcite, because aragonite is not stable: over time, it alters to calcite. Corals that have lived during the past 250 million years have secreted skeletons of aragonite, but most of these skeletons that have been preserved as fossils were long ago altered to calcite (Figure 2-8). This alteration is often brought about by watery solutions that destroy some of the original features of the coral skeleton.

Ion Substitution

By definition, a particular mineral can have a specified range of chemical compositions (p. 5). The reason for this latitude in the definition of a mineral is that many minerals have crystal lattices in which a small quantity of a particular ion can substitute for another ion without significantly altering the mineral's physical or chemical properties. Again, calcium carbonate serves to illustrate the point.

Two stable isotopes of strontium occur in Earth's crust and in the ocean. Because strontium ions are present in seawater, some of them find their way into skeletons of calcium carbonate secreted by marine organisms. The relative amounts of these two isotopes in seawater have changed during Earth's history. Therefore, as Chapter 6 will explain more fully, the isotopic ratio of strontium in a fossil can be used to establish the age of the fossil.

Figure 2-8 Coral skeletons. On the left is the cross section of a present-day coral skeleton that consists of aragonite. On the right is a piece of fossil coral of the same species that is only a few hundred thousand years old but that has been largely altered to calcite; secondary calcite also fills many of the pores of the original skeleton.

The Properties of Minerals

The molecular structure of a mineral determines many properties that influence its physical and chemical behavior in nature. Densities of minerals influence whether they are found in Earth's crust or mantle, for example, and crystal lattice structures influence how they break and how readily they abrade when they roll about as isolated grains. Table 2-1 shows the chemical and physical properties of the major mineral groups.

Hardness

Molecular bonds within minerals vary greatly in strength. The covalent bond structure of diamonds, in which each electron of the outer shell is shared by two atoms, is so strong that diamonds are the hardest of all minerals (see Figure 2-5). The powerful covalent bonding that forms diamonds can develop only under very high pressure; that is why diamonds form primarily within Earth's mantle and only rarely are elevated to its surface. Very small diamond crystals can also form when a large meteorite strikes

Table 2-1

Major mineral groups

	Chemical properties	Physical properties	Rock-forming contribution	Comments
Silicates	SiO_4 tetrahedra are the basic units	Mostly hard, except for mica and clay minerals; most have a glassy or pearly luster	Dominant mineral group in igneous, sedimentary, and metamorphic rocks	Most crystallize at high temperatures and occur in sediments only as detritus
Carbonates	Positive ions attached to CO_3	Soft, light-colored	Mostly sedimentary, but also form marble, a metamorphic rock	Include calcite, aragonite, and dolomite
Sulfates	Positive ions attached to SO_4	Soft, light-colored, water-soluble	Most rock-forming varieties are sedimentary	Form large sedimentary evaporite deposits, including gypsum and anhydrite
Halides	Positive ions attached to negative ions of elements such as chlorine (Cl) and bromine (Br)	Soft, light-colored, water-soluble	Most rock-forming varieties are sedimentary	Form large sedimentary deposits, including halite (rock salt)
Oxides	Metallic ions combined with oxygen	Soft to hard	Mostly sedimentary, but many varieties are present in igneous and metamorphic rocks	Some, including the iron materials magnetite and hematite, are major ore minerals
Sulfides	Metallic ions combined with sulfur	Soft to medium hard, often with a metallic luster	Have a minor role in rock forming	Many are important ore minerals that form at high temperatures

Earth from outer space, creating enormous pressures at impact.

Graphite, like diamond, is a mineral consisting of pure carbon, but some of the bonds within the crystal lattice of graphite are very weak. In fact, graphite is so soft that even paper can abrade it; that is why we can use it as what we call "lead" in our pencils. Graphite forms in Earth's crust, at much lower pressures than are required to produce diamonds in the mantle.

Density

Density is the weight of a given volume of any substance. The density of a mineral—often expressed as the number of grams per cubic centimeter (g/cm^3)—depends on two things: the atomic weights of the atoms that form the mineral and the degree to which those atoms are packed together. Iron, for example, with an atomic number of 26, has a greater atomic weight than those of many other rock-forming elements. For this reason, minerals that contain iron tend to be relatively dense. This is why such minerals are relatively abundant in Earth's mantle, on which the less dense crust floats (p. 15). Iron tended to sink deep within Earth early in its history, when the planet was in a hot, liquid state, and in fact it became the dominant element of Earth's core.

We can see how atomic packing increases a mineral's density by comparing diamond and graphite. The tight packing of diamond's carbon atoms, which results from its formation at high pressures deep within the Earth, gives it a density of 3.5 g/cm^3; graphite, which forms under lower pressures, has a density of only 2.1 g/cm^3.

Fracture Patterns

Weak bonding within a mineral's crystal lattice can create parallel planes of weakness along which the mineral tends to break. The relatively simple crystal structure of halite, for example, not only produces cubic crystals, but also creates planes of weakness known as cleavage planes (see Figure 2-6*C*). For this reason, when halite is shattered, it tends to break into fragments shaped like a typical crystal (see Figure 2-6*D*). If you observe several grains from a salt shaker through a strong magnifying glass, you will see that many of them are tiny cubes. These cubes were formed when larger chunks of halite were crushed.

The Origins of Minerals

The particular composition and internal structure of any mineral are determined by the conditions under which it was formed. We have seen that diamond, for example, forms at much higher pressures than graphite, although both consist solely of carbon. We have also noted that iron-rich minerals are rarer in Earth's crust than in its mantle, where elements of high atomic weight are concentrated. In lakes and the ocean, the only minerals that can precipitate are ones such as halite and aragonite that are soluble in water.

Tracing the Origins of Rocks

The constraints on mineral origins allow a geologist to identify the setting where a particular body of rock originated. Sometimes an individual mineral tells the story, and sometimes it is the entire suite of minerals within the rock. Minerals within rocks found at the top of a mountain, for example, may reveal that these rocks originated deep below Earth's surface at high temperatures and pressures; such rocks have obviously been uplifted to their present elevation by the forces that formed the mountain. As another example, the oldest sedimentary rocks that contain abundant iron minerals that are rich in oxygen have been taken to indicate when oxygen first built up to relatively high levels in Earth's atmosphere. These rocks are about 2 million years old.

Families of Rock-Forming Minerals

The **silicates** are the most abundant group of minerals in Earth's crust and mantle. In silicates, four negatively charged oxygen atoms form a tetrahedral structure around a smaller, positively charged silicon atom. Figure 2-9 illustrates how silicate tetrahedra unite in various ways, usually with other atoms, to form the most prominent silicate minerals of Earth's crust. Most silicate minerals form at high temperatures. Silicates, including feldspar and quartz, are the primary constituents of the igneous and metamorphic rocks of Earth's crust (p. 5). Because many silicate grains survive the weathering of these rocks and accumulate as sediments, most sedimentary rocks also consist primarily of silicate minerals.

Carbonate and **sulfate minerals** also play large roles in the formation of rocks, but, unlike silicates, most of these minerals form at low temperatures

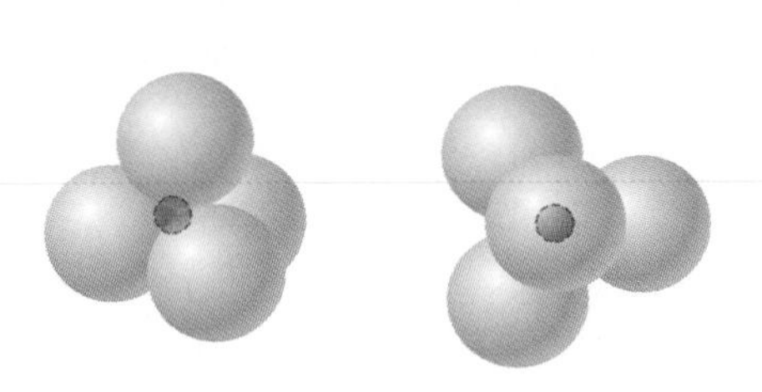

Two views of a tetrahedron of four oxygens with a silicon hidden in the center

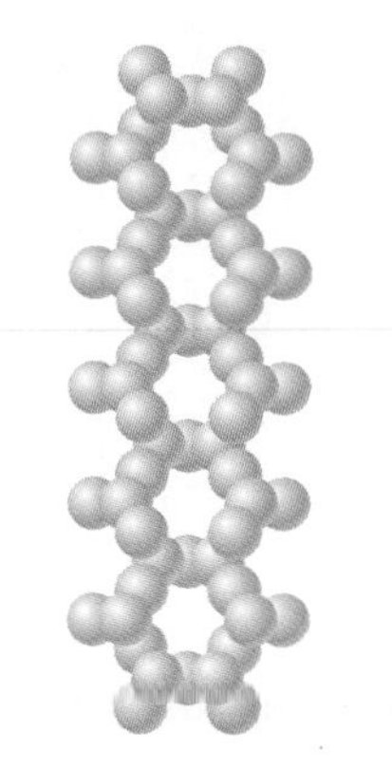

Amphibole (double chain)

Mica

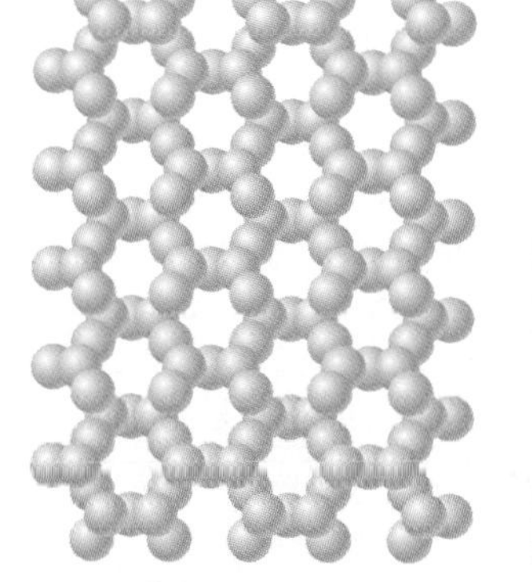

Clay

Very complex crystal lattices

Quartz

Pink feldspar

Figure 2-9 **Rock-forming silicate minerals.** Diagrams show the arrangements of tetrahedra in five rock-forming mineral groups. (Other ions are omitted for simplicity.) In amphiboles and pyroxenes, the silicate tetrahedra are assembled into long chains that are bonded together by ions of iron, calcium, or magnesium positioned between them. The iron and magnesium make these minerals dense and often dark. In mica and clay minerals, the silicate tetrahedra are more fully connected to form two-dimensional sheets that are bonded together by sheets of aluminum, iron, magnesium, or potassium. Because the bonds between these sheet silicates are weak, micas and clays cleave into thin flakes. Clay minerals are especially weak and almost always occur naturally as small flakes.

The feldspars are the most common group of minerals in Earth's crust, and quartz is second. Quartz is the simplest silicate mineral in chemical composition, consisting of nothing but interlocking silicate tetrahedra. Because each oxygen is shared by two adjacent tetrahedra, the mineral as a whole has only two times rather than four times as many oxygens as silicons (the ratio in a single tetrahedron). Hence the chemical formula of quartz is SiO_2. Quartz is very hard because its silicons and oxygens are tightly bonded.

Feldspars differ from quartz in that their structure includes both silicate tetrahedra and tetrahedra in which aluminum takes the place of silicon. Ions of one or more additional types (potassium, sodium, and calcium) also fit into the framework in varying proportions. Unlike quartz, feldspars display good cleavage. Feldspars are slightly softer than quartz, and they also weather chemically much more readily in nature.

near Earth's surface. Carbonates are constructed of one or more positive ions, such as calcium, magnesium, and iron, bonded to the complex ion CO_3^{2+}. We have already encountered the carbonate minerals calcite and aragonite, which are two forms of $CaCO_3$ with different crystal structures. **Dolomite** resembles calcite, but half of the calcium ions are replaced by magnesium. Unlike calcite and aragonite, dolomite is not secreted by any organism in the form of a skeleton.

Sulfates are formed of positive ions (such as calcium, iron, or strontium) that are attached to the complex ion SO_4^{2+}. As we will see later in this chapter, many sulfates also form at low temperatures near Earth's surface through the evaporation of ocean or lake water.

Although **oxides** make up only a small percentage of the large bodies of rock on Earth, these minerals form many important ore deposits. Rocks whose primary components are the oxides magnetite (Fe_3O_4) and hematite (Fe_2O_3), for example, yield most of the iron that is put to human use.

Types of Rocks

Rocks are classified on the basis of their size, their composition, and the arrangement of their constituent grains. Recall that a body of rock belonging to any of the three fundamental groups of rocks—igneous, sedimentary, and metamorphic—can be transformed into another body of rock of the same group or either of the other two (see Figures 1-7 and 1-9). Igneous, sedimentary, and metamorphic rocks can all weather and erode to produce sediment that accumulates to form a new body of sedimentary rock. Similarly, each of the three kinds of rock, when buried deep within the earth, can melt into magma that later cools to form igneous rock. Finally, each of the three kinds of rock can be altered by high pressures and temperatures to form metamorphic rock. When a metamorphic rock is remetamorphosed in this way, the result will usually be a new kind of metamorphic rock, because the new pressure and temperature conditions will almost always differ from the ones that formed the original rock.

Igneous Rocks

Igneous rocks are classified according to their chemical composition and grain size, both of which reflect a rock's mode of origin.

Composition and Density Recall from Chapter 1 that most igneous rocks of Earth's crust fall into either of two major groups—felsic or mafic. Felsic rocks, being rich in silicon and aluminum, are generally light-colored and of low density. For this reason, continental crust (see Figure 1-15) is predominantly felsic. Granite is the most abundant felsic igneous rock (Figure 2-10). Two kinds of feldspar constitute about 60 percent of its volume; one of them gives

Figure 2-10 **Common igneous rocks.** Basalt (upper left) and rhyolite (lower left) are mafic and felsic extrusive rocks, respectively. Gabbro (upper right) and granite (lower right) are the intrusive equivalents.

many bodies of granite a pink color (see Figures 1-2 and 2-9). Granite and other felsic rocks also contain large percentages of quartz. Quartz is the silicate mineral with the highest concentration of silica because, as SiO_2, it consists entirely of silica and oxygen. Feldspar is also relatively rich in silica, although it also contains aluminum and either potassium, sodium, or calcium, or a mixture of sodium and calcium. Mafic rocks, in contrast, are relatively low in silicon and contain no quartz. An abundance of magnesium and iron makes mafic rocks, such as **gabbro** and **basalt** (see Figure 2-10), darker and heavier than felsic rocks. Basalt forms most of the oceanic crust, while ultramafic rocks, which are even lower in silicon, form the mantle below the crust. Polished slabs of gabbro are sometimes sold commercially under the misleading name "black granite;" in fact, no granite is black.

Cooling Rate and Grain Size Igneous rocks are also classified according to grain size. This classification system is especially useful because grain size reflects the rate at which igneous rocks cooled from a molten state. If a rock cools slowly, deep within the crust, its crystals can grow large, thus producing a coarse-grained rock. In contrast, rapid cooling, which takes place at Earth's surface, "freezes" molten rock into small crystals that yield a fine-grained rock (see Figure 2-10).

Most molten rock that cools within Earth's crust or at its surface comes from the mantle; as this molten rock rises, in the form of a blob or plume, it melts crustal rock with which it comes in contact. Molten rock found within Earth is known as **magma.**

The coarse-grained bodies of rock that magma forms when it cools within Earth are referred to as **intrusions,** because they often displace or melt their way into preexisting rocks. They are also called **plutons. Sills** are sheetlike or tabular plutons that have been injected between sedimentary layers, and **dikes** are similarly shaped plutons that cut upward through sedimentary layers or crystalline rocks (Figure 2-11).

Molten rock that appears at Earth's surface through an opening, or **vent,** is called **lava.** Lava that has cooled to form solid rock often exhibits "frozen" flow structures on its surface (Figure 2-12). Some lavas erupt from tube-shaped vents to build cone-shaped volcanoes (Figure 2-11). One such structure is Mount St. Helens, which erupted in the state of

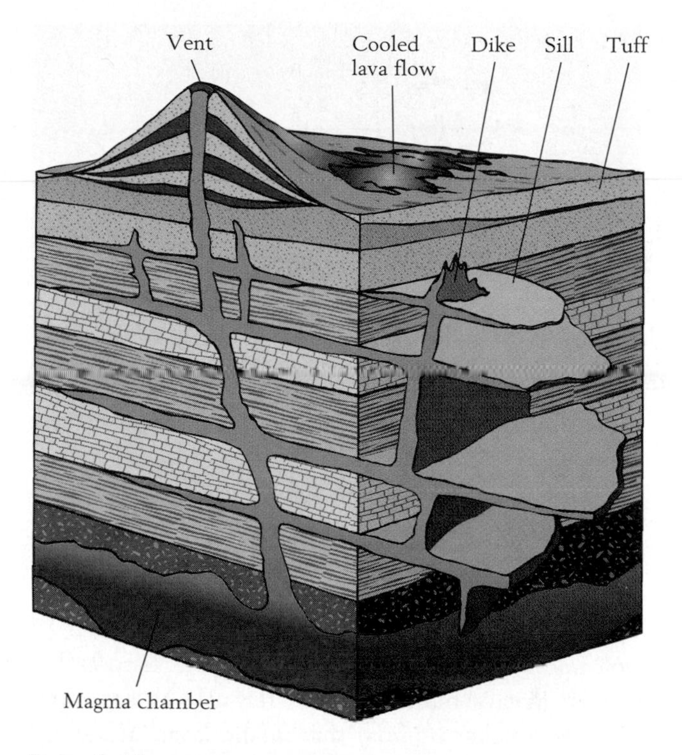

Figure 2-11 **Configurations of extrusive and intrusive bodies of igneous rock on a continent.** Magma in large chambers can cool to form massive plutons. Magma that rises through a steeply sloping crack cools to form a dike. Magma that is injected between sedimentary strata cools to form a sill.

Washington in 1980. A hollow crater forms at the summit of most volcanoes after an eruption, when the unerupted lava sinks back down into the vent and hardens. Volcanic rocks can also form in ways that do not involve the cooling of flowing lava. Some volcanic eruptions—including that of Mount St. Helens in 1980—are explosive, hurling solid fragments of previously formed volcanic rock great distances. These fragments range in size from dust to blocks several meters across. Loose debris of various sizes settles to form rock known as **tuff.** Although tuff is deposited in the same manner as sedimentary rock, it is usually classified as volcanic simply because it consists of volcanic particles. In fact, some tuffs form from hot grains that melt together as they settle, but others harden only after water percolating through them precipitates cement. Other volcanic rocks form simply by flowing out of cracks, or **fissures,** from which they spread over large areas. The

Figure 2-12 **Recently cooled lava in Hawaii.** The surface of the rock exhibits a ropy structure that reflects the pattern of the lava flow.

volcanic rocks that spread widely from fissures are almost always mafic, because felsic lavas are more viscous (thicker liquids) than mafic lavas. Mafic extrusive rocks that have flowed widely are often referred to as **flood basalts.** A flood basalt forms the broad Columbia Plateau in the northwestern United States (Figure 2-13). Another, in Siberia, was extruded at virtually the same time as the biggest global extinction of the Phanerozoic Eon. As we will see in Chapter 15, this volcanic event may have been associated with that extinction.

Lava that emerges from the crust beneath the sea cools rapidly in a way that gives its surface a hummocky configuration, creating rock known as **pillow basalt** (Figure 2-14). Thus pillow structures in ancient basalt indicate that it cooled under water rather than on the land.

Rocky material is not all that issues from volcanic vents. Gases of many kinds are also emitted from volcanoes, and in areas such as Yellowstone National Park, steamy geysers shoot up from sites where underground water is heated against magma and vaporized.

Sediments and Sedimentary Rocks

Most sedimentary rocks are formed of particles that belong to one of three categories: (1) fragments produced by the weathering and erosion of other rocks, (2) crystals precipitated from seawater, or (3) skeletal debris from organisms.

Fragments of rock produced by destructive processes are termed **clasts. Siliciclastic rocks,** then, are sedimentary rocks composed of clasts of silicate minerals.

Erosion and Sediment Production As we have seen, rocks at or near Earth's surface can be broken down by chemical reactions or physical forces. The products of this decay ultimately move away from the site of origin under the influence of gravity, wind, water, or ice. As we noted in Chapter 1, *erosion* is the group of processes that break down rock and transport the resulting products. Recall, too, that *weather-*

Figure 2-13 **Flows of Miocene basalt along the Columbia River near Vantage, Washington.** The cliff is about 330 meters (1100 feet) high.

Figure 2-14 Pillow basalts that formed on Española, one of the Galápagos islands, more than 3 million years ago, when lava cooled beneath the sea. Most of the "pillows" are less than 1 meter (3 feet) across.

ing is erosion that takes place before transport. There are physical as well as chemical weathering processes. Ice, snow, water, and earth movements are the primary agents of physical weathering. Water expands when it freezes, and when it freezes in cracks and crevices within rocks, it exerts such tremendous pressure that it can split the rocks apart.

Destructive chemical processes constitute the most pervasive kind of weathering, and water and watery solutions act as its primary agents. Water at Earth's surface readily converts feldspar to clay, for example, carrying away some ions in the process. Like micas, clay minerals are sheet silicates (see Figure 2-9). Clay differs from mica, however, in that the molecular structure of its sheets is weak. Thus clay minerals form only very small flakes, instead of large sheets like those that typify mica; this is why clay sediments have fine-grained textures.

Quartz, in contrast to feldspar, is quite resistant to weathering. This characteristic, and the resistance of quartz to abrasion, accounts for the abundance of sand on Earth's surface. Most sand consists of quartz grains that in size are similar to or slightly smaller than the quartz grains of granites and other crystalline rocks, from which most of them are derived.

When the feldspar grains of a crystalline rock such as granite weather to clay, the rock crumbles, releasing both flakes of clay and grains of quartz as sedimentary particles. Rainfall or the meltwaters of snow or ice wash many of these particles into streams, from which they are carried to larger bodies of water. Eventually many of the particles settle from the waters of rivers, lakes, or oceans as sediment, which in time may become hard sedimentary rock.

Both water and oxygen take part in the weathering of mafic rocks, converting these iron-rich minerals to iron oxide minerals that resemble rust. In the process, silica is carried away in solution. Mafic minerals are generally less stable at Earth's surface than felsic minerals and therefore undergo more rapid chemical weathering. Mafic minerals are not abundant in most beach sands that fringe oceans, because most of these minerals weather to crumbly iron oxide soon after they are exposed to air and water. Feldspars, too, are rarely found on sandy beaches, because most turn to clay either within or close to their parent rocks.

Siliciclastic Sedimentary Rocks Siliciclastic sedimentary particles and rocks are classified according to grain size (Figure 2-15). The term **clay** is applied to particles that are smaller than $\frac{1}{256}$ millimeter. Although this quantitative definition may seem redundant and contradictory, because clays also constitute a family of sheet silicate minerals, the use of the term *clay* in reference to both mineralogy and particle size seldom creates confusion: nearly all clay mineral particles are of clay size and nearly all particles of clay size are clay minerals. Silt includes particles in the next category, with diameters between $\frac{1}{256}$ and $\frac{1}{16}$ millimeters. **Mud** is a term that embraces aggregates of clay, silt, or some combination of clay and silt. Those that consist largely of clay exhibit a tendency to break along bedding surfaces; they are said to be **fissile** and are called **shale.** Fissility results from the tendency of flakes to align horizontally during their deposition (Figure 2-16). Rocks formed largely of mud are termed **mudstones.** The presence of silt along with clay makes mudstone less fissile than shale.

Material of **sand** size ranges from $\frac{1}{16}$ millimeter to 2 millimeters in diameter. When sand is cemented, it becomes sandstone (p. 6). Most sedimentary grains of silt and sand size are composed of quartz. Even though quartz is very hard, quartz grains suffer some abrasion as they bounce and slide downstream

Figure 2-15 Classification of sedimentary rocks according to grain size. Sediments range in size from clay to silt, sand, and gravel; gravel is divided into granules, pebbles, cobbles, and boulders. Gravelly rocks are called conglomerates when their pebbles and cobbles are rounded and breccias when they are angular; these rocks normally contain sand as well. Rocks in which sand dominates are called sandstones. Rocks in which silt dominates are called siltstones. Rocks formed of clay are called claystone if they are massive and shale if they are fissile (or platy). Siltstones, claystones, and shales are all varieties of mudstone. All of the specimens shown here would fit in the palm of your hand.

along the floors of rivers. In the process, they tend to become smaller and more rounded.

Gravel includes all particles larger than sand, including **granules, pebbles, cobbles,** and **boulders.** A rock that contains large amounts of gravel is termed a **conglomerate** if the gravel is rounded and a **breccia** if it is angular. In both cases, however, sand or granules nearly always fill the spaces between the pieces of gravel.

When sediment settles from water, coarse-grained particles settle faster than fine-grained particles, as can be seen when sediments of various grain sizes are mixed with water in a tall glass container and allowed to settle (Figure 2-17). Clay settles so slowly in water that in nature very little of it falls from rapidly moving water, such as that of a stream channel or a wave-ridden sandy shore of the ocean. Most of it is deposited in calm waters, such as those of lakes, quiet lagoons, and the deep sea. Sand tends to accumulate along beaches and on the bottoms of swiftly flowing streams.

Not only do coarse sediments settle more quickly from water than do fine sediments, but they are also less easily picked up or rolled along surfaces by moving water. Gravel, for example, tends to be deposited near its source area (the area in which it was originally produced by erosion); thus gravel usually accumulates along the flanks of the mountains from which it is eroded, seldom reaching the center of a large ocean.

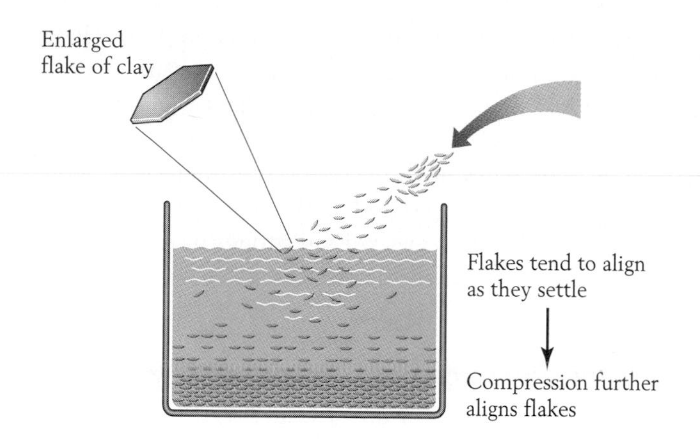

Figure 2-16 **Alignment of clay particles.** Flakes of clay tend to assume a horizontal orientation as they settle through a body of water and pile up on the bottom. Those that initially lie at an angle tend to be aligned by compaction of the sediment.

Differing degrees of sorting of sediment are seen in Figure 2-18. Grains are said to be *poorly sorted* when they are of mixed sizes; the implication is that moving water did not separate the grains well according to size before they were deposited. Sand along a beach, in contrast, has usually been washed and transported by water currents and waves, and thus it tends to be *well sorted*—that is, its particles tend to fall within a narrow size range. Most of the particles in a handful of beach sand are likely to be either medium- or fine-grained (see Figure 2-15).

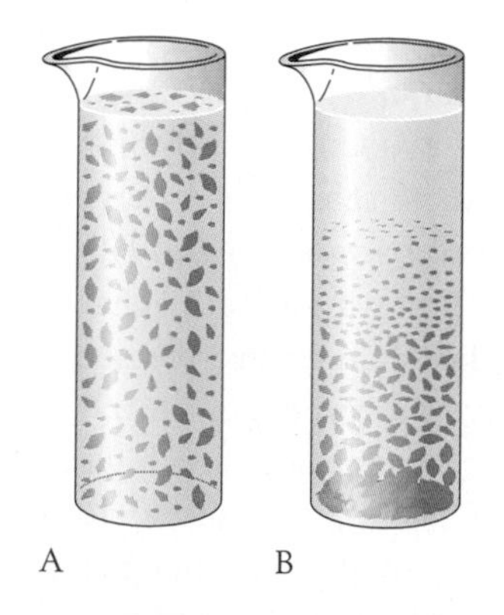

Figure 2-17 **The settling pattern of sediment after it is suspended in water.** *A.* The coarsest sediment settles most quickly and therefore ends up at the bottom of the deposit. *B.* The finest sediment settles last.

Siliciclastic rocks are classified according to composition as well as grain size. Because so many sand-sized grains are quartz, the word **sandstone** is sometimes automatically interpreted to mean quartz sandstone; but several other kinds of siliciclastic rocks also consist largely of sand-sized grains. One of them is **arkose,** whose primary constituents, grains of feldspar, often give the rock a pinkish color. Because feldspar tends to weather rapidly to clay, arkose usually accumulates only in proximity to its parent rock, soon after the feldspar grains are released by partial weathering and erosion.

Another rock in which sand-sized particles normally predominate is graywacke, which is so designated because it is usually dark gray. **Graywacke** consists of a variety of sedimentary particles, including sand-sized and silt-sized grains of feldspar and dark rock fragments and also substantial amounts of clay. Most of the clay in graywacke was not carried to the environment of deposition in its present state but

Figure 2-18 **Well-sorted (top) and poorly sorted grains (bottom) of sand.**

Figure 2-19 Cement bordering quartz grains in sandstone as seen in a thin section through a microscope. Here the rounded sand grains appear gray or white in polarized light, depending on their orientation, and granular calcite cement between the grains is iridescent.

was formed by the disintegration of larger grains within the rock.

Thus far we have discussed the nature of siliciclastic sediments, but not the ways in which these loose sediments become hard rock. A variety of processes lithify soft sediments, or transform them into rock. The primary physical process of **lithification** is *compaction*, a process in which grains of sediment are squeezed together beneath the weight of overlying sediment. Muddy sediments usually undergo a great deal of compaction after burial, as water is squeezed from them.

The most important chemical process of lithification is *cementation*. In this process, minerals crystallize from watery solutions that percolate through the pores between grains of sediment. The cement that is thus produced may or may not have the same chemical composition as the sediment. Sandstone, for example, is often cemented by quartz but more commonly by calcite. Cement is most effectively observed in slices of rock ground so thin that they transmit light and thus can be examined by microscope (see Figure 2-19). Cementation is less extensive in clayey sediments than in clean sands because after clays undergo compaction, they are relatively impermeable to mineral-bearing solutions. Clean sands are initially much more permeable than clays, but their porosity is reduced—and sometimes virtually eliminated—by pore-filling cement.

Cement sometimes gives sedimentary rocks their color. This is often the case for red siliciclastic sedimentary rocks called **red beds.** Some red beds are mudstones, and others are sandstones or conglomerates, but almost all derive their color from iron oxide, which acts as a cement.

Chemical and Biogenic Sedimentary Rocks The grains of some rocks are products of precipitation. **Chemical sediments** result from precipitation of inorganic material in natural waters, sometimes as a result of evaporation. **Biogenic sediments** consist of mineral grains that were once parts of organisms. Some of these grains are pieces of organic skeletons, such as snail shells or "heads" of coral, and others are the tiny, complete skeletons of single-celled creatures. Most biogenic sediments, however, consist of the skeletal remains of a variety of organisms rather than just one or two. Because it is not always possible to determine whether a sedimentary rock is of chemical or biogenic origin, we will consider both groups together.

The most common chemical sedimentary rocks are **evaporites,** which form from the evaporation of seawater or other natural water. Many evaporites are massive, well-bedded deposits that consist of vast numbers of crystals. Among the most abundant evaporites are **anhydrite** (calcium sulfate, $CaSO_4$) and **gypsum** (calcium sulfate with two water molecules attached, $CaSO_4 \cdot 2H_2O$). Note that the terms *anhydrite* and *gypsum* refer both to the minerals with these names and to rocks that are composed largely of these minerals. *Halite* is another term that refers both to a mineral and to an evaporite rock. Because it is the presence of sodium chloride in large amounts that makes seawater salty, it is no surprise that halite (Figure 2-6) should accumulate in large deposits that have great economic value. Recall that evaporites can form even in lakes where ions released from rocks by weathering have become concentrated, in regions where rates of evaporation are high (p. 22). Salty lakes in Death Valley, California, for example, lie atop salt deposits several kilometers deep that have accumulated over long millions of years.

Just as evaporites are readily precipitated from water, they are readily dissolved, so they do not survive long at Earth's surface except in arid climates.

Figure 2-20 Chert, a sedimentary rock composed of very small quartz crystals. Some chert is a chemical sedimentary rock and some is biogenic.

When evaporites are buried deep beneath younger deposits, however, and so are protected from potentially destructive groundwater, they can survive for long geologic intervals.

Other types of chemical sediments are less abundant than evaporites. Among the most common of these are chert, phosphate rocks, and iron formations. **Chert,** also called **flint,** is composed of extremely small quartz crystals that have been precipitated from watery solutions. Typically, impurities give chert a gray, brown, or black color. Chert breaks along curved, shell-like surfaces (Figure 2-20); Native Americans took advantage of this feature when they fashioned chert into arrowheads and other kinds of projectile points. Some cherts occur as scattered, irregular masses in other kinds of sedimentary rocks. These masses and some extensive beds of chert grow from silica-rich solutions that have moved through rock. Other bedded cherts are thought to have formed by direct precipitation of silica (SiO_2) from seawater. Still others are biogenic deposits that result from the accumulation on the seafloor of the microscopic skeletons of single-celled organisms. These skeletons consist of a type of silica that differs from quartz in that it is amorphous (or noncrystalline). Over long geologic intervals, water percolates through deposits of these skeletons, converting them to very hard chert (Figure 2-21). Cherts older than 100 million years or so have suffered such extensive chemical alteration that many are difficult to identify as biogenic or chemical.

Banded iron formations are complex rocks that consist of oxides, sulfides, or carbonates of iron interlayered with thin beds of chert. Iron formations are widespread only in very old Precambrian rocks, and many form large iron ore deposits.

Limestones include both chemical and biogenic bodies of rock. Because they are not as soluble in water as evaporites, limestones are much more common at Earth's surface, where they are quarried extensively for the production of building stone, gravel, and concrete (see Box 2-1). We have already noted

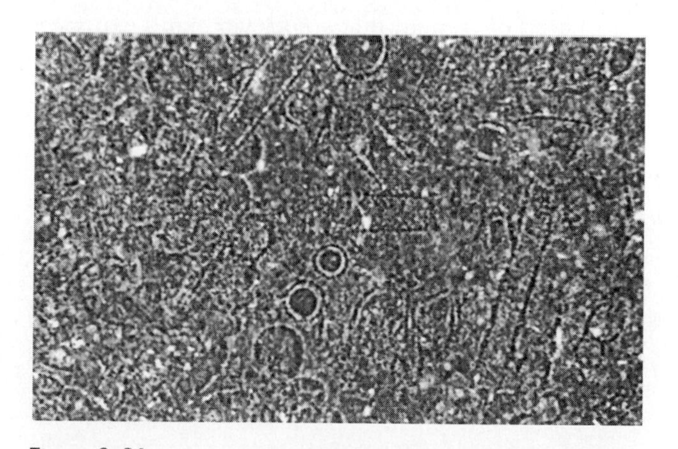

Figure 2-21 Photomicrograph of lithified siliceous ooze that forms the Caballos Formation of middle Paleozoic age. "Ghosts" of silicious skeletal elements of sponges and single-celled organisms called radiolarians are still visible in this thinly sliced rock.

For the Record 2-1

Why our buildings are made from fossils

Most buildings in the Western world would crumble if, through some act of magic, we removed from them all of the materials that are derived from fossils. Cement, mortar, and concrete are made from limestone, and fossils are major—often dominant—components of this rock. Limestone is the primary raw material of cement. Mortar, which is used to bond bricks and stone, is cement to which sand has been added to increase the strength and porosity of the hardened material. Concrete is simply mortar to which coarse material—usually gravel—is added for strength.

Ancient peoples formed cement by baking limestone ($CaCO_3$) to drive off carbon dioxide (CO_2) so as to form CaO, known as quicklime. They then "slaked" quicklime by adding water to form powdery $Ca(OH)_2$, or hydrated lime. When water containing CO_2 was mixed with the hydrated lime, the CO_2 reacted with the hydrated lime to form solid cement consisting of calcium carbonate ($CaCO_3$). Thus this cement had the same chemical composition as the original limestone, but it could be molded to any desirable shape before it hardened.

The ancient Greeks and Romans used this material, but they also invented a more durable kind of cement that could withstand exposure to water after it hardened. They made this more useful cement by mixing fine-grained silicate material such as clay or volcanic ash with ground limestone. When such a mixture is fired, a variety of compounds form, including calcium silicate and oxides of aluminum and iron; these compounds constitute the dry cement, to which water is added. As the silicate material for this cement, the Romans often used volcanic ash from the foot of Mount Vesuvius, the volcano that erupted in AD 79 and buried the city of Pompeii. The Romans became masters of concrete construction, erecting huge, enduring structures that they could never have built from stone. Still standing in Rome, for example, is the Pantheon, a temple with a hemispherical dome about 30 meters (~110 feet) in diameter. The Romans used pumice in place of gravel in the concrete near the top of this giant dome to reduce its weight; pumice is volcanic rock with so many gas-filled chambers that it can often float on water.

Building technology declined along with the Roman Empire. As a result, much of the concrete used during the Middle Ages was of low quality. Finally, in 1756, an Englishman named

A quarry from which limestone is being extracted along the Mississippi River to make cement.

John Smeaton rediscovered the value of including fine-grained silicate in cement. He made this discovery in the course of making cement from various kinds of limestone, some of which contained substantial amounts of clay. Early in the nineteenth century, a particular admixture of clay materials was found to produce cement that would harden under water. This product became known as Portland cement, because when it hardened it resembled the limestone that is still quarried as building stone on the island of Portland, along the southern coast of England. This was the limestone used to construct Westminster Abbey.

Portland cement quickly became the primary component of mortar and concrete in Europe and North America. Late in the nineteenth century many concrete sidewalks were built in the United States, and until about 1895 most Portland cement used for this purpose came from England. The great weight of this cargo was no problem because it moved across the Atlantic without any freight charge, as ballast on ships. When large-scale production of Portland cement finally got under way in North America late in the nineteenth century, the raw material was limestone that contained an abundance of clay. Early on, much of this rock was quarried in eastern Pennsylvania. In 1899 the great inventor Thomas Edison founded a company nearby in western New Jersey. Having conducted research on the production of Portland cement, Edison chose a site where both clay-rich limestone and pure limestone were accessible, so that the two kinds of rock could be quarried, ground to powder, and mixed in the proper proportions.

Since Edison's day, research has uncovered additives and production techniques that have endowed mortar and concrete with many useful new properties, such as greater strength, elasticity, and resistance to chemical destruction by acid rain. Of course, we use concrete to construct not only buildings and sidewalks but also bridges, tunnels, and superhighways, among many other things. In the process, we transform fragmentary fossils that were major components of limestone into the substance of these creations. In fact, we might view the marine organisms that built those skeletons millions of years ago as the initial laborers in the multistage construction of our buildings and the transportation networks that connect them.

Limestone. This particular rock, which consists of calcite ($CaCO_3$), is formed almost entirely of shells of organisms and cement that binds the shells together.

that although ancient limestones consist primarily of the mineral calcite, many of their grains were initially composed of the mineral aragonite. Recall that aragonite has the same chemical composition as calcite but differs in its crystal structure and is unstable at Earth's surface; in time, aragonite becomes transformed into calcite (p. 36).

Dolomite is a carbonate mineral that is relatively uncommon in modern marine environments but common in many ancient rocks. As mentioned earlier, it differs from calcite in that half of the calcium ions of the crystal lattice are replaced by magnesium ions. In fact, much dolomite has formed by the chemical alteration of calcite. When dolomite is the dominant mineral of an ancient rock, the rock, too, is called dolomite. Because limestones and dolomites are similar rocks and are often found closely associated, they are sometimes referred to collectively as **carbonate rocks.** Similarly, unconsolidated sediments consisting of aragonite, calcite, or both are often called **carbonate sediments.**

Carbonate sediments form in two ways: by direct precipitation from seawater and through accumulation of skeletal debris from organisms. Many types of marine life grow shells or other kinds of skeletons that consist of aragonite or calcite. These organisms contribute skeletal material to the seafloor as sedimentary particles. Some of these particles retain their original sizes, while others diminish through breakage or wear. The product of this biological contribution is an array of clastic particles that are similar to siliciclastic grains and, like siliciclastic grains, can be classified according to size. Thus we speak of carbonate sands and carbonate muds.

Most carbonate particles that are sand-sized or larger can be seen to be skeletal particles, but the origin of mud-sized material is more difficult to determine. The main constituents of carbonate muds in modern seas are **aragonite needles** (see Figure 2-7*B*). These needles are produced both by direct precipitation and by the collapse of carbonate skeletons, especially of algae. In ancient fine-grained limestones, aragonite needles have been transformed to tiny calcite grains. The resulting granular texture reveals little about the configuration or mode of origin of the original carbonate particles.

Oolites are sediments consisting of nearly spherical grains (*ooliths*) that grow by rolling about and accumulating aragonite needles in the way that a snowball becomes large enough to make a snowman when it is rolled about in the snow. Oolite forms in shallow water, where the seafloor is agitated by strong water movements, A cross section of an oolith displays a concentric structure, with thin dark bands representing breaks in deposition when the grain was at rest and ceased to grow for a time (Figure 2-22*B*).

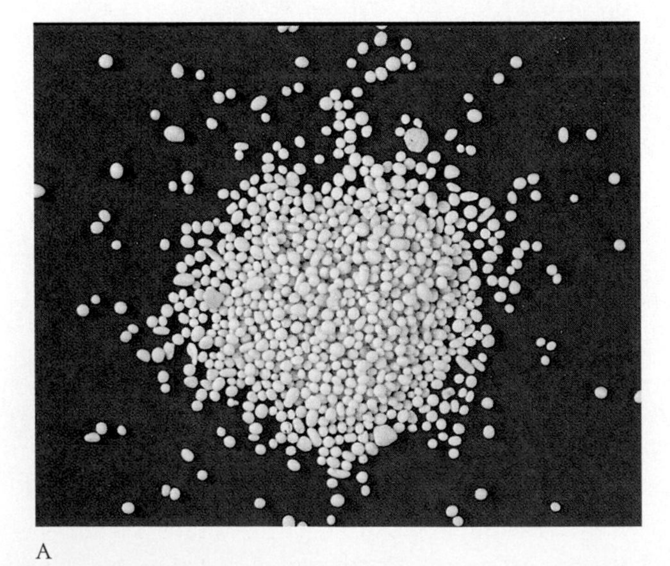

A

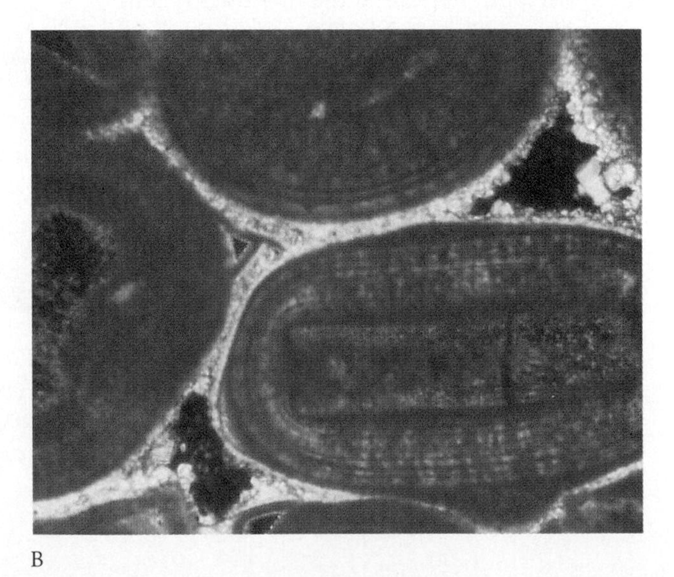

B

Figure 2-22 Oolite. *A*. Nearly spherical ooliths, which are the size of sand grains. *B*. A photomicrograph of thinly sliced oolite rock reveals the concentric structure of the ooliths.

Figure 2-23 **A graded bed.** When sediments consist of grains of mixed sizes, the coarser grains settle more rapidly than the finer ones, so that grain size diminishes from bottom to top.

Calcium carbonate is precipitated only from seawater that contains relatively little carbon dioxide. Carbon dioxide is less soluble in warm water than in cold water, so carbonate sediments accumulate primarily in tropical seas, where winter water temperatures seldom drop below 18°C (64°F). This condition not only explains the direct precipitation of carbonate minerals in warm seas but also accounts for the fact that few organisms that live in cold water secrete massive skeletons of calcium carbonate. Carbonate sediments can also form in freshwater habitats, usually as a result of the carbonate-secreting activities of certain algae.

Unlike siliciclastic clays, carbonate rocks do not compact greatly after burial. Instead, they harden primarily by cementation, in the same manner as siliciclastic sands. Carbonate sediments are nearly always cemented by carbonate minerals simply because rich sources of such cements are close at hand.

Sedimentary Structures Distinctive arrangements of grains in sedimentary rocks are termed **sedimentary structures.** Sedimentary structures reflect modes of deposition and provide useful tools for interpreting the environments in which ancient sediments were deposited.

A **graded bed** is a sedimentary structure in which grain size decreases from the bottom to the top (Figure 2-23). This pattern usually results from the normal settling process that characterizes sediments of mixed grain size, with the coarser sediment settling more rapidly than finer sediment (see Figure 2-17). Most graded beds form in nature when a strong current suddenly introduces a large volume of sediment to a quieter body of water, where it settles.

Figure 1-1 compared **ripples** forming today on a beach with ancient ripples preserved in sandstone. Such ripples have symmetrical cross sections because wave motion in water oscillates back and forth. Water currents and wind, in contrast, produce ripples and larger ridges—bars or dunes—with asymmetrical cross sections. Sediment accumulates along the lee (downstream) side of such a ridge (Figure 2-24). The accumulation of beds here produces what is termed **cross-bedding** or **cross-stratification** because sets of parallel beds slope at an angle to the horizontal. Although sets of cross-bedding in a single rock unit vary somewhat in orientation, their average direction of slope indicates the general direction of the prevailing winds when ancient beds were deposited to form a dune or of currents when the beds were deposited in a stream.

Mudcracks form as fine-grained, clay-rich sediments dry out and shrink. These structures, which often form hexagonal patterns like the tiles of a bathroom floor, are visible in ancient rocks even if sediment fills them. Mudcracks indicate deposition in shallow waters, such as those at the edge of a lagoon or lake, which receded and exposed the sediments to the air.

Metamorphic Rocks

Recall that metamorphic rocks form by the alteration of other rocks at temperatures and pressures that exceed those normally found at the Earth's surface. Metamorphism alters rocks without melting them, whereas igneous rocks, which also form at high temperatures, do so by the cooling of rock that has become hot enough to melt. Metamorphism alters both the composition and the texture of all kinds of rocks—igneous, sedimentary, and metamorphic. Mineral assemblages of metamorphic rocks serve as critical "thermometers" and "barometers," varying with the temperature and pressure of metamorphism. **Grade** is the word used to indicate the levels of temperature and pressure of metamorphism. High-grade assemblages of metamorphic minerals form at higher temperatures and pressures than low-grade assemblages.

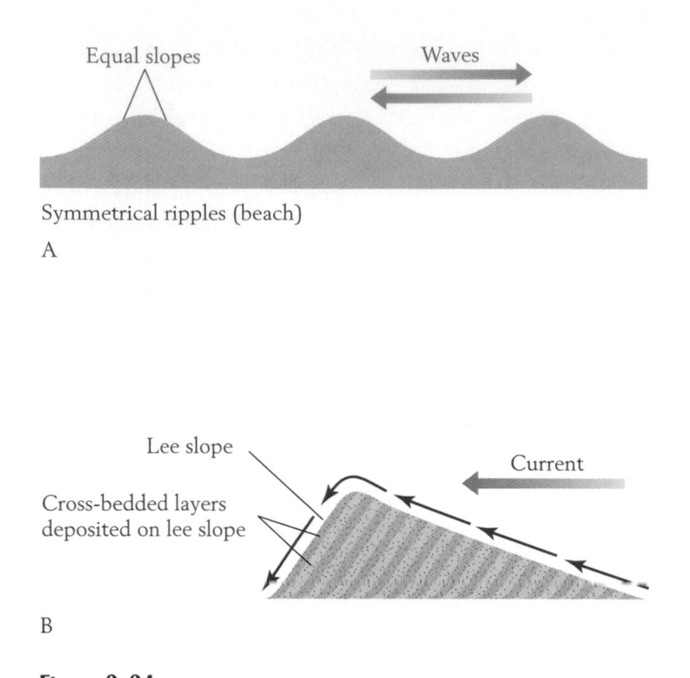

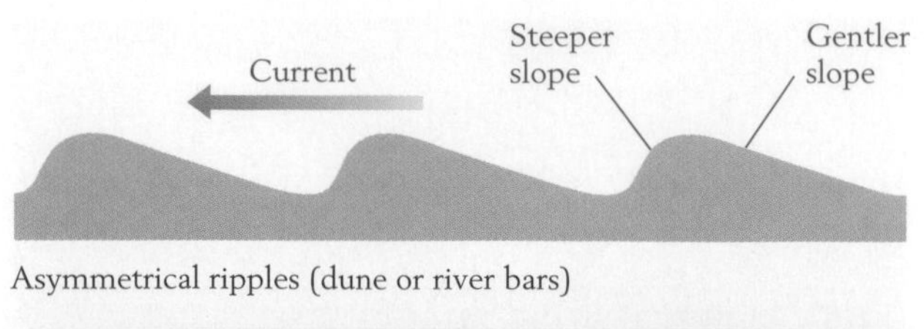

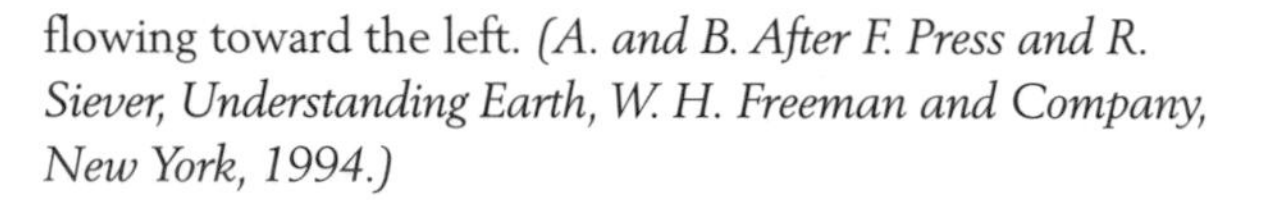

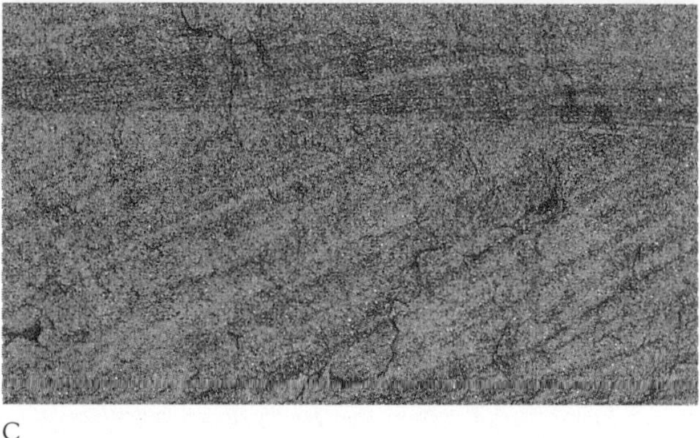

Figure 2-24 **Ripples and cross-bedding.** *A.* Profiles of ripples produced by waves and currents. *B.* The formation of cross-bedding on the lee slope of a ripple. C. Cross-bedding in coarse sandstone, formed as in *B,* by currents flowing toward the left. *(A. and B. After F. Press and R. Siever, Understanding Earth, W. H. Freeman and Company, New York, 1994.)*

Regional Metamorphism As its name implies, **regional metamorphism** transforms deeply buried rocks over areas whose dimensions are measured in hundreds of kilometers. Igneous activity usually extends along the length of an actively forming mountain chain, supplying heat for metamorphism. Along each side of such an igneous belt is a zone of regional metamorphism produced by high temperatures and pressures extending outward from the igneous belt. Most rocks in zones of regional metamorphism display a texture known as **foliation,** which is an alignment of platy minerals caused by the pressures applied during metamorphism.

The grade of metamorphism in a regional metamorphic zone typically declines away from the neighboring belt of igneous activity that supplied the heat for metamorphism. Let us examine three types of rocks that represent different grades of metamorphism. **Slate** is a fine-grained rock of very low metamorphic grade in which the pressure of metamorphism has aligned platy minerals to produce foliation that produces fissility much like that of shale (Figure 2-25*A*). Slate broken into rectangular plates covers the roofs of many houses. **Schist** is a low-to-medium-grade metamorphic rock that consists largely of platy grains of minerals, often including mica; because of its strong foliation, schist tends to break along parallel surfaces. **Chlorite** is a micalike green mineral that occurs primarily in schist. When abundant, it gives the rock the informal name *greenschist* (Figure 2-25*B*). **Gneiss** is a high-grade metamorphic rock whose intergrown crystals resemble those of igneous rock, being granular rather than platy, but whose minerals tend to be segregated into wavy layers (Figure 2-25C).

Not all rocks in regional metamorphic zones are foliated. Some have homogeneous granular textures, which indicate not only that their interlocking mosaics of crystals lack preferred orientations but also that certain minerals are not segregated into bands. **Marble** and **quartzite,** for example, are usually homogeneous granular metamorphic rocks (Figure 2-26). Marble consists of calcite, dolomite, or a mixture of the two minerals, and it forms from the metamorphism of sedimentary carbonates. Marble is popular as a decorative stone, not only because impurities often create attractive patterns within it but also because it is relatively soft and easy to cut and pol-

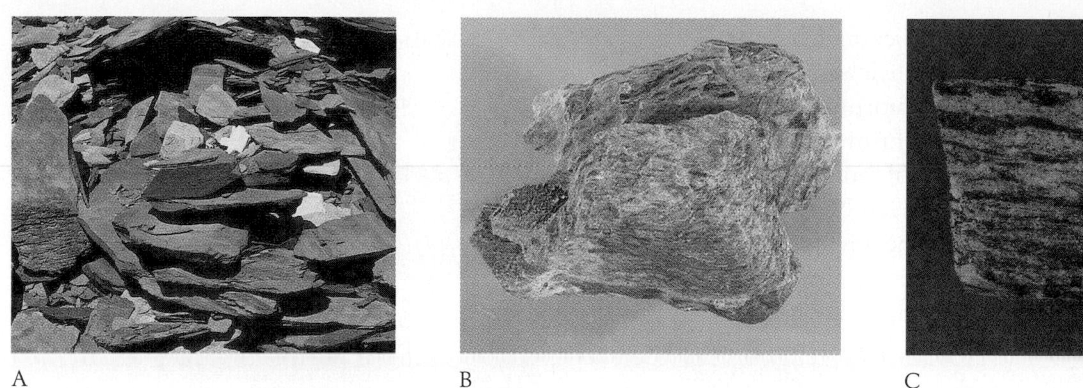

A B C

Figure 2-25 **Foliated metamorphic rocks.** Slate *(A)* is a low-grade metamorphic rock, schist *(B)* is a medium-grade rock, and gneiss *(C)* is a high-grade rock.

ish. Quartzite, consisting of nearly pure quartz, is much harder than marble; it forms from the metamorphism of quartz sandstone.

Contact Metamorphism When an igneous intrusion "bakes" surrounding rock (see Figure 1-9), the result is **contact metamorphism.** High temperature plays a larger role in this process than high pressure. The resulting rocks are usually fine-grained. Contact metamorphism may occur deep within Earth or near the surface. Though a more localized phenomenon, it resembles regional metamorphism in displaying a gradient: the grade of metamorphism declines away from the heat source.

Hydrothermal Metamorphism The percolation of hot, watery fluids through rocks can result in **hydrothermal metamorphism.** Fluids of this type escape when magma is intruded into continental crust, but most hydrothermal metamorphism takes place along mid-ocean ridges, where seawater circulates through the hot, newly formed lithosphere. Here basalt (see Figure 2-10) is extensively altered to greenschist (Figure 2-25*B*). Many valuable ore minerals, including gold deposits, have been emplaced from hydrothermal fluids.

Burial Metamorphism Rocks are altered when they are buried so deep that they are exposed to

A

B

Figure 2-26 **Nonfoliated metamorphic rocks.** *A.* A marble quarry in Ireland. *B.* Quartzite from Australia.

temperatures and pressures high enough to change their chemical composition. For siliciclastic rocks, the consequences of this **burial metamorphism** are similar to those of regional metamorphism.

When the original material of burial metamorphism is plant debris, the product can be coal (Figure 2-27). **Coal** is rock formed by the low-grade metamorphism of stratified plant debris. It can be burned because organic carbon compounds account for more than 50 percent of its composition. The starting material for the formation of coal is **peat,** which is leafy and woody plant tissue that accumulates in water that contains little oxygen, and therefore few bacteria that cause decay. The heat and pressure of burial turn peat into brown coal called *lignite* through compression and expulsion of water, hydrogen, and nitrogen. This metamorphism increases the carbon content of the organic material, and deeper burial continues the process, producing soft coal and eventually hard coal. The higher the carbon content of coal, the more energy it produces when it burns, because burning is rapid oxidation of carbon at a high temperature.

The Upper Limit of Metamorphism When conditions become so hot that a rock melts, metamorphism ceases. Rocks that form when the resulting molten material cools are then classified as igneous. Very high-grade metamorphic rocks that come close to a molten state before cooling resemble coarse-grained igneous rocks in being granular and lacking foliation. In fact, these metamorphic rocks are difficult to distinguish from igneous rocks.

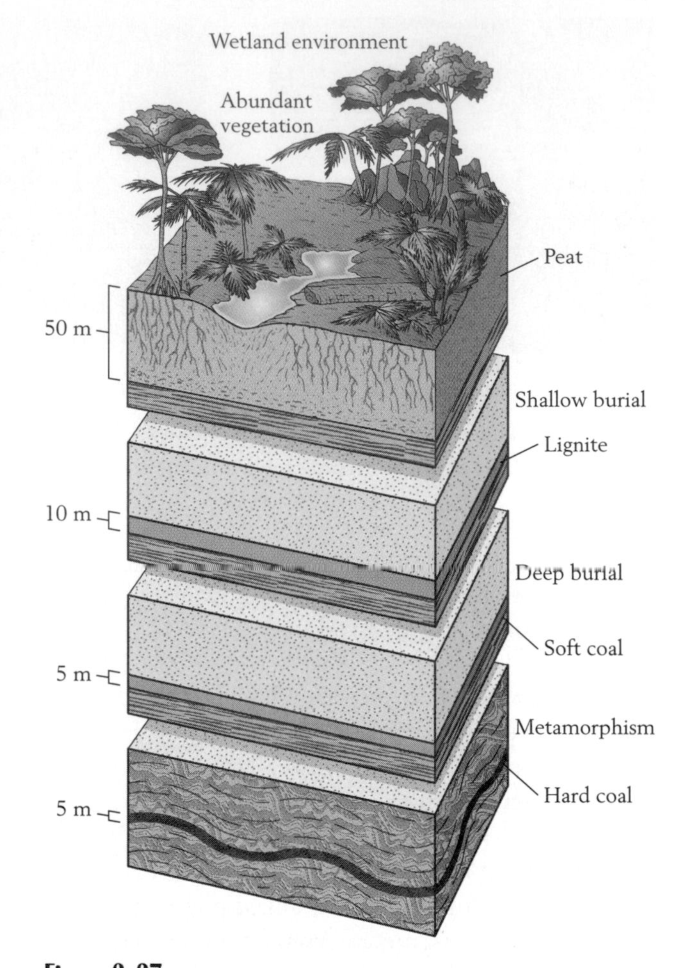

Figure 2-27 The origin of coal. Peat, which accumulates in water where little oxygen is present, becomes compressed and heated through burial, eventually becoming hard coal. *(After F. Press and R. Siever, Earth, W. H. Freeman and Company, New York, 1986.)*

Chapter Summary

1 Minerals are elements or chemical compounds that are the building blocks of rocks.

2 An isotope is a form of an element that differs in number of neutrons (and in atomic weight) from other isotopes of the same element.

3 Limestone is formed of calcium carbonate, which in some geologically young limestones consists of the chemically unstable mineral aragonite but in old limestones normally consists of the more stable mineral calcite.

4 A mineral's hardness is determined by the strength of the chemical bonds of its crystal lattice; its density is determined by the weight of its atoms and by their degree of packing within the crystal lattice.

5 Most rocks of Earth's crust and mantle are formed of silicate minerals, in which a silicon atom is surrounded by four oxygen atoms.

6 Mafic igneous rocks are denser than felsic rocks because they contain an abundance of minerals that are rich in iron, which has a high atomic weight.

7 Extrusive igneous rocks are fine-grained because the molten lava from which they form cools and solidifies quickly at Earth's surface.

8 Intrusive igneous rocks are coarse-grained because they form from magma that cools slowly below Earth's surface, allowing time for the growth of relatively large crystals.

9 Siliciclastic sedimentary rocks consist of silicate grains that are products of weathering and erosion. Clays are the most common minerals in fine-grained siliciclastic rocks, and quartz is the most common mineral in rocks formed of clasts of sand or gravel.

10 Chemical sediments are formed by precipitation of inorganic material from natural waters.

11 Biogenic sediments consist of mineral grains that were once parts of organisms. The most abundant biogenic rock is limestone, which consists largely of calcium carbonate.

12 Arrangements of grains, known as sedimentary structures, reflect the mode of deposition of sedimentary rocks.

13 Metamorphic rocks vary in composition, depending on the degrees of heat and pressure responsible for their origin. Regional metamorphism alters rocks over hundreds of kilometers, whereas contact metamorphism results when the heat of an igneous intrusion bakes local rock. Hydrothermal metamorphism results when hot, watery fluids percolate through rocks. Coal is a product of burial metamorphism, formed from plant debris.

Review Questions

1 How does an ion form from an uncharged atom?

2 What allows one element to substitute for another in the crystal lattice of a particular mineral?

3 How does the orientation of a dike of igneous rock differ from the orientation of a sill?

4 Under what circumstance does the eruption of lava produce a flood basalt instead of a volcano?

5 Why do most sediments that form from weathering consist of silicate minerals?

6 How does the origin of clay particles from igneous rock differ from the origin of grains of quartz sand from the same kinds of rock?

7 How does the process that turns clay into shale differ from the process that turns sand into sandstone?

8 Why do evaporite deposits weather quickly?

9 How do the following sedimentary structures form? (1) Graded beds. (2) Ripples. (3) Cross-bedding. (4) Mudcracks.

10 How does the origin of a metamorphic rock differ from the origin of an igneous rock?

11 Water plays a major role in the origin of sedimentary rocks. Use the Visual Overview on page 30 as a guide to explore the ways in which water serves as a medium for the production, transport, deposition, and lithification of the various kinds of mineral grains that form sediment.

Additional Reading

Erikson, J. *An Introduction to Fossils and Minerals: Seeking Clues to the Earth's Past,* Facts on File, New York, 1992.

Pellant, C., *Rocks, Minerals, and Fossils of the World,* Little, Brown & Co., Boston, 1990.

Press, F. and R. Siever, *Understanding Earth,* 2nd ed., W. H. Freeman and Company, New York, 1998.

CHAPTER 3

The Diversity of Life

In the course of geologic time many of the organisms that have inhabited Earth have left a partial record in rock of their presence and their activities. This record reveals that life has changed dramatically since it first arose and that its transformation has been intimately associated with changes in physical conditions on Earth—in climates or in the positions of continents, for example. At times in Earth's history, environmental conditions have changed suddenly and many forms of life have died out. After these crises, evolution has replenished life on our planet. This chapter introduces major groups of organisms that have evolved and become extinct in the course of Earth's history.

It is not easy to define *life* precisely, but two attributes that are generally regarded as essential to life are the capacity for self-replication and the capacity for self-regulation. Viruses are simple entities that can replicate themselves (or reproduce), but they do not regulate themselves—that is, they do not employ raw materials from the environment to sustain orderly internal chemical reactions. Thus viruses are not considered to be living things. On Earth today, all entities that are self-replicating and self-regulating are also cellular: they consist of one or more discrete units called cells. A living **cell** is a module with a variety of distinct features, including structures in which certain chemical reactions take place. The chemical "blueprint" for a cell's operation is coded in the chemical structure of genes. Essential to this blueprint is the cell's built-in ability to duplicate genes so that a replica of the blueprint can be passed on to another cell or to an entirely new organism.

A living chambered nautilus swimming close to the seafloor. It swims by squirting water from a tube that is visible beneath its tentacles.

Visual Overview

The Six Kingdoms

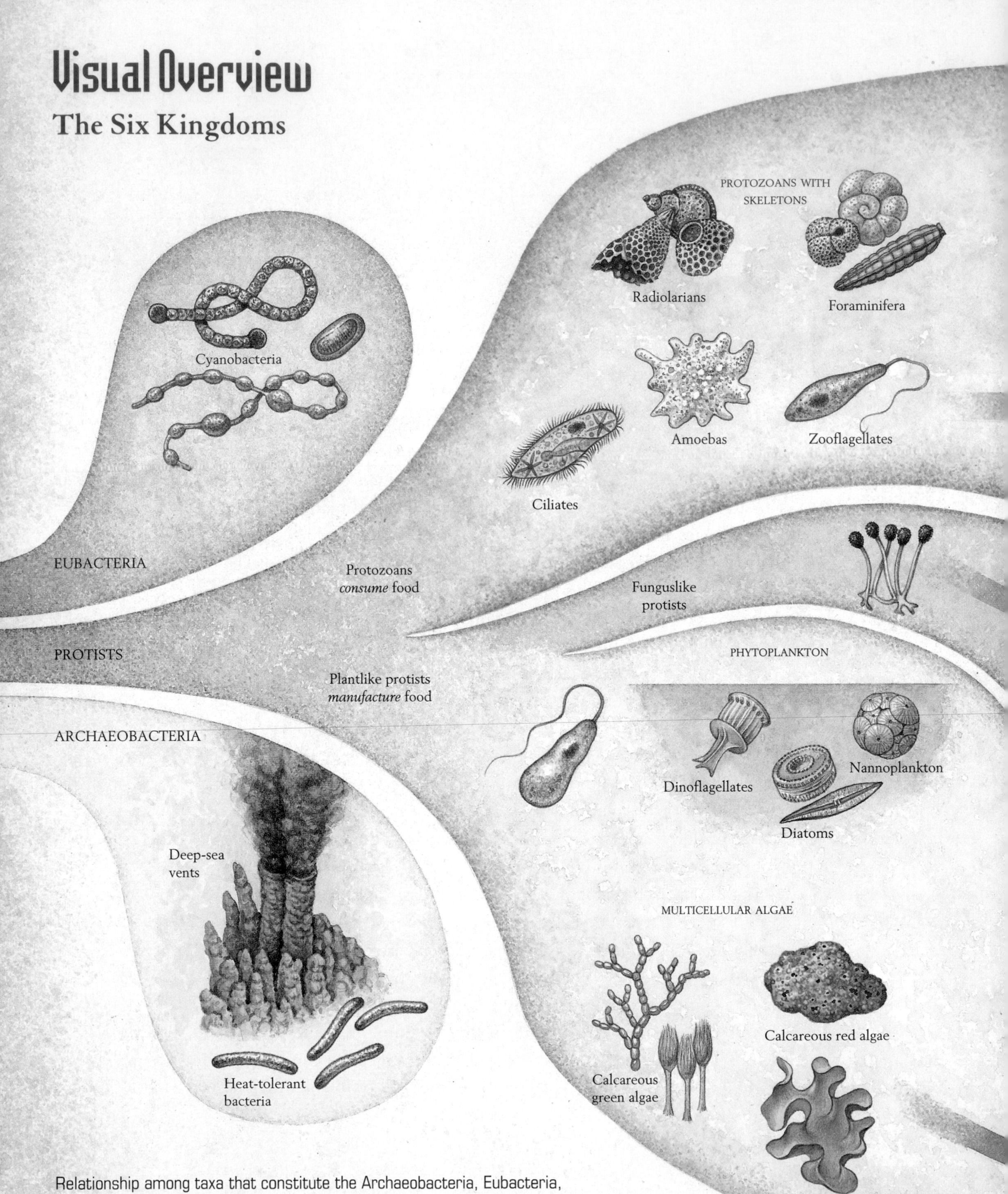

Relationship among taxa that constitute the Archaeobacteria, Eubacteria, Protists, Plants, and Animals.

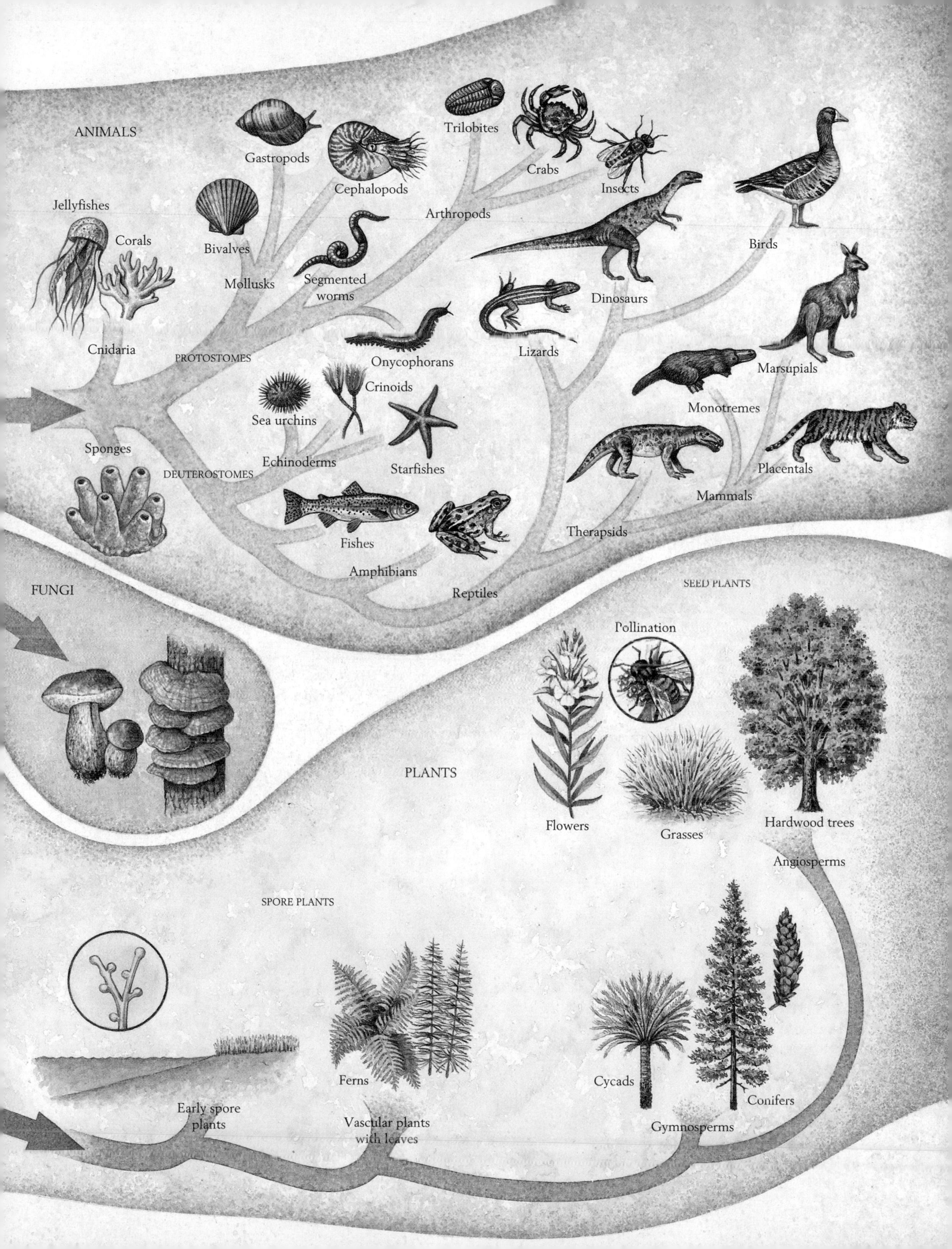

ANIMALS
Gastropods
Trilobites
Crabs
Insects
Cephalopods
Jellyfishes
Corals
Bivalves
Arthropods
Birds
Mollusks
Segmented
worms
Dinosaurs
Cnidaria
PROTOSTOMES
Onycophorans
Lizards
Crinoids
Marsupials
Sea urchins
Monotremes
Sponges
Echinoderms
Starfishes
Placentals
DEUTEROSTOMES
Mammals
Fishes
Therapsids
Amphibians
FUNGI
Reptiles
SEED PLANTS
Pollination
PLANTS
Flowers
Grasses
Hardwood trees
Angiosperms
SPORE PLANTS
Ferns
Cycads
Conifers
Early spore
plants
Vascular plants
with leaves
Gymnosperms

Fossils

Most of our knowledge about the life of past intervals of geologic time is derived from fossils. Recall that the term *fossil* is usually restricted to tangible remains or signs of ancient organisms that died thousands or millions of years ago (p. 11). **Fossilization** is the group of processes by which fossils form. Because few fossils consist of materials that can survive the high temperatures at which igneous and metamorphic rocks form, almost all fossils are found in sediments or sedimentary rocks.

Fossils are much more common in sedimentary rocks than most people realize. They are especially abundant in sedimentary rocks that were formed in the ocean, where animals with skeletons abound.

Hard Parts: Commonly Preserved Skeletal Features of Animals

The most readily preserved features of animals are the structures that are informally described as "hard parts"—teeth and bones of vertebrate animals and comparable solid structures of invertebrate animals. Many groups of invertebrates lack skeletons and therefore have left poor fossil records or none at all. Some invertebrate animals, however, have internal skeletons embedded in soft tissue; among them are relatives of starfishes called crinoids (informally called sea lilies) (Figure 3-1). Other invertebrates have protective external skeletons. Among these animals are the bivalve and gastropod mollusks, whose tissues are housed inside skeletons popularly known as seashells. Hard parts are often preserved with only a modest amount of chemical alteration, but at times they are completely replaced by minerals that are unrelated to the original skeletal material (Figure 3-1).

Preservation of Soft Parts of Animals

Fleshy parts of animals, or "soft parts," are occasionally found in the fossil record, but only in sediments that date back a few millions or tens of millions of years. Chemical residues of tissues and cells can be identified in much older rocks.

One deposit that is famous for preservation of soft parts is the Geiseltal Formation of Germany, which is more than 40 million years old. In the nearly impermeable Geiseltal sediments, which are rich in oily plant debris, the skin and blood vessels of long-extinct frogs can still be studied, fossil leaves are still green, and insects retain their iridescent color. Some animals are preserved here with their last meals

Figure 3-1 Fossil crinoids. In life each of these animals was attached to the seafloor by a flexible stalk. The calcium carbonate skeleton of the specimen on the right has been altered chemically to pyrite, a mineral that consists of iron sulfide and is known as "fool's gold."

Figure 3-2 Remarkable preservation of an Eocene mammal. This creature, which resembled a hedgehog, became buried in the bottom of a lake, where its flesh and fur were partially preserved because so little oxygen was present that scavenging animals and bacteria of the kind that cause decay were absent. This fossil is about 40 centimeters (16 inches) long.

remaining only partly decomposed in their stomachs, and some have fur still intact (Figure 3-2). Protection from oxygen is the secret for survival of soft tissue: it is most likely to be preserved when organisms are buried in fine-grained, relatively impermeable sediment, especially if oily water-repellent organic matter is also present. Bacteria that bring about the decay of soft tissues cannot live in the absence of oxygen.

Permineralization

Terrestrial plants do not generally have mineralized skeletal structures, but the cellulose walls of their cells are so rigid that woody tissue and even leaves are much more commonly preserved in the fossil record than is the flesh of animals. After plants are buried in sediment, the spaces left inside the cell walls of woody tissue may be filled by inorganic materials—most commonly by finely crystalline quartz, known as chert (p. 47). This process, known as **permineralization,** produces the fossils that are often called petrified wood.

Molds and Impressions

Sometimes solutions percolating through rock or sediment dissolve fossil skeletons, leaving a space within the rock that is a three-dimensional negative imprint of the organic structure. This type of imprint, called a **mold,** is sometimes filled secondarily with minerals. If this process has not occurred in nature, the mold can be filled with wax, clay, or liquid rubber in the laboratory to produce a replica of the original object (Figure 3-3).

Fossils called **impressions** might be viewed as squashed molds. Impressions usually preserve in flattened form the outlines and some of the surface features of soft or semihard organisms such as insects or

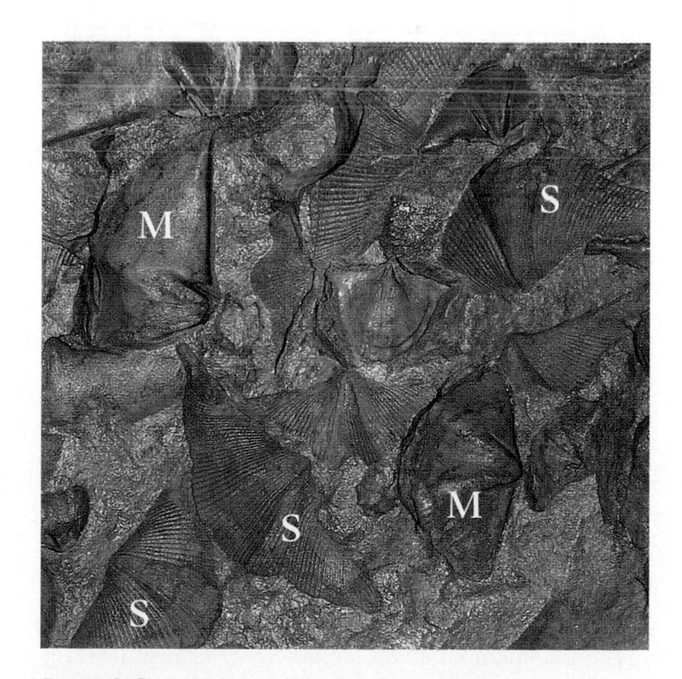

Figure 3-3 Preservation of brachiopods. This surface of a Paleozoic rock contains fossil shells of these animals (*S*) and also molds of the internal surfaces of shells (*M*).

Figure 3-4 **Carbonized fossil leaf impression from the Jurassic Period.** This leaf belongs to the group of plants known as cycads.

leaves (Figures 3-4). A residue of carbon remains on the surface of some impressions after other compounds have been lost through the escape of liquids and gases. This process of carbon concentration is known as carbonization.

Trace Fossils

Tracks, trails, burrows, and other marks left by animal activity are known as **trace fossils** (Figure 3-5). As the direct results of activity, trace fossils can reveal a great deal about the behavior of extinct animals—although the animal that made a particular kind of trace cannot always be identified with certainty.

Trace fossils reveal significant aspects of the behavior of extinct animals. Trackways of dinosaurs, for example, show that these mammals were very active. Unlike modern reptiles, they customarily moved about at a fast pace. Farther back in the geologic record, preserved tracks and burrows reveal how the earliest animals colonized the floors of ancient seas at a time when very few kinds of animals left fossils that revealed the shapes of their bodies.

Fossil Fuels

Organisms usually lose their identity in contributing to *fossil fuels.* As we have seen, coal is formed by metamorphism of plant debris that accumulates in water as peat (p. 54). Petroleum and natural gas form from smaller particles of organic matter, mostly the debris of small floating organisms that sinks to the seafloor and becomes buried there, thus escaping total destruction by bacteria. Deep burial exposes such material to elevated temperatures that transform it into liquid and gaseous compounds of carbon and hydrogen. These fluids sometimes accumulate in porous reservoirs within soft sediment or hard rock, from which they can be extracted for human use.

Figure 3-5 **Dinosaur tracks in Connecticut.** A dinosaur formed these tracks by walking across wet mud that hardened into rock.

The Quality of the Fossil Record

Although fossils are found with great frequency in many sedimentary rocks, fossils of most species of animals and plants that have existed in Earth's history have never been discovered. Rare species and those that lack skeletons are especially unlikely to be found in fossilized form. Even most species with skeletons have left no permanent fossil record. A variety of processes destroy skeletons. Animals that scavenge carcasses, for example, may splinter bones in the process. Also, many bones, teeth, and shells are abraded beyond recognition when they are transported by moving water before finally becoming buried. Even after burial, many fossils fail to survive metamorphism or erosion of the sedimentary rocks in which they are embedded. Finally, many fossil species remain entombed in rocks that have never been exposed at Earth's surface or sampled by drilling operations. Because sediments accumulate on nearly all parts of the seafloor, marine organisms are more frequently preserved than terrestrial plants and animals.

Kingdoms of Organisms

Most biologists divide life on Earth into six kingdoms (Figure 3-6). Two of these large groups contain bacteria and are known as **prokaryotes.** One is the **Archaeobacteria** ("old bacteria") and the other is the **Eubacteria** ("true bacteria"). Some workers still unite all bacteria in a single kingdom, but the two groups are so different that we will follow the new scheme of recognizing them as two distinct prokaryotic kingdoms. The four remaining kingdoms, which contain more advanced forms of life, are known as **eukaryotes.** The cells of prokaryotes differ from those of eukaryotes in lacking certain internal structures, including a nucleus to house their genetic material.

Three eukaryotic kingdoms—Plantae, Fungi, and Animalia—are distinguished by their nutrition. Members of the **Plantae** produce their own food by means of **photosynthesis.** This is a process by which an organism harnesses the sun's energy to produce its own food from water and carbon dioxide. Thus we

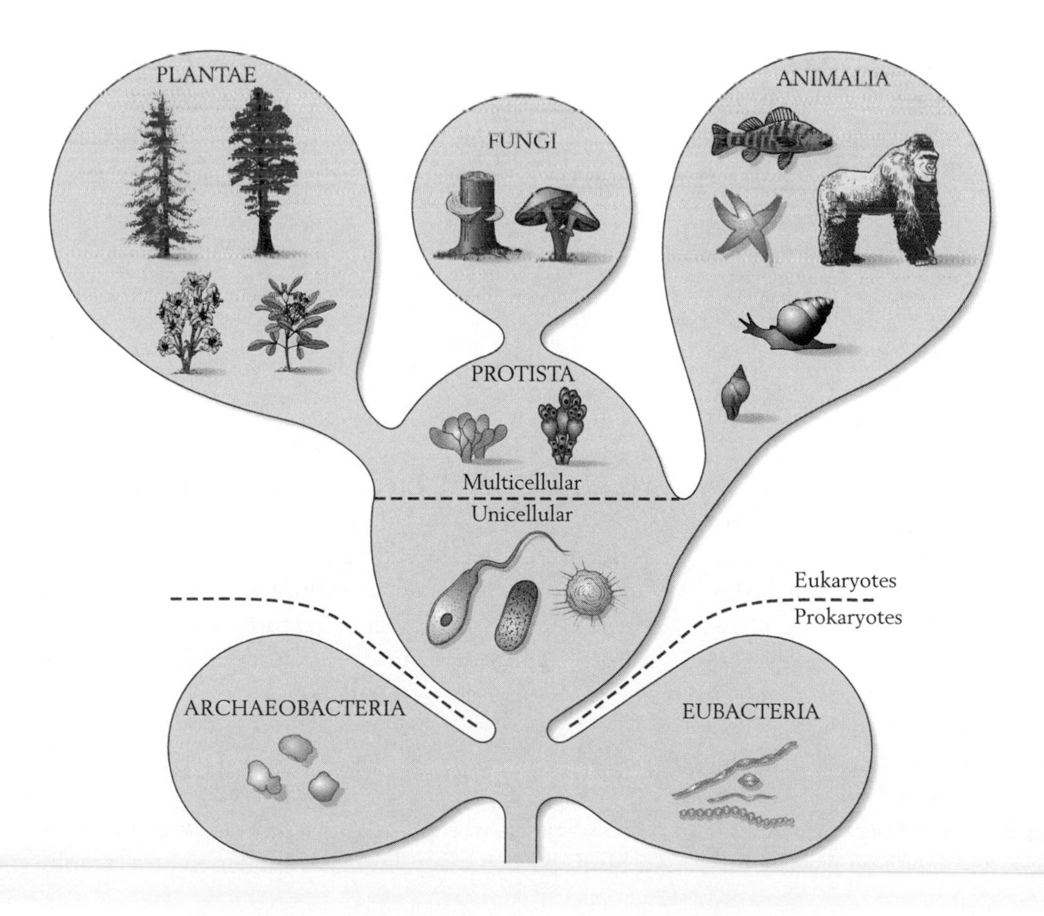

Figure 3-6 The six kingdoms of living things. Archaeobacteria and Eubacteria are termed prokaryotes because they lack a nucleus and other cellular features. These features are shared by the protists, plants, fungi, and animals, which are collectively termed eukaryotes.

speak of plants as **producers.** Members of the **Fungi** and the **Animalia,** in contrast, are **consumers:** they obtain nutrition by consuming the organic material of other forms of life. Whereas animals digest food outside their cells and then absorb the products, fungi absorb food materials into their cells and digest them there. Both plants and animals are *multicellular* (many-celled), and some fungi are too, but others are *unicellular* (single-celled).

The fourth eukaryotic kingdom is not easily defined, having been concocted as a receptacle for all forms of life that do not fit within any of the other three kingdoms. This hodgepodge, the **Protista,** is an assortment of simple, mostly unicellular species that include the groups that were ancestral to plants, fungi, and animals. Protists include both photosynthetic producers and consumers.

All six kingdoms—Archeobacteria, Eubacteria, Protista, Plantae, Fungi, and Animalia—are well represented in the fossil record.

Table 3-1

Major taxonomic categories within a kingdom, as illustrated by the classification of humans

Kingdom:	Animalia
Phylum:	Chordata
Class:	Mammalia
Order:	Primates
Family:	Hominidae
Genus:	*Homo*
Species:	*Homo sapiens*

Note: Between these categories, intermediate ones (e.g., superorders, suborders, superfamilies, and subfamilies) are sometimes recognized.

Taxonomic Groups

The six kingdoms and the numerous subordinate groups into which they are divided are known as **taxonomic groups,** or **taxa** (singular **taxon**), and the study of the composition and relationships of these groups is known as **taxonomy.**

Taxa within kingdoms range from the broad category known as the phylum (plural, phyla) to the narrowest category, the species (Table 3-1). A **species** is a group of individuals that interbreed or have the potential to interbreed in nature and that do not breed with other interbreeding groups. The basic categories of higher taxa—the kingdom, phylum, class, order, family, and genus (plural, genera)—are sometimes supplemented by categories such as the subfamily and the superfamily. Names of genera are printed in italics, as are species designations. Actually, the name of the species consists of two words, the first of which is the name of the genus to which the species belongs. This scheme of classification, introduced by the Swedish biologist Carolus Linnaeus in the eighteenth century, has the advantage of identifying every species as a member of a particular genus. For example, the first word in the name of *Panthera leo,* the species that we informally call the lion, connects this animal with closely related big cats, including *Panthera tigris,* the tiger, and *Panthera pardus,* the leopard. Thus, if we learn that an extinct animal belonged to the genus *Panthera* but was the size of a bear, we know that it was a huge cat. In fact, a lion much larger than the modern African lion inhabited North America in the recent geologic past, dying out just a few thousand years ago. As a very close relative of the modern lion, it is assigned to the genus *Panthera.*

Figure 3-7 illustrates how humans are classified within the order Primates of the class Mammalia. In general, the narrower the taxonomic category, the greater the biological similarity of its members. Humans and gorillas, for example, have enough in common to be assigned to the same superfamily. Monkeys, however, differ from these groups in several ways and are assigned to other superfamilies. All of these superfamilies are nonetheless similar enough to be united in a single suborder. Often one or a small number of biological features serve to distinguish one higher taxon from closely related taxa of the same rank.

Reconstructing the Tree of Life

The so-called tree of life is actually shaped more like a bush, having many branches but no central trunk;

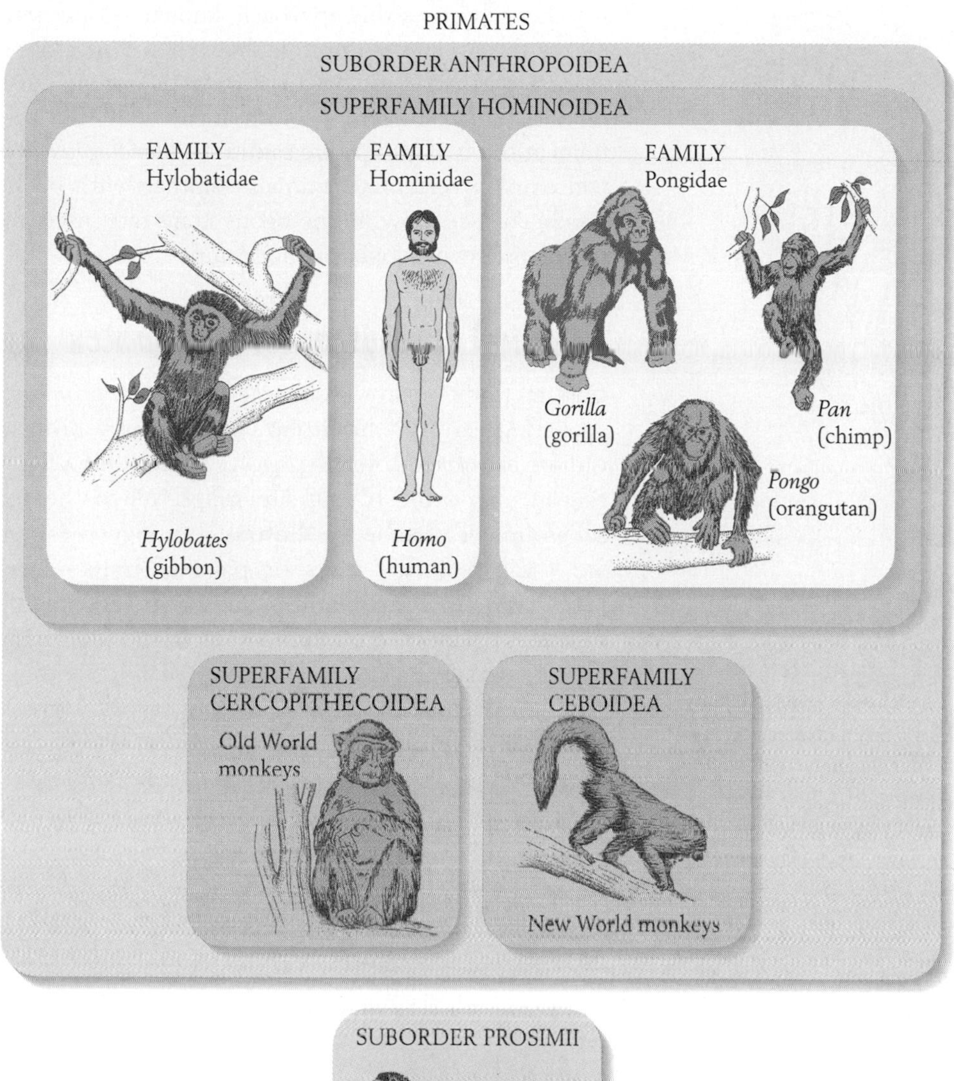

Figure 3-7 **The taxonomic position of the human genus, *Homo,* within the order Primates and the family Hominidae.** There are four other genera in the superfamily Hominoidea: three ape genera of the family Pongidae and one genus of the family Hylobatidae.

formally, it is known as the **phylogeny** of life. This structure grows as new species arise, each of them forming a separate branch. New species originate by branching off from others. Some of those species die out, but still the phylogeny grows because more species originate than become extinct.

Large phylogenies typically display a more complex structure. As Figure 3-8 indicates, their species tend to be clustered into groups that share particular traits. Researchers recognize such clusters as higher taxa. Small clusters constitute genera. Genera that share particular traits form larger clusters, which may be recognized as families. Even larger groups that are characterized by distinctive features are recognized as still higher taxa, such as orders and classes.

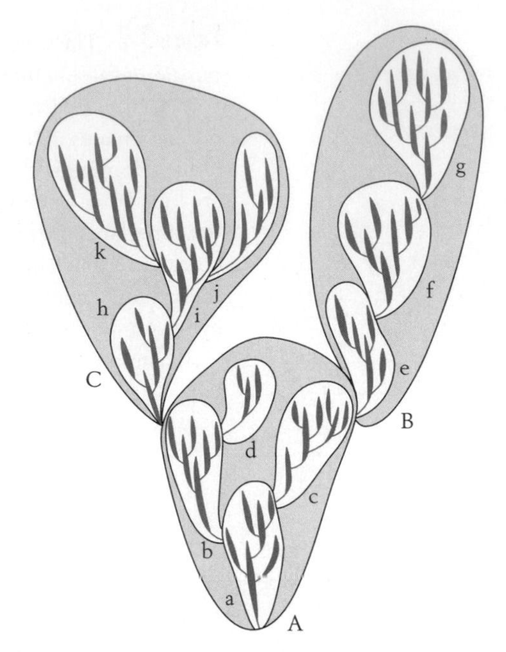

Figure 3-8 Natural clustering of species within phylogenies. Species that form clusters are closely related to one another and have a common ancestry. Such clusters form natural groupings that researchers often designate as higher taxa. The clusters labeled a–k, for example, might be recognized as genera. These clusters in turn form three larger clusters, A, B, and C, each of which might be designated as a family.

Identifying Clades and Their Relationships

Dividing a phylogeny into genera, families, and even higher taxa is a subjective matter, but one rule must be followed: to qualify as a higher taxon, any group of species must represent a coherent portion of a phylogeny. This means that all of the species must be traceable to a single branch, as illustrated in Figure 3-9. A cluster of species that share such an ancestry is termed a *clade*.

In the past, efforts to reconstruct phylogenies were based on degrees of resemblance between taxa and on evidence that some taxa were more primitive than others or existed earlier in geologic time. A relatively new technique, now widely employed, takes a different approach. It emphasizes branching events in phylogeny—events that form new clades. A researcher who uses this approach, known as *cladistics,* makes the initial assumption that when two groups share a particular biological trait, both groups have inherited the trait from a common ancestor. Thus the traits in the two groups are said to be *homologous*. We will employ vertebrate animals—animals with backbones—to see how homologous traits are used to reconstruct evolutionary relationships.

A General Cladogram for Vertebrates

Certain traits in any biological group are *primitive,* appearing early in the group's evolutionary history. Others are *derived,* having evolved later and occurring in only some of the subgroups. When the six subgroups of vertebrates illustrated in Figure 3-10*A* are compared, five traits or pairs of traits—jaws, lungs, claws or nails, feathers, and fur together with mammary glands—are found in one or more of them but are absent from the hagfishes. The five traits represent derived features, and the hagfishes are a primitive subgroup. Figure 3-10*A* depicts the relative

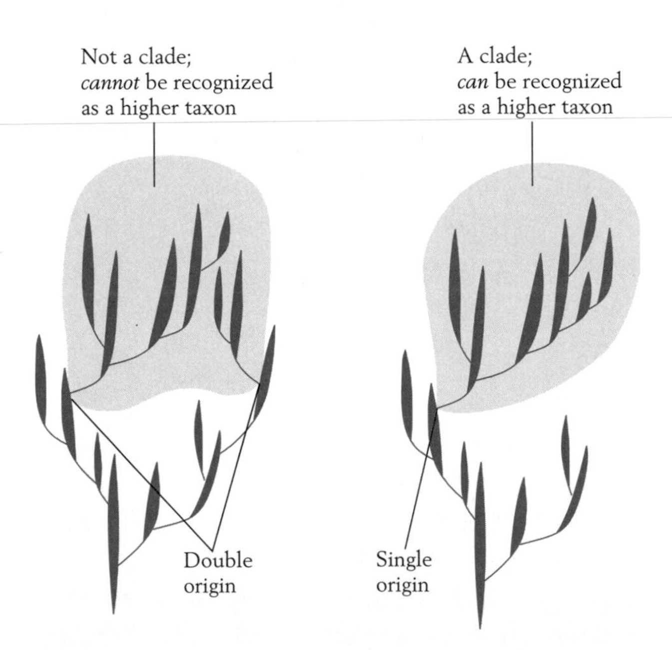

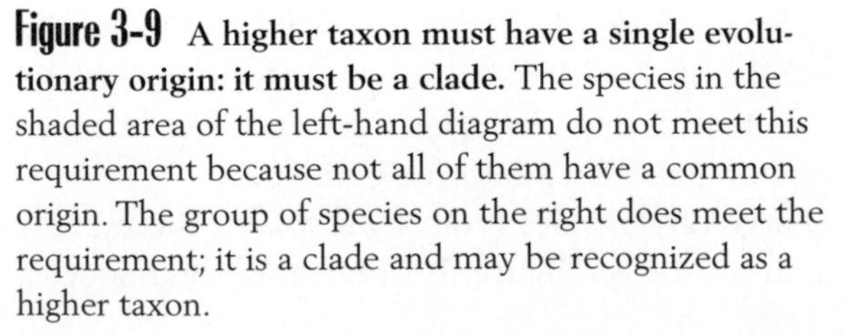
Figure 3-9 A higher taxon must have a single evolutionary origin: it must be a clade. The species in the shaded area of the left-hand diagram do not meet this requirement because not all of them have a common origin. The group of species on the right does meet the requirement; it is a clade and may be recognized as a higher taxon.

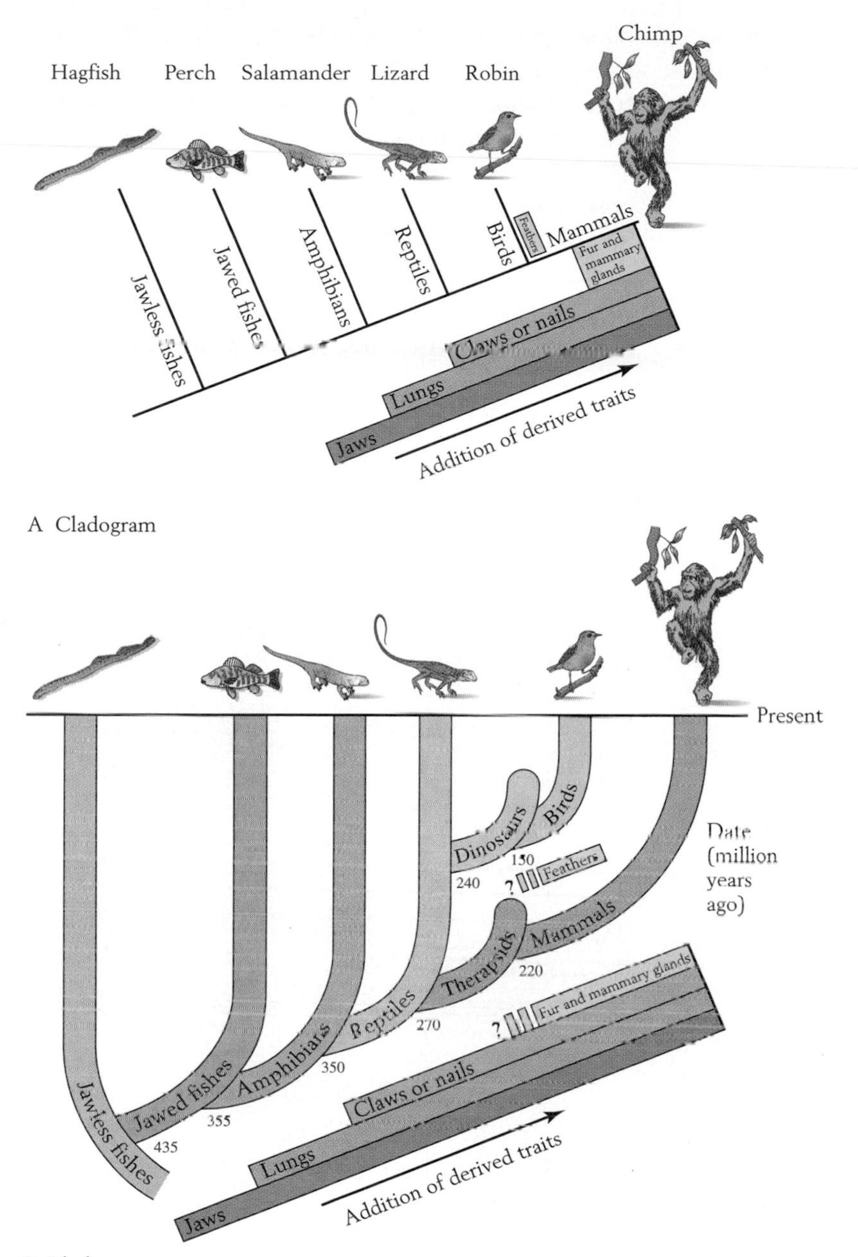

Figure 3-10 Relationships among groups of vertebrate animals. *A.* This cladogram, produced by cladistic analysis, shows major groups of present-day vertebrates arrayed along an axis on the basis of the presence or absence of derived traits. *B.* The actual, highly generalized phylogeny of mammals. The present-day groups are arranged in the order shown in the cladogram *(A)*, but this phylogeny also includes two extinct groups, therapsids and dinosaurs; cladistic analysis shows that these two groups occupy intermediate evolutionary positions between groups that survive today. Note that the approximate times of origin of the various vertebrate groups, based on fossil occurrences, form a sequence that corresponds to the branching sequence indicated by cladistic analysis.

phylogenetic positions of the various vertebrate subgroups, as reconstructed from the distribution of derived traits. This kind of figure is called a *cladogram*. Jaws are a derived trait that separates all five of the other subgroups from the hagfish. Lungs separate the other four subgroups with jaws from the jawed fishes. The origin of each derived trait, including jaws and lungs, marked a branching point in the evolution of vertebrates. Each of these branching events, by introducing the new trait, produced a new kind of animal.

A General Phylogeny for Vertebrates

Each of the five kinds of jawed animals shown in Figure 3-10*A* represents a particular class of vertebrates. A cladogram such as this does not depict a phylogeny; it simply shows what appear to be the relative evolutionary positions of selected taxa. As a result, a cladogram may not include a clade that is intermediate between two of the clades that it depicts. Figure 3-10*A*, for example, might give one the mistaken impression that mammals evolved from reptiles. The fossil record shows that an extinct group of animals known as **therapsids** actually occupied an intermediate evolutionary position between reptiles and mammals. Two conditions point to this status: therapsids share with mammals several derived features of the skull, teeth, and limbs that are absent from reptiles; yet they lack certain derived features of skull form that are present in mammals.

The evolutionary position of therapsids between reptiles and mammals is shown in Figure 3-10*B*, which is a general phylogeny of vertebrates. Anatomical evidence also indicates that birds did not evolve directly from reptiles, as one might conclude from Figure 3-10*A*. As Figure 3-10*B* illustrates, a group of dinosaurs was intermediate between reptiles and birds. Surprising as it may seem, birds evolved from small dinosaurs that apparently used feathers as insulation to maintain their body heat.

What we can conclude from the intermediate positions of therapsids and dinosaurs depicted in Figure 3-10*B* is that one can establish ancestral and descendant relationships only by including all biological groups in a cladistic analysis. Sometimes this task is impossible because a key group is unknown, never having been discovered in the fossil record.

Figure 3-10*B* indicates the geologic age of the oldest known fossil representatives of each of the large vertebrate groups that are included within it.

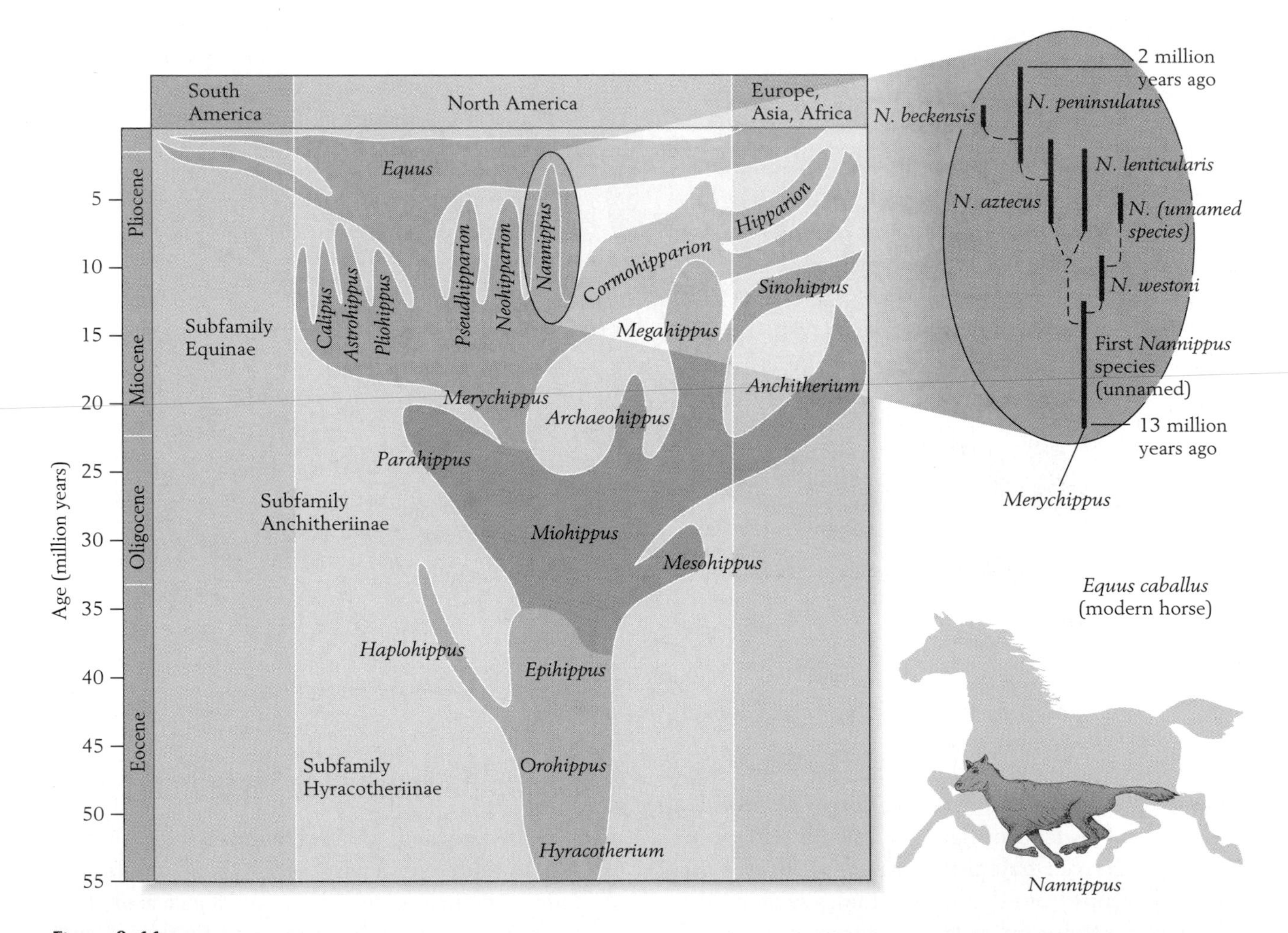

Figure 3-11 The phylogeny of *Nannippus,* a genus of small horses that lived in North America between about 13 million and 2 million years ago. On average, *Nannippus* was about half the height of a modern horse. It was a member of the Equinae, which is the subfamily that includes *Equus,* the genus that includes all living members of the horse family. The fossil record has yielded seven species of *Nannippus.* Cladistic analysis of the bones and teeth of *Nannippus* has produced the phylogeny shown here, in which the ancestry of two species remains uncertain. *(After B. J. MacFadden, Paleobiology, 11:245–257, 1985, and R. C. Hulbert, Paleobiology, 19:216–234, 1993.)*

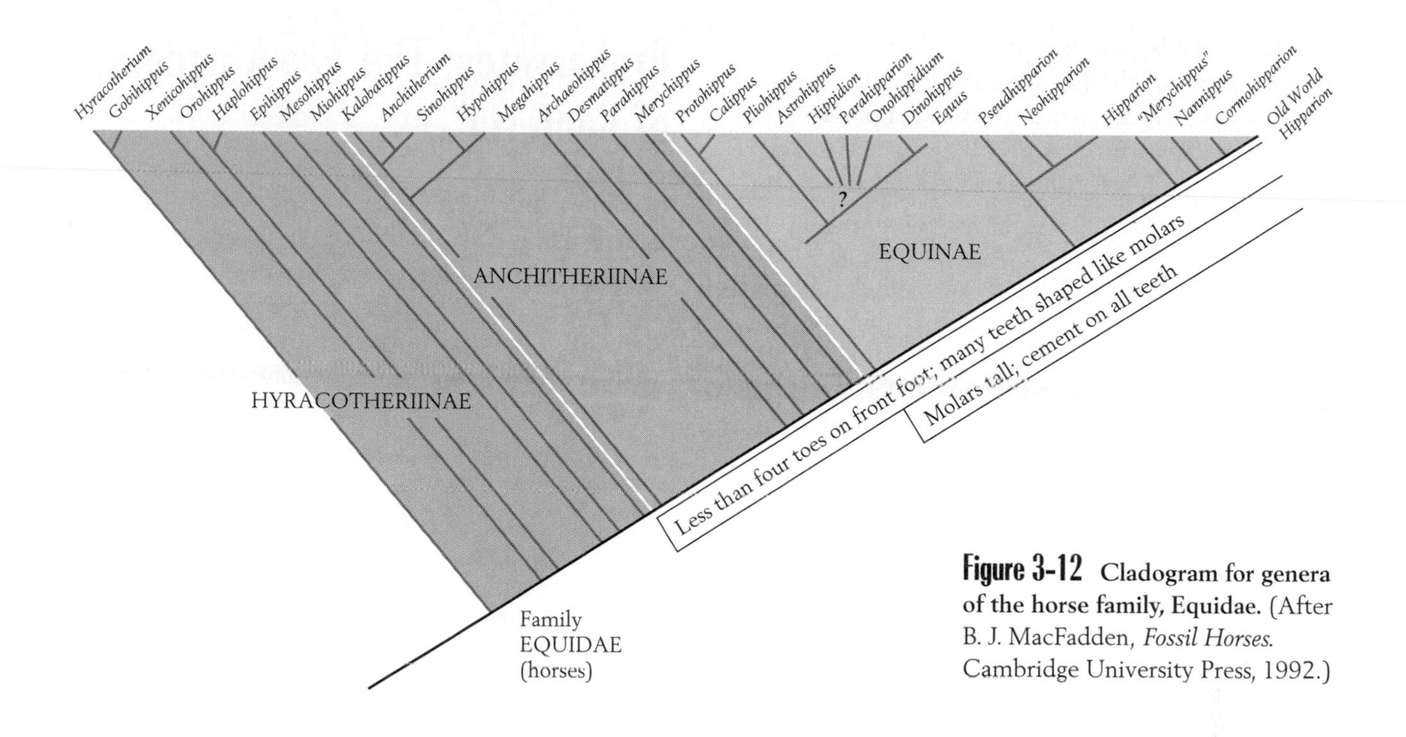

Figure 3-12 **Cladogram for genera of the horse family, Equidae.** (After B. J. MacFadden, *Fossil Horses.* Cambridge University Press, 1992.)

Note that the relative times of these first appearances are consistent with the relationships indicated by cladistics. Jawed fishes appear in the fossil record below amphibians, for example, amphibians appear below reptiles, and reptiles below mammals. This is a heartening result, supporting the cladistic methodology and suggesting that the fossil record is complete enough to be of value in the study of evolutionary relationships among major vertebrate groups.

Analyses at Lower Taxonomic Levels for Horses

Thus far we have examined relationships only among large groups of organisms, such as classes. Thus the phylogeny of vertebrates depicted in Figure 3-10*B* is very generalized. Paleontologists can begin this kind of analysis at lower taxonomic levels—even at the level of the species. Studies of the evolution of horses provide an illustration.

Horse Species Figure 3-11 shows the evolutionary relationships among species of the horse genus *Nannippus*. The name of this genus, from the Greek, means "extremely small horse," which is appropriate because typical members of *Nannippus* were only about half as tall as a modern horse. They also had three toes on each foot, rather than a single hoof like that of a modern horse.

At the level of the species, as at higher taxonomic levels, cladistic analysis does not always indicate certain ancestry. Thus the question mark beneath *Nannippus aztecus* and *Nannippus lenticulatus* indicates that the ancestry of each of these species within the genus is uncertain. In addition, an imperfect fossil record often leaves us with an underestimation of a species' stratigraphic range. One species may have descended from another even though the stratigraphic ranges of the two species that we recognize from fossils collected to date do not overlap. For some evolutionary relationships, the situation is even worse: fossils representing the ancestral species are never found.

Actually, the fossil record of horses is of relatively high quality and has been so well studied that it effectively illustrates how paleontologists reconstruct phylogenies. The phylogeny of *Nannippus* displayed in Figure 3-11 is part of a much larger phylogeny that has been reconstructed for species of fossil horses. This kind of analysis has clustered horse species into small clades that have been designated as genera, such as *Nannippus*.

Subfamilies of Horses Figure 3-12 is a cladogram for all horse genera, including *Nannippus*, that are recognized by one expert. This figure shows that the phylogeny of the horse family, Equidae, includes three clades that are recognized as subfamilies. The general phylogeny for the Equidae depicted in Figure 3-11 shows the relationships of these three subfamilies. This figure also shows the positions within the phylogeny of some of the genera, including *Nannippus*. Note that each of the first two subfamilies persisted for a time after giving rise to another subfamily. All present-day species of the horse family, including zebras, wild horses, and wild asses, belong to the genus *Equus*, the name of which is the Latin word for horse. Figure 3-11 shows that the horse family originated in North America and that most of its members have inhabited this continent. The last of the native horses of North America became extinct a few thousand years ago, however, and horses were absent from this continent until Spanish conquistadors introduced the domestic horse.

The Quality of Phylogenetic Reconstructions

The approach taken to produce cladograms such as those displayed in Figures 3-10*A* and 3-12 is not perfectly reliable. Not only does the imperfection of the fossil record produce gaps and uncertainties, but evolution sometimes follows unexpected paths that fool researchers. Some derived traits in many phylogenies, for example, have actually arisen more than once. In addition, traits have sometimes been lost secondarily and thus are absent from groups of species whose ancestors had possessed them. In general, researchers favor cladograms that minimize the numbers of multiple origins and disappearances of traits. Although the phylogenies of most other groups of animals and plants are less well known than that of the horses, we have an accurate picture of the general phylogenies—the relationships among major higher taxa—for many kinds of organisms.

In the following sections, which introduce various forms of life, we will not be concerned with the many formal taxonomic categories that scientists have erected. Rather, we will review the basic biological features and interrelationships of major groups of organisms—mainly phyla and classes—that have played large roles in the history of life and left conspicuous fossil records that allow us to assess those roles.

Prokaryotes: The Two Kingdoms of Bacteria

Bacteria gain nutrition in a great variety of ways. Some share with plants the ability to convert chemical compounds from sunlight into food. Others use light as a source of energy but absorb organic matter to obtain carbon for manufacturing food. Still others produce their own food by harnessing chemical energy rather than light. And then some are consumers, absorbing organic compounds from which they gain food and energy.

The fossil record of bacteria extends back more than 3 billion years, which is more than a billion years beyond the record of eukaryotes. Remarkably, many kinds of bacteria found in the modern world appear to differ little from forms that lived early in Earth's history.

Archaeobacteria

Several distinctive chemical compounds found within Archaeobacteria distinguish this group from Eubacteria, as do certain unique genetic features. Archaeobacteria are notable for their tolerance of extreme environmental conditions. Some forms thrive in hot springs in places such as Yellowstone National Park (Figure 3-13). One group tolerates only a combination of very high temperatures and extremely acidic conditions. Another lives only in the absence of oxygen. Still another occupies only very salty waters, such as those of the Dead Sea.

Eubacteria

Eubacteria are subdivided taxonomically on the basis of the structure of their cell walls. Some cause diseases in plants or animals. Others play a more positive role in the natural world, breaking down the cells and tissues of dead organisms and thus liberating nutrients for the use of other life forms. Many kinds of eubacteria are capable of locomotion.

The **cyanobacteria** are a group of photosynthetic eubacteria that have left an especially important fossil record. Some cyanobacteria are more or less spherical, and others are filamentous (threadlike) in form (Figure 3-14). Some filamentous types float as green-

Figure 3-13 Archaeobacteria that are adapted to high temperatures. A greenish mat of archaeobacteria surrounds a steaming hot spring in Yellowstone National Park.

Figure 3-14 *Oscillatoria,* a genus of eubacteria. This is a photosynthetic form that belongs to the group known as cyanobacteria. Its filaments often form sticky mats.

ish scum in lakes, streams, or the sea. Others form mats on the seafloor that can trap sediment to produce distinctive three-dimensional structures. The fossil record of these structures extends back more than 3 billion years.

The Protist and Fungus Kingdoms

Protists

Protists include many kinds of single-celled organisms and a few kinds of simple multicellular organisms. Protists are a varied assemblage of phyla, all of which contain simple organisms. Some scientists classify only unicellular organisms as protists, but we will include in this group the multicellular photosynthetic forms that are informally known as seaweeds. Seaweeds, together with single-celled protists that are also photosynthetic, are known as *algae* (singular, *alga*). Algae are the most prominent producer organisms in lakes and the ocean.

Animal-like protists are known as **protozoans** (Figure 3-15). Among them are amoebas, which have irregular, constantly changing shapes; flagellates, which employ a whiplike structure, called a *flagellum*, for feeding; and ciliates, which move by means of cilia—structures that resemble flagella but are shorter and more numerous, and that beat in unison.

Three groups of unicellular algae that float in natural bodies of water are well represented by abundant fossils: dinoflagellates, diatoms, and calcareous nannoplankton. These groups all originated during the Mesozoic Era, and together they are the most prominent producers in modern seas, serving as food for a great variety of animals.

Dinoflagellates employ two flagella for limited locomotion (Figure 3-16*A*) but are transported chiefly by movements of the water in which they drift. When conditions are unfavorable for survival, some dinoflagellates enter a state of dormancy, encasing themselves in a protective armor of tough organic matter, forming what is called a cyst. When conditions improve, a dinoflagellate can emerge from this state to resume an active life; some cysts, however, have become fossilized by sinking to the bottom and becoming buried in sediment.

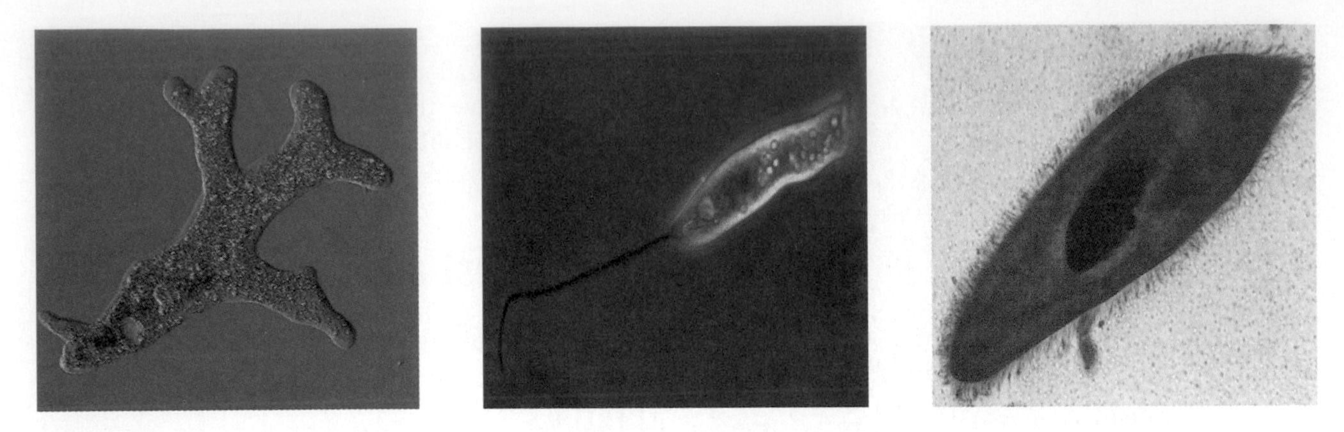

Figure 3-15 **Protozoans.** Pictured here are three types: *Amoeba*, which has a variable shape; a zooflagellate, with its flagellum extending to the left; and *Paramecium*, which is surrounded by cilia.

Diatoms are unicellular forms that secrete two-part skeletons of opal, a form of silicon dioxide that differs from quartz in lacking a crystalline structure (p. 38). The two halves of the skeleton fit together like the top and bottom of a petri dish (Figure 3-16*B*). Some diatom species live in lakes and others in the ocean. Most species float, but some lie on the bottom in shallow water where they receive enough light for photosynthesis. At certain times in the geologic past, diatoms have flourished so spectacularly that their skeletons have rained down on the seafloor to produce thick bodies of sediment. Some accumulations of diatom skeletons have eventually turned into chert (p. 47); others remain as soft accumulations that are mined to produce the scouring powder that we use in kitchens and bathrooms.

Calcareous nannoplankton are small, nearly spherical cells that secrete minute, shieldlike plates of calcium carbonate that overlap to serve as armor against small attackers (Figure 3-16C). Nearly all species of this group drift in the ocean but a few live in lakes. Some concentrated accumulations of plates

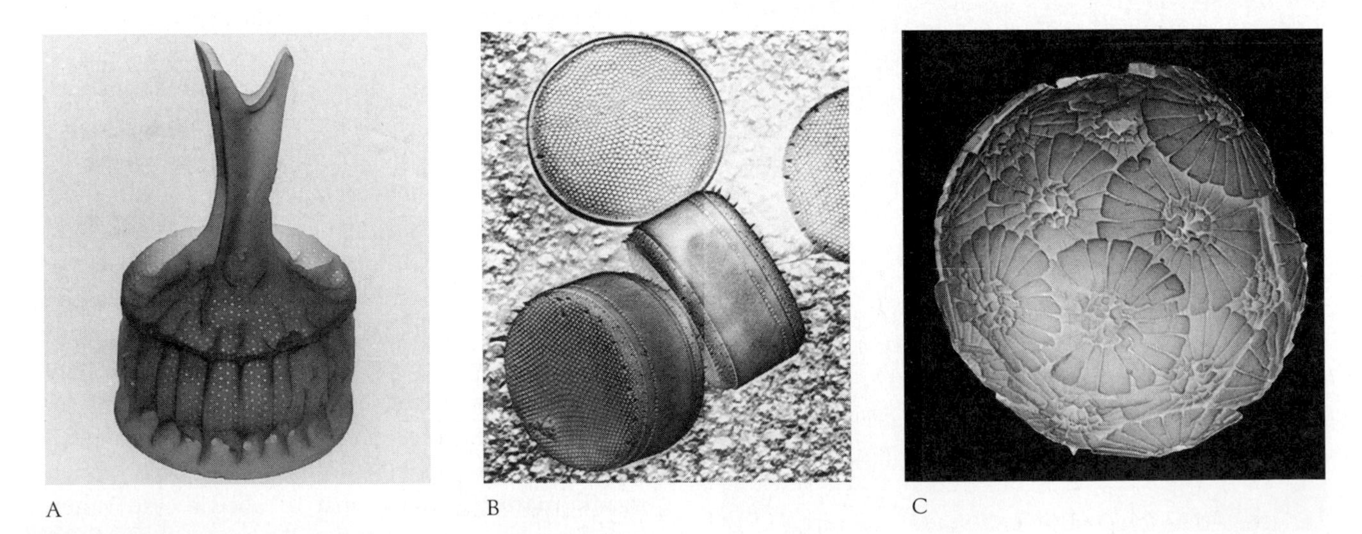

Figure 3-16 **Representatives of the major types of single-celled algae in natural bodies of water today: a dinoflagellate *(A)*, a diatom *(B)*, and a cell of calcareous nannoplankton *(C)*.** Their diameters are about 10, 10, and 8 μm, respectively. *(A. Michael Hoban, California Academy of Sciences. B. and C., L. Stein. C. Mitch Covington, Florida Geological Survey.)*

A

B

Figure 3-17 **A fleshy green alga *(Ulva) (A)* and a calcareous green alga *(Halimeda) (B).***

from calcareous nannoplankton have become weakly lithified to become the fine-grained form of limestone called chalk. We use this material, cut into cylindrical sticks, to write on chalkboards.

Multicellular algae include species that drift passively in bodies of water, but most kinds attach to the bottom. Most multicellular algae are soft, fleshy organisms, but some types that belong to the groups known as red algae and green algae secrete skeletons of calcium carbonate (Figure 3-17*B*). Fragments of such skeletons are major constituents of limestones.

Foraminifera and radiolarians are protozoan groups that secrete skeletons and have fossil records spanning the entire Phanerozoic Eon. **Foraminifera,** nicknamed "forams," form a chambered skeleton by secreting calcite or cementing grains of sand together. Long filaments of their protoplasm extend through pores in the skeleton and interconnect to form a sticky net in which they catch food (Figure 3-18). Some forams float in the ocean, and others live on the seafloor. So abundant are their skeletons on shallow seafloors that a few are likely to be present in a handful of beach sand. Foraminifers are widely used to date rocks. Because of their small size, they are especially useful when the samples to be dated are too small to contain many larger fossils. Thus forams are widely used in the search for petroleum, where the only available rock samples are cuttings from drill cores. Similarly, forams are frequently used to date layers in cores of sediment obtained by drilling into the floor of the deep sea.

Radiolarians are closely related to forams and, like them, capture food with threadlike extensions of protoplasm that radiate from their skeletons. These

Figure 3-18 **A living planktonic foraminifer.** Strands of protoplasm radiate from the skeleton, which is the size of a grain of sand.

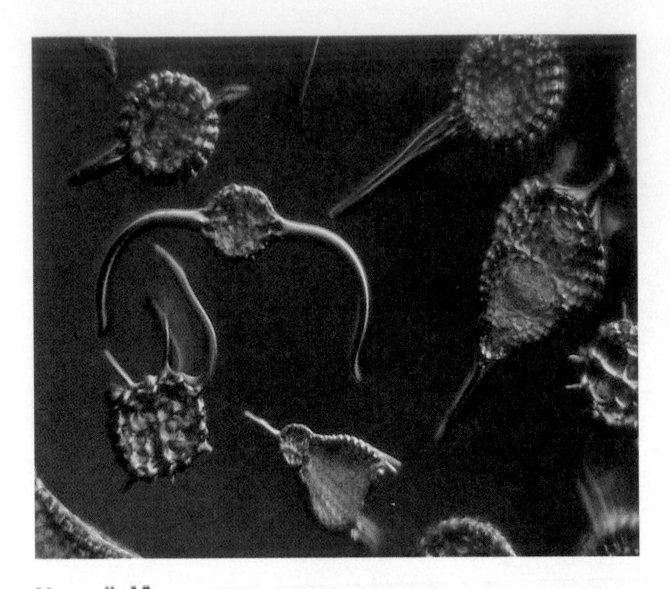

Figure 3-19 **Radiolarians.** The skeletons, which are made of silica, are the size of a grain of sand.

marine floaters secrete skeletons of opal that are perhaps the most beautiful organic structures in the sea (Figure 3-19).

Fungi

Fungi absorb most of their food from dead organisms. Most mushrooms, for example, feed on wood, bark, or dead leaves. Fungi typically have filamentous cells. Some groups, such as yeasts, are unicellular, but others, including mushrooms, have many cells that are packed tightly into bundles. Although fungi have a generally poor fossil record because they lack skeletons, those fossils that survive can be quite revealing. An abundance of fungal filaments in sedimentary layers just above those that mark the largest extinction of all time, at the end of the Paleozoic Era, points to a sudden killing off of life across broad regions of Earth: numerous fungi apparently were feasting on the victims.

The Plant Kingdom

Plants have a pattern of development that distinguishes them from multicellular green algae, the group from which they evolved (Figure 3-20). Algae shed their gametes into the water in which they live, so that fertilization is external and offspring develop independently in a watery environment. The egg of a plant, in contrast, is fertilized within the parent plant, and the embryo remains protected there, at least for a time, as what amounts to a parasite.

The early evolution of plants from green algae entailed a shift from life in the water to life on the land. Early plants were transitional, living in shallows along the margins of bodies of water and becoming first partly emergent, later entirely so. Many basic fea-

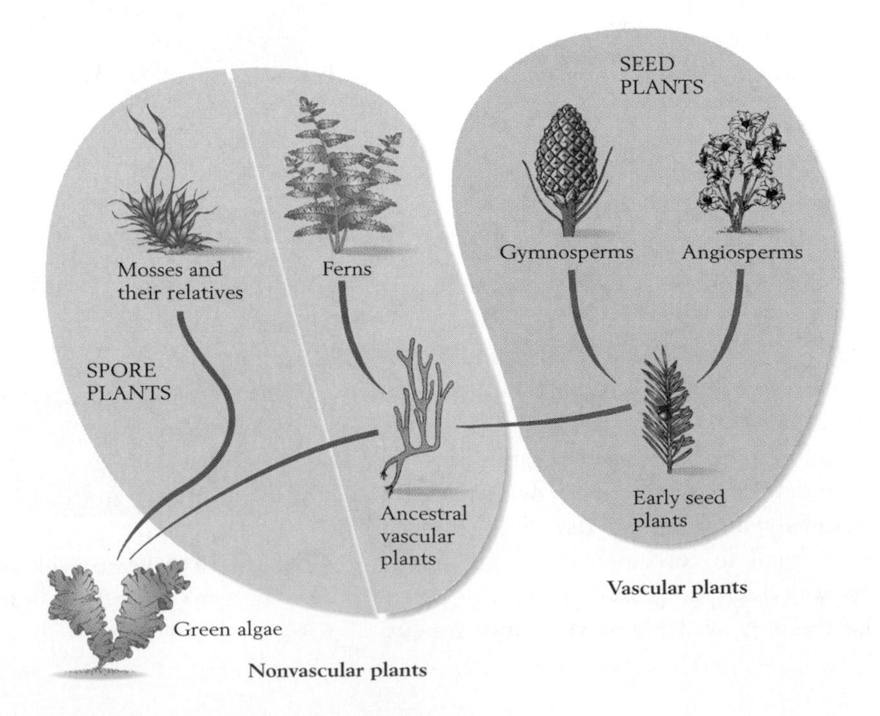

Figure 3-20 **Phylogeny of multicellular plants.**

tures of plants relate to problems of living on land. Early plants evolved rigid stems and roots that enabled them to stand upright without the support of water. Roots also provided water and nutrients from the soil, since the entire plant could no longer absorb them from a surrounding watery medium. Some early land plants also became **vascular:** they evolved vessels to transport water and dissolved nutrients and to distribute the food that they manufactured from these raw materials.

Mosses and a few other small, simple plants of the modern world are *nonvascular:* they lack conducting vessels and rely on diffusion for transport of materials from cell to cell. Lacking seeds, mosses reproduce by means of tiny spores. Mosses also lack multicellular roots; they soak up water and nutrients with tiny hairlike extensions of cells. Such forms have changed little from land plants that lived more than 400 million years ago.

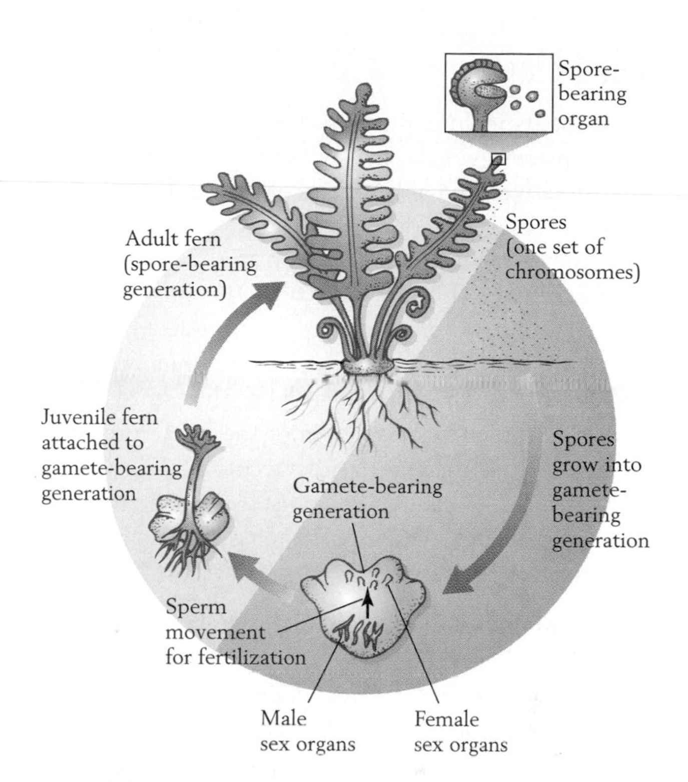

Figure 3-22 Life cycle of a fern. The large fern plant has two sets of chromosomes but produces spores that have only one. A spore grows into a tiny plant that also has only one pair of chromosomes but produces eggs and sperms. A sperm fertilizes an egg to produce another large plant with two sets of chromosomes that grows on top of the tiny parent plant.

Seedless Vascular Plants

The simplest vascular plant of the modern world is *Psilotum* (Figure 3-21) which lacks roots and leaves and resembles the earliest vascular plants known from the fossil record.

Ferns represent a more advanced level of evolution, having both roots and leaves. Ferns and other kinds of seedless plants, including mosses, reproduce by means of spores. **Spores** are tiny, durable structures that are shed by the conspicuous fern or moss plant that is familiar to us. As spores drift through the air, they disperse the offspring of the plant that produces them. Each spore grows into a tiny adult plant that is quite different from the large parent plant (Figure 3-22). This tiny plant produces eggs and sperms. A sperm crawls over the surface of the plant to fertilize an egg. This kind of reproductive cycle entails what is called an alternation of generations: a sperm-producing generation alternates with one that produces eggs and sperms. A sperm requires moisture to complete its journey to the egg; this is why ferns typically occupy moist habitats.

Figure 3-21 ***Psilotum,*** **the simplest vascular plant alive today.** This form lacks leaves and roots. The yellow knoblike structures at the ends of stalks are spore-bearing organs.

Most groups of spore plants in the modern world are only a few inches tall, but late in the Paleozoic Era several groups grew to the size of trees. They occupied broad swamps, where their remains accumulated to form peat (p. 54). Much of this peat turned into the coal that supplies a large part of the electric power that humans use today.

Seed Plants

Seed plants include most species of large land plants in the present world. Among them are pine trees and their relatives, and also species of plants that produce flowers. Seeds are durable structures that function to disperse plants. Some blow in the wind, and others have barbs that can attach to the fur of mobile animals. Still other seeds are surrounded by fruits that animals can eat, with the result that the seeds are deposited in nutrient-rich feces that serve as fertilizer. The evolutionary origin of the seed during the Paleozoic Era triggered a great ecological expansion of land plants beyond the moist environments that spore plants had required for their sperm to swim to eggs for reproduction.

Gymnosperms The plants known as **gymnosperms** ("naked seed" plants) produce seeds that are exposed to the environment. Gymnosperms include **conifers,** or cone-bearing plants (pine, spruce, and fir trees and their relatives), as well as less common groups such as cycads (see Figure 3-4). The seeds of conifers are exposed because they are located on the tops of the scales that together form a cone. After pollen forms on a cone, wind transports it to another cone that produces eggs, and there, after fertilization, seeds grow and are released. Gymnosperms were the dominant large plants of the Mesozoic Era, when dinosaurs roamed the land. Since early in the Cenozoic Era, however, flowering plants have been more abundant than gymnosperms in most terrestrial regions. Today gymnosperms are the dominant large plants only in cold regions and in some dry, sandy environments.

Angiosperms Flowering plants are more formally known as **angiosperms.** Their flowers endow them with special means of pollen transport. The colors and fragrances of flowers attract animals—especially insect and birds—to feed on nectar and carry away sticky pollen to other plants of the same species that the animals visit to obtain additional nectar. Many angiosperms, including hardwood trees, such as oaks and maples, have inconspicuous flowers. Others have showy flowers that attract particular species of pollinators. The seeds of angiosperms have a special advantage over those of gymnosperms: angiosperms engage in double fertilization, which places two eggs in close contact. One egg becomes the seed, and the second grows into a special tissue that nourishes the young plant that grows from the seed. Thus the seedling gets off to a good start—one of the reasons that angiosperms greatly outnumber gymnosperms in the modern world.

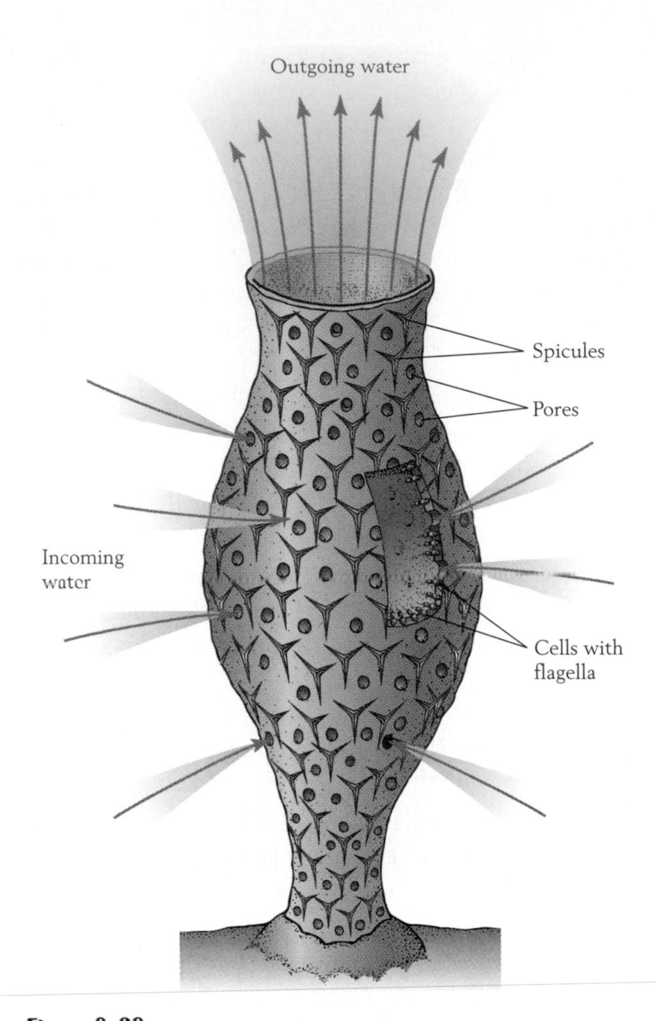

Figure 3-23 **Body plan of a simple sponge.** *(After W. K. Purves, G. H. Orians, and H. C. Heller, Life, Sinauer Associates, Sunderland, Mass., 1995.)*

The Animal Kingdom

Animals are conveniently divided into **vertebrates,** which possess a backbone, and **invertebrates,** which lack one. The bodies of most animals are formed of tissues. A **tissue** is a connected group of similar cells of a particular type that perform a particular function or group of functions. Only a few simple invertebrates lack tissues; the largest and most important of these groups is the sponges.

Sponges: Simple Invertebrates

The simplest animals that are conspicuous in the modern world are the *sponges.* These animals have irregular shapes and are formed of several cell types that are widely distributed throughout their bodies. Most species live in the ocean, but some live in lakes, and all are **suspension feeders;** that is, they strain small particles of food from water. Bacteria are the primary food that sponges obtain in this manner. Cells that bear flagella pump water into a sponge through numerous pores, and individual cells capture the bacteria from currents that pass through internal canals (Figure 3-23). The canals converge and exit through one or more central canals. Some sponges have supportive skeletons of tough organic material—the substance of bath sponges. Others secrete calcium carbonate or silica in the form of small, needlelike or many-pointed elements called *spicules* or as three-dimensional skeletons. Spicules and skeletons give sponges a conspicuous fossil record that extends back to the Cambrian Period.

Cnidarians

Jellyfishes, corals, and their relatives belong to the group known as **cnidarians.** Most members of this group live in the ocean, but a few occupy freshwater environments. Though simple animals, cnidarians exhibit a tissue grade of organization (Figure 3-24). Tissues that form an inner and an outer body layer change the shape of a cnidarian's cylindrical body by contracting or relaxing. Between these layers is a jellylike inner layer that stiffens the body. Cnidarians are carnivores that catch small animals with tentacles, which are armed with special stinging cells. The tentacles surround the mouth and pass food through it into a large digestive cavity. Cnidarians have nearly *radial symmetry*—that is, they have no left and right sides but face the environment with similar biological features on all sides. Radial symmetry is a typical trait of animals that are immobile or that have no preferred direction of movement. **Corals** are cnidarians that have left an excellent fossil record. Recall that they secrete skeletons of calcium carbonate (p. 36). Cnidarians reproduce not only sexually but also asexually, by budding. Budding can produce large colonies of interconnected individuals, such as the coral colonies that form reefs in modern seas.

Coral colonies grow in large aggregations that stand above modern seafloors as reefs. Reefs in modern seas serve as habitats for a great variety of plants and animals. Coral reefs that flourished long ago are conspicuous features of the geologic record. Some that are buried deep beneath Earth's surface are porous structures that serve as traps for petroleum.

A

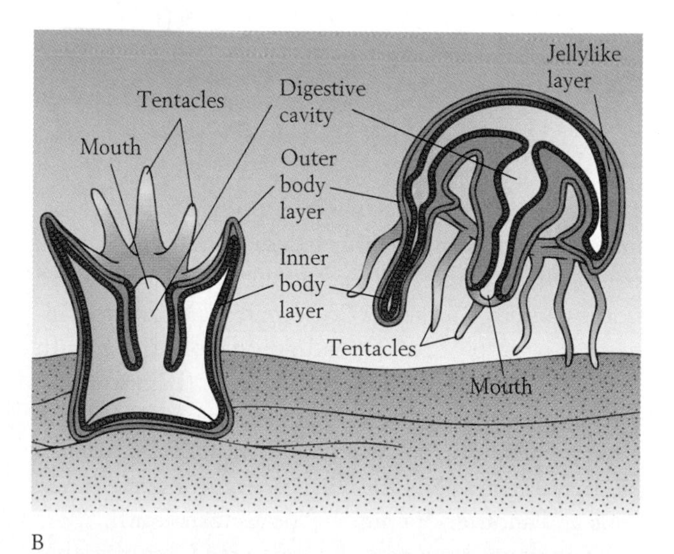

B

Figure 3-24 Body plans of cnidarians. *A.* A coral colony, some individuals of which are extending their tentacles from the skeleton. *B.* On the left is a cross section of a bottom-dwelling form that resembles a coral but lacks a skeleton. On the right is a cross section of a jellyfish.

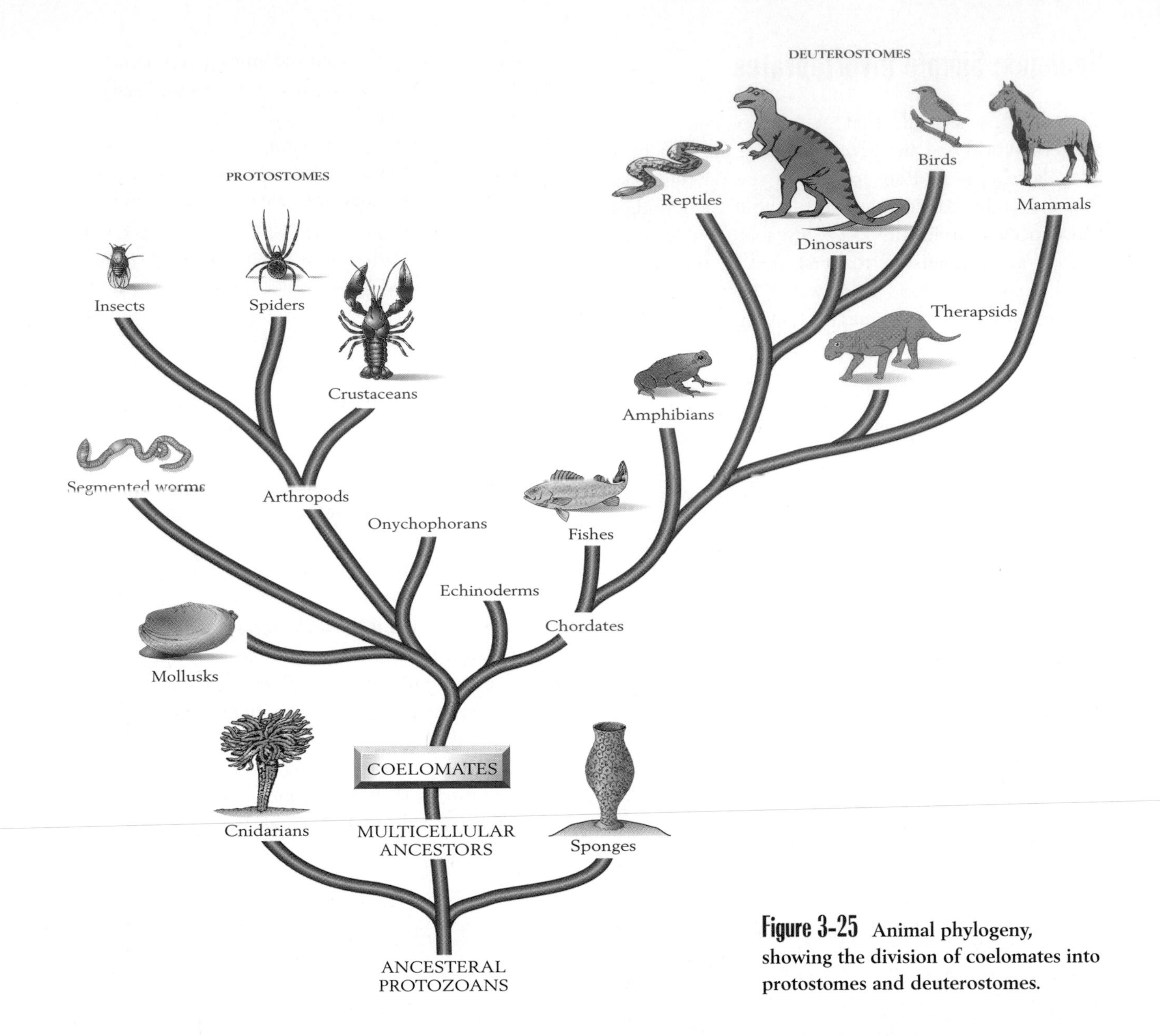

Figure 3-25 Animal phylogeny, showing the division of coelomates into protostomes and deuterostomes.

Coelomates: Animals with a Body Cavity **Coelomates,** animals more advanced than very simple worms, have a body cavity known as a *coelom*, which houses internal organs. Coelomates are divided into two large groups, depending on how the mouth forms during embryonic development (Figure 3-25). In *protostomes* the first opening to form in the embryo becomes the mouth. In most *deuterostomes*, on the other hand, the first opening becomes the anus and another opening becomes the mouth.

Segmented worms are advanced worms that have a fluid-filled coelom that serves as a primitive skeleton under the pressure of muscular contraction (Figure 3-26). Each segment of a segmented worm has its own coelomic cavity that can expand or contract independently, allowing the animal to move as bulging waves pass from end to end. Among the segmented worms are earthworms, which burrow in soil and feed on buried organic matter. Many other members of this group live in sand or mud at the bottom of rivers, lakes, or the sea; their burrows are conspicuous in sedimentary rocks hundreds of millions of years old.

Arthropods include a great variety of living animals, among them insects, spiders, crabs, and lobsters. Although their name means "jointed foot," most arthropods have appendages that serve as legs rather than feet. All arthropods have additional appendages

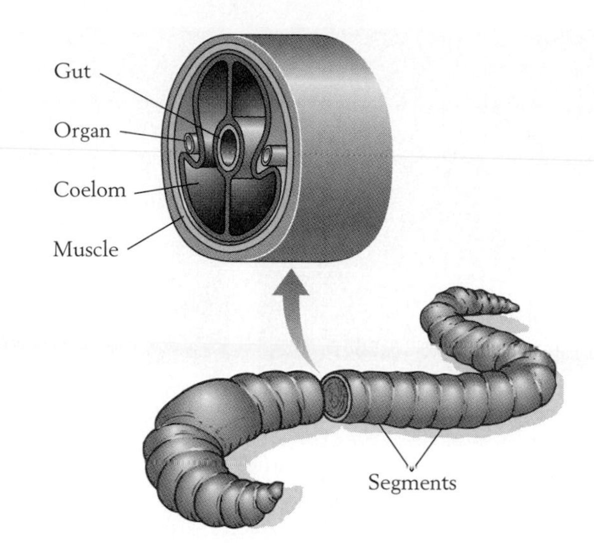

Figure 3-26 **The body plan of a segmented worm.** A fluid-filled coelom lies within the middle body layer. Each segment contains its own coelomic compartments. *(After W. K. Purves, G. H. Orians, and H. C. Heller, Life, Sinauer Associates, Sunderland, Mass., 1995.)*

that have become modified for such activities as manipulating food, sensing the environment, and swimming. Like segmented worms, which are their ancestors, arthropods have bodies that are divided into segments. The bodies of arthropods are not soft and wormlike, however, but are protected by an external skeleton of stiff organic material that in some forms is strengthened by impregnation with a mineral such as calcite. Like a suit of armor, this jointed skeleton allows for some flexibility of the body inside. Among the many groups of arthropods are the trilobites, crustaceans, and insects.

Trilobites are an extinct group of marine arthropods whose fossils are popular with the general public. They were especially common during the Cambrian Period, the interval during which a great variety of animals with hard parts first appeared on Earth. As their name indicates, trilobites had a body that was divided into three lobes, a central lobe and left and right lateral lobes (Figure 3-27). Several of their segments were fused to form a rigid head structure and several others were fused to form a rigid tail structure. The external skeleton of this animal was heavily calcified. Beneath the skeleton of most species were many pairs of appendages, each of which branched to form a gill-like structure for respiration and a leg for locomotion. Trilobites fed on small animals or particles of organic matter. A few types floated in the water or burrowed in sediment, but most crawled over the surface of the seafloor. Many species of trilobites had simple eyes.

Crustaceans are arthropods with a head formed of five fused segments, behind which are a thorax and an abdomen formed of additional segments. Among the many kinds of crustaceans are lobsters, shrimps, and crabs. Others are small floating creatures. The ocean contains the greatest variety of crustaceans, but many live in lakes or streams and a few are land dwellers. Because the external skeleton of most of these forms is uncalcified or only weakly calcified, crustaceans have a relatively poor fossil record. Even so, this record extends back to the lower Paleozoic.

Insects are a group of arthropods that includes most species of animals on Earth, yet almost none

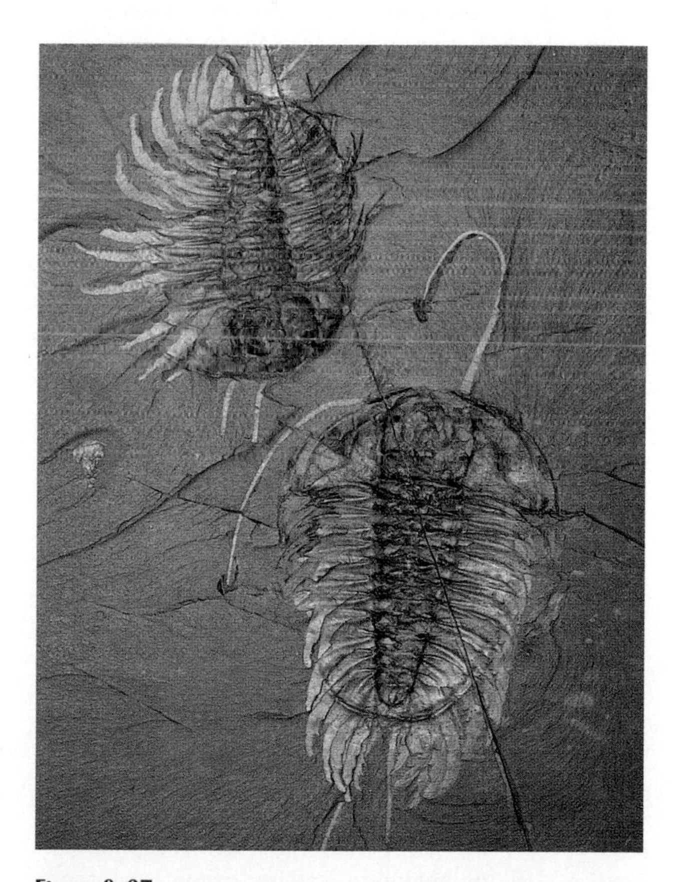

Figure 3-27 **Trilobites.** Not only are the segments and the three lobes of the skeleton visible in these exceptionally well-preserved specimens, but because the animals were buried in sediment that lacked free oxygen, their legs (one per segment) and antennae were also preserved. This species is about 10 centimeters (4 inches) long.

Figure 3-28 **Onychophorans, which are intermediate in form between segmented worms and arthropods.** *A.* A living animal, which is a terrestrial predator about the length of a pen. *B.* A fossil of a species belonging to this group, from Cambrian strata deposited in a marine environment more than 500 million years ago.

live in the ocean. Insects breath air through a system of tubes. Like crustaceans, they have bodies divided into a head, thorax, and abdomen, but nearly all have two pairs of wings (flies have only one). Insects play many roles in nature, but one of the most important is that of fertilizing plants by transporting pollen from flower to flower. Although insects are preserved in sediments only under unusual circumstances, their fossil record extends far back beyond the first appearance of flowering plants, revealing much about the group's general evolutionary history.

The **onychophorans** contribute greatly to our understanding of invertebrate evolution because they are intermediate in form between segmented worms and arthropods. They are wormlike in shape but, like arthropods, have a series of legs along their bodies (Figure 3-28). These legs are not jointed like those of arthropods, and thus represent an earlier stage of evolution. The fossil record of onychophorans extends back almost to the base of the Paleozoic Era. Early onychophorans lived in the sea. In contrast, modern representatives of the group, which reach about 15 centimeters (6 inches) in length, are terrestrial animals that occupy moist forests, where they feed on other small animals.

Mollusks Snails, clams, octapuses, and their relatives are familiar groups of **mollusks.** Most mollusks have a shell of aragonite, calcite, or both of these forms of $CaCO_3$ (p. 36). A *mantle,* which is a fleshy, sheetlike organ, secretes this shell. Most mollusks respire by means of featherlike gills, and many use a filelike structure called a *radula* to obtain food. The major groups of mollusks all have fossil records that extend back to the Cambrian Period. Figure 3-29 shows the anatomical relationships of the various classes of mollusks.

Monoplacophorans are the most primitive mollusks, with a fossil record extending back to the base of the Cambrian. This group was long considered to be extinct, but in 1952 marine biologists discovered living monoplacophorans on the floor of the deep ocean. Members of this class have cap-shaped shells and creep about on a broad foot. Monoplacophorans use their radula to graze on organic matter. It appears that all other mollusks ultimately trace back to monoplacophoran ancestors.

Snails, more formally termed **gastropods,** evolved from monoplacophorans at the beginning of the Paleozoic Era. A snail of the most basic type can be viewed as a monoplacophoran whose body has been twisted so that the digestive tract is U-shaped, with the anus positioned above the head. Snails constitute the largest and most varied class of mollusks. Most are marine, but some have come to live in freshwater environments. Through the evolution of a lung for respiring in air, others have become land dwellers. Some snails, including slugs, have lost their shells. A few types have become suspension feeders rather than grazers, and many marine groups feed on other animals.

Cephalopods include squids, octapuses, chambered nautiluses, and their relatives. All members of this molluscan group swim in the sea and feed on other animals. They have eyes and pursue their prey by jet propulsion, squirting water out through a small opening in the body. Cephalopods use a strong beak to eat other animals that they catch with tentacles.

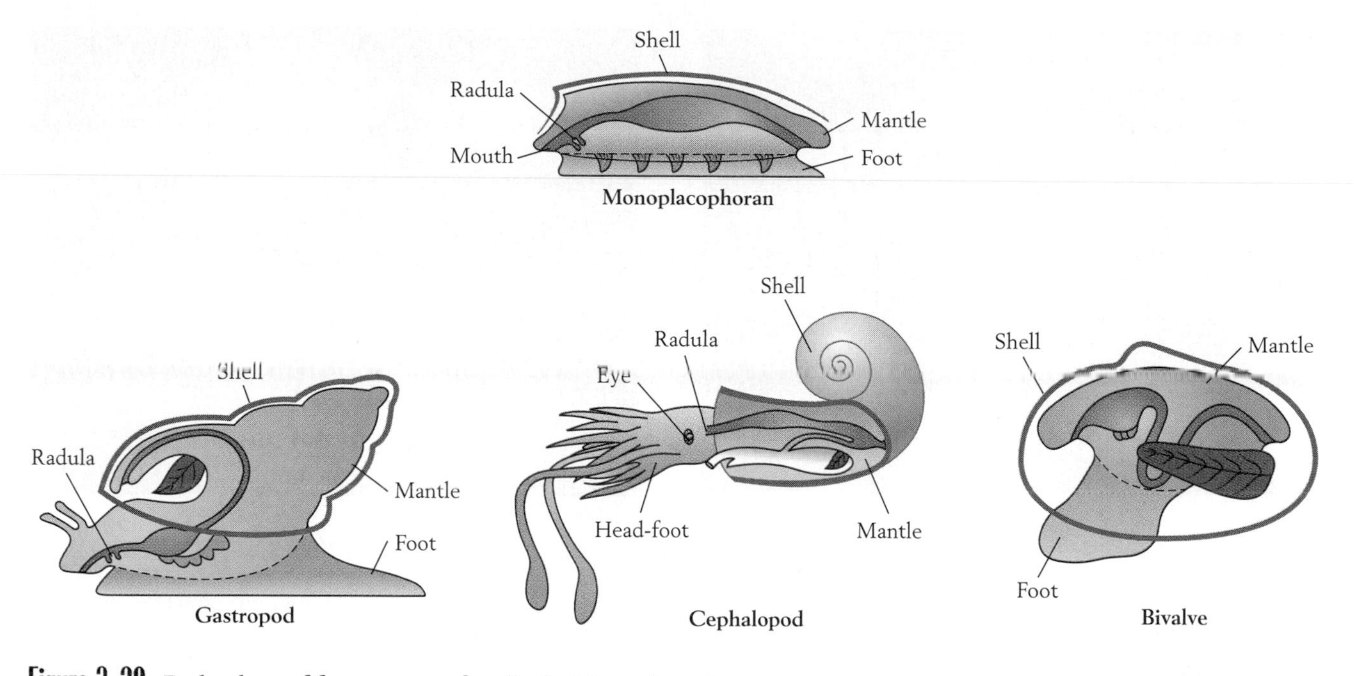

Figure 3-29 Body plans of four groups of mollusks. Monoplacophorans represent a primitive condition from which the other groups were derived.

The living chambered nautilus belongs to the most primitive cephalopod group (p. 56). The chambers of its shell are filled partly with water but also partly with gas, which prevents them from sinking to the seafloor. Cephalopods with chambered shells, though not well represented in modern seas, were a diverse group throughout most of the Phanerozoic Eon and their fossils are widely used to date rocks. Forms known as ammonoids secreted beautiful shells that had more complicated partitions between their chambers than those of the living chambered nautilus.

Bivalves, as their name suggests, have a shell that is divided into two halves, known as valves. This group includes clams, mussels, oysters, scallops, and their relatives (see Figure 1-12). The head and radula were lost in the evolution of this group, and the shell became folded along a hinge, with each of the two sides becoming one valve. One or two short cylindrical muscles pull the valves together. The single muscle of this type is the part of the scallop that we eat. Some bivalves use their foot to burrow in sediment. Others, including most species of oysters, cement their shell to a shell or rock or attach to these hard objects with threads that they secrete. Most bivalves are suspension feeders, but a few kinds of burrowers extract food from sediment.

Brachiopods Having shells that are divided into two halves, **brachiopods** look superficially like bivalve mollusks, but actually the two groups are not closely related (Figure 3-30; see also Figure 3-3). Brachiopods are sometimes called lampshells, because they resemble artists' portrayals of Aladdin's oil lamp. They employ a frilly loop-shaped structure to pump water and sieve small food particles from it. Brachiopods live in the ocean. They are uncommon today, but they played a major ecological role in ancient seas. In fact, they are the most conspicuous fossils in rocks of Paleozoic age. *Articulate brachiopods* are characterized by teeth that interlock along the hinge between the two valves. Most articulate brachiopods attach to the substratum by means of a fleshy stalk (Figure 3-30*B*), but some extinct forms lived free on the substratum or cemented their shells to hard objects. *Inarticulate brachiopods* lack hinge teeth. Among them is the living genus *Lingula*, which belongs to a group that originated early in the Paleozoic Era. Members of this group live in the sediment, anchored by their fleshy stalk (Figure 3-30*A*).

Bryozoans Although **bryozoans,** or moss animals, form colonies of tiny, interconnected individuals

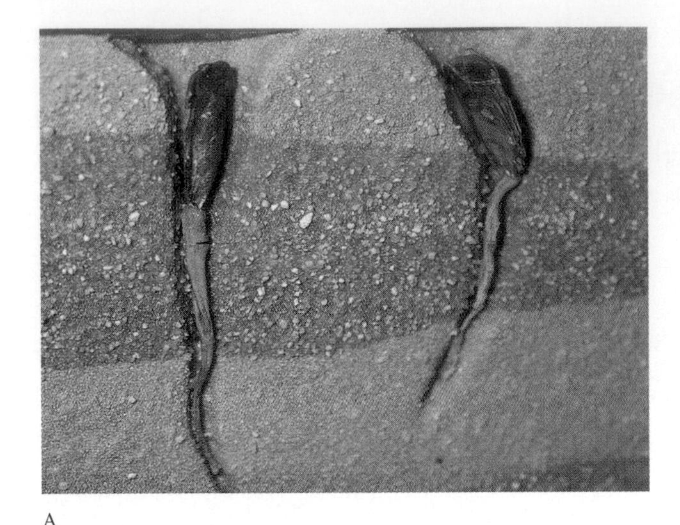
A

B

Figure 3-30 **Brachiopods.** These animals use a frilly loop-shaped structure to sieve food from the water. *Lingula (A)* is an inarticulate brachiopod that anchors itself in the substratum with its fleshy stalk; it is about the length of a pencil. Articulate brachiopods *(B)* attach to hard surfaces with their stalk; most species fall within the size range between a pea and a golf ball.

by budding, they are close relatives of brachiopods. Each individual is housed within a chamber in the colonial skeleton. In common with brachiopods, bryozoans employ a frilly loop-shaped structure for feeding. Unlike a brachiopod, however, a bryozoan animal extends this apparatus from its skeleton in order to feed (Figure 3-31). Many bryozoans have heavily calcified skeletons, and the group has an excellent fossil record that extends back to the Ordovician Period.

Echinoderms The name **echinoderm** means spiny-skinned form. The most conspicuous trait of echinoderms, however, is their fivefold radial symmetry: the five arms that radiate from the central body of a starfish, for example. Echinoderms also possess radial rows of tube feet, which are small appendages that terminate in suction cups. All echinoderms are ocean dwellers, and most have an internal skeleton formed of calcite plates that is readily fossilized in marine sediments.

With a few exceptions, **starfishes** are flexible animals whose internal plates are not locked rigidly together. Most starfishes are predators that use their tube feet to grasp their victims, such as bivalve mollusks. When the tension applied to a bivalve shell spreads the two valves apart slightly, the starfish extrudes its stomach and digests the victim with its shell. The fossil record of starfishes, though relatively poor, extends back to lower Paleozoic strata.

Sea urchins have plates that form a rigid skeleton to which numerous spines attach by ball-and-socket joints. There are two groups of these bottom dwellers. *Regular sea urchins* (Figure 3-32) have radially symmetrical bodies that typically are the size of a lemon or orange. By moving their spines, these animals crawl over the substratum with no preferred direction of movement; most graze on algae using a complex feeding apparatus that has five teeth. *Irregular sea urchins,* in contrast, are bilaterally symmetrical burrowers that have very short spines and feed on organic matter in the sediment. Flat irregular urchins that live along sandy beaches are known as sand dollars, and more inflated forms that live farther offshore are called heart urchins.

Crinoids sieve food from the water with featherlike arms and pass it to the centrally positioned mouth with tube feet. Most living species are free-living forms that can swim by waving their arms, but some that live in the deep ocean resemble fossil forms in being attached to the seafloor by a long, flexible stalk (Figure 3-33). Crinoid stalks are supported by disk-shaped, grooved plates that are stacked together like poker chips (see Figure 3-1).

Chordates

Vertebrate animals belong to the phylum known as chordates. **Chordates** are defined by their possession

Figure 3-31 **Bryozoans.** Most of these colonial animals cement their skeletons to hard objects. Individuals within the colony are no more than a few millimeters long.

of a *notochord*—a flexible, rodlike structure that runs nearly the length of the body and provides support—at least early in their life history, if not in adulthood. The *spinal cord,* a long stem of the nervous system, lies adjacent to the notochord. During the maturation of a vertebrate animal, the notochord develops into a vertebral column. This and other skeletal elements are bony in most vertebrates, but in a few groups, including sharks, they consist of cartilage.

Figure 3-33 **Stalked crinoid feeding in the deep sea.** The body is tilted, and the arms are spread to sieve food from a current. This animal is less than 1 meter (~3 feet) tall.

Figure 3-32 **A regular sea urchin.** Tube feet that terminate in suction cups extend between the spines. This animal is about the size of an orange.

Primitive Chordates *Amphioxus* is a primitive living chordate in which the notochord is the only skeletal feature (Figure 3-34). It swims like a fish, by flexing its tail back and forth with V-shaped muscles arrayed along its body. Most of the time, however, *Amphioxus* rests partly burrowed into the sediment, sieving small food particles from the water with its gills. In its basic anatomy, *Amphioxus* resembles the ancestors of all modern chordates. An extinct group of chordates called conodonts resembled *Amphioxus* in certain ways but apparently spent more time swimming (Box 3-1).

Vertebrates Because vertebrates are more familiar than many of the animal groups described earlier, they require a less extensive introduction. We have

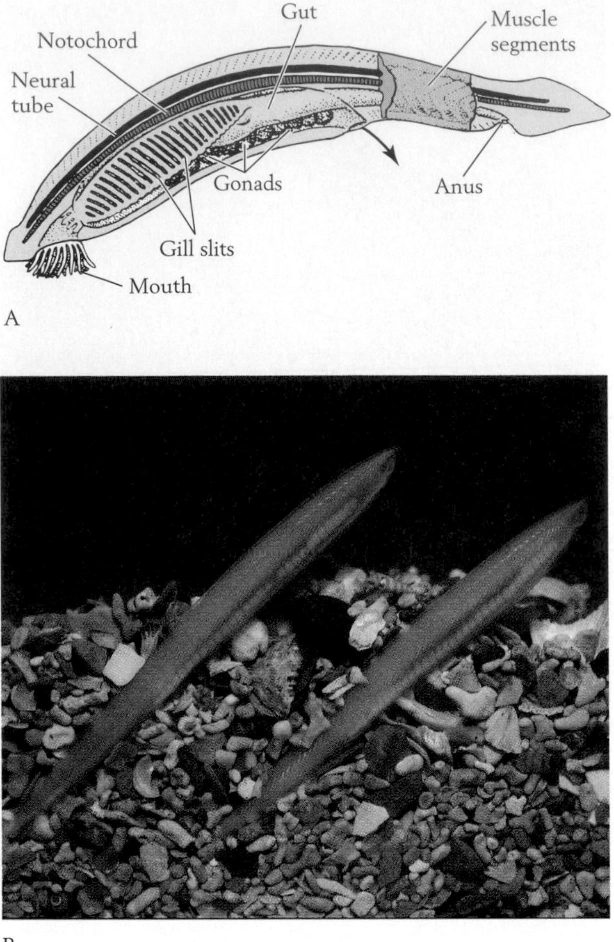

Figure 3-34 ***Amphioxus.*** *A.* The notochord, adjacent to the neural tube (which resembles a spinal cord), identifies this animal as a chordate. *B.* These two animals are suspension feeding while partly buried in sediment. *(Drawing from W. K. Purves, G. H. Orians, and H. C. Heller, Life, Sinauer Associates, Sunderland, Mass., 1995.)*

already examined relationships among major vertebrate groups (see Figure 3-10). The earliest vertebrates lacked jaws. They were primitive fishes or fishlike animals that evolved a supple structure to support the body—the jointed vertebral column, to which the skull and skeletal supports for appendages were attached. Partway through the Paleozoic Era, fishes became effective predators through the evolution of jaws. Early jawed fishes, including primitive sharks, had a cartilaginous skeleton. Other forms evolved bony external armor. Still others developed a bony internal skeleton, and their descendants include most fishes of the modern world.

Midway through the Paleozoic Era, two groups of bony fishes evolved, each with a distinctive kind of fin. As their name suggests, **ray-finned fishes** have fins that are supported by thin bones that radiate outward from the body. These are the dominant fishes of modern seas, lakes, and rivers; they include such groups as tuna, barracuda, salmon, and bass. **Lobe-finned fishes,** in contrast, have fleshy fins supported by a complex assembly of heavy bones. Only a few species of lobe-finned fishes survive in the modern world. One of the most remarkable of them is the coelacanth, a large, primitive creature that lives off the eastern coast of Africa and was discovered only in 1939 (Figure 3-35).

Lobe-finned fishes played a special role in the history of life. An early group of these fishes evolved into four-legged land animals. Each fin became a limb with toes. The lung, which terrestrial animals use to breath air, had evolved in the ancestral group of lobe-finned fishes, perhaps as a device for coping at times when the bodies of water in which they lived dried up.

Amphibians were the first four-legged vertebrates to spend their adult lives on land. They laid their eggs in water, however, and spent their early lives there, growing legs as they matured and then moving onto the land. We see this pattern of development, known as *metamorphosis,* in living amphibians such as frogs, whose juvenile forms are tadpoles that swim with fishlike tails. Most kinds of early amphibians were more similar to salamanders than

Figure 3-35 **A coelacanth.** This lobe-finned fish, which lives in moderately deep water, is about 1.5 meters (3 feet) long.

to frogs, but many were much larger than modern salamanders. Some grew to the size of a large pig.

Reptiles evolved from amphibians by way of another major evolutionary step: the origin of eggs that had protective shells and could survive on dry land. This biological innovation enabled reptiles to invade dry habitats—habitats that had been inaccessible to amphibians because of their dependence on water for reproduction. Surviving reptiles include turtles, lizards, snakes, and crocodiles. Like fishes and amphibians, reptiles are **ectothermic;** that is, the environment exerts control over their internal body temperature.

Dinosaurs were rather closely related to crocodiles, yet many experts do not consider them to have been reptiles. Certain groups of dinosaurs shared many additional anatomical features with birds. In fact, dinosaurs have recently been found with feathers attached to their bodies. There is much evidence that most dinosaurs were very active animals, not the slow, lumbering creatures they are often portrayed as being.

Birds evolved from a group of dinosaurs during the Mesozoic Era. Birds are **endothermic;** that is, they control their body temperatures internally. As we will see in Chapter 16, dinosaurs may have shared this feature with birds.

Mammals are separated from other vertebrates by a unique set of traits. Not only are mammals endothermic, but they have hair on their bodies for insulation and sweat glands for cooling when they become overheated. They also bear live young and suckle them with sweat glands modified to secrete milk. The teeth of mammals are more highly differentiated than those of reptiles, having a variety of shapes and functions. Mammals also exhibit advanced features for locomotion. In particular, their legs are positioned fully beneath their bodies rather than extending out from the sides as those of reptiles do (Figure 3-36). Three groups of mammals occupy the modern world, all with histories extending back to the Mesozoic Era.

Monotreme mammals are a small group today, including only the echidna and platypus. These primitive creatures are unusual in retaining the primitive trait of laying eggs, rather than giving birth to live young.

Marsupial mammals bear live young, but their offspring are tiny and immature at birth and grow for a time in a pouch on the mother's abdomen,

Figure 3-36 Differences between reptiles and mammals in tooth and limb structure. The alligator has simple spikelike teeth, whereas the lion has long "canine teeth" (the equivalent of "eyeteeth" in humans) near the front of its mouth, for slashing, and slicing teeth in the rear, for processing meat. The alligator's legs sprawl outward from its body, whereas the lion's legs are beneath its body.

For the Record 3-1

Bringing Some Very Old Vertebrates to Light

For more than a century, paleontologists used small toothlike fossils to date marine rocks of Paleozoic and Triassic age without knowing what kind of animal the fossils represented. These fossils were given the descriptive name *conodonts,* meaning "cone teeth." These toothlike fossils are typically less than a millimeter long, so small that they must be studied under a microscope. They can be extracted from shale by disaggregating it with special chemicals or from limestone by dissolving it in acid.

Long ago sets of various kinds of conodont fossils were found clustered together in consistent patterns. The clusters looked so much like grasping or chewing apparatuses that they were generally believed to be sets of teeth. Many conodont species were found to have broad geographic distributions and to occur in sedimentary rocks representing a wide variety of environments. This pattern of occurrence suggested that the teeth had belonged to animals that moved about effectively in ancient seas—probably a group of swimmers.

For decades paleontologists searched in vain for remains of the conodont animal. Some organic blobs were claimed to be outlines of the mystery creature because they contained its teeth, but they turned out to be nothing but remnants of the stomach contents or feces of some unknown predator that ate the conodont animal. Finally, in 1982, a worker who was examining fossil-bearing rocks of Lower Carboniferous age from Edinburgh, Scotland, came upon a fossil that represented an elongate, soft-bodied creature, about 4 centimeters (~2.5 inches) in length; a conodont tooth apparatus was embedded in one end. These were the fossil remains of an actual conodont animal! A few additional specimens have since been recovered from the same rock unit. Together these fossils portray an animal that had a pair of lobe-shaped structures protruding from its front end, presumably on either side of a mouth,

Conodont tooth (left) and the conodont animal (right).

just in front of the familiar conodont tooth apparatus. The body was formed of a series of V-shaped bundles of muscles, like those of *Amphioxus* (see Figure 3-34) or a modern fish. Two long fins supported by skeletal rays were attached along each side of the body, one near the center and one near the rear end. Spinelike rays provided skeletal support for these rays. With fins and teeth, the conodont animal fitted the description that scientists had previously pieced together for it on the basis of fragmentary and circumstantial evidence: it was a small predatory swimmer.

The question remained: How were conodonts related to other groups of animals? Were they chordate animals, or did they belong to a particular group of swimming worms with small teeth and long lateral fins—a group that survives today? The controversy was settled by detailed studies of the microstructure of the hard tissue that forms conodont teeth. One inner layer of the teeth was found to have a complex structure identical to one found in vertebrate bones. Another kind of tissue in conodont teeth turned out to resemble vertebrate enamel of the sort that covers human teeth, and a third was found to match a type of calcified cartilage found only in vertebrates. As a result of these discoveries, conodonts have joined the ranks of vertebrate animals.

The proper taxonomic assignment of conodonts to the proper group extended the known fossil record of internal hard tissues of vertebrates back some 40 million years, to Late Cambrian time. It also changed the identity of scientists who specialized in the study of conodonts. Previously these workers had identified themselves as invertebrate paleontologists, along with other workers who specialize in the study of groups such as trilobites, mollusks, and brachiopods. Suddenly veteran conodont specialists found themselves shifted to the field of vertebrate paleontology, in which scientists study teeth and bones rather than shells and other hard parts of so-called lower animals. At least the conodont specialists had the satisfaction of finally knowing what kind of animals they had been studying for so many years.

where they also obtain their milk. Today most marsupials, including kangaroos, inhabit Australia. An opossum is the only marsupial species native to North America.

Placental mammals include the vast majority of living mammal species, including humans. Their newborn offspring are much larger and more mature than those of marsupials and do not spend their infancy in a pouch. Not until the Cenozoic Era, when the dinosaurs were gone, did the placentals expand to become the dominant large animals on all continents except Australia, where marsupials came to prevail in the absence of placental competitors.

As we have seen, **therapsids** evolved from reptiles and were ancestral to mammals (see Figure 3-10*B*). They lived during late Paleozoic and early Mesozoic time. Therapsids were intermediate between reptiles and mammals in a variety of ways. They stood more upright than reptiles, for example, and their teeth were more highly differentiated, but in neither their posture nor their tooth patterns were they as advanced as mammals. Therapsids may have resembled mammals in being hairy animals that regulated their internal body temperature. Early in the Mesozoic Era, they shared the world with dinosaurs.

Chapter Summary

1 Fossils include "hard parts" of organisms and molds of these structures; they also include trace fossils, which are marks of activity, and impressions of soft parts.

2 The two groups of bacteria, Archaeobacteria and Eubacteria, are prokaryotic, differing from the other (eukaryotic) kingdoms of organisms in lacking a nucleus and other internal structures in their cells.

3 Taxonomy is the study of the composition and interrelationships of groups of closely related organisms called taxa. Taxa, ranging from species to kingdoms, are arranged into hierarchies.

4 A phylogeny is a tree of life produced by evolutionary branching.

5 Several groups of single-celled organisms called protists have important fossil records. Among them are three groups of floating algae that are major producers in the modern world (dinoflagellates, diatoms, and calcareous algae) and two groups of amoeba-like forms with skeletons (foraminifera and radiolarians).

6 Fungi are simple forms of life that absorb food from decaying organisms.

7 The earliest plants to invade the land evolved from algae and, like modern mosses, lacked vessels to conduct fluids through their tissues.

8 Like modern ferns, early vascular plants reproduced by means of spores.

9 Of plants that reproduce by means of seeds, the two largest groups are the gymnosperms, which include conifers (cone bearers) and angiosperms (flowering plants).

10 The simplest animals, including sponges, do not have their cells organized into tissues.

11 Cnidarians, which include corals, are among the simplest animals whose cells are organized into tissues.

12 Segmented animals include worms such as earthworms, arthropods, and also onychophorans, which are intermediate in form between the two other groups; of these groups, only the arthropods have an extensive fossil record (including the jointed skeletons of trilobites and even a wide range of insect remains).

13 The entire Phanerozoic displays a rich fossil record of mollusks—especially of gastropods (snails), cephalopods (relatives of the chambered nautilus), and bivalves (clams, mussels, oysters, scallops, and their relatives).

14 Brachiopods survive in relatively low diversity today but are the most conspicuous fossils in Paleozoic rocks.

15 Bryozoans, or moss animals, are colonial creatures that have been abundant since early in the Paleozoic Era.

16 Echinoderms—marine animals with radial symmetry—have an excellent fossil record that includes free-living forms (especially sea urchins) and attached forms such as crinoids (sea lilies).

17 A variety of fishes evolved during the first half of the Paleozoic Era, among them jawless groups and also jawed groups that included armored forms; ray-finned forms, which include most living fish species;

and lobe-finned forms, some of which gave rise to the first land-dwelling vertebrates (amphibians).

18 Reptiles, whose hard-shelled eggs liberated them from reproduction in water, evolved from amphibians late in the Paleozoic Era.

19 Before they died out, the dinosaurs gave rise to birds.

20 Mammals evolved from reptiles by way of the therapsids, which were intermediate in form between the two groups.

Review Questions

1 What conditions favor the preservation of soft parts as fossils within sediment?

2 What are the six kingdoms of organisms, and what traits characterize each one?

3 What is the value of derived traits for the reconstruction of phylogenies?

4 What kinds of organisms are grouped as algae?

5 How did the evolution of certain reproductive features allow early plants to invade the land? Answer the same question for animals.

6 In what ways do animals participate in the reproduction of seed plants?

7 What is a colonial animal?

8 What kinds of animals are included among the arthropods? What kinds are included among the mollusks?

9 How do lobe-finned fishes differ from ray-finned fishes, and why were the ray-finned forms unlikely to give rise to terrestrial animals?

10 How do mammals differ from reptiles?

11 The history of life has been highlighted by evolutionary innovations. Using the Visual Overview (p. 58) and what you have learned in this chapter, identify important evolutionary breakthroughs in the overall phylogeny of life that have distinguished new taxa from their predecessors, and explain the biological significance of each of the features that you identify.

Additional Reading

Cowen, R., *History of Life*, Blackwell Scientific Publications, Palo Alto, Calif., 1990.

Fortey, R., *Fossils: The Key to the Past*, Harvard University Press, Cambridge, Mass., 1991.

Margulis, L., and K. V. Schwartz, *Five Kingdoms: An Illustrated Guide to the Phyla of Life on Earth*, 3rd ed., W. H. Freeman and Company, New York, 1998.

Miller, S. A., and J. P. Harley, *Zoology: The Animal Kingdom*, Wm. C. Brown, Dubuque, Iowa, 1997.

Pellant, C., *Rocks, Minerals, and Fossils of the World*, Little, Brown & Co., Boston, 1990.

Raven, P. H., R. F. Evert, and S. E. Eichhorn, *Biology of Plants*, 6th ed., Worth Publishers, New York, 1998.

CHAPTER 4

Environments and Life

Organisms are able to live only in environments where they can find food, tolerate physical and chemical conditions, and avoid natural enemies. These were requirements for life in the past just as they are in the modern world. Climate is the environmental factor that exerts the greatest control over the distribution of species throughout the world, influencing conditions not only on land but also in bodies of water. Temperature and other climatic conditions strongly affect the distribution of plant species, and plant species in turn influence the distribution of many animal species. In this chapter, therefore, we will pay close attention to the mechanisms that create the prevailing weather patterns on Earth today. Our major focus, however, will be on the relationships between living things and the environments that they inhabit.

If the planet had undergone little change in the course of geologic time, we could directly apply what we know about life and environments today to ancient fossils and rocks. But Earth has changed dramatically over the course of its history: the planet's materials—including living mater—have changed continually both in composition and in location. Life, of course, has undergone vast evolutionary changes, and many conspicuous physical features of the modern world—including the polar ice caps of Greenland and Antarctica and the very cold body of water that now forms the deep ocean—were not present 100 million years ago. Although this situation does not violate the principle of uniformitarianism (p. 3), inasmuch as natural laws are not broken, it does require us to take changing environmental conditions into account when we interpret the rock record.

An African savannah. Zebras, springbok antelopes, and wildebeests gather at a water hole in this savannah in southwestern Africa.

Visual Overview

The Distribution of Environments and Life on Earth

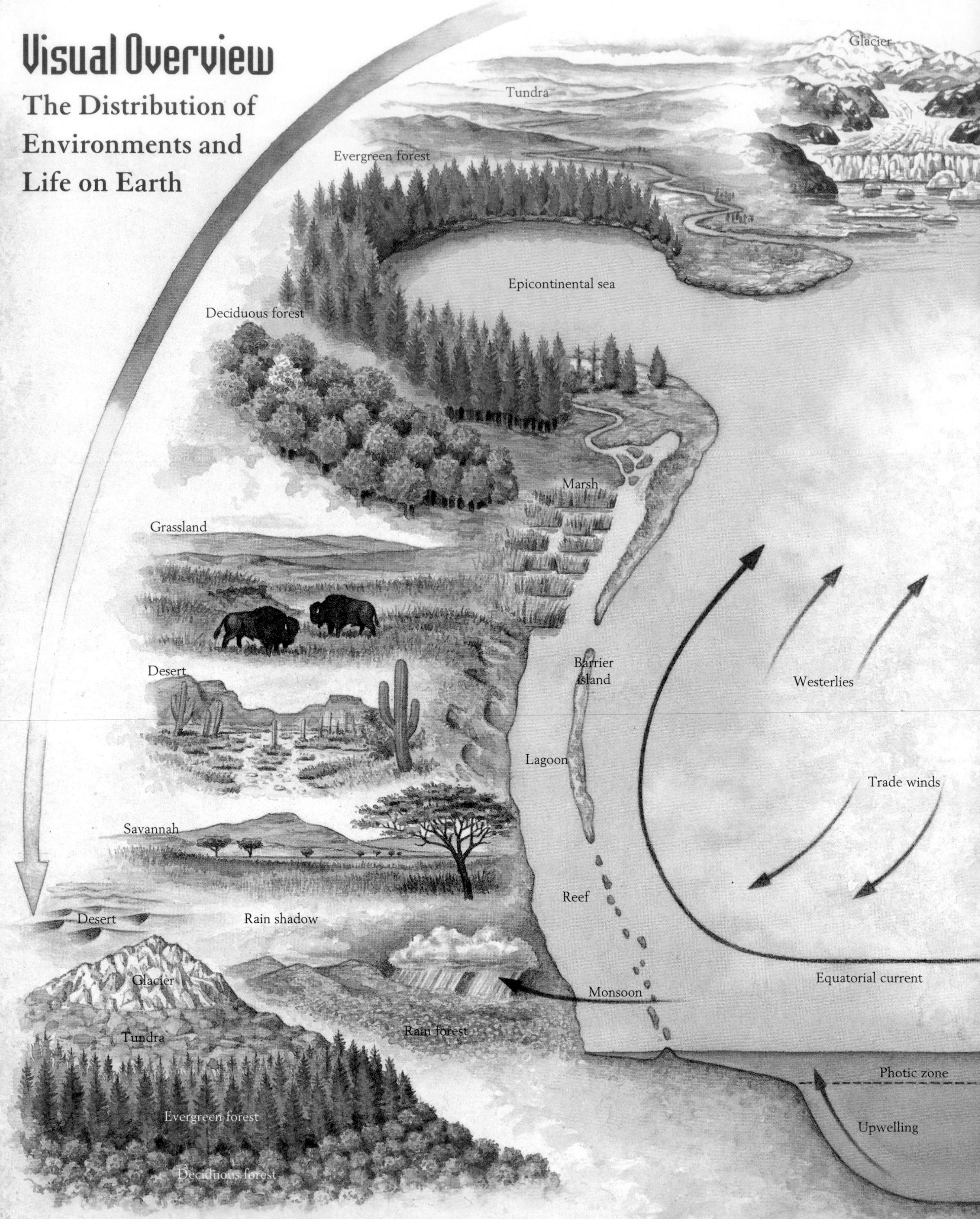

The distribution of life on Earth reflects physical conditions that vary from the tropics to polar regions, from lowlands to mountain tops, and from marginal marine environments to the deep sea.

EVERGREEN FOREST
Fir
Moose
Lynx
TUNDRA
Lichen
Lemming
Caribou or reindeer
GLACIAL ENVIRONMENT
Polar bear
Lichen
Penguin
DECIDUOUS FOREST
Maple
Squirrel
Wolf
GRASSLAND
Bison
Snake
Grass
DESERT
Owl
Saguaro
Scorpion
Dry, cool air descends
Cold air descends
Warm, moist air rises
Glacier
Tundra
Evergreen forest
Deciduous forest
Grassland
Desert
Savannah
Rain forest
Westerlies
Trade winds
Equatorial current
Summer
Winter
RAIN FOREST
Monkey
Gavial
Canopy-forming tree
SAVANNAH
Acacia
Zebra
Rhino
CORAL REEF
Palm
SEA
Fish
Killer whale
Shrimp
Organic matter
Abyssal plain

record. What we learn in this chapter will serve as a starting point for our exploration of environments and life through the eons, beginning with the planet's origins and moving forward until we reach the present.

One means of gaining an understanding of environments on Earth is to examine the configuration of the planet's surface. You will recall that Earth's crust is divided into the thin, dense oceanic crust and the thicker, less dense continental crust—a distinction that accounts for Earth's external shape, with continental surfaces standing above the seafloor. A **hypsometric curve** indicates the proportions of Earth's surface that lie at various altitudes above and depths below sea level (Figure 4-1).

Those settings on or close to Earth's surface that are inhabited by life are called **habitats.** Nearly all habitats can be classified as terrestrial or aquatic. Aquatic habitats are further divided into marine habitats (e.g., those within oceans and seas) and freshwater habitats (e.g., lakes, rivers, and streams). Birds, bats, and insects use the atmosphere above the ground as a part-time habitat, but virtually all of these creatures also conduct some activities, such as feeding, sleeping, and reproduction, at the surface. Later we will examine the nature of the various habitats on Earth together with the forms of life that occupy them. First, however, we will examine some of the principles that govern the distribution of species within habitats in general.

The Principles of Ecology

Ecology is the study of the factors that govern the distribution and abundance of organisms in natural environments. Some of these factors are conditions of the physical environment, and others are modes of interaction between species.

A Species' Position in Its Environment

The way a species relates to its environment defines its **ecologic niche.** The niche requirements of a species include particular nutrients or food resources and particular physical and chemical conditions. Some species have much broader niches than others. Before human interference, for example, the species that

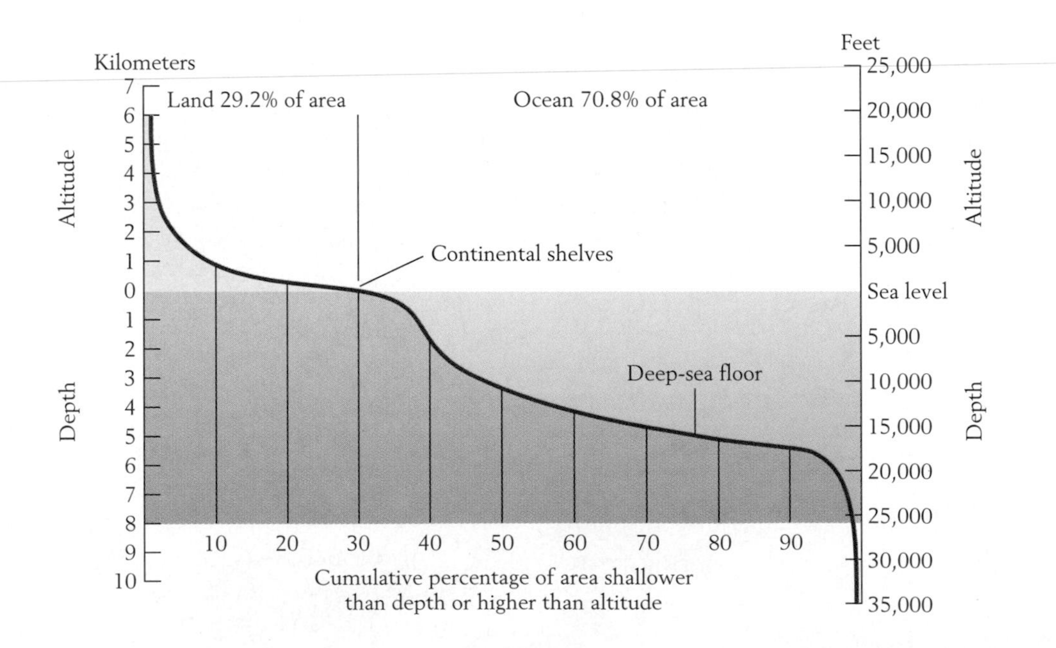

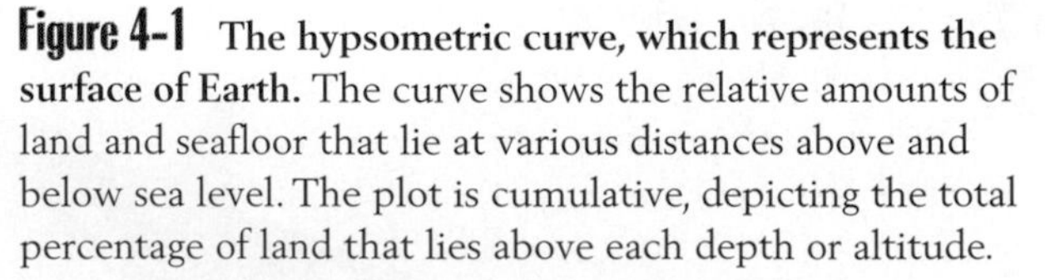

Figure 4-1 The hypsometric curve, which represents the surface of Earth. The curve shows the relative amounts of land and seafloor that lie at various distances above and below sea level. The plot is cumulative, depicting the total percentage of land that lies above each depth or altitude. About 70 percent of Earth's surface lies below sea level, and most of this area forms the deep-sea floor. Continental shelves are borders of continents flooded by shallow seas. The left side of the diagram shows that mountains account for relatively little of Earth's surface.

includes grizzlies and brown bears ranged over most of Europe, Asia, and western North America, eating everything from deer and rodents to fishes, insects, and berries. The sloth bear, in contrast, has a narrow niche. It is restricted to Southeast Asia, feeding mainly on insects, for which its peglike teeth are specialized, and on fruits. The ecologic niches of many other closely related species present similar contrasts.

We speak of the way a species lives within its niche as a **life habit**. A species' life habit is its mode of life—the way it obtains nutrients or food, the way it reproduces, and the way it stations itself within the environment or moves about.

Every species is restricted in its natural occurrence by certain environmental conditions. Among the most important of these **limiting factors** are physical and chemical conditions. Most ferns, for example, live only under moist conditions, whereas cactuses require dry habitats. The salt content of water is a major limiting factor for species that live in the ocean. Few starfishes or sea urchins, for example, can live in a lagoon or bay where normal ocean water is diluted by freshwater from a river.

Almost every species shares part of its environment with other species. Thus, for many species, **competition** with other species—or the process in which two or more species vie for an environmental resource that is in limited supply—is a limiting factor as well. Among the resources for which species commonly compete are food and living space. Often two species that live in similar ways cannot coexist in an environment because one species competes so much more effectively than the other that it monopolizes the available resources. Plants that grow in soil, for example, often compete for water and nutrients in that soil; as a result, plant species that live close together often have roots that penetrate the soil to different depths (Figure 4-2).

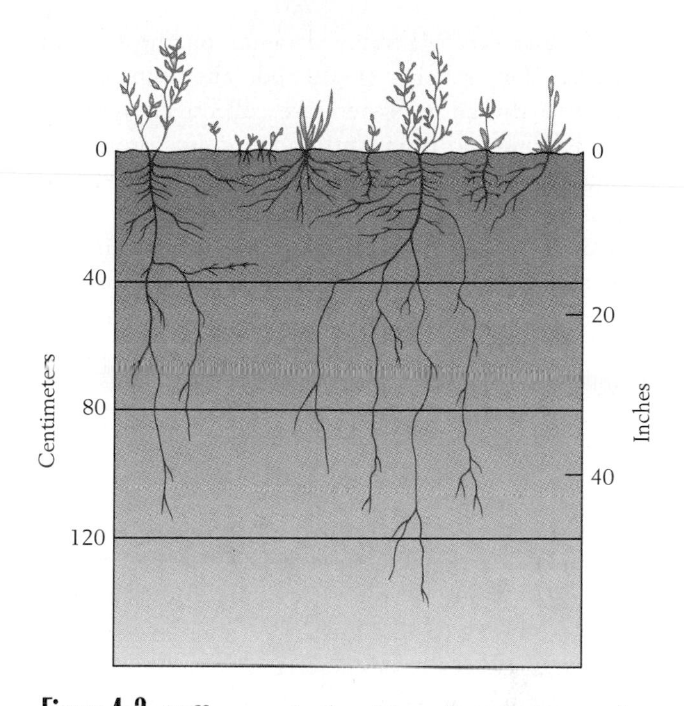

Figure 4-2 **Differences in the niches of coexisting species of plants.** The roots of different species occupy different depth zones of the soil and thus avoid competing for water and nutrients. *(After H. Walter, Vegetation of the Earth in Relation to Climate and Eco-Physiological Conditions, Springer-Verlag, Stuttgart, 1973.)*

Predation, or the eating of one species by another, is another limiting factor. An especially effective predator can prevent another species from occupying a habitat altogether.

Communities of Organisms

A **population** is a group of individuals that belong to a single species and live together in a particular area. Populations of several species living together in a habitat form an **ecologic community**. In most ecologic communities, some species feed on others. The foundation of such systems consists of producers—plants or plantlike organisms that manufacture their own food from raw materials in the environment. Consumers—animals and animal-like organisms—feed on other organisms (p. 64). Consumers that feed on producers are known as **herbivores,** and consumers that feed on other consumers are known as **carnivores.** Terrestrial herbivores include such diverse groups as rabbits, cows, pigeons, garden slugs, and leaf-chewing insects. Terrestrial carnivores include weasels, foxes, lions, and ladybugs.

The organisms of a community and the physical environment they occupy constitute an **ecosystem.** Ecosystems come in all sizes, and some encompass many communities. Earth and all the forms of life that inhabit it represent an ecosystem, but so does a tiny droplet of water that is inhabited by only a few microscopic organisms. Obviously, then, large ecosystems can be divided into many smaller ecosystems, and the size of the ecosystem that is treated in a particular ecologic study depends on the type of research that is being conducted. The animals and protozoans of an ecosystem are collectively referred to as a **fauna** and the plants and plantlike protists as a **flora.** A flora and a fauna living together constitute a **biota.**

One of the most important attributes of an ecosystem is the flow of energy and materials through it. When herbivores eat plants, they incorporate into their own tissue part of the food that these plants have synthesized. Carnivores assimilate the tissue of herbivores in much the same way. In most ecosystems, carnivores that eat herbivores are eaten in turn by other carnivores; in fact, several levels of carnivores are often present in an ecosystem. An entire sequence of this kind, from producer to **top carnivore,** constitutes a **food chain.** Because most carnivores feed on animals smaller than themselves, the body sizes of carnivores often increase toward the top of a food chain (Figure 4-3).

Simple food chains—sequences in which a single species occupies each level—are uncommon in nature. Most ecosystems are characterized by **food webs,** in which several species occupy each level. Most species below the top carnivore level serve as food for more than one consumer species. Similarly, most consumer species feed on more than one kind of prey.

Parasites and scavengers add further complexity to ecosystems. **Parasites** are organisms that derive nutrition from others without killing their victims, and **scavengers** feed on organisms that are already dead. A flea that feeds on the blood of a dog, for example, is a parasite, as is a tapeworm that lives within a human. A vulture, in contrast, is a scavenger, as is a maggot, which feeds on dead flesh.

Although material flows from one level of a food web to the next, it does not stop at the highest level. In fact, materials are cycled through the ecosystem continuously, with single-celled bacteria completing

Figure 4-3 A sequence of species forming a food chain within a stream. Single-celled algae are fed upon by nymphs (i.e., the juvenile stage) of mayfly insects. The nymphs, in turn, are eaten by sunfishes, which are preyed upon by large-mouthed bass. Otters eat the bass. (Organisms are not drawn to scale; the algae and nymphs are greatly enlarged.)

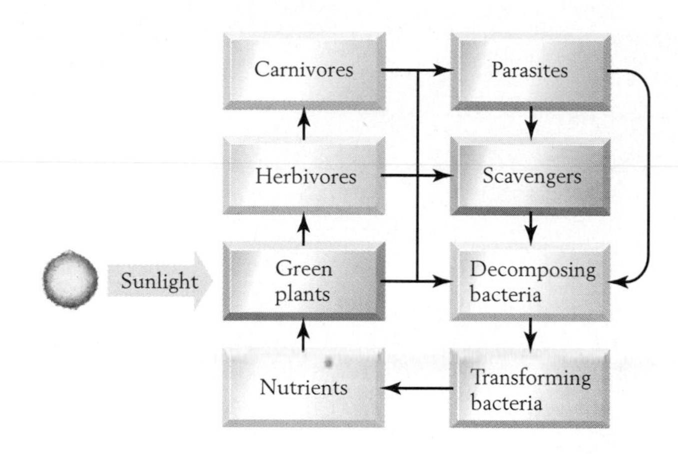

Figure 4-4 The cycle of materials through an ecosystem.

the cycle (Figure 4-4). Some of these bacteria decompose dead animals and plants of all types into simple chemical compounds, while others transform decomposed material, liberating nutrients to be reused by plants.

The term **diversity** is used to designate the variety of species that live together within a community. Diversity can be measured in several ways, some of which take into account the relative abundances of species, but the simplest measure of diversity is nothing more than a count of the species present. Diversity is normally low in habitats that present physical difficulties for life. Because plants require water to make food, for example, deserts contain fewer species of plants than do moist tropical forests. Only a few types of plants, such as cactuses, can sustain themselves in desert environments.

Predation is another factor that influences the diversity of a community. When a predator disappears from a region, a species that has served as its prey may become much more abundant. When a predator eliminates one species from a community, it may allow that species' competitors to survive there when they could not have done so before.

Certain types of physical disturbances can prevent strongly competitive species from excluding less competitive ones. Storm waves, for example, may tear animals and plants from rocky shores, leaving bare surfaces for the invasion of species that are weak competitors. Species that specialize in invading newly vacated habitats—land cleared by fire, say, or new shore areas formed along rivers that change course at flood stage—are aptly called **opportunistic species.** Populations of opportunistic species seldom survive for long in the face of better competitors, however. Because opportunistic species tend to be good invaders, while some of their populations are disappearing from one area, others are becoming established elsewhere. The plants that we call weeds are opportunistic species par excellence. In gardens and lawns, many types of weeds come and go in the course of a few seasons.

Biogeography

The distribution and abundance of organisms on a broad geographic scale are studied within the field known as **biogeography.** The limiting factor that plays the largest role here is temperature. Many species are restricted to polar regions, others to tropical regions near the equator, and still others to temperate regions in between.

Within most large, widespread taxonomic groups of animals and plants, more species are adapted to the tropics than to cold climates. Similarly, tropical communities are, on the average, more diverse than communities situated at high latitudes. In general, species increase in number toward the equator, because many more kinds of animals and plants can survive there than in the harsher conditions at higher latitudes.

Temperature, however, is not the only control of biogeographic patterns of occurrence, as evidenced by the fact that most species are not found in every habitat that meets their particular ecologic requirements. Dispersal of most species is also restricted by barriers, the most obvious of which are land barriers for aquatic forms of life and water barriers for terrestrial forms of life. Of course, these barriers change over time, and the geographic distributions of species shift accordingly. *Mammuthus,* the genus that includes the extinct members of the elephant family known as mammoths, evolved in Africa about 5 million years ago, during the early part of the Pliocene Epoch. Blocked by northern oceans, mammoths were unable to migrate to North America until the Pleistocene Epoch. During the Pleistocene Epoch, however, large volumes of water were locked up on the land as glaciers (or ice sheets), and sea level fell throughout the world. Consequently, a land bridge emerged between Siberia and Alaska, allowing mam-

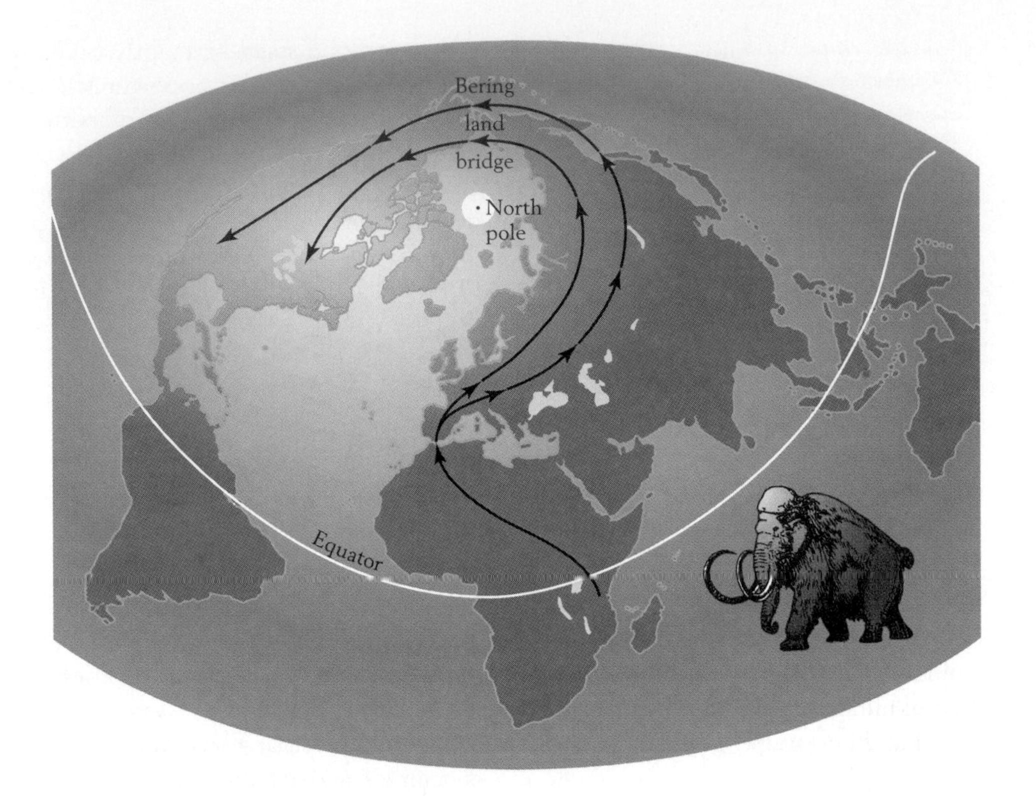

Figure 4-5 Changes in the geographic distribution of mammoths. Mammoths evolved in Africa during the Pliocene Epoch and spread to Eurasia. Later, during the Pleistocene interval, when sea level was lowered, they crossed the Bering land bridge, which emerged between Eurasia and North America. *(After V. J. Maglio, Amer. Philos. Soc. Trans. 63:1–149, 1986.)*

moths to invade the Americas, where they survived until several thousand years ago (Figure 4-5).

The survival of mammoths in the Americas represents an interesting biogeographic phenomenon—the development of a relict distribution, or the presence of a taxonomic group in one or two locations after it has died out elsewhere. By late in the Pleistocene Epoch, mammoths had disappeared from most of the areas they had previously occupied and remained in only a relatively small area of North America.

The Atmosphere

The atmosphere, or the envelope of gases that surrounds Earth, serves life primarily as a reservoir of chemical compounds that are used within living systems. The atmosphere has no outer boundaries; it merges gradually into interplanetary space. The dense part of the atmosphere, however, forms a thin envelope around Earth, so that more than 97 percent of the mass of the atmosphere lies within 30 kilometers (19 miles) of Earth's surface. (To place this figure in perspective, note that it resembles the average thickness of Earth's continental crust.)

The Chemical Composition of the Atmosphere

Nitrogen is the most abundant chemical component of the atmosphere, making up about 78 percent of the total volume of atmospheric gas. It is a major component of proteins, which stimulate chemical reactions and serve as building blocks of all living things. Second in abundance is oxygen, which forms about 21 percent of the volume of the atmosphere. Both nitrogen and oxygen exist in the atmosphere as molecules that consist of two atoms (N_2 and O_2); this is why we speak of them as compounds rather than elements.

Atmospheric nitrogen and oxygen are maintained at consistent levels by being cycled through living organisms continuously and returned to the air. Most oxygen enters the atmosphere from plants, which produce it through **photosynthesis**—the process by which water and carbon dioxide are converted to sugar and oxygen. The green compound chlorophyll acts as a catalyst in this process, while the energy that is necessary for the conversion is derived from the sun. A smaller amount of oxygen comes from the upper atmosphere, where sunlight breaks down water vapor (H_2O) into oxygen and hydrogen.

Carbon dioxide (CO_2), from which plants produce oxygen, is contributed to the atmosphere by respiration of animals and bacteria— and in the modern world by the burning activities of humans. It forms only about $\frac{3}{100}$ of 1 percent of the atmosphere's volume. In Chapter 10 we will see how oxygen and carbon dioxide are cycled through the atmosphere as part of large-scale geochemical cycles.

Temperatures and Circulation in the Atmosphere

Both the atmosphere and the ocean consist of fluid that is in constant motion in relation to the earth beneath them, and most of the energy that produces this motion comes ultimately from the sun. Thus solar radiation is responsible for much of what takes place in the atmosphere and in the ocean.

When solar energy reaches Earth, a good deal of it is absorbed and turned into heat energy. The amount of solar radiation that is absorbed varies from place to place according to the nature of Earth's surface. Sunlight generates less heat when it strikes ice, for example, than when it strikes water, soil, or vegetation, because ice reflects more radiation. The **albedo,** which is the percentage of solar radiation reflected from Earth's surface, ranges from 6 to 10 percent for the ocean; from 5 to 30 percent for forests, grassy surfaces, and bare soil; and from 45 to 95 percent for ice and snow.

Earth's polar regions receive just as many hours of sunlight as the equatorial region; the sunlight in polar regions is simply heavily concentrated in the summer. What makes polar regions cold even in summer is the low angle at which sunlight strikes them. A ray of sunlight that strikes at a low angle is spread over a broad area and supplies little heat (Figure 4-6).

The tilt of Earth's rotational axis creates Earth's seasons. When the north pole is tilted toward the sun, for example, it is summer in the northern hemisphere and winter in the southern hemisphere (Figure 4-6). There are two reasons for this pattern. First, at this time, the sun's rays strike most directly slightly to the north of the equator. Second, because the northern hemisphere is aimed toward the sun at this time, at every latitude in the northern hemisphere days are longer than in the southern hemisphere. When Earth is 180° from this position in its orbit and the north pole is tilted away from the sun, the seasons of the two hemispheres are reversed.

In Chapter 1 we saw how solar heating drives part of the planet's water cycle, causing H_20 to move from one portion of the cycle to another—promoting its transfer from the ocean to the atmosphere by

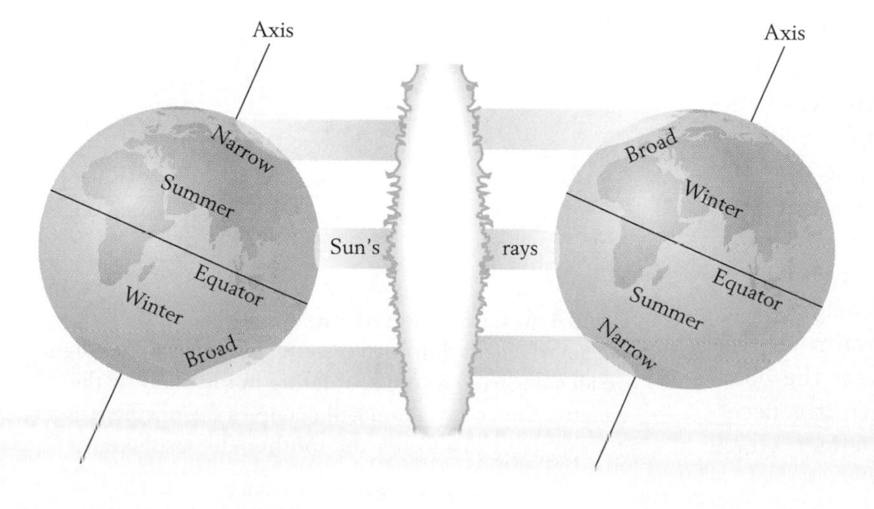

Figure 4-6 Sunlight and temperatures on Earth. A ray of sunlight of a given cross-sectional area strikes Earth at a lower angle near the poles than near the equator and is therefore spread over a broader area so that it heats the surface with less intensity. The tilt of Earth's axis creates the seasons by aiming one pole toward the sun when the other is aimed away from it. This orientation maximizes the angle at which the sun's rays strike near one pole (concentrating them for maximum heating) and minimizes this angle near the other pole.

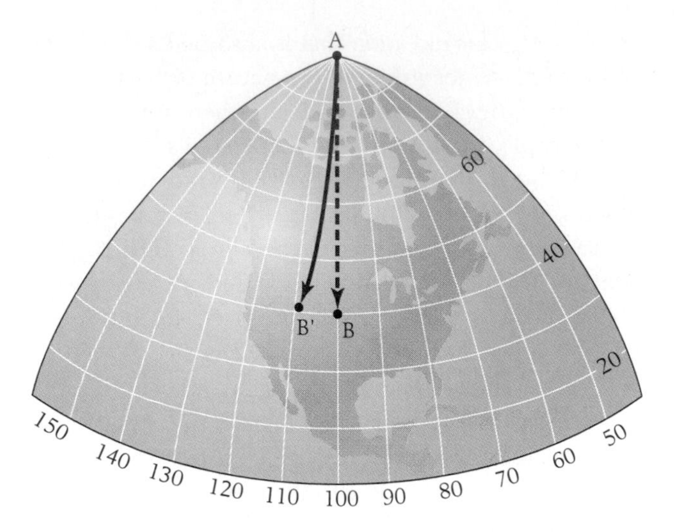

Figure 4-7 The Coriolis effect for an air current that flows from the north pole toward the equator. While the current moves, Earth rotates from west to east, and the current bends to the right with respect to features of Earth's surface. All currents of air and water in the northern hemisphere tend to bend toward the right.

evaporation, for example. The movements of water and water vapor through the water cycle (see Figure 1-21) are only a few of the many types of fluid movements driven by solar energy near Earth's surface. Solar heating also drives movements within large lakes, the atmosphere, and the ocean. Lakes and oceans are warmed by absorption of solar radiation, and the atmosphere is warmed primarily by heat that rises from land and water. The warming of fluids causes them to move.

Much of the transfer of heat from place to place in the ocean and atmosphere occurs by convection, which results from the fact that a liquid or gas is less dense when it is warm than when it is cool. In Chapter 2 we saw how this process produces large-scale motion within Earth's asthenosphere in the same way that it operates within a pot of water heated from below (see Figure 1-18).

Convection, operating in conjunction with other forces, produces the major movements within Earth's atmosphere. Let us first consider what happens where Earth is warmest and coldest: warm air rises near the equator while cool air sinks near the poles. To understand what happens in between it is necessary to take into account the **Coriolis effect,** which results from Earth's rotation. As Earth rotates, air currents are deflected clockwise in the northern hemisphere and counterclockwise in the southern hemisphere. The Coriolis effect is most easily envisioned for an air current moving from one of Earth's poles toward the equator (Figure 4-7). Instead of moving directly along a line of longitude toward the equator, such an air current will bend to the west with respect to landmarks on the surface of the rotating planet and will continue to do so because the surface speed of the planet increases toward the equator. More generally, the Coriolis effect causes an air current flowing in any direction in the northern hemisphere to flow toward the right. Similarly it bends all air currents in the southern hemisphere toward the left.

The Coriolis effect prevents Earth's atmosphere from circulating in the simple pattern shown in Figure 4-8. Consider what happens near the equator. Although air in this region rises from the warm surface of Earth and spreads in a general way to the north and south, the air that spreads poleward from the equator at high elevations does not move *directly* to the north and south because the Coriolis effect

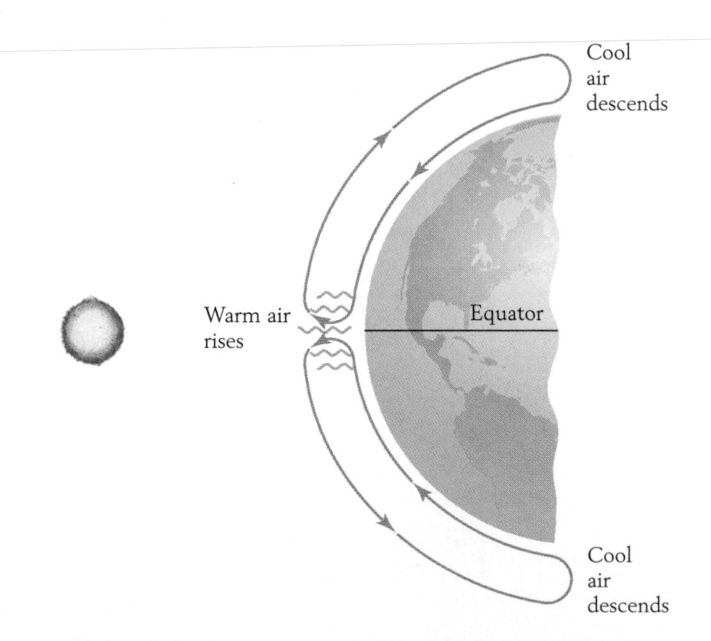

Figure 4-8 Circulation of atmosphere around an imaginary Earth that receives equal amounts of sunlight on all sides from a source rotating in the plane of the equator. One convection cell occupies the northern hemisphere and another the southern hemisphere.

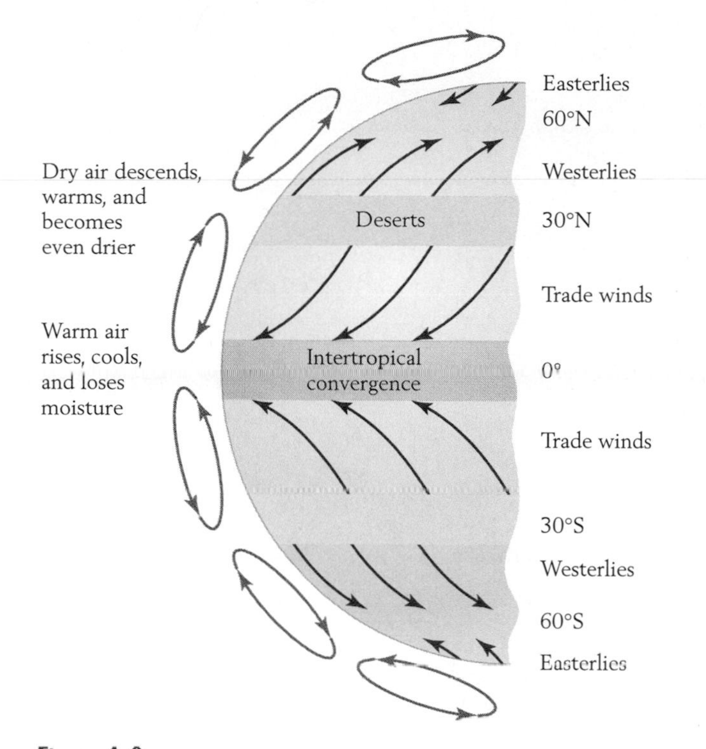

Figure 4-9 **The major gyres of Earth's atmosphere.** The lower segments of these gyres represent the major wind systems, labeled at the right.

deflects it toward the east (Figure 4-9). As a result, the air piles up north and south of the equator more rapidly than it can escape toward the poles, producing belts of high atmospheric pressure about 30° north and south of the equator. This high pressure bears down on Earth's surface. Air that rose near the equator and cooled aloft descends in this zone, pushing winds at the surface both poleward and equatorward. These winds are also deflected by the Coriolis effect. The ones that flow equatorward bend westward. These are the well-known trade winds (Figure 4-9). Note that the trade winds replace the warm air that rises near the equator, completing a cycle, or gyre, of airflow on either side of the equator.

The northern and southern trade winds converge in the **intertropical convergence zone.** Because of Earth's axial tilt, this warm, tropical zone is usually not positioned on the equator, but shifts seasonally with the location of maximum solar heating (see Figure 4-6): it lies a few degrees north of the equator during the northern summer and a few degrees south of the equator during the southern summer.

Another gyre is positioned poleward of the trade winds. Here **westerlies** flow toward the northeast. Like the trades, these winds originate from a high-pressure zone where air piles up between 20° and 30° from the equator. Because of the Coriolis effect, these winds bend toward the east instead of flowing directly northward. Still farther from the equator in each hemisphere is yet another gyre, which originates where cool air descends near the pole and, because of the Coriolis effect, twists westward to produce **easterlies.** These major air movements in the atmosphere influence the distribution of climates and organisms.

The Terrestrial Realm

At the present time, the continents of the world stand relatively high above sea level (see Figure 4-1). Thus the expanses of land are broader today than they were during most of Phanerozoic time. Another unusual feature of the modern world is a steep temperature gradient between each pole and the equator. Whereas tropical conditions prevail near the equator, the average summer temperature near the north and south poles is well below freezing. At certain times in the geologic past, polar regions were warmer than they are today and the pole-to-equator temperature gradient was gentler. Since climatic conditions have a profound effect on the distribution of organisms on the land, climates will be our first consideration when we discuss the distribution of life on the broad continental surfaces of the modern world.

Latitudinal Zones and Vegetation

It is remarkable how closely the distribution of terrestrial vegetation corresponds to the geographic pattern of climates. This correspondence, coupled with the fact that plants are the dominant producers of the food web and thus strongly affect the distribution and abundance of animals, makes climate an especially significant factor in terrestrial ecology. Plants, in fact, serve not only as sources of food but also as habitats for many animals; numerous insects, for example, spend their entire lives on certain types of trees, and insects account for most species of organisms living in the world today.

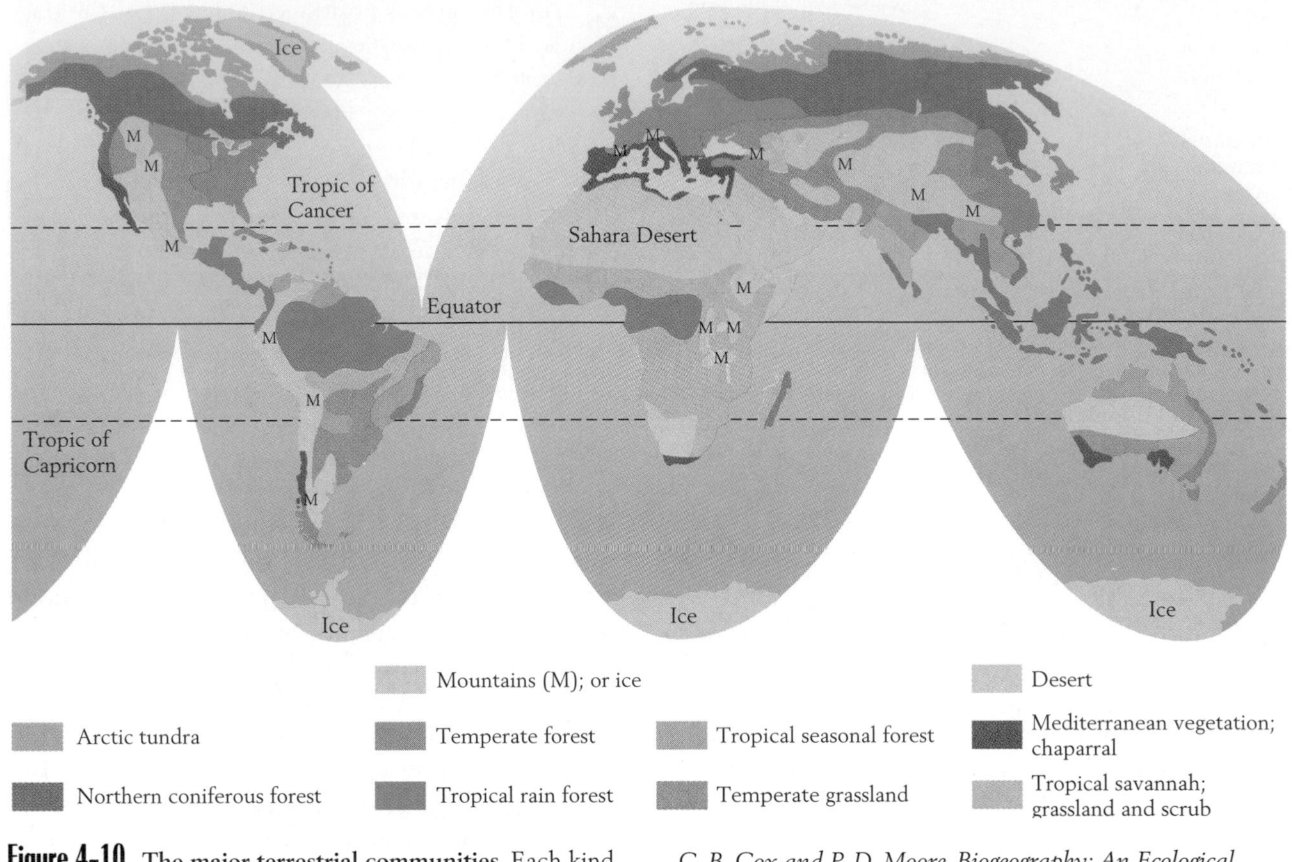

Figure 4-10 **The major terrestrial communities.** Each kind of community is characterized by a particular association of plants adapted to particular climatic conditions. *(After C. B. Cox and P. D. Moore, Biogeography: An Ecological and Evolutionary Approach, John Wiley & Sons, New York, 1980.)*

Terrestrial climates near the equator are not only very warm but often very moist as well. When air in this region rises after being heated at Earth's surface, it cools (see Figure 4-8); because cool air cannot hold as much water vapor as warm air, it loses moisture in the form of rain. In South America and Africa, the only large continents with equatorial regions (Figure 4-10), the warm, moist conditions that characterize **tropical rain forests** allow so many kinds of plants to thrive (Figure 4-11) that they form what are informally called jungles. These plants provide food and shelter for a wide variety of animals.

Tropical climates—those in which the average air temperature ranges from 18° to 20°C (64° to 68°F) or higher—are usually found at latitudes within 30° of the equator. You will recall that between 20° and 30° north and south of the equator, the air that rises from the equator piles up and, after cooling, descends to form trade winds (see Figure 4-9). The air of these winds is dry, having cooled at high altitudes and dropped much of its moisture as rain. As a result, the trade winds drop little rain; instead, they pick up moisture from the surface of Earth, leaving deserts in broad continental areas that lie about 30° north and south of the equator. The Sahara is the largest of these deserts, but broad deserts also occupy southern Africa, central Australia, and southwestern North America (see Figure 4-10).

Most deserts receive less than 25 centimeters (10 inches) of rain per year. Because only a few types of plants can live under such conditions, the desert environment is characterized by sand and bare rock rather than by dense vegetation (Figure 4-12). Some desert plants, such as cactuses, have the capacity to store water, and nearly all have small leaves so that a minimum of water is lost by evaporation. Few species of animals can survive in the desert environment, and a large percentage of those that do are nocturnal; many are small rodents that find refuge from the hot sun by remaining in burrows during the day.

Figure 4-11 A tropical rain forest in Peru.

Savannahs and **grasslands** form in areas where rainfall is sufficient for grasses to thrive, but not sufficient to allow the growth of forests. Many savannahs and grasslands are, in fact, positioned between dry deserts and wet woodlands. Savannahs differ from grasslands in containing scattered trees, which sometimes form small groves. The Great Plains of North America constitute a grassland, and savannahs are well represented in Africa.

In Africa and elsewhere, broad savannahs lie between the equatorial zone, where it is rainy year round, and deserts that occupy the trade wind belt and receive little rain throughout the year. In the northern hemisphere, such savannahs receive most of their rain during the northern summer, when the intertropical convergence zone has shifted north of the equator, bringing heavy tropical rainfall with it. Similarly, savannahs of the southern hemisphere receive most of their rain during the southern summer.

Savannahs are noted for their populations of large animals (see p. 90). Most of the herbivores found in savannahs and grasslands—including bisons, antelopes, zebras, and wildebeests—are relatively large animals that graze on grasses and have enough stamina to flee from carnivores such as jackals, lions, and cheetahs. In these regions the majority of carnivores, too, are large animals, with the ability to capture the large grazers. Many savannahs and grasslands also support scattered trees, and these habitats intergrade with open woodlands.

In sharp contrast to the rich biotas of tropical rain forests, warm savannahs, and woodlands are the

Figure 4-12 A desert landscape in Arizona.

meager biotas located near the north and south poles. Today large ice caps cover Greenland and Antarctica, the two large continents of polar regions. Such ice caps have been absent from Earth in past times, when no large continents have occupied polar regions or when the polar regions have been warmer. The ice caps of Greenland and Antarctica are now so heavy that they actually depress the continental crust (Figure 4-13). These ice caps are continental glaciers.

Glaciers are among the most impressive physical structures on Earth. They are not simply masses of ice; they are masses of ice in motion (Figure 4-14). Glaciers form from snow that accumulates until it is so thick that the pressure of its weight recrystallizes the individual flakes into a solid mass. Glaciers slide slowly downhill, but they also spread over horizontal surfaces because of internal deformation. This movement resembles the flowing of any solid material as it nears its melting point. Glaciers form not only at high latitudes but also at high elevations near the

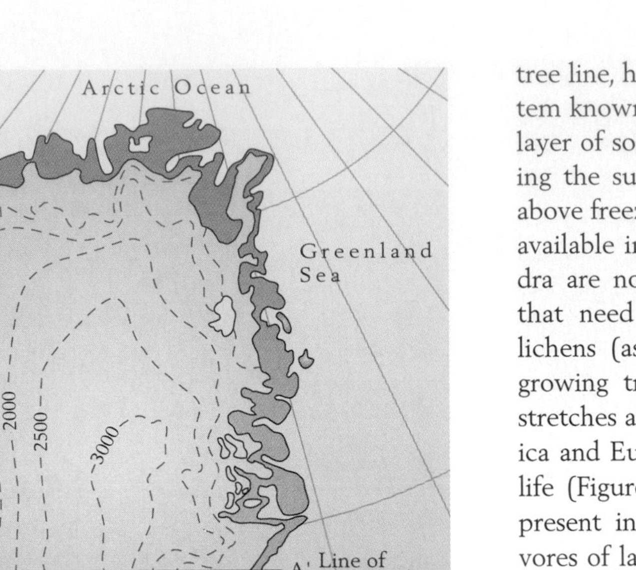

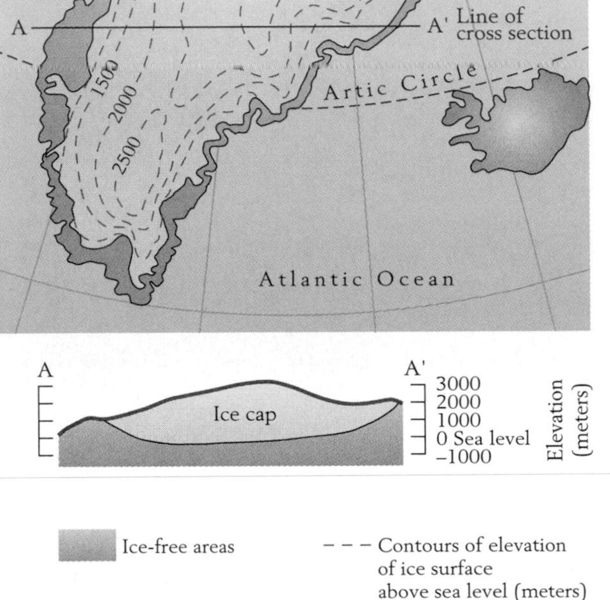

Figure 4-13 **The thick glacial ice cap of Greenland, which depresses the continental crust.** *(After F. Press and R. Siever, Earth, W. H. Freeman and Company, New York, 1986.)*

equator, where the atmosphere and surface of Earth are cool. Here they occupy mountain valleys, through which they flow downhill. When they encounter warmer temperatures closer to sea level, they usually melt, though near the poles they may reach the sea before they can do so. Large chunks of ice then break from their terminal portions and float off in the form of icebergs.

Nearly all of Antarctica is covered by ice, but no other large continent in the southern hemisphere has large areas close enough to the south pole to experience very cold conditions year round. In the northern hemisphere and in many tropical areas above the tree line, however, there is a type of subarctic ecosystem known as ***tundra.*** Tundra exists in areas where a layer of soil beneath the surface remains frozen during the summer, even though air temperatures rise above freezing. Under these conditions, water is never available in abundance. The dominant plants in tundra are not grasses and tall trees but rather plants that need little moisture, such as mosses, sedges, lichens (associations of algae and fungi), and low-growing trees and shrubs. A broad belt of tundra stretches across the northern margins of North America and Eurasia, supporting a low diversity of animal life (Figure 4-15). Rodents and snowshoe hares are present in tundras today, and the dominant herbivores of larger size are the caribou (which are called reindeer in Europe and Asia) and musk ox. Foxes and wolves are the primary hunters.

South of the tundra in the northern hemisphere, in areas where moisture is sufficient, forests rather than deserts or grasslands are found. The cold regions adjacent to tundra are cloaked in ***evergreen coniferous forests,*** dominated by such trees as spruce, pine, and fir. These trees are successful in very cold regions because they retain their needles year round. The summers in such regions are too short to permit deciduous trees to regrow their leaves and still have enough warm days ahead to produce the food they need. The variety of animals is also low, partly because relatively few plant species are present to support them and partly because many groups of animals simply cannot survive the cold winters.

Figure 4-14 **The Hubbard glacier of Alaska.** A large piece of the glacier is plunging into the water. Pieces that broke off earlier float in the foreground as icebergs.

Figure 4-15 A caribou grazing in North American tundra.

To the south, in slightly warmer climates with longer summers, ***temperate forests*** replace evergreen forests. Some evergreen trees can be found in such forests, but deciduous trees such as maples, oaks, and beeches are usually present in greater abundance. These deciduous trees are angiosperms, or flowering plants (p. 76). Where conditions are favorable for their growth, angiosperms defeat conifers, which are gymnosperms (p. 76), in competition for resources. Ground animals are more diverse in temperate forests than in cold evergreen forests, and birds are especially well represented.

Mediterranean climates, which are characterized by dry summers and wet winters, often prevail along coasts that lie about 40° from the equator. During the summer, the land in these areas is warmer than the ocean, so that moist air coming off the ocean is warmed over the land and retains its moisture. In the winter the land is cooler than the ocean, so that moist sea air cools over the land and drops its moisture. This type of climate characterizes much of California and southeastern Australia as well as the Mediterranean region. Mediterranean climates support chaparral vegetation, which consists primarily of shrubby plants with waxy leaves that retain moisture during summer droughts. Such climates have attracted large human populations, which have altered them greatly by decimating the native biotas.

Because oceans, continents, and continental topography are not distributed uniformly over Earth's surface, regional climates and vegetation vary from place to place even within a single broad climatic zone. Earth's surface features exert strong influence on climates and thus on forms of life.

The Effects of Altitude on Climate

The peaks of tall mountains are very cold even in the equatorial region because they project high into the atmosphere. The temperature gradient between the base of a mountain and the top resembles the latitudinal gradient between a warm climatic zone and a polar region. As a result, the zonation of vegetation on a mountain tends to resemble the broader geographic zonation between low and high latitudes (Figure 4-16). Thus a mountain in the temperate zone will typically have deciduous forest at its base,

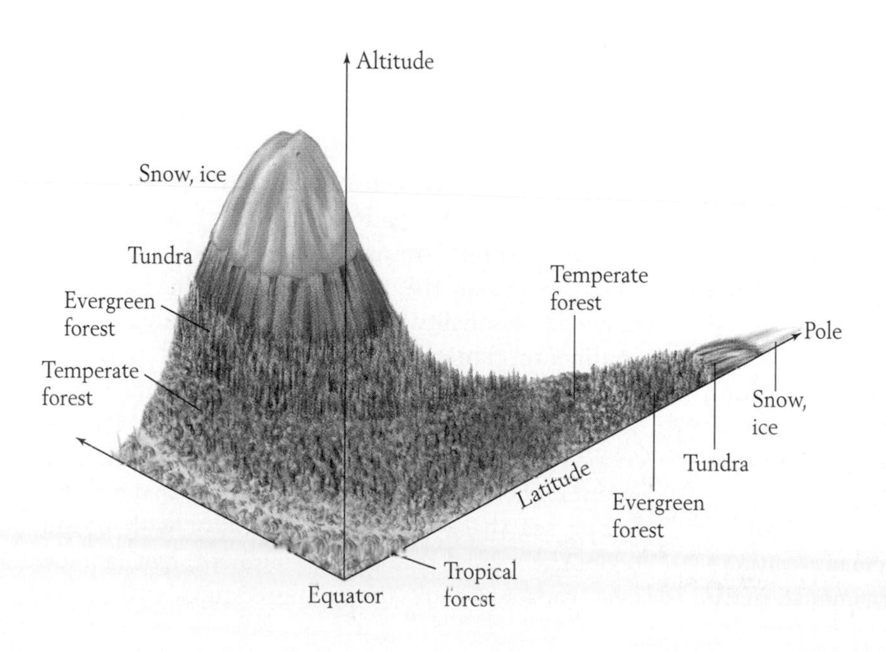

Figure 4-16 **The similarity between altitudinal and latitudinal zonation of terrestrial vegetation.** The zonation from the bottom to the top of a mountain is shown for a region near the equator where a tropical forest occupies the lowland in the foreground.

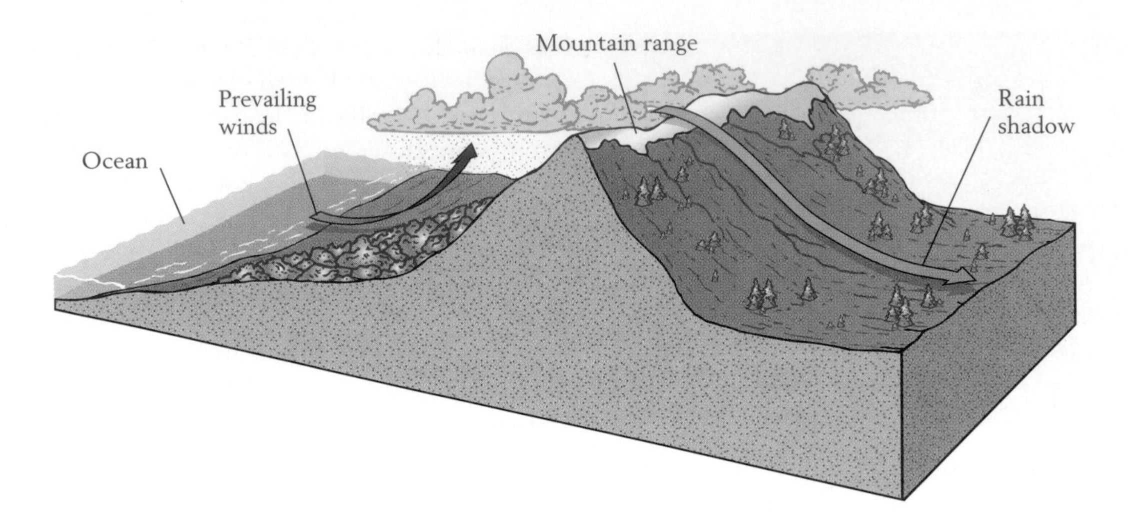

Figure 4-17 A mountain forming a rain shadow. The rising air loses its moisture as it cools, producing rain on the windward side of the mountain. The air becomes even drier as it descends and warms up on the other side, producing desert conditions. *(From W. K. Purves, G. H. Orians, and H. C. Heller, Life, Sinauer Associates, Sunderland, Mass., 1995.)*

evergreen forest higher on its slopes, and tundra above the so-called tree line. Snow and glaciers cap tall mountains, just as they cover a polar region. Near the equator, tropical forest may occupy the base of a mountain, below deciduous forest. At times in the past when global climates have been warmer than they are today, vegetation adapted to very cold climates may have been restricted to mountains.

Mountains not only offer plants cooler growing conditions than those of surrounding lowlands, they also alter rainfall patterns. A mountain range deflects prevailing winds upward, causing them to cool and drop their moisture. The result is a relatively large amount of rain on the windward side of the mountain, where the winds rise, and dry conditions on the leeward side, where the winds descend after losing most of their moisture (Figure 4-17). On the leeward side, in what is termed the **rain shadow,** the climate may be so arid as to produce a desert. For example, the Great Basin, a desert that includes most of the state of Nevada, lies in the rain shadow of the Sierra Nevada range. Winds carry moisture eastward from the nearby Pacific Ocean, but most of it falls on the western side of the Sierra Nevada, supporting abundant vegetation. Dry air descends into the Great Basin, which also receives little moisture from the trade wind belt to the south.

The Rocky Mountains, though much farther from the ocean, create a modest rain shadow immediately to their east, where there is only enough moisture to support short grasses. The so-called tall-grass prairie grows farther to the east, where winds bring more moisture northward from the Gulf of Mexico.

Land, Water, and Seasonal Temperature Change

Seas and large lakes tend to undergo less extreme temperature shifts between winter and summer than do interior regions of continents. These large bodies of water also tend to stabilize the climates of neighboring land areas. These relationships result from the high *heat capacity* of water, the ability to absorb or release a great deal of heat without changing temperature very much. Land (rock and sediment) has a much lower heat capacity than water, and air has a lower heat capacity still. Thus, outside the equatorial zone land areas and air masses tend to become warmer in summer and colder in winter than oceans at the same latitude. In fact, pronounced seasonality is the most conspicuous trait of continental climates.

At latitudes where temperatures change from season to season, temperatures tend to be less extreme near the margins of a continent than within its interior. The reason for this difference is that a neighboring sea shares its thermal stability with coastal regions. A coastal region of a continent, having a low heat capacity, cools down more quickly than the nearby ocean when winter comes. Winds transfer some heat from the ocean to the land, however, reducing the impact of winter. Similarly, the ocean remains relatively cool in summer and substantially lowers the temperature of nearby land areas through transfer of heat by winds.

Large lakes can stabilize the climate of a continental interior in the same way that an ocean can stabilize a terrestrial climate near the coast. Thus the Great Lakes of North America ameliorate climates within several kilometers of their shores. Lakes are so much smaller than the ocean, however, that winds from the land can significantly cool them in winter and warm them in summer. For the same reason, in regions with seasonal climates, bays and lagoons at the margin of the ocean display much more extreme summer and winter temperatures than those of the open ocean nearby. Coastal bays and lagoons are relatively small bodies of water that are only weakly connected to the ocean proper, so that winds can warm and cool them effectively.

Strong onshore and offshore winds near the margins of continents are termed **monsoons.** They are caused by the difference in heat capacity between land and water. When winter comes to a nonequatorial region, the land cools more rapidly than the neighboring ocean. The cool, dense air mass that forms over the land then pushes seaward, displacing warmer air above the ocean. The strong winds that carry the cooler air offshore are the monsoons. Summer monsoons blow in the opposite direction, because the land heats up more rapidly than the ocean and the air above it warms and rises, to be replaced by air that flows in from the ocean. Summer monsoons carry a great deal of moisture and release much of it as rain when they rise over the land and cool.

Today the most powerful monsoons occur in southern Asia (Figure 4-18). Here the effect of heat-

Winter

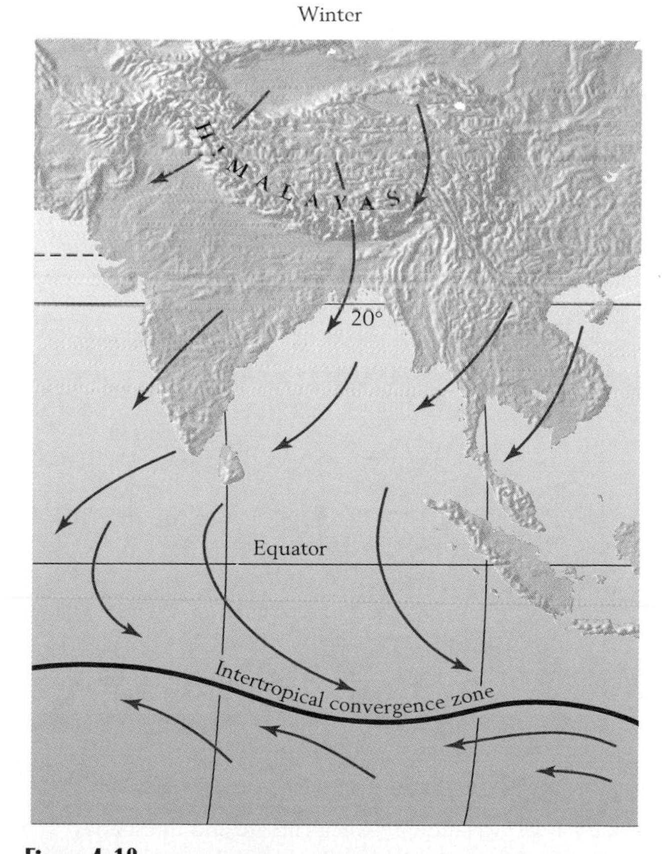

Summer

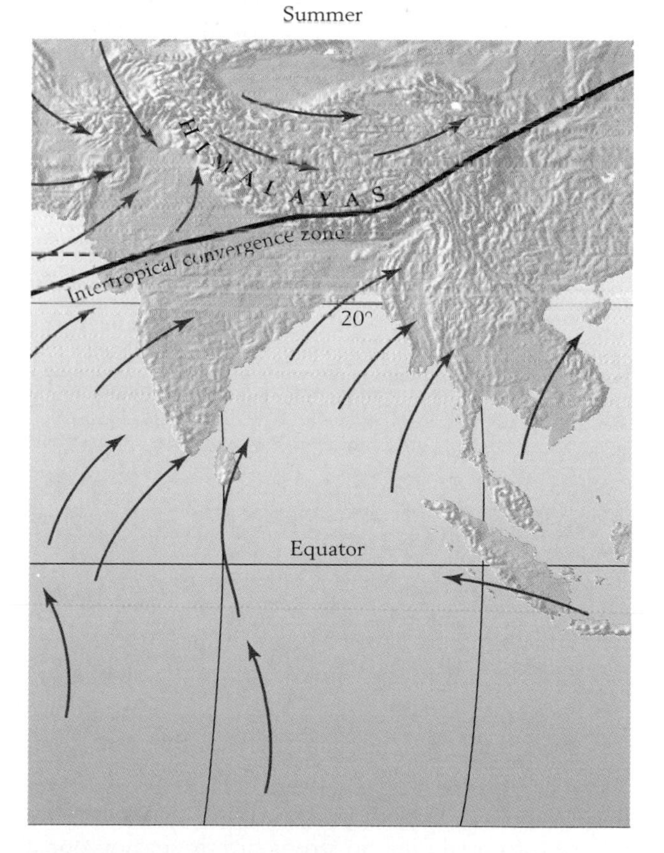

Figure 4-18 Monsoonal winds from the Indian Ocean. In wintertime the Himalaya mountains and the Tibetan plateau to their north become cold, and they cool the air above them. Increasing in density, this air descends and flows seaward as summer monsoons. In summertime the mountains and plateau warm up, and the air above them rises, twisting clockwise because of the Coriolis effect. The rising air is replaced by strong winds—summer monsoons—from the Indian Ocean. These moist winds produce heavy rains as the air rises over the land and cools.

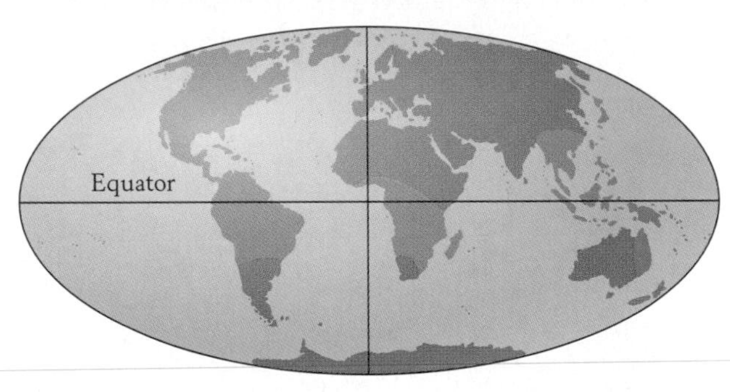

Figure 4-19 The cycad, a plant that was especially common during the Mesozoic Era. *A.* Living cycads. Today few cycads are found outside the tropics, and it appears that ancient cycads were also restricted to warm climates. *B.* The distribution of cycads today. *(B. After C. B. Cox and P. D. Moore, Biogeography: An Ecological and Evolutionary Approach, John Wiley & Sons, New York, 1980.)*

ing and cooling of the land is intensified by the huge mass of the Himalaya mountains and neighboring plateaus. A pronounced change in the temperature of this great body of uplands has a powerful effect on the atmosphere. The uplands also cause onshore summer winds to rise abruptly, cool markedly, and release torrential rain. So strong are these monsoonal winds that they cause the intertropical convergence zone in this region to shift northward and southward annually through a distance of about 5000 kilometers (~3000 miles).

Less powerful monsoons occur in other regions of the world, including the central United States, where summer heating brings winds laden with moisture inland from the Gulf of Mexico. Without these winds, summer droughts would be more severe in the American Midwest.

Fossil Plants as Indicators of Climate

Because plants are so sensitive to environmental conditions, the fossils of many plants can be used to infer climatic conditions of the past. The cycads, for example, are an ancient group of plants that are now found growing only in tropical and subtropical settings (Figure 4-19). Because this distribution seems to reflect a fundamental physiological limitation of the group, it is assumed that fossil cycads also lived in warm climates.

Angiosperm plants are valuable indicators of climates of the past 80 million or 90 million years, the interval during which they have been abundant on Earth. Angiosperms include not only plants with con-

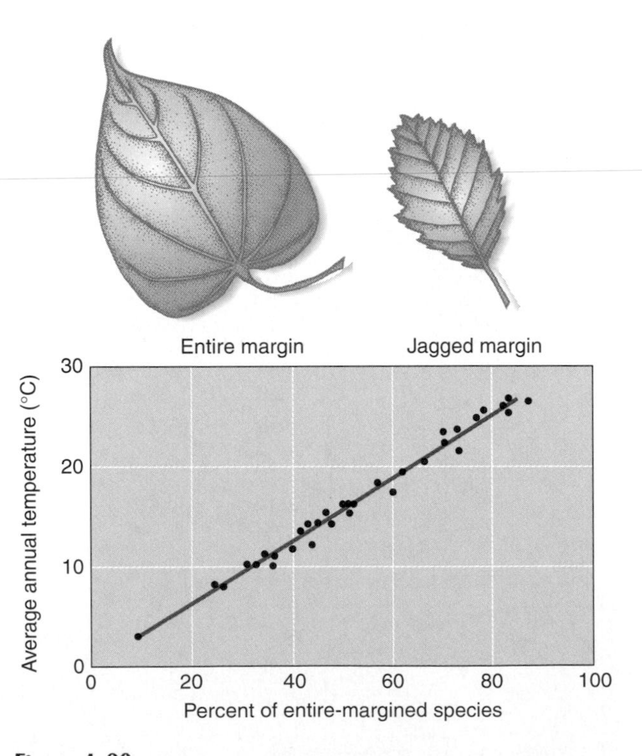

Figure 4-20 Relation between climate and the shapes of leaves of flowering plants. In the modern world the average annual temperature of a region is closely correlated with the percentage of plant species with entire (or smooth) margins. *(After J. A. Wolfe, American Scientist, 66:994–1003, 1978.)*

spicuous flowers and hardwood trees (e.g., maples, beeches, and oaks) but also grasses and their relatives. Thick, waxy leaves help some plants retain moisture. Fossil leaves of this type indicate that fossil hardwood plants lived in warm climates. The margins of leaves provide an even more useful means of assessing temperatures of the past. Leaves with smooth rather than jagged or toothed margins are especially common in the tropics. In fact, if a large flora occupies a region that is characterized by moderate or abundant precipitation, the percentage of plant species with smooth margins provides a remarkably good measure of the average annual temperature of that region (Figure 4-20). This relationship has revealed a considerable amount of information about temperatures on Earth during the past 100 million years.

Inasmuch as certain groups of animals also are restricted to warm regions, some animal fossils, too, serve as indicators of warm climates. Reptiles, which do not maintain constant warm body temperatures, are among the animals that cannot live in very cold climates.

The Marine Realm

The ocean floor is a vast basin in which most sediments have accumulated over the course of Earth's history. For this reason, and because so many types of organisms live in the ocean, the seafloor is also where most species of the fossil record have been preserved.

We have seen how the geographic distribution of terrestrial species reflects broad patterns of air movement in the atmosphere. In a similar way, the distribution of marine species reflects large-scale movements of water in the ocean.

Water Movements

Most ocean currents at Earth's surface owe their existence primarily to large-scale winds. The trade winds blow toward the equator from the northeast and southeast (see Figure 4-9), pushing equatorial water westward to form the north and south **equatorial currents** (Figure 4-21). These currents pile water

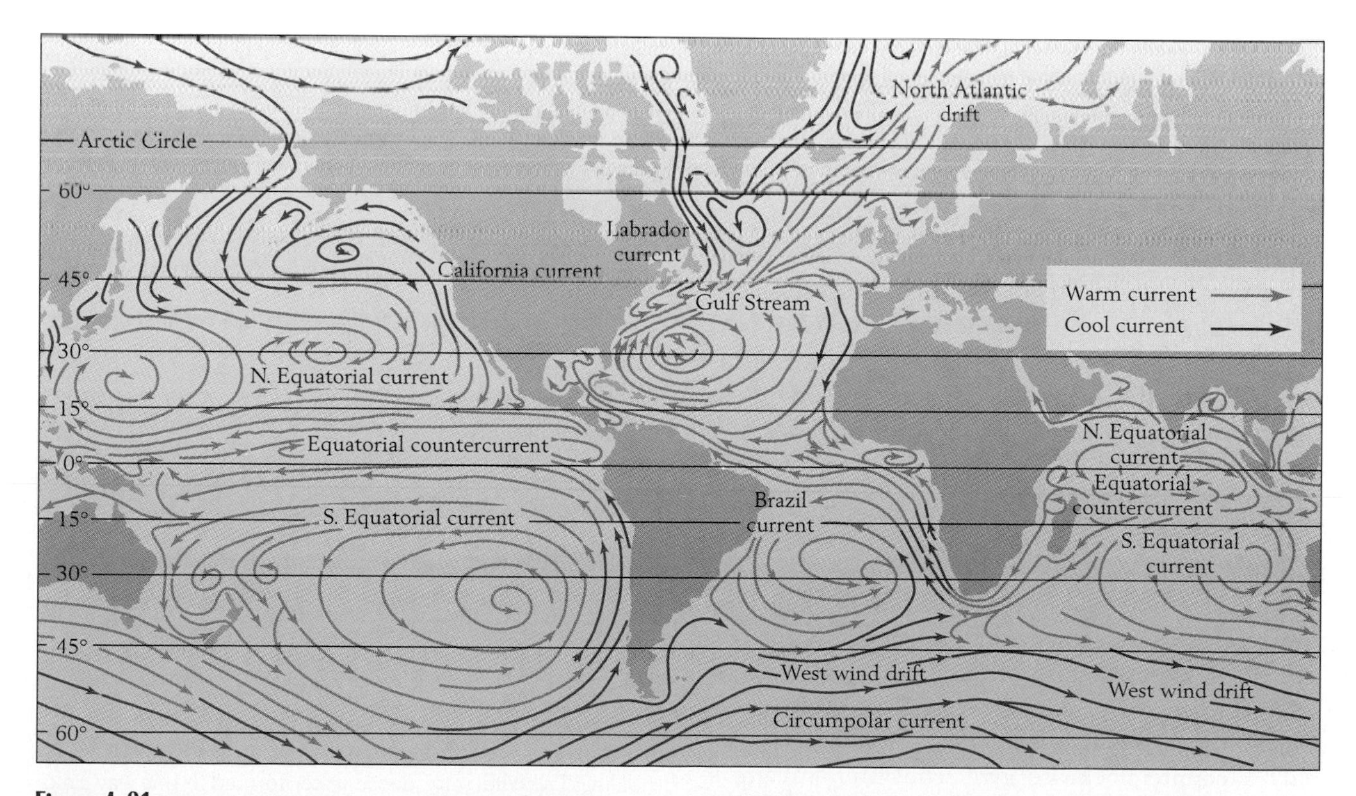

Figure 4-21 **Major surface currents of the ocean.** Note that large gyres north of the equator move clockwise, while those south of the equator move counterclockwise. *(After P. R. Ehrlich, A. H. Ehrlich, and J. P. Holdren, Ecoscience: Population, Resources, and Environment, W. H. Freeman and Company, New York, 1977.)*

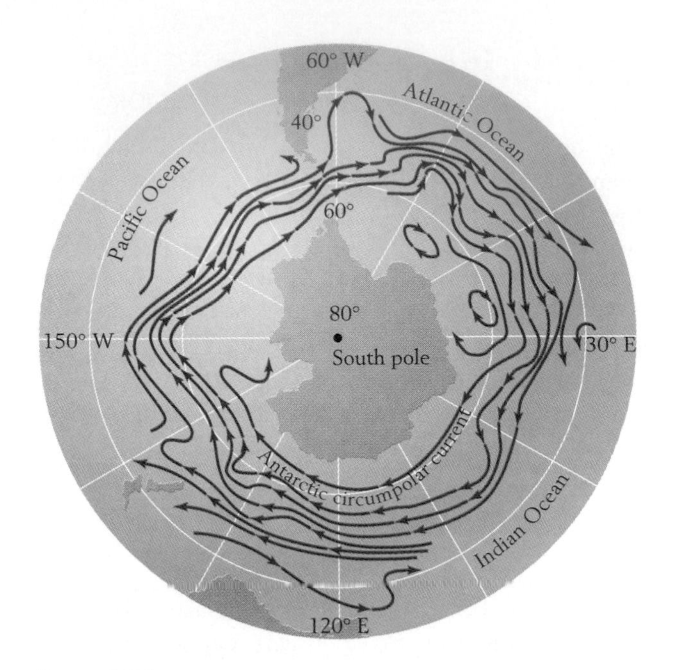

Figure 4-22 The eastward-flowing circumpolar current around Antarctica. Note in Figure 4-21 how this current is formed by the large counterclockwise gyres of southern oceans. *(After A. N. Strahler, The Earth Sciences, Harper & Row, New York, 1971.)*

up on the western sides of the major ocean basins, where some of the water flows backward under the influence of gravity as **equatorial countercurrents.**

Because equatorial currents are also affected by the Coriolis effect, they are deflected toward the poles as they approach the western boundaries of ocean basins. (Recall that the Coriolis effect bends a current in the northern hemisphere toward the right, or clockwise, and bends a current in the southern hemisphere toward the left, or counterclockwise.) When the western boundary current of an ocean reaches the zone of the westerly winds, between about 35° and 60° from the equator, it bends eastward. As the current then approaches the eastern margin of the ocean basin, it bends equatorward because of the Coriolis effect. Eventually the trade winds perpetuate the flow toward the equator. The result of these movements for each major ocean is a full clockwise gyre north of the equator and a counterclockwise gyre south of the equator. The Indian Ocean, most of which lies south of the equator, has only a counterclockwise gyre. The Gulf Stream, a famous segment of the North Atlantic gyre, carries warm water from low latitudes to the shores of Great Britain, where palm trees survive more than 50° north of the equator. An eastern segment of the Pacific gyre has the opposite effect, bringing cool water to the coast of California.

In the southern hemisphere, the southern segments of the three gyres are known as the ***west-wind drifts.*** Strengthened by westerly winds (see Figure 4-9), these drifts join to form the Antarctic ***circumpolar current*** (Figure 4-22). The landmasses of North America and Eurasia prevent the development of a comparable circumpolar current in the northern hemisphere.

Near the poles, water that is dense because it is frigid and more saline than normal sinks to great depths (Figure 4-23). When this water reaches the deep-sea floor, it spreads toward the equator. Antarctic water that descends in this manner is denser than the water from the north, so it hugs the bottom of the sea and flows well into the northern hemisphere. Above this water, which remains at near-freezing temperatures, is slightly warmer water that descends near the other pole, north of Iceland. These descending currents supply the deep sea with oxygen from Earth's atmosphere near the poles. This oxygen per-

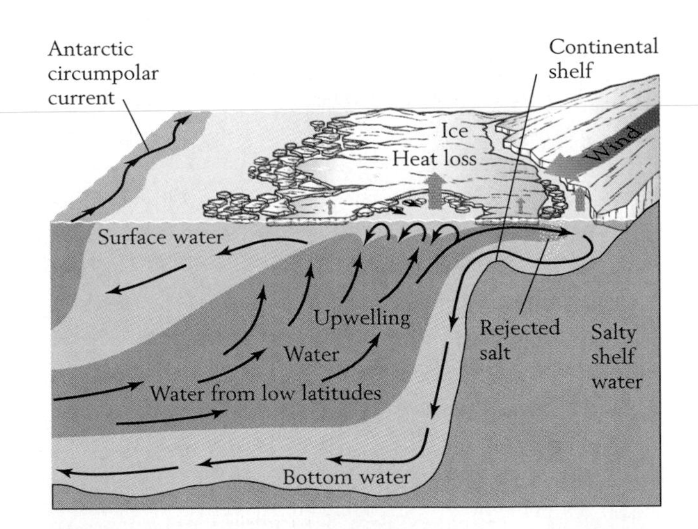

Figure 4-23 Formation of cold bottom water around Antarctica. Water flowing below the surface of the ocean from low latitudes rises up against the continental margin all around Antarctica. Here it becomes more dense, for two reasons. First, it loses heat to the cold atmosphere. Second, freezing of some of the water to form sea ice leaves excess salt behind, so that the remaining water is slightly hypersaline. The cold, salty, dense water formed in this way sinks to the deep-sea floor, where it spreads throughout the world. *(After A. L. Gordon and J. C. Comiso, Scientific American, June 1988.)*

mits a wide variety of animals to live in the bottom water despite the freezing temperatures. There have been times in the geologic past when the deep sea has been warmer and poorly supplied with oxygen.

Waves are yet another important form of water movement. Surface waves result from the circular movement of water particles under the influence of the wind. Because this movement decreases with depth, wave motion has no effect below a certain water depth, often several meters below the surface, depending on the size of the wave. Waves that are far from shore form swells that lack sharp crests. Very close to shore, the seafloor so greatly impedes the forward movement of waves that they steepen and break.

Tides, which also cause major movements of water in the oceans, result from the rotation of the solid Earth beneath bulges of water that are produced primarily by the gravitational attraction of the moon. As tides approach a coast, they often generate strong currents. A tide flows toward a coast and then ebbs again within a few hours as Earth rotates, moving the coast away from the tidal bulge. As tides ebb and flow at the edge of the sea, they cause the shoreline to shift back and forth across an ***intertidal zone.*** Landward of this zone is the ***supratidal zone,*** a belt that is dry except when it is flooded by storms or strong onshore winds that coincide with high tide.

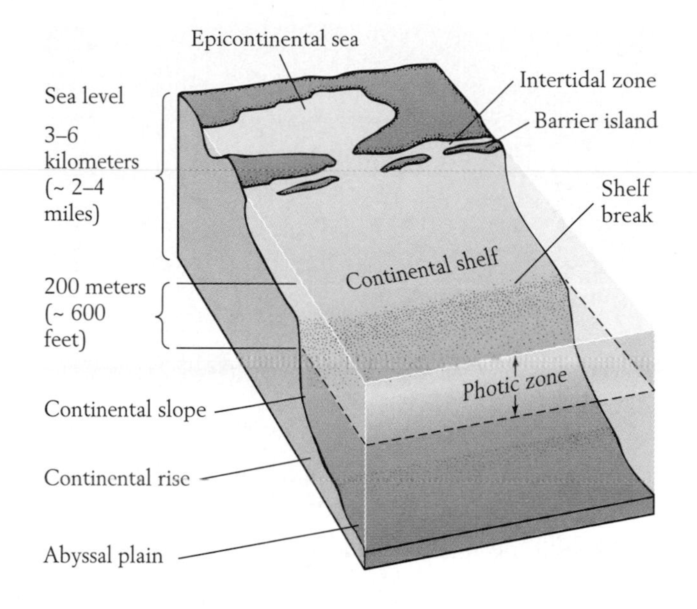

Figure 4-24 The aquatic environments at the edge of a continent. When the water is clear, the continental shelf—the submerged margin of a continent—usually lies within the photic zone, where enough sunlight penetrates to permit photosynthesis.

The Depth of the Sea

The depth of the sea varies from the thickness of a watery film at the shoreline to more than 10 kilometers (6 miles) in the deep sea. Large areas of seafloor lie between 3 and 6 kilometers (2 to 4 miles) below sea level (see Figure 4-1). These areas constitute the **abyssal plain.**

More details of the configuration of the seafloor are shown in Figure 4-24. A **continental shelf** is nothing more than the submarine extension of a continental landmass. The **shelf break** marks the edge of the shelf; seaward of it, the continent pinches out along the continental slope. Today, throughout the world, the shelf break of continents is about 200 meters (650 feet) below sea level. This consistency reflects the fact that all continents are roughly the same thickness, so that their surfaces stand at similar heights above the intervening oceanic crust (see Figure 1-15). Near the base of the **continental slope,** continental crust gives way to oceanic crust. Just seaward of this juncture is the **continental rise,** consisting of sediment that has been transported down the continental slope. Beyond the continental rise lies the abyssal plain, which is the surface of a layer of sediment resting on oceanic crust. When we speak of the **deep-sea floor,** we are usually referring to the region below the shelf break; the area above the deep-sea floor is often referred to as the **oceanic realm.**

Along the margin of the sea, **barrier islands** of sand heaped up by waves and wind often parallel the shoreline. In the protection of these elongate islands are relatively quiet lagoons or bays. **Marshes,** which are formed by low-growing plants that inhabit the intertidal zone, fringe the margins of these ponded bodies of water (Figure 4-25). Here plant remains accumulate as peat, which, if buried under the proper conditions of temperature and pressure, can turn to coal. In places the sea spreads farther inland over a continent, forming a broad, semi-isolated **epicontinental sea.** At the present time the seas happen to stand lower in relation to continental surfaces than they have done at most times during the past 600 million years. For this reason, epicontinental seas are not well developed. As we journey through

Figure 4-25 An intertidal marsh along the coast of Virginia.

Phanerozoic time in later chapters, we will examine many ancient epicontinental seas and the life they harbored.

How is life of the ocean related to the water's depth? Depth itself is probably a limiting factor for only a few species, but some significant limiting factors such as light (in the case of plants) and temperature are often closely related to depth. The upper layer of the ocean, where enough light penetrates the water to permit plants to conduct photosynthesis, is known as the **photic zone.** The base of this zone varies from place to place, depending on the clarity of the water and the angle of incidence of the sun's rays, but it usually lies between 100 and 200 meters (300 to 600 feet) below sea level. It is sheer coincidence that 200 meters is also the approximate depth of the shelf break in most areas.

For life of the seafloor, the most profound environmental change associated with depth takes place along the margin of the ocean. In contrast to life of the adjacent **subtidal** seafloor, which is never exposed to the air, the biota of the intertidal zone must endure large, often rapid fluctuations in environmental conditions. At some latitudes in this zone, hot dry conditions prevail at low tide during the summer, yet winter chills drop temperatures below freezing. Furthermore, in the surf zone, where waves break along a beach, the constant movement of the sand permits only a few species to survive. Species that live in this zone are exceptionally mobile and can quickly reestablish themselves in the sand if they are dislodged by a wave.

Conditions are unusual in the deep sea, too, but here the environment is more stable. As we have seen, the frigid water that descends over the abyssal plain from polar regions approaches the freezing point. In fact, all of the oceans' waters are cool at depths below about 500 meters (1600 feet). The environment formed by these waters is inhabited by unique groups of species that are adapted to cold conditions.

Marine Life Habits and Food Webs

Most of the photosynthesis that takes place in the ocean is conducted by single-celled drifting algae. For this reason, these algae are widely considered to be the most important producers at the base of the food web. Organisms that drift in water are known as **plankton,** and plantlike organisms that belong to this group constitute the **phytoplankton** (Figure 4-26). The most important groups of phytoplankton are *dinoflagellates, diatoms,* and (in warm regions) *calcareous nannoplankton* (see Figure 3-16). All three groups have extensive fossil records—the diatoms and calcareous nannoplankton because they have hard skeletons and the dinoflagellates because they form cysts that have durable cell walls (p. 71).

Feeding on the phytoplankton are drifting animals known as the **zooplankton,** among which are small shrimplike crustaceans and other animals that spend their full lives adrift. Also included in this group are the floating larvae of some invertebrate species that spend their adult lives on the seafloor. These larvae allow the seafloor species to disperse over large areas of the ocean; after spending time in the plankton, they settle to the sea bottom and develop into adult animals. Some members of the zooplankton are carnivores that feed on other zooplankton.

Although many types of plankton have the capacity to swim, planktonic species move through the ocean primarily by drifting passively along, going with the flow. Animals that move through the water primarily by swimming are termed **nekton.** The most important of these swimmers are fishes. Both the plankton and the nekton include not only herbivores, which feed on phytoplankton, but also carnivores, which feed on other animals. Planktonic and nek-

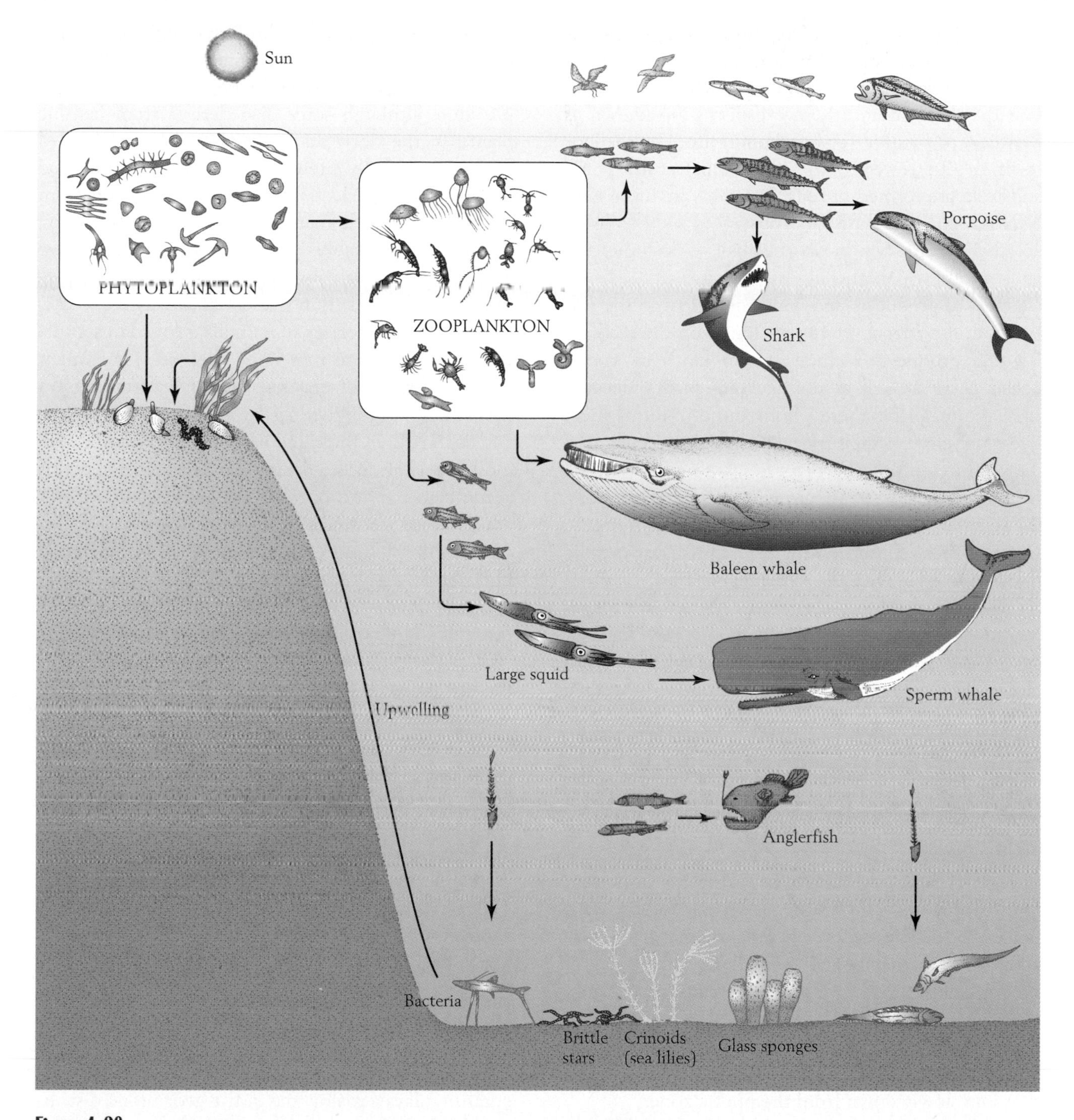

Figure 4-26 The food web in the ocean. (The various forms of life are not drawn to scale.) Phytoplankton occupy the photic zone of the ocean, and thus most zooplankton, which feed on phytoplankton, also live here. On continental shelves, especially near the shore, bottom-dwelling plants also contribute food to the marine ecosystem. Most species of large carnivores are fishes. Whales are warm-blooded mammals that include carnivorous porpoises and sperm whales, which feed on large animals, as well as baleen whales, which strain tiny zooplankton from the water. As the amount of plant material diminishes with depth, the abundance of animal life diminishes as well. A few herbivores that feed on plankton suspended in the water, such as sponges and crinoids (sea lilies), live on the deep-sea floor, but most herbivores there extract their food from sediment. Bacteria in the deep sea turn dead organic matter into nutrients that upwelling currents carry to the surface for use by phytoplankton and other photosynthetic life.

tonic organisms together constitute **pelagic life,** or oceanic life that exists above the seafloor.

Both immobile and mobile organisms also populate the seafloor, and these organisms are known as **benthos,** or **benthic** (or **benthonic**) life. The seafloor itself is often referred to as the **substratum.** Some substrata are formed of rock, but they are more likely to be composed of soft substances such as loose sediment. Some benthic organisms live on top of the substratum, while others live within it.

Just as producers and consumers float in the water, both nutritional groups are found on the seafloor. Benthic producers include certain kinds of single-celled algae as well as multicellular plants. Because they require light, these photosynthetic forms must live on or close to the surface of the substratum.

Some benthic herbivores graze on plantlike forms, especially algae growing on hard surfaces. Some strain phytoplankton and plant debris from the water. Others consume sediment and digest organic matter mixed in with the mineral grains.

Bottom-dwelling carnivores of modern seas include crabs and starfishes as well as several kinds of snails and worms. In addition, many fishes swim close to the sea bottom and feed on bottom-dwelling animals.

Figure 4-26 depicts the basic features of the marine food web. The phytoplankton occupy the photic zone above both the continental shelves and the deep sea. Joining them as producers on continental shelves, where the photic zone reaches the seafloor, are bottom-dwelling plants. Another important food supply in shallow water is organic debris transported to the ocean by rivers or washed in from marginal marine marshes. The high concentration of phytoplankton in the photic zone causes zooplankton to be concentrated there as well. Some herbivorous zooplankton can also be found at greater depths, where they feed on the algal cells and plant debris that rain slowly down from the photic zone.

Different kinds of swimmers occupy different depth zones in the ocean. Some, such as herring, feed on zooplankton. So do the great baleen whales, which strain zooplankton through a sievelike bony structure. Other fishes, including nearly all sharks, are carnivorous, as are many kinds of whales. Carnivores are found at all depths of the ocean, although some species are restricted to narrow depth zones.

A wide variety of benthos are found on the shallow seafloors of the photic zone. Here most of the food at the base of the food web comes from phytoplankton, although some also derives from benthic plants. In the deep sea, however, where suspended food is scarce, only a few kinds of benthos strain food from the water. Most herbivores extract food from sediment, and so do many types of carnivorous fishes. In fact, a remarkably large number of species live along the cold, dark abyssal plain. Organic debris arrives here from shallow waters at a very slow rate, however, so the density of animals is low. Thus a survey of 1000 square meters of deep-sea floor might uncover dozens of species, each represented by a small number of individuals.

Bacteria live throughout the open ocean but are most abundant in the deep sea, where organic debris accumulates. Some bacteria decompose this debris, while others transform some of the products of decay into simple nutrient compounds of nitrogen and phosphorus. Phytoplankton use these compounds to make food, thereby cycling the materials back through the ecosystem. A crucial step in this recycling is the physical process known as **upwelling,** or the movement of cold water upward from the deep sea to the photic zone. Upwelling tends to occur along the margins of continents, where the large oceanic gyres drag water away from the land (see Figure 4-21); this water is replaced by water that wells up from the deep sea. Upwelling often brings nutrients to the photic zone in large quantities, producing an unusually rich growth of phytoplankton. The phytoplankton support large populations of zooplankton, which in turn support large populations of fishes.

Marine Temperature and Biogeography

In the marine realm, as in the terrestrial realm, temperature plays a major role in the geographic distribution of species. A pattern can be seen in the distribution of planktonic life. The calcareous nannoplankton, for example, live primarily in warm waters; individual species are found in narrow latitudinal belts where water temperatures remain within certain limits (Figure 4-27). Most species of planktonic diatoms, in contrast, live in cool waters at high latitudes.

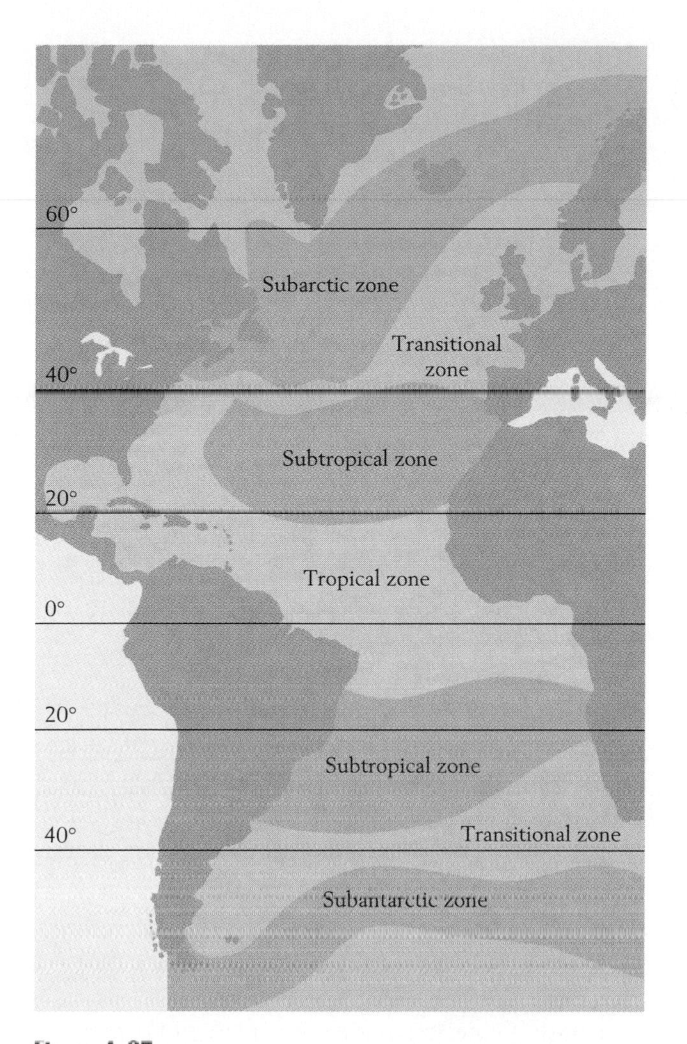

Figure 4-27 Latitudinal zones of calcareous nanno plankton in the modern ocean. Although these members of the phytoplankton are found in cold waters, fewer species live there than in warm-water zones. *(After A. McIntyre and A. W. H. Be, Deep Sea Research, 14:561–597, 1967.)*

The geographic distributions of species of the seafloor are also limited by temperature, and some large groups of animals and plants are restricted to certain latitudes. The reef-building corals are among the most prominent groups of this type; the massive reefs they construct are restricted largely to the tropics (Box 4-1).

Corals derive nutrition from smaller animals that they capture from the water by means of stinging cells on their tentacles. Reef-building corals are colonial animals that grow as clusters of connected individuals, or polyps. A colony forms from a single polyp that develops from a larva. The original polyp gives rise to a colony by budding off additional polyps, which in turn bud off others. Each polyp secretes a cup of calcium carbonate, and the adjacent cups are fused to form a large composite skeleton.

Some other types of reef dwellers join corals in contributing skeletons to the solid reef structure. Aiding corals in forming their large colonies are single-celled algae that live and multiply in the coral tissues. The corals supplement their diet by digesting some of these algae. Before they are digested, however, algal cells remove carbon dioxide from the corals for use in photosynthesis. In this way, the algae facilitate the corals' secretion of their calcium carbonate skeleton. Species that lack such algae cannot form reefs.

Reefs form their own fossil records as they grow. Beneath their living surface they consist of the remains of the dead animals and plants that were responsible for their construction.

Oxygen isotopes of fossil shells have been used in efforts to determine the temperatures at which animals lived millions of years ago. Oxygen occurs naturally in two isotopic forms, or forms of the element that have different atomic weights. Oxygen 18 has two more neutrons than the more common oxygen 16, and is thus the heavier isotope. The two isotopes have the same chemical properties, but marine organisms that secrete shells incorporate the isotopes in slightly different proportions, depending on the temperature of the environment. As temperature decreases, the percentage of oxygen 18 increases. (A difficulty encountered in attempts to analyze ancient ocean temperatures by means of oxygen isotopes is that some ancient shells have suffered chemical alteration after burial. As a result, temperature estimates based on oxygen isotopes are sometimes inaccurate.)

Salinity as a Limiting Factor

The saltiness of natural water is called **salinity.** Oceanic seawater contains about 35 parts of salt per 1000 parts of water, or is said to have a salinity of 35 parts per 1000 (3.5 percent). Salinities of 30 to 40 parts per 1000 are regarded as within the normal

For the Record 4-1

The Fragile Reef

Living coral reefs, with their colorful assemblages of animals and plants, are among the most beautiful biotic communities on Earth. Snorkelers flock to shallow tropical seas to marvel at these magnificent underwater kingdoms. Of all living communities, however, reefs are among the most vulnerable to environmental change, and today many coral reefs are under severe environmental stress because humans are altering their habitat. The fossil record alerts us to the alarming possibility that many modern reef ecosystems may collapse in the coming decades. During the past 600 million years a succession of communities have built reefs resembling those of the modern world. The reign of each group of reef builders ended abruptly, and reefs nearly disappeared from the world's oceans. After each crisis, reefs recovered their luxuriance only after millions of years of evolution.

The upper surface of a reef exposed at low tide.

The reef ecosystem is fragile because reef corals have very specific ecologic requirements. They can live only in waters of normal marine salinity; they cannot tolerate brackish or hypersaline conditions. Reef corals also require warm water, apparently because warmth facilitates the corals' secretion of their calcium carbonate skeletons. Few reefs grow more than 30° north or south of the equator. Yet temperatures only slightly higher than those of the warmest tropical seas in the world today are lethal to most coral species. Reefs can flourish only where water movements are strong enough to provide a rich supply of the zooplankton upon which corals feed. The water must also be clear, because sediment in suspension hinders corals' feeding and screens out sunlight needed by their symbiotic algae. Finally, reefs suffer if the water is overly rich in nutrients, especially phosphates, because such water favors the growth of mats of fleshy algae. These algae, which are quite unlike the tiny symbiotic algae in the corals' tissues, can smother corals and other members of the reef community.

Another factor that contributes to the instability of a coral reef community is the interdependence of its species; the disappearance of one kind of organism can devastate many others. In 1983, the sudden death of sea urchins wreaked havoc on reefs throughout the Caribbean. For unknown reasons, about 98 percent of the population of the sea urchin *Diadema* vanished. Some humans applauded the decline of this creature, whose poisonous, needlelike spines frequently cause injuries to swimmers, but the ecologic result was disastrous. Sea urchins graze on soft algae and prevent them from forming the mats that smother corals. After the decline of *Diadema*, mats of algae quickly spread over reefs, killing many corals and other reef builders. It was several years before populations of *Diadema* expanded to their former densities, and the damaged reefs are still in the process of recovering.

Modern coral reefs have held their own against natural disasters over vast stretches of time, but today human disturbances threaten to inflict more lasting damage on many reefs, or even total destruction. Nearly 2 million people a year visit the Pennekamp Coral Reef State Park in Florida, which includes a substantial fraction of the few well-developed shallow-water coral reefs in North American waters. About half of these enthusiastic visitors plunge into the water, and many bump into corals or trample them. Boats scrape the reefs. Coral colonies that have taken decades or even centuries to grow are being destroyed, and many areas of the reef are already dead. Reef dwellers are also being poisoned by sewage, toxic wastes, and petroleum compounds discharged by boaters and by the burgeoning population of the Florida Keys. Phosphates from fertilizers and human waste do their damage less directly. They promote the growth of smothering algal mats, as does the heavy fishing in surrounding waters, because many kinds of fish graze on algae.

Human activities may soon prove damaging to coral reefs in many areas of the world. Foremost among these activities is the burning of organic compounds, which releases carbon dioxide into the atmosphere, warming Earth's surface. Pulses of regional warming unrelated to human activities have killed some reef-building corals of the Pacific Ocean in recent years. These events have sounded an urgent warning. If human-induced greenhouse warming greatly elevates temperatures of tropical seas throughout the world during the next few decades, coral reefs may suffer widespread damage.

The fossil record, then, shows us that organic reefs have always tended to be ecologically fragile structures, and the ecology of modern reefs indicates that they are fragile still.

range for seawater; water of lower salinity is called **brackish,** while water of higher salinity is termed **hypersaline.**

Brackish and hypersaline conditions are most commonly found in bays and lagoons along the margins of the ocean. Brackish conditions result from an influx of water from rivers into bays or lagoons that are partially isolated from the open ocean. Hypersaline conditions also develop in bays and lagoons, but only in those whose waters evaporate rapidly—usually in hot, arid climates.

The salinity of brackish and hypersaline waters typically changes frequently in response to changes in rainfall and in evaporation rates. The salt content of the tissues of most animals is similar to that of normal seawater. Marine animals therefore tend to find it difficult to move into a habitat where the salinity is abnormal or fluctuating. It is hardly surprising that most bays and lagoons contain fewer species of animals than normal marine habitats. Many marine animals migrate into marshes to breed, however, because the scarcity of predators here and the abundance of organic matter from decaying vegetation improve the chances that their offspring will survive.

Freshwater Environments

The difficulty of living in water of low salinity is a major reason the faunas of rivers and lakes are not very diverse. Rivers and lakes are freshwater habitats, or habitats whose salinities remain below 5 parts per 1000 (0.5 percent). Most freshwater animals must have ways of excreting excess water that enters their body tissues from the environment. Phytoplankton and zooplankton similar to those of the ocean also inhabit lakes, but in reduced variety. Because streams and rivers are constantly in motion, they do not sustain a planktonic community as complex as that of the ocean. Most producers that occupy rivers live on the bottom, and a large proportion of consumers are immature growth stages of terrestrial insects (see Figure 4-3). Both lakes and rivers differ from the ocean also in their small variety of larger animal species. Unfortunately, relatively few kinds of these animals are readily preserved as fossils; fishes and shelled mollusks are the primary exceptions.

Chapter Summary

1 The way a species interacts with its environment defines its ecologic niche.

2 The distribution and abundance of any species are governed by a number of limiting factors: the availability of food, the nature of physical and chemical conditions in the environment, and the presence of other species that are potential predators or competitors.

3 Communities are groups of coexisting species that form food webs. Plants, which are at the base of most food webs, are fed upon by herbivores, which are fed upon in turn by carnivores. Communities and the environments they occupy constitute ecosystems.

4 The diversity of a community is the variety of species it encompasses.

5 Bacteria decompose dead organisms and transform the products of decay into compounds that serve as plant nutrients. Thus materials are cycled through the ecosystem continuously.

6 Physical barriers to dispersal and changes in environmental temperature are the most important factors that limit geographic distributions of species; many more species exist in warm climates than in cold climates.

7 Tropical rain forests develop near the equator, where warm air rises and loses its moisture. The dried air descends north and south of the equator and circles back toward the equator as trade winds. In many areas that lie between 20° and 30° from the equator, trade winds produce deserts and dry grasslands.

8 Continental glaciers (ice caps) exist near Earth's poles in Greenland and Antarctica.

9 Mountains provide cool habitats even in the tropics. They also alter climates in neighboring areas by intensifying monsoons and creating rain shadows.

10 Because land plants are highly sensitive to climatic conditions, fossil land plants are useful indicators of ancient climates.

11 In large oceans, prevailing winds and the Coriolis effect create huge gyres of water movement.

12 The dominant food producers in the ocean are photosynthetic single-celled algae that float in shallow waters.

13 The deep sea is cold because its waters come from near the north and south poles, where frigid water sinks to great depths and flows toward the equator. Many species of animals inhabit the deep sea, but their populations are quite small because little food reaches this environment.

14 The salinity of bays and lagoons near the margin of the ocean differs from that of normal seawater and fluctuates greatly; relatively few species are able to live in these environments.

15 Freshwater environments, such as lakes and rivers, usually harbor relatively few species because life in fresh water poses physiological problems for many kinds of animals.

Review Questions

1 Sometimes the species of a community are described as forming a food chain. Why is it usually more appropriate to speak instead of a food web?

2 Which terrestrial and marine environments characteristically contain few species? Explain why each of these environments is populated in this way.

3 How do the main kinds of producers (photosynthesizers) in the ocean differ in mode of life from those on the land?

4 What can fossil plants tell us about ancient environments?

5 What does the shape of the hypsometric curve tell us about the distribution of seafloor environments?

6 What produces the intertropical convergence zone?

7 What conditions create monsoons?

8 How do winds affect the ocean on a large scale?

9 Rain forests are sometimes likened to coral reefs, in that both support communities that include large numbers of species. Why are both communities restricted to the tropics?

10 What kinds of salinities characterize lagoons along the margin of the ocean? What causes these salinity conditions?

11 Identify the limiting factors that play important roles for various communities of marine organisms (from the deep sea to the intertidal zone and from the tropics to polar regions) and for various communities of nonmarine organisms (from lowland equatorial regions to mountain tops and polar regions). How do the compositions of various communities reflect the presence of the limiting factors? Use the Visual Overview on page 92 as a guide.

Additional Reading

Kormondy, E. J., *Concepts of Ecology,* Prentice-Hall, Inc., Englewood Cliffs, N.J., 1996.

Newton, C., and L. F. Laporte, *Ancient Environments,* Prentice-Hall, Inc., Englewood Cliffs, N.J., 1989.

Stiling, Peter D., *Ecology: Theories and Applications,* Prentice-Hall, Englewood Cliffs, N.J., 1992.

CHAPTER 5

Sedimentary Environments

The stratigraphic record reveals remarkably detailed pictures of ancient settings where sediments have accumulated. The most general reason for studying these depositional settings is to reconstruct geography of the past; that is, **paleogeography.** The goal is not only to learn about the distribution of land and sea at a particular time, but also to identify and reconstruct more localized environmental features, such as deserts, lakes, river valleys, lagoons, and submarine shelf breaks. In most instances, geologists can learn not only where a valley was located, for example, but also what kind of river occupied it—perhaps one formed by many small, intertwining channels choked with bars of gravel and sand, or one that flowed along a single broad, winding channel. Frequently geologists can also "read" from the sedimentary record whether the terrain that once bordered an ancient river was a dry, sparsely vegetated plain or a swamp densely populated by water-loving trees and undergrowth. Thus one aspect of paleogeography is the study of ancient climates.

The identification of ancient sedimentary environments also provides geologists with a framework within which to interpret life of the past. Although we can learn some aspects of how an animal or plant species lived by studying the configuration of its fossil remains alone, a fuller understanding of that species can come only when its habitat is taken into consideration. It was once widely believed, for example, that the largest dinosaurs were too big to be fully terrestrial and therefore must have spent much of their time in water, like hippopotamuses. The fossil record of these giant creatures contradicts this idea: their bones are frequently found in sedimentary rocks that represent nonaquatic environments.

Aerial view of oolite shoals along the Grand Bahama Bank, with deep water on the left.

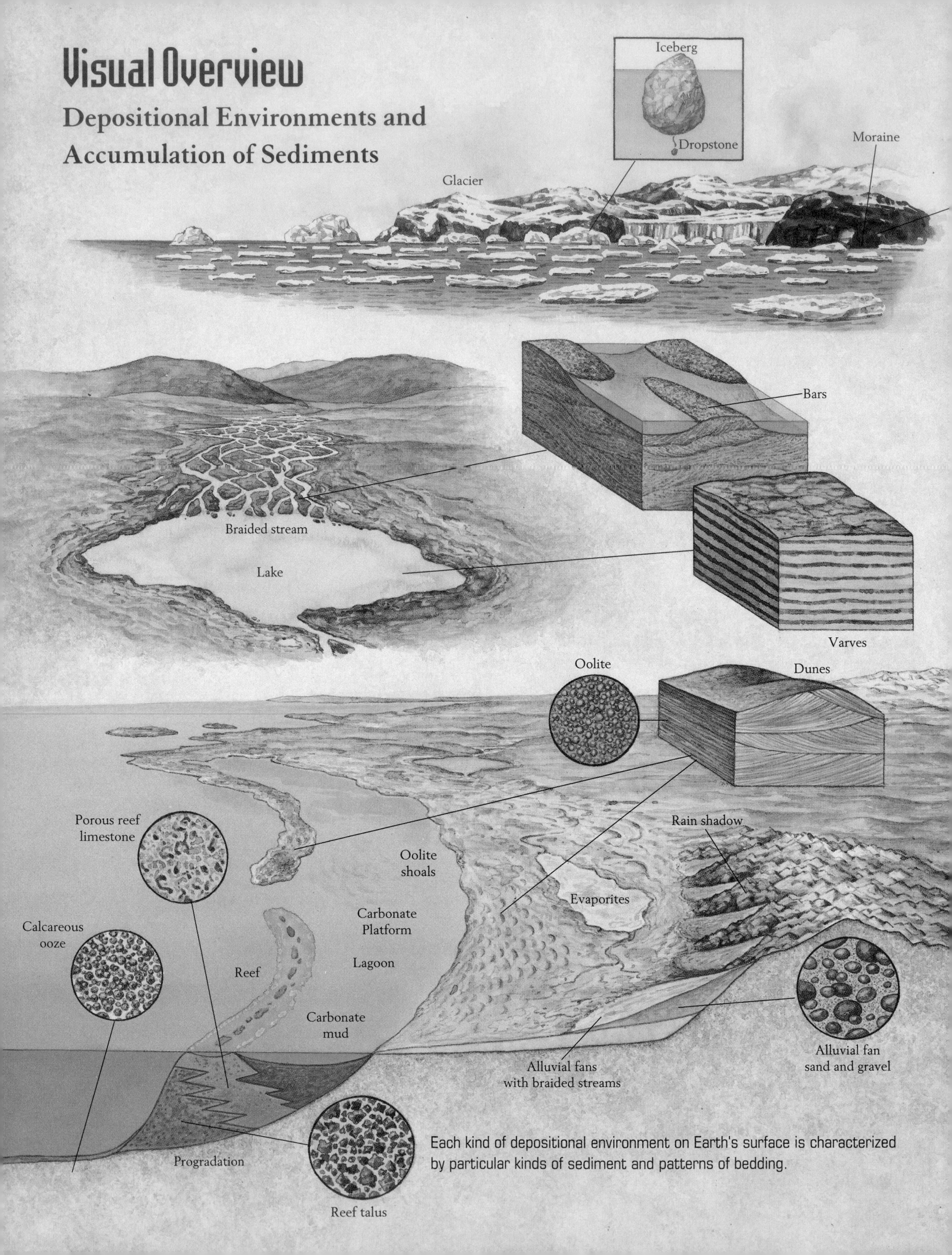

Visual Overview
Depositional Environments and
Accumulation of Sediments
Iceberg
Dropstone
Moraine
Glacier
Bars
Braided stream
Lake
Varves
Oolite
Dunes
Porous reef
limestone
Rain shadow
Oolite
shoals
Evaporites
Carbonate
Platform
Calcareous
ooze
Lagoon
Reef
Carbonate
mud
Alluvial fan
sand and gravel
Alluvial fans
with braided streams
Each kind of depositional environment on Earth's surface is characterized
by particular kinds of sediment and patterns of bedding.
Progradation
Reef talus

Glacial till
Delta
Coarse
Fine
Meandering river
Fine
Coarse
Fine
Coarse
Marsh
Inlet
Barrier
island
Turbidite
Lagoon
Regression
Transgression
Sandy mud
Beach sands
Turbidity current

The study of ancient sedimentary environments is also of considerable practical value in that many sedimentary deposits harbor natural resources, such as petroleum and natural gas (p. 62); other sedimentary bodies, such as deposits of halite and gypsum (p. 46), are themselves natural resources. By understanding the environmental relationships of sedimentary rocks, economic geologists can often predict where these resources will be found beneath Earth's surface. Coal is a resource whose location can be predicted on this basis, as are petroleum and natural gas, which tend to accumulate in porous rocks such as clean sands deposited along ancient shorelines or rivers and in ancient limestone reefs constructed in shallow seas by corals or other organisms.

Geologists rely heavily on actualism (p. 3) to reconstruct the environments in which sedimentary rocks have accumulated. In other words, they study patterns of deposition in modern settings in order to recognize ancient deposits that formed in similar environments. Because sediments that have been deposited in modern settings tend to be buried quickly beneath younger strata, geologists often have to excavate, tunnel, or core before they can observe them. *Coring* is a technique that can be used to examine a deposit at the center of a lake, lagoon, or deep ocean. A tube is driven into the bottom of the body of water and then withdrawn. The core of sediment thus extracted can be studied to determine the sequence of sediment deposition at the site. Coring at several locations provides a three-dimensional picture of sedimentary deposits. Similarly, geologists who wish to examine a meandering river's depositional record must either dig one or more pits in the valley floor adjacent to the channel or sample the floor by means of coring.

Some sedimentary features provide highly reliable information about the nature of the environments in which they formed. Many other sedimentary features, however, offer the geologist only ambiguous testimony. Most coal deposits, for example, represent swamps choked with vegetation∇but such swamps are typical of both the banks of rivers and the shores of marine lagoons. Geologists must therefore consider the nature of the beds that lie above or below coal deposits; if sediments containing fossils of marine animals lie above or below a coal bed, it is likely that a marine lagoon, not a river, lay adjacent to the swamp or marsh in which the coal formed (see Figure 4-25). Vertical stacking of sedimentary features often provides the key to recognizing ancient environments.

We will begin our exploration of depositional environments and their characteristic sediments with a discussion of soils. We will then move to freshwater systems, explore deserts and arid basins, and finally shift our focus to the marine realm.

Soil Environments

Soil can be defined as loose sediment that contains organic matter and has accumulated in contact with the atmosphere rather than under water. Soil rests either on sediment of some sort—sand or mud, for example—or on rock. It serves as a medium for the growth of plants by supplying essential nutrients and by providing a base for the physical support of roots and underground stems. Soils form in a variety of environments throughout the world—in tropical rain forests, in arid regions, and even on mountaintops. Moreover, layers of chemically altered soil can often be found buried within thick sequences of ancient sediment. Recognition of these ancient soils is of great geologic significance in that they can be used to define the configurations of ancient landscapes. Certain soils also offer evidence about the climatic conditions under which they formed.

How Soils Form

Soils develop in part by weathering processes and decay of plant material. The upper zone of many soils, referred to as **topsoil,** consists primarily of sand and clay mixed with organic matter called humus. **Humus,** which gives topsoil its dark color, is derived from the decay of plant debris and in turn supplies nutrients for other plants; thus it occupies an important position in the cycling of materials through the plants of terrestrial ecosystems. Bacteria and fungi produce most of the decay. Desert soils are poor in humus because vegetation in such environments is sparse.

The type of soil that forms depends in part on climatic conditions. In warm climates that are dry part of the year, calcium carbonate accumulates in the layer of soil below the topsoil, forming what is known as **caliche** (Figure 5-1). The accumulation of caliche, often as nodules, results from the evaporation of groundwater under hot, dry conditions.

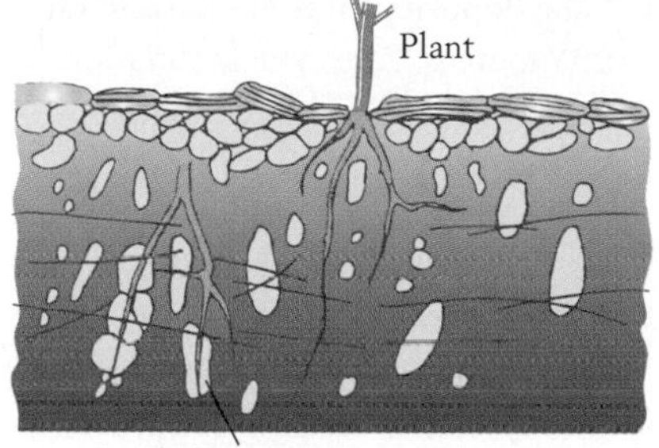

Figure 5-1 Nodules of caliche in soil. The white nodules (top) formed around plant roots, as shown in the drawing.

In moist, tropical climates warm waters percolate through the soil, destroying humus by oxidizing it. Silicate minerals in such areas also break down quickly, leaving the soil rich in aluminum oxides as well as in iron oxides, which give it a rusty red color. Tropical soils of this type are known as **laterites.**

Ancient Soils

Ancient buried soils can be exceedingly difficult to recognize and to interpret, partly because diagenesis often alters the chemical components of soils beyond recognition. One place where ancient soil is likely to be found, however, is beneath an unconformity. At such a site, soils will have formed while the rocks below the unconformity were exposed at Earth's surface—that is, before those rocks were buried beneath younger sediments.

Plant roots provide clues to the identity of ancient soils. Burrows made by animals such as insects and rodents are also diagnostic features. Certainly the most unusual of these excavations are the structures known as "devil's corkscrews," which are actually burrows that beavers of an extinct species dug with their teeth in the Oligocene and Miocene soils of Nebraska (Figure 5-2). Skeletons of those beavers have been found in the burrows, and scratches on the burrow walls match their front teeth! The fact that these animals lived as far as 10 meters (33 feet) below ground level indicates that the level of standing water in the ancient soil stood below this depth; had it been higher, the beavers would have drowned.

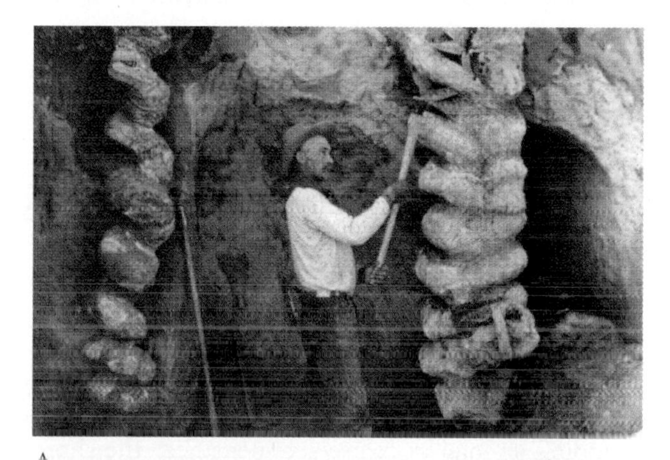

A

B

Figure 5-2 Fossil burrows made by beavers in soils more than 20 million years old. *(A)* Unlike modern beavers, which live near water, these animals lived in the grasslands of what is now Nebraska, where they formed colonies like those of prairie dogs. The burrows, which the beavers excavated with their front teeth *(B)*, extend downward as far as 10 meters (33 feet) into the soil. *(B. After L. D. Martin and D. K. Bennett, Palaeogeography, Palaeoclimatology, Palaeoecology 22:173–193, 1977.)*

The Depositional Environments of Freshwater Lakes and Glaciers

Freshwater lakes and glaciers serve as reservoirs of H_2O on the land, and each of these two kinds of reservoir leaves distinctive features in the geologic record.

Freshwater Lake Environments

At no time in geologic history have lakes occupied a large fraction of Earth's surface—but because lakes form in basins that lie at lower elevations than most soils, lake deposits are much more likely than soils to survive erosion.

Evidence of large freshwater lakes in the geologic past indicates abundant precipitation in the region, because a large freshwater lake must be fed by substantial runoff from the land. Second, as noted earlier, because of the high heat capacity of water, large lakes tend to stabilize the temperatures of nearby terrestrial areas (p. 107).

The sediments around the margins of a freshwater lake tend to be coarser than those that lie toward its center, partly because the current of a stream or a river slows down when it meets the waters of a lake, dropping its load of coarse sediment near the shore as it does so. Furthermore, the water movements related to wind-driven waves on the surface of the typical lake touch bottom only when they approach the shore, winnowing the sediment there and driving clay-sized particles into suspension. These particles later settle to the bottom toward the lake's center.

Fossils are valuable tools for distinguishing lake sediments from marine sediments. Although fish fossils are found in both kinds of deposits, the presence of exclusively marine fossils, such as corals, provides strong evidence that ancient sediments did not originate in a lake. Because burrowing animals are not as abundant in lakes as they are in many marine environments (and because waves and currents in lakes are generally weak), the fine-grained sediments that accumulate in the centers of lakes are likely to remain well layered, just as they were when they were laid down (Figure 5-3).

Figure 5-3 Lake deposits from the Eocene Green River Formation of Wyoming. The even laminations are typical of sediments that have accumulated in lakes.

Another clue in the identification of lake deposits is close association with other nonmarine deposits, such as river sediments. It would be highly unusual to find lake deposits directly above or below deep-sea deposits, or even above or below sediments deposited on the continental shelf, unless the two types of deposits were separated by an unconformity.

Glacial Environments

Even more useful than soils as clues to ancient climatic conditions are certain complex suites of depositional features that are known to develop only in particular climates. Sedimentary features associated with some types of glaciers, for example, are excellent indicators of cold climates. Glaciers that form in mountain valleys seldom leave enduring geologic records because the mountains through which they move, and on which they leave their mark, stand exposed above the surrounding terrain and therefore tend to be eroded rapidly on a geologic time scale. Continental glaciers, however, leave legible records that survive for hundreds of millions of years. As we will see in later chapters, the geologic record reveals that at several times during Earth's history ice sheets have spread over broad geographic areas. In fact, we are living during an interval of continental glaciation. As we have seen, one continental glacier now occupies most of Greenland (see Figure 4-13), and an even larger one occupies nearly all of Antarctica. Each of these glaciers is thickest in the center, tapering toward the continental margin.

As glaciers move, they erode rock and sediment, transporting both in the direction of flow. In this process, pieces of rock become embedded in the ice at the base of a glacier, where they become tools of further erosion, commonly leaving deep scratches in the bedrock that serve to record the direction in which the glacier is moving. When such scratches are found in an area that is now free of glaciers, it is safe to assume that a glacier passed that way in the distant past (Figure 5-4).

Glaciers leave records not only of erosion but also of deposition. The mixture of boulders, pebbles, sand, and mud that is scraped up by a moving glacier is deposited partly as the glacier plows along and partly when it melts. Some of the boulders bear scratches from abrasion during their journey at the base of a glacier. This heterogeneous material is called **till.** At the farthest reach of a glacier's advance, till plowed up in front of the glacier is left standing in ridges known as **moraines.** A considerable amount of till, however, after being transported for some distance, is deposited beneath the glacier; thus till may be deposited over a broad area even as a glacier continues to melt back. In front of a moraine, sediment from a melting glacier is often deposited by streams of **meltwater** issuing from the front of the retreating mass of ice. Here the sediment tends to be sorted by size into layers of gravel, cross-bedded sand, and mud, forming well-stratified glacial material known as **outwash.** Figure 5-5 shows till on the eastern side of the Sierra Nevada, which was glaciated during the recent (Pleistocene) Ice Age. Lithified till is known as **tillite.**

Figure 5-5 **Glacial till.** A mountain glacier deposited this bouldery sediment on the eastern side of the Sierra Nevada in California during the Pleistocene Epoch.

Figure 5-4 **A rocky surface in eastern Canada scoured by glaciers.**

In many instances, meltwaters converge to form a lake in front of the glacier. The alternating layers of coarse and fine sediment that typically accumulate in such lakes are called **varves** (Figure 5-6). Each coarse layer forms during the summer months, when streams of meltwater carry sand into the lake, whereas each fine layer is formed during the winter, when the surface of the lake is sealed by ice and all that accumulates on the lake bottom are clay and organic matter that settle slowly from suspension. Typically the abundant organic matter makes the winter layer darker than the summer layer. Each pair of layers of varved sediment thus represents a single

A

B

Figure 5-6 **Varved clays in Canada.** The varves are layers of clay and silt deposited in quiet lake waters. Varved clays produced during the Pleistocene Epoch at Toronto *(A)* are remarkably similar to Precambrian varves in southern Ontario that are nearly 2 billion years older *(B)*.

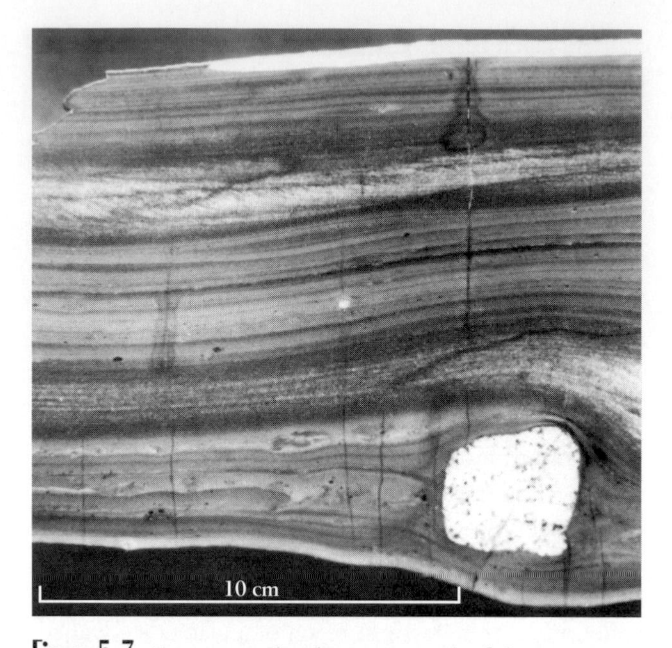

Figure 5-7 **Fine-grained sedimentary rock of the Proterozoic Gowganda Formation, which is about 2 billion years old, in southeastern Canada.** An igneous pebble apparently dropped from floating ice. The layers in the sedimentary rock are varves.

year's deposition, so geologists can count the number of years represented by a series of layers. In some areas, thousands of years of deposition have been tallied in this manner.

When a glacier encroaches on a lake or ocean, pieces of it break loose and float away as icebergs, some of which are immense (see Figure 4-14). As chunks of glacial ice melt in a lake or an ocean, their sediment sinks to the bottom, creating a highly unusual deposit in which pebbles or even boulders rest in a matrix of finer sediments (Figure 5-7). Unlike the tightly packed, coarse material that characterizes glacial till, these **dropstones** occur either singly or scattered throughout the matrix. Very few natural mechanisms other than this so-called ice rafting bring large stones to the middle of a lake or to a seafloor far from land. An uprooted tree that floats into a lake can release stones from its roots, but only in small numbers.

Deserts and Arid Basins

Geologists can identify dry environments of the distant past from diagnostic traits of sedimentary rocks. Recall, however, that dry climates are widespread in the trade wind belt but are also found outside this zone in the rain shadows of mountains and in inland regions, far from oceans. To determine which of these settings was the site of an arid environment recorded in ancient rocks, geologists must reconstruct the geographic features of neighboring regions.

Like frigid glacial environments, arid and semiarid basins are characterized by a unique suite of sedimentary deposits. Desert soils contain little organic matter, because dry conditions support little vegetation, the source of the organic matter in soils. The rain that occasionally falls in deserts leads to erosion

Figure 5-8 A playa lake in Death Valley, California.

and to deposition of sediment, often carrying the chemical products of weathering to desert basins. The subsequent precipitation of evaporite minerals in these basins sets arid regions apart from those with moist climates. In the latter, permanent streams flow great distances without being absorbed into the soil or evaporating into the air, so they are frequently capable of reaching the ocean by passing into large rivers. As a consequence, we speak of most humid regions as having **exterior drainage,** or drainage to areas beyond their borders. Runoff from arid regions, in contrast, is too sparse and intermittent to form permanent streams and rivers, and the result is **interior drainage**—a situation in which streams die out through evaporation and through seepage of water into the dry terrain.

Lakes in areas with interior drainage, known as **playa lakes** (Figure 5-8), also tend to be temporary. Temporary streams bring dissolved salts to these lakes along with suspended sediment. As the lakes shrink by evaporation, the salts precipitate as evaporites.

Death Valley: A Modern Example

Death Valley, seen in Figures 5-8, 5-9, 5-10, and 5-11, typifies arid basins in several ways. Temporary streams carry sediment down valleys incised into the naked rocks of nearby highlands to form low cone-

Figure 5-9 **Overlapping alluvial fans in Death Valley, California.** These alluvial fans have formed at the mouths of narrow valleys between mountains, where the flowing waters slow down and deposit the sediment they have carried. Braided streams can be seen on the surfaces of the fans.

shaped structures called **alluvial fans,** which spread out onto the floor of Death Valley (see Figure 5-9). These structures form where a mountain slope meets the valley floor, causing streams to slow down and drop much of their sediment. Alluvial fans consist of poorly sorted sedimentary particles that range from boulders to sand near the source area and from sand to mud on the lower, gentler slopes. Most alluvial fans include broad deposits of coarse, cross-stratified sediments laid down in a complex network of channels, or **braided streams,** that carry water and sediment during the infrequent rainy intervals. The braided streams of these channels lead to the center of the basin, where their occasional flow of water forms temporary playa lakes (see Figure 5-8). As the lake waters dry up, evaporite minerals accumulate. The same minerals also accumulate on those parts of the basin floor where groundwater seeps to the surface and evaporates. The evaporites of Death Valley are composed primarily of halite, gypsum, and anhydrite. Alternate wetting and drying in this basin produce large polygonal mudcracks in many areas (see Figure 5-10). Around the margins of these "salt pans," calcium carbonate deposits form caliche.

Sand Dunes

Sand dunes are hills of sand that have been piled up by the wind (p. 51). Dunes occupy less than 1 percent of some desert areas, including Death Valley, but where they are well developed, dunes form magnificent landscapes (see Figure 5-11). Dunes are familiar sights not only in desert terrains but also landward of the sandy beaches that border oceans and large lakes. Where the wind blows across loose sand, a dune can begin to form over any obstacle that creates a wind shadow in which sand can accumulate. Figure 5-12*A* shows how a dune tends to "crawl" downwind as sand from the upwind side moves over the top of the dune and accumulates on the downwind side. As the prevailing wind direction shifts back and forth, the direction in which the

Figure 5-10 Large mudcracks in Death Valley, California.

Figure 5-11 Sand dunes in Death Valley.

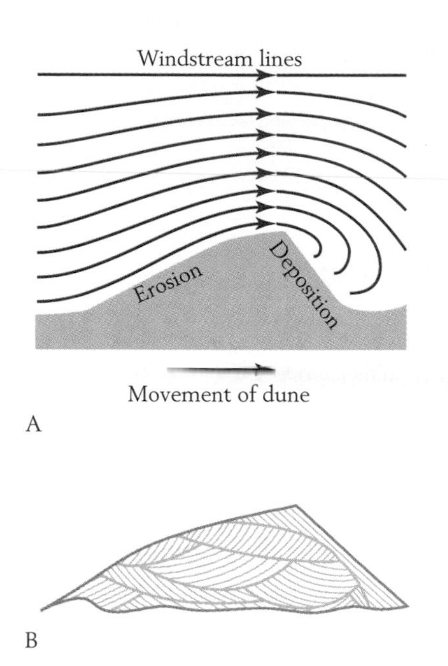

Figure 5-12 **The internal structure of a sand dune.** *A*. The windstream becomes compressed above a dune and consequently increases in velocity. The dune ceases to grow taller when its height becomes so great that it causes the windstream to move rapidly enough to transport sand. As sand passing over the dune accumulates on the steep leeward slope, the dune begins to "crawl" in that direction. *B*. This cross section of a dune shows the cross-bedding that results from shifts in the wind. As the wind direction changes, the shape of the dune is altered by removal of sand, and a new leeward slope forms. C. Cross-stratified dune deposits of the Jurassic Navajo Sandstone in Arizona.

dune migrates shifts as well. This shift in direction usually leads to the truncation of preexisting deposits, which often causes a new set of beds to accumulate on a curved surface cut through older sets. The result is called **trough cross-stratification.** Figure 5-12*B* and C show an idealized cross section through a dune and a real section through an enormous lithified dune that is more than 200 million years old.

River Systems as Depositional Environments

In areas that receive more rainfall than those we have been considering, the precipitation is more likely to create the features of exterior drainage—small streams that meet to form larger streams, which in turn flow into still larger rivers. Ultimately the waters of most large rivers reach the sea. Figure 5-13 summarizes the sequence of aquatic environments through which water typically passes as it moves from the headwaters of river systems in hills or mountains to the deep sea, transporting and depositing sediment in the process. Each of these depositional environments is represented by sedimentary deposits of many types. In the remainder of this chapter we will investigate the diagnostic features of ancient deposits formed in each of these environments—features that enable us to recognize each type of deposit as far back in the geologic record as a few hundred million or even 2 billion or 3 billion years.

Alluvial Fans and Braided-Stream Deposits in Moist Climates

Sediment in the form of alluvial fans accumulates at the feet of mountains and steep hills in moist regions, just as it does in arid regions. In moist climates, however, water often spreads these coarse sediments farther from their source, producing fans that slope more gently than those of dry basins. Runoff water moves rapidly over the relatively steep, poorly vegetated terrain that is commonly found on the lower reaches of an alluvial fan and in the region just beyond it. As a result, this water transports large volumes of coarse

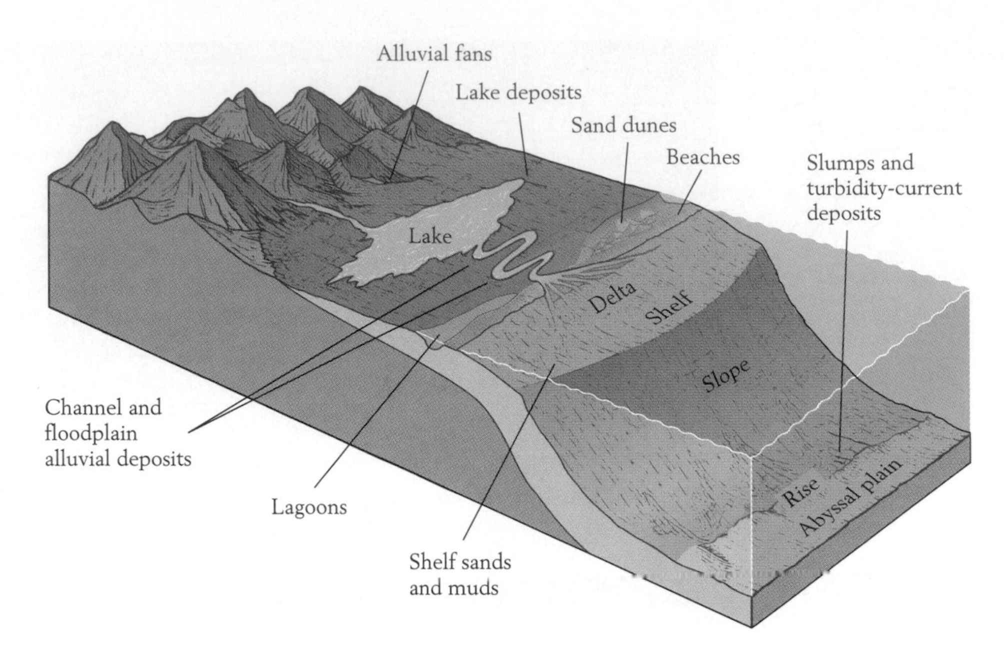

Figure 5-13 **The downhill course of sediment transported from mountains to the sea.** Along the way, sediment is trapped in a wide variety of depositional environments. *(After F. Press and R. Siever, Earth, W. H. Freeman and Company, New York, 1986.)*

sediment. In such an environment, a river does not flow as a simple winding water-way like the Mississippi or the Thames. Instead, it follows a complex network of interconnected channels, which give it the name *braided stream.* Braided streams also form in front of glaciers when large volumes of sediment plowed up by the glaciers choke the outflow of meltwater (Figure 5-14). The areas between the channels are elevated "bars" of coarse sediment.

Meandering Rivers

Many rivers occupy solitary channels that wind back and forth like ribbons (Figure 5-15). Unlike braided streams, these **meandering rivers** are not choked with sediment—that is, sediment is supplied to them slowly in relation to the rate at which the water flows. Rivers usually meander most actively in regions far from uplands and in gently sloping terrain. If there is any irregularity in this terrain, the river's path curves, and the water flows most rapidly near the outside of the bend and least rapidly near the inside. The river then tends to cut into the outer bank of the bend, where the current is swift. On the inside of the bend, the current is so weak that sediment is deposited rather than eroded, and it accumulates there to form what is known as a **point bar** (Figure 5-16*A*).

The sediment that forms point bars usually consists of sand. Most of this sand is cross-bedded, be-

Figure 5-14 **A braided stream in Alaska.** Bars of sediment divide the flow into numerous winding channels. The water and sediment emerge from the base of a melting glacier that is out of view.

Figure 5-15 A meandering river. Sand forms point bars on the insides of the meanders.

cause large ripples along the riverbed where it accumulates migrate in the course of time. In deeper water, where the current is stronger, the sediment in the river channel is coarser (Figure 5-16*B*). Gravel is often found along with coarse sand in the deepest part of the channel, and this gravel moves only when the river is flowing strongly.

Because mud tends to move downstream without settling, very little of it accumulates within a meandering river's channel. When the river overflows its banks, however, it carries fine sediment laterally to the lowlands adjacent to the river channel. In these areas, known as **backswamps,** the spreading floodwaters flow slowly, allowing mud to accumulate before they recede. In keeping with the normal pattern of sediment deposition, these floodwaters become progressively slower as they flow away from the channel, so they tend to drop the coarser portion of their suspended sediment before they spread far from the channel. Sand is therefore dropped first,

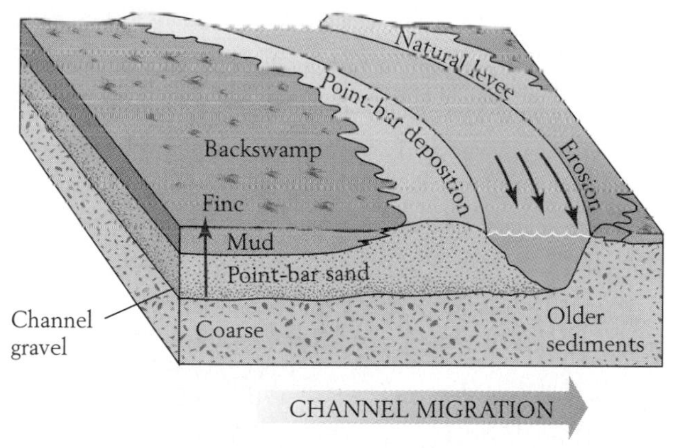

Figure 5-16 Deposition of sediment by a meandering river. *A*. At each bend the river migrates outward. The current flows most rapidly on the outside of a bend (long arrow), so the water cuts into the outer bank. On the inner side of the bend, where the current moves slowly, sand accumulates to form a point bar. As the channel migrates outward, the point-bar sands advance over the coarser, cross-bedded sands deposited within the original channel. Pebbles often accumulate at the base of the channel. Muddy backswamp deposits, which form when the river floods its banks, migrate in their turn over the point-bar sands. As a result of this shifting of environments, coarse sediments at the base grade upward into fine sediments at the top. *B*. Cyclical meandering-river deposits of Devonian age in the Catskill Formation, south of Harrisburg, Pennsylvania. In each cycle, coarse-grained sandstone (light-colored, thickly bedded deposits), which formed in the river channel, grade upward into fine-grained sediments, which formed in backswamps. *(B. From F. J. Pettijohn and P. E. Potter, Atlas and Glossary of Primary Sedimentary Structures, Springer-Verlag Publishing Company, New York, 1964.)*

followed by silt, and together they form a gentle ridge or **natural levee** alongside the channel. Because natural levees and backswamps are inundated only periodically, they tend to dry out and crack. Many such mudcracks have been preserved in the stratigraphic record. Levees and backswamps also become populated by moisture-loving plants, which may leave traces of their roots in the rock record. If after death these plants accumulate in large quantities, they may even form deposits that eventually turn into coal.

The vertical sequence in which sediments are deposited in a meandering river is seen in Figure 5-16—from coarse channel deposits at the base to cross-bedded point-bar sands in the middle to muddy backswamp deposits at the top. Levee sediments sometimes lie between the point-bar and backswamp deposits. In summary, the sequence passes from coarse-grained sediments at the bottom to fine-grained sediments at the top. This coarse-to-fine sequence forms because, as the channel migrates laterally, the point bar builds out over deeper gravels and the backswamp shifts over older point-bar deposits. As the channel migrates across a broad area, this sequence of deposits forms a broad composite depositional unit. The meandering-river sequence shown in Figure 5-16 illustrates **Walther's law,** which states that when depositional environments migrate laterally, sediments of one environment come to lie on top of sediments of an adjacent environment.

In a broad basin that happens to be subsiding, a river may migrate back and forth over a large area many times, piling one coarse-to-fine composite depositional unit on top of another as it goes (5-16*B*). Each of these composite units, or *sedimentary cycles,* lies unconformably on the one beneath it, because the channel in which the basal deposits accumulate removes the uppermost sediments of the preceding cycle as it migrates. Sometimes, however, a channel cuts deep into the sediments of the preceding cycle, removing not only the uppermost deposits but some lower ones as well. Thus many of the cycles that have been preserved in the geologic record of a migrating channel are really only partial cycles.

Two final points deserve mention. First, not every river can be assigned to the braided or meandering category. Some rivers have segments or branches that are braided and others that are meandering. Second, no river deposits its entire load of sediment in its valley. A river ultimately discharges its water and remaining sediment into a lake or the sea (see Figure 5-13).

Deltas

When a river empties into either a lake or the sea, its current dissipates, and it often drops its load of sediment in a fanlike pattern. The depositional body of sand, silt, and clay that is formed in this way is called a **delta** because of its resemblance to the Greek letter Δ. Most of the large deltas that have been well preserved in the geologic record formed in areas where sizable rivers emptied into ancient seas.

We have already seen that as moving water slows down, it drops first sand, then silt, and then clay. As river water mixes with standing water and begins to slow down, it, too, loses sand first. Silt, which is finer, spreads farther from the mouth of the river, and clay is carried even farther. The typical result is a delta structure that includes **delta plain, delta front,** and **prodelta** deposits (Figure 5-17).

Delta-plain beds, which consist largely of sand and silt, are nearly horizontal except where they are locally cross-bedded. Some delta-plain deposits accumulate within channels. As a river slows down on the surface of the delta, sand builds up on the bottom, causing the river channel to branch repeatedly into smaller channels that radiate out from the main-

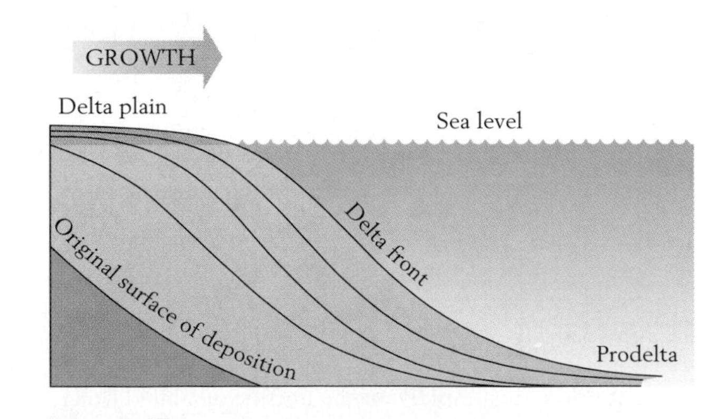

Figure 5-17 Cross section of a delta. As river water flows into the sea, it slows down. First sand drops from suspension in the delta plain; then silt and clay settle on the delta front (the slope is greatly exaggerated in the diagram); and finally clay settles in the prodelta. As the delta grows seaward, the sandy shallow-water sediments build out over the finer-grained sediments deposited in deeper water.

land. These **distributary channels** are floored by cross-bedded sands. Sand also spills out from the mouths of the channels, forming shoals and sheetlike sand bodies along the delta front. Between the distributaries and separated from them by levees are swamps, which are sometimes dotted by lakes. Here, as in the overbank areas of meandering rivers, muds accumulate and marsh plants often grow, contributing to future coal deposits.

Delta-front beds slope seaward from the delta plain, usually lying in waters that are deeper than those agitated by wind-driven surface waves. These beds consist largely of silt and clay, which can settle under these quiet conditions. Because they lie fully within the marine system, delta-front muds harbor marine faunas that often leave fossil records, but these muds usually contain fragments of waterlogged wood as well. In fact, the presence of abundant fossil wood in ancient marine muds testifies to the presence of both land and a river system near the site. In short, most ancient subtidal muds that harbor abundant fossil wood debris represent deltaic deposits.

Spreading seaward at a low angle from the lowermost delta-front deposits are the prodelta beds, which consist of clay. Even during floods, the freshwater that spreads from distributary channels slows down so abruptly that it loses its silt on the delta front. It is partly because freshwater is less dense than saline marine water that clay is carried far from the mouths of distributary channels. Because the freshwater floats on top of the more dense marine water, it does not mix in quickly; instead, it spreads seaward for some distance, carrying much of its clay.

As a delta **progrades,** or grows seaward, relatively coarse deposits on the delta plain build out over finer-grained foreset beds in accordance with Walther's law (see Figure 5-17), and these delta-front beds build out over still finer-grained bottom-set beds. The result is a sequence of deposits that coarsens toward the top. As we will now see, the stratigraphic expression of this upward-coarsening sequence varies with the nature of the delta.

The famous Mississippi River delta spills into the Gulf of Mexico in an area that is protected from strong wave action. As a result, the delta projects far out into the sea. Because construction of the delta from river-borne sediment prevailed decisively over the destructive forces of the sea, this type of delta is sometimes called a **river-dominated delta.**

That portion of the Mississippi delta which has grown at any given time has been much smaller than the delta as a whole. This growing portion, or **active lobe,** is the site of the functioning distributary channels (Figure 5-18). Many previously active lobes can be identified in the delta-plain portion of the Mississippi delta, and these lobes, which have been dated by the carbon 14 method (described in Chapter 6), provide a history of deltaic development during the

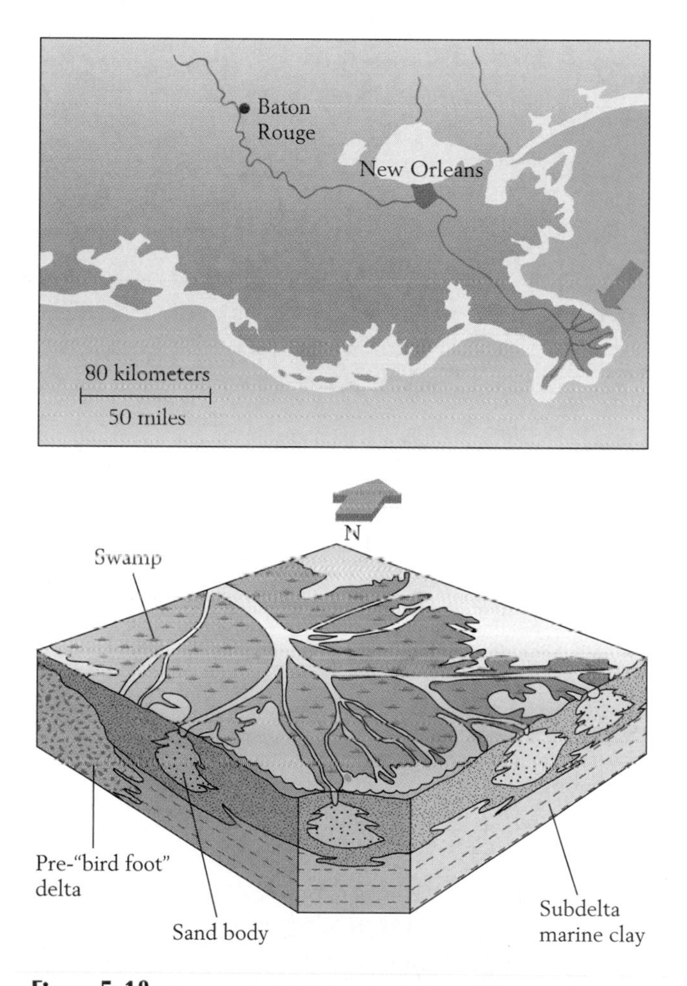

Figure 5-18 **Structure of the active lobe of the modern Mississippi delta.** At this lobe (identified by the arrow on the map) sediments are being deposited in the shallow water where the river meets the sea. Here, as shown in the block diagram below, the river channel divides into many distributary channels. Swamp deposits accumulate between the distributaries. Bodies of sand accumulate in front of the distributaries where the river water meets the ocean and slows down. As the active lobe builds seaward, the distributaries extend over these sand bodies. *(Modified from H. N. Fisk et al., Jour. Sedim. Petrol. 24:76–99, 1954.)*

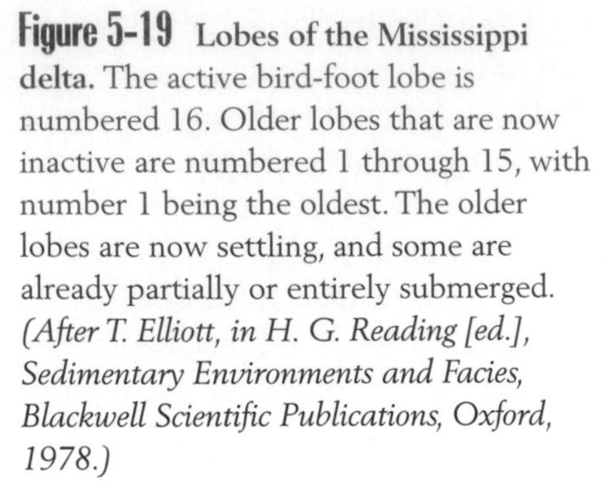

Figure 5-19 Lobes of the Mississippi delta. The active bird-foot lobe is numbered 16. Older lobes that are now inactive are numbered 1 through 15, with number 1 being the oldest. The older lobes are now settling, and some are already partially or entirely submerged. *(After T. Elliott, in H. G. Reading [ed.], Sedimentary Environments and Facies, Blackwell Scientific Publications, Oxford, 1978.)*

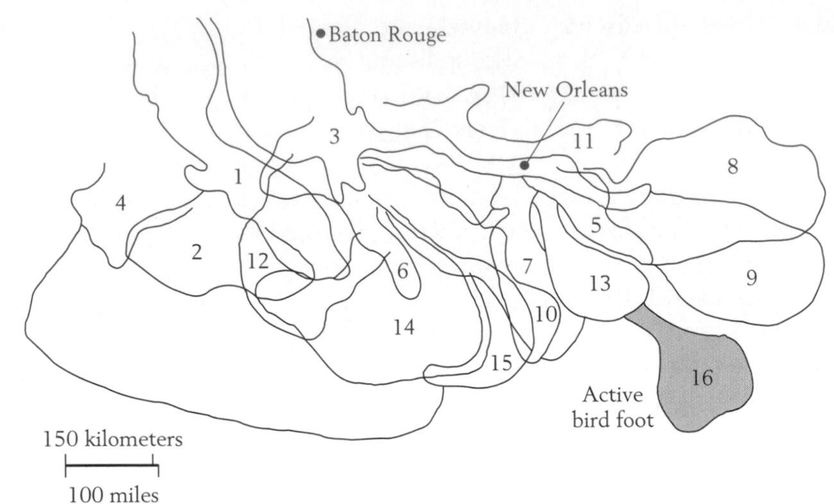

past few thousand years (Figure 5-19). Depositional activity (or lobe growth) periodically shifted in this delta when floods caused the river to cut a new channel and to abandon the previously active channel and its distributaries.

The fate of an abandoned delta lobe is, in fact, the key to the stratigraphic sequence produced by a river-dominated delta. In short, an abandoned lobe gradually sinks, for two reasons: first, the sediments of which it is formed compact under their own weight; and second, the lobe is part of the entire delta structure, which is constantly sinking as a result of the isostatic response of the underlying crust to the weight of the continually growing mass of sediment. After an abandoned lobe settles, a younger lobe eventually grows on top of it. Each lobe then consists of the typical upward-coarsening sequence. The result is an accumulation of cycles that differ markedly from those of meandering rivers, which, it will be recalled, become finer-grained toward the top.

Some deltaic cycles in the rock record lack tops because sediment was eroded away before another cycle was superimposed (Figure 5-20). Because the building of a river-dominated delta is limited to the active lobe at any given time, beds representing a delta can seldom be traced very far laterally in the rock record. When preserved, the porous sand bodies in the upper parts of deltaic cycles often serve as reservoirs for petroleum or natural gas.

Changes in the rate at which a delta sinks or the rate at which its river supplies it with sediment can alter the size of the active lobe. Today such changes are causing the Mississippi's bird-foot delta to shrink rapidly, with alarming consequences (Box 5-1).

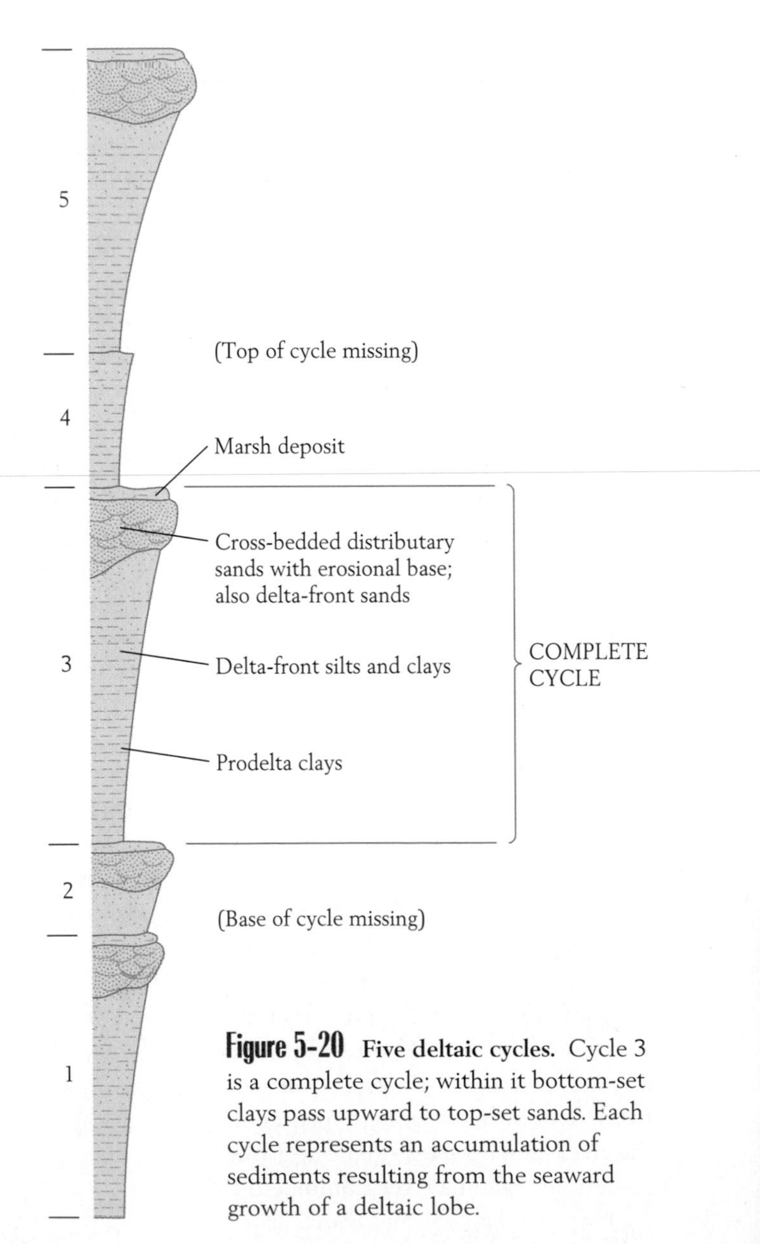

Figure 5-20 Five deltaic cycles. Cycle 3 is a complete cycle; within it bottom-set clays pass upward to top-set sands. Each cycle represents an accumulation of sediments resulting from the seaward growth of a deltaic lobe.

For the Record 5-1

The Shrinking Mississippi Delta

The coastal wetlands of Louisiana provide nearly 30 percent of the United States' annual seafood harvest. Every winter these wetlands support the nation's largest population of waterfowl. Although vitally important to the entire nation, the wetlands are now disappearing at an alarming pace. Land along the Louisiana coast, which is mostly wetlands, is shrinking at a rate of about 100 square kilometers (40 square miles) a year.

The Mississippi delta dominates the Louisiana coastline (pp. 135–136). In the past, sediment carried by the Mississippi River accumulated rapidly and built the delta out into the Gulf of Mexico. Today the river is no longer able to expand or even maintain its delta. Now the marine waters of the Gulf are drowning the delta, reducing the area available to plants and animals adapted to fresh and brackish water in marshes. This change has come about for two reasons. First, the Mississippi has been bringing less sediment to the delta every year. Second, the delta has been subsiding more rapidly than it once did.

The average volume of suspended sediment in the lower Mississippi was only half as large in 1980 as it was in 1950. Dams built on upstream tributaries of the river are largely responsible for this change. The dams have been beneficial in some ways—they have reduced flooding during times of high runoff from the land, and they have produced lakes for human recreation—but they have trapped sediment that would otherwise have traveled all the way to the Mississippi delta. At the same time, artificial levees built along the lower Mississippi to reduce local flooding have kept floodwaters from carrying sediment to some areas that otherwise would be floodplains.

An increase in the rate at which the massive delta has subsided has done even more than dams and artificial levees to prevent the buildup of sediments from keeping the delta above water. In fact, the shoreline of the Gulf of Mexico, from Florida to Texas, began to sink very rapidly just before the middle of the twentieth century. The positions of ancient shorelines indicate that during the past thousand years the average rate of sinking has been only about 1.55 millimeters (~1/16 inch) a year. In recent years the rate has been nearly five times greater. Again, human activities appear to have caused the change. So much water has been pumped out of the ground around the margin of the Gulf of Mexico in recent decades that the soft sediments of the region have become compacted, and the surface of the land has been sinking much more rapidly than it did earlier. As it sinks, the waters of the Gulf encroach farther and farther over the delta, drowning valuable wetlands.

Perhaps the best way to reduce the rate at which wetlands are disappearing from Louisiana would be to restrict the outflow of the river into the Gulf to a small number of distributaries, which would then deposit a large percentage of the sediment that the river carries. At least in these areas, the buildup of sediments might compensate for the sinking of the land. Wetlands would continue to shrink in other areas, but here, at least, they might be sustained. Unfortunately, even this strategy would not entirely solve the problem. The wetlands of Louisiana, which sustain such rich faunas of waterfowl and edible aquatic life, will continue to shrink for years to come.

A region of the Mississippi delta is being flooded as it subsides. The white objects are floats used to position a pipe that will carry petroleum from an offshore well.

Marine Depositional Environments

A series of distinctive marine environments extends from lagoons at the margin of the ocean across continental shelves and slopes to the floor of the deep sea. In addition to particular kinds of marine fossils, characteristic sedimentary features distinguish these environments.

The Barrier Island–Lagoon Complex

Because deltas form exclusively at river mouths, they occupy only a small percentage of the total shoreline of the world's oceans. Much longer stretches of shoreline are fringed by **barrier islands,** composed largely of clean sand that has been piled up by waves (see Figure 4-24). Although some barrier islands extend laterally from wave-dominated deltas, most such islands derive their sand not from neighboring rivers but from the marine realm. They are built up as waves and the shallow currents that flow along the coast, called **longshore currents,** winnow sediments and sweep sand parallel to the shoreline. In the beach zone that is washed by breaking waves, deposits tend to have nearly horizontal bedding but often dip gently seaward. Cross-bedding develops in areas where the beach surface is gently irregular and changes from time to time. Wind-blown sand often accumulates behind the beach as sand dunes, but in time these dunes are often eroded away.

Lagoons, shallow ponded bodies of water along a coastline, lie behind long barrier islands such as those that border the Texas coast of the Gulf of Mexico (Figure 5-21). Because they are protected from strong waves, these lagoons trap fine-grained sediment and are usually floored by muds and muddy sands. Rivers that empty into lagoons often build small deltas.

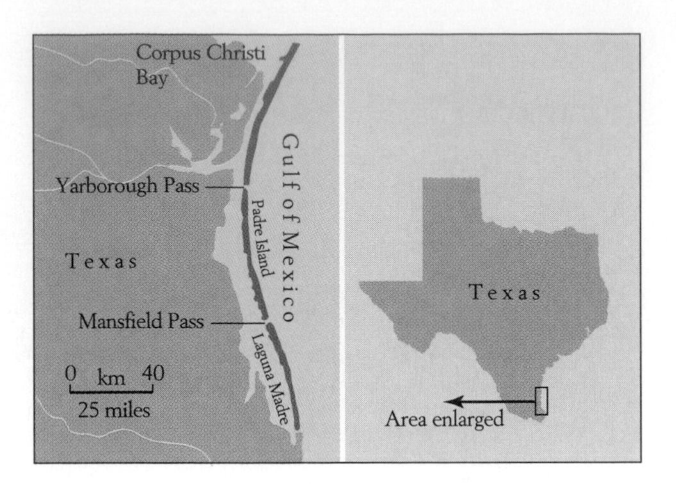

Figure 5-21 **Barrier islands along the Texas coast.** Tides here are weak, and there are few tidal channels or passes, so large lagoons lie behind these Texas barrier islands.

Barrier islands often form chains with tidal inlets separating adjacent islands. Tidal currents pass to and from the lagoon through these channels and deposit cross-bedded sand within them. Other depositional environments are also found along the shores of lagoons. Among them are **tidal flats** formed of sand or muddy sand whose surfaces are alternately exposed and flooded as the tide ebbs and flows. High in the intertidal zone, above barren tidal flats, marshes fringe one or both margins of many lagoons (see Figure 4-24). Here plant debris accumulates rapidly and decomposes to form peat, which can become coal after long burial.

Fresh water from rivers and streams tends to remain trapped in coastal lagoons for some time, so the waters of lagoons in moist climates are often brackish. The salinity of these waters at any given time depends on the rate of freshwater runoff from the land, which varies in the course of the year. Laguna Madre of Texas is typical of lagoons found in more arid climates, where temperatures are high (see Figure 5-21). Because they receive little fresh water from rivers and suffer a high rate of evaporation, the ponded waters of this long lagoon are hypersaline (p. 117). Whether lagoons are brackish or hypersaline, however, their abnormal and fluctuating salinity excludes many forms of life that require normal marine salinity. As a result, the fossil faunas in the ancient sediments of lagoons are not very diverse. Those species that are found in lagoons, however, often occur in large numbers. Usually some of them are burrowers that disturb the muddy sediments of the lagoonal floor, leaving these sediments either mottled or homogeneous and largely devoid of bedding structures.

When a **barrier island–lagoon complex** receives sediment at a sufficiently high rate, it progrades—that is, it migrates seaward—like the active lobe of a delta. Unlike the migration of a delta, however, this progradation takes place along a broad belt of shoreline (Figure 5-22). As the shoreline of a sea migrates landward, marsh and tidal-flat deposits prograde over sediments of the lagoon and associated tidal channels. These sediments in turn build out over deposits of the barrier beaches and over the tidal deltas and marshes behind them. Thus the horizontal sequence of environments (barrier beach, marsh or tidal delta, lagoon, tidal flat, and marsh) come to be represented by a corresponding vertical sequence of sedimentary deposits, in accordance with Walther's law.

The barrier island–lagoon system illustrates particularly well how the close vertical association of sediments that represent adjacent depositional environments can be used to identify ancient depositional systems. We have seen that vertical associations of rock types are also very useful in the identification of ancient meandering-river deposits (see Figure 5-16) and deltas (see Figure 5-20).

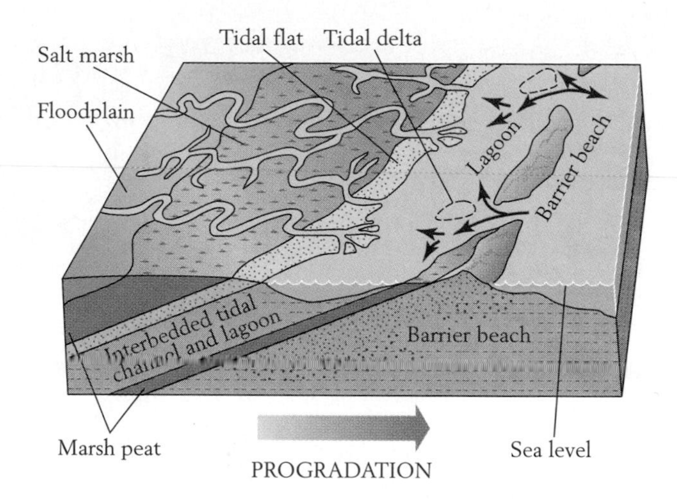

Figure 5-22 The stratigraphic sequence produced when a barrier island–lagoon complex progrades. Sediments of the lagoon and of the adjacent marshes and tidal flats are superimposed on beach sands of the barrier island.

Open Shelf Deposits

Open shelves, seaward of barrier islands, display a variety of physical conditions and therefore produce a variety of sedimentary deposits. On shelves where tides produce strong currents and sand is abundant, the currents may pile up the sand into large ridges or dunelike structures. On shelves where waves have a stronger effect than tidal currents, wave motion tends to flatten the bottom, and the sand spreads out in sheets. On quieter shelves, muddy sands accumulate. Animal burrows are abundant in most shelf sediments, and as we will soon see, skeletal fossils of particular kinds also point to an offshore, open marine environment.

Fossils as Marine Environmental Indicators

Ancient sediments deposited in barrier island–lagoon systems, together with the deposits that formed seaward of them, often yield fossils that help geologists to recognize particular environments of deposition. Figure 5-23 depicts an example in Wales, west of central England. Here, in rocks of mid-Silurian age, fossil communities of marine invertebrates are arrayed roughly parallel to the ancient shoreline. Adjacent to that shoreline is a narrow zone of fine-grained sediments, in which the inarticulate brachiopod *Lingula* is especially abundant, but only a few other species are present (Figure 5-23*A*). Presumably the water here was brackish. *Lingula* is a living fossil genus (see Figure 3-30*A*, *B*), which today tolerates nearshore environments where the salinity is brackish and variable and where few other species are able to live.

Seaward of the zone where *Lingula* predominates in the Silurian deposits is a more diverse fossil community, adapted to more stable conditions of the center and seaward margin of a lagoon (Figure 5-23*B*). Sandy deposits representing a barrier island on the outer margin of the lagoon are not preserved, but we can infer that a barrier was present because seaward of the lagoonal deposits are sandy, often cross-stratified marine deposits in which the most common fossil is a type of brachiopod that was adapted to agitated conditions (Figure 5-23C).

In finer-grained sediments deposited farther offshore is a fossil community that includes many species, none of which is very common. Many types of brachiopods are present, and trilobites are restricted to this belt. The high diversity of species reflects the stable conditions of an offshore shelf environment, beyond the influence of river water; the low abundance of species reflects a weak food supply, far from the algae and primitive higher plants that must have flourished in the vicinity of the lagoon and supplied its inhabitants with food. Where muddy sediments occur in this area, fragile planktonic graptolites are preserved, having settled to the quiet seafloor after death (Figure 5-23*D*).

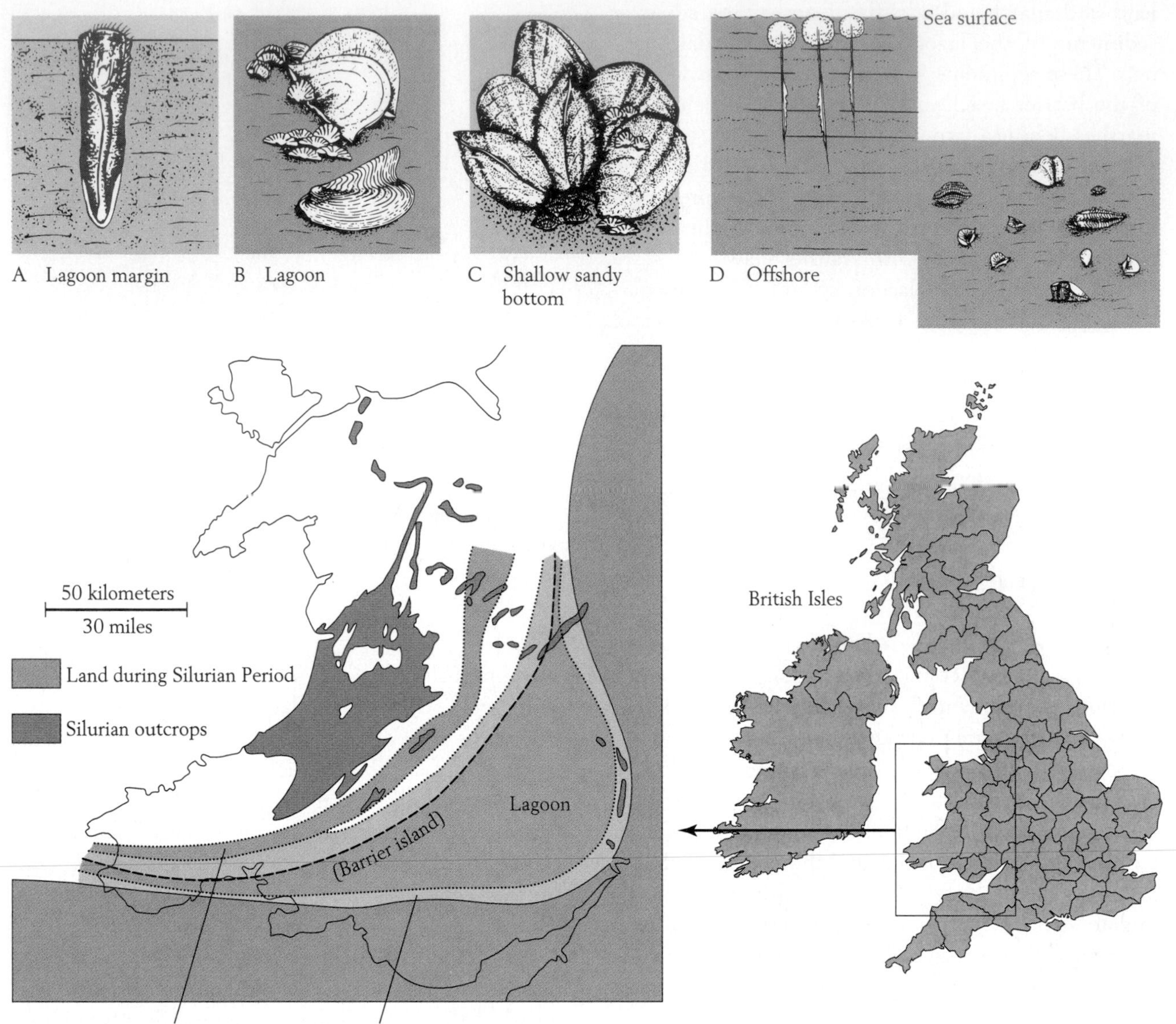

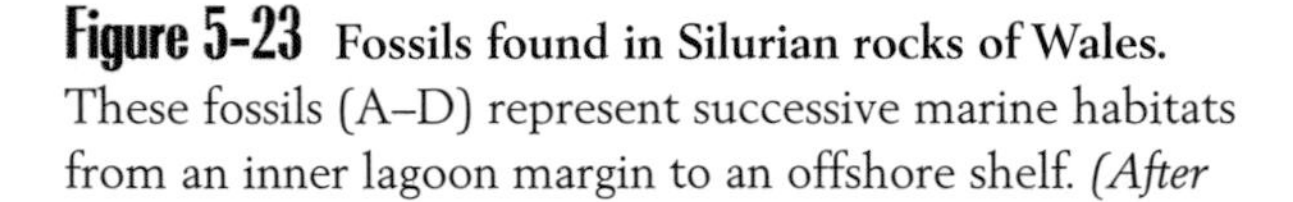

Figure 5-23 Fossils found in Silurian rocks of Wales. These fossils (A–D) represent successive marine habitats from an inner lagoon margin to an offshore shelf. *(After A. M. Ziegler, in W. S McKerrow [ed.], The Ecology of Fossils, MIT Press, Cambridge, Mass., 1978.)*

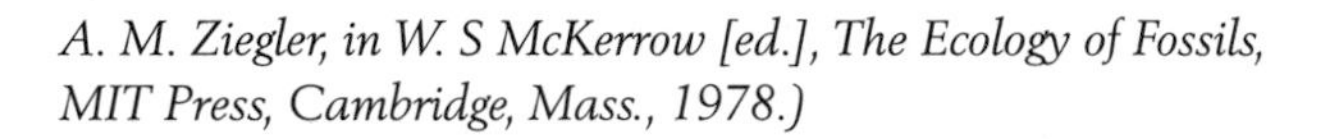

Fossils have played a key role in the reconstruction of this set of Silurian environments. In the next section, we will examine deposits that are composed entirely of recognizable fossils and fossil debris.

Reefs

In tropical shallow marine settings, if **siliciclastic sediments** are in short supply, carbonate sedimentation usually prevails. Here coral reefs are often prominent. Modern coral reefs are rigid structures that rise above the seafloor in shallow waters of high clarity and of normal marine salinity (p. 116). Because they are produced largely by organisms that secrete calcium carbonate, **organic reefs** form their own distinct depositional records—as bodies of limestone.

As we shall see in later chapters, large reefs at some times in the geologic past have been formed by organisms other than corals. Nonetheless, like their modern counterparts, ancient reefs grew in

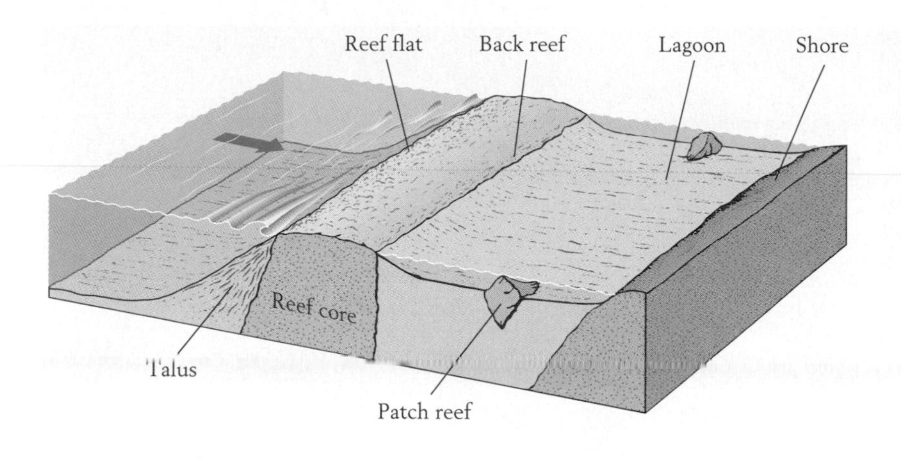

Figure 5-24 **A typical barrier reef.** The reef grows up to sea level, so the reef flat is exposed at low tide (see Box 4-1). Waves break across the reef flat, losing energy in the process and leaving a quiet lagoon behind. Sediment accumulates in the back-reef area and in the lagoon, and here and there patch (or pinnacle) reefs rise from the lagoon floor.

clear, shallow waters of normal salinity, so that the presence of an ancient reef in the stratigraphic record points to such conditions in the ancient environment.

A modern reef has several structural components that form a complex limestone deposit. The basic framework of a reef consists of the calcareous skeletons of organisms, primarily corals. This framework is strengthened by cementing organisms that encrust the surface of the reef. Carbonate sediment, composed of fragments of the skeletons of reef-dwelling organisms, is trapped within the porous framework, filling some voids. With their complex internal structure, reef limestones are typically either unbedded or only poorly bedded. Even with the presence of infilling debris, reef limestone is so porous that many ancient buried reefs serve as traps for petroleum, which migrates into them from sediments rich in organic matter.

Because living reefs stand above the neighboring seafloor, they alter patterns of sedimentation nearby. On the leeward side of an elongate reef there is often a relatively calm lagoon, especially if the reef has a typical **reef flat,** or horizontal upper surface, that stands very close to sea level (Figure 5-24). Below the living surface of the reef is a limestone core consisting of dead skeletal framework, dead skeletons of cementing organisms, and trapped sediment. A pile of rubble called **talus,** which has fallen from the steep, wave-ridden reef front, often extends seaward from the living surface.

Reefs build upward rapidly enough to remain near sea level even when the seafloor around them is becoming deeper. Many reefs, in fact, grow so rapidly and are so durable that they build seaward in the manner of a prograding delta (see Figure 5-17). Figure 5-25 shows a spectacularly exposed cross section

Figure 5-25 **Outcrop along Windjana Gorge, northwestern Australia, revealing the internal structure of a Devonian reef.** The reef core consists of unbedded limestone. The talus is crudely bedded, and the beds slope away from the reef core. The back-reef facies is also crudely bedded, but the beds are approximately horizontal.

Figure 5-26 A barrier reef. The open sea is on the right, the lagoon on the left.

of a Devonian reef in Australia. Although it was built by organisms that have been extinct for hundreds of millions of years, this reef closely resembles many modern reefs in its basic structure—it displays both a seaward talus slope and a leeward reef flat.

Isolated ***patch reefs*** are often found in lagoons behind elongate reefs (see Figure 5-24). Elongate reefs that face the open sea and have lagoons behind them are known as ***barrier reefs*** (Figure 5-26). Reefs that grow right along the coastline without a lagoon behind them are known as ***fringing reefs.*** Some fringing reefs grow seaward and eventually become barrier reefs.

Perhaps the most curious reefs in the modern world are the circular or horseshoe-shaped structures known as **atolls.** Atolls form on volcanic islands and thus are quite common in the tropical Pacific, which is dotted by many such islands. Charles Darwin's explanation for the origin of the Pacific atolls is still accepted today (Figure 5-27). According to Darwin, each atoll was formed when a cone-shaped volcanic island was colonized by a fringing reef. The island began to sink, turning the reef into a barrier reef, with a lagoon separating it from the remnant of the volcano (Figure 5-28). The island eventually sank beneath the sea, leaving a circular reef standing alone with a lagoon in the center, where limestone now accumulates in quiet water. Often the reef does not quite form a full circle but is broken by a channel on the leeward side, where reef-building organisms do not thrive. Horseshoe-shaped atolls range up to about

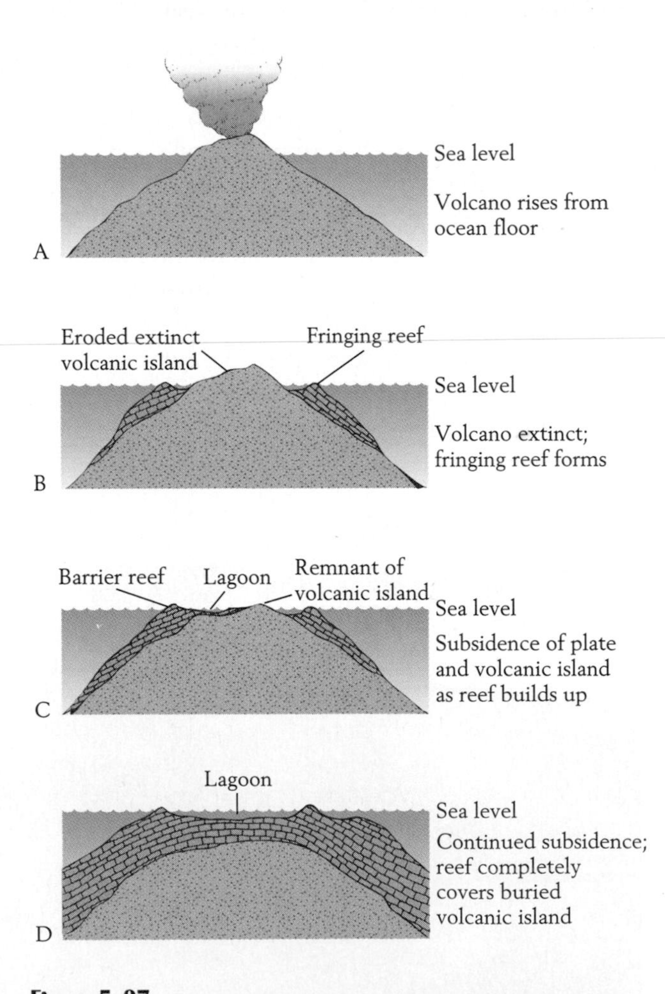

Figure 5-27 The development of a typical coral atoll in the Pacific as proposed by Charles Darwin. *(After F. Press and R. Siever, Earth, W. H. Freeman and Company, New York, 1982.)*

Figure 5-28 A coral reef encircling a volcanic island in the Pacific Ocean. This barrier reef will become an atoll when continued subsidence submerges the volcanic island. It represents the stage of atoll formation illustrated in Figure 5-27C.

65 kilometers (40 miles) in diameter; during World War II their lagoons served as natural harbors for ships.

Ancient atolls that lie buried beneath younger sediments can be identified by the study of cores of sediment brought up from drilling operations—and because porous reef rocks often serve as traps for petroleum, drilling in the vicinity of these atolls is often a profitable venture. Figure 5-29 shows the outline of a subsurface atoll of late Paleozoic age that has yielded considerable quantities of petroleum in the state of Texas.

Carbonate Platforms

A **carbonate platform** is a structure that is formed by the accretion of carbonate sediment and that stands above the neighboring seafloor on at least one of its sides. Like organic reefs, carbonate platforms consist largely of calcium carbonate that originated in shallow tropical waters at or near the site where they accumulated. In fact, organic reefs often grow along the windward margins of carbonate platforms and form parts of them.

In times past, carbonate platforms have stretched along most or all of the eastern margin of the United States; but because climates are cooler today than they have been during most of Earth's history, these structures are not well represented in the modern world. Tropical areas today do offer several examples, however. In the western Atlantic a large carbonate platform extends seaward from the Yucatan Peninsula of Mexico. Smaller platforms border the islands of the Antilles, and the platforms known as the Little Bahama Bank and Great Bahama Bank lie to the east and southeast of Florida (Figure 5-30).

The varied sediments that are now accumulating on the Bahama banks resemble those of many

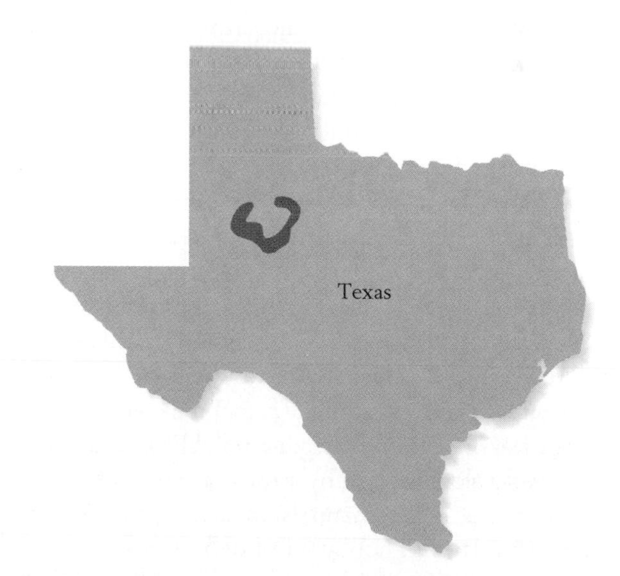

Figure 5-29 A Late Paleozoic horseshoe-shaped atoll that lies almost a kilometer (0.6 mile) below the surface of the land in Texas. The atoll was discovered when rocks of the region were drilled for petroleum. The reef appears to have faced prevailing winds from the south. *(After P. T. Stafford, U.S. Geol. Surv. Prof. Paper No. 315-A, 1959.)*

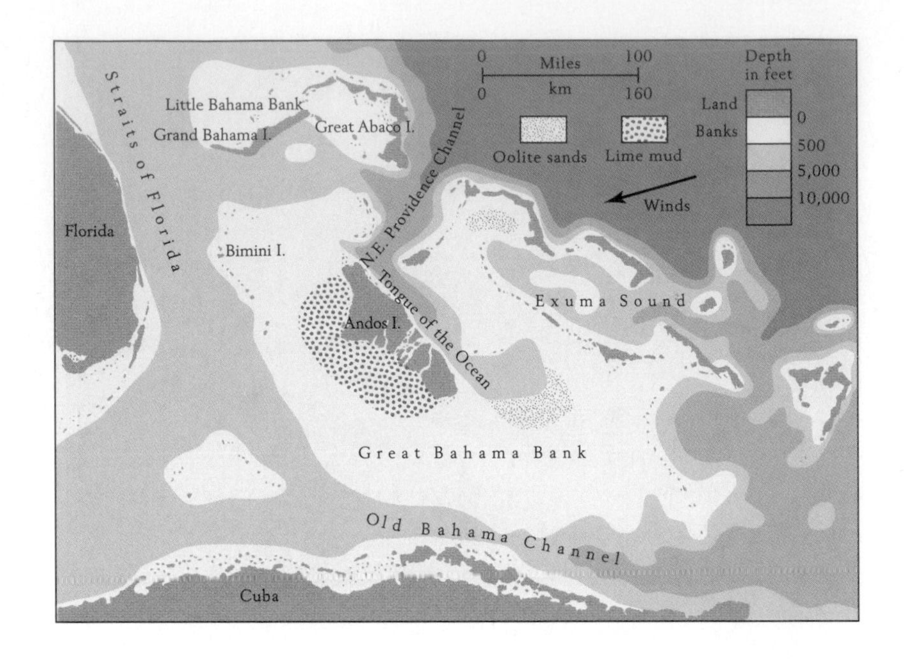

Figure 5-30 **The Bahama banks, which are now separated from Florida by the Florida Straits.** *(After N. D. Newell and J. K. Rigby, Soc. Econ. Paleont. and Mineral. Spec. Paper No. 5, 1957.)*

ancient carbonate platforms. Here, as on carbonate platforms generally, sediments accumulate rapidly. Since mid-Jurassic time, about 170 million years ago, some 10 kilometers (6 miles) of shallow-water carbonates have accumulated both on the Bahama banks and in southern Florida, which was part of the same carbonate platform during Cretaceous time and earlier. The heavy buildup of carbonate sediments in this region has caused the oceanic crust to bend downward, and shallow-water Jurassic deposits now lie up to 10 kilometers below sea level.

Among the accumulating carbonate sediments are oolites, which are composed of spherical grains that consist of aragonite needles precipitated from seawater (see Figure 2-7*B*). Strong currents pile oolites into spectacular shoals in some shallow subtidal areas of the Bahama banks (p. 120). Oolites formed in this manner display conspicuous cross-bedding. This feature and the need for individual ooliths to roll around in order to form (p. 50) makes cross-bedded oolites in the rock record reliable indicators of shallow-sea floors swept by strong currents.

Bordering tidal channels in some areas of the Bahama banks are knobby intertidal structures known as **stromatolites,** which are produced by threadlike organisms. As Figure 5-31 indicates, these organisms form sticky mats by trapping carbonate mud. They then grow through the mat thus formed to produce another mat. Repetition of this process on an irregular surface forms a cluster of stromatolites. Each is internally layered: organic-rich layers alternate with organic-poor layers. The threadlike organisms that form stromatolites are **cyanobacteria** (formerly known as blue-green algae), which are among the most primitive groups of organisms in the world (p. 70). The fossil record of stromatolites is unusually ancient, extending back through more than 3 billion years of geologic time.

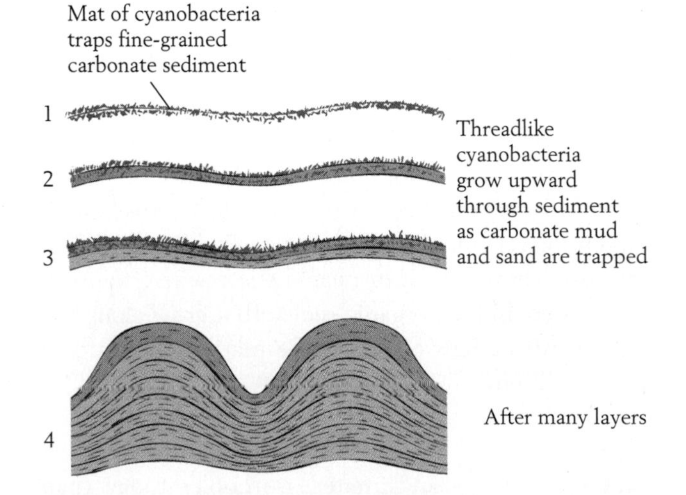

Figure 5-31 **The growth of stromatolites.** A mat of cyanobacteria traps mud, and the cyanobacteria grow through it to form another layer. The accumulation of several layers leads to the formation of a stromatolite.

There is a simple reason that stromatolites are found almost exclusively in supratidal and high intertidal settings. Because these environments are exposed above sea level much of the time, they become hot and dry, and relatively few marine animals can survive in them. Thus there is little to interfere with the tendency of the mats created by cyanobacteria to form layered structures. Such mats grow underwater here and there, but they are eaten by the many grazing marine animals and damaged by burrowers, so they seldom accumulate to form stromatolites or well-layered limestones. Exceptions are large column-shaped stromatolites that grow in subtidal channels in the Bahamas where tidal currents are very strong. Since few animals can survive in these current-swept areas, stromatolites flourish there. Stromatolites also flourish in Shark Bay, Western Australia, where the waters are hypersaline and animals are very rare (Figure 5-32). As we will see in later chapters, many very ancient stromatolites may have formed beneath the sea before the origin of marine animals that feed on cyanobacteria.

Ridges of sediment that build up along tidal channels in the Bahama banks dry out after occasional flooding by storm-driven seas. As a result, the surface of the sediment here is broken by mudcracks resembling those that often form when a mud puddle on the land dries up. Mudcracks associated with ancient marine deposits usually represent intertidal or supratidal environments that were alternately wetted by the sea and dried by the sun.

Figure 5-32 **Living stromatolites.** These forms, exposed at low tide, occupy the margin of Shark Bay, a lagoon in Western Australia where hypersaline waters allow few animals to survive.

Deep-Sea Sediments

To reconstruct the distribution of continents and ocean basins for any time interval in the past, geologists must identify not only nearshore deposits of the various kinds just described, but also deposits of deep-water environments beyond continental margins. Coarse clastic deposits derived from continental shelves accumulate near continental margins, but in the middle of the deep-sea floor, far from sources of clastic sediments, fine-grained sediments predominate, and they accumulate very slowly.

Submarine Slopes and Turbidites

One of the greatest advances in the study of sedimentary rocks took place in the middle of the twentieth century with the recognition that certain sedimentary rocks have been produced by turbidity currents. A **turbidity current** is a flow of dense, sediment-charged water moving down a slope under the influence of gravity.

Turbidity currents were first noticed in clear lakes, where flows were observed to form from muddy river water that hugged the lake floor. These currents flowed for a considerable distance toward the center of the lake, slowing down and dropping their sediment only when they reached gentler slopes and spread out. In the 1930s the Dutch geologist Philip Kuenen demonstrated in the laboratory that turbidity currents can attain great speed, especially when they are heavily laden with sediment and are moving down steep slopes. Sediment suspended in a turbidity current behaves as part of the moving fluid, and its presence increases the density of this fluid by as much as a factor of 2.

When the slope beneath a turbidity current begins to flatten out, the current slows and spreads out, dropping its sediment in the general sequence that we have seen again and again: first the coarse sediment falls from suspension, and then, much later, the fine material follows. The result is a graded bed of sediments, with poorly sorted sand and granules at its base and mud at the top. Such a graded bed is known as a **turbidite** (Figure 5-33*A*).

Large turbidity currents flow down continental slopes and deposit turbidites along continental rises and on the abyssal plain. Turbidity currents that originate near the edge of the continental shelf not only

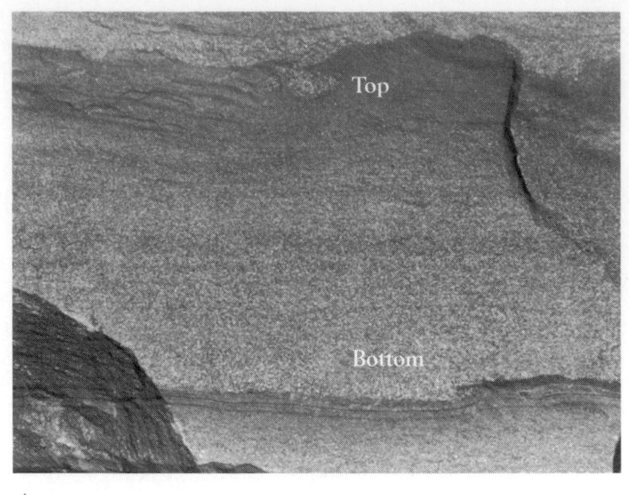

A

B

Figure 5-33 Ordovician turbidite beds in New York State. *A.* A turbidite bed that grades from coarse at the bottom to fine at the top; the upper surface is irregular because it was disturbed by the succeeding turbid flow and distorted by subsequent compaction. *B.* The bottom surface of a turbidite bed, showing "sole marks" produced when the sediment forming the bed filled depressions that the turbid flow depositing it scoured in the preexisting sediment surface; the current flowed toward the upper left.

carry sediment to the deep sea but also erode both the continental slope and part of the continental rise. Such currents are also largely responsible for carving the great submarine canyons that cut through many parts of the slope. The turbidity currents slowed down in front of these canyons, dropping part of their sedimentary load to form deep-sea fans that superficially resemble alluvial fans (Figure 5-34). In fact, much of the continental rise actually consists of coalescing submarine fans.

Ancient turbidites normally are stacked one on top of another in groups. The result is that the deposits they form are cyclical (see Figure 1-23). The bottom of each cycle consists of coarse material at the base of the turbidite, and the top consists of mud that accumulated in the quiet depths before the next turbidite formed.

The sandy portions of turbidites are typically graywackes (p. 45) that are quite unlike the clean sands of meandering-river deposits. This is only one of several ways in which a cyclical sequence of turbidites differs from the meandering-river cycle, although both show a grading of sediment from coarse to fine within each complete cycle. Another difference is that a single turbidite is usually only a few centimeters and seldom as much as a meter thick (see Figure 1-23), whereas most complete meandering-stream cycles measure several meters from bottom to top. In addition, turbidites lack the large-scale cross-bedding that characterizes the channels of a meandering river. Furthermore, the base of a turbidite is often irregular because the earliest, most rapidly moving waters of the turbid flow have scoured depressions in the sedimentary surface laid down earlier. These scours are then filled in by the first sediments

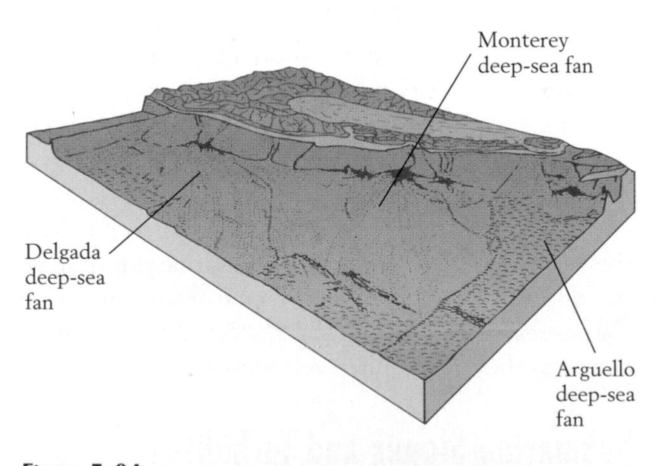

Figure 5-34 Deep-sea fans spreading from submarine canyons along the coast of California. The fans consist primarily of turbidite deposits. *(After H. W. Menard, Geol. Soc. Amer. Bull. 71:1271–1278, 1960.)*

that settle as the current slows. When the base of a lithified turbidite is turned over for inspection, its irregularities, or "sole marks," can reveal the direction of water flow (see Figure 5-33*B*).

Pelagic Sediments

Turbidity currents and other bottom flows carry mud to the abyssal plain of the ocean well beyond the continental rise. None of these flows, however, contributes sediment to the deep sea at a high rate; indeed, sediment in most areas of the abyssal plain accumulates at a rate of about 1 millimeter per 1000 years! In the deep sea, sparse sediments from turbidity currents are joined by clays from two other sources. One source is the weathering of rocks that have been produced by oceanic volcanoes. Such clays are less abundant in the Atlantic than in the central Pacific, where volcanoes are common. Clays also reach the deep sea by settling from the water above, which they reach after traveling through the air as wind-blown dust or moving seaward from the land at very low concentrations in surface waters of the ocean. Just as organisms that occupy the water above the deep-sea floor are termed pelagic forms of life (p. 114), fine-grained sediment that settles from these waters to the deep-sea floor is called **pelagic sediment.**

Clays that settle from the upper part of the water column are one type of pelagic sediment. Other types are contributed by small pelagic organisms whose skeletons become sedimentary particles when the organisms die. Whether clays predominate in any area of the deep sea depends on the extent to which they are diluted by the more rapid accumulation of biologically produced sediments. Some of the latter consist of calcium carbonate, while others consist of silica.

Where deposition of calcium carbonate predominates in the deep sea, its fine grain size has led oceanographers to refer to it as **ooze** (Figure 5-35). This calcareous ooze consists of skeletons of single-celled planktonic organisms. Among them are the globular skeletons of planktonic foraminifera, which are amoeba- like creatures. Other important constituents are the armorlike plates that surround calcareous nannoplankton, the single-celled floating algae that are major components of tropical phytoplankton (see Figure 3-16C; see also Figure 4-26).

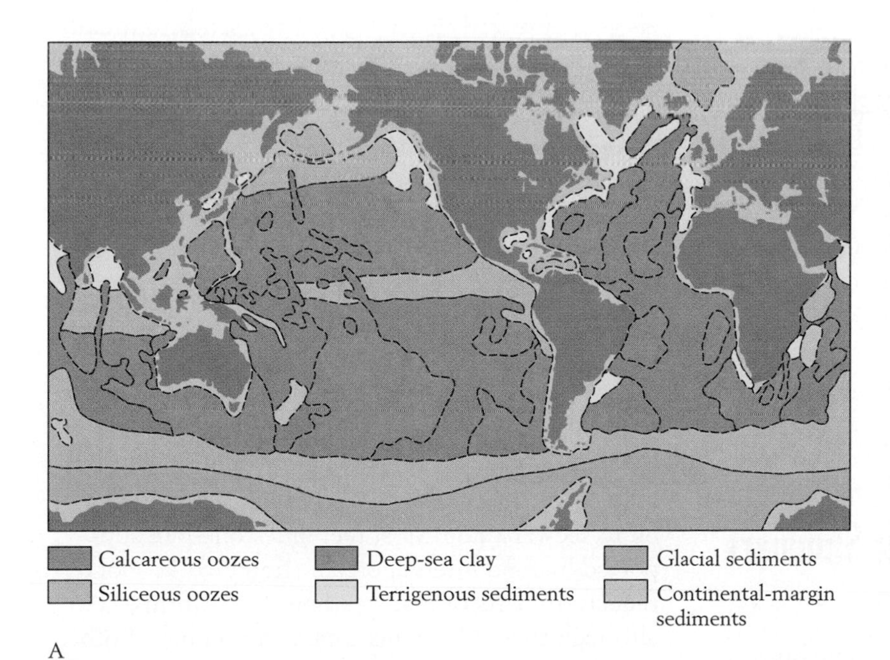

Figure 5-35 The global pattern of deep-sea sediments. *A.* Calcareous oozes are restricted to low latitudes. Most areas of siliceous ooze lie closer to the poles, although some occur close to the equator in areas of the Pacific and Indian oceans. *B.* Globigerina ooze. *(A. After T. A. Davies and D. S. Gorsline, in J. P. Riley and R. Chester [eds.], Chemical Oceanography, Academic Press, Inc., London, 1976.)*

Calcareous ooze is abundant only at depths of less than about 4000 meters (13,000 feet). The reason for this limitation is that calcium carbonate tends to dissolve as it settles through the water column. As pressure increases and temperature declines with depth, the concentration of carbon dioxide increases; as a result, the water becomes undersaturated with respect to calcium carbonate and thus tends to dissolve it.

In many regions at high latitudes as well as in tropical Pacific regions characterized by strong upwelling, biologically produced siliceous ooze carpets the deep-sea floor. This sediment consists of the skeletons of two groups of organisms that thrive where upwelling supplies nutrients in abundance: the diatoms, a highly productive phytoplankton group found in nontropical waters (see Figure 3-16*B*), and radiolarians, which are single-celled planktonic protozoans related to foraminifera (see Figure 3-18). Recall that the skeletons of both diatoms and radiolarians consist of a soft form of silica called opal, which tends to recrystallize so that they lose their identity (see Figure 2-21). In the process, the rock that they form becomes a dense, hard chert (see Figure 2-20). Before recrystallization, the soft sediment, consisting largely of diatoms, is known as diatomaceous earth. This is the abrasive component of many scouring powders used in kitchens.

Thick diatomaceous bodies of sediment have formed in marine areas of strong upwelling, where diatoms have been unusually abundant. Diatoms did not exist until late in Mesozoic time, however, and here we come to an important point: the composition of pelagic sediments has changed markedly in the course of geologic time as groups of sediment-contributing organisms have waxed and waned within the pelagic realm.

Chapter Summary

1 Modern environments where sediment is deposited offer valuable means to determine how ancient sedimentary rocks formed. Sometimes one kind of rock alone serves to identify an ancient environment. Usually, however, suites of closely associated rock types are required for this purpose, and these are commonly organized into cycles in which one kind of sediment tends to lie above another.

2 Ancient soils are sometimes found beneath unconformities, although they may be hard to identify because of chemical alteration.

3 Lake deposits, which are much less common than marine deposits, are characterized by thin horizontal layers, by few burrows, and by an absence of marine fossils.

4 Glaciers, which plow over the surface of the land, often leave a diagnostic suite of features, including scoured and scratched rock surfaces, poorly sorted gravelly sediment, and associated lake deposits that exhibit annual layers.

5 In hot, arid basins on the land, erosion of the surrounding highlands creates gravelly alluvial fans. Braided streams flowing from the fans toward the basin center deposit cross-bedded gravels and sands. Beyond these deposits there may be shallow lakes and salt flats where evaporites accumulate. Some arid basins also contain dunes of clean, cross-bedded, wind-blown sand.

6 In moist climates, meandering rivers leave characteristic deposits in which channel sands and gravels grade upward through point-bar sands to muddy backswamp sediments.

7 Where a river meets a lake or an ocean, it drops its sedimentary load to form a delta; deltaic deposition typically produces an upward-coarsening sequence as shallow-water sands build out over deeper-water muds.

8 More widespread than deltas along the margin of the ocean are muddy lagoons bounded by barrier islands formed of clean sand.

9 Coral reefs border many tropical shorelines. A typical reef stands above the surrounding seafloor, growing close to sea level and leaving a quiet lagoon on its leeward side. Most reef limestones are supported by rigid internal organic frameworks. Coral reefs form parts of many carbonate platforms, although these platforms contain a number of other deposits as well.

10 Beyond the edge of the continental shelf, turbidity currents intermittently sweep down continental slopes to the continental rise and deep-sea floor,

where they spread out, slow down, and deposit graded beds of sediment.

11 Still farther from continental shelves, only fine-grained sediments accumulate. Clay reaches these deep-sea areas very slowly. In some areas the deposition of clay is far surpassed by the accumulation of minute skeletons of planktonic marine life that settle to the seafloor to form calcareous or siliceous oozes.

Review Questions

1 What kinds of nonmarine sedimentary deposits reflect arid environmental conditions?

2 What kinds of nonmarine sedimentary deposits reflect cold environmental conditions?

3 What kinds of deposits indicate the presence of rugged terrain in the vicinity of a nonmarine depositional basin?

4 In what nonmarine settings do gravelly sediments often accumulate?

5 Contrast the patterns of occurrence of sediments and sedimentary structures in the following three kinds of depositional cycles: the kind produced by meandering rivers, the kind produced by deltas, and the kind produced by turbidity currents.

6 Draw a profile of a barrier island–lagoon complex, and label the various depositional environments.

7 What features typify sediments that accumulate in the centers of lakes?

8 How do stromatolites form?

9 Describe the kind of rock found in a typical organic reef.

10 Where is a lagoon in relation to a barrier reef? Where is it in relation to an atoll?

11 Which features of carbonate rocks suggest intertidal or supratidal deposition? Which features suggest subtidal deposition?

12 What types of sediments and sedimentary structures usually reflect deposition in a deep-sea setting?

13 What are the important depositional environments of the basic kinds of sediment (such as mud, well-sorted sand, gravel, evaporites, and various kinds of limestone) and of particular sedimentary structures (such as cross-bedding, graded beds, mudcracks, and ripples)? How are the kinds of sediment and sedimentary structures found within each environment related to processes operating within it? Use the Visual Overview on page 122 for reference.

Additional Reading

Blatt, H., and R. Tracy, *Petrology: Igneous, Sedimentary, and Metamorphic*, 2nd ed., W. H. Freeman and Company, New York, 1996.

Miall, A., *Principles of Sedimentary Basin Analysis*, Springer-Verlag, New York, 1990.

Prothero, D., *Interpreting the Stratigraphic Record*, W. H. Freeman and Company, New York, 1990.

Reading, H. G., *Sedimentary Environments: Processes, Facies, and Stratigraphy*, Blackwell Publications, Palo Alto, Calif., 1996.

Selley, R. C., *Ancient Sedimentary Environments and Their Subsurface Diagnosis*, Chapman and Hall, London, 1996.

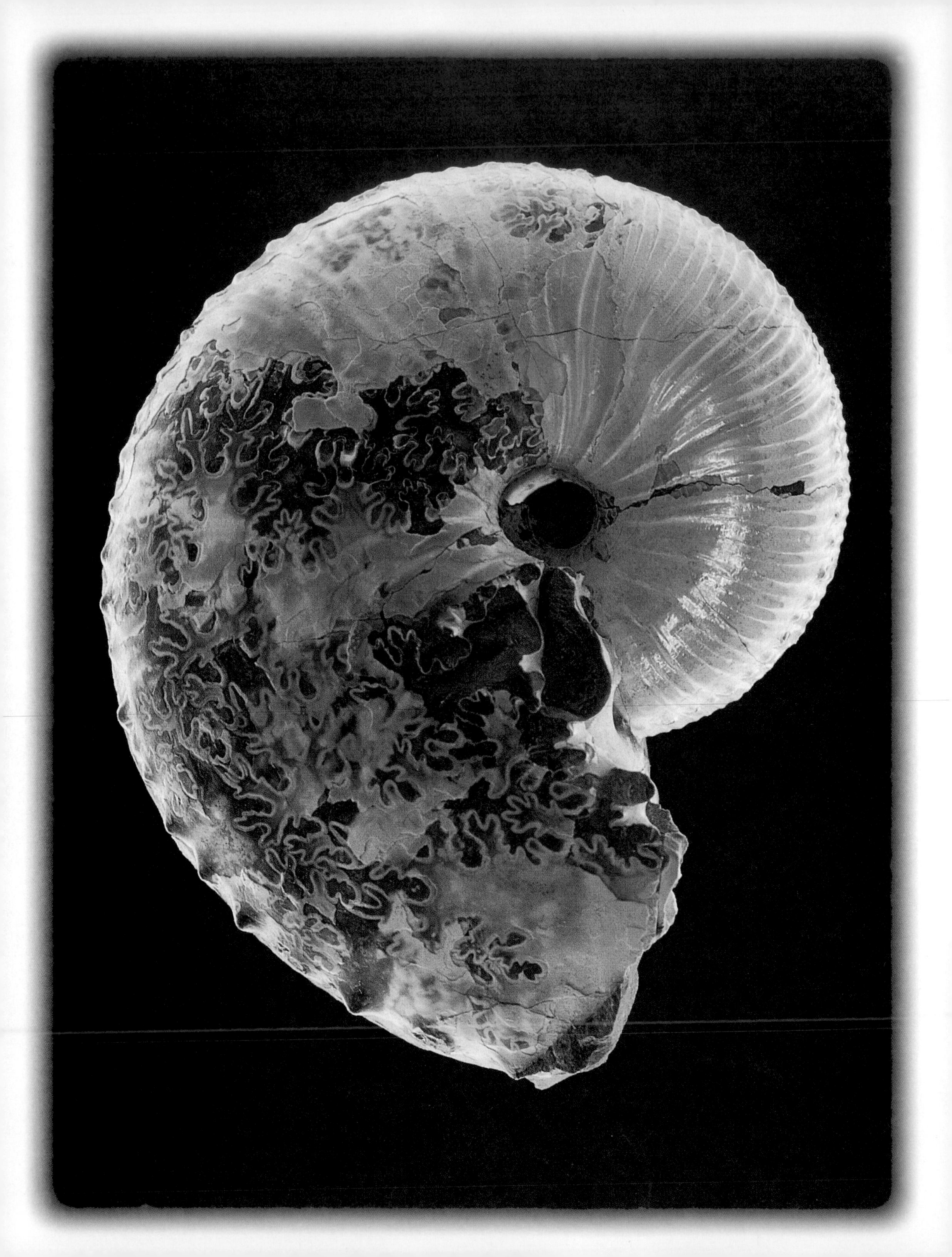

CHAPTER 6

Correlation and Dating of the Rock Record

During the 1830s the arguments of James Hutton and Charles Lyell convinced most scientists of the immensity of geologic time (p. 5). The challenge of piecing together the chronological details of Earth's history, however, remained. First, the *relative* ages of bodies of rock needed to be established: was a body of rock younger, older, or the same age as another? Second, the *absolute* ages of bodies of rock needed to be established: how long ago, in years, or thousands or millions of years, did they form?

The Geologic Time Scale

Between the early eighteenth century and the present, geologists have made great progress in determining relative and absolute ages of rocks throughout the world. This progress has been crucial to all endeavors to unravel the history of Earth and its life. Only after placing rocks and fossils in their relative positions in geologic time can scientists trace out the history of environmental change and reconstruct the comings and goings of species within this environmental context. Absolute dating provides a true time scale for the chronological framework, revealing just how rapidly ice ages have begun and ended, how suddenly mass extinctions have occurred, and how quickly mountains have risen and eroded away.

A distinctive Upper Cretaceous ammonite, displaying some original brownish shell material on the upper left, pearly inner shell material on the right. Dark sediment below fills chambers between convoluted interior partitions of the shell called septa.

Visual Overview

Methods of Stratigraphic Correlation

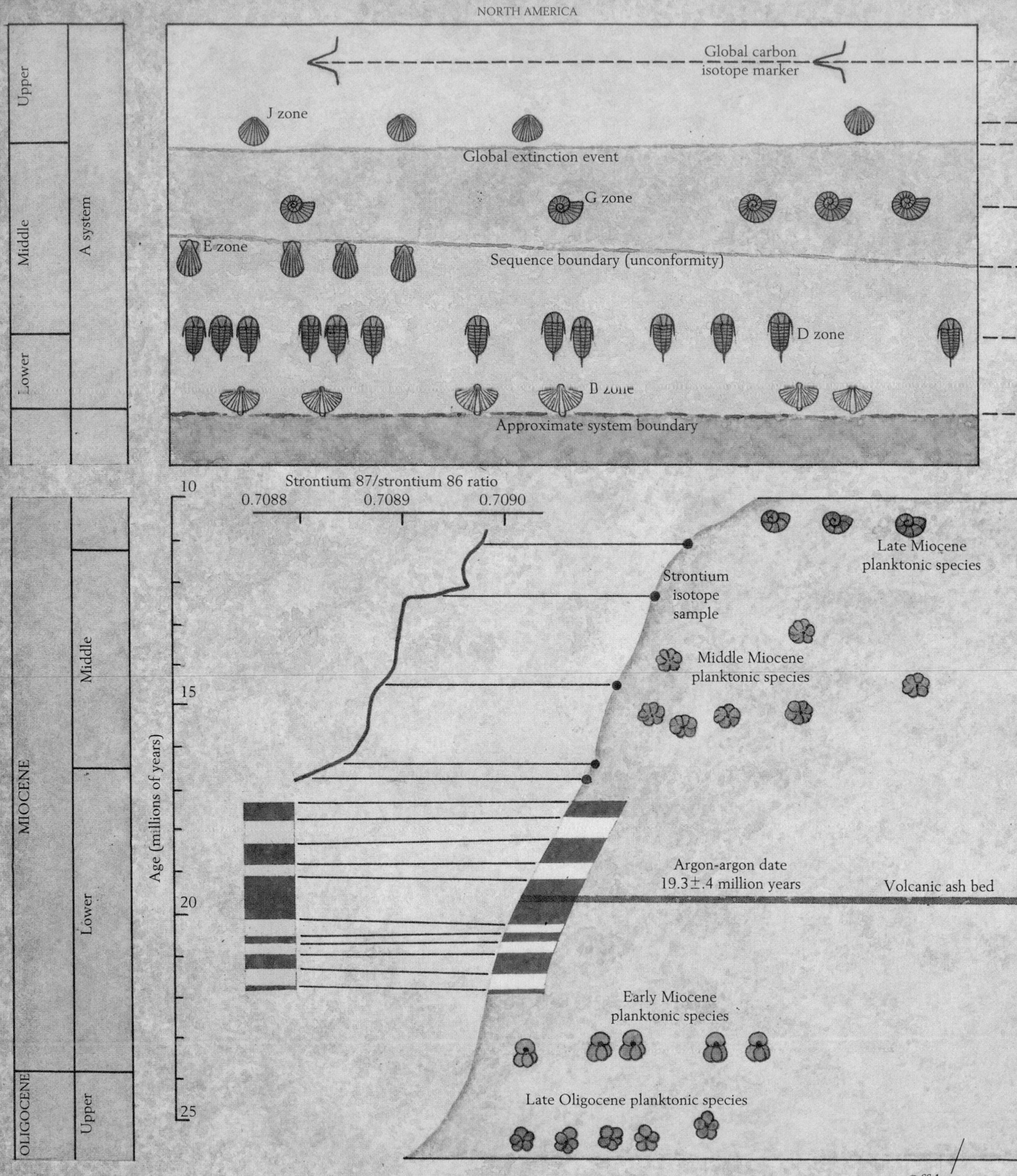

Scientists use fossils, radioactive dating, unconformities, isotope stratigraphy, and patterns of transgression and regression to correlate rocks with respect to time.

EUROPE

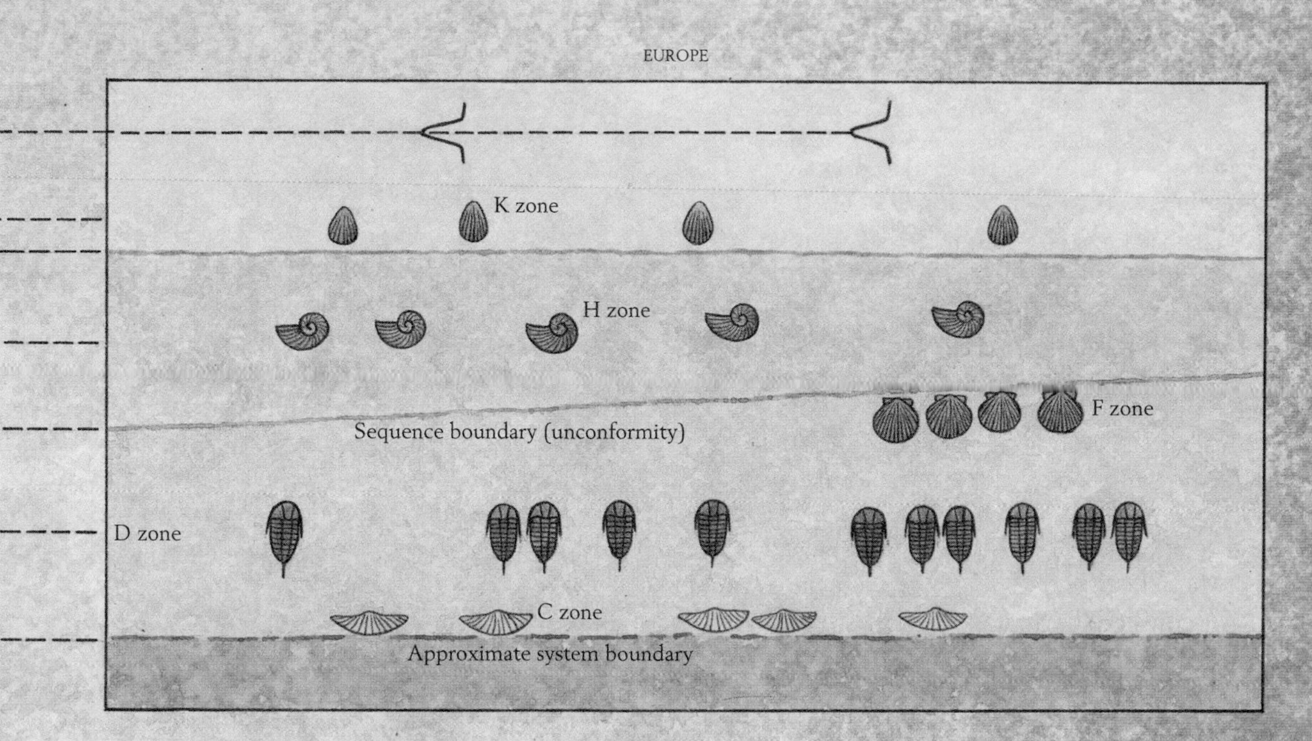

K zone
H zone
F zone
Sequence boundary (unconformity)
D zone
C zone
Approximate system boundary

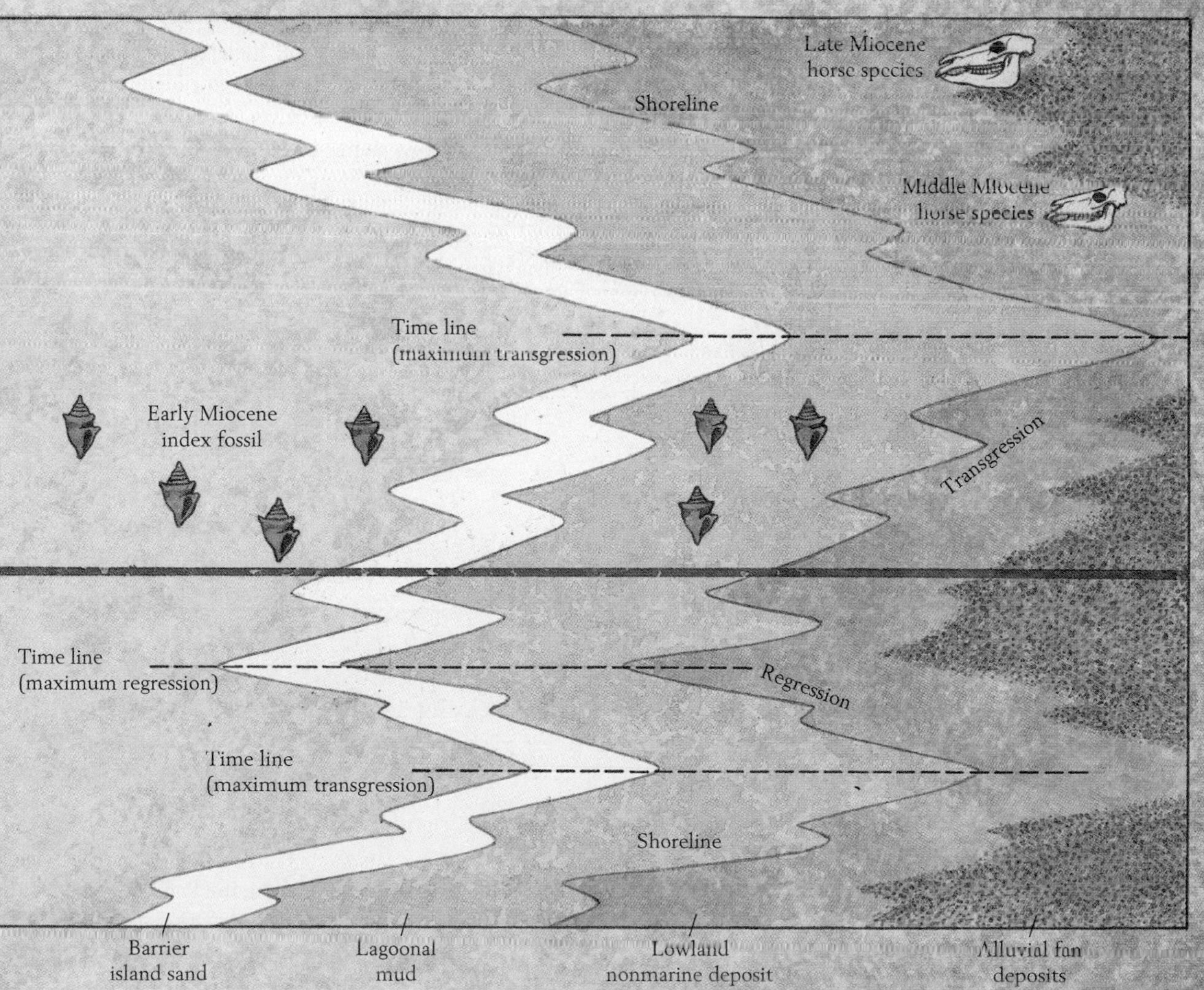

Late Miocene
horse species
Shoreline
Middle Miocene
horse species
Time line
(maximum transgression)
Early Miocene
index fossil
Transgression
Time line
(maximum regression)
Regression
Time line
(maximum transgression)
Shoreline
Barrier
island sand
Lagoonal
mud
Lowland
nonmarine deposit
Alluvial fan
deposits

Fossil Succession

In the seventeenth century Nicolaus Steno established ways of showing that some rocks were older than others, but these laws applied only to bodies of rock that were in contact with one another (p. 9). Early in the nineteenth century scientists discovered that fossils could often be used to establish the relative ages of strata that lay far apart. The pioneer here was William Smith, who introduced the concept of fossil succession (p. 12). Smith showed that fossil mollusks could be used to determine the relative ages of certain bodies of rock in England. Having established the sequence in which various kinds of fossils occurred, from lowest to highest, he showed that he could place an isolated outcrop of rock in its proper position in this stratigraphic sequence. In addition, he concluded that the entire sequence represented a very long interval of time. William Smith was a man of modest background and limited education. His studies of fossil succession served a practical purpose. He was a surveyor and engineer who planned the siting of canals, so he had much to gain by learning how to recognize particular strata easily, on the basis of their fossils, and to predict what other strata would be found below them at localities he visited for the first time.

Baron Georges Cuvier, of the National Museum of Natural History in France, was a very different kind of scientist. He was a highly educated aristocrat dedicated to academic pursuits. Cuvier was the first scientist to conclude that species have become extinct in the course of Earth's history. From the Paris Basin, where sediments were being quarried extensively, Cuvier collected the fossils of many species of mammals unknown in the present world. We now know that these sediments were deposited early in the Cenozoic Era (Figure 6-1). Cuvier identified a succession of seven distinctive fossil faunas, each of which disappeared abruptly from the stratigraphic record, to be followed by another. In between any two successive fossil faunas, however, were strata that yielded marine mollusks. Cuvier concluded that cataclysmic environmental changes had punctuated the history of life in this region. The sea had occasionally swept inland from the northeast, wiping out terrestrial mammals and introducing marine life. When the waters receded again, a new terrestrial fauna appeared. Cuvier believed that species of each new mammalian fauna had been living elsewhere before the crisis and

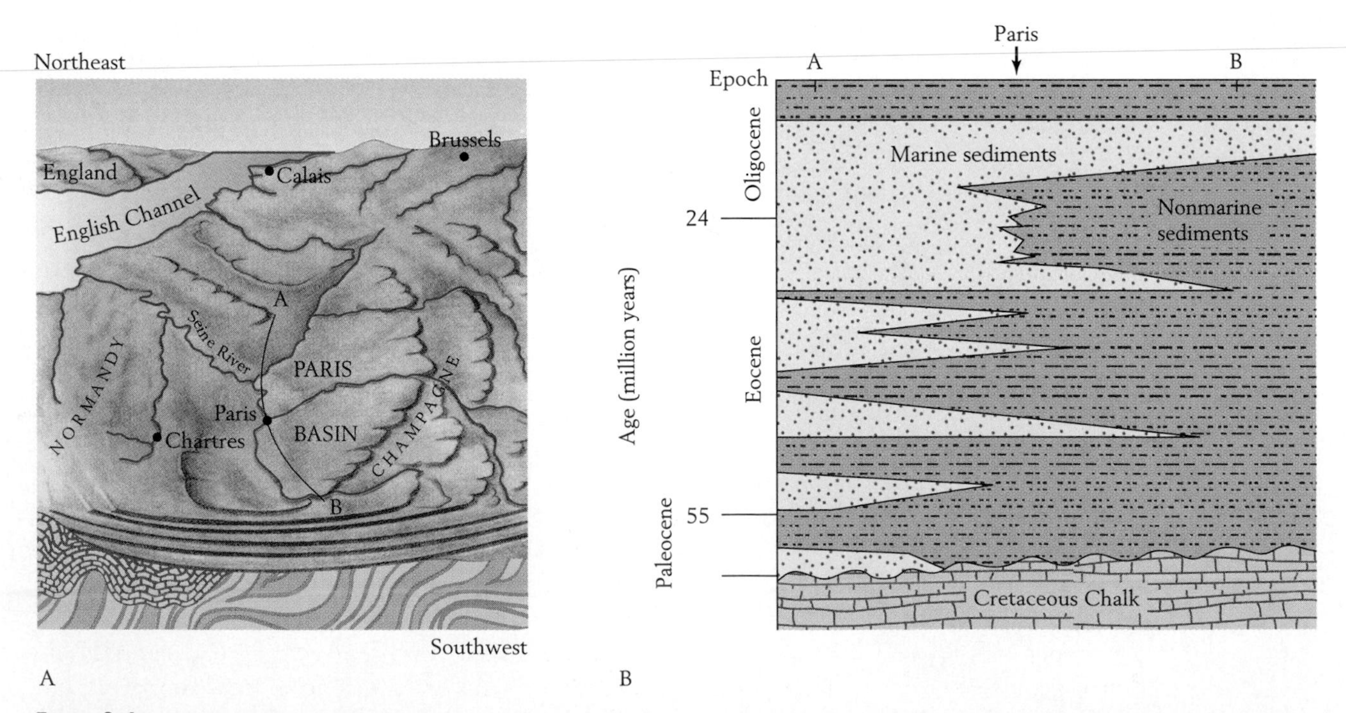

Figure 6-1 **Cenozoic strata of the Paris Basin, where Cuvier discovered successive faunas of mammals.** *A.* A schematic view of the basin structure; erosion of weak strata around the basin margins has left durable strata standing as ridges. *B.* Schematic cross section along the line *AB* shown in *A*. Marine strata interfinger with nonmarine strata, having been laid down during episodic incursions of the sea from the northeast; the nonmarine strata yield mammalian fossils.

had survived because their habitat underwent no major change. They had then migrated to the Paris Basin when conditions became favorable there.

Certainly, the first members of each new fauna to occupy the Paris Basin immigrated to this region from other regions, but we now know that many other species arose in this region or elsewhere by evolving from other species. Cuvier never accepted the idea that new species have arisen on Earth by evolving from others. He died in 1832, long before Charles Darwin made his great contribution. What Cuvier established, however, was that entire biotas had become extinct because of environmental change and had then been replaced by other biotas. Thus Cuvier added new understanding to William Smith's discovery of fossil succession within strata.

Since Cuvier's time, scientists have established the relative positions of many kinds of fossils in strata, and these fossil occurrences have provided the most important evidence of the relative ages of strata around the world. Before examining the ways in which geologists establish the relative and absolute ages of rocks today, let us review how early geologists came to divide the rock record into major intervals.

Geologic Systems

The modern science of geology was born in Great Britain, and it was here and in nearby areas of Europe that early geologists divided the bodies of rock that contain fossils of diverse animal life into what we now call **geologic systems** (see Figure 1-13). The oldest of these systems is the Cambrian, and thus the label Precambrian was applied to all older rocks, which we now know represent the first 4 billion years or so of Earth's history. Because early scientists had no way of knowing the relative ages of rocks separated by great distances, they founded the various systems haphazardly. They could establish the relative ages of systems with certainty only by finding two of them in the same area and observing that one was positioned above the other.

The British geologists Adam Sedgwick and Roderick Murchison named the Cambrian and Silurian systems in a joint paper published in 1835, principally on the basis of geologic studies in Wales. (The distribution of Silurian rocks in Wales is shown in Figure 5-23.) Sedgwick, a professor at Cambridge University, derived the name Cambrian from Cambria, the Roman name for Wales, and Murchison, a wealthy landowner who made geology his full-time hobby, named the Silurian for the Silures, an ancient Welsh tribe. Murchison defined the Silurian primarily on the basis of its fossils—including taxa of trilobites, brachiopods, and crinoids—and also corals that formed small reefs. Sedgwick, finding no diagnostic fossils in his Cambrian System, simply defined the Cambrian as a body of sedimentary rocks that rested directly on ancient crystalline rocks.

By 1839, Silurian fossils had been found throughout Europe and even in North and South America and South Africa. As a result, Murchison proclaimed the Silurian System to be a global entity. In time, he went too far, claiming that the Silurian included the earliest fossil-bearing rocks on Earth. He argued that the Welsh rocks on which Sedgwick had founded the Cambrian System actually belonged to the lower portion of the Silurian and simply happened to contain few fossils. Murchison's expanded version of the Silurian System incurred Sedgwick's bitter enmity. Eventually Sedgwick was vindicated: distinctive Cambrian fossils came to light not only in Wales but throughout the world. The Cambrian is now universally recognized as the oldest geologic system to contain a great variety of fossil shells and other skeletal remains of invertebrate animals.

Even after the conflict between Murchison and Sedgwick was resolved, their Cambrian and Silurian systems failed to remain intact. In 1879 Charles Lapworth, a British schoolmaster, showed that in many parts of the world the rocks that had been assigned to these systems actually displayed a succession of three distinctive groups of fossils. He proposed that the Cambrian be retained as the system harboring the oldest group and the Silurian be retained as the system harboring the youngest group. For the intervening rocks, with their own distinctive fossils, he erected the Ordovician System.

In the twentieth century, geologists came to recognize that many taxa fossilized in uppermost Cambrian rocks became extinct abruptly, before the earliest Ordovician strata were laid down. Thus the Ordovician System records the evolution of many new forms of life. Uppermost Ordovician rocks document another episode of widespread extinction—and also an expansion of glaciers near the south pole that we now know was related to that extinction. Silurian rocks, in turn, contain many new kinds of fossils that represent another evolutionary expansion of life.

Unlike the Cambrian, Ordovician, and Silurian, some other systems were founded to embody distinctive strata. The Cretaceous System, for example, was erected for a body of rocks whose upper portions include great thicknesses of chalk in Britain, France, Texas, Kansas, and many other areas. The Belgian Omalius d'Holloy established the Cretaceous System in France in 1822, aptly deriving its name from *creta*, the Latin word for chalk. Not until later in the century did geologists learn that Cretaceous chalk is formed of very small plates that encrusted calcareous nannoplankton and sank to the seafloor after the death of these spherical, single-celled floating algae (p. 72). These algae were more productive when they produced Cretaceous chalk than at any other time in their history.

Stratigraphic Units

Broadly defined, **stratigraphy** is the study of stratified rocks, especially their geometric relations, compositions, origins, and age relations. **Stratigraphic units** include strata, or groups of adjacent strata, that are distinguished by some physical, chemical, or paleontological property; they also include units of time that are based on the ages of such strata. Geologic systems, such as the Cambrian, Silurian, and Cretaceous, represent one kind of stratigraphic unit.

Correlation is the procedure of demonstrating correspondence between geographically separated parts of a stratigraphic unit. Recall that a distinctive body of rock may be formally recognized as a *formation* on the basis of its composition and appearance (p. 8). After a formation is established in one region, outcrops at other locations may be correlated with it and assigned to it. This or any other correlation based on rock type is termed **lithologic correlation.** In correlating separate portions of a formation in this way, geologists are implicitly following Steno's principle of original continuity (p. 9).

The term "correlation" is more commonly used to indicate that widely separated bodies of rock are the same *age*. In fact, much of this chapter is devoted to discussion of methods used to correlate strata with respect to time, a procedure known as **temporal correlation.**

Time-Rock Units and Time Units

A **time-rock unit,** formally termed a *chronostratigraphic unit*, includes all the strata in the world that were deposited during a particular interval of time. A system is a time-rock unit. A **time unit,** formally termed a *geochronologic unit*, is the interval during which a time-rock unit formed. Thus a time-rock unit, such as the Silurian System, is defined in the field, and we refer to the time interval that this system represents as the Silurian Period.

Systems are grouped into **erathems.** For the Phanerozoic Eon, these are the Paleozoic, Mesozoic, and Cenozoic erathems. Eras are the time units that correspond to erathems. Systems are also subdivided into **series,** which are divided into **stages; epochs** and

Table 6-1

Geologic Time Units and Time-Rock Units

Time unit	Example	Time-rock unit	Example
Era	Paleozoic	Erathem	Paleozoic
Period	Devonian	System	Devonian
Epoch	Late Devonian	Series	Upper Devonian
Age	Famennian	Stage	Famennian

Note: Time-rock units are bodies of rock that represent time units bearing the same formal name. When an epoch (the time unit) is designated by the term *Early, Middle,* or *Late,* the corresponding series (the time-rock unit) is identified by the adjective *Lower, Middle,* or *Upper.* For example, the Lower Devonian series of rocks represents Early Devonian time.

ages are the corresponding time units for these smaller time-rock units (Table 6-1).

A boundary between two systems, series, or stages is formally defined at a single locality, known as a **boundary stratotype.** Often the lower boundary of a time-rock unit is defined in one region and the upper boundary in another. The challenge is to extend each boundary throughout the world by means of temporal correlation. As we shall see shortly, this task can be undertaken by a variety of methods.

Biostratigraphic Units

Fossil occurrences are the most widely used means of extending boundaries of systems, series, and stages around the globe, The results are far from perfect, however, for reasons that will soon be evident.

Stratigraphic units of the rock record that are defined and characterized by their fossil content are termed **biostratigraphic units.** These units are based on the stratigraphic ranges of fossil taxa. The *stratigraphic range,* usually just called the range, of a species is the total vertical interval through which a taxon occurs in strata, from lowermost to uppermost occurrence.

Zones The most fundamental biostratigraphic unit is the zone, more formally termed a *biozone.* A **zone** is a body of rock whose lower and upper boundaries are based on the ranges of one or more taxa—usually species—in the stratigraphic record. A zone can be defined to coincide with the range of a single taxon. Most zones are more complex than this, however. Many have a lower boundary defined by the lowermost or uppermost occurrence of one taxon and an upper boundary defined by the lowermost or uppermost occurrence of another taxon. Other zones are defined to include the stratigraphic interval within which three or more taxa occur together. Every zone is named for a taxon that occurs within it. Figure 6-2 shows how the vertical ranges of a number of graptolite species in the Silurian System of Britain relate to two graptolite zones. One of these zones is defined by the range of a single species, *Monograptus convolutus,* and is named for this species. The other is named for *Monograptus sedgwickii,* a species named in honor of Adam Sedgwick, but this zone is defined by the co-occurrence of several graptolite species.

Unfortunately, no zone represents exactly the same time interval everywhere it occurs. For one

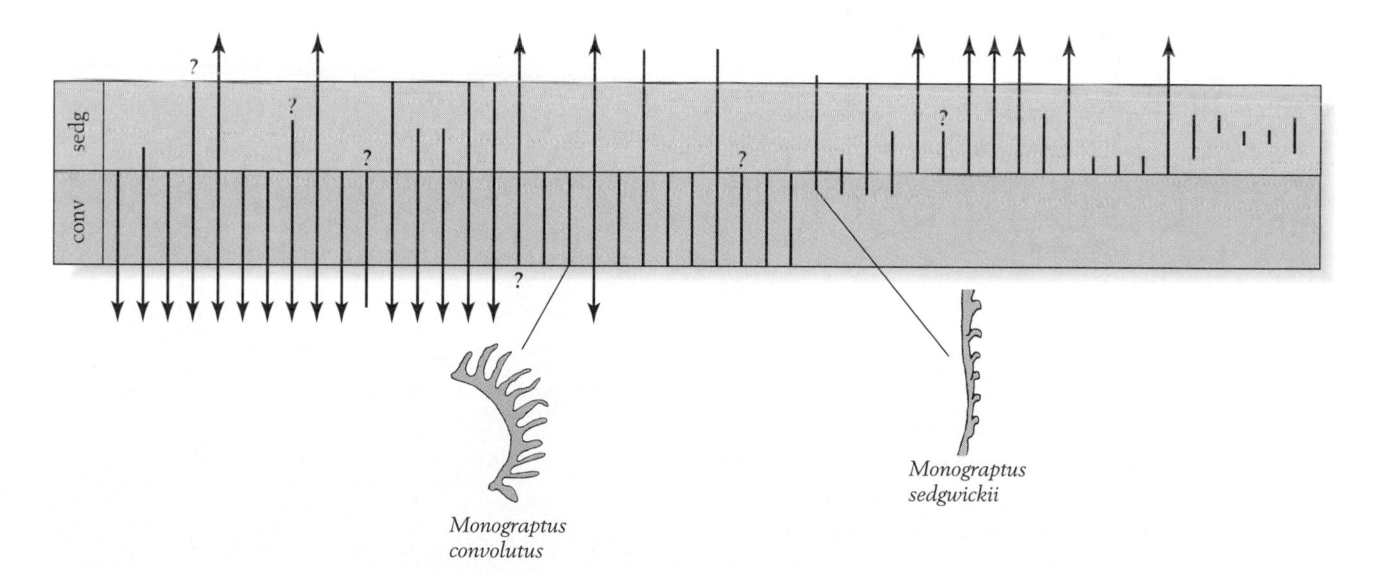

Figure 6-2 Zones based on the presence of fossils of planktonic graptolites, an extinct group of animals whose fossils are extremely useful for correlation. The two zones represented in this bar graph are situated in the lower part of the Silurian System in the British Isles. Vertical bars represent the known ranges of species; the species for which the two zones are named are illustrated. The stalks of these animals are about 2 or 3 millimeters ($\frac{1}{8}$ inch) wide. All ranges shown are for species that first appear in the *convolutus* or *sedgwickii* zone. An arrow indicates that a range continues beyond the figured interval; a question mark indicates that a range may be longer than that shown here. *(Adapted from R. B. Rickards, Geological Journal, 11:153–188, 1976.)*

thing, the members of an extinct taxon will not have appeared or disappeared simultaneously in all of the areas the taxon inhabited. A species will commonly have originated in a small area and later greatly expanded its geographic range. Furthermore, a species or genus—the mammoth, for example—often persisted in a restricted region after it died out in most of the areas it once inhabited (see Figure 4-5). In fact, many taxa have had complex histories of migration, largely as a result of changing environmental conditions. Another problem is that the fossil record is incomplete. A taxon may have existed at a given time and place without leaving a fossil record—or its fossils may remain undiscovered.

Although fossil species and genera do not provide for perfect correlation, some are reliable enough to be designated **index fossils,** or **guide fossils.** Such a taxon has some or all of the following desirable characteristics:

1 It is abundant enough in the stratigraphic record to be found easily.

2 It is easily distinguished from other taxa.

3 It is geographically widespread and thus can be used to correlate rocks over a large area.

4 It occurs in many kinds of sedimentary rocks and therefore can be found in many places.

5 It has a narrow stratigraphic range, which allows for precise correlation if its mere presence, rather than its lowermost or uppermost occurrence, is to be used to define a zone.

Unfortunately, few taxa exhibit all of these traits as strongly as we might wish. Consider the planktonic foraminiferal fossils found in Late Mesozoic and Cenozoic sediments (p. 73; see also Figure 5-35). They meet two criteria for index fossils: they are easily identified under the microscope, and they floated across large areas of the sea and settled in a wide variety of sedimentary environments. They are not ideal index fossils, however, because they lived in offshore areas and are therefore seldom found in sediments deposited near the shore. Moreover, some extinct species of planktonic foraminifera lived over much longer intervals than others. Those that survived for as long as 15 or 20 percent of the Cenozoic Era make poor index fossils; only the earliest and latest appearances of such long-lived species provide useful information for correlation. Other species, however, lived for such short intervals that their mere occurrence permits quite precise correlation.

Magnetic Stratigraphy and Polarity Time-Rock Units

Some rocks acquire magnetic properties as they form, and the pattern of magnetization changes abruptly throughout the world from time to time. Changes of this kind provide a means of temporal correlation that is usually much more accurate than the use of biostratigraphic zones. This use of rock magnetism constitutes **magnetic stratigraphy.**

As we have seen, Earth's core consists of dense material made up of iron and other heavy substances. In the outer part of the core, this material is in a liquid state, and its motion generates a magnetic field. As a result, the planet behaves like a giant bar magnet, with a north and south pole (Figure 6-3). Reversal in the polarity of Earth's **magnetic field** provides for accurate correlation throughout the world.

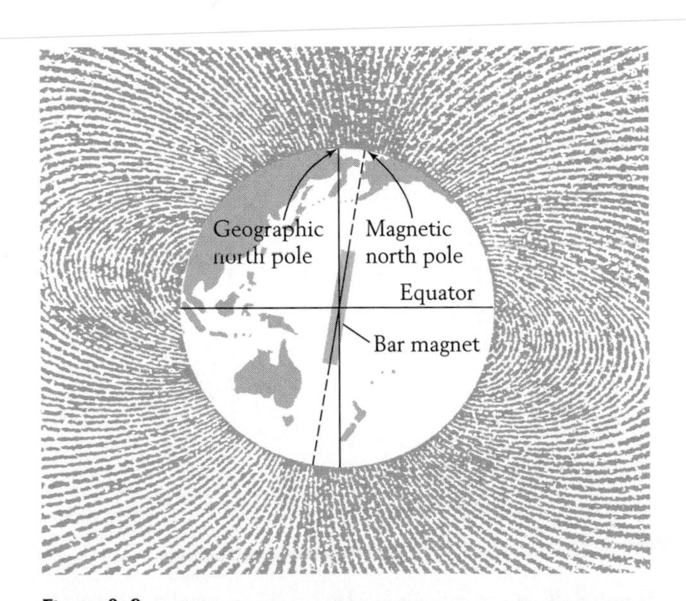

Figure 6-3 **Earth's magnetic field.** The magnetic field, represented by the lines of force surrounding the planet, resembles the one that would be produced by a bar magnet located within the planet with its long axis inclined slightly (11°) from Earth's axis of rotation. *(After F. Press and R. Siever, Earth, W. H. Freeman and Company, New York, 1986.)*

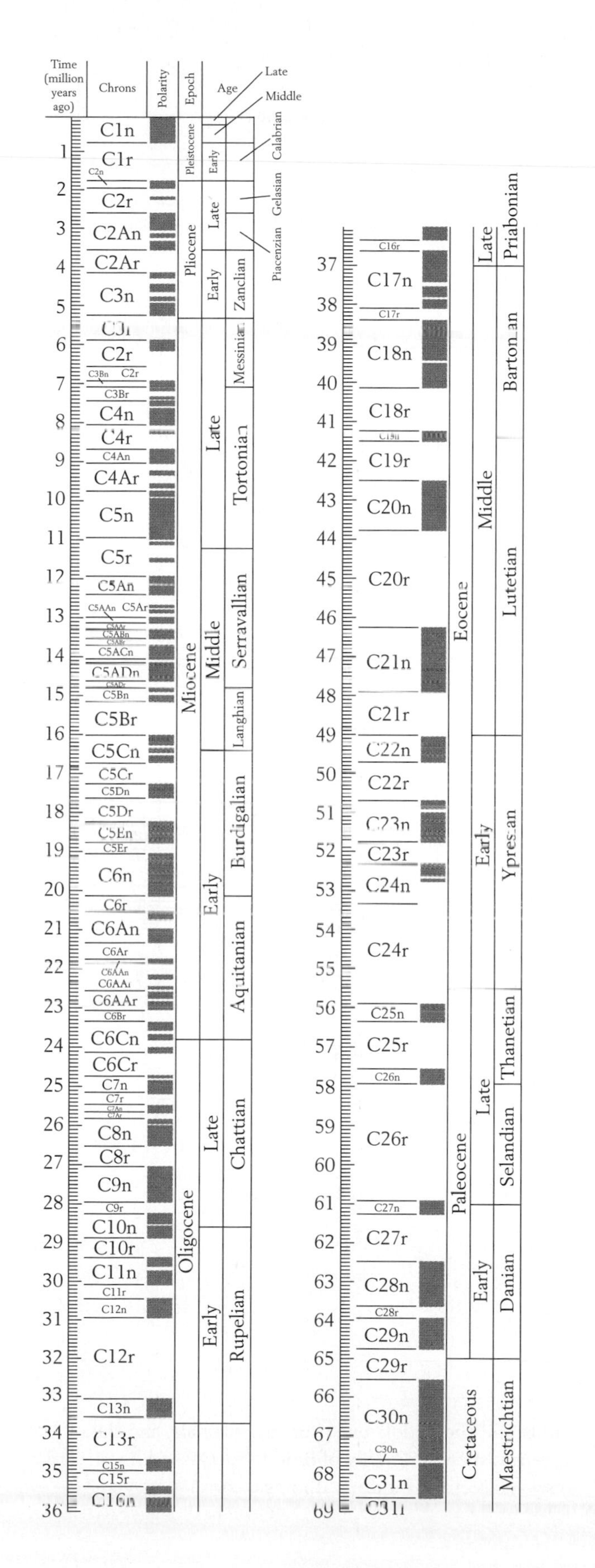

Figure 6-4 **Magnetic polarity scale for the Cenozoic Era.** Numbers represent millions of years. About half of the time the magnetic field has had normal polarity (polarity like that of the present), as indicated by the colored segments of the scale, and about half of the time the polarity has been reversed, as indicated by white space. *(After W. B. Berggren et al., Geol. Soc. Amer. Bull. 96:1407–1418, 1985.)*

When iron-containing minerals form sedimentary or igneous rocks at or near Earth's surface, they often become aligned with Earth's magnetic field just as a compass needle does when it is allowed to rotate freely. As small grains of iron minerals settle from water to become parts of sedimentary rocks, they often rotate so that their magnetism becomes aligned with that of the planet. Also, iron minerals that crystallize from lava or magma automatically become magnetized by Earth's magnetic field as they cool.

It is a startling fact that Earth's north and south magnetic poles occasionally switch positions. No one knows why. Intervals between such magnetic reversals vary considerably, but during the Cenozoic Era they have averaged about a half-million years. Sequences of magnetized rocks that can be dated radiometrically have revealed the history of magnetic reversals during the Cenozoic Era (Figure 6-4) and during much of the Mesozoic Era as well. Periods when the polarity was the same as it is today are known as *normal intervals*, and periods when the polarity was the opposite of what it is today are called *reversed intervals*. Each of these intervals can be formally recognized as a **polarity time-rock unit** and either assigned a number or formally named for a geographic locality where it is well represented by magnetized rocks.

The primary difficulty encountered in efforts to use the magnetic record for correlation lies in assigning strata known to have normal or reversed magnetism to a particular polarity time-rock unit. Especially helpful in this respect is a "signature;" that is, a distinctive sequence of reversals that, when plotted on a stratigraphic column, produces a pattern resembling the bar code for an item in a store. In rocks known to be of Eocene age, for example, a pattern that indicates a long normal interval flanked by two long reversed intervals can only be early middle Eocene in age; furthermore, the base of the first long reversal

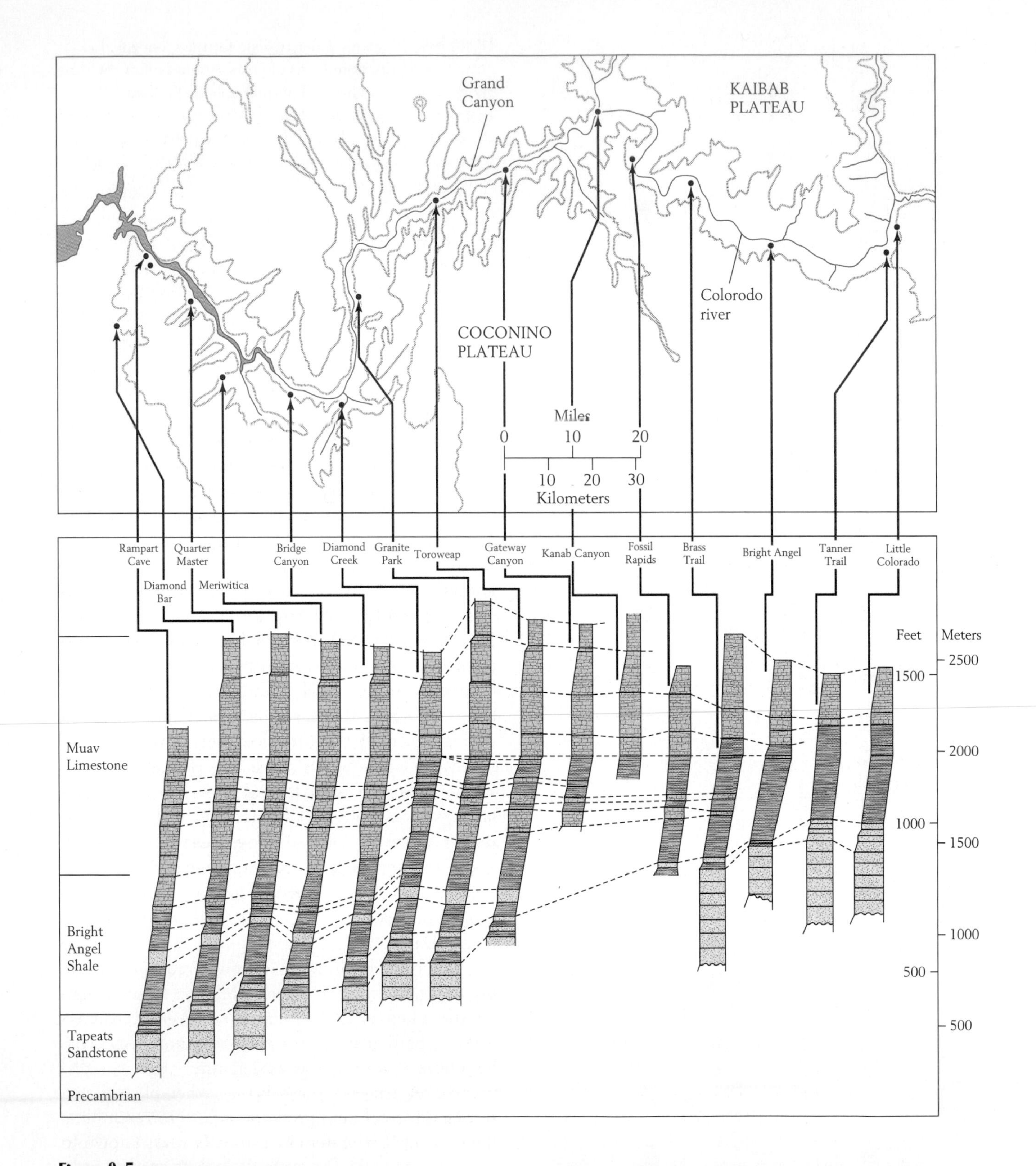

Figure 6-5 Stratigraphic sections measured along the walls of the Grand Canyon where Cambrian strata are exposed. The strata here are divided into three formations; the lowermost one consists primarily of sandstone, the middle one primarily of shale, and the uppermost one primarily of limestone. Dashed lines between sections indicate lithologic correlation.

interval coincides with the base of the middle Eocene (see Figure 6-4).

Unfortunately, magnetic stratigraphy has not yet been extensively employed for temporal correlation of Paleozoic and Mesozoic strata.

Rock Units, Stratigraphic Sections, and Facies

As we saw in Chapter 1, geologists divide the stratigraphic record into local three-dimensional bodies of rock known as *formations*. Recall that formations are sometimes united into larger divisions known as *groups* and sometimes include smaller units called *members*. Groups, in turn, may be united into *supergroups*. All these entities are known as **rock units,** or more formally, *lithostratigraphic units*. Formations are delineated on the basis of **lithology,** or the physical and chemical characteristics of rock. Many are relatively homogeneous bodies, consisting of a single rock type. Others, as we have seen, consist of two or more rock types in alternating layers. Formations of this sort include the sedimentary rock cycles produced by deltas (see Figure 5-20) and by meandering rivers (see Figure 5-16). Every rock unit is assigned a **type section** at a particular locality where it is well exposed. Rock units are named for local geographic features, such as rivers and towns. The name of the geographic feature is followed by the word "Formation," "Group," or "Member." or by the name of a specific rock type, such as Sandstone, Shale, or Limestone.

Geologists construct cross sections of strata to establish their geometric relationships and interpret their modes of origin Such cross sections are constructed from local stratigraphic sections (Figure 6-5). A **stratigraphic section** is a local outcrop that displays a vertical sequence of strata. Typically the section is measured from bottom to top, and the positions of various types of rock are recorded, as are the locations of all the kinds of fossils that can be collected and identified. The rocks of two or more sections in a given region can then be correlated, either as formations or as informal bodies of rock, as can fossil occurrences (or biostratigraphic zones) and polarity time-rock units. Such correlations produce a cross section of the regional stratigraphy (Figure 6-6).

A key aspect of rock units is that their boundaries are defined without reference to biostratigraphic units or time. In fact, the ages of the upper and lower boundaries of many rock units vary widely from place to place. This point is well illustrated by three Cambrian formations of marine origin exposed along the walls of the Grand Canyon (Figure 6-6*A*). To early geologists it appeared that the Bright Angel Shale rested on the Tapeats Sandstone, and so was the younger unit, and the Muav Limestone appeared to bear a similar relation to the Bright Angel Shale. In the 1940s, however, a geologist named Edwin McKee showed that this idea of a layer-cake pattern was in error. By carefully measuring the sections illustrated in Figure 6-5 and collecting fossils from them, McKee showed that a trilobite zone that belongs to the Early Cambrian Series passes through the Tapeats Sandstone and into the lower part of the Bright Angel Shale (Figure 6-6*B*). He further discovered that a second trilobite zone, belonging to the *Middle* Cambrian Series, passes through the *upper* part of the Bright Angel Shale and into the Muav Limestone. This pattern shows that the eastern portion of each formation is younger than the western portion.

The strata represented in Figures 6-5 and 6-6 accumulated along the western margin of North America, which in Cambrian time lay far to the east of its present location. The Tapeats Sandstone was deposited above an unconformity on crystalline rocks of Precambrian age. This means that it was deposited along the shoreline. The eastward migration of the base of the Tapeats Sandstone during Cambrian time indicates that the shoreline shifted in this direction. This shift occurred because of a global rise in sea level that caused a corresponding shift of the shoreline inland along the east coast of North America.

We observed in Chapter 5 that when a sandy beach progrades over muddy offshore deposits, the process of progradation pushes the shoreline seaward (see Figure 5-22). This migration of a shoreline is known as **regression.** The Cambrian strata depicted in Figure 6-6 display the opposite pattern, termed **transgression,** in which the sea spreads over the land. Not only does the shoreline migrate inland as sea level rises, but so do the environments seaward of it, in accordance with Walther's Law (p. 134). Thus, during Cambrian time in the Grand Canyon region, muddy sediments came to rest on top of nearshore sands (Figure 6-6*B*). Similarly, carbonate sediments deposited even farther offshore came to rest on the muddy sediments.

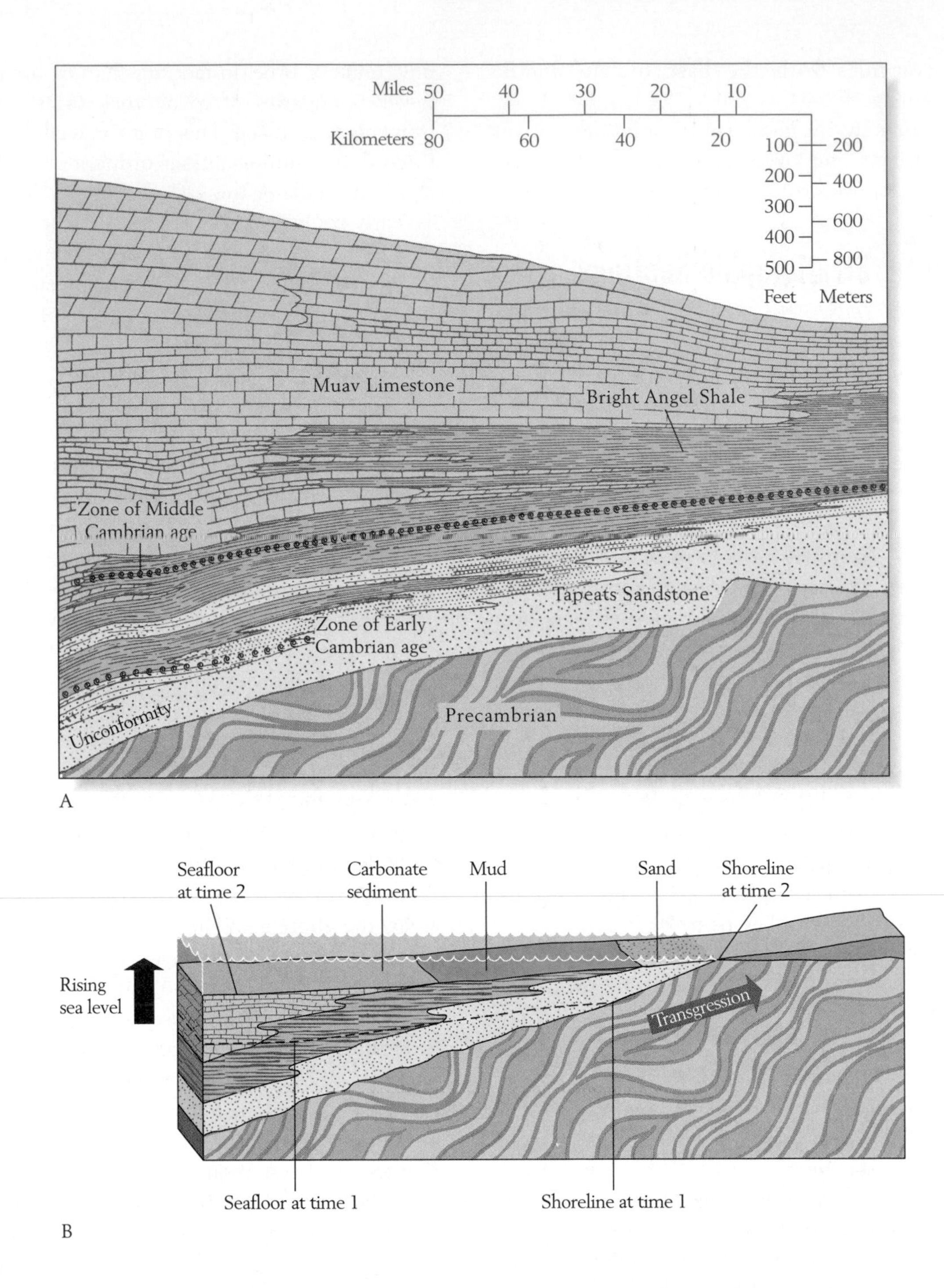

Figure 6-6 Stratigraphic pattern of Cambrian rocks exposed along the walls of the Grand Canyon. *A.* A cross section constructed from the stratigraphic sections seen in Figure 6-5. Fossils collected from the various sections show that biostratigraphic zones do not parallel formation boundaries, but intersect them, passing from one formation into another. Thus the formations represent facies that were forming simultaneously within neighboring environments that shifted through time. The basal formation, the Tapeats Sandstone, was deposited along the shoreline that bordered western North America, and this shoreline shifted eastward through time. *B.* The eastward shifting of the shoreline as sea level rose between two "moments" of Cambrian time. The result was a transgressive pattern of deposition, in which offshore muddy sediments came to overlie nearshore sand deposits, and carbonate sediments deposited even farther offshore came to overlie the muddy sediments. *(After E. D. McKee, Carnegie Inst. of Washington Publ. 563, 1945.)*

The pattern of sediment deposition illustrated in Figure 6-6 reflects the fact that no particular type of environment stretches infinitely far in any direction. One environment of deposition inevitably gives way to another, and when it does so, the transition between the two modes of deposition is either abrupt or gradual. The set of characteristics of a body of rock that represents a particular environment is called a **facies.** Accordingly, lateral changes in the characteristics of ancient strata, which reflect lateral changes in the depositional environment, are known as facies changes. The three formations displayed in Figure 6-6A represent distinct facies.

In another portion of the geologic record, a porous reef limestone full of fossil reef-building organisms might constitute one facies. This facies might pass laterally into steeply dipping talus deposits of poorly sorted rubble that slid down the front of the reef; in the other direction, the reef facies might pass into a facies of bedded limestone that formed in a lagoon (see Figures 5-24 and 5-25). All of these limestone facies might be assigned to a single formation, whereas the three facies illustrated in Figure 6-6, being lithologically very different, are assigned to separate formations.

Early Estimates of Earth's Absolute Age

Until late in the nineteenth century, when radioactivity was discovered, geologists could make only crude estimates of the absolute ages of bodies of rock or of Earth itself. Most attempts to establish Earth's age led to erroneous suggestions that the planet was younger than the hundreds of millions of years that the conventional uniformitarian view seemed to demand. Recall that dating of radioactive materials now shows that Earth is about 4.6 billion years old, even older than many uniformitarian geologists of the nineteenth century believed.

Salts in the Ocean

In the eighteenth and nineteenth centuries, some scientists estimated Earth's age by calculating how long it should have taken for the ocean to accumulate its salts. They assumed that the ocean's waters had originally been fresh and that runoff from the land had progressively increased their salt content, and was still doing so. In 1899, the Irish geologist John Joly estimated that at the rate at which rivers were then contributing salts to the ocean, about 90 million years would have been required to produce its current salinity. Joly greatly underestimated the age of the ocean by this method because precipitation of evaporites, including halite and gypsum (p. 46), constantly removes salts from the ocean. In fact, we now recognize that the salinity of the ocean may not have changed greatly since early in Earth's history. During this time of little change, the removal of salts by deposition of evaporites has approximately balanced the addition of salts by rivers and other sources.

Rates of Accumulation of Sediment

In the late eighteenth and early nineteenth centuries, some geologists attempted to use rates of accumulation of sediment to estimate Earth's age. First, they estimated rates of deposition of sediment in various modern settings. Then they estimated the thickness of sedimentary rocks in Earth's crust. Multiplying these two numbers together gave estimates for the total time of accumulation. These estimates were typically 100 million years or less, a small fraction of Earth's actual age. We now know that the estimates were inaccurate for several reasons:

1. The stratigraphic record is full of gaps. In many depositional settings, sediments are deposited in pulses. Within just a few seconds, a current may deposit a bed of sediment a few centimeters thick, but then there may be no net accumulation on top of this bed for a few years or even hundreds of years. Moreover, during this longer interval, scour by currents may remove the bed that was laid down so rapidly, along with other beds below it (p. 24).

2. Unconformities in the rock record represent even larger breaks in sediment accumulation—times when sediments were exposed to erosion on the land instead of continuing to accumulate (p. 10). Some unconformities represent millions of years when sediment failed to accumulate in a given region and when strata representing millions of years were eroded away.

3 Early geologists failed to recognize that many metamorphic rocks had once been sedimentary rocks. Thus they failed to recognize that these rocks once formed a significant part of the stratigraphic record.

Earth's Temperature

The most formidable challenge to the uniformitarian view that Earth must be very old came from Lord Kelvin, the British physicist who in 1865 presented an address with the presumptuous title "The Doctrine of Uniformity in Geology Briefly Refuted." By "uniformity" Kelvin meant uniformitarianism. His argument was simply that Earth had been very hot when it formed and had been cooling ever since, yet it was still warm; had it been very old, it would have been much cooler than it was. The temperature was known to rise as one descended a mine shaft, and calculations showed that the interior of the planet must remain very hot. Kelvin argued that Earth retained so much of its original heat that it could be only about 20 million or at most 40 million years old. This seemed much too brief a history for the uniformitarian interpretation of Earth's history—for processes such as erosion, deposition, igneous activity, and mountain building to have produced the complex array of rocks displayed at Earth's surface. Thus Lord Kelvin and his followers went so far as to challenge uniformitarianism, which had become the foundation of geological science. The discovery of radioactivity finally beat back this challenge, providing an explanation for the persistence of high temperatures within Earth (p. 19). Soon thereafter, radioactivity came to play a different and much larger role in the evaluation of geologic time. Scientists developed methods of using radioactive materials to determine absolute ages of rocks and of Earth itself.

Radioactivity and Absolute Ages

In 1895 Antoine-Henri Becquerel discovered that the element uranium undergoes spontaneous radioactive decay; that is, its atoms change to those of another element by releasing subatomic particles and energy. Geologists soon recognized that radioactive elements and the products of their decay could be used as geologic clocks to measure the ages of rocks.

Radiometric Dating

Only a few naturally occurring chemical elements are useful for dating rocks by means of **radioactive decay.** Recall that isotopes are forms of an element that differ in the number of neutrons in their nuclei; furthermore, some isotopes are stable and others are unstable, or radioactive (p. 32). Atoms of radioactive isotopes decay spontaneously, changing to atoms of a different element. The isotope that undergoes decay is known as the *parent isotope,* and the product is known as the *daughter isotope.* There are three modes of decay:

Loss of an alpha particle. An alpha particle consists of two protons and two neutrons. In other words, it represents an atom of helium. Loss of an alpha particle converts the parent isotope into the element whose nucleus contains two fewer protons (Figure 6-7).

Loss of a beta particle. A beta particle is an electron, which has a negative charge but no mass. Its loss turns a neutron into a proton, changing the parent isotope into the element whose nucleus contains one more proton.

Capture of a beta particle. Addition of a beta particle turns a proton into a neutron, changing the parent isotope into the element whose nucleus has one less proton.

Some radioactive isotopes decay into other isotopes that are also radioactive. In fact, several steps of decay are required to yield a stable isotope from some parent isotopes.

	Uranium 238		Alpha particle		Thorium 234
Protons	92	–	2	=	90
Neutrons	146	–	2	=	144

Figure 6-7 Decay of an atom of uranium 238 to thorium 234 by loss of an alpha particle. This loss reduces the atomic number by 1 and the atomic weight by 4.

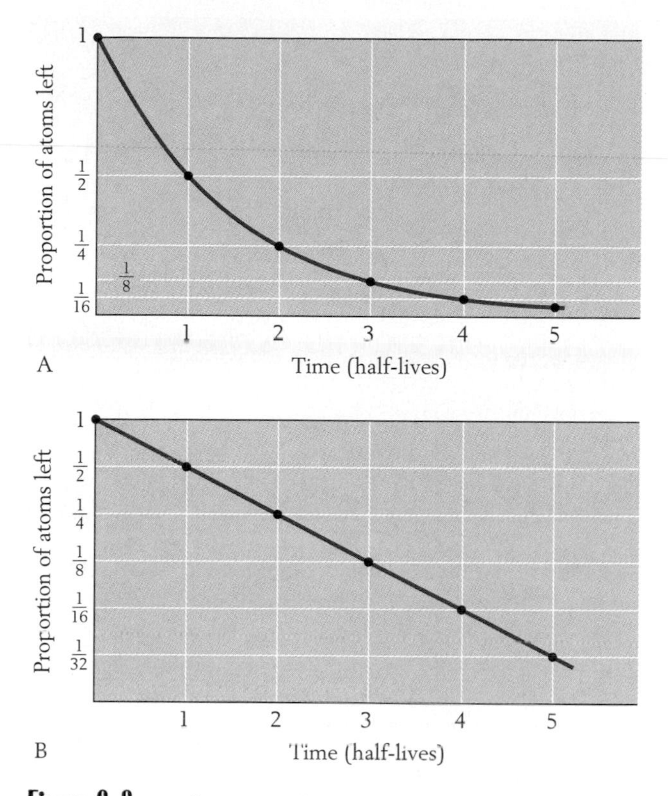

Figure 6-8 Arithmetic and geometric patterns formed by loss of atoms through radioactive decay. When plotted on a standard arithmetic scale *(A)*, the number of atoms can be seen to decrease more slowly with each successive interval of time. When the number of atoms is scaled as a geometric progression *(B)*, the plot forms a straight line. Half of the atoms present at the beginning of each interval (or half-life) survive to the beginning of the next interval.

Radioactive elements are useful for dating rocks because each radioactive element decays at its own nearly constant rate. Once geologists have measured this rate in the laboratory, they can calculate the length of time over which decay in a natural system has been proceeding by measuring the amounts of both the radioactive parent isotope and the daughter isotope that remain in the rock. This procedure is known as **radiometric dating.** Radioactive isotopes decay at a constant exponential or geometric rate (not at a constant arithmetic rate), as Figure 6-8 indicates. Thus, no matter how much of the parent element is present when it begins to decay, after a certain amount of time half of that amount will survive. After another interval of the same duration, half of the surviving amount (one-fourth of the original amount) will remain, and so on. This characteristic interval is known as the **half-life** of a radioactive element. Thus, in the course of four successive half-lives, the number of atoms of a radioactive element will decrease to one-half, one-fourth, one-eighth, and one-sixteenth of the original number of atoms. The number of atoms of the daughter product will increase correspondingly.

Useful Isotopes

Many elements occur naturally as both radioactive and nonradioactive (or stable) isotopes. As Table 6-2 indicates, several radioactive isotopes are abundant in rocks and therefore are useful for geologic dating. Most of these isotopes occur in igneous rocks; thus, if we know the amounts of parent and daughter elements currently present in an igneous rock, we can calculate the time that has elapsed since the parent element was trapped—the date when magma cooled to form the rock. A few minerals on the seafloor incorporate radioactive elements as well, and radiometric dates obtained for these minerals represent the interval of time that has elapsed since they formed. Unfortunately, radiometric clocks are reset by metamorphism. Metamorphic processes reposition radioactive isotopes in new minerals that contain

Table 6-2

Properties of Some Radiometric Isotopes That Are Commonly Used to Date Rocks

Radioactive isotope	Approximate half life (years)	Product of decay
Rubidium 87	48.6 billion	Strontium 87
Thorium 232	14.0 billion	Lead 208
Potassium 40	1.3 billion	Argon 40
Uranium 238	4.5 billion	Lead 206
Uranium 235	0.7 billion	Lead 207
Carbon 14	5730 billion	Nitrogen 14

Note: The number after the name of each element signifies the atomic weight of that element and serves to identify the isotope. Carbon 14, which has a very short half-life (that is, a high rate of decay), is used for dating materials younger than about 70,000 years. The other radioactive isotopes are employed for dating much older rocks.

none of their decay products. As a result, subsequent decay gives the age of the metamorphic event, not that of the original rock.

Half-lives of naturally occurring radioactive isotopes vary greatly (see Table 6-2), and these differences have a bearing on the ultimate use of various isotopes. More specifically, isotopes with short half-lives are useful for dating only very young rocks, while those with long half-lives are best used to date very old rocks. These limitations are related to measurement problems associated with the various isotopes. In essence, isotopes that have short half-lives, such as carbon 14, decay so quickly that their quantities in old rocks are too small to be measured. By the same token, isotopes that have long half-lives, such as rubidium 87, decay so slowly that the quantities of their daughter elements in very young rocks are too small to be measured accurately. Let us examine the radiometric utility of the various decay systems listed in Table 6-2.

Rubidium-Strontium Rubidium occurs as a trace element in many igneous and metamorphic rocks and even in a few sedimentary rocks. Nonetheless, because of the long half-life of rubidium 87, the parent isotope, the rubidium-strontium system is generally useful only for dating rocks older than about 100 million years.

Uranium and Thorium The useful radioactive isotopes uranium 235, uranium 238, and thorium 232 decay to different isotopes of lead. The most useful mineral for dating by means of uranium and thorium is the silicate zircon because it is widespread in low concentrations in igneous and metamorphic rocks as well as in detrital sediments derived from them. By artificially abrading the surfaces of zircon grains to obtain unaltered material and by employing several grains to study a single rock unit, geologists are now able to date even very old (Precambrian) rocks quite precisely. Some calculated ages in the range of 2 billion to 3 billion years are considered to be within just a few million years of the actual ages! The uranium-lead and thorium-lead systems have also been used to date rocks from the moon. The ages of the oldest dated moon rocks—slightly more than 4.6 billion years—closely approximate the age of Earth and its solar system, estimated from other evidence.

Several radiometric techniques have been developed to employ uranium and its daughter products

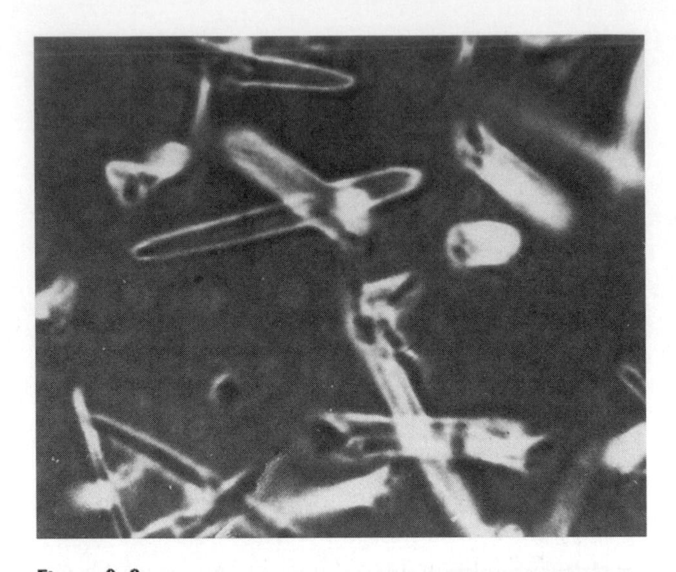

Figure 6-9 **Fission tracks in a mineral grain enlarged by etching.** The complete track oriented almost horizontally in the center is 13.5 micrometers long.

to determine the ages of rocks or fossils a few million years old or younger. One of these techniques is **fission-track dating,** a method for measuring the decay of uranium 238 that usually yields more accurate age estimates than conventional measurements of the daughter product (lead 206) when a rock is too young to have accumulated much lead. When uranium 238 decays, it emits subatomic particles that fly apart with so much energy that they penetrate the surrounding crystal lattice, producing fission tracks (Figure 6-9). These tracks can be enlarged in the laboratory by acid etching and can then be counted under a microscope. After these tracks have been counted, the remainder of the uranium can be subjected to a neutron field, which causes it to decay completely. The number of tracks thus produced can be compared to the number that formed naturally, and the resulting numerical ratio reveals the age of the mineral.

Additional methods of dating apply to reef-building corals (see Box 4-1), which incorporate a small amount of uranium in their skeletons. The fact that uranium 234 decays rapidly to thorium 230 allows accurate dating of corals that range in age from a few thousand years to about 300,000 years. Other radioactive isotopes of uranium yield helium as one of their final daughter products. Thus, by measuring the amount of helium and the amount of undecayed uranium trapped in well-preserved coral skeletons, geochemists can date corals several million years old.

Potassium-Argon and Argon-Argon Argon, the decay product of potassium 40, is an inert gas—that is, a gas that does not combine chemically with other elements. Argon becomes trapped in the crystal lattice of some minerals that form in igneous and metamorphic rocks. Generally, more precise dating results from a procedure called argon-argon dating: a sample is bombarded with neutrons in a nuclear reactor to convert potassium 40 to argon. The amount of argon gas produced in this way can be measured more precisely than the amount of potassium in a rock. One deficiency of the potassium-argon and argon-argon methods is that argon can leak from the lattice of a crystal, making it appear that less than the actual amount of decay has occurred. An error of the opposite kind will occur if a sample has absorbed argon from the atmosphere. This method has provided the dates of important events in the evolution of humans in Africa.

Radiocarbon Dating Radiometric dating that makes use of carbon 14, the radioactive isotope of carbon, is known as **radiocarbon dating.** This is the best known of all radiometric techniques, but because the half-life of carbon 14 is only 5730 years, this method can be used only with materials that are less than about 70,000 years old. Objects of biological origin, such as bones, teeth, and pieces of wood, make up the bulk of the materials dated by this method. Despite its limitations, radiocarbon dating is of great value for dating materials from the latter part of the Pleistocene Epoch—an interval so recent that most other radioactive materials found in its sediments have not decayed sufficiently to permit their products to be measured accurately. Fortunately, the useful range of carbon 14 encompasses the entire time interval during which modern humans have existed, as well as the interval during which glaciers most recently withdrew from North America and Europe at the close of the recent Ice Age (Pleistocene Epoch). Thus radiocarbon dating plays a valuable role in the study of human culture; it is sometimes used to date materials that are no more than a few hundred years old.

Carbon 14 is a rare isotope of carbon that forms in the upper atmosphere, about 16 kilometers (10 miles) above Earth's surface, as a result of the bombardment of nitrogen by cosmic rays. Both carbon 14 and the stable isotope carbon 12 are assimilated by plants, which turn them into tissue. Once a plant dies, however, carbon is no longer incorporated in its tissues, and the carbon 14 that was present when the plant died decays back into nitrogen 14. Thus the percentage of carbon 14 in the tissues of plants declines in relation to the percentage of carbon 12, and the ratio of the two can be used to determine when the tissue died.

Radioactivity vs. Fossils: The Accuracy of Correlation

The fact that the half-lives of radioactive isotopes are well established does not imply that radioactive decay permits more accurate correlation of sedimentary rocks than fossils. For one thing, most minerals that can be dated radiometrically are of igneous origin, and the dating of igneous rocks often yields only a maximum or a minimum age for associated sedimentary rocks (Figure 6-10). Clasts of igneous rocks or minerals found in sedimentary rocks can also be dated radiometrically, but only estimates of the maximum ages of sedimentary rocks can be derived in this way.

Even more fundamental uncertainties are inherent in radiometric dating. Most published radiometric

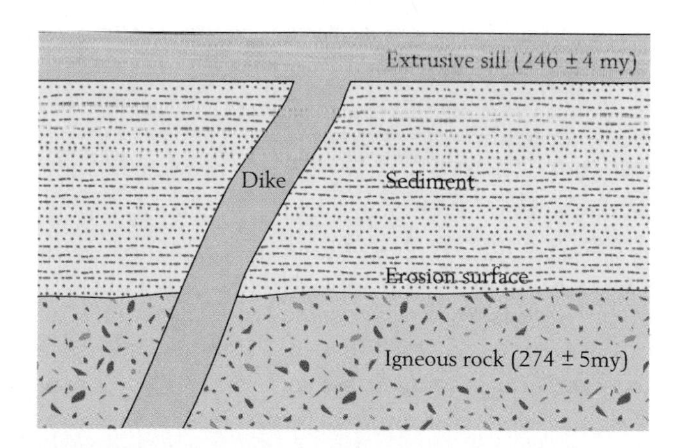

Figure 6-10 A cross section of a group of rocks showing how the age of a sedimentary unit can be bracketed by the radioactive dating of associated igneous rocks. Here the sediments lie on top of a body of igneous rock dated at 274 ± 5 million years, so they must be younger than that. The sediments are also cut by a dike and covered by a sill of igneous rock dated at 246 ± 4 million years; they are therefore older than that. We can conclude that this sedimentary unit may be as old as 279 million years or as young as 242 million years.

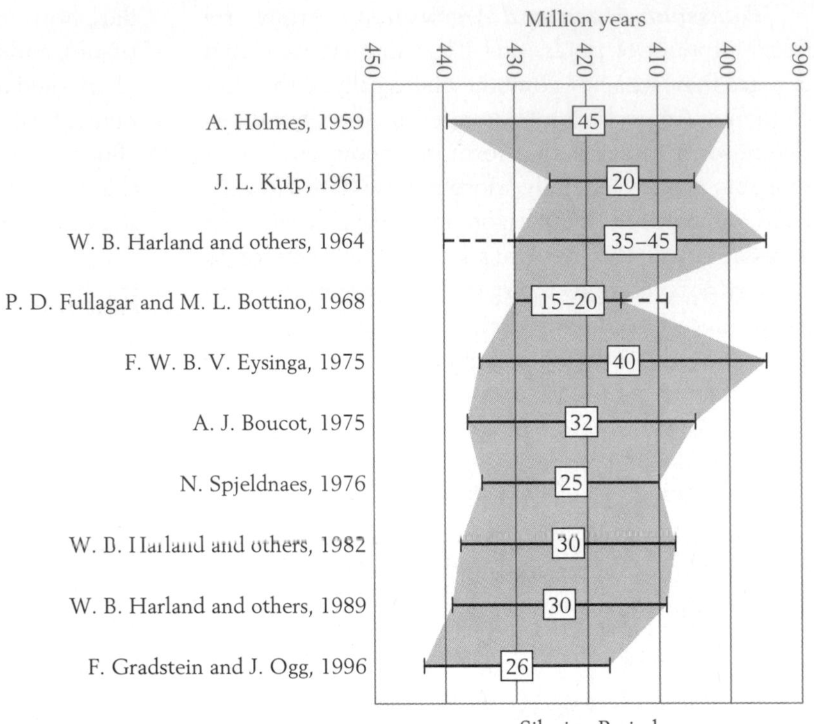

Figure 6-11 **Estimates of the interval of time represented by the Silurian Period.** The date each estimate was made and the author or authors responsible for each estimate are listed. *(Adapted from N. Spjeldnaes, Amer. Assoc. Petrol. Geol., Studies in Geol. 6:341–345, 1978.)*

dates, for example, are followed by a symbol and a number representing a smaller interval of time (for example, 47 ± 3 million years). The plus-or-minus sign indicates uncertainty that is attributable to possible errors in the measurement of the quantities of parent and daughter elements.

Another problem is that rocks that are dated radiometrically have not necessarily remained closed systems in nature: they may have gained or lost atoms of the parent or the daughter isotope. This is a particularly common source of error for the potassium-argon decay system, whose daughter element frequently seeps from rock, leading to underestimation of the amount of decay that has occurred and hence to an underestimation of the rock's age.

Still another source of error in dating stratigraphic boundaries is the absence of appropriate radioactive isotopes in rocks positioned close to the boundaries.

These types of errors sometimes add up to sizable total errors, especially when very old rocks are being dated. This point is well illustrated by past estimates made for the beginning and end of the Silurian Period (Figure 6-11). In evaluations made between 1959 and 1968 alone, the duration of the Silurian was halved and then doubled and then halved again. It now seems likely that the Silurian began about 443 million years ago and that it ended about 417 million years ago, but each of these dates may be adjusted slightly in the future.

Fossil graptolites (see Figure 6-2) of Silurian age allow for more accurate correlation of sedimentary rocks within particular regions than do radiometric dates. Many kinds of graptolites floated in the sea and thus spread quickly over large areas. In addition, graptolites evolved rapidly. These two factors together have made graptolites highly useful index fossils. Many individual species of graptolites existed for about 1 million or 2 million years; thus correlations based on the occurrences of such species cannot be inaccurate by a larger interval than this.

Although radiometric dating provides a special kind of geologic time scale—specifically, an absolute scale, or one based on years—most geologic correlations are still based on fossils. Not only are fossils more common than radioactive elements in sedimentary rocks, but the analysis of fossils usually allows for greater accuracy. However, the geologic intervals at the upper and lower ends of the geologic scale of time represent striking exceptions to this general rule. At the upper end, most rocks that are older than about 1.4 billion or 1.5 billion years contain few fossils that are well enough preserved and easily enough identified to serve as guides. Thus radiometric dates serve as a primary basis for correlation of these early rocks, especially from continent to continent. And at the lower end of the time scale, the interval extending from the present back to about 70,000 years ago, radiocarbon dating offers estimates of age that commonly have plus-or-minus values of only a few percent. Because relatively few species of animals or plants have appeared or disappeared during this brief interval, fossils found in the corresponding rocks have much less value for correlation.

Dating by Other Stratigraphic Features

Several other stratigraphic features often permit more precise correlation than either fossils or radioactive isotopes.

Isotopic Stratigraphy

Within a chemical reservoir such as the ocean, the ratio between two isotopes of an element may vary through time. If organisms incorporate the two isotopes into their skeletons, the fossilized skeletons will record changes in the ratio of the two isotopes in the ocean and distinctive patterns of change can be used for correlation. One correlation method of this kind employs the element strontium. Two stable isotopes of strontium that occur in all modern seas provide a special opportunity to date relatively young strata containing fossils that have not been altered appreciably since their burial. These two isotopes, strontium 86 and strontium 87, occur in the same relative abundance in all modern seas. The ratio of the two isotopes' abundances has changed through time, however, for many reasons, including changes in the rates at which rocks of various isotopic composition have yielded their strontium to the ocean as a result of exposure and erosion. During the past 25 million years, the relative abundance of strontium 87 has increased slightly. Small amounts of strontium take the place of calcium in the crystal structure of the calcareous skeletons of marine organisms. This strontium has the same isotopic ratio as the seawater in which the animal lives. Thus well-dated fossils provide a record of changes in the ratio of strontium 87 to strontium 86 in seawater (Figure 6-12). Once a record of such changes has been established, geologists can date fossil skeletons by measuring their precise isotopic composition. This valuable dating method, known as **isotope stratigraphy,** has only recently been put to use. It promises to resolve many stratigraphic problems, especially with respect to Cenozoic fossils, which often have undergone relatively little diagenetic alteration and retain their original strontium isotope ratios.

Event Stratigraphy

Imagine that someone with supernatural powers were in an instant to spray paint over many areas of Earth—ocean floors, lake bottoms, and terrestrial lowlands. In keeping with the law of superposition, this layer would become a time-parallel surface—that is, it would separate deposits that had formed

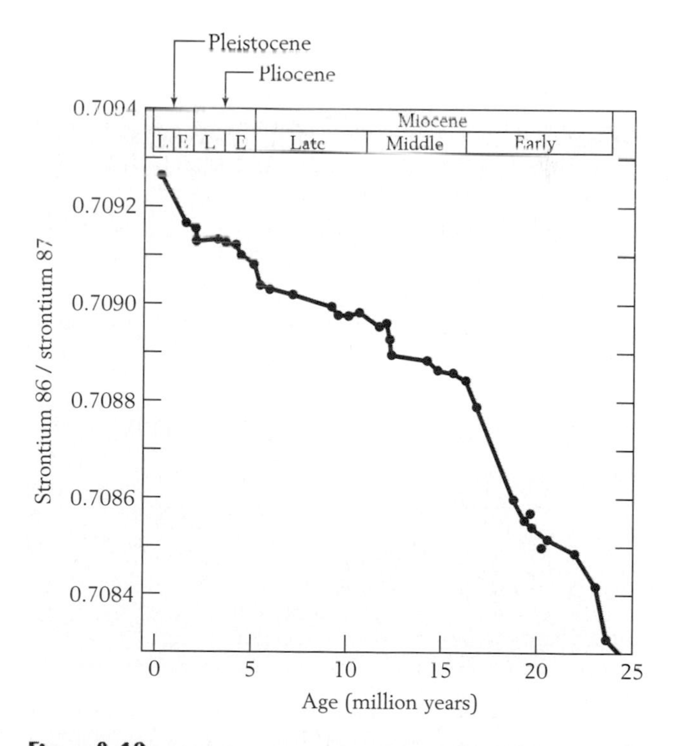

Figure 6-12 The changing ratio of strontium isotopes in marine fossils composed of calcium carbonate during the past 25 million years. *(After D. J. DePaolo, Geology 14:103–106, 1986.)*

before it was laid down from accumulations that were laid down afterward. Similarly, if a powerful force were suddenly to lower sea level throughout the world by 200 meters, the deposition of sediments at the shoreline would shift seaward for great distances, and the first of the new shoreline deposits thus created would be of nearly the same age on all continents, so that these deposits could be accurately correlated throughout the world. These two hypothetical scenarios—the formation of a time-parallel surface and a sudden relocation of shorelines—are not far removed from actual sudden events whose geologic records have enabled us to correlate rocks of widely separated regions. The use of geologic records of this kind for correlation is termed **event stratigraphy.**

Key Beds A **key bed,** or **marker bed,** is a bed of sediment that resembles our hypothetical layer of paint: all parts of it are virtually time-parallel; that is, of identical age. Widespread layers of volcanic ash, for example, often function as key beds. Sometimes an ancient ash fall, which may represent either a single volcanic eruption or a series of nearly simultaneous eruptions, can be traced for thousands of square kilometers, and so provides a time-parallel surface in the stratigraphic record. The Bishop Tuff, for example, is a bed of volcanic ash that was emitted from a volcano in eastern California slightly more than 700,000 years ago and spread halfway across the United States (Figure 6-13). Figure 6-14 shows a much older ash bed positioned within nonvolcanic marine strata. This bed formed during Cretaceous time in Colorado, when a volcano erupted in a region of mountain building to the west and emitted ash that fell into a large inland sea.

Some key beds are global in distribution. For example, the top of the Cretaceous System can be identified by a thin layer of sediment that is enriched in iridium, an element that is generally rare on Earth. The iridium at the top of the Cretaceous System was contributed by fallout from the explosion of an asteroid that struck Earth, resulting in the extinction of the dinosaurs and many other forms of life. Occurring with the high concentration of iridium are sand-sized grains that (as will be described more fully in Chapter 17) exhibit features that reflect the

A

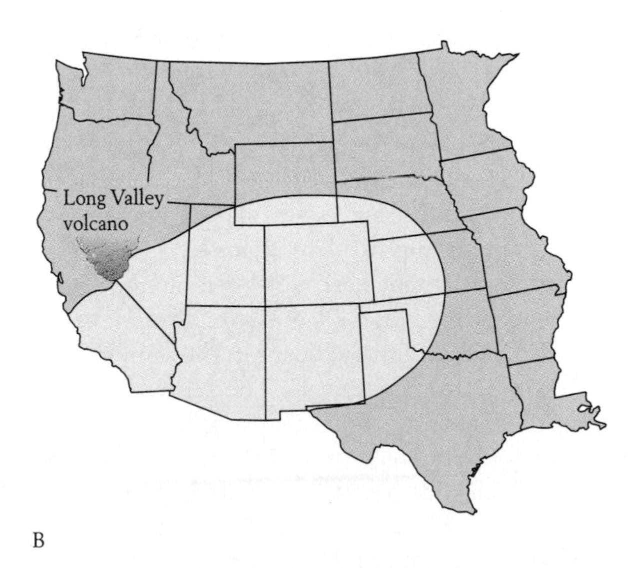

B

Figure 6-13 **The Bishop Tuff, which formed when the Long Valley volcano in eastern California erupted slightly more than 700,000 years ago.** *A.* A geologist examines the ash bed at Shoshone, where it is very thick, having been deposited close to the source. *B.* Areal extent of recognized occurrences of the tuff; most of the ash spread eastward, blown by the prevailing winds. *(After G. A. Izett and C. W. Naeser, Geology 4:587–590, 1976.)*

heat and pressure of the asteroid's impact. These grains were blown high into the atmosphere and spread throughout the world.

Glacial tills such as those described in Chapter 5 also serve as useful marker beds; even tills formed by glaciers in different parts of the world can be useful if they represent only a brief interval of global cooling.

Certain events produce key beds that allow for correlation within particular basins of deposition. The character of deep-water evaporite deposits in a basin, for example, can reflect aquatic conditions throughout the basin. Because deep-water evaporites (p. 46) typically form widespread horizontal beds (Figure 6-15), an individual bed that differs in chemical composition from beds above and below can often be traced over thousands of square kilometers.

Back-and-Forth Shifting of Depositional Boundaries When a transgression is followed by a regression, the resulting stratigraphic pattern is useful for correlation (Figure 6-16C). A point along the boundary between two facies marks the time of maximum transgression. Two or more such points that represent the same time of maximum transgression can therefore be connected to form a line of temporal correlation. Most time-parallel surfaces of this type are useful only for correlating stratigraphic sections that represent single depositional basins—for example, sections that represent different parts of a lake or shallow sea. The unconformities discussed in the next section can provide evidence for correlation over longer distances.

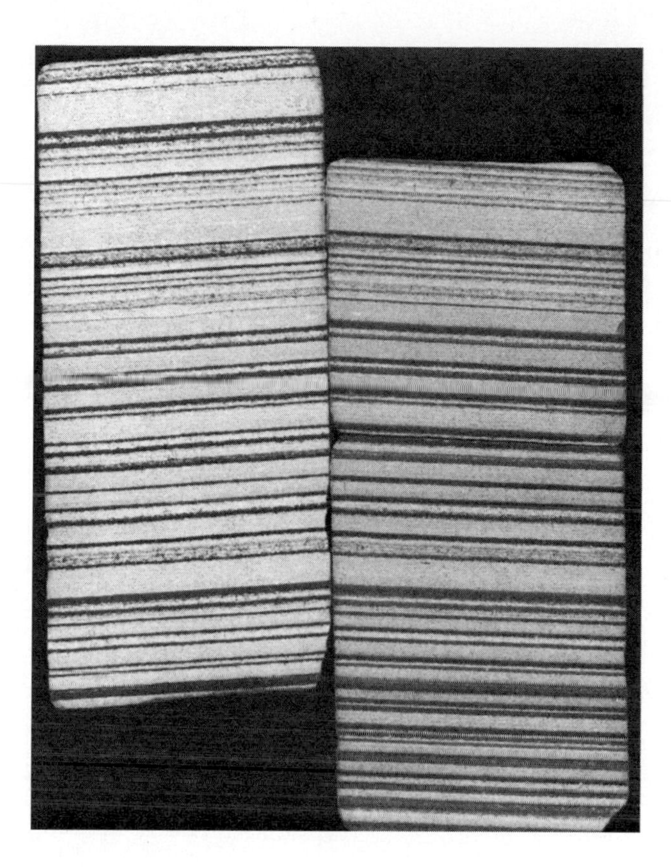

Figure 6-15 Two cores taken from the Castile Evaporites, which were precipitated in western Texas near the end of the Permian Period. The cores are from localities 14.5 kilometers (9 miles) apart, yet their laminations match almost perfectly, allowing for precise correlation. The alternating dark and light bands, which range up to a few millimeters in thickness, probably represent seasonal organic-rich (winter) and organic-poor (summer) layers. If this is the case, each pair of bands represents one year, and the 200,000 or so paired bands of the Castile Formation represent about 200,000 years of deposition. *(R. Y. Anderson et al., Geol. Soc. Amer. Bull. 83:59–86, 1972.)*

Figure 6-14 A thick band of volcanic ash (above the hammer head) overlain by two similar but thinner ash beds. This conspicuous group of Cretaceous beds permits correlation over a large area of Colorado.

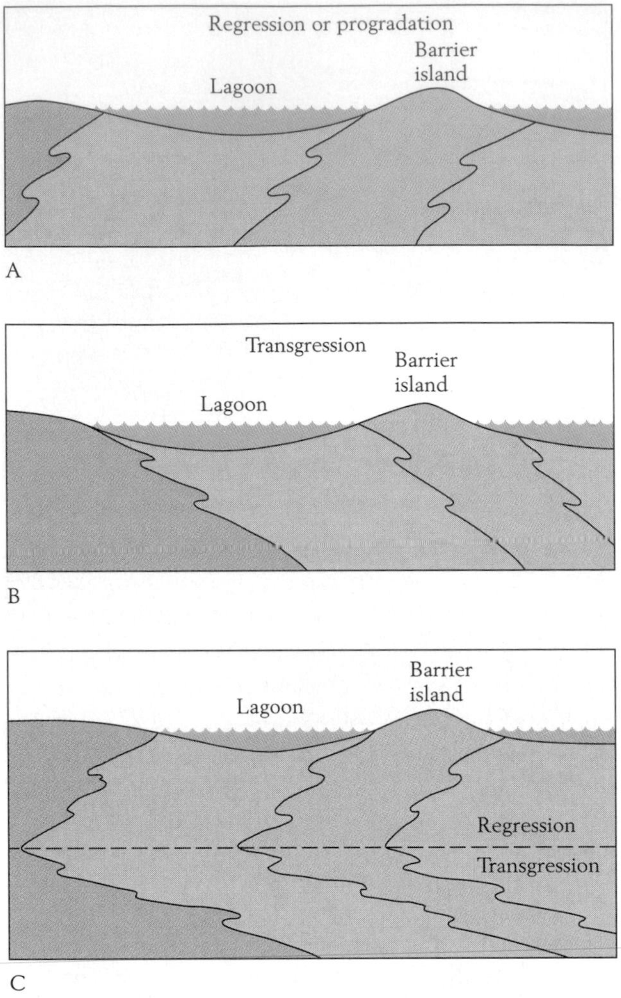

Figure 6-16 Correlation based on a stratigraphic pattern in which a regressive depositional sequence follows a transgressive sequence. *A.* The pattern of a regressive (progradational) sequence. *B.* The pattern of a transgressive sequence. C. When a regression follows a transgression, the points of maximum transgression of various facies can be connected to form a line of correlation, as indicated by the broken line in the diagram. This was the "moment" in geologic time when the sea shifted farthest inland. A similar line can be constructed for a stratigraphic pattern in which a transgression follows a regression.

Unconformities, Bedding Surfaces, and Seismic Stratigraphy

Unconformities that formed during the same interval of tectonic uplift or nondeposition sometimes constitute nearly time-parallel surfaces. Such unconformities may truncate rocks of many ages, but the sediments resting directly on top of the erosional surface are often nearly the same age in all parts of a depositional basin. The deposits resting on the surface of an unconformity are rarely precisely the same age, however, and the ages of some of these deposits vary greatly from place to place. A sea that has deserted an area, for example, may later invade it again slowly, after a period of erosion, reaching different parts of the area at different times. On the other hand, if sea level drops suddenly throughout the world and then rises again rapidly, the resulting global unconformities may represent fairly accurate time markers.

Global unconformities that occur within Mesozoic and Cenozoic sediments lying along continental shelves have been used with great success as time markers. Most continental shelves have remained be-

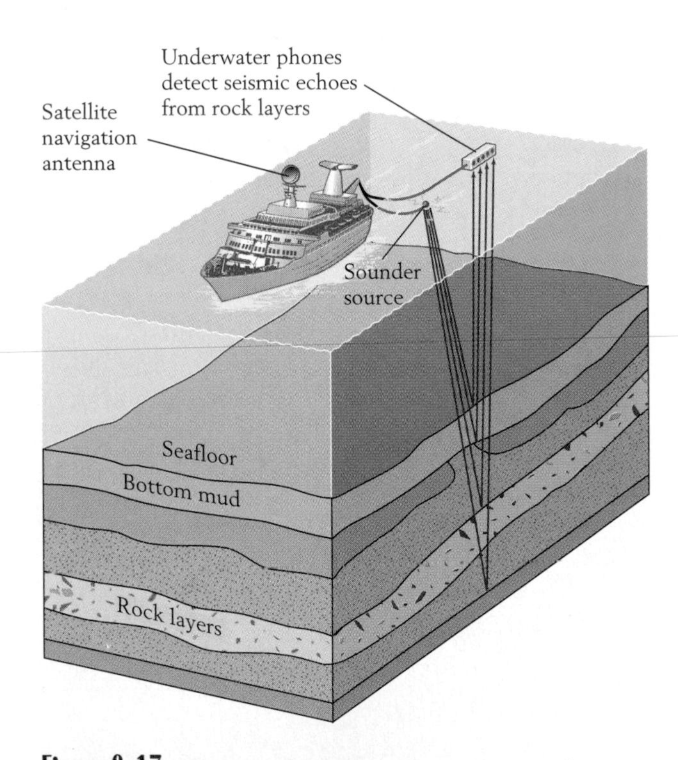

Figure 6-17 The use of seismic reflections to study sediments and rocks buried beneath the seafloor. Sound waves are generated by a sounder that makes a pneumatic explosion like that of a bursting balloon. The sound waves bounce off surfaces of discontinuity, which include bedding surfaces and unconformities. Underwater phones then pick up the reflections, allowing marine geologists to determine the configurations of buried features. Fossils recovered from sediment cores reveal the ages of individual sedimentary beds. *(After F. Press and R. Siever, Earth, W. H. Freeman and Company, New York, 1986.)*

low sea level since their formation and thus have not been destroyed. Such sediments and unconformities have been examined by **seismic stratigraphy**—the study of sedimentary rocks by means of seismic reflections generated when artificially produced seismic waves bounce off physical discontinuities within buried sediments (Figure 6-17). Some of these discontinuities have been found to be time-parallel bedding surfaces. Others have been identified as unconformities (Figure 6-18).

When compilations of many seismic profiles are used together with fossil evidence to date local changes in sea level, global patterns of sea-level change may be revealed. Evidence that sea level rose or fell in many widely separated areas at the same time indicates that the change occurred on a global scale. Such evidence has yielded a global curve of sea-level changes during the Cenozoic Era (Figure 6-19). Less precise information for earlier intervals, based in part on rocks, fossils, and unconformities visible on the continents, has yielded a less detailed curve for the entire Phanerozoic interval.

The changes in sea level shown in Figure 6-19 are based on seismic evidence of simultaneous shifts in many parts of the world and are thus assumed to reflect global changes in sea level known as **eustatic events.** However, eustatic changes in sea level are not reflected in seismic profiles where there is an offsetting tectonic change in the position of the land (Figure 6-20). If the land in one area rises in pace with a eustatic rise in sea level, for instance, the eustatic rise will not be expressed as it would be if the land were to remain stationary (Figure 6-21*A*). Moreover, if the tectonic uplift exceeds the eustatic rise in sea level, sea level in that area actually falls in relation to the land (Figure 6-21*B*) regardless of what is happening in the rest of the world. Box 6-1 illustrates how, in a similar way, a continental margin can sink so as to lower rocks that originated in shallow water far below the position at which they formed.

In addition, vertical sea-level changes have often been mistaken for transgressions and regressions, which are lateral shifts in the position of the shoreline. Of course, a strong correlation exists both between transgressions and eustatic rises and between regressions and eustatic falls, but there is a complicating factor that weakens this correlation, making transgressions and regressions unreliable indicators of eustatic changes. As Figure 6-21C indicates, the rate at which sediment accumulates strongly influences whether transgression or regression will occur in a particular area. Thus regressions have often occurred locally during global intervals of rising sea level simply because sediment that has been supplied from the land at a very high rate has pushed the sea back from the land.

Sequences

The widespread unconformities that represent eustatic falls have been used to divide the stratigraphic record into units called sequences (Figure 6-22). **Sequences** are large bodies of marine sediment deposited on continents when the ocean rose in relation

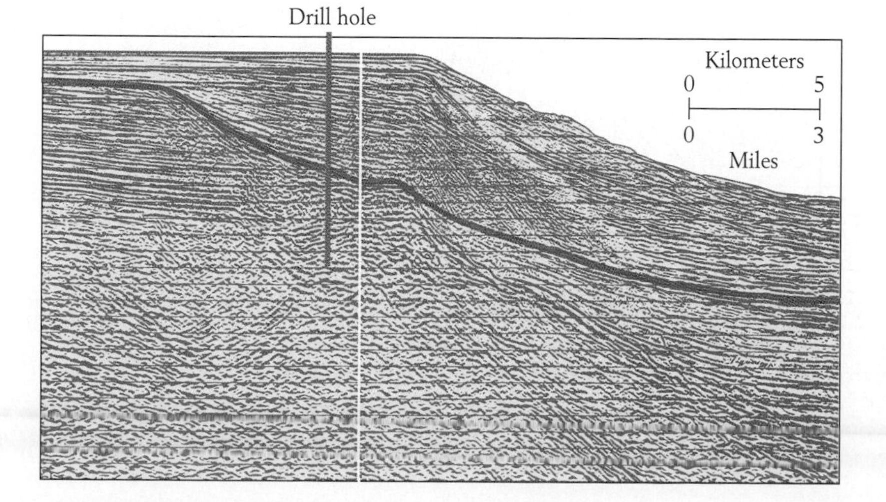

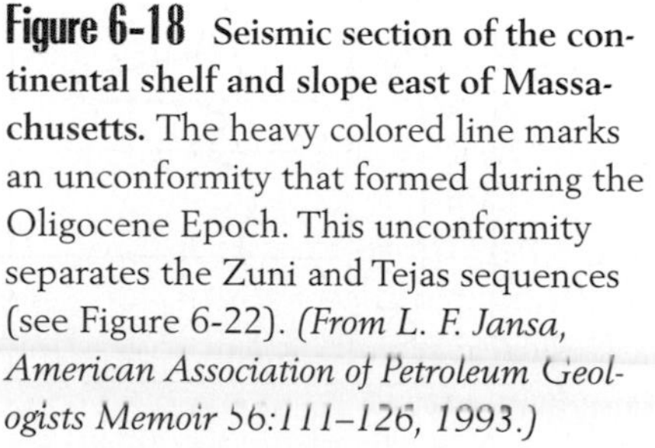

Figure 6-18 **Seismic section of the continental shelf and slope east of Massachusetts.** The heavy colored line marks an unconformity that formed during the Oligocene Epoch. This unconformity separates the Zuni and Tejas sequences (see Figure 6-22). *(From L. F. Jansa, American Association of Petroleum Geologists Memoir 56:111–126, 1993.)*

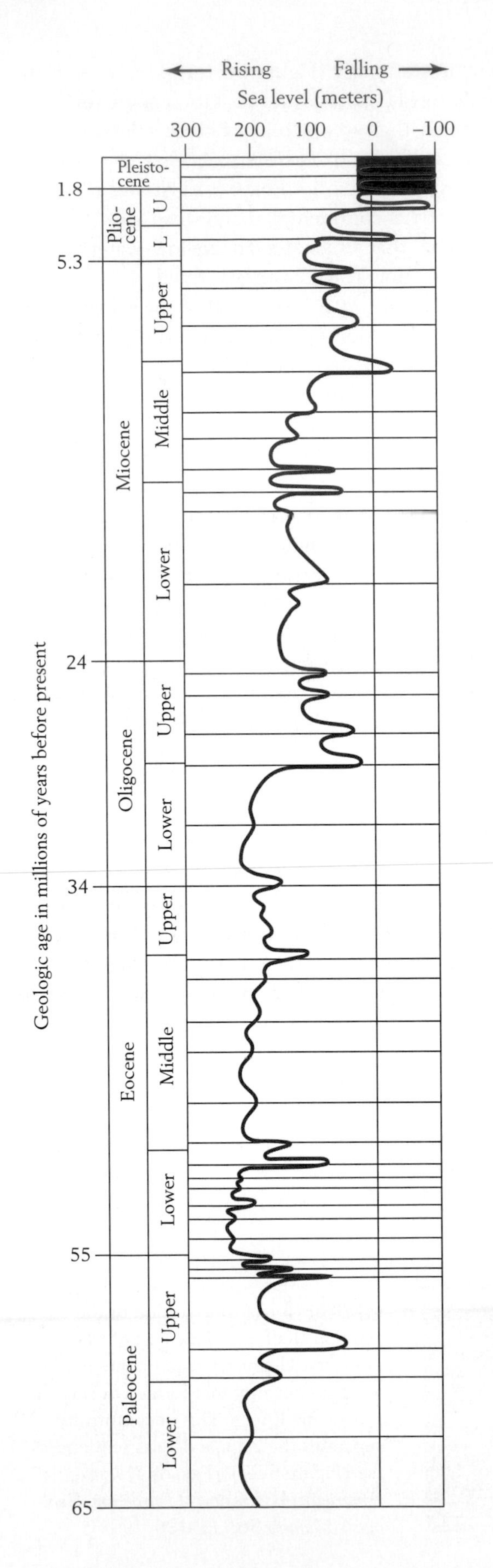

Figure 6-19 Estimates, based primarily on seismic stratigraphy, of relative changes in sea level during the Cenozoic Era. Horizontal segments of the sea-level curve represent sudden drops in sea level. The horizontal scale is in arbitrary units. *(After B. Haq et al., Science 235:1156–1167, 1987.)*

to the level of continental surfaces and formed extensive epicontinental seas (p. 111). Individual sequences represent tens of millions of years of time during which sea level rose quite high and then receded again. During most of the time when sequences have been deposited—which means during most of Phanerozoic time—the sea has stood higher than it does today.

Most long-term changes in sea level that have produced sequences separated by unconformities have resulted from changes in the rate at which new lithosphere has formed along mid-ocean ridges. Mid-ocean ridges, as we know, are great swellings of the seafloor where new lithosphere rises up and remains swollen from heat rising within it. The lithosphere cools and shrinks as it moves away from a ridge axis. Consequently, the seafloor descends on either side of

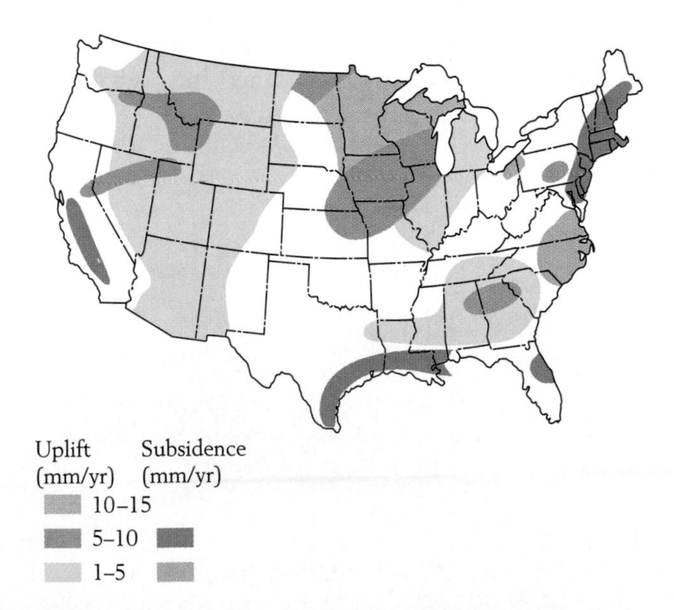

Figure 6-20 Rates of uplift and subsidence of Earth's crust in the United States today. *(After S. P. Hand, National Oceanic and Atmospheric Administration.)*

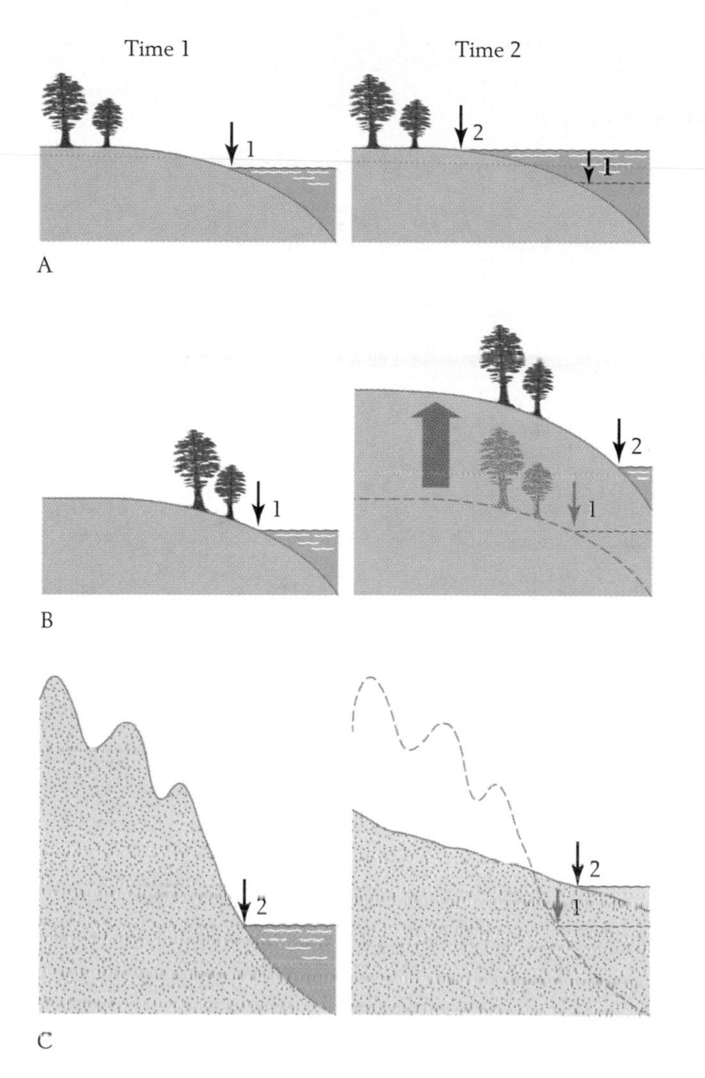

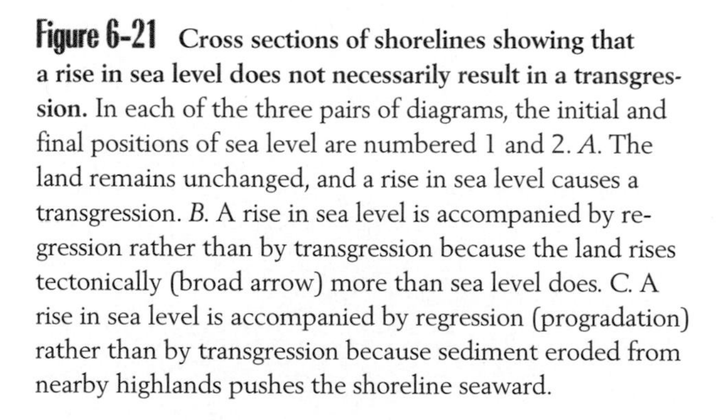

Figure 6-21 Cross sections of shorelines showing that a rise in sea level does not necessarily result in a transgression. In each of the three pairs of diagrams, the initial and final positions of sea level are numbered 1 and 2. *A.* The land remains unchanged, and a rise in sea level causes a transgression. *B.* A rise in sea level is accompanied by regression rather than by transgression because the land rises tectonically (broad arrow) more than sea level does. C. A rise in sea level is accompanied by regression (progradation) rather than by transgression because sediment eroded from nearby highlands pushes the shoreline seaward.

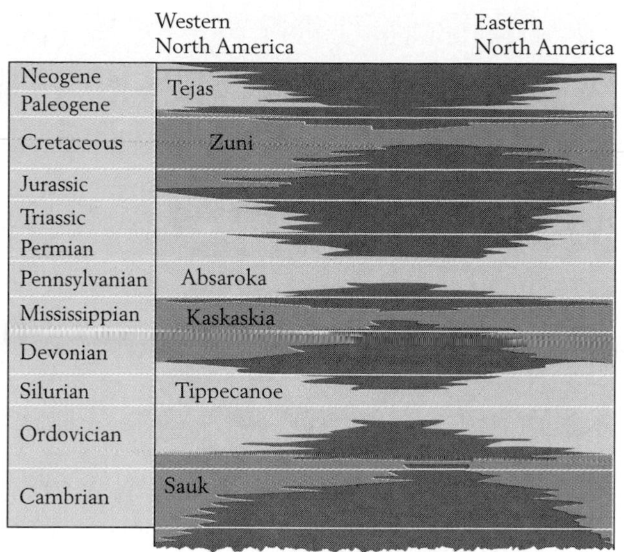

Figure 6-22 Phanerozoic sequences on the North American continent. These sequences resulted from transgressions on both sides as global sea level rose, and they ended with regressions as global sea level fell. The black areas are gaps represented by unconformities. *(From L. L. Sloss, Geol. Soc. Amer. Bull. 74:93–114, 1963.)*

a ridge, eventually leveling out to form the abyssal plain (see Figure 4-24). Lithospheric plates have been more active at some times than at others in the course of Earth's history. At times of intense plate tectonic activity, the total length of mid-ocean ridges has been relatively great. Rates of spreading have tended also to be high at these times, and so much heat has flowed to the ridges that individual ridges have stood relatively tall. The large total volume of ridges has displaced ocean water, pushing sea level upward and causing broad continental areas to flood. At such times, marine deposition on continents has formed sequences.

Then, when plate tectonic activity has become less intense again, total ridge volume has shrunk over millions of years and sea level has declined correspondingly. Unconformities have then formed as seas have receded from continents, producing sequence boundaries.

Changes in the volume of mid-ocean ridges have moved sea level up and down at rates on the order of 10 meters (33 feet) per million years. Expansion and contraction of continental glaciers have caused much more rapid and dramatic changes in sea level. Many times during the modern Ice Age, sea level has fallen by as much as 100 meters (330 feet) within just a few thousand years when glaciers have expanded over the land, "locking up" water and thus removing

For the Record 6-1

Searching for Oil Off Southern New Jersey

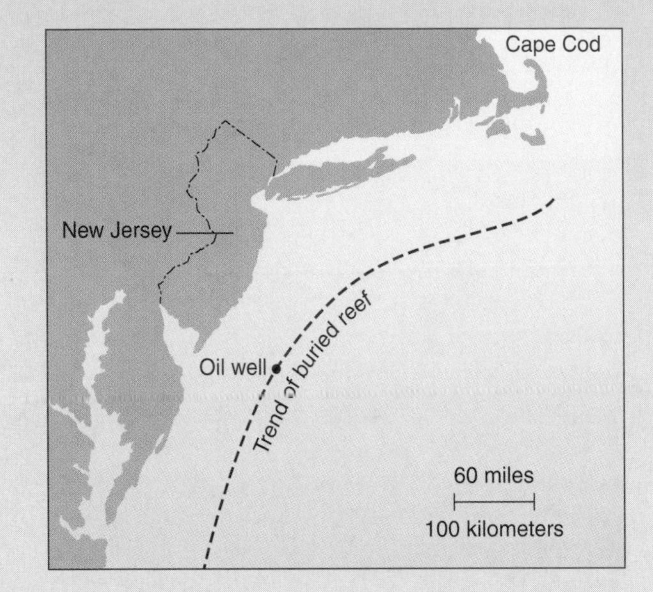

Seismic stratigraphy is a major tool in the search for petroleum and natural gas. Seismic cross sections reveal that a large ridge parallels the Atlantic coast of the United States for hundreds of kilometers, deeply buried in sediment below the continental slope. The seismic sections display broad surfaces within the pile of sediments encasing the ridge. These surfaces represent brief intervals when patterns of deposition changed or deposition ceased altogether. Thus the surfaces allow for correlation, and those that are deflected upward in the vicinity of the ridge indicate that it once stood above the surrounding seafloor. When the ridge was discovered, its location and configuration suggested that it might be an ancient

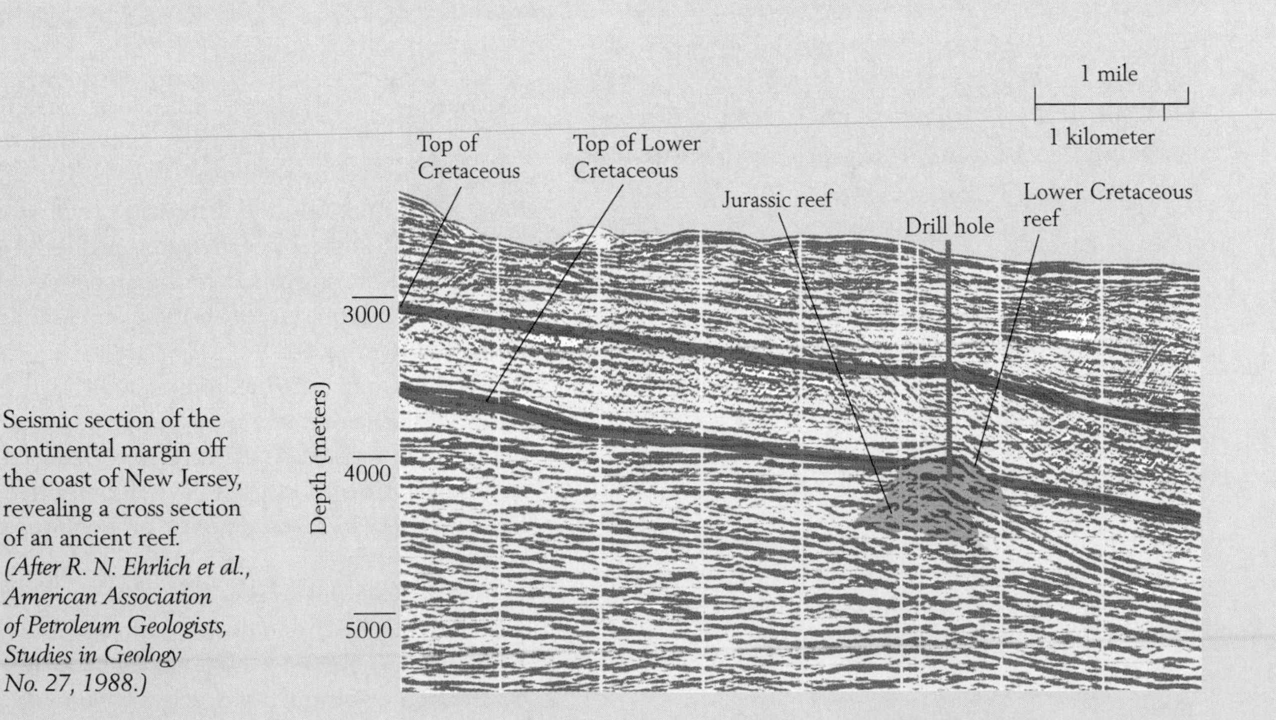

Seismic section of the continental margin off the coast of New Jersey, revealing a cross section of an ancient reef. *(After R. N. Ehrlich et al., American Association of Petroleum Geologists, Studies in Geology No. 27, 1988.)*

barrier reef that long ago grew along the edge of the continental shelf. This possibility aroused the interest of the petroleum industry, because buried reefs are porous structures that frequently trap oil and gas.

Barrier reefs grow near the surface of the ocean, but the ridge beneath the continental slope of the eastern United States lies about 4 kilometers (~2.5 miles) below present sea level—much lower than sea level has ever stood during Phanerozoic time. In fact, the entire continental margin has subsided greatly since it formed early in the Mesozoic Era when a huge continent broke apart, separating the landmasses that are now North and South America from Europe and Africa. This event marked the birth of the Atlantic Ocean, which has been widening ever since as the continents to the east and west have moved farther and farther apart. As the continents separated, the newly formed continental margins subsided, and sediments accumulated to great thickness along eastern North America. Thus it was reasonable to suppose that a reef had once occupied the shallow waters off the newly formed North American coast and that after it died, it sank and was buried beneath younger deposits.

The likelihood of striking oil led three corporations to cooperate in constructing an offshore rig and drilling into the buried ridge in 1984. The drilling produced samples containing fossils that revealed the ages of the ridge and the deposits above it. The results were informative: the ridge turned out to be a barrier reef of Late Jurassic and Early Cretaceous age. Unfortunately, the results were negative from a practical standpoint: the reef contained no commercially valuable oil or gas.

it from the global water cycle. There were several other intervals of continental glaciation earlier in Earth's history, and they also featured rapid and dramatic eustatic events. One of these episodes took place briefly at the end of the Ordovician Period, during deposition of the Tippecanoe sequence (see Figure 6-22). Glaciers grew rapidly at this time, causing a dramatic eustatic fall, but soon they contracted again so that sea level rose and widespread deposition resumed on continental surfaces.

Because sea level happens to be lower in relation to continental surfaces today than it has been during most of Earth's history, sequences are well displayed in outcrops on modern continents, as are the unconformities between them. Young sequences also include bodies of sediment that now lie below the sea, having accumulated along continental borders relatively recently. Geologists use seismic stratigraphy to study these bodies of sediment, which underlie continental shelves and slopes. The seismic section of the continental shelf and slope offshore from Massachusetts, for example (see Figure 6-18), depicts the unconformity between the Zuni and Tejas sequences (see Figure 6-22). This unconformity shows that sea level fell substantially throughout the world during the Oligocene Epoch, about 30 million years ago. Actually, the drop in sea level was not so great as one might conclude from the fact that the unconformity extends far below modern sea level. Since the unconformity formed, the weight of sediments deposited above it has depressed the strata that contain it.

Chapter Summary

1 Systems were erected haphazardly in the nineteenth century and pieced together to form the geologic time scale. Along with erathems, which are larger units, and series and stages, which are smaller units, systems are considered time-rock units.

2 A time-rock unit is defined by upper and lower boundaries, which are defined at localities called boundary stratotypes.

3 Time units—eras, periods, epochs, and ages—correspond to erathems, systems, series, and stages.

4 Biostratigraphic units are defined by fossil occurrences; the zone is the basic unit of this type.

5 The most useful species for correlation are called index or guide fossils. Ideally, these species are easily identified, widely distributed, abundant in many kinds of rock, and restricted to narrow vertical stratigraphic intervals.

6 Periodic reversals in the polarity of Earth's magnetic field provide excellent markers for correlation of magnetized rocks.

7 A body of rock that is characterized by a particular lithology or group of lithologies is often recognized as a formal rock unit (a member, formation, or group). Rock units do not have time-parallel upper or lower boundaries, so they are often transsected obliquely by biostratigraphic units.

8 A facies is the set of characteristics of strata that formed in a particular environment. A formation may consist of a single facies, or of two or more adjacent facies.

9 The relative scale of geologic time is based on superposition and fossil succession. Bodies of rock are simply ordered from oldest to youngest. The scale based on radiometric dating, in contrast, is an absolute scale in which events are measured in years.

10 Correlations that are based on radiometric dates are often less accurate than ones based on index fossils.

11 Global shifts in the isotopic ratios of certain elements preserved in sediments provide methods for correlating strata.

12 Key beds such as ash falls, glacial tills, and evaporite beds are sedimentary layers that form almost simultaneously over large areas, and thus serve as useful time markers.

13 Some unconformities are useful for identifying times in Earth's history when sea level has suddenly dropped throughout the world. The analysis of seismic reflections often allows geologists to recognize these unconformities and their associated bedding surfaces deep within Earth.

Review Questions

1 What kinds of stratigraphic correlation do geologists undertake?

2 What is the difference between the relative and absolute ages of rocks?

3 For what interval of geologic time is radiocarbon dating applicable?

4 How are strontium isotopes useful in dating rocks?

5 Why does a geologic formation not necessarily have upper and lower boundaries that are time-parallel?

6 What factors prevent biostratigraphic zones from having boundaries that are perfectly time-parallel?

7 What factors lead to imperfections in radiometric dating?

8 Construct a diagram resembling Figure 6-16C to depict the depositional history of a barrier island–lagoon complex that has undergone a regression followed by a transgression.

9 Construct a diagram resembling Figure 6-21*B* showing that a lowering of sea level need not always lead to regression.

10 How are seismic profiles used to study regional stratigraphy?

11 What is event stratigraphy?

12 Using the Visual Overview on page 152 as your guide, evaluate the various methods of correlating rocks with regard to age. Which methods are especially useful for intercontinental correlation?

Additional Reading

Ager, D. V., *The Nature of the Stratigraphical Record,* John Wiley, New York 1993.

Blatt, H., W. B. N. Berry, and S. Brande, *Principles of Stratigraphic Analysis,* Blackwell Scientific Publications, Oxford, 1991.

Davis, R., *Depositional Systems: An Introduction to Sedimentology and Stratigraphy,* Prentice-Hall, Englewood Cliffs, N.J., 1983.

Miall, A. M., *Principles of Sedimentary Basin Analysis,* Springer-Verlag, New York, 1990.

Prothero, D., *Interpreting the Stratigraphic Record,* W. H. Freeman and Company, New York, 1990.

CHAPTER 7

Evolution and the Fossil Record

The central concept of modern biology is that living species have come into being as a result of the evolutionary transformation of quite different forms of life that lived long ago. Indeed, it is often maintained that very little of what is now known about life would make sense in any context other than that of organic evolution. It is important to remember, however, that while the broad definition of evolution is "change," organic evolution does not encompass every kind of biological change; the term refers only to changes in populations, which consist of groups of individuals that live together and belong to the same species.

When we examine the broad spectrum of organisms that inhabit our planet, we cannot help being impressed by the success with which each form functions in its own particular circumstances. Members of the cat family, for example, have sharp fangs at the front of the mouth for puncturing the flesh of prey and bladelike molars in the rear for slicing meat. Horses, for their part, are equipped with chisel-like front teeth for nipping grass and broad molars for grinding it up (Figure 7-1). Plants, too, exhibit a variety of forms and features that vary with the plants' ways of life. The leaves of most tree species that are native to tropical rain forests, for example, are waxier than those of plants found in cool regions, and a typical tropical leaf terminates in an elongate tip called a drip point.

Megatherium americanum was the largest species of ground sloths, all of which are long extinct. This Pleistocene species, which was about 6 meters (20 feet) long from head to tail, migrated to North America from South America before it died out.

Visual Overview

The Evolution of Life

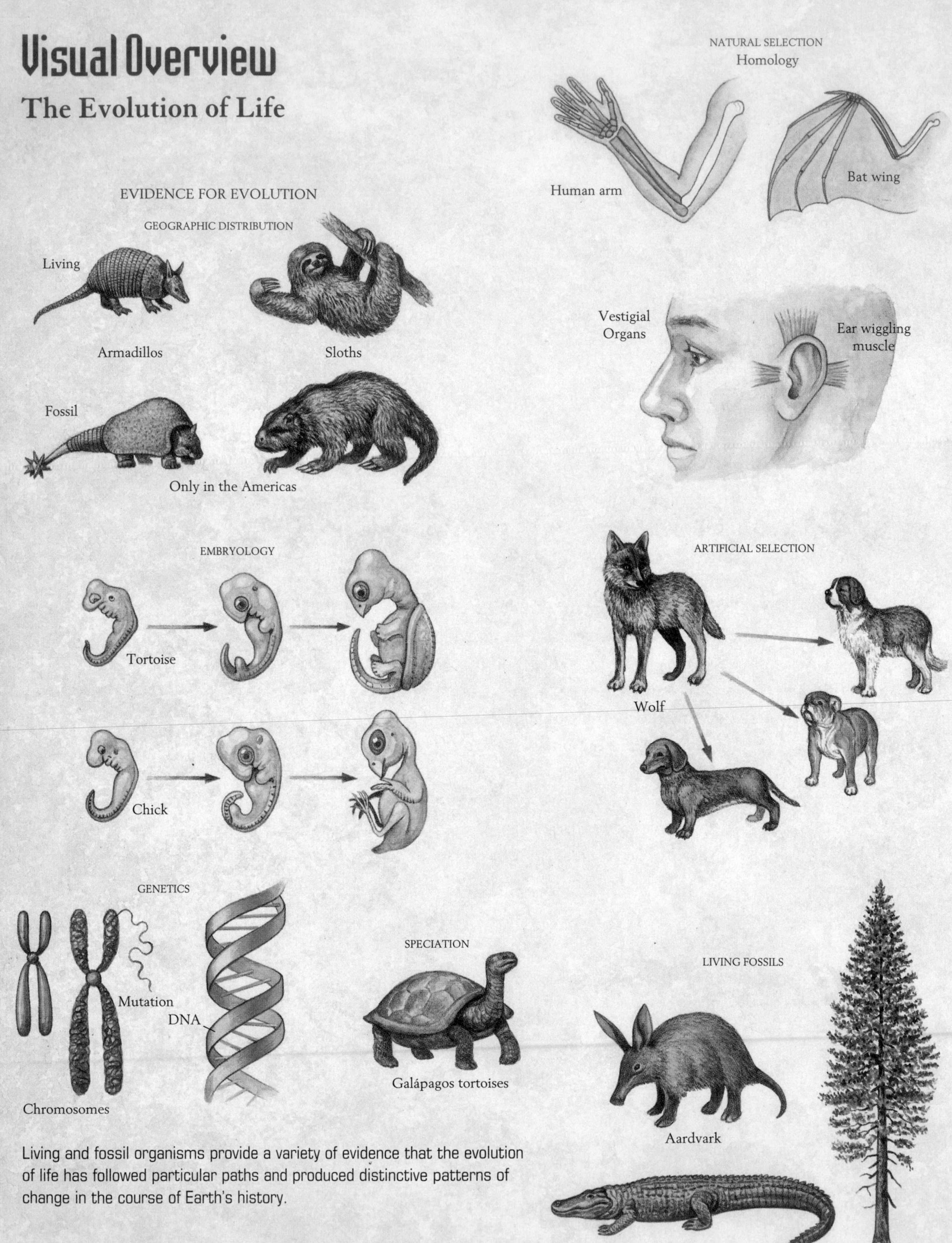

Living and fossil organisms provide a variety of evidence that the evolution of life has followed particular paths and produced distinctive patterns of change in the course of Earth's history.

TRENDS AND PATTERNS
RAPID CHANGE
Axolotl
Salamander
(Adult)
(Larval stage)
GRADUAL CHANGE
Coiled oysters
COPE'S RULE
EVOLUTIONARY RADIATION
Following Appearance of New Habitat
Cichlid fishes
Following Mass Extinction
Mammals
Dinosaurs
Following Adaptive Breakthrough
Early amphibian
Lobe-finned fish
Fins
Legs
CONVERGENCE
Thylacine (marsupial)
Wolf (placental)

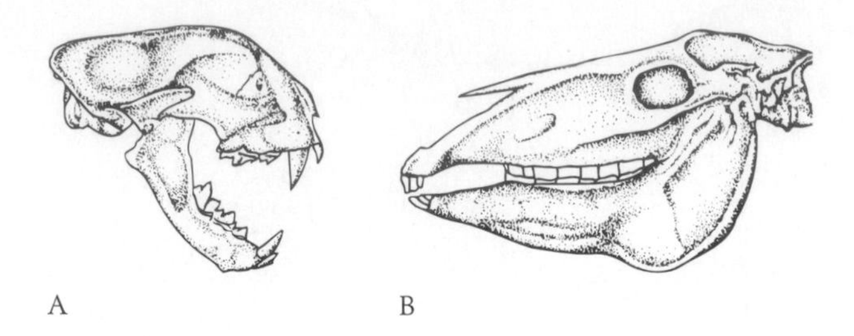

Figure 7-1 The teeth of a cat *(A)* and a horse *(B)*, both of Pliocene age.

The drip point and the waxy surfaces help the leaves shed the rainwater that falls on them daily in a rain forest; the waxy surfaces also keep them from drying out in the tropical heat. In contrast, the leaves of another rain-forest plant, the bromeliad, form a cup that acts as a private reservoir for rain. Without this feature, the bromeliad would dry up and die, because it lives high above the moist forest floor, attached to a tree. Yet another rain-forest plant, the Venus's-flytrap, secretes a sweet nectar that lures insects to the midrib of its leaf (Figure 7-2). On the margins of these leaves are rows of spines that mesh when the leaves snap shut around an unsuspecting insect. In a reversal of the normal roles of plant and animal, the Venus's-flytrap then devours the insect.

These specialized features, which allow animals and plants to perform one or more functions that are useful to them, are known as **adaptations.** Each individual organism possesses many adaptations that function together to equip it for its particular way of life. Before the middle of the nineteenth century, adaptations were not well understood, or even recognized. It was assumed that all features of a species were perfect mechanisms that had been specially designed to allow the species to function optimally within its ecologic niche. Since then it has become widely acknowledged that the features we now recognize as adaptations are fraught with imperfections, many of which stem from evolutionary heritage. An animal or plant, in other words, may develop a useful new feature with which to perform a function, but the evolution of this feature will sometimes be constrained by the structure of the ancestral organism. Evolution can operate only by changing what is already present; it cannot work with the freedom of an engineer who is designing a new device from raw materials. The business of evolution, in other words, is and always has been remodeling rather than new construction.

It becomes obvious that evolution is a remodeling process when we observe that certain organs, such as the cheek teeth of mammals and the leaves of higher plants, serve different functions in different species but nonetheless have a common "ground plan," or fundamental biological architecture. All mammals, for example, possess teeth that are rooted in bone and consist of both dentin and enamel. Similarly, certain types of cells and tissues form the leaves of nearly all flowering plants. Common ground plans suggest common origins, and this is one of the many pieces of evidence that indicate that groups of species of the modern world have a common evolutionary heritage: no matter how greatly the species of a given order or class may differ, they share certain basic features that reflect their common ancestry.

We will begin by reviewing the ideas of Charles Darwin, the man who popularized the idea of evolution. Next we will examine natural selection and discuss current knowledge and theories of how new species come into being. Finally, we will learn more of what the fossil record has revealed about extinction and about rates, trends, and patterns of evolution.

Figure 7-2 A Venus's-flytrap luring a bee onto a gaping pair of leaves, which will snap shut.

Figure 7-3 Charles Darwin in old age.

Charles Darwin's Contribution

Few biologists gave serious consideration to the idea of organic evolution until 1859, when Charles Darwin (Figure 7-3) published his great work, *On the Origin of Species by Natural Selection*. You can appreciate the power of the basic evidence that evolution has occurred by putting yourself in Darwin's position when, in 1831, at the age of 23, he set sail aboard the *Beagle* on the ocean voyage that took him around the world. In the course of this trip Darwin became convinced of the workings of evolution and also accumulated much of the evidence that later enabled him to convince others of the validity of his ideas.

The Voyage of the *Beagle*

Recall from Chapter 1 that Darwin read Charles Lyell's *Principles of Geology* during the voyage and became convinced that the uniformitarian approach to geology was valid.

While Darwin's adherence to the uniformitarian view of Earth's history provided a framework for his acceptance of evolution, it was his observation of the geographic distributions of living things that ultimately led him to theorize that many different forms of life possess a common biological heritage. Darwin was surprised to find South America inhabited by animals that differed substantially from those of Europe, Asia, and Africa. The large flightless birds of South America, for example, were species of rheas (Figure 7-4), which belonged to a different family from the superficially similar birds of other continents—the ostriches of Africa and the emus of Australia. Among other unique South American creatures were the sloths and the armadillos. Not only was South America the home of living representatives of these groups, but it was there that Darwin dug up the fossil remains of extinct giant relatives of the living forms (see p. 180 and Figure 7-5). Why, he asked, were all rhea birds as well as all living and extinct members of the sloth and armadillo families found nowhere but in the Americas?

Darwin was also intrigued to find that species of marine life on the Atlantic side of the Isthmus of

Figure 7-4 The rare flightless bird *Rhea darwinii,* named after Charles Darwin.

A

B

Figure 7-5 Unusual South American mammals. *A.* A living three-toed sloth, hanging upside down in its normal mode of life. This animal is the size of a small dog. *B.* A reconstruction of two of the Pleistocene mammals of South America that Darwin unearthed as fossils. *Megatherium*, the giant ground sloth, was more than 6 meters (~ 20 feet) in length—larger than an elephant. It ranged northward into the United States. The giant armadillo is *Glyptodon*.

Panama differed from those on the Pacific side. In places the isthmus is only a few miles wide, and it struck Darwin as strange that the marine creatures on opposite sides of this narrow neck of land should differ from each other—unless the various species had somehow come into being where they now lived. If the species had instead been scattered over the planet by an external agent, he reasoned, many should have landed on both sides of the isthmus.

Perhaps Darwin's most striking observations concerned life forms on oceanic islands. Darwin noted that no small island situated more than 5000 kilometers (~3000 miles) from a continent or from a larger island was inhabited by frogs, toads, or land mammals unless they had been introduced by human visitors. The only mammals native to such islands were bats, which could originally have flown there. This observation led Darwin to suspect that species could originate only from other species. Otherwise, why would isolated areas of land be left without forms of life prominent elsewhere?

The Galápagos Islands, which lie astride the equator about 1100 kilometers (700 miles) from South America, played an especially large part in the development of Darwin's new ideas. Darwin found the Galápagos to be inhabited by huge tortoises, and he thought it curious that the people who lived on the islands could look at a tortoise shell and immediately identify the island from which it had come. The fact that different species of giant tortoises occupied different islands (Figure 7-6) led Darwin to suspect that these distinctive populations of tortoises had a common ancestry but had somehow become differentiated in form as a result of living separately in different environments.

Even more striking were the various kinds of finches that Darwin found in the Galápagos; some types had slender beaks, others had somewhat stur-

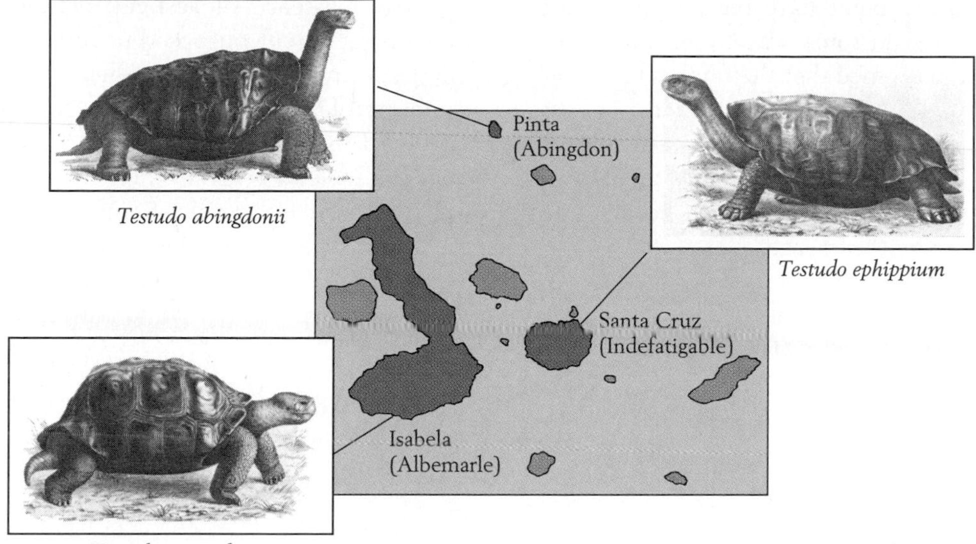

Figure 7-6 Three tortoises, each of which inhabits a different island of the Galápagos. *Testudo abingdonii,* which inhabits Pinta Island, has a long neck and a shell that is elevated in the neck region; these features represent adaptations for reaching tall vegetation. The shells of these animals can exceed 1 meter (3 feet) in length. *(Adapted from T. Dobzhansky et al., Evolution, W. H. Freeman and Company, New York, 1977.)*

dier ones, and still others had very heavy beaks, which served the function of breaking seeds (Figure 7-7). One kind of finch behaved like a woodpecker, using a cactus spine as a woodpecker uses its long beak to probe for insects in wood. Furthermore, all of the finches in the Galápagos resembled a species of finch on the South American mainland (the closest large landmass) rather than the finches found in other regions of the world. Darwin began to ponder whether a population of finches from the South

C

Figure 7-7 Three of the finch species that Darwin observed in the Galápagos Islands. *A.* The large tree finch's parrotlike beak operates like heavy pliers to crush fruits and buds. *B.* The warbler finch's beak operates like needle-nose pliers to catch insects. C. The woodpecker finch, which excavates tree bark with its chisellike beak, uses a cactus needle as a tool to probe for insects. *(A and B. From P. R. Grant, "Natural Selection and Darwin's Finches," Scientific American. Copyright ©1991 by Scientific American, Inc. C. From D. Lack, "Darwin's Finches," Scientific American. Copyright ©1953 by Scientific American, Inc. All rights reserved.)*

American mainland might have reached the islands and become altered in some way to assume a wide variety of forms. It seemed that the finches had somehow differentiated in such a manner that they were able to pursue ways of life that on the mainland were divided among several different families of birds. As Darwin put it in the journal in which he described his scientific work on the voyage:

Seeing this gradation and diversity of structure in one small, intimately related group of birds, one might really fancy that from an original paucity of birds in this Archipelago, one species had been taken and modified for different ends.

Anatomical Evidence

When Darwin returned to England and weighed other evidence indicating that one type of organism evolved from another, he found that certain anatomical relationships seemed to build an especially compelling case. One such piece of evidence was the remarkable similarity of the embryos of all vertebrate animals. Darwin was intrigued by the admission of Louis Agassiz, a noted American scientist, that he could not distinguish an early embryo of a mammal from that of a bird or a reptile. This, Darwin reasoned, was exactly what could be expected if all vertebrate animals had a common ancestry: although adult animals might become modified in shape as they adapted to different ways of life, early embryos were sheltered from the outer world and would thus undergo less change.

Equally convincing to Darwin was the evidence of **homology**—the presence in two different groups of animals or plants of organs that have the same ancestral origin but serve different functions. The principle of homology is illustrated by the variations in teeth and leaves discussed earlier in this chapter. Another example is the common origin of the toes of land-dwelling mammals and the wings of bats. Bats' wings are actually formed of four toes whose external appearance and bone configuration resemble those of walking mammals. If bats' wings did not share the biological origin of the feet of walking mammals, why should the two types of organs have similar bone configurations? Such evidence of common origins abounds in both the animal world and the plant world.

The existence of **vestigial organs**—organs that serve no apparent purpose but resemble organs that do perform functions in other creatures—further supported Darwin's argument in favor of evolution. We humans, for example, retain muscles that other mammals use to prick up their ears in order to catch sounds more effectively. Some people can wiggle their ears slightly with these muscles, but this ability serves no function: in humans the muscles are vestigial structures.

Natural Selection

Darwin also recognized a different type of evidence that pointed to the validity of biological transformation in nature: animal breeders were known to have produced major changes in domesticated animals by means of selective breeding. If wild dogs could be modified into greyhounds, Saint Bernards, and chihuahuas under domestic conditions, Darwin saw no reason why animals should be anatomically straitjacketed in nature. The question was: What could bring about biological changes under natural conditions?

This question led to Darwin's second great contribution. The first, of course, had been his amassing of an enormous amount of evidence indicating that species had evolved in nature. The second was his conception of a mechanism through which evolution was likely to have taken place. The mechanism Darwin proposed was **natural selection**—a process that operates in nature but parallels the artificial selection by which breeders develop new varieties of domestic animals and plants for human use.

Essentially, artificial selection in domestic breeding involves the preservation of certain biological features and the elimination of others. A breeder simply chooses certain individuals of one generation to be the parents of members of the succeeding generation. Darwin recognized that in nature, many more individuals of a species are born than can survive. Accordingly, he reasoned that success or failure here, as in artificial breeding, would not be determined by accident. In nature it would be determined by advantages that certain individuals had over others—greater ability to find food, for example, or avoid enemies, or resist disease, or deal with any of a number of environmental conditions. By virtue of their longevity, these individuals would tend to produce more offspring than others.

Darwin also recognized, however, that survival was not the only factor influencing success in natural selection, because some individuals with only average life spans were capable of producing more total offspring than others simply because they bore large litters or shed large numbers of seeds. Thus, as long as the members of a breeding population varied substantially in either longevity or rate of reproduction, certain individuals would pass on their traits to an unusually large number of members of the next generation. The kinds of individuals that came to predominate as generation followed generation could then be said to be favored by natural selection.

Alfred Russel Wallace, another British scientist who had traveled extensively in far-off lands, conceived of natural selection after Darwin had but before Darwin had published his ideas. If these two men had not brought the concept to light, some other scientist would have done so before long, because the evidence for evolution by natural selection is so vividly displayed in nature.

Genes, DNA, and Chromosomes

Darwin faced a major obstacle in his efforts to convince others that natural selection could operate effectively to produce evolution. Because he lived before the birth of modern genetics, Darwin was not familiar with the mechanisms of inheritance and thus could not explain how an organism could pass along a favorable genetic trait to its offspring. Although the Austrian monk Gregor Mendel had outlined the basic elements of modern genetics only a few years after Darwin published *On the Origin of Species*, Mendel's work was not acknowledged until the turn of the century, two decades after Darwin's death.

Mendel's most significant contribution to modern genetics was the concept of **particulate inheritance,** which explains how certain hereditary factors, which we now call **genes,** retain their identities while being passed on from parents to offspring. Mendel's experiments with pea plants demonstrated that individuals possess genes in pairs, with one gene of each pair coming from each parent. In one of his experiments, Mendel employed a true-breeding white-flowered strain of pea plants and a true-breeding red-flowered strain. (*True breeding* signifies that within the strain, descendants resemble parents throughout a long series of generations.) Mendel's first step was to cross plants of the white-flowered strain with those of the red-flowered strain. The surprising result was that all of the daughter plants had red flowers. When these red plants were crossed with each other, however, they produced both red-flowered and white-flowered descendants.

The most significant aspect of Mendel's work was his discovery that the effect of the gene for white color could surface in the third generation even after its presence had been masked in the second generation. Preservation of genes in this manner—that is, as discrete entities that maintain their identity from generation to generation constitutes the basis of particulate inheritance, the cornerstone of modern genetics.

Another discovery that was made during the emergence of modern genetics was that genes can be altered. It is now understood that genes are, in fact, chemical structures that can undergo chemical changes, and these changes, or **mutations,** provide much of the variability on which natural selection operates. Genes are now known to be segments of long molecules of deoxyribonucleic acid, or **DNA**—a compound that carries chemically coded information from generation to generation, providing instructions for the growth, development, and functioning of organisms.

Changes in one or more of the hundreds of nucleotides that form genes along a strand of DNA are known as point mutations. These mutations can result from imperfect replication of the DNA strand during cell division. They can also occur when an already-existing DNA strand is chemically altered by an external agent, which may be a chemical substance or a dose of radiation such as cosmic radiation or ultraviolet light. In any case, point mutations usually produce changes in the structure of the proteins that are coded by the mutated segments of DNA.

In organisms other than bacteria and their close relatives (see p. 70), DNA is concentrated within **chromosomes,** which are elongate bodies found in the nucleus of the cell. Most organisms have chromosomes that are paired, one having been inherited from each parent (Figure 7-8). Chromosomal mutations take the form of changes in the number of chromosomes or in the positions of segments of individual chromosomes. Such changes cause segments of DNA to move with respect to one another,

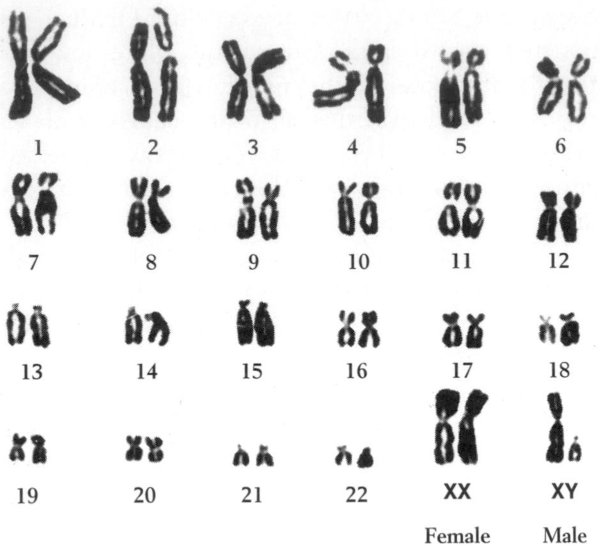

Figure 7-8 **The complete set of human chromosomes.** One member of each of 23 pairs comes from each parent. The presence of two X chromosomes in pair 23 indicates that this set of chromosomes represents a female; the alternative, male condition, is determined by the presence of one X and one Y chromosome in pair 23.

and such a change in relative positions sometimes alters the way one segment affects the operation of another segment. The result can be a change in the way an organism develops and functions.

Populations, Species, and Speciation

When two animals breed, each normally contributes half of its chromosomes to each offspring by way of a **gamete,** a special reproductive cell that contains only one (rather than a pair) of each type of chromosome. The female transmits this set of chromosomes by way of an egg cell, which is the female gamete, and the male by way of a sperm cell, the male gamete. Similarly, the offspring, if it mates, combines half of its chromosomes with half of those of a member of the opposite sex in order to produce still another generation. The mixing that takes place in this manner, which is known as **sexual recombination,** continually yields new combinations of chromosomes and hence of genes. The process of sexual recombination, in conjunction with occasional mutations, is responsible for the variability among organisms that provides the raw material for natural selection. Unfortunately, Darwin and his contemporaries had no knowledge of these sources of variability.

In the study of evolution today, the sum total of genetic components of a population, or group of interbreeding individuals, is referred to as a **gene pool.** And as we have seen (p. 64), populations form a species if their members can interbreed. Reproductive barriers between species keep gene pools separate and thus prevent interbreeding. These barriers include differences in mating behavior or habitat preference, incompatibility of egg and sperm, and failure of offspring to develop into fully functioning adults.

Not only can a species as a whole evolve in the course of time, but it can also give rise to one or more additional species. The origin of a new species from two or more individuals of a preexisting one is called **speciation.** Because species are kept separate from one another by reproductive barriers, speciation by its very definition entails evolutionary change that produces such barriers. It is widely believed that most events of speciation involve the geographic isolation of one population from the remaining populations of the parent species. This isolated population then follows an evolutionary course that causes it to diverge from the parent species in both physical form and way of life. Its divergence may result from such phenomena as the occurrence of unique mutations and the guidance of natural selection by unusual environmental conditions. The development of distinct species of finches on the various Galápagos islands (see Figure 7-7) illustrates this principle.

Extinction

Fossils provide the only direct evidence that life has changed substantially over long spans of geologic time. They also offer the only concrete evidence that millions of species have disappeared from Earth, or suffered **extinction.**

The idea that a species could become extinct was not widely accepted until late in the eighteenth century. Before that time, fossil forms that seemed no longer to inhabit Earth were thought to live in unexplored regions. In 1786, however, Georges Cuvier, a French naturalist, pointed out that fossil mammoths were so large that any living mammoths could not possibly have been overlooked. Cuvier thus concluded that mammoths were extinct. His argument was well received, and soon the extinction of many species was accepted as fact.

In general, extinction results from particularly extreme impacts of the limiting factors that normally hold populations in check. A limiting factor may be predation on a population, disease, competitive interaction with one or more other species, a restrictive condition of the physical environment, or chance fluctuation in the number of individuals in the population (p. 95). Changes resulting from one or more of these factors have led to the extinction of most of the species of animals and plants that have inhabited Earth; in fact, of all the species that have existed in the course of Earth's history, only a tiny fraction remain alive today.

Species have also disappeared by evolving to the point at which they have been formally recognized as different species. In this process, known as **pseudoextinction,** a species' evolutionary line of descent continues, but its members are given a new name. The point at which the new species comes into being is often arbitrarily designated, because there is no way of determining precisely when members of an evolving group lost the ability to interbreed with its original members.

Rates of Origination and Extinction of Taxa

One unique contribution of fossils to biological science is the ability they afford us to assess rates of evolution and extinction. It is only through data derived from the fossil record, for example, that we have been able to measure the rates at which new species, genera, and families have appeared within large groups of animals and plants. We may use the number of species, genera, or families living today as a final number in some calculations, but the geologic record provides essential information about the events that have produced the numbers of these living taxa.

Evolutionary Radiation

At many times in Earth's history, groups of animals or plants have undergone remarkably rapid evolutionary expansion—that is, one or more phyla, classes, orders, or families have produced many new genera or species during brief intervals of time. Rapid expansions of this kind are known as **evolutionary radiations.** The word *radiation* refers to the pattern of expansion from some ancestral adaptive condition to the many new adaptive conditions represented by descendant taxa. Figure 7-9 shows how the fossil record permits us to measure the rate at which evolutionary radiation has taken place. Here we can see that the number of families of corals increased rapidly during the Jurassic Period. The coral families depicted belong to the group known as the hexacorals, which first appeared in Middle Triassic time, about 215 million years ago. Living species of hexacorals form the beautiful coral reefs of the modern world (see Box 4-1).

Evolutionary radiation often occurs in groups of plants or animals within just a few million years of their origin. This pattern is common because the modes of life of recently formed groups often differ from those of groups that originated earlier. Since the old and new groups occupy different niches, ecologic competition does not restrain the diversification of the new group. In addition, when a group first evolves, predatory animals may not yet have developed efficient methods for attacking its members. Until they do, the new group is free to form many new species in a short period of time.

Sometimes the extinction of one biological group has allowed for the evolutionary radiation of another even though the radiating group was not a new one on Earth. Mammals, for example, inhabited Earth during almost all of the Mesozoic Era, but they remained small and relatively inconspicuous until the close of that era, when the dinosaurs suffered extinction. Unrestrained by competition or predation by dinosaurs, mammals then underwent a spectacular radiation. Their rise to dominance on the land has led paleontologists to label the Cenozoic Era the Age of Mammals. Most of the living orders of mammals, including the order that comprises bats and the one that comprises whales, came into existence within only about 12 million years of the start of the Cenozoic Era. This interval represents only about 2 percent of all Phanerozoic time.

Many episodes of evolutionary radiation have been preceded by **adaptive breakthroughs**—the appearance of key features that, along with ecologic opportunities, have allowed the radiation to take place. The rapid growth of the hexacorals' skeletons, for example, has often allowed these creatures to crowd out other animals that inhabit hard surfaces in

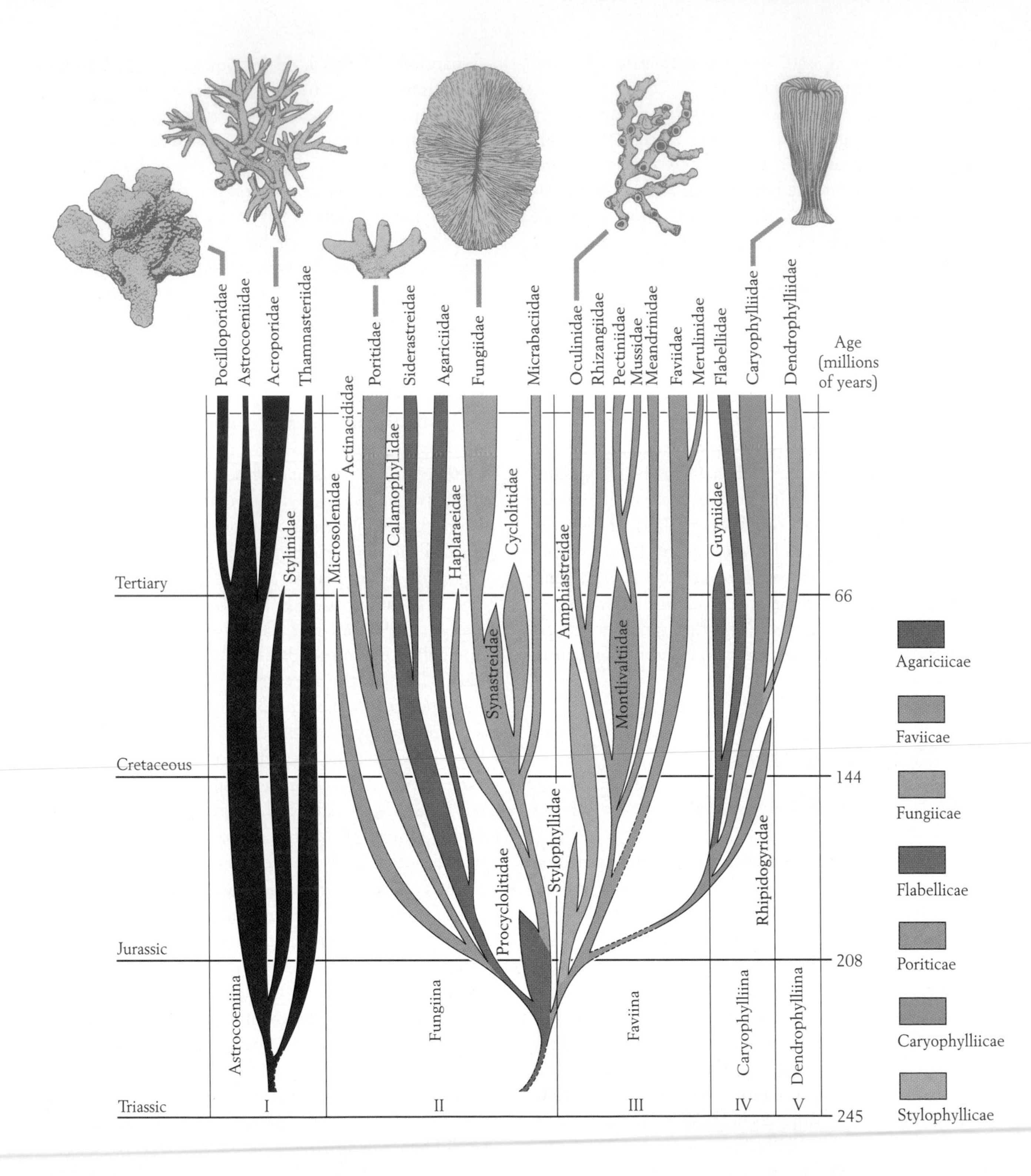

Figure 7-9 The pattern of adaptive radiation of the hexacorals since the group originated in Middle Triassic time. This group builds modern coral reefs. Typical representatives of six families are illustrated. All of these colonies are between about 10 and 20 centimeters (4–8 inches) in maximum dimension. Four of the five orders of hexacorals (roman numerals), and nearly all superfamilies (patterns), were already present by Middle Jurassic time, but families (names printed in color) continued to proliferate during the Cretaceous Period.

shallow marine environments (p. 115). Because their skeletons are porous, hexacorals can quickly assume large proportions without the need for large volumes of calcium carbonate. Thus they have had an edge over slowly growing organisms with which they compete for space on the seafloor. Another adaptive breakthrough for the hexacorals was the development of a symbiotic relationship with algae that live in the tissues of the reef-building species and help the hexacorals develop their skeletons (p. 115).

A variety of adaptive breakthroughs played major roles in the early Cenozoic evolutionary radiation of mammals. The key feature for rodents, for example, was gnawing front teeth, which most of these animals use for eating nuts and other hard seeds. Beavers, however, use them to cut down trees, and recall that one extinct species of beaver used them to gnaw deep burrows in the ground (see Figure 5-2). The development of grinding cheek teeth was a major adaptive breakthrough for horses.

The pattern of evolutionary radiation seen in Figure 7-9 is typical: early expansion produced large-scale evolutionary divergence at a very early stage. Note that four of the five modern orders of hexacorals were already present halfway through the Jurassic Period, shortly after the adaptive radiation of hexacorals began. After the new orders of hexacorals became established, however, evolutionary change was more restricted. New families continued to develop at a high rate, but new orders did not. Later, during the Cenozoic Era, few new families evolved, but divergence continued at the genus and species level. This pattern seems to indicate that as a group of animals or plants begins to expand, it quickly exploits any adaptations that its body plan allows it to develop with ease. Later, however, evolution is restricted to the development of variations on the basic adaptive themes that evolved early on. In time, few new families evolve, and eventually only new genera and species evolve.

Local evolutionary radiations of the recent past offer special insights into larger adaptive radiations that took place earlier. Many of these local events occurred in isolated environments such as islands or lakes, whose well-defined boundaries prevent the escape of the species within them. When these sites of evolution are of recent origin, they can provide evidence of the remarkably rapid diversification of life. This diversification, in turn, usually reflects the fact that the site of adaptive radiation was uninhabited territory and thus lacked the predators and competitors that might have inhibited the evolutionary diversification.

The Galápagos, where Darwin studied the unique groups of tortoises and finches (see Figures 7-6 and 7-7), are the islands most famous as sites of recent evolutionary radiation. The Galápagos originated as a result of volcanic eruptions a few million years ago. Many large lakes have also been sites of radiations in the recent past. Cores taken from the center of Lake Victoria in Uganda have yielded a layer of fossil grass that radiocarbon dating has shown to be only about 13,000 years old. This is the maximum age for the lake, which evidently formed on top of a broad grassland. Lake Victoria now contains about 170 species of fishes of the cichlid group, and all but three of them can be found nowhere else in the world. The brevity of the lake's history indicates that the radiation was spectacularly rapid.

Despite the relative youth of the lake, many of the cichlid fishes have highly distinctive adaptations; some are specialized for eating insects, others for attacking other fishes, and still others for crushing shelled mollusks (Figure 7-10). Several species of cichlids in Lake Victoria look very much like the fishes that seem to have given rise to the great evolutionary radiation that has occurred in the lake. In other words, the original species, or descendants very much like them, remain in the lake along with much more distinctive products of the evolutionary radiation.

In view of such dramatic evolutionary radiations in lakes and on islands, it is interesting to consider what may have happened in the aftermath of major extinctions of the past. When the dinosaurs disappeared from Earth at the end of the Mesozoic Era, for example, the great continents of the world must have been the equivalents of large vacant islands that mammals could colonize on a grander scale than had been possible when dinosaurs ruled. For the mammals, which had remained small and inconspicuous in the shadow of the dinosaurs, the opportunity to undergo evolutionary radiation must have resembled the evolutionary opportunity that was available to the first cichlids that arrived in Lake Victoria or the first finches that landed on a Galápagos island. The entire world, not just a small area, became available for mammals to occupy.

Haplochromis chilotes,
a specialized insect eater (46% of actual size)

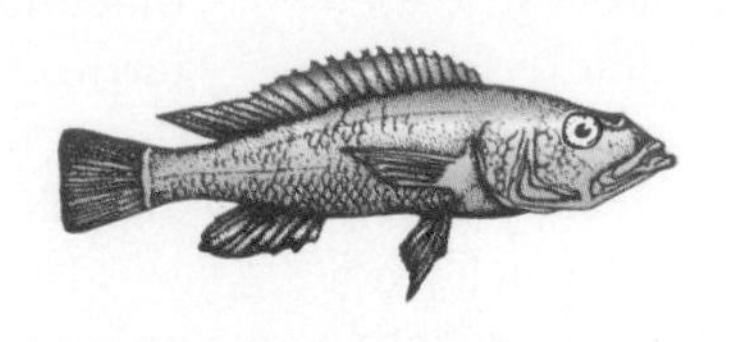

Haplochromis estor,
a fish eater (16% of actual size)

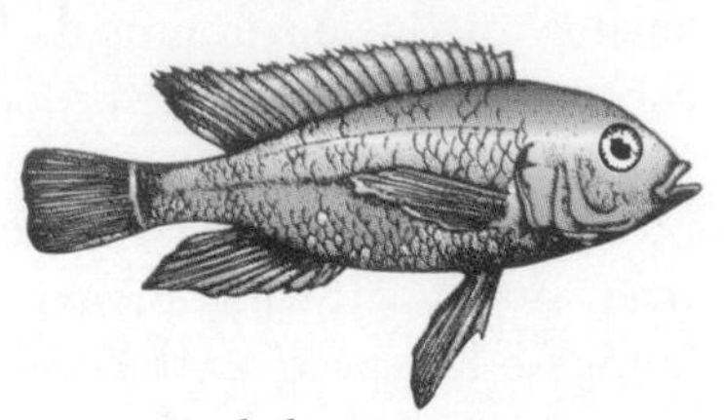

Haplochromis sauvagei,
a mollusk eater (44% of actual size)

Figure 7-10 Three of the more than 170 species of cichlid fishes that evolved in Lake Victoria, Uganda, within the last few hundred thousand years. The fishes exhibit a wide variety of adaptations. *(After P. H. Greenwood, Brit. Mus. [Nat. Hist.] Bull. Suppl. 6, 1974.)*

Thus small-scale evolutionary radiations of the recent past seem to offer useful models for understanding larger evolutionary radiations of the more distant past. The most basic lesson is that when preexisting species do not interfere, a small number of founder species can rapidly produce many new species, some of which differ substantially from the original forms.

Rates of Extinction

Extinction rates have varied greatly among most large groups of animals and plants in the course of geologic time, and they have varied just as greatly from taxon to taxon. An average mammalian species, for example, has survived for just 1 million to 2 million years, which means that the extinction rate for mammals has exceeded 50 percent per million years. In contrast, within many groups of marine life an average species has existed for 10 million years or more; among these groups are bivalve mollusks (clams, scallops, oysters, and their relatives). Under ordinary circumstances, then, only a small fraction of the species within such a group has disappeared every million years.

Groups of animals and plants that are well represented in the fossil record and that have experienced high extinction rates tend to serve well as index or guide fossils (p. 158). The ammonoids, an order of swimming mollusks with coiled shells, meet both requirements (p. 150). As Figure 7-11 indicates, few ammonoid genera found in any stage of the Mesozoic Erathem are also found in the next stage—and yet these genera are usually succeeded by a large number of new ones. Thus the rate at which genera of ammonoids died out was high, but so was the rate at which new genera formed.

Different groups of organisms may exhibit different rates of extinction, but when many groups are viewed together, global trends in rates of extinction become evident. Figure 7-12 shows that the average rate of extinction for animals in the ocean has declined since the Cambrian Period. One explanation is that taxa whose genera have tended to suffer high rates of extinction have tended to disappear altogether, while groups characterized by low rates of extinction have tended to accumulate in the course of the Phanerozoic. These resistant taxa make up a relatively large percentage of the taxa of the modern world.

Figure 7-12 also shows that extinctions of species have sometimes been clustered within brief intervals of time. During several intervals of a few million years or less, large numbers of species and larger taxonomic groups have vanished in **mass extinctions.** The fossil record of species is so incomplete that geologists use fossil genera to tally rates of extinction; even if only a modest percentage of the species that belonged to any genus are known from the fossil record, they can provide a good estimate of the time that genus became extinct. In Figure 7-12 the largest mass extinctions are represented by peaks that rise above the 40 percent level for extinction of genera. There is no convention specifying how severe a crisis must be to constitute a mass extinction, but many relatively small extinction events have been described as mass extinctions—events represented by small peaks (loss of 10 to 20 percent of genera).

Geologists are now studying mass extinctions very intensively. Although few of these biotic crises are well understood, it is clear that they did not all result from a single kind of cause. As noted earlier, it now appears almost certain that the one that swept away the dinosaurs resulted from the impact on

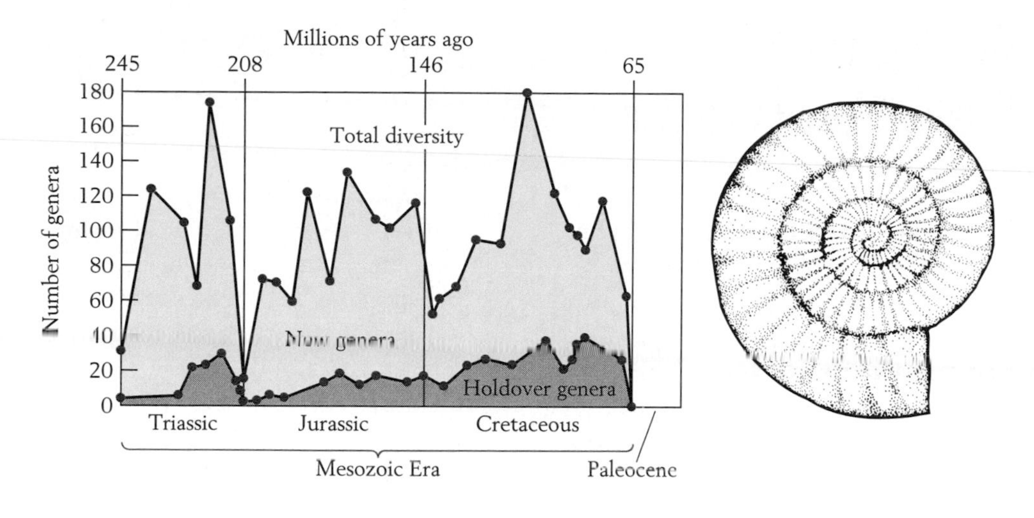

Figure 7-11 The appearance and disappearance of ammonoid genera through time. The ammonoids were relatives of the chambered nautilus. The turnover rate of genera was high throughout the ammonoids' history. Data plotted for the many ages of the Mesozoic Era show that few genera present during one age were still present in the next. *(After W. J. Kennedy, in A. Hallam [ed.], Patterns of Evolution, Elsevier, Amsterdam, 1977.)*

Earth of a large asteroid. As we will see in later chapters, other mass extinctions had earthly causes.

After every mass extinction, the number of species on Earth has increased again. A mass extinction affects some groups of living things more severely than others, and some taxonomic groups that have suffered huge losses have failed to regain their previous diversity. Conversely, some taxa that previously were not very diverse have come to flourish in the aftermath of a biotic crisis.

Unfortunately, it appears today that the world is entering a unique interval of mass extinction: one that we humans are bringing about both by exploiting species and by modifying our environment. As we confront this crisis, lessons from the geologic record may serve us well. Study of past biotic crises gives us insight into the fundamental nature of mass extinction, suggesting how we may modify our behavior in order to reduce the impact of the impending crisis (Box 7-1).

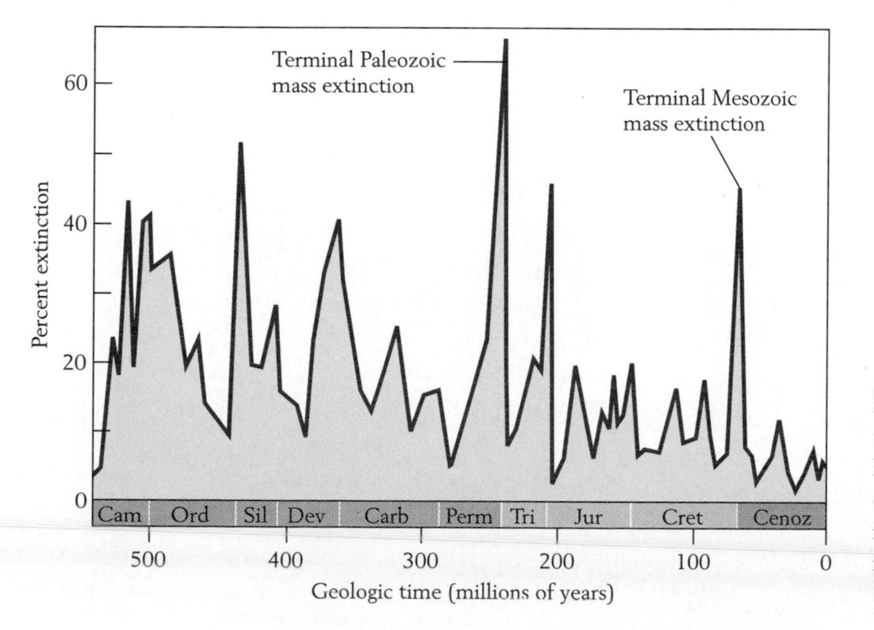

Figure 7-12 Percentages of extinction for genera of marine animals throughout Phanerozoic time. Each point represents data for a single stage of the geologic record. Conspicuous peaks represent mass extinctions. *(After J. J. Sepkoski, Geotimes, March 1994.)*

For the Record 7-1

The Coming Mass Extinction

The great extinctions of the geologic past provide a unique perspective on the extinctions that humans are now precipitating throughout the world. Comparisons indicate that our activities are triggering a mass extinction that will rival any that the world has known in the past 600 million years.

The fossil record alerts us to look for particular patterns in the crisis we are entering. Large animals, for example, have always tended to suffer relatively high rates of extinction, apparently because their typically small populations leave them constantly at risk. The dinosaurs and mammoths spring most readily to mind.

The fossil record also shows that when the environment deteriorates on a global scale, the percentage of species that are lost tends to be especially great in tropical regions. Tropical species are particularly vulnerable because the great diversity of life in the tropics is packed into complex communities in which many species have specialized ecologic requirements and small populations. A large proportion of these species can exist only in association with certain other species, which provide their habitat or their food. Thus when one species goes, others are sure to follow. The patterns seen in the fossil record are already apparent in the modern-day crisis: large-bodied and ecologically specialized species have been disappearing most rapidly. (Recall Box 4-1 and the ecologic fragility of the species of coral that build tropical reefs.)

Rates of extinction have varied throughout geologic time, but calculations suggest that during most intervals, only about one species died out each year on the average. These days several species must be dying out somewhere every day. Experts believe that the rate of extinction may climb to several hundred species a day within 20 or 30 years.

The largest number of extinctions can be traced to our destruction of habitats. Of all our depredations, the destruction of tropical forests—most of them rain forests—has the most dire consequences, for two reasons. First, even though these lush habitats occupy less than 10 percent of Earth's land area, they contain most of the world's species. Second, the total expanse of tropical forests being destroyed every year would make an area about the size of West Virginia; and only about 5 percent of the remaining area is protected in parks and preserves. Because their populations fall below critical levels, many species die out in a shrinking forest long before the forest has disappeared altogether. Rain forests can grow again in areas from which they have disappeared, but recovery requires more than a century. The forests are disappearing so rapidly that even a massive restoration program would be too late to preserve vast numbers of species of great beauty, scientific interest, and possible value to humankind as sources of medical drugs and other useful products. Evolution would require several million years to restore the lost diversity through the natural speciation process. Even then the process would not precisely duplicate any lost species, or even produce species remotely resembling those that belonged to genera or families that had vanished altogether.

Even when we do not attack a habitat with fire, chain saws, and bulldozers, we can still manage to alter it in ways that make it hostile to many forms of life. Acid rain has devastated vast areas of North America and Europe. The acid forms in the atmosphere from oxides of nitrogen and sulfur emitted by automobiles and power plants. Precipitation and runoff carry it to lakes and rivers. The resulting pollution has killed all the fishes in hundreds of lakes in New York State, and in Nova Scotia it has eliminated the Atlantic salmon from numerous rivers where it once came every year to spawn.

Direct exploitation of animals and plants is another cause of extinction. The African black rhinoceros will be lucky to survive hunters catering to people in all parts of the world who cling to the old fantasy that its horn enhances sexual prowess.

After earlier mass extinctions, a few kinds of surviving organisms inherited a world all but free of competitors and predators, and their populations exploded. These were ecologic opportunists—species capable of invading vacant terrain readily and then multiplying rapidly (p. 97). Weeds are not the only opportunists in today's world; so are noisy, aggressive birds such as starlings, the cyanobacteria that form scummy masses in polluted water and then decay to rob their environment of oxygen, and many other forms of life inimical to ours. Thus we not only face the prospect of losing vast numbers of species important to us and many entire ecologic communities in the next few decades; we must also expect their places to be filled by species that impair the quality of human life.

The Siberian tiger, a subspecies, is on the verge of extinction.

Evolutionary Convergence

Convincing evidence that biological form is adaptive is seen in instances of **evolutionary convergence**—that is, the evolution of similar forms in two or more different biological groups. This principle is strikingly illustrated by the similarity between many of the **marsupial mammals** of Australia and the other kinds of mammals that live in similar ways on other continents (Figure 7-13). Marsupial mammals, which carry their immature offspring in a pouch, are the products of a radiation that took place on this isolated island continent during the Cenozoic Era. That this radiation has been adaptive is indicated by the fact that these marsupials have *diverged* from each other but simultaneously have *converged,* both in way of life and in body form, with one or more groups of **placental mammals** living elsewhere. (Nearly all nonmarsupial mammals are placentals.) The strong similarities between many Australian marsupial mammals and mammals of other regions must partially reflect the basic evolutionary limitations of the mammals in general. It would appear, in other words, that certain adaptations are likely to develop under a variety of circumstances, while others are highly unlikely to evolve.

Almost all adaptive radiations, however, have produced some surprises as well. Judging from what we see on other continents, we might have predicted, for example, that hoofed, four-footed herbivores resembling deer, cattle, and antelopes would populate the continent of Australia. As it turns out, the Australian equivalents of these large galloping herbivores are kangaroos—animals that hop around on two legs. Apparently it just happened that the breakthrough represented by the kangaroos' hopping adaptation evolved in Australian herbivorous marsupials before an adaptive breakthrough could produce an efficient running apparatus. Once the kangaroos had occupied many regions of Australia, there was simply little opportunity for animals resembling deer, cattle, or antelopes to evolve.

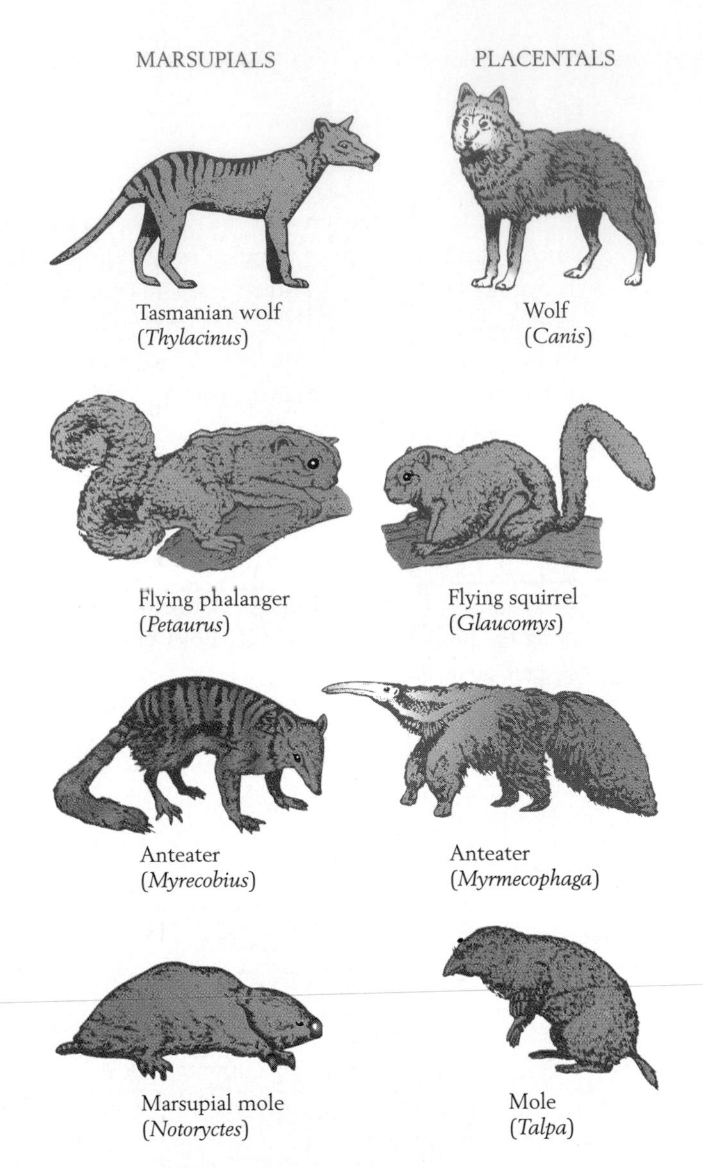

Figure 7-13 Evolutionary convergence between marsupial mammals of Australia and nonmarsupial mammals of other continents. Each of the marsupials is more closely related to a kangaroo than to its counterpart in the other column. *(After G. G. Simpson and W. S. Beck, Life, Harcourt, Brace & World, Inc., New York, 1965.)*

Evolutionary Trends

By examining the evolutionary history of any higher taxon that has left an extensive fossil record, we can observe long-term evolutionary trends—general changes that developed over the course of millions of years. Some of these changes affected form, but others simply affected body size.

Change of Body Size

A general tendency for body size to increase during the evolution of a group of animals is known as **Cope's rule,** after Edward Drinker Cope, a nineteenth-century American paleontologist who observed this phenomenon in his studies of ancient vertebrate animals.

Numerous factors may cause a group of animals to become larger as the group evolves, but all have to do with the tendency of large individuals to produce more offspring than smaller ones. Within species in which males fight for females, for example, larger males tend to win battles and hence to produce a disproportionate number of offspring. Within other species, larger animals may survive to produce more offspring because they are better equipped to obtain food or to avoid predators.

This evolutionary trend cannot continue indefinitely in any animal group, however, because at some point a further increase in body size will inevitably cease to be advantageous. A four-legged animal the size of a large building, for example, could not run or even stand, because its weight would greatly exceed the strength of its limbs. Indeed, many animals could not gather sufficient food or move efficiently if they were appreciably larger than they are.

Given the fact that increases in body size are advantageous only within limits, the great number of animal groups that have evolved toward larger size seems to indicate that most animal orders and families have evolved from relatively small ancestors. Large size tends to impose many adaptive problems for animals, and the specialized adaptations associated with these problems are not easily altered to produce entirely new adaptations. Thus large, highly specialized animals tend to represent evolutionary dead ends.

Some of the problems associated with large size can be seen in the physical adaptations of the elephant. An elephant has such a huge body to feed that it must spend most of its time grinding up coarse food with its molars. The need for constant chewing dictates that an elephant's teeth and jaws must be quite large in relation to its overall size. The elephant's head is therefore large as well—so large that the neck must be quite short to support it (Figure 7-14). To compensate for their consequent short reach, members of the elephant family evolved an enormous trunk from an originally short nose, together with long tusks from originally short teeth. These and other unusual features make it unlikely that modern elephants will ever evolve into very different types of animals.

Manatees (or sea cows), which are blubbery ocean swimmers, are relatives of the elephants (see Figure 7-14). Like elephants, manatees are highly specialized animals with limited potential to give rise to substantially different types of animals. The two groups have common ancestors of early Cenozoic age that were small and rodentlike in their general form and adaptations. These small, relatively unspecialized forms easily evolved in a variety of directions, most of which led to larger animals. Elephants and manatees are among the largest.

Of course, whales are the largest of all mammals; the water that surrounds them supports their massive bodies. The fossil record of whales is not complete enough to document the details of their evolution, but it confirms that whales evolved from four-legged animals (Figure 7-15). Ultimately, whales trace back to mesonychids, which were predatory terrestrial mammals that evolved early in the Cenozoic Era. The earliest whale fossils—jawbones and teeth that closely resemble those of certain mesonychids—come from sediments that were deposited in a shallow river in what is now Pakistan during early Eocene time, about 50 million years ago. From slightly younger marine deposits have come more

Figure 7-14 The elephant and the manatee. These animal groups differ in form but have a common ancestry within the class Mammalia.

complete skeletons, which show that whales had invaded shallow seas. Even these marine forms retained hind legs that they flapped in order to swim but also used for awkward locomotion on land, in the manner of seals (Figure 7-15*B*). Their skeletal structure indicates that they were about 3 meters (10 feet) long. Later in the Eocene, only about ten million years after the earliest whales appeared, more advanced forms existed that had no hind limbs and only small pelvic bones. Some of the most advanced Eocene whales, measuring more than 20 meters, were as long as relatively large modern whales, though they were more serpentlike in form (Figure 7-15*D*).

Figure 7-15 depicts a general trend, showing stages of evolution in the transition from four-legged, land-dwelling mesonychids to fully modern whales. We cannot be sure that any of the whales depicted here actually descended from any of the earlier forms pictured below them—only that they descended from these *kinds* of animals. Now let us examine how fossil data depict more precise trends for some taxonomic groups.

The Structure of Evolutionary Trends

Trends in evolution occur on both small and large scales. A transition from one species to another, for example, represents a simple trend on a small scale. A large-scale trend is one that occurs within a branching limb of the tree of life—a **phylogeny** (p. 65). The change from the oldest known kind of horse to the modern horse represents a large-scale trend in the phylogeny of the horse family (Figure 7-16). The oldest known horse had four toes on each forefoot and three on each hind foot, had relatively simple molars, and was the size of a small dog. The modern horse, in

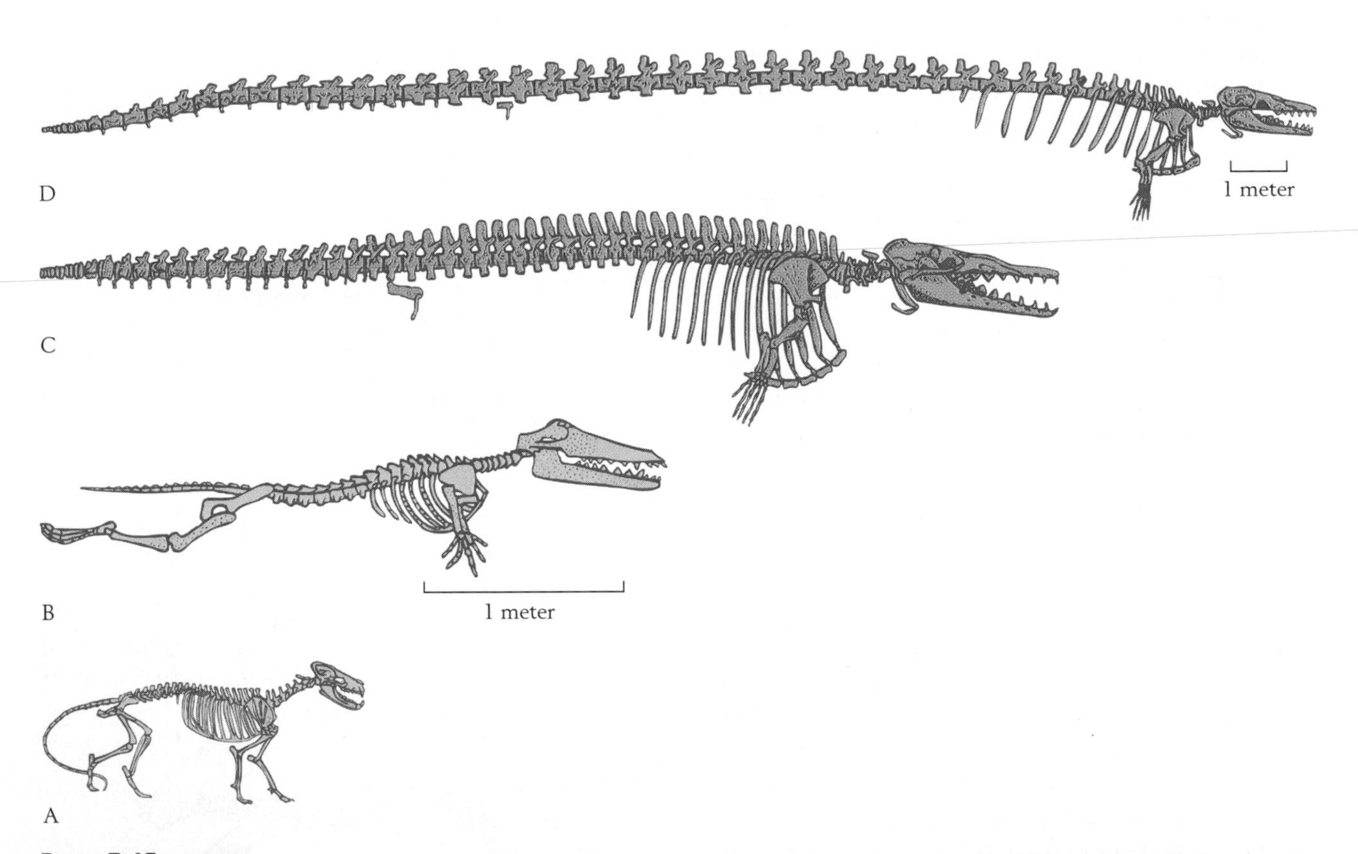

Figure 7-15 Stages in the evolution of whales. *A.* A member of the mesonychid group of terrestrial mammals, from which whales evolved. *B. Ambulocetus*, an early whale that lived about 50 million years ago and probably spent some time on land at the water's edge. C and *D.* Fully aquatic whales that lived about 35 million years ago. (Lower scale applies to *A, B,* and C.) *(A, B, and D. From R. L. Carroll, Vertebrate Paleontology and Evolution, W. H. Freeman and Company, New York, 1987; C. from G. M. Thewissen et al., Science 263:210–212, 1994.)*

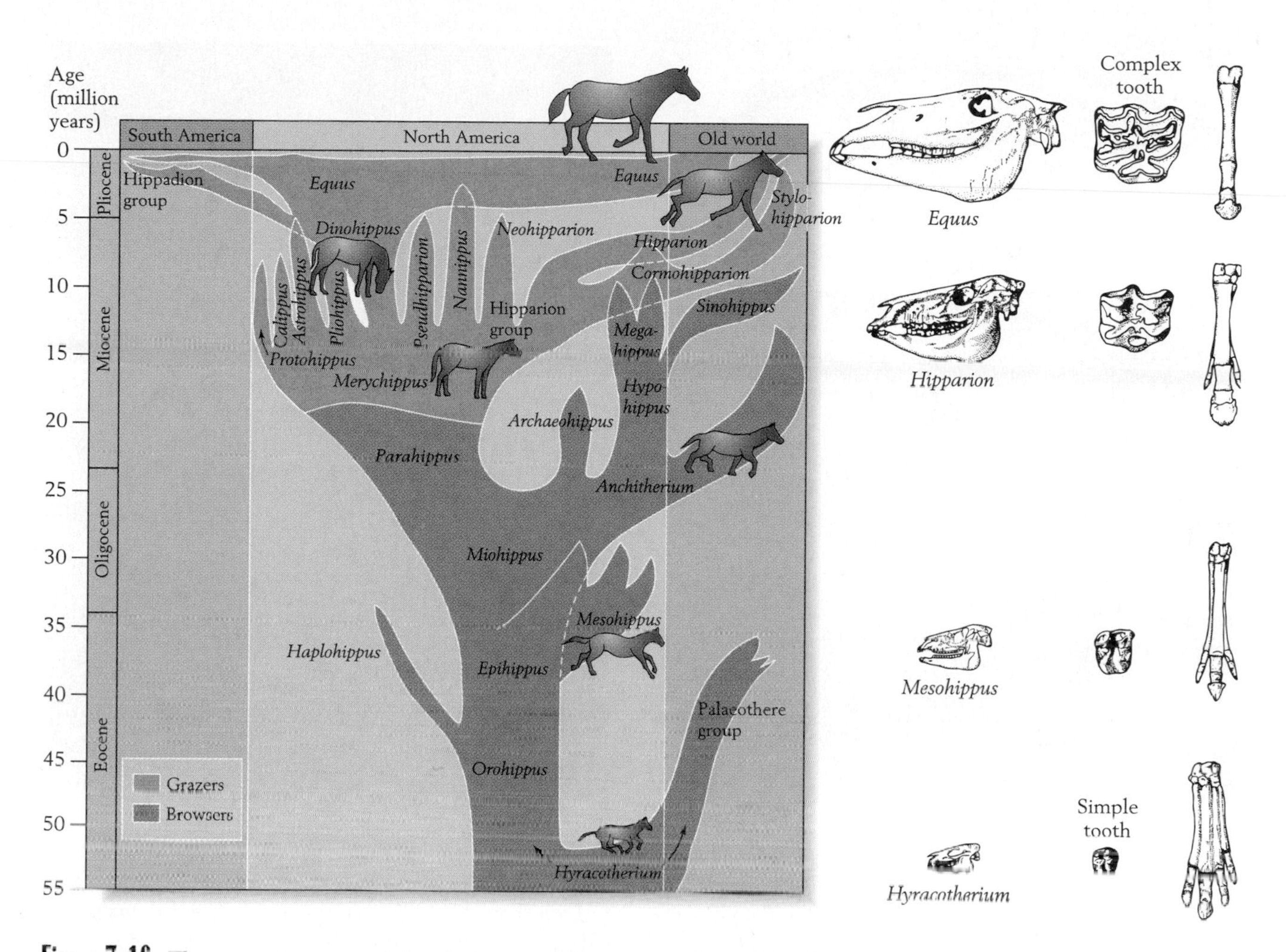

Figure 7-16 The general pattern of the phylogeny of the horse family. The surfaces of the molars of some members developed complex cusps that were associated with a transition from browsing on soft leaves to grazing on tough grasses. The number of toes on the front foot was reduced from four to one. Heads and teeth (but not feet) are drawn to the same scale, to show the general increase in size. *(After B. J. MacFadden, Paleobiology, 11:245–257, with modifications by D. R. Prothero; and G. G. Simpson, Horses, Oxford University Press, New York, 1951.)*

contrast, has a single hoofed toe on each foot, has complex molars, and is a relatively large mammal. Many speciation events separated the ancestral kind of horse from the modern kind. Both the ancestral genus and the modern one have included several species, as have most intermediate genera, and so the transition from one kind to the other represents a complex, large-scale evolutionary trend.

In considering the structure of evolutionary trends, we will focus first on simple trends that cause one species to be transformed into another by a single thread of evolution, and we will then examine complex trends involving many species—trends that consist of many threads of evolution and can thus be viewed as having a fabric.

Some changes from one species to another have occurred in rapid steps of speciation, with the descendant species evolving from the parent species in a relatively short span of time. A probable example is the axolotl, a species whose members remain aquatic throughout life (Figure 7-17). The axolotl evolved from a normal species of salamander, one that is still extant—an amphibian that underwent metamorphosis from an aquatic juvenile to a terrestrial adult form. The axolotl becomes reproductively mature even though it retains the juvenile body form of its

Figure 7-17 **Aquatic and terrestrial amphibians.** *A*. The axolotl, which lives its entire life in fresh water. *B* and C. Two stages in the ontogeny of a typical salamander. The aquatic larval stage *(B)* closely resembles the adult axolotl *(A)*, which in effect never grows up. *(A. After J. Z. Young, Life of Vertebrates, Oxford University Press, London, 1962.)*

ancestors. The evolutionary transition that produced the axolotl was genetically simple. An axolotl that would normally remain aquatic throughout life can be artificially forced to metamorphose into a terrestrial animal if it is injected with thyroxine, a substance normally produced by the thyroid gland but missing in the axolotl. It thus appears that the axolotl was produced by a speciation event consisting of a simple genetic change that impeded the normal development of the thyroid gland. This change probably occurred quite rapidly on a geologic scale of time. The simple genetic change that produced the axolotl probably spread throughout a single population of the ancestral species. In this respect, however, the axolotl appears to be unusual, because most species that form rapidly from a small ancestral population do not evolve by means of just one genetic change; it appears that several such changes are often entailed.

Other species have evolved slowly by the gradual transformation of an entire species—that is, an entire species has changed sufficiently in the course of many generations to be regarded as a new species. Figure 7-18 illustrates this type of evolutionary change in a group of coiled oysters during the Jurassic Period. Recall that the "disappearance" of a species because of evolutionary change is termed pseudoextinction (p. 191).

Like many biologists and paleontologists of the twentieth century, Darwin believed that gradual trends such as that of the coiled Jurassic oysters produced most large-scale evolutionary trends, including those involved in the origin of the modern horse (see Figure 7-16). Recently, however, it has been suggested that gradual trends, such as those evident in the evolution of the Jurassic coiled oysters, are relatively rare. Oysters are bivalve mollusks, and more than 300 species of bivalve mollusks have been identified from Jurassic rocks of Europe—yet very few of these species exhibit gradual trends like the one illustrated in Figure 7-18. In fact, it is estimated that an average bivalve species living in Europe during the Jurassic Period existed without appreciable change for about 15 million years, or for about one-quarter of the entire Jurassic Period.

Many other animal and plant species have also survived for long geologic intervals. Benthic foraminifera, for example, are estimated to have survived for 30 million years, diatoms 25 million years, mosses and their relatives more than 15 million years, seed plant species 6 million years, freshwater fishes 6 million years, beetles more than 2 million years, and mammals about 2 million years. These wide-ranging estimates suggest that most species evolve very slowly; they indicate that an average animal or plant species is not likely to evolve sufficiently to be regarded as a new species even after it has passed through about a million generations. Note that the oyster species shown in Figure 7-18 changed enough to be regarded as a new species within 2 million to 4 million years; even though an individual oyster became reproductively mature in 2 or 3 years, evolution required a million generations or so to produce a new species. All of these longevities must be viewed in the light of the length of time it has taken for new higher taxa of the same group to develop. Recall that early in the Cenozoic Era, whales evolved from small, rodentlike mammals in no more than 12 million years. A typical survival time of 2 million years for a single mammal species seems quite sizable in comparison.

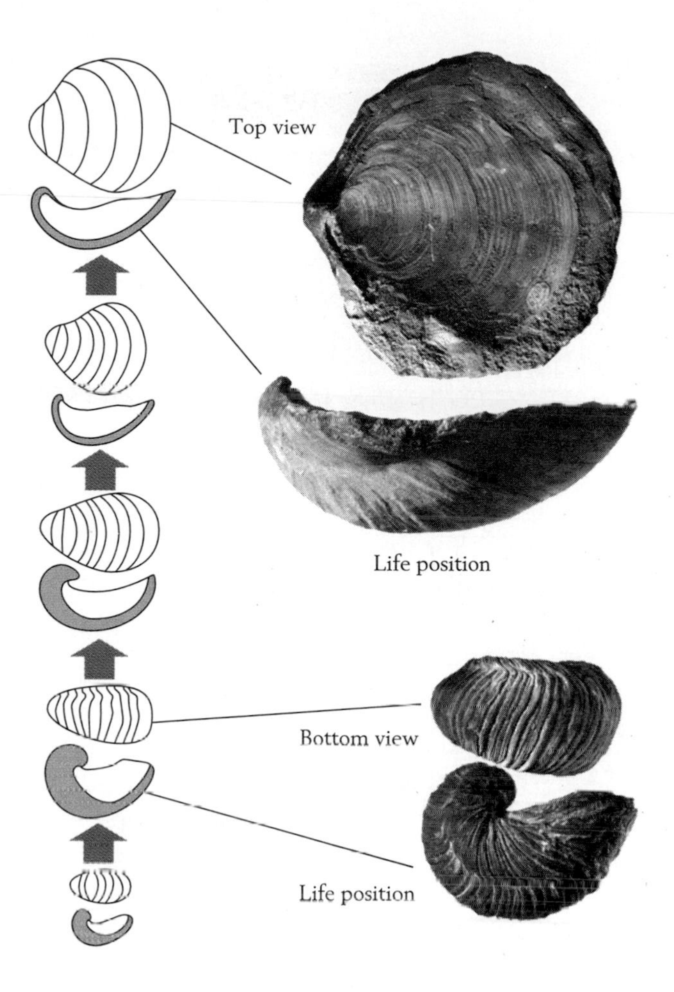

Figure 7-18 **An apparently gradual trend in a lineage of coiled oysters of the genus *Gryphaea* during an Early Jurassic interval of about 12 million years.** During the interval shown here, the shell became larger, attaining the diameter of a small saucer, but it also became thinner and flatter. These animals rested on the seafloor in the orientation labeled "life position." Perhaps the flatter shell was more stable against potentially disruptive water movements. *(After A. Hallam, Philos. Trans. Roy. Soc. London 254B:91–128, 1968.)*

According to the traditional, gradualistic model of evolution, most evolutionary change takes place in small steps within well-established species. Some paleontologists have come to oppose the gradualistic model on the basis of the very slow rates of evolution that have characterized many well-established species. They believe that such rates have been too slow to account for many large evolutionary changes that occurred quite rapidly on a geologic scale of time. These paleontologists conclude that such evolutionary changes must be associated with speciation—that is, with the rapid evolution of new species from others. This is the **punctuational model** of evolution.

Another line of evidence cited in favor of this idea is the evolutionary history that typifies long, narrow segments of phylogeny—segments that undergo little branching but span long intervals of geologic time. If speciation were indeed the site of most evolution, such segments of phylogeny would be expected to exhibit little evolution for the simple reason that they have experienced very little speciation. This is exactly the pattern exhibited by the bowfin fishes (Figure 7-19), a group that has experienced very little speciation and very little evolution during the last 60 million years. The single living bowfin species so closely resembles those of early Cenozoic

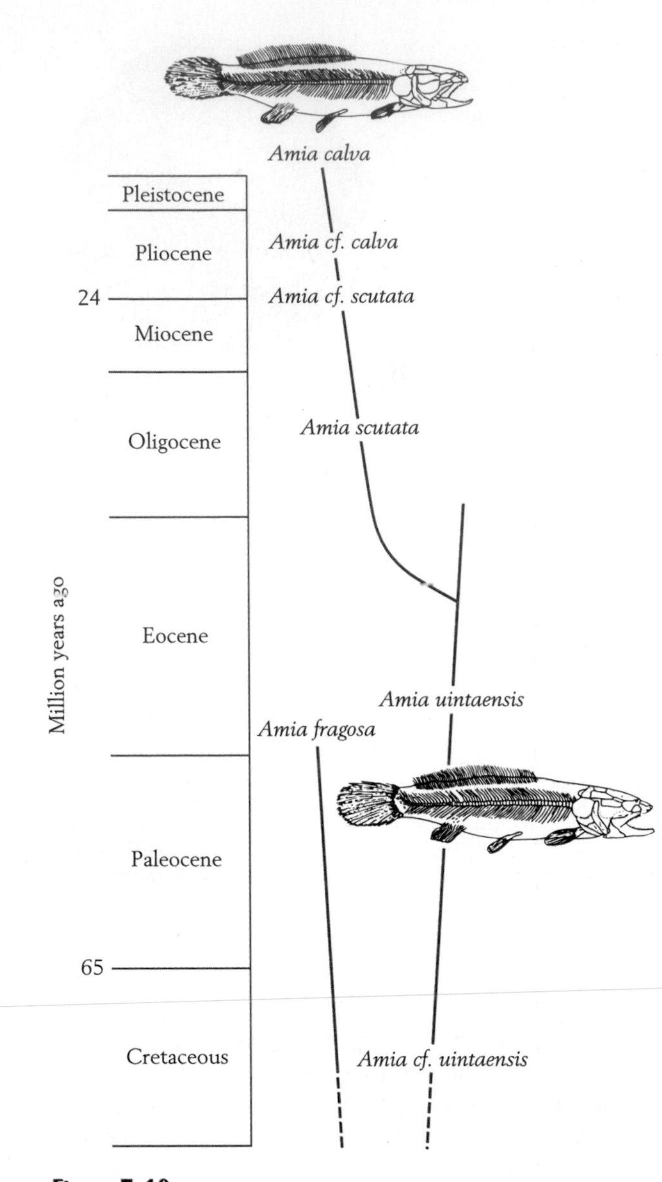

Figure 7-19 Living and fossil bowfin fishes. The living bowfin, *Amia calva*, is the product of a long history in which there has been little speciation (branching) and also little evolutionary change. *Amia* grows to a length of more than a meter (3 feet). *(After J. R. Boreske, Museum of Compar. Zool. Bull. 146:1–87, 1974.)*

time that it has been labeled a living fossil. As it turns out, all of the living species that we know to be at the end of long, narrow segments of phylogeny are living fossils. Among other living fossils are the alligator, the snapping turtle, and the aardvark. A well-publicized living fossil plant is the dawn redwood, which was thought to be extinct until it was discovered living in a small area of China in the 1940s.

Even when individual species have evolved slowly, large-scale trends have sometimes developed through steps of change in particular evolutionary directions that have led to distinctive new species. Trends have also developed when species of one type have survived for long intervals while related species of other types have suffered high rates of extinction. The result has been a shift in the biological composition of the genus or family to which the species have belonged.

The Irreversibility of Evolution

An evolutionary trend that has resulted from at least several genetic changes is highly unlikely to be reversed by subsequent evolution. This principle is called **Dollo's law,** for Louis Dollo, the Belgian paleontologist who proposed it early in the twentieth century. Dollo's law reflects the fact that it is extremely unlikely that a long sequence of genetic changes in a population will be repeated in reverse order. Thus evolution occasionally produces a species of animals or plants that crudely resembles an ancestor, but it never perfectly duplicates a species that has disappeared. In other words, once a species has been changed by evolution or eliminated by extinction, it is gone forever.

Chapter Summary

1 Several lines of evidence convinced Charles Darwin that organic evolution produced the multitudinous species that inhabit the modern world. Among these pieces of evidence were:

- The restriction of many closely related groups of species to discrete geographic regions separated by barriers.
- Embryological similarities between groups of animals that are dissimilar as adults.
- The presence of similar anatomical "ground plans" in animals that live in quite different ways.
- The existence in certain animals of vestigial organs that serve no apparent function but resemble functioning organs in other species.

- The ability of humans to alter domestic animals and cultivated plants by artificial selection in breeding.

2 Natural selection is a process in which certain kinds of individuals become more numerous in a population because they produce an unusually large proportion of the total offspring. They manage to do so either by surviving a long time or by reproducing at a high rate.

3 The variability that is the basis of natural selection is generated by two mechanisms: genetic mutation and the generation of new gene combinations by sexual reproduction.

4 Speciation is the process by which an existing species gives rise to an additional species; it is believed that, in most cases, the population that becomes the new species is geographically isolated from the remainder of the parent species.

5 Extinction is the dying out of a species. The primary agents of extinction are the ecological factors that normally govern the sizes of populations in nature. Pseudoextinction is the disappearance of a species through its evolutionary transformation into another species.

6 Mass extinction is the disappearance of many species during a geologically brief interval of time.

7 Evolutionary radiation is the proliferation of many species from a small ancestral group. Evolutionary radiation has usually followed new access to ecological opportunities.

8 Under normal circumstances, the rates of speciation and extinction among groups of animals and plants vary greatly.

9 One of the most convincing kinds of evidence that evolutionary changes are adaptive is evolutionary convergence, or the evolution within two or more higher taxa of species that resemble one another in form and also live in the same way.

10 Long-term evolutionary trends are evolutionary changes that have developed in the course of millions of years. Such trends can develop in many ways, and the relative importance of these ways depends on what percentage of evolutionary change is associated with rapid speciation events and what percentage is represented by the gradual transformation of well-established species.

Review Questions

1 What geographic patterns suggested to Charles Darwin that certain kinds of species descended from others?

2 What characteristics make a particular kind of individual successful in the process of natural selection?

3 What conditions make it likely that a small group of closely related species will increase to a large number of species by means of rapid speciation?

4 How can evolution proceed by a change in the growth and development of a species?

5 In what ways can an evolutionary trend develop during the history of a genus or a family?

6 What kinds of environmental change can lead to the extinction of a species?

7 What is a mass extinction?

8 How is pseudoextinction related to gradual evolutionary change?

9 Give an example of evolutionary convergence.

10 What general trends has the evolution of horses displayed?

11 Using the Visual Overview on p. 182 and what you have learned in this chapter, compare the kinds of evidence that living organisms on the one hand and fossils on the other hand contribute to our understanding of the evolution of life. What kinds of evidence does each of these two bodies of evidence contribute that the other does not?

Additional Reading

Fox, R. F., *Energy and the Evolution of Life,* W. H. Freeman and Company, New York 1988.

Gould, S. J., *Dinosaurs in a Haystack,* Crown Publishers, New York, 1996.

Grant, P. R., "Natural Selection and Darwin's Finches," *Scientific American,* October 1991.

McKinney, M. L., *Evolution of Life: Processes, Patterns, and Prospects,* Prentice-Hall, Englewood Cliffs, N.J., 1993.

CHAPTER

The Theory of Plate Tectonics

The emergence of the theory of plate tectonics during the 1960s fostered a revolution in the science of geology. **Tectonics** is a term that has long been used to describe movements of Earth's crust. Accordingly, **plate tectonics** refers to the movements of discrete segments of Earth's crust in relation to one another. Whereas continents were once thought to be locked in place by the oceanic crust that surrounds them, the theory of plate tectonics holds that continents move over the surface of Earth because they represent parts of moving plates (see Figure 1-17). Moreover, continents occasionally break apart or, alternatively, fuse together to form larger continents. The theory of plate tectonics explains why most volcanoes and earthquakes occur along curved belts of seafloor, why mountain belts tend to develop along the edges of continents, and why the present ocean basins are very young from a geologic perspective. Most kinds of large-scale rock deformation also result from the movements of plates.

The History of Opinion about Continental Drift

When the concept of plate tectonics emerged quite suddenly in the 1960s, it resolved many long-standing disputes. For many years, the idea that continents move horizontally over Earth's surface, an idea labeled **continental drift,** failed to receive general support in Europe or North America. In 1944 one prominent geologist went so far as to assert that the idea of continental drift

Thingvellir Graben in Iceland. This portion of the Mid-Atlantic Ridge is dramatically exposed above sea level. As the rift separates, lava squeezes upward to form new basaltic crust.

Visual Overview
Elements of Plate Tectonics
Gondwanaland
Lystrosaurus
Mesosaurus
Glossopteris flora
Thrust fault
Trench
Mélange
Thin layer of sediment
Drag of convection
Pull of slab
Earthquakes
Plate tectonics provides a unifying picture of the dynamic features of Earth's lithosphere, accounting for the distributions of plants and animals, volcanoes, earthquakes, midocean ridges, and the three basic kinds of faults.

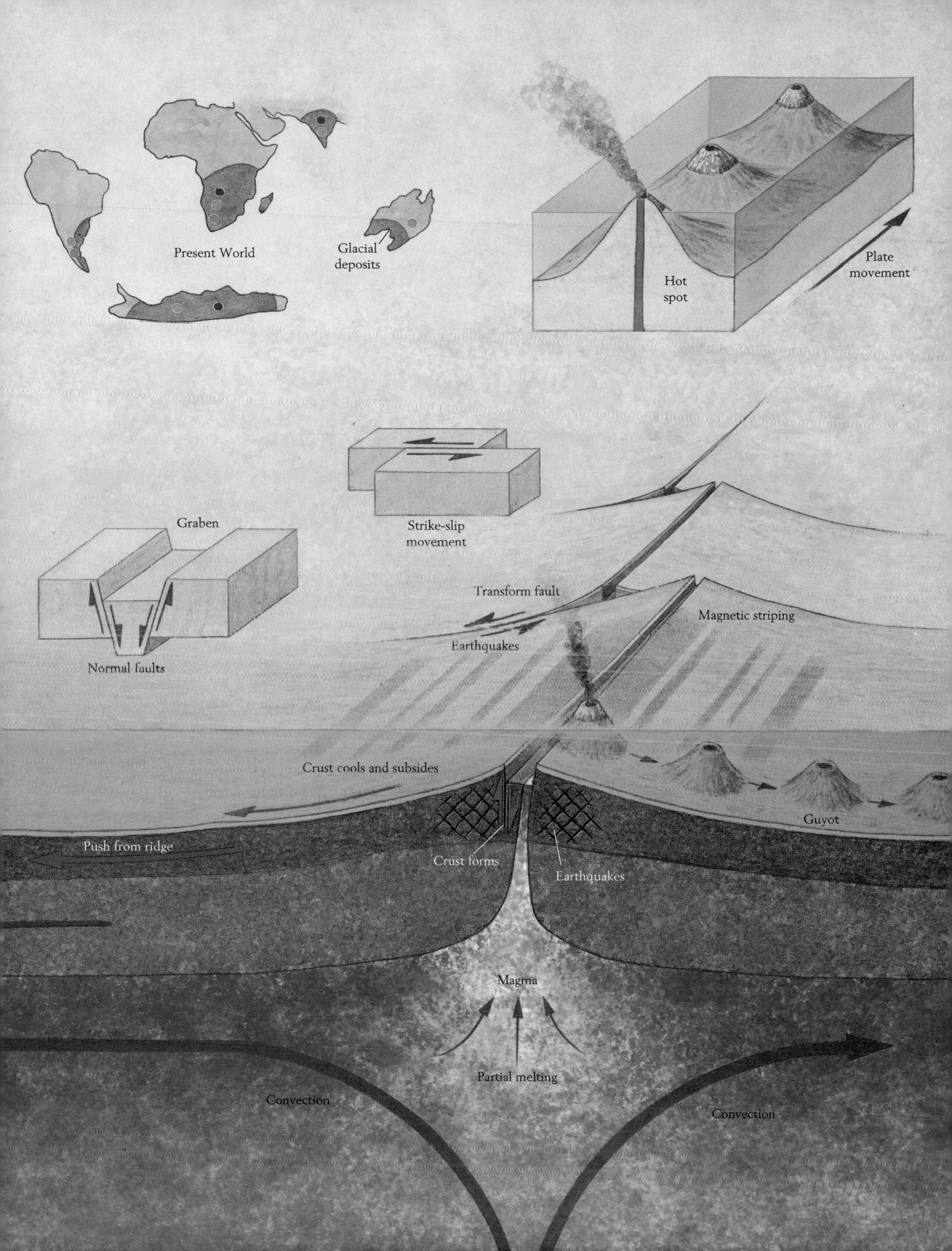
Present World
Glacial
deposits
Hot
spot
Plate
movement
Strike-slip
movement
Graben
Normal faults
Transform fault
Earthquakes
Magnetic striping
Crust cools and subsides
Guyot
Push from ridge
Crust forms
Earthquakes
Magma
Partial melting
Convection
Convection

should be abandoned outright because "further discussion of it merely encumbers the literature and befogs the minds of students." Although many geologists may not have read this comment, most agreed with its spirit, and during the 1950s, little attention was given to the possibility that continental drift was a real phenomenon.

Earlier in the twentieth century, however, when the idea of continental drift first emerged, it had attracted considerable attention, primarily as a result of the arguments of two scientists: Alfred Wegener of Germany and Alexander Du Toit of South Africa. We will briefly examine the case that these two men and their followers made and the reasons their arguments were rejected by most of their contemporaries.

The observation that the coasts on the two sides of the Atlantic Ocean fit together like separated parts of a jigsaw puzzle (Figure 8-1) provided the first evidence that continents might once have broken apart and moved across Earth's surface. Centuries ago, map readers noted with curiosity that the outline of the west coast of Africa seemed to match that of the east coast of South America.

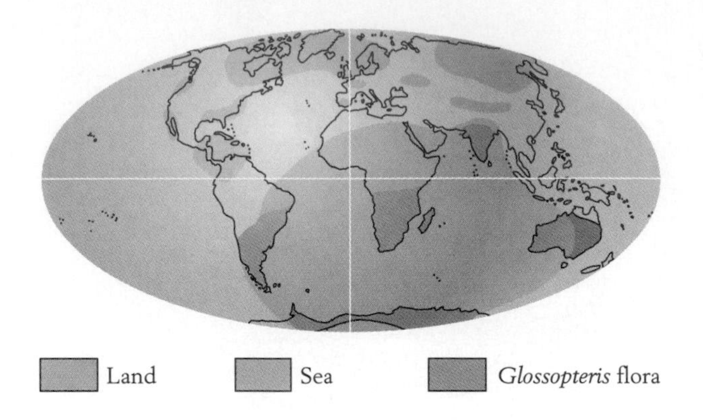

Figure 8-2 Distribution of land and seas during the Carboniferous Period as perceived by Alfred Wegener's predecessors, who believed that large areas of the present-day ocean floor then stood above sea level. The distribution of the *Glossopteris* flora is also consistent with the opposing idea that the Gondwanaland continents were once united.

Figure 8-1 Computer-generated "best-fit" union of continents that now lie on opposite sides of the Atlantic. This fit was calculated by Sir Edward Bullard and his co-workers at the University of Cambridge. The fit was made along the 500-fathom line of each continental slope. Continental shelves are shown in light green. *(After P. M. Hurley, Scientific American, April 1968.)*

Nonetheless, nineteenth-century geologists clung to the idea that large blocks of continental crust could not move over Earth's surface. Thus, when the distribution of certain living and extinct animals and plants began to suggest former connections between landmasses now separated, most geologists tended to assume that great corridors of felsic rock (the most abundant material in continental crust) had once formed land bridges that connected continents but later subsided to form portions of the modern seafloor (Figure 8-2). Today it is recognized that this scenario is not realistic, because felsic crust is of such low density that it cannot possibly sink into the mafic rocks that underlie the oceans. Nonetheless, many prominent geologists of the late nineteenth century presented schemes of Earth history that included the concept of felsic corridors.

One phenomenon that led scientists to speculate about ancient land bridges was the similarity between the fauna of the island of Madagascar and that of India, a land that is separated from Madagascar by nearly 4000 kilometers (2500 miles) of ocean. Madagascar's mammals are primitive; altogether missing are the zebras, lions, leopards, gazelles, apes, rhinoceroses, giraffes, and elephants that inhabit nearby Africa. In contrast, some of the native animals of India closely resemble those of Madagascar. Some scientists consequently favored the idea that a now-sunken land bridge had once spanned the western part of the Indian Ocean, connecting Madagascar to India.

A second line of evidence for ancient land connections was found in the fossil record. During the nineteenth century, late Paleozoic coal deposits of India, South Africa, Australia, and South America were found to contain a group of fossil plants that were collectively designated the *Glossopteris* flora, after their most conspicuous genus, a variety of seed fern (Figure 8-3). After the turn of the century, the *Glossopteris* flora was discovered in Antarctica as well. The presence of this fossil flora on widely separated landmasses was one of the facts that led to the idea that land bridges had once connected all of these continents. The name Gondwanaland came to connote the hypothetical continent that consisted of these landmasses and the land bridges that were believed to have connected them. (Gondwana is a region of India where seams of coal yield fossils of the *Glossopteris* flora.)

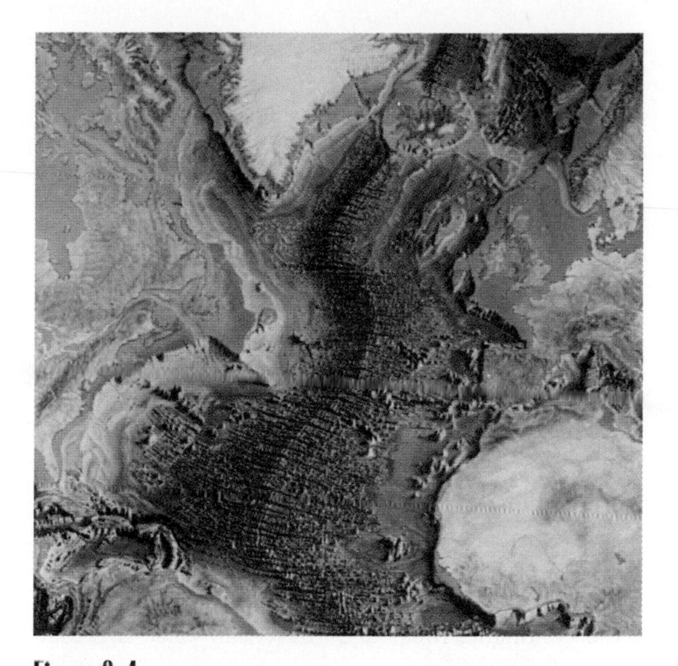

Figure 8-4 Artist's representation of the Mid-Atlantic Ridge in the North Atlantic region. Transform faults are conspicuous features oriented at right angles to the ridge axis; they produce offsets along the ridge.

Figure 8-3 A fossil *Glossopteris* leaf from the Permian of India. The leaf is about 12 centimeters (5 inches) long.

Not until early in the twentieth century was it hypothesized that the continents had once lain side by side as components of very large landmasses that eventually broke apart and moved across Earth's surface to their present positions. This general idea is central to modern plate-tectonic theory.

One early surmise, now generally accepted, was that the immense submarine mountain chain called the Mid-Atlantic Ridge, which is one of several ridges of the present-day seafloor, marks the line along which one ancient landmass ruptured to form the Atlantic Ocean (Figure 8-4).

A Twentieth-Century Pioneer: Alfred Wegener

In 1915 Alfred Wegener, a German meteorologist, presented evidence that virtually all of the large continental areas of the modern world were united late in the Paleozoic Era as a single supercontinent, which he labeled Pangea. Wegener reasoned that Pangea had broken apart and that the fragments had drifted about. He cited the great rift valleys of Africa as possible newly forming or failed rifts. Wegener's insight was again correct. As we will see, the African rift valleys are now regarded as instances of continental rifting at an early stage.

Wegener supported his theory with several additional arguments. He noted numerous geologic similarities between eastern South America and western Africa, for example, and he also called attention to the many similarities between the fossil biotas of these two widely separated continents. Several extinct groups of animals and plants, in addition to members of the *Glossopteris* flora, had been found in the fossil records of two or more Gondwanaland continents, and this discovery led Wegener to argue that these continents must once have lain close together (Figure 8-5).

Today twenty of the twenty-seven species of fossil land plants recognized within the *Glossopteris* flora of Antarctica have been found as far away as India. It might be suggested that winds could have spread the plants this far by carrying their seeds, but in fact the seeds of the genus *Glossopteris* are several millimeters in diameter—much too large to have been blown across wide oceans. Today, after the development of plate tectonic theory, professional geologists accept the former existence of Pangea—although, as we will see, they find major errors in Wegener's proposed chronology for the fragmentation of this supercontinent (see Figure 8-5).

Alexander Du Toit and the Gondwanaland Sequence

Wegener's arguments were more fully developed by the South African geologist Alexander Du Toit. Du Toit and others introduced a wealth of circumstantial evidence in support of the idea of continental drift—evidence that was publicized both before Wegener's death in 1930 and during the three decades of controversy that followed. Du Toit noted, for example, that fossils of the small reptile *Mesosaurus* (Figure 8-6) occurred at or near the position of the Carboniferous-Permian boundary in both Brazil and South Africa. On both continents, fossils of *Mesosaurus* occur in dark shales along with fossil insects and crustaceans. *Mesosaurus* occupied freshwater and perhaps brackish habitats, and most paleontologists found it difficult to imagine that the animal had somehow made its way across an ocean as broad as the present Atlantic and had then found freshwater depositional settings that were nearly identical to its former habitat.

Even *living* animal and plant groups were shown to exhibit a "Gondwanaland" pattern: a number of individual species and genera were found to be distributed among the southern continents. One genus of earthworm, for example, was found to live only at the southern tips of South America and Africa, which lay close together in Wegener's Gondwanaland reconstruction. Another genus was encountered only in southern India and southern Australia.

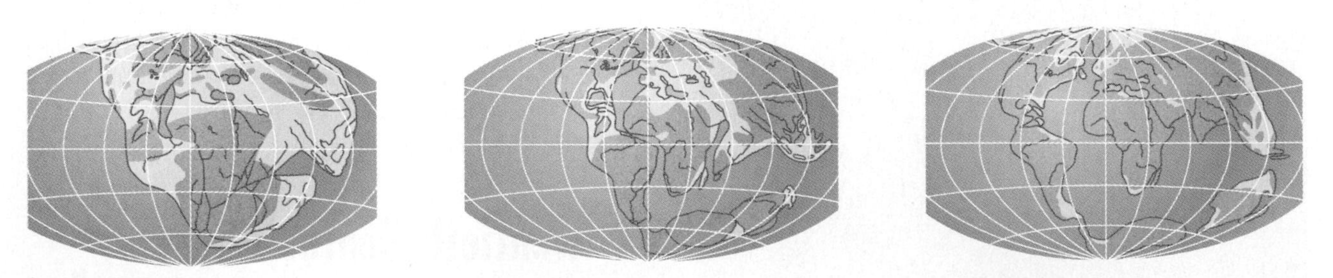

Figure 8-5 Alfred Wegener's reconstruction of the map of the world for three past times. Africa is placed in its present position as a point of reference. Heavy shading represents shallow seas. Wegener erred in suggesting that Pangea, the supercontinent shown in the left-hand map, did not break apart until the Cenozoic Era (center and right-hand maps). *(After A. Wegener, Die Entstehung der Kontinents und Ozeane, Friedrich Vieweg und Sohn, Brunswick, Germany, 1915.)*

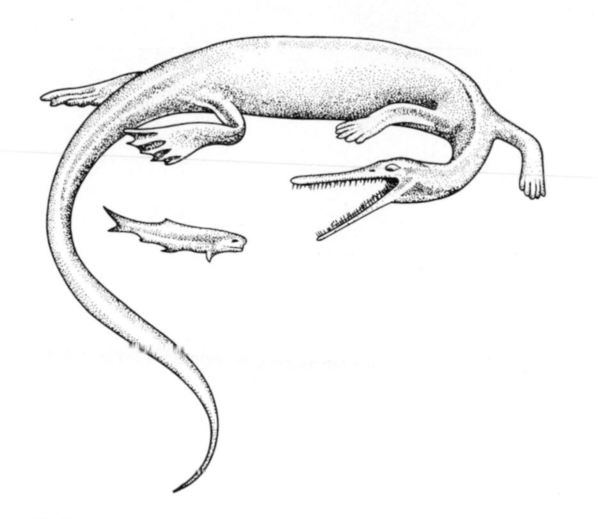

Figure 8-6 ***Mesosaurus,*** **a small Early Permian reptile found in freshwater deposits in both South Africa and southern Brazil.** The animal was about 0.6 meter (2 feet) long. *(After H. F. Osborn, in A. S. Romer, Proc. Amer. Philos. Soc., 1968, pp. 335–343.)*

The general stratigraphic context of the *Glossopteris* flora and *Mesosaurus* offered further support for the existence of Gondwanaland. Specifically, Carboniferous and Permian rock units that yield the *Glossopteris* flora form what is known as the Gondwana sequence, which occurs with remarkable similarity in South America, South Africa, India, and Antarctica.

The Gondwana sequence of Brazil bears an uncanny resemblance to a sequence representing the same geologic interval in South Africa. At the bases of both sequences (Figure 8-7) are glacial tillites that are coarsest at the base and alternate with interglacial sediments, including coals that yield members of the *Glossopteris* flora. As in South Africa, *Mesosaurus* is found near the base of the Permian in dark shales. Much of the Triassic record consists of dune deposits, which, like similar dune deposits of South Africa, are succeeded by Jurassic lava flows. Similar Gondwana sequences occur in Antarctica and India.

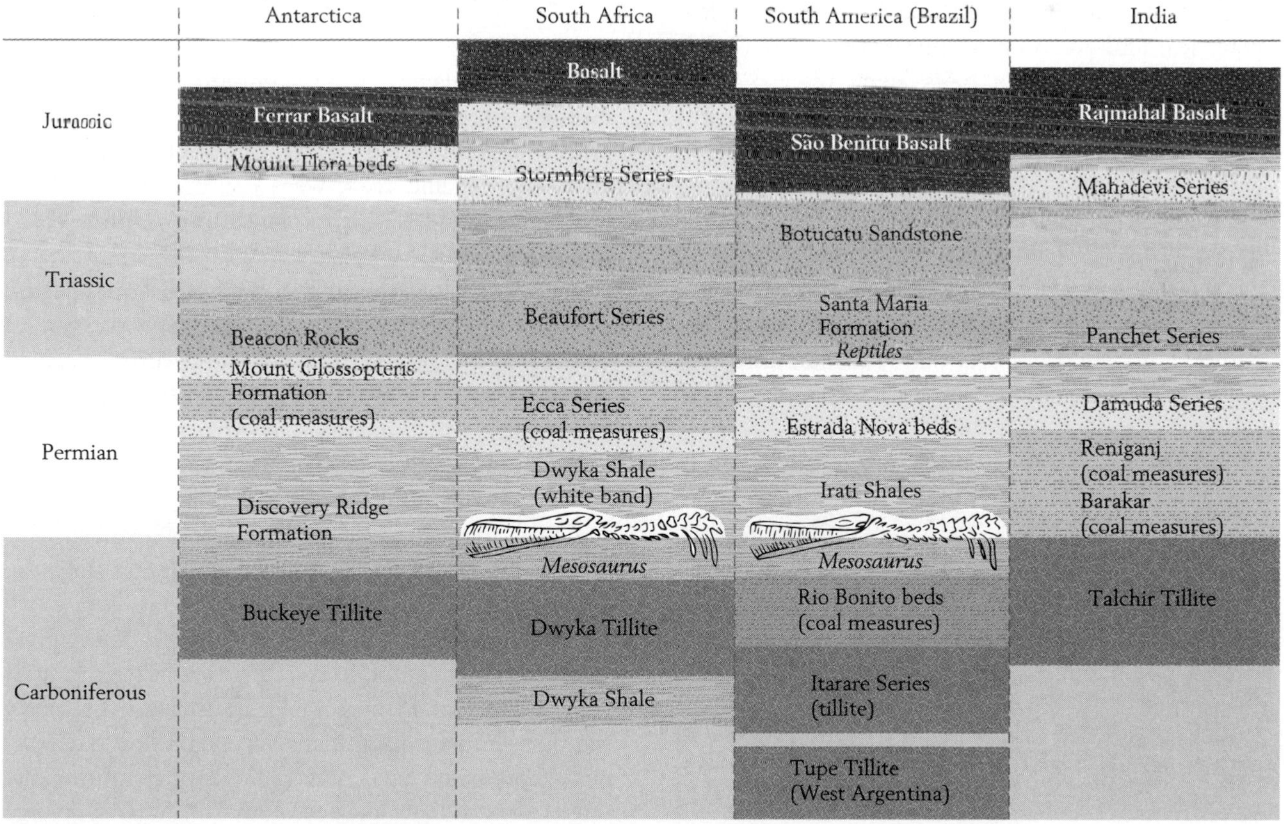

	Antarctica	South Africa	South America (Brazil)	India
Jurassic	Ferrar Basalt Mount Flora beds	Basalt Stormberg Series	São Benitu Basalt	Rajmahal Basalt Mahadevi Series
Triassic	Beacon Rocks	Beaufort Series	Botucatu Sandstone Santa Maria Formation *Reptiles*	Panchet Series
Permian	Mount Glossopteris Formation (coal measures) Discovery Ridge Formation	Ecca Series (coal measures) Dwyka Shale (white band) *Mesosaurus*	Estrada Nova beds Irati Shales *Mesosaurus*	Damuda Series Reniganj (coal measures) Barakar (coal measures)
Carboniferous	Buckeye Tillite	Dwyka Tillite Dwyka Shale	Rio Bonito beds (coal measures) Itarare Series (tillite) Tupe Tillite (West Argentina)	Talchir Tillite

Figure 8-7 **Correlation of the stratigraphic sequences of four continents.** In each sequence, glacial tillites are followed by shales and coal beds containing the *Glossopteris* flora. *(Modified from G. A. Duamani and W. E. Long, Scientific American, September 1962.)*

When Du Toit and other followers of Wegener measured the orientations of features scoured into underlying bedrock by glaciers, they found telltale patterns (Figure 8-8). The glacial movement in eastern South America, for example, had been primarily from the southeast, where today no landmass exists that might support large glaciers. Likewise, in southern Australia there was evidence of glacial flow from the south, where there is now only ocean. Obviously, it would not be at all difficult to account for such movement if the continents had been united at the time the glaciers were flowing. Ice flow would then have radiated from the center of a large continent that could have supported large glaciers under cold climatic conditions.

Du Toit correctly deduced from geologic evidence that Pangea did not form until late in the Paleozoic Era. Before Pangea was formed, Gondwanaland existed as a distinct supercontinent, and the northern continents were united as a second supercontinent called Laurasia.

Du Toit also recognized that if South America, Antarctica, and Australia were assembled as Gondwanaland, the mountain belts along their margins would line up, as would regional trends of rock deformation (Figure 8-9).

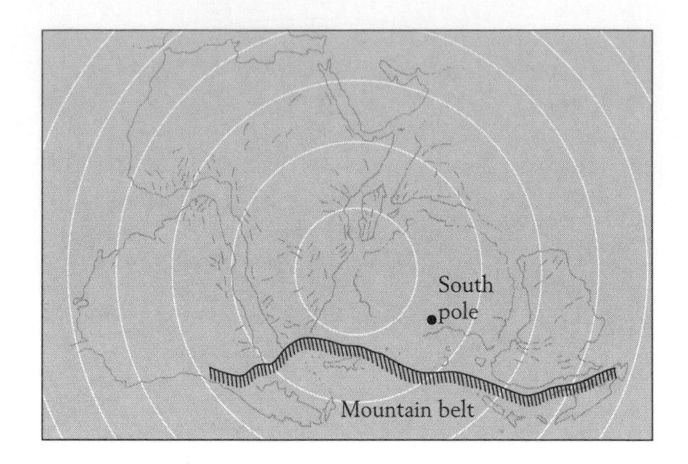

Figure 8-9 Alexander Du Toit's reconstruction of Gondwanaland. The short lines show that regional patterns of faulting and other kinds of rock deformation align well when the continents are assembled to form Gondwanaland. The Andes mountain chain of South America aligns with mountain systems of South Africa, Antarctica, and Australia. *(After A. L. Du Toit, The Geology of South Africa, Oliver & Boyd Ltd., Edinburgh, 1937.)*

The Rejection of Continental Drift

Despite mounting evidence supporting Wegener's and Du Toit's ideas, geologists of the northern hemisphere continued to view the theory of continental drift with considerable skepticism. The primary source of their dissatisfaction lay in the apparent absence of a mechanism by which continents could move over long distances. Geophysicists knew that both continental crust and oceanic crust were continuous above the Mohorovičić (or Moho) discontinuity, so they could not imagine how continents could be made to move laterally—to plow through oceanic crust (see Figure 1-15). In addition, some fossil evidence seemed to contradict the notion of continental drift. Wegener had suggested a brief timetable for drift: he proposed that Pangea, which incorporated virtually all the modern continents, had survived into the Cenozoic Era (see Figure 8-5). But when paleontologists looked for evidence that the world's biotas had evolved into increasingly distinctive geographic groupings since the start of the Cenozoic Era, they found none.

We now know that Wegener made a dating error that misled paleontologists of his time. The rifting of Pangea had actually begun near the start of the Mesozoic Era—much earlier than Wegener believed. Continents could not have moved far enough during the brief Cenozoic Era to have allowed biotas to diverge greatly; instead, continents have been relatively widely dispersed since the very beginning of this era.

Burdened by so many apparent problems, the idea of continental drift remained highly unpopular in the United States and Europe for decades.

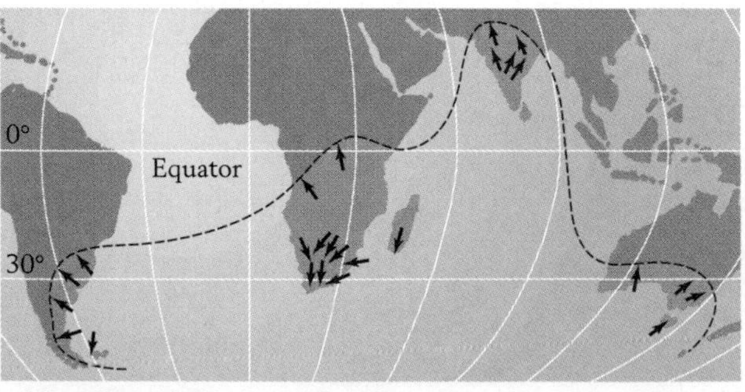

Figure 8-8 Locations of late Paleozoic glaciation and the directions in which glaciers flowed. *(After A. Holmes, Principles of Physical Geology, Ronald Press Company, New York, 1965.)*

Ironically, after new data supporting plate tectonics had finally brought continental drift into favor, an exciting fossil find was made in Antarctica. This was the discovery in 1969 of the genus *Lystrosaurus*, an animal classified as a member of the therapsids, the group that gave rise to mammals (p. 68). *Lystrosaurus* was a heavyset herbivorous animal with beaklike jaws (Figure 8-10). Antarctica is so distant from other fragments of Gondwanaland that discovery of *Lystrosaurus* fossils there a decade earlier might have revived enthusiasm for continental drift before the advent of plate tectonic theory.

The Puzzle of Paleomagnetism

Interest in continental movements was renewed in the late 1950s as a result of new evidence derived from **paleomagnetism,** or the magnetization of ancient rocks at the time of their formation. We have already seen that Earth's magnetic field has reversed its polarity on many occasions. During the 1950s, geophysicists attempted to ascertain whether the north and south magnetic poles not only reversed their positions but also wandered about periodically. To explore this possibility, these researchers attempted to determine the previous positions of the magnetic poles by using magnetized rocks as compasses for the past.

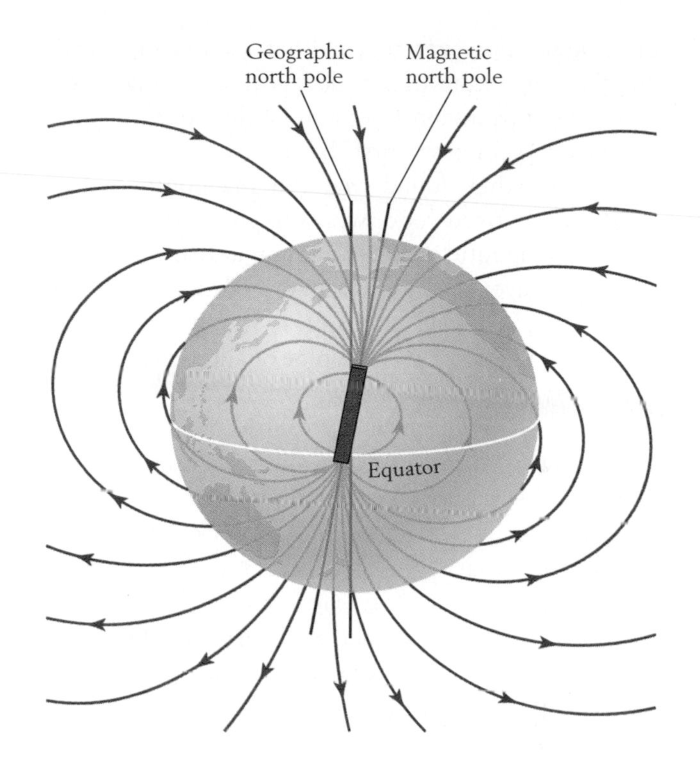

Figure 8-11 **The structure of Earth's magnetic field.** The core has north and south poles and thus behaves like a bar magnet. At the present time, the north–south axis has a declination of 15 degrees from Earth's north–south geographic axis. Curved lines represent magnetic lines of force. These lines of force intersect Earth's surface at high angles near the poles and low angles near the equator.

Figure 8-10 ***Lystrosaurus*, the mammallike reptile now known from Antarctica as well as from Africa and southeast Asia.** *Lystrosaurus* was a herbivorous animal about a meter (3 feet) long, with short legs, beaklike jaws, and a pair of short tusks. *(Painting by Mark Hallett.)*

As we learned in Chapter 6, a magnetic field frozen into a rock is similar to the magnetic field that is "read" by a compass. The angle that a compass needle makes with the line running to the geographic north pole is called the **declination.** Today the magnetic pole lies about 15 degrees from the geographic pole, so the declination is small (Figure 8-11). A compass needle not only points at the north magnetic pole but, if allowed to tilt in a vertical plane, also dips at a particular angle. As Figure 8-11 shows, the dip of a compass needle varies with the distance of the compass from the magnetic pole. The dip is lowest near the equator, where the lines of force of Earth's magnetic field intersect Earth's surface at a low angle.

Paleomagnetism in a rock also has both a declination and a dip, which indicate the apparent direction of the north magnetic pole at the time when the rock was first magnetized, as well as the distance between the rock and the pole. It is important to understand, however, that neither a compass needle nor the magnetism of a rock reveals anything about longitude (position in an east–west direction).

When geologists first began to measure rock magnetism, they found that the magnetism of recently magnetized rocks was consistent with Earth's current magnetic field. The magnetism in older rocks, however, had different orientations. As data accumulated, it began to appear that Earth's magnetic pole had wandered. A plot of the pole's apparent positions, indicated by rocks of various ages in North America and in Europe, showed that the pole seemed to have moved to its present position from much farther south, in the Pacific Ocean. However, the path obtained from European rocks differed in detail from that obtained from North American rocks (Figure 8-12*A*). It was recognized that this pattern might actually reflect a history in which the north pole did not wander at all; instead, as Wegener had suggested, the continents of Europe and North America might have moved in relation to the pole and to one another, carrying with them rocks that had been magnetized when the continents were in different positions. This possibility led to the use of the cautious term *apparent polar wander* to describe the pathways that geophysicists plotted.

Tests were conducted to examine the possibility that continents rather than poles had moved. It was hypothesized, for example, that if North America and Europe had once been united and had drifted over Earth's surface together, they should have developed identical paths of apparent polar wander during their joint voyages. The test, then, was to fit the outlines of North America and Europe together along the Mid-Atlantic Ridge to determine whether, with the continents in this position, their paths of apparent polar wander coincided. As Figure 8-12*B* shows, the apparent polar-wander paths of North America and Europe did coincide almost exactly in both Paleozoic and early Mesozoic time. This evidence strongly suggested that the continents had indeed drifted apart, carrying their magnetized rocks with them.

The Rise of Plate Tectonics

During the late 1950s, these new paleomagnetic data generated widespread discussion of continental drift, but most geologists continued to doubt its validity. There were two reasons for this continuing skepticism: first, many paleomagnetic measurements were imprecise; and second, the belief persisted that no

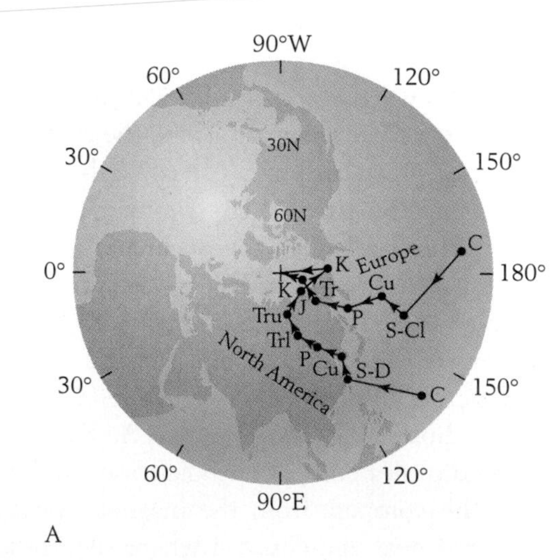

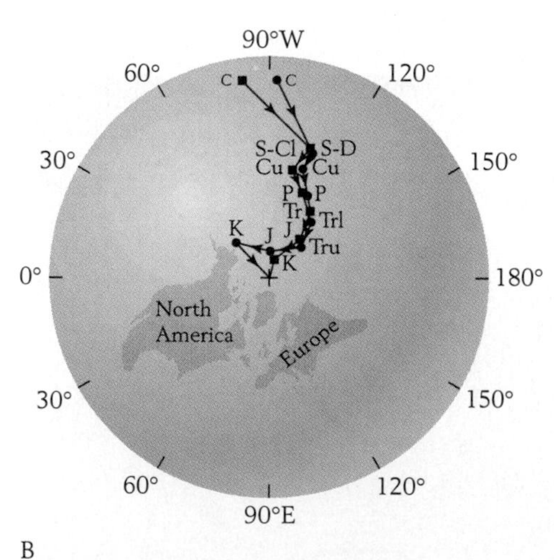

Figure 8-12 **Apparent polar-wander paths for North America (circles) and Europe (squares).** *A.* Plot of polar-wander paths based on the assumption that the continents have remained in their present positions. *B.* Plot for North America and Europe juxtaposed, as postulated for Paleozoic time by Wegener and his followers. Here the Paleozoic and Mesozoic apparent polar-wander paths of the two continents nearly coincide, suggesting that the continents were united during the Paleozoic Era. The time-rock units represented are Cretaceous (K), Triassic (Tr), Upper Triassic (Tru), Lower Triassic (Trl), Permian (P), Upper Carboniferous (Cu), Siluro-Devonian (S-D), Silurian to Lower Carboniferous (S-Cl), and Cambrian (C). *(After M. W. McElhinny, Paleomagnetism and Plate Tectonics, Cambridge University Press, London, 1973.)*

natural mechanism could move continents against oceanic crust. Then, in 1962, the American geologist Harry H. Hess published a landmark paper proposing a novel solution to this problem.

Seafloor Spreading

In essence, Hess suggested that the felsic continents had not plowed through the dense mafic crust of the ocean at all but that instead the entire crust had moved. Hess's ideas were highly unconventional (he labeled his contribution "geopoetry"), but the manner in which he compiled his facts exemplifies the way geologists assemble circumstantial evidence to construct theories. In the following summary of Hess's paper, the critical facts and inferences appear in italics.

During World War II, Hess commanded an American naval vessel in the Pacific, and he seized upon this opportunity to pursue his interest in geology. In order to study the configuration of the ocean floor, for example, Hess kept his ship's echo-sounding equipment operating for long stretches of time. While profiling the bottom in this way, he discovered curious flat-topped seamounts rising from the floor of the deep sea, which he named **guyots** after the nineteenth-century geographer Arnold Guyot. On the basis of their size and shape, Hess concluded that *guyots were volcanic islands that had been eroded by the action of waves near sea level.* Two decades later, shallow-water fossils of Cretaceous age were recovered from the tops of some guyots, proving that the guyots had indeed once stood near sea level. How the ocean floor on which they sat had subsided to such great depths, however, remained a mystery.

Another piece of evidence that Hess pondered was the apparent youth of the ocean basins. At the time, it was estimated that sediment was being deposited in the deep sea at a rate of about 1 centimeter ($\frac{1}{2}$ inch) per thousand years. At that rate, 4 billion years of Earth history would theoretically produce a layer of deep-sea sediment 40 kilometers (25 miles) thick. In fact, *the average thickness of sediment in the deep sea today is only 1.3 kilometers (less than 1 mile). Thus, allowing for some compaction, Hess estimated that the layer of sediment existing in the deep sea represented only about 260 million years of accumulation—a figure that might therefore approximate the average age of the seafloor.* (Hess's calculation was of the right order of magnitude, but we now know that the average age of the seafloor is even younger than 260 million years; in fact, little or none of the seafloor is as old as 200 million years.)

Hess found support for the idea of a youthful seafloor in his observation that there are only about 10,000 volcanic seamounts (volcanic cones and guyots) in all the world's oceans. Hess knew that when a volcano has been eroded to the level at which waves can act on it, it withstands further erosion very effectively; so he assumed that if the oceans were permanent features, their oldest volcanic seamounts should still be extant. Given the fact that there were only 10,000 volcanic seamounts in modern oceans, Hess further reasoned that if the oceans were nearly as old as Earth—say, 4 billion years old—an average of only one volcano would have formed every 400,000 years or so. The existence of so many obviously young volcanoes indicated to Hess that new volcanoes appear much more frequently, perhaps at a rate of one every 10,000 years. Thus *the relatively small number of volcanic seamounts in modern oceans suggested to Hess that current ocean basins are much younger than Earth.*

Like many earlier workers, Hess noted the central location of the Mid-Atlantic Ridge. He also noted that other mid-ocean ridges tend to be centrally located within ocean basins. (A "best fit" restoration of continents along the Mid-Atlantic Ridge, calculated after Hess's paper was written, is shown in Figure 8-1.) Four other curious facts about these ridges seemed significant to Hess:

1 *They are characterized by a high rate of upward heat flow from the mantle to neighboring segments of seafloor.*

2 *Seismic waves from earthquakes move through the ridges at unusually low velocities.*

3 *A deep furrow runs along the crest of each ridge.*

4 *Volcanoes frequently rise up from mid-ocean ridges.*

Hess developed a hypothesis that seemed consistent with all of these observations. Essentially, he suggested that mid-ocean ridges represent narrow zones where oceanic crust forms as material from the mantle moves upward and undergoes chemical changes. Hess further maintained that as this material rises, it carries heat from the mantle to the surface of the seafloor. *The expanded condition of the warm, newly forming crust thus accounts for the swollen condition of the crust there—that is, for the presence of a ridge.*

Hess then revived a geophysical concept that other researchers had discussed earlier—that the mushy material that makes up Earth's mantle rotates by means of large-scale thermal convection (see Figure 1-18). Hess proposed that Earth's liquidlike mantle is divided into **convective cells** (Figure 8-13) whose low-density material forms crust as it rises and cools. This crust then bends laterally to become one flank of a ridge (Figure 8-14). *The furrow down the center of many ridges could then be explained as the site at which newly formed crust separates and flows laterally in two directions. Similarly, volcanoes along mid-ocean ridges would represent the rapid escape of mantle material at certain sites, while the low velocity of earthquake waves passing through a ridge would result from the fact that the rocks of the ridge are extremely hot and are extensively fractured where they bend laterally to form the basaltic seafloor.*

How do the guyots that Hess discovered fit into this theory? According to his scheme, the seafloor adjacent to a mid-ocean ridge (together with anything attached to this seafloor) moves laterally, away from the spreading center. The volcanoes that frequently form along mid-ocean ridges sometimes grow upward to sea level, as is the case with Ascension Island in the Atlantic. As a volcano moves laterally from the ridge along with the crust on which it stands, it moves away from the source of its lava. It then becomes an inactive seamount, and its tip is quickly planed off by erosion. Recall that the seafloor gradually deepens away from a mid-oceanic ridge, because newly formed material of the crust cools and therefore shrinks as it moves laterally away from the ridge (see Figure 1-19). Thus *a truncated seamount is gradually transported out into deep water as if it were on a descending conveyor belt, and it then becomes a guyot* (Figure 8-15). Assuming that the Atlantic Ocean has developed by seafloor spreading since the end of the Paleozoic, Hess calculated a spreading rate of about 1 centimeter per year.

Continents can be viewed as enormous bodies that float in oceanic crust by virtue of their low density. They would be expected to ride passively along like guyots. Here, then, was Hess's explanation for the fragmentation of continents: he reasoned that when convective cells in the mantle change their locations, the upwelling limbs of two adjacent cells must sometimes come to be positioned beneath a continent. Convective spreading should then rift the continent into two fragments and move them apart from the newly formed spreading center. New ocean floor should subsequently form at the same rate on each side of the spreading center. Hess further maintained that the spreading center would continue to

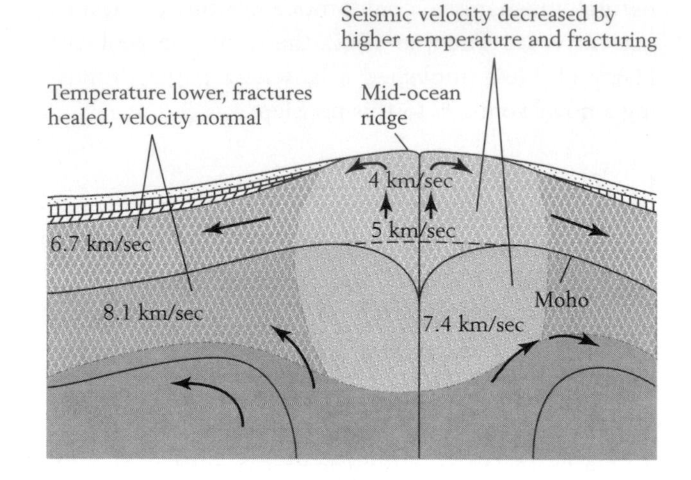

Figure 8-14 **Hess's model of the structure of a mid-ocean ridge.** Arrows show the flow of new crust derived from the convecting mantle. The newly formed crust carries heat from the mantle. This factor, along with fracturing of the rock as it bends laterally, results in low velocities (shown in kilometers per second) for seismic waves passing through the ridge. The elevation of the ridge results from the hot, swollen condition of the newly formed crust. *(After H. H. Hess, in Petrologic Studies: A Volume in Honor of A. F. Buddington, Geol. Soc. Amer., 1962.)*

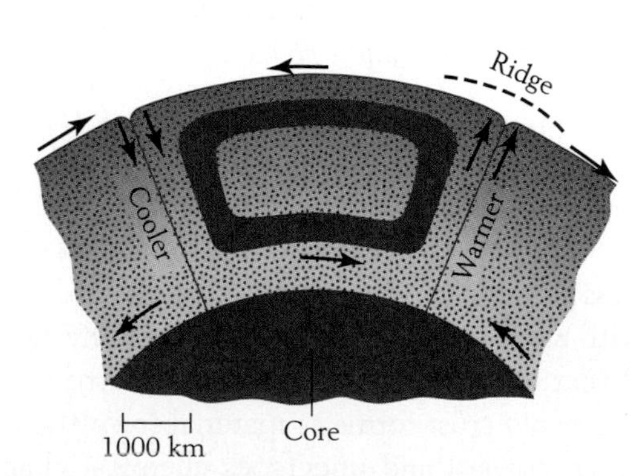

Figure 8-13 **Convective motion within the mantle as envisioned by Harry H. Hess.** Mid-ocean ridges form where the upward-flowing limbs of two adjacent convection cells approach the surface. *(After H. H. Hess, in Petrologic Studies: A Volume in Honor of A. F. Buddington, Geol. Soc. Amer., 1962.)*

operate along the midline of the new ocean basin—and thus persist as a mid-ocean ridge—as long as the convective cells remained in their new location.

If oceanic crust forms and flows laterally without an enormous change in thickness, however, it must disappear somewhere. Hess postulated that it must be swallowed up again by the mantle along the great **deep-sea trenches** that exist at certain places in the ocean floor (Figure 8-16). Movement of crust into the mantle along one side of a trench provided a ready explanation for the fact that *Earth's gravitational field here is unusually weak;* the presence of low-density crustal rock in deep-sea trenches in place of dense mantle rock would be expected to weaken the gravitational force exerted by Earth on objects at or above its surface. Hess estimated that *the formation of new crust along mid-ocean ridges and the simultaneous disappearance of crust into the deep sea trenches would produce an entirely new body of crust for the world's oceans every 300 million or 400 million years.*

Hess's hypothesis of seafloor spreading had two great strengths. First, by asserting that continents move along *with* oceanic crust, it overcame the objection that continents could not move *through* that crust. Second, the hypothesis was consistent with a variety of facts, the most important of which are italicized above. Most of these facts had not previously seemed to make sense.

The Triumph of Paleomagnetism

Despite the strong circumstantial evidence in support of Hess's hypothesis, its publication in 1962 created no great stir within the geologic profession. What was needed was a really convincing test of the basic idea. Such a test was soon found. It was based on the well-known fact that Earth's magnetic field has periodically reversed its polarity (p. 159). In 1963 the British geophysicists Fred Vine and Drummond Matthews reported that newly formed rocks lying along the axis of the central ridge of the Indian Ocean were magnetized while Earth's magnetic field was polarized as it is now. This finding came as no surprise, because it was known that other mid-ocean ridges also exhibited "normal" magnetization. It turned out, however, that seamounts on the flanks of the Indian Ocean ridge were magnetized in the reverse way. Vine and Matthews concluded that this pattern might confirm Hess's seafloor-spreading model. They reasoned that if crust is now forming along the axis of any mid-ocean ridge, it must become magnetized with the magnetic field's present polarity as it crystallizes from the molten mantle. In older crust lying at some distance from the ridge, however, reversed polarity should be encountered, and in even older crust farther from the ridge, the polarity should be normal again (Figure 8-17).

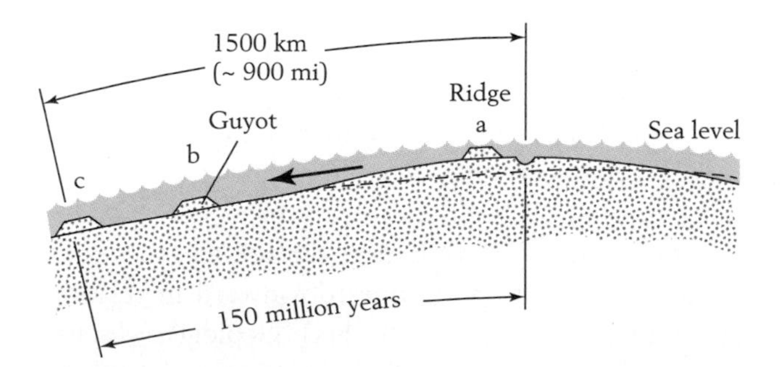

Figure 8-15 Hess's interpretation of the way a typical guyot is formed. First, a volcano builds a cone along a mid-ocean ridge. The cone initially stands partly above sea level, and its tip is later planed off by wave erosion (a). The resulting flat-topped structure moves laterally with the spreading crust and is carried gradually downward (b and c), because the newly formed crust beneath it cools and therefore shrinks as it moves away from the ridge. *(After H. H. Hess, in Petrologic Studies: A Volume in Honor of A. F. Buddington, Geol. Soc. Amer., 1962.)*

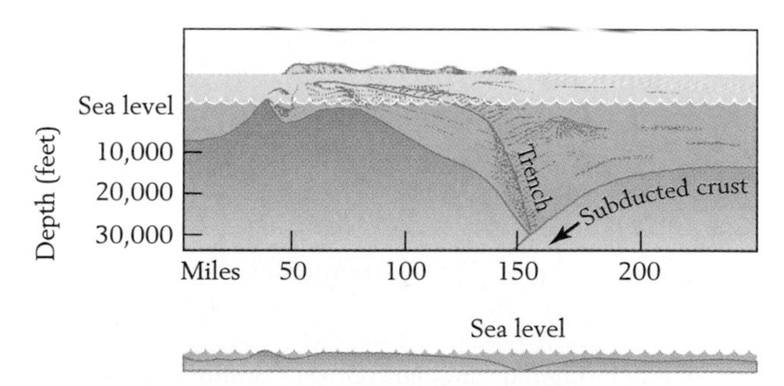

Figure 8-16 Section through the Tonga trench in the Pacific Ocean. The view is northward. In the upper picture, vertical distances are exaggerated by a factor of 10. The lower diagram is drawn without vertical exaggeration. The island toward the left is Kao, a dormant volcano. The oceanic crust to the right (or east) of the trench is moving down into the mantle beneath the oceanic crust to the left. *(Modified from R. L. Fisher and R. Revelle, Scientific American, November 1955.)*

Magnetic "striping" had, in fact, recently been observed on many parts of the seafloor, but striping patterns were soon put to a more rigorous test. During the 1960s a time scale was developed for late Cenozoic magnetic reversals. This scale was based on measurements of the magnetic polarity of terrestrial rocks of known age. It was assumed that the spreading rate for each mid-ocean ridge had remained reasonably constant over the past 4 million or 5 million years. The relative widths of seafloor stripes then turned out to be proportional to the time intervals that these stripes were thought to represent—that is, long intervals of normal polarity were represented by broad stripes and short intervals by narrow stripes. Thus the detailed patterns of striping were found to match the known timing of magnetic reversals (see Figure 8-17).

An interesting story concerns the misfortunes of a Canadian geologist named L. W. Morley. Morley developed the same model for magnetic anomalies that Vine and Matthews published, but the manuscript in which he outlined his model was rejected by the two journals to which he submitted it in 1963. One reviewer of the manuscript cynically commented that "such speculation makes interesting talk at cocktail parties." Because Vine and Matthews were fortunate enough to have had their paper accepted for publication, they were the ones who ultimately received recognition in the scientific world. Radically new ideas are not easily established.

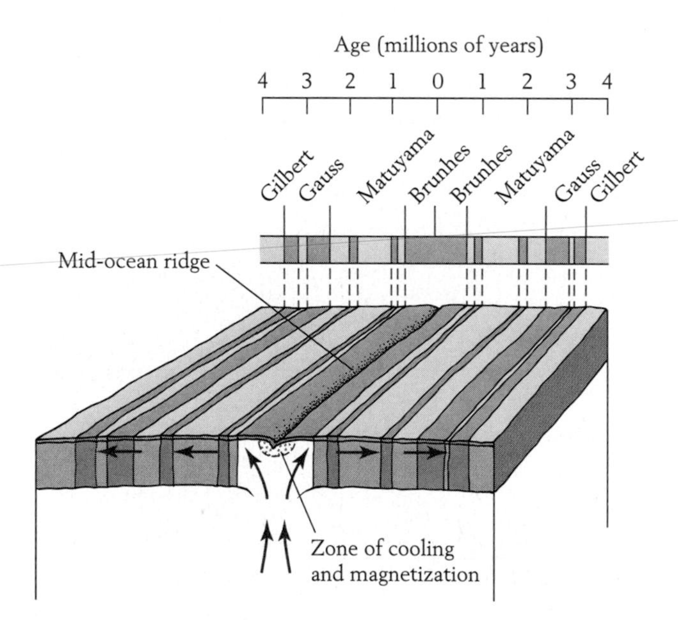

Figure 8-17 **Magnetic anomaly patterns of the seafloor fit the prediction that they represent magnetic reversals.** The time scale shows known magnetic reversals of the past 4 million years. The labels (Gilbert through Brunhes) represent intervals that are characterized by either normal or reversed polarity, dated by the polarity of terrestrial rocks whose ages are known. The relative lengths of these intervals are remarkably similar to those of the magnetic-anomaly stripes on either side of a mid-ocean ridge. *(After A. Cox et al., Scientific American, February 1967.)*

Faulting and Volcanism along Plate Boundaries

Since the advent of plate tectonics, geologists have learned how plates move in relation to one another along their mutual boundaries—and how they fracture and experience igneous activity there as well. To comprehend these dynamic features, we must first understand the basic kinds of faults along which rocks move.

Kinds of Faults

Recall that faults are surfaces along which bodies of rock break and move past each other (p. 15). Faults are classified according to their orientation and the directions in which rocks move along them (Figure 8-18).

Normal faults take their name from the fact that usually nothing more than gravity accounts for the direction of movement along them. They result from tension: in effect, the two blocks are pulled apart along the fault. The plane of a typical normal fault lies at more than 45° to Earth's surface, and the block of rock above the fault slides downward in relation to the one below (Figure 8-18*A*). Geologists discuss motion along faults as "relative movement" because any type of fault can result from movement of the rocks on both sides of the fault plane or from movement of the rocks on only one side. A normal fault, for example, occasionally entails uplift of the block of rock below the fault without downward movement of the upper block.

Thrust faults display movement that is opposite in direction to movement along normal faults: the

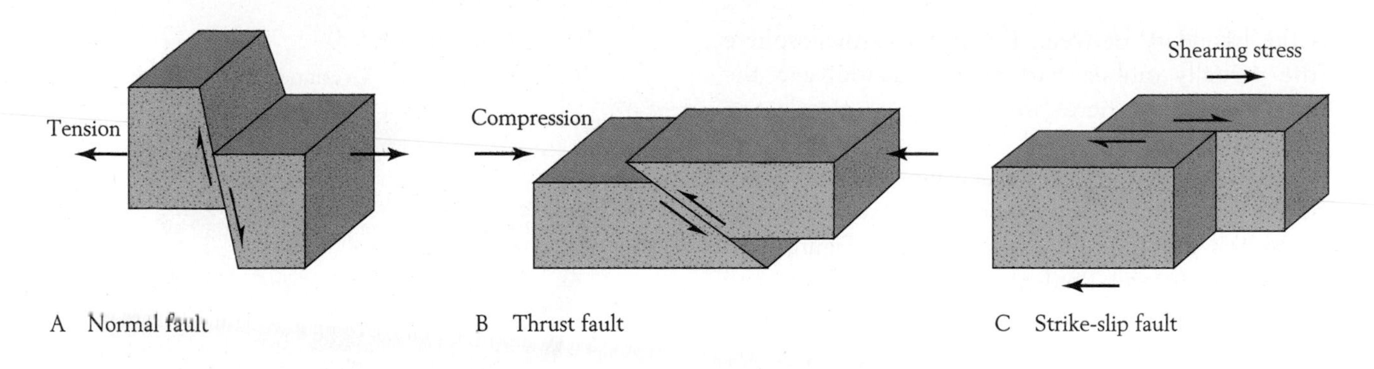

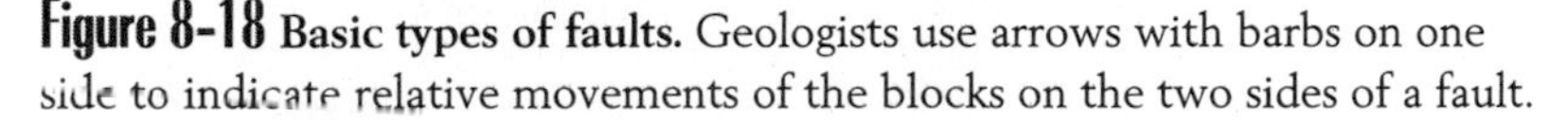

Figure 8-18 Basic types of faults. Geologists use arrows with barbs on one side to indicate relative movements of the blocks on the two sides of a fault.

relative movement of the upper block is uphill along the fault surface (Figure 8-18*B*). Thrust faults occur in areas where opposing horizontal forces compress the lithosphere enough to fracture it and to cause the rocks on the two sides of the fracture to slide past each other.

Strike-slip faults are nearly vertical, and movement along them is nearly horizontal. This movement results from shearing stress that causes the rocks on opposite sides of the fault to move in opposite directions (Figure 8-18C). The most famous strike-slip fault is the San Andreas fault of California. Earthquakes caused by movement along the San Andreas have caused considerable damage in California. Los Angeles and San Francisco lie on opposite sides of this fault. At the present rate of movement along it, the two cities will lie alongside each other in about 30 million years.

What Happens at Ridges

At places in Iceland, the furrow down the center of the Mid-Atlantic Ridge can be seen to be a structural graben. A **graben** is a valley bounded by normal faults along which a central block has slipped downward (Figure 8-19). Grabens form where the crust is extending—where it is forming and moving laterally along a mid-ocean ridge. As the crust periodically breaks apart along a mid-ocean ridge, lava moves upward to fill the space thus vacated, producing new oceanic crust. Extruded lavas have also been recorded along submarine mid-ocean ridges. In some places the lavas spread out to form broad flows, but occasionally they build volcanoes (see Figure 8-19). When the lavas cool underwater, they form **pillow basalt** (see Figure 2-14).

It is now recognized that the boundary between the crust and the mantle—the Moho (see Figure 1-15)—is not the surface along which Earth's "skin" moves. This surface, which lies well below the Moho,

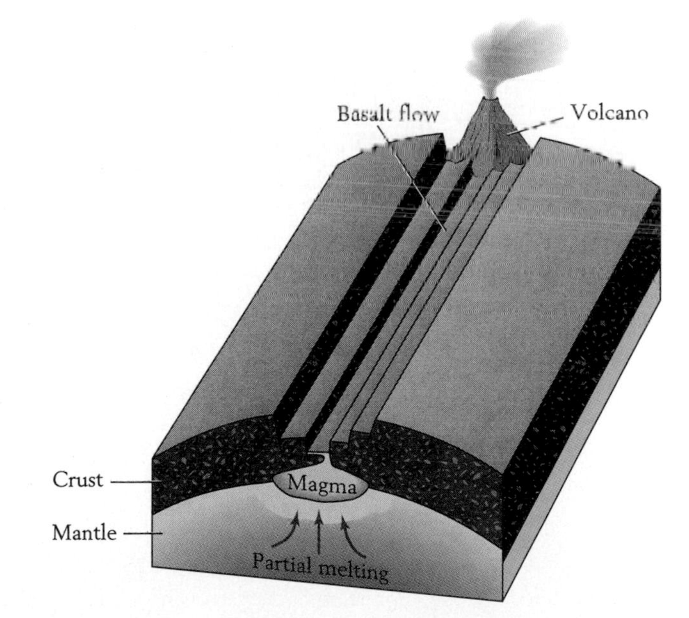

Figure 8-19 A graben of the sort that a mid-ocean rift represents. As tension breaks the crust and spreads it laterally, a block of crust sinks between normal faults, forming a graben. Lavas that emerge along faults fill in the space produced by rifting and also flow laterally and solidify on the floor of the graben. In places, the lavas erupt through small vents to form volcanoes.

is the boundary between the plastic asthenosphere (the partially molten part of the mantle) and the more rigid lithosphere (the uppermost mantle and the crust). The asthenosphere-lithosphere boundary is situated closer to the surface beneath mid-ocean ridges, where high temperatures keep mantle material molten even at very shallow depths. Figure 1-19 shows the current configuration of the crust and upper mantle in the vicinity of the Atlantic Ocean, which is still growing by seafloor spreading.

Transform Faults

Ridges are frequently offset along **transform faults,** which are enormous strike-slip faults (Figures 8-4 and 8-20). Transform faults form because pressures are uneven along ridges, and some segments of newly formed crust break away from others that move less rapidly away from the ridge axis. Earthquakes emanate from transform faults episodically, when movement takes place along them. The San Andreas fault is a transform fault that happens to cut across the edge of the North American continent.

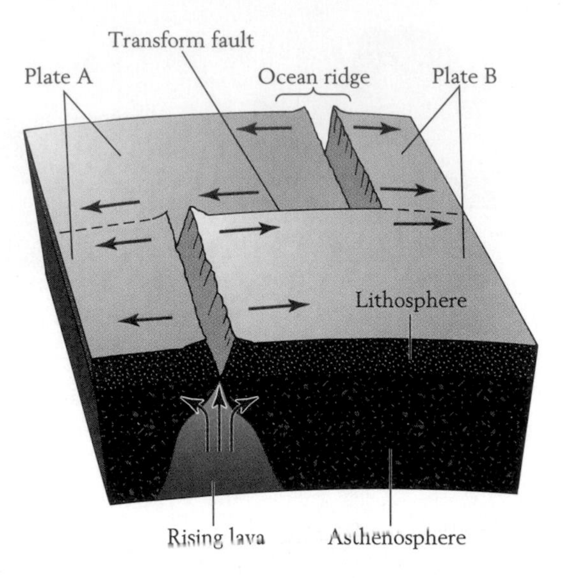

Figure 8-20 **A transform fault.** The central part of this fault is a plate boundary along which two plates slide past each other. Here plate A and plate B are separating at a rift, and the transform fault offsets this ridge. Arrows indicate the opposite directions of plate motion between the two segments of the ridge. *(After F. Press and R. Siever, Earth, W. H. Freeman and Company, New York, 1986.)*

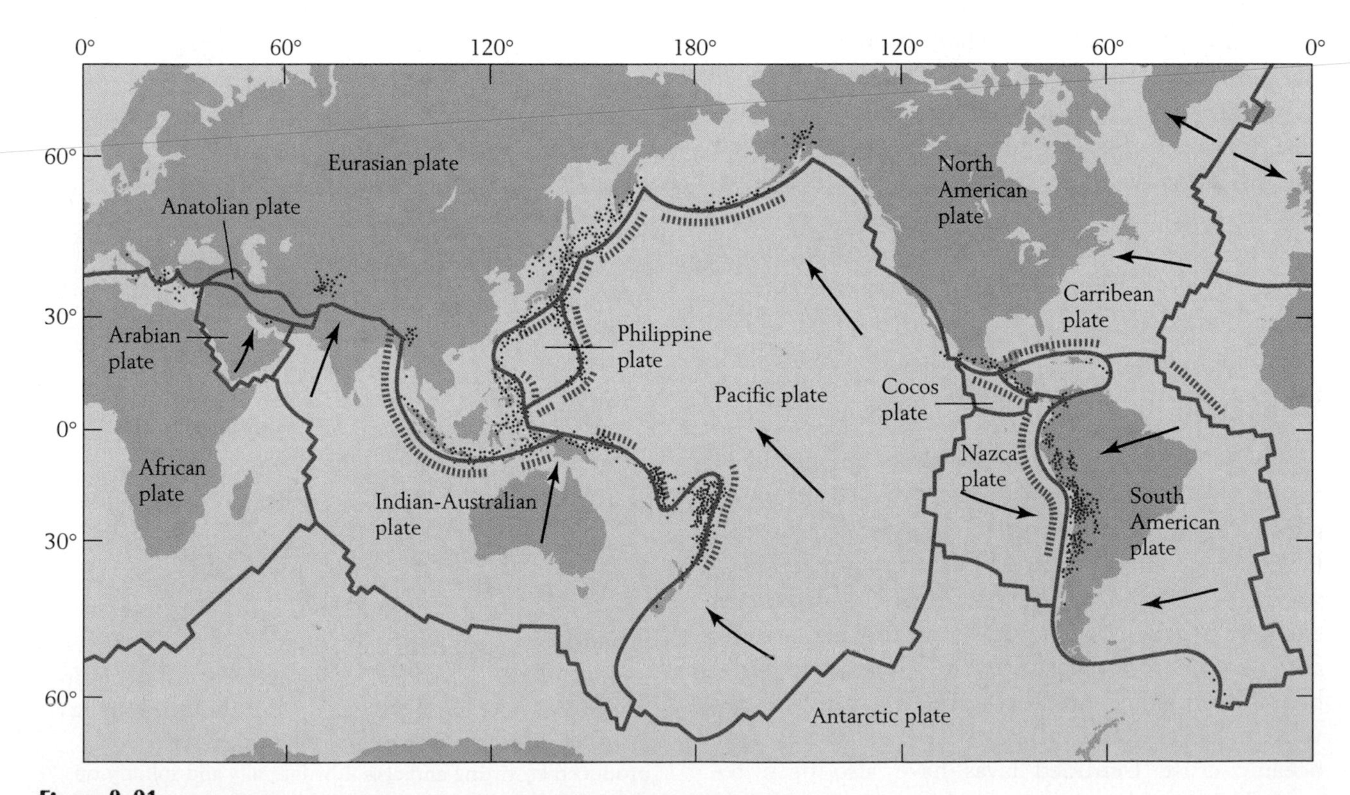

Figure 8-21 **Distribution of deep-sea trenches and deep-focus earthquakes in the Pacific.** Note that when trenches (blue dashes) are viewed from above, many can be seen to curve, and that deep-focus earthquake centers (red dots) are concentrated along trenches.

Subduction at Deep-Sea Trenches

Deep-sea trenches are the sites where slabs of lithosphere descend into the asthenosphere—a process called **subduction.** Most of the trenches of the modern world encircle the Pacific Ocean (Figure 8-21). In the decade before Hess developed his ideas, geologists had noted that trenches are associated with two other geologic features: volcanoes and **deep-focus earthquakes.** The latter are earthquakes that originate more than 300 kilometers (190 miles) below Earth's surface (p. 15). In areas far from deep-sea trenches, deep-focus earthquakes are rare. Near trenches, however, both shallow- and deep-focus earthquakes are frequent.

The typical spatial relationship between trenches, volcanoes, and deep earthquake foci is shown in Figure 8-22. The earthquake foci fall along a narrow, nearly planar zone that lies along or within a descending slab. The slab descends because it is cooler and therefore denser than the partially molten asthenosphere, and it produces earthquakes because it occasionally takes a sudden step downward.

Chains of volcanic islands often parallel deep-sea trenches (see Figure 8-21). They are positioned in this way because the descending slab undergoes partial melting (see Figure 8-22), and any of the molten material that is less dense than the asthenosphere rises toward the surface as magma. Some of this magma solidifies within the crust to form intrusive igneous bodies. The rest reaches the surface to emerge in volcanic eruptions.

A band of subducted lithosphere demarcates a **subduction zone.** Subduction zones border much of the Pacific Ocean (see Figures 1-17 and 8-21). The volcanoes associated with these zones form what is known as the "ring of fire" around the Pacific; Box 8-1 describes cataclysmic volcanism within the ring. When deep-sea trenches and the chains of volcanoes associated with them are viewed from above, many of them can be seen to have the shape of an arc. The volcanoes that rise above sea level form curved arrays of volcanoes termed **island arcs** (see Figure 1-17).

Plate convergence typically creates a zone of intensely deformed rocks in the belt between the zone called the **forensic basin** and the deep-sea trench (see Figure 8-22). Most of the rocks in this deformed belt are deep-ocean sediments such as

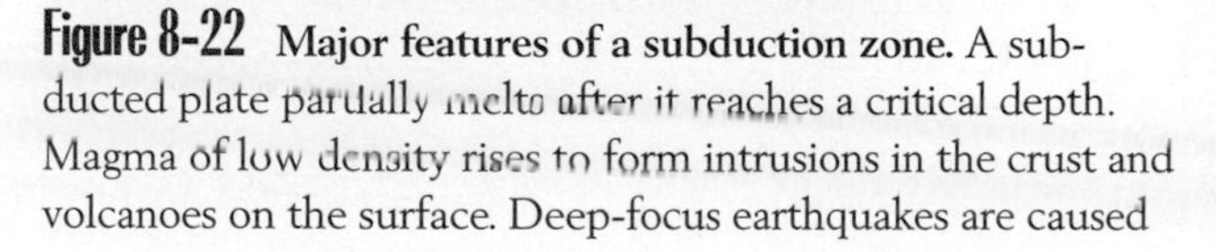

Figure 8-22 Major features of a subduction zone. A subducted plate partially melts after it reaches a critical depth. Magma of low density rises to form intrusions in the crust and volcanoes on the surface. Deep-focus earthquakes are caused by episodic downward movement of the slab. Compression along the trench forms a mélange and piles up slices of seafloor along thrust faults. *(After S. Uyeda, The New View of the Earth, W. H. Freeman and Company, New York, 1978.)*

For the Record 8-1

The Ring of Fire

In 1928 a small volcano rose above the surface of the sea between the large Indonesian islands of Sumatra and Java. The islanders named it Anak Krakatau, or "child of Krakatau." Nearly half a century before it was born, its parent, a much larger volcanic island, had all but self-destructed in a series of volcanic eruptions that affected the entire world.

Early in 1883, earthquakes repeatedly shook islands near Krakatau, which until then had been quiescent for about 200 years. On May 20, Krakatau's pointed summit began to spew volcanic ash and steam. They foreshadowed one of the greatest natural catastrophes of recorded history. On August 26, 27, and 28, a series of violent explosions audible to people across 10 percent of Earth's surface sent volcanic ash several tens of kilometers above Krakatau. The resulting airwaves disturbed barographs throughout the world. During these convulsions, Krakatau belched out an estimated 20 cubic kilometers (5 cubic miles) of lava, rock, and ash. By the time the event was over, most of the volcano had collapsed into a huge chamber that had formed beneath it as the volcanic material was expelled.

With each explosion of Krakatau, an avalanche of volcanic debris poured into the sea and huge waves surged thousands of kilometers across the ocean, some swelling to heights of 40 meters (130 feet). Altogether they destroyed 165 coastal villages and took more than 36,000 human lives.

When the eruptions of 1883 were over, only a third of the original volume of Krakatau remained above sea level, and steaming debris that it had expelled formed new islands nearby where the sea had previously been 36 meters (100 feet) deep. In the months after the eruptions, volcanic dust in the atmosphere caused such deep-red sunsets that frightened citizens as far away as New York City mistakenly called out fire engines. The effects on climates throughout the world were longer lasting. Ash that spread throughout the atmosphere screened out sunlight, reducing average global temperatures by as much as 1–2°C. Only five years later did climates return to normal.

Indonesia, the home of Krakatau, is a veritable nation of volcanoes. With a total of 132, it ranks first in the world in number of volcanoes that have been active within the last 10,000 years. Indonesia lies within the "ring of fire" around the Pacific Ocean, where lithosphere is descending into hot asthenosphere along subduction zones and melting to produce magma. More than 20 countries that border the Pacific are vulnerable to tidal waves produced by

Anak Krakatau.

volcanic eruptions. Most of these countries also face the more direct dangers of volcanism: exposure to flowing lava, flying debris, spreading poisonous gases, and smothering mudslides.

Directly or indirectly, volcanoes have claimed more than a quarter-million human lives during the past 400 years. It is sobering to note that the eruption of Krakatau in 1883 was not an especially large volcanic event. When Tambora, another Indonesian volcano, came to life in 1815, its activity went largely unstudied because the science of geology was still in its infancy, but it is clear that Tambora's eruption dwarfed Krakatau's. The size of Tambora's crater shows that it spewed forth about five times as much material, and its dust screened out so much sunlight that 1816 came to be known as "the year without a summer."

How severely a volcanic event affects humans actually depends more on the proximity of large populations than on the size of the eruption. The famous eruption of Mount Vesuvius in AD 79 was relatively small, yet it totally destroyed the nearby towns of Pompeii and Herculaneum, trapping thousands of unsuspecting Romans. Ironically, the people of Pompeii had worshipped the magnificent smoldering mountain that loomed impressively over them, and they had built their main street to align with its summit.

We can take various measures to reduce the damage that future volcanic eruptions wreak on human populations. We can construct coastal barriers to tidal waves, for example, or move dwellings inland. More generally, we can establish plans of evacuation for frequently endangered areas, and we can monitor volcanoes to sense movements of subterranean magma so that potential victims can be alerted to the danger of imminent eruptions. Unfortunately, millions of humans inhabit risky locations and geologists are monitoring only a handful of menacing volcanoes. Unless efforts are made to improve safety, eruptions will undoubtedly claim thousands of lives during the next several decades.

Figure 8-23 **A mélange of the Franciscan sequence in California.** Large blocks of exotic material are visible in the dark, metamorphosed deep-sea sediment. This mélange formed during the Mesozoic Era, when deep-sea sediments were pushed against the margin of the continent along a subduction zone.

dark muds and graywackes, with bits of ocean crust mixed in. Some have been scraped from the descending plate. Rocks of subduction zones are characteristically metamorphosed at low temperatures (because the depth at which they are deformed is not great) and at high pressures (because the plates converge with great force). This chaotic, deformed mixture of rocks is called a **mélange** (the French word for mixture). The mélange shown in Figure 8-23 formed near San Francisco when a subduction zone passed beneath the western margin of North America.

We have seen that spreading zones display normal faults and that the transform faults that offset spreading zones are strike-slip faults. In contrast, subduction zones exhibit thrust faults (see Figure 8-18*B*). Under enormous compressive forces, huge slices of a mélange and adjacent seafloor break from a downgoing lithospheric plate and pile up along thrust faults (see Figure 8-22). The entire body of rocks formed in this way constitutes what is descriptively termed an **accretionary wedge.** Between the accretionary wedge and the igneous arc lies a basin where turbidites and other sediments accumulate in moderately deep water.

Why Plates Move

Three driving forces cause plates of lithosphere to move away from spreading zones toward subduction zones (Figure 8-24):

1 Convective motion in the asthenosphere applies drag to the base of a plate.

2 The ascent of magma at a spreading zone pushes the lithosphere upward, and the weight of the elevated ridge then causes the lithosphere to spread laterally in both directions, pushing the plate on each side ahead of it.

3 At the other end of a plate, the cold, relatively dense slab sinks into the hot asthenosphere, dragging the rest of the plate toward the subduction zone.

The relative importance of each of these forces that keep plates in motion remains to be determined. Nonetheless, their combined strength moves not only the relatively thin oceanic crust but also immense continents that float on Earth's mantle.

Where Slabs Go

A slab eventually breaks away from the plate that is being subducted. The free slab absorbs heat very slowly from the hot asthenosphere around it. Remaining cold and dense, it sinks deep into the asthenosphere. In fact, the behavior of earthquake waves indicates that some slabs sink all the way to the boundary between Earth's core and mantle (see Figure 1-18). Eventually, perhaps only after hundreds of millions of years, a slab at this depth must melt. The mafic magma thus formed, being less dense than the ultramafic asthenosphere, must then rise through convective movements. The particulars of this process remain unclear, but the overall pattern amounts to an enormous version of the rock cycle (p. 9).

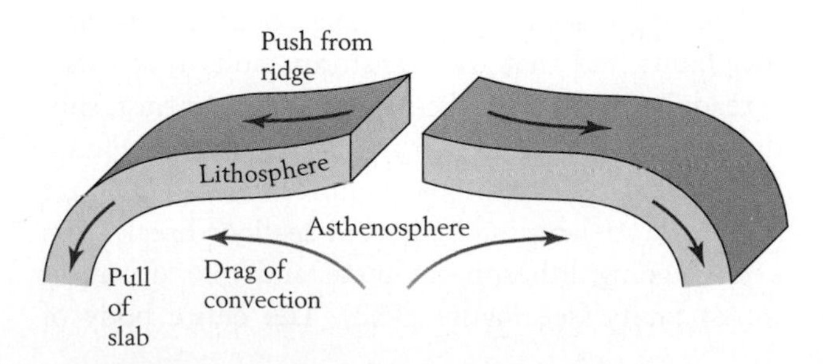

Figure 8-24 Forces that drive plate movements.

Relative Plate Movements

Because all plates are moving, no piece of lithosphere represents a perfectly immobile block against which the movement of all others can be assessed. When two plates are in contact, their relative movements determine whether the boundary between them is a spreading zone, a subduction zone, or a transform fault.

Today Earth's lithosphere is divided into eight large plates and several small ones (see Figure 1-17). The configuration of lithospheric plates and plate boundaries has changed dramatically throughout Earth's history. From time to time, new ridges and subduction zones have formed and old ones have disappeared. Although we cannot reconstruct the history of plate movements in detail for all of geologic time, this history is moderately well known for the Paleozoic Era and very well known for the Mesozoic and Cenozoic eras.

Cores obtained by drilling the floor of the deep sea from ships have shown that all segments of the deep-sea floor are of Mesozoic or Cenozoic age. Fossils of planktonic organisms provide relative ages of deep-sea sediments, and radiometric dating gives absolute ages for the oceanic crust beneath the sediments. The resulting patterns reveal rates of seafloor spreading. Australia and Antarctica have been moving apart much more rapidly, for example, than the Americas and Africa (Figure 8-25).

The complex history of continental movement described by plate tectonics differs in an important way from the pattern that Wegener envisioned. Wegener thought that the enormous landmass of Pangea existed as a stable crustal feature for hundreds of millions of years and that it fragmented in a single event. Plate tectonics, however, entails continuous movement of most landmasses in relation to one another.

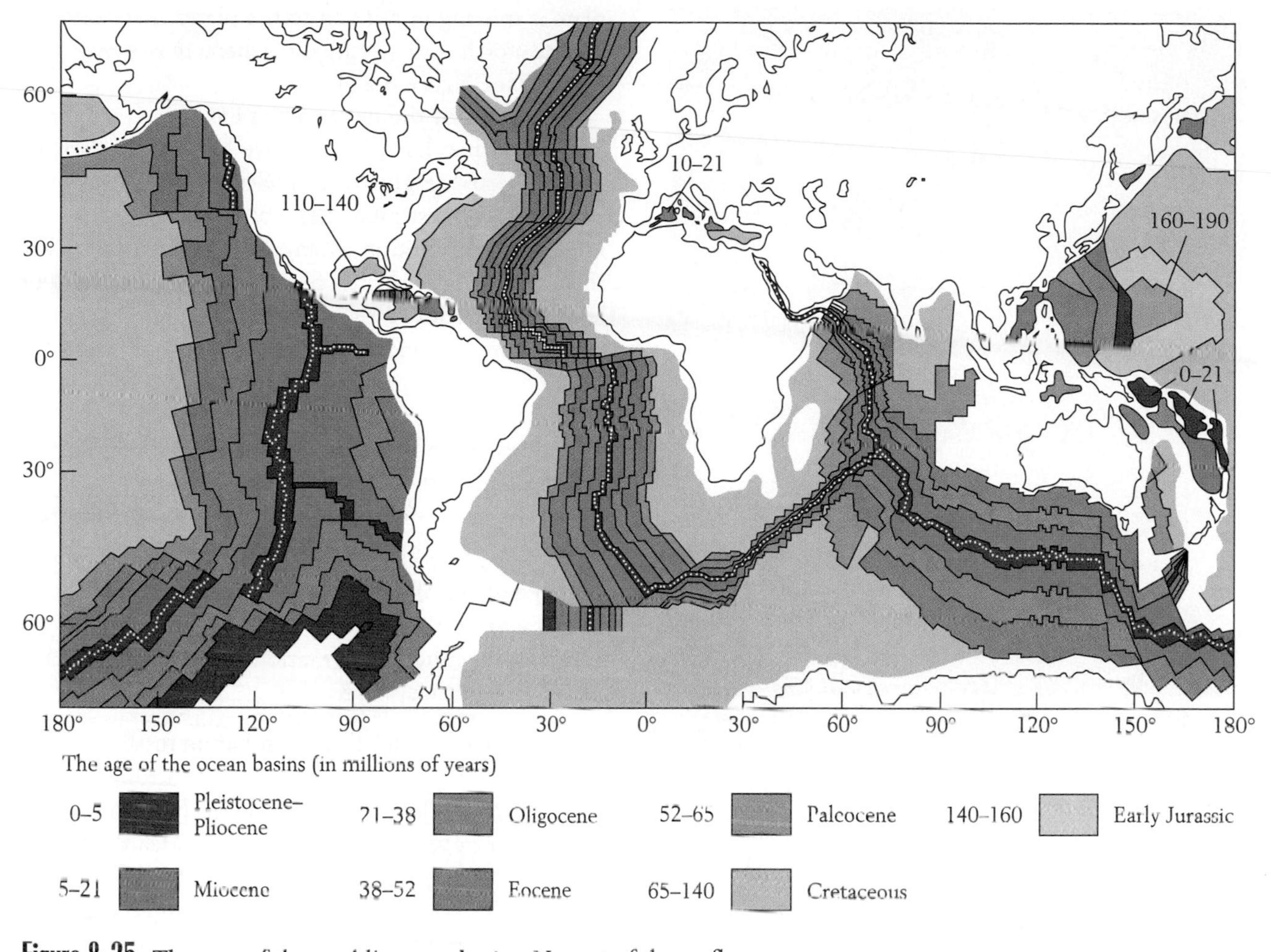

Figure 8-25 **The ages of the world's ocean basins.** No part of the seafloor is older than Mesozoic. *(After W. C. Pitman et al., Geol. Soc. Amer. Map and Chart MC-6, 1974.)*

Absolute Plate Movements

Today the enormous Pacific plate is moving to the northwest *in relation to* the plates that neighbor it on the north and northeast (see Figure 8-21). How can we measure the *absolute* movement of a plate? Absolute movement is defined as any movement in relation to a fixed feature, such as an immobile point at the surface of Earth's mantle. Nearly immobile points appear to have been discovered in the form of hot spots. A **hot spot** is a small geographic area where heating and igneous activity occur within the crust. A hot spot is located at Yellowstone National Park in Wyoming, where geysers and volcanoes have been present for millions of years. Some hot spots result from the arrival at Earth's surface of a **thermal plume,** or a column of molten material that rises from the mantle. Plumes appear to arise near the border between Earth's core and mantle and melt their way upward through the convecting mantle. Often a large volcano forms at the surface above a plume that rises through thin oceanic crust. As a plate moves over a plume, its successive positions are commonly recorded as a chain of volcanoes such as the Hawaiian Islands (Figure 8-26).

Radiometric dating tells us that Hawaii, the largest and easternmost of the islands, is less than 1 million years old, while the small northwestern island of Kauai is about 5.6 million years old, and a long train of even older submarine seamounts extends northwestward beyond Kauai. This age pattern seems to indicate that the Pacific plate is moving in a west-northwest direction over a stationary hot spot. This direction approximates the one in which the Pacific

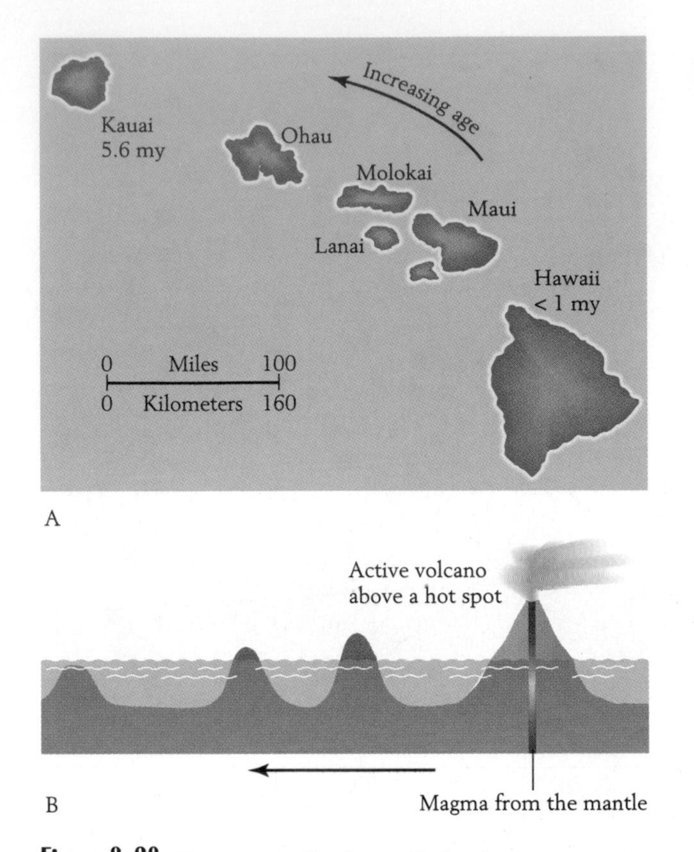

Figure 8-26 Formation of a chain of islands as an oceanic plate moves over a hot spot. *A.* The major Hawaiian islands. The islands increase in age toward the northwest. They have formed, one after the other, as the Pacific plate has moved northwestward over a hot spot in the mantle. *B.* Volcanic islands form, one after the other, as an oceanic plate moves over a hot spot.

plate is moving in relation to the plates that border it on the north and northwest, where it is being subducted (see Figure 8-21).

A search of the entire globe has turned up many hot spots that have been active within the last 10 million years (Figure 8-27). Several of them have been used to estimate the absolute directions and rates of plate motion. Many lithospheric plates are moving at a rate of only 5 centimeters (2 inches) or less per year, but rates of more than 10 centimeters per year are not uncommon.

Chapter Summary

1 Large plates of lithosphere cover Earth. The movement of plates over the asthenosphere accounts for much deformation and breakage of rocks in the lithosphere.

2 Similarities of rocks and nonmarine fossils on opposite sides of ocean basins provide strong circumstantial evidence that continents have moved horizontally over Earth's surface. The idea that continents actually drift was strongly opposed for several decades, partly because it was assumed that continents would have to move over or through the oceanic crust that lies between them.

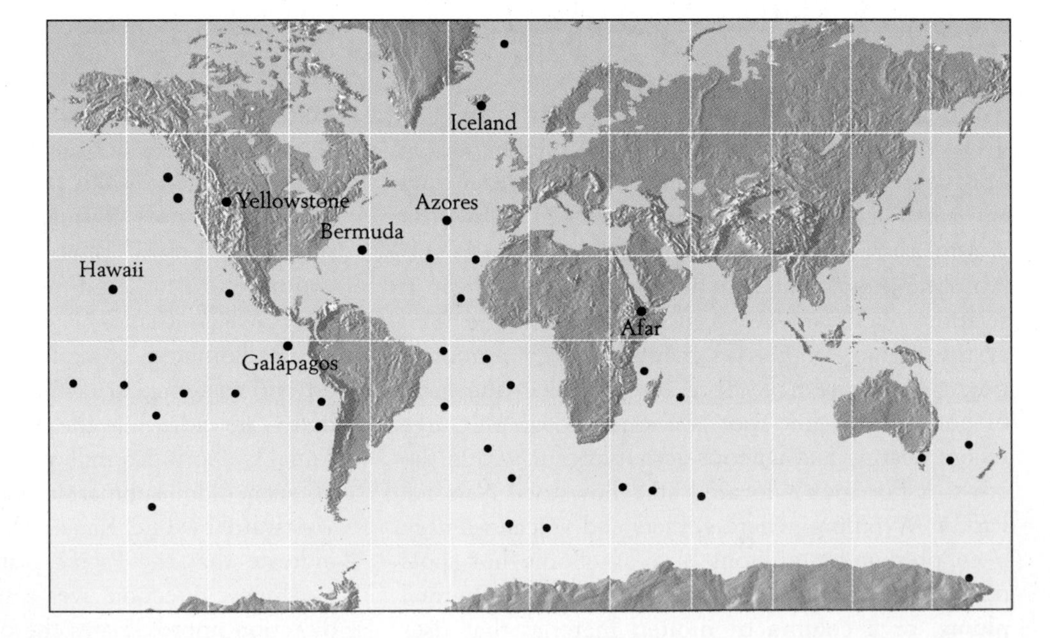

Figure 8-27 All of the world's identified hot spots. Many hot spots are positioned on or close to mid-ocean ridges. Note that hot spots are located at the Hawaiian Islands, Yellowstone National Park, Iceland, and the Afar region of northeastern Africa. *(After S. T. Crough, Ann. Rev. Earth Planet. Sci. 11:165–193, 1983.)*

3 It is now recognized that oceanic crust moves right along with the continents.

4 Material derived from the mantle is extruded along mid-ocean ridges, spreading laterally in both directions to form the oceanic crust. Oceanic crust is recycled back to the mantle by means of subduction along deep-sea trenches.

5 Geologists can measure the rate of seafloor spreading by observing how rapidly the age of the seafloor increases away from a mid-ocean ridge. The rates of spreading are high enough so that all segments of the modern seafloor are of Mesozoic and Cenozoic age.

6 At spreading zones the lithosphere swells to form a ridge and undergoes normal faulting and volcanism.

7 At subduction zones, where plates converge, compression contorts sediments and basalt into a mélange, and thrust faulting forms an accretionary wedge. The plate that descends into the lithosphere partly melts, releasing magma that rises to form an igneous arc.

8 Plates that are not converging or diverging move past each other along strike-slip faults called transform faults.

Review Questions

1 List as many pieces of evidence as you can in support of the idea that continents have moved over Earth's surface.

2 What is the geographic extent of the lithospheric plate on which you live?

3 What is apparent polar wander? Draw pictures of Earth showing how the movement of a continent can produce apparent wander of the north magnetic pole.

4 Why are most volcanoes that have been active in the last few million years positioned in or near the Pacific Ocean?

5 How can a hot spot indicate the absolute direction of movement of a plate?

6 Why do mid-ocean ridges form?

7 Why do deep-focus earthquakes occur along subduction zones?

8 Describe three kinds of faults.

9 What are the multiple driving forces of plate movement?

10 What happens to slabs of subducted lithosphere?

11 Suppose that you were to encounter a well-trained geologist who was unfairly imprisoned in 1955 and was deprived of reading materials until he was released last week. He entered prison firmly opposed to the idea that continents have moved large distances across Earth's surface. Given an hour of time, how would you convince this unfortunate geologist that continents have actually moved thousands of kilometers? Use the Visual Overview on p. 208 as a guide to develop your argument.

Additional Reading

Allegre, C. *The Behavior of the Earth: Continental and Seafloor Mobility*, Harvard University Press, Cambridge, 1988.

Erikson, J., *Plate Tectonics: Unraveling the Mysteries of the Earth*, Facts on File, New York, 1992.

Keller, E. A., and Pinter, N., *Active Tectonics: Earthquakes, Uplift, and Landscape*, Prentice-Hall, Upper Saddle River, N.J., 1996.

MacDonald, K. C., and P. J. Fox, "The Mid-Ocean Ridge," *Scientific American*, June 1990.

Murphy, J. B., and Nance, R. D., "Mountain Belts and the Supercontinent Cycle," *Scientific American*, April, 1992.

Nance, R. D., T. R. Worsley, and J. B. Moody, "The Supercontinent Cycle," *Scientific American*, July 1988.

Sullivan, W., *Continents in Motion*, American Institute of Physics, New York, 1991.

CHAPTER 9

Continental Tectonics and Mountain Chains

Plate tectonic forces modify the thick, felsic crust of continents as well as the thin, mafic crust beneath the deep ocean. These forces break continents apart, weld them together, and build mountain chains along their margins. Continental crust undergoes changes along all three kinds of plate boundaries: transform faults, spreading zones, and subduction zones.

Sometimes transform faults—strike-slip faults that offset spreading zones (p. 222)—intersect the margins of continents. The San Andreas fault, for example, slices through western California, where in some places movements along it have shattered rocks and accelerated erosion to create a narrow, scarlike valley (Figure 9-1). San Francisco sits at the western edge of the North American plate, whereas Los Angeles occupies a sliver of continental crust that is part of the Pacific plate. Movements along the San Andreas fault are bringing these two cities closer together at the rate of 5.5 centimeters (~2 inches) per year. In several tens of millions of years, the sliver that Los Angeles occupies may move beyond San Francisco to end up as a slender island in the Pacific Ocean.

We will see in later chapters that movements along transform faults in the distant past have juxtaposed dissimilar geologic **terranes**—that is, geologically distinctive regions of Earth's crust, each of which has behaved as a coherent crustal block. The offset along these faults is commonly very large—tens or

Mount Everest, at 8853.5 meters (29,028 feet), is the tallest mountain on Earth. Standing along the border between Tibet and Nepal, it is part of the immense Himalaya chain, which formed when a fragment of Gondwanaland collided with Asia to become the peninsula of India.

Visual Overview

Formation and Deformation of Continental Margins

Domes and Three-Armed Rifts

Rift Valley

New Ocean

Passive Margin

CONTINENTAL RIFTING

Continent Stalls at Subduction Zone

Subduction Reverses

Continents Collide

MOUNTAIN BUILDING BY CONTINENTAL COLLISION

Continent Stalls at Subduction Zone

Slow movement

High angle

Rapid movement

Low angle

MOUNTAIN BUILDING WITHOUT CONTINENTAL COLLISION

Mountain

Erosion

Uplift

Erosion

Uplift

ISOSTASY

Continents rift apart, forming passive margins that become sites of mountain building along subduction zones, with or without collision with other continents.

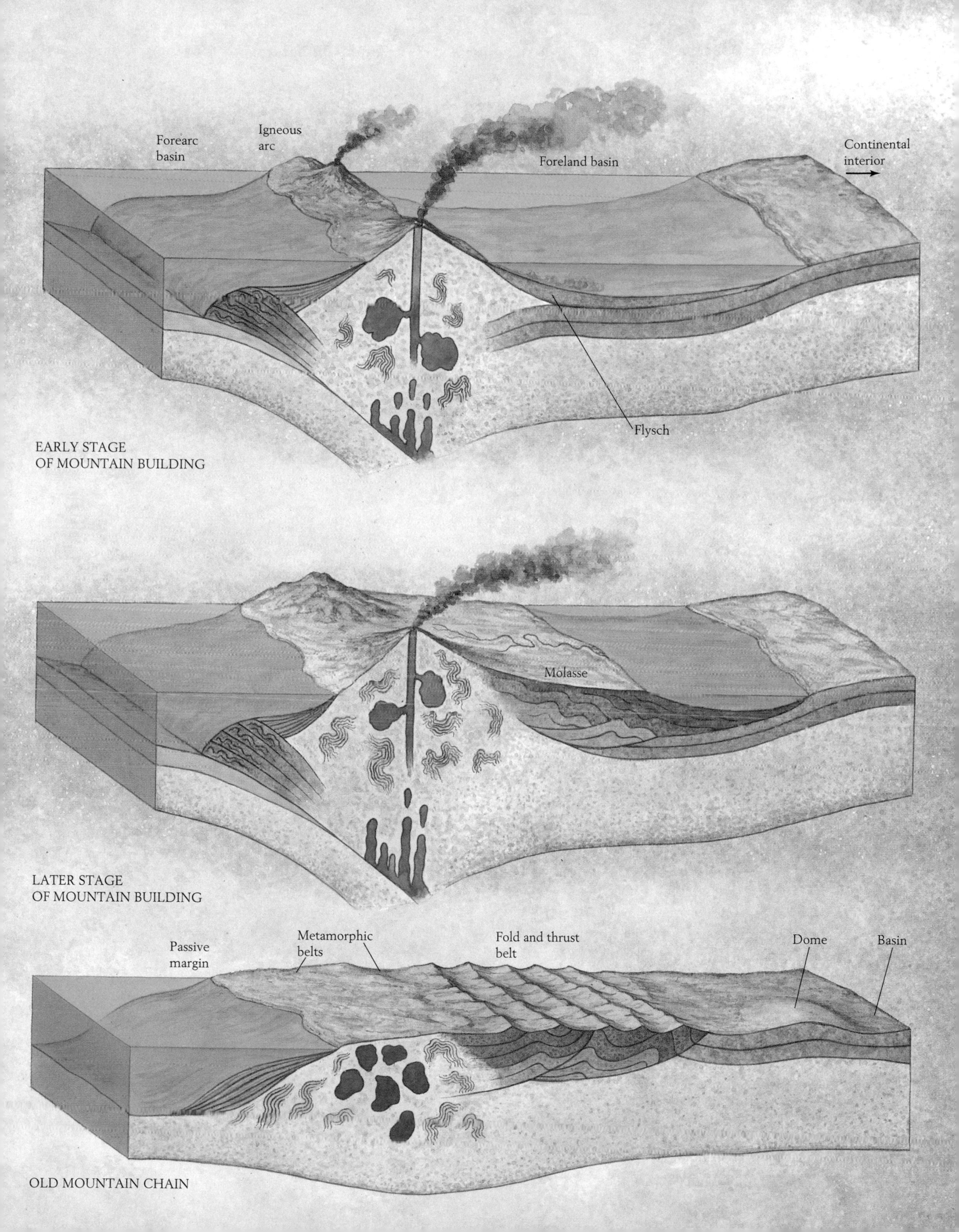
Forearc
basin
Igneous
arc
Foreland basin
Continental
interior
Flysch
EARLY STAGE
OF MOUNTAIN BUILDING
Molasse
LATER STAGE
OF MOUNTAIN BUILDING
Passive
margin
Metamorphic
belts
Fold and thrust
belt
Dome
Basin
OLD MOUNTAIN CHAIN

Figure 9-1 The San Andreas fault in California, a great break within Earth's lithosphere that separates large segments, or plates, of the lithosphere. The Pacific plate is on the left, and the North American plate is on the right. The Pacific plate periodically slides northwestward in relation to the North American plate.

hundreds of kilometers—and this kind of strike-slip displacement has brought together terranes that developed in different regions and under quite different circumstances.

The Rifting of Continents

We saw in Chapter 8 that spreading zones occupy the central portions of ocean basins, where oceanic lithosphere forms along a medial ridge and spreads in either direction. Continental crust, too, can spread apart. A new ocean then forms between the two continental fragments that remain. As continental crust is about five times thicker than oceanic crust, it does not break apart so easily. In fact, a *rift* may begin to fracture a continent and then fail to complete the break, leaving telltale scars of its activity. Africa and adjacent seas provide the best example of continental rifting in progress in the modern world.

Three-Armed Rifts and Hot Spots

A system of rift valleys, formed during the Cenozoic Era, extends southward through Africa from the Red Sea and the Gulf of Aden (Figure 9-2). The central rift valleys and the broader basins harboring the Red

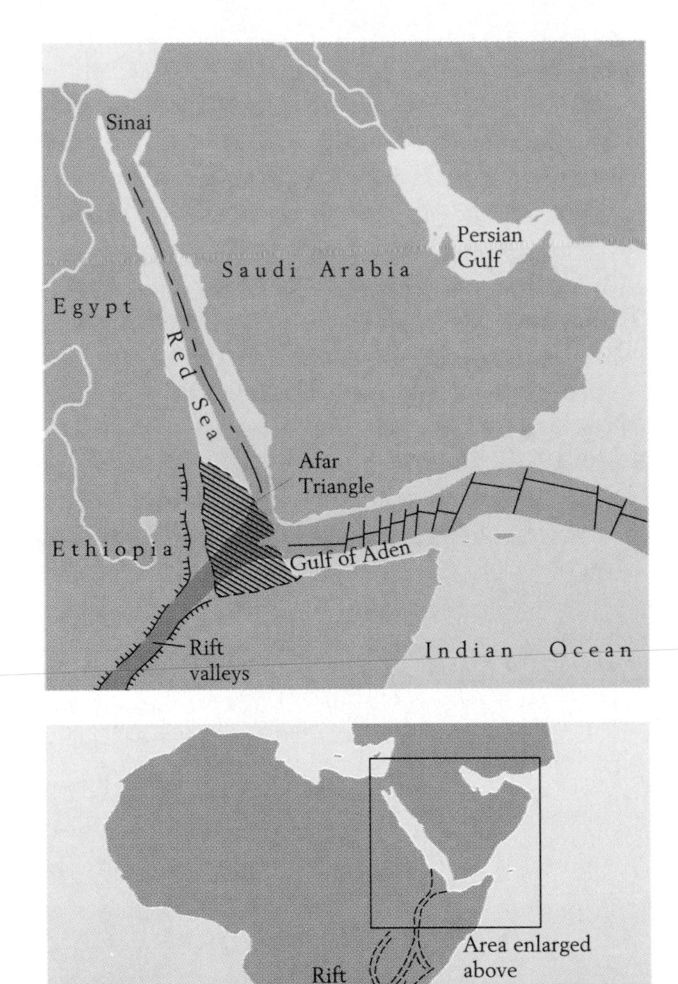

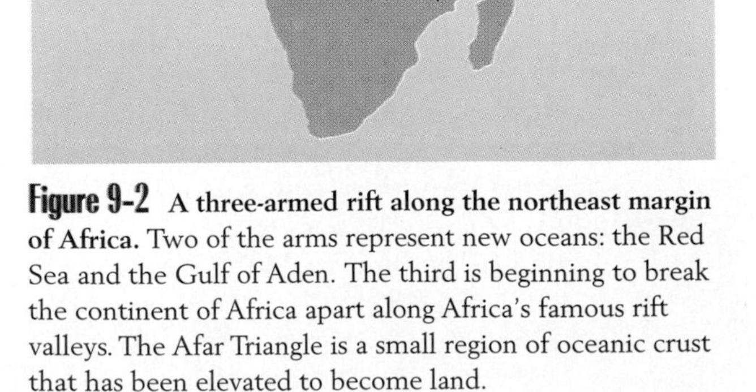

Figure 9-2 A three-armed rift along the northeast margin of Africa. Two of the arms represent new oceans: the Red Sea and the Gulf of Aden. The third is beginning to break the continent of Africa apart along Africa's famous rift valleys. The Afar Triangle is a small region of oceanic crust that has been elevated to become land.

Sea and Gulf of Aden are grabens formed by extension and breaking of the continental lithosphere.

When rifts develop, they often begin as three-armed grabens at plate boundaries known as *triple junctions.* Long before the advent of plate tectonics, geologists noticed that at the locations of three-armed grabens, continental crust is frequently elevated into a dome. In the context of plate tectonics, it seems evident that this doming represents the development of a hot spot. In the area of Ethiopia called the Afar Triangle, the Red Sea, the Gulf of Aden, and the north end of the African rift system form a triple junction. Such junctions are common features of Earth's crust (see Figure 1-17). More than one kind of plate boundary can meet at a triple junction. Each boundary may consist of a spreading zone, a subduction zone, or a transform fault. At the Afar Triangle, the junction happens to involve three spreading zones.

It appears that when a large continent is rifted apart, the jagged line along which it divides often represents a composite structure formed from arms of several three-armed rifts. This phenomenon is seen in the rifting that formed the Atlantic Ocean (Figure 9-3). A three-armed rift usually contributes two of its arms to the composite rift, while the third arm becomes a **failed rift**—a plate tectonic dead end. Before it ceases to be active, this third arm forms a graben or a system of grabens that projects inland from the new continental margin formed by the other two arms. Some of the world's largest rivers, including the Mississippi and the Amazon, flow through valleys located in failed rifts that border the Atlantic basin.

It is not uncommon, however, for all three arms of a three-armed rift to develop into segments of plate boundaries. Note that the Mid-Atlantic Ridge, for example, terminates at a triple junction of ridges in the South Atlantic. Similarly, although the rift arm that projects into Africa has not yet divided the continent, it may do so in the future.

It is not surprising that many hot spots are situated on or very near mid-ocean ridges (see Figure 8-27). These may be the only surviving hot spots of a larger number that produced three-armed rifts that contributed one or two arms to the formation of active spreading ridges. The southern portion of the Mid-Atlantic Ridge has shifted to a position slightly to the west of three surviving hot spots that appear to have played a role in its origin.

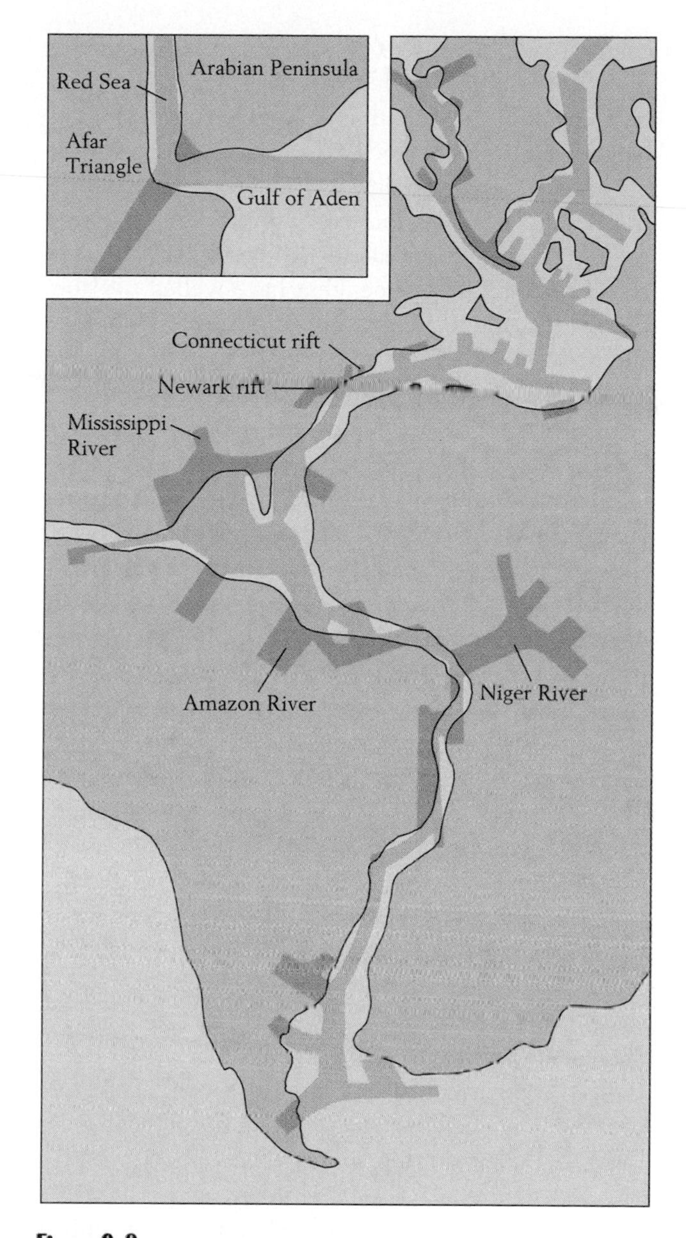

Figure 9-3 Ancient three-armed rifts become apparent when we reassemble continents now bordering the Atlantic Ocean. Many of these rifts contributed two arms to the fracture zone that became the Mid-Atlantic Ridge. *(After K. C. Burke and J. T. Wilson, Scientific American, August 1976.)*

Geologic Features of Continental Rift Valleys

As we have seen, a mid-ocean ridge is often associated with block faulting that is expressed by a graben running along the ridge's midline (see Figures 8-4 and 8-19). When a spreading zone first encounters continental crust, however, it seldom produces faults that cut cleanly through. Instead, the extension tends to break the thick continental crust into a complex band of fault blocks. Today this process is under way in Africa, where a system of rift valleys passes southward from the Red Sea (see Figure 9-2). Each valley is a long, narrow, downfaulted block of crust associated with mafic volcanoes that have welled up from the mantle; the rifting has also produced mafic dikes and flood basalts. Some of the rift valleys cradle great lakes such as Lake Tanganyika. These rift valleys have been in existence only since early in Miocene time (less than 20 million years).

The rapid subsidence of a downfaulted basin creates a rugged landscape that is subject to rapid erosion, so that clastic sediments often accumulate quickly to great thicknesses. The sediments typically include conglomerates derived from the steep valley walls and also red beds—sediments whose reddish color is usually attributable to iron oxide cement—and alluvial-plain deposits. Lakes that form within the elongate valleys also leave a sedimentary record. In arid climates, temporary lakes leave accumulations of nonmarine evaporites.

If rifting continues long enough, a rift valley becomes so wide and so extended that it opens up to the sea. Because inflow from the sea tends at first to be restricted or sporadic, saline waters within the rift valley evaporate more rapidly than they are renewed. Under such circumstances, marine evaporites form. Waters from the Indian Ocean, for example, only recently gained full access to the currently widening Red Sea. Beneath a thin veneer of marine sediment in the Red Sea, geologists have found evaporites that formed during an earlier time, when there was only a weak connection to the larger ocean, or perhaps even earlier, when the basins were nonmarine but arid.

The margins of the Red Sea also exhibit geologic features that typify the early stages of continental rifting. The Afar Triangle (see Figure 9-2) was once part of the Red Sea floor. Most of the rocks in this area are basalts similar to those that form oceanic crust. Much of the topography is the product of block faulting and uplift produced by the high rate of heat flow from the mantle, and in places there are great thicknesses of evaporite deposits.

In summary, then, regions of continental rifting are characterized by normal faults, mafic dikes and sills, and thick sedimentary sequences within fault block basins that often include lake deposits, coarse terrestrial deposits, and evaporites followed by oceanic sediments. This suite of features can be found in fault basins of the eastern United States. Two of the basins are labeled in Figure 9-3 as the Newark and Connecticut rifts. These are actually failed rifts—rift arms that never became part of the composite rift that formed the Atlantic Ocean. These rifts extended far inland and never opened wide enough to allow the sea to invade, although they contain other sedimentary sequences that typify incipient continental rifts.

Passive Margins of Continents

When continental rifting does not fail, one continent becomes two, and a narrow ocean forms between them. Eventually the two new continental margins move far from the spreading zone where they formed. Soon they are likely to be flooded by shallow seas, because they move laterally away from the ridge axis, down the slope of the asthenosphere's surface, to regions where heat flow is lower and the asthenosphere is not so swollen (Figure 9-4).

Thus the continental borders, which were tectonically active when they were still close to the spreading zone, become what are termed **passive margins.** Having descended below sea level, these tectonically inactive areas of continental crust accumulate sediment along shallow shelves. Thus, after Pangea broke up early in the Mesozoic Era, the newly formed Atlantic margin of the United States migrated away from Africa and soon began to accumulate great thicknesses of sediment (see Figure 9-4). As later chapters will describe in greater detail, to the present day this passive margin has continuously subsided under the weight of added sediment, making way for still more to be laid down.

As the term "passive margin" suggests, continental margins can also be tectonically active. Indeed, **active margins** are zones of tectonic deformation and igneous activity. Simply put, they are sites of mountain building. Before we can investigate how tectonic forces and igneous activity form mountain chains, however, we must consider how bodies of rock bend and flow under stress from tectonic forces, in addition to moving past one another along faults.

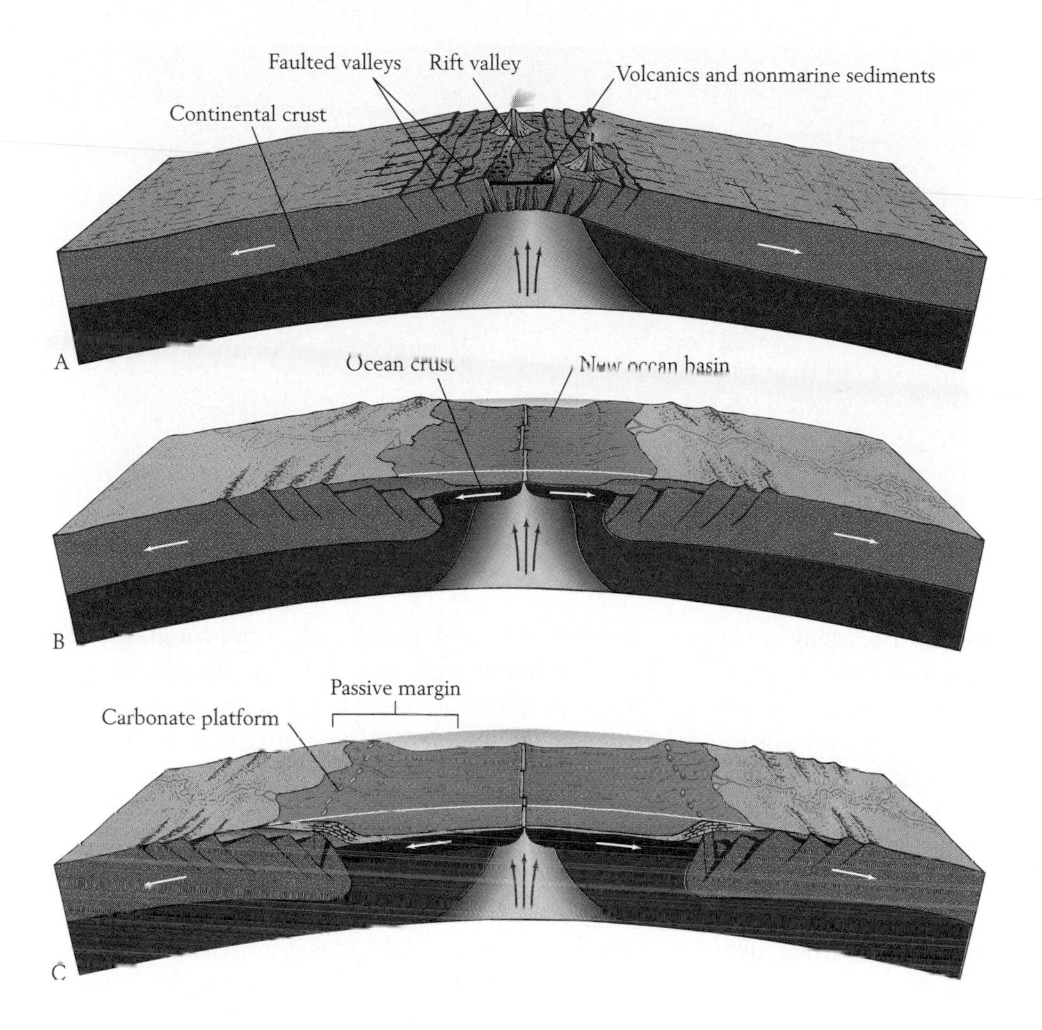

Figure 9-4 The rifting of Pangea that produced the Atlantic Ocean. Rifting began early in the Triassic Period, forming nonmarine basins along block faults *(A)*. The sea spread into the zone of rifting during the Jurassic, and passive margins formed on either side of the new Atlantic ocean *(B)*. The Atlantic has continued to expand, and the passive margins here persisted to the present day. *(After F. Press and R. Siever, Understanding Earth, W. H. Freeman and Company, New York, 1994.)*

Bending and Flowing of Rocks

Geologic outcrops reveal that some rocks have been warped, twisted, and folded; they have even flowed. When forces are applied to a rock, the component grains may be affected in various ways; they may slide past one another, change shape, or break along parallel planes. (In the last case, a grain deforms in the way a deck of cards on a table changes shape when we push on one end of the deck near the top.) Internal deformation of a large body of rock by any of these mechanisms takes place very slowly, but when many of the grains are affected, the entire body of rock can undergo radical changes in shape in the course of millions of years. Such changes usually take place at great depths within Earth's crust.

When rocks are deformed at high temperatures and pressures, they also undergo high-grade metamorphism (p. 51). Under less extreme conditions, sedimentary rocks and the fossils they contain may be deformed with little or no metamorphism.

Folding

One common type of large-scale rock deformation is referred to as **folding.** Compressive forces can shorten Earth's crust, creating folds (Figure 9-5). Folds come in many sizes. When folded sedimentary rocks are viewed with their oldest beds at the bottom and their youngest beds on top, they display two kinds of folds. **Synclines** are concave in an upward direction, with their vertexes at the bottom. **Anticlines,** in contrast, are concave in a downward direction, with

Figure 9-5 Folded rocks in the Andes.

their vertexes at the top (Figure 9-6). Many rocks do not simply bend when they are folded; instead, material is displaced from one part of a bed toward another. When a large, complex body of rock is subjected to an external force, weak beds tend to become more intensely folded than durable beds. Shales, for example, tend to be weak and to deform much more severely than massive sandstones and limestones that are subjected to the same forces. Igneous and metamorphic rocks are relatively durable, but they, too, can be folded.

The Terminology of Folds

Geologists have developed special terminology to describe shapes of folds in detail. A tilted bed, for example, is said to have a **dip**—a term that describes the angle that the bed forms with the horizontal plane. In other words, the dip is the direction in which water would run down the surface of the bed. The **strike** of a bed, in contrast, is the compass direction that lies at right angles to the dip (Figure 9-7); strikes are always horizontal. It is sometimes said that the **regional strike** of a given area is in a particular geographic orientation—north–south, say. This does not mean that every strike in this area has the same orientation, only that most of the fold axes trend north–south, so that the strikes of most beds do too.

A fold is said to have an **axial plane,** which is an imaginary plane that cuts through the fold and divides it as symmetrically as possible. In actuality, many folds are asymmetrical, with one limb (or flank) dipping more steeply than the other. If either limb is rotated more than 90° from its original position, the fold is said to be **overturned** (Figure 9-8).

The **axis of a fold** is the line of intersection between the axial plane and the beds of folded rock. Often the axis plunges—that is, it lies at an angle to the horizontal (Figure 9-9*A*). When a **plunging fold** is truncated by erosion, its beds form a curved outcrop

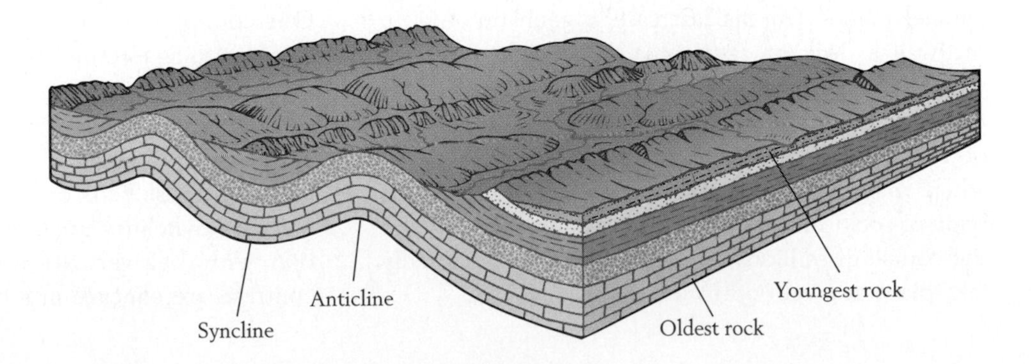

Figure 9-6 Humplike folds, or anticlines, and troughlike folds, or synclines. *(After F. Press and R. Siever, Earth, W. H. Freeman and Company, New York, 1986.)*

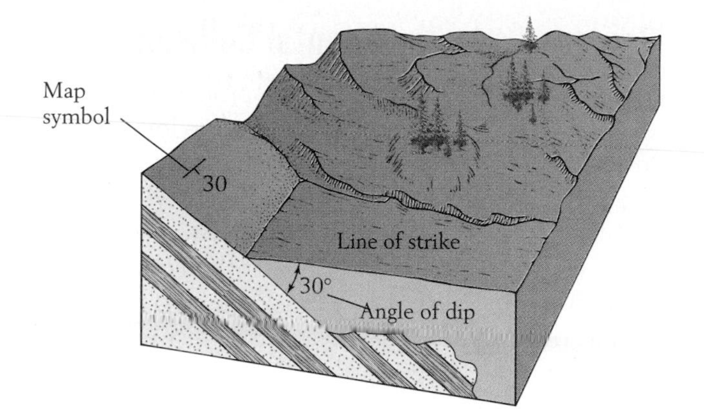

Figure 9-7 **The strike and dip of inclined beds.**

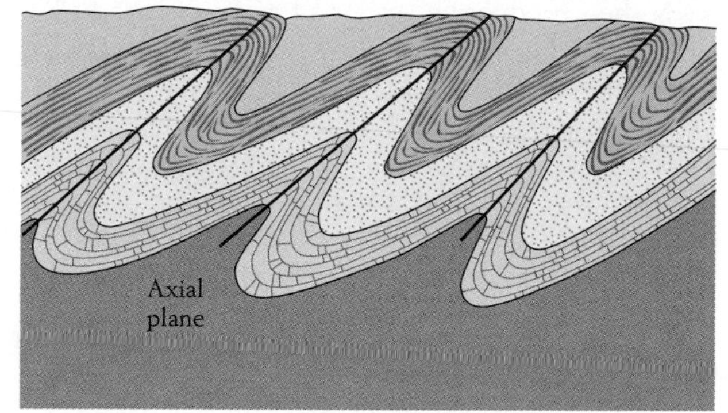

Figure 9-8 **An overturned fold in cross section.** Both limbs of an overturned fold dip in the same direction. One limb of a recumbent fold is upside down. *(After F. Press and R. Siever, Earth, W. H. Freeman and Company, New York, 1986.)*

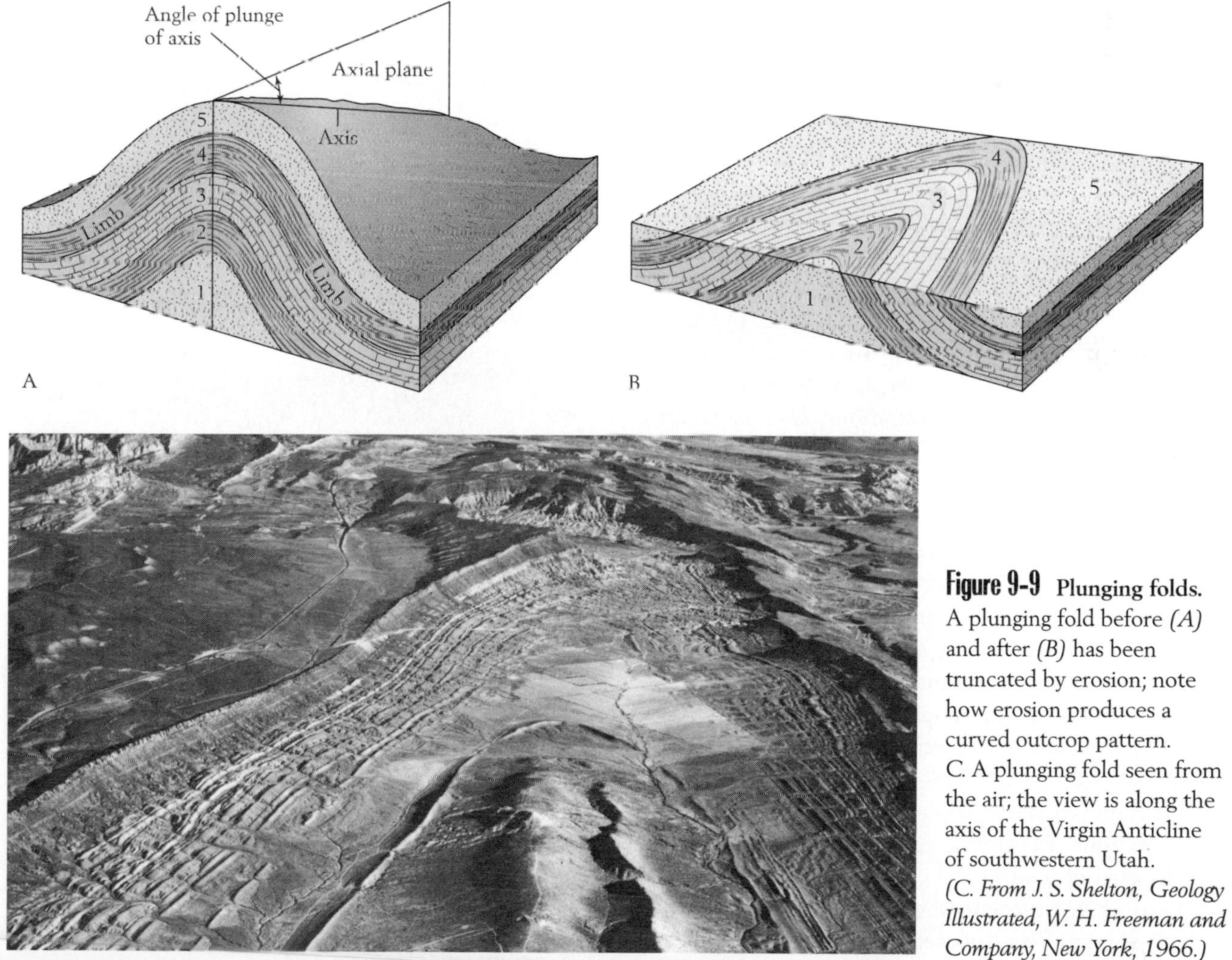

Figure 9-9 **Plunging folds.** A plunging fold before *(A)* and after *(B)* has been truncated by erosion; note how erosion produces a curved outcrop pattern. C. A plunging fold seen from the air; the view is along the axis of the Virgin Anticline of southwestern Utah. *(C. From J. S. Shelton, Geology Illustrated, W. H. Freeman and Company, New York, 1966.)*

pattern (Figures 9-9*B* and C). A series of plunging folds then produces a scalloped surface pattern. Because most folds plunge (we could not expect many to be perfectly horizontal), this scalloped pattern is characteristic of regions where sedimentary rocks have been extensively folded. Most large folds in continental crust form where two plates converge, applying compressive forces to the rocks adjacent to the juncture. Folding in such regions thickens the crust and contributes to the growth of mountain chains.

Mountain Building

Before the development of plate tectonic theory, the process of mountain building, or **orogenesis,** was a subject of debate. North American geologists had long been struck by the fact that one chain of mountains, the Cordilleran system, parallels the west coast of their continent and another, older chain, the Appalachian system, parallels the east coast. Before the 1960s, this symmetrical pattern led some North American geologists to conclude that mountains tend to form along the margins of relatively stable continental masses known as **cratons.** Most European geologists, however, pointed to the Ural mountains, standing between Europe and Asia within the largest landmass on earth, as evidence that long mountain chains could indeed rise up in the center of a continent. Plate tectonic theory explains both situations. It is true that continuous long mountain chains form only along continental margins. Continents can unite, however, and when they do, a mountain chain forms where the margins are welded together within the newly formed large landmass.

The unification of two continents along a subduction zone is termed **suturing.** According to the theory of plate tectonics, subduction zones are the key to the origin of mountain chains on continents. But not all mountain-building events—called **orogenies**—result from continental suturing; a mountain chain also forms when an oceanic plate descends beneath the margin of a solitary continent along a subduction zone. Furthermore, not all mountain chains are formed of continental crust. Rifting and swelling of oceanic crust create the great mid-ocean ridges, which are, in effect, submarine mountain chains (see Figure 8-4).

Orogenesis by Continental Collision

When two continents collide, two properties of continental crust—its great thickness and its low density in relation to the density of the asthenosphere—lead to mountain building. Continents are simply too thick and too buoyant to be subducted. When a continent encounters a deep-sea trench, its resistance to subduction forces a reversal in the direction of subduction (Figure 9-10). As a result, the oceanic plate

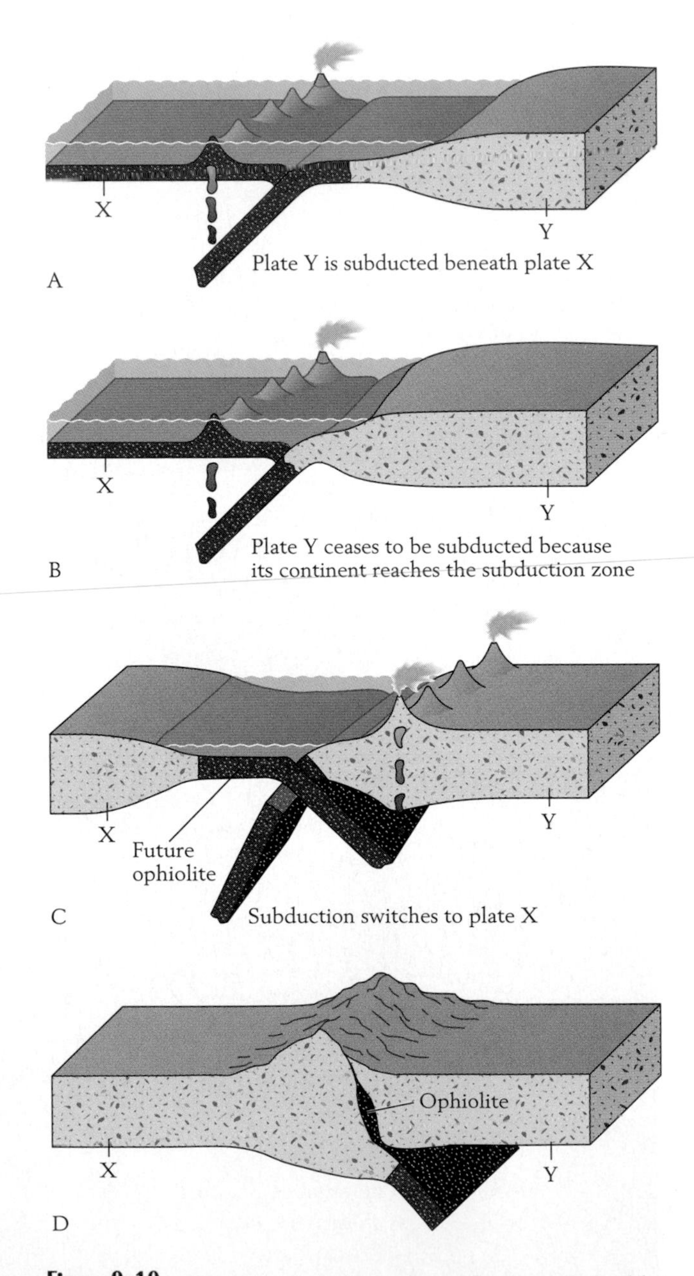

Figure 9-10 The processes of suturing and mountain building where two continents meet along a subduction zone.

opposite the continent is forced to descend into the mantle. If a second continent is riding on the newly subducting plate, it will eventually collide with the first continent along the subduction zone, and the two continents will be welded together. The juncture formed in this way is called a **suture.**

Continental suturing creates mountain chains because subduction along a trench causes the margin of one continent to wedge beneath the margin of the other (Figure 9-10*D*). The forces of collision cause both continental margins to thicken, and a mountain chain is uplifted along the suture. Often remnants of seafloor are pinched up along the suture, forming an ophiolite. **Ophiolites** consist of pillow lavas that formed oceanic crust and ultramafic rocks that formed the upper mantle. Often present as well are deep-sea sediments, such as turbidites, black shales, and cherts, that originally rested on the seafloor above the lavas and ultramafic rocks. In effect, ophiolites are samples of ancient ocean basins that have been conveniently elevated for our study. Because ophiolites often mark the positions of vanished oceans that once lay between continents, they are key features in our recognition of plate convergence along subduction zones.

Orogenesis without Continental Collision

Mountains can form along the margin of a continent that is resting against a subduction zone even when that continent does not collide with another. Figure 9-11 shows what happens when a continent encounters a subduction zone. Magma rises into the margin of the continent positioned above a subducted slab. Some of the magma reaches the surface and forms a chain of volcanoes that elevate the crust, forming mountain peaks. Some magma also cools within the crust, forming plutons, or massive intrusions of igneous rock.

In keeping with the principle of isostasy (p. 16), the addition of large volumes of low-density igneous rock to the base of the crust causes the crust

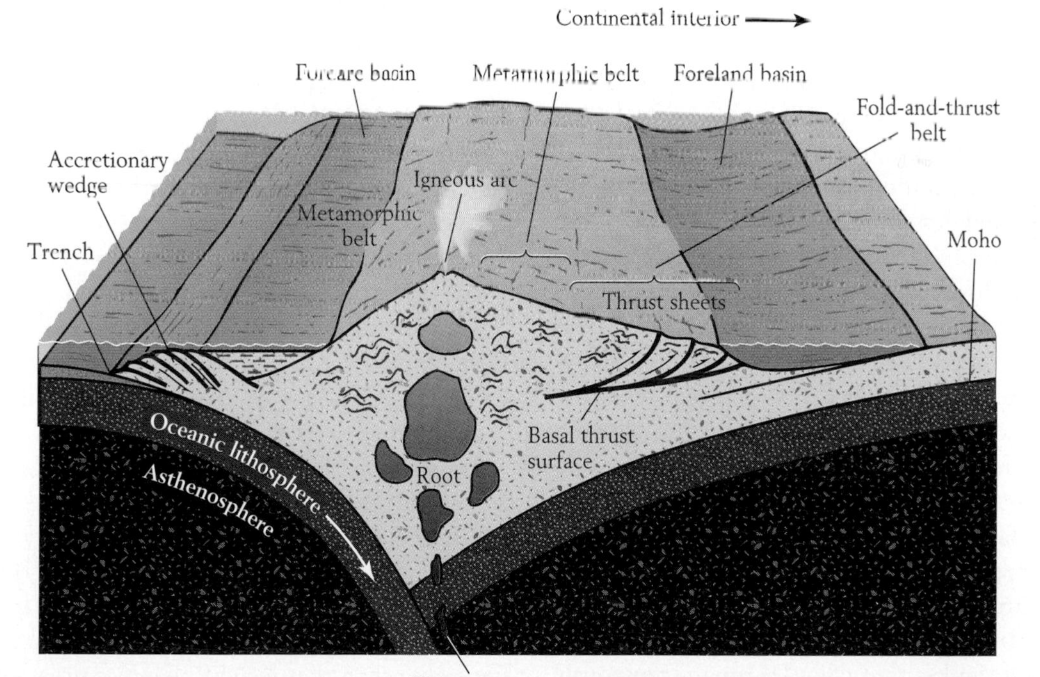

Figure 9-11 An idealized mountain chain forming where an oceanic plate is being subducted beneath the edge of a continent. This cross section illustrates the general symmetry of the mountain chain. Metamorphism dies out both toward the sea and toward the land from the central igneous arc, and beyond the metamorphic belt, in the direction of the continental interior, is a fold-and-thrust belt. Beyond the inland fold-and-thrust belt, the crust is warped downward to form a foreland basin, where sediments from the mountain system accumulate.

along the igneous arc to bob upward. This vertical movement contributes to the elevation of the mountain chain. At the same time, a root of crustal rock produced by the igneous arc extends downward beneath the mountain chain, balancing the weight of the mountains by displacing dense rocks in the asthenosphere.

An igneous arc like the one just described forms the core of a typical mountain chain. Extending along either side of this core is a belt of regional metamorphism (see p. 52). This belt consists of crustal rocks that have been metamorphosed by heat from the core and locally intruded by igneous activity issuing from the core. Rocks of this **metamorphic belt** (see Figure 9-11) are also deformed by processes that will be described shortly.

Metamorphism dies out in both directions from the igneous core. Toward the continental interior, the metamorphic belt gives way to a **fold-and-thrust belt** (see Figure 9-11). The folds of the fold-and-thrust belt are typically overturned away from the core of the mountain, reflecting the fact that the prevailing forces of deformation come from the direction of the core. Because of their distance from the igneous core, the preexisting sedimentary rocks in the fold-and-thrust area are largely unaffected by metamorphism and are folded less severely than the rocks in the metamorphic belt. As a result, the behavior of these sedimentary rocks during the deformation process is more brittle and less plastic than that of the rocks in the metamorphic belt. Thus the folds are broken by enormous thrust faults along which large slices of crust, known as **thrust sheets,** have moved away from the core. The thrust sheets usually slide along a basal thrust surface. A look at Figure 9-11 reveals how the telescoping action of folding and thrusting above a basal thrust shortens and thickens the crust.

Figure 9-12 shows the relationship between folds and thrust faults. After an overturned fold forms, continued stress along it can break the folded rock, forming a thrust fault. Figure 9-13, a cross section of the fold-and-thrust belt of the Rocky Mountains, shows thrust faults that slice through previously folded rocks.

Seaward of the igneous arc, along the subduction zone, rocks are also deformed by folds and thrust faults. This is the position of the accretionary wedge, in which thrust sheets are piled up along the continental margin (see also Figure 8-22).

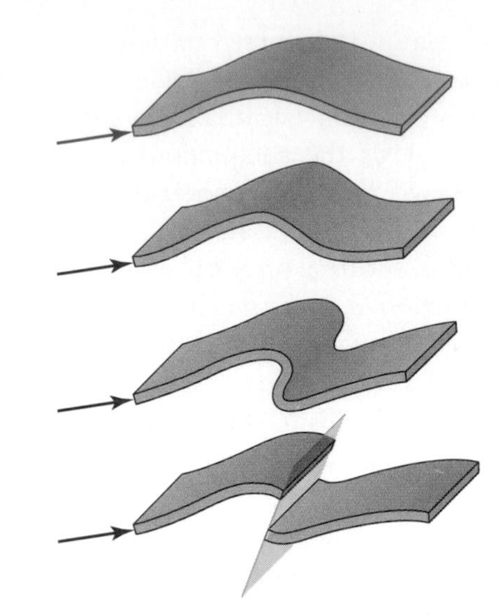

Figure 9-12 **An overturned fold can give rise to a thrust fault as force continues to be applied.** *(After F. Press and R. Siever, Earth, W. H. Freeman and Company, New York, 1986.)*

Mechanisms of Deformation

It is easy to understand how deformation occurs in the accretionary wedge along the trench (see Figure 9-11). Here compression breaks off slices of material and piles them up along thrust faults. The compression also folds material within individual thrust slices. Deformation within the metamorphic belts and the fold-and-thrust belt is more complex, having two primary causes. The first cause is simply pressure that the subducted plate applies to the mountain chain, pushing it laterally toward the interior of the continent. The resulting compression takes the form of folding near the igneous arc and folding and thrusting in the more brittle terrain farther toward the continental interior.

The second cause is less intuitively obvious. Its name, **gravity spreading,** is aptly descriptive, however. This mechanism depends on the fact that rock, although seemingly rigid, can deform under its own weight when that weight becomes great enough. The necessary weight can develop as the igneous arc builds the core of the mountain to a high elevation, while a root develops by isostatic adjustment. When the mountain chain becomes tall enough, it tends to spread laterally, like a mound of pudding heaped too high to remain stable. A mountain chain spreads by

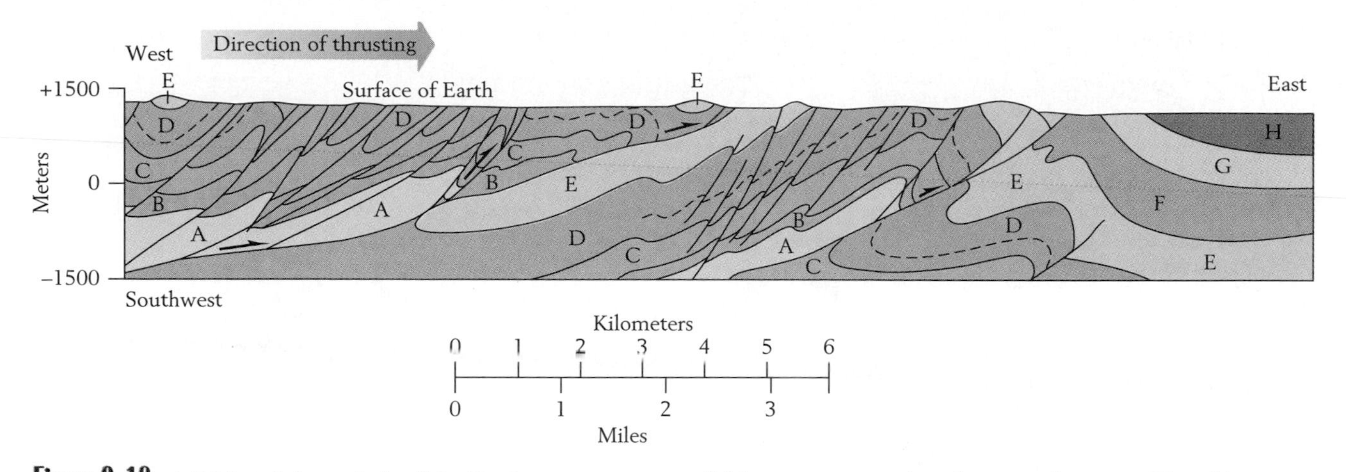

Figure 9-13 A fold-and-thrust belt of the Rocky Mountains southwest of Calgary, Alberta, Canada. Thrust faults, shown as dark lines, are intimately associated with overturned folds. Thrusting has been toward the east, and folds are overturned in the same direction. The oldest beds (A) are Lower Carboniferous, and the youngest (H) are Paleogene. *(After P. B. King, The Evolution of North America, Princeton University Press, Princeton, N.J., 1977.)*

deformation along folds and thrust faults. This process can cause thrust sheets to move uphill along the basal thrust, just as pudding that is heaped too high can spread up the gently sloping sides of a dish.

Foreland Basin Deposition

The downwarping of the lithosphere beneath an actively forming mountain chain continues for some distance beyond the fold-and-thrust belt. This activity produces an elongate foreland basin whose long axis lies parallel to the mountain chain (Figures 9-11 and 9-14). The foreland basin forms rapidly and is usually so deep initially that the sea floods it either through a gap in the mountain chain or through a passage around one end of the chain (Figure 9-14C).

The foreland basin typically subsides so quickly that the first sediments to accumulate in it are deep-water deposits, especially muds. Turbidites also form if the slope from the foot of the mountain is steep enough to send turbidity flows out into the basin. Figure 1-23, for example, displays turbidite deposits that accumulated in front of Cretaceous mountains in northern Alaska. When both shales and turbidites accumulate in foreland basins, they are collectively known as **flysch.**

As a mountain system evolves, folding and thrust faulting move progressively farther inland, and mountain building proceeds toward the continental interior. In the process, the shale or flysch deposits become folded and faulted. At the same time, the mountain core rises and sheds sediments more and more rapidly. Eventually sediment chokes the foreland basin, pushing marine waters out and leaving nonmarine depositional settings in their place (see Figure 9-14*D* and *E*). These settings comprise alluvial fans along the mountain chain and also river beds, floodplains, and other lowland environments. Collectively, these nonmarine sediments are termed **molasse.** Molasse deposits can accumulate to great thicknesses; as deformation continues, they too can sometimes be folded and faulted. During molasse deposition, the foreland may no longer be a topographic basin but may appear instead as a broad depositional surface sloping away from the mountain front (Figure 9-14*E*). As the foreland subsides beneath the accumulating sediments, however, it remains a *structural* basin in which the older strata have the form of a syncline (see Figure 9-6). Molasse deposits pinch out from the margin of a mountain belt toward the interior of the craton. Because of its prismlike configuration, a thick package of molasse is sometimes referred to as a **clastic wedge.**

The depositional transition from the deep-water sediments to nonmarine sediments occurs during the evolution of most foreland basins. Even molasse deposition comes to an end when orogenic activity stops. Not only does igneous activity cease in the core of the mountain chain, but so do folding and

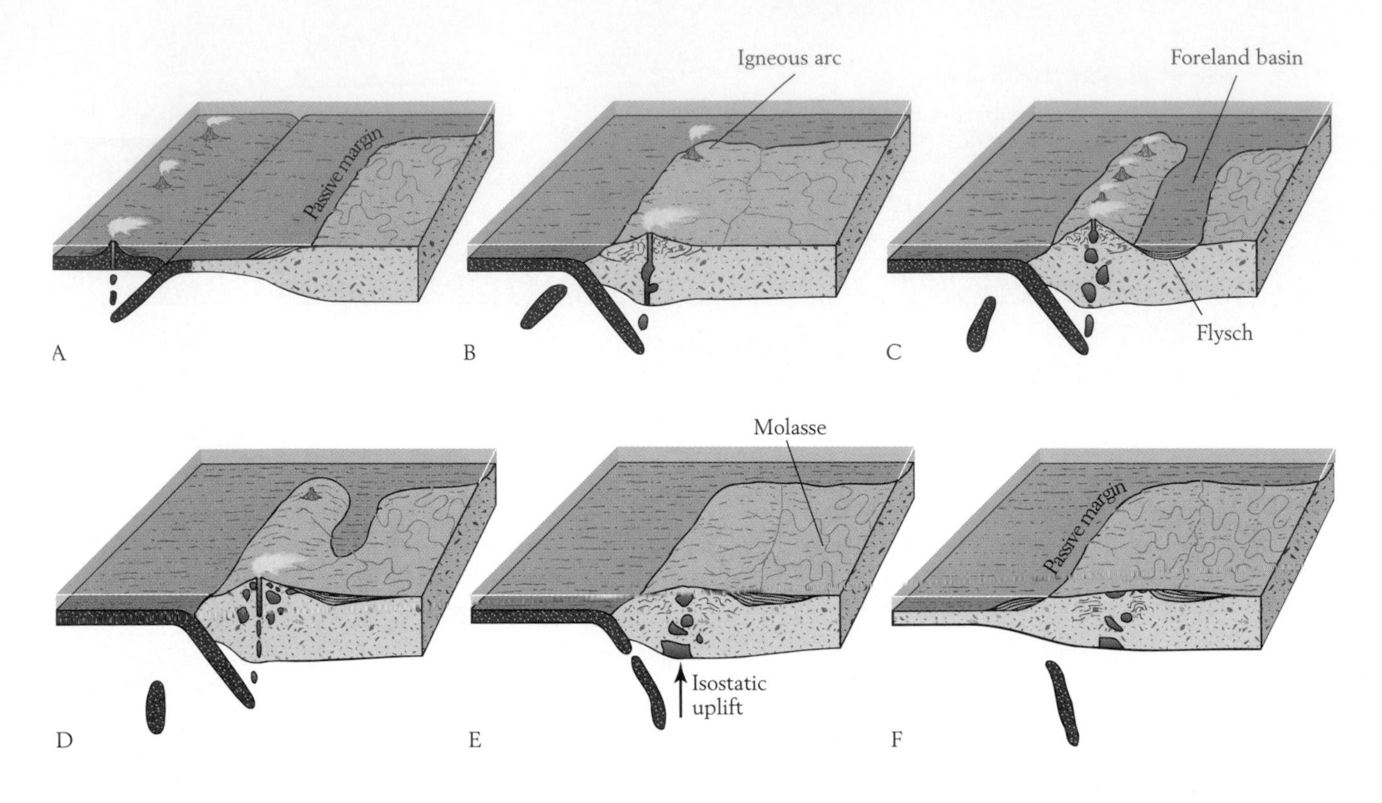

Figure 9-14 History of a typical mountain belt and foreland basin. *A.* A continent bounded by a passive margin approaches a subduction zone. *B.* The continent stalls at the subduction zone, forcing a reversal in the direction of subduction, and mountain building begins. C. A deep foreland basin forms and accumulates flysch. *D.* As mountain building progresses, the accumulation of sediment pushes the sea from the foreland basin. *E.* The foreland basin accumulates nonmarine sediment (molasse) until the mountain chain wears down, but its root remains and isostasy causes the mountain belt to bob up again. *F.* Isostatic uplift and erosion continue until no root remains, but geologic features of the mountain system remain in the bedrock of the region.

thrusting along the margin. Soon erosion subdues the mountainous terrain and the source of the molasse sediment disappears (Figure 9-14*F*).

After a mountain system is initially leveled by erosion, some of its root remains. That remnant of a root creates gravitational instability, so that the thick mass of felsic rocks in the mountain belt tends to rise up, just as a block of wood floating in water will do if its top is sliced off (see Figure 1-16). Orogenic belts bob up sporadically long after subduction has ceased, and each time they rise, erosion temporarily subdues them. This process can continue over hundreds of millions of years, until no root is left (Figure 9-14*E*). Even after the root is gone and the land is level, folds and faults remain in the bedrock of the region, as do igneous and metamorphic rocks—and often flysch and molasse of the foreland basin. These are the marks of an ancient mountain belt.

The Andes: Mountain Building without Continental Collision

The Andes mountains of South America are still in the process of being formed. In fact, Figure 9-11 may be viewed as an idealized representation of the Andean belt as it is today.

The Andes are associated with the ring of fire. This composite feature, which encircles much of the Pacific Ocean, is in most areas a product of the subduction zones where oceanic plates are colliding (see Figure 8-21 and Box 8-1). In certain segments of the ring, however, subduction is occurring instead along blocks of continental crust. One of these segments is the coast of South America, where the Andes continue to form.

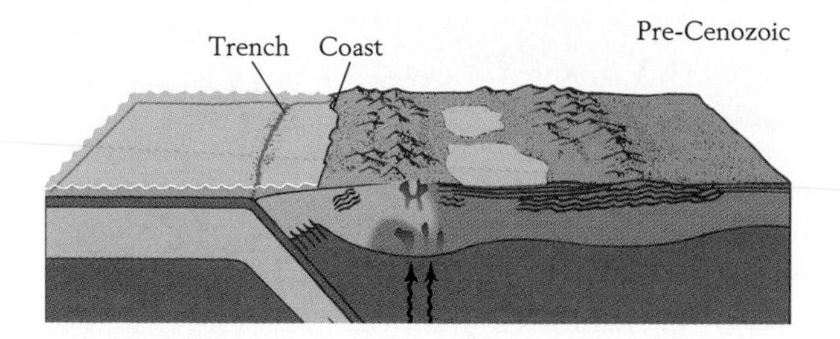

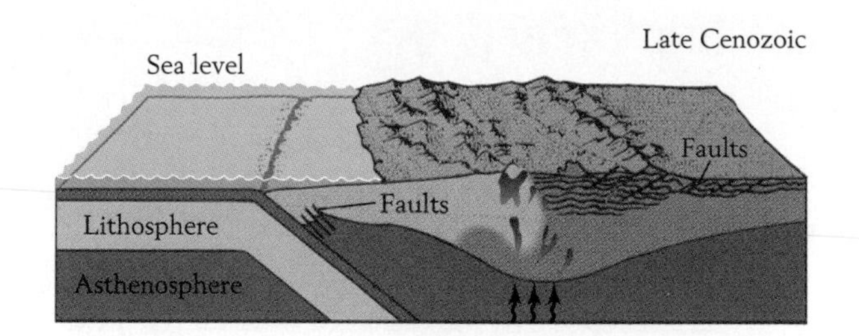

Figure 9-15 Formation of the Andes. During pre-Cenozoic time, igneous material was added to the crust from the oceanic plate descending along the marginal trench. During Cenozoic time, igneous activity shifted farther east. Thrust faulting has occurred both east and west of the area of igneous activity. *(After D. E. James, Scientific American, August 1973.)*

The Andean system is the longest continuous mountain chain in the world. Its history extends well back into the Paleozoic Era, but the present pattern of mountain building began early in the Mesozoic Era, when a subduction zone came to lie along the margin of South America (Figure 9-15). Enormous volumes of igneous rock have since risen from the subducted oceanic plate and have been added to the Andean crust, thickening it in some places to more than 70 kilometers (45 miles). When Charles Darwin sailed around the world on the *Beagle*, he noted the presence of Cenozoic marine fossils at high altitudes in the Andes. These fossils offered proof that the Andes had been elevated greatly during recent geologic time. Darwin also saw at firsthand that the movements occurred in pulses. He witnessed earthquakes during which land along the seacoast was suddenly raised several feet. From a distance Darwin also observed the eruption of Andean volcanoes. We now know that for about the last 200 million years the Andean crust has been thickened by the addition of igneous material below and also has been bobbing up isostatically. At the same time, volcanic rocks have been piled on top. We also understand why most major earthquakes originate in major orogenic belts such as the Andes (Box 9-1).

Igneous activity has steadily shifted toward the interior of South America during Mesozoic and Cenozoic time; in other words, magma has ascended at positions farther and farther inland (see Figure 9-15). Probably because today the subducted plate descends at a low angle, the zone where volcanoes are formed and magma cools below the surface to form intrusive rocks is now centered about 200 kilometers (125 miles) inland from the coast. Earlier the subducted plate probably descended at a steeper angle, thus reaching the depth of partial melting nearer the coast.

Sometimes a change in the angle of subduction reflects a change in the rate of plate movement. When the plate that is not being subducted is moving rapidly toward the subduction zone, it overrides and bears down on the upper part of the subducted slab. This pressure "rolls back" the plate, much as you might roll back a wrinkle in a carpet by pushing on it with a stick (Figure 9-16). Migration of the

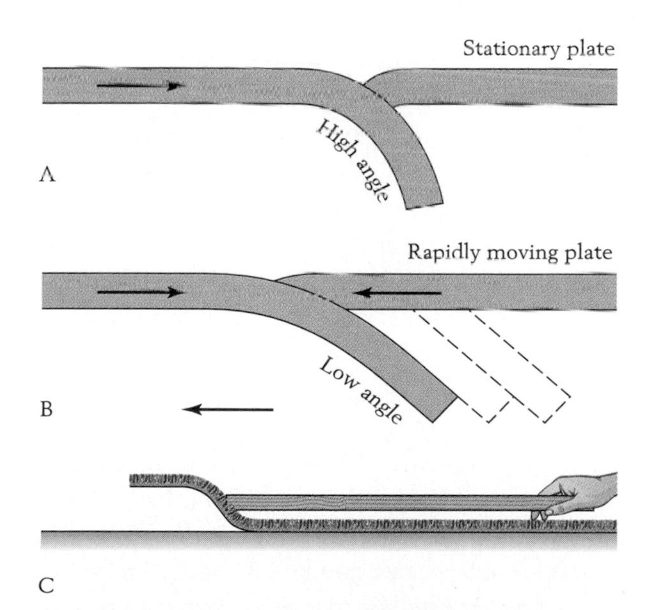

Figure 9-16 The rate of plate movement and the angle of subduction. *A.* The plate not being subducted moves slowly, and the weight of a subducted slab rotates the slab to a high angle. *B.* The plate not being subducted moves rapidly against the subducted slab, rolling it back and forcing it to maintain a low angle of descent. C. A situation analogous to the "roll back" phenomenon in *B;* the stick rolls back the wrinkle in the carpet.

For the Record 9-1

Where Earth Shakes

The Richter scale is the universal yardstick by which geologists measure the magnitude of earthquakes. An earthquake that scores 3 on the scale will be felt indoors by some people but will do no damage to buildings; one that reaches 6 will topple chimneys and weak walls; and one of magnitude 8 will destroy virtually all human structures. Most major earthquakes are sudden movements of earth at the boundaries of lithospheric plates. Since 1906, all continental earthquakes that have exceeded 8 on the Richter scale have originated in mountain belts produced by the collision of plates. A subducted plate moves downward in sudden steps. When the massive plate takes an unusually large step, it produces a very large earthquake, and where the subduction is beneath a continent, the continent shakes violently. Similarly, collision of continents along a subduction zone causes episodic earthquakes until the subduction zone shifts to a new location.

Two regions have been the sites of all large continental earthquakes of recent decades: mountain belts that form segments of the ring of fire around the Pacific Ocean and the chain of mountain belts that stretches from southern Europe to China. Individual earthquakes have taken tens of thousands of lives in densely populated sectors of these zones, especially in China and in Central and South America. The 1976 earthquake centered beneath the city of Tangshan, China, stands as the most disastrous of modern times. It leveled the city and killed about 240,000 people. In contrast, the much larger 1964 Alaskan earthquake, which originated beneath a bay near Anchorage, caused only 131 deaths. Eleven of the victims lost their lives as far away as Crescent City, California, where a huge wave crashed ashore. This wave had spread from the coast of Alaska, where a pulse of subduction had elevated the continental margin and disturbed shallow seas. Few lives were lost in Alaska because the population near the earthquake's center was sparse.

The famous San Francisco earthquake of 1906 occurred when the Pacific plate suddenly moved northwestward against the North American plate along the San Andreas fault (p. 231). Although this earthquake was of magnitude 8.3, it caused little damage outside California because its shock waves were damped by the soft sediments that form much of the terrane nearby. Nonetheless, the San Francisco earthquake and the fires it triggered claimed about 700 lives. If an earthquake of similar magnitude struck the same area today, it would take an even heavier toll because the population of the region has exploded since 1906. The Loma Prieta earthquake, caused by movement along the San Andreas fault in 1989, was only of magnitude 7.0, yet it was more costly than any previous natural disaster in the history of the United States, causing more than $10 billion in property damage in California as well as more than 2400 human injuries.

Only rarely does an earthquake that registers higher than 8 on the Richter scale originate in the middle of a continent. In 1811 and 1812, however, earthquakes of this magnitude shook a vast area of the United States. They were named the New Madrid earthquakes, after a town close to their place of origin in Missouri. Because these earthquakes shook the rigid craton, their shock waves traveled great distances, ringing church bells and cracking pavement as far away as the District of Columbia. Fortunately, so few towns had been built west of the Appalachians by 1811 that the New Madrid earthquakes caused no more than about ten human deaths. Even though the New Madrid earthquakes originated far from any plate boundary, they also resulted from plate movements. They were caused by movement along faults that formed long ago, in Proterozoic time, along a failed rift that sliced into the southern flank of North America. The Mississippi River follows the axis of the ancient rift, where Earth's crust remains thin and weak even though the faults lie buried beneath Phanerozoic sediments. Every year many tiny earthquakes vibrate the land near New Madrid without attracting the average resident's notice. An earthquake that rivals those of 1811 and 1812 would damage human constructions across a large region of the United States. Such an event may not occur for many decades or centuries, but long-range predictions are impossible. Episodic movements along the ancient faults, some small and some large, apparently result from stresses produced by North America's movement over the asthenosphere.

The Borah Peak earthquake in 1983 produced this fault scarp in the Rocky Mountains.

Only early warnings, safety instruction, and construction of resilient buildings can protect populations against earthquakes. Less than two years before the catastrophic 1976 earthquake in China, earth movements damaged hundreds of buildings in the Chinese city of Haicheng. Weeks before that crisis had begun, however, sensitive instruments had detected subtle but ominous geologic changes, including a slight tilting of the land. The instruments' warning led to evacuation before Earth began to shake, and casualties were relatively light.

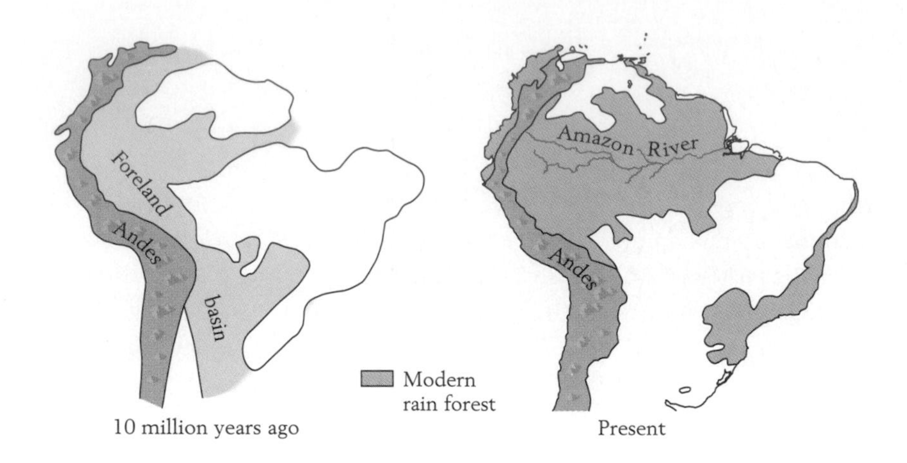

Figure 9-17 The shallow sea that occupied the foreland basin of the Andes during the Miocene Epoch. *(After S. D. Webb, Science 269:361–362, 1995.)*

subduction zone causes the subducted slab to dip at a low angle. In contrast, when the plate that is not being subducted is moving too slowly to roll the subduction zone back rapidly, the subducted slab is free to rotate toward a vertical position (Figure 9-16*A*).

As recently as 10 million years ago, during the Miocene Epoch, a long seaway occupied the foreland basin to the east of the Andes (Figure 9-17). Though relatively shallow, it stretched along most of the length of the continent. Interestingly, a smaller seaway connected this inland sea to the Atlantic Ocean along the axis of a failed rift that today marks the path of the lower Amazon River (see Figure 9-3). As the foreland basin filled in and its marine waters receded, freshwater runoff that had flowed into the shallow inland sea coalesced to form the Amazon River. The Amazon is still home to dolphins, stingrays, and sea cows—freshwater species descended from marine ancestors that occupied the foreland basin sea millions of years ago.

The Andes orogenic belt has matured to the point where large volumes of sediment shed from the mountains now keep the sea from flooding the zone to the east of the Andes. Today only nonmarine molassic sediments accumulate here.

The Andes exemplify mountain building along a single continent bordered by a subduction zone. In contrast, the Himalayas—the second actively forming mountain range we will examine—are the product of collision between two continents.

The Himalayas

The Himalayas, having formed primarily during the Neogene Period, are a relatively youthful mountain system. Partly because of their youth, the Himalayas are the tallest mountain range on Earth (p. 230). The Himalayan front rises abruptly from the flat Ganges Plain; not far from the front, Mount Everest, the tallest mountain on Earth, towers to 8848 meters (5.5 miles) above sea level. Even the broad Tibetan plateau, which lies to the north of the Himalayan front (Figure 9-18), stands at an average elevation of about 5 kilometers (3 miles) above sea level—higher than any mountain peak in the 48 contiguous United States.

Plate Movements

The Himalayas are part of a great series of mountain chains of Cenozoic origin that stretch from Spain and North Africa to Indochina (Figure 9-19). All of these chains formed as a result of the northward movement of fragments of Gondwanaland. Recall that Gondwanaland was the immense southern continent that became part of Pangea at the end of the Paleozoic Era and then broke apart during the Mesozoic Era (p. 211). The Alps and other Cenozoic mountains of the Mediterranean region formed as the African plate moved northward against the Eurasian plate. The Indian peninsula, which projects southward from the Himalayas, was originally a fragment of Gondwanaland. By late in the Mesozoic Era, this fragment was moving northward as an island continent within the large Australian plate (Figure 9-20). The collision of this Indian craton with Eurasia resulted in the uplifting of the Himalayas.

When did the Indian craton arrive? During Eocene time, shallow seas covered much of the Indian craton, and limestones were laid down over large areas. Coarse sediment derived from mountainous

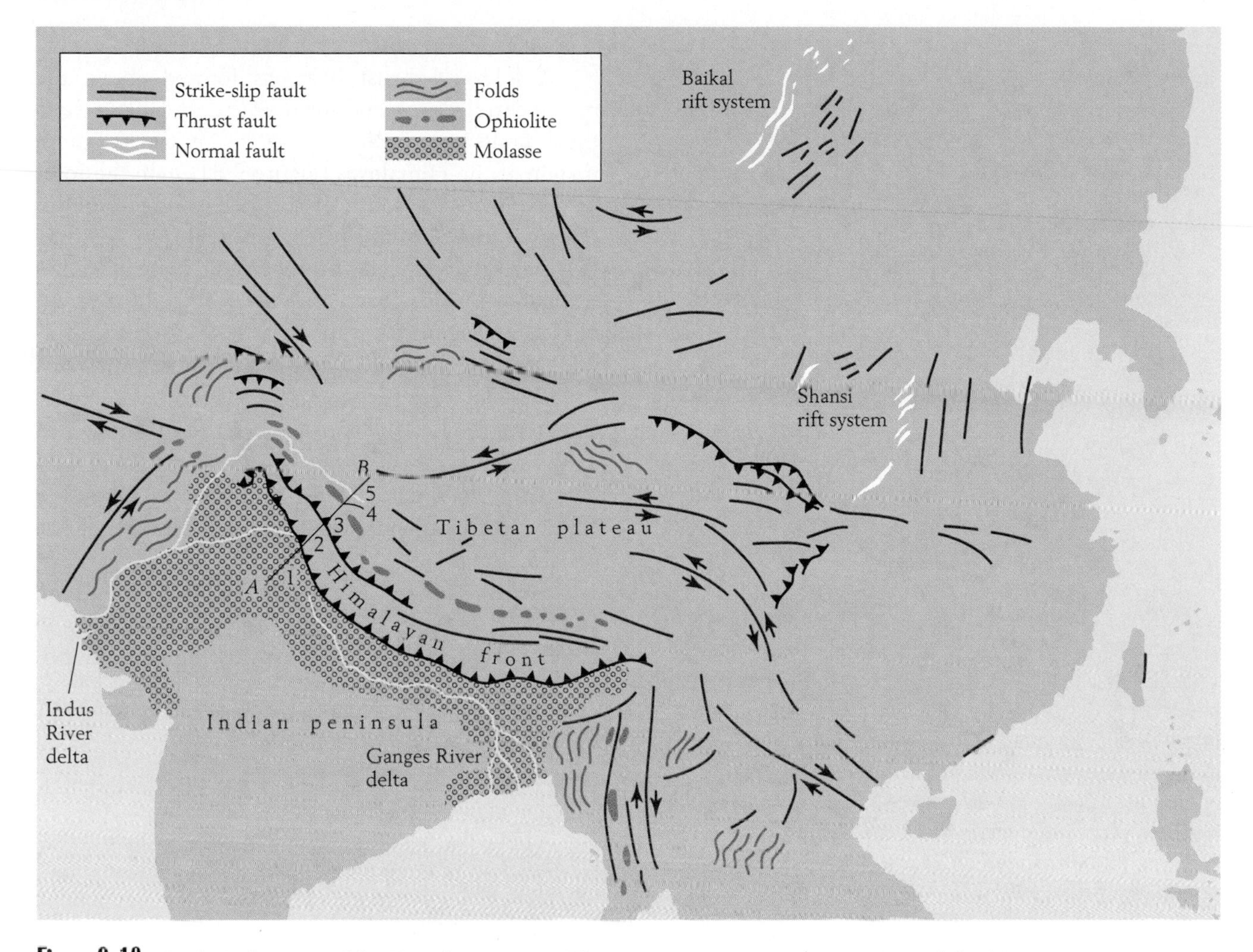

Figure 9-18 **Geologic features of the Himalayan region.** The high-standing Tibetan plateau is bounded by thrust faults, especially in the south, and molasse is being shed southward from the plateau. Numerous strike-slip faults throughout the region seem to have permitted the Asian crust to squeeze eastward, as the Indian peninsula has pushed northward. *(Modified from P. Molnar and P. Tapponier, Scientific American, April 1977.)*

terrain was first deposited on top of the limestones in late Miocene time. Apparently it was not until shortly before this time that mountain building began. Sediments in the Indian Ocean provide additional evidence of the timing of orogenic activity. The oldest deep-sea turbidites deposited offshore from the Indus and Ganges rivers (see Figure 9-18) date to the middle Miocene. The rivers themselves cannot be much older, and they came into being when the Himalayas began to form, perhaps 20 million years ago. Indeed, much of the Himalayan chain has been uplifted during the last 15 million years.

The Pattern of Orogenesis

Figure 9-21 shows in greater detail how the Himalayas formed. When India was approaching Eurasia, riding on the Australian plate, the northern margin of this plate was being subducted beneath Eurasia (see Figure 9-20). When India arrived, being a continental mass, it could not be subducted. As a result, subduction ceased, and so did the associated igneous activity along the southern margin of Tibet. Convergence of the Australian and Eurasian plates continued, however, and about 20 million years ago India began to wedge beneath the southern margin of Tibet without descending into the asthenosphere (Figure 9-21*B*).

Sediments of the forearc basin that had bordered Tibet were squeezed up along the suture, along with material of the accretionary wedge and solid oceanic crust, to form an ophiolite. At some unknown time another dramatic event took place: the northern margin of India, consisting of sediments and underlying continental crust, broke away from the rest of the

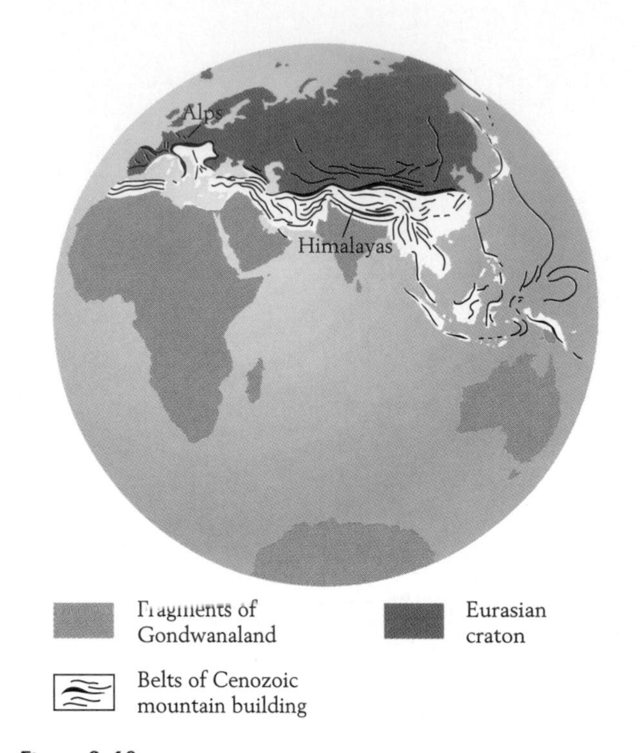

Figure 9-19 The series of mountain chains that formed along the southern margin of Eurasia when fragments of Gondwanaland moved northward against the large northern continent during the Cenozoic Era. *(After H. Cloos, Einfuhrung an die Geologie, Verlag von Gebrüder Borntraeger, Berlin, 1936.)*

Indian continent. The remaining Indian continent then slid beneath the margin of Eurasia for at least 100 kilometers (about 60 miles) along a huge thrust fault, now known as the main central thrust. This fault can be seen today in many areas of the Himalayas, where valleys have cut deep into the mountains. Movement along the main central thrust ceased sometime before 10 million years ago and a new fault, the main boundary fault, developed below it (Figure 9-21C). Movement along this fault has continued to the present day. The result of the movement along the two faults has been a great thickening of the Indian crust, as the margin of India has underthrust the slices of crust that have broken from it.

Presumably, as the Indian subcontinent continues to press against Tibet, a new thrust fault will form sometime in the future so that yet another slice of crust will be added to the Himalayan region. In any event, it is doubling of the crust that has made the Himalayas the tallest mountain chain of the modern world.

A fold-and-thrust belt has formed above the main central thrust and main boundary faults where these faults approach the surface at the southern margin of the Himalayas (Figure 9-21C). In the foreland basin to the south of this belt, a huge body of sediment, the Siwalik beds, has formed from material that has eroded from the mountains. The Siwaliks, which have yielded large numbers of fossil mammals of Neogene age, constitute molasse that has accumulated in a foreland basin that formed where the crust has been depressed by the adjacent mountain chain. This foreland basin has never been deep enough to admit the ocean, however, so it has received only nonmarine sediments. As the fold-and-thrust belt has continued to advance, it has deformed some of the Siwalik strata.

Earthquakes still rumble through the Himalayan region as a result of movement along faults, and there is every reason to believe that mountain building here is far from over.

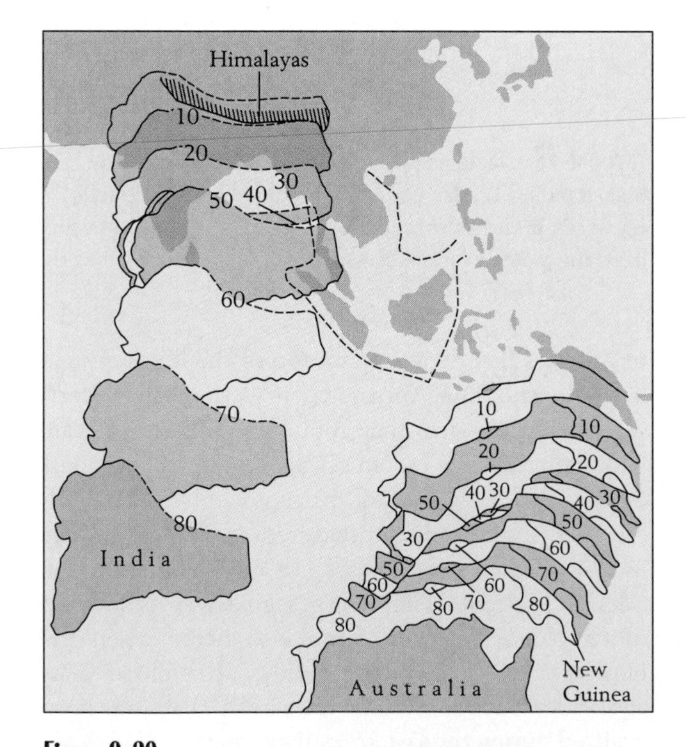

Figure 9-20 Northward movement of the Indian craton between 80 and 10 million years ago. Numbers represent the millions of years before the present when geographic boundaries reached various positions. *(After C. McA. Powell and B. D. Johnson, Tectonophysics 63:91–109, 1980.)*

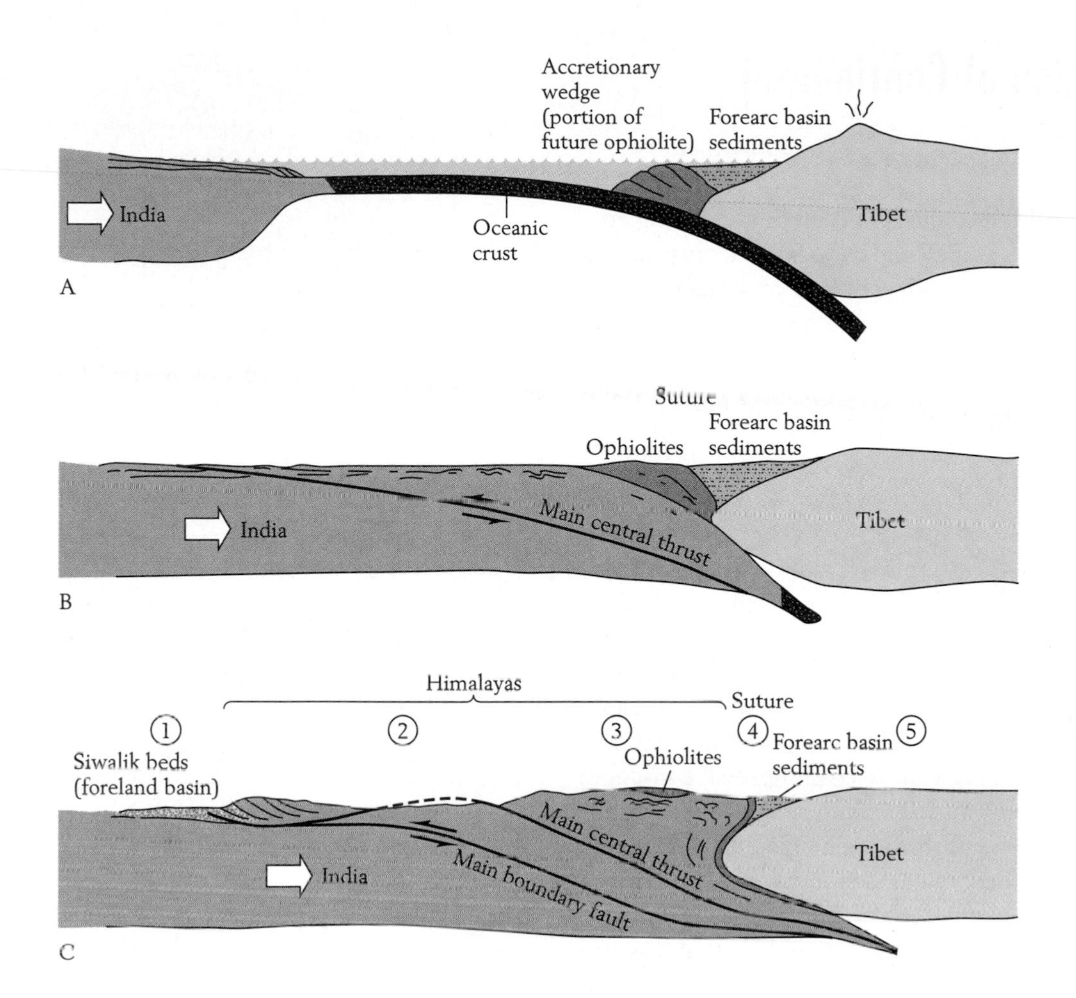

Figure 9-21 **Cross section showing how the Himalayas may have formed when the Indian craton wedged beneath the margin of Eurasia.** *A.* Slightly before 20 million years ago the Indian craton was rafted toward Eurasia as part of a plate being subducted beneath Tibet. *B.* About 20 million years ago the Indian craton began to wedge beneath Tibet and fractured along the main central thrust. Movement along this fault thickened the crust, forming mountainous terrain. Compression in the suture zone uplifted and deformed the accretionary wedge that had bordered Tibet, producing ophiolites. *C.* Today movement has shifted to a new thrust fault, the main boundary fault, and the crust has further thickened. Molasse is being shed southward, and older molasse deposits are being deformed in the vicinity of the main boundary fault. (Numbers refer to zones identified in Figure 9-18 along line *AB*.) *(Adapted from P. Molnar, American Scientist 74:144–154, 1986.)*

Suturing of Small Landmasses to Continents

Peninsular India, which became attached to Eurasia early in the Himalayan orogeny, is a relatively large landmass. Smaller pieces of continental lithosphere, known as *microcontinents,* also become sutured to large continents, as do island arcs that have ceased to be active but remain as large bodies of volcanic rock and sediment.

Later chapters will describe the attachment of several small landmasses to the eastern and western margins of North America during the past 500 million years. The addition of these **exotic terranes** has expanded the size of the continent considerably, especially in the west, where they constitute large areas of the United States and Canada, from Alaska to California.

Tectonics of Continental Interiors

The powerful forces that build mountains along continental margins can even buckle Earth's crust or create faults far inland from the zone of orogenesis. The causes of tectonic activity in continental interiors are poorly understood, but it is evident that the building of both the Appalachian and Rocky mountains caused mild deformation in the interior of the North American craton.

Mountain building in eastern and western North America has caused stresses to build up in neighboring regions of the continental interior, warping the crust in irregular patterns. The result was minor faulting and gentle folding that produced structural basins and domes. A *structural basin* is a circular or oval depression of stratified rock, and a *structural dome* is a comparable uplift—a blisterlike structure (Figure 9-22). Both result from compressive forces of the sort that form anticlines and synclines when they are applied along a broader zone (see Figure 9-6). In fact, an extremely elongate dome amounts to an anticline that plunges in two directions. Similarly, a very long basin amounts to a syncline that has upturned ends.

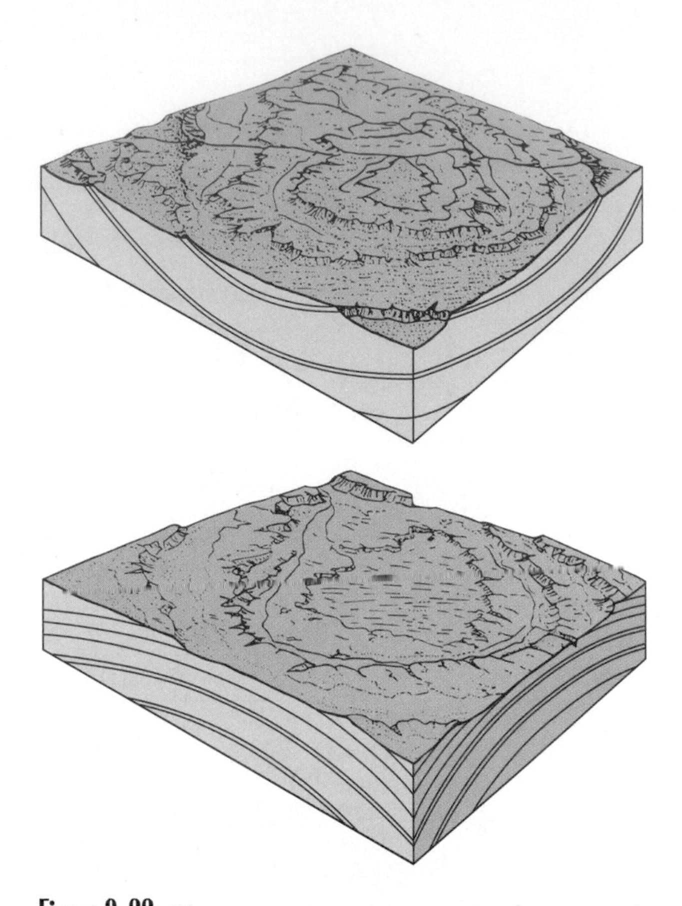

Figure 9-22 The concentric outcrop pattern of a structural basin (below) and of a structural dome (above). The rocks in the center of the structural basin are relatively durable and thus have remained at a high elevation; those in the center of the structural dome are relatively weak and have been deeply eroded to form a topographic basin. *(After W. K. Hamblin and J. D. Howard, Exercises in Physical Geology, Burgess Publishing Company, Minneapolis, 1975.)*

Both basins and domes, when eroded, yield concentric, circular bands of outcrop for stratified rocks. There is a simple way to distinguish the outcrop pattern of a dome from that of a basin (see Figure 9-22). In a dome, the oldest beds lie in the center, whereas in a basin, it is the youngest beds that are centrally positioned. Figure 9-22 illustrates how rocks within a structural basin that are resistant to erosion can form topographic ridges; similarly, erosion of weak rocks in the center of a structural dome can create a central basin.

The Black Hills of South Dakota are the surface expression of an oblong dome (Figure 9-23). Although they lie about 250 kilometers (150 miles) from the Rockies, they rose up with these mountains early in the Cenozoic Era, and erosion has since exposed ancient rocks in the center of the dome.

Large domes and basins also originate deeper within the interiors of cratons. The entire state of Michigan, for example, constitutes the central part of a structural basin that began to form during the Paleozoic Era (Figure 9-24). It remains a mystery why this basin originated far from plate tectonic activity along the margins of North America.

Occasionally rocks of continental interiors undergo small movements along faults that lie far from zones of rifting or mountain building. Most movements of this kind probably result from slight amounts of compression or extension of the continental lithosphere as it moves over the asthenosphere. Less than two centuries ago, earthquakes with epicenters in Missouri issued from movements of this kind along faults that formed long ago, during Proterozoic time (see Box 9-1).

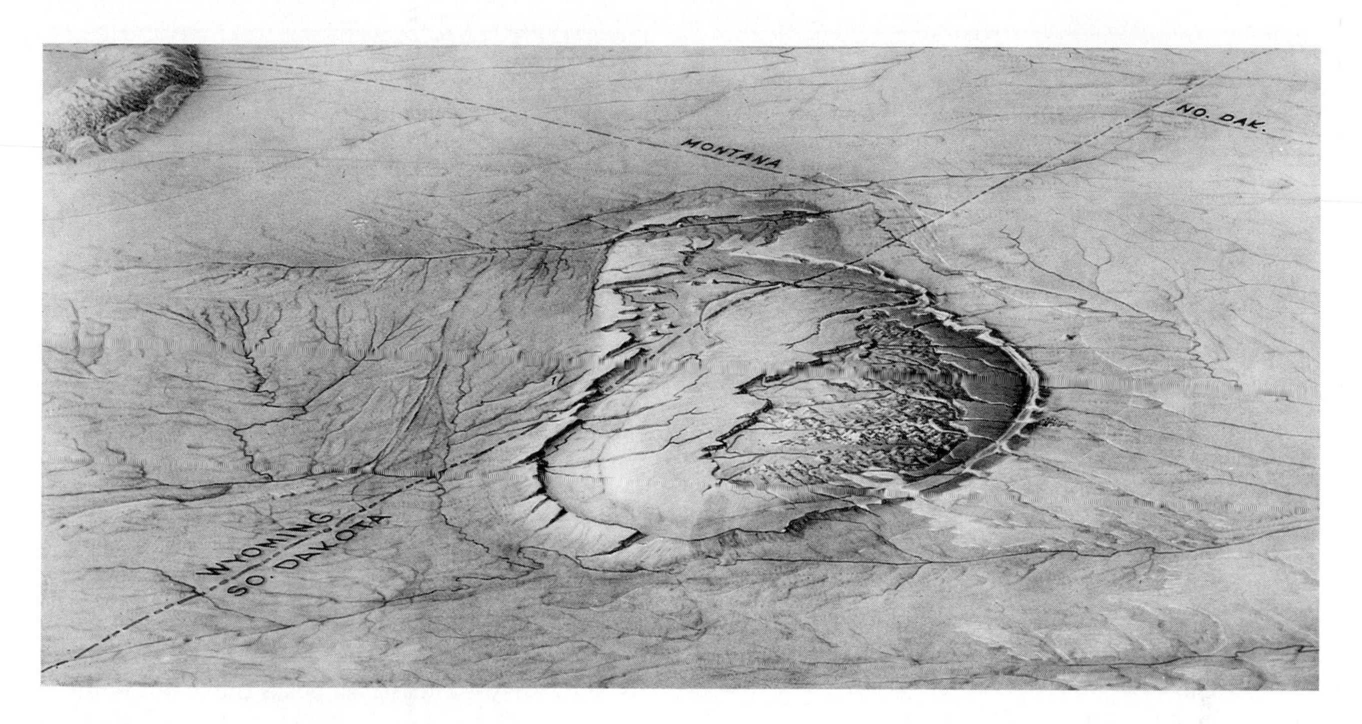

Figure 9-23 The Black Hills, a blisterlike structural dome to the east of the Rocky Mountains. Paleozoic and Mesozoic strata flank crystalline Archean rocks that erosion has exposed in the center of the dome. *(From J. S. Shelton, Geology Illustrated, W. H. Freeman and Company, New York, 1966.)*

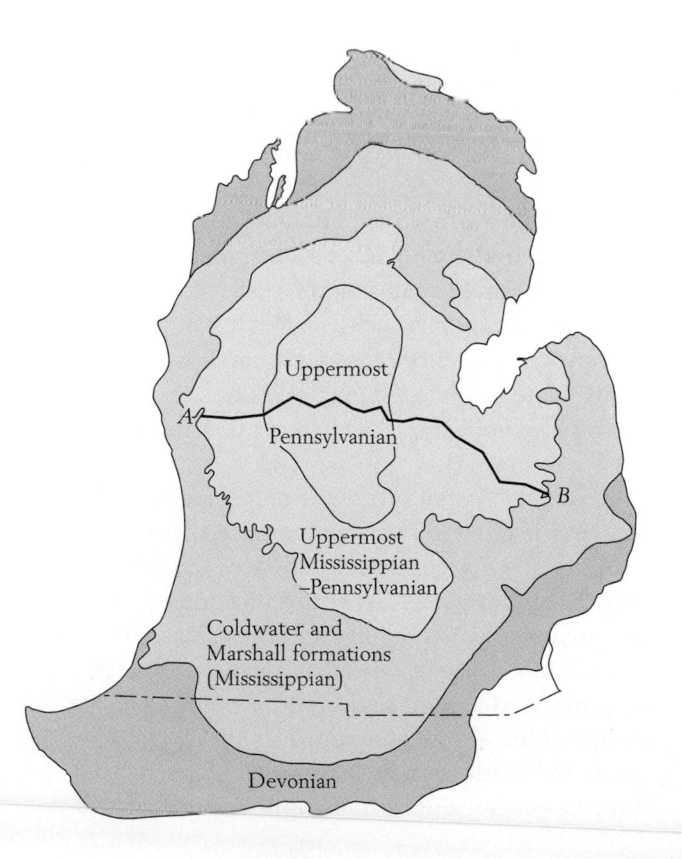

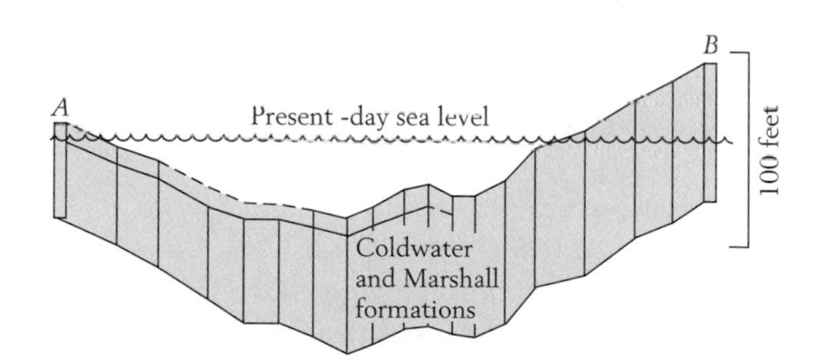

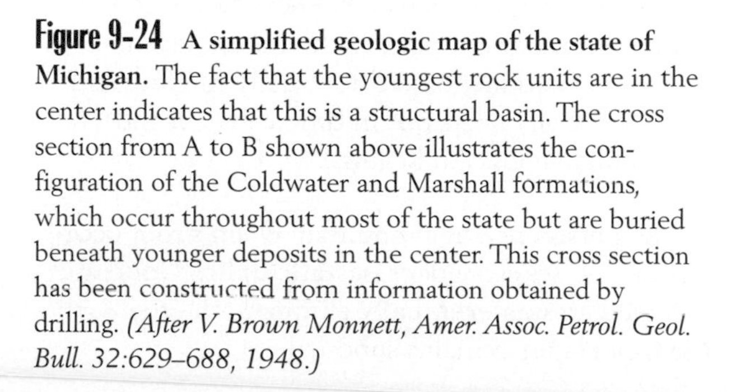

Figure 9-24 A simplified geologic map of the state of Michigan. The fact that the youngest rock units are in the center indicates that this is a structural basin. The cross section from A to B shown above illustrates the configuration of the Coldwater and Marshall formations, which occur throughout most of the state but are buried beneath younger deposits in the center. This cross section has been constructed from information obtained by drilling. *(After V. Brown Monnett, Amer. Assoc. Petrol. Geol. Bull. 32:629–688, 1948.)*

Chapter Summary

1 Fracturing of continents often begins with the doming of continental crust in several places. Each dome then fractures to form a three-armed rift system. The joining of some of the rift arms may produce a fracture that cuts across the entire continent.

2 Block faulting and deposition of thick siliciclastic sequences and evaporites often mark the beginning of continental fracturing.

3 A mountain chain also forms when the edge of a moving continent encounters a subduction zone. This situation is causing the elevation of the Andes today.

4 Continental material is of such low density that it cannot be consumed by the mantle. As a result, when two continents converge along a deep-sea trench, neither becomes subducted; rather, the two become sutured together. The Himalayas formed in this way, when the crustal block that now forms peninsular India was forced beneath the margin of Asia, doubling the thickness of the crust.

5 When a seafloor remnant, called an ophiolite, is found within a modern continent, it marks the position of an ancient ocean that disappeared when two continents were united.

6 An igneous arc occupies the central axis of a typical mountain chain, and it is bordered by a metamorphic belt, beyond which lies a less intensely deformed fold-and-thrust belt.

7 A foreland basin forms inland of the mountain chain, because the weight of the mountains depresses the continental crust. Usually the foreland basin is initially deep, but in time it fills in and nonmarine deposition begins.

8 The persistence of a mountain chain's root occasionally causes secondary isostatic uplift. Uplift and renewed erosion eventually eliminate the root, and the topography remains subdued.

Review Questions

1 What geologic features enable us to recognize ancient continental rifting? What features enable us to recognize ancient subduction zones?

2 What are failed rifts, and how are they important to our understanding of the breakup of continents? (Hint: Refer to Figure 9-3.)

3 What is flysch and where does it form?

4 What is molasse? Why does it normally accumulate after flysch?

5 Why are the Andes taller than the Appalachians?

6 Why are the Himalayan peaks the tallest in the world today?

7 How can mountain chains form without continental collision?

8 Why do mountains have roots?

9 Are the rocks that become ophiolites within a mountain chain older or younger than molasse deposits that form along the mountain chain?

10 Examine a world map or globe to locate mountain chains that are not discussed in this chapter. Then locate these chains on the plate tectonic map of the world (see Figure 8-21). See if you can figure out how the presence of each mountain system might relate to plate tectonic processes. (Some of the answers appear in the chapters that follow.)

11 Using the Visual Overview on p. 232 and what you have learned in this chapter, trace the history of a continental margin that experiences the following events: (a) it originates by rifting, becomes a passive margin along which sediment accumulates, (b) stalls at a subduction zone, where subduction reverses, (c) grows a mountain chain, (d) becomes a passive margin again when the igneous arc that formed it ceases to function, and (e) eventually loses its root through erosion and isostatic uplift.

Additional Reading

Allegre, C., *The Behavior of Continental and Seafloor Mobility*, Cambridge University Press, New York, 1988.

Bonati, E., "The Rifting of Continents," *Scientific American*, March 1987.

Molnar, P., "The Geologic History and Structure of the Himalaya," *American Scientist* 74:144–154, 1986.

Murphy, J. B., and Nance, R. D., "Mountain Belts and the Supercontinent Cycle," *Scientific American*, April 1992.

Russo, R. M., and Silver, P. G., "The Andes' Deep Origins," *Natural History*, February 1995.

Chapter 10

Major Chemical Cycles

Many environmental changes that affect broad regions of Earth are of a chemical nature—or they are the physical results of chemical changes. The most widely publicized of these environmental changes are the increases in so-called Greenhouse gases—atmospheric gases that trap warming solar radiation near Earth's surface—caused by human activities. This chapter describes evidence that carbon dioxide, the most important greenhouse gas, has undergone major changes during the Phanerozoic Eon. During long stretches of geologic time, Earth's climate has been even warmer than it will become during the next few decades as a result of human induced greenhouse warming. These ancient "greenhouse" intervals serve as models for the future Earth system. Humans evolved in a much cooler, "icehouse" world, more than 100,000 years before we began to burn fossil fuels at a rate that now threatens our habitat. Will human activities eventually turn this icehouse world into a greenhouse world? Some chemical cycles influence the chemistry of the ocean in ways that affect marine organisms and the sediments they produce. This chapter explores how changes in rates of plate tectonic activity have altered the chemistry of the ocean, with profound effects on the kinds of organisms that have produced reefs and carbonate sediments.

Reservoirs

The chemical changes that we will review are quite simple. Most are changes in the rates at which key chemical elements and compounds of the Earth system move in huge cycles that carry them through two or more vast reservoirs. These **reservoirs** are bodies of chemical

A moist temperate forest in Oregon, in which tall, deeply rooted Douglas fir and rapid recycling of water produce rapid weathering.

Visual Overview

Key Chemical Cycles in Earth System History

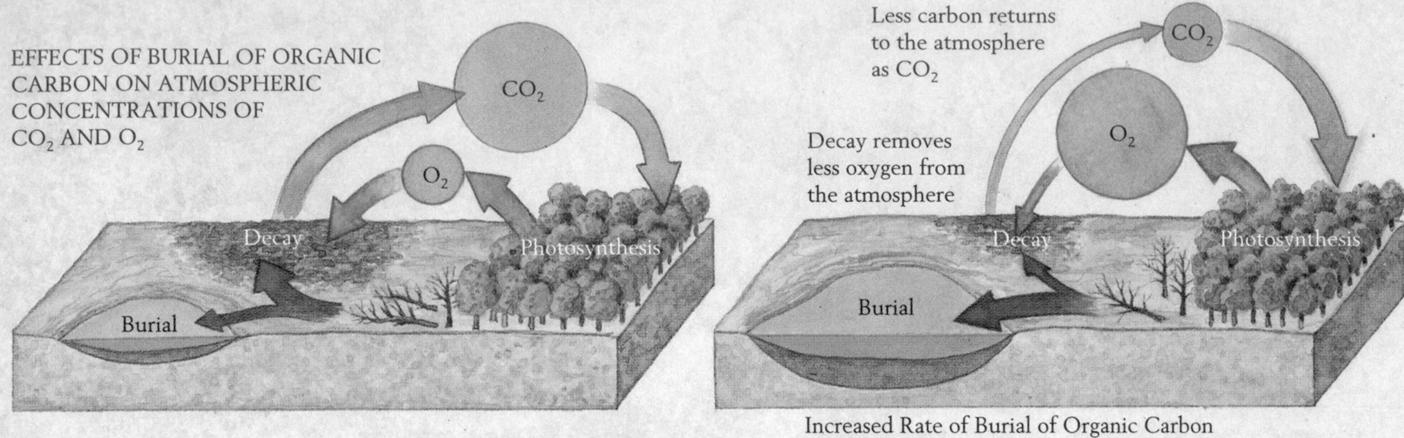

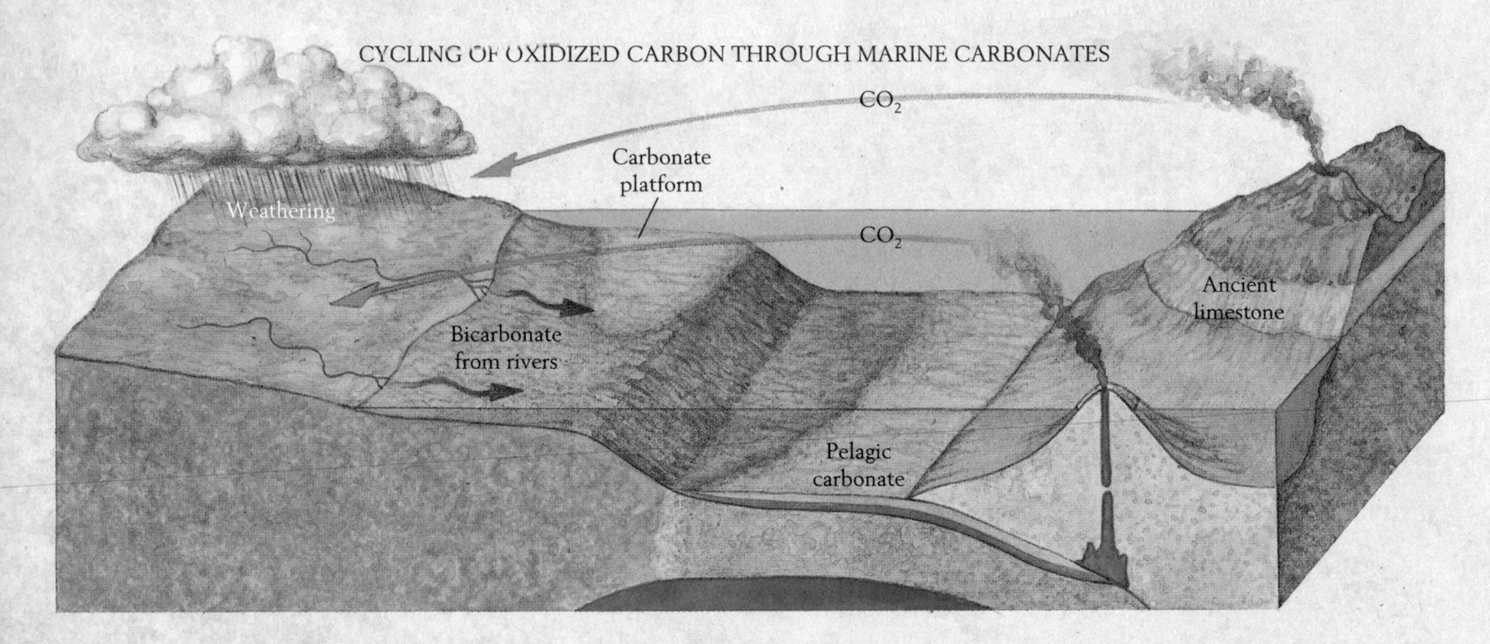

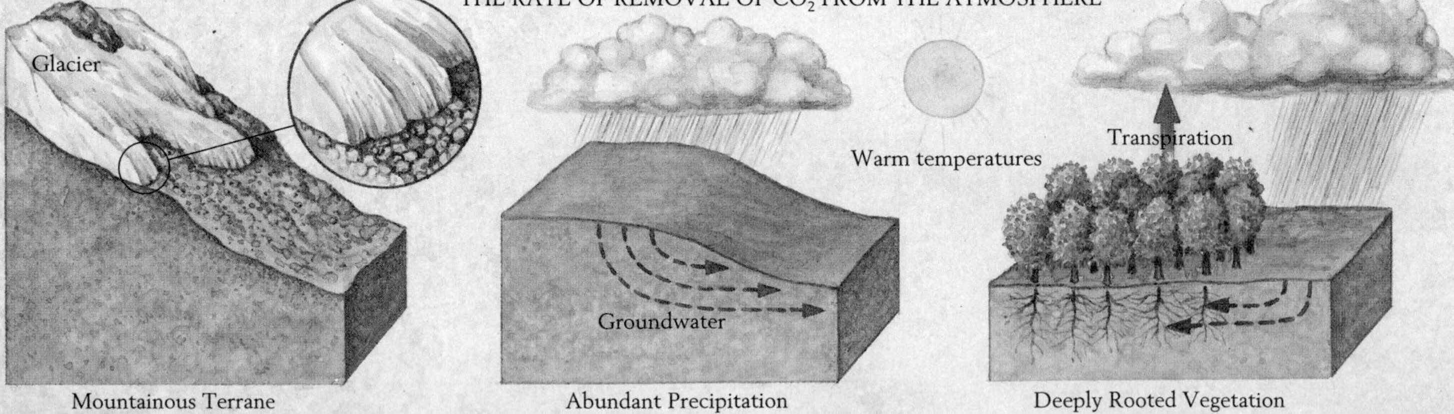

Chemical cycles that influence Earth's climate, the composition of its atmosphere, and the ratios of carbon isotopes in its atmosphere and natural waters.

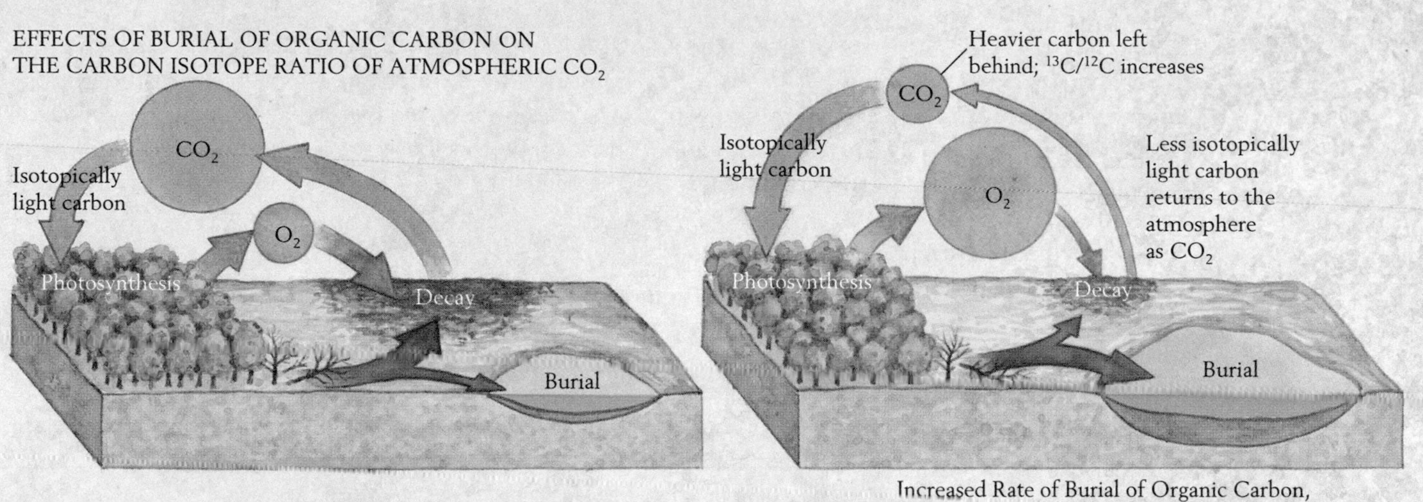
EFFECTS OF BURIAL OF ORGANIC CARBON ON THE CARBON ISOTOPE RATIO OF ATMOSPHERIC CO_2
CO_2
Isotopically light carbon
O_2
Photosynthesis
Decay
Burial
Heavier carbon left behind; $^{13}C/^{12}C$ increases
CO_2
Isotopically light carbon
O_2
Less isotopically light carbon returns to the atmosphere as CO_2
Photosynthesis
Decay
Burial
Increased Rate of Burial of Organic Carbon, Which Is Isotopically Light

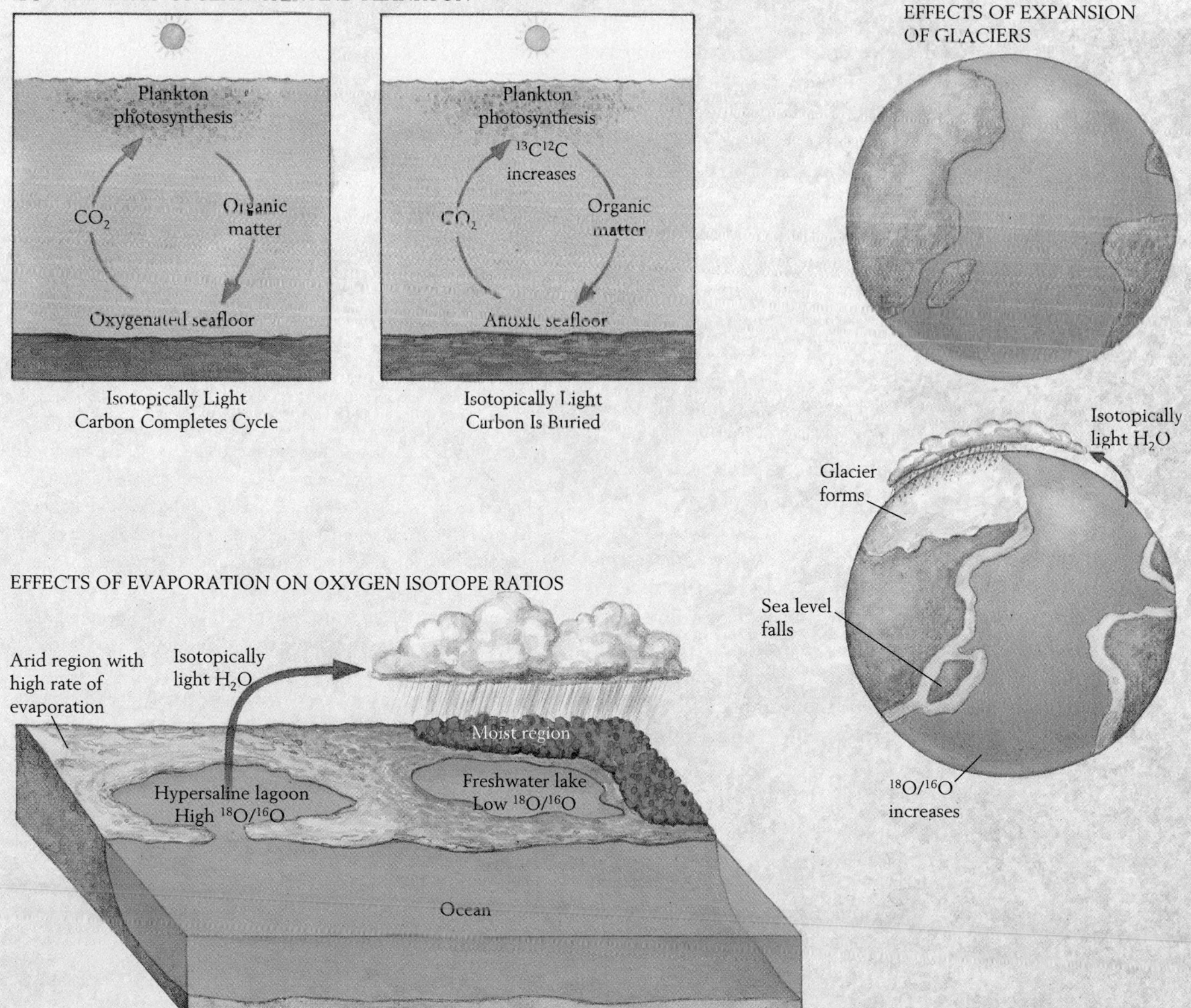
EFFECT OF BURIAL OF ORGANIC CARBON ON THE CARBON ISOTOPE RATIO OF SEAWATER AND PLANKTON
Plankton photosynthesis
CO_2
Organic matter
Oxygenated seafloor
Isotopically Light Carbon Completes Cycle
Plankton photosynthesis
$^{13}C^{12}C$ increases
CO_2
Organic matter
Anoxic seafloor
Isotopically Light Carbon Is Buried
EFFECTS OF EXPANSION OF GLACIERS
Isotopically light H_2O
Glacier forms
Sea level falls
$^{18}O/^{16}O$ increases
EFFECTS OF EVAPORATION ON OXYGEN ISOTOPE RATIOS
Arid region with high rate of evaporation
Isotopically light H_2O
Moist region
Hypersaline lagoon High $^{18}O/^{16}O$
Freshwater lake Low $^{18}O/^{16}O$
Ocean

entities that occupy particular spaces. They include Earth's atmosphere, its oceans, portions of its crust, and the biomasses of its biological communities. These reservoirs expand and contract through changes in the rates at which elements or compounds flow to or from them.

Chapter 4 described the cycling of nitrogen and phosphorus—key nutrients for producer organisms—through ecosystems. Here, instead, we focus especially on cyclical pathways of two other elements essential to life: carbon and oxygen. All of these large-scale chemical cycles move materials through portions of the water cycle (discussed in Chapter 1).

We will also examine how two isotopes of a particular element can differ in their movements through global chemical cycles. Scientists use these differences as indicators of past conditions and events. The distribution of carbon isotopes in sedimentary rocks and of the fossils they contain, for example, gives evidence of changes in the abundance of carbon dioxide and oxygen in Earth's atmosphere over the course of hundreds of millions of years. The distribution of oxygen isotopes in marine fossils offers testimony about the temperatures and salinities of ancient seas.

The fundamental aspects of chemical cycles presented in this chapter provide a framework for understanding many environmental and biological events described in later chapters.

Fluxes

When one reservoir for chemical materials increases in size, one or more other reservoirs must shrink. In the water cycle, for example, the expansion of glaciers on the land robs the ocean of water, and sea level drops (p. 21). Reservoirs expand or contract because of changes in the rate at which they gain or lose their contents. For a global cycle, a rate of this kind is termed a **flux**. An everyday equivalent of a flux is the number of gallons of water per minute that a pump draws from a well and sends through a hose to a swimming pool. Another is the number of bushels of corn per year harvested from a field and fed to a herd of farm animals.

Feedback

To envision how reservoirs and fluxes operate, imagine a balloon with two openings in it, so that water can flow in at one end and out the other (Figure 10-1). The water in the balloon increases in volume until its pressure forces water out as rapidly as it is flowing in. The pressure exerted by the balloon operates as a **feedback,** opposing the expansion of the reservoir within the balloon more and more strongly, until the fluxes to

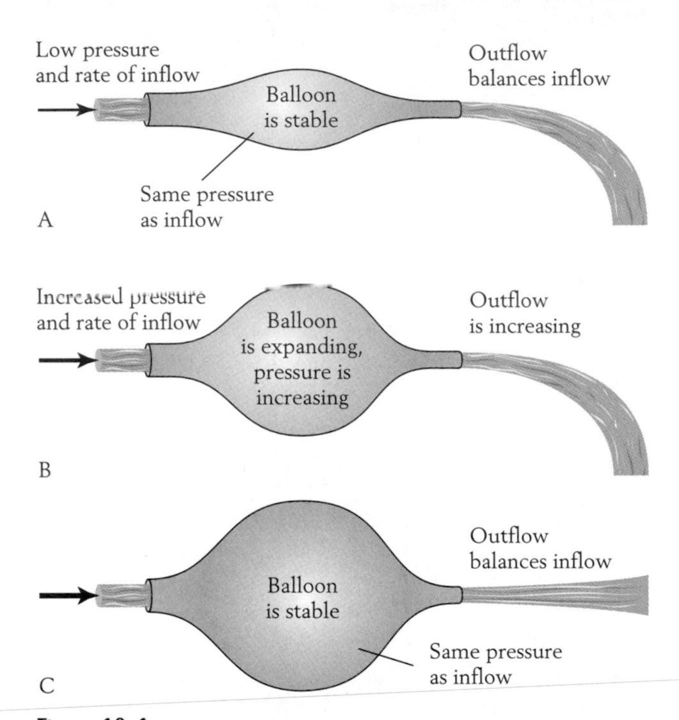

Figure 10-1 How fluxes control the volume of a reservoir. In this imaginary laboratory experiment, the reservoir is a balloon with a tube-shaped opening at each end. Water flows into the balloon through one pipe and out of it through another, smaller one. Initially the pressure and rate of inflow are low. Even so, the water cannot escape from the small exit opening as rapidly as it enters the balloon, so the balloon begins to fill with water and expand. As the balloon expands, its rubber stretches and applies increasing backpressure to the water it contains. The balloon continues to expand until its constantly increasing backpressure balances the pressure of the inflowing water. At this point *(A)* the balloon stabilizes, and the outflow from it balances the inflow. (It is as if the balloon simply became an extension of the inflow tube.) *B* and *C* show what happens when the pressure and rate of inflow are then increased. At first, the newly applied higher pressure is greater than the opposing pressure from the balloon. The balloon therefore begins to expand *(B)*, and it continues to expand until its backpressure again equals that of the water inside *(C)*. Then the balloon (reservoir) is stabilized at a larger volume than in *A*, and the inflow (flux) to it and the outflow (flux) from it are again equal but larger than in *A*. Backpressure from the balloon is the feedback that has stabilized the system twice *(A* and *C)*.

and from this reservoir are in balance. At this point the volume of the reservoir is stabilized.

A variety of factors operate as feedbacks in global chemical cycles. Forests may be providing an important negative feedback today. During recent decades, humans have increased the concentration of carbon dioxide in the atmosphere by burning wood and fossil fuels. Plants use carbon dioxide in photosynthesis, the process by which they produce their own food. An increase in the concentration of atmospheric carbon dioxide in effect fertilizes plants. If they have not already done so, forests of middle and high latitudes may soon increase their biomass as a result of rising levels of atmospheric carbon dioxide. The biomass of plants is a reservoir for carbon, and an increase in the size of this reservoir will reduce the rate of buildup of carbon dioxide in the atmosphere. Thus the fertilization effect of carbon dioxide on plants is a negative feedback against the buildup of this gas in the atmosphere.

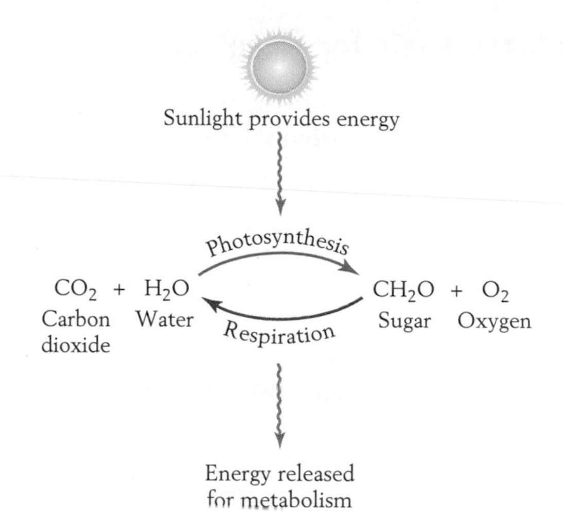

Figure 10-2 **The photosynthesis-respiration cycle in plants.** This cycle, which requires no input of carbon, hydrogen, or oxygen, turns energy from sunlight into energy that can be used in metabolism.

Carbon Dioxide, Oxygen, and Biological Processes

Living things play major roles in chemical cycles within the Earth system. We have already seen how land plants form an integral part of the water cycle, soaking up moisture with their roots and transpiring large volumes of water vapor to the atmosphere (p. 21). Water is also one of the two raw materials for photosynthesis, the most important process that provides food for both plants and animals. The other raw material for photosynthesis is carbon dioxide, whose two constituent elements, carbon and oxygen, move cyclically through the Earth system. First we will examine the factors that control the abundances of oxygen and carbon dioxide in the atmosphere. Then we will consider how these abundances may have changed in the course of geologic time.

The Photosynthesis-Respiration Cycle in Plants

Carbon and oxygen move back and forth between plants and animals by way of the *photosynthesis-respiration cycle.* The biological processes of photosynthesis and respiration are opposites (Figure 10-2). In photosynthesis plants combine carbon dioxide and water to form what we can loosely call sugars—compounds of carbon, hydrogen, and oxygen. Having consumed plenty of sugars, we are all familiar with the large amount of energy they contain. When plants conduct photosynthesis, sunlight provides the energy for the formation of sugars, and this energy ends up stored within the sugars. Oxygen is a by-product of photosynthesis.

Respiration moves in the opposite direction, combining sugar and oxygen to release the sugars' energy. Organisms use this energy to fuel their metabolism. Carbon dioxide and water, the raw materials of photosynthesis, are the products of respiration. Because oxygen is added to the elements of sugar in respiration, the process is an example of *oxidation.* Fire, representing very rapid oxidation, illustrates how this process releases energy. Carbon compounds that contain relatively little or no oxygen are referred to as "reduced carbon" compounds, and the carbon atoms within them are described as reduced carbon. Sugar is such a compound.

Plants employ the photosynthesis-respiration cycle because they are unable to use sunlight directly for their metabolism. They are able to store the energy of sunlight in sugars, however, and, through respiration, they can later release it in a usable form. Although energy is released less rapidly here than in fire, we can refer to it as *burning* of sugars.

Photosynthesis for Tissue Growth

Note that the photosynthesis-respiration cycle in a plant harnesses energy from the sun and releases it for the plant's use without gaining or losing chemical components (see Figure 10-2). Thus there is no net exchange of CO_2 or O_2 between the plant and the atmosphere. The plant has to conduct more photosynthesis than it needs for energy, however. It needs sugar to build tissue while growing from a spore or seed and to produce its own spores or seeds. To manufacture sugar that becomes tissue—a reservoir for carbon—a plant cannot destroy it through respiration. When the plant builds sugars and does not burn them, the CO_2 and water that it uses in forming them are not returned to the atmosphere but are stored within the plant's tissues as sugars and other compounds that contain carbon, hydrogen, and oxygen and are derived from sugars. In addition, because plants do not use O_2, the by-product of photosynthesis, to burn the stored sugars, they release it to the atmosphere. Thus it is because plants grow leaves, stems, roots, and reproductive structures that they remove CO_2 from the atmosphere and contribute O_2 to it.

Almost every bit of plant tissue has one of three final destinies. It can be eaten by an animal, it can decay, or it can be buried in sediment.

Animal Respiration

For humans, as for all other animals, respiration is a process in which gases are exchanged with the environment. We release CO_2 when we exhale and use up O_2 that we inhale.

Animals employ respiration to gain energy from the sugars of the plants they eat. Thus, as Figure 10-3 illustrates, animals form a photosynthesis-respiration cycle with plants like the one that plants employ in manufacturing and burning sugars. Plants use as much CO_2 and H_2O to make sugars eaten by animals as the animals release in the respiration that liberates the sugars' energy. Similarly, animals use the same amount of O_2 to burn the sugars that the plants liberate in producing them. In this cycle, unlike the comparable one that occurs entirely within plants, CO_2 and O_2 pass through the atmospheric reservoir as they move between plants and animals. Even so, because the exchange between plants and animals is in balance, the volume of O_2 and CO_2 in the atmospheric reservoir is unaffected by the cycle.

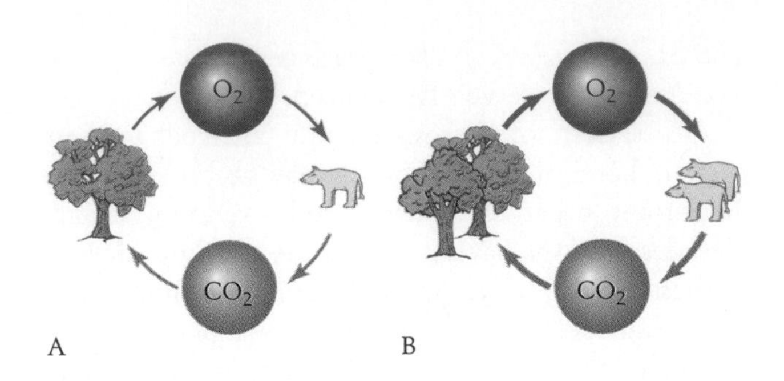

Figure 10-3 The photosynthesis-respiration cycle involving both plants and animals. In this cycle oxygen and carbon dioxide move through atmospheric reservoirs (balloons) between plants and animals. Doubling the biomass of both plants and animals (*B* versus *A*) doubles the volume of all of the fluxes (arrows) both to and from each reservoir (balloon), leaving the volume of each reservoir unchanged.

If the number of plants increases, providing more food for animals, the number of animals will increase in proportion—as will all the fluxes of O_2 and CO_2 between the organisms and atmosphere. If the biomasses of plants and animals double, for example, so will the various fluxes of O_2 and CO_2 to and from the atmosphere. The sizes of the atmospheric reservoirs of these gases will therefore remain unchanged (Figure 10-3*B*).

Of course, the food web extends beyond herbivores to carnivores. Carnivores consume mostly the protein and fat of other animals rather than sugars, but these compounds have ultimately been manufactured from sugars, and carnivores oxidize them to obtain metabolic energy in the same way that herbivores oxidize sugars. Thus the photosynthesis-respiration cycle extends to the top of the food world without essential modification.

Bear in mind that some animals—endothermic forms (p. 85)—normally maintain a body temperature higher than the temperature of their environment. To accomplish this feat, mammals and birds must consume more food and respire at higher rates than the normal rates of endothermic animals, such as fishes, amphibians, and reptiles.

Respiration by Decomposers

To this point we have ignored an important component of ecosystems: the decomposers, which break down dead organic debris not consumed by animals.

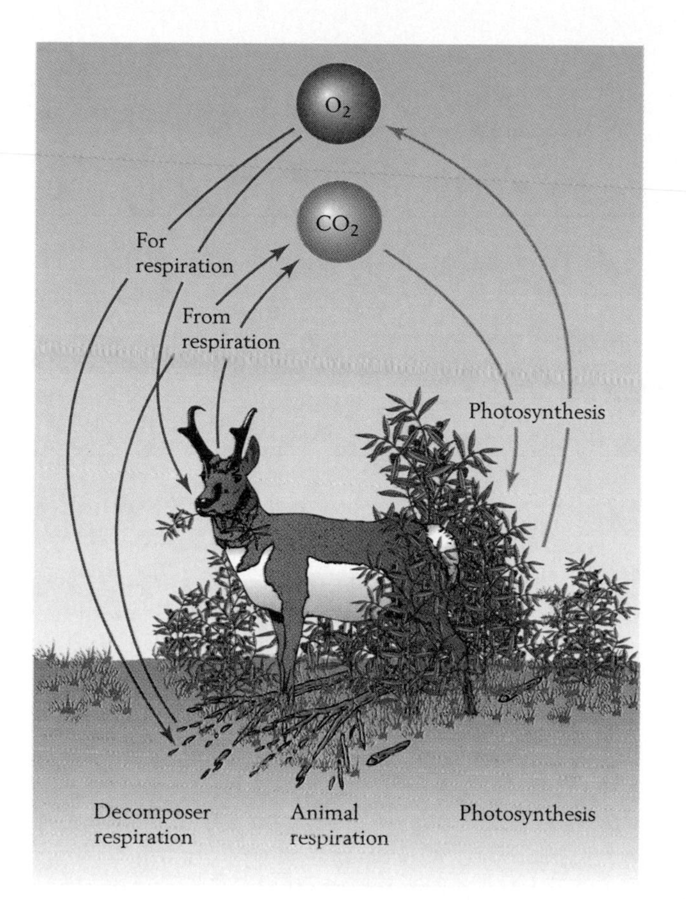

Figure 10-4 Respiration of plant material. The total amount of consumption and respiration in nature is divided in variable proportions between animals and decomposers.

The most important decomposers are bacteria and fungi. Decomposers, like animals, use respiration to break down tissues. They simply consume dead rather than living organic matter. Thus decomposers, too, extract O_2 from the atmosphere and release CO_2 (Figure 10-4).

There is plenty of room in the ecosystem for decomposers of dead plant tissue because animals do not harvest the entire biomass of living plants. Animals' consumption is limited by their lack of access to certain plant structures and by their inability to eat and digest some parts of plants. In addition, predation or limiting factors imposed by the physical environment often keep animal populations below the levels that the biomass of plants can potentially support.

Simply put, some plant tissues are consumed by animals and others are consumed by decomposers when the plants die. Both kinds of consumption entail respiration. If, then, populations of herbivorous mammals decline, more tissue is left for decomposers, and these organisms contribute more total respiration. Conversely, if herbivore populations expand, they assume a larger portion of the total amount of respiration. All of the plant material *that is not somehow removed from the ecosystem* ends up as part of the photosynthesis-respiration cycle that includes plants and the organisms that destroy their tissues.

Burial of Plant Debris and Atmospheric Chemistry

The italicized portion of the previous sentence provides the key to understanding long-term changes in the abundances of CO_2 and O_2 in Earth's atmosphere. Some dead plant debris escapes from the photosynthesis-respiration cycle through burial—burial in swamps, for example, or in sediments along the margins of continents. Buried organic matter, including that contained in coal, constitutes a reservoir for reduced carbon compounds (Figure 10-5). Just as sedimentation is always burying reduced carbon, however, erosion is always exposing it to the atmosphere, where it is soon oxidized by decomposers or inorganic processes. Over long stretches of geologic time, the rate of burial of organic carbon has nearly balanced its rate of exposure by erosion, and the reservoir of buried carbon has undergone little net change in volume. In other words, the cycle of burial and erosion has been in balance, like the photosynthesis-respiration cycle of plants and animals. As a result, levels of CO_2 and O_2 in the atmosphere have been relatively stable.

Occasionally the system is thrown out of balance, however, when the overall rate of carbon burial increases. Atmospheric CO_2 was used to produce any reduced carbon compounds that are buried. Thus, in effect, excess burial of these compounds removes CO_2 from the photosynthesis-respiration cycle—specifically from the atmospheric reservoir—and stores it in a subterranean reservoir. As a result, the concentration of CO_2 in Earth's atmosphere declines.

What happens to oxygen when the overall rate of carbon burial increases? It also becomes more abundant in the atmosphere (Figure 10-6). Had the large volume of reduced carbon compounds not been buried, decomposers would have oxidized it through respiration. Because these carbon compounds have been buried, however, the oxygen that would have oxidized them remains in the atmosphere.

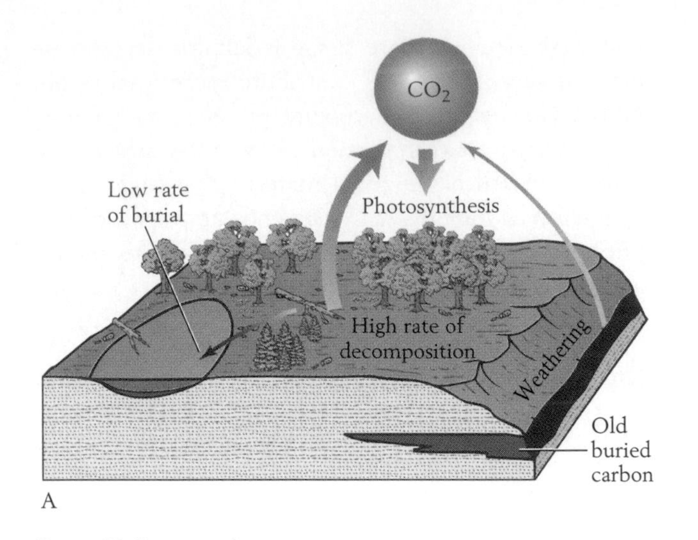

Figure 10-5 The effect of the rate of burial of organic carbon on the size of the atmospheric reservoir of carbon dioxide. Initially *(A)* the rate of carbon burial (left) balances the rate of weathering of buried carbon (right). Later *(B)* swamps expand and vegetation adapted to them flourishes. The increased rate of burial of dead plant material in swamps leaves less carbon to be returned to the atmosphere through decomposition, as CO_2. As a result, the atmospheric reservoir of CO_2 shrinks.

A large change in the rate of burial of organic matter for the entire planet powerfully alters the concentration of atmospheric CO_2 and O_2. Where does carbon accumulate in large quantities? The most important sites are anoxic bodies of water. An *anoxic* condition (or *anoxia*) is the virtual absence of O_2. Anoxia allows debris from dead plants to survive on the floor of a body of water and eventually to become deeply buried without having fully decayed. The reason is quite simple. The organisms—mostly bacteria—that cause decay by respiration require oxygen to do so. In environments where these organisms cannot live, organic matter survives and accumulates with silt and clay. Conditions of this kind have provided for exceptional preservation of fossils, sometimes allowing remnants of soft tissue to become fossilized (p. 60). Today debris from sphagnum moss and other plants accumulate in bogs in cold climates. The bottom waters of these bogs, being not only anoxic but also highly acidic, are inhospitable to the kinds of bacteria that cause decay. If deeply buried, the peat that accumulated within them would eventually turn to coal. Remaining near the surface, they are nonetheless readily exploited for use as fuel and, to a lesser extent, for enrichment of garden soils.

When anoxic conditions become widespread on Earth, burial of plant debris can form large reservoirs of organic compounds within Earth's crust. At times when deep portions of the oceans became anoxic or when swamps with stagnant, anoxic bottom waters expanded to occupy large areas of continents, these swamps became burial sites for large volumes of organic matter from trees, some of which has become coal (p. 54).

The Carboniferous Period was a time when primitive trees were buried extensively in what have come to be called *coal swamps.* The remains of the trees accumulated on the anoxic bottoms of the swamps, where they turned into peat. This peat eventually became coal, which gave the Carboniferous Period its name and which provides humans with large quantities of fossil fuel.

At times when organic matter is buried rapidly, there is no reason why the rate at which older buried organic carbon is exposed should increase. In fact, because the erosion process occurs throughout the world and uncovers buried carbon of many ages, different rates in different areas tend to average out: there is little overall change through time. The rate of burial is more unstable because a global change in climate can alter floras or thereby expand swampy areas and drastically change rates of burial of organic matter throughout the world. Thus, given little change in the rate at which buried carbon is exposed to oxidation, a large, long-term increase in the flux of organic debris to the buried carbon reservoir increases the size of this reser-

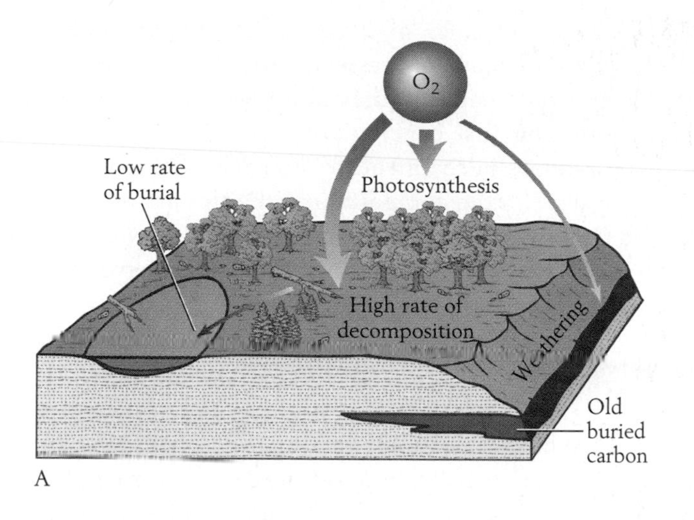

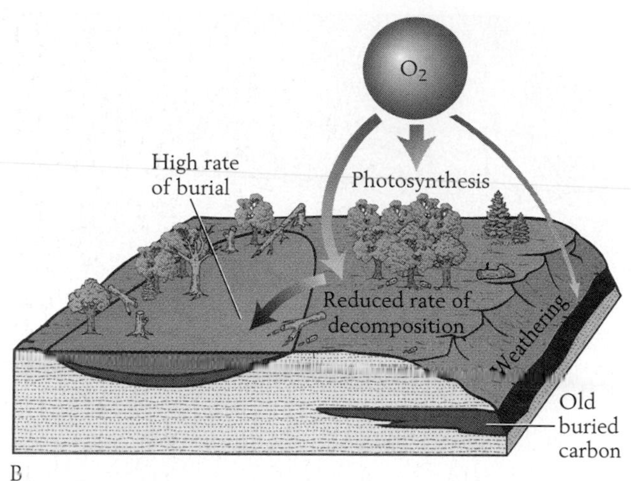

Figure 10-6 The effect of the global rate of burial of organic carbon on the size of the atmospheric reservoir of O_2. Initially *(A)* the rate of carbon burial (left) balances the rate of weathering of buried carbon (right). Later *(B)*, as in Figure 10-5, swamps expand and vegetation adapted to them flourishes. An increased rate of burial leaves behind less dead plant material to use up oxygen by decomposing. As a result, the atmospheric reservoir grows.

voir substantially. The result is a substantial reduction of the CO_2 in Earth's atmosphere and a corresponding increase in O_2 (see Figures 10-5 and 10-6).

Aquatic Ecosystems

Our discussion thus far has focused on nonmarine plants and animals, but the same principles hold for the oceans, where dissolved O_2 is available for photosynthesis. Of course, the main producers of sugars in these bodies of the ocean are single-celled planktonic algae.

Animals consume a larger percentage of the biomass of marine plankton than the percentage of plant biomass that animals are able to consume on the land. Thus a small proportion of the sugars produced by marine photosynthesis is available for either decay or burial. Terrestrial plants, however, contribute additional dead tissue to marine environments. Rivers carry large amounts of partly decomposed sugars from the land to lagoons and deltas, where those that are not eaten by marine animals are decomposed by bacteria and accumulate in muddy sediments.

Organic matter can also become buried farther from shore. In the modern oceans cold, dense surface water sinks near both of Earth's poles, spreading throughout the deep sea and supplying it with oxygen. At certain times in the past, however, polar regions were relatively warm, and water in their vicinity did not descend. At these times, the deep sea was relatively stagnant and anoxic. With few bacteria to cause decay, much of the organic matter from dead phytoplankton that settled on the deep-sea floor was buried. This burial, like that of terrestrial plant debris in swamps, transferred carbon from the atmosphere to the reservoir of buried carbon.

In fact, the abyssal plane of the ocean is never the site of rapid carbon burial. It lies far from the zones of upwelling that lie along continental margins, where productivity of phytoplankton is highest and where plant debris from the land also accumulates in the sediments of lagoons and deltas. At certain times, however, low-oxygen conditions have extended upward to such shallow depths that the floors of large epicontinental seas have become largely anoxic. Organic-rich muds accumulated in these settings, which received plant debris from nearby land areas and from plankton in overlying waters. These muds eventually became shales that were black because they contained a large amount of carbon. Figure 10-7 shows the locations of deposits of this type in mid-Cretaceous time, slightly more than 100 million years ago, and the distribution of the epicontinental seas in which they accumulated. High temperatures and pressures later altered some of the organic matter buried in these Cretaceous settings to liquid and gaseous

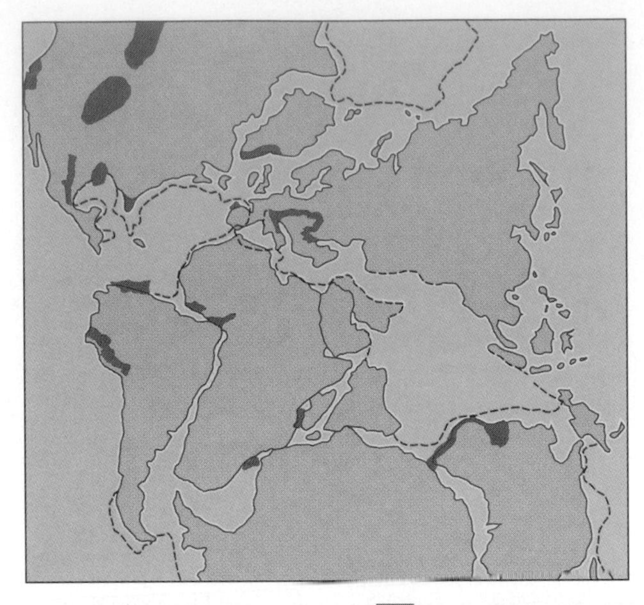

Figure 10-7 Locations of black shales and muds deposited in anoxic marine waters of mid-Cretaceous epicontinental seas. *(After A. G. Fischer and M. A. Arthur, Soc. of Econ. Paleont. and Mineral. Spec. Publ. 25, 19–50, 1977.)*

compounds. These fluids migrated through porous rocks and became trapped in underground reservoirs that have provided humans with large quantities of petroleum and natural gas.

The Use of Isotopes to Study Global Cycles

Carbon isotopes serve as tools for tracing the history of some aspects of atmospheric chemistry. Recall that isotopes are varieties of chemical elements whose atoms differ only in the number of neutrons they contain. Thus an atom of carbon 13 has one more neutron and is heavier than an atom of carbon 12. These are stable isotopes—they do not spontaneously decay in the manner of radioactive isotopes, such as those used to date rocks (p. 32). The relative proportion of one isotope, such as carbon 13 (abbreviated ^{13}C), in a specimen is usually specified by the symbol δ, which relates the isotopic composition of this specimen to that of a standard specimen to which all others are compared. This relative proportion is expressed as parts per thousand (‰).

Two molecules of a particular compound that contain different isotopes of a given element behave slightly differently as they flow through the Earth system. Our first concern will be with carbon. Carbon isotopes in sedimentary materials offer evidence of changes in the composition of Earth's atmosphere. To understand why, we must understand how carbon 12 and carbon 13 move through carbon reservoirs in slightly different ways.

Carbon Isotopes

Most molecules of CO_2 in the atmosphere contain carbon 12; a smaller proportion contain carbon 13. When organisms of any kind conduct photosynthesis, they employ a slightly larger proportion of carbon 12 than the proportion found in the atmosphere because this isotope is relatively light. As a result, all plant tissue contains a disproportionately large percentage of carbon 12. The distribution of carbon isotopes in sedimentary materials allows scientists to identify past changes in the concentration of CO_2 and O_2 in Earth's atmosphere.

Isotopes and the Organic Carbon Cycle Isotopically light carbon that plants preferentially extract from the atmosphere is returned to the atmosphere through the respiration of animals and decomposers (as in Figure 10-4). Thus the photosynthesis-respiration cycle has little effect on the isotopic ratio of CO_2 in the atmosphere. The same is true for burial of isotopically light organic carbon, as long as the carbon is returned to the atmosphere by weathering of organic matter as rapidly as it is added to the underground reservoir by burial—as long as the system is in balance.

The system is *not* always in balance, however. At certain times carbon can be buried rapidly even though weathering rates remain unchanged (see Figure 10-5). At these times a relatively large proportion of carbon 12 from the atmosphere becomes locked up in the reservoir of buried organic matter, leaving the atmosphere with an elevated ratio of carbon 13 to carbon 12 (Figure 10-8).

At any time in Earth's history when the rate of carbon burial increases on continents or beneath the sea, both CO_2 in the atmosphere and CO_2 dissolved in the ocean become depleted of carbon 12 and therefore enriched in carbon 13. Both reservoirs are

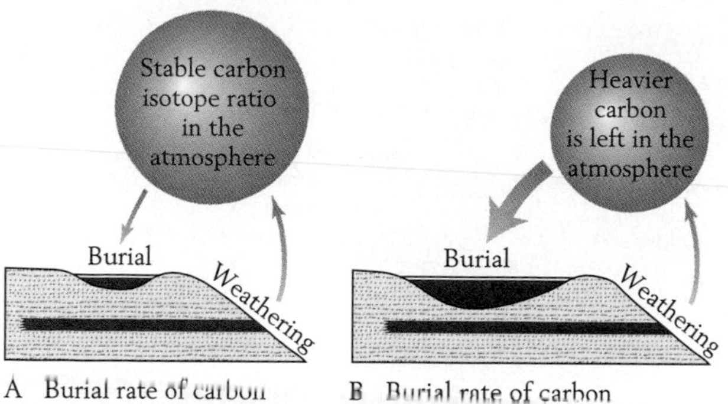

Figure 10-8 The effect of the rate of carbon burial on the isotope ratio of carbon in atmospheric CO_2. In situation *A*, the rate at which organic carbon is buried equals the rate at which weathering returns organic carbon to the atmosphere as CO_2; burial of carbon therefore does not alter the isotope ratio of carbon in the atmospheric CO_2. In situation *B*, organic carbon is buried at a much higher rate, expanding the total reservoir of buried organic carbon, which is isotopically light; conversely, the atmospheric reservoir of CO_2 has shrunk and become isotopically heavier.

affected because the movement of carbon between them is rapid on a geologic scale of time. Exchange between the reservoirs results largely from movement of CO_2 back and forth between the atmosphere and surface waters of the ocean. Changes in the abundance and isotopic composition of CO_2 are spread rapidly through the atmosphere by winds and vertical air movements and through the upper ocean by large-scale water currents.

Isotopes in Limestones and Organic Matter in Sediments Study of carbon isotopes allows scientists to identify changes in the global rate of burial of carbon. Because of the rapid mixing between carbon reservoirs, carbon isotope ratios in shallow-water limestones provide a record of changes for carbon isotope ratios in the atmosphere as well as the ocean.

Figure 10-9*A* is a plot of the carbon isotope ratios in the $CaCO_3$ of limestones over the entire Phanerozoic Eon. The most conspicuous feature of this plot is the large increase in the relative percentage of carbon 13 in the latter part of the Paleozoic. Limestones with the highest values of carbon 13 formed in late Carboniferous time, when large volumes of carbon accumulated in coal swamps. Because this carbon was isotopically light, an excess of carbon 13 was left in

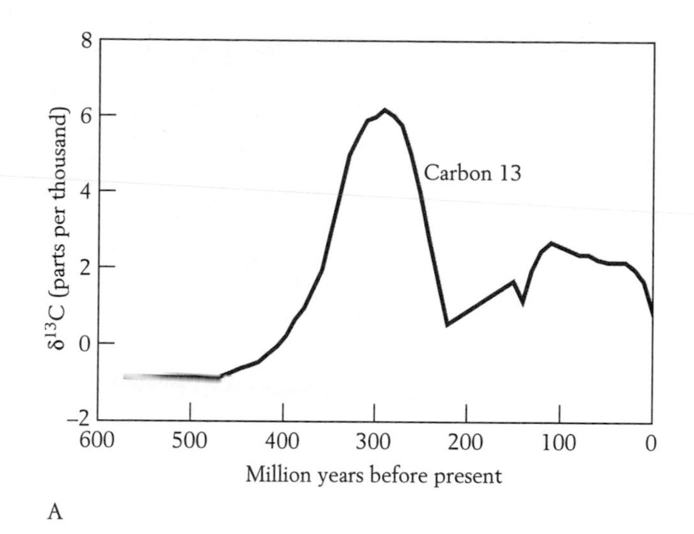

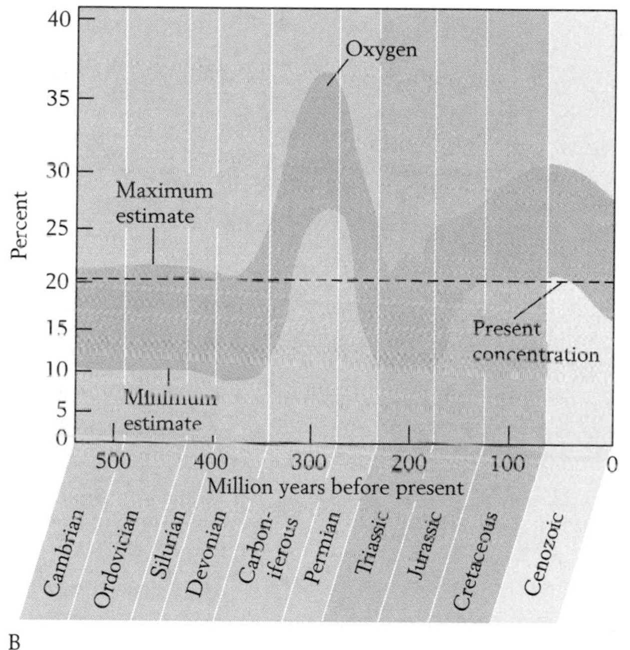

Figure 10-9 Estimated changes in the size of the atmospheric reservoir of oxygen during the Phanerozoic. *A*. Changes in the relative percentage of carbon 13 in seawater, estimated from the isotopic composition of limestones of various ages. *B*. Estimated changes in the fraction of Earth's atmosphere that has consisted of free oxygen. This estimate is based on the carbon content of sediments; an increase in the rate of carbon burial causes oxygen to build up in the atmosphere. The broad band plotted in *B* depicts uncertainties in calculations. *(A. From R. A. Berner, Amer. Jour. Sci. 287:177–196, 1987. B. From R. A. Berner and D. E. Canfield, Amer. Jour. Sci. 289:333–361, 1989.)*

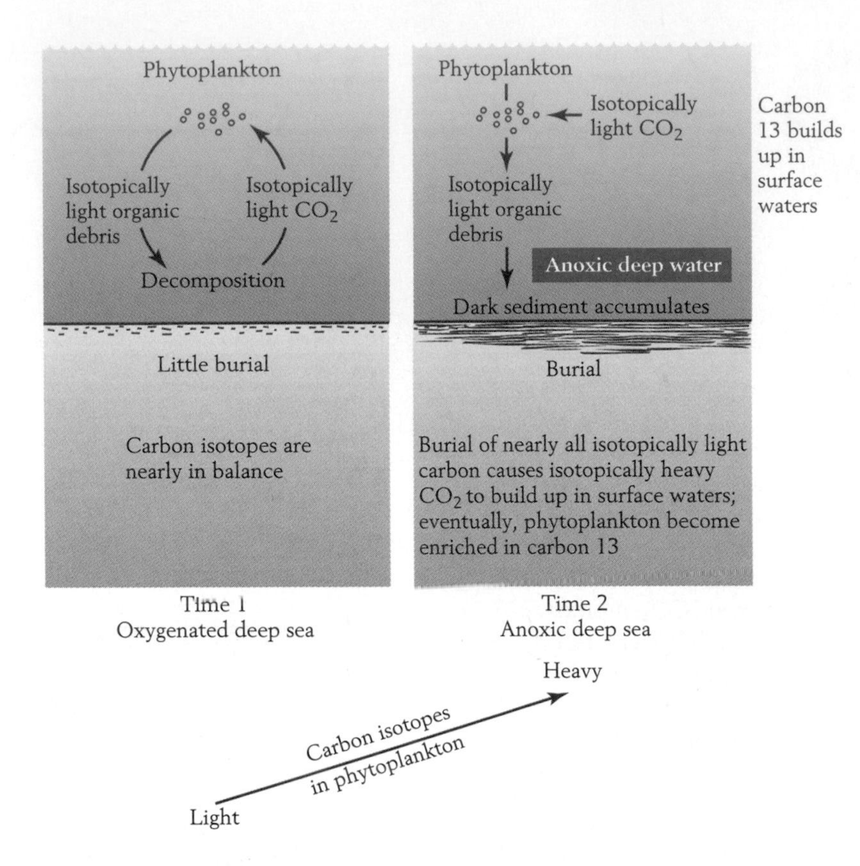

Figure 10-10 Influence of the rate of burial of organic carbon on carbon isotope ratios in phytoplankton and sedimentary organic matter. Debris from phytoplankton is isotopically light. When deep water becomes anoxic (time 2), most of the organic debris that sinks is buried. As a result, decomposition and upwelling no longer return its light carbon to the upper ocean, as they did at time 1. Thus the CO_2 that remains in the upper ocean becomes enriched in carbon 13. Because phytoplankton conduct their photosynthesis there, they become isotopically heavier, as does the organic debris that sinks to accumulate in sediments when the phytoplankton die. Thus an increase in the isotope ratio of organic carbon in the geologic record may reflect an increased rate of carbon burial.

the atmosphere. As a result, $CaCO_3$ precipitated in the ocean became isotopically heavy.

Phytoplankton, the primary photosynthesizers of the open ocean, follow a similar pattern. When high burial rates of organic matter, which is isotopically light, leave behind an excess of carbon 13 in the CO_2 of the atmosphere and ocean, marine phytoplankton that use this CO_2 for photosynthesis assimilate isotopically heavy carbon. When waters of the deep sea become anoxic, for example (Figure 10-10), the carbon in organic matter accumulating on the seafloor becomes increasingly heavy. In other words, a sudden pulse in the carbon isotope ratio toward heavy values in the marine stratigraphic record can indicate an episode of increased burial of organic carbon.

Atmospheric Oxygen

We have seen that an increase in the rate of burial of organic carbon causes oxygen to build up in the atmosphere (see Figure 10-6). Conversely, a decrease in the rate of burial leaves more carbon behind to be oxidized by composers; increased respiration then pulls oxygen out of the atmosphere.

Because changes in carbon isotope ratios in marine carbonates provide a continuous record of overall rates of carbon burial, these ratios provide a general picture of rising and falling oxygen in the atmosphere (see Figure 10-9*B*). A shift of the carbon in the $CaCO_3$ toward heavier values, for example, reflects burial of more organic carbon, which is isotopically light. The fact that this carbon was buried rather than decomposed must have caused oxygen to accumulate in the atmosphere.

We cannot reconstruct oxygen levels in the atmosphere precisely from the graph displayed as Figure 10-9*A* because secondary factors also come into play. We can nonetheless conclude that atmospheric oxygen reached its highest Phanerozoic level during the late Paleozoic interval, when so much carbon was buried in coal swamps (see Figure 10-9*B*). Some estimates suggest that the level of oxygen then was about twice what it is now.

Atmospheric Carbon Dioxide

Of course, burial of organic carbon not only enlarges the atmospheric reservoir of O_2 but also shrinks the atmospheric reservoir of CO_2. It does so by shifting reduced carbon to the reservoir of buried organic carbon (see Figure 10-5). But whereas the level of O_2 in the atmosphere depends largely on the rate of burial of organic carbon, other factors exert a stronger influence over levels of atmospheric CO_2. Especially large changes in the concentration of CO_2 result from chemical reactions that create and destroy minerals in soils and rocks near Earth's surface.

Figure 10-11 shows how chemical processes apart from photosynthesis and respiration (p. 261) cycle oxidized carbon through the atmosphere, ocean, and solid Earth. These processes form part of the rock cycle (outlined in Chapter 1). Weathering processes on the land use up CO_2 from the atmosphere to break down rocks. This is CO_2 that has

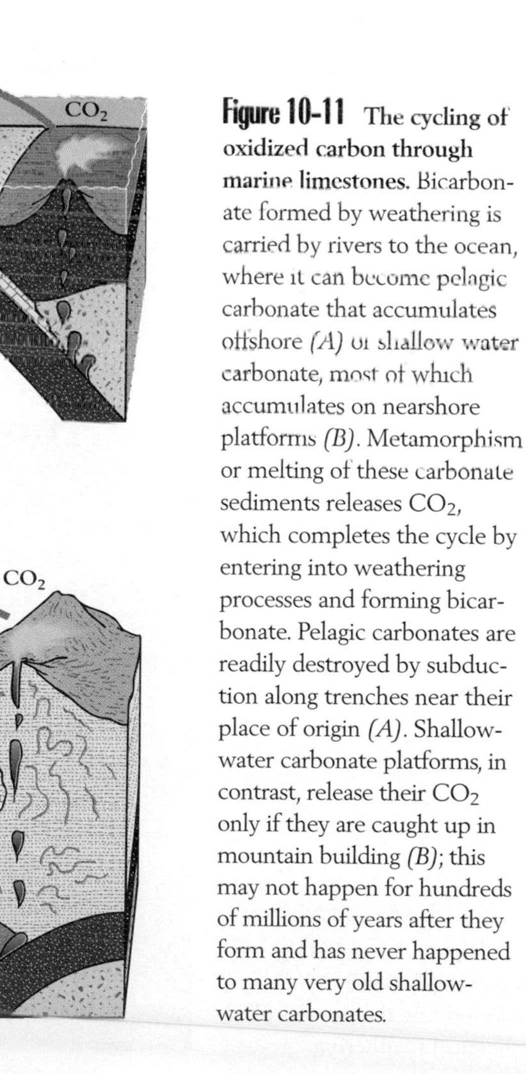

Figure 10-11 **The cycling of oxidized carbon through marine limestones.** Bicarbonate formed by weathering is carried by rivers to the ocean, where it can become pelagic carbonate that accumulates offshore *(A)* or shallow water carbonate, most of which accumulates on nearshore platforms *(B)*. Metamorphism or melting of these carbonate sediments releases CO_2, which completes the cycle by entering into weathering processes and forming bicarbonate. Pelagic carbonates are readily destroyed by subduction along trenches near their place of origin *(A)*. Shallow-water carbonate platforms, in contrast, release their CO_2 only if they are caught up in mountain building *(B)*; this may not happen for hundreds of millions of years after they form and has never happened to many very old shallow-water carbonates.

combined with water in Earth's atmosphere to form carbonic acid (H_2CO_3). Carbonic acid attacks both limestone ($CaCO_3$) and silicate rocks, releasing both positively charged ions of calcium (Ca^{++}) and negatively charged combinations of carbonate and hydrogen known as bicarbonate ions (HCO_3^-). Dissolved in water, these chemical species travel in rivers to the ocean, where they recombine to form $CaCO_3$. As we have seen, some of this $CaCO_3$ is precipitated by inorganic processes in warm seas, and some is secreted by organisms in the form of skeletons. Thus some limestones are bioclastic sediments and some are chemical sediments (p. 46).

Eventually metamorphism and melting complete the cycle by breaking down limestone and soft carbonate sediment, extracting CO_2 and releasing it to the atmosphere. Two distinct settings are the primary sites where calcium carbonate sediments accumulate: carbonate platforms, such as the Great Bahama Bank (p. 143), and deep-sea settings, where minute skeletons of plankton accumulate as fine-grained pelagic carbonates (p. 147). The two separate reservoirs of carbonate sediment that form in this way—one in deep water and one in shallow water—are subject to different kinds of metamorphism. Many pelagic carbonates are eventually subducted into Earth's upper mantle, riding on the surface of oceanic crust. Their destruction liberates CO_2, which escapes into the atmosphere through volcanoes that form along subducted zones (see Figure 10-11*A*). Carbonate platforms, in contrast, are thick bodies of low density that cannot be subducted. Only by becoming caught up in mountain building are they likely to undergo metamorphism and contribute CO_2 to the atmosphere (Figure 10-11*B*).

In short, carbonate sediments and rocks most commonly undergo metamorphism and return their CO_2 to the atmosphere along subduction zones. This liberated CO_2 is available to take part in weathering, which produces HCO_3^- that returns to the ocean. There $CaCO_3$ forms from the HCO_3^- through direct precipitation and skeletal secretion by organisms, completing the global cycle for oxidized carbon. In contrast to bodies of pelagic carbonate, shallow-water carbonate platforms amount to reservoirs where carbon extracted from Earth's atmosphere is stored for long geologic intervals.

The Importance of Weathering

We have traced the movement of oxidized carbon through a great cycle, but the question remains: What can cause one key segment of this cycle—the atmospheric reservoir—to expand and contract? The greenhouse effect of CO_2 makes changes in the abundance of this gas a prime factor in the warming and cooling of Earth's climate. Weathering of calcium and magnesium silicate rocks is the primary

Figure 10-12 A glacier in the Alps that is grinding up the rock beneath it, accelerating the process of chemical weathering.

process that removes CO_2 from the atmosphere. As we shall see shortly, changes in the global rate of weathering appear to have been the most important factor in the largest drop in concentration of atmospheric CO_2 of the entire Phanerozoic Eon.

Mountain Building and Weathering The elevation of a large mountain range is one geological change that accelerates weathering. The steep slopes of a mountain range undergo rapid weathering and erosion because of the strong influence of gravity and the activity of glaciers (Figure 10-12). Glaciers accelerate chemical weathering by grinding up rock at Earth's surface. Such weathering extracts a large volume of CO_2 from the atmosphere. As we have seen, most of this CO_2, along with calcium released by the same weathering process, ends up forming marine carbonate sediments. Tens or hundreds of millions of years must pass before these sediments reach the metamorphic settings of mountain belts where their gases can be released. Thus CO_2 from Earth's atmosphere remains bound for a considerable time in vast reservoirs of shallow-water carbonate sediments. Furthermore, mountain building can substantially reduce the concentration of CO_2 in the atmosphere by transferring it to these sediments.

Temperature and Weathering Like many other chemical processes, weathering speeds up as the temperature rises. As a result, chemical weathering occurs more rapidly on average in tropical climates than in cold climates.

Precipitation and Weathering Water flowing through soils and porous rock is responsible for most chemical weathering. In any region, then, the rate of chemical weathering will vary with the amount of precipitation that falls, including snow that melts. The dry climate of the Colorado Plateau, for example, produces such low rates of chemical weathering that limestone forms broad upland surfaces and steep cliffs, such as those that rim the Grand Canyon (see Figure 1-5). In the moist Appalachian region of the eastern United States, in contrast, bodies of limestone tend to weather preferentially and floor valleys rather than uplands.

At some times in the geologic past moist conditions have been widespread on Earth and arid environments have been confined to small areas. At other times, moist climates have been more restricted and arid environments more widespread. The overall rate of weathering of Earth's surface has been rapid when moist conditions have extended over much of the globe.

What factors influence global patterns on so large a scale? One is the distribution of continents. When large continental areas occupy the central tropics, for example, moist terrestrial climates are relatively widespread (p. 102). On the other hand, when large continental areas lie in the trade wind belt, arid terrestrial climates are relatively widespread.

The size of continents also plays an important role in the weathering process. Much of the precipitation that falls on continents returns to the ocean by way of rivers, and continents rely on wind-borne moisture from the ocean to replenish this runoff. For this reason, regions of large continents that lie far inland tend to be dry. It follows that a large portion of a huge continent is likely to be arid. When a continent is so small that all portions of it lie close to the ocean, it tends to receive a large amount of precipitation for its size.

At certain times in Earth's history, most of its continental lithosphere was united to form one or a very few large continents. When such continents were not extensively invaded by shallow seas, arid conditions were widespread and global rates of weathering relatively low.

Vegetation and Weathering The effects of plentiful precipitation on weathering are greatly amplified by a key biological factor. Abundant moisture allows forests to occupy the land, and the roots of large plants accelerate chemical weathering. Through their roots, land plants secrete acids and other compounds that break down minerals. Furthermore, forests trap water in a local cycle that moves it through the soil repeatedly. This cycle results from plants' transpiration of water into the atmosphere—photosynthesis uses only a small percentage of the water that their roots absorb—and much of the transpired moisture forms clouds that hover above the forest. Eventually the clouds release the moisture as rain, which soaks into the forest soil, completing the local cycle. This cycling results in the rapid solution of soil minerals. Studies of the chemical products of weathering in modern streams reveal that the rate of weathering is typically about seven times higher in a forested area than in a nearby barren area.

Extensive vegetation on Earth's surface has accelerated weathering at some times in the geologic past. At other times, the aridity of much of the terrain has reduced rates of weathering.

Phanerozoic Trends in Atmospheric CO_2

By taking into account relevant factors of the kinds we have discussed, scientists have created computer models depicting the history of CO_2 in Earth's atmosphere. Figure 10-13 depicts such a model. Let us examine the major features of this model, recognizing that it may require changes as new information appears. By assessing the net impact of a variety of factors that have influenced atmospheric CO_2, the model provides an estimate of the CO_2 level in Earth's atmosphere since the start of the Cambrian.

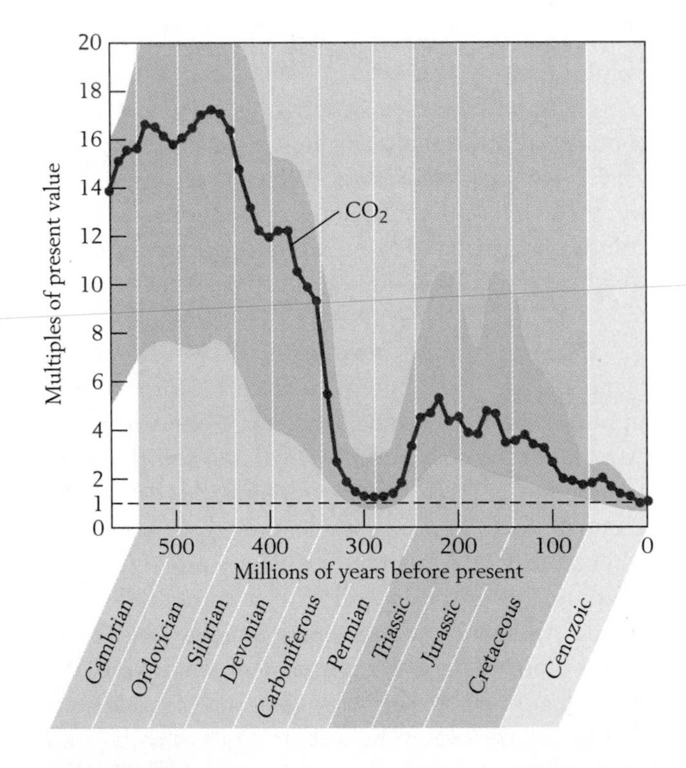

Figure 10-13 Plot of results of a computer model created to estimate changes in the concentration of CO_2 in the atmosphere during Phanerozoic time. The plotted values are multiples of the present value, which is designated as 1. The shaded area shows the estimated range of uncertainty. The severe decline during Devonian and Carboniferous time is the most conspicuous feature of the plot. *(After R. A. Berner, Amer. Jour. Sci. 294: 56–91, 1994.)*

The most conspicuous feature of Figure 10-13 is a sharp decline in CO_2 during the latter part of the Paleozoic Era, beginning in the Devonian Period. This decline results largely from a substantial increase in the rate of weathering estimated for terrestrial environments. Weathering intensified during Devonian time because of evolutionary changes in land plants that allowed them to inhabit upland environments, rather than simply occupying swamps and fringes of lakes and rivers. This evolutionary step, which we will examine in greater detail in Chapter 14, allowed forests to spread over broad areas of Earth's surface for the first time. In newly forest-clad regions, weathering rates should have increased to about seven times their previous levels (p. 271). Incorporation of this change into the model for a time when other factors underwent little change produced the dramatic drop in atmospheric CO_2 estimated for the Devonian Period.

Note, however, that Figure 10-13 shows the late Paleozoic decline in atmospheric CO_2 continuing into the Carboniferous Period, after upland forests had become widespread. In this model, the continued decline results from another change that we have already discussed: an increase in the rate of burial of organic carbon when coal swamps became widespread (p. 265). Burial of organic material that was produced by photosynthesis depleted the atmospheric reservoir of CO_2 by transferring the carbon to a buried reservoir rather than allowing respiration to return it to the atmosphere as CO_2 (see Figure 10-5). Recall that the resulting shift toward heavier isotope ratios for CO_2 in the atmosphere and oceans is reflected in the $CaCO_3$ of marine limestones (see Figure 10-9*A*).

The large total decline in atmospheric CO_2 calculated for the Devonian and Carboniferous would have greatly weakened the greenhouse warming of Earth's surface and lower atmosphere. Reduced temperatures from this change probably contributed to another phenomenon of the Carboniferous: the expansion of glaciers across broad areas of the southern hemisphere.

Let us step back for a moment and note the very high estimate of atmospheric CO_2 for the early portion of the Paleozoic Era. Might such a high concentration of CO_2, through its powerful greenhouse effect, have produced exceedingly warm temperatures on Earth? Indeed, climates may have been relatively warm, on average, but an opposing factor was the progressive increase in the sun's output of radiation since early in the history of the solar system. Weaker solar warming during the first half of the Paleozoic may have more or less offset the powerful greenhouse effect operating at that time. As a result, the mean annual temperature of the early Paleozoic Earth may not have differed greatly from that of the Mesozoic Earth.

The model depicted in Figure 10-13 produces a modest rise in atmospheric CO_2 after the late Paleozoic decline. This rise results largely from two components of the model. One is the rate of mountain building. Since mountain building was not very extensive on a global scale during the Mesozoic Era, the overall rate of weathering on Earth was low—and therefore so was the rate at which weathering removed CO_2 from the atmosphere.

The second factor that elevates the level of Mesozoic atmospheric CO_2 in the model is an increase in the amount of pelagic $CaCO_3$ that accumulated on the deep-sea floor relative to the amount of $CaCO_3$ that accumulated on nearshore carbonate platforms (see Figure 10-11). This increase resulted from the evolutionary origin of calcareous nannoplankton and planktonic foraminifera, which have formed pelagic oozes in the deep sea only during the past 150 million years or so. Once these oozes became plentiful, their descent along subduction zones greatly increased the rates at which metamorphism within these zones released CO_2 to the atmosphere.

Intensified weathering resulting from mountain building produces the modest decline in atmospheric CO_2 that Figure 10-13 shows for the Cenozoic Era. One factor not incorporated into the model may have made the Cenozoic decline in CO_2 less severe than the model suggests. This is a change in Earth's climate. During the past 35 million years or so, broad areas of the Earth have become drier. Fossil plants and other climatic indicators show that deserts and grasslands have expanded greatly during this interval, while forests adapted to moist conditions have contracted. The resulting reduction of water flow through soils and rocks must have slowed weathering and reduced the rate at which this process removed CO_2 from the atmosphere. By reducing the greenhouse effect, a substantial decline of atmospheric CO_2 during the Cenozoic Era may have set the stage for the modern ice age of the northern hemisphere, which began more than 3 million years ago and continues to the present day. To gain a better picture of the history of atmospheric CO_2 concentrations in the Cenozoic Era, scientists will need to gain more information about the relative significance of mountain building and climatic change.

Negative Feedbacks: How CO2 Levels Are Held in Check

Could the atmospheric reservoir CO_2 grow to the point where severe greenhouse warming would elevate temperatures at Earth's surface far above present levels? Similarly, could greenhouse cooling plunge the entire world into frigid conditions? In fact, neither drastic heating nor drastic cooling of Earth's climate appears to have been possible in the Phanerozoic world. The stabilizing factors here are examples of feedbacks (p. 260). A *negative feedback* is a consequence of a particular kind of change that applies a brake to this change (see Figure 10-1). The effect of a negative feedback intensifies as the change that it opposes moves the system farther toward an extreme condition. In other words, the warmer or colder global climates become, the more strongly brakes are applied. Negative feedbacks on weathering rates appear to be the most important factor in stabilizing Earth's climate. Both temperature and precipitation exert strong negative feedbacks on Earth's climate.

Temperature and Rates of Weathering

Recall that chemical weathering speeds up as the temperature rises. Consider, then, what must happen when for some reason CO_2 begins to build up in the atmosphere and, through the greenhouse effect, increases the average temperature of Earth's surface (Figure 10-14). The warming effect accelerates chemical weathering, increasing the rate at which this process extracts CO_2 from the atmosphere. The more the CO_2 that builds up, the more rapid the weathering. This increase in weathering slows the rate of CO_2 buildup until finally the system is in balance. In this way, barring additional changes, the atmospheric reservoir is stabilized.

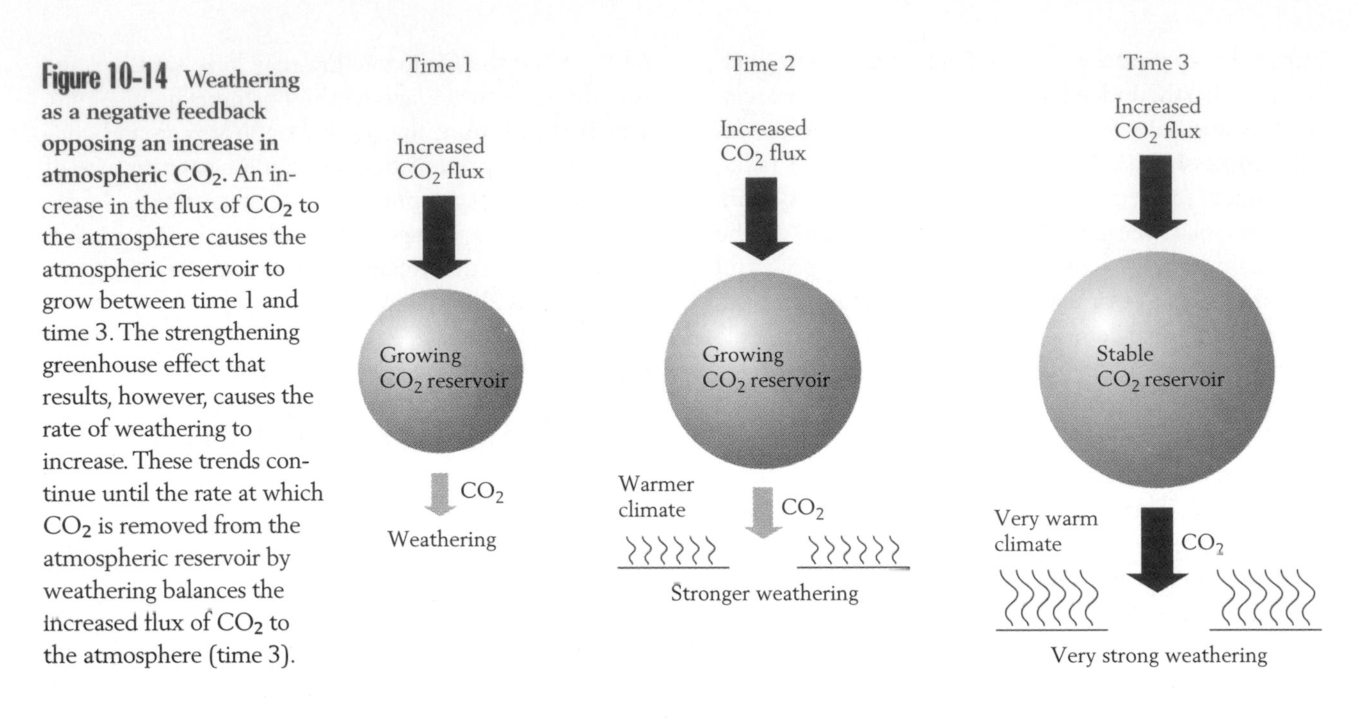

Figure 10-14 Weathering as a negative feedback opposing an increase in atmospheric CO_2. An increase in the flux of CO_2 to the atmosphere causes the atmospheric reservoir to grow between time 1 and time 3. The strengthening greenhouse effect that results, however, causes the rate of weathering to increase. These trends continue until the rate at which CO_2 is removed from the atmospheric reservoir by weathering balances the increased flux of CO_2 to the atmosphere (time 3).

Chemical weathering operates as a negative feedback in the opposite way when the atmospheric reservoir of CO_2 begins to shrink for some reason. Then global temperatures decline and greenhouse warming weakens. As temperatures fall, the overall rate of weathering declines and so does the rate at which weathering extracts CO_2 from the atmosphere. The result is that the atmospheric reservoir of CO_2 shrinks less rapidly. The more slowly weathering proceeds, the more slowly the concentration falls, until a stable level is reached.

Precipitation and Rates of Weathering

We have seen that the configuration of continents influences patterns of precipitation. The temperature of the ocean also affects rates of precipitation on a global scale. The surface waters of warm oceans evaporate more rapidly than those of cold oceans and tend to supply nearby land areas with abundant moisture. This relationship between temperature and precipitation gives us a second negative feedback for changes in atmospheric CO_2, and it works in the same direction as the first.

Consider what happens when increasing atmospheric CO_2 warms Earth considerably. As we have seen, the first negative feedback is acceleration of the chemical reactions of weathering by the higher temperatures. But, in addition, the climatic warming increases precipitation on a global scale. As a result, more watery fluids—the agents of weathering—move through soil and rock every year. Forests expand as a result of the increase in both of the materials they employ for photosynthesis: water in the soil and CO_2 in the atmosphere (p. 261). This expansion of forests further accelerates weathering, for reasons described earlier. All of this weathering uses up CO_2 from the atmosphere (see Figure 10-11). The result is less CO_2 in the atmosphere and a reduced greenhouse effect. Thus the increased global precipitation, like the increased temperatures that produce it, acts as a negative feedback for greenhouse warming.

Presumably, increases in weathering as climates have become warmer and moister have prevented Earth from undergoing what might be called a runaway greenhouse effect: persistent buildup of atmospheric CO_2 leading to extraordinarily high temperatures on Earth.

Oxygen Isotopes, Climate, and the Water Cycle

Oxygen, like carbon, exists in two stable isotopic forms. Oxygen 16 is more common than oxygen 18. The slightly different behavior of these two isotopes as oxygen moves through the Earth's system provides information about the temperatures of ancient environments, the volumes of glacial ice on Earth, and the salinities of ancient oceans.

Temperatures and Isotope Ratios in Skeletons

Just as photosynthesizing plants slightly favor CO_2 molecules that contain carbon 12 as opposed to carbon 13, organisms that secrete skeletons of calcium carbonate ($CaCO_3$) incorporate oxygen 18 and oxygen 16 in ratios that differ from those of the environment. In addition, the ratio of the two isotopes incorporated into skeletons varies with temperature. At low temperatures, organisms incorporate a relatively large proportion of oxygen 18 because ions containing this heavier isotope tend to become more sluggish than ions containing oxygen 16 and readily combine with other ions. Once scientists have determined how the oxygen isotope ratio in the skeletons of a particular group of organisms varies with temperature, they can use the fossil skeletons of these organisms as what amount to paleothermometers.

Although the fossil record of planktonic foraminifera extends back only to the Cretaceous Period, these fossils have been widely used to assess temperatures of ancient oceans. The utility of these organisms stems from their widespread occurrence in the offshore waters of the ocean and from their relative abundance in sediments of the deep-sea floor (p. 147).

The oxygen isotope technique for determining temperatures can be applied to fossils older than Cretaceous, but the older the fossil, the greater the possibility that some of the material within it has been altered after burial. Watery solutions may have recrystallized the skeletal material, removing some oxygen atoms and replacing them with others of different isotope ratios. The skeletal material in question has to be examined through a microscope and by other means to establish that it has undergone little recrystallization.

Figure 10-15 depicts a remarkably detailed record of oxygen isotope ratios for a rudist bivalve from Late Cretaceous deposits in Greece. Rudists are an extinct group of large bivalve mollusks that built reefs during the Cretaceous Period. This particular specimen comes from the Tethyan Seaway, which at that time connected the Pacific and the Atlantic oceans by way of the Mediterranean region. Figure 10-15 is based on samples from a vertical section through a rudist shell. As would be expected, these

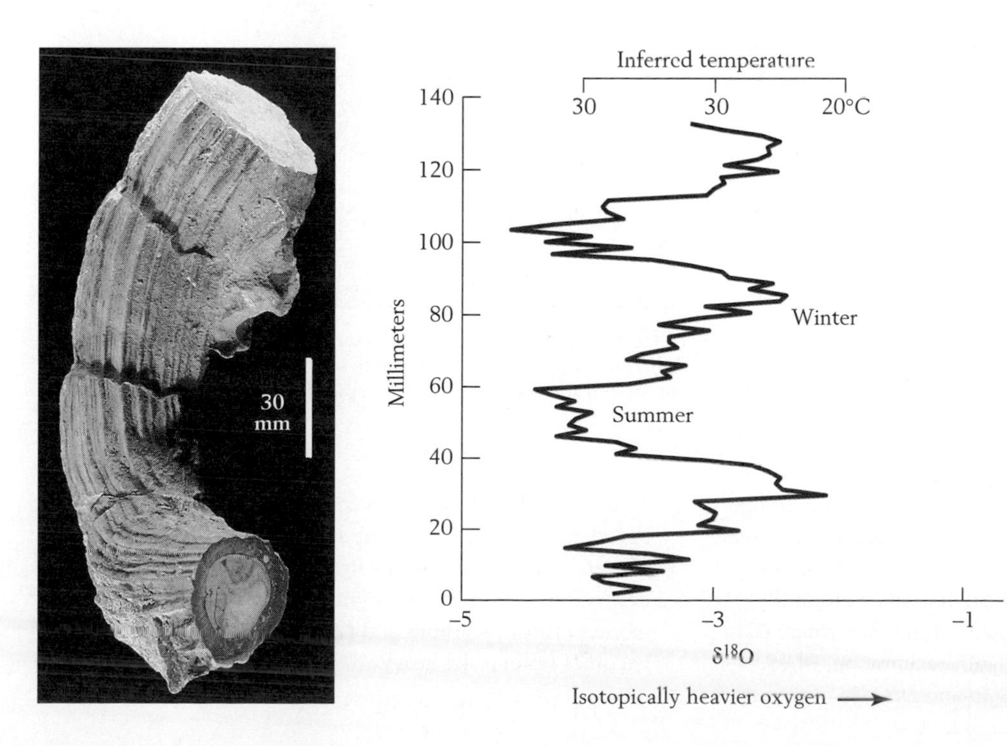

Figure 10-15 **Seasonal temperature shifts estimated from oxygen isotope ratios in a reef-building rudist bivalve from Greece.** Samples for isotopic analysis were taken along the length of a shell similar to the cucumber-sized one on the left. The plotted data depict about three years of upward growth. *(From T. Steuber, Geology 24:315–318, 1996.)*

samples reveal that the oxygen isotope composition of the calcium carbonate secreted by the animal was heavier in winter than in summer. Comparison of isotope ratios of living species indicate an annual temperature range between about 22° and 32°C (72° and 88°F) for the rudist's environment. These temperatures are much warmer than those of the Mediterranean waters near Greece today, resembling the present temperatures of seas off Miami, Florida. In other words, the isotope ratios show that the Mediterranean region was remarkably warm during the Late Cretaceous: warm waters were flowing

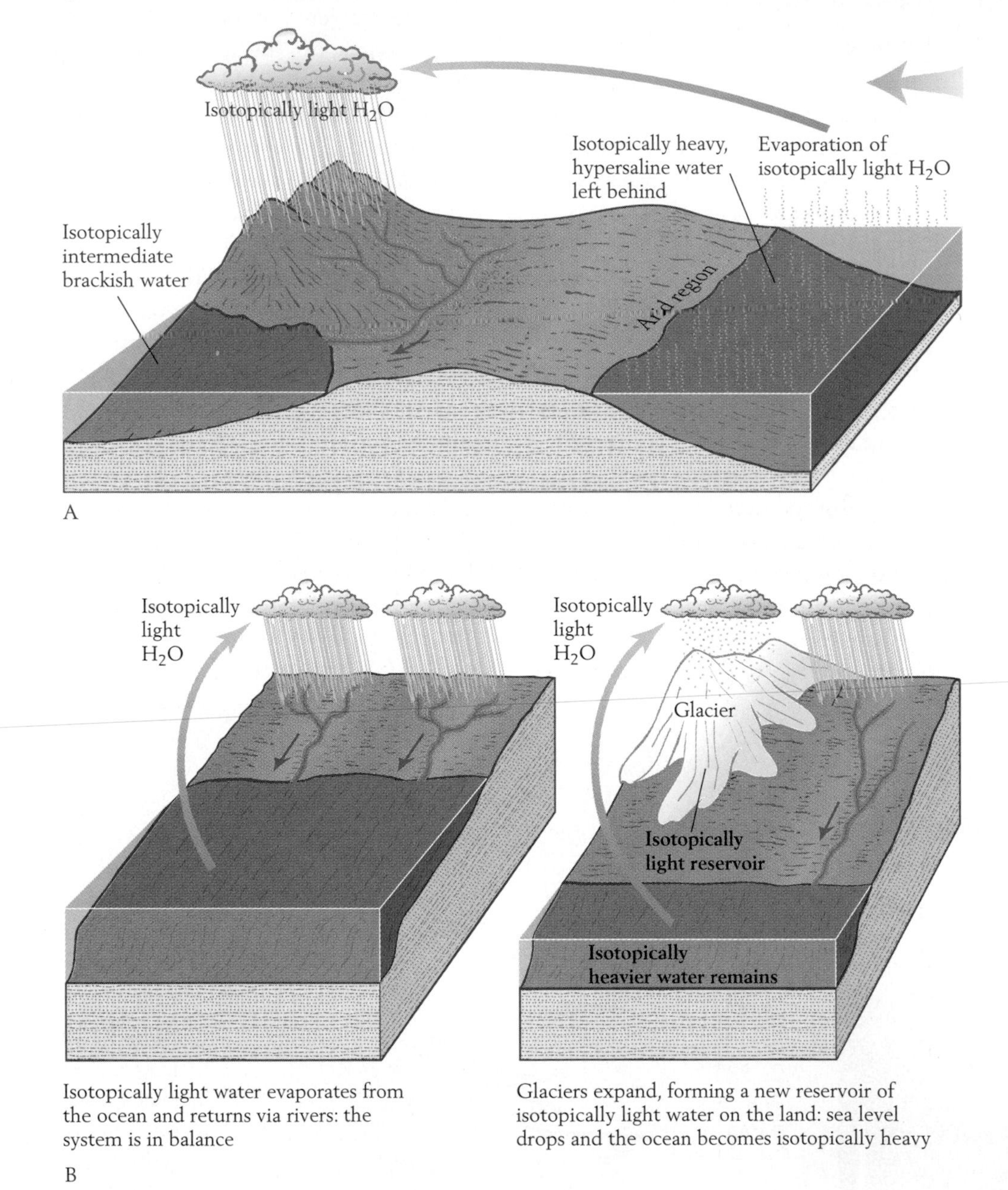

Figure 10-16 **Effects of evaporation on the oxygen isotope ratio of seawater.** *A.* A high rate of evaporation preferentially removes isotopically light H_2O from one arm of the sea, leaving the remaining water hypersaline and isotopically heavy. The water vapor thus produced moves through the atmosphere to an area of abundant precipitation, where it moves via rainfall and river flow to another arm of the sea. As a result, the waters of this second body of water are brackish and isotopically light. *B.* Growth of continental glaciers locks up isotopically light atmospheric H_2O in a new reservoir on the land (right). As a result, throughout the world (1) sea level drops and (2) the H_2O of the ocean becomes isotopically heavy.

through the Tethyan Seaway from the tropical Pacific. The isotopes also show that under these conditions rudists grew about 3.0–4.5 centimeters (~1.2–1.8 inches) per year—more rapidly than most modern reef corals.

Salinity and Isotope Ratios

Additional factors complicate the use of oxygen isotopes to estimate past temperatures. One is the fact that the oxygen isotope ratio of ocean water varies from time to time and from place to place. The most important cause of this phenomenon is variation in the rate at which surface waters evaporate. Because H_2O molecules containing oxygen 16 are lighter than those containing oxygen 18, they evaporate more readily. As a result, high rates of evaporation leave behind seawater that is enriched in oxygen 18 (Figure 10-16). This pattern is evident in modern seas, in which hypersaline waters are isotopically heavier than waters of normal salinity (~35 parts per thousand). It is easy to see why brackish waters are isotopically lighter than waters of normal salinity. Because H_2O molecules that contain oxygen 16 evaporate more readily than those containing oxygen 18, atmospheric moisture is isotopically light. When this moisture reaches Earth's surface as precipitation and returns to the sea in rivers, the brackish waters that result are isotopically lighter than normal seawater.

The variation of oxygen isotope ratios with salinity complicates efforts to derive accurate paleotemperatures from these ratios. Sometimes, however, salinity can be shown to be a relatively minor source of error. This is the case with the rudist depicted in Figure 10-15, because reefs grow under a narrow range of salinity conditions. The waters in which this rudist grew probably underwent only minor isotopic shifts as a result of high evaporation rates in summer or seasonal increase in freshwater runoff to the sea.

Volume of Glacial Ice and Isotopes

Since atmospheric moisture is isotopically light, glaciers are too, because they form from snow that precipitates from clouds. For this reason, the expansion and contraction of large glaciers affects the oxygen isotope ratio of seawater (see Figure 10-16*B*). Recall that glaciers have at times locked up a large portion the H_2O of the water cycle (p. 21). At many times during the past two and a half million years, expansion of glaciers has temporarily lowered sea level by about 100 meters (~330 feet). At such times the storage of relatively large amounts of oxygen 16 in glacial ice has left the oceans enriched in oxygen 18. During the intervening intervals, when glaciers have shrunk, the ratio of oxygen 18 to oxygen 16 has fallen again. We live at such a time today. Figure 10-17 shows a plot of oxygen isotope ratios for bottom-dwelling deep-sea foraminifera over the past 2.5 million years.

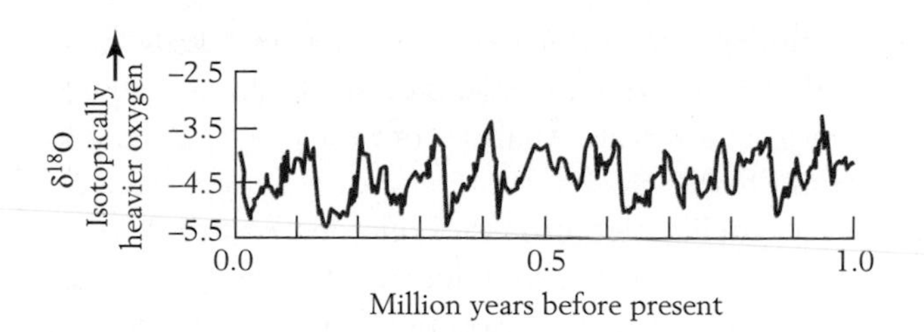

Figure 10-17 Fluctuations of oxygen isotope ratios in the skeletons of foraminifera that resulted from the expansion and contraction of continental glaciers during the past 2.5 million years. The foraminifera sampled here lived on the Pacific deep-sea floor. Peaks on the graph depict relatively high values that represent times when glaciers expanded, locking up large volumes of isotopically light H_2O. The lowest isotope values represent times when glaciers melted back, returning isotopically light H_2O to the oceans. *(After R. Tiedmann, M. Sarnthein, and N. J. Shackelton, Paleoceanography 9:619–638, 1994.)*

Like changes in salinity, changes in the volume of ice create problems for the scientist who wants to use oxygen isotopes to estimate ancient ocean temperatures. Nonetheless, the effects of ice volume and temperature during a glacial interval combine to give a clear picture of the timing of expansion and contraction of large glaciers. When glaciers have expanded, foraminifera in most regions have secreted isotopically heavy skeletons because of the effects of both ice voume and temperature. The oceans in which they lived have been enriched in oxygen 18. In addition, the organisms have preferentially taken up this isotope in secreting their skeletons because the seawater is cooler. Similarly, when glaciers have contracted, the effects of ice volume and temperature have both contributed to lower oxygen 18/oxygen 16 ratios.

It is estimated that changes in ice volume contributed a substantially larger portion of each isotopic shift shown in Figure 10-17 than the portion contributed by temperature change.

Shifts in oxygen isotope ratios in fossils have also been related to episodes of glaciation in pre-Cenozoic intervals. Analyses of ratios are less reliable for old fossils than for young ones, however, because of recrystallization of fossil skeletons. At times when few glaciers have existed, ice volume can be ignored in the study of oxygen isotopes. Isotope ratios reflect temperature and salinity alone. The Late Cretaceous, when the reef-building rudist depicted in Figure 10-15 lived, was such a time.

Ocean Chemistry and Skeletal Mineralogy

A variety of evidence suggests that the general chemical composition of seawater has not changed greatly for hundreds of millions of years. Sodium and chloride, which combine to form halite when seawater evaporates (p. 46), have remained the most abundant dissolved ions, and other components have also varied only modestly. The abundances of some chemical components of seawater have nonetheless fluctuated enough to bring about changes in the types of minerals that have precipitated as sediments.

For our purposes, the fluctuation of two ions, Ca^{++} and Mg^{++}, in seawater has been especially important. These are chemically similar ions, having the same charge and differing only modestly in diameter. Magnesium can substitute for calcium in the crystal lattice of calcite but is too small to lodge in the lattice of aragonite, the other form of $CaCO_3$ that precipitates from seawater (see p. 36).

Today both aragonite and high-magnesium calcite precipitate from seawater. The latter is calcite with magnesium substituting for several percent of the calcium ions. Most oolites are composed of aragonite, but both minerals fill voids in coral reefs and cement sediment here and there along tropical seashores. Aragonite and high-magnesium calcite also formed oolites and cements on the seafloor during two earlier geologic intervals—early in the Cambrian and again from late in the Paleozoic Era until well into the Mesozoic Era. Between these intervals were ones during which oolites and seafloor cement were composed of ordinary calcite—calcite containing little magnesium. Thus geologists speak of Earth as three times having had "aragonite seas" and twice having had "calcite seas."

It turns out that the relative proportions of Ca^{++} and Mg^{++} in seawater have determined which minerals have precipitated from warm seas. Laboratory experiments show that relatively high Mg^{++}/Ca^{++} ratios result in precipitation of aragonite and high-magnesium calcite. Relatively low Mg^{++}/Ca^{++} ratios result in precipitation of normal calcite.

What causes the Mg^{++}/Ca^{++} ratio of seawater to rise and fall in the course of geologic time? The primary cause is change in the volume of mid-ocean ridges. Seawater circulates through the sediments and fractured rocks that flank mid-ocean ridges. In the process, the water heats up and reacts chemically with the rock of the newly forming lithosphere. The chemical reactions transfer Ca^{++} from the rocks to the seawater, while also extracting Mg^{++} from the sea water and locking it up in rocks of the newly forming oceanic crust.

In other words, the mid-ocean ridges amount to ion-exchange systems, extracting Mg^{++} from seawater and releasing Ca^{++} to it. Variations in the volume of ridges throughout geologic time have therefore exerted strong influence on seawater chemistry. When the total volume of ridges is great, causing a rise in global sea level (p. 175), the Mg^{++}/Ca^{++} ratio falls because the ion exchange system speeds up (Figure 10-18). The result is precipitation of the mineral calcite from seawater. At other times, such as the present, when mid-ocean ridges are smaller and less numerous and sea level stands lower in relation to continental surfaces, the Mg^{++}/Ca^{++} ratio is higher and high-magnesium calcite and aragonite precipitate.

More important than the changes in the mineralogy of inorganically precipitated carbonates during transitions between calcite and aragonite seas are corresponding changes in the kinds of marine organisms that secrete large volumes of calcium carbonate. Box 10-1 describes how the fine-grained limestone called chalk (p. 73) accumulated in large quantities during a brief interval when the Mg^{++}/Ca^{++} ratio of seawater was lower than at any other time in the Phanerozoic and chalk-producing plankton flourished as never before or after.

Seawater chemistry has also influenced which groups of organisms have formed reefs at particular times and which have contributed large volumes of carbonate sediment. Because organisms need ideal conditions to form large reefs (p. 116), we might expect the success of reef builders to reflect seawater

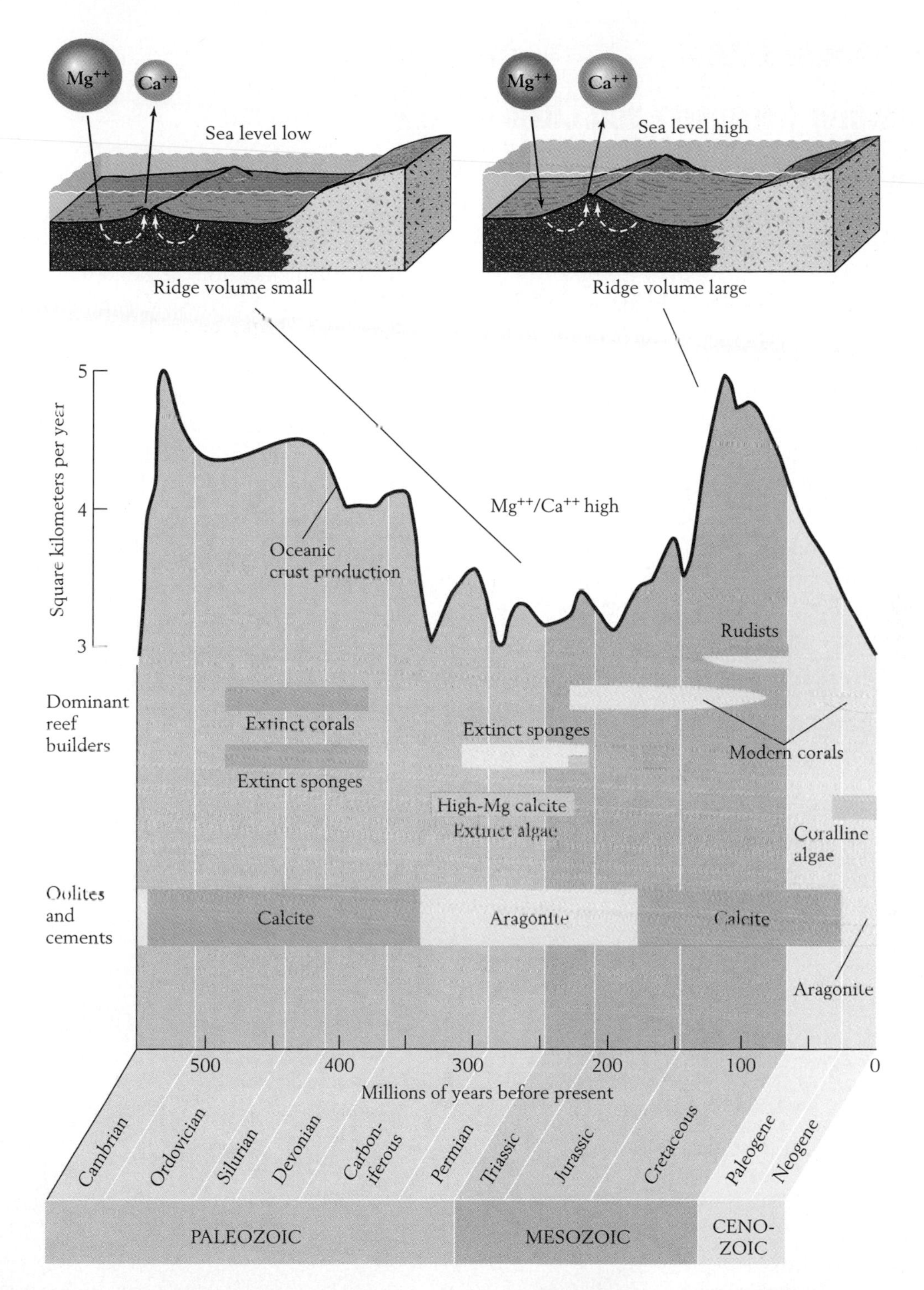

Figure 10-18 Changes in spreading rates along mid-ocean ridges and corresponding changes in the mineralogy of oolites and carbonate cements and of dominant reef-building organisms. The curve shows changes in the rate of oceanic crust production during Phanerozoic time. This curve is based on evidence of long-term changes in global sea level. Seawater that circulates through mid-ocean ridges gives up Mg^{++} and carries away Ca^{++}. Thus an increase in the volume of mid-ocean ridges causes not only a global rise in sea level but also a decrease in the Mg^{++}/Ca^{++} ratio of seawater. Low Mg^{++}/Ca^{++} ratios produce calcite seas: calcite forms oolites, marine cements, and skeletons of dominant reef builders. Aragonite and high-magnesium calcite play these roles instead when the total volume of mid-ocean ridges is low and the Mg^{++}/Ca^{++} ratio of seawater is high. *(After S. M. Stanley and L. A. Hardie, Palaeogeography, Palaeoclimatology, Palaeoecology, 1998.)*

For the Record 10-1

Seawater Chemistry and Chalk

Chalk is a soft, white, fine-grained limestone that we use to write on blackboards, but it is also the formal name for a large Upper Cretaceous body of rock in western Europe. Along the coast of France the Chalk forms cliffs that Allied troops scaled during the invasion of Normandy in 1944, and it rises up as the famous White Cliffs of Dover on the British side of the English Channel. The Cretaceous contains a vastly larger volume of chalk than any other geologic system. As it turns out, this unusual abundance probably reflects the chemical composition of Cretaceous seas.

The most massive deposits of Cretaceous chalk are those of western Europe, which are typically about 200 meters (~660 feet) thick, but substantial bodies of chalk also accumulated in an epicontinental sea in North America. Between Alabama and Kansas, Upper Cretaceous chalks have yielded spectacularly well preserved fossils, including skeletons of huge marine reptiles that might be described as sea monsters. Fossils are exceptionally well preserved in chalk for the same reason that the chalk is soft: it is formed of minute grains of calcite. The small size of these grains makes the rock they form so nearly impermeable to water that they dissolve very slowly. Furthermore, at temperatures and pressures near Earth's surface calcite is less soluble than aragonite, the other common calcium carbonate mineral. Much of the calcium carbonate dissolved from limestones is quickly reprecipitated as cement, and the weak dissolution of its grains has prevented chalk from becoming cemented into harder limestone.

It is because grains of Cretaceous chalk are predominantly armor plates of calcareous nannoplankton (p. 72) that they are calcitic and minute. (Recall that "nanno" means very

Cliffs of the Chalk in Denmark. The darker, upper portion was deposited in earliest Cenozoic time, after the extinction that killed off the dinosaurs.

small.) The plates encrust these nearly spherical floating cells and protect them from tiny predators. Calcareous nannoplankton were so prolific when Chalk was deposited that their minuscule plates often accumulated at a rate of about a millimeter per year (~ an inch every 25 years). At no other time during the group's history has it approached such productivity, despite the fact that it has remained taxonomically diverse since soon after its Jurassic origin (today it includes nearly 400 species). Here is where seawater chemistry comes in. During the Cretaceous interval of massive chalk deposition the ratio of Mg^{++} to Ca^{++} in seawater was lower than at any other time in the Phanerozoic. Thus calcite precipitated from seawater more easily at this time than at any other.

Two additional patterns support the idea that this condition was responsible for the remarkable productivity of phytoplankton: First, calcareous nannoplankton suffered heavy extinction at the end of the Cretaceous, when the dinosaurs died out, yet they recovered to flourish and form chalk at the beginning of the Cenozoic Era, when the Mg^{++} to Ca^{++} ratio remained low. Only when the ratio then rose substantially did chalk deposition cease, even though warm global climates should have favored the growth of calcareous nannoplankton.

Second, individual plates of calcareous nannoplankton became small and spindly as calcite seas gave way to aragonite seas during the Cenozoic. Plates were broad and thick in Cretaceous and earliest Cenozoic species, but became much smaller as the era progressed. In modern seas they encrust cells very thinly. One genus, *Discoaster*, at first had robust circular plates but then evolved embayments that gave them a starlike shape. When *Discoaster* became extinct about 3 million years ago, all of its species had scrawny plates that, instead of providing a complete armor shield, formed a cagelike structure that left more than three-quarters of the spherical cell exposed. It appears that during the last 60 million years calcareous nannoplankton suffered the evolutionary equivalent of osteoporosis, the disease that strikes many humans in middle age, weakening their bones because of a deficiency of calcium in their diet!

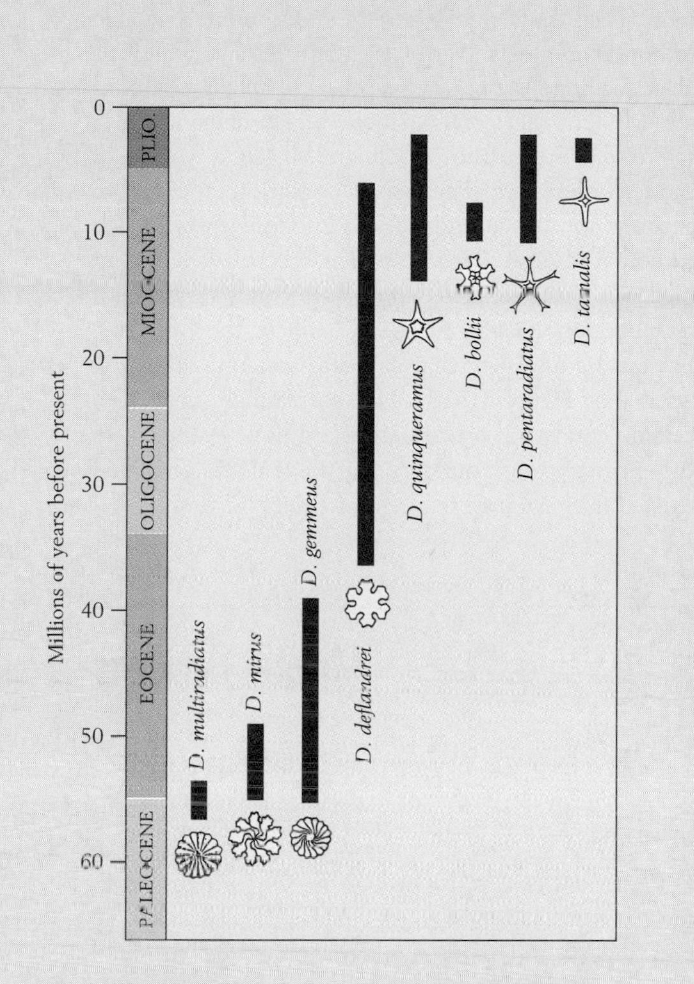

Typical forms of skeletal plates of *Discoaster* during various Cenozoic intervals. Species with robust plates were replaced by species with delicate, star-shaped plates.

chemistry. Modern taxa of reef-building corals can exist outside the tropics and can survive without symbiotic algae, but under neither condition are they able to build large reefs. Warm temperatures and symbiotic algae facilitate the corals' secretion of calcium carbonate. Modern reef corals, which secrete aragonite, also seem to benefit from the high Mg^{++}/Ca^{++} ratio of modern seawater. The same condition appears to account for the abundance on modern reefs of calcareous red algae (p. 73), which secrete skeletons of high-magnesium calcite. Aragonitic corals were major reef builders in aragonite seas early in the Mesozoic Era but relinquished that role after calcite seas developed (see Figure 10-18). Not surprisingly, groups of calcitic corals and sponges that are now extinct were the primary reef builders in the calcite seas that existed during a long segment of Paleozoic time.

Chapter Summary

1 Reservoirs of key elements and compounds in the Earth system can shrink or expand as fluxes between them change.

2 Photosynthesis is the process by which plants use the energy of sunlight to produce sugars from carbon dioxide and water; oxygen is a by-product of this process. Respiration entails the opposite chemical reaction and is used by organisms to oxidize sugars in order to release their energy.

3 Sugars that plants produce but do not use in respiration form plant tissue, which animals and decomposers consume for their respiration.

4 If no dead plant tissue were buried, the photosynthesis-respiration cycle involving plants, animals, and decomposers would be roughly in balance. Both O_2 and CO_2 would cycle through these organisms and the atmosphere without much change in the volume of the atmospheric reservoir of either gas.

5 Burial of dead plant tissue, which contains reduced carbon, upsets the balance of the global photosynthesis-respiration cycle. Because it prevents some reduced carbon from decomposing and returning to the atmosphere as CO_2, the atmospheric reservoir of CO_2 shrinks. At the same time, oxygen that would have been used to decompose the organic matter if it had not been buried remains in the atmosphere, and the atmospheric reservoir of O_2 expands.

6 Most burial of organic carbon takes place in anoxic swamps or marine environments that exclude most kinds of bacteria that decompose organic matter. The spread of these environments in the geologic past has caused the atmospheric reservoir of CO_2 to shrink, reducing greenhouse warming and simultaneously expanding the atmospheric reservoir of O_2.

7 Photosynthesis employs a disproportionate percentage of relatively light carbon 12. Therefore, at times in the geologic past when much organic carbon has been buried, the relative abundance of carbon 13 has increased in the atmosphere and ocean. Shifts of this kind toward heavier carbon are reflected in the carbon isotope composition of limestone and of sedimentary organic matter produced by photosynthesis.

8 Weathering of minerals removes CO_2 from the atmosphere, so that factors that intensify weathering can deplete the atmospheric reservoir of CO_2. These factors include mountainous topography, warm climates, high rates of precipitation, and deeply rooted vegetation.

9 During the latter part of the Paleozoic Era, the initial spread of forests on Earth intensified weathering and depleted the atmospheric reservoir of CO_2. This depletion continued as coal swamps became widespread sites of carbon burial, and it contributed to the cooler climatic conditions of a major ice age.

10 Negative feedbacks have prevented extreme changes in the atmospheric reservoirs of CO_2 and O_2.

11 Oxygen isotopes in skeletons of marine organisms provide a record of past ocean temperatures.

12 As water molecules containing oxygen 16 are lighter than those containing oxygen 18, they evaporate more readily. This pattern permits geologists to detect salinity changes in ancient oceans and also the buildup of continental glaciers, which lowered sea level and left ocean water isotopically heavy.

13 Changes in spreading rates along mid-ocean ridges alter the magnesium-calcium ratio of seawater. Shifts in this ratio have caused different kinds of minerals to precipitate from seawater at different times in the geologic past, and have strongly influenced which kinds of organisms have been reef builders and sediment producers.

Review Questions

1 What are the two possible fates of plant material not eaten by animals?

2 What is a negative feedback? What is a positive feedback?

3 How can carbon isotopes in limestones provide evidence about the history of atmospheric oxygen? (Hint: Refer to Figure 10-9.)

4 Draw sketches illustrating how increased burial of carbon reduces the atmospheric reservoir of CO_2 and enlarges the atmospheric reservoir of O_2.

5 In what kinds of marine environments can carbon be buried in large quantities?

6 Draw a diagram depicting the cycle of oxidized carbon that includes limestone, atmospheric CO_2, and weathering.

7 Why are pelagic carbonates more likely than shallow-water carbonates to melt and return CO_2 to the atmosphere?

8 How do glaciers promote chemical weathering?

9 Why does the influence of moist climates on vegetation accelerate weathering?

10 What can the study of oxygen isotopes tell us about ancient oceans?

11 What controls the ratio of magnesium to calcium in the ocean? How have changes in this ratio influenced the mineralogical composition of limestone during Phanerozoic time?

12 Using the Visual Overview on page 258 and what you have learned in this chapter, (a) summarize how burial of organic carbon, alteration of carbonates at high temperatures, and changes in rates of weathering alter greenhouse warming by Earth's atmosphere and (b) explain how carbon isotopes are used to assess rates of burial of organic carbon.

Additional Reading

Berner, E. K., and R. A. Berner, *Global Environment: Water, Air, and Geochemical Cycles,* Prentice Hall, Englewood Cliffs, N.J., 1996.

Mackenzie, F. T., and J. A. Mackenzie, *Our Changing Planet: An Introduction to Earth System Science and Global Environmental Change,* Prentice Hall, Englewood Cliffs, N.J., 1995.

Smill, V., *Cycles of Life: Civilization and the Biosphere,* Scientific American Library, New York, 1997.

Turekian, K. K., *Global Environmental Change: Past, Present, and Future,* Prentice Hall, Englewood Cliffs, N.J., 1996.

CHAPTER 11

The Archean Eon of Precambrian Time

Since the nineteenth century, the interval of Earth history that preceded the Phanerozoic Eon has been known as the Precambrian. Although the term "Precambrian" has no formal status in the geologic time scale, it has traditionally been employed as though it did. The Precambrian includes nearly 90 percent of geologic time, ranging from 4.6 billion years ago, when Earth formed, to the start of the Cambrian Period, about 4 billion years later.

Two eons are formally recognized within the Precambrian: the Archean and the Proterozoic. The Archean Eon includes about 45 percent of Earth's history—the interval from about 4.6 billion to 2.5 billion years ago. During the Archean Eon, Earth underwent enormous physical changes and life developed on its surface. Many details of Archean history, however, remain unknown or poorly understood.

One reason for our limited understanding of Precambrian history is that, although this history spans about a billion years, Precambrian rocks form less than 20 percent of the total area of rocks exposed at Earth's surface (Figure 11-1). Erosion has destroyed many Precambrian rocks,

In northwestern Canada the Canadian Shield is a largely barren Precambrian terrane that was scoured by glaciers during the recent ice age and is now dotted by lakes. Outcrops of numerous dikes formed when magma forced its way into giant cracks in the Precambrian crust.

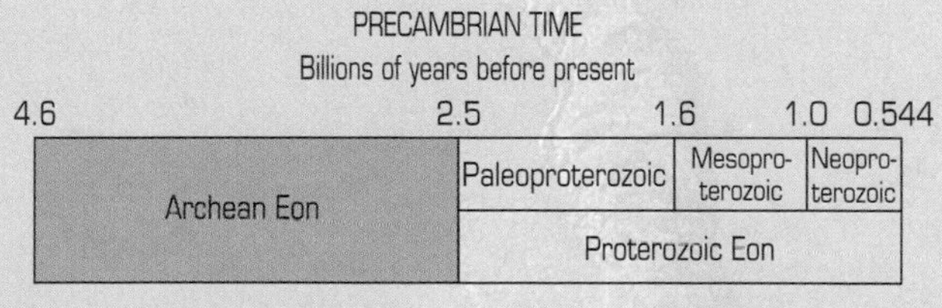

PRECAMBRIAN TIME

Billions of years before present

4.6 – 2.5	2.5 – 1.6	1.6 – 1.0	1.0 – 0.544
Archean Eon	Paleoproterozoic	Mesoproterozoic	Neoproterozoic
	Proterozoic Eon		

Visual Overview

Major Events of the Archean Eon

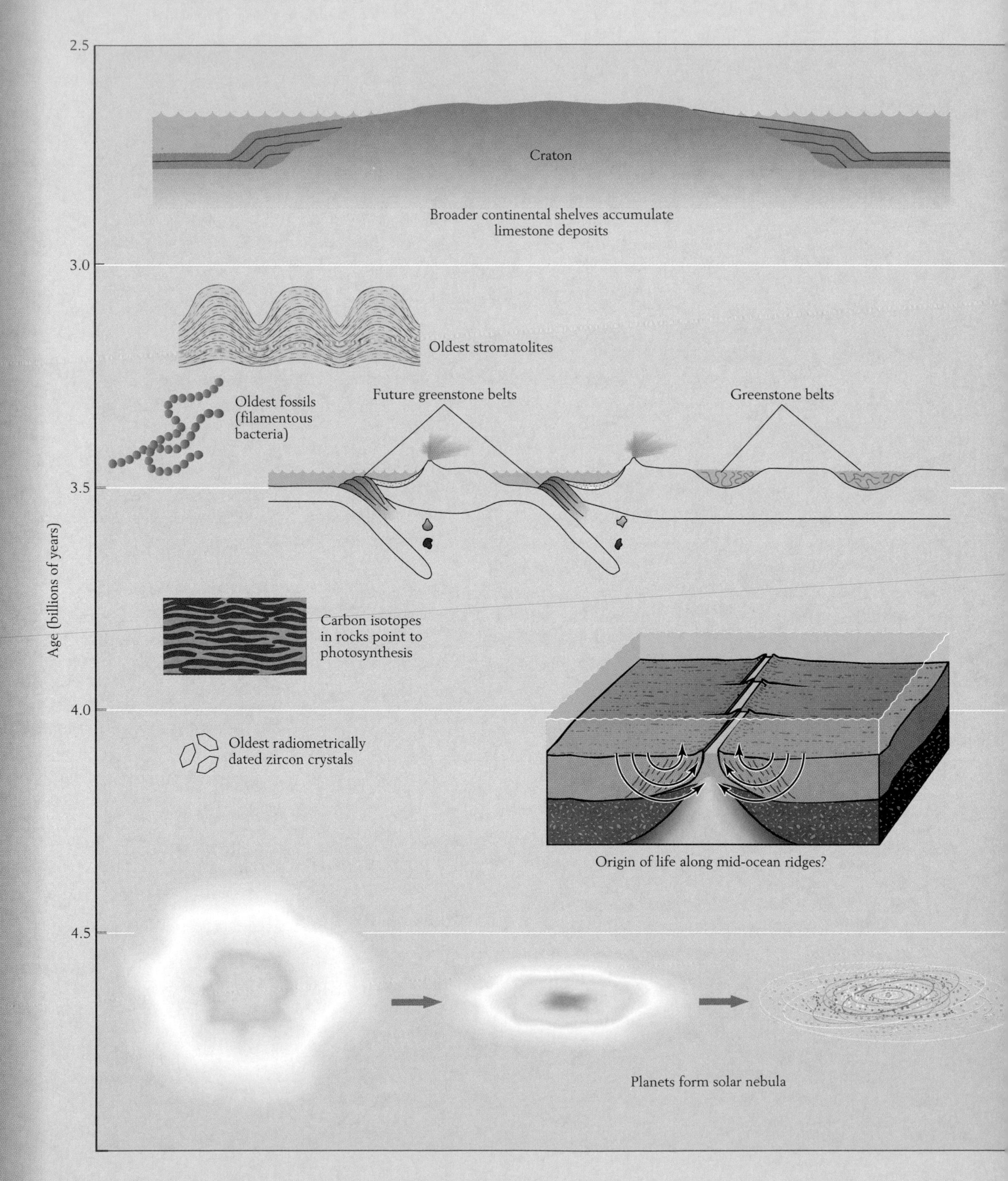

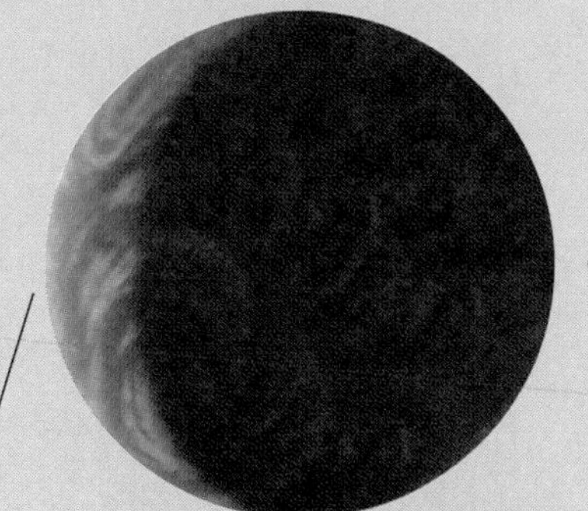
Origin of the early atmosphere

Larger
continents
form
Huge impact creates Earth's moon
Rate of
impacting
declines
Earth's
interior
continues
to cool

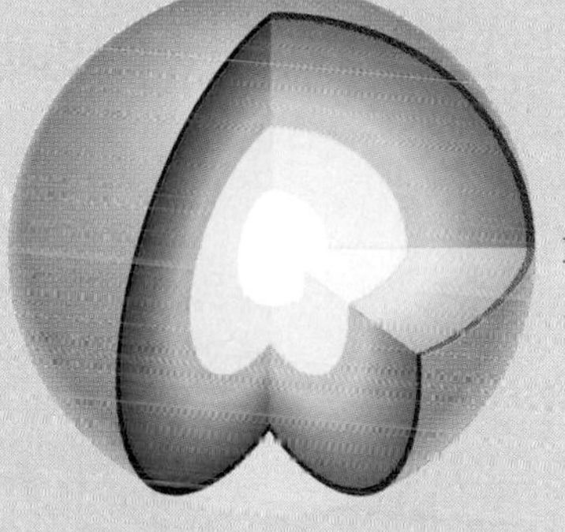
Formation of Earth's layers

Differentiation of materials
according to density
The first
100 hundred
million years
Molten Earth

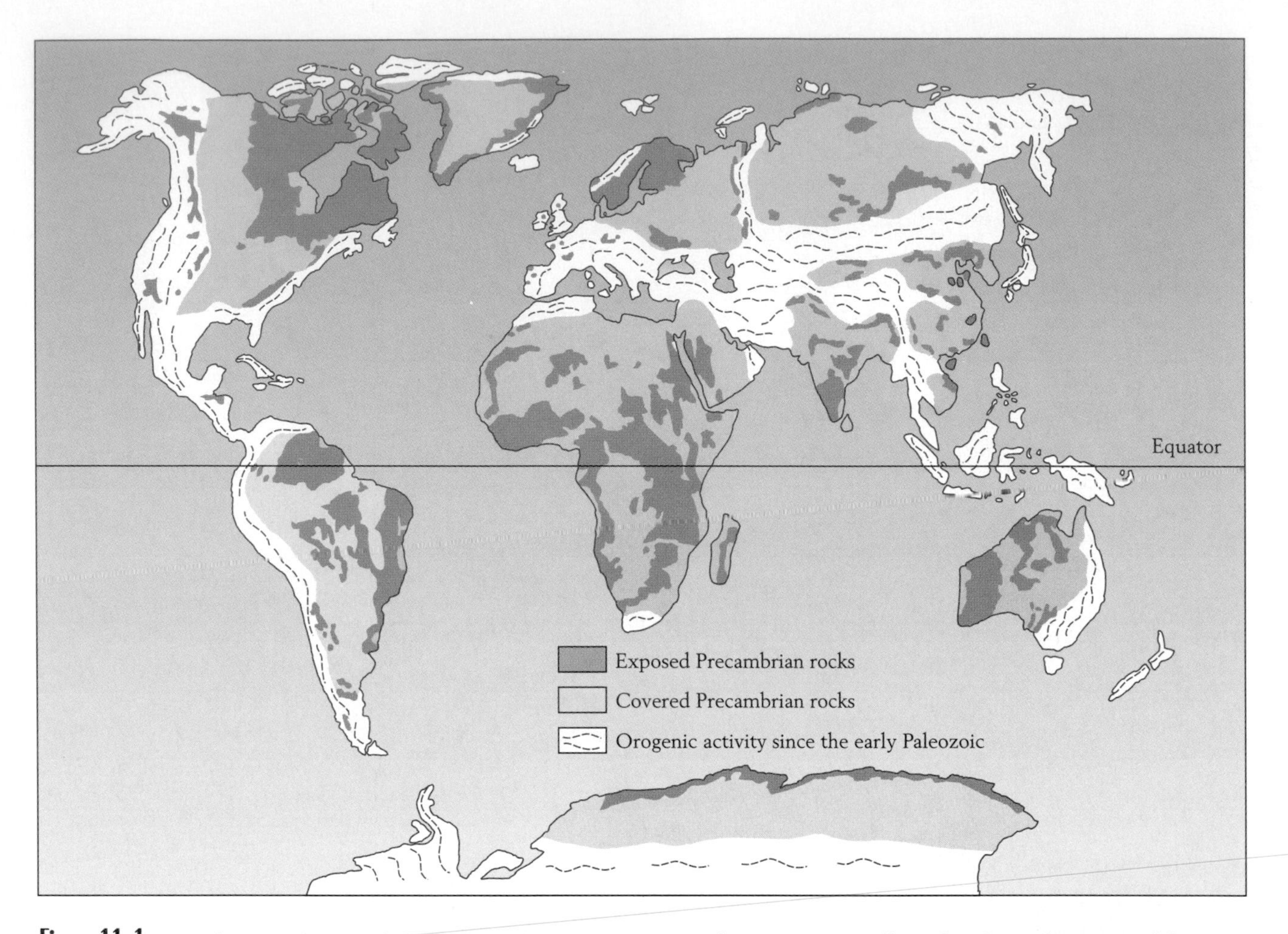

Figure 11-1 Distribution of Precambrian rocks in the modern world. Note that these rocks form or underlie most of the cratonic area of the modern world. *(After A. M. Goodwin, in B. F. Windley [ed.], The Early History of the Earth, John Wiley & Sons, New York, 1976.)*

and metamorphism has so altered others that they can no longer be dated and therefore cannot be recognized as Precambrian. Still other Precambrian rocks lie buried beneath younger sedimentary and volcanic rocks.

The simple unicellular fossils of Archean rocks are uncommon and difficult to assign to species and genera, so that few are recognized as index fossils. Stratigraphic correlation of these rocks has therefore been based largely on radiometric dating. The richer fossil record of later Proterozoic strata contains more advanced life forms and provides many guide fossils for dating rocks.

Most geologic information about the Precambrian is derived from **cratons,** those large portions of continents that have not undergone tectonic deformation since Precambrian or early Paleozoic time. All the continents of the present world include cratons that consist primarily of Precambrian rocks (Figure 11-2). A **Precambrian shield** is a largely Precambrian portion of a craton that is exposed at Earth's surface. The largest is the vast Canadian Shield, which has recently (geologically speaking) become more fully exposed by the action of Pleistocene glaciers (p. 127). Although shields contain some sedimentary rocks, they consist primarily of crystalline (igneous and metamorphic) rocks. As we will see, mountain belts that formed during Precambrian time left traces that can be recognized within Precambrian shields, but erosion long ago destroyed their elevated topography. Today the only Precambrian rocks that stand at high elevations in mountain ranges are those that have been uplifted by Phanerozoic orogenies.

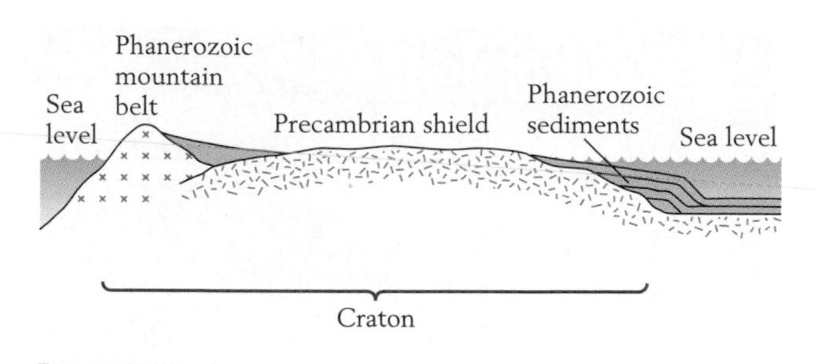

Figure 11-2 A Precambrian shield flanked by younger (Phanerozoic) mountain belts and sediments. A shield and the solid rocks that surround it constitute a craton.

It was during the Archean Eon that Earth acquired its basic configuration, with the mantle and crust surrounding the core (p. 15). Early in Archean time, Earth's crust seems to have differed from what it is today. The planet was much hotter than it is now, and apparently there were no large felsic cratons. By the end of Archean time, however, large cratons had begun to form, and plate tectonic processes were modifying these cratons in the same way that they do now. The transition to this modern style of tectonism ushered in the Proterozoic interval of Precambrian time (discussed in Chapter 12). In the absence of useful biostratigraphic data, the boundary between the Archean and Proterozoic intervals is defined by its absolute age of 2.5 billion years before the present.

We will begin our review of Archean events by considering when and how Earth and other planets of the solar system may have formed. We will then review evidence that suggests how Earth changed during the remainder of Archean time.

The Ages of the Planets and the Universe

The planets of our solar system all rotate around the sun in the same direction, and in orbits that lie in nearly the same plane. This is strong evidence that the planets formed simultaneously from a single disk of material that rotated in the same direction as the modern planets.

Precisely when the planets came into being has been a difficult issue to resolve. Astronauts report that the most beautiful object they see from space is Earth, whose surface is partially blanketed by swirling white clouds set against the blue background of extensive oceans (Figure 11-3). But while Earth's water is aesthetically pleasing and necessary for life, its abundance near the planet's surface makes rapid erosion inevitable. Continuous alteration of the crust by erosion and also by igneous and metamorphic processes makes unlikely any discovery of rocks nearly as old as Earth. Thus geologists have had to look beyond this planet in their efforts to date Earth's origin. Fortunately, we do have samples of rock that appear to represent the primitive material of the solar system. These samples are **meteorites**—extraterrestrial objects that have been captured in Earth's gravitational field and have then crashed into our planet.

Some meteorites consist of rocky material and, accordingly, are called **stony meteorites** (Figure 11-4). Others are metallic and have been designated **iron meteorites** even though they contain lesser amounts of elements other than iron. Still others consist of mixtures of rocky and metallic material and thus are called **stony-iron meteorites**. Meteorites come in all sizes, from small particles to the small planets known as asteroids; no asteroid, however, has struck Earth during recorded human history. Many meteorites appear to be fragments of larger bodies that have undergone collisions and broken into pieces. Iron

Figure 11-3 Earth viewed from space. The blue regions are oceans and the swirling white masses are clouds.

Figure 11-4 A stony meteorite. This particular stone fell in a meteorite shower near Plainville, Texas.

Figure 11-5 The configuration of a spiral galaxy. This is galaxy NGC 6744, its arms spiraling outward from the central bulge.

meteorites are fragments of interiors of these bodies, comparable to Earth's core, and stony meteorites are from outer portions of these bodies, comparable to Earth's mantle.

Meteorites have been radiometrically dated by means of several decay systems, including rubidium-strontium, potassium-argon, and uranium-thorium (p. 166). The dates thus derived tend to cluster around 4.6 billion years, which suggests that this is the approximate age of the solar system. After many meteorites had been dated, it was gratifying to find that the oldest ages obtained for rocks gathered on the surface of the moon also approximated 4.6 billion years. This must, indeed, be the age of the solar system. Ancient rocks can be found on the moon because the lunar surface, unlike that of Earth, has no water to weather and erode rocks and is characterized by only weak tectonic activity.

Determining the age of the universe, which turns out to be more than three times as old as our solar system, has been more complicated. Most stars in the universe are clustered into enormous disk-like galaxies (Figure 11-5). The distance between our galaxy, known as the Milky Way, and all others is increasing. In fact, all galaxies are moving away from one another, evidence that the universe is expanding. It is not the galaxies themselves that are expanding, but the space between them. What is happening is analogous to inflation of a balloon with small coins attached to its surface (Figure 11-6). The coins behave like galaxies: although they do not expand, the space between them does. Before the galaxies formed, matter that they contain was concentrated with infinite density at a single point, from which it exploded in an event irreverently called the **big bang.** Even after it became assembled into galaxies, matter continued to spread in all directions from the site of the big bang.

The evidence that the universe is expanding makes it possible to estimate its age. This evidence, called the **redshift,** is an increase in the wavelengths of light waves traveling through space—a shift toward the end of the spectrum of wavelengths where visible light is red. Expansion of the space between galaxies

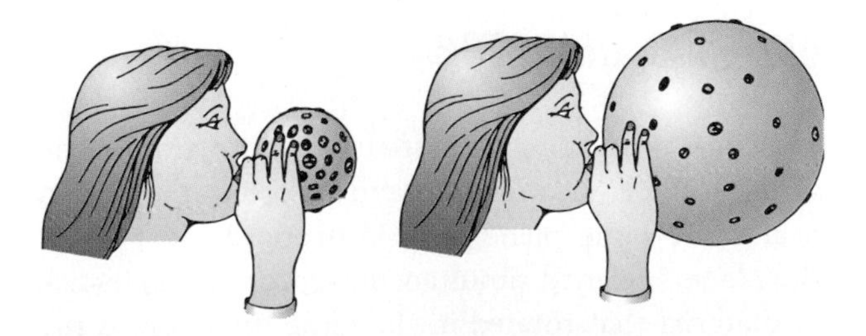

Figure 11-6 An analogy for the expansion of the universe. The space between galaxies expands like the balloon, but the galaxies themselves, like the coins, do not expand but only move farther apart.

causes this shift by stretching light waves as they pass through it. The farther these light waves have traveled through space, the greater the redshift they have undergone. For this reason, light waves that reach Earth from distant galaxies have larger redshifts than those from nearby galaxies. Calculations based on these redshifts indicate that between about 15 billion and 18 billion years ago all of the galaxies would have been at one spot, the site of the big bang. This, then, is the approximate date of the big bang and the age of the universe.

The Origin of the Solar System

We know more about the origin of distant stars than we do about the origin of our own solar system. Our solar system formed long ago, and we have no younger solar systems to observe at close range. Scientists can train their telescopes on multitudes of large stars in various stages of development, and sometimes they can detect planets. Unfortunately, these stars are too far away to allow us to observe how planets may be forming in association with them.

Galaxies form by the gravitational collapse of dense clouds of gas (mainly hydrogen) into stars. Our galaxy, which originated less than 10 billion years ago, is made up of approximately 250 billion stars. And even after a galaxy is formed, secondary stars are continually born within its spiral arms, where galactic matter is concentrated as matter contracts and expands (Figure 11-7). The sun is a star that formed in such a setting.

The Sun

Our sun formed from material remaining from another star that collapsed violently, to form heavy elements. After this collapse, a **supernova**—an exploding star that casts off matter of low density—was formed. What remained was a dense cloud that condensed as it cooled. This dense cloud, or **solar nebula**

Figure 11-7 The birth of stars. In this photograph, taken by the Hubble Space Telescope, stars emerge from bright terminations of huge columns of hydrogen and dust in the constellation Serpens.

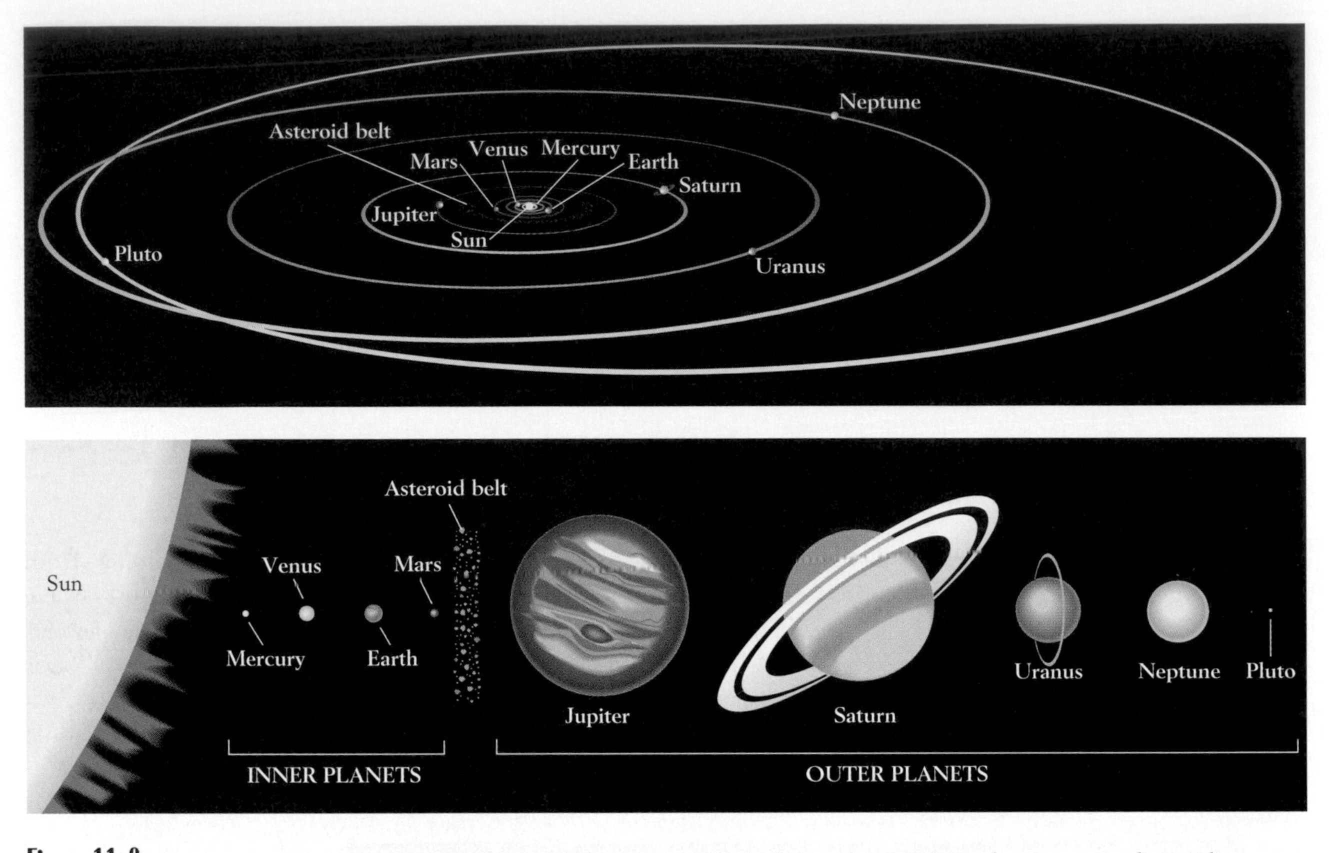

Figure 11-8 **The solar system.** The inner planets, including Earth, are small rocky bodies. The asteroid belt lies between them and the outer planets, which consist largely of volatiles. *(After F. Press and R. Siever, Understanding Earth, 2nd ed., W. H. Freeman and Company, New York, 1998.)*

(see Figure 11-7), is assumed to have had some rotational motion when it formed and must then have rotated more and more rapidly as it contracted, just as ice skaters automatically spin more rapidly (conserving angular momentum) as they pull in their arms.

The Planets

The planets formed either during or soon after the birth of the sun. In fact, the sun and the planets originated during an interval no longer than 50 million to 100 million years, roughly the length of the Cenozoic Era (the Age of Mammals). This has been deduced from the excessive concentration in some meteorites of the stable isotopes xenon 129 and plutonium 244. (In this case "excessive" means present in an amount greater than is typical for meteorites.) Excessive amounts of these isotopes in a meteorite indicate that some of the short-lived parent isotopes were originally present in the meteorite material and then decayed to xenon 129 and plutonium 244. Thus the planet or planetlike body of which such a meteorite is a fragment must have formed soon after the elements of the solar system came into being; otherwise, the short-lived parent isotopes of xenon 129 and plutonium 244 could not have been incorporated in the material of meteorites in a sufficiently high concentration to leave excessive amounts of their daughter products. The conclusion is that the sun cannot be very much older than the meteorites, the moon, and the planets, which are about 4.6 billion years old.

Planets that are positioned far from the sun formed largely from volatile (i.e., easily vaporized) elements. These elements were expelled from the hot inner region of the nebula to solidify in colder regions far from the sun. Materials that are denser and less volatile tended to be left behind, and these materials formed the inner planets, including Earth (Figure 11-8).

Several steps led to the formation of the planets. When the rotating dust cloud reached a certain density and rate of rotation, it flattened into a disk, and the material of the disk then segregated into rings, which later condensed into the planets (Figure 11-9). Each planet began to form by the aggregation of material within one of these rings. The aggregates eventually reached the proportions of *asteroids,* which are commonly about 40 kilometers (25 miles) in diameter, and these coalesced to form planets.

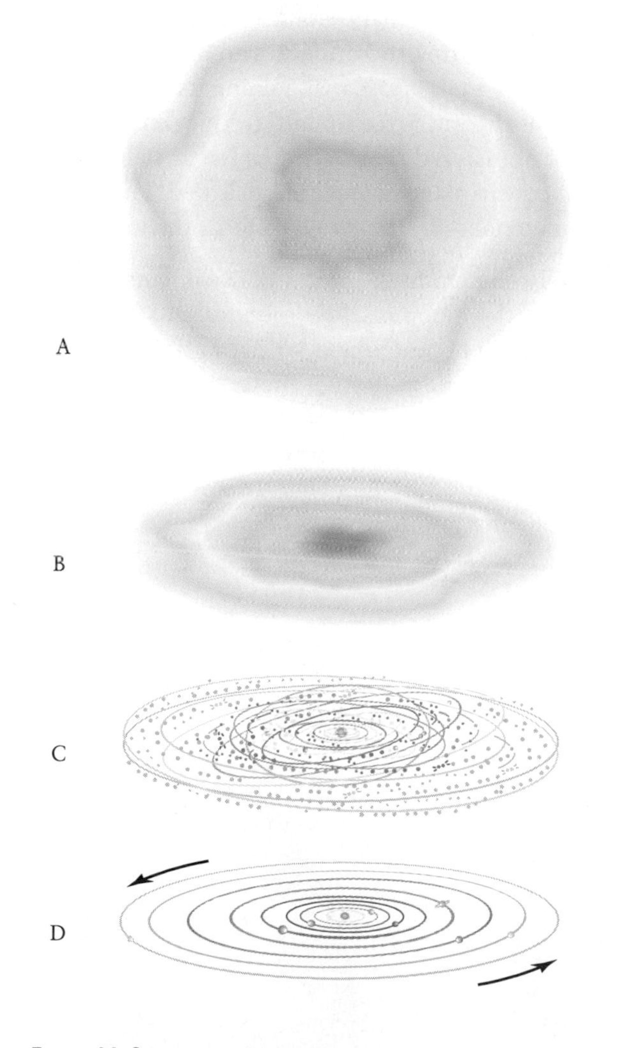

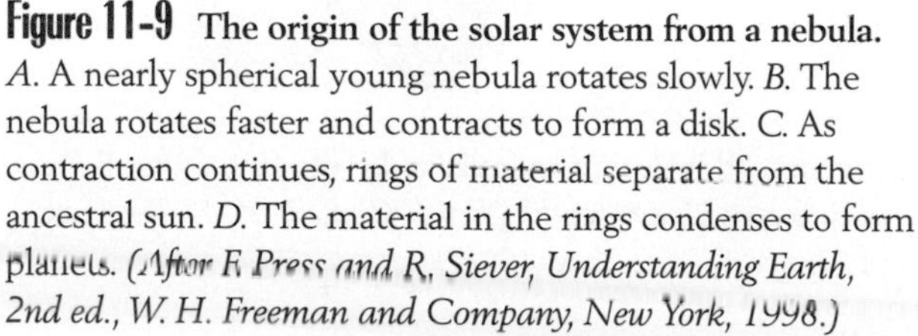

Figure 11-9 The origin of the solar system from a nebula. *A.* A nearly spherical young nebula rotates slowly. *B.* The nebula rotates faster and contracts to form a disk. C. As contraction continues, rings of material separate from the ancestral sun. *D.* The material in the rings condenses to form planets. *(After F. Press and R. Siever, Understanding Earth, 2nd ed., W. H. Freeman and Company, New York, 1998.)*

Figure 11-10 An asteroid. This false-color picture of 243 IDA, taken by the Galileo spacecraft, reveals many craters.

After the planets formed, the rocky debris that we refer to as asteroids remained in orbit around the sun (Figure 11-10). Although some of these asteroids survive to this day, most have collided with larger planets to become part of them. Others have had their motions so severely disturbed by near collisions that they have passed out of the solar system. A large swarm of asteroids is in orbit between Mars and Jupiter, not far from Earth in the solar system (see Figure 11-8). The motions of these asteroids are perturbed by the gravitational attraction of nearby planets—especially Jupiter because of its large size—and occasionally one leaves its orbit. Most asteroids that have struck Earth during the Phanerozoic Eon have escaped in this way.

The Origin of Earth and Its Moon

As Earth accreted, the impacts of giant bodies, the size of Mercury or Mars (see Figure 11-8), are thought to have contributed between half and three-quarters of its mass. These giant bodies must have been quite hot initially and would have become even hotter as their energy of motion turned into heat at impact.

Earth's Layers

Decay of radioactive isotopes within early Earth generated additional heat. The result of the various sources of heat was a molten planet, in which the most dense material sank toward the center and the least dense material rose toward the surface (Figure 11-11). The result was a predominantly iron core and a mantle of dense silicate minerals. Less dense silicates must have floated to the surface to form what has been termed a **magma ocean.** Eventually this liquid surface layer cooled to form a basaltic crust resembling the oceanic crust of the modern world. As we shall see shortly, the continental crust originated later.

The Catastrophic Birth of the Moon

For many decades, scientists debated the origin of Earth's moon. Was it an asteroid captured by Earth's gravitational field, a body that accreted separately from numerous small bodies that once orbited Earth, or a chunk of Earth knocked loose by collision with an asteroid? A variation of the third alternative is now widely accepted.

It now appears that a body about the size of Mars (one-tenth of Earth's mass) formed the moon by striking Earth a glancing blow shortly after our planet's initial accretion (Figure 11-12). The moon is not a chunk of Earth, however; it formed almost entirely from the mantle of the impacting body. This origin accounts for the fact that the proportions of two elements abundant in lunar rocks, iron and magnesium, differ from their proportions in Earth's mantle, from which the moon would have been largely derived if it had broken loose from Earth.

A computer simulation of the kind of glancing impact now thought to have formed the moon has predicted features that the moon actually displays:

1 A virtual absence of water. Water is absent even from the crystal lattices of minerals in lunar rocks. In the computer simulation, volatile elements and compounds, including water, were expelled from the moon as it formed.

2 A small metallic core. The simulation showed the metallic core and mantle of the impacting body separating during the impact. The impactor's dense core sank into Earth, joining its core, whereas the impactor's mantle exploded to form a disk of material that encircled Earth, held in orbit by the planet's gravitational field. This orbiting material soon coalesced to form the moon.

3 A feldspar-rich outer layer. The astronauts who reached the moon in 1969 found such a layer everywhere except where volcanoes and asteroid impacts had allowed other material to rise from the moon's mantle. This outer layer resulted from the heat of the moon's formation, which in the simulation produced a magma ocean resembling that of early Earth.

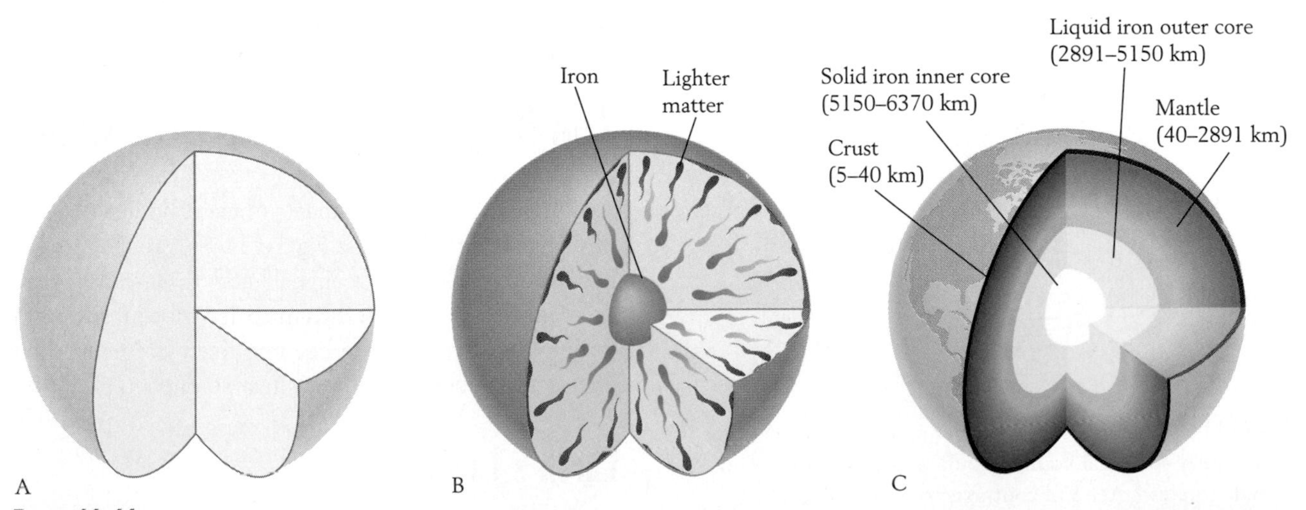

Figure 11-11 **The origin of Earth's layers.** Early Earth accreted as a generally homogeneous body *(A)*. Iron sank to the center of the liquid Earth, and less dense material rose to the surface *(B)*. Eventually the planet became fully differentiated into core, mantle, and crust *(C)*. *(After F. Press and R. Siever, Understanding Earth, 2nd ed., W. H. Freeman and Company, New York, 1998.)*

Figure 11-12 The origin of the moon. This artist's rendition shows a Mars-sized object striking Earth a glancing blow. The moon is thought to have formed from the impactor's mantle, which broke free.

The model also indicated that the heat generated by the moon-forming impact should have been intense enough to melt newly formed Earth and disrupt its layering. The result would have been a new magma ocean—and when it cooled, an entirely new crust on Earth, the one that survives today.

The impact of a Mars-sized object also explains why Earth rotates more rapidly than it should for its size and position in the solar system. The glancing blow that formed the moon knocked Earth into a faster rate of spin.

Earth's Early Atmosphere

The asteroids that coalesced to form Earth were too small to permit their gravitational fields to hold gases around them to form atmospheres. We can thus conclude that Earth did not inherit its atmosphere from these ancestral bodies; instead, the gases that the primitive planet retained as an atmosphere must have been emitted from within Earth after it formed, while it was in a liquid state, so that they could easily escape to the surface. As the core, mantle, and crust became differentiated when Earth was first liquefied, extensive **degassing**—or loss of gases to Earth's surface—would have accompanied this differentiation.

Weaker degassing has continued to the present by way of volcanic emissions. The chemical composition of gases released from modern volcanoes indicates what the early atmosphere contained: primarily water vapor, hydrogen, hydrogen chloride, carbon monoxide, carbon dioxide, and nitrogen. In the modern world, photosynthesis is responsible for most of the oxygen in the atmosphere. Less oxygen was present in the early Archean atmosphere, before the advent of photosynthesis. This atmosphere would therefore have been inhospitable to most forms of modern-day life.

The Oceans

Scientists traditionally have assumed that our planet's oceans formed by the volcanic emission of water vapor, which cooled and condensed at Earth's surface to form liquid water. Recently, however, astronomers have found that comets, which come from outside the solar system, are likely to have contributed much of this water. Comets consist of ice and cosmic dust, and they melt as they pass into Earth's upper atmosphere and produce a continual rain of water to our planet's surface. Comets several kilometers wide are readily seen in the sky, but as many as 15 million smaller ones, less than 12 meters (~40 feet) in diameter, pelt Earth's atmosphere every year. It now seems that comets must have contributed much of the water that formed Earth's early ocean; degassing of water vapor from the planet's interior played a secondary role.

The rain that fell to the early Earth would have contained almost no salts because it was derived from water vapor that comets and volcanic emissions contributed to the atmosphere. Salts were brought to early seawater by rivers that carried the products of weathering on the land. Calculations show that seawater should have become approximately as saline early in Archean time as it is now. Since then, salts have been precipitated from seawater as evaporite sediments about as rapidly as they have been added to the oceans, so the salinity of seawater has not varied greatly. At the same time, the global water cycle has moved water but not salts from the oceans to the atmosphere and back again (see Figure 1-21).

When the World Was Young

Even after a giant impact formed the moon, additional large impacts must have liquefied Earth's crust and mantle repeatedly. The frequent occurrence of such impacts early in the history of the solar system explains why Earth and other planets rotate on tilted axes. Earth's axis tilts at an angle of 23.5 degrees from the plane of its orbit. The planet Uranus, which must have been struck a more powerful blow, lies on its side, with its axis aimed almost directly at the sun (see Figure 11-8). Venus rotates about a more upright axis but has a retrograde motion, an indication that a large asteroid struck it off center and reversed its direction of spin from that of the other planets.

For about 500 million years after Earth accreted, large bodies continued to strike it and the other planets of the solar system at a high though declining rate. The planets were sweeping up debris left over from the formation of the solar system. Given the rarity of early Archean rocks, our only evidence that this extraterrestrial shower took place comes from other planets and the moon, which have not undergone the kind of weathering and erosion that constantly alter Earth's surface.

Figure 11-13 The crescent-shaped arrangement of the moon's maria on the side that faces Earth.

The Craters of the Moon

Earthbound observers have long commented on the moon's pockmarked appearance (Figure 11-13). The moon's enormous craters, which are known as *maria* (singular, mare), the Latin word for "seas," were first sketched by Galileo. It is only from manned lunar exploration and from photographs provided by artificial satellites that we have gained detailed knowledge of these maria. The maria that face Earth have an average diameter of about 200 kilometers (125 miles) and are distributed in a crescent-shaped pattern. At first it was not known whether the maria were craters produced by volcanoes or were the impact scars of huge meteorites. Detailed study of the moon's surface, which is pockmarked with craters, established that meteorite impacts formed the maria. The lunar highlands surrounding the maria consist of rock fragments that testify to the pulverization of the lunar crust by the impacts of falling meteorites. The maria facing Earth are floored by immense flows of dark volcanic basalts, which give the maria their dusky appearance. These basalts, which formed as a result of the heat generated by impacts, are themselves scarred by smaller craters. The density of cratering indicates their relative ages: the older the basalt, the more cratering it has undergone.

Dating of associated rocks has shown that most large lunar craters are quite old, ranging in age from slightly less than 4.0 billion years to about 4.6 billion years. Estimates indicate that during the early cataclysmic interval of lunar history, meteorite impacts were more than a thousand times more frequent than they are now. Craters of all sizes are also abundant on planets of the solar system whose surfaces have not been as heavily altered as Earth's. The conclusion seems inescapable that Earth was subjected to the same kind of meteorite bombardment as its neighboring moon and nearby planets.

During that cataclysmic period before 4 billion years ago and ever since, the huge planet Jupiter has actually afforded our planet considerable protection from bombardment. Acting as an immense shield, Jupiter's powerful field of gravity has pulled in many asteroids that would otherwise have struck Earth. Without Jupiter's presence, our planet would have suffered about a thousand times more impacts than it has actually received, and life on Earth would have been devastated by frequent major impacts even after 4 billion years ago. Despite the presence of Jupiter, one such crisis did occur 65 million years ago, when the arrival of a large asteroid wiped out the dinosaurs and many other forms of life on Earth (Box 11-1).

A Hotter Earth and Smaller Plates

During Archean time, heat must have flowed upward through Earth's lithosphere more rapidly than it does today, because Earth's radioactive "furnace" was hotter. Recall from Chapter 1 that Earth's heat source is constantly diminishing as radioactive isotopes decay without being renewed (p. 23). Because particular isotopes decay at constant rates, geologists can calculate the approximate difference between the rate at which Earth produces heat today and the rates of times past. The total rate of heat production was perhaps twice as high near the end of Archean time as it is now (Figure 11-14), and earlier it was even higher. As a consequence, hot spots should have been numerous during the Archean and the lithosphere should have been fragmented into small plates separated by numerous rifts, subduction zones, and transform faults. As we shall see, the nature of Archean rocks confirms this prediction.

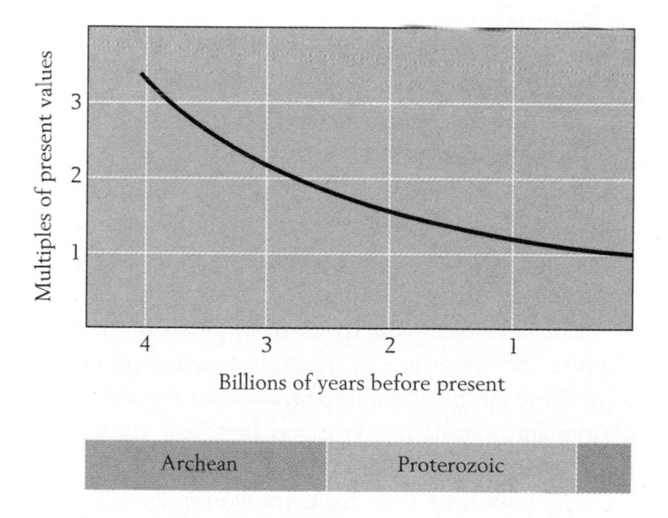

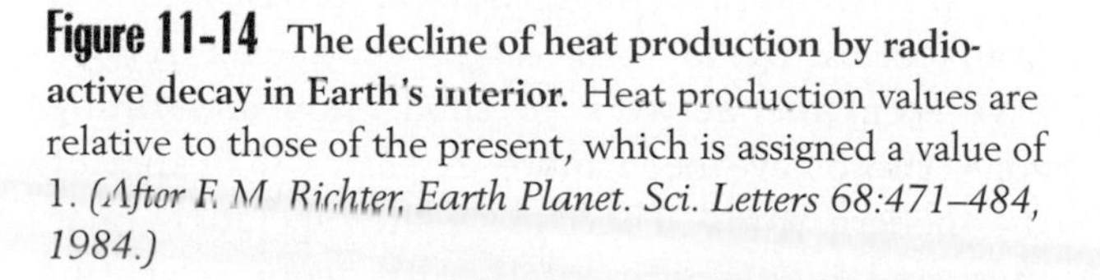

Figure 11-14 The decline of heat production by radioactive decay in Earth's interior. Heat production values are relative to those of the present, which is assigned a value of 1. *(After F. M. Richter, Earth Planet. Sci. Letters 68:471–484, 1984.)*

Geologists still debate how rapidly the total volume of Earth's continental crust increased during Archean time. There is no question, however, that even late in Archean time Earth's felsic crust was divided into relatively small continents, sometimes termed **protocontinents.**

The Origin of Continents

Only after the final magma ocean on Earth had cooled to form a basaltic crust, like that of modern ocean basins, did felsic material begin to segregate from this mafic crust and the mantle to form nuclei of continental crust. Igneous activity in the modern world reveals how the earliest continental crust probably formed.

Hot Spots: Where Felsic Crust Forms

For continental crust to form, felsic components must be extracted from mafic rocks. Although igneous material is extracted from such rocks along subduction zones, the resulting volcanic rocks—those extruded along island arcs—are also mafic or intermediate in composition between mafic and felsic. In contrast, igneous processes associated with hot spots in oceanic crust can produce felsic volcanic rocks—rocks with the chemical composition of granite.

Iceland is an island formed entirely by volcanic activity at a hot spot along the Mid-Atlantic Ridge (see Figure 8-27). Here the generation of large bodies of felsic igneous rock begins in the lower crust, where small pods and lenses of felsic composition segregate from igneous material derived from the mantle (Figure 11-15). Recall that numerous normal faults run parallel to a mid-ocean ridge, where two plates are diverging (see Figure 8-20). Mafic magma flows along the normal faults beneath Iceland, and its great heat melts the felsic bodies that the faults intersect. Felsic magma rises readily because of its low density, and it forms volcanoes that have the chemical composition of granite. Basaltic volcanoes are more common, but at least 10 percent of Iceland's upper crust is felsic.

As volcanic rocks build up on the surface of Iceland, a second process produces additional felsic magmas. This is the isostatic sinking of the pile of volcanics to form a root, like that of a mountain

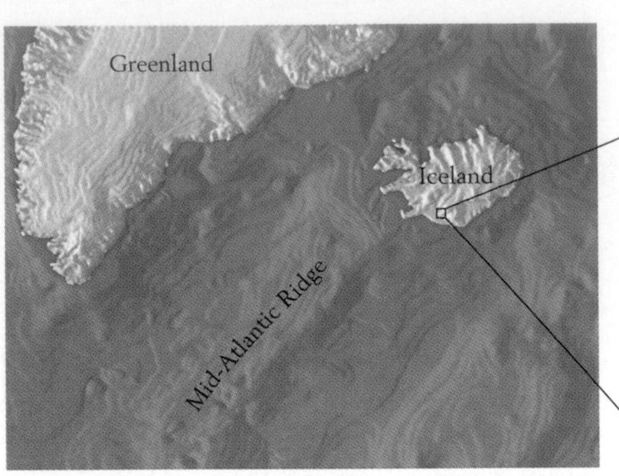

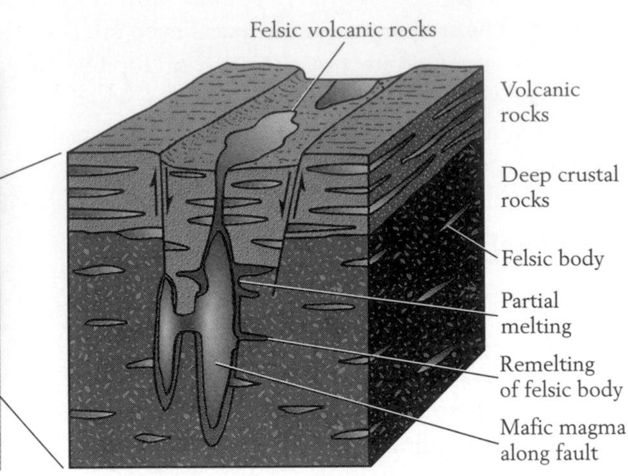

Figure 11-15 The origin of felsic volcanic rocks in Iceland, which is situated over a hot spot along the Mid-Atlantic Ridge. Small felsic bodies that have formed in deep ocean crust are melted by mafic magma moving along faults. The resulting felsic magma rises to the surface. Additional felsic magma forms by the isostatic sinking and partial melting of the pile of volcanics. *(After B. Gunnarsson and B. D. Marsh, Jour. Volcanology and Geothermal Res., 1998.)*

(p. 16). Partial melting of the volcanics extracts additional felsic components, forming magma that later cools to form felsic rocks. A special feature of Iceland allows felsic material to form rapidly: the low rate of crustal spreading here—about 2 centimeters per year—allows many areas of crust to undergo repeated melting and concentration of felsic components before they move away from the zone of igneous activity.

Today, because of its high rate of formation, the crust of Iceland is 8 to 10 kilometers thick—about twice the average thickness of oceanic crust. In fact, Iceland can be viewed as a modern-day protocontinent. It has been forming for only about 16 million years and is still accumulating felsic volcanics.

Presumably, numerous protocontinents resembling Iceland emerged during the last billion and a half years of Archean time and grew larger than Iceland before the hot spots that formed them died out when plate tectonic patterns shifted. Weathering and metamorphism would have generated additional felsic crystalline rocks. First, weathering would have removed iron and magnesium from the mafic rocks of protocontinents, leaving behind clays of felsic composition that eventually formed shales. Then metamorphism would have converted some of these shales into more durable felsic metamorphic rocks. In addition, small protocontinents would have become sutured together to form larger protocontinents.

Why Archean Continents Remained Small

During Archean time, episodes of felsic crust production at hot spots—and also plate tectonic suturing—must have produced numerous protocontinents. Nonetheless, it appears that for a long interval no huge continents formed on Earth, because broad blocks of crust that date to 3.5 billion years or earlier are absent from Precambrian shields. Even younger Archean blocks of crust—ones dating to between 3.5 billion and 2.5 billion years—are relatively small. Archean crustal areas embedded in modern continents are shown in Figure 11-16. Some of them have been fragmented by post-Archean rifting, some have had their borders obscured by metamorphism, and all have been reduced by erosion at their margins. Despite the operation of these destructive processes, some large Archean shield areas would remain today if Archean cratons had been as large, on average, as those of the present world.

The high heat flow from the young Earth's interior was what prevented Archean protocontinents from coalescing to form larger continents. As we have seen, the network of subduction and rifting zones must have been more extensive than that of the modern world. Therefore, plates, including ones that contained continental crust, were relatively

For the Record 11-1

The Threat from Outer Space

During the past few years, geologists have gathered evidence that a large extraterrestrial object struck Earth about 65 million years ago and killed off the dinosaurs. This evidence has heightened interest in the threat that asteroids and comets may pose to life on Earth.

Nearly all of the meteorites that have struck Earth in recent times have come from the belt of asteroids that circles the sun between Mars and Jupiter. Jupiter is such a large planet that its movement disturbs these bodies, and occasionally one is tossed into an orbit that crosses that of Earth. Sometimes the result is collision. Comets that strike Earth have traveled farther than meteorites—from far beyond the solar system, where they are concentrated in two huge clouds. A few, however, have orbits that bring them close to the sun, and the sun's intense heat vaporizes some of the ice that forms them. The tail we observe when a comet streaks across the night sky as a "shooting star" is the water vapor streaming away behind it. Some comets that have lost all of their ice remain in solar orbits as solid masses; most of these comets are indistinguishable from asteroids. When a comet loses mass as a result of vaporization, its orbit changes, and occasionally one crashes to Earth.

Just how likely is it that an asteroid or comet will strike Earth in the near future? We have two ways of estimating how often extraterrestrial objects of various sizes crash into our planet. One is to count the objects of various sizes whose orbits cross that of Earth. Knowing the configurations of the orbits as well, scientists can calculate the probability of collision. A second approach is to measure and date craters excavated during the past 3 billion years. Radiometric dating of metamorphic rocks formed during impact gives the age of a crater.

An object must weigh about 350 tons or more to form an impact crater. Smaller objects simply break apart and lodge in Earth's surface. Stony meteorites are so brittle that they break into small pieces when they encounter Earth's atmosphere. In general, only metallic meteorites are still large enough to form craters when they strike Earth.

An artist's conception of the marine landing of a large bolide as it collides with Earth. *(Painting by William K. Hartmann.)*

(continued)

For the Record 11-1 *(continued)*

Studies of orbiting bodies and craters indicate that extraterrestrial objects have struck Earth at a roughly constant rate during the past 3 billion years. By the beginning of this interval, the planets had swept up most of the debris left over from the formation of the solar system. During the past 3 billion years, the deaths of comets have supplied some of the new bodies that have threatened Earth; the rest have been asteroids that have escaped from the belt between Mars and Jupiter.

Telescopes reveal that today there are about 1000 bodies larger than 1 kilometer (0.6 mile) in diameter whose orbits cross that of Earth. Few of these bodies exceed 20 kilometers (12 miles) in diameter, and the sizes of craters discovered on Earth suggest that few objects larger than that have in fact struck our planet during the past 3 billion years. On the other hand, it is estimated that an object as large as 10 kilometers in diameter has struck Earth about once every 40 million years.

Most extraterrestrial bodies that have collided with Earth have made marine landings because oceans have always covered most of the planet. Unfortunately, geologic processes have destroyed many craters that extraterrestrial bodies have left on continents. Even so, craters recognized on continents tell interesting stories. The two largest, each about 140 kilometers (85 miles) in diameter, are the Sudbury crater in Ontario, Canada, and the Vredefort crater in South Africa. Each was formed nearly 2 billion years ago by the impact of an object that must have been in the neighborhood of 10 kilometers in diameter. The Popigai crater in Siberia is the third largest yet discovered, with a diameter of about 100 kilometers (60 miles). Its age, somewhere between 50 million and 30 million years, places the time of impact within the Age of Mammals, and the object that landed was probably slightly less than 10 kilometers in diameter.

Collision with a very large extraterrestrial object—one slightly larger than 10 kilometers—should have serious consequences for environments on Earth. The cloud of particles blasted into the atmosphere should plunge the planet into total darkness for at least a few months. Lack of sunlight would produce subfreezing temperatures and perhaps accumulation of deep snow on continents, even at low latitudes. Even after the dust began to settle and the sun peeked through the haze, the reflectance of light from widespread snow would probably leave Earth cool for years. The energy of the explosion would convert atmospheric nitrogen to acidic nitrous oxides, and atmospheric moisture would fall to Earth as acid rain. Acid rain could damage many forms of life. It would also attack carbonate rocks and liberate carbon dioxide, producing a greenhouse effect and eventually reversing the initial climatic trend, perhaps warming the planet to unusually high temperatures.

What forms of life would survive such changes? How many humans would die? We have no certain answers; we can only wonder whether the arrival of an extraterrestrial object 10 or 15 or 20 kilometers in diameter may someday bring an end to our species' reign on Earth.

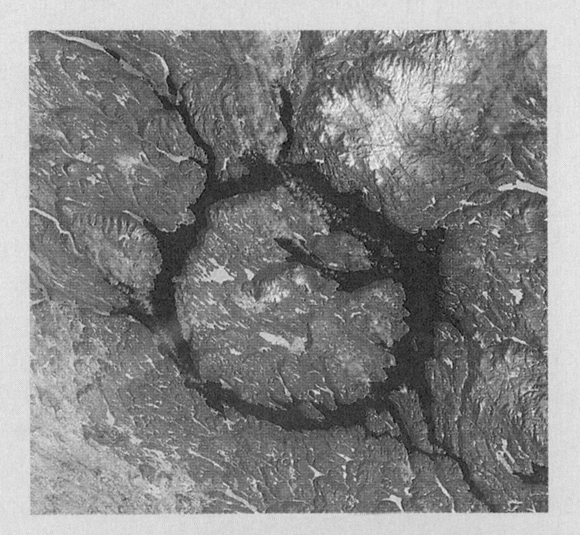

Manicouagan crater, an impact structure in the Canadian Shield of Ontario, is about 70 kilometers (~43 miles) in diameter and is partly occupied by lakes. The impact occurred in Late Triassic time, more than 200 million years ago.

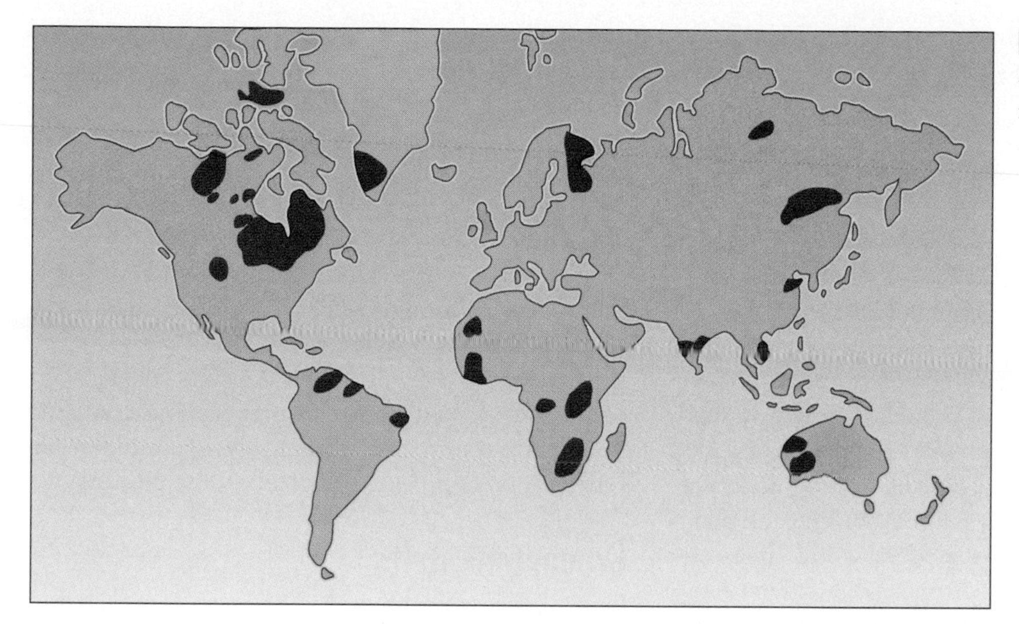

Figure 11-16 The locations of Archean shields in the modern world.

small. Only after Earth's plate tectonic engine slowed down considerably did protocontinents begin to coalesce into large continents.

The Oldest Continental Crust

Small grains of the mineral zircon are the only identifiable remnants of very early Archean felsic rocks. Zircon crystals form when molten rock cools to form granite or felsic volcanic rock. Clastic grains of zircon are so resistant to heat and weathering that they often survive unaltered in metamorphic and sedimentary rocks formed from the igneous rocks in which the zircons crystallized. Fortunately, zircon grains also contain small quantities of uranium, so their ages can be measured. The oldest zircon grains yet dated, from weakly metamorphosed sandstone in Western Australia, give ages of 4.1 billion to 4.2 billion years. These grains are the oldest known remnants of continental crust, having weathered from felsic igneous rock before lodging in sandstone.

The oldest body of actual continental crust yet discovered is in the northwestern corner of the Canadian Shield. Here radiometric dating of felsic rocks of the Acosta Formation gives ages that range from 3.8 billion to 4.0 billion years. A block of continental crust in Western Australia is also older than 3.5 billion years. How much additional continental crust existed before 3.5 billion years ago but has been destroyed by erosion or metamorphosed beyond recognition? And how rapidly did continental crust grow to its present total volume?

The Overall Growth of Continental Crust

Although individual continents remained small during Archean time, some experts concluded that the total volume of continental crust approached its present size by about 3.5 billion years ago. Today, however, the consensus is that the volume grew more slowly than this. To understand this issue, it is useful to consider what is happening to continental crust today.

In fact, the total volume of continental crust is changing very little today. Igneous contributions from the oceanic crust and mantle continually add to the total volume of felsic crust, but felsic material is lost at about the same rate through erosion and subduction of continental crust. These processes of addition and subtraction slowly replace old continental crust with new continental crust. Through the ages, they have obviously removed a considerable amount of Archean crust, replacing it with younger crust. What remains from the Archean represents about 7 percent of the modern continental crust (see Figure 11-16). It is difficult to assess just how much crust existed at the end of the Archean Eon, however, so it is difficult to estimate the rate at which the volume of continental crust expanded during Archean time. Recent estimates of the volume of continental crust 3.8 billion years ago generally range from 5 to 40 percent of the present volume, and most estimates for 2.5 billion years ago range from 60 to 100 percent of the present volume.

Archean Rocks

Archean rocks reveal a world that differed in interesting ways from that of the Proterozoic and Phanerozoic time. The rocks themselves differ in average composition from younger rocks. During Archean time, numerous volcanic arcs produced large volumes of dark igneous rocks. In addition, large bodies of dark sedimentary rocks formed from the erosion of these volcanic rocks. The thinness of the Archean continental crust is preserved in the modern world; areas of existing cratons that are of Archean age extend less far downward from the surface, on the average, than do younger portions of cratons.

General Features of Sedimentary Rocks

It is a striking fact that most Archean sediments are of deep-water origin. They include graywackes, mudstones, iron formations, and sediments derived from volcanic activity. Sediments deposited in terrestrial and shallow marine environments are relatively uncommon in the Archean record. These sediments include quartz sandstones and also carbonates (Figure 11-17). We have seen that in more recent times, crustal subsidence has caused shallow marine sediments such as those of the Bahamas (p. 143) and nonmarine sediments such as those of rift valleys (p. 236) to be buried quite deep quite rapidly. Why, then, were no similar sequences preserved in Archean time? The apparent answer is that these are continental and continental-shelf deposits, and no large continents existed in Archean time. The configuration of Archean rocks within modern continents also reflects a predominance of small protocontinents.

Greenstone Belts

The characteristic configuration of Archean terranes is evident in satellite photographs of shield areas (Figure 11-18). Over broad areas, podlike bodies known as **greenstone belts** sit in masses of high-grade metamorphic rocks of felsic composition (gneisses, for example). Rocks of the greenstone belts themselves are generally weakly metamorphosed, and the green metamorphic mineral chlorite (p. 52) gives them their name.

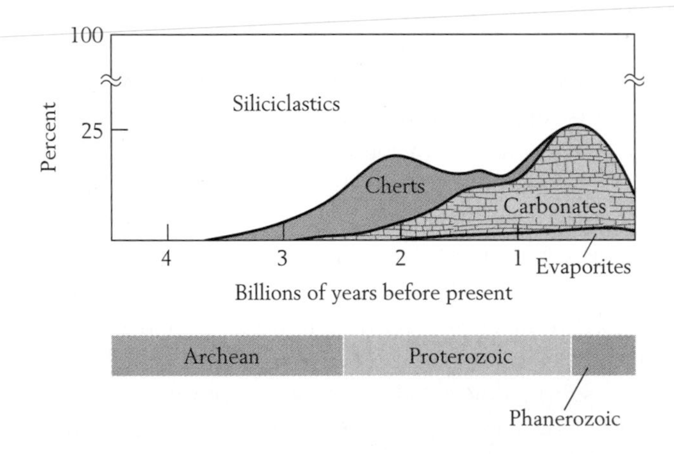

Figure 11-17 Changes in the relative percentages of sedimentary rocks in the course of Earth's history. Cherts were relatively abundant in late Archean and early Proterozoic rocks, including banded iron formations. Carbonates became much more abundant during the Proterozoic, as continents and continental shelves expanded. Evaporites appear to have been rare before mid-Proterozoic time, but this is partly a matter of preservation because these rocks dissolve readily in water. *(After A. B. Ronov, Geochim. Int. 4: 13–737.)*

Figure 11-18 A satellite photograph of greenstone belts in the Pilbara Shield of Western Australia. These belts are dark bodies of rock between circular bodies of light-colored crystalline rock that represent felsic crust of Archean protocontinents. The felsic body at the top is about 40 kilometers (25 miles) across.

Figure 11-19 **Archean pillow basalts in the Yellowknife region of Canada.** The pillows have been planed off by erosion.

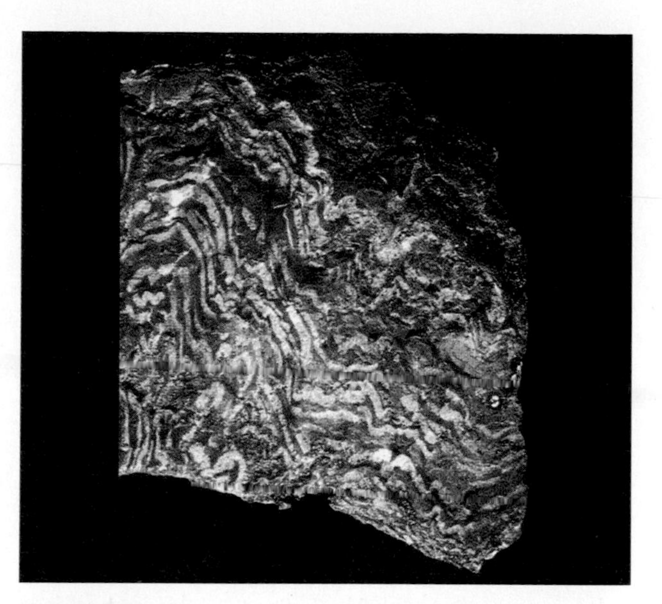

Figure 11-20 **Some of the oldest known rocks on Earth.** These banded iron formations from the Isua area of southern Greenland were deposited as sediments at least 3.8 billion years ago. Individual layers, which are deformed and weakly metamorphosed, are a few millimeters thick. Layers of chert (fine-grained quartz) alternate with darker layers that contain iron minerals.

Igneous rocks of greenstone belts, before they were metamorphosed, were mostly mafic and ultramafic volcanic rocks of the kind extruded along volcanic arcs, with felsic volcanic rocks present in smaller volumes. Many of the volcanics display pillow structures, which indicate that the lava that formed them was extruded underwater (Figure 11-19; see also Figure 2-14). Not surprisingly, many of the sedimentary rocks of greenstone belts, though now metamorphosed, can be seen to have originally formed from detritus eroded from dark volcanic rocks. Metamorphosed turbidites are common, as are dark mudstones, now mostly metamorphosed to slate. These sediments were apparently deposited in forearc basins and other environments situated along subduction zones (p. 223).

Cherts (p. 47) and the iron-rich sedimentary rocks known as **banded iron formations** are also found in the Archean sedimentary belts. The banded iron formation at Isua, in southern Greenland, represents one of the oldest known bodies of rocks on Earth (Figure 11-20). Like other banded ironstones of the Precambrian, it consists of iron-rich layers alternating with quartz layers. The Isua rocks are believed to have originated by chemical precipitation in marine basins, and the quartz within them is thought to have existed initially as chert precipitated from seawater. Submarine volcanic eruptions were the source of the dissolved silica that was precipitated as chert.

The rocks that today form greenstone belts came to lie within Archean protocontinents through continental accretion (Figure 11-21). Having formed along subduction zones, they were wedged between protocontinents that collided and became sutured together. Since the protocontinents were so small and the subduction zones so abundant, greenstone belts formed frequently and came to constitute a large proportion of Archean terranes.

Large Cratons Appear

Major metamorphic episodes occurred in many parts of the world between about 2.7 billion to 2.3 billion years ago. Why this widespread metamorphism occurred is unclear, but it reset many radioactive clocks and consolidated many small crustal elements into sizable cratons.

There is evidence that "cratonization" did not occur simultaneously throughout the world, however.

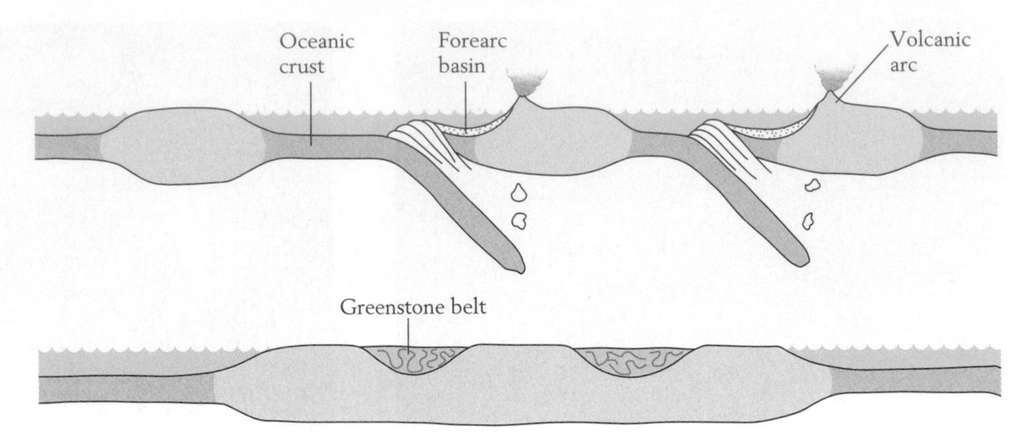

Figure 11-21 The formation of greenstone belts. Forearc basin sediments, deformed oceanic crust, and arc volcanics along the margins of protocontinents (top) become squeezed between protocontinents during suturing to become podlike greenstone belts (bottom) in a larger protocontinent.

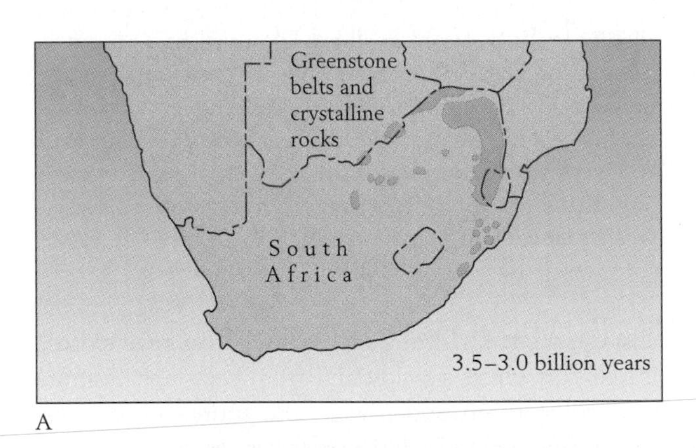

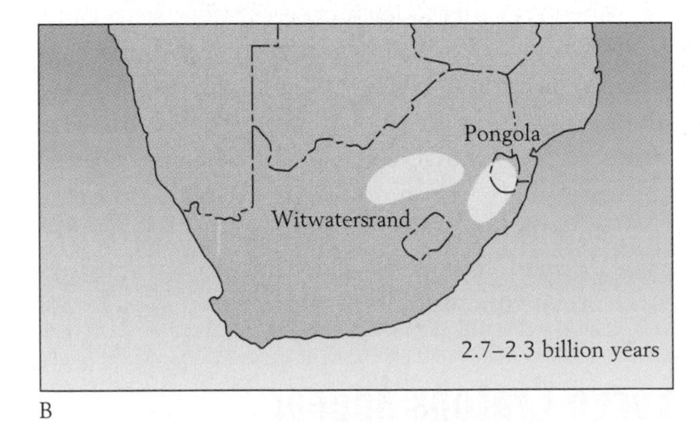

Figure 11-22 The distribution of some of the world's oldest known extensive cratonic rocks, in the Archean of southern Africa. Siliciclastic deposits of the Witwatersrand sequence and Pongola Supergroup *(B)* accumulated in a basin that occupied a broad area of continental crust composed partly of greenstone sequences *(A)*. *(Adapted from C. R. Anhaeusser, Philos. Trans. Roy. Soc. London A273:359–388, 1973.)*

In most areas, typical Archean greenstone associations formed until approximately 2.5 billion years ago, but in southern Africa a large craton was already present about a half billion years earlier (Figure 11-22). Here a greenstone sequence more than 3 billion years old is immediately overlain by a substantial body of clastic sediments known as the Pongola Supergroup. The Pongola Supergroup, which ranges in age from about 2.8 billion to 2.5 billion years, consists of deposits that are strikingly similar to intertidal sequences of younger portions of the stratigraphic record (Figure 11-23). The Witwatersrand sequence accumulated in nonmarine environments to the west of the Pongola sediments, on the surface of the same craton. Witwatersrand sediments have yielded abundant small clasts of detrital gold, whose great density caused it to accumulate with larger silicate pebbles in braided-steam deposits that now form conglomerates. Thus South Africa long ago became a major gold mining region. The Witwatersrand sediments cover about 40,000 square kilometers, with a maximum thickness of nearly 8 kilometers (5 miles). Together with the adjacent Pongola strata, they indicate that a sizable landmass existed 3 billion years ago in southern Africa.

Even at the close of Precambrian time, large continents remained unusual in one important respect: they were barren of advanced forms of life. Long before the first large cratons existed, however, living cells had begun to populate the marine realm, where they remained at a primitive stage of development for a billion years or more.

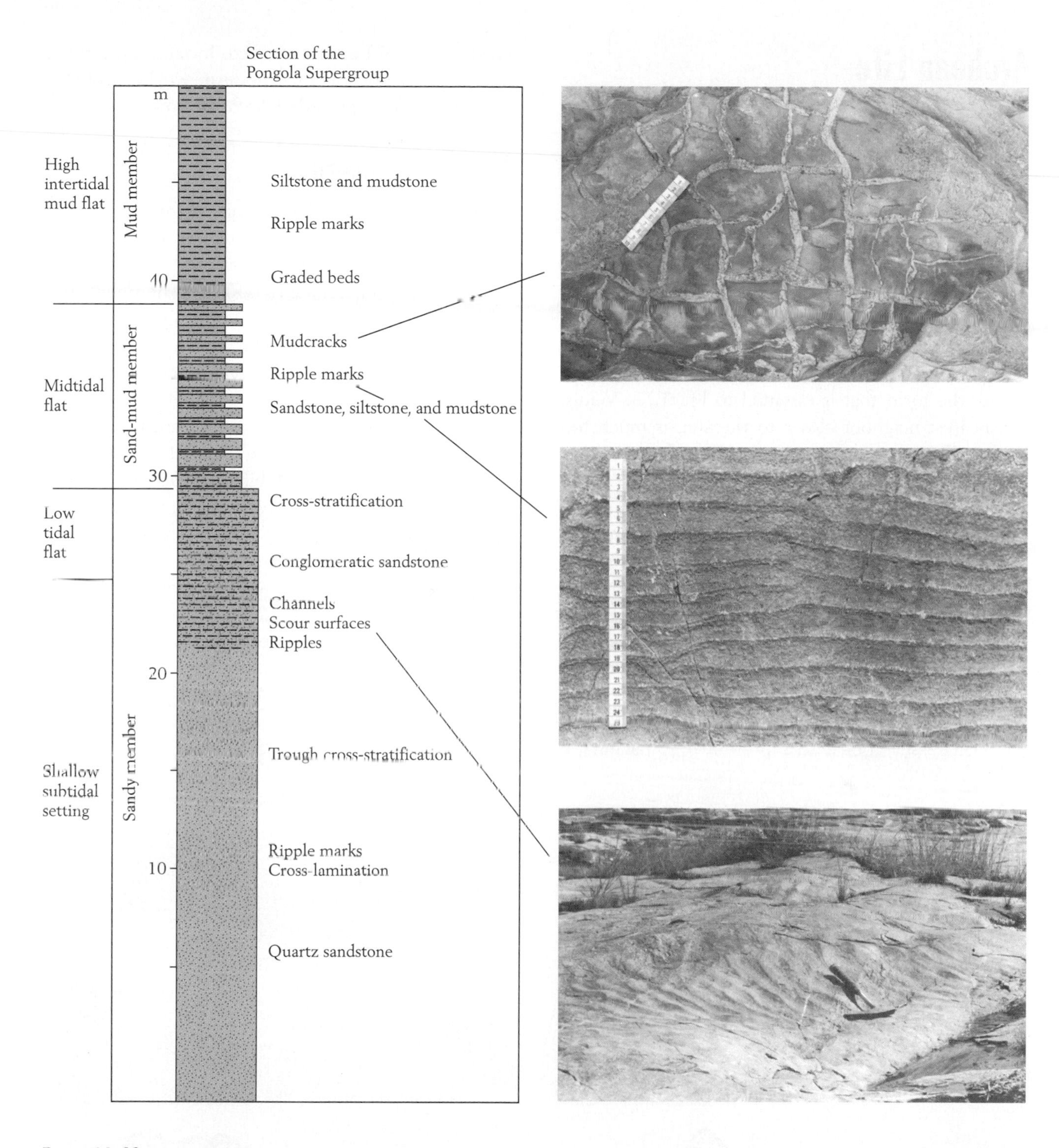

Figure 11-23 Depositional environments represented within a stratigraphic section of the Pongola Supergroup of southern Africa. The section includes a Late Archean regressive sequence in which a tidal flat prograded seaward over subtidal environments. In the lower, sandy member, cross-stratification and symmetrical ripples are common. Some of these ripples are double-crested, reflecting the ebb and flow of tides. Tidal channel deposits floored by pebbles are present, particularly in the upper part. Above them, the sand-mud member bears ripples and mudcracks, which suggest a shallower, midtidal environment. The uppermost mud member bears smaller mudcracks as well as mud chips and appears to represent a high intertidal mud flat that lay landward of the zones where sand settled from tidal waters. *(After V. von Brunn and D. K. Hobday, Jour. Sedim. Petrol. 46:670–679, 1976.)*

Archean Life

Of all the planets in our solar system, only Earth is well suited to life as we know it. One of the reasons is that its size is right. On a much larger planet, the gravitational pull on the atmosphere would be so great that the atmosphere would be too dense to admit sunlight, which is the fundamental source of energy for life. A much smaller planet would lack sufficient gravitational attraction to retain an atmosphere with life-giving oxygen. In addition, Earth's temperatures are such that most of its free water is liquid, the form that is essential to life. Even Venus, our nearest neighbor closer to the sun, is much too hot to allow water to survive in a liquid state. Mars, our nearest neighbor farther from the sun, has a cooler surface, but its atmosphere is so thin that liquid water would evaporate from the planet's surface almost immediately. Evidence indicates, however, that water once flowed over the surface of Mars, and it is speculated that life may have evolved independently there. If so, there may be fossils in Martian rocks.

Archean rocks have yielded no fossils that appear to represent organisms with **eukaryotic cells**—cells that contain chromosomes and nuclei. Although Archean fossils appear to represent only bacteria, a great variety of bacteria evolved. In fact, most major bacterial groups of the modern world probably appeared before the end of Archean time.

The Fossil Record

In the 1950s paleontologists were astounded by the discovery of molds of individual bacterial cells in Precambrian cherts. Chert forms by hardening of gelatinous silicon dioxide (p. 47), and the finely crystalline quartz thus formed can faithfully preserve the shapes of individual cells. This kind of preservation has shed much light on the Precambrian evolution of unicellular organisms.

The oldest fossils yet discovered are filamentous cells from chert in Western Australia and South Africa that are about 3.4 billion years old (Figure 11-24). These forms resemble photosynthetic eubacterial forms of the modern world and may have been cyanobacteria (p. 70).

Photosynthesis was apparently well established on Earth by about 3.8 billion years ago. This is the age of the Isua Formation of Greenland (see Figure 11-20). Organic carbon compounds extracted from the Isua are isotopically light: like carbon compounds produced today by photosynthesis, they contain a relatively large percentage of carbon 12 (p. 266). The

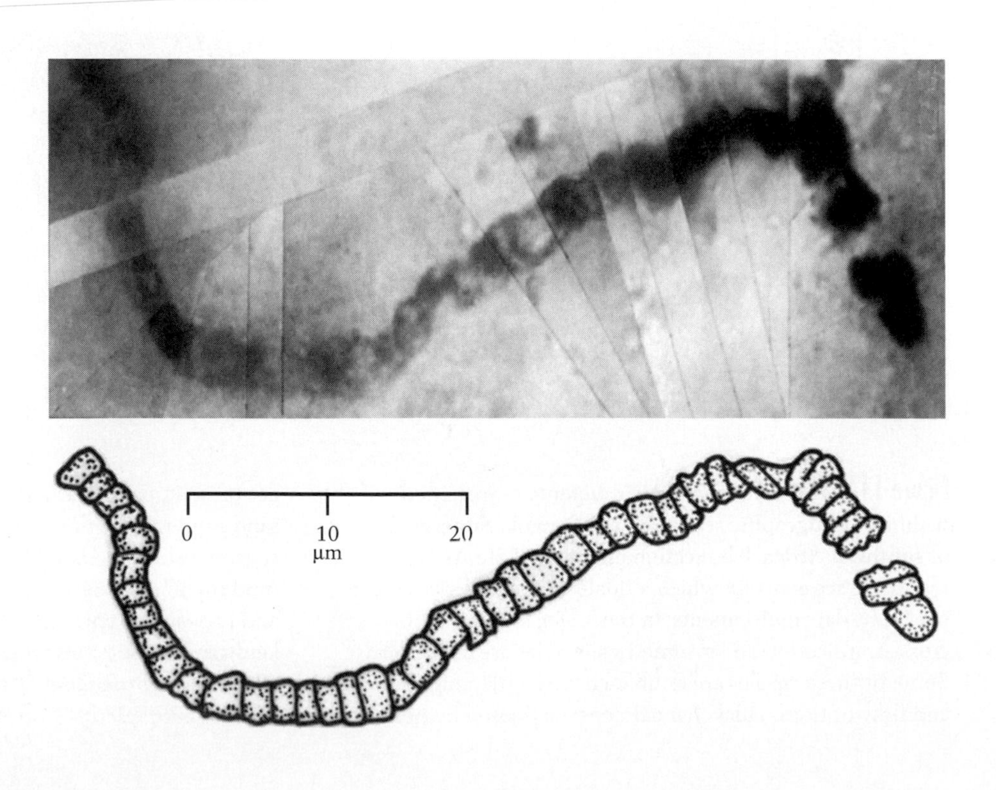

Figure 11-24 **One of the oldest known fossils.** This and other cyanobacteria-like filaments that are nearly 3.5 billion years old have been found in cherts in Western Australia.

Figure 11-25 One of the oldest known stromatolites. This form, from the Fig Tree Group of South Africa, is about 3.2 billion years old.

Isua's organic material, then, appears to be a product of photosynthesis.

Stromatolites provide more tangible evidence that photosynthesis was widespread on Archean seafloors as early as 3.2 billion years ago (Figure 11-25). Presumably cyanobacteria played a major role in the production of these distinctive layered structures, just as they do today (p. 144). Stromatolites become increasingly abundant toward the top of the Archean. This change may in part reflect an increase in the area of shallow seafloor, the habitat where stromatolites grow.

Amino Acids

Recall that two essential attributes of life are self-replication, or the ability to reproduce, and self-regulation, or the ability to sustain orderly internal chemical reactions (p. 57). Sustaining chemical reactions requires energy, such as that provided by respiration (p. 261).

Many of the compounds that life requires for self-replication and self-regulation are proteins. Some kinds of proteins form physical structures, and others enable particular chemical reactions to take place within cells. The building blocks of proteins are twenty amino acids, which are compounds of carbon, hydrogen, oxygen, and nitrogen.

In 1953 Stanley Miller and Harold Urey reported on a simple laboratory experiment in which they produced nearly all of the amino acids found in proteins. The experiment was designed to mimic the conditions under which life arose on Earth. In a closed vessel, above an "ocean" of water, the researchers created a primitive "atmosphere" of hydrogen, water vapor, methane (CH_4), and ammonia (NH_3) (Figure 11-26). To trigger chemical reactions, as lightning might have done on early Earth, they caused a spark to discharge continuously through

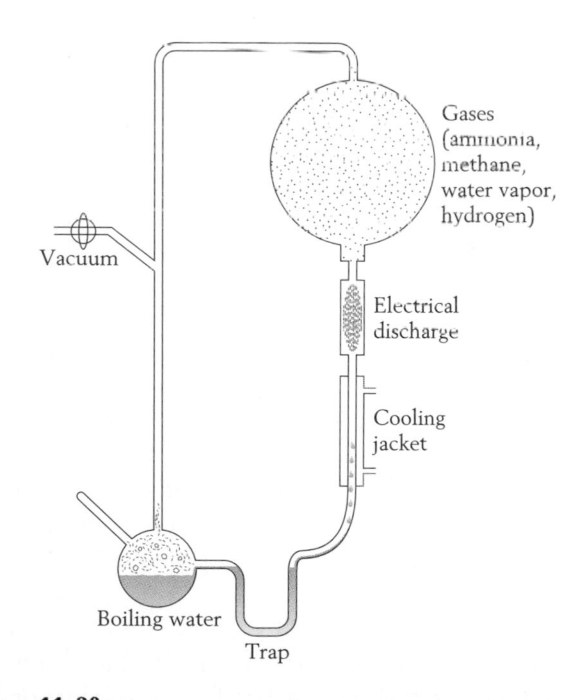

Figure 11-26 The laboratory experiment in which Miller and Urey produced amino acids by circulating ammonia (NH_3), methane (CH_4), water vapor (H_2O), and hydrogen past an electrical discharge. Amino acids accumulated in the trap. *(After G. Wald, "The Origin of Life." Copyright 1954 by Scientific American, Inc. All rights reserved.)*

the atmosphere above the vessel. A series of chemical reactions soon formed numerous amino acids.

As we will see shortly, Miller and Urey turned out to be unrealistic in assuming that Earth's early atmosphere contained no free oxygen. Nonetheless, their experiment showed that amino acids can readily form from simple compounds. Amino acids have obviously formed on other bodies of the solar system as well. The carbonaceous Murchison meteorite, which fell in Australia in 1969, was found to contain the same amino acids in the same relative proportions as those produced by Miller and Urey's experiment. This discovery showed that some amino acids incorporated into proteins on Earth could have been delivered from outer space by meteorites and comets.

An RNA World?

Nucleic acids are other essential compounds for life as we know it. They include one type known as DNA and another type known as RNA (Figure 11-27). DNA carries the genetic code for an organism, providing information for growth and regulation. It also has the ability to replicate itself in order to pass this critical information on to subsequent generations.

RNA can also replicate itself, but it plays a larger number of roles than DNA. One kind of RNA, called messenger RNA, carries the message of DNA to sites where it provides information for the formation of particular proteins. Another kind of RNA, called transfer RNA, ferries appropriate amino acids to sites where they are assembled into these proteins. RNA can also act as a catalyst, enabling certain kinds of proteins to form.

To produce life as we know it, evolution had to establish a chemical structure for the construction of particular proteins that could replicate itself in order to pass its code on to future generations. Nucleic acids must have been the original compounds to perform this function because they perform it in all living organisms.

Because of its versatility, RNA is likely to have been the nucleic acid of the earliest life forms. It could have served as a catalyst for the production of key proteins, and it could also have replicated itself in order to pass its coded message on to descendants. Thus most experts now envision an early global ecosystem known as the *RNA world*. Once the RNA system was in place, Darwinian evolution was possible, with natural selection operating on occasional mutations of the RNA structure. Eventually DNA, a more stable molecule, evolved to replace RNA as the genetic code.

At a very early stage in the history of life, organisms must have evolved a protective external structure. This must have been a semipermeable membrane like the one that bounds modern cells. Such a membrane would have protected the chemical system of the primitive organism, while allowing only a few kinds of compounds to pass in and out.

Where Did Life Arise?

It is possible that all life on Earth traces back to microscopic aliens simple forms of life that arrived from outer space and flourished on our planet. If, as seems more likely, life arose on Earth, it could have come into being only after about 4 billion years ago, when bombardment by asteroids had subsided substantially. Even after Earth's surface had stabilized, certain conditions were required for the origin of life.

Where Miller and Urey's experiment produced amino acids, they concluded that precursor compounds and ultimately life itself arose in small, ponded bodies of water that were struck by lightning and turned into what is sometimes referred to as the "primordial soup." Even Charles Darwin speculated that life may have first arisen in "a warm little pond." The problem with that idea is that it would have required an atmosphere lacking free oxygen, because even a small amount of O_2 would have oxidized, and thereby destroyed, chemical raw materials necessary for the production of essential organic compounds.

Knowing that photosynthesis produces the preponderance of oxygen in Earth's atmosphere today, scientists once assumed that the atmosphere lacked free oxygen before the origin of photosynthetic organisms. We now know, however, that ultraviolet light breaks down water vapor in Earth's upper atmosphere, slowly liberating oxygen, which spreads in small quantities throughout the atmosphere. This process would have contributed a small amount of oxygen to the Archean atmosphere—enough to destroy chemical compounds essential to life.

Life must have originated not in a small pond that was exposed to atmospheric oxygen but in some environment that was isolated from Earth's atmosphere. The most likely setting was a warm area

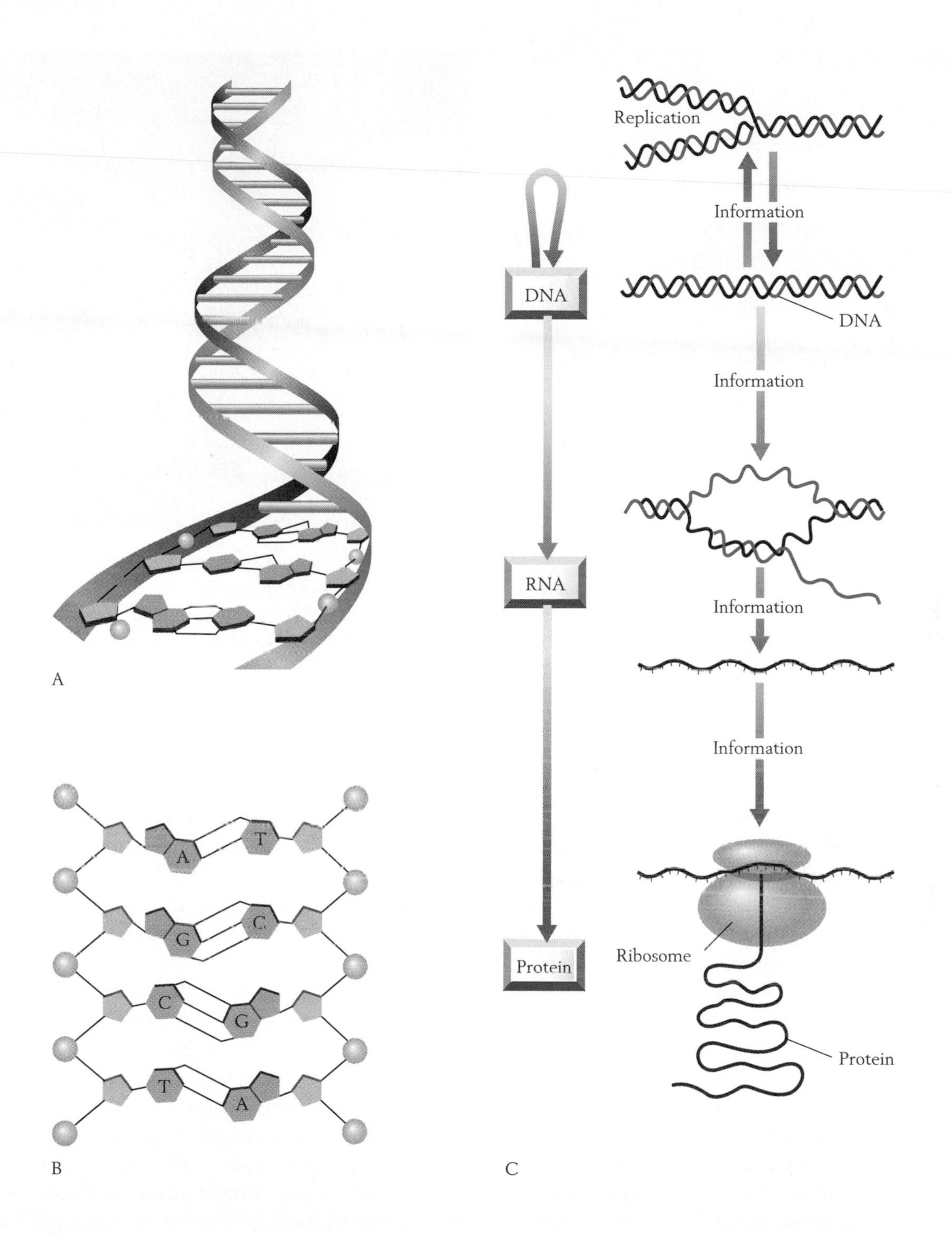

Figure 11-27 The structure and function of nucleic acids. *A.* Each strand of the double helix of DNA consists of a chain of nucleotide units. A nucleotide unit includes a phosphate group (sphere), a sugar (pentagon), and a nitrogenous base (blue). The nucleotide bases bond the two strands of the helix together. *B.* Of the four kinds of bases, adenine (A) and thymine (T) are mutually attached by a double bond, and cytocene (C) and guanine (G) are mutually attached by a triple bond. C. For replication, the two strands of DNA separate, and each then duplicates itself. The double helix also separates to allow messenger RNA to translate a portion of the code carried within the bases of DNA. Messenger RNA carries this translation to a site called a ribosome, where it specifies a sequence of amino acids that will form a particular protein. Transfer RNA brings the appropriate amino acids to assemble the protein.

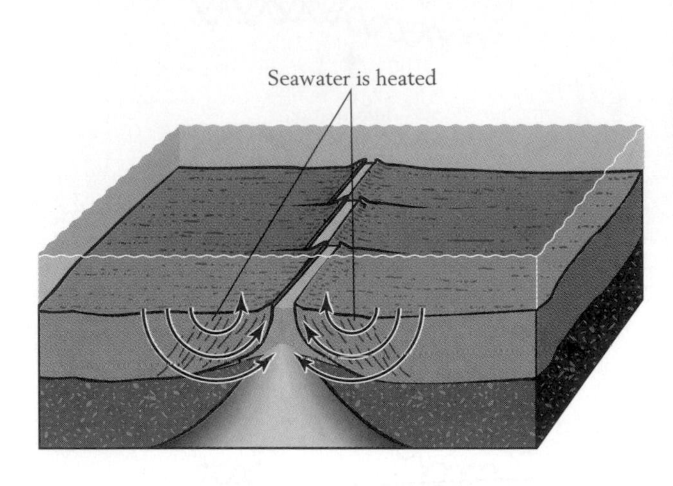

Figure 11-28 The setting where bacteria flourish along a mid-ocean ridge. Water that penetrates the crust is heated, dissolves minerals, and rises through pores, cracks, and vents. Some of this warmed water simply seeps out of the seafloor, but some of it spouts out of chimneylike structures formed by sulfides and other minerals that precipitate as the rising water cools. A wide range of thermal and chemical conditions beneath the sea floor offers varied habitats for life.

beneath the seafloor in the vicinity of a mid-ocean ridge.

The heat that rises from Earth's mantle along mid-ocean ridges warms seawater that has percolated into the crust through pores and cracks. Because heating reduces its density, this water rises back to the ocean. In some areas it flows from the seafloor through large vents as columns of very hot water (Figure 11-28). Bacteria of many kinds inhabit the warm water of ridge environments, occupying pores, cracks, and vents. They live in a variety of ways, but most of them make use of chemicals that the hot water has dissolved while moving through the ridge system. Some of these bacteria live in water warmer than 100°C, which remains in a liquid state because of the great pressure applied by the ocean above. Others live in lukewarm water farther from ridge axes. In general, these high-temperature bacteria may be inhabiting the kind of setting where life originated.

The principle that most warm-adapted bacteria put into practice to obtain energy is quite simple: they harness the energy of naturally occurring chemical reactions. Many of the chemical compounds that emerge from deep within mid-ocean ridges are not in chemical equilibrium after the rising water in which they are dissolved cools and mixes with seawater; and so they will enter into chemical reactions. Many such reactions do not occur quickly, however, and the bacteria take advantage of this situation. The bacteria consume the chemical compounds and simply allow chemical reactions that would have occurred in seawater to take place inside their cells. These reactions release energy. The bacteria harness this energy for their metabolism and excrete the chemical products of the reactions.

Some of the warm-adapted bacteria are producers, but unlike photosynthetic organisms, they do not use light as an energy source. The processes they employ are collectively termed **chemosynthesis.** Some chemosynthetic bacteria combine hydrogen and sulfur to produce hydrogen sulfide, for example; others combine hydrogen and carbon dioxide to form methane and water. Other kinds of bacteria are consumers, oxidizing an element such as hydrogen or sulfur in the way that higher organisms oxidize sugars.

There is evidence that the warm-adapted bacteria that live in the vicinity of mid-ocean ridges are actually the most primitive living bacteria. Biologists have studied the DNA and RNA of all kinds of bacteria to reconstruct their evolutionary tree. Features of DNA and RNA shared by many bacterial groups are regarded as primitive features that were inherited from very ancient ancestors. Derived traits shared by few groups identify younger branches of the tree (see p. 66). As it turns out, the most primitive living Achaeobacteria (p. 70) are adapted to warm habitats. This finding suggests that bacteria evolved in such habitats, perhaps in the vicinity of mid-ocean ridges.

Mid-ocean ridges exhibit features that would also have made them likely sites for prebacterial evolution, even for the origin of life:

1 Their enormous size offered a large range of temperatures, which provided many opportunities for key evolutionary events to take place.

2 Organic compounds of the kind required for the origin of early life readily dissolve in their warm waters. Furthermore, many of these waters are anoxic, so that they can protect those chemicals essential to life that are destroyed by free oxygen.

3 They are unusual environments in offering an abundance of phosphorus, an element that all organisms require in substantial quantities.

4 They contain metals, such as nickel and zinc, that all organisms require in trace quantities.

5 They are well supplied with clays, which are known to serve as useful substrates for the assembly of large organic molecules.

6 As we have seen, they provide simple organisms with the opportunity to harness a variety of naturally occurring chemical reactions that release energy.

There is a good chance, then, that life evolved in warm, anoxic waters that circulated through oceanic crust in the vicinity of mid-ocean ridges. This may be where the RNA world began, and in all likelihood it was where bacteria later came into being.

We will probably never have any idea where eukaryotes later arose. Nonetheless, remarkable evidence concerning the timing of this group's origin has emerged from studies of DNA and RNA, such as the studies that have established the general phylogeny of bacteria. The genetic structure of primitive eukaryotes indicates that they diverged from bacteria close to the time when Archaeobacteria and Eubacteria diverged—well back in Archean time. As we will learn in Chapter 12, however, eukaryotes do not appear in the fossil record until the Proterozoic Eon. Apparently for hundreds of millions of years before this time they had remained simple, inconspicuous, unicellular forms that in many ways resembled bacteria.

Chapter Summary

1 Precambrian shields form the cores of all modern cratons.

2 Radiometric dating has revealed that stony meteorites, which represent the primitive material of early bodies of the solar systems, are 4.6 billion years old, as are the most ancient moon rocks. This, then, is the apparent age of Earth and the other planets of the solar system.

3 Earth originated by condensation of material that had been part of a rotating dust cloud.

4 The moon originated when a body the size of Mars struck Earth a glancing blow; the moon formed from the mantle of the impacting body.

5 Between the time it formed and slightly later than 4 billion years ago, Earth was pelted by large numbers of meteorites. During the same interval, meteorites produced most of the large craters that are still visible on the moon, whose surface is less active than Earth's.

6 Earth was liquefied by the impact that formed the moon, and perhaps by later giant impacts. Earth became stratified into core, mantle, and oceanic crust because material of high density sank toward the center of the young, liquid Earth.

7 The formation of large continents was inhibited in early and middle Archean time by the abundance of radioactive elements, whose decay

produced heat at a high rate. Under conditions of high heat flow, Earth's crust was divided into small protocontinents, which probably formed by the accumulation of felsic material above hot spots.

8 The most readily studied Archean rocks occur in greenstone belts. Greenstones consist of metamorphosed dark volcanic and sedimentary rocks. They formed along subduction zones adjacent to small continents, from which were derived dark mudstones and graywackes associated with the volcanics.

9 Large continental landmasses apparently did not form until late in Archean time. The oldest of these landmasses now recognized are in South Africa, where shallow marine and nonmarine siliciclastic sediments were spread over sizable areas between 3 billion and 2 billion years ago.

10 Life may have arisen along mid-ocean ridges, where temperatures were warm, free oxygen was nonexistent, and other conditions were favorable.

11 All known Archean fossils appear to represent bacteria, which are the most primitive forms of cellular life on Earth today, lacking cell nuclei and chromosomes. Bacteria are preserved as outlines of fossil cells, and cyanobacteria also formed stromatolites.

Review Questions

1 What is a Precambrian shield? Where is one located in North America?

2 What reasons are there to believe that Earth was pelted by vast numbers of meteorites early in its history?

3 Why might we expect Earth to be nearly the same age as its moon and the material that forms meteorites?

4 What geologic features characterize greenstone belts and how did these belts form?

5 What types of sedimentary rocks were rare in the Archean Eon? What does this suggest about the nature of cratons during Archean time?

6 Why did magma rise from the mantle to Earth's surface at a higher rate during Archean time than it does today?

7 What features make Earth a more hospitable place than other planets for life as we know it?

8 Why is it likely that life arose in the vicinity of mid-ocean ridges?

9 What are stromatolites? From what we know of their formation today, why might we expect them to have been present early in Earth's history?

10 How may the modes of life of certain living bacteria shed light on early evolution?

11 The composition and configuration of Earth's crust changed more profoundly in the course of Archean time than during any later interval of Earth's history. Using the Visual Overview on page 286 and what you have learned in this chapter, describe major changes in the Archean crust and explain how they relate to each other and to changes in Earth's deep interior.

Additional Reading

deDuve, C., "The Beginnings of Life on Earth," *American Scientist* 93:428–437, 1995.

Grieve, R. A. F., "Impact Cratering on the Earth," *Scientific American*, April 1990.

Nisbet, E. G., *The Young Earth: An Introduction to Archean Geology*, Allen & Unwin, Boston, 1987.

Orgel, L. E., "The Origin of Life on Earth," *Scientific American*, October 1994.

Silk, J., *The Big Bang: The Creation and Evolution of the Universe*, W. H. Freeman and Company, New York, 1988.

Taylor, G. J., "The Scientific Legacy of Apollo," *Scientific American*, July 1994.

Taylor, S. R., "The Evolution of Continental Crust," *Scientific American*, January 1986.

Taylor, S. R., "The Origin of the Moon," *American Scientist* 75:469–471, 1987.

Taylor, S. R., and McLennen, S., "The Evolution of Continental Crust," *Scientific American*, January 1996.

Windley, B. F., *The Evolving Continents*, John Wiley & Sons, New York, 1995.

CHAPTER 12

The Proterozoic Eon of Precambrian Time

The Proterozoic Eon, which succeeded the Archean Eon 2.5 billion years ago, was in many ways more like the Phanerozoic Eon, in which we live. We have already seen a foreshadowing of this difference between the Proterozoic and Archean eons in the origin of large cratons late in Archean time. The persistence of large cratons throughout the Proterozoic Eon produced an extensive record of deposition in broad, shallow seas—a pattern that differed substantially from the Archean record of deep-water deposition, which is now confined largely to greenstone belts and adjacent areas. In addition, more Proterozoic than Archean sedimentary rocks remain unmetamorphosed and are therefore accessible for study.

The extensive deposits of Proterozoic age document ancient mountain-building events that are strikingly similar to those of the Appalachians and other younger orogenic belts, and they reveal records of major intervals of glaciation, at least one of which seems to have affected most of the world. Proterozoic rocks also bear a fossil record of organic evolution that reveals a transition from the simplest kinds of single-celled organisms at the start of the Proterozoic Eon to more advanced single-celled forms and finally to multicellular plants and animals, some of which belonged to modern phyla. This fossil record provides one of the methods by which geologists divide the

Billions of years before present

2.5	1.6	1.0	0.544
Paleoproterozoic	Mesoproterozoic	Neoproterozoic	
Proterozoic Eon			

Highly oxidized iron-rich sediments of the Hamersley Group in Western Australia.

Visual Overview

Major Events of the Proterozoic Eon

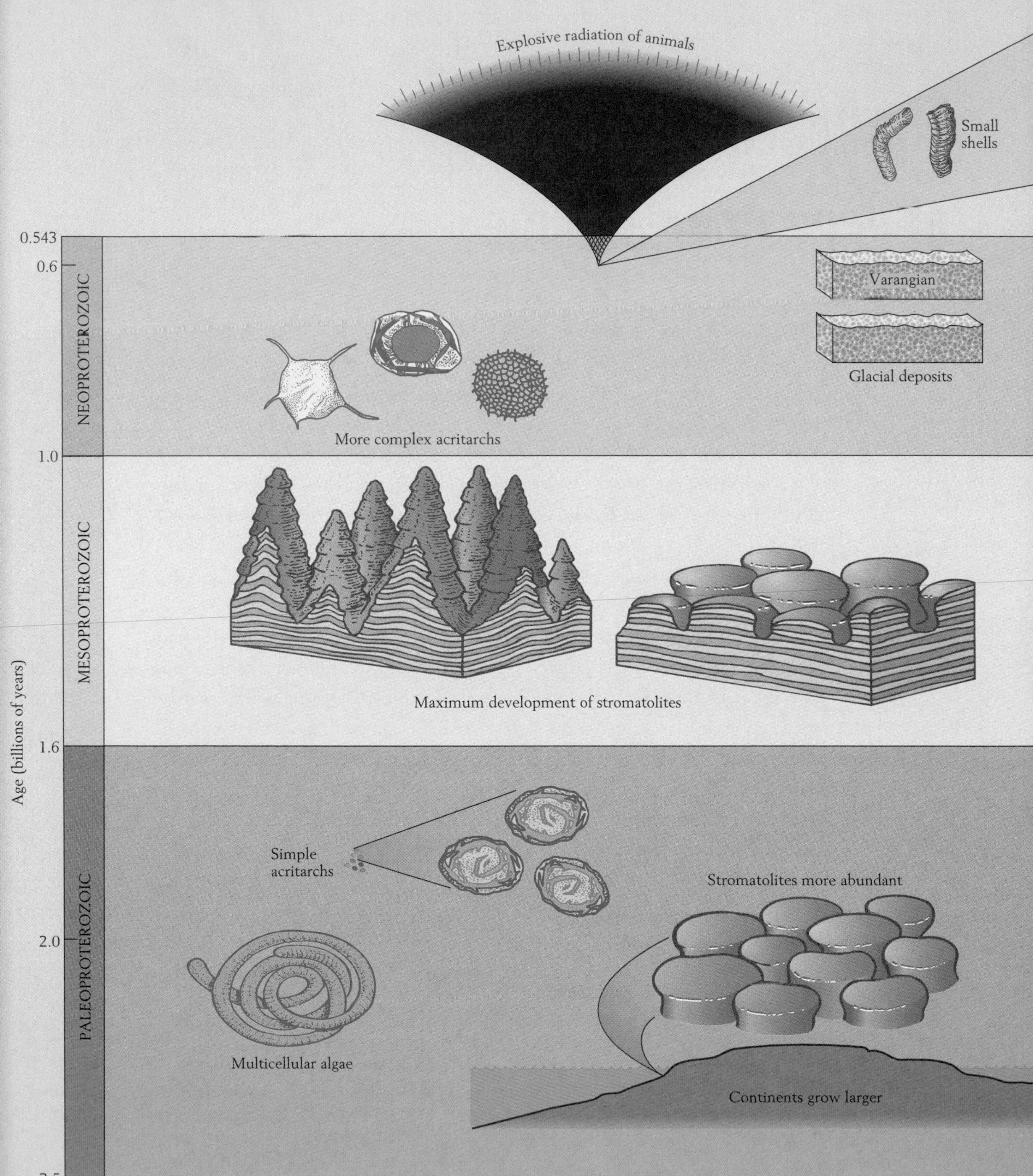

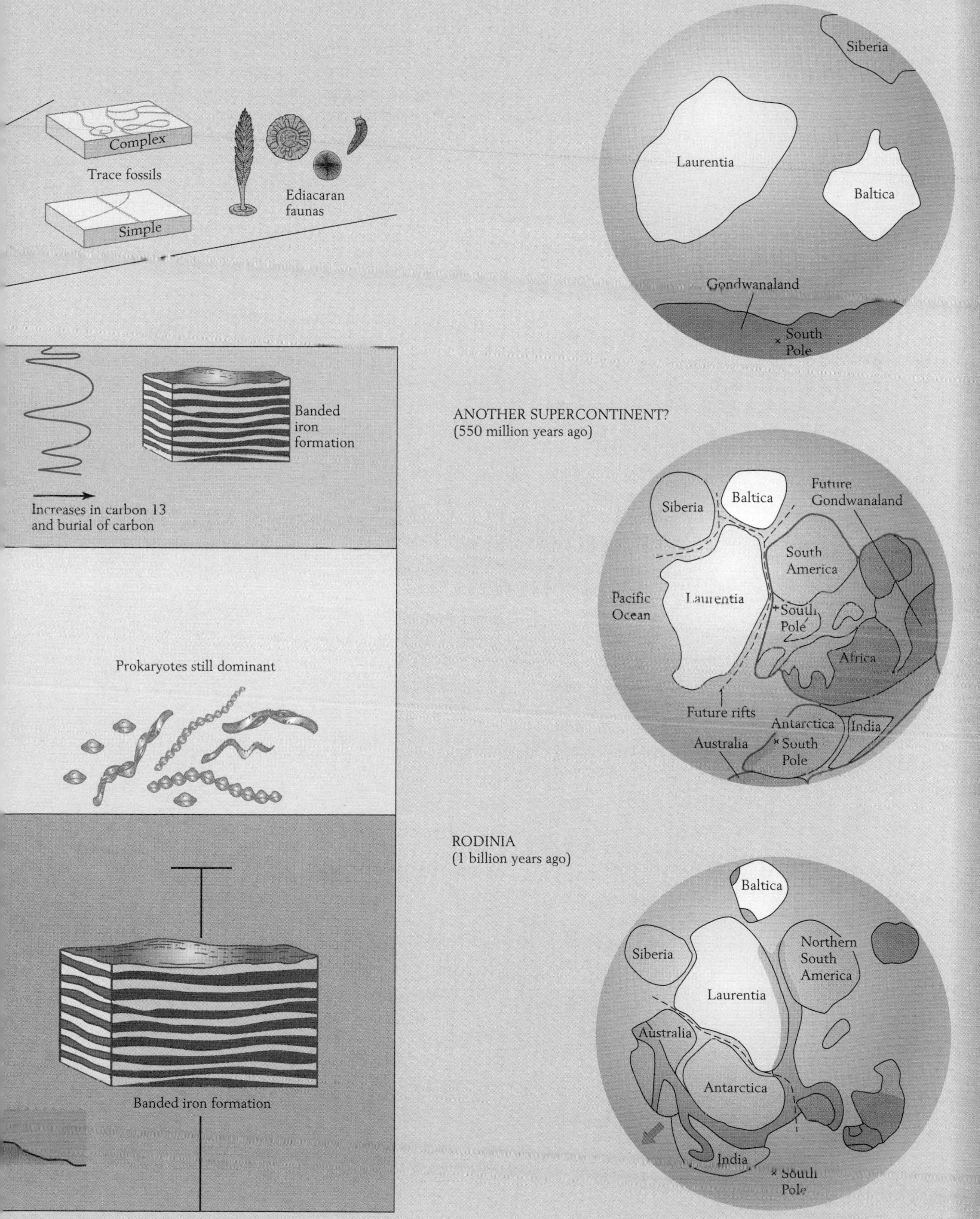

EARLY ORDOVICIAN
Siberia
Laurentia
Baltica
Gondwanaland
South Pole
Complex
Trace fossils
Simple
Ediacaran faunas
Banded iron formation
Increases in carbon 13 and burial of carbon
ANOTHER SUPERCONTINENT?
(550 million years ago)
Siberia
Baltica
Future Gondwanaland
South America
Pacific Ocean
Laurentia
South Pole
Africa
Future rifts
Antarctica
India
Australia
South Pole
Prokaryotes still dominant
RODINIA
(1 billion years ago)
Baltica
Siberia
Northern South America
Laurentia
Australia
Antarctica
India
South Pole
Banded iron formation

Proterozoic into three eras, the Paleoproterozoic, Mesoproterozoic, and Neoproterozoic.

Global events are one of the subjects of this chapter, but we will also view the Proterozoic world on a regional scale and learn how the modern continents began to take shape.

A Modern Style of Orogeny

As we saw in Chapter 11, cratons of modern proportions first began to form about 3 billion years ago, late in Archean time. This was when the oldest sedimentary deposits to be laid down over broad continental areas accumulated in southern Africa (p. 304). Although mountain-building processes resembling those of the Phanerozoic world were undoubtedly in operation by this time, it is in rocks about 1 billion years younger that geologists have found the oldest well-displayed remains of a mountain system that is thoroughly modern in character. The Wopmay orogen of Canada, which formed along the margin of an early continent, developed slightly after 2 billion years ago, over a large area that is now approximately 100 kilometers (60 miles) to the west of Hudson Bay. Today remarkably well preserved sedimentary rocks of this orogen are exposed along the low-lying surface of the Canadian Shield as a result of continental glaciation that has repeatedly scoured the orogenic belt over the past 3 million years or so.

The Wopmay orogen lies along the western margin of the geologic region known as the Slave Province and displays an ancient fold-and-thrust belt (Figure 12-1). Although it has long been planed off

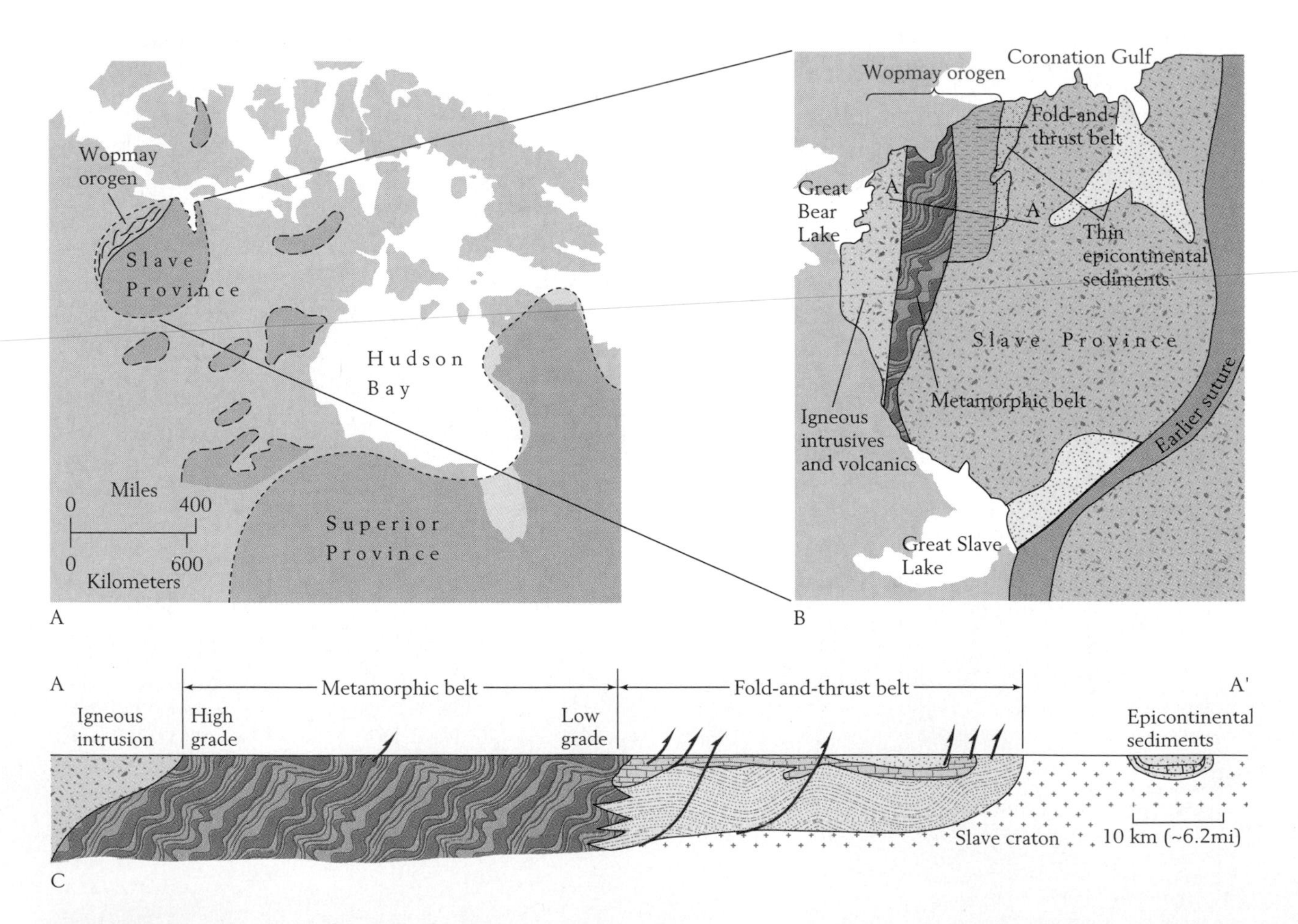

Figure 12-1 The Wopmay orogen, which formed nearly 2 billion years ago along the margin of the Slave Province of northwestern Canada. *A.* The Slave Province and other regions of Archean terrane are shown in blue. *B.* The igneous, metamorphic, and fold-and-thrust belts of the Wopmay orogen. C. Cross section of the Wopmay orogen (depicted with vertical exaggeration). *(After P. F. Hoffman, Philos. Trans. Roy. Soc. London A233:547–581, 1973.)*

by erosion, this zone of deformed rocks bears a striking resemblance to the younger belts described in Chapter 9. In the Wopmay orogen, thrusting was toward the east, and igneous intrusions associated with the deformation now lie primarily within the Bear Province, to the west. A belt of metamorphism lies between the igneous belt and the fold-and-thrust belt. To the east, epicontinental sedimentary rocks continuous with those of the fold-and-thrust belt are relatively undeformed. Like sedimentary deposits of younger fold-and-thrust belts, those of the Wopmay belt show a clear relation to tectonic history. Near the end of Archean time, before the orogen was formed, most of what is now called the Slave Province existed as a discrete craton. Early in the Proterozoic Eon, this craton was sutured to another craton to its east, and thick shelf deposits were deposited along its western margin. As in younger mountain belts, the shelf deposits were succeeded by flysch and then by molasse deposits (p. 243). Wopmay sequence has the following characteristics:

1 The first thick deposit, which formed along the passive margin of the Slave craton, is a quartz sandstone that prograded toward the basin (Figure 12-2). This quartz sandstone grades westward into deep-water mudstones and turbidites that now lie within the metamorphic belt.

2 Carbonate rocks that contain abundant stromatolites accumulated along the passive margin on top of the quartz sandstone. These rocks formed a carbonate platform. Sedimentary cycles in these platform deposits record repeated progradation of tidal flats across a shallow lagoon. Laminated dolomite that formed in the lagoonal environment is at the base of each cycle, while at the top are oolitic or stromatolitic deposits that must have formed in environments fringing the lagoon on its landward side (Figure 12-3). Enormous stromatolite mounds grew to the west, along the shelf margin. The fine-grained deposits of the lagoon were trapped behind the persistent barrier formed by these mounds. Thus stromatolites bounded the lagoon on both its landward and seaward margins. The present metamorphic zone consists of a thinner sequence of mudstones that represent deeper environments beyond the shelf edge, together with beds of dolomite breccia that contain blocks as long as 50 meters (165 feet). These blocks were transported down the steep slope in front of the shelf edge by catastrophic flows of submarine debris.

3 The carbonate platform deposits give way to transitional mudstones, which reflect a downwarping of the platform as a foreland basin was formed.

4 As is typical of foreland basin sequences of Phanerozoic age, flysch deposits (turbidites)

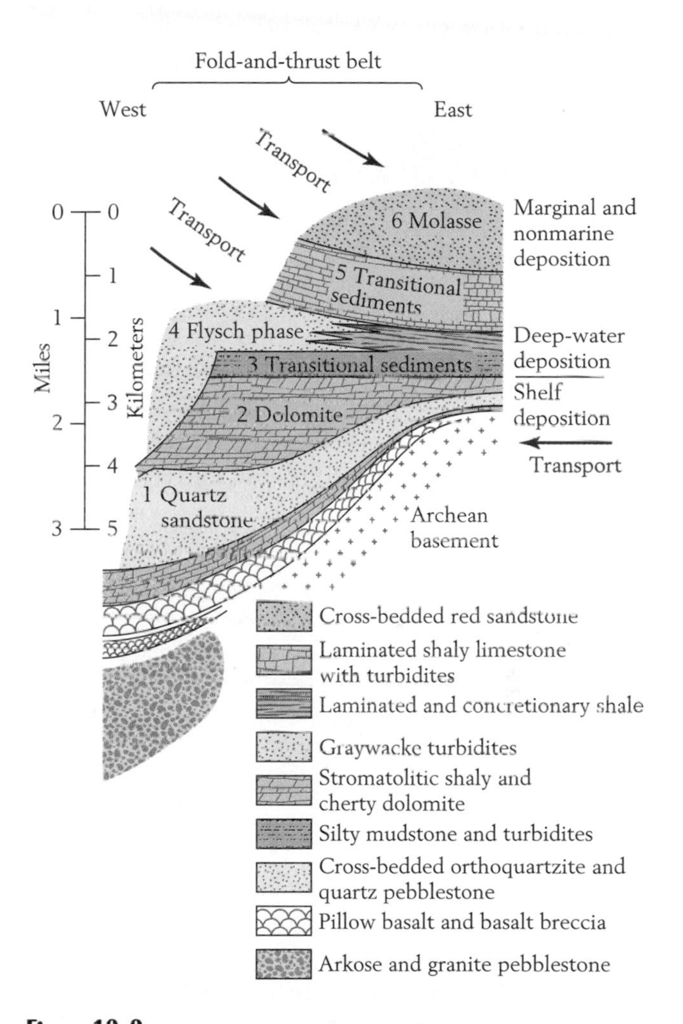

Figure 12-2 The sequence of deposition of sediments in the fold-and-thrust belt of the Wopmay orogen. The units numbered 1 and 2 represent marine deposition along a shallow continental shelf. Units 3 and 4 are deep-water deposits, including flysch, that accumulated when the shelf foundered as mountain building began to the west. Unit 5 consists of shallow-water deposits transitional between flysch below and molasse above. Unit 6, the molasse phase of deposition, followed the exclusion of marine waters by a heavy influx of sediment from the west. *(After P. F. Hoffman, in M. R. Walter [ed.], Stromatolites, Elsevier, Amsterdam, 1976.)*

Figure 12-3 Stromatolites within shelf deposits of the Wopmay orogen.

follow the mudstones. The Wopmay flysch thickens to the west, and includes particles derived from uplifted plutonic rocks to the west. This pattern is also typical of Phanerozoic foreland basins: the source area of siliciclastics was seaward of the foreland basin (see Figure 9-11).

5 The deep-water turbidites grade upward into beds containing mudcracks and stromatolites, both of which formed in shallow-water environments and thus point to a shallowing of the foreland basin.

6 The influx of sediments eventually pushed marine waters from the Wopmay foreland basin, and the Wopmay cycle, like tectonic cycles of the Phanerozoic, ended with an interval of molasse deposition (p. 243). The Wopmay molasse consists largely of river deposits in which cross-bedding is conspicuous.

In summary, two kinds of evidence suggest that the Proterozoic Wopmay orogen had the same pattern of formation as a modern orogenic system. First, the parallel igneous, metamorphic, and fold-and-thrust belts resemble similarly arranged belts of younger mountain ranges (see Figure 12-1). Second, within the fold-and-thrust belts, shallow-water shelf deposits are succeeded by flysch deposits that give way to molasse deposits. The rocks to the west of the Wopmay orogen are those of an island arc. The Wopmay orogeny occurred when the Slave craton collided with this island arc slightly after 1.9 billion years ago.

Global Events between 2.5 Billion and 1 Billion Years Ago

The fact that the Slave terrane along the Wopmay orogen behaved like rigid continental crust when it was rifted and deformed indicates that by 2 billion years ago Earth was markedly cooler than it had been a billion years earlier, when magmas were pushing up from the mantle to the surface throughout much of what is now the Canadian Shield. That climates in this region were also quite cool approximately 2 billion years ago is shown by evidence that glaciers spread over the land. After reviewing the evidence for this glacial interval, we will examine other remarkable changes that took place in Earth's environments and life early in the Proterozoic Eon.

Early Proterozoic Glaciation

Just to the north of Lake Huron in southern Canada are some of the most spectacularly exposed ancient glacial deposits: those of the Gowganda Formation. Well-laminated mudstones in this formation consist of varves that formed in the standing water of a lake or ocean in front of glaciers. In Chapter 5 these ancient deposits were compared to the strikingly similar glacial varves that formed nearby, where Toronto is now located, just a few thousand years ago (see Figure 5-6). Some of the laminated Gowganda mudstones contain dropstones—pebbles and cobbles that appear to have fallen from ice that melted as it floated out from a glacial front (see Figure 5-7). These mudstones alternate with tillites, which were deposited when glaciers encroached on the body of water. Some of the pebbles and cobbles of these tillites are faceted or scratched from having slid along at the bases of moving glaciers.

The Gowganda deposits cannot be dated directly, but the best estimate of their age is about 2.3 billion years, because they rest on 2.6-billion-year-old crystalline rocks and are intruded by igneous rocks that are 2.1 billion years old. Tillites of similar age are found elsewhere in Canada and in Wyoming, Finland, southern Africa, and India. These widespread glacial deposits testify to extensive continental glaciation not long after the transition from Archean to Proterozoic time.

Early Proterozoic Life

No abrupt change in life on Earth marked the Archean-Proterozoic transition. There is good evidence that eukaryotic algae existed early in Proterozoic time, but bacteria, including cyanobacteria, remained more abundant than these algae in Earth's oceans.

Stromatolites Beginning in strata about 2.2 billion years old, stromatolites become increasingly abundant in Proterozoic rocks. This proliferation probably reflects an increase in the size of continents and therefore in the breadth of continental shelves, where stromatolites flourished.

Stromatolites also began to grow into a greater range of shapes, attaining their greatest variety about 1.2 billion years ago (Figure 12-4). Distinctive cone-shaped stromatolites, which must have been produced by a particular kind of cyanobacteria, lived primarily on offshore continental shelves. Stromatolites became less diverse during the Neoproterozoic Era, and cone-shaped forms were the first to decline.

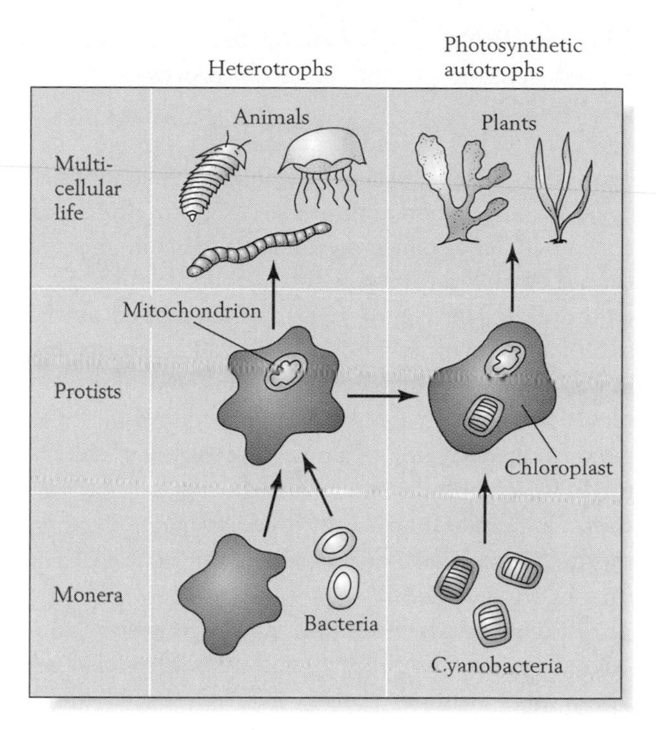

Figure 12-5 The probable sequence of major events leading from Monera to multicellular animals and plants. The first protist apparently evolved when one bacterium engulfed but failed to digest another, which then became a mitochondrion. A plantlike protist evolved when an animal-like protist engulfed but failed to ingest a cyanobacterium cell, which then became a chloroplast.

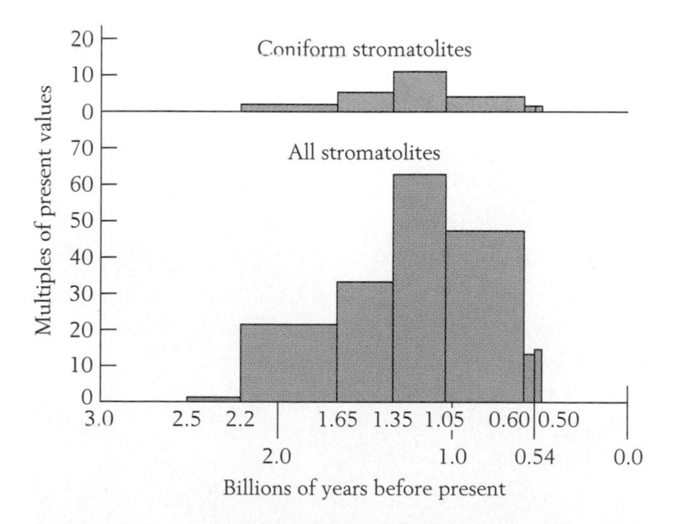

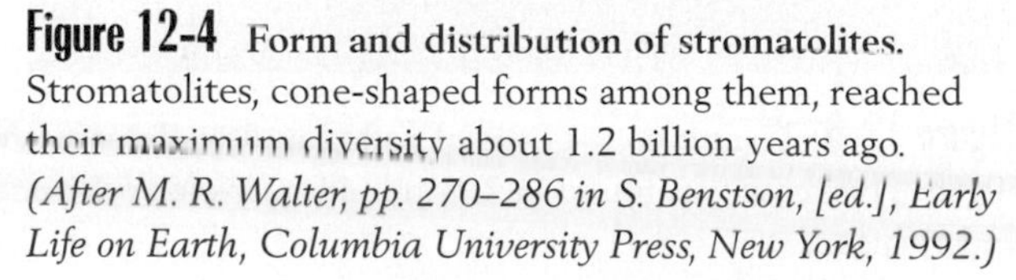

Figure 12-4 Form and distribution of stromatolites. Stromatolites, cone-shaped forms among them, reached their maximum diversity about 1.2 billion years ago. *(After M. R. Walter, pp. 270–286 in S. Benstson, [ed.], Early Life on Earth, Columbia University Press, New York, 1992.)*

Early Eukaryotes Recall from Chapter 11 that the genetic structure of simple eukaryotes reveals that they evolved well back in Archean time. Presumably these primordial forms lacked important features of advanced eukaryotes, however. To understand what they probably lacked, let us review the steps by which the advanced eukaryotic cell arose.

It now appears that the eukaryotic cell arose from the union of two prokaryotic cells, one of which came to reside within the other. The cell that lived within the other was altered in minor ways to form a structure called a **mitochondrion,** one or more of which is present in nearly all eukaryotic cells (Figure 12-5). Mitochondria are the structures that allow cells to derive energy from their food by means of respiration (p. 261). Evidence of this strange origin of mitochondria is the presence within them of both DNA and RNA that differ from those of the surrounding cell. It is assumed that the smaller cell that became a mitochondrion, complete with its own

DNA and RNA, was eaten by the larger one but proved resistant to the digestive processes of the predator cell.

It is widely agreed that plantlike protists later evolved as a result of another union of two kinds of cells. In this major evolutionary step, a protozoan consumed and retained a cyanobacteria cell. This cell then became an intracellular body known as a **chloroplast** (see Figure 12-5), which served as the site of photosynthesis both in plantlike protists and in higher plants, which evolved from them. The similarity between cyanobacteria and chloroplasts is striking. In both, for example, the pigment chlorophyll, which absorbs sunlight and permits photosynthesis, is located on layered membranes. Furthermore, chloroplasts, like mitochondria, contain their own DNA and RNA.

In many kinds of protists, photosynthesis is conducted within chloroplasts, and it is generally believed that plantlike protists evolved several times, when protozoans retained within their cells the cyanobacteria they had eaten. Thus mobile protozoans may have evolved into mobile photosynthetic forms, while immobile protozoans evolved into immobile photosynthetic forms.

Those living eukaryotes that, according to genetic data, represent the lowermost branches of the eukaryotic family tree are parasites in animals. They live without oxygen and lack mitochondria, obtaining energy directly from their hosts. These forms seem to represent an early stage of eukaryotic evolution. Possibly, like them, all Archean eukaryotes had nuclei but in other ways resembled bacteria.

Perhaps eukaryotes are not found in the Archean fossil record because they continued to resemble bacteria, which were rarely preserved except in the form of stromatolites. Eukaryotes may have continued to resemble bacteria because they had not yet taken one key evolutionary step that was necessary for the origin of mitochondria. Before some of them could develop the habit of eating prokaryotic cells, including the cell that became the first mitochondrian, they had to evolve a structure that bacteria lack. This is the *cytoskeleton*—a dynamic set of fibers that underlie the outer membrane of the cell and allow the cell to change its shape for various purposes. The presence of a cytoskeleton allows one cell to engulf another. The origin of this structure may have triggered an evolutionary expansion of eukaryotes to include a variety of new species, some of which were larger and more readily preserved as fossils than most kinds of bacteria.

Algae Having formed by the ingestion of one cell by another, the earliest eukaryotes must have been unicellular. Multicellular protists—seaweedlike algae—may have arisen soon after the evolution of fully developed eukaryotic cells with mitochondria and chloroplasts. In fact, the oldest fossil eukaryotes now recognized are algal ribbons, commonly wound into loose coils, that date to about 2.1 billion years ago (Figure 12-6). Even after 2 billion years ago, however, prokaryotes greatly outnumbered eukaryotes in floras of single-celled organisms. The Gunflint flora of the Lake Superior region, for example, includes only prokaryotic forms (Figure 12-7*A*).

Nonetheless, single-celled algae termed **acritarchs** become increasingly conspicuous in Proterozoic rocks younger than 2 billion years (Figure 12-7*B*). These nearly spherical or many-pointed forms are the dominant group of algal plankton found in Paleozoic as well as Precambrian strata. Some acritarchs are believed to have been the resting stages (or cysts) of dinoflagellates, which are one of the most prominent groups of planktonic algae today (p. 71). Although some Proterozoic acritarchs probably belonged to other groups of algae, the size and complexity of many Proterozoic acritarchs, together with the chemical composition of their cell walls, indicate a eukaryotic level of organization. All living bacteria are smaller than Proterozoic acritarchs, and many have simpler wall patterns.

Figure 12-6 ***Grypania*, a genus of coiled multicellular algae.** This fossil was found in 2.1-million-year-old rocks in Michigan.

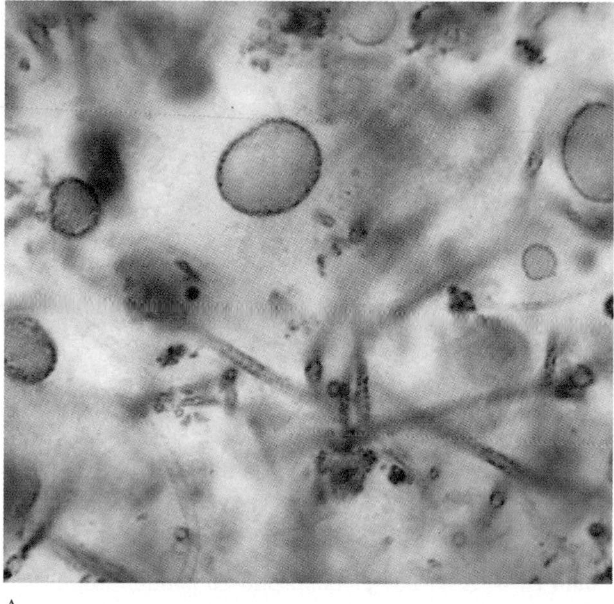
A

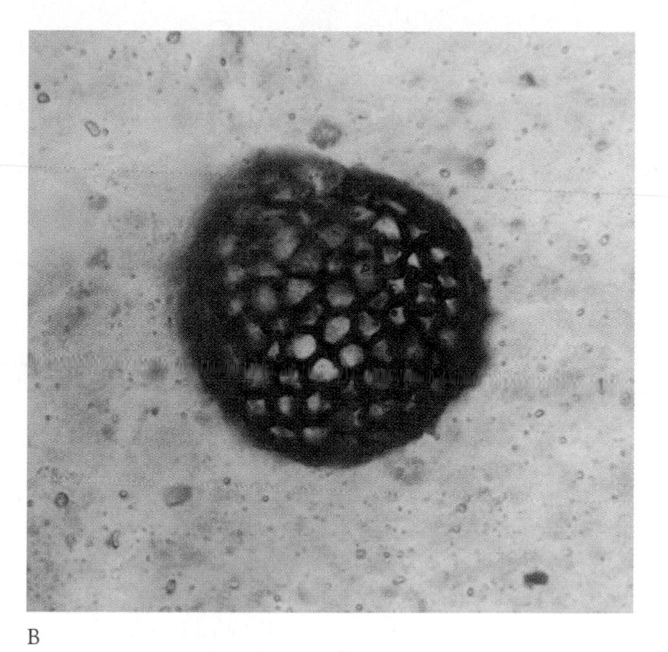
B

Figure 12-7 Fossil prokaryotic and eukaryotic cells of Proterozoic age. Eukaryotic cells were larger and more complex in form. *A*. Fossil cells of the Gunflint Formation, which is about 1.9 billion years old. Both filamentous and spheroidal forms are present. The large spherical forms are about 10 microns in diameter. *B*. The acritarch *Dictyotidium*, which has a thick, complex wall structure that is unknown in prokaryotes. This fossil is about 750 million years old and has a diameter of about 35 microns.

Distinctive organic compounds known as *biomarkers* also indicate the presence of eukaryotes early in Proterozoic time. For example, compounds called steranes, which are present in eukaryotic but not prokaryotic cell membranes, have been detected in rocks about 1.7 billion years old.

Fossil microbiotas indicate that bacteria, including cyanobacteria, continued to play a more important role in marine ecosystems than acritarchs or other eukaryotes until after 2 million years ago. As we shall see, however, changes in the physical environment may have given eukaryotes a special advantage.

The Buildup of Atmospheric Oxygen

Recall that from early in Earth's history, radiation from the sun has broken down water vapor in the upper atmosphere, releasing small amounts of oxygen (p. 308). Sometime before 3.5 billion years ago, photosynthetic prokaryotes also began releasing oxygen to Earth's atmosphere.

It is widely agreed that photosynthesis caused atmospheric oxygen to build up during Precambrian time. Before oxygen could build up in the atmosphere, however, natural reservoirs known as chemical sinks had to be filled. *Chemical sinks* are chemical elements and compounds that combine readily with oxygen. Sulfur and iron were two of the most important oxygen sinks present in Earth's crust and oceans immediately after Earth formed. (Note how iron that we extract from naturally occurring compounds rusts when it is exposed to the oxygen in the atmosphere.)

Evidence from Minerals There is abundant evidence that until about 2.3 billion years ago chemical sinks were soaking up oxygen so effectively that the concentration of oxygen in the atmosphere remained at only 1 or 2 percent of its modern level. This evidence comes in the form of uranium and iron minerals in early Proterozoic rocks.

Under the atmospheric conditions of the modern world, the uranium oxide mineral uraninite (UO_2) is readily oxidized further and quickly dissolves from rocks. The iron sulfide mineral pyrite (FeS_2), known as "fool's gold," also disintegrates readily by oxidation when exposed to the modern atmosphere. Today uraninite and pyrite are seldom found in the sands of rivers and beaches, and they are similarly rare in sandstones younger than about 2.3 billion years. These minerals accumulated in much greater concentrations in nonmarine and shallow marine deposits before 2.3 billion years ago, an indication that at this time little oxygen was present in bodies of water that were in contact with the atmosphere.

The distribution of banded iron formations may also reflect an increase in the concentration of atmospheric oxygen slightly before 2 billion years ago. Banded iron formations are among the oldest known rocks on Earth (see Figure 11-20) and are quite common in Archean terranes. They are absent from the Phanerozoic record, however, and most of them accumulated between about 3.5 billion and 1.9 billion years ago. The term "banded iron formation" refers to a bedding configuration in which layers of chert that is sometimes contaminated by iron alternate with layers of other minerals that are richer or poorer in iron than the chert (Figure 12-8). The iron in these formations may occur in a variety of minerals, and in many cases the mineralogy of the iron has altered over time in ways that cannot be reconstructed. Banded iron formations account for most of the iron ore that is mined in the world today. Those with great economic value contain iron in the form of magnetite (Fe_3O_4), whose oxygen-to-iron ratio is lower than that of hematite (Fe_2O_3). In most of these rocks the average composition of iron oxide is intermediate between hematite and magnetite.

Banded iron formations accumulated in offshore waters. Many are associated with turbidites. Both the iron and the silica in these sediments appear to have come from hot, watery emissions from the seafloor associated with igneous activity. The kind of layer that was deposited at any time probably depended on the chemical composition of nearby watery emissions. The weak oxidation of the iron indicates that deep and even moderately deep waters of the ocean were poorly supplied with oxygen. Iron formations apparently ceased to form about 1.9 billion years ago because the concentration of oxygen built up in the waters of the deep ocean. An increase in the concentration of atmospheric O_2 may have ended this episode, but it is possible that the deep ocean simply became better mixed so that more atmospheric oxygen reached deep levels.

Significantly, red beds (p. 46) display the opposite pattern: they are never found in terrains much older than 2 billion years. Hematite, a highly oxidized iron mineral, gives red beds their color (p. 314). Often the hematite found in red beds has formed secondarily by oxidation of other iron minerals that

Figure 12-8 A weakly metamorphosed banded iron formation, about 2 billion years old, in northern Michigan.

accumulated with the sediments. Oxygen has been plentiful in Earth's atmosphere during Phanerozoic time, so that this secondary oxidation has often occurred within a few millions or tens of millions of years after the sediments were deposited. It would appear that oxidation of this type did not occur early in Earth's history.

The rare units of Precambrian soils that are well preserved offer even stronger testimony about the history of atmospheric oxygen. These thin units reveal the chemical nature of weathering during the time when the soil was formed. The concentration of oxygen in the atmosphere influences this weathering. In moist soils, as in the ocean, iron exposed to abundant oxygen precipitates as hematite and other highly oxidized minerals. When little oxygen is present, iron that weathers from rocks remains in solution and is carried away by moving water. It is striking that soils that formed on basaltic rocks before about 1.9 billion years ago lost nearly all of their abundant iron. Thus there was still not enough oxygen in the atmosphere to precipitate the iron in these soils as hematite and other highly oxidized minerals. In contrast, all Proterozoic soils younger than 1.9 billion years accumulated highly oxidized iron. The heavy oxidation of one well-preserved soil that formed 1.9 billion years ago in South Africa indicates that by this time atmospheric oxygen had built up to at least 15 percent of its present level.

Why the Atmospheric Reservoir of Oxygen Expanded Presumably part of the reason for the buildup of atmospheric oxygen about 2 billion years ago was that chemical sinks, including reduced iron compounds, were filling up. As they did so, more of the oxygen produced by photosynthesis accumulated in the atmosphere. In addition, however, a large amount of organic carbon was apparently buried in marine sediments. Oxygen that would have been used up in the decay of this organic matter was left to accumulate in the atmosphere (see Figure 10-6). The evidence for this burial of organic carbon is a marked shift toward heavier carbon isotope ratios in limestones throughout the world between about 2.2 billion and 2.0 billion years ago. Recall that the burial of large volumes of organic carbon, which is isotopically light, leaves isotopically heavy carbon behind in the ocean. This heavy carbon ends up in limestones that are precipitated from the seawater (see Figure 10-9). As we shall see shortly, the buildup of atmospheric oxygen had major ramifications for life on Earth.

The Impact on Life As oxygen built up in the early Proterozoic atmosphere, the concentration of dissolved oxygen inevitably increased in the upper ocean. The amount of nitrogen that was oxidized to form nitrate (NO_3^-), which is an important nutrient for eukaryotic algae (p. 114), must then have increased also. Cyanobacteria, in contrast, do not require nitrate from their environment, as they can use pure nitrogen (N_2), which is abundant in the atmosphere and waters of shallow seas. As cyanobacteria could prosper even before nitrates became abundant in their environment, they had a temporary advantage over eukaryotic algae. The early Proterozoic oxygen buildup must have fertilized eukaryotic algae, however, by increasing their supply of nitrate. This may partly explain the Proterozoic expansion of acritarchs and multicellular algae.

The Neoproterozoic Era

The Neoproterozoic Era, which began 1 billion years ago, was marked by a variety of profound global changes, at least some of which were probably causally connected to one another. The animal kingdom apparently originated during Neoproterozoic time, and the most spectacular event of this era was the initial radiation of well-fossilized animals during the last 50 million years of the Neoproterozoic, some 4 billion years after Earth's origin.

Ice Ages

During the interval between about 850 million and 600 million years ago, Earth went through at least four ice ages. These episodes are difficult to correlate stratigraphically with one another, but the last one, which took place about 600 million years ago, left glacial deposits on all major continents of the modern world except Antarctica. It is difficult to comprehend how ice could have spread over so many continents during this glaciation, which is known as the Varangian. Indeed, the limited paleomagnetic data now available suggest that even regions lying close to the equator experienced some degree of continental glaciation. Almost all of Australia, for example, seems to have lain within 30 degrees of the equator in Neoproterozoic time, and yet glaciation in Australia was extensive. It is possible that most of Earth became quite cold.

The Beginnings of Modern Life

Momentous evolutionary changes took place near the end of the Neoproterozoic Era, but even early in the era there were stirrings of change. Fossil seaweeds reveal that multicellular green and red algae became abundant in Neoproterozoic ecosystems, and distinctive biomarkers confirm this conclusion. In the planktonic realm, acritarch species evolved that were larger and had more complex shapes than species of earlier times. Thus it appears that a large adaptive radiation of eukaryotes began slightly before a billion years ago.

Immediately after the enormous Varangian glacial episode about 600 million years ago, the acritarchs experienced major changes: many species became extinct and were replaced by others. At the same time, stromatolites declined in both abundance and variety. Slightly later, the oldest unquestioned fossils of animals were preserved in many parts of the world, marking the start of the great evolutionary radiation of animals that continued into the Phanerozoic Eon.

The Explosive Evolution of Animals

During the final 30 million years or so of the Neoproterozoic Era, animal life underwent an evolutionary radiation that has been aptly described as explosive.

Three kinds of fossils contribute to our understanding of this spectacular radiation: trace fossils, soft-bodied fossils, and skeletal fossils. The earliest unquestioned fossils in each of these categories are about 570 million years old. Possibly very simple animals evolved tens or even hundreds of millions of years earlier, but we have no certain evidence of their existence. Structures that may represent simple soft-bodied animals occur in rocks about 600 million years old. These simple, disk-shaped forms may actually be multicellular protists, possibly even fleshy algae.

Trace Fossils Tracks, trails, and burrows—what are called trace fossils (p. 62)—have provided special evidence about the early diversification of animal life. Even during the nineteenth century it was acknowledged that fossil skeletons appear in the stratigraphic record quite suddenly near the base of the Cambrian System. The complexity and variety of fossilized Cambrian life gave rise to speculation that multicellular animals had a long Precambrian history, during which they lacked hard parts and therefore left no fossil record. One effective way to test this hypothesis was based on the assumption that soft-bodied multicellular animals—those that lack hard parts—have crawled over the seafloor or burrowed into it throughout their existence, and in doing so have often produced trace fossils in sedimentary rocks. If soft-bodied invertebrate animals had existed for a long interval of Proterozoic time, scientists reasoned, some of them would have left such trace fossils.

A search for trace fossils in Precambrian rocks has since turned up a striking pattern: such fossils have been found only in rocks about 570 million years old or younger. For example, 1.3-billion-year-old sedimentary rocks of the Belt Supergroup of Montana exhibit no tracks or trails. As Figure 12-9 indicates, Belt mudstones are often strikingly well layered in comparison with younger deposits, in which burrowing animals have often disrupted or destroyed layers of sediment.

Neoproterozoic trace fossils display a general evolutionary pattern. The oldest ones are simple tubes made by wormlike animals that burrowed through the sediment. In several regions of the world, as stratigraphic sections progress up toward and into the Cambrian System, trace fossils become increasingly complex and varied (Figure 12-10). This increase in both complexity and variety seems to represent the initial evolutionary diversification of mobile animals in the world's oceans.

Imprints of Soft-Bodied Animals Bodies of animals trapped beneath sediment also left conspicuous imprints in Neoproterozoic sediments younger than about 570 million years. Many of these fossils have

Figure 12-9 Laminated siltstone from the 1.3-billion-year-old Greyson Shale (Belt Supergroup) in Montana. Here, as in other rocks older than late Proterozoic, we see no evidence of burrowing by invertebrate animals.

Figure 12-10 Undersurfaces of sandstone with fillings of relatively complex Proterozoic burrows in Norway. *A.* Filling of a feeding burrow. *B.* Filling of a shallow burrow on which can be seen scratch marks left by the legs of the animal that dug it.

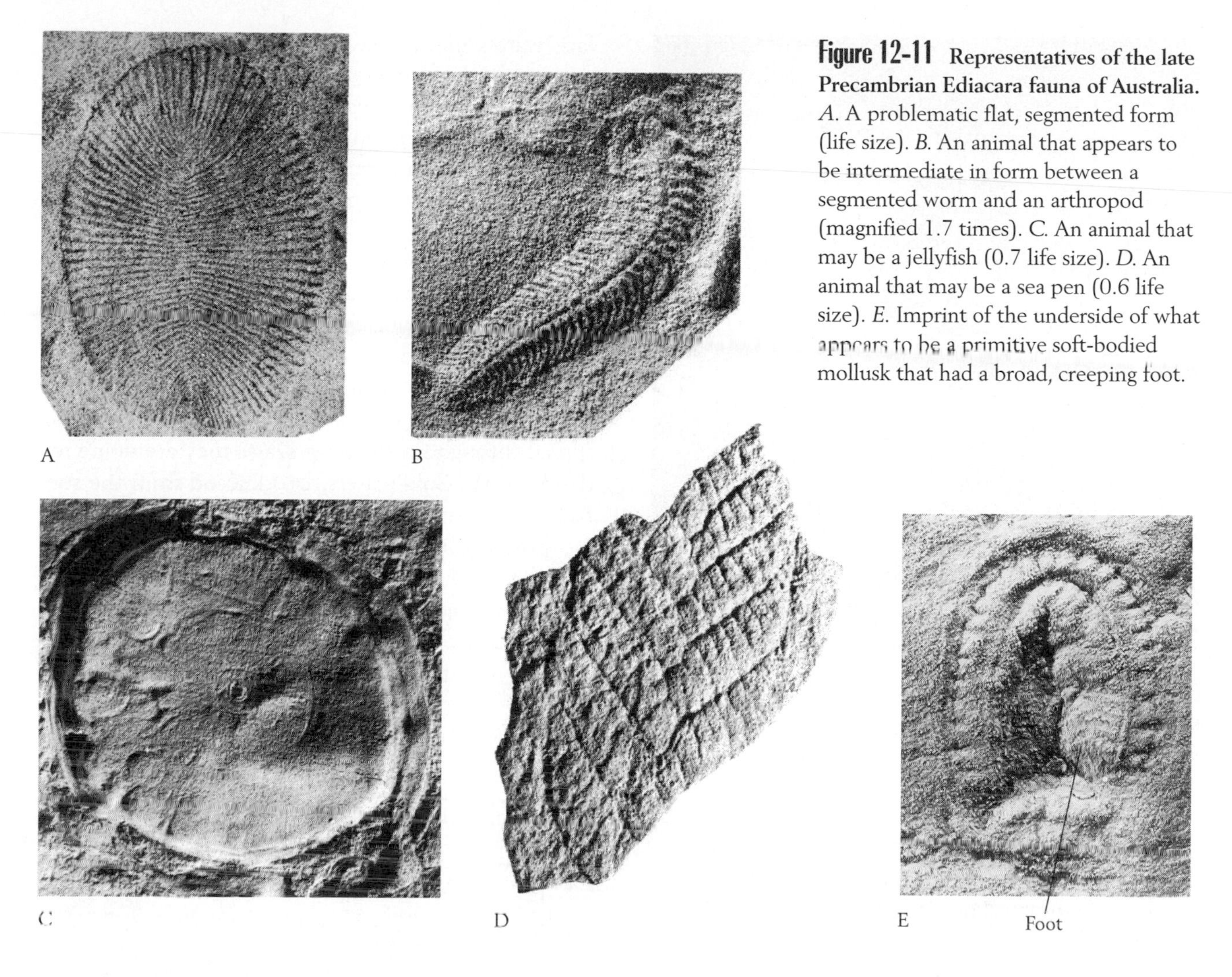

Figure 12-11 **Representatives of the late Precambrian Ediacara fauna of Australia.** *A.* A problematic flat, segmented form (life size). *B.* An animal that appears to be intermediate in form between a segmented worm and an arthropod (magnified 1.7 times). C. An animal that may be a jellyfish (0.7 life size). *D.* An animal that may be a sea pen (0.6 life size). *E.* Imprint of the underside of what appears to be a primitive soft-bodied mollusk that had a broad, creeping foot.

been thought to represent jellyfishes or sea pens, both of which belong to the phylum **Cnidaria,** which contains modern corals (p. 77). Whereas jellyfishes float in the water, sea pens are stalked creatures that stand upright on the seafloor. The assignment of the Proterozoic fossils to the Cnidaria has recently been questioned, however, and they may actually represent groups without living representatives. Because the Ediacara fauna of Australia is the most famous of late Precambrian "soft-bodied" faunas, all of the Neoproterozoic faunas that resemble it are termed Ediacaran faunas (Figure 12-11).

Assemblages of soft-bodied animals were never preserved in a similar way on sandy, well-oxygenated seafloors of Phanerozoic age. Probably the Ediacaran animals were preserved only because they lived before the evolution of predators or scavengers capable of readily devouring their carcasses.

Whatever kinds of animals the Ediacaran fossils represent, it is clear that animals more advanced than cnidarians were well established before the end of Proterozoic time. Among these advanced groups were the segmented worms known as annelids, which today include not only modern earthworms but also many kinds of marine and freshwater species (see Figure 3-26). Annelids undoubtedly formed many of the tubelike fossil burrows of Neoproterozoic age. Also present were early members of the phylum Arthropoda, which includes modern crabs, lobsters, and insects. Arthropods have external skeletons and jointed legs, a set of which presumably made the scratches visible in the burrow shown in Figure 12-10*B*. One Ediacaran form appears to have been a primitive mollusk that crept over the seafloor like a snail (see Figure 12-11*E*).

Skeletal Fossils For many decades, the oldest known fossil shells and other hard parts of animals were from the base of the Cambrian System. Although abundant and varied shelly faunas make their

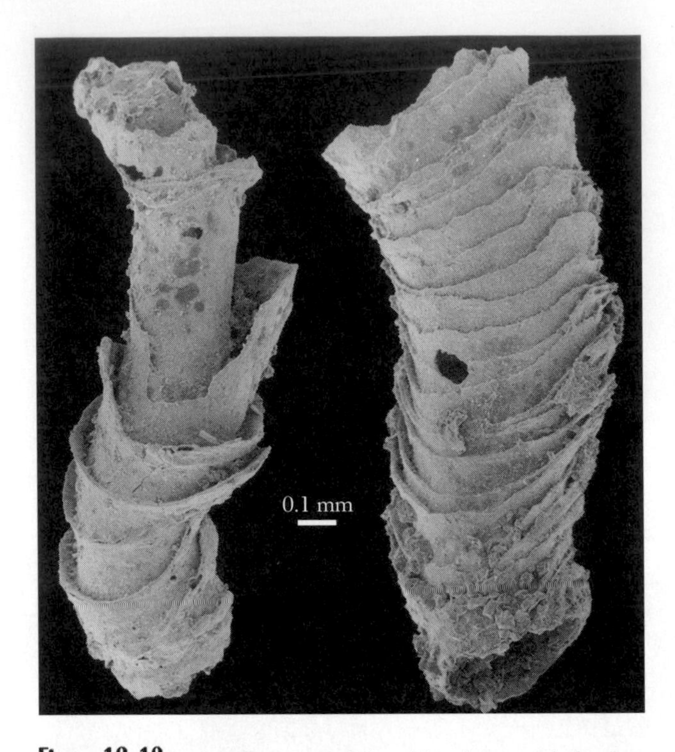

Figure 12-12 ***Cloudina*, one of the oldest known skeletal fossils.** This tubelike form was slightly less than 1 millimeter in diameter.

earliest appearance at this level, scientists have recently found small calcium carbonate shells throughout the Neoproterozoic interval that contain fossils left by soft-bodied animals. These shells are vase-shaped and tubular structures made of calcium carbonate (Figure 12-12). Although these fossils cannot be assigned with assurance to any previously recognized taxonomic group, they provide further evidence of the broad scope of the Proterozoic radiation of animals.

What Triggered the Initial Radiation of Animals? We have seen that acritarchs experienced a pulse of evolutionary change shortly after 600 million years ago, while stromatolites declined, and soon after this time animals began their rapid Neoproterozoic adaptive radiation. It seems likely that these episodes were interrelated in some way. They may also be connected to physical events that left their marks in the geologic record. These events included the Varangian glacial episode, deposition of banded iron formations, and oscillations of carbon isotope ratios in shallow seas.

The Neoproterozoic glacial episodes appear to be related to oscillations in carbon isotope ratios in surface waters of the ocean (Figure 12-13). These isotope ratios have been measured in carbonate sediments. At times they rose to very high levels, probably because a large amount of light organic carbon was being buried in marine sediment (see Figure 10-8). Some experts have proposed that these were times when the deep waters of the ocean were anoxic, so that much isotopically light organic matter was removed from the ecosystem by burial.

Note that during the two glacial episodes of the late Neoproterozoic, carbon isotope ratios declined dramatically (see Figure 12-13). It may be that at these times cold waters sank in frigid regions and spread throughout the deep sea as they are doing today (p. 110). Cold waters that descend from the surface are oxygenated by the atmosphere. During the Neoproterozoic Era, as today, oxygenation of deep waters would have allowed bacterial decay and return of isotopically light carbon to surface waters through upwelling. Perhaps such oceanographic mix-

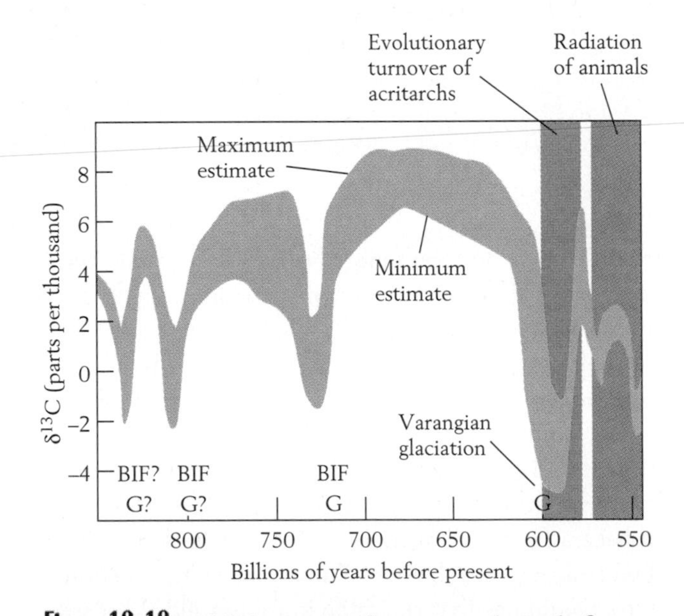

Figure 12-13 Global events in the sea near the end of Proterozoic time. Carbon isotope ratios in limestones are very high for two long intervals. These ratios declined at times of continental glaciation (G). Banded iron formations (BIF) are also associated with glacial deposits. After the final, Varangian glaciation, major biological changes occurred. *(After A. T. Kaufman and A. H. Knoll, Precambrian Research 73:27–49, 1995; and A. H. Knoll and H. D. Holland, p. 21–33 in Effects of Past Global Change on Life, National Academy Press, Washington, D.C., 1995.)*

ing is what caused the carbon of shallow seas to become isotopically light during glacial episodes.

Remarkably, deposition of banded iron formations resumed during the Neoproterozoic, at times when glacial deposits accumulated. It is unlikely to be coincidental that two such unusual kinds of deposition took place at the same time. Perhaps iron from volcanic emissions accumulated in deep waters of the ocean between glacial episodes, when no cold waters were descending and deep waters remained relatively stagnant and deprived of oxygen. Then, during glacial episodes, these waters were mixed with oxygenated waters, which caused iron to precipitate and form sedimentary iron deposits.

How might global changes in the physical environment have triggered the biological changes that took place near the end of Proterozoic time? The enormous expansion of glaciers 600 million years ago may have caused the extinction of nearly all preexisting acritarch species and led to the evolution of new species adapted to new climatic conditions. The diversification of conspicuous animals requires a separate explanation, however, because it began perhaps 30 million years later.

A buildup of atmospheric oxygen to a critical level may have triggered the evolutionary radiation of conspicuous animals. Perhaps before late in Proterozoic time there was only enough oxygen in the atmosphere to support tiny, simple animals—creatures so small that oxygen could diffuse throughout their bodies even though they lacked circulatory systems to carry it efficiently in bloodlike fluids.

There is independent evidence that oxygen may have built up in the Neoproterozoic atmosphere. Recall that massive burial of organic carbon probably produced the episodes in which carbon isotopes in shallow Neoproterozoic seas became isotopically light (see Figure 12-13). Burial of large volumes of organic carbon would have left behind oxygen—oxygen that would have oxidized this organic carbon if the carbon had not been buried (p. 264). We do not know for certain that any of this oxygen accumulated in the atmosphere, however. The problem is that so much oxygen would have been released that if all of it had remained in the atmosphere, its concentration would have become impossibly high. If so much oxygen was released, most of it must have been absorbed by chemical sinks, such as unoxidized and weakly oxidized iron. With so much oxygen necessarily having been absorbed by sinks, we cannot be certain that sinks did not absorb all of it, leaving none to accumulate in the atmosphere.

Thus it remains unclear whether a Neoproterozoic buildup of atmospheric oxygen permitted the evolution of sizable animals—ones several centimeters in length, height, or breadth. The issue remains unresolved: oxygen may have reached this critical level at an earlier time. It is possible that the late Neoproterozoic radiation was triggered instead by the initial evolution of certain key adaptive features, such as muscle or nerve cells. Unfortunately, we do not know the time of origin or biological nature of the earliest animals, which were small, simple multicellular forms that left no fossil record. We know much more about the diversification of the larger animals that left fossil records, but even the timing of this very important episode in the history of life remains unexplained.

The Expansion and Contraction of Continents

Although geologists have long attempted to determine how the continents of the modern world originated, they have managed to trace back the histories of these continents only into Neoproterozoic time. Uncertainties remain about the histories of older Proterozoic cratons, and the configurations and relative positions of Archean microcontinents will probably never be known. The difficulty is that depositional patterns and structural trends of rocks more than half a billion years old are often obscured by erosion, metamorphism, or burial; paleomagnetic data are also sparse and difficult to interpret. Nonetheless, geologists have reconstructed partial histories for most large blocks of Proterozoic crust and have thus gained some knowledge about how they became part of the Phanerozoic world.

In the next section we will learn how North America grew episodically during Precambrian time and how this continent became part of a vast supercontinent that fragmented late in the Proterozoic Eon. We will also review the origins of the landmasses that came to constitute the continents of the Paleozoic Era, including Gondwanaland.

Before we discuss the histories of individual cratons, let us consider how cratons in general become larger or smaller. As we have seen, cratons increase greatly in size when they become sutured together along a subduction zone, and this process is usually accompanied by mountain building in the vicinity of the suture (p. 240).

Cratonic growth on a smaller scale, which is known as **continental accretion,** also entails mountain building, but this process occurs at the margin of a single large craton. As we noted earlier, marginal accretion can result either through the suturing of a **microplate** to a large craton along a marginal subduction zone (p. 251) or from the compression and metamorphism of sediments that have accumulated along a continental shelf. The latter process is sometimes referred to as **orogenic stabilization,** because it thickens the crust and hardens both unconsolidated sediments and soft sedimentary rocks (Figure 12-14). The Wopmay orogenic episode enlarged the Slave craton by both suturing of a microplate and orogenic stabilization (see Figure 12-1).

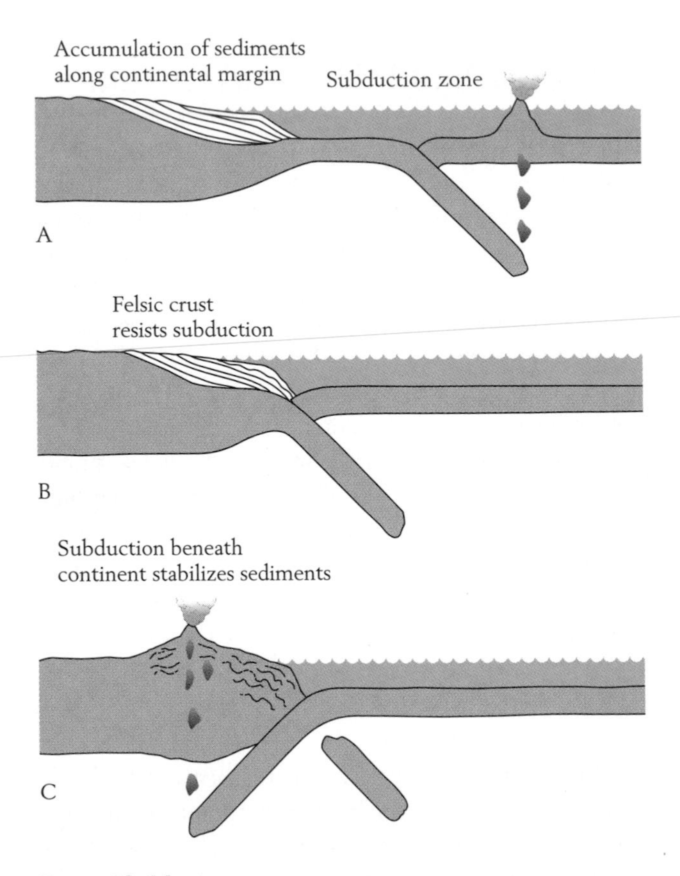

Figure 12-14 Stabilization of sediments that have accumulated along the margin of a continent *(A)*. The continent comes to rest along a subduction zone. Because the continent is of low density and resists subduction, the direction of subduction is reversed *(B)*. Igneous activity then adds rock to the continental margin and metamorphoses the sediments that have collected there *(C)*.

Stabilization is a cannibalistic process inasmuch as some of the sediment that is deposited and stabilized along a continental margin is derived from the interior of the continent by erosion. On the other hand, limestone that accumulates along continental margins is precipitated from seawater or is secreted by organisms and thus represents an external contribution to the mass of the continent—as do the igneous rocks and oceanic crust that become welded to a continental margin resting along a subduction zone.

Orogenic processes do not simply add material to continents; they also alter preexisting crust. Regional metamorphism, for example—which sometimes operates in conjunction with structural deformation—remobilizes continental crust. Metamorphism often alters the character of preexisting rocks beyond recognition and resets their radiometric clocks so that the age of the crust can no longer be determined (p. 165). In reviewing the Precambrian history of individual Proterozoic cratons, we will encounter many examples of crustal **remobilization.**

How do continents decrease in size? They can shrink by erosion, but this process operates so slowly that it has little overall significance. Far more important is the process of continental rifting, which operates on many scales. It can remove a small sliver of crust, or it can divide a large craton in half. Although continental rifting that took place more than half a billion years ago is difficult to document, evidence suggests that major rifting occurred late in Proterozoic time.

The Assembly of North America

Greenland today is a continent in its own right, but during Proterozoic and most of Phanerozoic time it was attached to North America. Laurentia is the name given to the combined landmass. The core of Laurentia was the crustal block that now forms most of the North American craton. This ancient block is well exposed today as the largest Precambrian shield in the world: the Canadian Shield (p. 284).

Continental Accretion

The Canadian Shield constitutes a large portion of the North American craton, including a small part of

Figure 12-15 Major geologic features of North America. The Canadian Shield ends where sedimentary rocks of the interior lowlands lap over it on the south and west. The Cordilleran, Ouachitas, and Appalachian orogens flank the North American craton on the west, south, and east.

the northern United States (Figure 12-15). Precambrian rocks also underlie the interior of the continent to the south, where they are overlain by a relatively thin veneer of Phanerozoic sedimentary rocks. Rocks obtained from wells that penetrate the Phanerozoic cover have provided a good picture of the general distribution of buried Precambrian rocks. These rocks, together with the exposed rocks of the Canadian Shield and Precambrian rocks that have been elevated by Phanerozoic mountain building in the American West, reveal that in the course of Proterozoic time, Laurentia gained territory by continental accretion.

Evidence that Laurentia was growing by accretion during Proterozoic time began to appear decades ago, when the recognition of structural trends and regional occurrences of rock units permitted geologists to recognize natural geologic provinces within the Canadian Shield. More recently, reliable radiometric dates for rocks of the Canadian Shield and for subsurface rocks bordering the shield have yielded a much more detailed picture. Uranium-lead techniques now yield dates with precision within about 10 million years for rocks that are about 2 billion years old. These dates are obtained from crystalline rocks, and thus they represent episodes of igneous and metamorphic activity.

North America also grew rapidly during Proterozoic time, when it became sutured to other cratons. In fact, as we will see shortly, near the end of the Proterozoic Eon it was united with nearly all of Earth's other landmasses to form a vast supercontinent only slightly smaller than Pangea.

The first stage in the formation of North America, before it became part of a supercontinent, was the assembly of at least six microcontinents into a sizable craton (Figure 12-16). This amalgamation took place within only about 100 million years, between 1.95 billion and 1.85 billion years ago. Each of the microcontinents that were combined had formed during Archean time. Today the former microcontinents represent Archean terranes, which lie mostly within the Canadian Shield. The largest of these Archean terranes is the Superior Province, which crops out as far south as Minnesota. The Wyoming and Hearne provinces may actually have formed a single microplate; in the narrow zone of contact between the two, the geologic evidence is inconclusive. The Wyoming Province is exposed south of the Canadian Shield in mountainous uplifts in Wyoming (Figure 12-17) and also in the Black Hills, a blisterlike structure in South Dakota (see Figure 9-23), whose gold deposits are discussed in Box 12-1.

Most of the Archean terranes were sutured directly together, but the Superior Province is separated from the Wyoming and Hearne provinces by a broad zone of rocks that formed about the time of the suturing, 1.9 billion to 1.8 billion years ago (see Figure 12-16). This zone comprises both deep-sea sediments squeezed up between the converging cratons and crystalline rocks produced by an igneous arc.

South of the Archean terranes that were sutured together between 1.95 billion and 1.85 billion years ago is a broad zone of crust that formed shortly thereafter, between about 1.8 billion and 1.6 billion years ago. The rocks of this zone are exposed in uplifts from southern Wyoming to northern Mexico, and geologists have sampled them by drilling through the sedimentary rocks that blanket most of the central and western United States. The composition of these

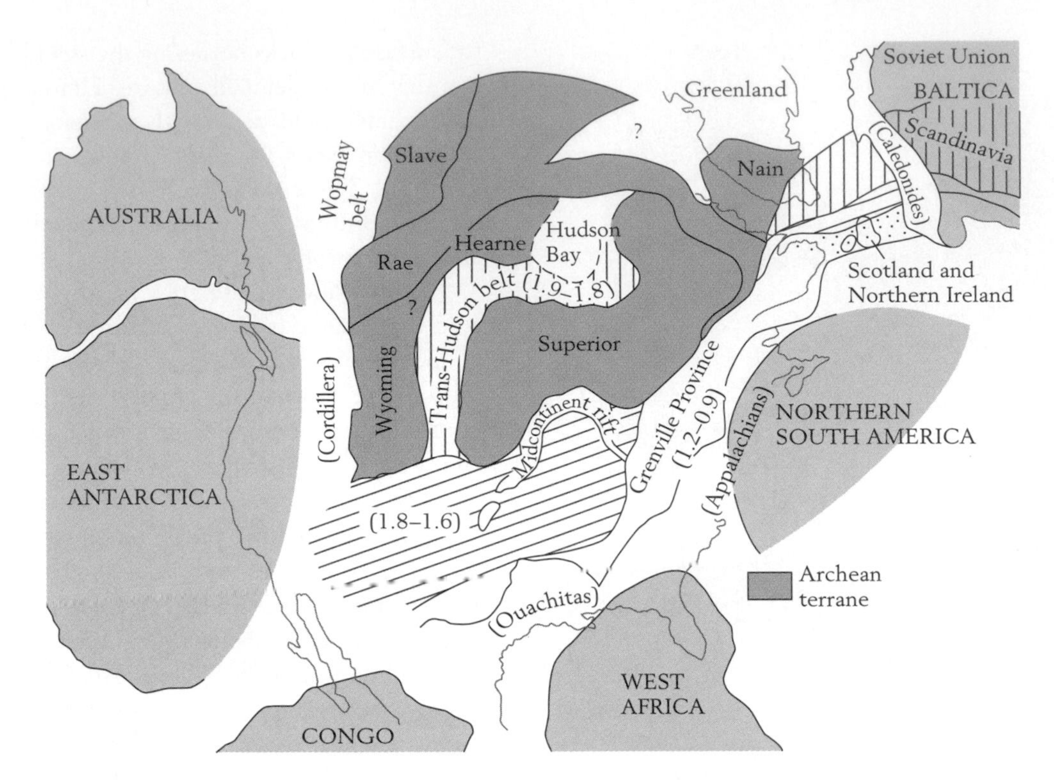

Figure 12-16 Geologic provinces of North America late in Proterozoic time, when this continent was attached to other landmasses (see Figure 12-20). Numbers in parentheses represent times of origin in billions of years before the present. Provinces of Archean terrane, in the north, represent Archean microcontinents that were amalgamated 1.95–1.85 billion years ago. The Wyoming and Hearne provinces may constitute a single terrane. The Trans-Hudson Belt consists of newly formed crust that was caught between the Superior terrane and the Archean terrane to the west. The origin of the broad province of the central United States 1.8–1.6 billion years ago resulted in substantial continental accretion toward the south. The Grenville Province formed when North America was sutured to Baltica and landmasses that later became portions of Gondwanaland. *(Derived from P. F. Hoffman, Ann. Rev. Earth and Planet. Sci. 16:543–603, 1988.)*

rocks suggests that they formed by igneous activity and by sedimentation along an island arc. The rate of continental accretion was very rapid by Phanerozoic standards. In the central and western United States, the continental margin expanded southward about 800 kilometers (500 miles) in 200 million years.

A Larger Continent

Thus far we have considered only regions that remain parts of North America and Greenland. In fact,

Figure 12-17 A U-shaped glaciated valley in the Beartooth mountains of Wyoming. Archean rocks form the core of the Beartooth uplift. The glacial scouring took place within the last 2 million years, during Earth's most recent ice age.

For the Record 12-1

A Mountain of Gold

After more than a century of operation, the Homestake mine in the Black Hills of South Dakota ranks first among all gold mines, with total profits in excess of $1 billion. The forested Black Hills, a dome-shaped outlier of the Rocky Mountains, stand conspicuously above the surrounding prairie. Their gold comes from Archean rocks that were metamorphosed early in Proterozoic time and are now exposed in the center of the dome. The gold was probably emplaced there by hot watery solutions, perhaps along an Archean mid-ocean rift. Then, about 1.6 billion years ago, hot fluids from regional metamorphism concentrated much of the gold in veins, along with quartz and other minerals. The age of the deposits is not unusual. Most of the gold in Earth's crust is in Precambrian rocks. Perhaps this heavy metal has come largely from the dense mantle. We know that early in the planet's history mantle material moved upward in great volumes along numerous mid ocean ridges

The quest for gold in the Black Hills in many ways epitomizes the nineteenth-century conquest of the American West. Meriwether Lewis and William Clark heard of the Black Hills in 1804, during their epic journey to the Pacific, but did not visit them. Beginning in the 1820s, other explorers and prospectors ventured into the mountains, but at great risk. Ezra Kind was the last survivor of a party of six who rode into the Black Hills in 1833, and before he too died, he scraped a final message on a slab of sandstone: "Got all of the gold we could carry our ponys all got by the Indians I have lost my gun and nothing to eat and Indians hunting me."

Gold excites prospectors not only because of its value but also because even in nature it is usually as shiny as a wedding ring. As a so-called noble metal, it does not form compounds with other elements, and it flashes its purity to the naked eye. Sporadic reports of gold in the Black Hills tantalized adventurers for years. Congress nonetheless ratified a treaty in 1868 that included the Black Hills in a new reservation for the Sioux. The treaty banned prospectors from the region, but dreams of wealth led many adventurers to violate the law, and the actions of a reckless general named George Armstrong Custer only added to the incentive. Custer was ordered to lead a combined force of soldiers and civilians to reconnoiter the Black Hills region in 1874. With gold clearly on his mind, Custer hired a geologist

The craggy peaks of the Black Hills stand high above the Great Plains.

(continued)

For the Record 12-1 *(continued)*

to join the expedition. Once in the Black Hills, Custer diverted the group's activities from exploring to prospecting. His party found only a few particles of the noble metal, but apparently he wanted recognition as the man who first discovered the rich gold deposits that were widely anticipated. Newspapers throughout the country reported on Custer's expedition, and when the general failed to correct exaggerated stories about an abundance of gold, thousands of prospectors flocked to the Black Hills. The treaty with the Sioux became a worthless piece of paper.

The flurry of activity quickly bore fruit. The first rich deposits to be found were stream sediments, but soon prospectors tracked down the veins of quartz that supplied this detrital gold, and mining of these veins yielded even greater riches. The great Homestake mine opened in 1876. That year General Custer, still seeking fame, led his soldiers into the famous massacre at the Little Big Horn, in which he and 269 of his men lost their lives. The Sioux's victory did them more harm than good; increased hostility toward them added to the economic incentives to terminate the treaty of 1868. In 1877, faced with the government's threat to cut off supplies to their reservation, the Sioux reluctantly signed a new agreement that removed them from the Black Hills, a region that for generations they had regarded as sacred.

Legitimate at last, mining camps in the Black Hills expanded into towns. Deadwood, which sprang up close to the Homestake mine, became the town most infamous for lawlessness. Calamity Jane earned her nickname there, and a bullet took Wild Bill Hickock's life in a Deadwood saloon. Civilization has subdued the town, but the Homestake mine continues to churn out thousands of tons of ore every day and still has larger gold reserves than any other American mine.

early in Proterozoic time the combined landmass called Laurentia was part of a larger craton. Geologic similarities between the Wyoming Province and both eastern Antarctica and eastern Australia suggest that these regions were attached to one another well before a billion years ago (see Figure 12-16). Thus western North America was connected to cratons that today are positioned in the southern hemisphere. Similar evidence points to a connection between the terranes that now constitute Siberia and the northern Canadian region of Laurentia. Exactly when Laurentia became attached to these other landmasses is not yet known.

Middle Proterozoic Rifting in Central and Eastern North America

The growth of the landmass that included North America was threatened in mid-Proterozoic time by the greatest disturbance of the central North American craton during the last 1.4 billion years. In this episode of continental rifting, between about 1.2 billion and 1.0 billion years ago, lavas poured into downwarped basins along a belt that extended from the Great Lakes region to Kansas (Figures 12-16 and 12-18). Had the crescent-shaped zone of rifting extended to the margins of the craton, the eastern United States would have drifted away as a separate small craton. But the rifting failed.

The rocks that formed within the failed midcontinent rift include hardened lavas known as Keweenawan basalts, which are exposed near the southern border of the Canadian Shield (Figure 12-19). They contain ore deposits of native copper, a mineral consisting of elemental copper uncombined with other elements. Similar basin basalts lie to the southwest beneath the sedimentary cover of the Midwest, as has been revealed both by examination of rock cuttings taken from deep wells and by detection of strong magnetism from Earth's surface. Because these basalts are rich in iron and magnesium and therefore are very dense, their presence is also associated with a feature known as the "midcontinental gravity high," which is a local increase in Earth's gravitational field as measured from the surface. While the basalts were forming, numerous basic dikes were also emplaced across the Canadian Shield to the north (see Figure 12-18 and p. 284)

The configuration and composition of Keweenawan rocks and their subsurface counterparts indi-

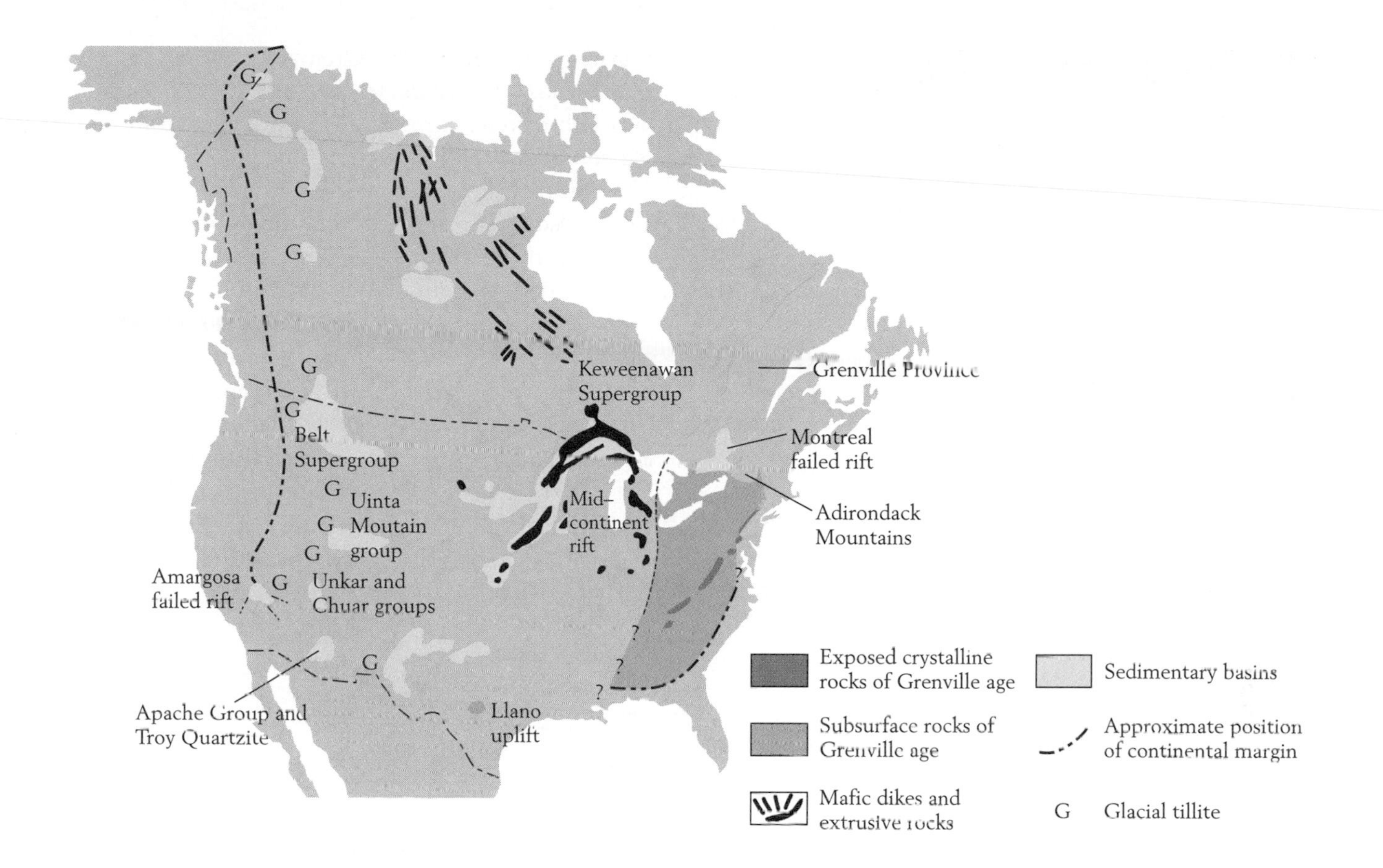

Figure 12-18 The North American craton about 850 million years ago, with major geologic features that developed after 1.4 billion years ago. Many of the glacial tillites (G) are actually slightly younger than 850 million years. The Keweenawan Supergroup, which includes mafic volcanics, accumulated in a rift that ultimately failed. Some of the sedimentary basins of western North America were failed rifts. The Grenville Province represents an orogenic belt that was active slightly before a billion years ago. *(Partly after J. H. Stewart, Geology 4:11–15, 1976.)*

Figure 12-19 Columnlike joints in the Edwards Island flow, a body of volcanic rock in the Keweenawan Supergroup of northern Michigan.

cate the presence of a failed rift in the eastern United States (p. 235). The Keweenawan volcanics, for example, are associated with red siliciclastic rocks and alluvial-fan conglomerates in what appear to be downfaulted troughs—configurations that tend to occur in newly forming continental rifts. Thus it would appear that about 1.3 billion years ago, a spreading center formed beneath the late Precambrian craton and began to rift it apart. It is possible that this spreading center intercepted the eastern margin of the ancient craton. Whatever the reason, obviously the Keweenawan rifting was abortive. Rifting ceased before the continent was split but left its mark in the enormous volumes of mantle-derived lavas that were disgorged along a belt more than 1500 kilometers (900 miles) long and 100 kilometers (60 miles) wide.

The Grenville Orogenic Belt

While the midcontinent rifting was in progress, an episode of mountain building took place along the east coast of North America. The Grenville orogeny, which spanned the interval from about 1.2 billion to 1.0 billion years ago, was another step in the accretion of the North American continent, adding a belt of terrane that stretched from northern Canada to the southeastern United States (see Figures 12-16 and 12-18). The Grenville orogeny stabilized a large volume of sediments that had accumulated along the margin of eastern North America before about 1.2 billion years ago.

The igneous and metamorphic activity of the Grenville orogeny ended about 1 billion years ago. The resulting crystalline rocks are best exposed in the Canadian portion of the Grenville Province. To the south, most crystalline rocks of Grenville age are buried, but some crop out here and there—for example, in the Adirondack uplift of New York State, the Blue Ridge Mountains of the central Appalachians, and an isolated prominence in Texas known as the Llano uplift (see Figure 12-18).

The relationship between the Grenville orogeny and the midcontinent rift remains a puzzle. What is clear is that the Grenville event entailed the collision of eastern North America with a landmass that would later become northern South America, where there are remnants of mountain systems that are the same age as the Grenville orogenic belt (Figures 12-16 and 12-20*A*). Forming another segment of the combined landmass was Baltica, which today forms northern Europe.

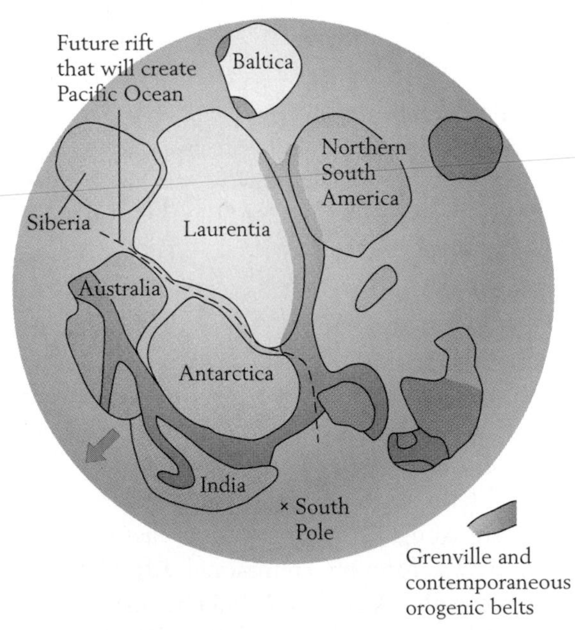

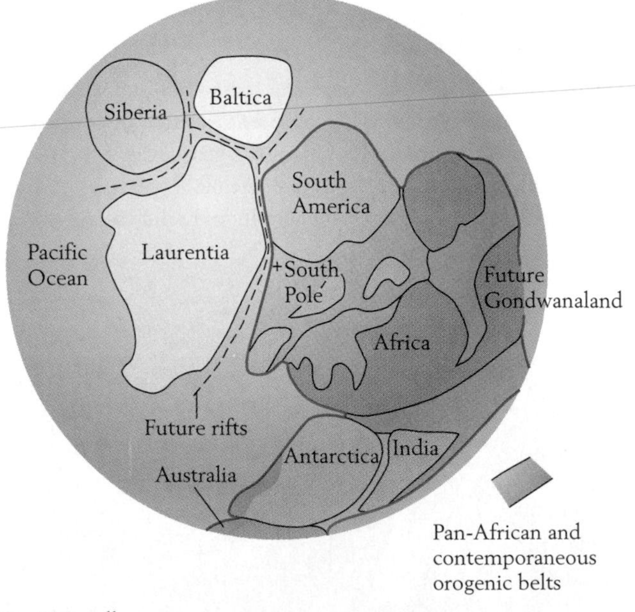

Figure 12-20 Changing paleogeographic patterns in late Proterozoic and early Paleozoic time. *A.* The likely configuration of the late Proterozoic supercontinent about a billion years ago, at the end of the Grenville orogeny. Laurentia occupied the central position. *B.* The clustering of continents in a new configuration about 550 million years ago. By this time, landmasses had broken away from the western margin of Laurentia, forming the Pacific Ocean, and had clustered on the eastern side of Laurentia; there, partly as a result of the Pan-African orogeny, they formed the landmass that would soon break away to become Gondwanaland. *(Based on I. W. D. Dalziel, GSA Today 2:240–241, 1992; and R. Unrug, GSA Today 7:1–6, 1997.)*

The Assembly and Breakup of Supercontinents

Between the time of the Grenville orogeny, 1.2 billion to 1.0 billion years ago, and 500 million years ago, Earth underwent major episodes of continental suturing and fragmentation. At least one supercontinent—possibly two—formed and broke apart during this interval, which spanned the boundary between the Proterozoic and Phanerozoic eons.

The Origin of the Supercontinent Rodinia

The Grenville collision was actually part of a long zone of tectonic suturing that encircled much of the continent to which North America belonged. Orogenic belts in southern Africa, the Indian peninsula, and Australia suggest that these regions, along with Laurentia, became attached to eastern Antarctica at this time. As a result of this extensive suturing, the landmasses that would later become Gondwanaland wrapped around most of Laurentia (see Figure 12-20*A*). The combined landmass thus formed, known as Rodinia, rivaled the Phanerozoic supercontinent Pangea in total size. Rodinia was fully assembled by about 1 billion years ago.

The Birth of the Pacific Ocean and America's West Coast

Between about 800 million and 700 million years ago, Rodinia split in half. This was one of the most significant rifting events of all time, because it created the Pacific Ocean to the west of Laurentia. In western North America, tectonic episodes that preceded this rifting produced failed rifts that harbored large depositional basins in western Laurentia. From northern Canada to southern Arizona, these basins (shown in Figure 12-18) received large volumes of sediment. In the northern United States, the largest of the basin sequences is the Belt Supergroup (Figure 12-21), which ranges in age from about 0.9 billion to 1.5 billion years. In general, the Belt thickens toward the west, sometimes reaching a thickness of 16,000 meters (53,000 feet).

The Belt formed in a northwesterly trending failed rift. Sandstones increase in abundance toward the western part of the sequence, while limestones increase toward the east, where sediments accumulated

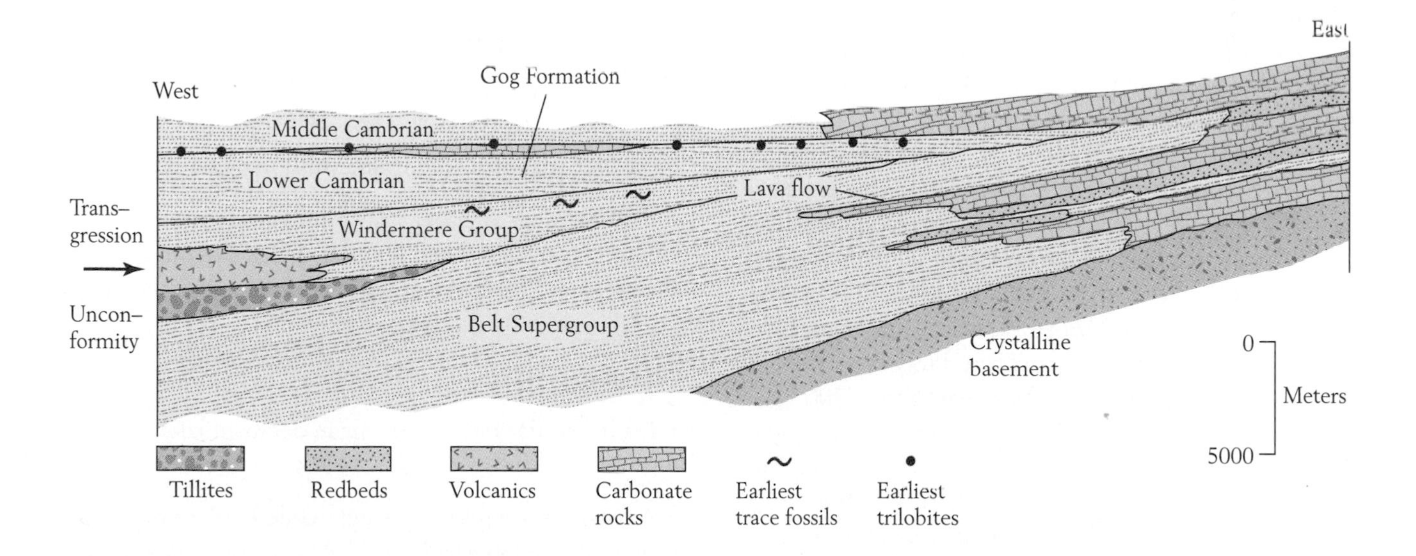

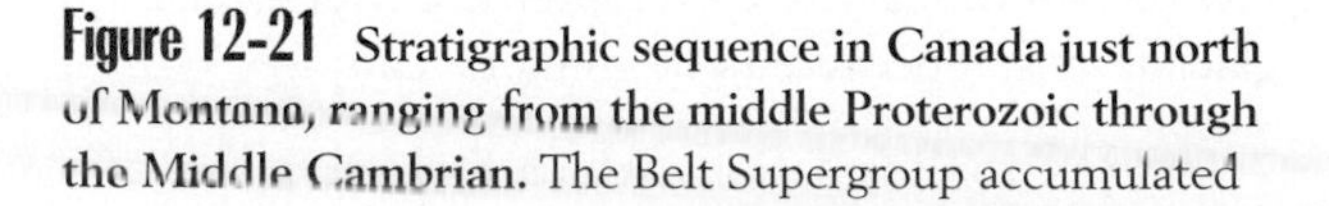

Figure 12-21 Stratigraphic sequence in Canada just north of Montana, ranging from the middle Proterozoic through the Middle Cambrian. The Belt Supergroup accumulated in a failed rift trending to the northwest (see Figure 12-18). *(Modified from P. B. King, The Evolution of North America, Princeton University Press, Princeton, N.J., 1977.)*

in shallower water. In general, however, mudstones predominate. The Belt apparently formed as a result of the accumulation of sediments in very shallow water during rapid subsidence. Salt crystals and mudcracks, both of which indicate a drying up of shallow bodies of water, are present in Belt sediments, and shallow-water stromatolites are also abundant in the limestones.

The final global glaciation of the Proterozoic took place about 600 million years ago, shortly after the great rifting event (see Figure 12-13). In the region occupied by the Belt Supergroup, and in fact from one end of western North America to the other, tillites accumulated. After the rifting event, the western border of North America remained a passive margin for hundreds of millions of years. In fact, it has never since been sutured to another large craton. As we will see, however, it has grown westward through suturing events that have added small bodies of crust. It has also, of course, been the scene of extensive mountain building and igneous activity that have added crust during the latter part of Phanerozoic time. Thus, at the start of the Phanerozoic Eon, the passive western margin of North America lay well to the east of its present position.

Another Supercontinent?

The large block that separated from Laurentia as the Pacific Ocean formed was eventually to become the eastern segment of Gondwanaland. This block and Laurentia continued to separate, and the Pacific Ocean continued to grow, until their leading edges collided with opposites sides of the newly forming African craton. In this way a new supercontinent may have formed, with Africa in its center (see Figure 12-20*B*). Whether such a supercontinent actually formed is uncertain, however, because some of the crustal blocks that would have been part of it may have broken away before others were sutured together.

We know for sure that between about 700 million and 500 million years ago numerous small continental blocks were sutured together to form the large body of crust that today constitutes Africa. The orogenies that sutured these blocks together are collectively termed the Pan-African orogeny. This orogeny ended about 500 million years ago, about 40 million years after the start of the Paleozoic Era. By the start of the Paleozoic, however, both Laurentia and Baltica had broken away from the supercontinent (see Figure 12-20*B*). If they did so before the African crust was fully assembled, then a supercontinent never completely formed near the end of the Proterozoic Eon. At least, we know that nearly all of Earth's continental crust was clustered together at this time. This clustering of broad land areas, by keeping large portions of them far from oceans that would have warmed them in winter (p. 106), presumably contributed to the widespread Varangian glaciation.

The Birth of Paleozoic Continents

The Pan-African suturing, together with the separation of Laurentia and Baltica as distinct continents, left more than half of Earth's continental crust locked up in the supercontinent Gondwanaland. Siberia, another large landmass, had rifted away from Laurentia slightly earlier, before 600 million years ago (see Figure 12-20*B*). The chapters that follow will trace the history of these landmasses throughout the Paleozoic Era and show how they eventually were combined to form the supercontinent Pangea.

Chapter Summary

1 At least as early as 2 billion years ago, plate tectonic processes formed mountain belts with characteristics similar to those of Phanerozoic age.

2 Continental glaciers spread over parts of Canada and other regions more than 2 billion years ago.

3 Stromatolites first became abundant about 2 billion years ago. Their success at this time may have resulted from the growth of continental shelves.

4 Eukaryotes evolved through the ingestion of one kind of bacterial cell by another.

5 Multicellular eukaryotic algae occur in rocks about 2.1 billion years old.

6 Acritarchs are fossils of single-celled eukaryotic algae that lived as plankton. They experienced a high rate of evolutionary turnover after the widespread glaciation that occurred 600 million years ago.

7 Banded iron formations, which formed in the presence of oxygen, accumulated in great abundance in marine basins between about 3.0 billion and 1.9 billion years ago. They may all have formed when the concentration of oxygen in the atmosphere was lower than it is today.

8 The presence of red beds in sedimentary sequences younger than about 2 billion years, together with the rarity of the easily oxidized minerals pyrite and uraninite in such sequences, suggests that atmospheric oxygen had reached a moderate level early in Proterozoic time.

9 Episodes of continental glaciation occurred between about 850 million and 600 million years ago. In the last of these events, glaciers spread even to low latitudes.

10 The oldest unquestioned fossils of multicellular animals are younger than about 570 million years. They include trace fossils, imprints of soft-bodied fossils, and small tube- and vase-shaped calcareous skeletons.

11 Burial of organic carbon late in Proterozoic time may have caused oxygen to build up in the atmosphere, allowing animals to increase in size and diversity.

12 The modern North American craton was assembled from smaller Archean cratons early in Proterozoic time. By about a billion years ago, it had become part of a huge supercontinent that contained nearly all of Earth's landmasses.

13 Near the time of the Proterozoic-Phanerozoic transition, rifting events formed Gondwanaland, Laurentia, and Baltica.

Review Questions

1 Compare the basic features of the Wopmay orogenic belt with those of the Appalachian orogenic belt described in Chapter 9.

2 What kinds of geologic evidence suggest that glaciers were present on Earth more than 2 billion years ago?

3 List as many differences as you can between the Archean world and the world as it existed 1 billion years ago.

4 What arguments favor the idea that little atmospheric oxygen existed on Earth until slightly before 2 billion years ago?

5 What evidence is there that eukaryotic organisms existed 2 billion years ago?

6 How does the history of North America illustrate continental accretion?

7 How does the Appalachian orogenic belt of North America relate to the Grenville orogenic belt?

8 How were the crustal elements of Gondwanaland assembled?

9 Using a world map, locate the modern positions of the various landmasses that make up the supercontinents shown in Figure 12-20*B*.

10 What global physical and biological events occurred near the end of Proterozoic time?

11 Life underwent extraordinary changes in the course of Proterozoic time. Using the Visual Overview on page 316 and what you have learned in this chapter, describe these changes and explain how some of them may have been related to changes in the chemistry of the atmosphere.

Additional Reading

Bengtson, S. (ed.), *Early Life on Earth*, Columbia University Press, New York, 1994.

Dalziel, I. W. D., "Earth Before Pangaea," *Scientific American*, January 1995.

Knoll, A. H., "End of the Proterozoic Eon," *Scientific American*, October 1991.

Vidal, G., "The Oldest Eukaryotic Cells," *Scientific American*, February 1984.

Windley, B. F., *The Evolving Continents*, John Wiley & Sons, New York, 1984.

CHAPTER 13

The Early Paleozoic World

The establishment of the Cambrian and Ordovician systems in Britain more than a century ago illustrates how early geologists formally divided the stratigraphic record into useful intervals (p. 14). Early Paleozoic rocks of marine origin were later found to be well displayed on the broad surfaces of many large cratons, reflecting the fact that, with brief interruptions, sea level rose in the course of the Cambrian Period and stood high during most of Ordovician time. During the latter part of the Ordovician Period, mountains rose up along eastern North America when it collided with several small landmasses. Life in the oceans diversified rapidly at the end of Proterozoic time, in what has been termed the Cambrian explosion of life; brief episodes of mass extinction also occurred during the Cambrian Period. Then during the Ordovician Period a greater variety of animal life evolved than had ever existed before. The Ordovician ended, however, in widespread global extinctions that were linked to a brief episode of glaciation near the south pole.

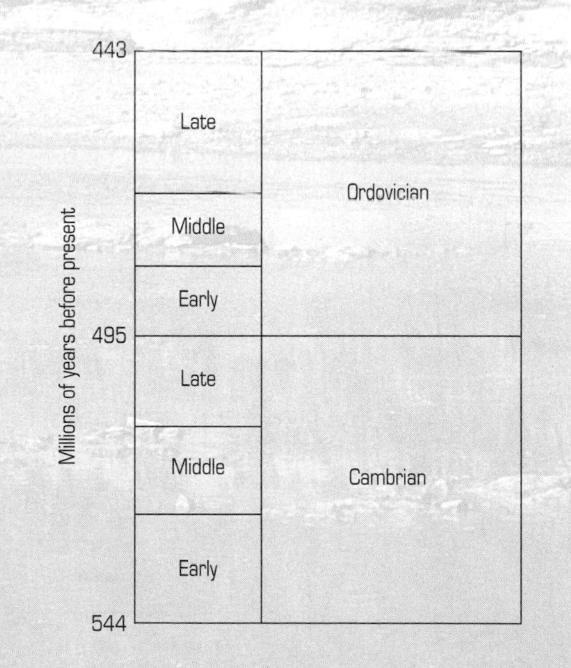

The boundary between Ordovician and Silurian systems on Anticosti Island in eastern Canada. The light gray mounds along the beach are Upper Ordovician reefs; darker Lower Silurian strata stand above them.

Visual Overview

Major Events of the Early Paleozoic

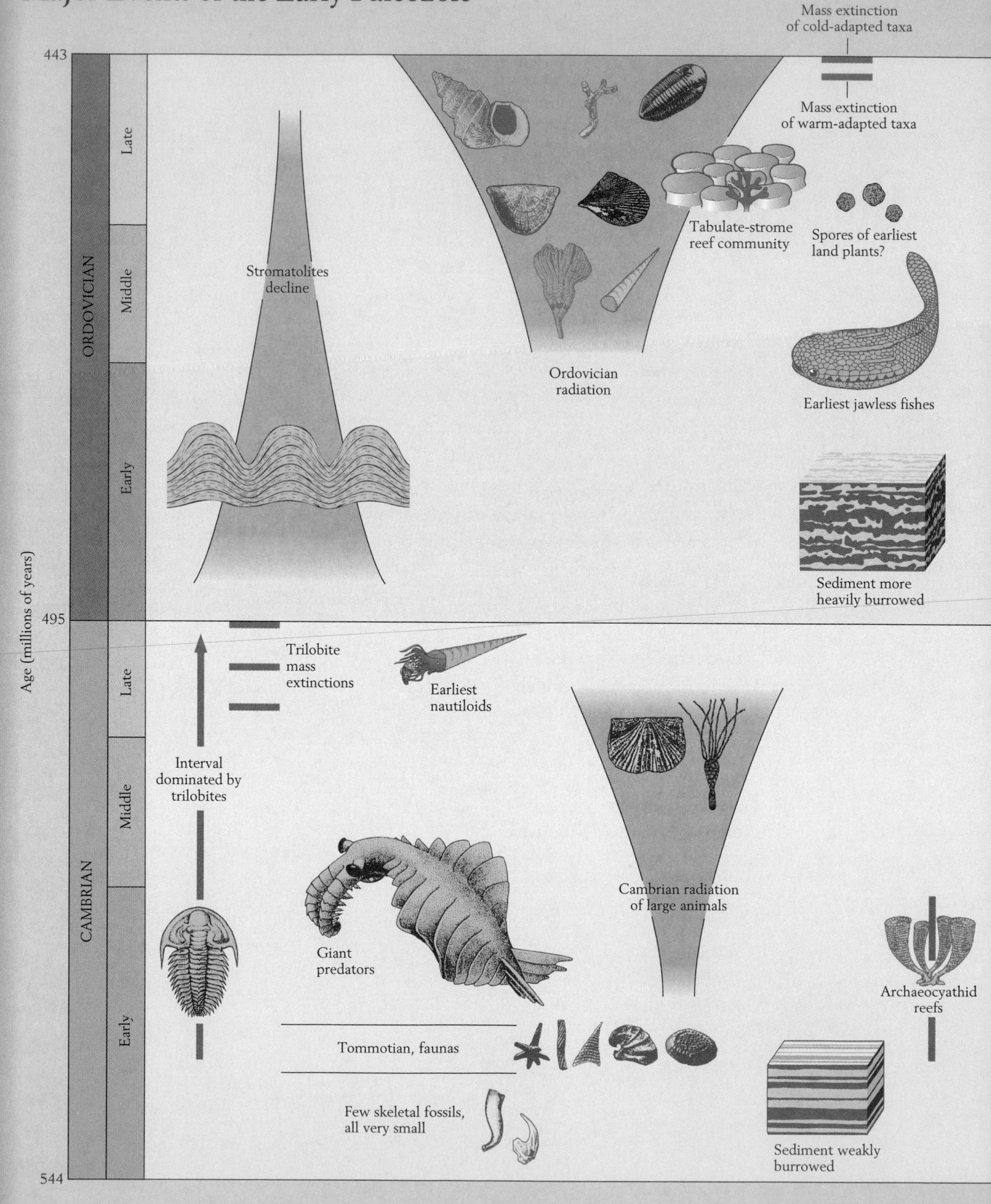

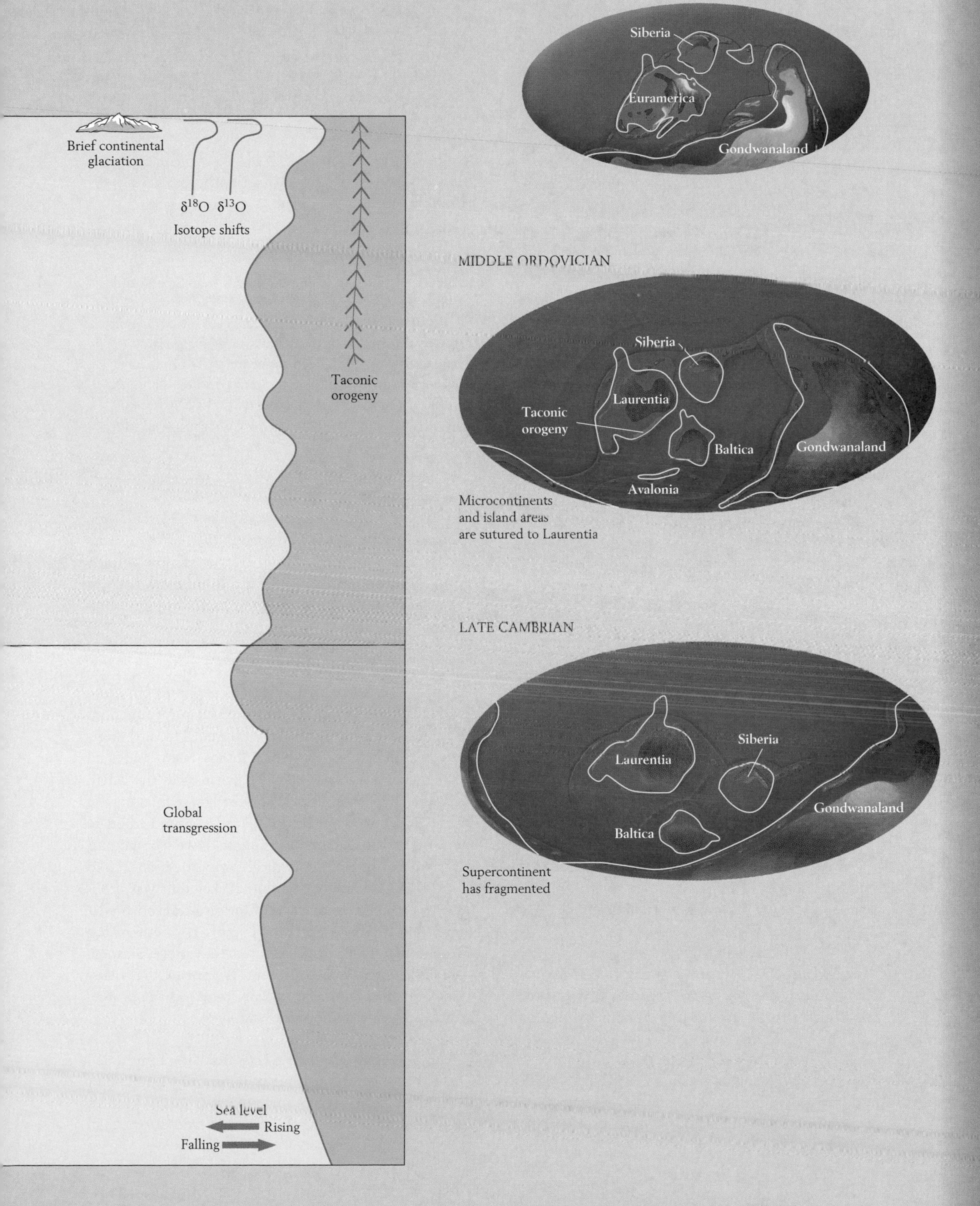
MIDDLE SILURIAN
Siberia
Euramerica
Gondwanaland
MIDDLE ORDOVICIAN
Siberia
Laurentia
Taconic orogeny
Baltica
Gondwanaland
Avalonia
Microcontinents and island areas are sutured to Laurentia
LATE CAMBRIAN
Laurentia
Siberia
Gondwanaland
Baltica
Supercontinent has fragmented
Brief continental glaciation
δ18O δ13O
Isotope shifts
Taconic orogeny
Global transgression
Sea level
Rising
Falling

The Cambrian Explosion of Life

The story of early Paleozoic biotas is essentially a story of life in the sea. It is presumed that certain simple kinds of protists and fungi had made their way into freshwater habitats by this time, but no fossil record of early Paleozoic freshwater life is known. The terrestrial realm, too, was barren of all but the simplest living things. Neither insects nor vertebrate animals occupied the land before middle Paleozoic time.

The organisms that we will discuss in this chapter were the first biotas on Earth to leave a conspicuous fossil record—one that is plainly visible even to a casual observer in many areas, because it includes an abundance of skeletons composed of durable minerals.

As we learned in Chapter 12, all but a few of the creatures that emerged during the initial radiation of animal life in the Late Proterozoic were soft-bodied. During the earliest segment of Cambrian time, the seas became populated by a different kind of fauna, consisting of a great variety of animals with skeletons.

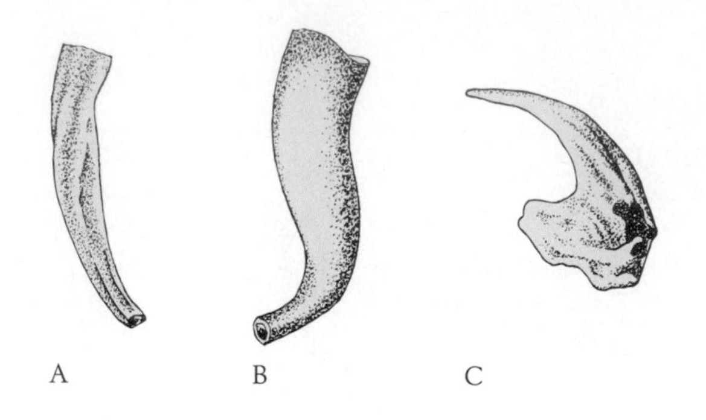

Figure 13-1 **Earliest Cambrian fossils from Siberia and China.** *Anabarites (A)* and *Cambrotubulus (B)* are tubular fossils of unknown relationships; they are ~3–5 millimeters long. C. A conodontomorph, the tooth of a predatory animal that may have been closely related to conodonts; this fossil is about 1 millimeter long. *(After A. Y. Rozanov and A. Y. Zhuravlev in J. H. Lipps and P. W. Signor [eds.], Origin and Early Evolution of the Metazoa, Plenum Press, New York, 1992.)*

Early Cambrian Life

The Early Cambrian was the first time in Earth's history when many modern phyla of animals appeared. This interval, which spanned about two-thirds of Cambrian time, was not marked by a uniform radiation of animal life, however. The Early Cambrian can be divided into three intervals, each characterized by a distinctive fauna.

The Lowermost Cambrian The lowermost portion of the Lower Cambrian record has yielded only simple skeletal fossils. Most of them are tube- or vase-shaped, but one group, which appeared near the end of the interval, consisted of teeth (Figure 13-1). This limited variety of forms represented only a slight advance over the modest variety of animals that existed in latest Proterozoic time.

The Tommotian Fauna A much richer fauna appears abruptly in the central portion of the Lower Cambrian record. This so-called *Tommotian fauna,* named for the Tommotian Stage of Early Cambrian time, was first discovered in Siberia. It includes a host of small skeletal elements that cannot be assigned to any living phylum and that show no relation to any group of fossils found in post-Cambrian rocks (Figure 13-2). The Tommotian fauna also contains the oldest known members of a few groups that survive to the present day—sponges, which are very simple animals (p. 76); monoplacophorans, which were ancestral to all present-day groups of mollusks (p. 80); and brachiopods (p. 81).

The origin of the many kinds of skeletons found in Tommotian faunas was a major evolutionary development. We know that skeletons support soft tissue and facilitate locomotion, but such adaptive functions cannot explain why so many different kinds of skeletons developed so suddenly. It has been suggested that a chemical change in the oceans was the cause, but this hypothesis does not explain why some of these skeletons were composed of calcium carbonate and others of calcium phosphate—two compounds with quite different chemical properties. The rapid evolution of a variety of external skeletons was probably a response to the evolution of advanced predators. The presence of several kinds of teeth in the Tommotian fauna indicates that a variety of small predators were indeed present. It appears that many kinds of small animals suddenly required the protection of a skeleton in order to survive.

Large Animals with Skeletons The Tommotian fauna occupied the seas for perhaps only 3 million or 4 million years. Then, for unknown reasons, most members of this fauna disappeared, and a variety of

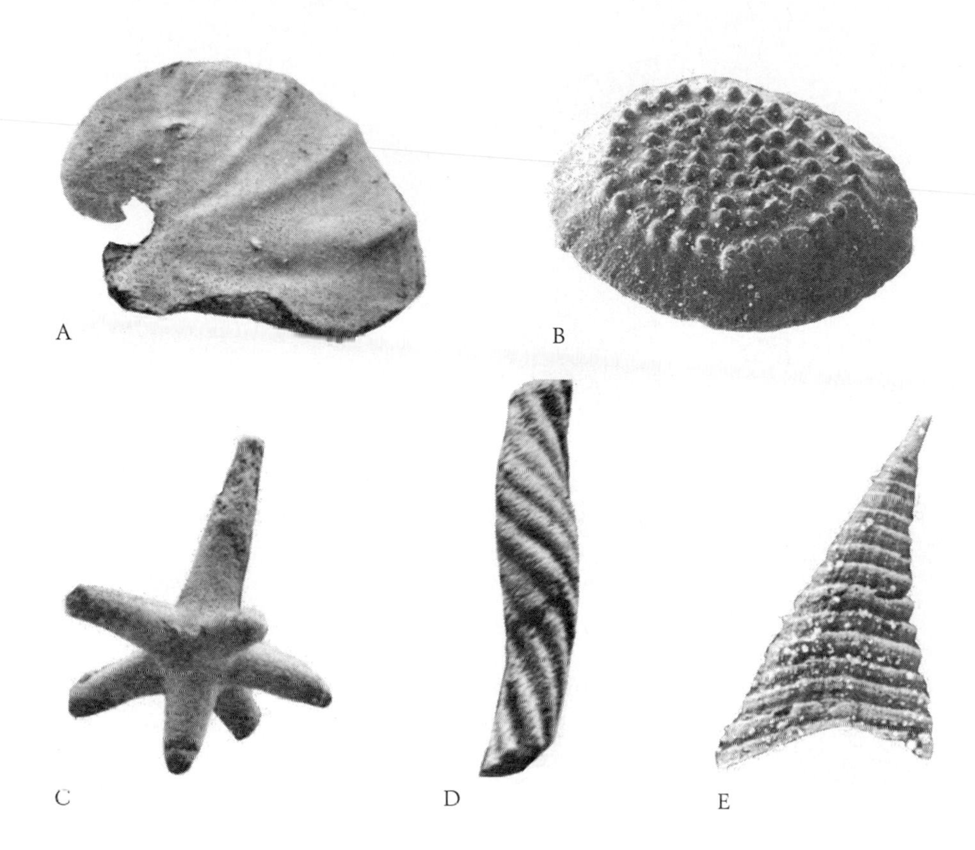

Figure 13-2 Fossils that represent the Tommotian fauna, the oldest diverse skeletonized fauna on Earth. All of the specimens shown here are small; none exceeds a few millimeters in length. *A.* A fossil that appears to represent primitive mollusks with coiled shells. *B* to *E.* None of these specimens can be assigned to a familiar group of animals. *(A, C, and D. From S. C. Matthews and V. V. Missarzhevsky, Jour. Geol. Soc. London 131: 289–304, 1975. B and E. S. Bengston.)*

new skeletonized animals emerged during the final few million years of Early Cambrian time. This new group of animals differed from the Tommotian fauna in two important ways: many of them were much larger, and most of them belonged to phyla that have survived to the present time.

Among the new animals were the trilobites, a group that survived to the end of the Paleozoic Era. Popular with fossil collectors, trilobites were arthropods whose segmented skeletons were heavily calcified (p. 78). Lacking mouth parts for chewing, most trilobites were deposit feeders: they extracted small particles of organic matter from sediment. Some lived on the surface of the sediment, and others were shallow burrowers. A few kinds of trilobites, however, were small, planktonic forms that must have been suspension feeders (Figure 13-3). Some tracks of bottom-dwelling trilobites show scratch marks that their legs produced when they were digging in sediment (Figure 13-4). These trace fossils closely resemble some tracks of Neoproterozoic age. This similarity suggests that trilobite-like arthropods existed millions of years before trilobites but failed to leave a skeletal fossil record because their skeletons were not mineralized with calcium carbonate.

Once present, trilobites diversified quickly, producing such a conspicuous fossil record throughout the remainder of the Cambrian that geologists commonly refer to the faunas that preceded them as the pretrilobite Cambrian faunas. The most abundant animal groups with skeletons that shared the late Early Cambrian seafloor with trilobites were monoplacophoran mollusks, inarticulate brachiopods (Figure 13-5), and a variety of echinoderms.

Perhaps the appearance of new kinds of large predators led to the demise of the soft-bodied Ediacara taxa and at the same time promoted the evolution of skeletons by animals larger than the minute Tommotion creatures. Large predators are, in fact, well represented in an unusually well preserved Early Cambrian fauna of soft-bodied animals recently discovered in China. Members of this fauna were rapidly buried by mud, which inhibited the decay of soft tissues. This fauna is only about 30 million years younger than the soft-bodied Ediacaran faunas of late Proterozoic age but contains a very different group of soft-bodied taxa. Apparently the globally distributed Ediacaran fauna largely disappeared at the end of Proterozoic time or very early in Cambrian time. Conspicuous in the remarkable Chinese fauna is a

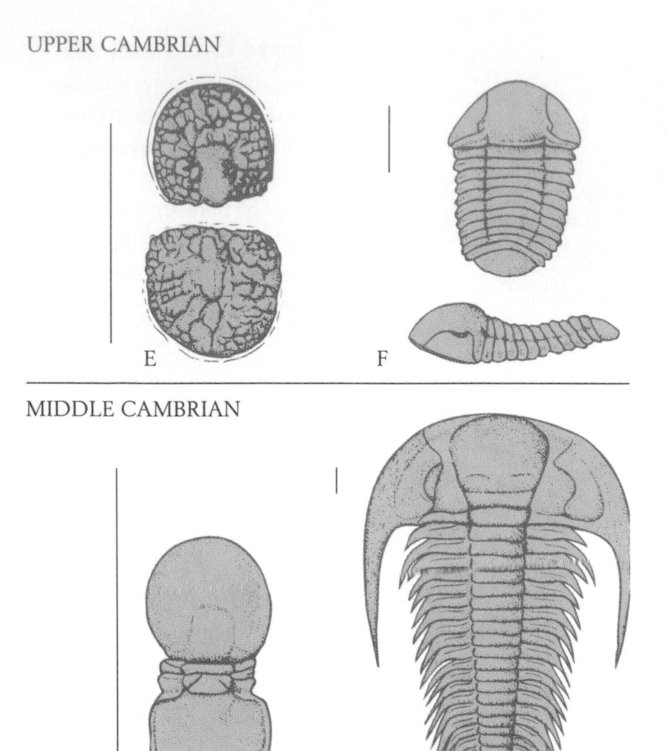

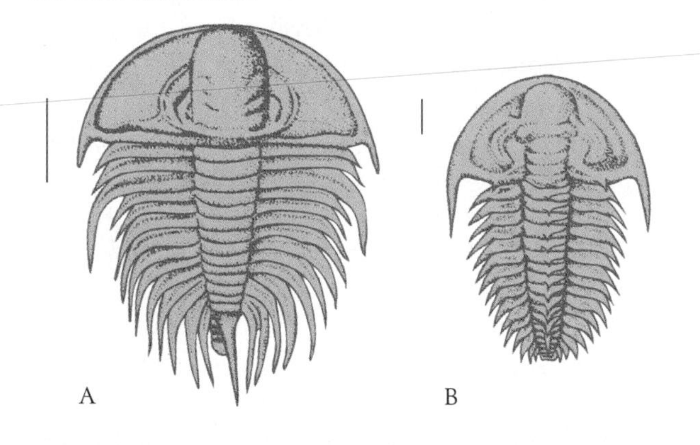

Figure 13-3 **Typical Cambrian trilobites.** Trilobites were arthropods (invertebrate animals with segmented bodies and jointed legs). The soft body and many legs were positioned beneath the flexible, jointed skeleton. Trilobites had mouth parts for chewing small pieces of food. Most species crawled over the seafloor, but some burrowed in sediment, and a few small species, including those labeled C and E, were planktonic. *A. Olenellus. B. Holmia. C. Lejopyge. D. Paradoxides. E. Glyptagnostus. F. Illaenurus.* (Scale bars represent 1 centimeter [3/8 inch].) *(After R. C. Moore [ed.], Treatise on Invertebrate Paleontology, pt. O, Geological Society of America and University of Kansas Press, Lawrence, 1959.)*

Figure 13-4 **A Cambrian trilobite track preserved as a cast on the underside of a bed of sediment.** Scratches made by the appendages of the trilobite are clearly visible.

group of predatory worms known as *priapulids.* This group survives in modern seas but includes only eight living species, all of which feed on other worms.

Arthropods are the most diverse animal group in the Chinese fauna. Some of the arthropod species had segmented bodies but most had a flexible carapace that was folded along a central axis so that each half protected one side of the animal's body. The most spectacular Early Cambrian arthropods were huge carnivores known as **anomalocarids.** These forms resembled arthropods, but only the appendages in front of the mouth were jointed. They were swimmers that propelled themselves with flaps positioned along

Figure 13-5 **Cambrian brachiopods (close to life size).** The articulate genus *Eorthis* (left) lived on the surface of the sediment; the inarticulate genus *Lingula,* a burrowing form (right), survives in modern seas.

their bodies and impaled prey on daggerlike spines along their frontal appendages. One of their species, which must have been a fearsome predator, reached a length of about 2 meters (~6 feet) (Figure 13-6).

Modes of Life of Late Early Cambrian Animals Although many animal taxa evolved during Early Cambrian time, they encompassed a narrower range of life habits than later Paleozoic faunas. Most Early Cambrian seafloor dwellers, for example, lived close to the surface of the sediment. Most of those that burrowed lived only slightly below the surface of the sediment, and those that attached to the seafloor did not stand high above it.

Most free-living animals of the Early Cambrian were **deposit feeders,** extracting organic matter from sediment; these forms included trilobites and other arthropods, as well as some echinoderms. Monoplacophorans, in contrast, grazed on algae growing on the seafloor. Suspension feeders, which collected organic matter and algal cells from the water, included brachiopods and attached echinoderms (p. 82). The most abundant of these echinoderms were the **eocrinoids,** whose name means "dawn crinoid." Recall that crinoids, or sea lilies, survive today (see Figure 3-33); eocrinoids were their evolutionary ancestors. Eocrinoids were obviously abundant in Cambrian seas, because their plates are the principal components of many Cambrian limestones. They must have formed simple communities because all of them attached to the seafloor by very short stalks (Figure 13-7).

Figure 13-7 **An eocrinoid from the Cambrian of Utah.** This animal, whose stem was only about 2.5 centimeters (1 inch) long, was buried suddenly by sediment that washed over it from the left, in this view, toppling the animal to the right.

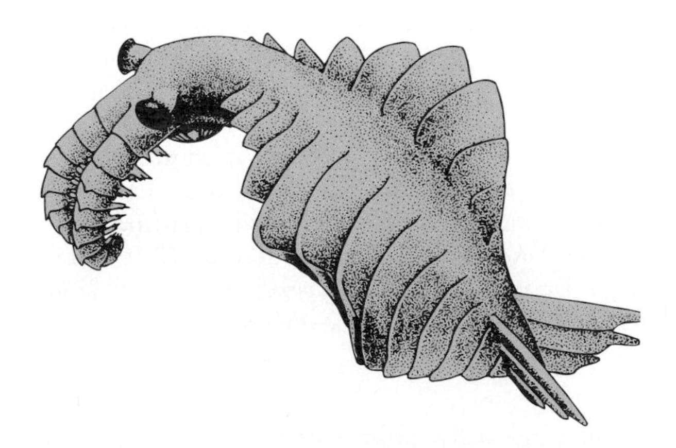

Figure 13-6 ***Anomalocaris,* a giant Early Cambrian anomalocarid.** This animal, which was about 2 meters (6 feet) long, swam by undulating lobes along its body and captured prey with appendages that bore sharp spines. *(After D. Collins, Jour. of Paleontology 70:280–293, 1996.)*

Crinoids that lived later in the Paleozoic Era, in contrast, formed complex communities of species that stood at various heights.

We have limited knowledge of life in the water above Early Cambrian seafloors, but we know that acritarchs persisted from the Proterozoic as important members of the phytoplankton and that zooplankton included not only agnostid trilobites (see Figure 13-3C and *E*) but also jellyfishes.

Stromatolites and Animals Stromatolites are less abundant in Cambrian rocks than in Proterozoic rocks and more restricted in their occurrence. Whereas Proterozoic stromatolites commonly grew in subtidal settings, Cambrian forms were largely confined to the intertidal zone. This Cambrian restriction probably resulted from the effects of grazing animals, including monoplacophoran mollusks. In modern seas, relatively few kinds of organisms can tolerate the unstable temperatures and salinities of the intertidal zone (p. 117), and presumably even fewer animals occupied this hostile environment during the Cambrian. Weak grazing pressure in the intertidal zone apparently allowed stromatolites to grow there during the Cambrian Period. Even so, Cambrian stromatolites are so riddled with holes made by animals that they display little layering and therefore have a very different appearance from Proterozoic stromatolites.

Reefs The oldest organic reefs with skeletal frameworks are low mounds that formed in Early Cambrian time, beginning in the Tommotian. The main builders of these reefs were **archaeocyathids,** which apparently were suspension feeders that pumped water through holes in their vase-shaped and bowl-shaped skeletons (Figure 13-8). Archaeocyathids were probably sponges, the simplest of which resemble them in general body plan (p. 76). Although archaeocyathids were the primary frame builders of Early Cambrian reefs, organisms of unknown taxonomic relationships actually contributed a larger volume of calcium carbonate to these reefs by encrusting archaeocyathid skeletons and binding them together.

At the end of Early Cambrian time, nearly all archaeocyathids became extinct. From then until mid-Ordovician time, all that remained were small, inconspicuous reeflike structures formed by the encrusting organisms that had previously lived with the archaeocyathids.

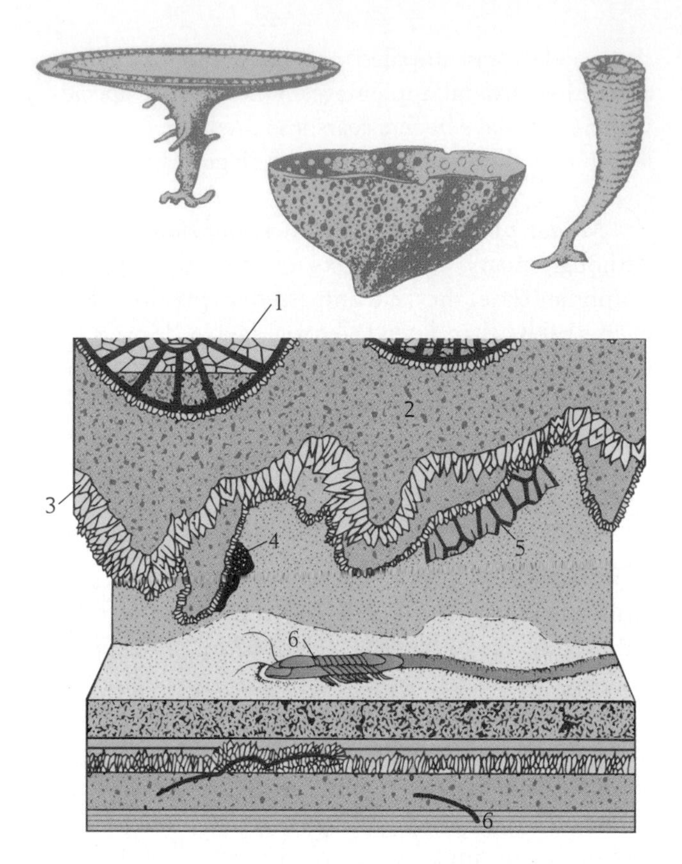

Figure 13-8 Archaeocyathid species (top) and one of the world's oldest organic reefs, located in the Lower Cambrian Series of Labrador (bottom). All three archaeocyathid species were vase- or bowl-shaped. It seems likely that these animals were similar to sponges in that they pumped water through their porous walls, but it is not certain whether they were closely related to any group of modern organisms. The diagram shows the composition of the reef, which was constructed by several kinds of organisms, the most important of which were archaeocyathids (1) and calcareous algae (2). Cavities were encrusted with crystals of the mineral calcite (3), which were precipitated from seawater, and by organisms whose biological relationships are uncertain (4, 5). Trilobites (6) left tracks on sediment flooring cavities in the reef and also left fossil remains within the sediment. *(Archaeocyathid drawings from D. R. Kobluck and N. P. James, Lethaia 12:193–218, 1979. Diagram after I. T. Zhuravleva, Akad. Nauk. USSR, Geol. Geofiz. Novosibirsk 2:42–46, 1960.)*

Evolutionary Experimentation Evolution during the Cambrian Period produced many groups of animals that included only a few genera and species; indeed, some are classified as discrete classes or even phyla. This phenomenon is seen in the early Paleozoic history of the phylum Echinodermata. Today this phylum includes only a few large groups, such as starfishes and sea urchins, but quite a number of bizarre echinoderm classes evolved during Cambrian and Ordovician time (Figure 13-9). None included many species or genera, and most survived only a short time. Many body plans were "tried out" in this manner, but only a few succeeded (see Box 13-1). This pattern is sometimes referred to as evolutionary "experimentation," with the understanding that what happens is not a planned event but rather a development produced blindly by nature. Of the many groups of invertebrates that appeared during Early Cambrian time, only a few—such as sponges, snails, brachiopods, and trilobites—flourished long afterward.

Later Cambrian Life

The Middle and Late Cambrian together spanned only about 15 million years. This interval was marked by the expansion of several preexisting animal groups, especially the trilobites. Remarkably, of some 140 families of trilobites recognized in Paleozoic rocks, more than 90 have been found in Cambrian strata. Echinoderms also continued to diversify, as did articulate brachiopods, which would become the most conspicuous fossils in younger Paleozoic rocks.

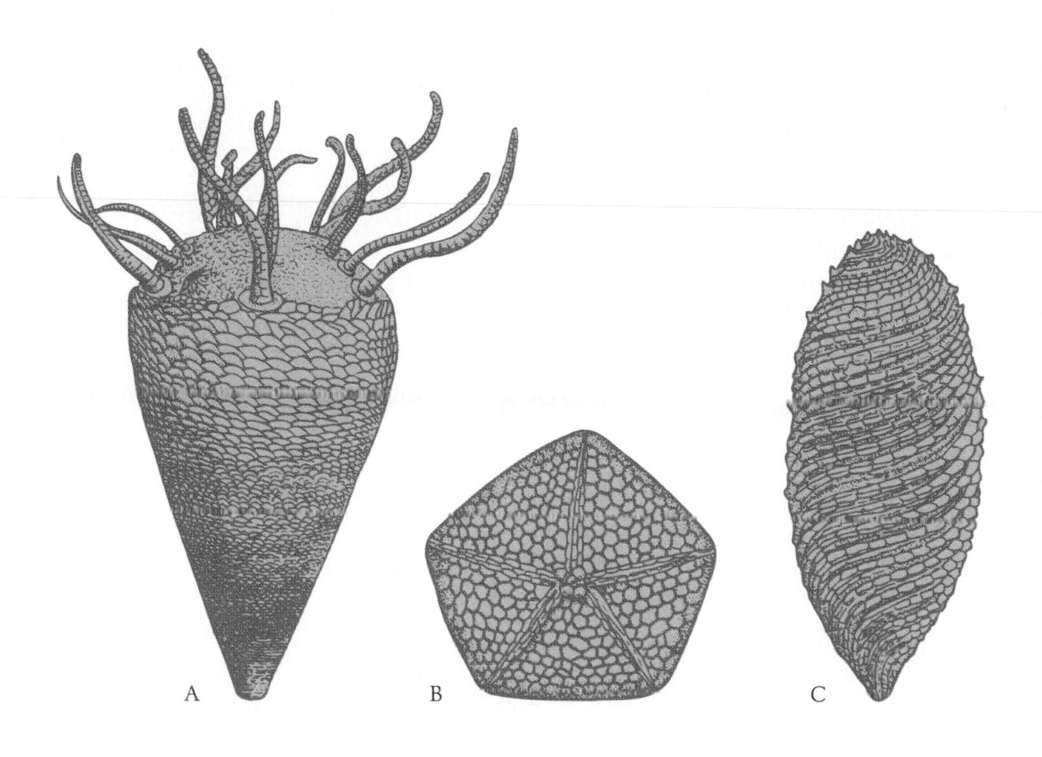

Figure 13-9 **Strange Cambrian echinoderms that show no close relation to any younger group.** *A* and *B.* Attached forms that apparently fed on organic matter suspended in the water. C. A flexible form that probably burrowed in sediment. Most living echinoderm groups, including starfishes and sea urchins, display a fivefold radial symmetry similar to that which can be seen in *B.* *(After R. C. Moore [ed.], Treatise on Invertebrate Paleontology, pt. U, Geological Society of America and University of Kansas Press, Lawrence, 1966.)*

The Earliest Vertebrates Conodonts also diversified in the course of the Cambrian Period. Their teeth are abundant in the fossil record, but the recent discovery of fossils of their soft bodies has shown them to have been small, swimming animals; the teeth themselves reveal that conodonts were the earliest known vertebrate animals (see Box 3-1). Similar small teeth in very early Cambrian faunas (see Figure 13-1C) may represent conodont ancestors.

Fishes also evolved during the Cambrian, but we know of their existence only through the preservation of isolated bony external plates (Figure 13-10). These early vertebrates were probably deposit feeders, like more fully preserved fishes of Silurian and Early Devonian age.

The Burgess Shale Fauna Strata in western North America have yielded a spectacular fauna of Middle Cambrian soft-bodied animals that invites comparison with the Early Cambrian Chinese fauna described earlier. The largest group of species in the North American soft-bodied fauna comes from the Rocky Mountains of British Columbia (Figure 13-11; see Box-13-1). Later we will examine the environment in which the Burgess Shale formed, but for now we can simply note that it accumulated in a deep-water setting where soft-bodied animals were buried in the absence of oxygen and bacterial decay. Arthropods are the most abundant of these fossils, and some of them resemble certain of the Chinese taxa. Also present were anomalocarids smaller than the one depicted in Figure 13-6. In addition, both the Chinese and American faunas include onychophorans. Elongate animals with jointed legs, onychophorans are generally intermediate in form between segmented worms and arthropods. Today members of this group live as predators on moist forest floors, having somehow invaded the land (see Figure 3-28). Priapulid

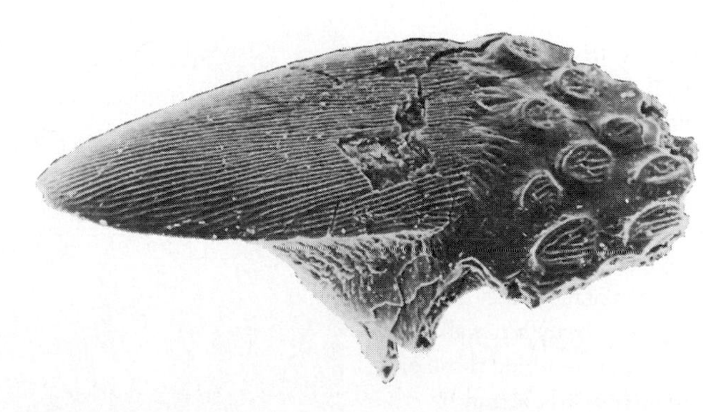

Figure 13-10 **A tiny, bony plate of a small jawless fish from the Cambrian System of Wyoming.**

For the Record 13-1

What Does It Take to Survive?

The diversification of animal life in Late Proterozoic and Cambrian time was a veritable evolutionary explosion. When humans reflect on this episode, they tend to focus on the many successful groups of animals it produced—the groups that have survived—and to ignore the groups of very early animals that cannot be assigned to living phyla. Many of these animals, such as the strange-looking one shown below on the left, died out soon after they appeared.

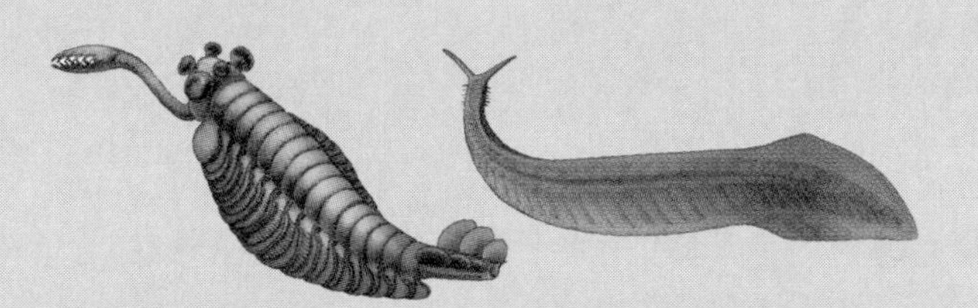

Opabina (left) had five eyes and food-gathering pincers at the end of a strange nozzle. *Pikaia* (right) was a swimming chordate. *(Drawing by Marianne Collins; reprinted by permission from S. J. Gould, Wonderful Life, W. W. Norton & Company, New York, 1989.)*

When we assemble the full cast of characters found in these early faunas, we can hardly avoid wondering why some groups survived long after Cambrian time whereas others soon vanished. Were the victims of early extinction biologically inferior to the groups that survived? Were they less effective in competing with other taxa or in avoiding predators? Or were they simply the unlucky victims of catastrophic extinctions that struck down species without regard to their ecological abilities in a healthy environment?

Such accidental death might have come in the mass extinctions that decimated the trilobites during the latter part of Cambrian time, for example, or the global crisis that brought the Ordovician Period to a close.

Not only did the very ancient groups of animals that have survived to the present escape these catastrophes, but for hundreds of millions of years thereafter they coped successfully with predators and with taxa that vied with them for food and other resources. Some of the taxa that

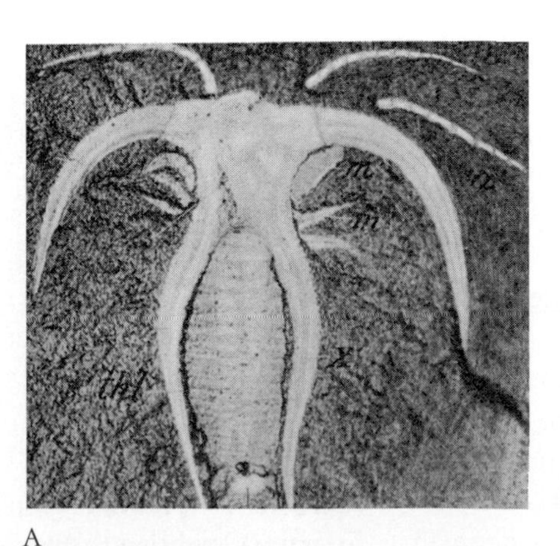

A

B

C

Figure 13-11 **Animals without durable shells from the Burgess Shale of British Columbia.** *A.* An arthropod related to the trilobites. *B.* A polychaete worm. C. An onycophoran, which had a wormlike body but walking legs that resembled those of the arthropods. *A* and *B* magnified 3 times; C magnified 4 times.

failed to survive appear to have been ecologically deficient. Some, for example, were weakly armored against predators and probably could no longer flourish after advanced predators evolved early in the Paleozoic Era. It is clear that external skeletons, which evolved in many taxa in earliest Cambrian time, were particularly effective in thwarting predators. Some of these skeletons, however, were weak and flexible. A multitude of fossil plates from the skeletons of eocrinoids shows that these stalked ancestors of modern sea lilies were very abundant, yet their fossil skeletons are rarely intact; they normally fell to pieces soon after the animals died. Some of their relatives, however—cystoids, blastoids, and crinoids—had robust skeletons that abound intact in the fossil record, and these animals flourished long after Cambrian time. Although we have an incomplete picture of Paleozoic predators, we know that one important group, the nautiloids, evolved at the end of Cambrian time. Rapid adaptive radiation of these beaked predators produced numerous families early in Ordovician time, and the impact on other marine animals, including some of the kinds found in the Burgess Shale, must have been devastating.

All of the early groups of animals were at risk soon after they evolved simply because they included few species. Diversification increases the chances of survival, and many groups of animals that failed to survive beyond Cambrian time are known from only a few fossil species.

Among the many strange animals of the Burgess Shale fauna was a rather ordinary-looking creature that belonged to a group with a great future. This was the genus *Pikaia*, which had a notochord. The notochord evolved into the vertebrate backbone, and it identifies the animal as a chordate—a member of the phylum to which humans and all other vertebrates belong. Probably *Pikaia* itself was not our ancestor, but one of its Cambrian relatives certainly was—some form undiscovered in the fossil record. It is sobering to observe that if accidental extinction had swept away the early chordates before some of them evolved into fishes, no four-legged vertebrate, to say nothing of a human being, would ever have walked the land.

worms occur in the Burgess Shale fauna, along with several types of segmented worms. An overall comparison of the Chinese fauna with the younger American fauna indicates that evolutionary changes between Early and Middle Cambrian time were relatively minor for soft-bodied invertebrate animals.

Ordovician Life

The Ordovician Period was marked by a great evolutionary radiation of life in the seas. Many of the new animal groups remained successful for most of the Paleozoic Era. This radiation did not get under way until Middle Ordovician time, however. The Early Ordovician was a time of only modest evolutionary expansion.

The Early Ordovician

Later in this chapter, we will learn that trilobites suffered a major extinction at the end of Cambrian time. Trilobites recovered from this crisis, however, and remained the most abundant members of many marine communities throughout Early Ordovician time. Other groups, such as brachiopods and snails, that had originated on Cambrian seafloors became increasingly well represented early in Early Ordovician time.

In the waters above Ordovician seafloors two groups expanded dramatically from modest evolutionary beginnings in the Cambrian: graptolites and nautiloids. Most graptolite species floated as zooplankton and were preserved in offshore settings, where their fragile colonies settled in muddy sediment (see Figure 5-23) Widespread occurrence in the resulting black shales has made them useful for dating these rocks (see Box 6-1). Nautiloids were swimmers that probably rested on the seafloor at times and also fed on bottom-dwelling animals. Like the living chambered nautilus (p. 56) and other cephalopod mollusks, they pursued their prey by means of jet propulsion and caught them with tentacles. Cambrian nautiloids are known only from China, where they arose shortly before the beginning of the Ordovician Period. Nautiloids diversified rapidly early in Ordovician time, however, and spread throughout the world.

The Great Ordovician Radiation of Life

Marine life underwent a spectacular adaptive radiation during Middle and Late Ordovician time. Many new forms of life came to live in and on the sediment of the seafloor and in the overlying waters (Figures 13-12 and 13-13).

Life in Sediment It is not easy for animals to evolve adaptations for living in sediment. The primary difficulty is in maintaining access to oxygen-bearing water. Some burrowers manage by pumping water down from above the surface of the sediment. Trace fossils reveal that an increasing variety of worms came to live in this way near the end of Proterozoic time (p. 326). Fabrics of sedimentary rocks reveal that animals burrowed through subtidal seafloors with increasing intensity between Cambrian and Late Ordovician time (Figure 13-14). Presumably this increase reflected a continued diversification of worms and other soft-bodied burrowers. Shelled burrowers diversified as well. As the Ordovician Period progressed, burrowing bivalve mollusks attained the position that they hold today: they became the most successful group of burrowing suspension feeders with shells. Some new kinds of trilobites were also smoothly contoured shallow burrowers (see Figure 13-12).

Life on the Seafloor A large variety of trilobites continued to scurry over the seafloor, and in the latter part of Ordovician time they were joined by an increasing array of other animals (see Figure 13-12C). Jawless fishes continued to grub for food in the sediment. Snails grazed on algae that grew on the seafloor, and a few types were stationary forms that strained food from the water.

Articulate brachiopods diversified markedly; beginning in the Middle Ordovician, they became the most conspicuous Paleozoic fossils. Some attached by a fleshy stalk, but others rested freely on the sediment. Crinoids also expanded, forming complex communities in which species stood at various levels above the seafloor.

Rugose corals, known as horn corals because of their typical shape, also became well represented on

Figure 13-12 Ordovician invertebrate fossils. *A.* A straight-shelled nautiloid about 15 centimeters (6 inches) long. *B.* A spiny epifaunal trilobite. C. A smooth-shelled burrowing trilobite. *D.* A snail (gastropod). *E* and *F.* Two kinds of articulate brachiopods. G. An epifaunal bivalve mollusk. *H.* A branched bryozoan colony. *I.* A rugose coral. *J.* A stromatoporoid colony. *K.* A tabulate coral colony.

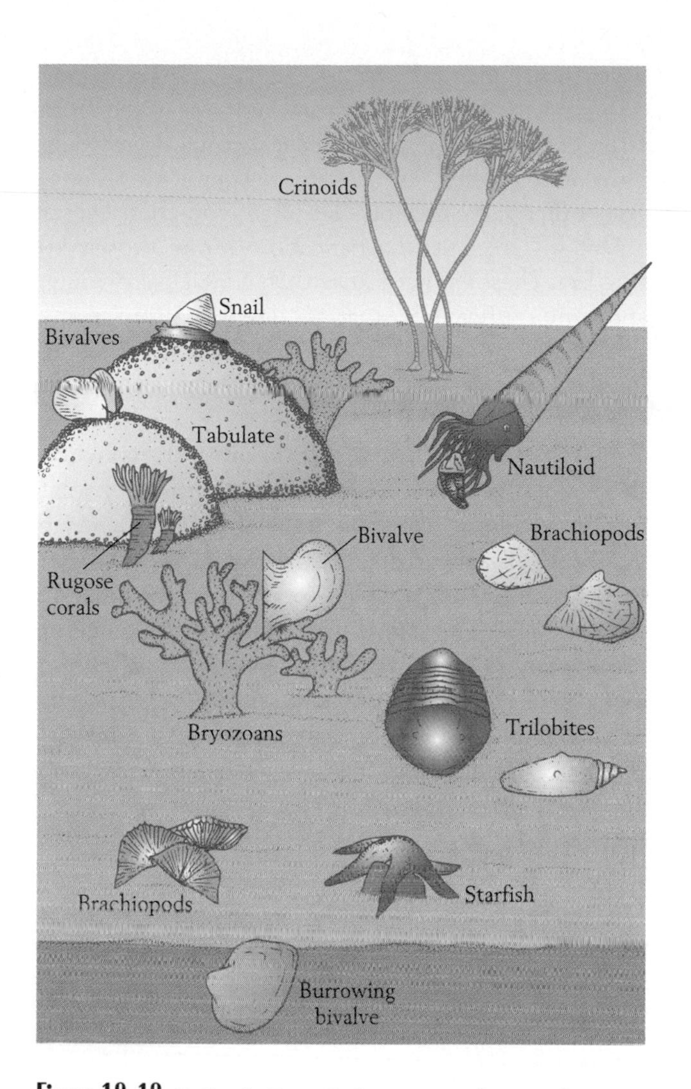

Figure 13-13 Life of a Late Ordovician seafloor in the area of Cincinnati, Ohio. Fossils of many of the groups of animals represented here can be seen in Figure 13-12. Note that at this early stage of Phanerozoic evolution, relatively few animals lived within the sediment. At the left a snail crawls over a large tabulate colony, and two bivalve mollusks are attached to another tabulate colony by threads that give the bivalves stability. Another bivalve is similarly attached to the branch of a bryozoan colony. Two solitary rugose corals, lodged alongside colonies, have their tentacles outstretched for food. Stalked crinoids are waving about at the top of the picture, feeding on suspended matter with their arms. To their right, a large nautiloid prepares to eat a trilobite that it has trapped in its tentacles; below the nautiloid's eye is a spoutlike siphon that the animal uses to expel water for jet propulsion. Two kinds of suspension-feeding brachiopods live on the seafloor. In the right foreground are trilobites of a type that left trace fossils, indicating a burrowing mode of life. In the central foreground a starfish prepares to devour a bivalve by prying apart the shell halves with its sucker-covered arms; then, by extruding its stomach, the starfish can digest the bivalve within its opened shell.

the seafloor during the Ordovician Period. Some formed large colonies by budding, and certain colonial forms contributed to the growth of organic reefs. In fact, the most dramatic ecological development on Ordovician seafloors was the ascendancy of a new reef community, which produced large tropical reefs throughout the world. The most important reef formers were an extinct group of corals called **tabulate corals** and a group of sponges called **stromatoporoids;** both of these groups secreted massive skeletons of calcite. There had been no major reef-building community since the archaeocyathids all but disappeared at the end of Early Cambrian time. Many Ordovician reefs were larger than those built by archaeocyathids, and the new community continued to thrive well beyond the Ordovician Period.

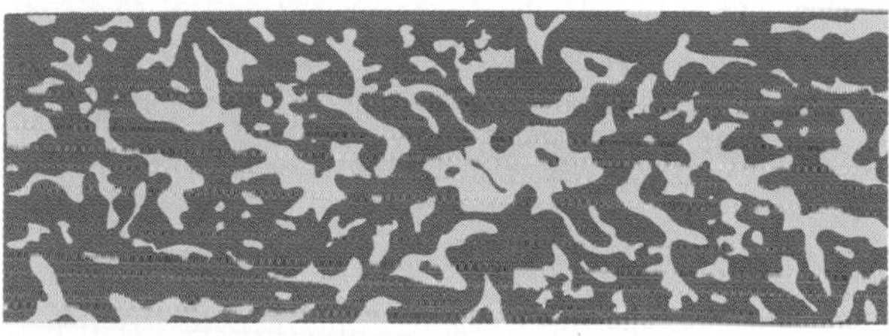

Upper Ordovician

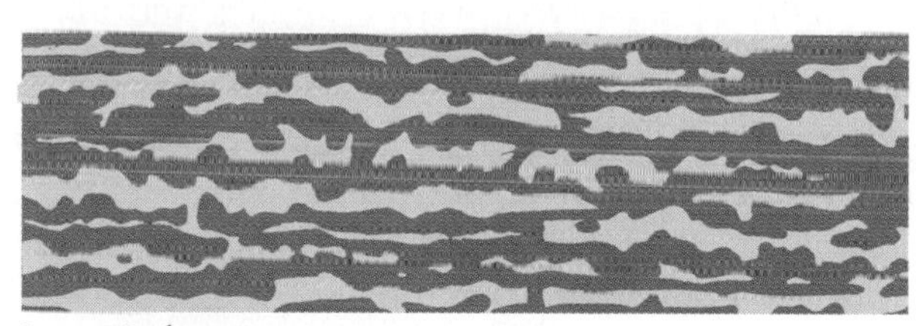

Lower Ordovician

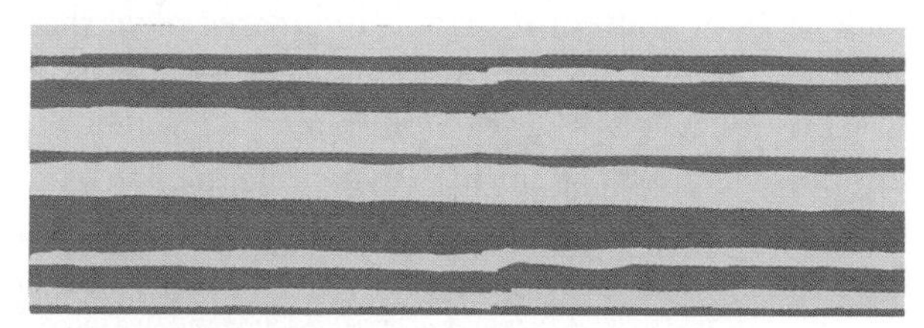

Pretrilobite Lower Cambrian

Figure 13-14 Fabrics produced by burrowing in lower Paleozoic sediments of the Great Basin of the western United States. Each figure illustrates the degree of burrowing most commonly observed in shallow-water limestones for a particular interval. The intensity of burrowing increased markedly between earliest Cambrian and Early Ordovician time and between Early Ordovician and Late Ordovician time. *(After M. L. Droser and D. J. Bottjer, Geology 16:233–236, 1988; 17:850–852, 1989.)*

Predators The evolutionary expansion of nautiloids was the most conspicuous advance in predation during the Ordovician Period. Some straight-shelled nautiloids reached lengths of about 3 meters (10 feet). On the seafloor starfishes appeared and assumed their role of feeding on other animals (see Figure 13-13).

Animal Life and the Decline of Stromatolites

Of all the Phanerozoic periods, only the Cambrian and Ordovician were characterized by abundant stromatolites. This abundance was carried over from late Precambrian time, but by the end of the Ordovician interval large stromatolites were rare. As we have seen, even in Cambrian time grazing animals had largely restricted stromatolites to intertidal areas.

The few areas where stromatolites grow in the modern world offer some clues as to what happened to stromatolites during Ordovician time. The types of cyanobacteria that form stromatolites occur widely in modern seas, but cyanobacteria prosper well enough to form conspicuous stromatolites only in environments that are hostile to nearly all animals: the fringes of land along the ocean that seas only occasionally flood, when tides are very high; subtidal channels in which very strong water movements exclude animals; and hypersaline lagoons (see Figure 5-32). In more normal marine environments, animals burrow through algal mats and also eat them; their ravages prevent the mats from forming stromatolites. Experiments have shown that when animals are excluded from small areas of seafloor in tropical climates, algal mats flourish, just as they did long ago. It seems evident that the great adaptive radiation of Ordovician life produced a variety of animals that prevented stromatolites from developing except in habitats where conditions excluded these animals. The severe restriction of stromatolites early in the Paleozoic Era permanently altered the nature of shallow seafloors.

Extinction and Diversity at Sea

Slightly more than 400 families of marine invertebrates have been recognized in Upper Ordovician rocks. As Figure 13-15 illustrates, the number of known Late Ordovician genera—about 1300—was close to the Paleozoic maximum. Some scientists have concluded that the leveling off of taxonomic diversity late in the Ordovician Period resulted from saturation of marine ecosystems: ecological crowding prevented further diversification. Figure 13-15 suggests otherwise, showing that large extinction events repeatedly set back diversification during the Paleozoic Era. These events, rather than ecologic crowding, apparently prevented marine diversity from rising appreciably above the Late Ordovician level before the end of Paleozoic time. Fewer large extinctions occurred during the Mesozoic and Cenozoic eras. The primary reason for the decline in rates of extinction as the Phanerozoic Eon progressed was that taxa that were particularly vulnerable to environmental change tended to disappear in major extinctions. Taxa that were more resilient survived and gave rise to similarly resilient taxa. Thus, through a kind of weeding-out process, animal life became increasingly resistant to extinction. The rarity of major extinction events during the Mesozoic and Cenozoic eras allowed animal life to become markedly more diverse.

Did Plants Invade the Land?

The fossil record shows that small multicellular plants were well established in moist terrestrial envi-

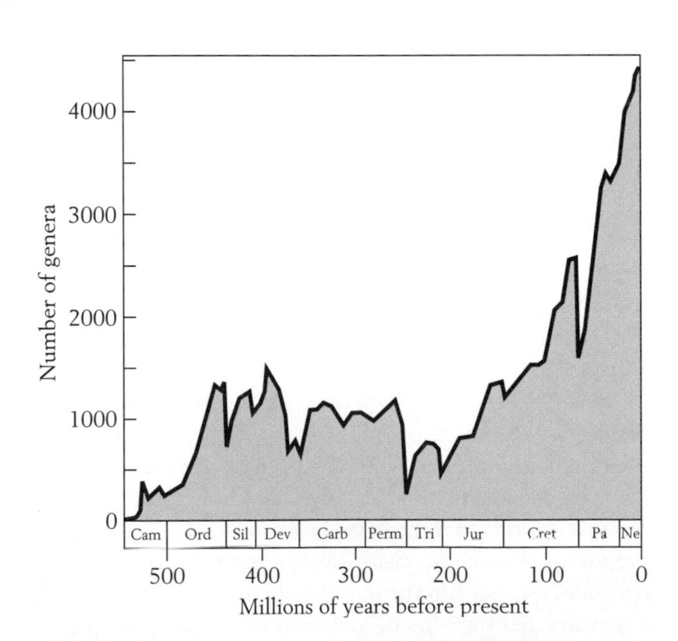

Figure 13-15 **The number of genera of marine animals with skeletons that have existed at various times during the Phanerozoic Eon.** *(J. J. Sepkoski, Geotimes, March 1994.)*

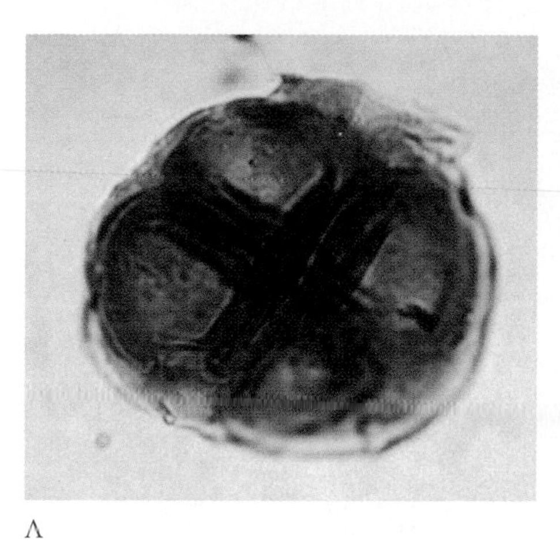

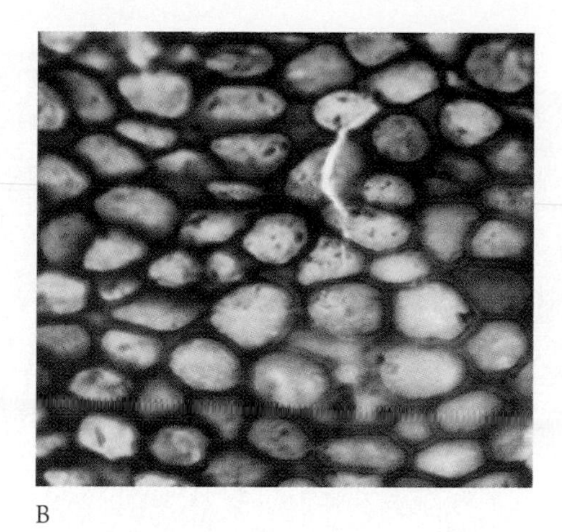

Figure 13-16 Late Ordovician fossils that may represent plants that lived on the land. *A.* Spores that resemble those of modern land plants. *B.* A sheet of fossil cells that resemble those covering the surfaces of some land plants. *(From J. Gray et al., Geology 10:197–201, 1982.)*

ronments during the Silurian Period. Although similar plants probably occupied Ordovician landscapes, the evidence is not yet conclusive. It consists of fossilized sheets of cells similar to those that cover the surfaces of modern land plants as well as structures that resemble the spores released by primitive (nonseed) land plants of the modern world (Figure 13-16). If land plants did evolve during the Ordovician Period, they were probably restricted to moist habitats, as mosses are in modern times.

The Paleogeography of the Cambrian World

As we saw in Chapter 12, rock magnetism and other geologic evidence strongly suggest that most cratons were clustered together near the end of Precambrian time and may have formed one giant supercontinent. The arrangement of continents late in Cambrian time was strikingly different. By that time, broad continental surfaces were positioned at low latitudes (Figure 13-17) and accumulated shallow-water limestones.

The Cambrian Period was notable for the progressive flooding of continents. The stage for this trend was set near the end of Precambrian time, when most of Earth's cratons stood largely exposed above sea level. As a result, only scattered local areas on modern continents yield a continuous record of shallow-water deposition across the Precambrian-Cambrian boundary. In Chapter 12 we discussed one of these areas: the Rocky Mountain region of southern Canada (see Figure 12-21).

As the Cambrian Period progressed, many parts of Gondwanaland remained above sea level, partly as a result of regional uplifts caused by orogenic activity between about 800 million and 500 million years ago. Other cratons, however, show evidence of continued encroachment of Cambrian seas until little of their total area remained exposed late in Cambrian time (see Figure 13-17). This flooding represented one of the largest and most persistent sea-level rises of the entire Phanerozoic Eon. It was interrupted in North America only by a modest regression in Middle Cambrian and another during Late Cambrian time. As the seas began to encroach on broadly exposed continents slightly before the beginning of the Cambrian, siliciclastic sediments were eroded from the continents and accumulated around the continental margin. Figure 13-18 shows this pattern in North America. When the seas encroached farther over most continents during Middle and Late Cambrian times, a characteristic sedimentary pattern emerged. To understand the nature of this pattern, let us examine the geography of Laurentia, the landmass that included North America, Greenland, and Scotland (shown in Figure 13-17).

At all times during the Middle and Late Cambrian, some part of central Laurentia stood above sea level (see Figure 13-18). Around the margin of the land, belts of marine deposition were arranged in concentric patterns. Siliciclastic sediments derived from the craton were deposited in the innermost belt. This belt was essentially the same as the marginal siliciclastic belt that surrounded the continent during earliest Cambrian time, but it had shifted inland along with the shoreline. Seaward of this belt were broad carbonate platforms that were sometimes fringed by

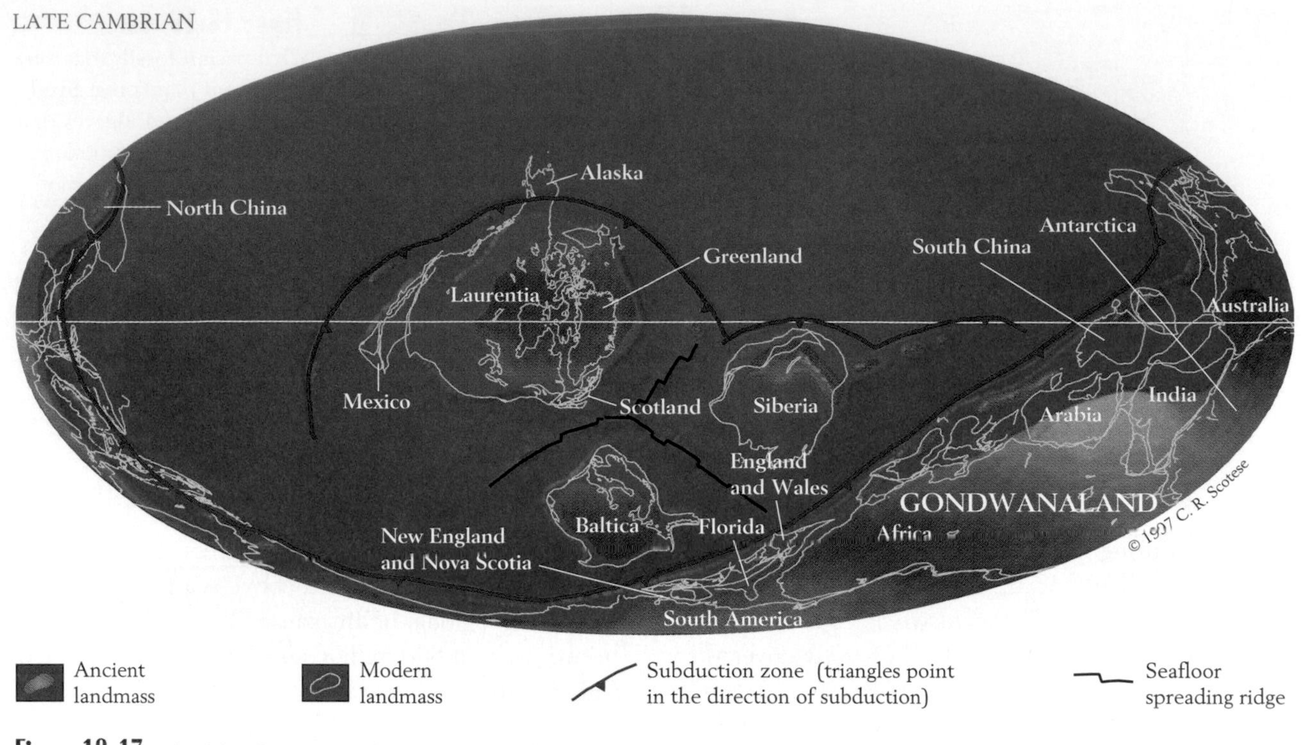

Figure 13-17 **World paleogeography in Late Cambrian time.** Laurentia was positioned at the equator and inundated by shallow seas. *(Adapted from paleogeographic map by C. R. Scotese, PALEOMAP Project, University of Texas at Arlington, 1997.)*

reefs. As the Cambrian progressed and seas shifted inland, limestones came to lie on top of nearshore sands. We viewed this pattern in the Grand Canyon region as an example of the concept of facies (see Figure 6-6). Muds and breccias derived from the platform accumulated in deep water near the base of its steep slope, and the nearby subduction zone contributed volcanic sediments to the deep-water belt from offshore.

Trilobites, the dominant skeletonized animals of Middle and Late Cambrian oceans, were distributed around continents in a pattern corresponding to the arrangement of the sedimentary belts. Certain groups of these trilobites, including small forms that lived as plankton (see Figure 13-3C and *E*), are found primarily in deep-water deposits.

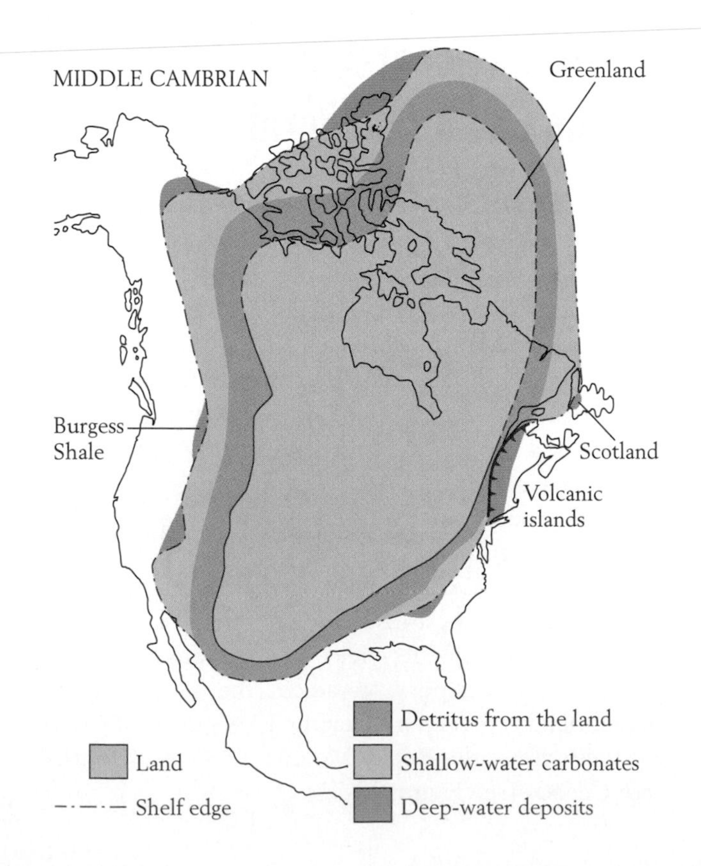

Figure 13-18 **Concentric pattern of sediment deposition around the margin of Laurentia during Middle Cambrian time.** Note the location of the Burgess Shale, renowned for its fauna of soft-bodied invertebrates, at the base of the Middle Cambrian continental shelf in western Canada. *(After A. R. Palmer, American Scientist 62:216–224, 1974.)*

Episodic Mass Extinctions of Trilobites

Trilobite species that inhabited tropical seas—including small planktonic forms—were the ones that suffered in the repeated mass extinctions of the Cambrian Period. Each mass extinction was followed by an adaptive radiation that restored the diversity of shallow-water trilobites to a high level (Figure 13-19). These events are well documented in North America and thus far have been recorded elsewhere only in Australia.

Each adaptive radiation of Cambrian trilobites proceeded for several million years, but each extinction was quite sudden. The fossil record reveals that each extinction took place during the deposition of a layer of sediment just a few centimeters thick and thus must have lasted no more than a few thousand years. The transition from one adaptive radiation to another followed a characteristic pattern that is illustrated in Figure 13-20. Above the thin layer that records the extinction, in beds just a meter or so thick, a variety of new trilobite genera join species that survived the mass extinction. These beds accumulated during a brief time of biotic adjustment in which opportunistic species flourished for a short time and then dwindled in the face of competition from new, more successful forms. The beds above contain a different fauna that included only about half as many species. From this fauna there issued a new adaptive radiation that lasted several million years—until another mass extinction started the cycle again.

What led to the periodic mass extinction of trilobites? Elimination of shallow-water habitats by lowering of sea level can be ruled out, because the extinctions did not all coincide with widespread regression of seas. In fact, a sizable regression occurred between two of the mass extinctions in Late Cambrian time. Because the adaptive radiation that preceded each mass extinction was associated with the deposition of tropical limestones, it has been suggested that a sudden, temporary cooling of the seas was the agent of the trilobites' periodic massive death. This idea gains support from evidence that the adaptive radiation that followed each mass extinction issued from a group of trilobites that lived offshore in cool, deep waters marginal to the continent. These offshore trilobites did not suffer in the mass extinction. Such evidence is only circumstantial, however, and the case for temperature as an agent of mass extinction remains unproved.

Ordovician Paleogeography

Early in Ordovician time, Baltica had been centered midway between the equator and the south pole (Figure 13-21), but then it moved northward. By the end of the Ordovician it had moved into the tropics

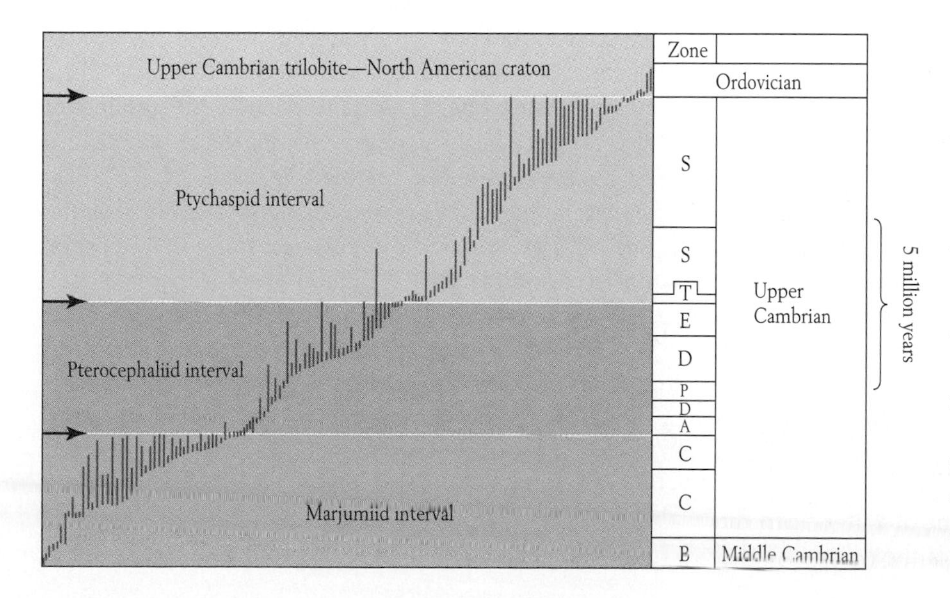

Figure 13-19 Repeated adaptive radiation and mass extinction of Cambrian trilobites. The vertical bars indicate the stratigraphic ranges of prominent species in Middle and Upper Cambrian deposits of North America. The ranges form clusters and thus delineate three successive adaptive radiations, each of which was terminated by a mass extinction (see arrows at left). A stratigraphic interval representing approximately 5 million years is shown on the right. Note that many trilobite species survived less than 1 million years. *(After J. H. Stitt, Oklahoma Geol. Surv. Bull. 124:1–79, 1977.)*

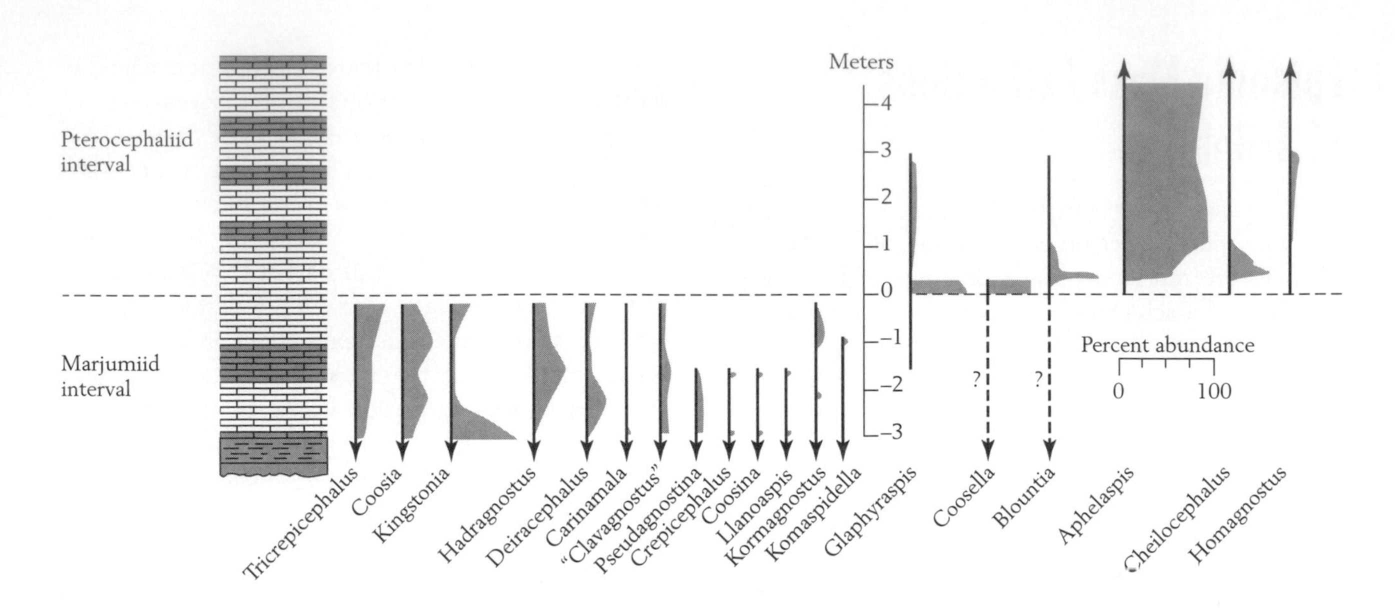

Figure 13-20 The pattern of trilobite mass extinction recorded in limestones at the top of the Cambrian Bonanza King Formation of Nevada. The gray band above each genus shows how the relative abundance of the genus changed through time. Many genera of the Marjumiid interval (see Figure 13-19) disappear at the 0 meter level. Just above this level, only a few forms are found and, in the absence of other genera, these forms were very abundant. Arrows indicate that species are also found below or above the occurrences displayed in the diagram. *(After A. R. Palmer, Alcheringa 3:33–41, 1979.)*

and tropical limestones accumulated in the area of the present Baltic Sea. Some of these limestones, including oolite deposits, resembled those that are forming today in the Bahama Banks (p. 143). In contrast, Gondwanaland shifted so that its southern margin encroached on the south pole. Near the end of Ordovician time, a large glacier grew on this southern region, and associated changes in the world's oceans caused a profound global extinction event.

Glaciation and Sea-Level Lowering

The ice cap that grew on Gondwanaland was centered near the south pole, in what is now northern Africa (see Figure 13-21). Evidence of this glacial episode comes in many forms: tillites, scratches on bedrock, and dropstones in marine sediments (p. 127).

Near the end of the Ordovician Period, a global drop in sea level caused an unconformity to form on top of shallow-water strata throughout the world. Sea level fell because the ice cap that grew in Gondwanaland removed a significant amount of water from the global water cycle. In some areas of the central United States, rivers cut deep valleys near the end of the Ordovician Period, eroding rapidly downward to reach the new, suddenly lower level of the sea. The depth of their canyonlike valleys—about 50 meters (165 feet)—indicates that global sea level fell by at least this amount.

It once seemed that the Ordovician glacial episode might have lasted several million years, but oxygen isotope ratios contradict this idea. Recall that when large glaciers grow on the land, isotopically light H_2O accumulates in them. As a result, seawater becomes correspondingly enriched in oxygen 18, the heavier isotope. This condition and also cooling of seas produce heavier oxygen isotope values in calcium carbonate secreted by marine organisms (p. 275). Fossil brachiopod shells reveal this kind of shift near the end of the Ordovician Period and show that it lasted only 0.5 million to 1.0 million years (Figure 13-22). This, then, was the length of the glacial episode.

The brevity of the drop in sea level explains why it does not appear in Figure 6-22, which displays sequences of deposition in North America that are bounded by unconformities. The sequences depicted there are separated by widespread unconformities that formed while sea level declined over millions of years. In general, such large but gradual changes of sea

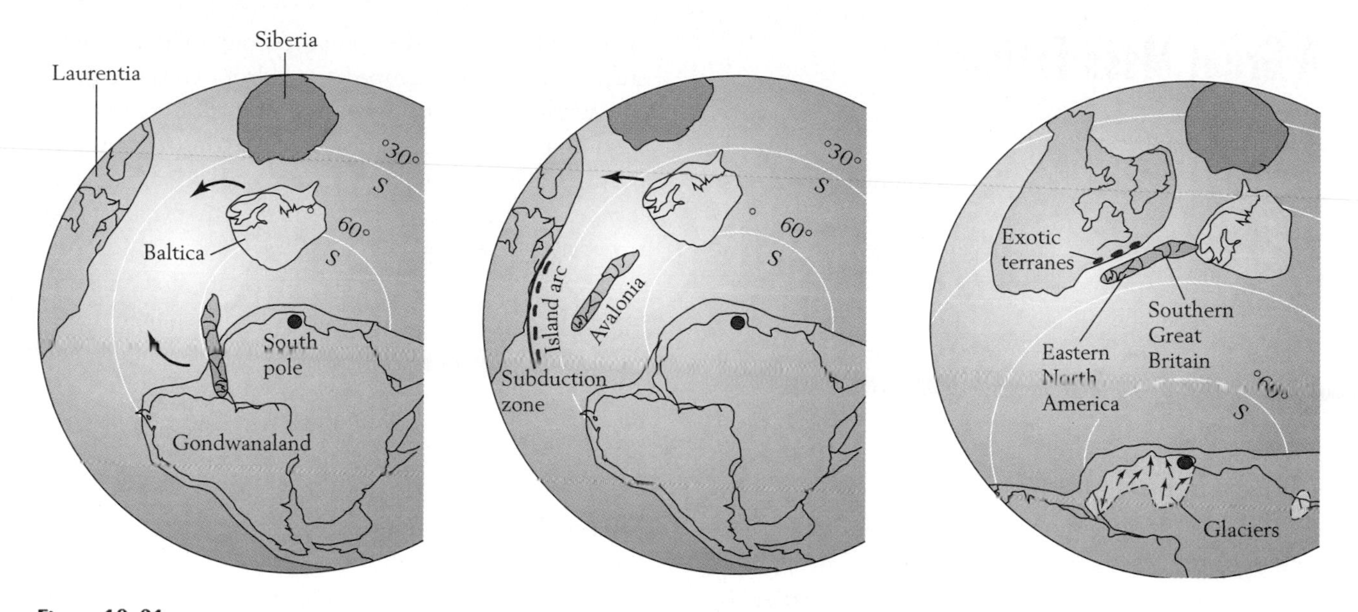

Figure 13-21 Movement of landmasses during Ordovician time. Avalonia was a fragment of Gondwanaland that moved toward Baltica, and islands to its north collided with Baltica, forming exotic terranes now located in Maine and eastern Canada. Near the end to the Ordovician, glaciers expanded over Gondwanaland near the south pole. *(Partly after B. A. van der Pluijm and R. Van der Voo, Geol. Soc. Canada Special Paper 41, 127–136, 1995.)*

level have accompanied global changes in the rates of seafloor spreading: sea level has risen or fallen along with the seafloor (p. 175). One such unconformity that formed earlier in the Ordovician Period marks the upper boundary of the first sequence of the Phanerozoic record (see Figure 6-22).

What caused glaciers to expand abruptly near the end of the Ordovician Period? Certainly the movement of Gondwanaland over the south pole was an essential element, but the supercontinent remained in this position for millions of years. Some additional factor must have come into play to trigger the glacial episode and then quickly bring it to an end. Carbon isotopes suggest that this factor was a reduction of Earth's greenhouse warming. Figure 13-22 shows how the carbon isotope ratios in seawater, as recorded in brachiopod shells, shifted toward much heavier values during the glacial interval. Possibly an unusually large amount of organic matter was buried to produce this shift. Recall that organic carbon is isotopically light, so that its excess burial leaves isotopically heavy CO_2 in the ocean and atmosphere (p. 266). Perhaps, then, burial of carbon reduced the greenhouse effect in the Ordovician world, and the cooling that resulted caused glaciers to expand in the region of the south pole. If this is the case, something must quickly have ended the greenhouse cooling. In any event, more recent ice ages, including the one in which we live today, have lasted much longer than the Late Ordovician glacial episode.

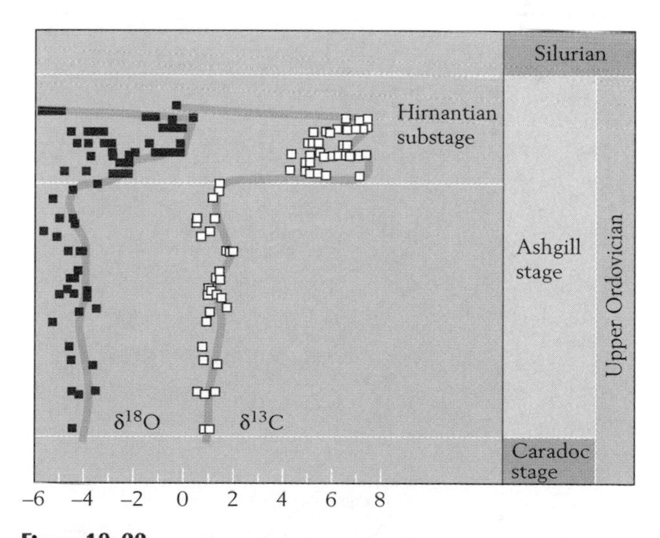

Figure 13-22 Shifts of oxygen and carbon isotope ratios toward more positive values late in the Ordovician Period. The data shown here, for brachiopod shells from the Baltic region, reveal a return to lower values near the end of the Ordovician. Data from other regions show the same pattern. *(After P. J. Brenchley et al., Geology 22:295–298, 1994.)*

A Great Mass Extinction

During the Ordovician glacial event, marine life suffered one of the largest mass extinctions in Earth's history. Many groups of brachiopods, trilobites, bryozoans, and corals died out on the seafloor. In the waters above, many species of acritarchs, graptolites, conodonts, and nautiloids disappeared.

There were actually two pulses of extinction in the Late Ordovician crisis. The first pulse coincided with the onset of glaciation and the second with the end of the glacial interval (see Figure 13-22). The first pulse is marked in the stratigraphic record by the disappearance of many fossil groups at the level of the unconformity that formed when sea level dropped as glaciers first expanded. The second pulse is marked by the disappearance of many fossil groups higher in the stratigraphic record, where changes in the nature of sedimentary rocks indicate that sea level was rising.

Tropical forms of life died out preferentially in the first episode of extinction. The reef community, for example, was devastated by the loss of numerous species of corals and stromatoporoids. This is hardly surprising because temperatures dropped throughout the world. Shifts in the distribution of marine life accompanied this climatic change: animals that had occupied seafloors in cool regions expanded into shallow seas nearer the equator. Apparently even shallow tropical seas became cooler.

Another distinctive pattern of the first pulse of extinction was the almost total disappearance of species that had been restricted to a single epicontinental sea on a single landmass. These species died out when sea level fell and eliminated the particular environments to which they had been adapted. Some species that lived in the open ocean also died out, apparently succumbing to changes in the temperature of the ocean.

The second pulse of extinction, which occurred when sea level rose at the end of the brief glacial interval, eliminated many of the cool-water species that had spread toward the equator when the glaciers expanded. Apparently the climatic warming that ended the glacial interval was responsible for their disappearance.

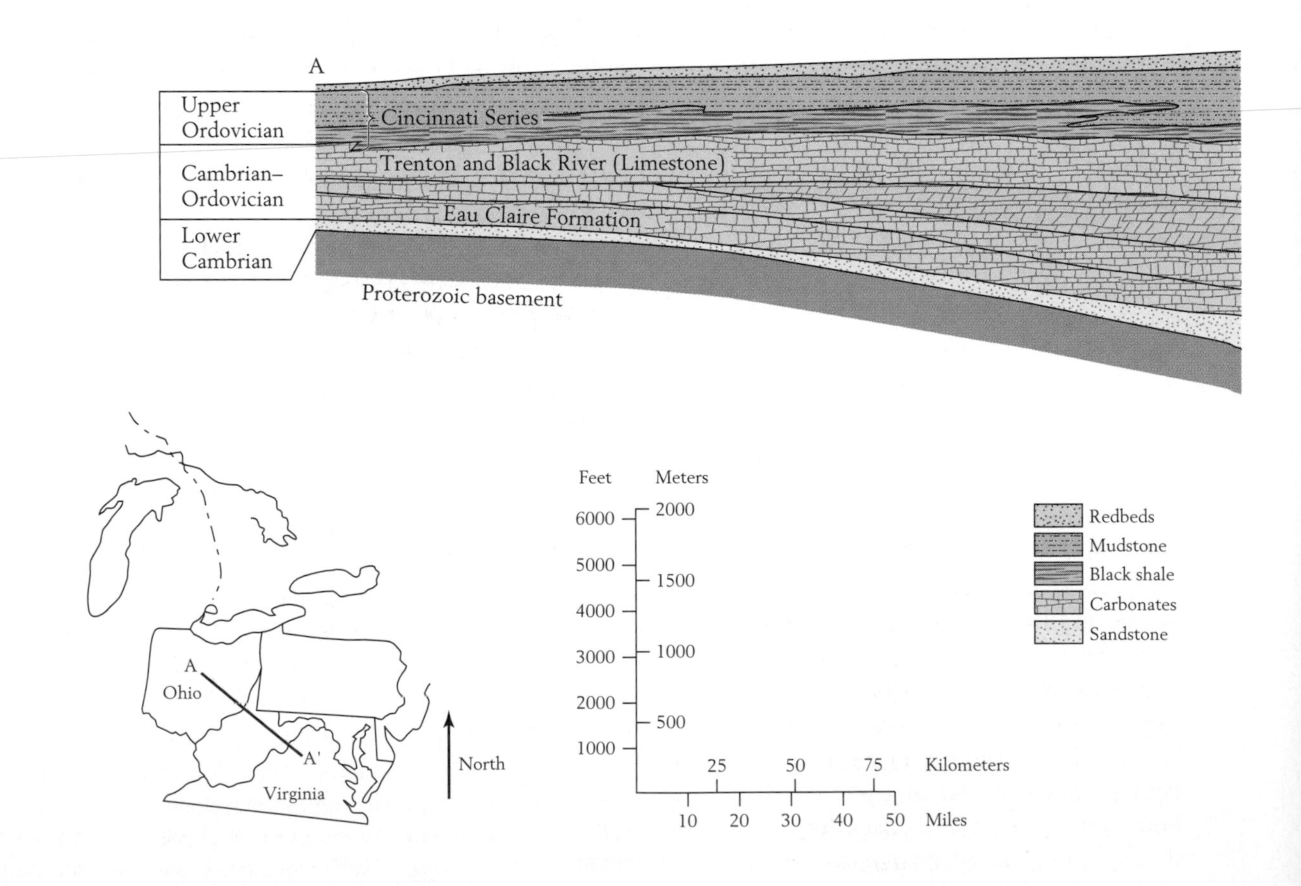

Regional Examples

Some interesting regional events took place along the margins of Laurentia early in the Paleozoic Era. We will focus first on the active eastern margin of what is now North America, and then on its passive western margin.

The Taconic Orogeny in Eastern Laurentia

Ordovician mountain building events in eastern North America are collectively termed the *Taconic orogeny.* This event was the first of three orogenic episodes in what is now the Appalachian mountain belt. Chapters 14 and 15 discuss the second and third Appalachian orogenies.

The Taconic orogeny did not result from collision of two large landmasses. Rather, it entailed collisions between Laurentia and several islands that had occupied the ocean between Laurentia, to the north, and Baltica and Gondwanaland, to the south (see Figure 13-21). Before we examine the evidence for these collisions, let us examine sedimentary rocks now located in the central Appalachian mountains. These rocks record a change in the pattern of deposition that offers evidence of the onset of orogenic activity.

The carbonate platform that bordered the eastern margin of Laurentia during the latter part of the Cambrian Period (see Figure 13-18) persisted through Early Ordovician time. Then, in mid-Ordovician time, the depositional framework changed drastically. Carbonate deposition ceased, and flysch deposits were laid down in deep water. Deposition of black shale predominated at the outset of this activity; later, turbidites became prevalent (Figure 13-23). Sole markings in turbidites of the Martinsburg Formation (see Figure 5-33) indicate that these deposits were derived from upland regions to the east. This change represents the classic pattern for the onset of mountain building along a continental margin (p. 243).

Clearly, the shallow carbonate platform foundered and dark sediment accumulated on top of it at considerable water depths. Eventually sediment was

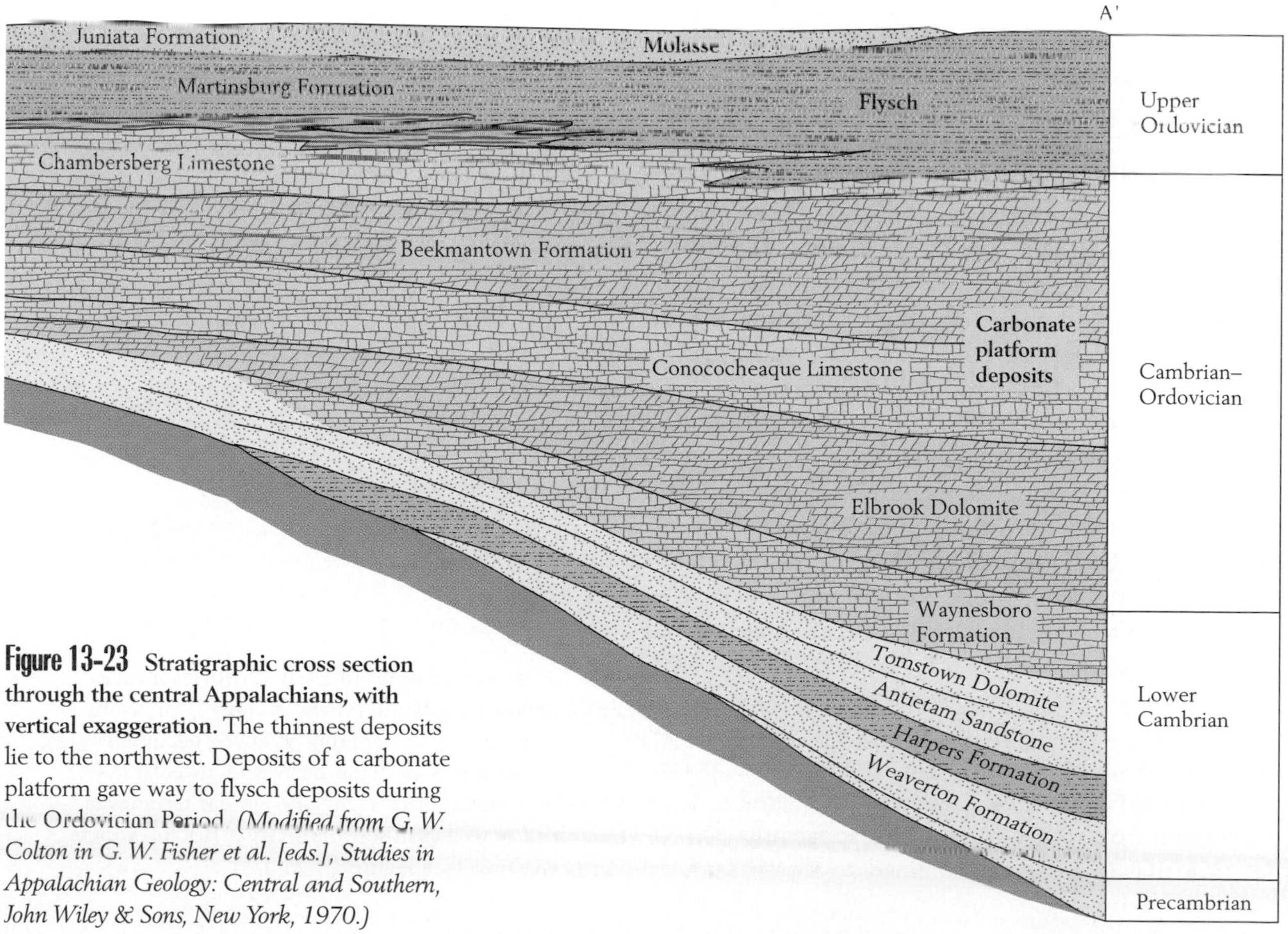

Figure 13-23 Stratigraphic cross section through the central Appalachians, with vertical exaggeration. The thinnest deposits lie to the northwest. Deposits of a carbonate platform gave way to flysch deposits during the Ordovician Period. *(Modified from G. W. Colton in G. W. Fisher et al. [eds.], Studies in Appalachian Geology: Central and Southern, John Wiley & Sons, New York, 1970.)*

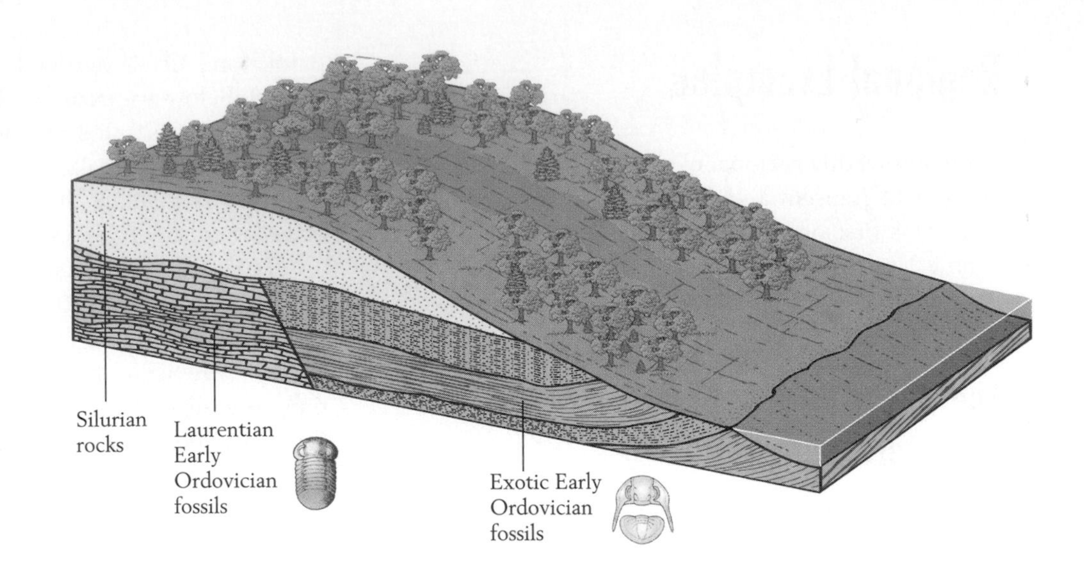

Figure 13-24 Exotic terranes in eastern North America. Both sedimentary rocks and fossils in these terranes contrast with those in adjacent Laurentian terranes.

being supplied by the eastern area of uplift faster than the mountains were subsiding. Near the end of the Ordovician Period, flysch gave way to molasse in the form of shallow marine and nonmarine clastics, some of which were red beds. Thus the Ordovician and Silurian Juniata and Tuscarora formations of the central Appalachians consist of coarse shallow marine and nonmarine deposits in a clastic wedge or molasse sequence that tapers out toward the northwest (see Figure 13-23). Cross-bedding and other features reveal that this was indeed the primary direction of transport.

Orogenies commonly result from the subduction of oceanic crust along a continental margin. The orogeny that is currently uplifting the Andes is an example (see Figure 9-15). In the Taconic orogeny, however, subduction was in the opposite direction: the margin of the carbonate platform that bordered Laurentia was wedged into a subduction zone (see Figure 13-21). Several islands of an igneous arc collided with Laurentia during the Taconic orogeny. The rocks of these small landmasses form exotic terranes that are now embedded in the eastern margin of North America, forming parts of Newfoundland and New Brunswick in Canada and Maine in the United States. As Figure 13-24 indicates, both the rocks and the fossils of these terranes testify to their origin outside of Laurentia. Many of the terranes are formed of rocks that differ from the cratonic rocks of the same age that now lie immediately to their west. Some of these terranes also yield Cambrian or Early Ordovician fossils that differ from fossils of the same age in other parts of North America (Figure 13-25). Some of these fossils are unknown from other regions and apparently represent taxa that in life were confined to islands. Others belong to taxa that occur in southern Great Britain.

Before the advent of plate tectonics, the foreign-looking fossils in sections of eastern North America were quite puzzling. These fossils make sense only in the context of plate tectonics. Plate tectonic reconstructions reveal that southern Great Britain was part

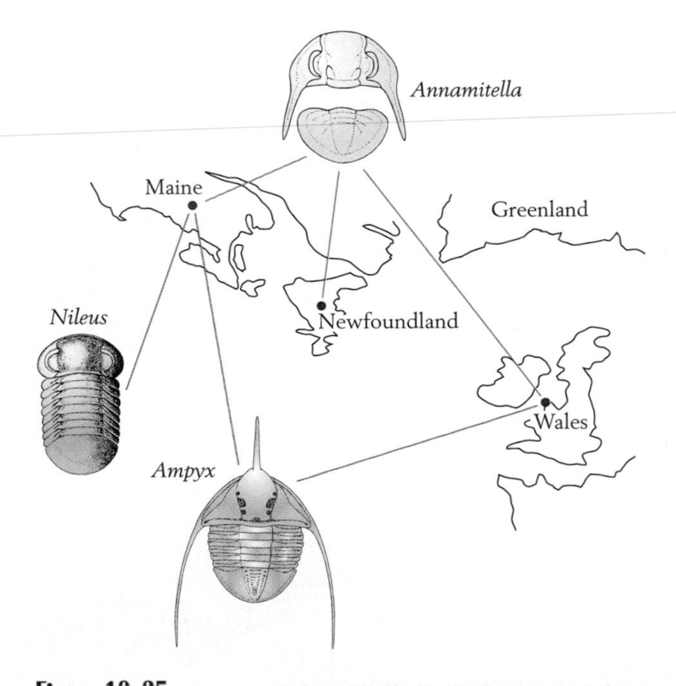

Figure 13-25 Exotic trilobites in Early Ordovician rocks of Maine and Newfoundland. *Nileus* is known only from an exotic terrane in Maine, apparently having occupied this terrane when it was an island far from Laurentia. Other species occur in this terrane and similar terranes in Newfoundland as well as in southern Great Britain, which was part of Avalonia (see Figure 13-21.)

of a sliver of continental crust, called Avalonia, that rifted away from Gondwanaland during the Ordovician Period and encroached on Baltica (see Figure 13-21). The exotic terranes that collided with Laurentia were islands of igneous arcs that had been positioned near Avalonia. We can no longer be surprised that they contain fossils resembling those of the British Isles.

Figure 13-26 shows how the eastern margin of Laurentia was deformed as small landmasses collided with it, and how the sedimentary record in eastern New York State reflects the collision. In this region oceanic crust and the accretionary wedge adjacent to the neighboring island arc rode up over the carbonate platform. The depressed continental margin became a foreland basin, where dark sediment shed

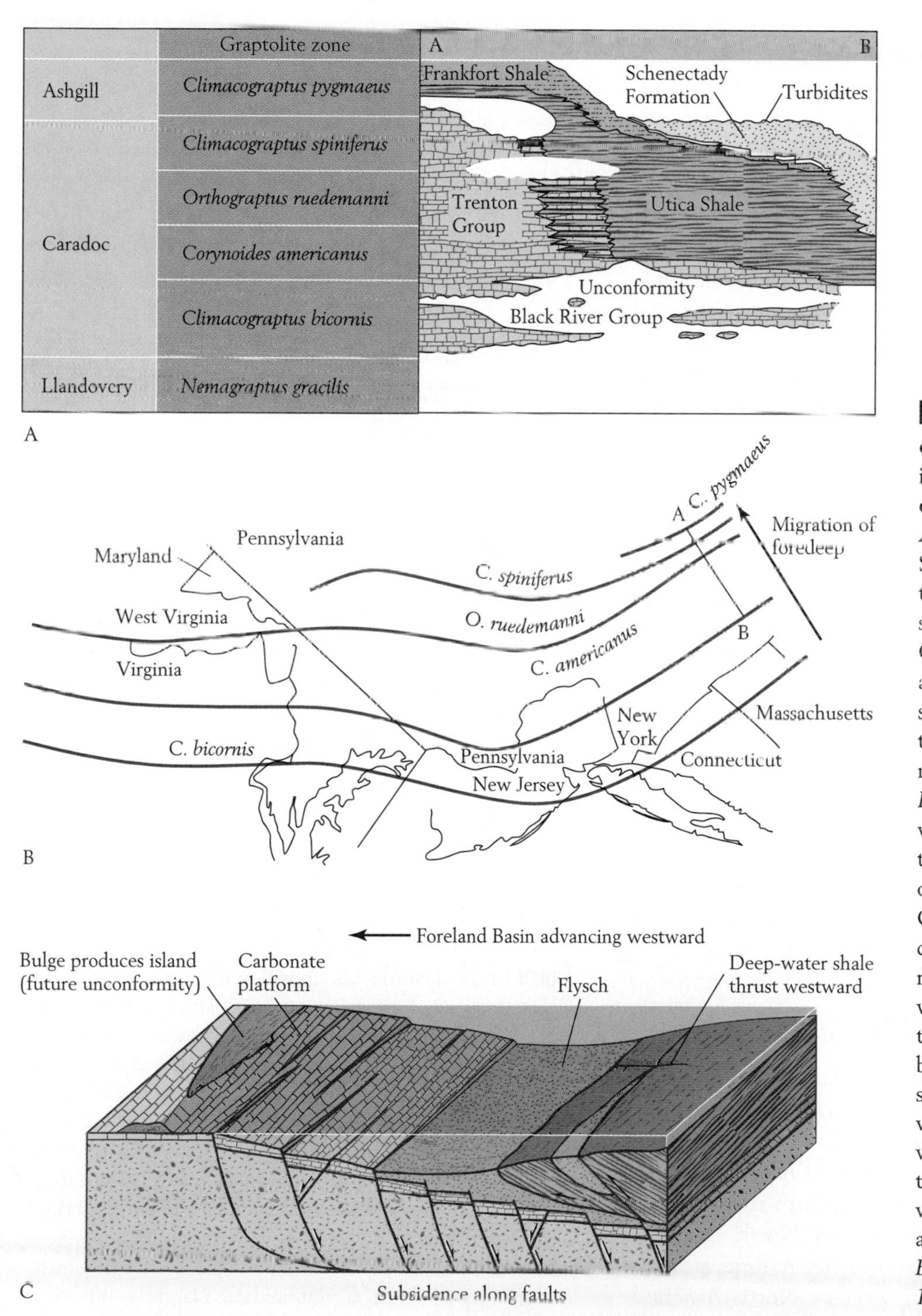

Figure 13-26 Development of a foreland basin in what is now eastern New York during the Taconic orogeny. *A.* The deep-water Utica Shale was deposited above the shallow-water carbonate sediments of the Trenton Group; flysch, derived from an island arc to the east, then spread over the Utica Shale to form the Schenectady Formation and Frankfort Shale. *B.* A heavy line marks the western boundary of graptolite occurrences in each of the zones shown in *A.* Graptolites, which occupied deep-water environments, migrated progressively westward during Ordovician time, along with the foreland basin. C. The foreland basin subsided along faults as deep-water sediments were thrust westward. Upward bulging of the carbonate bank to the west elevated some areas above sea level. *(After D. C. Bradley, Tectonics 8:1037–1049, 1989.)*

from the island arc accumulated as flysch in deep water (p. 243).

As the collision proceeded, the foreland basin migrated inland, and large wedges of flysch that formed within it were thrust farther and farther westward, forming a thick pile of deep-water deposits on top of shallow-water carbonates. In places, ophiolites—slices of oceanic crust (p. 241)—were thrust up onto the craton along with turbidites. In advance of the foreland basin, the carbonate platform bulged up above sea level, creating unconformities in the record of carbonate deposition. The bulge formed because lithospheric material squeezed westward from beneath the foredeep. Normal faults formed along the seaward margin of the bulge, and the foreland basin subsided in advance of the thrust sheets.

Graptolite species that floated above deep water spread farther and farther west as the orogeny progressed, providing evidence of the basin's westward migration (see Figure 13-26*B*). Radiometric dating of igneous rocks to the east of the zone occupied by the foreland basin provide absolute ages for igneous activity during the Taconic orogeny. The orogeny ended near the close of the Ordovician Period, apparently when the continental margin of Laurentia ultimately offered so much resistance that continental convergence could not continue.

Near the end of the Taconic orogeny, bodies of molasse sometimes referred to as *clastic wedges* (p. 243) spread far to the west (Figure 13-27; see also Figure 13-23). Beyond the clastic belt, in the interior of the United States (southern Laurentia), limestones accumulated in shallow seas. Stratigraphic unconformities reveal that several low islands stood along the Transcontinental Arch in what is now the west-central United States. This arch was a gentle ridge that persisted throughout much of the Paleozoic Era; at various times during this era, one or more segments of it stood above shallow epicontinental seas.

Stability in Western Laurentia

Recall that a marine carbonate platform rimmed the craton of Laurentia during Cambrian time. Of course, the Taconic orogeny ended deposition along the carbonate platform in eastern North America during mid-Ordovician time, but the platform survived into middle Paleozoic time in western North America. Thus throughout Cambro-Ordovician time a stable continental shelf bounded western North America, passing diagonally across what is now the southeastern corner of California (see Figure 13-27). Coarse, poorly sorted sediments derived from shallow-water environments accumulated at the base of a steep platform edge. In many places there are great thicknesses of these limestones, composed in part of debris from shallow-water stromatolites and invertebrates (Figure 13-28). Deposits that accumulated on deep seafloors beyond the platform include black limey mudstones and limestones.

To the north, in British Columbia, the continental rise at the base of the carbonate bank was the depositional setting for the Burgess Shale, where a remarkable fauna of soft-bodied animals was preserved. In 1909 Charles Walcott, the secretary of the Smithsonian Institution and an expert on Cambrian fossils, discovered a spectacularly well preserved fossil along a mountain trail. Careful examination of the strata

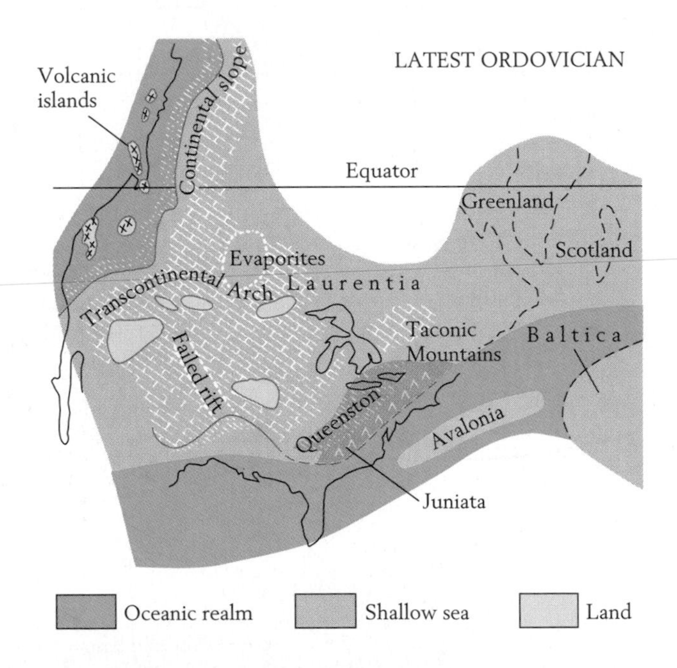

Figure 13-27 Patterns of deposition in North America during latest Ordovician time, when Baltica and Avalonia lay close to Laurentia. The Juniata and Queenston formations formed as clastic wedges of sediment shed inland from newly formed mountains to the east (see Figure 13-23). Carbonate sediments accumulated in the shallow seas that inundated almost all of the craton of North America. These deposits blanketed a failed rift in southern Oklahoma that had extended inland from the Gulf Coast in the early Paleozoic. Several low islands remained along the Transcontinental Arch. The continental margin bordering western North America remained stable, but volcanic islands lay offshore.

Figure 13-28 **An outcrop of the Cambro-Ordovician Hales Limestone in Nevada, representing the ancient continental slope of western North America (see Figure 13-27).** The coarse fragments of shallow-water limestone in this graded bed were transported down the continental slope.

above the trail turned up the layer, 2 meters (7 feet) thick, from which the fossil-bearing block had fallen. Walcott organized a quarrying operation and removed nearly all of the fossil-bearing material.

Extensive study of the Burgess assemblage since Walcott's day has revealed that the fauna is so well preserved because it was entombed in an oxygen-free environment from which destructive bacteria and scavenging animals were excluded. Stratigraphic evidence further indicates that the Burgess Shale was deposited at the foot of the steep carbonate shelf. In fact, the escarpment is still preserved in cross section in the mountainside some 200 meters (650 feet) above the beds that preserve the fossils (Figure 13-29). Presumably the carbonate bank stood close to sea level, so this figure of 200 meters approximates the depth at which the Burgess fauna was preserved. The Burgess fauna was collected from a series of turbidite beds. Within each bed, calcareous siltstone grades upward into fine-grained mudstone. The beds

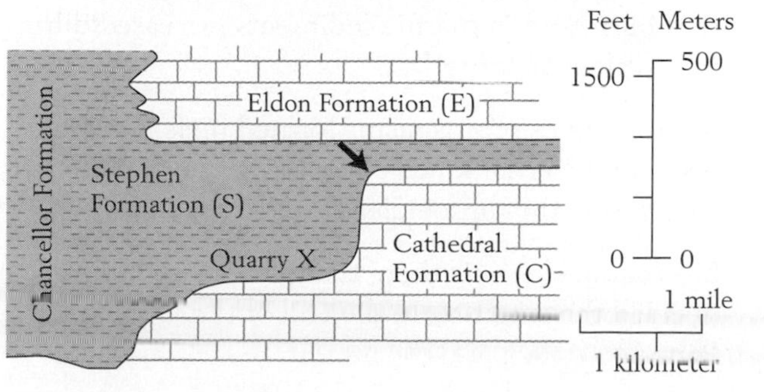

Figure 13-29 **Location of the Burgess Shale quarry within the Stephen Formation in British Columbia.** The formations labeled in the diagram are identified in the photograph by letters. The arrow in the diagram points to the edge of the ancient continental shelf. The Burgess Shale accumulated at the foot of the steep slope below this shelf edge. *(Photograph from W. H. Fritz, Proc. N. Amer. Paleont. Conv. I:1155–1170, Allen Press, Lawrence, Kans., 1969.)*

apparently formed when turbid flows descended the escarpment from one or more channels in the carbonate bank. Most of the animals found in the Burgess Shale probably lived along the continental margin and were swept farther down the steep continental slope by the turbid flows. Possibly they were preserved in the absence of oxygen because they were buried very rapidly. Because several flows produced the same result, however, it seems more likely that the entire site was an oxygen-free basin near the foot of the continental slope—a depression filled with stagnant water from which oxygen had become depleted. The Santa Barbara Basin off the coast of California may be a modern analog. In any case, we must be grateful for this spectacular glimpse of the soft-bodied marine life of Middle Cambrian time.

Chapter Summary

1 The earliest Cambrian skeletal forms included a small variety of very small tubes and teeth. Above them is the more diverse Tommotian fauna of small animals, many of which are unknown from later intervals; later Cambrian shelly faunas were dominated by trilobites but included the earliest vertebrates and large invertebrate predators. A great evolutionary radiation late in the Ordovician Period produced a much more diverse invertebrate fauna resembling that of later Paleozoic time. This fauna included many kinds of corals, bryozoans, brachiopods, mollusks, echinoderms, and graptolites.

2 Stromatolites declined during the Cambrian Period, apparently because of the grazing and burrowing activities of new groups of animals.

3 A highly successful reef community developed during the Ordovician Period. This community, which was dominated by corals and stromatoporoid sponges, went on to thrive throughout almost all of middle Paleozoic time.

4 Early in the Cambrian Period, many continents stood unusually high above sea level; but as the period progressed, the continents were increasingly flooded. Siliciclastic deposits fringed the land, and carbonates were laid down across the expanding continental shelves and along marginal platforms.

5 During the Cambrian Period, trilobites suffered periodic mass extinctions, the last of which occurred at the close of the period.

6 Near the end of the Ordovician Period, continental glaciers spread over the region of Gondwanaland positioned over the south pole. Oxygen isotopes in fossils indicate that this glacial event was relatively brief, and carbon isotopes suggest that it may have resulted from global warming.

7 Two pulses of marine extinction near the end of the Ordovician Period constituted one of the largest mass extinctions of all time. The first pulse appears to have resulted from cooling and lowering of sea level when glaciers expanded, and the second pulse from warming when the glacial episode came to an end.

8 The Taconic orogeny occurred when the eastern margin of Laurentia was wedged into a subduction zone and islands of an igneous arc were attached to it, forming what are now exotic terranes in Maine and eastern Canada.

9 Whereas the carbonate platform bordering eastern North America was destroyed by Late Ordovician mountain building, the carbonate platform that bordered western North America remained intact into middle Paleozoic time.

Review Questions

1 Why do geologists know more about the life that colonized early Paleozoic seafloors than about the life that floated and swam above those seafloors?

2 What fossil evidence suggests that distinctive new kinds of predatory animals evolved during Cambrian time?

3 What evidence is there that the variety of animals that burrowed in marine sediments increased during early Paleozoic time?

4 What kinds of organisms formed reefs in Cambrian time? What kinds of organisms performed this role during the Ordovician?

5 What evidence is there that plants may have invaded the land before the end of the Ordovician Period?

6 What evidence is there that major extinctions of trilobites occurred very suddenly during the Cambrian?

7 On which landmasses is the climate likely to have been warmer in Late Cambrian time than it is today? (Compare Figure 13-17 with a map of the modern world.)

8 Review the history of sediment deposition along the eastern margin of North America and relate this history to plate movements.

9 What is the significance of the Burgess Shale? In what geographic region and environmental setting did it form?

10 Why do some Lower Ordovician rocks in Maine share some trilobite taxa with Great Britain but share no trilobite taxa with neighboring areas of the United States?

11 The Cambrian and Ordovician periods differed from one another in many ways. Using the Visual Overview on page 342 and what you have learned in this chapter, compare these two periods with respect to sea level, the distribution of landmasses on Earth, and the nature of life in the oceans.

Additional Reading

Droser, M. L., R. A. Fortey, and X. Li, "The Ordovician Radiation," *American Scientist* 84:122–131, 1996.

Erwin, D., J. Valentine, and D. Jablonski, "The Origin of Animal Body Plans," *American Scientist* 85:126–137, 1997.

Lipps, J. H., and P. W. Signor, *Origin and Early Evolution of the Metazoa*, Plenum Press, New York, 1992.

Palmer, A. R., "Search for the Cambrian World," *American Scientist* 62:216–224, 1974.

Simonetta, A. M., and S. Conway Morris, *The Early Evolution of Metazoa and the Significance of Problematic Taxa*, Cambridge University Press, Cambridge, 1991.

CHAPTER 14

The Middle Paleozoic World

The oceans of the world stood high during most of Silurian and Devonian time, leaving a widespread sedimentary record on every continent. Marine deposition was interrupted in one region, however, by the most far-reaching plate tectonic event of middle Paleozoic time: the suturing of Baltica to Laurentia along a zone of mountain building.

In the northern British Isles, Silurian rocks were tilted by the Caledonian orogeny, and thus an angular unconformity separates them from the overlying Devonian sediments. It was farther south, in Wales, however, that in 1835 Roderick Murchison founded the Silurian System, along with the Cambrian. Five years later, Murchison and Adam Sedgwick formally recognized the Devonian System, naming it for the county of Devon, along the southern coast of England. They recognized that the fossils of this system were intermediate in character (we would now say intermediate in evolutionary position) between those of the Silurian System below and those of the Carboniferous System above. (The Carboniferous System, though younger than the Devonian, had been recognized earlier in the century.)

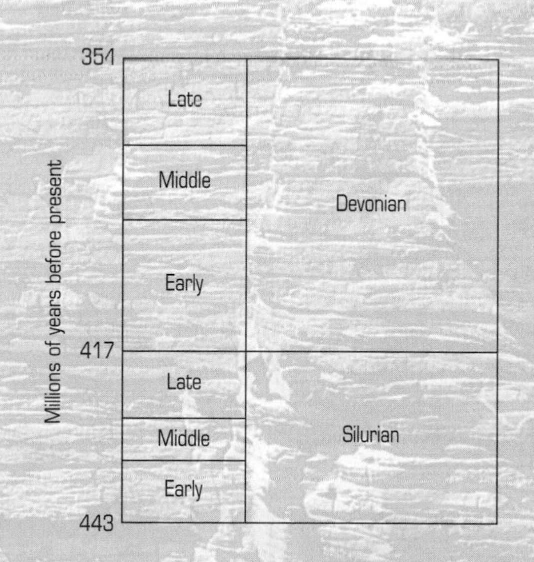

The Old Red Sandstone forms cliffs along the coast of the Orkney Islands off northern Scotland.

Visual Overview

Major Events of the Middle Paleozoic

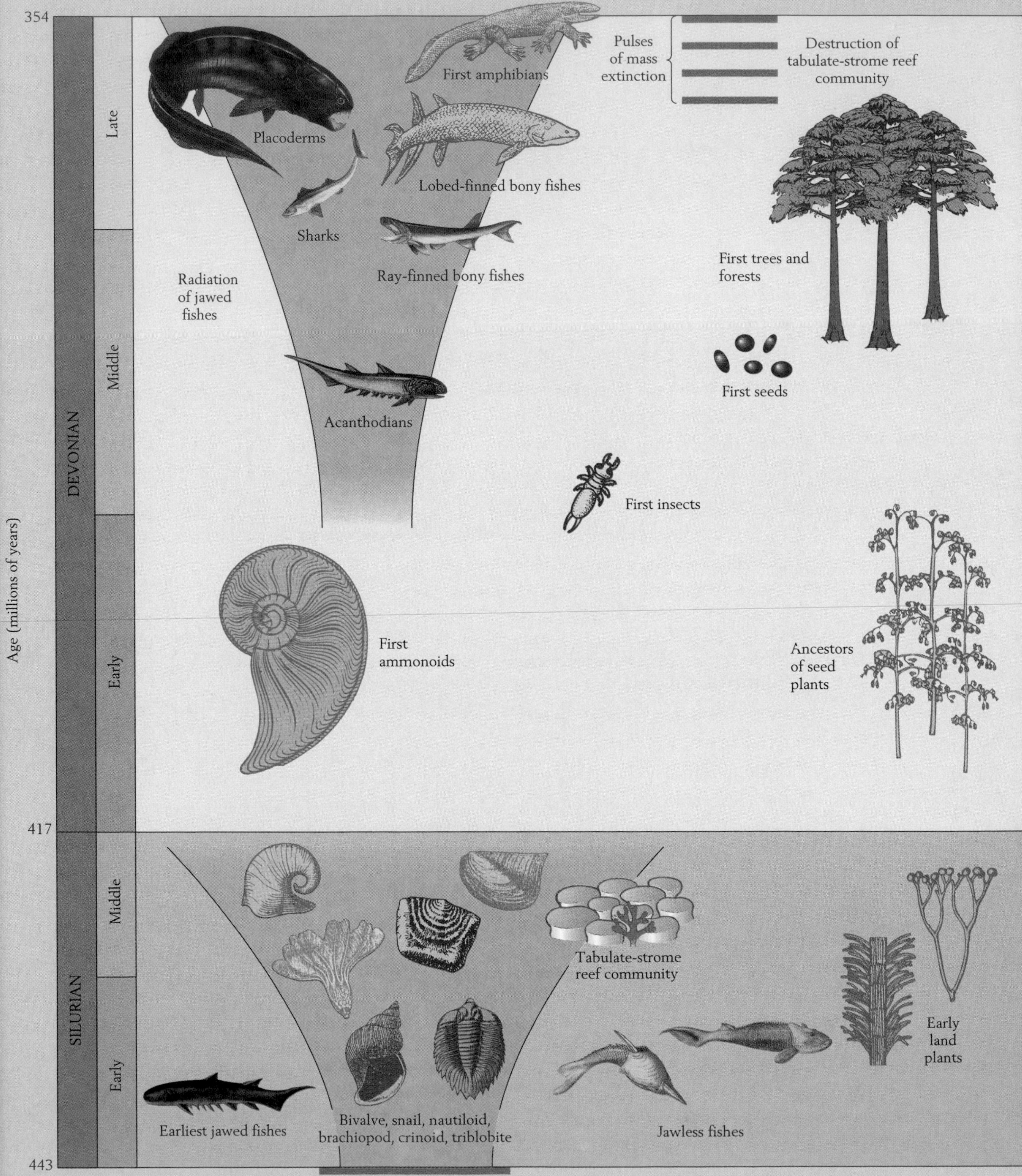

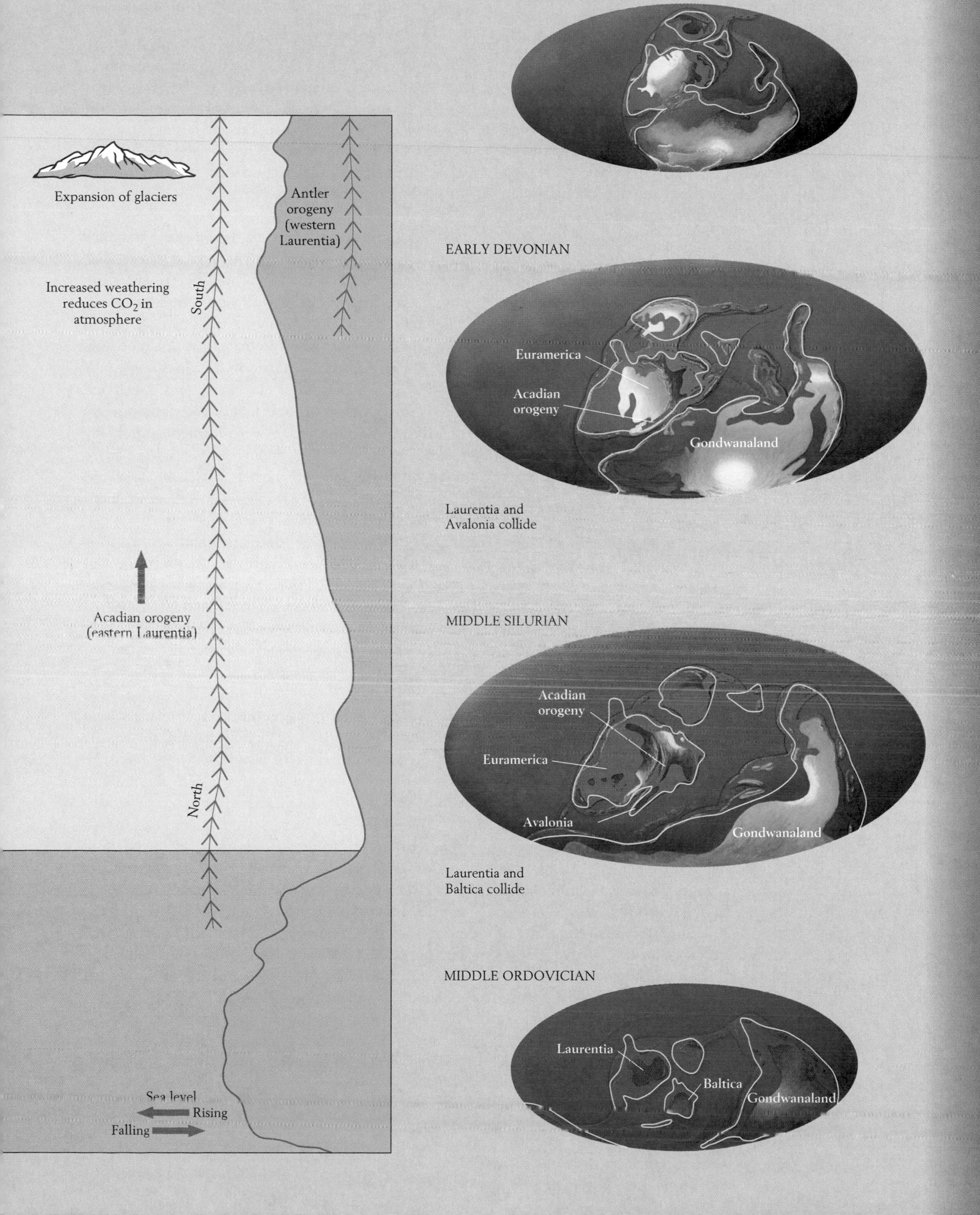

EARLY CARBONIFEROUS
EARLY DEVONIAN
Euramerica
Acadian
orogeny
Gondwanaland
Laurentia and
Avalonia collide
MIDDLE SILURIAN
Acadian
orogeny
Euramerica
Avalonia
Gondwanaland
Laurentia and
Baltica collide
MIDDLE ORDOVICIAN
Laurentia
Baltica
Gondwanaland
Expansion of glaciers
Antler
orogeny
(western
Laurentia)
Increased weathering
reduces CO_2 in
atmosphere
South
Acadian orogeny
(eastern Laurentia)
North
Sea level
Rising
Falling

The broad, shallow epicontinental seas of Silurian and Devonian time teemed with life. In the tropical zone, a diverse community of organisms built reefs larger than any that had formed during early Paleozoic time. More advanced predators were also on the scene, including the first jawed fishes—a few of which were the size of large modern-day sharks. The Devonian Period was also distinguished by the progressive colonization of land habitats by new forms of life. Plants were restricted to marshy environments in Silurian time but were forming large forests by Late Devonian time. The oldest known insects are also of Early Devonian age; and near the end of the Devonian Period, the first vertebrate animals crawled up onto the land, the fins of their ancestors having been transformed into legs. Shortly before the end of the Devonian Period, however, a wave of mass extinction swept away large numbers of aquatic taxa, leaving an impoverished biota during the final 3 million or 4 million years of the period.

A New Expansion of Life

After the great mass extinction at the close of the Ordovician Period, many of the decimated taxa diversified once again. Their recovery surpassed the Ordovician adaptive radiation, yielding superior reef builders and swimming predators. Meanwhile, plants spread over the land, and near the end of the Devonian Period animals invaded the terrestrial realm.

Aquatic Recovery

Most of the marine taxa that had flourished during the Ordovician Period rediversified after the terminal Ordovician mass extinction to become prominent members of the Silurian and Devonian marine biota. The trilobites failed to recover fully, however, and were less conspicuous in middle Paleozoic than in early Paleozoic seas. For other groups, recovery took the form of renewed adaptive radiation. As Figure 14-1 shows, all of the important Paleozoic articulate brachiopods were well represented in middle Paleozoic seas.

The bivalve mollusks expanded their ecologic role by invading nonmarine habitats; some of the oldest known freshwater bivalves are found in the Upper Devonian strata of New York State. One of the most spectacular Early Silurian adaptive radiations was that of the graptolites, which had nearly disappeared at the end of the Ordovician Period. During just the first 5 million years or so of Silurian time, the number of species of graptolites known in the British Isles increased from about 12 to nearly 60.

Luxuriant Reefs Most of the Silurian radiations of marine life did not vastly alter marine ecosystems; instead they refilled niches vacated by the mass extinction at the end of the Ordovician. Builders of organic reefs did diversify in new ways, however, and in some shallow waters they produced much larger reefs than any of Cambro-Ordovician age. In Chapter 13 we saw that a new kind of organic reef developed in mid-Ordovician time. Whereas the earliest reefs of the Middle Ordovician were formed entirely by bryozoans, later Middle Ordovician reefs were more complex, with tabulate corals and stromatoporoid sponges playing dominant roles in their construction. Reef communities of this second type, which we can call *tabulate-strome reefs,* diversified and persisted for about 120 million years, until late in the Devonian Period. The success of these reefs was a result of mid-Paleozoic adaptive radiations of tabulates, colonial rugose corals, and stromatoporoid sponges.

Tabulate-strome reefs occasionally attained heights of 10 meters (35 feet) above the seafloor during the Silurian Period, but during Devonian time they assumed enormous proportions. The largest one grew near the equator in Gondwanaland, in what is now Western Australia (see Figure 5-25). In areas subjected to strong wave action, the growth of tabulate-strome reefs followed a characteristic ecologic succession (Figure 14-2). First, sticklike tabulates and rugose corals colonized an area of subtidal seafloor. A low mound was then formed when these fragile forms were encrusted by platy and hemispherical tabulates and colonial rugose corals. Finally, as the mound grew up toward the sea surface, stromatoporoids and algae encrusted the seaward side, forming a durable ridge. Tabulates and colonial rugose corals occupied a zone of quieter water behind the ridge, and beyond them was a lagoon in which mud-sized sedimentary grains accumulated along with coarser skeletal debris from the reef. Pockets of fossils preserved in reef rock reveal that a wide variety of invertebrates inhabited tabulate-strome reefs; brachiopods and bivalve mol-

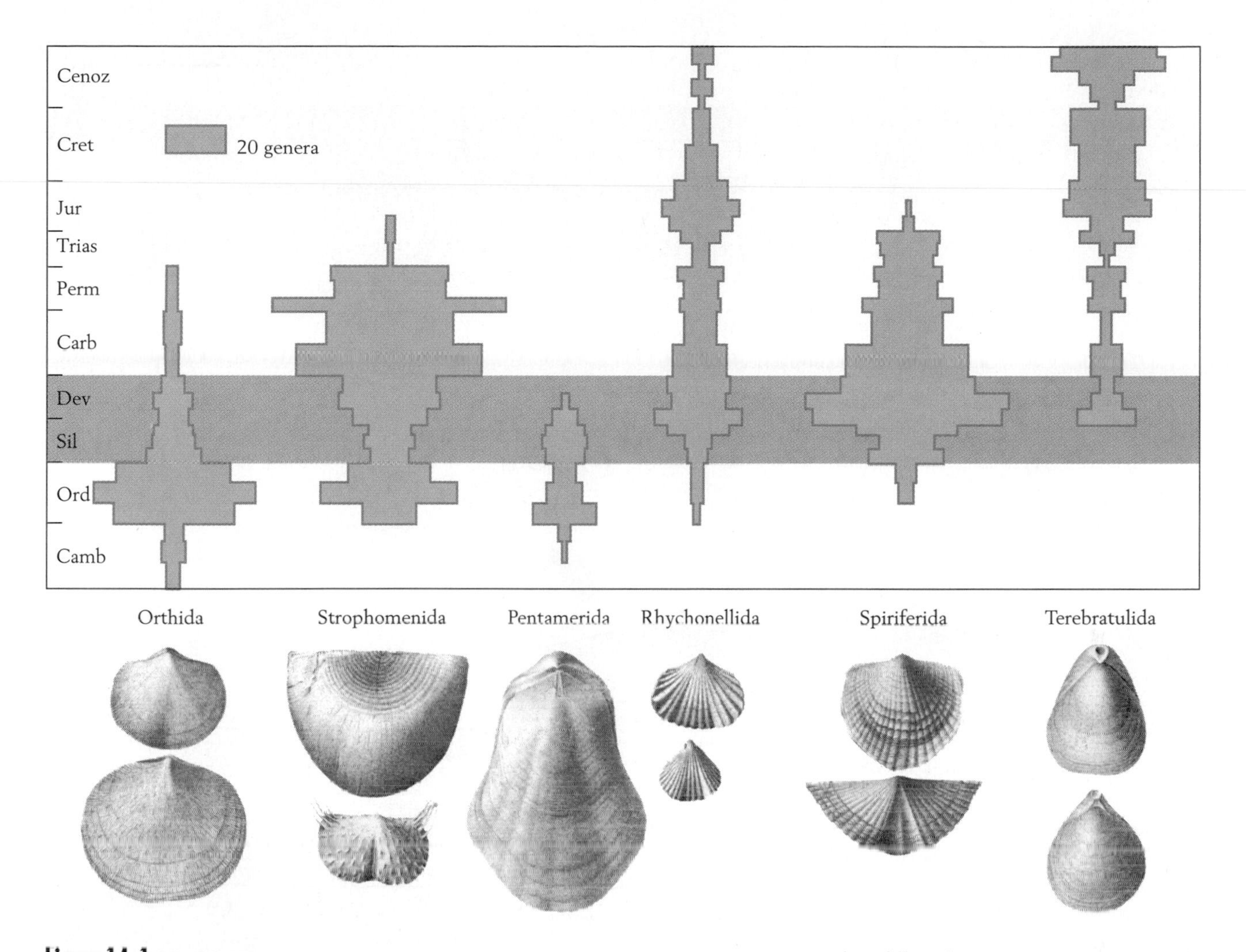

Figure 14-1 Middle Paleozoic articulate brachiopods. Top: The diversity (numbers of genera) of the six major orders of Paleozoic brachiopods, all of which were well represented in middle Paleozoic (Silurian and Devonian) time. Bottom: Typical middle Paleozoic representatives of each brachiopod order. *(Brachiopod illustrations from James Hall's volumes of the New York State Natural History Survey [1862–1894].)*

lusks attached themselves to a typical reef, snails grazed over it, and crinoids and lacy bryozoans reached upward from its craggy surface. Although its fauna would look unusual to us today, a living, fully developed tabulate-strome reef (Figure 14-3) would certainly seem as colorful and spectacularly beautiful as the coral reefs that flourish in the modern tropics (see Box 4-1).

New Swimming Invertebrates Perhaps the greatest change in the nature of aquatic ecosystems during middle Paleozoic time resulted from the origin of new kinds of nektonic (swimming) animals, many of which were predators. The most prominent new invertebrate swimmers were the ammonoids. These coiled cephalopod mollusks evolved from a group of straight-shelled nautiloids during Early Devonian time (Figure 14-4). After giving rise to the ammonoids, the nautiloids persisted at low diversity. The ammonoids, in contrast, diversified rapidly, and because their species were distinctive, widespread, and relatively short-lived, they serve as guide fossils in rocks ranging in age from Devonian to latest Mesozoic. (Ammonoids died out along with the dinosaurs at the end of the Mesozoic Era.)

Another group of invertebrate predators that proliferated during middle Paleozoic time were the eurypterid arthropods. These distant relatives of scorpions

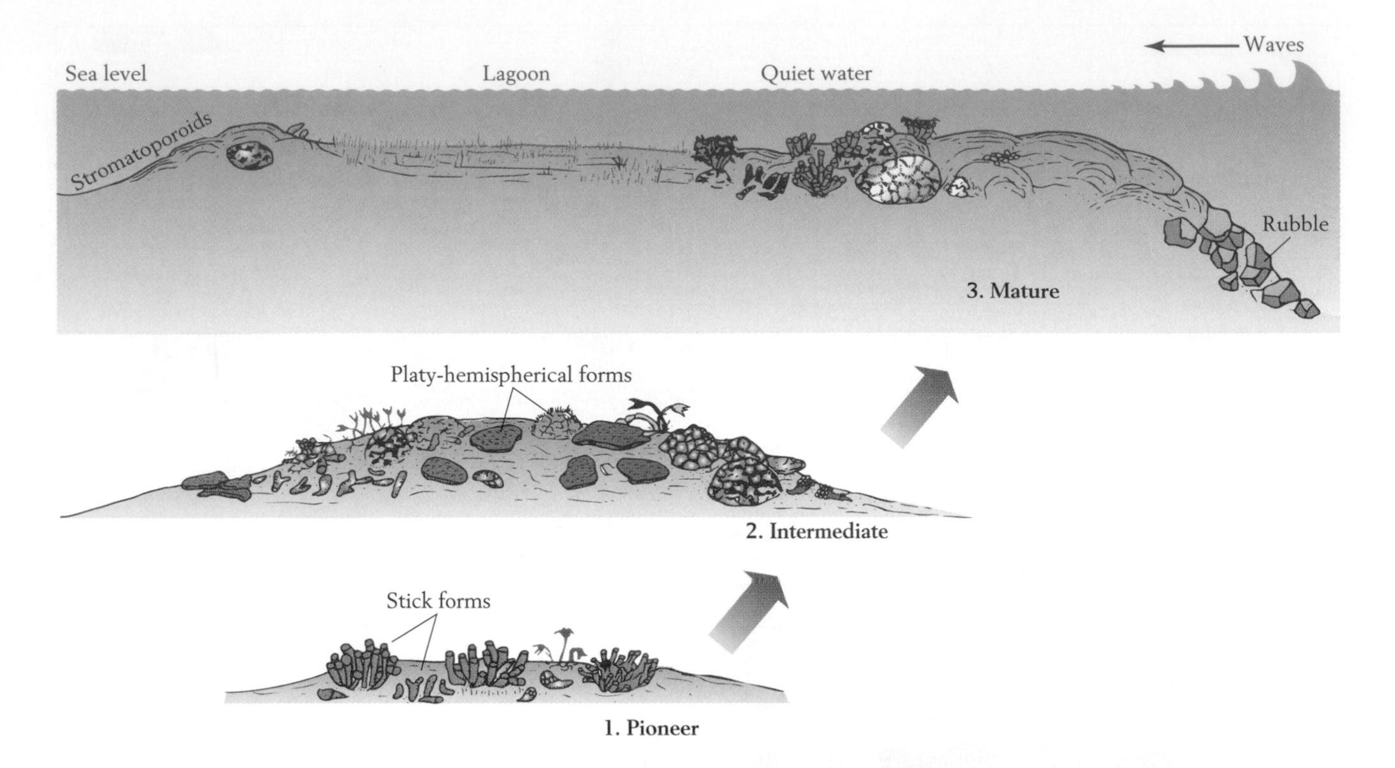

Figure 14-2 Ecologic succession of a typical Devonian reef. (1) The pioneer community consisted of fragile, twiglike rugose corals and tabulates. (2) Broad and moundlike tabulate colonies were dominant during the intermediate stage of development. (3) In the mature stage, the reef grew close to sea level, and waves broke against a ridge of massive, encrusting stromatoporoids; behind them were species that were adapted to quieter water, and leeward of the reef was a lagoon populated by fragile, twiglike species. The leeward side of the lagoon was bounded by a small stromatoporoid ridge. At left is a moundlike reef formed by tabulates and rugose corals. The fossils were collected in Michigan (see Figure 14-23) and were then reassembled at the Smithsonian Institution to recreate the reef. This moundlike reef represents the intermediate stage (2). *(Diagram after P. Copper, Proc. Second Internat. Coral Reef Symp. 1:365–386, 1975.)*

were swimmers, and many had claws (Figure 14-5). Although the eurypterids appeared in the Ordovician Period and survived until Permian time, their most conspicuous fossil record is in middle Paleozoic rocks. Unlike ammonoids, eurypterids ranged into brackish and freshwater habitats.

Jawless Fishes Other swimmers that were adapted to both marine and freshwater conditions were the fishes. The major groups are seen in Figure 14-6. Whereas only fragments of fish skeletons have been found in early Paleozoic sediments (see Figure 13-10), the Silurian and Devonian systems have yielded diverse, fully preserved fish skeletons, many of which are found in freshwater deposits of lakes and rivers.

We do not know when fishes first occupied freshwater habitats. The fact that all known Cambro-Ordovician fish remains have been found in marine deposits supports the idea that fishes evolved in the ocean. Most Silurian fish remains, unlike most

Figure 14-3 Reconstruction of an Upper Devonian reef in New York State. Numerous kinds of corals are present. In the right foreground is a huge, spiny trilobite that measured about 45 centimeters (18 inches) in length.

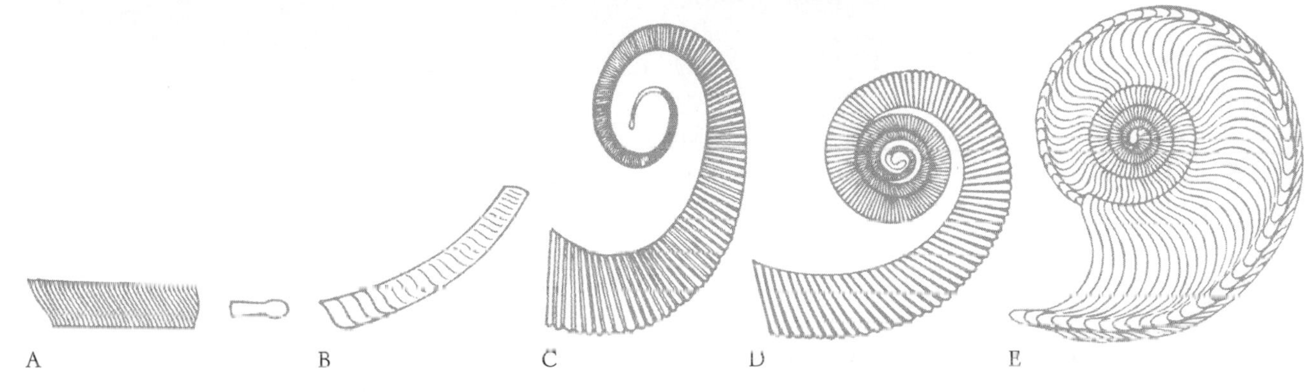

Figure 14-4 Shells of Lower Devonian cephalopod mollusks from the Hunsruck Shale of Germany that reveal the apparent evolutionary sequence from nautiloids to early ammonoids. *A* and *B*. Fragments of nautiloids of the group that evolved into ammonoids. C through *E*. Early ammonoid species representing various degrees of coiling. The bulblike shape of the earliest part (tip) of each shell, together with other shared features, suggests that these species are closely related; the coiling sequence displayed here apparently represents the evolutionary sequence. *(After H. K. Erben, Biol. Rev. 41:641–658, 1966.)*

Figure 14-5 Reconstruction of a Late Silurian eurypterid. This animal was about 0.5 meters (20 inches) long. The appendages beneath the head of this species bore sharp spikes for stabbing prey.

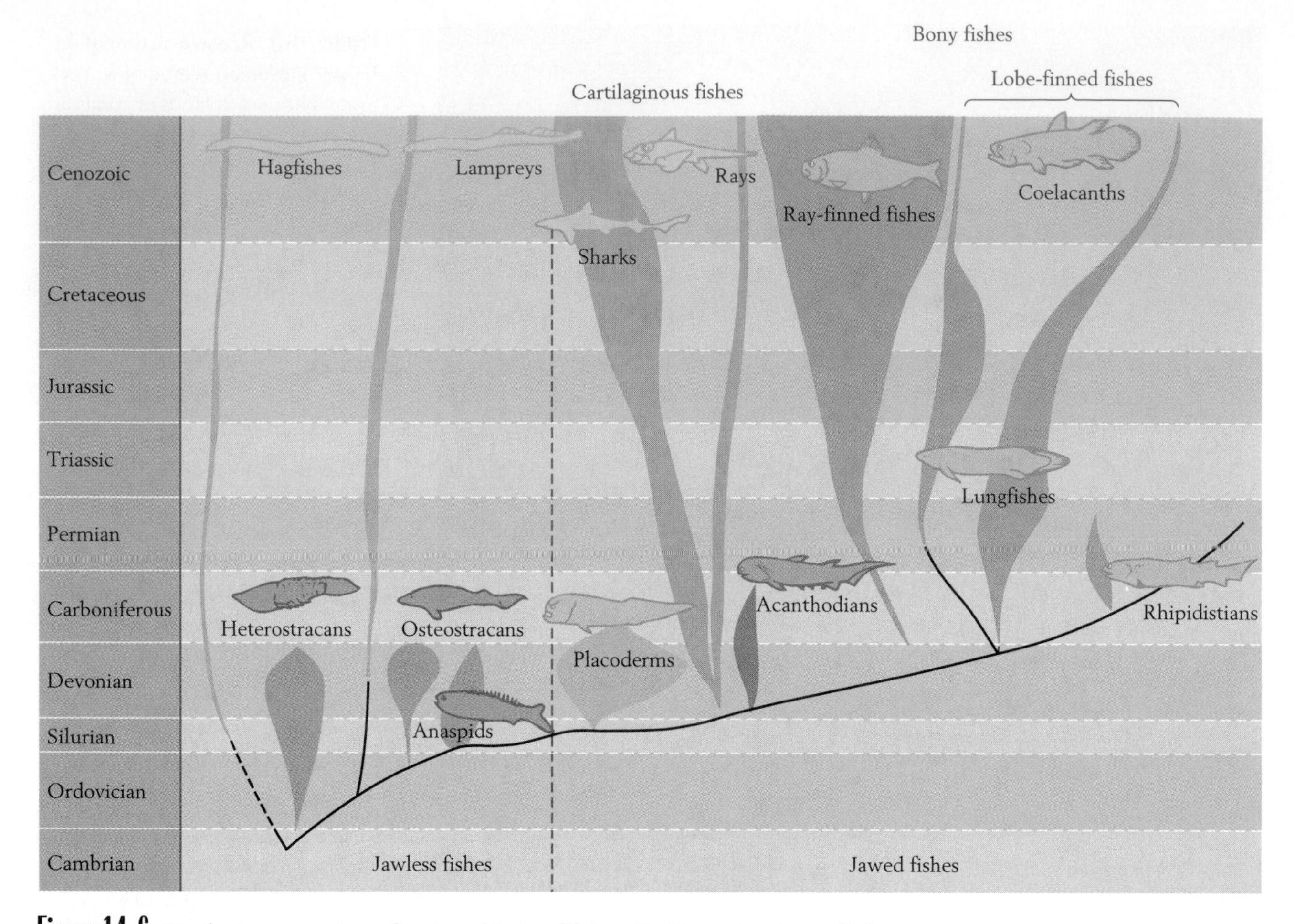

Figure 14-6 **Geologic occurrence of various kinds of fishes.** By Devonian time, all the major groups living today were in existence; no ostracoderms and few placoderms survived beyond the Devonian Period. *(Modified from E. H. Colbert, Evolution of the Vertebrates, John Wiley & Sons, New York, 1980.)*

Cambro-Ordovician fish fossils, come from freshwater deposits. One of the most conspicuous new groups of fishes was the **ostracoderms,** whose name means "bony skin." Ostracoderms were small animals with paired eyes like those of higher vertebrates. Lacking jaws and covered by bony armor, ostracoderms differed from modern fishes. Their small mouths allowed them to consume only small items of food. Many of these fishes, such as *Hemicyclaspis* (Figure 14-7*B*), also had flattened bellies that apparently were adapted to a life of scurrying along the bottoms of lakes and rivers. The upper fin of the asymmetrical tail of *Hemicyclaspis* was elongate; when the animal wagged it back and forth for swimming, this structure would have pushed its head downward rather than upward. In contrast, the ostracoderm known as *Pteraspis* (Figure 14-7*A*) had a curved belly and an elongate lower fin in its tail that would have lifted it upward, suggesting a life of more active swimming well above lake floors and river bottoms. Ostracoderms not only lacked jaws for chewing, but they also lacked bony internal skeletons as well as highly mobile fins, which provide more advanced fishes with stability and control of their movements; it is assumed that ostracoderms had cartilaginous internal skeletons. These animals continued to thrive throughout most of the Devonian Period but disappeared at the end of that interval.

Fishes with Jaws Late in Silurian time a quite different group of small marine and freshwater fishes made their appearance. These were the **acanthodians**—elongate animals with numerous fins supported by sharp spines (Figure 14-8; Box 14-1). The

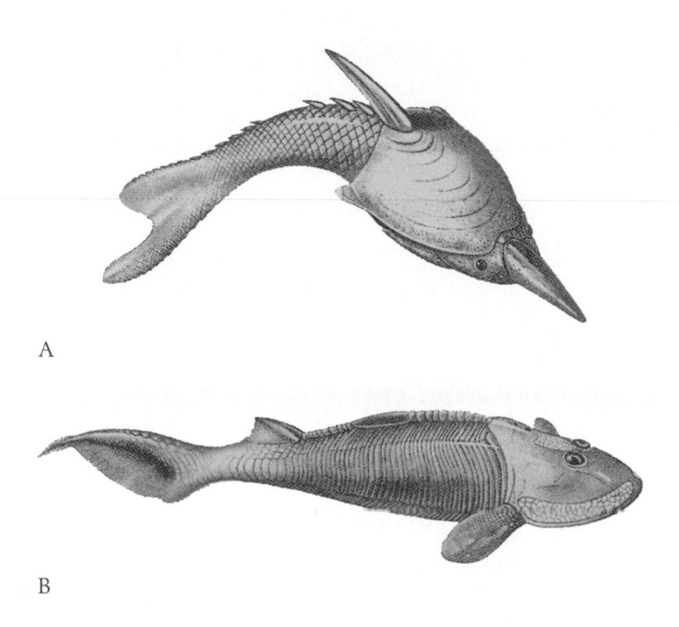

Figure 14-7 Reconstruction of Devonian ostracoderms (jawless fishes). *Pteraspis (A)* was about 20 centimeters (8 inches) long, and *Hemicyclaspis (B)* was about 13 centimeters (5 inches) long. *(After D. Dixon et al., The Macmillan Illustrated Encyclopedia of Dinosaurs and Prehistoric Animals, Macmillan Publishing Company, New York, 1988.)*

acanthodians appear to have been the first fishes to possess several features that were passed on to more advanced, modern fishes: Their fins were paired; scales rather than bony plates covered their bodies; and, most important, they had jaws. With the origin of jaws, a wide variety of new ecologic possibilities opened up for vertebrate life—possibilities that related primarily to the ability to prey on other animals. Unlike ostracoderms, many acanthodians must have been predators that fed on small aquatic animals. As Box 14-1 indicates, the jaws of acanthodians evolved from the gill supports of ancestral fishes.

Acanthodians declined near the end of Devonian time, but they left an evolutionary legacy of great ecologic significance. We do not know precisely how more advanced groups of fishes were related to acanthodians, but during the Devonian Period, a great adaptive radiation of descendant jawed fishes added new levels to the food webs of both freshwater and marine habitats. Soon very large fishes were feeding on smaller fishes, which in turn fed on still smaller fishes. At the top of this food web were the largest members of the group, the **placoderms.** A few of these heavily armored jawed fishes are known from uppermost Silurian and Lower Devonian freshwater deposits, and a wide variety of freshwater species existed by mid-Devonian time. Only secondarily did placoderms make their way into the oceans, and they were not highly diversified there until Late Devonian time. *Dunkleosteus,* a Late Devonian marine genus, attained a length of some 7 meters (23 feet). Like other placoderms, it had armorlike bone protecting the front half of its body (Figure 14-9), but its unarmored

Figure 14-8 Restoration of *Climatius,* a member of the most primitive group of jawed fishes, known as acanthodians. This animal was about 7.5 centimeters (3 inches) in length. *(After D. Dixon et al., The Macmillan Illustrated Encyclopedia of Dinosaurs and Prehistoric Animals, Macmillan Publishing Company, New York, 1988.)*

Figure 14-9 The massive armored skull of *Dunkleosteus,* a placoderm fish of Late Devonian age. This skull was more than a meter (~3 feet) long. Note the bony teeth and the armor protecting the eye.

For the Record 14-1

Jaws, Evolution, and Genetic Engineering

The origin of jaws teaches us an important lesson about evolution. The process of natural selection must work with the materials at hand. Genetically based variability within populations is the raw material on which natural selection must work, and this variability is limited. New biological structures must evolve from old ones, not from raw materials derived from sources outside the organism. Even some potentially useful modifications of animals and plants that simple genetic changes might produce have never actually occurred simply because, by chance, the genetic changes have never taken place. Mutations, after all, are genetic accidents.

The evolution of jaws transformed the marine ecosystem. It also added a new dimension to life on land when jawed vertebrates emerged from aquatic environments near the end of the Devonian Period. Both in the sea and on the land, the existence of jaws permitted sophisticated predators to evolve. Ultimately, evolution produced human jaws by remodeling those of fishes. Of course, there were many stages of development along the way, and other lines of evolution produced jaws of various other types.

Where did fishes get their jaws? Evolution did not grow them from tiny beginnings. Instead, as evolution has so often done in "building" new biological structures, it produced jaws from other features that were already well developed. Jaws evolved during the Devonian Period from bars that supported the gills of primitive fishes. Fossil sharks provide much of the evidence.

In sharks, as in other primitive vertebrates, skeletal bars lie on either side of the throat between the gill slits—the openings that allow water to pass through the gills. Each bar has an upper and lower part that connect to form a backward-pointing V. The jaws of some primitive Devonian sharks resemble these bars in both shape and orientation. Unlike our jaws, those of primitive sharks were not attached to the skull: they were positioned directly in front of the gill bars and were aligned with them. In fact, the jaws of some fossil sharks give the appearance of being the first gill bars in the series, although they differ slightly in shape from the bars and they bear teeth. It is difficult to avoid the conclusion that the jaws amount to modified gill bars. A strong similarity between the muscles that operate the jaws and those that close the gill slits reinforces this conclusion. Fossils of primitive sharks found in black shales in northeastern Ohio have been magnificently preserved under anoxic conditions similar to those that produced the Burgess Shale fauna of soft-bodied animals (p. 349). These fossils display traces of the jaw muscles of the sharks, along with many other anatomical features.

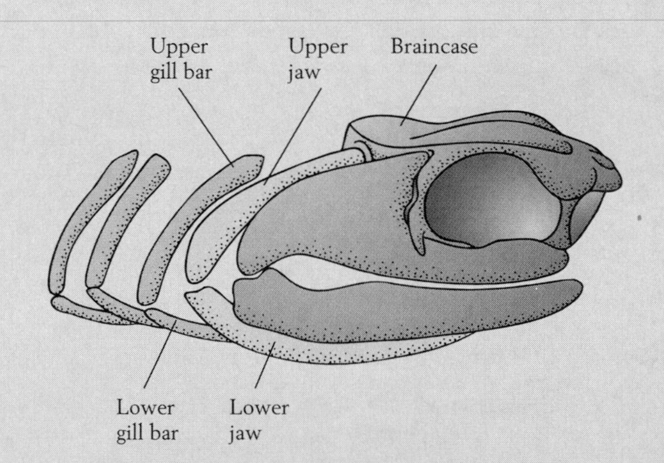

The braincase, jaws, and gill supports of a Carboniferous shark illustrate the primitive configuration of jaws in fishes. *(After R. Zangrel and M. E. Williams, Paleont. 18:333–341, 1975.)*

The similarity of gill bars and primitive jaws extends even to the teeth. Small, pointed structures called denticles toughen the skin of sharks, giving it the texture of sandpaper. The teeth that line the jaws of primitive sharks resemble the denticles of their skin. Thus it appears that evolution produced primitive teeth along the jaws simply by enlarging the denticles in the skin that overlay the ancestral gill bars.

An engineer with the power to design and build a shark would produce a better jaw than the primitive one that evolved from the gill bar. The skeleton of the early jaw consisted of soft cartilage rather than bone, and it floated loosely in soft tissue, without connection to the braincase. The primitive teeth also remained too much like denticles to function as effectively as those that an engineer might design or those that evolution eventually produced.

When humans interfere with the natural process of evolution by breeding domestic animals and plants selectively to improve their value to society, they are at the mercy of the variability that appears in the populations they breed. They cannot produce any kind of cow, chicken, or corn plant they want. Genetic engineering—the process in which biologists artificially redesign organisms by restructuring their genetic codes—carries human interference further, but even genetic engineering has limitations. Only certain genetic recombinations are possible, and many of them produce organisms that will not develop properly or function well.

It is easy to see why evolution has produced many organisms that seem highly imperfect and why it has failed to produce all kinds of organisms that our imaginations can conjure up. Chance has limited its productions, and so has the requirement that it create new forms of life from ones that already exist.

tail, which remained flexible for locomotion, was exposed to attack. *Cladoselache*, a small shark, is commonly found with the giant *Dunkleosteus* in black shales of northern Ohio (Figure 14-10).

Sharks were, in fact, among the most important groups of fishes in Devonian seas. Unknown in rocks older than mid-Devonian, the sharks may have been the last major group of fishes to evolve. Devonian sharks were primitive forms, and few grew much longer than 1 meter (3 feet).

Also making an appearance in Devonian time were the **ray-finned fishes.** These jawed forms, which attained only modest success during the Devonian Period, went on to dominate Mesozoic and Cenozoic seas. They include most of the familiar modern marine and freshwater fishes, such as trout, bass, herring, and tuna. The term *ray-finned* refers to the fact that thin bones that radiate from the body support the fins of these fishes; these bones are visible through the transparent fins of living fishes. The oldest mid-Devonian ray-finned fishes, such as *Cheirolepis* (Figure 14-11), differed from modern representatives in having asymmetrical tails and diamond-shaped scales that did not overlap.

The origin of the ray-finned fishes was an event of great significance, but so was the origin of a related group of jawed fishes—one that included the lungfishes and the lobe-finned fishes.

Fishes with Lungs The Devonian Period was the time of greatest success for the **lungfishes,** only three genera of which survive today—one in South America, one in Africa, and one in Australia. (Presumably this fragmented distribution reflects the Mesozoic breakup of Gondwanaland.) The Australian genus, *Neoceratodus*, so closely resembles the Triassic genus *Ceratodus* that it is commonly referred to as a "living fossil." The surviving lungfishes are named for the lungs that allow them to gulp air when they are trapped in stagnant pools during the dry season. Such lungs presumably served a similar function in Devonian time.

Lungfishes belong to a group known as **lobe-finned fishes** (Figure 14-12). These fishes derive their name from their paired fins, whose bones are not radially arranged, as in ray-finned fishes, but instead attach to their bodies by a single shaft. Most lobe-finned fishes occupied freshwater habitats, but one unusual group, the coelacanths, invaded the oceans. A

Figure 14-10 **The giant placoderm *Dunkleosteus* in pursuit of the Late Devonian shark *Cladoselache.*** These creatures were, respectively, about 7 meters (23 feet) and 2 meters (6 feet) long. *(After D. Dixon et al., The Macmillan Illustrated Encyclopedia of Dinosaurs and Prehistoric Animals, Macmillan Publishing Company, New York, 1988.)*

single coelacanth genus survives today in deep waters off the southeastern coast of Africa (see Figure 3-35). Lobe-finned fishes declined after the Devonian Period but left a rich evolutionary legacy. As we will see, lobe-finned fishes are the ancestors of all terrestrial vertebrates, including humans; their lungs were the predecessors of our own.

The Impact of Swimming Predators The great diversification of jawed fishes and, to a lesser extent, the expansion of ammonoids and eurypterids must have had a profound effect on many relatively defenseless aquatic animals. These predators may have contributed to the decline in the trilobites' diversity in middle Paleozoic time. About 80 families of trilobites are known from the Ordovician, but only 23 families have been found in Silurian deposits. It seems likely that the weakly calcified external skeletons of these animals offered little resistance to the jaws of fishes, and certainly trilobites had no mechanism for rapid locomotion. Advanced cephalopods and jawed fishes were probably responsible for the mid-Paleozoic decline of trilobites. The small, apparently defenseless ostracoderms, which died out late in the Devonian Period, must also have been easy prey for jawed fishes. Ostracoderms even lacked the

Figure 14-11 ***Cheirolepis,* a primitive ray-finned fish of mid-Devonian age.** The tail of this animal was strongly asymmetrical, and the small, diamond-shaped scales did not overlap. The animal was about 55 centimeters (22 inches) long. *(After D. Dixon et al., The Macmillan Illustrated Encyclopedia of Dinosaurs and Prehistoric Animals, Macmillan Publishing Company, New York, 1988.)*

Figure 14-12 **The lobe-finned fish *Eusthenopteron,* a large animal that exceeded 50 centimeters (20 inches) in length.** This unusually well preserved specimen is from Upper Devonian deposits at Scaumenac Bay, Canada. (See Figure 11-23.)

ability to burrow in sediment, which at least some of the trilobites could do (see Figures 13-4 and 13-13).

Plants Invade the Land

It is difficult to imagine how the landscape looked in Precambrian and early Paleozoic times, before terrestrial plants became widespread. Moist terrestrial environments must have been populated by algae, cyanobacteria, and fungi, but forests and meadows were absent, and there must have been large areas of barren rock and soil with little or no humus (decayed organic matter). Thus one of the most important events revealed by the fossil record of Silurian and Devonian life was the invasion of terrestrial habitats by higher plants.

The basic requirements for the terrestrial existence of large multicellular plants are quite different from those of plants that live in water. Unlike water, air is much less dense than the tissues of a plant, so if a plant is to stand upright in air, it must have a rigid stalk or stem. A tall plant must also be anchored by a root system or by a buried horizontal stem, which serves the further indispensable function of collecting water and nutrients from the soil.

The first upright plants to make their way onto the land lacked the roots, leaves, and efficient means to transport nutrients that made their descendants so successful. Essentially, these plants were simple rigid stems. Fragments of such early plants have been found in Silurian rocks. Silurian plants seem to have been pioneers that lived near bodies of water, and they may actually have been semiaquatic marsh dwellers rather than fully terrestrial plants.

Vascular Plants Most large plants of the modern world are **vascular;** that is, their stems have one set of special tubes to carry water and nutrients upward from their roots and another to distribute the food that the plants manufacture for themselves. Most large modern plants also bear leaves, which serve to capture the sunlight necessary for photosynthesis.

A major adaptive breakthrough for life on land, before the evolution of roots and leaves, was the origin of vascular tissue. Two kinds of tubes developed—one to transport water and nutrients and another to transport manufactured food. Figure 14-13 shows the tubes within a stem of the Early Devonian genus *Rhynia*. A few kinds of vascular plants are found in nonmarine deposits of latest Silurian age. These plants had

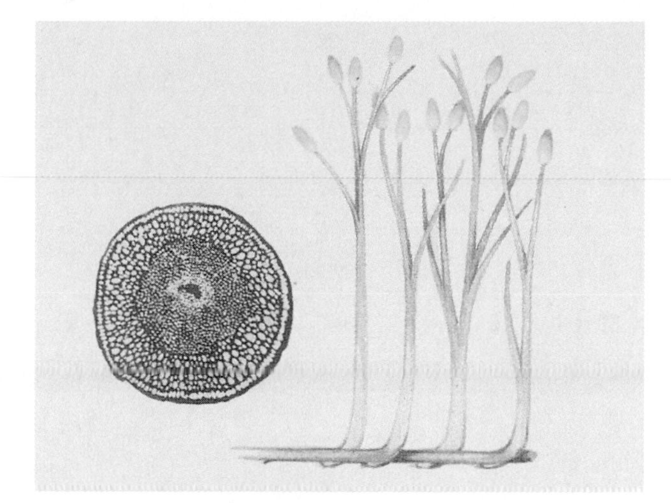

Figure 14-13 The Early Devonian vascular plant *Rhynia.* The reconstruction shows that a simple horizontal stem served the function of a root system, and this primitive plant bore no leaves. The yellow structures are spore organs. The vascular tissue that transported water is seen as a small dark area in the cross section of the stem; around it, and visible as a narrow, light-colored ring, is vascular tissue that transported food. *(From M. E. White, The Flowering of Gondwana, Reed Books Pty. Ltd., 1990.)*

branched leaves as well as bulbous organs that shed spores. *Baragwanathia*, the largest such plant yet discovered, grew to a height of about 1 meter (3 feet) (Figure 14-14).

Spore Plants **Spores** are reproductive structures that can grow into new adult plants when they are released into the environment. Ferns are familiar spore plants in the modern world. The fossil record of spores resembling modern ones extends well back into the Ordovician System (see Figure 13-16), but while these older spores suggest that upright land plants existed much earlier than the Late Silurian, they may in fact represent aquatic or semiaquatic species. In some Early Devonian forms, solitary spore organs stood atop upright stalks (see Figure 14-13), while other species displayed clusters of spore organs in similar positions; and in still other species, spore organs were arrayed along the upright stalks.

Whether or not land plants existed much before latest Silurian time, it was apparently near the end of the Silurian Period that vascular tissues evolved. As a result of this physiological breakthrough, a great adaptive radiation took place in Early Devonian time. The plants that resulted were still relatively low, creeping

Figure 14-14 **One of the oldest known vascular plants, a specimen of the genus *Baragwanathia* from the Upper Silurian of Victoria, Australia.** This plant was about 2.5 centimeters (1 inch) in diameter.

forms that lacked well-developed roots and leaves, but during Early and Middle Devonian time, more complex plants evolved. The vascular tissues of early vascular plants such as *Rhynia* were confined to a narrow zone of the stem (see Figure 14-13) and so were mechanically weak and inefficient at conducting liquid. By Late Devonian time, however, some plants had developed vascular tissues that occupied a larger volume within the stem and were therefore mechanically stronger and also more efficient transporters of nutrients. Plant groups with these useful traits also evolved roots for support and effective absorption of nutrients, as well as leaves for capturing sunlight. These plants seem to have competitively displaced such plants as *Rhynia*, which were less efficient at obtaining nutrients and synthesizing food.

Certain of the small plants that arose during Early and Middle Devonian times are classified as **lycopods.** This group includes the tiny club mosses of the modern world (Figure 14-15). Some late Paleozoic lycopods grew to the proportions of trees, and their petrified remains supply much of modern soci-

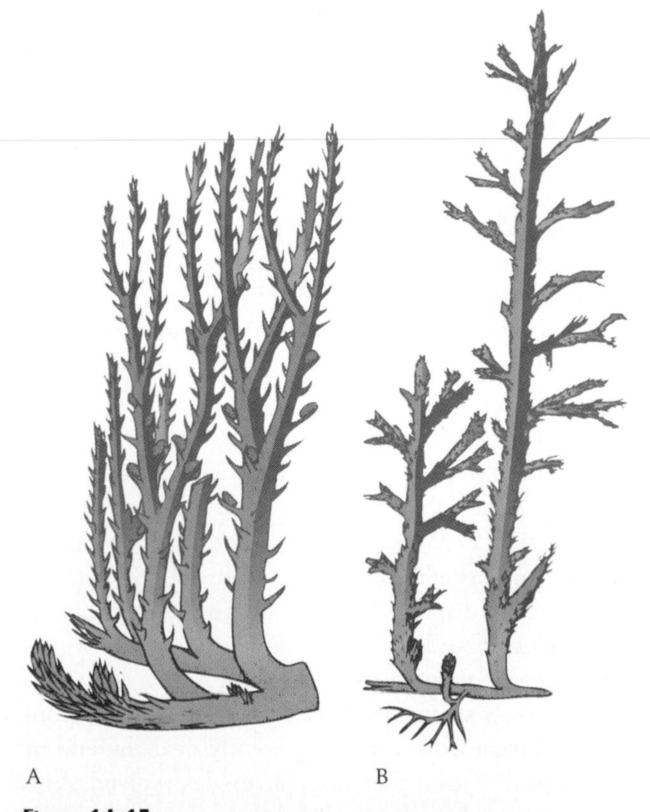

Figure 14-15 **Reconstructions of *Protolepidodendron (A)* and *Asteroxylon (B)*, primitive Devonian forms of lycopods.** *Lycopodium* (C) is a small form like the Devonian fossils, growing only a few centimeters tall. In contrast to these oldest and youngest representatives, many lycopods of late Paleozoic time were large trees.

A

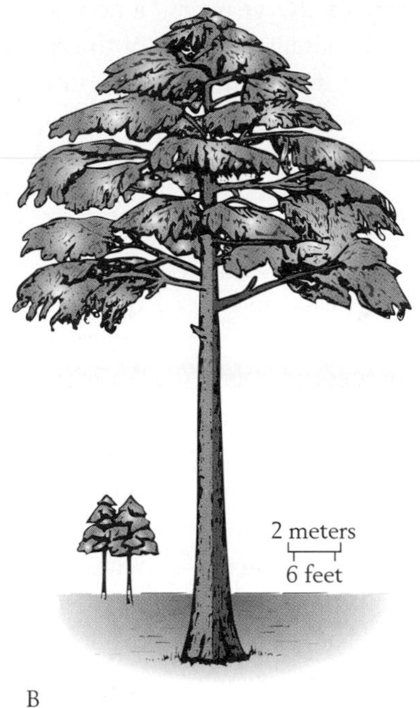

B

Figure 14-16 ***Archaeopteris*****, one of the very oldest kinds of large tree.** *A.* A segment of a branch with leaves. *B.* The entire tree. *(After C. R. Beck, Biol. Rev. 45:379–400, 1970.)*

ety's coal, but only tiny creeping lycopods resembling the primitive types of the Early Devonian have survived to the present.

During Late Devonian time, trees occupied the land for the first time. Quite abruptly, large trees with strong, woody stems changed the face of Earth, forming the world's first forests where there had been only barren land before. The spore plant *Archaeopteris* was the first genus to assume the proportions of a tree, reaching a height of about 30 meters (100 feet). Probably the adaptive breakthrough responsible for its large size was the evolution of broad leaves, which captured sunlight effectively (Figure 14-16).

Because of the limitations imposed by their mode of reproduction, low-growing Early Devonian spore plants must have formed marshes along bodies of water (Figure 14-17). Even *Archaeopteris* trees must have inhabited swamps. All of these spore plants, like those of the present day, had a complex reproductive cycle that would have restricted them

Figure 14-17 **Reconstruction of an Early Devonian landscape, where many of the first land plants still bordered bodies of water.** These plants were generally less than 1 meter (3 feet) tall.

to habitats that were damp at least part of the year (see Figure 3-22). This reproductive cycle entails not only a conspicuous spore plant, such as a fern, but also a tiny, inconspicuous plant over whose surface a sperm must travel to fertilize an egg. The sperm requires moist conditions to make its journey. As the Devonian Period progressed, however, the appearance of a second adaptive innovation, the **seed,** liberated land plants from their dependence on moist conditions and allowed them to invade many habitats.

Since fertilization is an internal process in seed plants, environmental moisture is not necessary. The seed, which results from fertilization, is released as a durable structure that can sprout into a plant when conditions become favorable. Pollen, which represents a different part of the cycle in seed plants, also tolerates a wide variety of environmental conditions. Pollen travels through the air to fertilize eggs so that seeds can form. Today most large land plants grow from seeds.

Flowerless seed plants originated in Late Devonian time and soon became important elements of late Paleozoic terrestrial floras (Figure 14-18), opening up a new world to the plant kingdom. One consequence of the spread of terrestrial vegetation was that, for the first time in Earth's history, plants carpeted the soil and gripped it with their roots, thereby stabilizing it against erosion. Braided-river deposits, which reflect rapid erosion (p. 132), characterize the clastic wedges of Precambrian and early Paleozoic age. Only in middle Paleozoic time, when vegetation first stabilized the land, did rivers begin to meander and to deposit sediment in orderly cycles (p. 134) on a large scale. An increase in the intensity of weathering was a second consequence of the spread of trees. Tree roots and the fungi associated with them release chemicals that weather rocky soil deeply and rapidly (p. 271). Fossils of so-called *root fungi* of the kind associated with modern tree roots have, in fact, been identified in Devonian rocks.

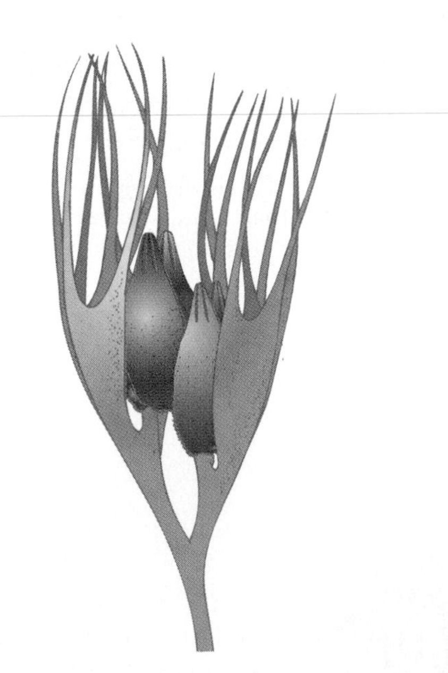

Figure 14-18 Reconstruction of Upper Devonian plants that bore some of the oldest known seeds at the ends of their branches. Individual seeds, four of which are shown, were about 1 centimeter (~$\frac{3}{8}$ inch) in length. *(J. M. Pettitt and C. B. Beck, University of Michigan Paleont. Contrib. 22:139–154, 1968.)*

Animals Move Ashore

The Lower Devonian Rhynie Chert of Scotland has yielded not only a variety of beautiful fossil plants but also some of the oldest known nonmarine arthropods, including scorpions and flightless insects.

Arthropods probably invaded dry land in Late Silurian time, before some of their terrestrial representatives were preserved in the Rhynie Chert, but it was not until the Late Devonian interval that vertebrate animals made a similar transition. Anatomical evidence indicates that the four-legged vertebrates most closely related to fishes are the **amphibians**—frogs, toads, salamanders, and their relatives (see Figure 3-10). In fact, amphibians return to the water to lay their eggs and spend their juvenile period there. Then most kinds of amphibians metamorphose into air-breathing, land-dwelling adults. Living amphibians are small animals that differ substantially from the large fossil amphibians found in upper Paleozoic rocks. Their evolutionary history began late in the Devonian Period.

In eastern Greenland, the remains of unusual vertebrate animals have been found in uppermost Devonian rocks of the Old Red Sandstone continent. These fossils, some of which are assigned to the genus *Ichthyostega*, represent creatures that are strikingly intermediate in form between lobe-finned fishes and amphibians. The lobe fin itself is formed of an array of bones resembling that found in amphibians; similarly, the complicated teeth of lobe-

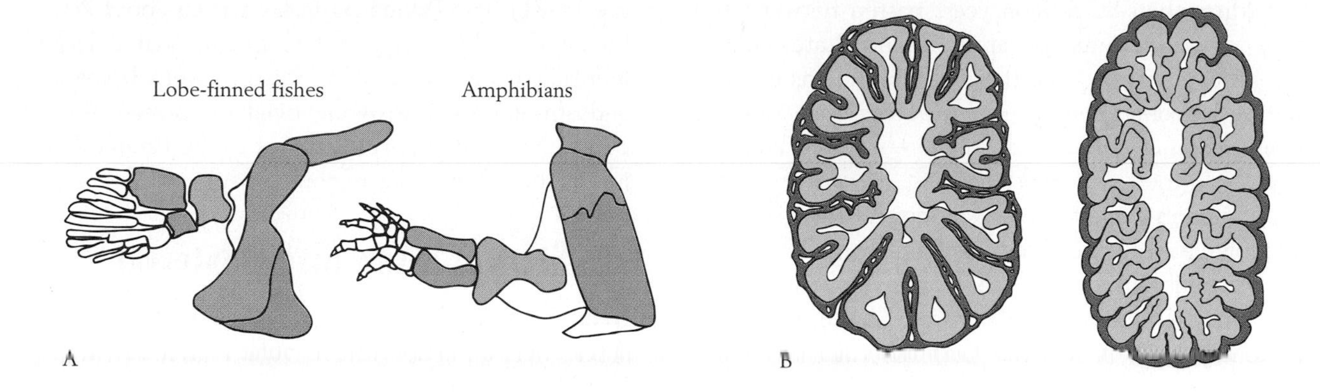

Figure 14-19 **Lobe-finned fishes (left) and amphibians (right).** *A.* The shoulder and limb bones; shading identifies particular bones that the two groups have in common. *B.* Cross sections of teeth, showing the unusual, complex structure found in both groups.

finned fishes closely resemble the teeth of early amphibians (Figure 14-19). These features alone strongly suggest that amphibians evolved from lobe-finned fishes, and additional features make the derivation a certainty. *Ichthyostega* had four legs, as do amphibians, but its skull structure was remarkably like that of a lobe-finned fish. The creature also had a fishlike tail—a feature of its ancestors that was probably not useful on land (Figure 14-20). Because of this intriguing combination of features, *Ichthyostega*, which was not discovered until the twentieth century, represents what is commonly termed a "missing link."

The lung, which early fishes occasionally put to use to breathe air, was available for exploitation long before amphibians evolved. In the way the lung became a full-time supplier of oxygen we can see yet another example of the "opportunism" of evolution. Unlike the gill supports that evolved into jaws earlier in vertebrate evolution, the structure that developed into lungs required very little evolutionary modification to open up an entirely new mode of life.

Figure 14-20 **The primitive amphibian *Ichthyostega*.** The legs of the *Ichthyostega* contrast with the fins of the approximately contemporary lobe-finned fish *Eusthenopteron* (left; see also Figure 14-12) and the lungfish *Rhynchodipterus* (right). The trunk and branches belong to the tree fern *Eospermatopteris*. *(Drawing by Gregory S. Paul.)*

More than 80 million years passed between the time vascular plants appeared on land (Late Silurian or earlier) and the time the first amphibians evolved (latest Devonian). It is not surprising that vascular plants colonized the land before vertebrate animals, because a food web must be built upward from the base; animals cannot live on land in the absence of an adequate supply of edible vegetation there.

Amphibians evolved so late in Devonian time that they played no significant role in the ecosystem of that period. It was the Carboniferous and Early Permian that might be called the Age of Amphibians.

The Paleogeography of the Middle Paleozoic World

Figure 14-21 displays the positions of continental areas during Middle Silurian time and during late Early Devonian time, about 65 million years later. An important new geographic feature to appear during Devonian time was the continent of Euramerica, which was formed by the union of Laurentia, Baltica, and Avalonia. In general, the Silurian and Devonian were periods when sea level stood high in relation to the surfaces of major cratons. Early in Silurian time, sea level rose from its low position at the end of the Ordovician Period—a rise that is thought to have resulted from continued melting of the extensive polar glaciers that had formed late in Ordovician time. Simultaneously, many marine invertebrate taxa underwent the radiations discussed earlier in this chapter.

The wide distribution of organic reefs strongly suggests that middle Paleozoic climates were relatively warm. Climates were also relatively dry in many areas, as evidenced by the accumulations of large volumes of evaporite deposits. Most evaporites of Silurian age formed within 30° or so of the ancient equator, but some Devonian evaporites accumulated farther north and south, apparently reflecting the broad distribution of warm climates during the Devonian Period (see Figure 14-21).

Early in Devonian time, perhaps because of cooling in polar regions, a discrete marine province formed along the margin of southern Gondwanaland. This southern realm, called the Paraná Basin, was populated by a fauna adapted to cool water (see Figure 14-21). The Paraná Basin lay within about 20° of the south pole, so it is not surprising that it lacked tabulate-strome reefs. Also missing were bryozoans and ammonites. Burrowing bivalves formed a large percentage of the marine species in the Paraná Basin, just as they do in polar regions today.

The Start of a New Glacial Interval

Near the end of the Devonian Period, glaciers again spread over a portion of this polar region, leaving behind tills and scratches on bedrock in South America. An interesting hypothesis relates the Late Devonian glaciation to the initial expansion of trees across landscapes throughout the world. We have noted that this expansion dramatically accelerated chemical weathering (p. 272). *Archaeopteris*, the first genus to form trees (see Figure 14-16), spread throughout the world but was quickly replaced by newly evolving groups of seed plants, which were able to invade drier terrain and expand the overall distribution of forests. Recall that the initial expansion of forests may explain the onset of the global decline in atmospheric carbon dioxide that seems to have begun in Devonian time (see Figure 10-13). Atmospheric carbon dioxide is thought to have declined when increased amounts of carbon were buried as wood and leaves decayed. In addition, accelerated chemical weathering extracted carbon dioxide from the atmosphere (p. 272).

Certainly, then, we must consider the possibility that a reduced greenhouse effect resulting from the growth of Earth's first forests caused climates to cool throughout the world. Such global cooling may explain the expansion of glaciers in Gondwanaland.

Late Devonian Mass Extinction

One of the most devastating mass extinctions of marine life in all of Phanerozoic time took place near the end of the Devonian Period, when more than 40 percent of all marine genera disappeared (see Figure 7-12). Geologists divide the Upper Devonian Series into two stages—the Frasnian Stage and the Famennian Stage. The crisis spanned much of late Frasnian and Famennian time.

Offshore deposits in several parts of the world point to two intervals when deep waters of the ocean were anoxic. These organic-rich deposits, known as Kellwasser beds, typically occur as two separate

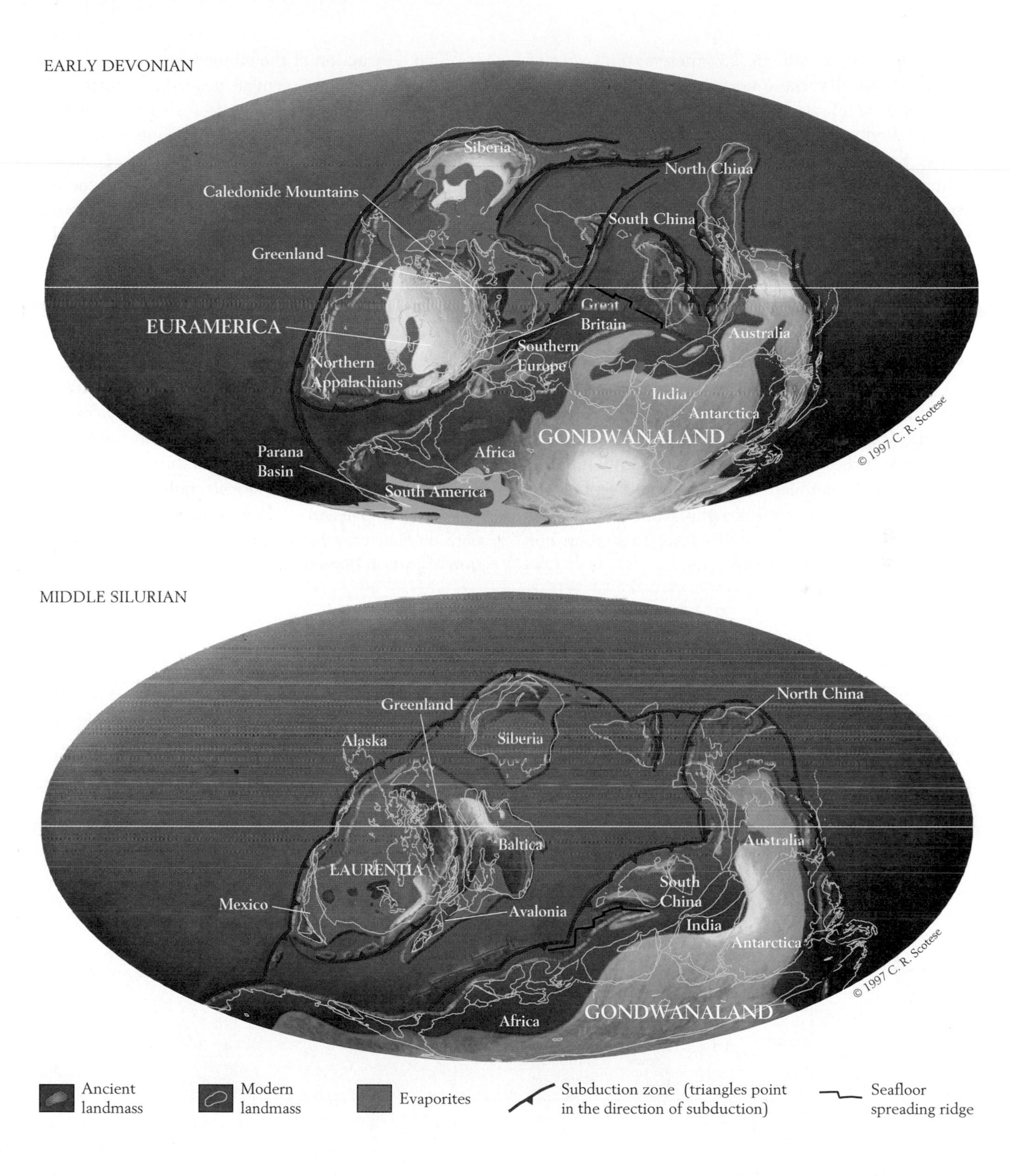

Figure 14-21 World geography during Middle Paleozoic time. During this interval, Baltica and Avalonia collided with Laurentia to form Eurasia; this collision connected the landmasses that now form the British Isles. During the Devonian Period, shallow seas near the south pole formed the Paraná Basin. *(Adapted from paleogeographic maps by C. R. Scotese, PALEOMAP Project, University of Texas at Arlington, 1997.)*

units, each about 0.5 to 2.0 meters thick. Facies shifts indicate that these units accumulated at times when sea level was high (Figure 14-22). Many species died out about the times when the Kellwasser beds formed. Reduced oxygen cannot be the primary cause of these extinctions, however, because waves constantly stir and oxygenate the upper waters of the ocean; yet many shallow-water taxa died out in the Late Devonian crisis. The tabulate-strome reef community, which occupied shallow seas, was so severely affected that it never recovered. In fact, it declined gradually during the interval between the two times of Kellwasser deposition.

A final pulse of extinction took place at the end of the Famennian age (the very end of the Devonian Period). At this time the planktonic acritarchs and the placoderm fishes—two groups that had flourished throughout Famennian time—nearly disappeared. Only a few freshwater species of placoderms survived into Carboniferous time, and neither this group nor the acritarchs ever diversified again.

The destruction of the tabulate-strome reef community exemplifies a general geographic pattern of the Late Devonian extinction: losses of tropical taxa were especially heavy, whereas inhabitants of the cold Paraná Basin were almost unscathed (see Figure 14-21). This geographic pattern suggests that cooling of seas played a major role in the extinction. Whereas nontropical species could have migrated toward the equator to find suitably warm waters as temperatures declined, no region would have remained warm enough to serve as a refuge for equatorial forms.

A rapid global decline of sea level followed each of the two Kellwasser intervals (see Figure 14-22). These two eustatic events may well have resulted from the expansion of glaciers in Gondwanaland at about this time. It would be remarkable if the extinction of many forms of life in the ocean could eventually be connected to greenhouse cooling traceable to the initial expansion of forests on the land—an expansion that may have caused a considerable reduction of carbon dioxide in the atmosphere (p. 386).

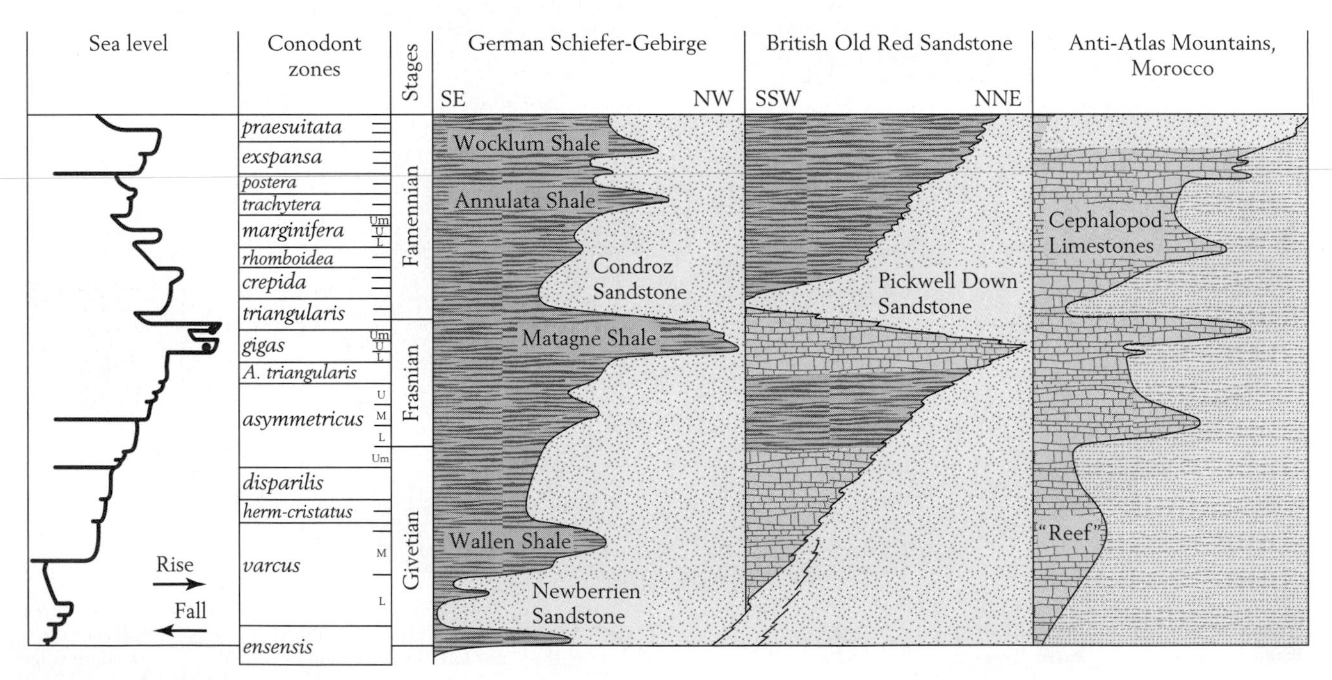

Figure 14-22 Evidence of declines in sea level about the times of the Late Devonian extinctions. Each of the Kellwasser events was followed by a eustatic fall. The sea-level curve (right) is a general one for Euramerica. The stratigraphic sections are for three separate regions; each shows a general regression in the uppermost Frasnian. *(Stratigraphic sections after W. Buggisch, Geol. Rundschau 80:49–72, 1991; sea-level curve after J. G. Johnson and C. A. Sandberg, Canadian Soc. Petrol. Geol. Mem. 14:171–178, 1988.)*

Regional Examples

The Silurian and Devonian periods were times of widespread reef development and carbonate deposition, but they were also times of orogeny. While eastern North America was being transformed from an Early Silurian highland to a Middle Silurian carbonate shelf, reefs and evaporite deposits were forming farther to the west. Later in the Devonian Period, Laurentia and Baltica united to form the continent of Euramerica, and mountains rose up in the Appalachian region. Mountains also formed in western North America, but reefs continued to grow there as well.

Eastern North America

The Taconic orogeny in eastern Laurentia ended with the deposition of clastic wedges of sediment shed westward from the Taconic Mountains, which formed along eastern Laurentia late in the Ordovician Period (see Figure 13-27). The Silurian Period began with a continuation of this pattern. As the eastern highlands were subdued by erosion, however, the site of clastic deposition became more broadly flooded by shallow seas, and late in Silurian time shallow-water carbonates accumulated along a new passive margin.

To the west, tabulate-strome reefs dotted shallow epicontinental seas (Figure 14-23). Here, however, the pattern of sedimentation and reef development changed drastically from Middle to Late Silurian time. First, during the Middle Silurian, two basins accumulated muddy carbonates. One was the Michigan Basin (see Figure 9-24), and the other lay in what is now north-central Ohio. These basins were bounded by large barrier reefs and were populated by scattered pinnacle reefs. At this time siliciclastic muds were still accumulating on broad tidal flats to the east.

As the Silurian Period progressed, this pattern changed. To the east, in Pennsylvania and neighboring areas, the deposition of siliciclastic mud gave way to carbonate sedimentation. At the same time, the Michigan and central Ohio basins came to be only weakly supplied with seawater and thus turned into evaporite pans in which dolomite, anhydrite, and halite were precipitated. The resulting deposits are a major source of rock salt today. It appears that marginal reefs in the Michigan Basin grew so high during Middle Silurian time that they restricted the flow of water into the basin. In time, evaporation and possibly a slight lowering of sea level led to the exposure and consequent death of the reefs. Although a weak flow of seawater into the basin replenished the water that had been lost by evaporation, the rate of evaporation was so high that evaporite minerals were precipitated around the margins of the basin and even at considerable depths within it. At first the center of the basin was moderately deep, but as evaporites accumulated, the water there grew progressively shallower until eventually the sea was excluded altogether.

Because the conditions within the evaporite basins were so inhospitable during the Late Silurian, reefs grew only to their southwest, in Indiana and Illinois. The most famous of them is the Thornton Reef of northern Illinois (Figure 14-24). The structure of this reef indicates the direction of the prevailing winds

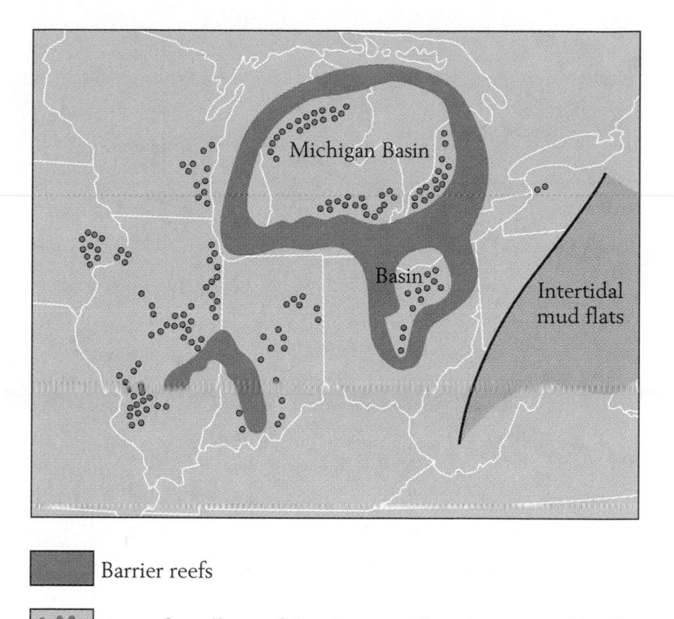

Figure 14-23 Middle Silurian reefs of the Great Lakes region. Barrier reefs encircled the Michigan Basin and a smaller basin in Ohio. They also flourished in southern Indiana and Illinois. Extensive mud flats now lay to the east in the Pennsylvania region, in contrast to the environments of coarse clastic deposition that occupied this area in Early Silurian time. *(Modified from K. J. Mesolella, Amer. Assoc. Petrol. Geol. Bull. 62:1607–1644, 1978.)*

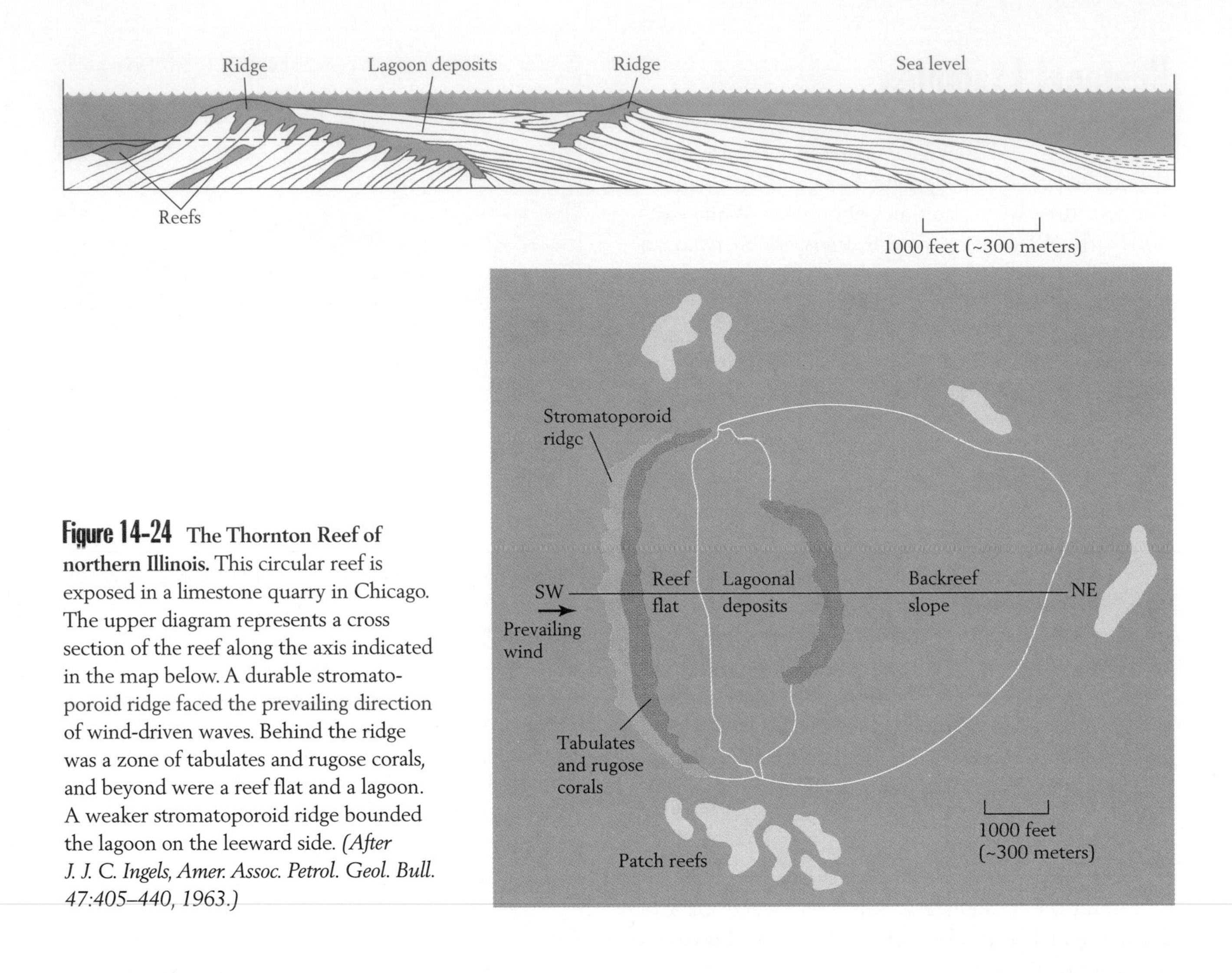

Figure 14-24 The Thornton Reef of northern Illinois. This circular reef is exposed in a limestone quarry in Chicago. The upper diagram represents a cross section of the reef along the axis indicated in the map below. A durable stromatoporoid ridge faced the prevailing direction of wind-driven waves. Behind the ridge was a zone of tabulates and rugose corals, and beyond were a reef flat and a lagoon. A weaker stromatoporoid ridge bounded the lagoon on the leeward side. *(After J. J. C. Ingels, Amer. Assoc. Petrol. Geol. Bull. 47:405–440, 1963.)*

at the time it was growing: the stromatoporoid ridge obviously faced waves advancing from the southwest.

Euramerica

One of the classic angular unconformities in the geologic record occurs in Scotland between Devonian beds of the nonmarine Old Red Sandstone and the nearly vertical Silurian marine beds on which they rest. It was at the locality shown in Figure 14-25 that James Hutton recognized the meaning of stratigraphic unconformity in 1788 (p. 10). Thus it came to be understood that the Old Red Sandstone was deposited after a Silurian episode of mountain building. The Old Red crops out over large areas of Scotland. Chunks of its distinctive red sandstones (see p. 368) can be found in parts of Hadrian's Wall, built by the Roman emperor Hadrian across northern England in the second century.

The Old Red includes not only rocks of Early, Middle, and Late Devonian age but also rocks representing Late Silurian and earliest Carboniferous times. For a long time geologists found it puzzling that such a large volume of sediment could have been shed from highlands in the British Isles when most of the islands' area formed a depositional basin. Now the puzzle has been solved within the framework of plate tectonics by the reassembly of the landmass that formed when Laurentia and Baltica were united during mid-Paleozoic time.

Continental Suturing and the Acadian Orogeny The second Phanerozoic interval of mountain building in the Appalachian region is known as the *Aca-*

Figure 14-25 Angular unconformity between the Old Red Sandstone (top left) and Silurian rocks (bottom left) at Siccar Point, Berwickshire, Scotland. The Silurian rocks were tilted and folded when Baltica collided with Laurentia to form Euramerica (see Figure 11-21). The Old Red Sandstone was subsequently deposited near the margin of this continent.

dian orogeny. Like the Taconic orogeny before it, the Acadian entailed collision of landmasses, but this time the landmasses were larger. In fact, the Acadian event resulted from a double collision along the eastern margin of Laurentia. In the north, the collision was with Baltica; in the south it was with the microcontinent Avalonia (see Figure 14-21).

The Acadian orogeny began in the north, in mid-Silurian time. Here the suturing of Laurentia to Baltica produced an orogenic belt now located in northeastern Laurentia and Greenland. This suturing also created the Caledonide Mountains, which now lie along the Atlantic coast of Norway.

Orogenic activity then progressed southward as Avalonia, which was evidently part of the same lithospheric plate as Baltica, collided with eastern Canada and the northeastern United States. Recall that Avalonia was an elongate microcontinent that broke away from Gondwanaland and moved close to Laurentia during the Ordovician Period. Small islands moving ahead of it were attached to Laurentia during the Taconic orogeny (see Figure 13-21). Seaward of these exotic terranes are others, known as Avalon terranes, that represent segments of Avalonia attached during the Acadian orogeny, along with elements of oceanic crust (Figure 14-26). The Avalon terranes share fossil taxa that are unknown from North American rocks immediately to the west. These foreign taxa, which also occur in larger fragments of Gondwanaland, include the distinctive Early Cambrian trilobite *Paradoxides*.

Farther south, another elongate microcontinent became sutured to the eastern United States to form the Carolina terrane (see Figure 14-26). Metamorphism has obscured the original nature of this terrane, but a few surviving fossil remains, including those of *Paradoxides*, point to an origin in or close to Gondwanaland.

Paradoxides also occurs in southern Great Britain, which formed the northern segment of Avalonia and became attached to Scotland and Northern Ireland during the Acadian orogeny. Thus the British Isles were assembled. The assembly of continental Europe would come later. Early in the Devonian Period a microcontinent destined to become southern Europe was encroaching on the newly forming Euramerica (see Figure 14-21). This microcontinent had rifted away from Gondwanaland after the departure of Avalonia. It would attach to the former landmass of Baltica (northern Europe) to create continental Europe late in the Paleozoic Era, when Gondwanaland also was sutured to Euramerica. Thus Great Britain and continental Europe became locked within the supercontinent Pangea (see Figure 8-5); their Atlantic and Mediterranean coastlines did not form until this supercontinent began to fragment early in the Mesozoic Era.

The Second Great Depositional Cycle in the Appalachian Region Let us now return to the eastern margin of Laurentia to examine the rocks that provide a record of Avalonian mountain building. During the Silurian Period, erosion subdued the mountains formed during the Taconic orogeny, and a new passive margin formed along Laurentia's east coast; an immense carbonate platform soon extended

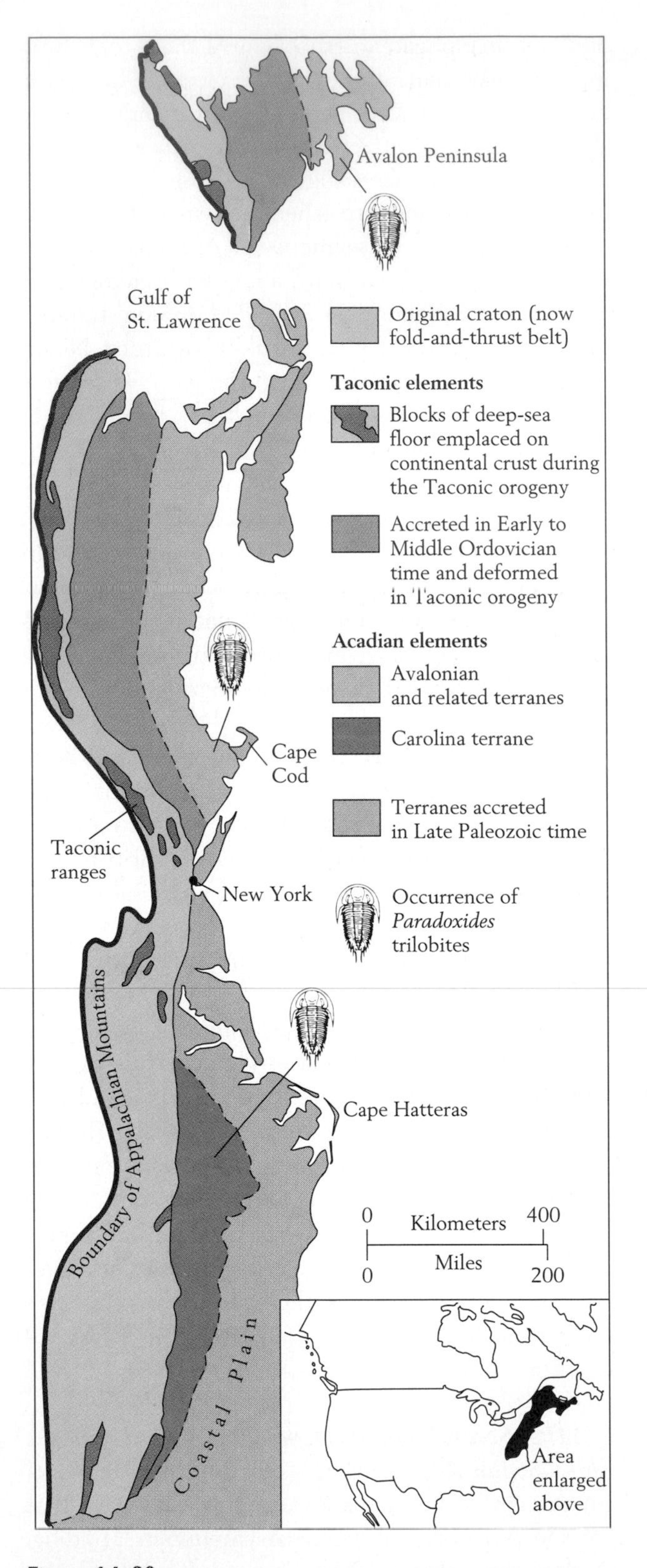

Figure 14-26 **Exotic terranes of eastern North America.** Distinctive fossils, including those of the Early Cambrian trilobite *Paradoxides,* characterize terranes accreted during the Acadian orogeny. *(Modified from H. Williams and R. D. Hatcher, Geology 10:530–536, 1982.)*

along much of its length. Reflecting the global prosperity of the tabulate-strome community, it included more extensive reefs than the Cambro-Ordovician carbonate platform of the same region (see Figure 13-18). Like the earlier platform, however, this one suddenly subsided to be replaced by a foredeep as mountains began to rise up to the east. The Oriskany Sandstone, a sandy beach deposit, spread over the uppermost carbonate rocks of the Helderberg Group (Figure 14-27). Then shallow-water sedimentation suddenly gave way to deposition of deep-water flysch. In New York State black muds of the Marcellus Formation came first, and they were followed by turbidites and shales of the Hamilton Group. A similar transition took place farther south, in Maryland (Figure 14-28). In accordance with the typical pattern of foreland basin deposition (p. 243), an increased supply of sediment from the adjacent mountain belt then pushed back waters of the basin, and deep-water deposits gave way to shallow marine and nonmarine molasse.

The enormous Catskill clastic wedge is a regressive body of molasse that records a westward progradation of sedimentary environments during the Acadian orogeny (see Figure 14-27). Nonmarine red beds of the Catskill wedge include sandstones and conglomerates that accumulated in braided streams near highlands (see Figure 5-14); at lower elevations meandering rivers formed cycles with coarse sediments at the base and muds at the top (see Figure 5-16). In some areas coarsening-upward deltaic cycles formed along the coastline (see Figure 5-20).

The cycle of deposition associated with the Acadian orogeny was much like the one associated with the Taconic orogeny. And in both cycles passive margin deposition was followed by accumulation of flysch and then molasse (compare Figures 13-23 and 14-27).

Interior and Eastern North America Figure 14-29 depicts general environments of Euramerica late in the Devonian Period. During much of Devonian time, an arm of land may have extended southwestward across what is now the western interior of North America, where the Transcontinental Arch persisted from early Paleozoic time (see Figure 13-27) and little or no sediment accumulated. The Devonian equator passed through Euramerica, so that prevailing winds, blowing from the east, must have supplied abundant moisture to the continent's east-

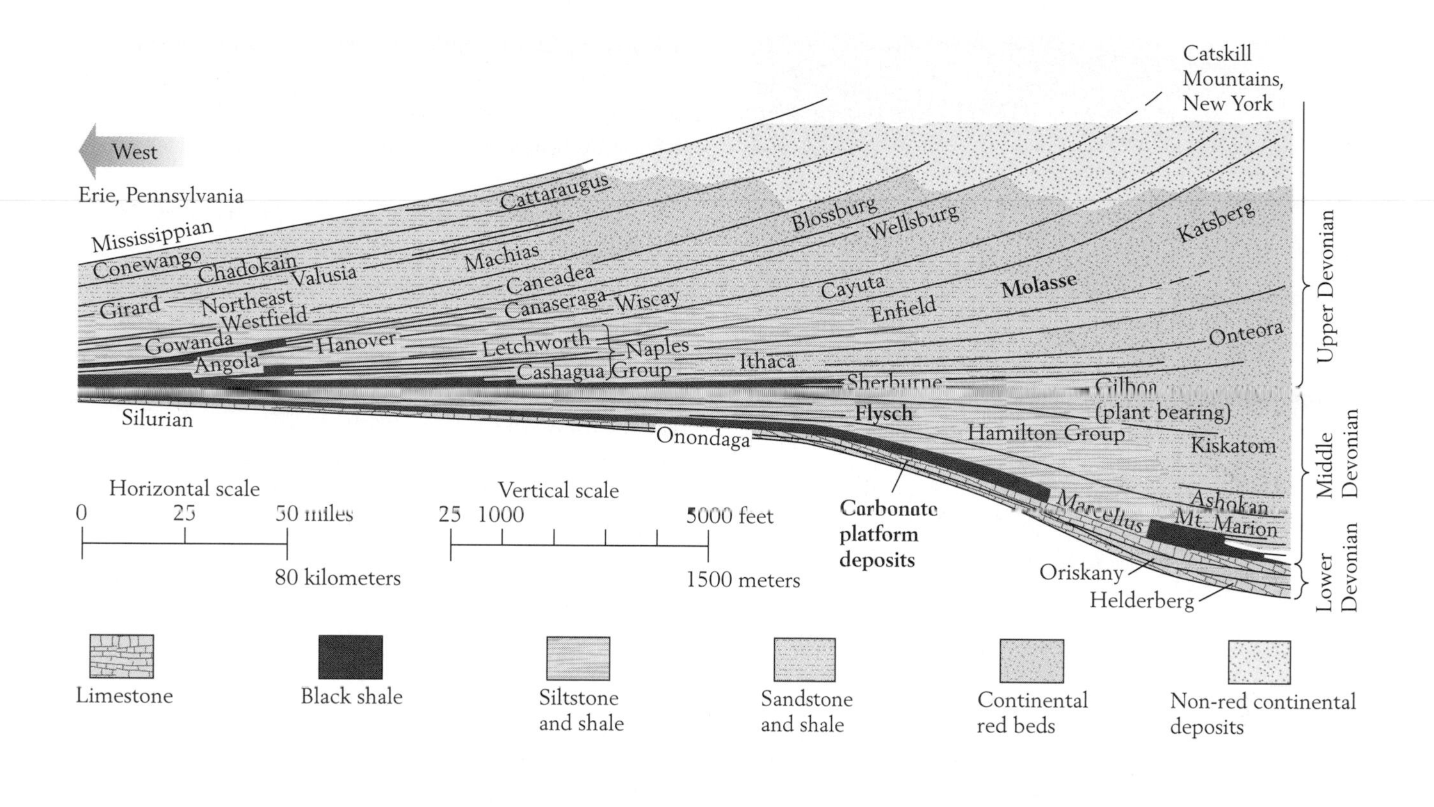

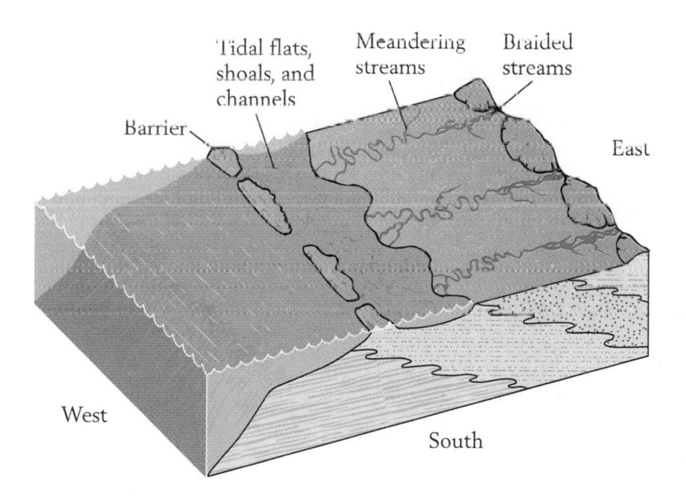

Figure 14-27 **Devonian rocks of New York State that record a regression caused by the Acadian orogeny.** (Top) Shallow-water carbonates (the Helderberg Group and Oriskany Sandstone) gave way to deep-water deposits (the Marcellus Shale and Hamilton Group). Next, coarse molasse deposits shed from the mountains formed a clastic wedge that spread westward. (Bottom) Environments of deposition of the Catskill clastic wedge and associated deposits. Braided streams meander seaward from the feet of the mountains to the east. Eventually they empty into lagoons behind barrier islands. Muds are deposited offshore. *(Top, after P. B. King, The Geological Evolution of North America, Princeton University Press, Princeton, N.J., 1977; bottom, after J. R. L. Allen and P. F. Friend, Geol. Soc. Amer. Spec. Paper 106, 21–74, 1968.)*

ern margin. It is therefore not surprising that coal deposits, formed from early land plants, are found here. Although evaporites are found here and there near the eastern margin of the continent, they are best developed in the rain shadow to the west, in the area of North America where the Rocky Mountains now stand. Along the east and west coasts of Euramerica, the climate was at least intermittently hot and dry enough for caliche nodules (p. 124) to form in abundance in low-lying areas that are now represented by ancient soils.

Late in the Devonian Period a mud-floored seaway lay to the west of the coastal mountains and the area of molasse deposition, extending northward to the Hudson Bay area (see Figure 14-29). The giant placoderm *Dunkleosteus* flourished in this sea, together with other fishes now preserved in the black shales of northern Ohio (see Figures 14-9 and

Figure 14-28 The Oriskany Sandstone in western Maryland. It was tilted late in the Paleozoic and is now weathering to loose sand. This unit formed along a beach, but it is overlain by deep-water shales that mark the onset of the Acadian orogeny.

14-10). Near the end of Devonian time, the deposition of black muds extended farther to the west, leaving a vast area of eastern and central North America blanketed with these sediments.

Reef Building and Orogeny in Western North America Along the continental shelf west of Eurasia, in what is now western Canada, tabulate-strome reef complexes developed during the latter half of the Devonian Period (Figure 14-30). Reefs here took the form of elongate barriers, atolls with central lagoons, and platforms. Many of these reefs now lie deeply buried, and their porous textures have created traps for petroleum. South of the belt of reef growth, carbonates were deposited in shallow seas, just as they had been early in Paleozoic time (see Figure 14-29).

Through Silurian and Devonian time, the western margin of North America remained approxi-

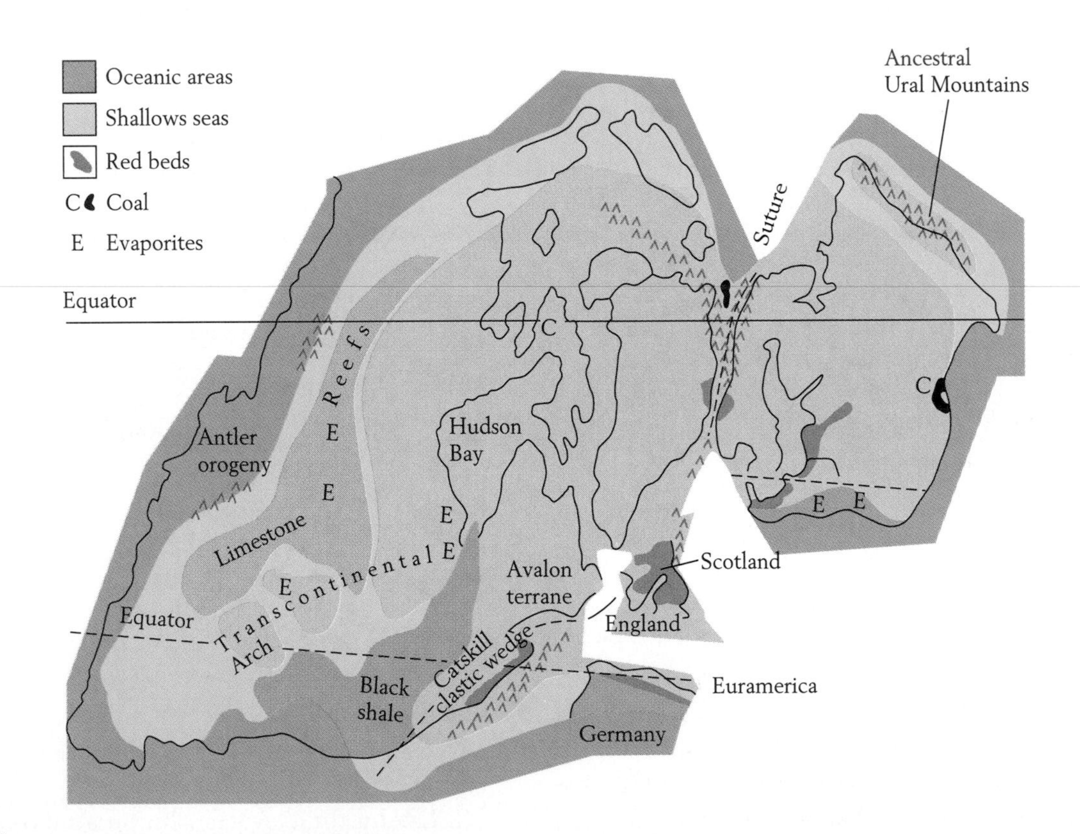

Figure 14-29 Euramerica during Late Devonian time. The central mountain belt running from north to south formed as Baltica converged with Laurentia (see Figure 14-21). Red beds were concentrated in the south, near the equator. Deep-water deposits accumulated in what is now central Germany. Shales accumulated in North America west of the Catskill clastic wedge, and limestones, reefs, and evaporites formed farther west. The Antler orogeny affected the western margin of the continent.

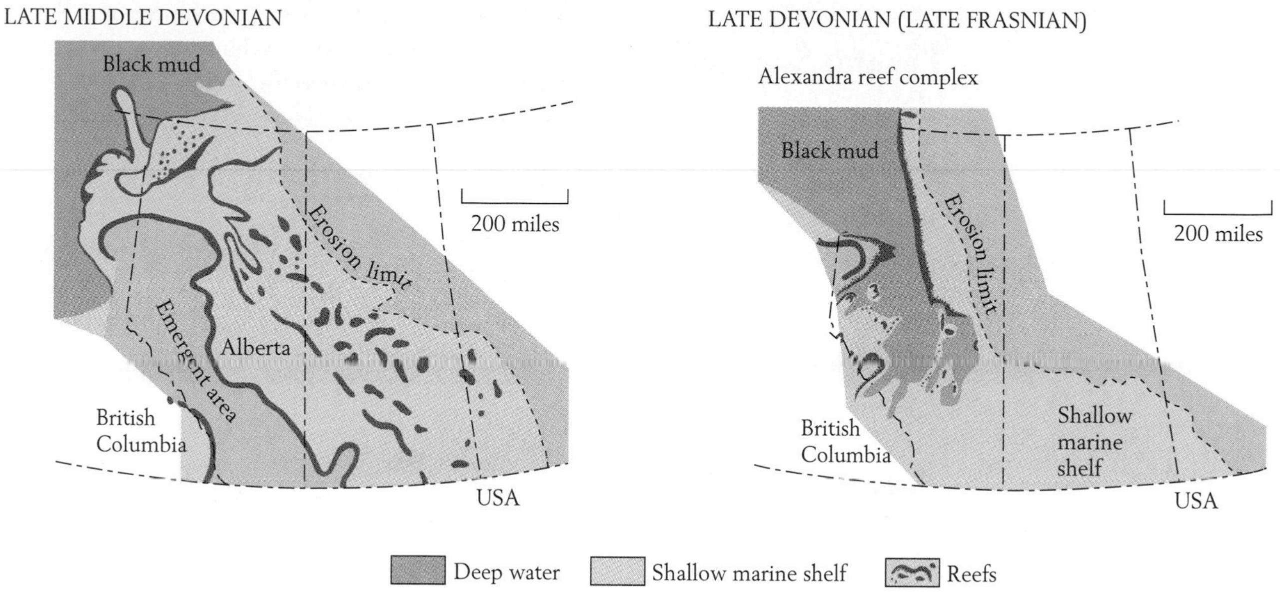

Figure 14-30 The distribution of reefs in western Canada during the latter part of Devonian time. Late in Frasnian time, reefs ceased to grow and black mud spread onto the continental shelf. Northeast of the erosion limit, Middle and Late Devonian rocks have been eroded away. *(After E. R. Jamieson, Proc. N. Amer. Paleont. Conv. J:1300–1340, 1969.)*

mately where it had been during the Ordovician Period. In middle Paleozoic time, however, an island arc stood offshore. In the Klamath Mountains and in the Sierra Nevada of present-day northern California, ophiolite sequences, which include graywackes, shales, cherts, and volcanics, record the presence of this Klamath Arc (Figure 14-31*A*). Rocks in Nevada show that this simple geographic picture became more complex between Middle Devonian and Early Mississippian time; they reveal closure of the basin between the Klamath Arc and the craton. In central Nevada, deep-sea deposits like those of northern California can be seen to have been thrust as far as 160 kilometers (100 miles) onto the craton (Figure 14-31*B*). The principal thrust fault along which this movement occurred is called the Roberts Mountains Thrust.

The collision of the arc and continental margin that produced the Roberts Mountains Thrust is known as the *Antler orogeny* (see Figure 14-29). This was the first sizable episode of mountain building in the Cordilleran region of North America during Phanerozoic time. The remainder of the Cordilleran story, which is told in the following chapters, has mountain building as its dominant theme.

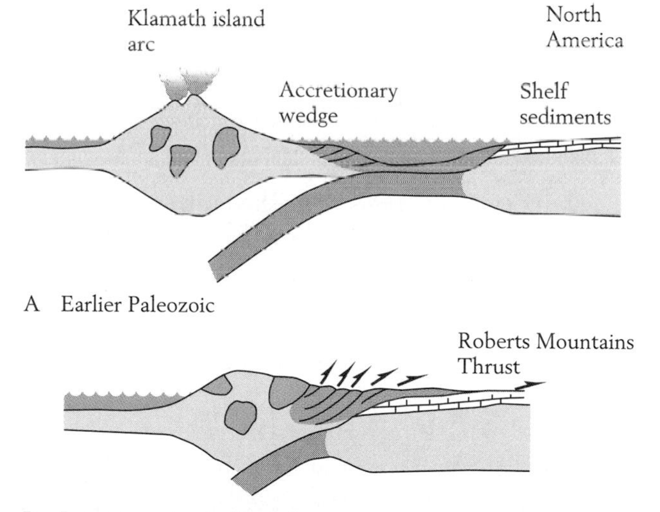

Figure 14-31 The likely mechanism by which the Klamath Arc was added to the North American continent by the Antler orogeny during Late Devonian and Mississippian time. The basin between the craton and the Klamath Arc *(A)* closed. As the continental crust was thrust beneath the volcanic crust of the Klamath Arc, deep-sea sediments slid onto shallow-water carbonates along the Roberts Mountains Thrust *(B)*.

Chapter Summary

1 Many important middle Paleozoic groups of marine life, depleted during the great mass extinction at the end of the Ordovician Period, expanded again at the start of the Silurian Period. Other forms, such as the ammonoids and the jawed fishes, were new. Jawed fishes and mollusks occupied freshwater as well as marine habitats.

2 The Silurian Period witnessed the invasion of the land by vascular plants, followed in Devonian time by the invasion of arthropods (scorpions and insects) and vertebrate animals (amphibians). During the Devonian Period, spore plants were joined by seed plants, which did not require moist habitats for reproduction and thus were not restricted to the fringes of aquatic habitats.

3 During most of middle Paleozoic time, world climates were relatively warm, and the seas stood high in relation to the surfaces of continents.

4 The Acadian orogeny occurred when Laurentia, Baltica, and the microcontinent Avalonia united to form Euramerica. As a result, exotic terranes were attached to eastern North America. A thick clastic wedge spread westward from the mountain chain that formed along the zone of suturing.

5 Tabulate-strome reefs flourished throughout middle Paleozoic time and were especially abundant in the Great Lakes region and near the western continental margin of North America.

6 Shortly before the end of the Devonian Period, a great mass extinction eliminated many forms of marine life, including nearly all members of the tabulate-strome reef community and nearly all placoderm fishes. Species that occupied cold regions seem to have survived better, suggesting that changing climatic conditions may have been the cause of the extinction.

7 The evolution of trees in Late Devonian time led to the initial spread of forests over terrestrial terrains. The deep roots of forest trees increased rates of weathering, and forests' production of wood and debris increased rates of carbon burial. Both these changes reduced the concentration of carbon dioxide in the atmosphere, perhaps triggering the glacial episode that occurred near the end of Devonian time.

Review Questions

1 In what important ways did invertebrate life change between Ordovician time and Devonian time?

2 What animals have the oldest extensive fossil record in freshwater sediments?

3 In what way did terrestrial environments of the Late Devonian Period look different from those of Early Devonian time?

4 What evidence do fossil bones and teeth provide that amphibians evolved from fishes?

5 Where was the landmass that now forms southern Europe toward the end of the Devonian Period?

6 How did the landmass that now forms Great Britain come into being?

7 How do reefs of middle Paleozoic age illustrate ecological succession?

8 Reefs are commonly porous structures that serve as traps for petroleum. If you wanted to drill for oil in Devonian reefs, what geographic regions would seem most promising?

9 What caused large quantities of sediment to accumulate in the south-central part of Euramerica during the Devonian Period?

10 What evidence is there of a decline in the concentration of atmospheric carbon dioxide during Devonian time? (Hint: Refer to Figure 10-9*A* and p. 272.)

11 Land areas changed dramatically in the course of middle Paleozoic time. Using the Visual Overview on page 370 and what you have learned in this chapter, describe how continental surfaces changed during Silurian and Devonian time with respect to their distribution on Earth, their topography, and their colonization by terrestrial life.

Additional Reading

Dineley, D. L., *Aspects of a Stratigraphic System: The Devonian,* John Wiley & Sons, New York, 1984.

Gensel, F. G., and H. N. Andrews, "The Evolution of Early Land Plants," *American Scientist* 75:478–489, 1987.

Gray, J. and W. Shear, "Early Life of the Land," *American Scientist* 80:444–456, 1992.

Long, J. A., *The Rise of Fishes,* Johns Hopkins University Press, Baltimore, 1995.

McGhee, G. R., *The Late Devonian Mass Extinctions: The Frasnian-Famennian Crisis,* Columbia University Press, New York, 1996.

CHAPTER

The Late Paleozoic World

The late Paleozoic interval of geologic time included the Carboniferous Period, when coal formed from the remains of new kinds of plants, and the subsequent Permian Period, when many organisms died out in the greatest mass extinction in all of Phanerozoic time.

The late Paleozoic world was marked by major climatic changes. Glaciers spread over the south polar region of Gondwanaland during the Carboniferous Period and then receded during the Permian. A general drying of climates at low latitudes during the Permian Period led to widespread accumulation of evaporites and to a contraction of coal swamps. Increased aridity also led to the extinction of spore plants and amphibians, both of which required moist conditions. Seed plants and mammal-like reptiles inherited the Earth.

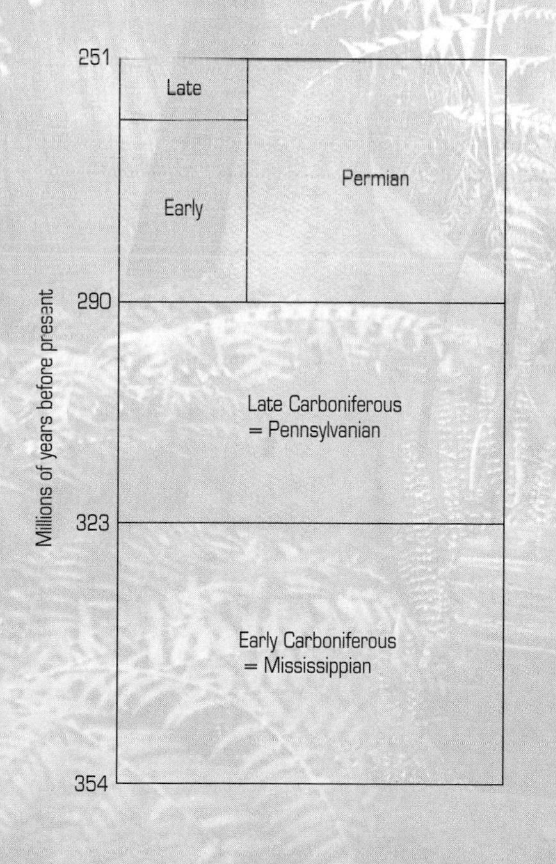

During Late Carboniferous time, logs of spore trees accumulated in broad swamps to form coal deposits that are now exploited by modern societies. Here ferns and seed ferns form the undergrowth beneath lycopod trees. On the fallen log on the right is a cockroach, one of the many kinds of insects that evolved during Late Carboniferous time.

Visual Overview

Major Events of the Late Paleozoic

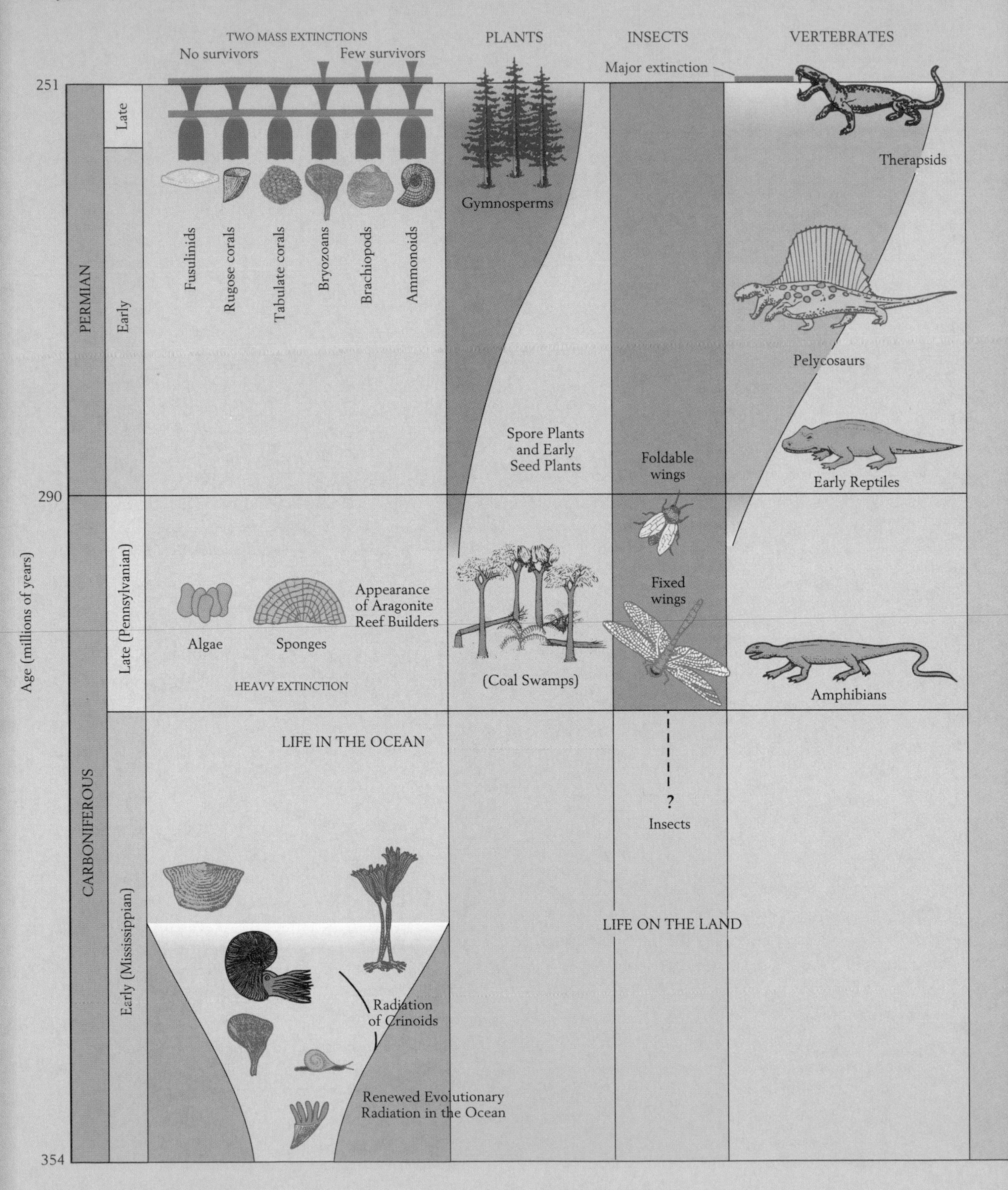

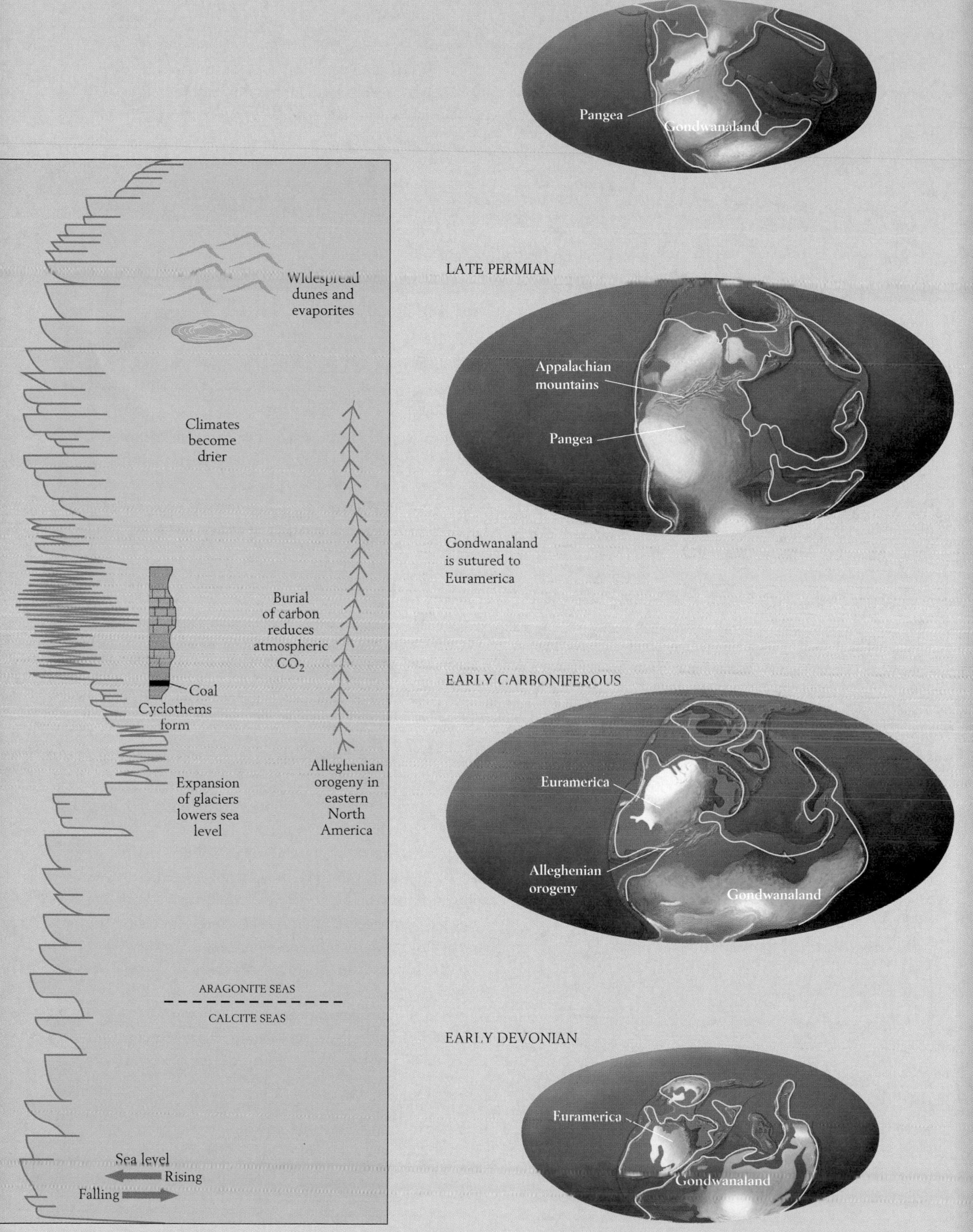

EARLY TRIASSIC
Pangea
Gondwanaland
LATE PERMIAN
Appalachian mountains
Pangea
Gondwanaland is sutured to Euramerica
EARLY CARBONIFEROUS
Euramerica
Alleghenian orogeny
Gondwanaland
EARLY DEVONIAN
Euramerica
Gondwanaland
Widespread dunes and evaporites
Climates become drier
Burial of carbon reduces atmospheric CO_2
Coal
Cyclothems form
Expansion of glaciers lowers sea level
Alleghenian orogeny in eastern North America
ARAGONITE SEAS
CALCITE SEAS
Sea level
Rising
Falling

In addition to the great mass extinction, another major event took place near the end of the Paleozoic Era. This was the attachment of Gondwanaland to Eurasia, accompanied by mountain building in Europe and in eastern North America. By the time this major suturing event was completed, almost all of the supercontinent of Pangea was in place.

The Carboniferous System was formally recognized in Britain in 1822, early in the history of modern geology. Its name was chosen to reflect the system's vast coal deposits, which had long been mined for fuel. Actually, only the upper part of the Carboniferous System harbors enormous volumes of coal; the lower part contains an unusually large percentage of limestone. Recognizing this distinction, American geologists, late in the nineteenth century, began to refer to the lower, limestone-rich Carboniferous interval as Mississippian because of its excellent exposure along the upper Mississippi Valley and to the upper, coal-rich interval as Pennsylvanian because of its widespread occurrence in the state of Pennsylvania. Soon the Mississippian and Pennsylvanian were informally recognized as separate systems in North America, and in 1953 the United States Geological Survey officially granted them this status.

Although this difference between the upper and lower parts of the Carboniferous System is evident elsewhere in the world, most geologists in Europe continue to recognize just one system: the Carboniferous. Because this book covers world geology, we will follow the European practice, but occasionally it will be helpful to reiterate that the Lower and Upper Carboniferous systems are equivalent to the Mississippian and Pennsylvanian, respectively.

Roderick Murchison, who established the Silurian System and coestablished the Devonian, recognized and named the Permian System in 1841 for rocks in the town of Perm in Russia. Permian rocks were later identified throughout the world.

Life

Marine life of the late Paleozoic interval did not differ markedly from that of Late Devonian time except for the appearance of new reef builders. Changes on land were far more profound. Numerous insects of remarkably modern appearance evolved during this time, and many new kinds of spore trees colonized broad swamps, where their remains formed coal deposits. During Permian time, these plants were replaced by seed trees. In parallel fashion, amphibians, which were tied to water for reproduction—like spore plants—initially dominated terrestrial habitats but were later replaced by more fully terrestrial reptile groups. By the end of Permian time, the terrestrial reptiles displayed a variety of adaptations for feeding and locomotion, many of which resembled those of mammals.

New Forms of Marine Life

Some forms of marine life never recovered from the mass extinction of Late Devonian time. Tabulates and stromatoporoids, for example, never again played a major ecologic role. The ammonoids, on the other hand, rediversified quickly and once again assumed a prominent ecologic position; indeed, ammonoid fossils are widely employed to date late Paleozoic rocks (see Figure 14-4). Also persisting from Devonian time as mobile predators were diverse groups of sharks and ray-finned bony fishes. Gone shortly after the start of Carboniferous time, however, were the armored placoderms that had ruled Devonian seas. The absence of armored placoderms and of similar fishes after earliest Carboniferous time reflected a general trend: although the late Paleozoic is not known for vast changes in the composition of marine life, heavily armored taxa of nektonic (swimming) animals tended to give way to more mobile forms. Apparently, as the Paleozoic Era progressed, the ability to swim rapidly became a near necessity, probably because of the increasingly effective predators that inhabited the seas. After Devonian time, armored fishes never again dominated marine habitats, and heavy-shelled nautiloids also declined in number. In contrast to these heavy, awkward forms, the swimmers that thrived in late Paleozoic time—the ammonoids and especially the sharks and ray-finned fishes—were highly mobile.

We know little about the groups of algae that floated in late Paleozoic seas alongside fishes and ammonoids. Phytoplankton are not well represented in the late Paleozoic fossil record, although some groups must have prospered without leaving recognizable fossils.

Brachiopods rebounded from the Late Devonian mass extinction to resume a prominent ecologic role. A group of spiny brachiopods known as productids enjoyed particular success. Some immobile productids employed their spines to anchor or support them-

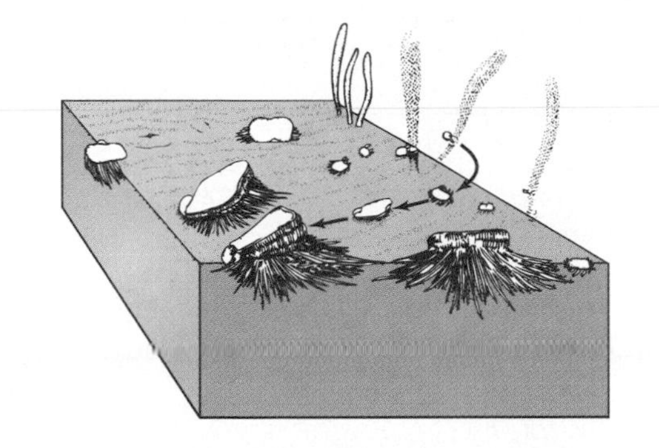

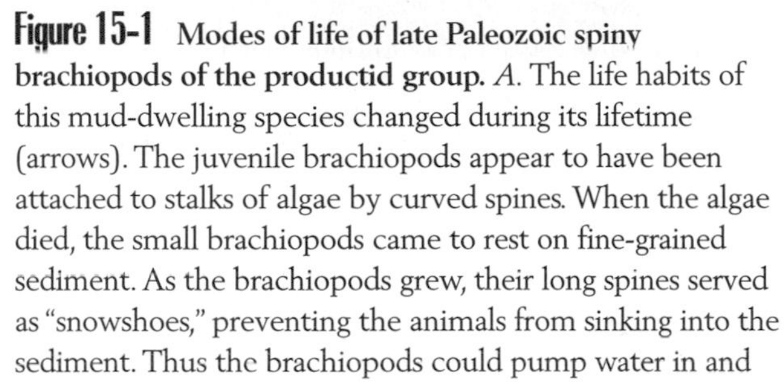

Figure 15-1 Modes of life of late Paleozoic spiny brachiopods of the productid group. *A.* The life habits of this mud-dwelling species changed during its lifetime (arrows). The juvenile brachiopods appear to have been attached to stalks of algae by curved spines. When the algae died, the small brachiopods came to rest on fine-grained sediment. As the brachiopods grew, their long spines served as "snowshoes," preventing the animals from sinking into the sediment. Thus the brachiopods could pump water in and out between the two halves of their shells to obtain food and oxygen without being clogged by mud. *B.* A group of Permian brachiopods of the genus *Prorichthofenia.* The lower halves of the shells of these coral-like animals were cone-shaped rather than cup-shaped, and throughout their lives their spines were attached to hard objects—in this case, the shells of neighboring brachiopods. The upper halves of the shells were flattened lids. *(A. After R. E. Grant, Jour. Paleont. 40:1063–1069, 1966.)*

selves in sediment, and a group of Permian productids developed cone-shaped shells that were attached by spines to the frameworks of solid reefs (Figure 15-1). Like the brachiopods, burrowing and surface-dwelling bivalves continued to thrive in late Paleozoic time, as did gastropods.

Crinoids—animals that were attached to the seafloor and captured floating food that came within reach of their waving arms as it floated by—expanded to their greatest diversity early in the Carboniferous Period, forming meadows on many areas of the seafloor (Figure 15-2). During this time, these organisms contributed vast quantities of carbonate debris to the rock record (Figure 15-3), leading to widespread limestone deposition during the Early Carboniferous (Mississippian) Period. Other animal groups also contributed to the formation of Carboniferous limestones.

Figure 15-2 Reconstruction of an Early Carboniferous (Mississippian) meadow of crinoids (sea lilies). Sharks, which also were well represented in Early Carboniferous seas, cruise above the crinoids.

Figure 15-3 **Early Carboniferous limestone composed largely of skeletal debris from crinoids.** The cylindrical fossils are segments of crinoid stems, which are corrugated so that in life they stacked one on top of another like poker chips.

Figure 15-4 **A lacy, fanlike bryozoan of late Paleozoic age.** In life this colony stood upright and was the size of a small fern.

Lacy bryozoans were sheetlike colonial animals that stood above the seafloor and fed on suspended organic matter (Figure 15-4). These organisms not only contributed skeletal debris to limestones but also trapped sediment to form mound-shaped structures.

The **fusulinids,** a group of large foraminifera that lived on shallow seafloors, included only a few genera in Early Carboniferous time but underwent an enormous adaptive radiation during the Late Carboniferous and Permian (Figure 15-5). Some 5000 species have been found in Permian rocks alone. Although they were single-celled, amoeba-like creatures with shells, some fusulinid species exceeded 10 centimeters (4 inches) in length. Their abundance and rapid evolution make fusulinids useful guide fossils for Upper Carboniferous and Permian strata. They also became major constituents of limestone.

Reefs and Aragonite Seas

After the Late Devonian collapse of the tabulate-strome community, reefs were poorly developed. Tabulate-strome reefs had been built by organisms that secreted calcite skeletons in calcite seas—seas with a low magnesium–calcium ratio (p. 278). The magnesium–calcium ratio rose early in the Carboniferous Period, and calcite seas gave way to aragonite seas (see Figure 10-18). It appears that few kinds of organisms can build massive reefs if the magnesium–calcium ratio of seawater does not favor their skeletal mineralogy. Thus it was only after aragonite-secreting reef builders evolved in mid-Carboniferous time that reefs were able to flourish once again. Aragonitic algae built Late Carboniferous reefs, and aragonitic sponges assumed a larger role in the growth of Permian reefs (Figure 15-6).

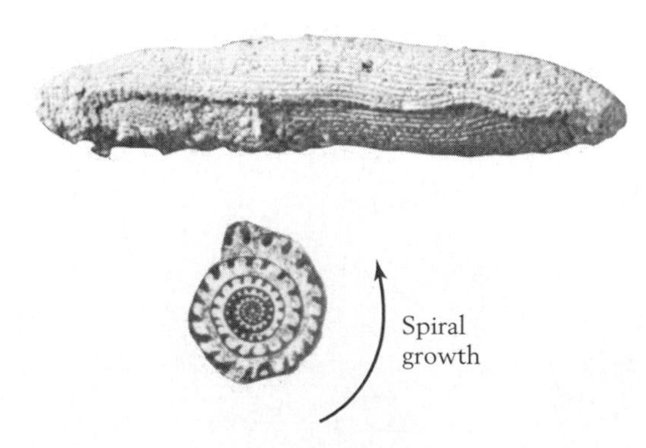

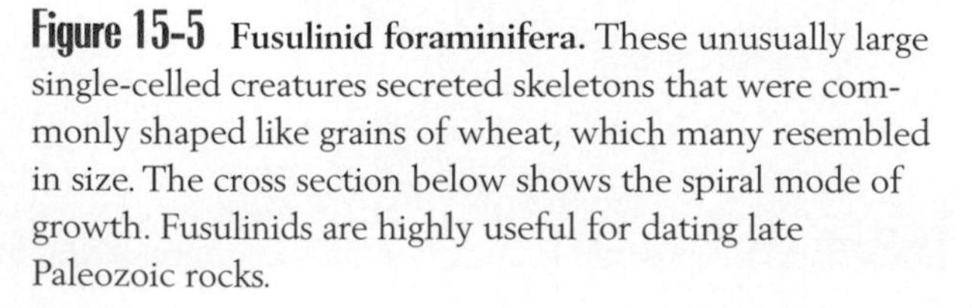

Figure 15-5 **Fusulinid foraminifera.** These unusually large single-celled creatures secreted skeletons that were commonly shaped like grains of wheat, which many resembled in size. The cross section below shows the spiral mode of growth. Fusulinids are highly useful for dating late Paleozoic rocks.

Figure 15-6 A Permian calcareous sponge of the genus *Girtyocoelia*. This reef builder had pea-sized chambers that were interconnected. Water that passed through the chambers was expelled through holes after food particles were filtered from it.

Plant Life on Land

Plants gave the Carboniferous Period its name, and in no other geologic interval are plant fossils more conspicuous; soft coal (p. 54) from this period typically contains recognizable stems and leaves. Coal deposits developed chiefly in lowland swamps, where fallen tree trunks accumulated in large numbers. Because it takes several cubic meters of wood to make one cubic meter of coal, it is evident that the vast coal beds of Late Carboniferous age represent an enormous biomass of original plant material. Wetlands were far more extensive than they are today (Box 15-1).

A small number of genera, each represented by large numbers of species, emerged as the dominant late Paleozoic flora of the coal swamps and adjacent habitats. The most important coal-swamp genera were *Lepidodendron* and *Sigillaria*, two types of lycopod trees that contributed many of the logs that were buried and compressed to form coal (Figure 15-7). As we have seen, the lycopods present during Early and

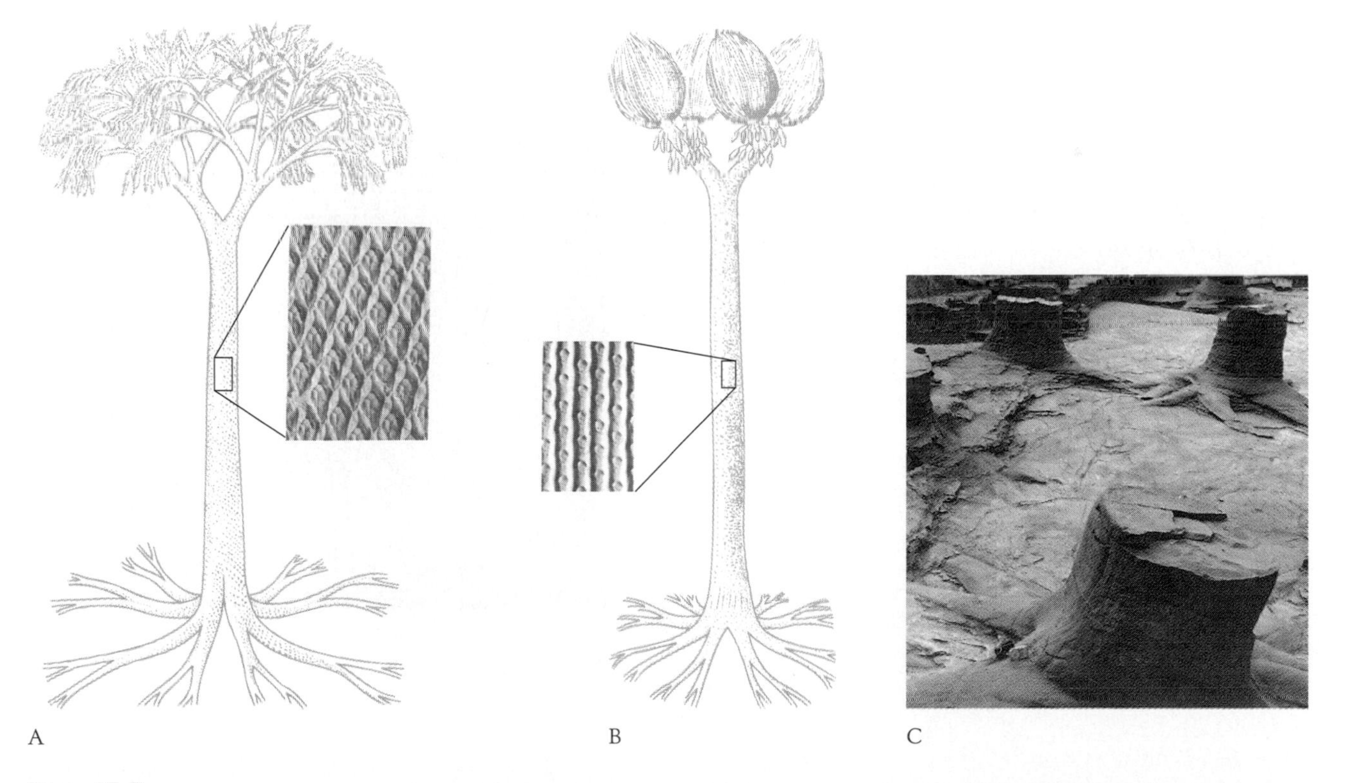

Figure 15-7 The dominant trees of Carboniferous coal swamps, belonging to the lycopod group. The genus *Lepidodendron (A)* has a trunk with a spiral pattern of scars where leaves were formerly attached. Note the similar spiral arrangement of branches in the early vascular plant *Baragwanathia* (Figure 14-14); this small, simple plant may have been an ancestor of the treelike lycopods. The trunk of *Sigillaria (B)* has vertical columns of leaf scars. The roots of lycopod trees are often preserved as fossils. Those shown here *(C)* are in Scotland.

For the Record 15-1

Wetlands, Then and Now

Coal swamps of the Carboniferous Period were among the most extensive wetlands of all time. Wetlands are continental areas that are normally moist at least part of the year and that support luxuriant growths of vegetation. Some have standing water throughout the year; the soil of others is constantly moist; and still others are flooded only about once a year. Wetlands were widespread during Carboniferous time because the seas stood high enough to spread across large areas of cratons. Along the margins of these epicontinental seas were broad, nearly flat lowlands where brackish and fresh water tended to accumulate and coal swamp floras flourished. Today, because continents stand higher above sea level, wetlands form only narrow fringes along the coasts and occupy small areas of the interiors. Altogether, wetlands cover only 6 percent of the surfaces of modern continents, and they are shrinking rapidly under the impact of human activities. When the United States was founded in 1776, its wetlands spread across 215 million acres, but only about 100 million of these acres remain.

Most of the wetlands of the modern world are informally termed marshes or swamps. Tall, grasslike plants that can tolerate brackish conditions form coastal marshes in relatively cool climates. In frostfree areas these plants give way to mangroves, which are shrublike trees that are able to grow in standing salt water. Plants such as grasses and cattails also grow in freshwater marshes, many of which have open water in the center or stretch along rivers. In contrast to marshes, swamps support abundant trees and shrubs, such as willows and cypresses, that tolerate moist soil or even permanent standing water.

Among the most interesting freshwater wetlands are peat bogs, where low-growing plants, including mosses, produce an environment that is so acidic and so depleted of oxygen that it is hostile to the kinds of microorganisms that would ordinarily decompose dead plant material. In the absence of these organisms, the plant

A cypress swamp in the southeastern United States.

material accumulates as peat. Peat bogs are concentrated in moist regions at high latitudes—the northern Great Lakes area and New England in North America, for example, as well as large areas of Canada and Alaska. Peat is widely used as fuel. Peat harvesters in eastern Europe have recovered about 2000 almost perfectly preserved human bodies from ancient bogs. The absence of oxygen in the soggy peat that floors bogs preserves the flesh of any animal that sinks into it: no bacteria, no decay. One of the oldest bodies found, that of a man who lived in Denmark about the time of Christ, was found with a rope around the neck.

Wetlands perform such important functions in the global ecosystem that their recent decline is alarming. They replenish groundwater, for example, by trapping fresh water and allowing it to percolate into the soil instead of flowing quickly to the sea. The plants that flourish in wetlands trap sediment and organic matter rich in nutrients. Many of these nutrients would otherwise be lost to the sea, or they would pollute rivers, lakes, and lagoons by promoting the growth of cyanobacteria. When scummy masses of cyanobacteria die, their decay robs water of oxygen and excludes other forms of life. The abundant vegetation of wetlands releases a large quantity of oxygen to the atmosphere every year. It also yields organic detritus that supports dense populations of animals. About 90 percent of the commercial fish species of Florida, for example, depend on the mangrove ecosystem for food or shelter.

Humans are currently destroying about one-third of a million acres of wetlands every year. The Everglades of Florida, the largest and most famous freshwater marsh of North America, are now less than half their original size. The Everglades are known as "the River of Grass" because their broad waters actually flow southward very slowly from Lake Okeechobee to the sea. Artificial dikes and levees now restrict the flow of water from the lake, and some of it is diverted for human use. As a result of human interference, many parts of the Everglades have dried up or contain so little water during the dry season that their biotas are shrinking. About 60,000 wading birds occupied this great marsh in 1940. Today there are only about one-quarter as many.

Prairie potholes—small basins scoured by glaciers—contain inland marshes that are also disappearing at a rate that threatens many forms of life. Prairie potholes are scattered over a large area of the northern Great Plains, where they support a great variety of animals and serve as breeding grounds for half of the ducks of North America. Farmers have drained or filled in many of these small marshes, vastly reducing the total area available to waterfowl and other animals.

Unfortunately, humans have tended to regard wetlands as nuisances—places where alligators lurk or mosquitoes breed. These soggy areas have seemed so unattractive that "swamp" has become a metaphor for a place not to be. Swamps, it has seemed, are best drained or filled in to produce solid land. Only recently have humans begun to value wetlands even though they are not the sort of place where most of us would want to live. And today our valuable bogs, swamps, and marshes look fragile indeed when we compare their shrunken remains with the vast coal swamps that cloaked lowlands during Carboniferous time.

Middle Devonian time had all been small plants (see Figure 14-15). Like smaller lycopods, *Lepidodendron* and *Sigillaria* were spore plants that were confined to swampy areas. *Lepidodendron* was the more successful genus; some of its species grew more than 30 meters (100 feet) tall and measured 1 meter (3 feet) across at the base.

At the feet of the treelike plants of the Upper Carboniferous was an undergrowth that consisted primarily of a wide variety of ferns and fernlike plants. Although some of them were spore plants like the modern ferns, many others were so-called **seed ferns,** which, as their name suggests, reproduced by means of seeds (Figure 15-8). Seed ferns are difficult to distinguish from spore ferns on the basis of their foliage alone, and they were not recognized as a separate group until 1904. Many seed ferns were small, bushy plants, but others were large and treelike. *Glossopteris,* the famous plant so abundant in Gondwanaland, was a seed fern. Its species were trees with tonguelike leaves (see Figure 8-3).

Not all Late Carboniferous vegetation occupied coal swamps. In fact, seed ferns and spore plants, called **sphenopsids,** were more abundant on higher ground (Figure 15-9). Fossils of these groups are more common in sands and muds that accumulated along levees and floodplains of rivers than in coals that accumulated in permanently wet coal swamps. Late Carboniferous sphenopsids, though often tree-

Figure 15-8 A frond and seeds of a Late Carboniferous seed fern.

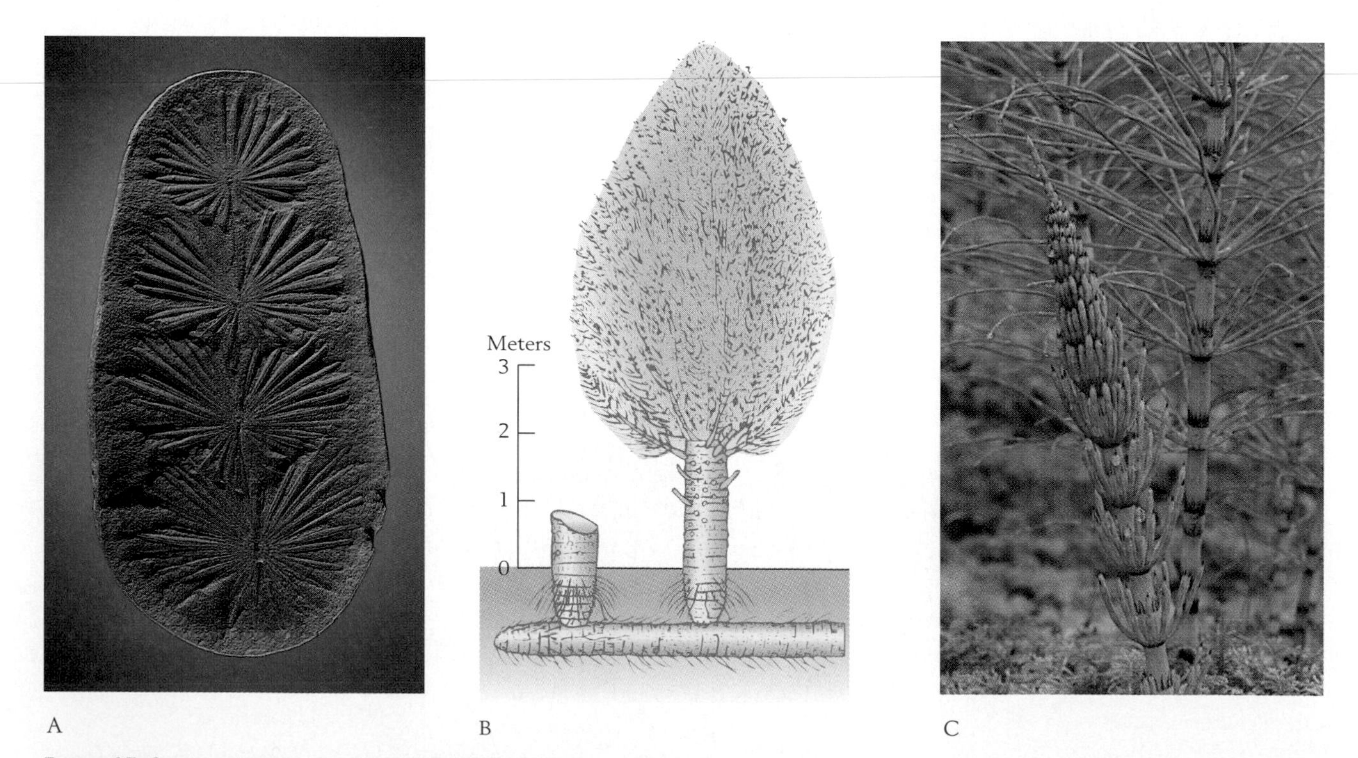

Figure 15-9 The Late Carboniferous sphenopsid plant ***Calamites.*** Branches such as the one preserved in the fossil *(A)* were positioned at intervals along the segmented tree trunk *(B)*, just as they are on the segmented stalk of living horsetails *(C)*, which grow to only about a meter (3 feet) in height.

Figure 15-10 **Reconstruction of a tall cordaite tree of Late Carboniferous time.** Cordaites were seed plants that formed large forests on dry ground.

sized, were similar in general form to horsetails, which are small sphenopsids of the modern world: they were characterized by branches that radiated from discrete nodes along the vertical stem and by horizontal underground stems that bore roots.

Another important group of Late Carboniferous plants that occupied high ground was the **cordaites,** tall trees that often reached 30 meters (100 feet) in height (Figure 15-10). As seed plants, cordaites were liberated from moist habitats and formed large woodlands resembling modern pine forests. In fact, cordaites belonged to the group known as **gymnosperms** ("naked-seed plants"), which include the living **conifers,** or cone-bearing plants (pines, spruces, redwoods, and their relatives). The seeds of these plants are lodged in exposed positions on cones or on other reproductive organs. Gymnosperm seeds thus differ from the covered seeds of flowering plants, a group that did not emerge until the Mesozoic Era.

During the Permian Period, gymnosperms, including conifers, took over terrestrial environments. Figure 15-11 shows the foliage of one of these conifers that resembled the needled branches of certain living conifers. In Paleozoic conifers, as in modern ones, seeds were borne naked on cones. Gymnosperm floras, having expanded in Late Permian time, prevailed throughout Triassic, Jurassic, and Early Cretaceous time (Figure 15-12) and thus are often thought of as representing Mesozoic vegetation. Mesozoic vegetation, however, had a head start on other life of the new era.

Freshwater and Terrestrial Animals

In late Paleozoic freshwater habitats, aquatic ray-finned fishes continued to diversify and were joined by freshwater sharks that have no close modern relatives. For the first time, mollusks also became conspicuous in freshwater environments; shells of many species of clams are found in freshwater and brackish sediments associated with coal deposits.

On land, a group of invertebrate animals, the insects, assumed a very important ecologic role—one

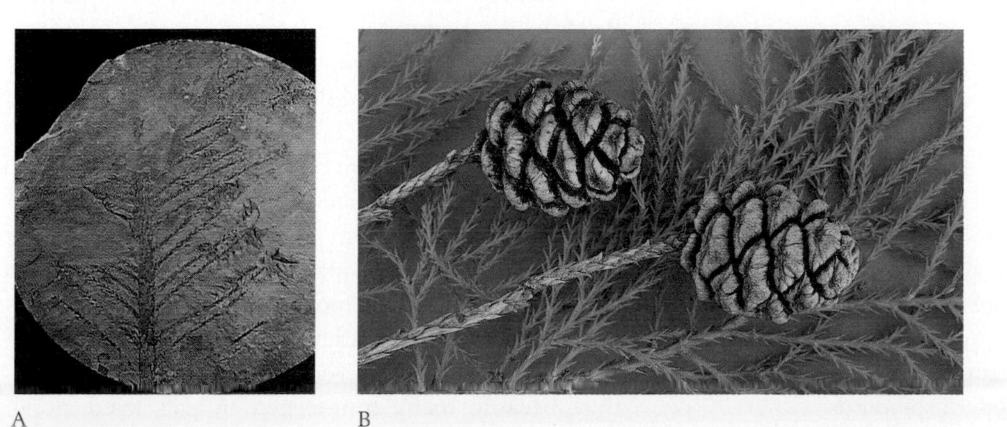

A B

Figure 15-11 **An early conifer of Late Carboniferous age *(A)* and a living species of conifer that is related to the redwoods *(B)*.** Like living conifers, *Walchia* had needled branches and reproduced by means of cones.

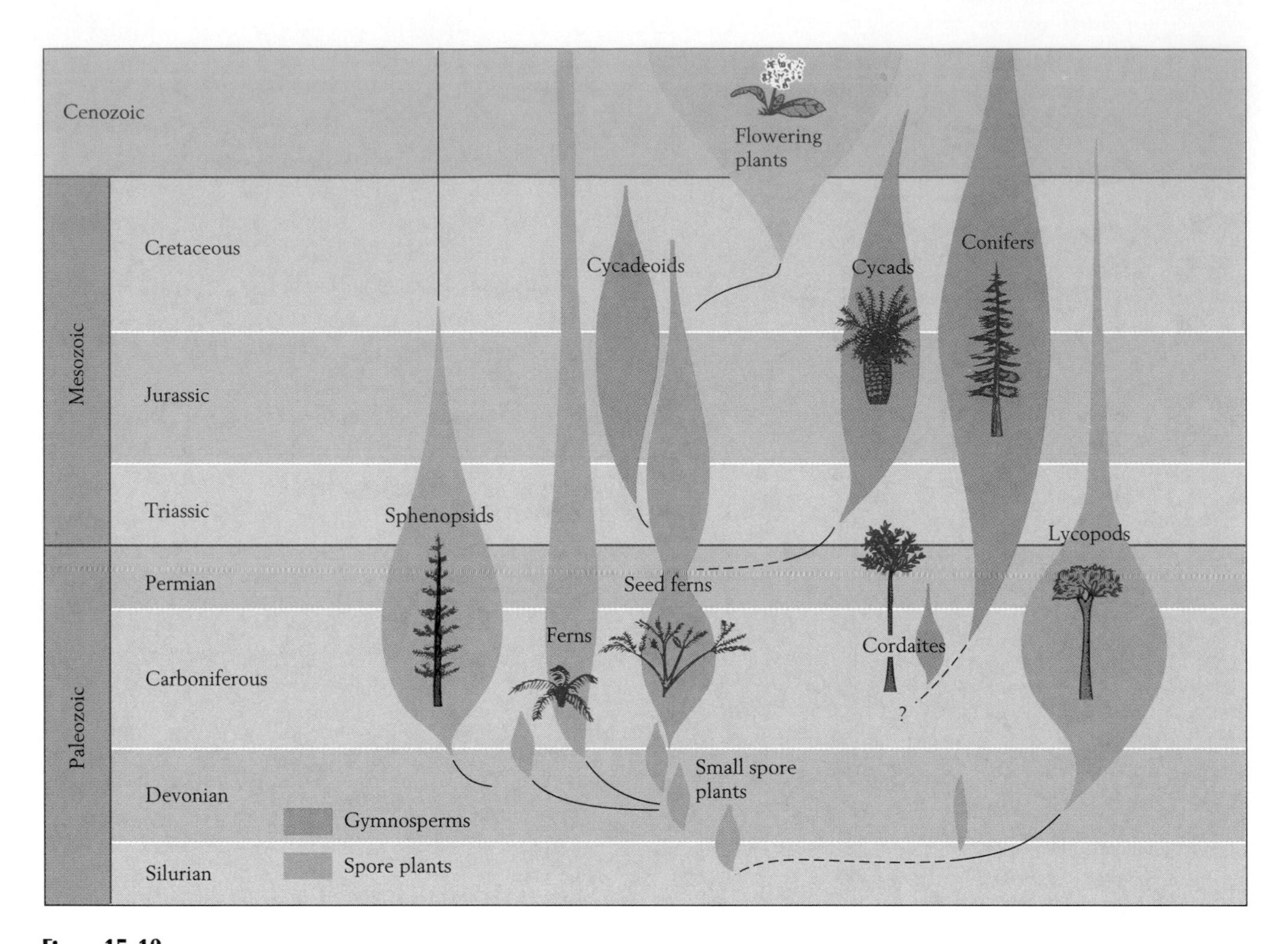

Figure 15-12 The history of major groups of swamp- and land-dwelling plants. Spore plants dominated Silurian, Devonian, and Carboniferous floras. Seed ferns are the oldest known seed plants. Gymnosperms (naked seed plants) dominated Mesozoic floras, but the early conifers, which belonged to this group, diversified greatly during the Permian Period, while spore groups declined. *(Modified from A. H. Knoll and G. W. Rothwell, Paleobiology 7:7–35, 1981.)*

that they have never relinquished. The oldest known insects are of Early Devonian age, but they were wingless forms. By Late Carboniferous time, however, many kinds of insects had wings (Figure 15-13). The earliest flying insects differed from most modern species in that they could not fold their wings back over their bodies; the only two living orders of insects that lack this ability are the dragonflies and the mayflies, both of which occur in Upper Carboniferous deposits. Giant dragonflies—animals with wingspans of nearly half a meter (18 inches)—found in the Carboniferous of France have given rise to the false impression that Carboniferous landscapes were populated with many kinds of huge insects. In fact, only one giant Carboniferous species is known. The rest were of normal size by modern standards.

The fact that advanced insects with foldable wings are found in younger Upper Carboniferous deposits indicates that the insects underwent an extensive radiation before the beginning of the Permian Period. Like many modern insect species, some of these Carboniferous forms had eggs that hatched into caterpillar-like larvae, while others possessed specialized egg-laying organs or mouth parts adapted to sucking juices from plants. The legs of still other species were highly modified for grasping prey, leaping, or running. Indeed, many of these insects appear to have been as highly adapted for particular modes of life as are insects of the modern world.

It is sometimes difficult to distinguish between aquatic and terrestrial vertebrates of late Paleozoic time because many four-legged animals lived along

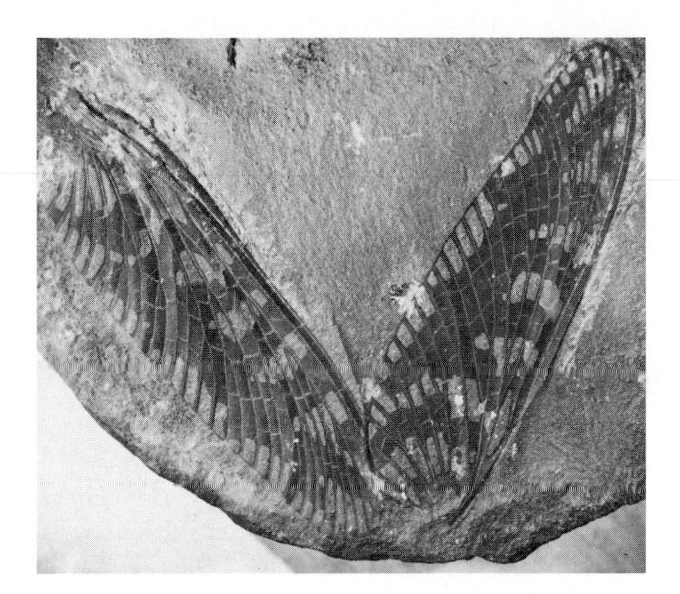

Figure 15-13 Wings of a Late Carboniferous insect from the Mazon Creek Formation of Illinois. The color pattern is beautifully preserved in this remarkable specimen. *(F. M. Carpenter, Proc. N. Amer. Paleont. Conv. I:1236–1251, 1969.)*

the shores of lakes, rivers, shallow seas, or swamps and divided their time between the land and the water. The only vertebrates to populate the Early Carboniferous landscape were amphibians, many of which retained aquatic or semiaquatic habits throughout their lives. Carboniferous amphibians, however, did not closely resemble their modern relatives. The frogs, toads, and salamanders that comprise most living species of Amphibia are small, inconspicuous creatures; they seem to be the only kind of animals belonging to this class that can thrive in the modern world in the face of competition and predation by advanced mammals and reptiles. Carboniferous and Early Permian amphibians, in contrast, had the world largely to themselves, and thus displayed a much broader spectrum of shapes, sizes, and modes of life; some superficially resembled stubby alligators (Figure 15-14), others were small and snakelike, and a few were lumbering plant eaters. Some Carboniferous amphibians measured 6 meters (20 feet) from the ends

Figure 15-14 An Early Permian scene beside a body of water. *Dimetrodon* (see Figure 15-15) threatens *Eryops*, an amphibian that was about 2 meters (6 feet) long. Early insects are in the foreground, and the small reptile *Araeoscelis* is climbing a tree of the genus *Cordaites*. The vine is *Gigantopteris* and the small plants are *Lobatannularia*. *(Drawing by Gregory S. Paul.)*

of their snouts to the tips of their tails. The species that were fully terrestrial as adults were covered by protective scales.

The Rise of the Reptiles The oldest known reptiles are found in deposits near the base of the Upper Carboniferous (Pennsylvanian) System. Most of the skeletal differences between the earliest reptiles and their amphibian ancestors were minor. The most important way in which reptiles differ from amphibians is in their mode of reproduction. The key feature in the origin of the reptiles was the **amniote egg,** which is also employed by modern reptiles and birds. This egg provides the embryo with a nutritious yolk and two sacs: one (the amnion) to contain the embryo and the other to collect waste products. A durable outer shell protects the developing embryo. The amniote egg allowed vertebrates for the first time to live and reproduce away from bodies of water. The amniote egg apparently originated in Carboniferous time, when reptiles evolved.

Because the amniote egg was in essence a self-contained pond, it eliminated the need for the young animal to live in water and thus enabled reptiles to exploit the land more fully. There is an interesting parallel here with the evolution of the seed in plants. As we have seen, spore plants, like amphibians, require environmental moisture during part of their life cycle. The origin of the more advanced groups—the seed plants and reptiles—represented a transition to a fully terrestrial existence.

Later reptiles developed yet another feature of great importance: an advanced jaw structure that could apply heavy pressure upon closing and could slice food by means of bladelike teeth. Carboniferous amphibians and early reptiles had jaws that could snap closed quickly but could apply little pressure. Moreover, they had pointed teeth that could kill prey by puncturing it but that could not slice or tear food apart, so these animals were forced to swallow their meals whole.

Despite the origin of reptiles in Late Carboniferous time, amphibians continued to prosper into Early Permian time. During the Permian Period, however, reptiles diversified and apparently began to replace amphibians in various ecologic roles, probably because the reptiles had more advanced jaws and teeth as well as greater speed and agility. Permian rocks of Texas have yielded large faunas of amphibians and reptiles that reveal this pattern. By Early Permian time, the **pelycosaurs,** fin-backed reptiles and their relatives, had become the top carnivores of widespread ecosystems (Figure 15-15). Their stratigraphic occurrence suggests that many lived in swamps and that some may have been semiaquatic. *Dimetrodon,* one such carnivore (see Figures 15-14 and Figure 15-15), was

Figure 15-15 Skeletons of pelycosaurs from the Lower Permian Series of Texas. The fin of the herbivore *Edaphosaurus* (left) and the carnivore *Dimetrodon* (right), which was supported by long vertebral spines, served an uncertain function. Some workers believe that skin stretched between the spines was used to catch the sun's rays, allowing the animal to raise its temperature to a level above that of its surroundings. From snout to tail, *Dimetrodon* exceeded 2 meters (6 feet) in length.

Figure 15-16 A Late Permian scene in the South African part of Gondwanaland. Therapsids move along an ice-covered stream in a snowy environment, where they may have been able to live by virtue of being endothermic. In this reconstruction, they are shown as having hair, which is associated with endothermy in mammals. The largest animal is *Jonkeria;* in the background is *Trochosaurus;* the animals at lower right are *Dicynodon;* and the very small form is *Blattoidealestes. (Drawing by Gregory S. Paul.)*

about the size of a jaguar and had sharp, serrated teeth. Whereas even the Permian carnivorous amphibians, such as the alligator-like *Eryops,* were forced to swallow small prey whole, *Dimetrodon* could tear large animals to pieces.

Dimetrodon and other pelycosaurs had a skull structure that in some ways resembled that of mammals, which evolved from them. Their descendants, the **therapsids,** were especially similar to mammals (Figure 15-16). Therapsids' legs were positioned more vertically beneath their bodies than were the sprawling legs of primitive reptiles or even pelycosaurs. In addition, the jaws of therapsids were complex and powerful, and the teeth of many species were differentiated into frontal incisors for nipping, large lateral fangs for puncturing and tearing, and molars for shearing and chopping food.

A New Level of Metabolism Many experts believe that the therapsids were **endothermic,** or warm-blooded: by virtue of a high metabolic rate, they maintained their body temperatures at relatively constant levels that were usually above those of their surroundings. Hair similar to that of modern mammals may have insulated therapsids' bodies (see Figure 15-16). Even if they were endothermic, however, therapsids may not have kept their body temperatures at levels as constant as those of mammals. In any case, the upright postures and complex chewing apparatuses of advanced Permian therapsids show that these active animals approached the mammalian level of evolution in anatomy and behavior.

The endothermic condition allows animals to maintain a sustained level of activity—to hunt prey or to flee from predators with considerable endurance. **Ectothermic** (or cold-blooded) reptiles, in contrast, must rest frequently in order to soak up heat from their environment. Endothermic metabolism, along with advanced jaws, teeth, and limbs, may account not only for the success of the therapsids during Permian time but also for the decline of the pelycosaurs, which were probably ectothermic. In fact, while pelycosaurs were declining to extinction during Late Permian time, therapsids were undergoing a spectacular adaptive radiation. More than 20 families of these advanced animals seem to have evolved in just 5 million or 10 million years, and they were the dominant groups of large animals in Late Permian terrestrial habitats. Therapsids seem to have represented an entirely new kind of animal—one so advanced that it was able to diversify very quickly.

The Paleogeography of the Late Paleozoic World

During late Paleozoic time the major continents moved closer and closer together until by early Mesozoic time nearly all of them were fused together as the supercontinent Pangea. Even early in the Carboniferous Period, however, the continents were rather tightly clustered (Figure 15-17). As the period progressed, the sector of Gondwanaland that lay over the south pole became covered by a large continental glacier that persisted into the Permian Period. Meanwhile, hot conditions prevailed in equatorial regions. Coal deposits formed most extensively during Late Carboniferous time, accumulating at both low and high latitudes. Global climates changed dramatically early in the Permian Period, when glaciers melted in Gondwanaland and the spread of arid conditions caused coal swamps to contract throughout the world.

The Early Carboniferous Period: Limestone and Glaciers

Sea level, which had declined near the end of the Devonian Period, rose at the start of Early Carboniferous time, so that warm, shallow seas spread across broad continental surfaces at low latitudes. As a result, limestones accumulated over large areas, often with crinoid debris as their major component.

South polar temperatures became strikingly cooler early in Carboniferous time. Tillites reveal that throughout all or nearly all of Carboniferous time, large areas of Gondwanaland were blanketed by ice sheets. The earliest of the ice sheets may have been the same ones that formed in Late Devonian time (p. 386), or they may have developed after the Devonian glaciers melted. Uncertainty remains because preserved tillites provide an incomplete record of glaciation; also, many tillites are poorly dated.

Warm, moist conditions prevailed in some continental areas nearer the equator. Thus coal-swamp floras, which first became established early in Carboniferous time, flourished along the northeastern margin of Euramerica (see Figure 15-17). Trade winds must have continued to bring this continent moisture from the oceans to the northeast, but the western part of the continent was in the rain shadow of the Appalachian Mountains. Here, across what is now central and western North America, evaporites and limestones accumulated in broad, shallow seas.

Events at the Mid-Carboniferous Boundary

The transition from Early to Late Carboniferous time was marked by two important events: a global decline in sea level and heavy extinction of marine life. In many parts of the world, the drop in sea level is evidenced by a disconformity in shallow marine deposits. In North America, the marine records of the Mississippian and Pennsylvanian systems are separated by a disconformity that is estimated to represent more than 4 million years in some areas. Among the marine groups that suffered heavy extinction during this interval were the crinoids and ammonoids, which lost more than 40 and 80 percent of their genera, respectively. Presumably sea level fell during this crisis because glaciers expanded in Gondwanaland, locking water up on the land. Presumably the mid-Carboniferous extinctions resulted from cooling of the seas that was associated with expansion of glaciers and from loss of some shallow-water habitats when sea level fell.

The Later Carboniferous Period: Continental Collision and Temperature Contrasts

During the middle portion of Carboniferous time, the northward movement of Gondwanaland caused that continent to collide with Eurasia. The mountains thus formed in southern Europe are known collectively as the Hercynides, and the orogeny as a whole is known

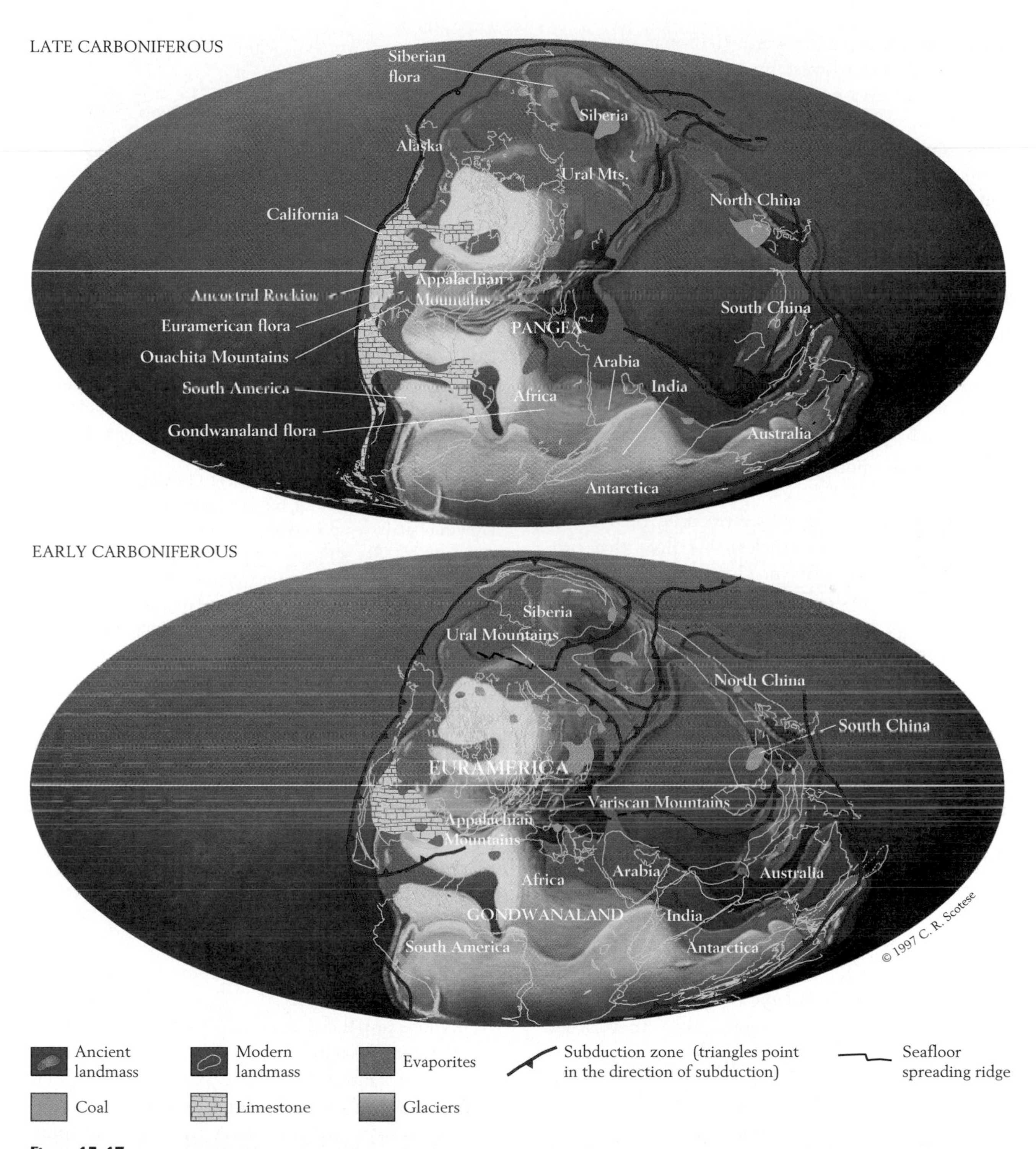

Figure 15-17 World geography in Carboniferous time. The major continents were rather tightly clustered on one side of the globe in Early Carboniferous (Mississippian) time. Coal deposits formed near the sea that bordered eastern Euramerica, and limestones and evaporites accumulated in the epicontinental sea that flooded western Euramerica. Glaciers spread over Gondwanaland near the South Pole. In Late Carboniferous time Gondwanaland and Euramerica collided. Coal deposits formed over a larger total area at this time than at any other period in Earth's history. In Gondwanaland, continental glaciers spread to remarkably low latitudes and were separated from tropical coal swamps (formed by the Euramerican flora) by steep temperature gradients. The Gondwanaland and Siberian floras flourished under cooler conditions. *(Adapted from paleogeographic maps by C. R. Scotese, PALEOMAP Project, University of Texas at Arlington, 1997.)*

as the Hercynian (or Variscan). Hercynian mountains also formed in northwestern Africa. In North America the Hercynian orogeny, known here as the Alleghenian, in effect continued where the Caledonian orogeny left off, extending the Appalachian mountain chain southwestward and forming the adjacent Ouachita Belt in Oklahoma and Texas.

On the land, latitudinal temperature gradients steepened during Late Carboniferous time—that is, there were extreme differences in temperature between the equator and the poles. Continental glaciers pushed northward to within nearly 30° of the equator, a latitude where subtropical conditions have prevailed during most of Phanerozoic time. It seems amazing that tropical coal swamps flourished in North America and western Europe not much farther north than the northernmost Carboniferous glaciers (see Figure 15-17).

Recall that coal deposits formed in frigid Gondwanaland as well. Nonetheless, the flora that produced the coal deposits in Gondwanaland differed substantially from the equatorial Euramerican flora. *Lepidodendron* and *Sigillaria*, the dominant Euramerican elements, were present in Gondwanaland, but many of the plants of Gondwanaland are unknown from northern continents. The Gondwanaland flora was adapted to the cool climates of the glacial regime in the south (see Figures 8-2 and 15-17). Siberia, which lay near Earth's other pole, also had a distinctive flora adapted to cold conditions.

Many cold climates are strongly seasonal, and seasonal growth of wood produces distinctive rings visible in the cross sections of tree trunks. The Upper Carboniferous floras of Gondwanaland in the south and of Siberia in the north are known for their distinctive tree rings (Figure 15-18). In contrast, the Euramerican fossil trees that grew near the Carboniferous equator were of the tropical type: they lacked seasonal rings.

The composition of fossil floras in North America and Europe reveals that tropical climates changed significantly in the course of Late Carboniferous time. About two-thirds of the way through Late Carboniferous time, lycopods, sphenopsids, and seed ferns suddenly declined in terrestrial plant communities, while spore ferns assumed a correspondingly larger role. This floral turnover occurred at a time when glaciers shrank in Gondwanaland. Coal continued to form from decaying vegetation in swamps, but lycopods were no longer the primary contributors. This change foreshadowed even greater climatic changes during Permian time.

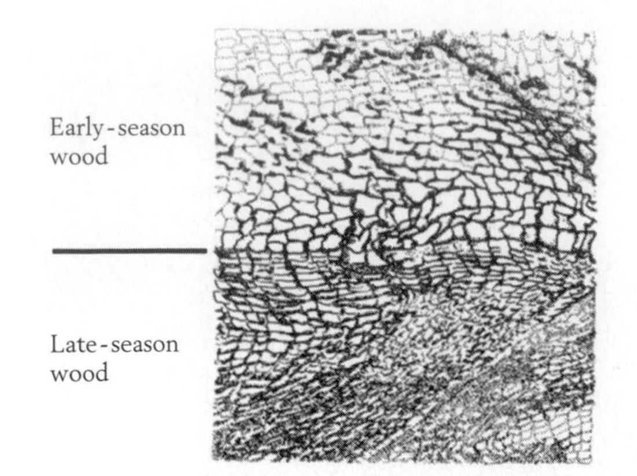

Figure 15-18 A section of fossil wood of late Paleozoic or Triassic age from Antarctica, showing the details of the boundary between two growth rings. Each space in the woody tissue was occupied by a single cell. When the late-season wood of the earlier (lower) layer formed, the cells were small and the tree grew slowly. Growth was interrupted during winter, but spring then stimulated the growth of early-season wood, resulting in large cells and rapid growth. The number of growth rings indicates the age of a tree, but rings are not well developed in tropical climates. In late Paleozoic time, high-latitude climates were strongly seasonal, producing growth rings in Siberia near the north pole and in such regions as Antarctica near the south pole.

The Permian Period: Climatic Complexity

As a result of complex topographic conditions and steep climatic gradients, the floras of Permian time were more provincial than those of any other Phanerozoic period—with the possible exception of the most recent interval, when continents have been widely dispersed and the waxing and waning of continental glaciers have produced much geographic differentiation.

The Permian floras remained distinct even though they were not separated by vast oceans. In Permian time, the suturing of Siberia to eastern Europe along the Ural Mountains resulted in the nearly complete assembly of Pangea (Figure 15-19). Southeast Asia remained as the only separate landmass of large size, and it would be attached early in the Mesozoic Era.

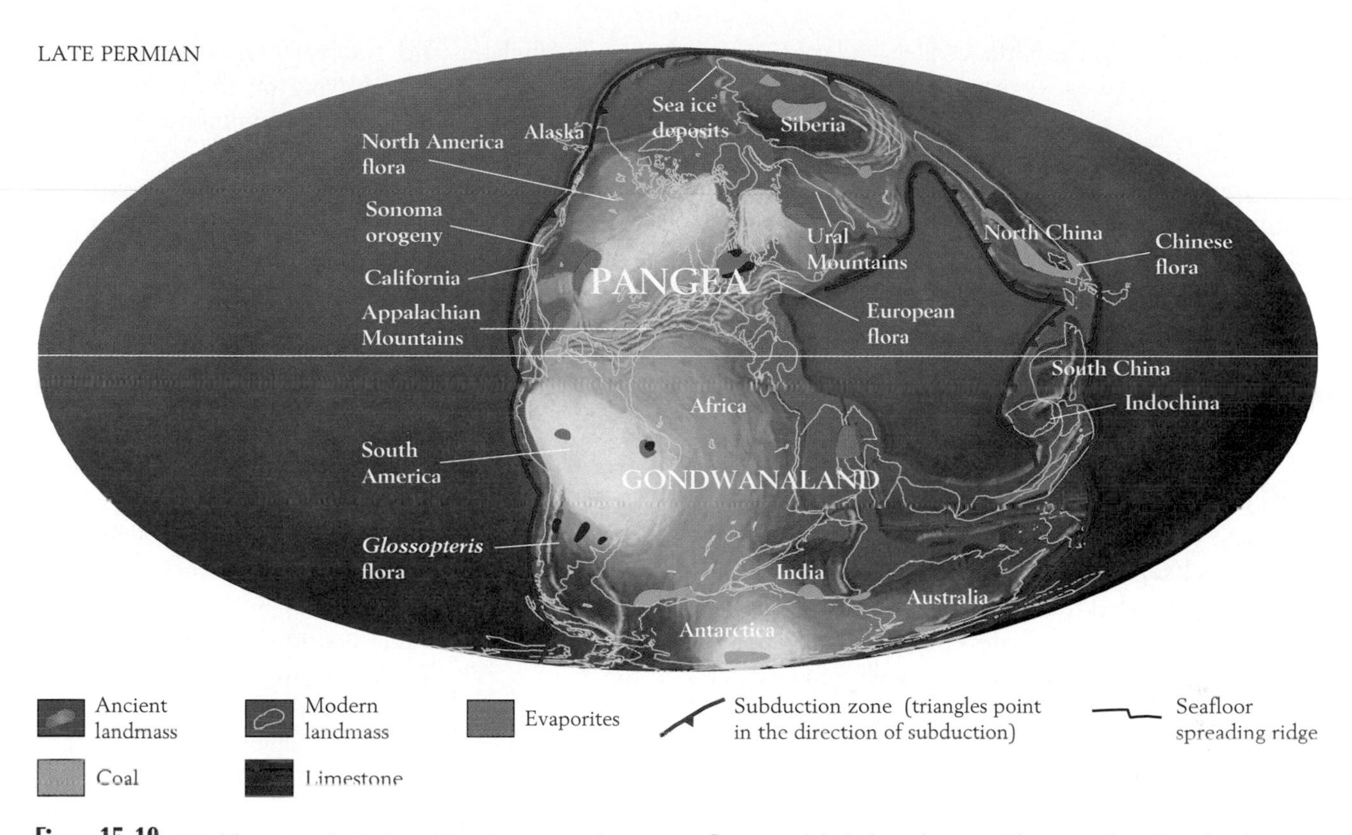

Figure 15-19 World geography in Late Permian time, when the ocean separating Europe from Asia was closing along the Ural Mountains to form the supercontinent Pangea. The Permian landmasses had a complex geography. Many mountain ranges stood high above lowlands. Five distinctive floras are labeled on the map. Floras produced coal only at high latitudes, while equatorial regions were dry. Small glaciers formed in Siberia, which now encroached on the north pole. *(Adapted from paleogeographic maps by C. R. Scotese, PALEOMAP Project, University of Texas at Arlington, 1997.)*

Contributing to the climatic contrasts of the late Paleozoic globe were several mountain chains, including those formed during the suturing of Gondwanaland to Euramerica.

The various Permian floras had one thing in common: they were distinct from Carboniferous floras, except in China, where lycopods persisted because China was still a warm, moist island surrounded by tropical seas (see Figure 15-19). Coal continued to form only in China and at high latitudes in Pangea. Most of Pangea was too dry to support coal swamps. Instead, dune deposits and evaporites accumulated, especially at low latitudes, in the dry trade wind belt (Figure 15-20). The Permian, in fact, has a greater concentration of salt deposits than any other geologic system.

At the beginning of the Permian, *Glossopteris* and related plants came to dominate the landscape in Gondwanaland. As the Permian progressed, plants adapted to moist conditions gave way to ones favored

Figure 15-20 Cross-bedded dune deposits of the Permian Coconino Sandstone, which crops out in the vicinity of the Grand Canyon in Arizona.

by drier habitats. In the north, conifers and other gymnosperms replaced coal-swamp floras early in the Permian (see Figure 15-11). Many late Paleozoic conifers apparently resembled conifers of the modern world: they could thrive under relatively dry conditions. For the most part, the dry conditions must have resulted from the union of Euramerica and Gondwanaland, which formed a continent so huge that many sections of it lay far from any ocean.

Atmospheric Carbon Dioxide and Carboniferous-Permian Climate

It appears likely that changes in greenhouse warming initiated and ended the massive Carboniferous glaciation. Extensive burial of reduced carbon in Carboniferous coal swamps must have markedly reduced the concentration of carbon dioxide in the atmosphere, weakening greenhouse warming of Earth's surface (see Figure 10-5). Thus it appears to be no accident that the glaciation more or less coincided with the interval of widespread coal swamps. Recall that the spread of forests in Late Devonian time had already reduced the atmospheric concentration of CO_2 by intensifying the weathering of continental surfaces (p. 386).

When climates became drier in Late Carboniferous and Permian time, levels of atmospheric carbon dioxide must have risen, for two reasons. First, dry conditions would have reduced rates of weathering. Second, as coal swamps dried up, rates of carbon burial declined. Through greenhouse warming, the resulting buildup of atmospheric carbon dioxide presumably ended the late Paleozoic ice age.

The Late Permian Extinctions

The Paleozoic Era ended with the greatest mass extinction of animal life of all time. This was the first great mass extinction to strike Earth's biota after vertebrate animals had invaded the land on a grand scale, and terrestrial vertebrates were among its primary victims. Nearly 20 families of Permian therapsids failed to survive into Triassic time.

In the marine realm, the Permian crisis entirely swept away the fusulinids, which had been highly successful in mid-Permian seas, and also the rugose corals, tabulates, and trilobites—although the latter two groups were already very much on the decline long before the crisis began. The ammonoids hung on by a thread: only a handful of their species survived into the Triassic Period. The brachiopods, bryozoans, and stalked echinoderms suffered heavy losses, and the bivalve and gastropod mollusks were struck moderately hard.

The First Extinction

What was long considered a single Late Permian crisis appears actually to have been a double event: a mass extinction at the end of the second-to-last age of the Permian, known as the Guadalupian Age, followed by a greater one at the very end of the period. Most reef-building taxa died out abruptly in the first crisis, and so did about three-quarters of all genera of fusulinid foraminifera. The fusulinid species that disappeared included all species longer than 6 millimeters ($\frac{1}{4}$ inch) and all species whose skeletons had complex wall structures. Symbiotic algae probably lived within these fusulinids, as they do in large living foraminifera. Perhaps, then, the large fusulinid species became extinct because their symbiotic algae died out.

The total number of marine genera that disappeared at the end of the Guadalupian suggests that as many as 70 percent of marine species died out. The number may have been slightly smaller because some of the genera thought to have died out may actually have survived into latest Permian time without having been detected in the post-Guadalupian fossil record.

Dropstones from icebergs are abundant in Guadalupian rocks of northern Siberia, indicating that glaciers expanded near the north pole. Possibly this glacial event was related to climatic changes that caused the extinction, but the precise timing of the glacial episode has not been established.

The Terminal Extinction

In the past, researchers who failed to recognize that two extinctions occurred in Late Permian time estimated that the terminal Permian crisis eliminated between 90 and 95 percent of marine species. It now appears that only about 80 percent of the species alive near the end of the Permian died out in the terminal Permian crisis, but that number still makes this the largest of all Phanerozoic mass extinctions. The Guadalupian event was itself among the largest.

Several patterns may shed light on the cause of the terminal Permian crisis:

1. Extinction was sudden, with most losses probably occurring within less than a million years.
2. Tropical taxa died out in especially large numbers. The reef community, which had rebounded after the Guadalupian extinction, totally collapsed.
3. Carbon isotopes in marine sediments became markedly enriched in carbon 12 at the expense of carbon 13 close to the end of the Permian (Figure 15-21).
4. Sea level began a general decline in mid-Permian time and rapidly dropped about 100 meters during the final 2 million years or so of the Permian. At the end of the period shallow seas were confined to narrow continental margins.
5. The *Dicroidium* flora, which was adapted to warm, dry conditions, migrated from low latitudes to high southern latitudes, replacing the *Glossopteris* flora, which was adapted to cooler, moister conditions. This shift at the very end of the Permian, along with a simultaneous shift in many regions from deposition of coal to deposition of red beds, indicates that major climatic changes occurred at the time of the mass extinction.
6. Spores and fossil remains of fungi are very abundant in shallow marine rocks of the uppermost Permian; fungi appear to have flourished by deriving nutrition from large amounts of dead vegetation in coastal areas.
7. In what may have been the largest continental fissure eruption of the entire Phanerozoic, flood basalts spread across a broad area of Siberia about 251 million years ago, very close to the time of the terminal Permian extinction. Emissions from this volcanic event may have altered global climates by screening out sunlight or enhancing the greenhouse effect.
8. The terminal Permian extinction occurred during a brief interval of anoxia in the deep sea.

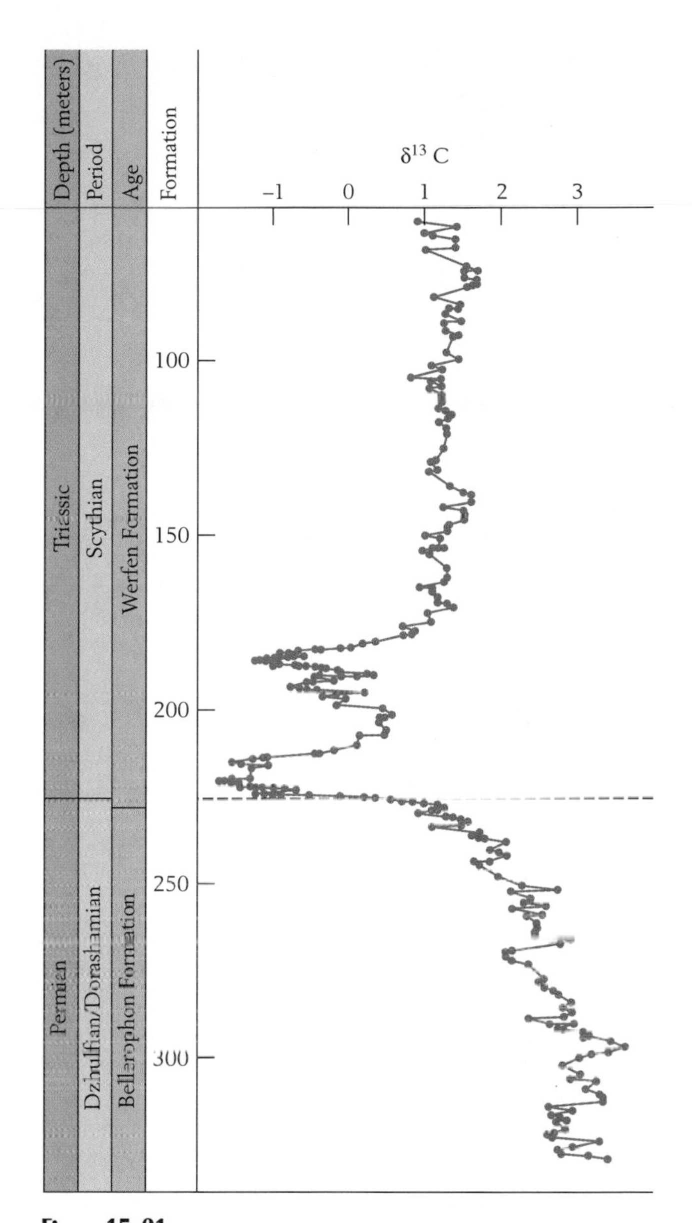

Figure 15-21 **Graph showing the strong shift of carbon isotope ratios in strata spanning the boundary between the Permian and Triassic systems in the Austrian Alps.** *(After W. T. Holser et al., Abj. Geol. Bundesanstalt 45:213–232, 1991.)*

The eighth point is especially intriguing. Uplifted deep-sea strata in Japan reveal a striking pattern of change in the state of the ocean (Figure 15-22). The deep sea was well oxidized until the very end of Guadalupian time. Guadalupian cherts in the uplifted rocks consist of the remains of radiolarians, whose abundance on the deep-sea floor indicates that upwelling supplied abundant nutrients to the shallow waters where the radiolarians lived. Surface waters must move downward to replace upwelling waters; therefore, when strong upwelling was allowing the Permian radiolarians to flourish, downwelling waters must have been oxygenating the deep sea. In fact, the cherts formed by the radiolarians are stained red by hematite, a highly oxidized iron mineral (p. 40).

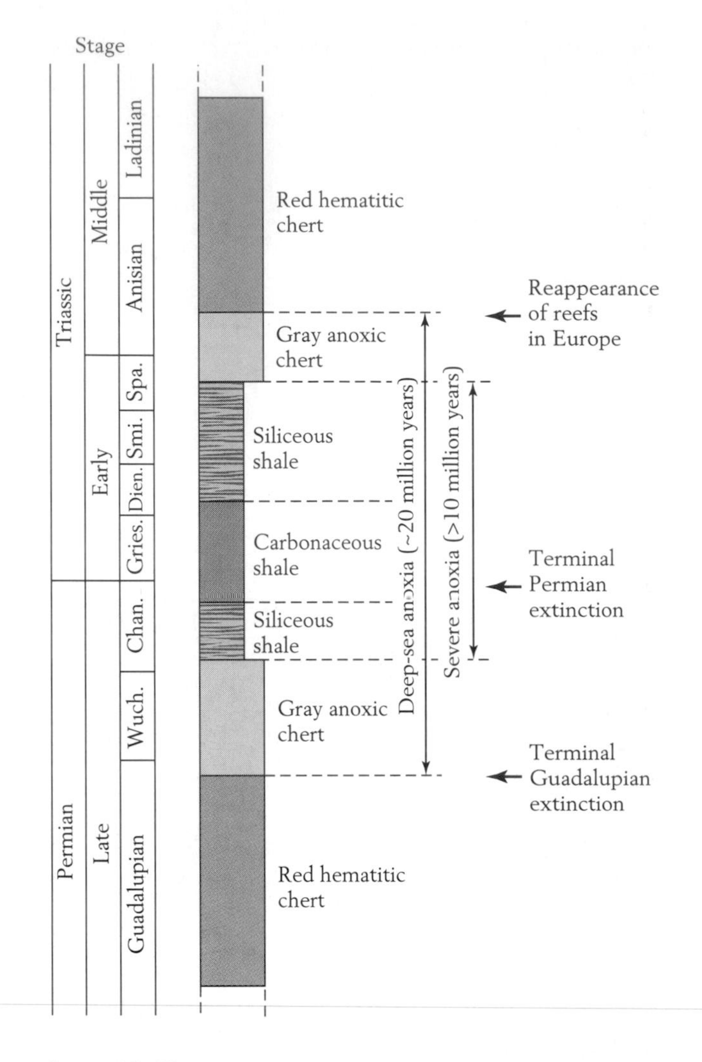

Figure 15-22 A stratigraphic section of uplifted rocks in Japan that document an episode of deep-water anoxia in Late Permian time. When anoxia began, at the time of the terminal Guadalupian extinctions, gray chert replaced hematitic (highly oxidized) red chert. An interval of severe anoxia began at the time of the terminal Permian extinction. Deposition of hematitic chert resumed at the time when reefs began to grow again in Europe. *(After Y. Isosaki, Science 276:235–238, 1997.)*

Significantly, the onset of anoxic conditions in the deep sea coincided with the extinction at the end of Guadalupian time, when gray sediments began to replace the red cherts (see Figure 15-22). Higher up in the Japanese sequence the deep-sea sediments become very rich in organic carbon, reflecting stagnation of the deep ocean. Furthermore, when this severe stagnation began, anoxic sediments also began to accumulate in shallower waters around the margins of the Pacific Ocean. Apparently for 3 million or 4 million years only the upper waters of the ocean were oxygenated. Early Triassic sediments show the opposite trend, recording the return of fully oxygenated conditions to the deep sea over several million years.

Changes in ocean circulation must have been related, directly or indirectly, to the two Late Permian extinctions: the first event coincided with the onset of deep-sea anoxia, and the second occurred at the time of maximum anoxia in the oceans. Some workers have suggested that brief episodes of mixing in the ocean carried large quantities of carbon dioxide to surface waters and the atmosphere, poisoning many forms of life. The sudden replacement of the *Glossopteris* flora by *Dicroidium* flora presumably reflected a sudden, widespread shift to warmer or drier conditions. Possibly this shift resulted from greenhouse warming by carbon dioxide that rose from the deep sea. This carbon dioxide would originally have been released by bacterial decay of organic matter on the deep-sea floor. Because organic matter is enriched in carbon 12 (p. 266), movement of carbon dioxide from deep waters to shallow seas might also explain the abrupt decline of the carbon 13/carbon 12 ratio in marine carbonates at the end of the Permian (see Figure 15-21).

As yet we have no certain explanation for either of the two Late Permian mass extinctions. It seems unlikely that the occurrence of the terminal Permian crisis at the same time as the massive outpouring of basalt in Siberia was coincidental, but we have no evidence of a connection. Much of the most intriguing evidence bearing on the Late Permian extinctions has appeared in recent years, so there is reason to expect that solutions will also emerge before long.

Regional Examples

The Paleozoic Era witnessed two orogenies in what is now the United States: the final episode of mountain building in the Appalachian region and a related orogeny in the southwestern region. Both orogenies resulted from the collision of Euramerica with Gondwanaland. Coal deposits formed in western Europe as well as in North America, and tectonic events produced mountains in the Rocky Mountain region. An enormous reef complex formed and came to an end in west Texas (the site of large petroleum reservoirs), and an orogenic episode visited the Pacific margin of North America.

The Alleghenian Orogeny and the Appalachian Mountains

The Late Carboniferous collision of Euramerica with Gondwanaland shifted the region of mountain building southward along the eastern margin of North America. The conspicuous fold-and-thrust belt of the central and southern Appalachians formed at this time. This still-mountainous region is a zone of low-temperature deformation called the Valley and Ridge Province (Figure 15-23). In this zone large slices of crust slid westward along thrust faults and were crumpled by pressure from the east. To the east of the Valley and Ridge Province, separated from it by the Blue Ridge Mountains, lies the Piedmont Province. Rocks of the Piedmont were metamorphosed and highly deformed because of their proximity to the zone of suturing. The Blue Ridge is a band of Proterozoic rocks that were metamorphosed about a billion

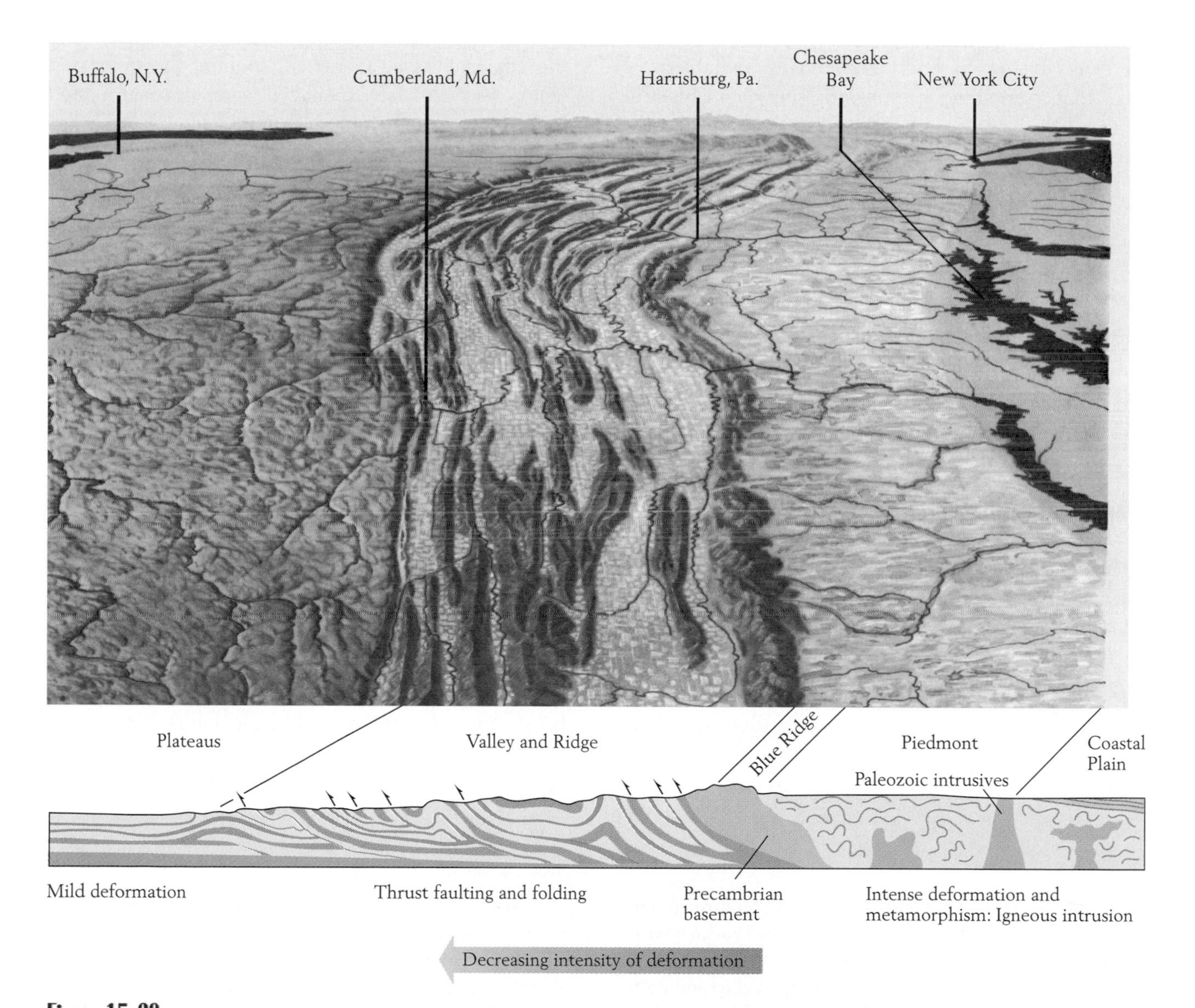

Figure 15-23 Aerial view and idealized cross section of the Appalachian region. The aerial view is toward the northeast. Sediments of the Coastal Plain lap up on the worn-down eastern portion of the Appalachian system. To the west of the Coastal Plain, the low-lying Piedmont Province is separated from the Valley and Ridge Province by the conspicuous Blue Ridge Province. *(Modified from J. S. Shelton, Geology Illustrated, W. H. Freeman and Company, New York, 1966.)*

years ago, during the Grenville orogeny (p. 335). These rocks were elevated along a large thrust fault during the late Carboniferous *Alleghenian orogeny.*

The Valley and Ridge Province displays the passive margin/flysch/molasse cycle that formed during the Taconic orogeny (see Figure 13-23) and the similar one that formed during the Acadian orogeny (see Figure 14-27). The Alleghenian Orogeny followed quickly after the Acadian orogeny, at a time when the continental margin was still an upland from which molasse was spreading westward. The Alleghenian orogeny simply built new mountains and perpetuated the accumulation of molasse (Figure 15-24). Some of the molasse deposits, such as the Pottsville Formation of Pennsylvania, harbor commercially valuable coal beds that formed in swamps bordering rivers. Most of the coal beds are positioned in the upper portions of meandering-river cycles (see Figure 5-16).

The Alleghenian orogeny culminated in the folding and thrust faulting of Paleozoic rocks of all ages, including those that had accumulated as flysch and molasse during the Taconic, Acadian, and Alleghenian orogenies. The ridges of the Valley and Ridge Province still stand as the Appalachian Mountains because they are underlain by ancient roots of lower density than the mantle. Recent isostatic uplift has created uplands in this old orogenic belt.

Earth Movements in the Southwestern United States

The late Paleozoic was also a time of mountain building along a zone extending from Utah across Oklahoma and Texas to Mississippi (Figure 15-25). Here the Ouachita Mountains formed as a westward continuation of the Appalachians when Gondwanaland and Euramerica collided (see Figure 15-17). Today traces of the two mountain chains meet at right angles beneath flat-lying younger deposits, and although the zone of contact is not well understood, the exposed segment of the Ouachitas is a fold-and-thrust belt resembling the Appalachian Valley and Ridge Province (Figure 15-26). One difference is that the folded rocks of the Ouachitas, which range in age from Ordovician to Middle Pennsylvanian, consist of deep-water black shale and flysch deposits that have been thrust northward against shelf-edge carbonates of similar age. In other words, deformation took place offshore from the continental margin (see Figure 15-25), and after it began, the rate of deposition in the adjacent basin increased. In fact, the deformed region seems to have behaved as an unusually deep foreland basin in which enormous volumes of deep-water Carboniferous deposits continued to accumulate on top of already-deformed older deposits. These younger, thicker deposits were, in turn, folded and thrust northward. Most of this deformation was completed before the start of the Permian Period. Although plate tectonic events relating to the origin of the Ouachita system were complex and remain poorly understood, it is known that several microplates were involved in the collision. Some microplates south of the Ouachita system eventually became parts of Central America (Figure 15-27).

The craton to the north and west of the Ouachita deformation also underwent tectonic movements. It is not clear just how these cratonic movements were related to the Ouachita orogeny, but they were largely vertical; enormous areas in what is now the southwestern United States became transformed into a series of uplifts and basins (see Figure 15-25). Many of these structural features are bounded by high-angle faults. The basins

Figure 15-24 *(right)* **Stratigraphic cross section through the central Appalachians west of the Blue Ridge Province, with vertical exaggeration.** Folds and faults are not shown. The thickest deposits lie to the southeast, in the Valley and Ridge Province. The thinnest deposits lie to the northwest, in the plateau region west of the Appalachians. This package of Paleozoic strata represents three cycles of deposition that relate to orogenic activity to the east. The Taconic and Acadian orogenies produced complete tectonic cycles of deposition, including passive margin deposition, flysch, and molasse. The Alleghenian orogeny, however, occurred when mountains produced during the Acadian orogeny were still shedding sediment to the east. Therefore, the Alleghenian orogeny simply piled Carboniferous molasse on top of the Devonian molasse produced by the Acadian event. *(Modified from G. W. Colton, in G. W. Fisher et al. [eds.], Studies of Appalachian Geology: Central and Southern, John Wiley & Sons, New York, 1970.)*

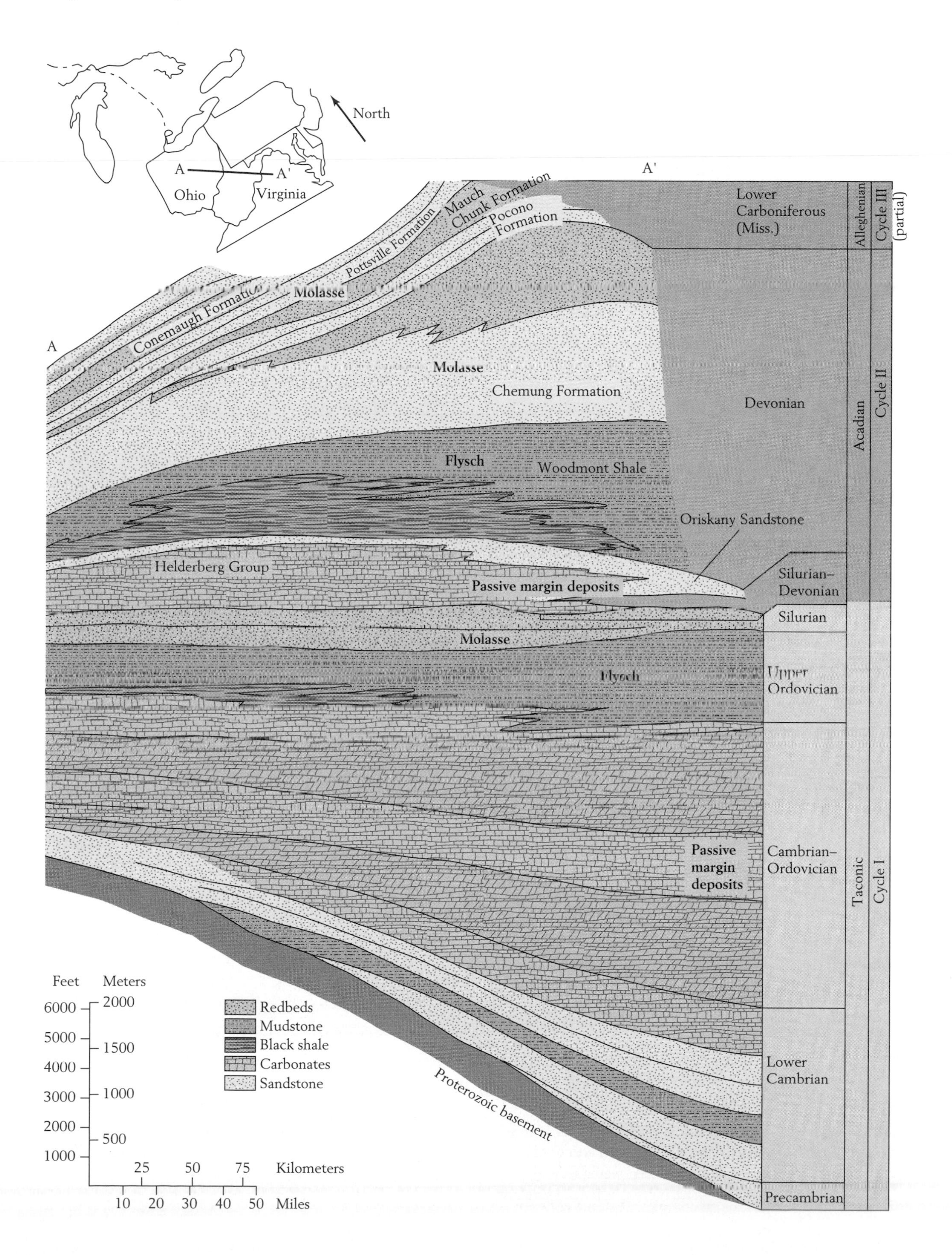

North
A
A'
Ohio
Virginia
Lower Carboniferous (Miss.)
Devonian
Silurian–Devonian
Silurian
Upper Ordovician
Cambrian–Ordovician
Lower Cambrian
Precambrian
Alleghenian
Acadian
Taconic
Cycle III (partial)
Cycle II
Cycle I
Mauch Chunk Formation
Pocono Formation
Pottsville Formation
Conemaugh Formation
Molasse
Molasse
Chemung Formation
Flysch
Woodmont Shale
Oriskany Sandstone
Helderberg Group
Passive margin deposits
Molasse
Flysch
Passive margin deposits
Proterozoic basement
Feet
Meters
6000
5000
4000
3000
2000
1000
2000
1500
1000
500
25 50 75 Kilometers
10 20 30 40 50 Miles
Redbeds
Mudstone
Black shale
Carbonates
Sandstone

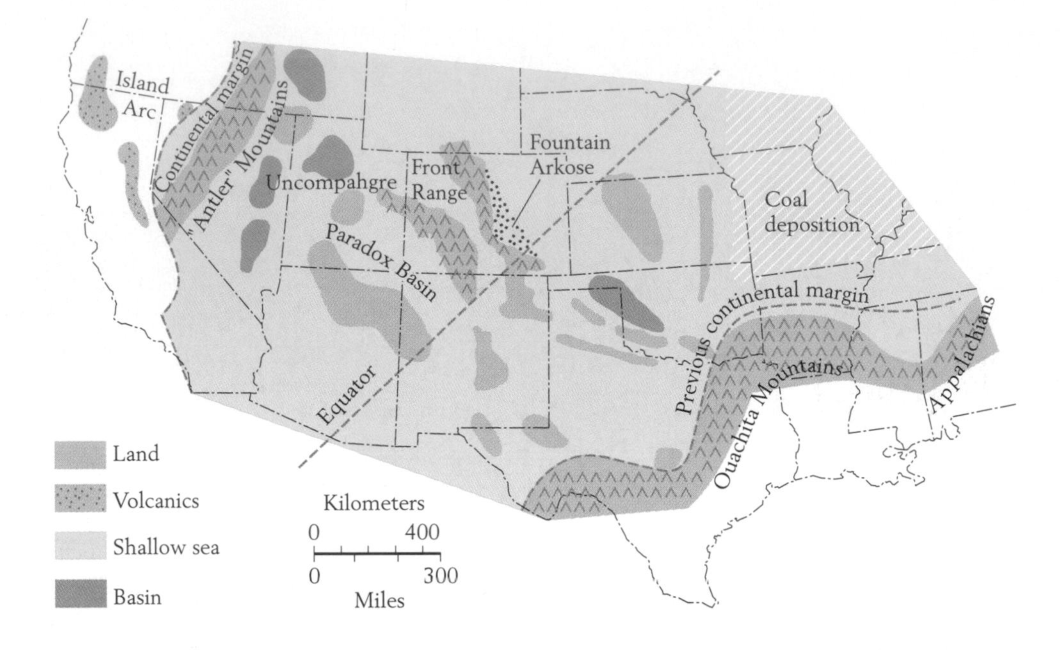

Figure 15-25 Paleogeography of the southwestern United States during Late Carboniferous (Pennsylvanian) time. Partway through the Late Carboniferous interval, the Ouachita Mountains began to form at the southern end of the Appalachians. Most of this deformation occurred offshore, uplifting oceanic sediments and welding them onto the previous continental margin. Coal-bearing cyclothems formed in marginal marine environments in the midcontinent region. To the west, shallow seas covered most of the craton, but many uplifts and basins developed from Texas to eastern Nevada, apparently in association with the Ouachita orogeny. The highest uplifts, the Front Range and the Uncompahgre, are together known as the Ancestral Rocky Mountains. Farther west, mountains produced by the earlier Antler orogeny still bordered the continent, and an island arc now lay along a subduction zone offshore in what is now California and Oregon.

accumulated Late Carboniferous (Pennsylvanian) and, in some cases, Permian deposits. Clastic debris shed from the uplifts was deposited rapidly in nearby basins as coarse arkose. Arkose is a sedimentary rock that consists largely of feldspar, a mineral that weathers to clay if it is not buried rapidly (p. 45).

Two of the uplifts, the Front Range and Uncompahgre uplifts, are commonly referred to as the Ancestral Rocky Mountains. The Ancestral Rockies developed during Late Carboniferous time; they were elevated and then subdued by erosion in an area where portions of the Rocky Mountains stand

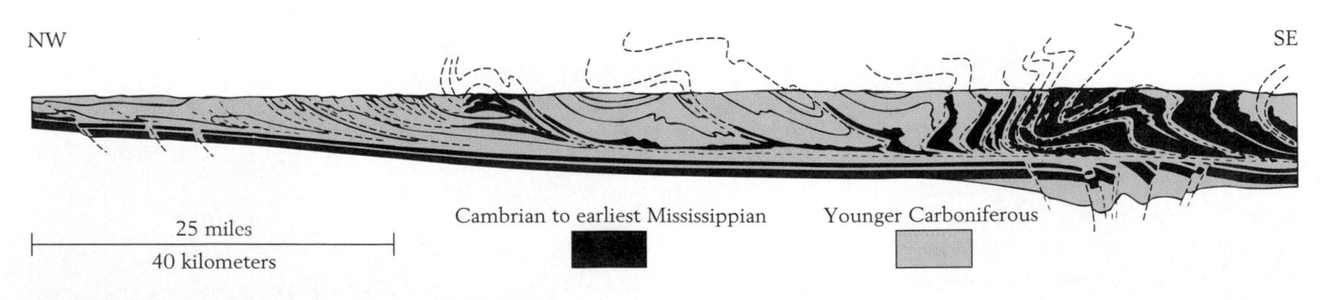

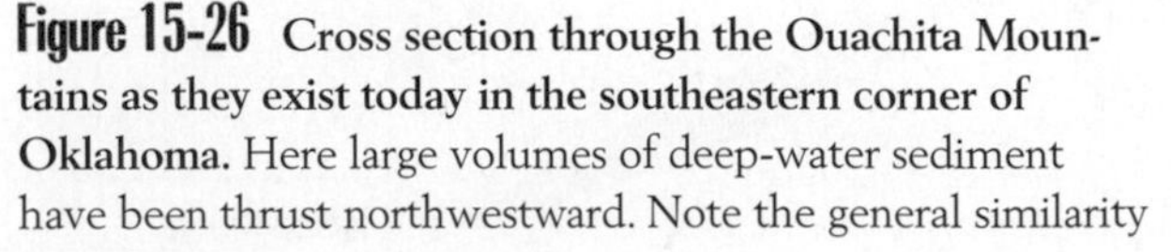

Figure 15-26 Cross section through the Ouachita Mountains as they exist today in the southeastern corner of Oklahoma. Here large volumes of deep-water sediment have been thrust northwestward. Note the general similarity between the style of deformation here and in the Appalachian fold-and-thrust belt (see Figure 7-10). *(After J. Wickham et al., Geology 4:173–180, 1976.)*

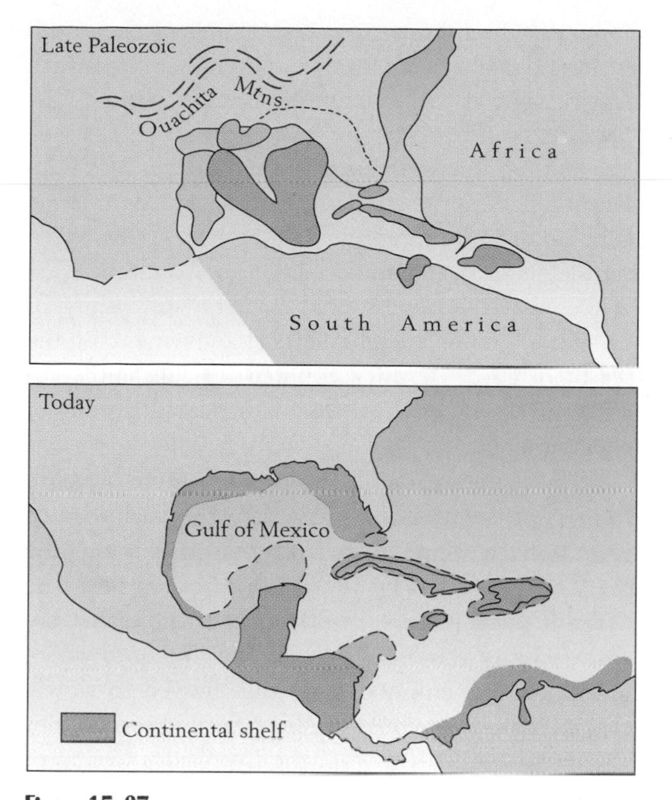

Figure 15-27 Approximate positions of microplates south and east of the Ouachita fold-and-thrust belt late in Paleozoic time (above) and today (below). Since late Paleozoic time some microplates have shifted southward, leaving the Gulf of Mexico in their place. *(Modified from A. G. Smith and J. C. Briden, Mesozoic and Cenozoic Paleocontinental Maps, Cambridge University Press, Cambridge, 1977.)*

today. It is estimated that the Uncompahgre uplift rose to between 1.5 and 3.0 kilometers (1 or 2 miles) above the surrounding seas, which flooded much of western North America. This elevation is comparable to that of the modern Rockies above the Great Plains to the east. The Front Range uplift is named for the Front Range of the modern Rockies, which now extends slightly farther east than the late Paleozoic uplift. Growth of the Ancestral Rockies elevated Precambrian rocks beneath, which later were leveled by erosion. The Precambrian roots of the Ancestral Rockies can still be observed where more recent secondary uplift has caused rivers to cut deep gorges.

At places in the basin between the Uncompahgre and Front Range uplifts, arkosic sands and conglomerates accumulated to thicknesses exceeding 3 kilometers (2 miles). The Fountain Arkose, which formed along the eastern flank of the Front Range uplift, was later upturned along the front of the modern Rockies when they were uplifted. Here, through differential erosion, the Fountain stands out in central Colorado as a series of spectacular ridges (Figure 15-28).

The Ancestral Rockies lay close to the Late Carboniferous equator, where easterly equatorial winds must have prevailed. It is therefore understandable that the ancient mountains seem to have produced a rain shadow to their west. Here, in the Paradox Basin (see Figure 15-25), great thicknesses of evaporites—primarily halite (p. 46)—accumulated.

Figure 15-28 Northward view along the Front Range of the Rocky Mountains at Boulder, Colorado. The core of the Rockies lies to the right. Tilted upward along its margin are the so-called Flatirons. The Flatirons consist of the Fountain Arkose, formed of sediment shed from the Ancestral Rocky Mountains, which lay slightly to the west in Late Carboniferous time. The Fountain Arkose was later tilted upward when the modern Rockies formed.

Coal within Cyclothems

In Late Carboniferous (Pennsylvanian) time, while rivers flowing from the Appalachians continued to form molasse deposits in eastern North America, coal swamps spread over the floodplains of these rivers and over the margins of epicontinental seas. In North America these swamps extended far to the west of the mountains over much of the nearly flat midcontinent. Some coal basins that are now separate were probably connected at the time they were formed. The Michigan Basin, however, formed in isolation (see Figure 9-24), and the basins in New England and eastern Canada may have done so as well (Figure 15-29).

In the nearly flat midcontinent region of the United States, Late Carboniferous (Pennsylvanian) cycles of a special kind formed. The coal beds of these cycles formed in swamps that bordered shallow seas. Here and in similar settings on other continents, coal beds are thin but widespread, occurring within cycles that include marine deposits. Dozens of similar cycles are commonly found superimposed on one another. Such cycles in coal beds are known as **cyclothems** in North America and as coal measures in Britain.

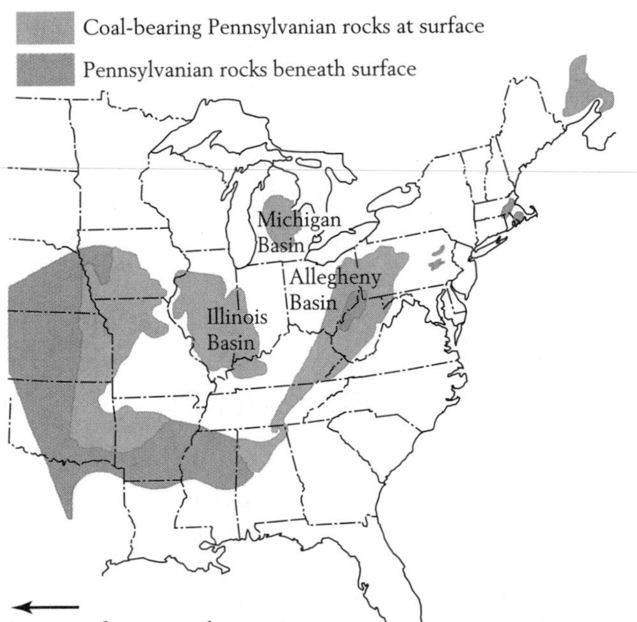

Figure 15-29 Distribution of coal-bearing cyclothems of Pennsylvanian (Upper Carboniferous) age in eastern North America. Before undergoing erosion, the cyclothems were even more extensive in the areas east and south of the Illinois Basin. The main area of coal formation extended from the margin of the early Appalachian Mountains on the east to an area of fully marine deposition to the west.

Shifting Environments and the Origin of Cyclothems The fact that many marine and nonmarine habitats are often represented in just a few vertical meters of stratigraphic section in a cyclothem indicates that the depositional gradient was very gentle. Clearly, only a slight vertical movement of the sea or of Earth's crust accounted for substantial advance or retreat of the water with related shifting of environments.

The coal now found within a cyclothem began to form as peat within a coal swamp. The coal swamps seem to have occupied lowland areas neighboring the sea. They were fed by the rivers whose deposits lie beneath them (Figure 15-30). It is possible that the entire swamp was, in effect, a broad river into which inland streams emptied—one that flowed so slowly that its movement could not have been observed with the naked eye. This is what the Everglades swamp of Florida is today (p. 407). The Everglades "river" also flows over a very flat region. The southern Florida peninsula is being partially drowned by the rising sea. The water of the Everglades remains fresh except near the edge of the sea.

Cyclothems were formed by alternating transgressions and regressions of shallow seas. A transgression resulted in the deposition of marginal marine peat (future coal) on top of nonmarine deposits and of marine sediment on top of the peat (see Figure 15-30). Regression reversed the sequence, burying marine deposits beneath peat and then nonmarine sediments (Figure 15-31).

Glaciers and Sea Level The oscillations in sea level responsible for the cyclothems were so rapid that they must have resulted from the repeated expansion and contraction of glaciers in Gondwanaland. Why, then, do we not find similar cycles representing the Pleistocene interval of the past 1.6 million years, when continental glaciers have expanded and retreated many times? The explanation seems to be that in recent times—even during interglacial periods of high sea level such as the one in which we live—the continents have remained relatively emergent. The seas are rising and falling over steeply sloping continental margins; they are not invading

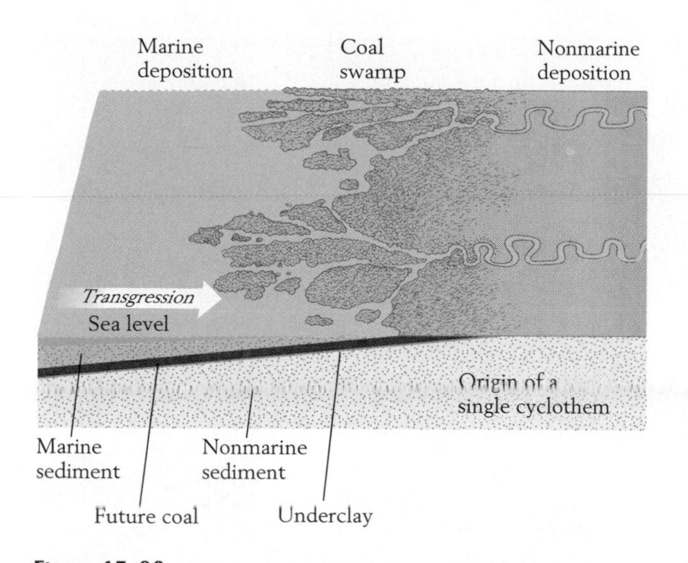

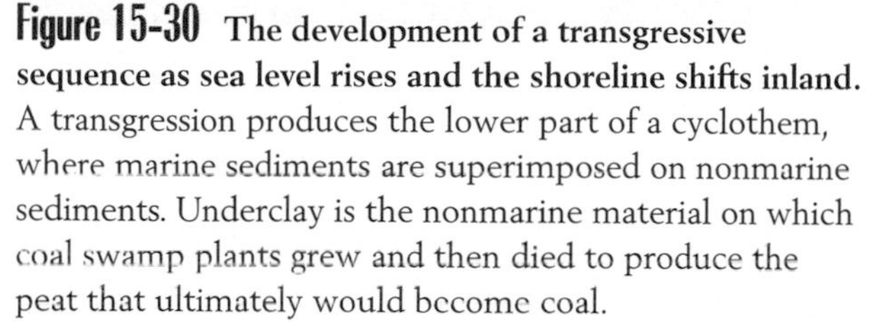

Figure 15-30 The development of a transgressive sequence as sea level rises and the shoreline shifts inland. A transgression produces the lower part of a cyclothem, where marine sediments are superimposed on nonmarine sediments. Underclay is the nonmarine material on which coal swamp plants grew and then died to produce the peat that ultimately would become coal.

and receding from vast, almost flat interior lowlands as they did when they formed the cyclothems of Kansas, Illinois, and neighboring regions.

Sedimentary records show that glaciation ceased altogether in South America long before glaciers disappeared from Australia. This observation suggests that glaciers did not wax and wane simultaneously in all parts of Gondwanaland. Thus the rise or fall of the world's oceans at any time must have depended on the averaging of glacial events within the entire glaciated area.

The Permian System of West Texas

The Delaware Basin of Texas and New Mexico is one of the most famous geologic structures in the world, both because of its economic importance and because it offers spectacular scenery. Although it has not been occupied by the sea for more than 200 million years, it remains a topographic basin. A person can stand in its center today and view ancient carbonates that formed as banks or reefs around the margin during the Permian Period (Figure 15-32). Earlier, during the latter part of Late Carboniferous (Pennsylvanian)

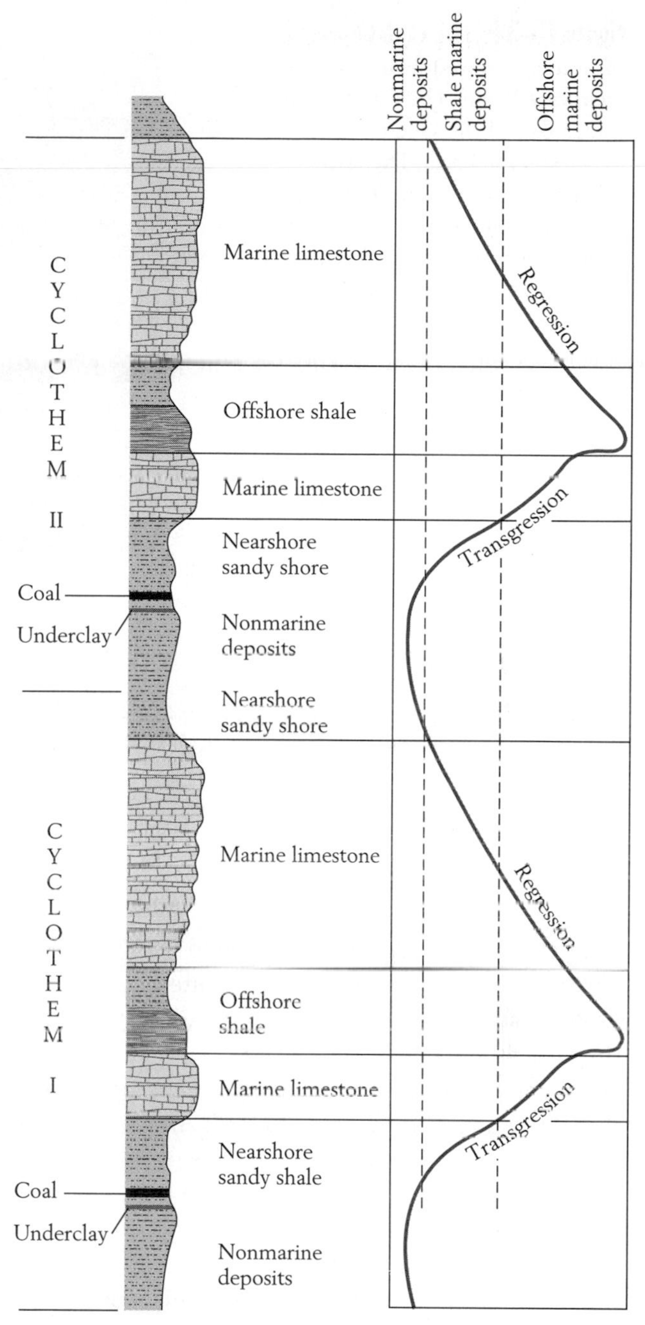

Figure 15-31 Two idealized cyclothems. The coal swamps migrated over these nonmarine deposits as the transgression proceeded. Above the coal are deltaic and other marginal marine sediments, which were deposited above the coal as the sea spread inland. The marginal marine deposits are succeeded, in turn, by marine limestones that represent fully marine conditions and then by black shales that represent deep environments that were present in the region at the time of maximum transgression. In time, sea level began to fall again, and the depositional sequence was reversed.

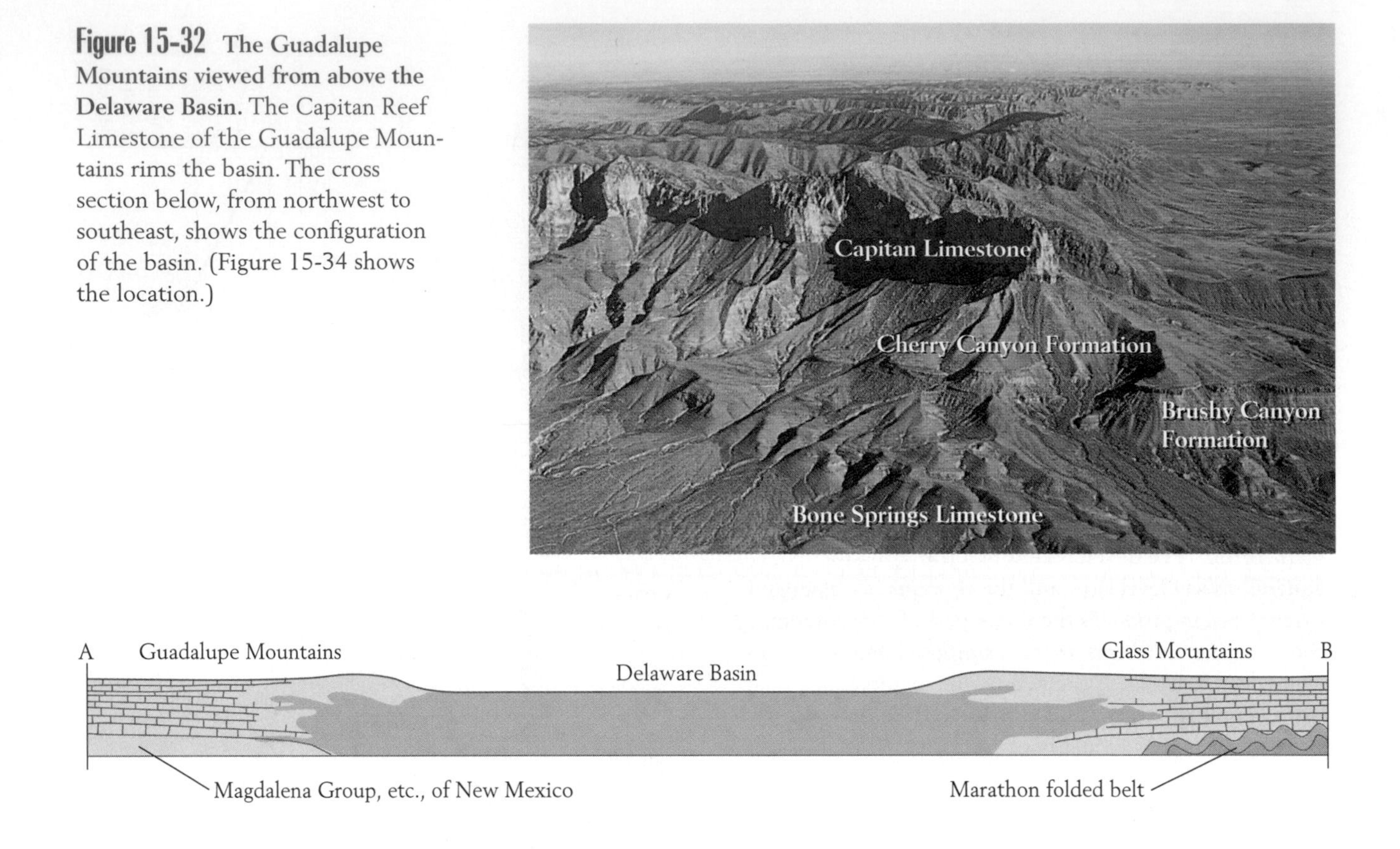

Figure 15-32 The Guadalupe Mountains viewed from above the Delaware Basin. The Capitan Reef Limestone of the Guadalupe Mountains rims the basin. The cross section below, from northwest to southeast, shows the configuration of the basin. (Figure 15-34 shows the location.)

time, the shallow seas that had shifted back and forth over the coal basins of the central United States withdrew westward, never to return. In earliest Permian time they remained only in Texas and neighboring areas, where they were connected to the seas that still flooded the western margin of North America (Figure 15-33). The Ancestral Rockies still stood high, as a mountainous island, and the young Ouachita Mountains bordered the southern margin of North America as what must have been a rugged and imposing mountain range. To the northwest of this range, detrital material came to rest in a marginal foreland basin.

During early and middle Paleozoic time, marine deposits accumulated in the area that now forms west Texas, which was a broad, shallow basin on the continental shelf. During the uplift of the Ouachita range and other Carboniferous uplands, a small fault block rose up within the west Texas basin, dividing it into the Delaware Basin and the Midland Basin (Figure 15-34). Both of these basins subsequently received large thicknesses of sediment that have yielded enormous quantities of petroleum.

Reef Growth While reefs grew upward around the Delaware Basin, the Midland Basin to the east became filled with sediment (see Figure 15-34). Along with surrounding areas, it was then flooded by shallow seas. Although by this time the Ancestral Rockies had been lowered by erosion, they still formed a large island to the northwest. The Delaware Basin lay very close to the Permian equator, and the Ouachita chain must have left the basin in the rain shadow of equatorial winds blowing from the east. The shallow seas that surrounded the basin were the sites of carbonate and evaporite deposition in what was obviously an arid climate.

As time passed and sea level rose, the reef grew upward more rapidly than the Delaware Basin filled in, and eventually the reef stood high above a basin that was some 600 meters (2000 feet) deep (Figure 15-35). Although the waters have long since withdrawn, the configuration remains today (see Figure 15-32). Early in its history, when the Delaware Basin was relatively shallow, its floor was inhabited by snails, deposit-feeding bivalves, sponges, and brachi-

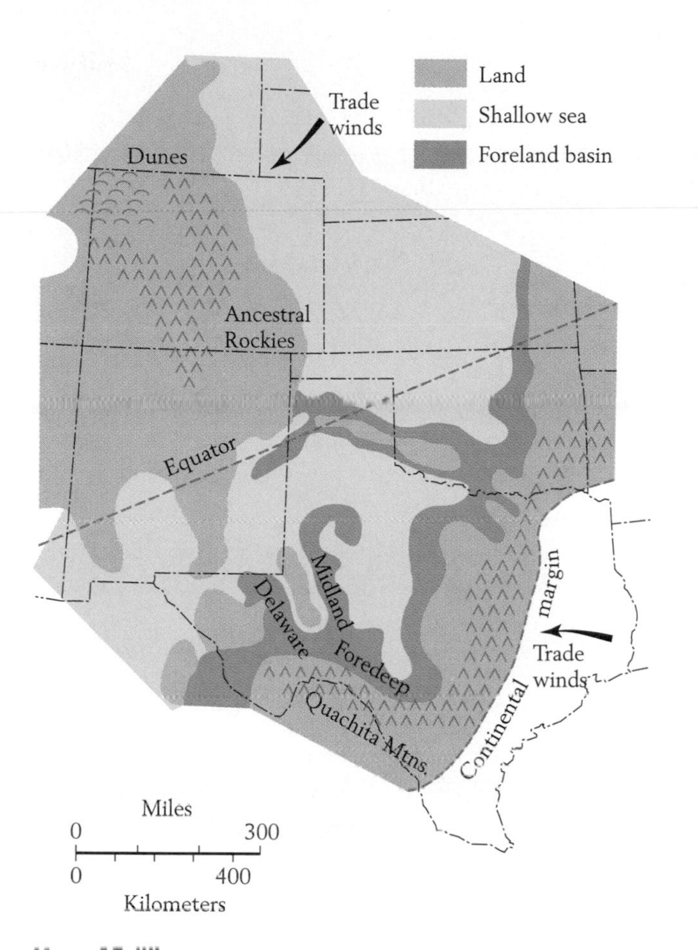

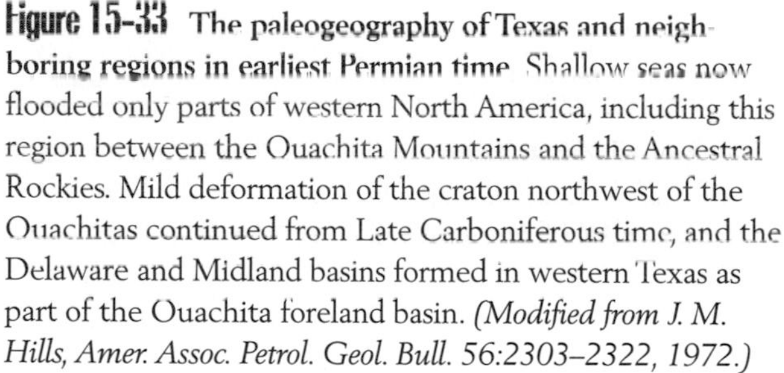
Figure 15-33 **The paleogeography of Texas and neighboring regions in earliest Permian time.** Shallow seas now flooded only parts of western North America, including this region between the Ouachita Mountains and the Ancestral Rockies. Mild deformation of the craton northwest of the Ouachitas continued from Late Carboniferous time, and the Delaware and Midland basins formed in western Texas as part of the Ouachita foreland basin. *(Modified from J. M. Hills, Amer. Assoc. Petrol. Geol. Bull. 56:2303–2322, 1972.)*

Figure 15-34 **The paleogeography of Texas and neighboring regions when reefs encircled the Delaware Basin in Late Permian time.** During this interval a narrow passageway (Hovey Channel) connected the Delaware Basin with the open ocean to the west, but the Midland Basin was eventually filled with sediment. Line *A–B* shows the location of the cross section in Figure 15-33.

opods. Later, when the basin had deepened, these animals decreased in number; the only abundant marine fossil remains in the younger basin deposits are conodonts (see Box 3-1), radiolarians (see Figure 3-19), and ammonoids (see Figure 14-4), all of which lived high in the water column and sank to the bottom when they died. Also present are plant spores that were blown into the basin from the land. We can conclude that when the Capitan Limestone formed (see Figure 15-35), the floor of the deep basin below was poorly oxygenated; oxygen used up in the decay of organic matter was not replenished, and few bottom-dwelling animals could survive.

The reeflike structure that rims the Delaware Basin was built during the Guadalupian Age, the second-to-last age of the Upper Permian. The reef was formed primarily by sponges (see Figure 15-6), algae, and lacy bryozoans (see Figure 15-4). The crest of the reef was covered by shallow water that also bathed an extensive back-reef flat (see Figure 15-35). Rubble from the reef periodically tumbled down the forereef slope into the basin. The ancient talus slope is present

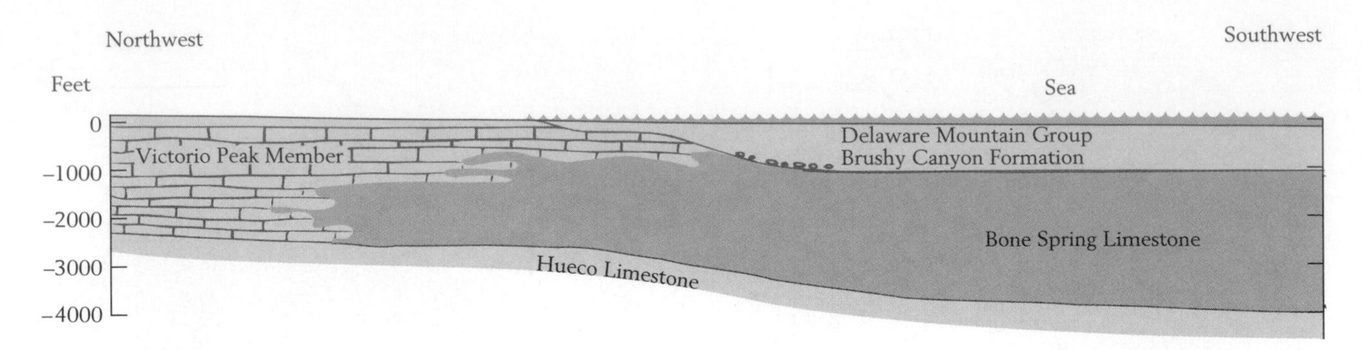

A At close of lower Guadalupian time

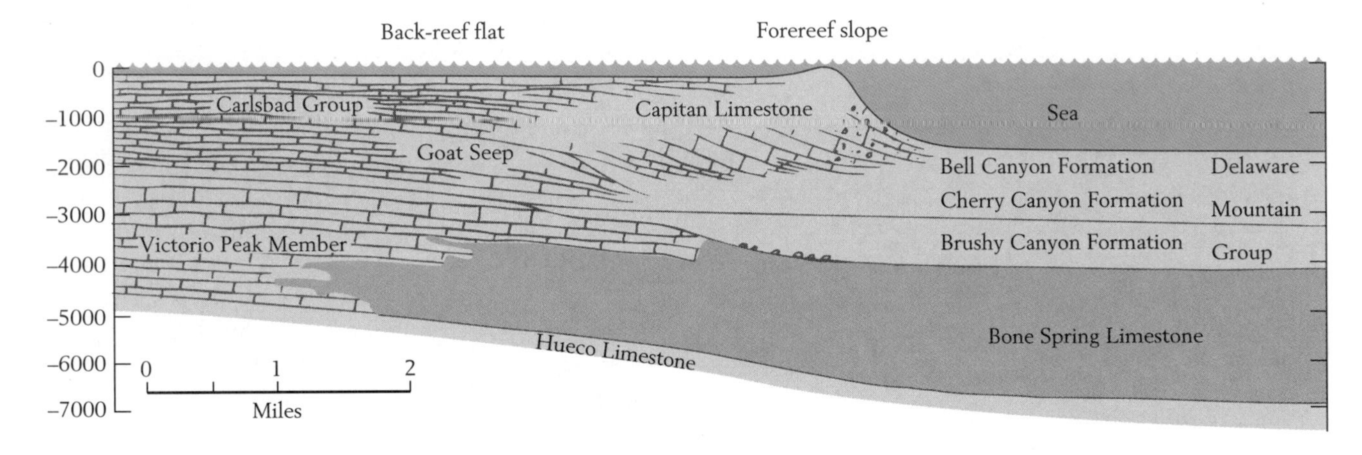

B At close of upper Guadalupian time

Figure 15-35 Profiles of the reef that now forms the Guadalupe Mountains early in Guadalupian time *(A)* and later in Guadalupian time *(B)*, when the reef had grown rapidly upward while the basin deepened. (The Guadalupian Age was the next-to-last age of the Permian Period.) Note that the reef advanced toward the basin center as it grew upward. *(After N. D. Newell et al., The Permian Reef Complex of the Guadalupe Mountains Region, Texas and New Mexico, W. H. Freeman and Company, New York, 1953.)*

today in bedding that dips at angles as high as 40°. In the rubble of the forereef slope are many beautifully preserved fossils whose originally calcareous hard parts have been replaced by durable silica; among them are sponges (see Figure 15-6) and shells of fusulinid foraminifera that lived in shallow habitats but periodically washed down to lodge in the slope rubble. Some were swept into the basin by turbidity flows that left conspicuous graded beds in the rocks of the basin; these beds constitute the Delaware Mountain Group. Most sediments of this unit are dark sands and silts that periodically washed into the basin, apparently when sea level was low and the reef surface was exposed to erosion.

When the older bedding surfaces of the carbonates that ring the Delaware Basin are traced laterally, a different configuration becomes apparent. These older bedding surfaces show that the early reefs, known as the Goat Seep Formation, stood in much lower relief above the basin (see Figure 15-35). From its earliest days until late in the Permian, the Delaware Basin was connected with the open sea to the southwest through what is called Hovey Channel (see Figure 15-34). Early in the evolution of the basin, when the reefs were low, the connection and the resulting pattern of water circulation permitted oxygen-rich waters to reach the basin floor so that animals could live there (Figure 15-36). Later the basin deepened,

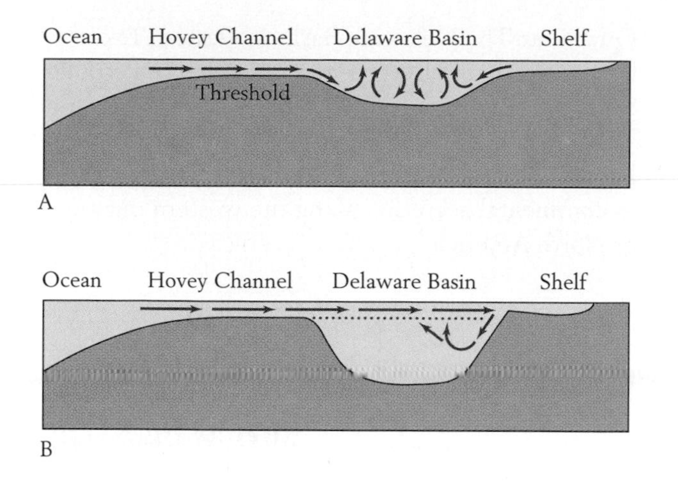

Figure 15-36 Patterns of water circulation in the Delaware Basin. Early in its history, the basin was shallow enough that well-oxygenated surface water reached its floor *(A)*. Later, when the basin had deepened, good circulation was restricted to the upper waters, and the bottom waters thus became stagnant *(B)*. *(After N. D. Newell et al., The Permian Reef Complex of the Guadalupe Mountains Region, Texas and New Mexico, W. H. Freeman and Company, New York, 1953.)*

but Hovey Channel remained shallow, a conformation that caused the bottom waters of the basin to become stagnant and poor in oxygen, excluding almost all forms of life.

Death of the Reef Eventually, near the end of Permian time, the Delaware Basin filled with evaporites. As we have seen, the climate of Texas and neighboring areas became arid toward the end of the Permian, and the rate at which waters evaporated from the Delaware Basin may have increased. Hovey Channel may also have become so constricted that the rate of evaporation there occasionally exceeded the rate at which new water was supplied. In any case, the reef stopped growing and the basin ultimately filled with evaporites; distinct layers in some of these evaporite deposits extend over many hundreds of square kilometers (see Figure 6-15). It is possible that this layering reflects seasonal changes similar to those responsible for glacial varves.

The Delaware Basin evaporites remained in place for a long period, protecting the magnificent geologic record along the walls and floor of the basin. Fresh water later dissolved the evaporites in many areas, exposing the ancient reef-encircled structure (see Figure 15-32).

The Western Margin of North America

What was happening to the west of the Delaware Basin and the Ancestral Rocky Mountains? During late Paleozoic time the western margin of the North American craton passed through what is now northwestern Nevada (see Figure 15-25). Hundreds of kilometers from the margin of the craton, volcanoes were active again in the area that is now California and slightly to the north. Here coarse clastic deposits were shed from volcanic highlands into the surrounding seas. An orogenic episode in Nevada in latest Permian and Early Triassic times was remarkably similar to the Antler orogeny (see Figure 14-31). In this second, Sonoma orogeny, as in the Antler, marine deposits were thrust upward over the continental margin. The Sonoma orogeny was of great significance in that it entailed the complete closure of the basin between the volcanic arc and the North American continent. While some of the deep-sea deposits of the basin were thrust onto the continent, others were welded onto the continental margin along with the volcanic terrane of the arc. The result was a considerable westward growth of the North American crust.

Chapter Summary

1 Marine life of late Paleozoic time in many ways resembled life of the middle Paleozoic, but the tabulate-strome reef community was gone. In addition, four groups that expanded enormously contributed large volumes of skeletal debris to limestones: first crinoids and lacy bryozoans, later flakelike algae and fusulinid foraminifera.

2 In Carboniferous time the coal-swamp floras, which were dominated by trees of the genera *Lepidodendron* and *Sigillaria*, played a major ecologic role, as did seed ferns and, on drier land, sphenopsids and cordaites. During the Permian Period, however, climates in the northern hemisphere became warmer and drier, and these plant groups gave way to conifers and other gymnosperms.

3 Carboniferous coal swamps produced coal beds, which occur within depositional cycles. Some coal beds represent meandering rivers and others alternating transgressions and regressions of shallow seas.

4 Throughout nearly all of late Paleozoic time, continental glaciers blanketed the south polar region of Gondwanaland and large areas of this great southern continent were populated by the cold-adapted *Glossopteris* flora. Greenhouse cooling resulting from burial of carbon in coastal swamps may have initiated this glacial episode.

5 Insects originated during the Carboniferous Period and underwent a great adaptive radiation. During the Permian Period, amphibians were displaced from terrestrial habitats by early mammal-like reptiles, including finbacks, but these soon gave way to more advanced mammal-like reptiles, the therapsids.

6 In mid-Carboniferous time Gondwanaland became sutured to Euramerica, causing widespread orogenies, including the Alleghenian of the Appalechian region. Because of its great size, the landmass formed by this collision developed widespread aridity and coal swamps shrank.

7 The latter part of the Permian Period was a time of hot, dry conditions and widespread evaporite deposition in equatorial regions

8 The Permian Period—and thus the Paleozoic Era as well—ended with two major extinction events separated by just a few million years. The second of these events, at the very end of the Permian, was the largest mass extinction of animals that has ever occurred.

9 In the United States west of the Appalachians, the Alleghenian orogenic episode affected the central and southern Appalachians and also created the Ouachita Mountains in Oklahoma and Texas. Uplifts and basins formed north and west of the Ouachitas. The Delaware Basin in western Texas became encircled by a reef complex in the Permian System, and near the end of the Permian this basin was filled by evaporite deposits. Beginning in the latest Permian time, the Sonoma orogeny resulted in continental accretion along the western margin of North America.

Review Questions

1 What caused shallow seas to expand and contract over the midcontinental United States many times during Late Carboniferous time?

2 If you could examine Late Carboniferous (Pennsylvanian) coal-bearing cycles in the field, how would you determine whether the coal deposits formed along a river or a shallow sea? (Hint: Refer to p. 134 and Figures 5-16 and 15-31.

3 What groups of terrestrial plants that existed in late Paleozoic time survive today?

4 What justification is there for dividing the Carboniferous interval into the Mississippian and Pennsylvanian periods?

5 How did the history of glacial activity in Late Carboniferous time relate to the deposition of coal?

6 In eastern North America, mountain building progressed from New England to Texas during Paleozoic time. How does this pattern relate to continental movements? (Review the relevant parts of Chapters 9, 13, and 14, as well as the relevant part of this chapter.)

7 In what way may therapsids have been superior to the amphibians and reptiles that preceded them?

8 Why were evaporite deposits widespread in Europe and North America during Late Permian time?

9 What apparently caused greenhouse warming of Earth to weaken during the Carboniferous and then strengthen again during the Permian?

10 What important changes occurred in the deep sea during Late Permian time and how may these changes have related to major extinctions?

11 The formation of Pangea during the Carboniferous Period was a major geologic event. Using the Visual Overview on page 400 and what you have learned in this chapter, describe how sea level, climates, and life changed late in the Paleozoic Era, after the formation of Pangea. How did some of these changes relate to each other and to the existence of Pangea?

Additional Reading

Benton, M. J., *Vertebrate Palaeontology*, Chapman & Hall, London, 1997.

Erwin, D. H., *The Great Paleozoic Crisis: Life and Death in the Permian*, Columbia University Press, New York, 1993.

Thomas, B. A., and R. A. Spicer, *The Evolution and Palaeobiology of Land Plants*, Croom Helm, London, 1987.

CHAPTER 16

The Early Mesozoic Era

The Mesozoic Era, or "interval of middle life," began with the Triassic Period. The Triassic and the subsequent Jurassic Period together constitute slightly more than half of the era. Rocks representing these periods are especially well exposed and well studied in Europe.

Near the transition from the Paleozoic Era to the Mesozoic Era, the great supercontinent Pangea took its final form, encompassing virtually all the major segments of Earth's continental crust. Pangea was so large that much of its terrain lay far from any ocean, and as a result it became arid. During Jurassic time, however, sea level rose and marine waters spread rapidly over the land, leaving a more extensive record of shallow marine deposition than that of the Triassic System. Then, later in early Mesozoic time, Pangea began to fragment, and before the end of the Jurassic Period, Gondwanaland was once again separate from the northern landmasses.

Life of early Mesozoic time differed substantially from that of the Paleozoic Era. For many groups of animals, recovery from the Late Permian biotic crisis was sluggish, but by the end of the Triassic Period, mollusks had reexpanded to become more diverse than they had been during the Paleozoic Era. Their success has continued to the

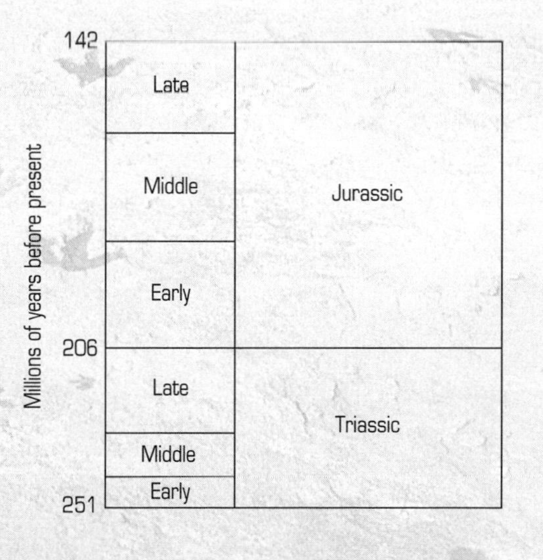

Dinosaur tracks of Jurassic age in the Painted Desert in Arizona.

Visual Overview

Major Events of the Early Mesozoic

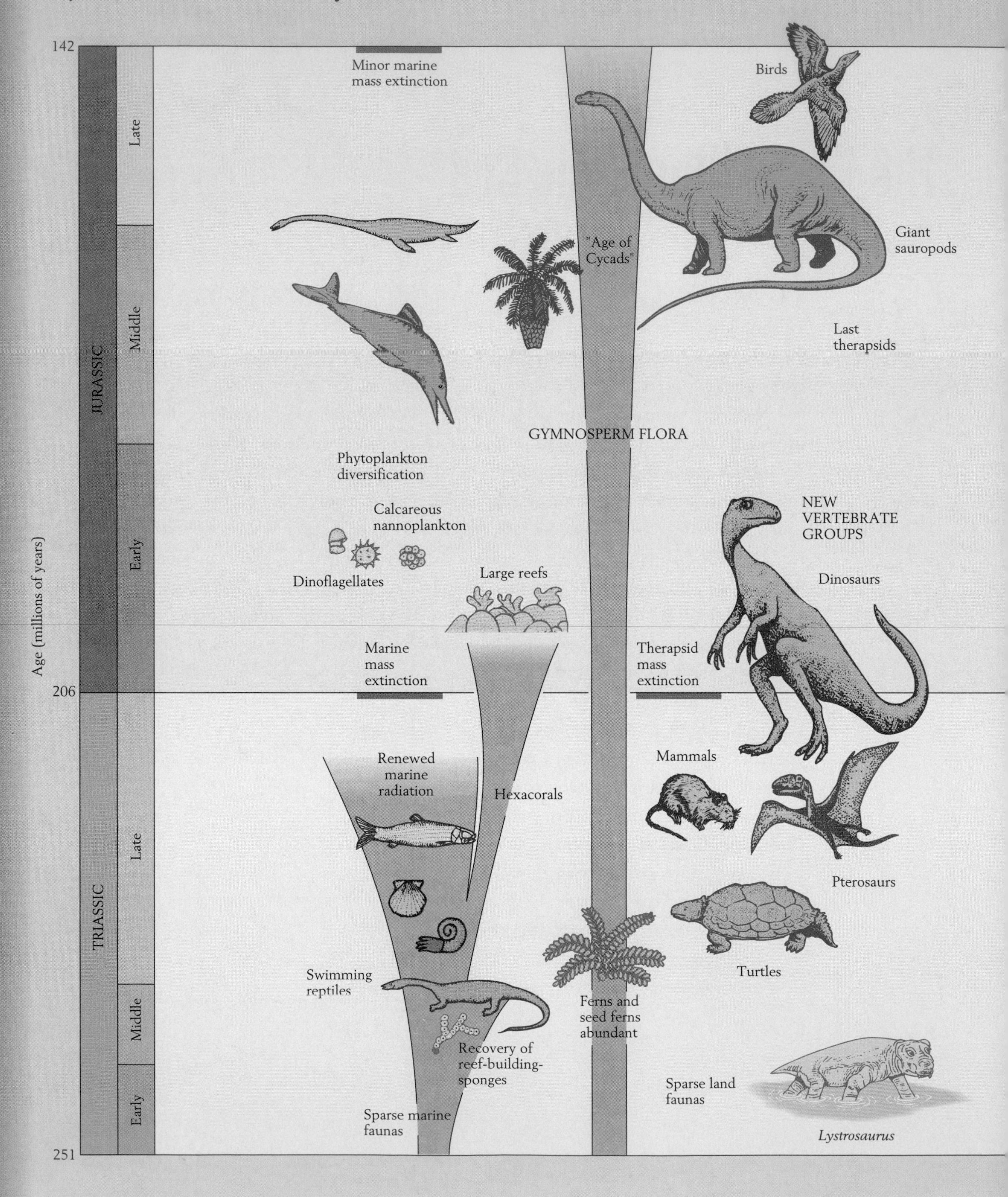

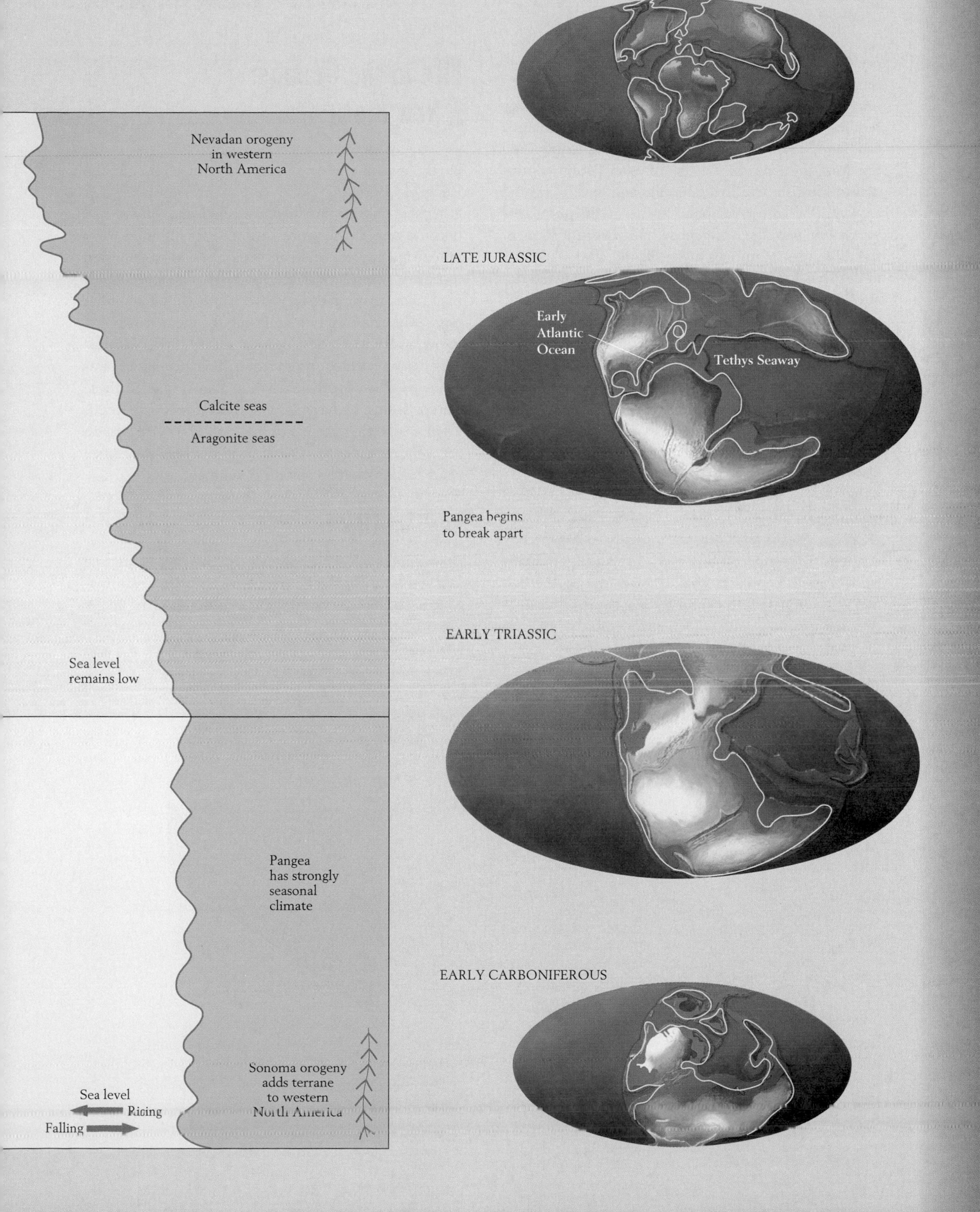

LATE CRETACEOUS
Nevadan orogeny
in western
North America
LATE JURASSIC
Early
Atlantic
Ocean
Tethys Seaway
Calcite seas
Aragonite seas
Pangea begins
to break apart
EARLY TRIASSIC
Sea level
remains low
Pangea
has strongly
seasonal
climate
EARLY CARBONIFEROUS
Sonoma orogeny
adds terrane
to western
North America
Sea level
Rising
Falling

present time. The marine ecosystem was also transformed during Triassic and Jurassic time by the addition of both modern reef-building corals and large reptiles, which joined fishes as swimming predators. On land, gymnosperm floras that had conquered much of the land during the Permian Period continued to flourish, and flying reptiles and birds appeared as well. The most dramatic event in the terrestrial ecosystem was the emergence and diversification of the dinosaurs. Mammals arose slightly after dinosaurs in the Triassic Period, but they remained small and relatively inconspicuous throughout the Mesozoic Era.

The Triassic System is bounded by the terminal Permian extinction below and by another extinction above. It was the unique fauna of this system that led Friedrich August von Alberti to distinguish the Triassic in 1834. Alberti originally named the system the Trias for its natural division in Germany into three distinctive stratigraphic units.

At first the Jurassic System also had a shorter name, Jura, a label that was borrowed from a portion of the Alps in which the system is especially well exposed. The Jurassic was not formally established by a published proposal, however; instead, it gradually came to be accepted as a valid system during the first half of the nineteenth century, when its many distinctive marine fossils were widely investigated.

Life in the Oceans: A New Biota

By the end of the great extinction that brought the Paleozoic Era to a close, several previously diverse groups of marine life had vanished and others had become rare. Gone were fusulinid foraminifera, lacy bryozoans, rugose corals, and trilobites. Most common in Lower Triassic rocks are mollusks. The ammonoids made a dramatic recovery after almost total annihilation; although only two ammonoid genera are known to have survived the Permian crisis, Lower Triassic rocks have yielded more than 100 genera of ammonoids. The adaptive radiation that produced these genera seems to have issued from the single genus *Ophiceras*. Other groups of marine life were slower to recover, but by Late Triassic time the seas were once again teeming with a variety of animals.

Seafloor Life

One remarkable consequence of the loss of marine life at the end of the Permian was a return of stromatolites to shallow subtidal environments in many parts of the world. These cyanobacterial structures

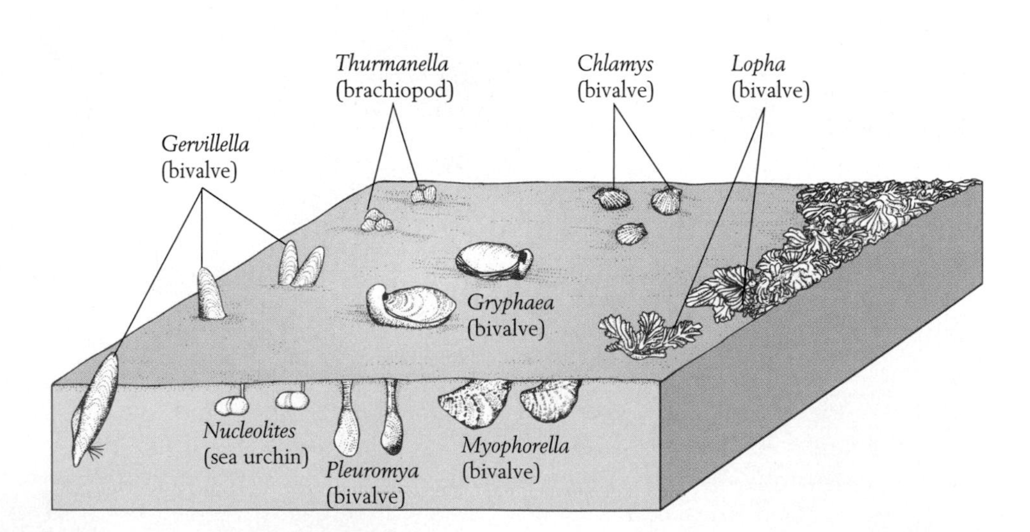

Figure 16-1 **Life of a Late Jurassic seafloor.** As in Paleozoic time, many animals lay exposed on the seafloor, but some of them were of new types, such as the irregularly shaped oyster *Lopha*, which cemented itself to other shells, and the coiled oyster *Gryphaea*. Other new animals, including sea urchins such as *Nucleolites*, lived within the sediment. *(Modified from F. T. Fursich, Palaeontology 20:337–385, 1977.)*

flourished briefly in the Early Triassic, just as they had done for a longer interval before the evolution of animals that ate and burrowed through them (p. 347). Bivalve mollusks, like ammonoids, are frequently found in Lower Triassic rocks, although their diversity is somewhat limited. Both bivalves and gastropods expanded in number and in variety to become among the most prominent groups of early Mesozoic marine animals. As in the Paleozoic Era, some of the bivalves burrowed in the seafloor, while others rested on the sediment surface (Figure 16-1).

Other than ammonoids and bivalves, only brachiopods make a modest showing in Lower Triassic rocks; all other marine invertebrates are rare. The brachiopods also diversified during the Triassic and Jurassic periods, but they subsequently declined until today they are very sparse.

Sea urchins, which had existed in limited variety during the Paleozoic Era, diversified greatly during the first half of the Mesozoic Era. Some of the new forms that emerged at this time were surface dwellers, like most of the Paleozoic sea urchins, but others lived within the sediment as actively burrowing deposit feeders (see Figure 16-1).

Reefs did not recover from the terminal Permian extinction until Middle Triassic time, when vertical mixing in the oceans increased, as shown by the return of oxygenated conditions to the deep sea (see Figure 15-22). The earliest of the new reefs were built by sponges and algae resembling the ones that had built Late Permian reefs. As the Triassic progressed, however, the dominant force in reef building shifted to a newly evolved group that is still successful today– the **hexacorals** (Figure 16-2). This group includes not only colonial reef builders, but also solitary species that resemble the solitary rugose corals of the Paleozoic Era (see Figure 13-13). During the Middle Triassic a few species of hexacorals built small mounds that stood no more than 3 meters (10 feet) above the seafloor. By latest Triassic time, reefs were larger, some having been constructed by more than 20 species.

Some of the early coral mounds grew in relatively deep waters, so it appears that the earliest hexacorals, unlike the corals that form large tropical reefs today, did not live in association with symbiotic algae, which require strong sunlight (see Box 4-1). Perhaps it was not until latest Triassic or Early Jurassic time, when hexacorals began to form large reefs, that this symbiotic relationship was established.

Figure 16-2 **Triassic hexacorals.** The fragments of colonies shown here are 3 to 4 centimeters (1.5 inches) across.

Because of the success of bivalve and gastropod mollusks, sea urchins, and reef-building hexacorals, seafloor life of the Late Jurassic world looked much more like that of today than had seafloor life of the Paleozoic Era. Still missing were many kinds of modern arthropods, but the group that includes crabs and lobsters got off to a modest evolutionary start during the Jurassic Period.

Pelagic Life

Presumably many kinds of planktonic organisms in Triassic and Jurassic seas left no fossil record. The **dinoflagellates** produced many fossils in the form of cysts, which were durable resting stages that these algae formed when the environment became inhospitable. Dinoflagellates underwent extensive diversification during mid-Jurassic time and remain an important group of producers in modern seas (see Figure 3-16). The **calcareous nannoplankton,** another important group of living algae (see Figure 3-16), made their first appearance in earliest Jurassic time. Today these floating algae are most diverse in tropical seas (see Figure 4-27), and their armor plates rain down on the deep sea to become prominent constituents of deep-sea sediments (see Box 10-1).

Figure 16-3 The Jurassic ammonoid *Phylloceras.* The suture pattern is shown below the shell. (The suture is the juncture between the convoluted internal partitions, or septa, and the coiled outer shell.)

Higher in the food chain, the ammonoids and the belemnoids played major roles as swimming predators. The ammonoids' evolutionary recovery after the Permian crisis led to great success throughout the Mesozoic Era. Individual ammonoid species, however, survived for relatively brief intervals, often a million years or less, and so they are extremely useful as index fossils for Mesozoic rocks (Figure 16-3). The belemnoids, which were squidlike relatives of the ammonoids, also pursued prey by jet propulsion (Figure 16-4). They evolved in late Paleozoic time but remained inconspicuous until the Mesozoic Era, at which time many types evolved. Conodonts, the toothlike structures that are now known to have belonged to eel-like animals (see Box 3-1), have also proved useful in the correlation of Triassic rocks, but by Jurassic time, conodont-bearing animals no longer existed.

Paleozoic ray-finned bony fishes gave rise to forms that were successful in early Mesozoic time but were still more primitive than most of their modern-day descendants. The scales that covered the bodies of these fishes were diamond-shaped structures that overlapped slightly or not at all (Figure 16-5), in sharp contrast to the circular, strongly overlapping scales of nearly all modern bony fishes. Presumably the primitive diamond-shaped scales were less protective than the modern kind. Other features that distinguished early Mesozoic bony fishes from their modern counterparts were skeletons that consisted partly of cartilage rather than entirely of bone; relatively simple, primitive jaws; and tails that were highly asymmetrical, like those of all Paleozoic bony

A

B

Figure 16-4 **Belemnoids, squidlike cephalopod mollusks that were related to ammonoids but lacked external shells.** *A.* Like ammonoids, belemnoids were predators that swam by jet propulsion. *B.* The most commonly preserved part of a belemnoid is the cigar-shaped counterweight. This heavy structure, shown here on a rock surface displaying numerous ammonoids, acted to offset the buoyant effect of gas within the shell, thereby maintaining balance.

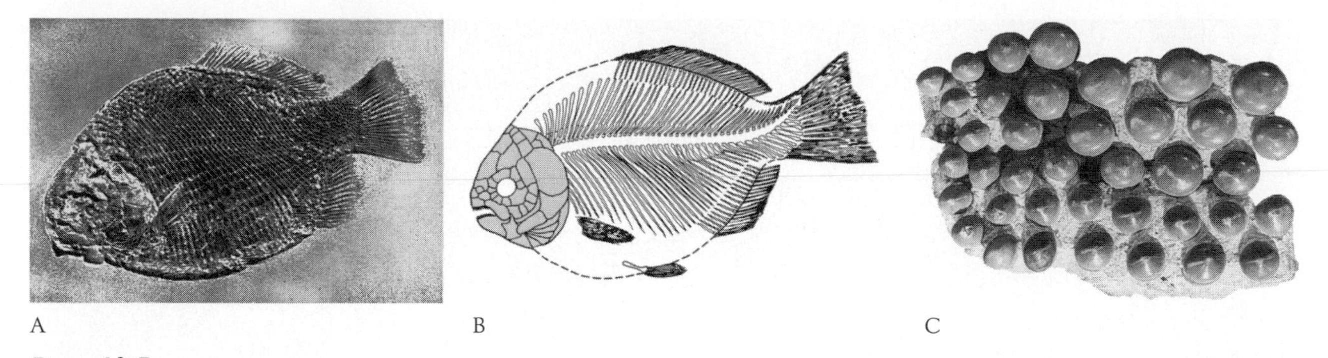

Figure 16-5 The Jurassic fish *Dapedius*. *A.* Unlike modern fishes, this early form had scales that barely overlapped one another. *B.* It also had asymmetrical skeleton supports within its tail (note the upturned spinal column). C. This particular genus had knobby teeth for crushing shellfish.

fishes. Some early Mesozoic bony fishes had teeth shaped like rounded pegs, which served to crush durable items of food– probably small shellfish (see Figure 16-5C). Bony fishes underwent many changes during the Mesozoic Era, and few species with these primitive traits survived. One especially useful feature that developed during this time was the swim bladder, a sac of gas that allows advanced fishes to regulate their buoyancy. The swim bladder evolved from the lung, which was present in some primitive fishes.

Sharks were also well represented in early Mesozoic seas. Some had teeth that were adapted for crushing shellfish, like those of the bony fish shown in Figure 16-5C. Some modern groups of sharks appeared during the Jurassic Period, among them the family that includes the modern tiger shark.

Many reptiles that resembled the popular conception of sea monsters emerged in early Mesozoic seas. Among them were the **placodonts,** which, like many early Mesozoic fishes, were blunt-toothed shell crushers (Figure 16-6). The placodonts' broad, armored bodies gave them the appearance of enormous turtles. Cousins of the placodonts were the **nothosaurs** (Figure 16-7), which have been found in Early Triassic deposits and seem to have been the first reptiles to invade the marine realm. Nothosaurs had paddlelike limbs resembling those of modern seals. It seems likely that, like modern seals, they were not fully marine, but lived along the seashore and periodically plunged into the water to feed on fishes. Although the placodonts and nothosaurs did not survive the Triassic Period, a group of more fully aquatic reptiles evolved from the nothosaurs in mid-Triassic time, and these reptiles, known as **plesiosaurs,** played an important ecologic role for the remainder of the Mesozoic Era. Plesiosaurs apparently fed on fishes and, in Cretaceous time, attained the proportions of modern predatory whales, reaching some 12 meters

A

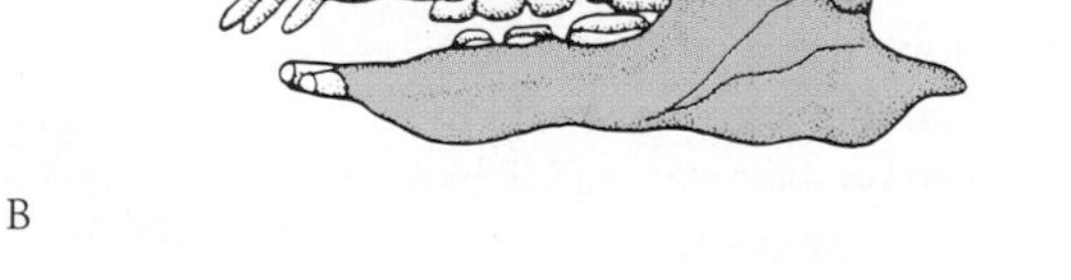
B

Figure 16-6 Reconstruction of a Triassic placodont. This aquatic reptile *(A)*, which crushed shelled marine invertebrates of the seafloor with large, rounded teeth *(B)*, was about 1.5 meters (5 feet) long.

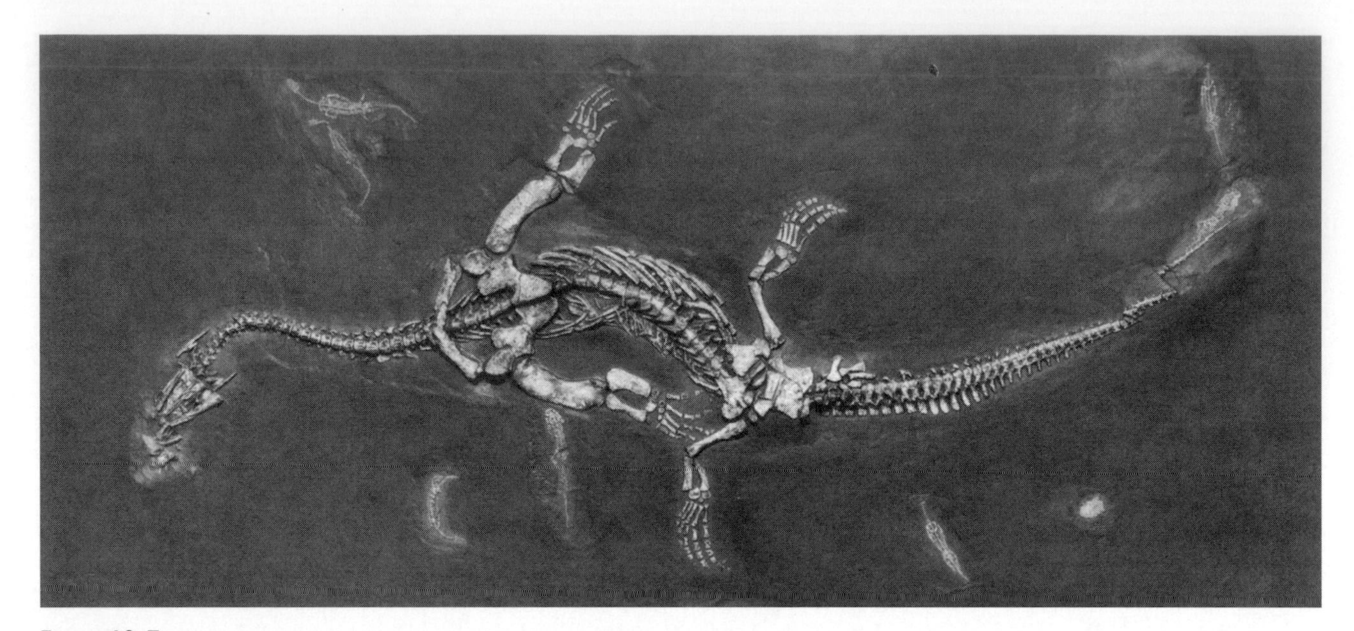

Figure 16-7 The nothosaur *Ceresiosaurus*, which was about 2.2 meters (7 feet) long, preserved with small nothosaurs of a different family in the Middle Triassic Muschelkalk of Germany. Like modern seals, these animals probably fished along the shore.

(40 feet) in length. The limbs of plesiosaurs were winglike paddles that propelled these animals through the water in much the same way that birds fly through the air (Figure 16-8).

By far the most fishlike reptiles of Mesozoic seas were the **ichthyosaurs,** or "fish-lizards," many of whose species must have been top predators in marine food webs. Superficially, ichthyosaurs bear a closer resemblance to modern dolphins, which are marine mammals, than to fishes; outlines of skin preserved in black shales under low-oxygen conditions show the dolphinlike profiles of some ichthyosaurs (Figure 16-9). The ichthyosaurs, however, had upright tail fins rather than the horizontal pair of rear flukes

Figure 16-8 Late Jurassic (Oxfordian) plesiosaurs from England mounted in swimming position. Note the paddlelike limbs. These two animals illustrate the two body types of plesiosaurs. *Cryptoclidus,* above, has a long neck and a short head, whereas *Peloneustes,* below, has a short neck and a long head. *Cryptoclidus* is about 3 meters (10 feet) long.

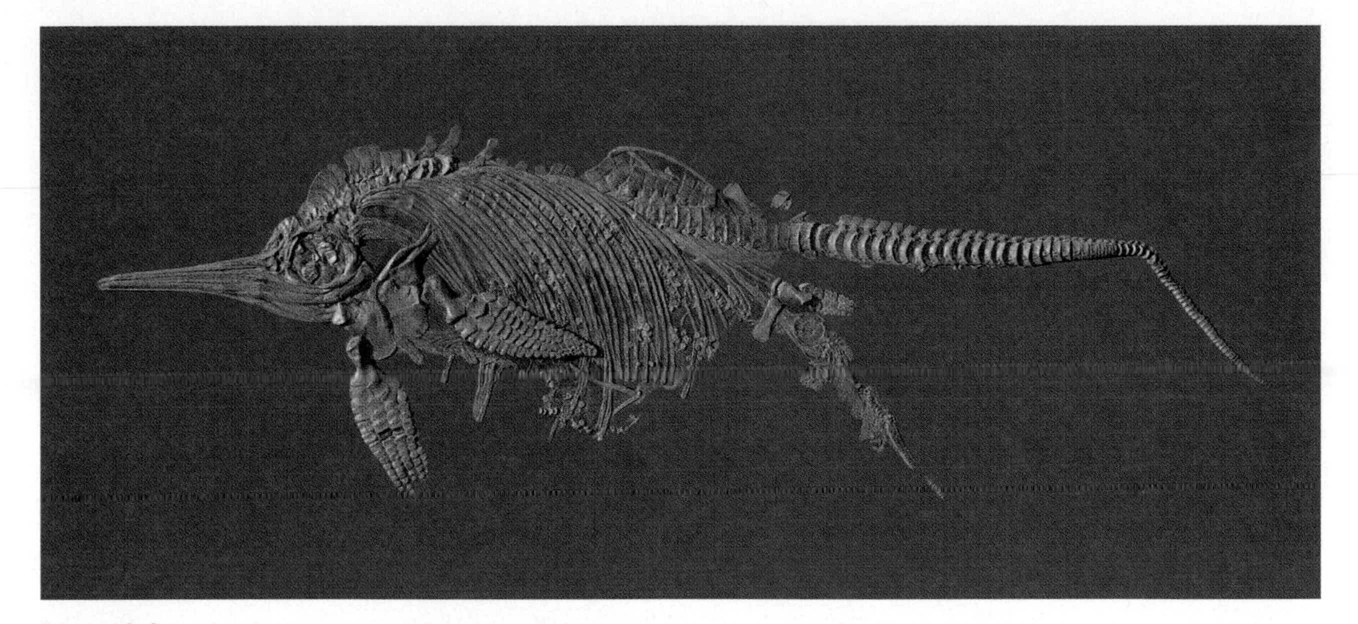

Figure 16-9 An ichthyosaur that died in the act of giving birth. The infant's head apparently stuck in the mother's birth canal and both animals died. These swimming reptiles were preserved in a Jurassic deposit in Germany. The mother was about 2 meters (6 feet) long.

that propel dolphins through the water. In addition, the extension of the backbone into the ichthyosaur tail bent downward, in contrast to the upward curve that characterized early Mesozoic bony fishes (see Figure 16-5). Large eyes supplemented other adaptations of ichthyosaurs for rapid swimming in the pursuit of prey. Ichthyosaurs were fully marine and thus could not easily have laid eggs; instead, they bore live young. In fact, skeletons of ichthyosaur embryos have been found within the skeletons of adult females.

Surprising as it may seem, the last important group of early Mesozoic marine reptiles to evolve were the early **crocodiles,** which, as we will see, were related to the dinosaurs. Although crocodiles evolved in Triassic time as terrestrial animals, some were adapted to the marine environment by earliest Jurassic time. In fact, some crocodiles became formidable oceangoing predators whose finlike tails were well adapted for rapid swimming.

Life on the Land

The presence of dinosaurs during the Mesozoic Era gave the biotas of large continents an entirely new character, but Mesozoic land plants were also distinctive. Because these plants were positioned at the bases of the food webs to which dinosaurs belonged, we will review them first.

Land Plants: The Mesozoic Gymnosperm Flora

Unlike terrestrial animals, land plants do not appear to have undergone a dramatic mass extinction at the close of the Paleozoic Era. As we learned in Chapter 15, the decline of the late Paleozoic floras began long before the end of the Permian Period. In effect, the transition from the late Paleozoic kind of flora to the Mesozoic kind of flora began before the start of the Mesozoic Era.

Among the groups that decreased in diversity long before the end of Permian time were the lycopod trees, which formed coal swamps, and the sphenopsid and cordaite trees, which inhabited higher ground. Persisting into the Mesozoic Era in greater numbers were ferns and seed ferns. Seed ferns, however, were reduced in abundance and apparently failed to survive into Jurassic time. Ferns, of course, survived, but diverse and abundant as they are today, they are nowhere near as prevalent as they were in

A

B

Figure 16-10 Leaves of the living ginkgo, *Ginkgo biloba (A)*, and similar leaves of Jurassic age *(B)*.

Triassic time. Ferns, in fact, dominate Triassic fossil floras.

Most of the trees that stood above Triassic ferns belonged to three groups of gymnosperms that had already become established during the Permian Period. The most diverse of these three groups was the one comprising the **cycadeoids** and **cycads** (see Figure 3-4 and p. 76); they were followed by the conifers, which we have already discussed (p. 108), and the ginkgos. All three of these groups survive to this day, but the cycads are rare, and only one living species of ginkgo remains on Earth (Figure 16-10).

The trees that belonged to these three dominant groups are united as gymnosperms because they were all characterized by exposed seeds. The seeds of pines and other conifers, for example, rest on the projecting scales of their cones. There is a reason for this configuration. Whereas flowering plants, which did not evolve until Cretaceous time, can attract insect and bird pollinators, most gymnosperms rely primarily on wind to carry their pollen from tree to tree. With the possible exception of the pine family, all of the modern conifer families were present in early Mesozoic time. The few modern species of the less familiar cycads are tropical trees that superficially resemble palms (see Figures 3-4 and 4-19). Cycadeoids, which were similar in form and closely related to the cycads, are extinct. The trunks of these plants are well known as early Mesozoic fossils. The single living species of ginkgo looks more like a hardwood tree (that is, like an oak or a maple) than a conifer, and, like hardwoods, it sheds its leaves seasonally. This surviving species of ginkgo is a true living fossil form whose record extends back some 60 million years to the Paleocene Epoch (see Figure 16-10), early in the Cenozoic Era.

Cycads, cycadeoids, conifers, and ginkgos formed the forests of the Jurassic Period, but the cycads were so dominant that the Jurassic interval has been called the Age of Cycads. Both Triassic and Jurassic landscapes, however, would have looked more familiar to us than Paleozoic landscapes, largely because of the presence of conifers that closely resembled modern evergreens (see Figure 15-11). Even so, the absence of flowering plants such as grasses and hardwood trees would have made Mesozoic floras appear monotonous to a modern observer (Figure 16-11).

Terrestrial Animals: The Age of Dinosaurs Begins

Local fossil records for land animals have been found to span the boundary between the Permian and Triassic systems in only two regions: the Karroo Basin of South Africa and an area of Russia near the Ural Mountains. Significantly, the fossil records of these two regions tell the same story. Just below the Permian-Triassic boundary in both regions, most of the dozens of genera of Late Permian mammal-like reptiles disappear suddenly from the fossil record, marking a major mass extinction. What remained at the start of Triassic time were a few predatory genera and the large herbivore *Lystrosaurus* (see Figure 8-10).

Early Mammals Although therapsids rediversified during the Triassic Period to play an important ecologic role once more, they barely survived into the

Figure 16-11 Reconstruction of a Mesozoic landscape. Cycads, cycadeoids, and ferns appear in the foreground and conifers along the horizon. *(Drawing by Z. Burian under the supervision of Professor J. Augusta.)*

Jurassic Period. Nonetheless, they left an important legacy in the form of the true mammals, which evolved from them near the end of the Triassic Period. **Mammals** remained small and peripheral throughout the Mesozoic Era; apparently no species grew larger than a house cat (Figure 16-12). Their problem seems to have been that the **dinosaurs** evolved slightly earlier in Late Triassic time and quickly rose to dominance.

Dinosaur Origins The dinosaurs inherited an advanced locomotory ability from their ancestors, the **thecodonts,** which evolved earlier in the Triassic Period (see Figure 16-12). Some thecodonts were adapted for speedy two-legged running in the fashion of ostriches and other flightless birds, but all thecodonts probably spent much time standing or walking on all fours. The upper portion of the legs of many thecodonts stood beneath their bodies rather than sprawling slightly out to the side as they did in mammal-like reptiles. This feature, which facilitated running, was passed on to the dinosaurs and seems to have been a key to the dinosaurs' success.

The first dinosaurs resembled bipedal thecodonts (that is, thecodonts that traveled on their hind legs), but the dinosaurs' skulls were differently formed and their teeth were more highly developed. Dinosaurs (formally termed Dinosauria) did not become gigantic before the end of the Triassic, but during Triassic

Figure 16-12 **Triassic animals of the genus *Lagosuchus* intimidating a smaller mammal.** *Lagosuchus,* which was about 30 centimeters (1 foot) tall, was a thecodont that closely resembled the earliest dinosaurs. Thecodonts were the ancestors of dinosaurs. *(Drawing by Gregory S. Paul.)*

Figure 16-13 **Terrestrial life of Late Triassic time in Argentina.** The plants are of the widespread genus *Dicroidium*. The largest animals depicted here are thecodonts of the genus *Saurosuchus*, which were about 7 meters (25 feet) long. Confronting them are three small, primitive dinosaurs of the genus *Herrerasaurus*. The dead animal is the rynchosaurian reptile *Scaphonyx*. Two small thecodonts of the genus *Ornithosuchus* are scampering off in the foreground. In the left foreground is the long-legged primitive crocodile *Trialestes*. *(Drawing by Gregory S. Paul.)*

time they did reach lengths of more than 6 meters (20 feet). Figure 16-13 displays a small species along with thecodonts that stood on all fours, large mammal-like reptiles, and a primitive long-legged crocodile. The crocodiles, like the dinosaurs, evolved from thecodonts in Late Triassic time.

The Fall of the Therapsids and the Rise of Dinosaurs By Late Triassic time, therapsids lived alongside increasing numbers of dinosaurs, the still-diverse thecodonts, and other, mainly smaller amphibians and reptiles. A few kinds of large amphibians persisted as well. Near the end of the Triassic, however, a mass extinction that took a heavy toll on marine life also eliminated all but a few groups of therapsids. The stage was thereby set for the ascendancy of the dinosaurs to a dominant position for the remainder of the Mesozoic Era.

The known fossil record of Early Jurassic time is too poor to permit us to piece together the details of the dinosaurs' great rise to dominance. Nonetheless, fossil remains of huge dinosaurs even in Lower Jurassic rocks indicate that these animals evolved rapidly; the oldest dinosaur giants are found in Australia.

Dinosaurs fall into two groups, which are characterized by different pelvic structures (Figure 16-14). All of the "bird-hipped" (ornithischian) dinosaurs were herbivores, whereas the "lizard-hipped" (saurischian) group included both herbivores and carnivores. In both groups there were species that traveled on two legs and others that moved about on all fours. Largest of all the dinosaurs were the **sauropods,** lizard-hipped herbivores that moved about on all fours. Some interesting aspects of dinosaur biology are presented in Box 16-1.

By Late Jurassic time, both bird-hipped and lizard-hipped dinosaurs were quite diverse. The most spectacular Jurassic assemblage of fossil dinosaurs in the world is found in the Upper Jurassic Morrison Formation, which extends from Montana to New Mexico. At Como Bluff, Wyoming, dinosaur bones are so common that a local sheep herder constructed a cabin of them because they were the most readily available building materials. The dinosaurs of the Morrison Formation, which include more than a dozen genera, are representative of the kinds of dinosaurs that lived throughout the world during Late Jurassic time. Several of the common Morrison species are shown in Figure 16-15. The skeleton of *Allosaurus,* the large carnivore in this scene, is seen in Figure 16-16.

Frogs and Turtles Two groups of small reptiles that remain successful in the modern world also became established in Triassic time. One was the frogs,

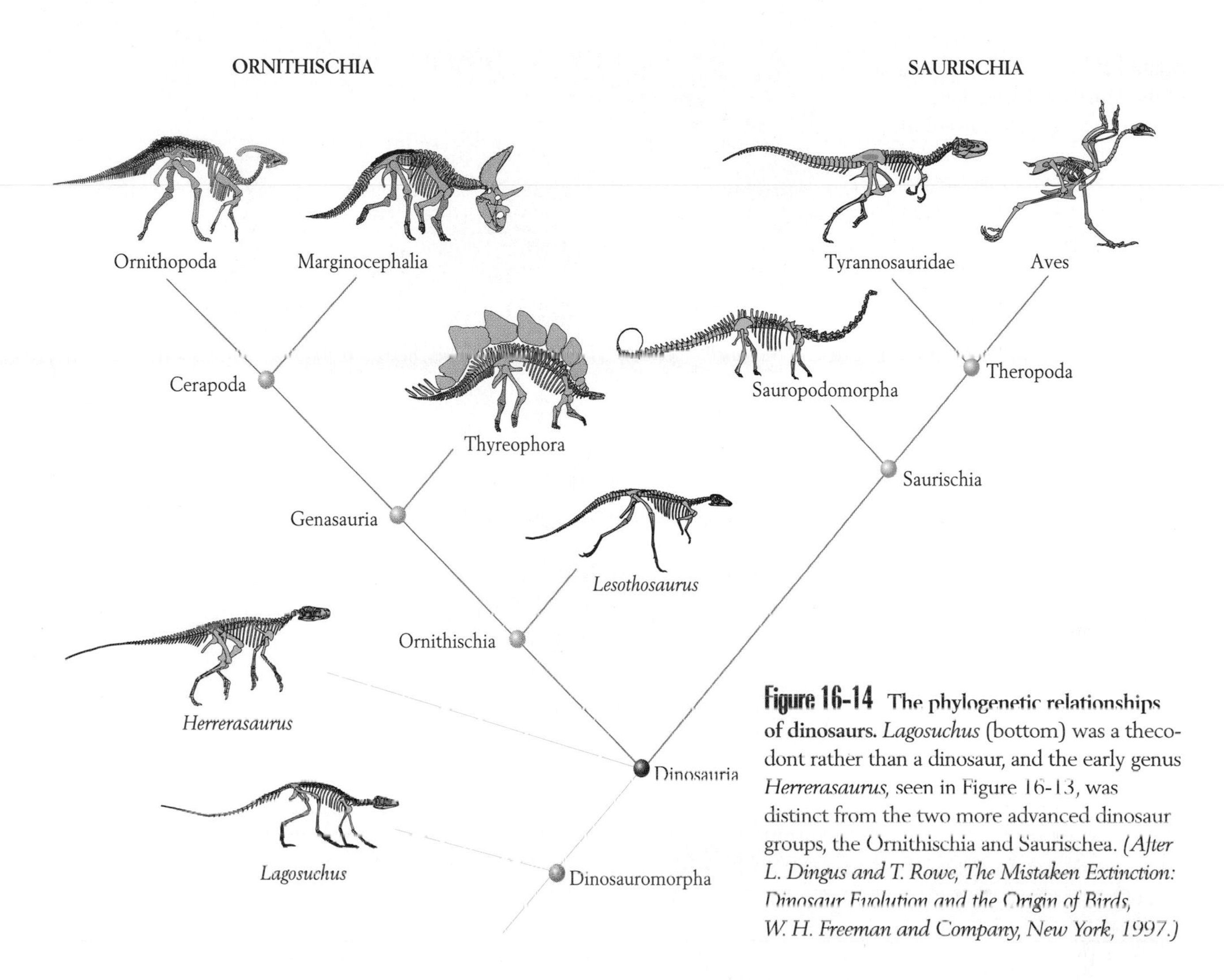

Figure 16-14 The phylogenetic relationships of dinosaurs. *Lagosuchus* (bottom) was a thecodont rather than a dinosaur, and the early genus *Herrerasaurus*, seen in Figure 16-13, was distinct from the two more advanced dinosaur groups, the Ornithischia and Saurischea. *(After L. Dingus and T. Rowe, The Mistaken Extinction: Dinosaur Evolution and the Origin of Birds, W. H. Freeman and Company, New York, 1997.)*

which were and remain amphibians of small body size. The oldest known fossil displaying the form of a modern frog is of earliest Jurassic age, but froglike skeletons have also been found in Triassic rocks. The other modern group was the turtles, although the earliest turtles lacked the ability to pull their heads and tails fully into their protective shells.

Creatures That Took to the Air Late in the Triassic Period, vertebrate animals invaded the air for the first time as the pterosaurs came into being. These animals had long wings and hollow bones that served to facilitate flight. Some species had long tails as well (Figure 16-17). The structure of the pterosaur skeleton reveals that these reptiles were capable of flying, but the great length of the wings suggests that most species flapped their wings primarily when taking off and then, once airborne, soared on air currents without much flapping. The behavior of pterosaurs when not in flight has been widely debated. Most species appear to have been able to walk and also to climb adeptly with the aid of their hooklike claws.

The oldest known fossil birds are of Late Jurassic age. The first clue to their existence was a feather discovered in 1861 in the fine-grained Solnhofen Limestone of Germany, followed a few months later by the discovery of an entire skeleton of the species to which the feather belonged (Figure 16-18). This feathered animal was given the name *Archaeopteryx*, which means "ancient wing." *Archaeopteryx* had a skeleton so much like that of a dinosaur that it would be regarded as one were it not for its plumage. This animal is a classic missing link– in this case, the link between birds and their flightless ancestors. The teeth, large tail, and clawed forelimbs of *Archaeopteryx*, which are absent from advanced birds, reflect

Figure 16-15 **The dinosaur fauna of the Morrison Formation.** *Camarasaurus* walks toward the right in the background; *Diplodocus* is drinking water; at the far right is *Stegosaurus.* Reclining on the left is *Allosaurus,* a large carnivorous dinosaur, about 6 to 7 meters (20 to 25 feet) long. In the foreground are dragonflies, pterosaurs, turtles, and a crocodile. The trees are conifers. *(Drawing by Gregory S. Paul.)*

Figure 16-16 ***Allosaurus,* a huge carnivorous dinosaur of Jurassic age that roamed the American West, reached a length of about 12 meters (40 feet) and its skull was nearly a meter (3 feet) long.**

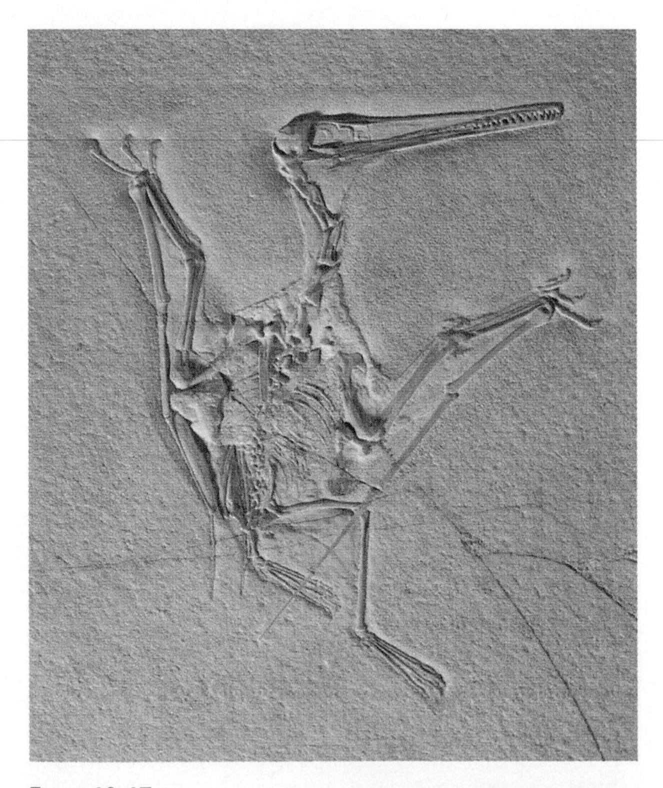

Figure 16-17 **The pterosaur, or "flying lizard," *Pterodactylus.*** This skeleton was preserved intact in the Solnhofen Limestone of Germany; it measures about 60 centimeters (2 feet) long. Many of the bones are hollow, like those of a bird.

its dinosaur ancestry. *Archaeopteryx* lacks a breastbone and so is assumed to have possessed weak flying muscles. It was probably a clumsy flier by the standards of modern birds. Unfortunately, the hollow bones of birds are fragile, and no other bird bones have been found within the Jurassic System. There have been claims that bird bones occur in Upper Triassic rocks. If verified, these finds would place the origin of birds much farther back in geologic time than is now believed. *Archaeopteryx* would then represent a primitive type of bird that happened to survive for a long interval of time.

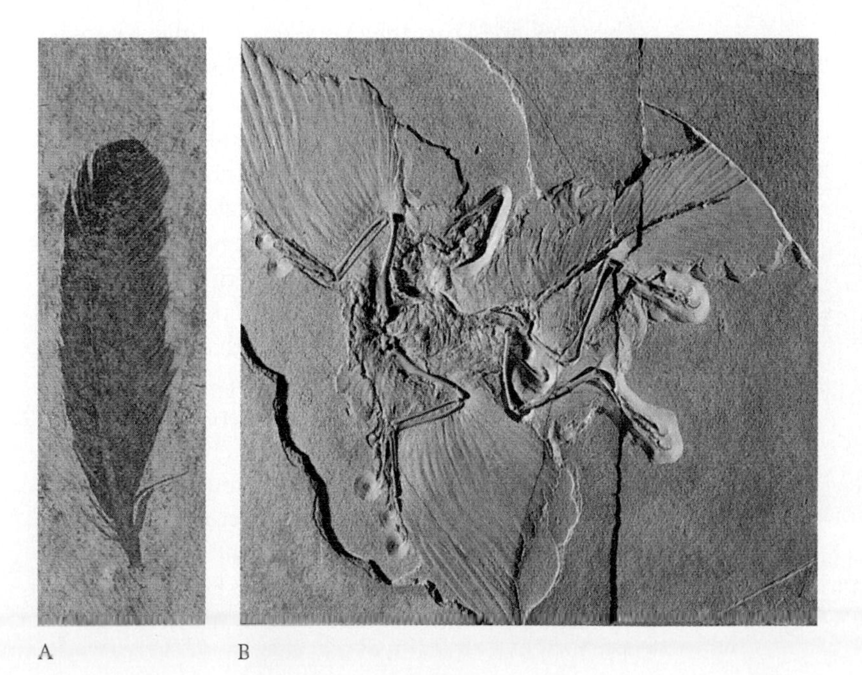

Figure 16-18 Fossil remains of *Archaeopteryx lithographica*, the oldest known bird, from the Upper Jurassic Solnhofen Limestone of Germany. This animal was about the size of a crow. The existence of a bird during Solnhofen deposition was first indicated by the discovery of a feather *(A)*. The asymmetry of the feather suggests that it aided in flight; flightless living birds have feathers that are symmetrical about the central shaft. The bird itself was soon found, and impressions of long feathers are clearly visible around it in the fine-grained limestone *(B)*. Despite the feathers, *Archaeopteryx* had a skeleton and teeth similar to those of dinosaurs.

For the Record 16-1

Who Were the Dinosaurs?

The average person has many misconceptions about dinosaurs. Not all dinosaurs were of massive proportions. Dinosaurs have often been portrayed as hulking, lumbering creatures, but actually many were less than 1 meter (3 feet) long. Furthermore, some dinosaurs were as agile as ostriches, which are famous for their great speed. The orientations of dinosaur limbs in their sockets indicate that the legs were positioned almost vertically beneath the body, and fossil trackways of dinosaurs confirm that this posture was typical. The left and right tracks are nearly in line, signifying that both feet were positioned almost beneath the center of the body. Most fossil trackways also reveal long strides for the size of the individual tracks, suggesting that dinosaurs tended to move rapidly.

Dinosaurs also appear to have been social animals. Each of the various duck-billed dinosaurs, for example, had a tall, crested skull, which functioned like the resonating chambers of a trumpet. Not only did many species of dinosaurs probably communicate by sound, but some of their well-preserved trackways show that large groups of animals sometimes traveled together.

Many dinosaur eggs have been found, some preserved in the circular pattern in which they were carefully laid; often several circles of eggs were buried on top of one another. The eggs were tapered toward one end, which the mother thrust into the soil. In Upper Cretaceous rocks of Mongolia paleontologists have found a parent dinosaur preserved in the act of brooding her eggs. Nests of baby dinosaurs found in the Upper Cretaceous of Montana—clusters of juvenile skeletons with broken eggshells in a depression—show that dinosaurs also cared for their young, which grew rapidly. Having hatched at a length of just a few inches, the offspring reached about 1.5 meters (5 feet) in length before leaving the nest at the end of the warm season, perhaps only 3 or 4 months after hatching.

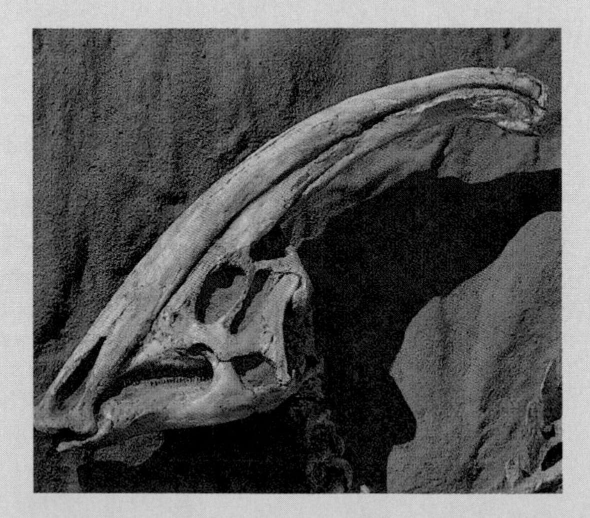

The skull crest of *Parasauralophus,* which was about one meter (3 feet) long, probably served as a resonating chamber for trumpeting.

One might ask how the largest dinosaurs, which had relatively small jaws, chewed up enough food to survive. The answer is that these giant herbivores used their mouths and jaws only for gathering and swallowing plant food. In the animals' intestinal tracts were "gizzard stones," like those of birds but much larger, which helped the animals grind up coarse material after they had swallowed it. Birds evolved from dinosaurs, and remnants of feathers have been found preserved on a small dinosaur discovered in China.

The greatest controversy about dinosaurs has related to their metabolism. Dinosaurs have traditionally been classified as reptiles, and it has thus been assumed that they were ectothermic (or cold-blooded). There is much evidence, however, that dinosaurs were actually endothermic, or warm-blooded. Part of this evidence comes from fossil trackways showing that dinosaurs moved around so rapidly that they must have had high energy levels. Otherwise, they might not have been able to maintain their dominance over endothermic mammals throughout the Mesozoic Era. Ectothermic animals, including modern reptiles, have little endurance (p. 413). Additional evidence derives from the fact that in communities of dinosaurs, predators usually made up less than 10 percent of the volume of living species, as is the case in living and fossil mammal communities. In communities of ectothermic animals, predators commonly represent 40 percent or so of the volume of living tissue. Having low metabolisms, they need little food, and many can

This *Maiosaura* hatchling, about 50 centimeters (20 inches) long, was found in a nest.

sustain themselves on small populations of prey animals. The low percentage of predators in many dinosaur communities argues for a closer resemblance to mammal communities than to reptile communities. The microscopic structure of their bones has also been debated. Endothermic animals differ from ectothermic animals in bone structure, and it appears that the dinosaurs' bone structure was of the endothermic type. The discovery that dinosaurs brooded their eggs is a strong indication of endothermy. Ectothermic animals do not have to warm their eggs by sitting on them. In all likelihood the feathers possessed by at least some dinosaurs served as insulation to retain internally generated body heat. Finally, the rapid growth of juvenile dinosaurs points to a high metabolism. Modern reptiles grow more slowly than dinosaurs did.

Whatever the details of dinosaurs' metabolism may have been, we at least recognize that even large dinosaurs were active, highly adapted animals rather than simple, lumbering beasts. For all we know, many dinosaurs might have fared well in the modern world had they been given the chance. The sudden extinction of the dinosaurs at the end of the Cretaceous Period was an accident, not a result of biological inferiority.

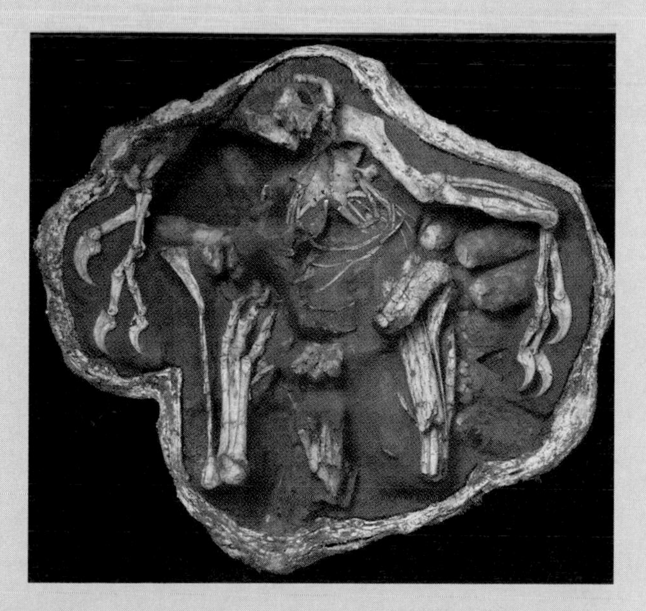

Bones of *Oviraptor,* whose name ironically means "egg stealer," preserved in a brooding posture on top of her eggs in the Upper Cretaceous of China. The skeleton is about 0.7 meters (28 inches) across.

A cluster of feathers preserved as part of the remains of *Protarchaeopteryx* from Chinese rocks close to the Jurassic-Cretaceous boundary.

The Paleogeography of the Early Mesozoic World

At the start of the Mesozoic Era, all of the major landmasses of the world were united as the supercontinent Pangea (see Figure 15-17). Near the end of Triassic time, Pangea began to break apart, but continental movement is so slow that even by the end of Jurassic time, the newly forming continental fragments were hardly separated. Thus, throughout the early Mesozoic Era, Earth's continental crust was concentrated on one side of the globe. Sea level rose slightly at the start of Triassic time (see p. 437). As in Late Permian time, however, the bulk of the continental crust during the Triassic Period stood above sea level, forming one vast continent. At the start of Triassic time, the Tethys Seaway was an embayment of the deep sea projecting into the portion of equatorial Pangea that today constitutes the Mediterranean. Later in Triassic and Jurassic time, rifting extended the Tethys between Eurasia and Africa and all the way westward between North and South America to the Pacific.

Pangea during the Triassic Period

Although the dominant land plants of the Triassic Period differed from those of the Permian, the distributional pattern of floras on Pangea remained much the same; a Gondwana flora existed in the south and a Siberian flora in the north (Figure 16-19). The Euramerican flora grew under warmer, drier conditions at low latitudes; in fact, unusually extensive deposition of evaporites attests to the presence of arid climates far from the equator. This condition resulted in part from the sheer size of Pangea, which was so large that

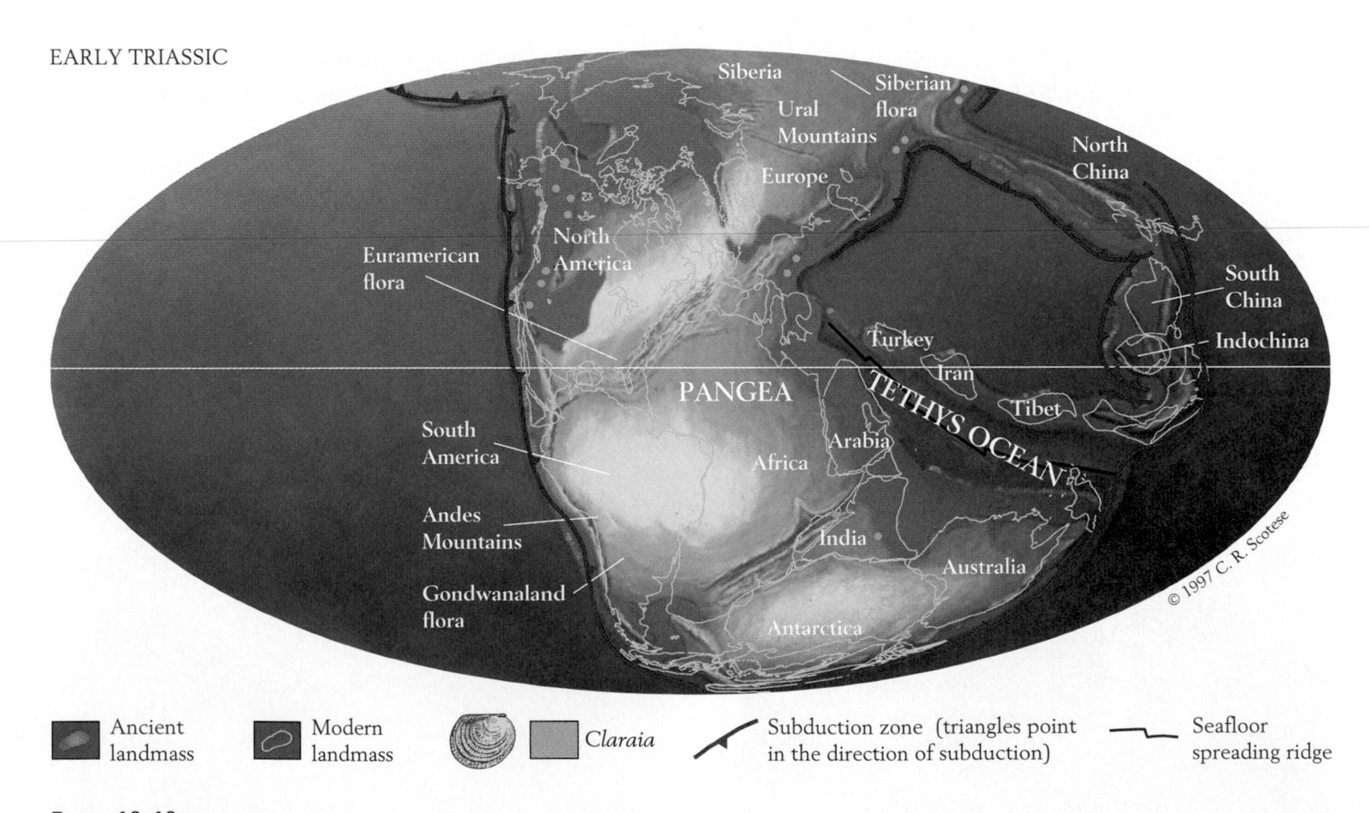

Figure 16-19 **World geography of Early Triassic time.** The Euramerican flora occupied a broad, warm belt across the middle of Pangea, and the Siberian and Gondwanaland floras occupied regions to the north and south. The bivalve mollusk genus *Claraia* was broadly distributed along both the eastern and western borders of Pangea. The widespread evaporites illustrated here represent the entire Triassic Period, which was relatively warm and dry even at high latitudes. *(Adapted from paleogeographic maps by C. R. Scotese, PALEOMAP Project, University of Texas at Arlington, 1997.)*

many of its regions lay far from the low-standing oceans.

Pangea also had strongly seasonal climates: because of its great size, the ocean only weakly affected its temperature, and the low heat capacity of land therefore caused it to become very hot in summer and cold in winter (p. 106). When summer came to the northern half of Pangea, hot air rising from the land must have drawn strong monsoonal winds from the southern half (p. 107); similarly, powerful winds must have flowed southward across the equator when it was summer in the southern hemisphere.

The survival of only a small percentage of Permian species into the Triassic Period produced some striking biogeographic distributions. In the oceans, the scalloplike bivalve genus *Claraia* ranged over an enormous area and is assumed to have occupied the seafloor even in deep water. On land, the large herbivorous therapsid *Lystrosaurus* ranged over large areas of the globe; *Lystrosaurus* has been found in the fossil records of several continents that represent fragments of Gondwanaland (see Figure 8-10). It appears that at the start of the Triassic Period, few therapsids preyed on *Lystrosaurus,* so its populations were huge. As we saw earlier, the therapsids underwent renewed radiation during the Triassic Period, and partway through the period thecodonts and dinosaurs began their expansion. Even when these vertebrate groups radiated to high diversities, most of their species were also wide-ranging. Many families of Triassic terrestrial vertebrates are found as fossils on several modern continents. In fact, the Triassic is the only period whose fossil land vertebrates clearly indicate that all of Earth's continents were connected.

The Breakup of Pangea

The most spectacular geographic development of the Mesozoic Era was the fragmentation of Pangea, an event that began in the Tethyan region. As the Triassic Period progressed, the Tethys Seaway spread farther and farther inland, and eventually the craton began to rift apart. The Tethys subsequently became a deep, narrow arm of the ocean separating what is now southern Europe from Africa. During the Jurassic Period, this rifting spread westward, ultimately separating North and South America. South America and Africa, however, did not separate to form the South Atlantic until the Cretaceous Period; in fact, all of the Gondwanaland continents remained attached to one another until Cretaceous time. North America began to break away from Africa in mid-Jurassic time. Interestingly, this rifting generally followed the old Hercynian suture. Rifting occurred as some of the arms of a series of triple junctions joined, tearing Pangea in two (see Figure 9-3).

The rifting that formed the Atlantic had another important consequence. When continental fragmentation begins in an arid region near the ocean, evaporite deposits often form there (p. 106). Thus, as rifting began in Pangea, extension produced normal faults between Africa and the northern continents; zones bounded by such faults sank, and water from the Tethys to the east periodically spilled into the trough and evaporated. Evaporites that were precipitated in this trough are now located on opposite sides of the Atlantic, both in Morocco and offshore from Nova Scotia and Newfoundland (Figure 16-20).

During Middle and Late Jurassic time, one arm of rifting passed westward between North and South America, giving rise to the Gulf of Mexico. Early intermittent influxes of seawater into this rift, apparently from the Pacific Ocean, caused great thicknesses

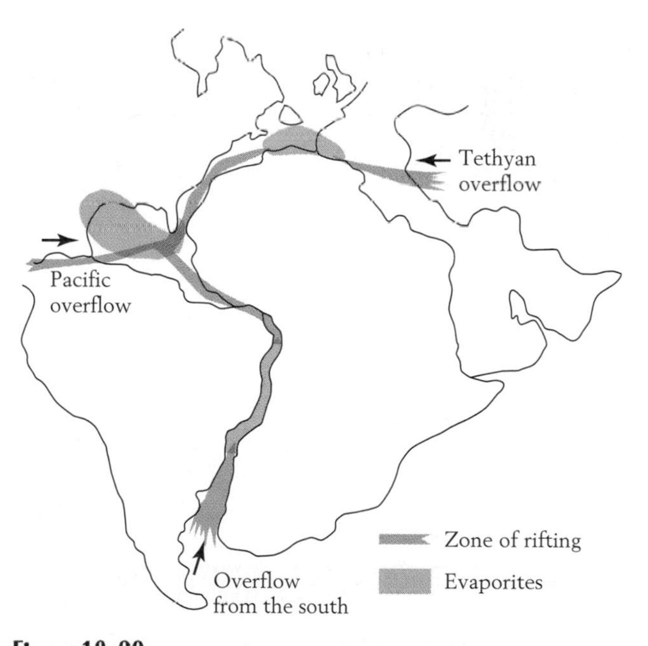

Figure 16-20 **Early Mesozoic evaporites.** Evaporites accumulated during the early stages of rifting that formed the Atlantic Ocean as oceans overflowed intermittently into newly forming fault basins. *(After K. Burke, Geology 3:613–616, 1975.)*

of evaporites to accumulate. Today these evaporites, which are known as the Louann Salt, lie beneath the Gulf of Mexico and in the subsurface of Texas. Because its density is low, the Louann Salt has in some places pushed up through younger sediments to form salt domes (Figure 16-21), many of which are associated with reservoirs of petroleum and sulfur. The rifting that forms the South Atlantic did not begin until Early Cretaceous time, when the record shows that salt deposits formed after seawater spilled inland from the south (see Figure 16-20).

The Jurassic World

Although sea level underwent only minor changes during Late Triassic and Early Jurassic time, it subsequently rose, with minor oscillations, until Late Jurassic time. Then, very late in the Jurassic Period, it underwent more rapid oscillations but remained at a high level, causing epicontinental seas to flood large areas of North America and Europe (see Figure 6-22). Long before the advent of plate tectonic theory it was recognized that there were two biogeographic provinces of marine life in Europe during the Jurassic Period: the southern province, which was centered in the Tethys and is designated the Tethyan realm, and the northern province, which is labeled the Boreal realm.

Coral reefs were largely restricted to the Tethyan realm, as were limestones and certain groups of mollusks, indicating that the Tethyan was largely a tropical province. The transition from the Tethyan to the Boreal Province resembled what we see today between tropical southern Florida, with its carbonates and coral reefs, and subtropical northern Florida, where siliciclastic sediments prevail and reefs are lacking.

There is no doubt that temperature gradients from equator to pole were gentle throughout the Jurassic Period. Plants that appear to have required warmth (the Euramerican flora of Figure 16-22) occupied a broad belt extending to about 60° north latitude. Even the Gondwana flora to the south and the Siberian flora to the north included groups of ferns whose modern relatives cannot tolerate frost. The high-latitude floras do not seem to have been

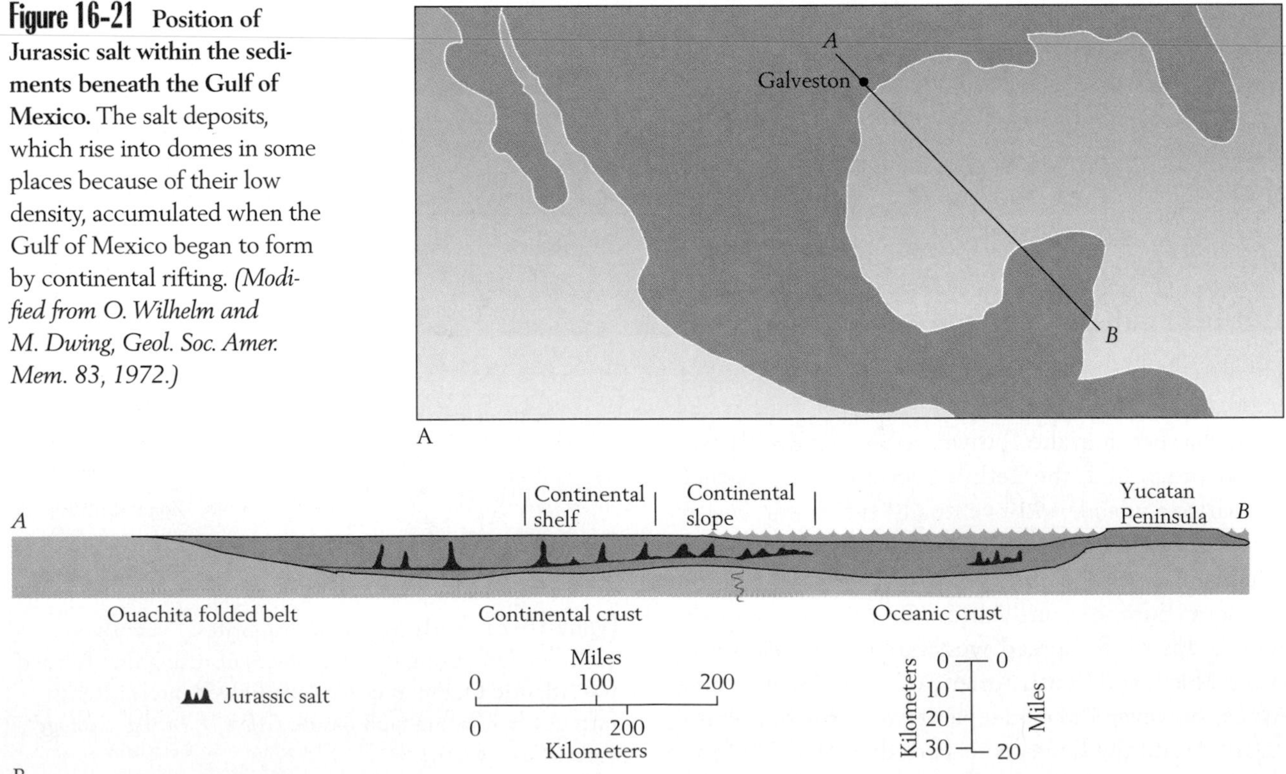

Figure 16-21 **Position of Jurassic salt within the sediments beneath the Gulf of Mexico.** The salt deposits, which rise into domes in some places because of their low density, accumulated when the Gulf of Mexico began to form by continental rifting. *(Modified from O. Wilhelm and M. Dwing, Geol. Soc. Amer. Mem. 83, 1972.)*

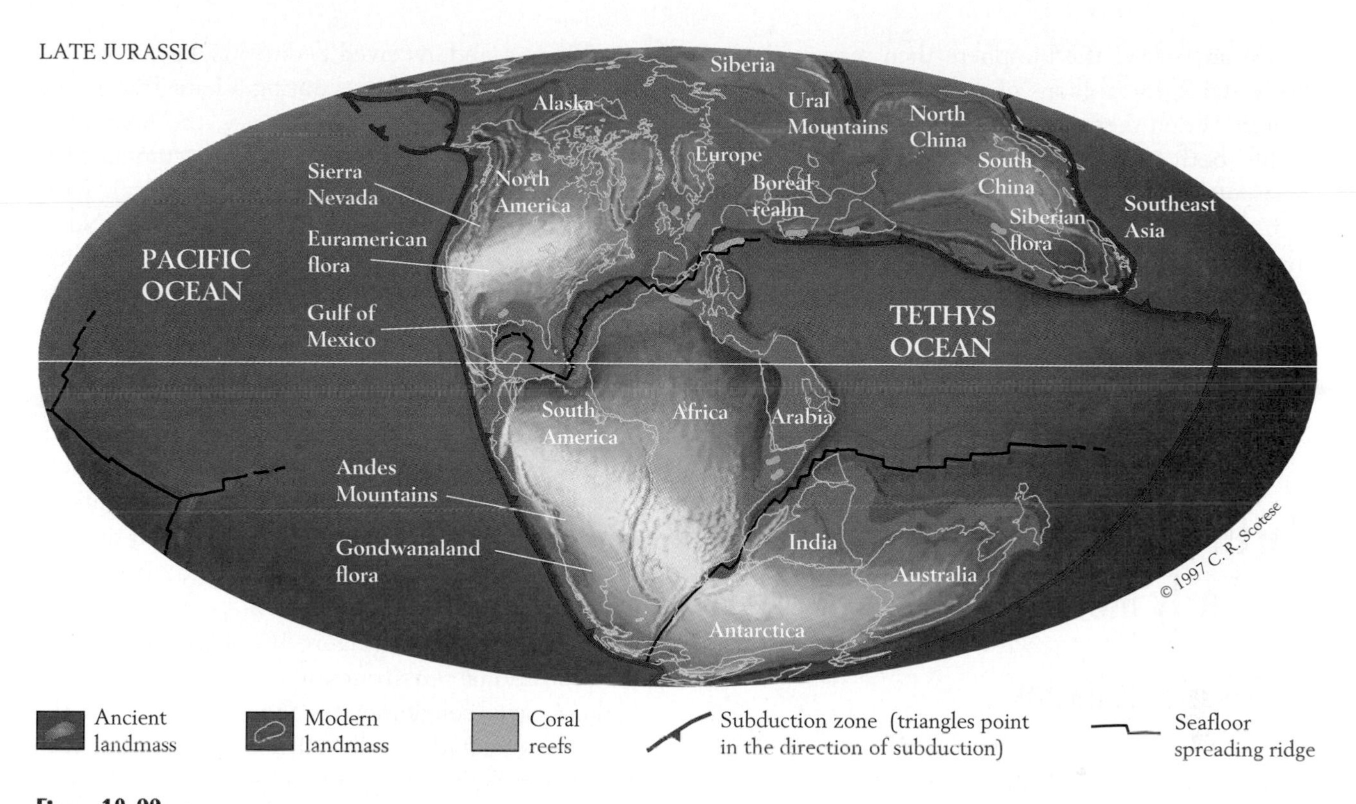

Figure 16-22 Geography of the Late Jurassic world. The three floras persisted from Triassic time. The Tethys marine realm, which was characterized by tropical life, including reef corals, extended from the eastern Pacific across the newly forming Mediterranean and Gulf of Mexico to the western Pacific. The cooler Boreal realm lay to the north. *(Adapted from paleogeographic maps by C. R. Scotese, PALEOMAP Project, University of Texas at Arlington, 1997.)*

tropical, however; they contained few cycads, and modern cycads are restricted to warm regions (see Figure 4-19). Thus it appears that the large landmass that began to fragment during the Jurassic Period was bathed in tropical climates well to the north and south of the equator, even though temperatures were somewhat cooler toward the poles.

Mass Extinctions

The Triassic Period ended with one of the largest mass extinctions of all time. This crisis struck both on the land and in the sea. In the marine realm, about 20 percent of all families of animals suffered extinction. Conodonts and placodont reptiles (see Figure 16-6) died out altogether. So did most species of bivalves, ammonoids, plesiosaurs (see Figure 16-8), and ichthyosaurs (see Figure 16-9), although all of these groups recovered in Jurassic time. The terrestrial victims included most genera of mammal-like reptiles and large amphibians.

The primary beneficiaries of the extinction on the land were the dinosaurs, which radiated rapidly during the Jurassic and then continued to dominate terrestrial habitats throughout the remainder of the Mesozoic Era. Two pulses of Late Triassic extinction occurred on the land—one at the end of the Norian Age, the final stage of the period, and one at the end of the preceding Carnian Age. The timing of extinction in the seas is less clear, but many genera died out during the final few million years of the Norian Age. The cause of the Late Triassic extinctions remains unknown. Perhaps it is significant that late in Norian time, conifers and other groups of gymnosperms replaced the *Dicroidium* flora (see Figure 16-13), which had prevailed in lowland habitats of Gondwanaland since early in the Triassic Period. Some workers believe that this floral transition signaled a climatic change, perhaps an increase in aridity.

In any event, the biosphere then remained relatively stable for millions of years. At the end of the Jurassic Period there was moderately heavy extinction of life both in the oceans and on the land. Many dinosaurs died out late in Jurassic time, but the fossil record is not complete enough to indicate whether the extinction was sudden. In any event, with the dawning of the Cretaceous Era animal life on the land had a new aspect. The stegosaurian dinosaurs failed to make the transition, as did the larger sauropods (see Figures 16-14 and 16-15). New kinds of dinosaurs populated the Cretaceous world.

North America in the Early Mesozoic Era

Eastern Fault Basins

During Early and Middle Triassic time, erosion subdued the Appalachian Mountains, which were centrally located in Pangea. In Late Triassic time, long, narrow depositional basins bounded by faults developed on the gentle Appalachian terrain (Figure 16-23). These basins formed when Pangea was splintered by normal faults on either side of the great rift that began to divide the continent, forming the Atlantic Ocean (see Figure 9-3). One of the largest of these basins extended from New York City to northern Virginia and received sediments known as the Newark Supergroup. Here, during a Late Triassic and Early Jurassic interval of subsidence, the nonmarine sediments of the Newark Supergroup accumulated to a thickness of nearly 6 kilometers (4 miles). Early Mesozoic basins resembling those of eastern North America are also found in Africa and South America, but these contain thick evaporite deposits. It was in the early phase of rifting that water from the Tethys periodically spilled into these southern basins to form vast salt deposits (see Figure 16-20).

The best locations for investigating the basin sediments are in the eastern United States. One particularly well studied basin passed through present-day Connecticut and Massachusetts, bounded on the east by a large normal fault along which the basin subsided continually while sediments accumulated from an eastern source area (Figure 16-24*A*). Several types of depositional environments existed within this basin. Coarse conglomerates that wedge out to the west accumulated on alluvial fans that spread from the eastern fault margin. Many sand-sized sediments of the basin are stream deposits. The fact that most of these deposits are composed of red arkose suggests that deposition in this area was rapid, since apparently there was little time for feldspars to disintegrate to clay. Well-laminated mudstones floored the lakes in the basin center. Cycles now visible in the sediments reflect expansion and contraction of the lakes, which for the most part must have been quite shal-

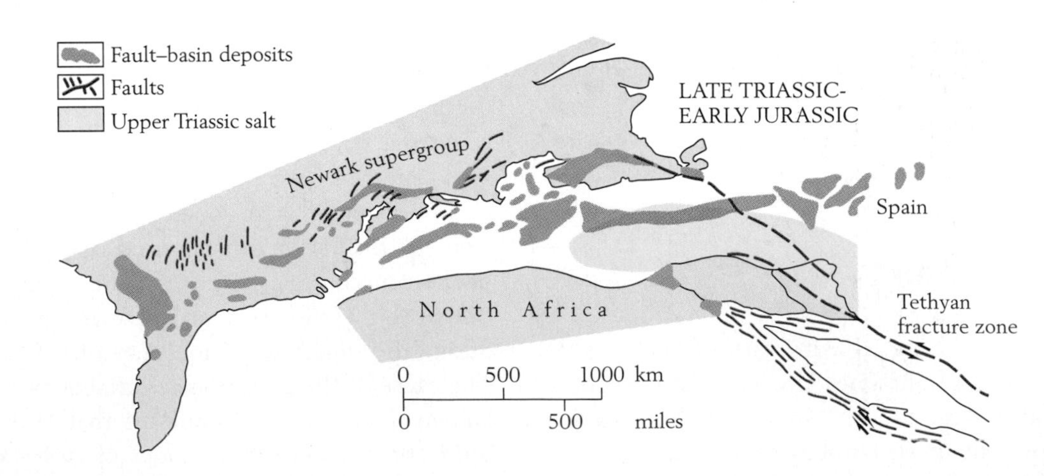

Figure 16-23 Geologic features of eastern North America and nearby regions during Late Triassic and Early Jurassic time. In eastern North America, block faulting produced elongate depositional basins, most of which paralleled the enormous rift that eventually formed the Atlantic Ocean. Salt deposits accumulated from the sporadic westward spilling of seawater from the Tethys Seaway, where the Mediterranean was forming as a result of Africa's movement in relation to Europe. *(After W. Manspeizer et al., Geol. Soc. Amer.* Bull. *89:901–920, 1978.)*

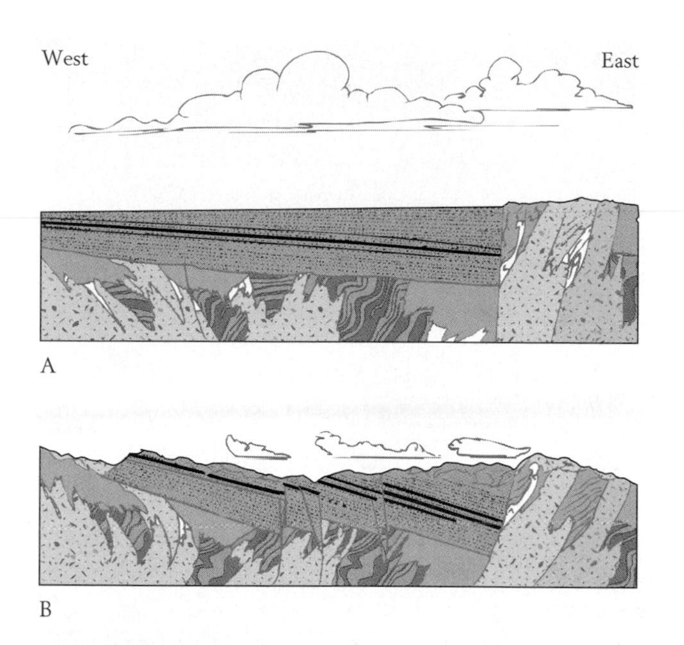

Figure 16-24 Cross sections of the large early Mesozoic fault basin that passes through central Connecticut, where part of the Newark Supergroup was deposited (Figure 16-23). *A*. The basin late in its depositional history, when great thicknesses of sediment had accumulated. As the basin subsided, lavas welled up periodically, forming dikes and sills (heavy black lines), and gravels from the uplands to the east spread into the basin as alluvial fans. *B*. The eventual destruction of the basin by extensive faulting.

low. During some dry intervals, evaporite minerals were precipitated from the shrinking waters, but abundant fossil fish remains indicate that at other times the waters were hospitable to life. In fact, freshwater fishes underwent spectacular adaptive radiations in some of the larger lakes. These radiations resembled those that have occurred very recently in the great African lakes (see Figure 7-10). It is interesting to note that these modern African lakes occupy rift valleys much like those in which the Newark Supergroup was deposited (see Figures 9-2 and 9-3).

Although dinosaur tracks are common in rocks representing lake margins, conditions in the basins seldom favored the preservation of dinosaur skeletons, except in eastern Canada, where a number of bones have been found. Some of the ancient soils in the basin that extends through Connecticut and Massachusetts contain caliche nodules, indicating that the climate here was warm and seasonally arid (see Figure 5-1). Apparently bones decayed rapidly under these conditions, so that relatively few were preserved as fossils.

Periodically mafic magmas welled up through faults, forming dikes and widespread sills within the basin. One of the largest of these sills forms the Palisades along the Hudson River near New York City (Figure 16-25). At least some of the North American

Figure 16-25 The Palisades sill, exposed along the Hudson river across from New York City.

basins continued to subside until Early Jurassic time, when deposition ended with a final episode of faulting. After this time, the basins apparently moved so far westward along with the North American plate that they were no longer affected by the mid-Atlantic rifting. The fact that some of the basins are located several hundred kilometers from the present margin of North America (see Figure 16-23) indicates how extensive the fracturing of a large continent can be; in most such instances, many small breaks and ruptures occur rather than a clean separation from sea to sea.

Western North America

Throughout the Triassic Period, much of the American West was the site of nonmarine deposition. Shallow seas expanded and contracted along the margin of the craton but for the most part remained west of Colorado. During Middle and Late Triassic time, western North America was especially free of marine influence.

Terrestrial and Marine Environments As in the Permian Period, the climate here remained largely arid. At times, however, there was sufficient moisture to permit the growth of large trees belonging to the Euramerican flora. The river and lake sediments in Utah and Arizona that are collectively known as the Chinle Formation, for example, erode spectacularly in some places to reveal the well-known Petrified Forest of Arizona (Figure 16-26). In southwestern Utah the Chinle is overlain by the Wingate Sandstone, a desert dune deposit. Above it lies a river deposit called the Kayenta Formation, on top of which rests the Navajo Sandstone. The Navajo, also a desert dune deposit, ranges upward in the stratigraphic sequence from approximately the position of the Triassic-Jurassic boundary. The Navajo is famous for its large-scale cross-bedding in the neighborhood of Zion National Park (see Figure 5-12C).

During Middle and Late Jurassic time, as sea level rose throughout the globe (see p. 437), waters from the Pacific Ocean spread farther inland in a series of four transgressions, each more extensive than the last. The first such transgression went no farther than British Columbia and northern Montana, but the last, which is known as the Sundance Sea, spread eastward to the Dakotas and southward to New Mexico (Figure 16-27). Eventually, as mountain building progressed along the Pacific coast in Late Jurassic time, the Sundance Sea retreated.

Figure 16-26 Silicified logs that have weathered out of the Triassic Chinle Formation in the Petrified Forest of Arizona.

Subduction and the Accretion of New Terranes During the Mesozoic Era, the western margin of North America expanded by the addition of numerous island arc terranes and other microplates (Figure 16-28), in a manner analogous to the addition of exotic terranes to eastern North America in Paleozoic time, during episodes of mountain building in the Appalachian region (p. 391). This mode of continental accretion actually began earlier. The Antler orogeny of Devonian and Early Carboniferous time had entailed the collision of the Klamath island arc with the western margin of North America (see Figure 15-25). This event added a sliver of exotic terrane,

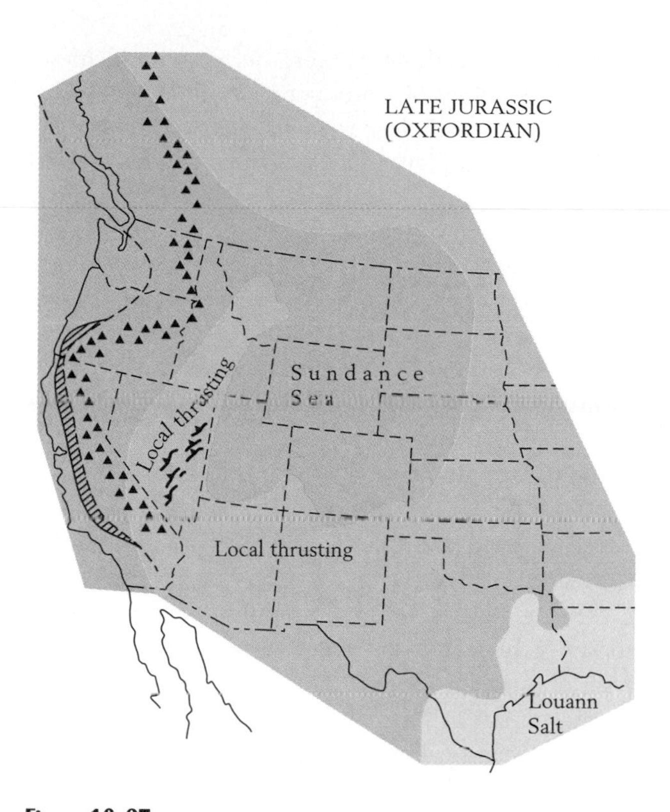

Figure 16-27 Geologic features of western North America during Late Jurassic (Oxfordian) time, when the Sundance Sea flooded a large interior region from southern Canada to northern Arizona and New Mexico. In mid-Triassic time, the western margin of the continent had ridden up against a subduction zone. As a result, during the Jurassic Period a belt of igneous activity extended for hundreds of kilometers parallel to the Pacific coast. At this time, thrust faulting was limited largely to the state of Nevada.

the Roberts Mountains terrane, to the western margin of North America.

In late Paleozoic time, after the Antler episode of accretion, the Golconda arc approached the Pacific margin of North America. Early in the Triassic Period this arc collided with the North American continent, just as the Klamath arc had done in the earlier Antler orogeny. The suturing of the Golconda arc, during what is known as the Sonoma orogeny, differed from the Antler orogeny in one important way, however: rather than adding a narrow slice of island arc terrane to North America, it attached a broad microcontinent, known as Sonomia (Figure 16-29). Today Sonomia comprises southeastern Oregon and northern California and Nevada (see Figure 16-28). Squeezed in between Sonomia and the Roberts Mountains terrane was the Golconda terrane, formed largely of the accretionary wedge associated with the Golconda arc (see Figure 16-29).

After the Sonoma orogeny ended, early in the Triassic Period, there was a brief interlude of tectonic quiescence along the west coast of North America. Then, in mid-Triassic time, the continental margin once again came to rest against a subduction zone and began an orogenic episode that extended from Alaska

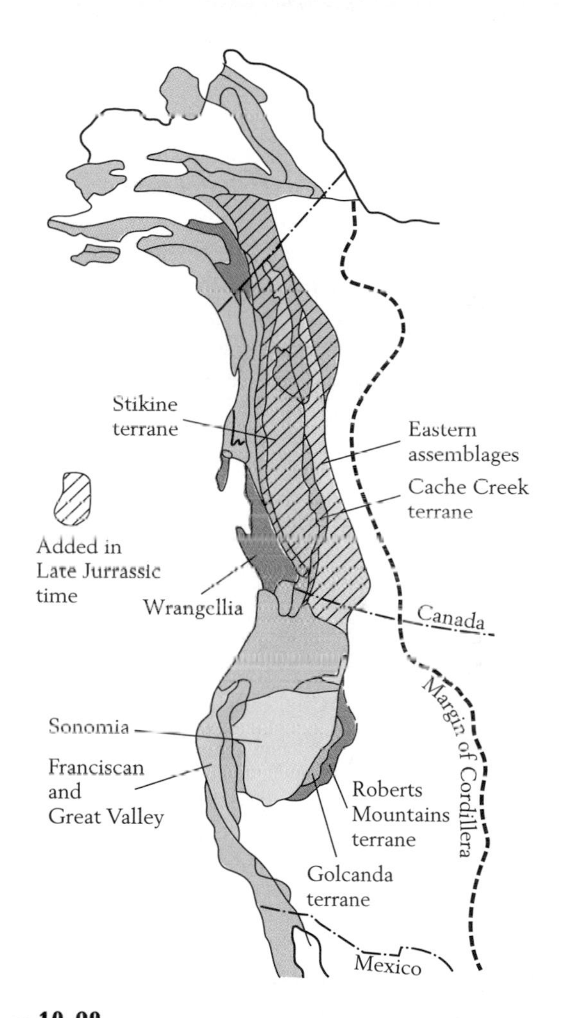

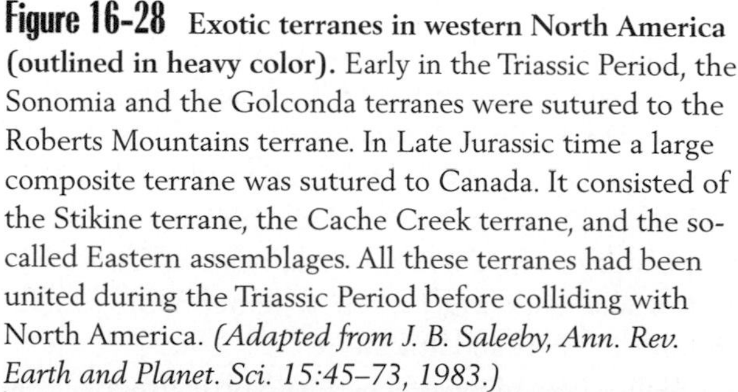

Figure 16-28 Exotic terranes in western North America (outlined in heavy color). Early in the Triassic Period, the Sonomia and the Golconda terranes were sutured to the Roberts Mountains terrane. In Late Jurassic time a large composite terrane was sutured to Canada. It consisted of the Stikine terrane, the Cache Creek terrane, and the so-called Eastern assemblages. All these terranes had been united during the Triassic Period before colliding with North America. *(Adapted from J. B. Saleeby, Ann. Rev. Earth and Planet. Sci. 15:45–73, 1983.)*

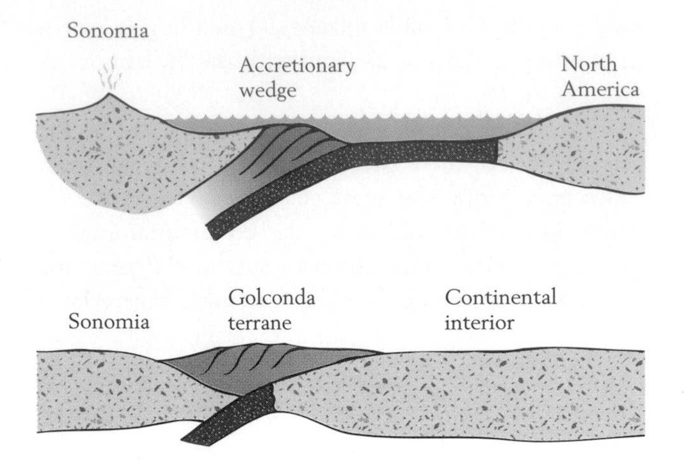

Figure 16-29 The accretion of Sonomia and the Golconda terrane to the western margin of North America. The eastern portion of the microcontinent of Sonomia was formed by the eruptions of an island arc. The Golconda terrane formed from an accretionary wedge that was squeezed between Sonomia and North America as the western margin of North America became wedged against the island arc that bordered Sonomia. (Figure 16-28 shows the location of Sonomia and the Golconda terrane in North America today.)

all the way to Chile. Mountain building along the Pacific coast of North America during the Mesozoic Era resembled the growth of the Andes to the south, which has continued to the present day (p. 244).

Subduction of the oceanic plate beneath the margin of North America thickened the continental crust by leading to the accumulation of intrusive and extrusive igneous rocks. The oldest intrusives of the Sierra Nevada were emplaced during Jurassic time (Figure 16-30); larger volumes were added later in the Mesozoic Era.

The Mesozoic history of the Pacific coast of North America is highly complex. At times more than one subduction zone lay offshore, and exotic slivers of crust were added to the continental margin. Near the end of the Jurassic Period the continent accreted westward when the Franciscan sequence of deep-water sediments and volcanics was forced against the craton along a subduction zone after having been metamorphosed at high pressures and low temperatures. The Franciscan sediments include graywackes and dark mudstones, together with smaller amounts of chert and limestone.

Before becoming attached to North America, the Franciscan sequence constituted an accretionary wedge, whose sediments were deformed and metamorphosed along the subduction zone at high pressures and relatively low temperatures; they represent a mélange (see Figure 8-23). When the continental margin eventually collided with the accretionary wedge, the Franciscan rocks were piled up against the continent, along with the Great Valley sequence of deep-sea turbidites, which accumulated in the forearc basin (Figure 16-31). This Late Jurassic event coincided approximately with eastward folding and thrusting from the Sierra Nevada uplift. These tectonic events of Jurassic age are collectively known as the Nevadan orogeny. Orogenic activity that is related to the Nevadan orogeny, although it is not generally assigned this name, continued well into the Cretaceous Period.

Farther north, from northern Washington State to southern Alaska, a large exotic terrane collided with the margin of North America, resulting in substantial westward accretion. This exotic terrane was actually a composite block, formed of several smaller terranes (see Figure 16-28). These smaller terranes include quite diverse suites of Paleozoic rocks and fossils, an indication that the terranes were once separate entities. They do, however, share rock units of Triassic age, an indication that they were a single unit during the Triassic Period. The entire composite terrane was then accreted to North America late in Jurassic time, along the subduction zone that bordered the continent.

Figure 16-30 The Sierra Nevada at Yosemite National Park, California. Many of the granitic rocks that formed the Sierra Nevada were emplaced during the Jurassic Period.

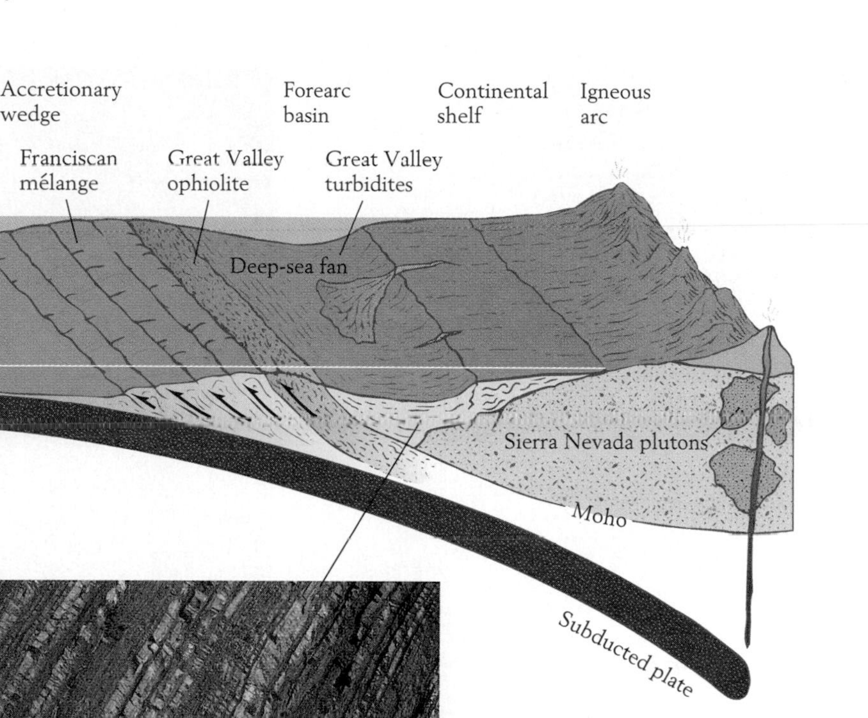

B

Figure 16-31 The Pacific margin of northern California in Jurassic time. The Franciscan mélange formed an accretionary wedge along the marginal subduction zone (see Figure 8-23). The Great Valley ophiolite was a zone of seafloor that was squeezed up along the eastern margin of the accretionary wedge, and the Great Valley sequence formed in Late Jurassic time as turbidites on deep-sea fans and in adjacent environments. The photograph shows turbidites that now lie along the western margin of the Sacramento Valley in California, where they have been tilted to a high angle by tectonic activity. The accretion of the Franciscan and Great Valley terranes to the continental margin during Late Jurassic and Cretaceous time extended North America westward (Figure 16-28). Today the Great Valley sequence still occupies a low region, the Central Valley of California. West of the Central Valley, portions of the Franciscan mélange have been elevated as part of the Coast Ranges. *(A. Adapted from R. K. Suchecki, Jour. Sedim. Petrol. 54:170–191, 1984.)*

Deposition in a Foreland Basin To the south, in the western United States, the eastward thrusting and folding of Late Jurassic time greatly altered patterns of deposition as far east as Colorado and Wyoming. The Sundance Sea spread over a broad foreland basin east of the mountains. This was the most extensive marine incursion since late Paleozoic time (see Figure 16-27). In latest Jurassic time, however, the folding and thrust faulting that extended over Nevada, Utah, and Idaho produced a large mountain chain. The shedding of large volumes of clastics eastward from the mountains drove back the waters of the Sundance Sea, leaving only a small inland sea to the north (Figure 16-32). What remained in Colorado, Wyoming, and adjacent regions was a nonmarine foreland basin in which molasse deposits accumulated. Apparently, on the gentle profile of the foreland basin, even the lowest depositional environments were above sea level, because there was no initial deposition of marine flysch. The molasse of the

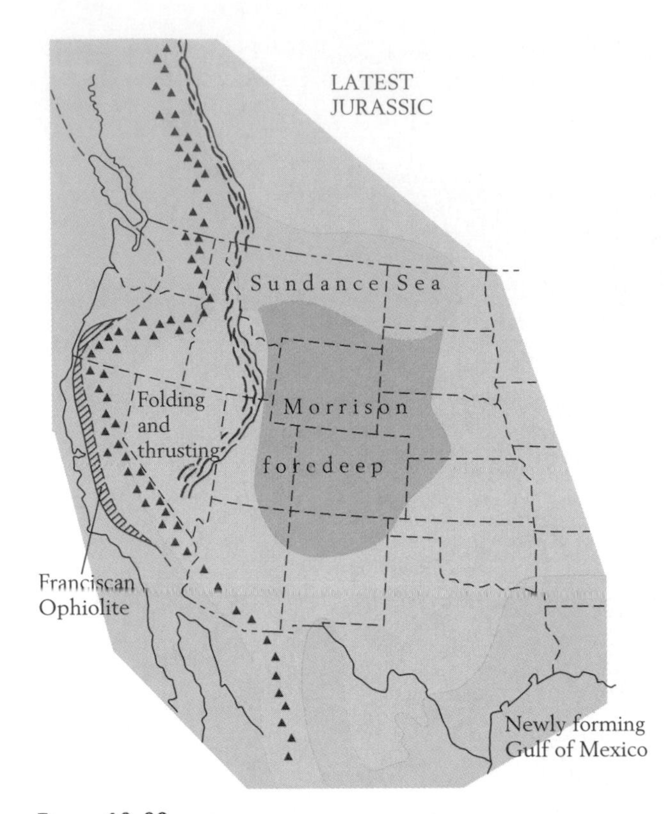

Figure 16-32 Geologic features of western North America during latest Jurassic time. A fold-and-thrust belt now extended for hundreds of kilometers roughly parallel to the coastline but far inland. Tectonic activity had driven the Sundance Sea from the western interior, leaving a foreland basin where the nonmarine Morrison Formation accumulated.

Figure 16-33 Excavation of dinosaur fossils in the Morrison Formation. These partly intact skeletons remain embedded in the rock at Dinosaur National Monument, Utah.

foreland basin was deposited in rivers, lakes, and swamps, creating the famous Morrison Formation, which has yielded the world's most spectacular dinosaur faunas (Figure 16-33; see also Figures 16-15 and 16-16). The dinosaur skeletons in this formation are usually disarticulated, but often as many as 50 or 60 individuals are found together in one small area, an indication that these fossils may have accumulated during floods.

The Morrison Formation consists of sandstones and multicolored mudstones deposited over an area of about 1 million square kilometers. Caliche soil deposits indicate that the climate was seasonally dry during at least part of the Morrison depositional interval, while the scarcity of crocodiles, turtles, and fishes suggests that many of the lakes may have been saline at this time. The dinosaurs are found in deposits representing all of the Morrison environments—rivers, lakes, and swamps. This broad environmental distribution suggests that none of the species, not even the huge sauropods (see Figure 16-15), were adapted specifically for a life of wading in large bodies of water. The Morrison Formation spans the last 10 million years or so of the Jurassic Period and is overlain by the nonmarine Cloverly Formation of Early Cretaceous age, which contains a completely different fauna of dinosaurs, apparently because of major extinctions at the end of the Jurassic Period.

Chapter Summary

1 Triassic and Jurassic seas lacked taxa that had been prominent in Paleozoic time, such as fusulinid foraminifera, rugose corals, and trilobites. Important groups of marine life that were present during this time included bivalve gastropods, ammonoid mollusks, brachiopods, sea urchins, hexacorals, bony fishes, sharks, and swimming reptiles.

2 Ferns were abundant during the Triassic Period, but gymnosperm plants dominated Jurassic landscapes.

3 Although mammals originated in Triassic time, dinosaurs were much more successful than mammals during the Mesozoic Era. Flying reptiles evolved in the Triassic Period, and birds later evolved from dinosaurs.

4 A mass extinction that was less severe than the terminal Permian event marked the end of the Triassic Period, and moderately heavy extinction occurred at the end of Jurassic time.

5 Very early in the Triassic Period, nearly all of Earth's continental crust was consolidated in the supercontinent Pangea, and even at the end of Jurassic time, all of the continents remained close together. Evaporites mark zones where Pangea began to rift apart early in the Mesozoic Era.

6 Fault-block basins that formed in eastern North America received thick deposits of sediment during the rifting episode that eventually formed the Atlantic Ocean between this continent and Africa.

7 The Sundance Sea invaded western North America during the Jurassic Period but was expelled by uplifting and sediment influx associated with the Nevadan orogeny. Dinosaur fossils are abundantly preserved in the molasse sediments that were then deposited in the vicinity of Utah.

Review Questions

1 What important groups of Paleozoic marine animals were absent from Triassic seas?

2 Why should we not be surprised that stromatolites spread over Early Triassic seafloors?

3 What two kinds of flying vertebrates evolved during early Mesozoic time?

4 How did reefs formed by hexacorals during the Jurassic Period differ from those formed during Triassic time?

5 What suggests that dinosaurs had relatively advanced behavior of the sort that might explain their evolutionary success?

6 What evidence is there that dinosaurs brooded their eggs and tended their young long after they hatched?

7 What was the geographic setting in which the most spectacular known assemblage of Jurassic dinosaurs was preserved?

8 In what areas is there evidence that oceans started to form in Triassic and Jurassic time? What is that evidence?

9 What plate tectonic change might explain the eastward migration of the zone of igneous activity in western North America during the Jurassic Period?

10 By what mechanism did western North America expand westward during early Mesozoic time?

11 Vertebrate life underwent spectacular evolutionary changes early in the Mesozoic Era. Using the Visual Overview on page 436 and what you have learned in this chapter, review the ways in which vertebrate animals expanded their ecological role in the ocean, on the land, and in the air during the Triassic and Jurassic periods.

Additional Reading

Alexander, R. M., "How Dinosaurs Ran," *Scientific American*, April 1991.

Dingus, L., and T. Rowe, *The Mistaken Extinction: Dinosaur Evolution and the Origin of Birds*, W. H. Freeman and Company, New York, 1997.

Padian, K., and L. M. Chiappe, "The Origin of Birds and Their Flight," *Scientific American*, February 1998.

Thomas, B. A., and R. A. Spicer, *The Evolution and Palaeobiology of Land Plants*, Croom Helm, London, 1987.

CHAPTER 17

The Cretaceous World

The Cretaceous Period was in many ways an interval of transition. Some Cretaceous sediments are lithified, like nearly all those of older systems; many others, however, consist of soft muds and sands, like most deposits of the younger Cenozoic Era. Fossil biotas of the Cretaceous Period also display a mixture of archaic and modern features. They include members of diverse taxa that failed to survive the Cretaceous Period—among them the dinosaurs and ammonoids—as well as diverse modern taxa, such as flowering plants and the largest subclass of fishes in the world today. It was during the Cretaceous Period that continents moved toward their modern configuration. At the start of the period the continents were tightly clustered. By the end of Cretaceous time, however, the Atlantic Ocean had widened and the southern portion of Pangea—the former Gondwanaland—had fragmented into most of its daughter continents.

The Cretaceous System was first described formally in 1822. Its name derives from *creta*, the Latin word for chalk, which is a soft, fine-grained kind of limestone that accumulated over broad areas of the Late Cretaceous seafloor.

Life

Life of the Cretaceous Period, both in the seas and on land, was a curious mixture of modern and archaic forms. In the marine realm, strikingly

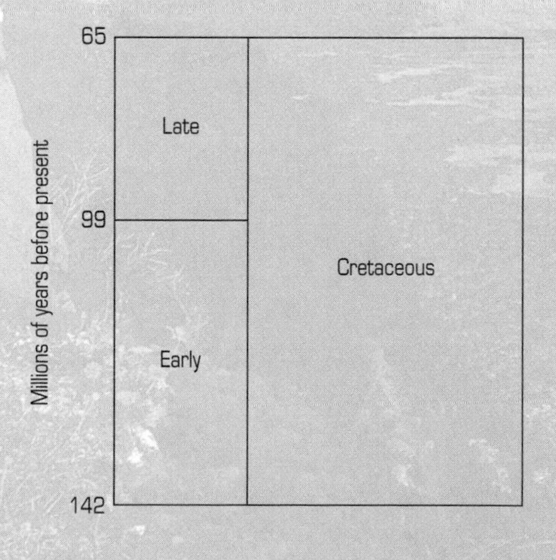

Chalk, the soft, powdery rock that is unusually abundant in the Upper Cretaceous Series in many areas. The chalk deposits shown here stand above the coastline of southeastern England, where they form the famous White Cliffs of Dover.

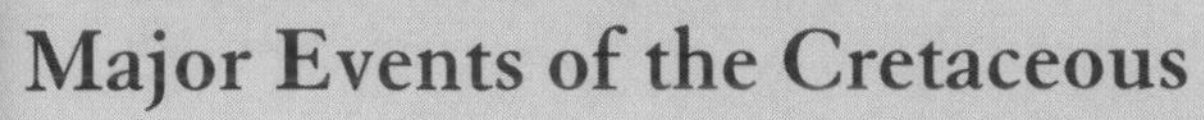

Visual Overview

Major Events of the Cretaceous

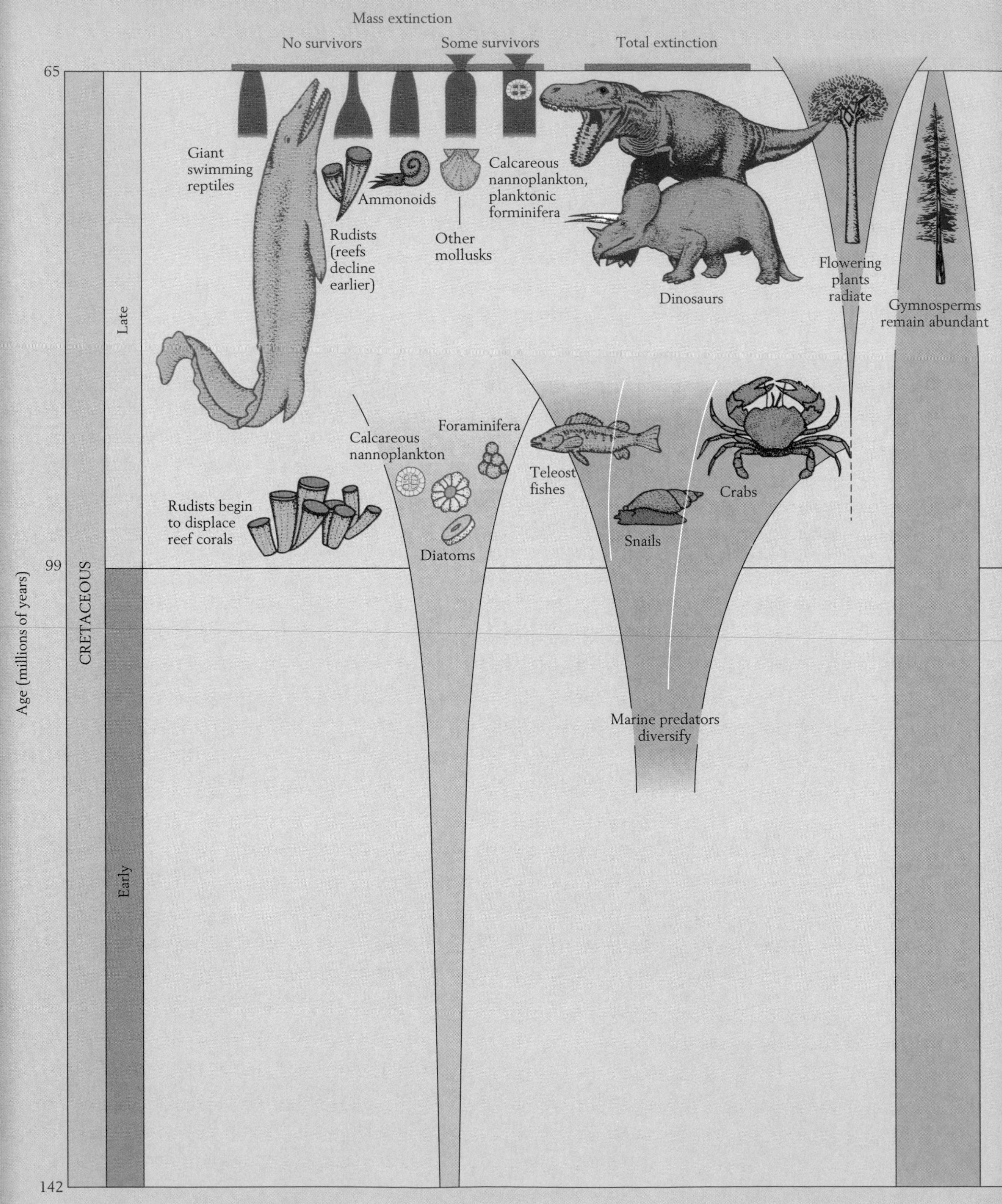

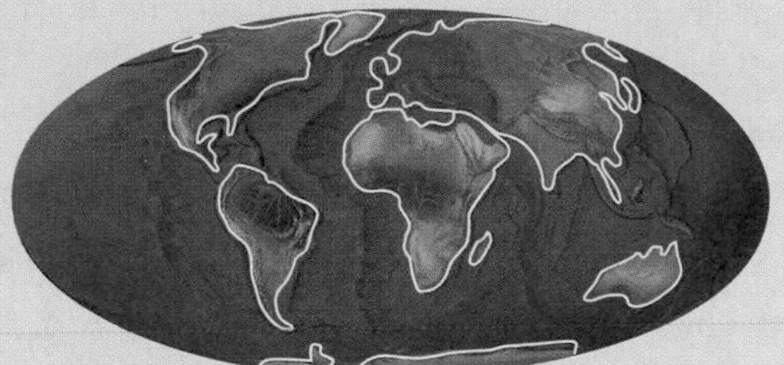
MIDDLE MIOCENE

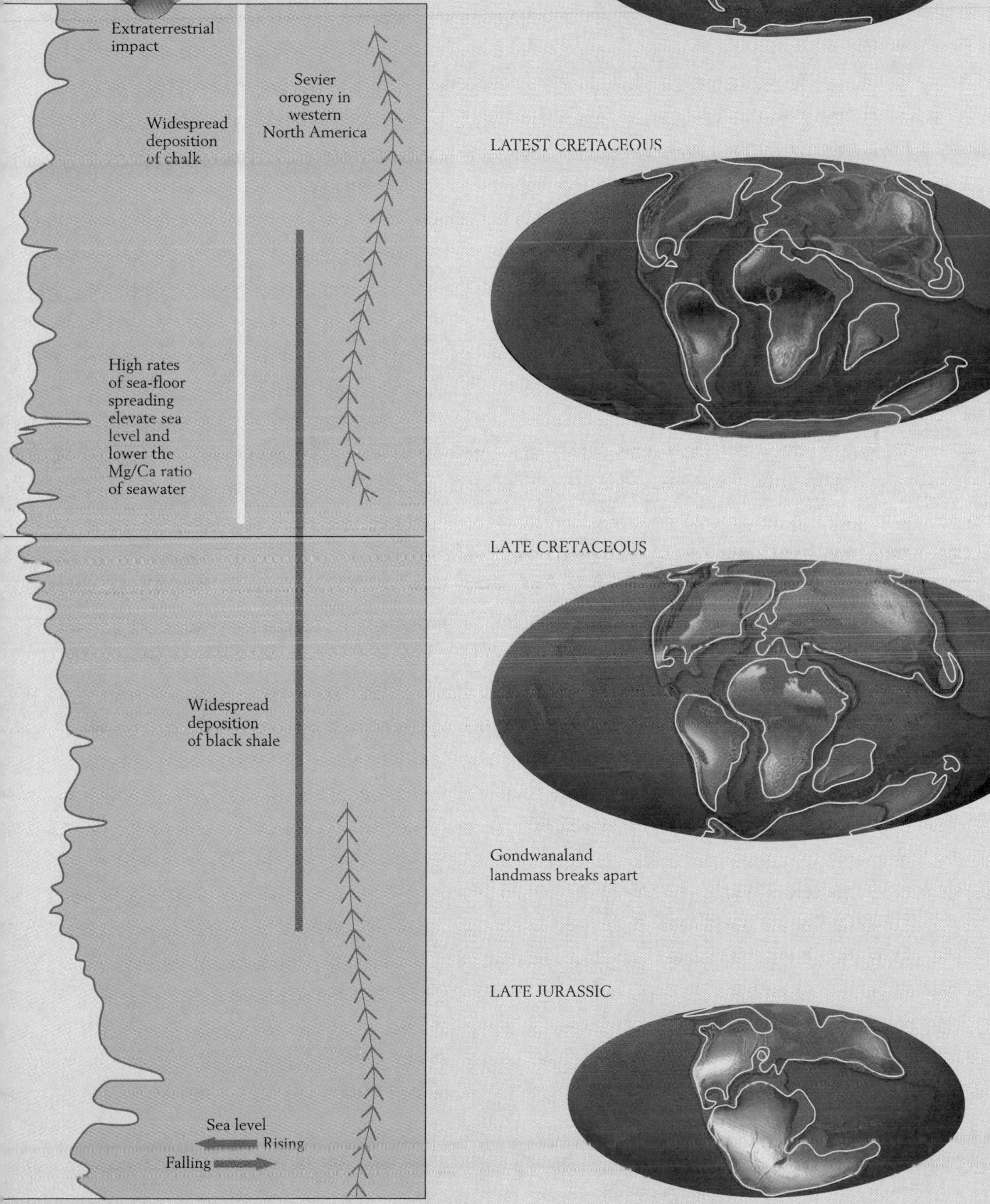
Extraterrestrial impact
Sevier orogeny in western North America
Widespread deposition of chalk
High rates of sea-floor spreading elevate sea level and lower the Mg/Ca ratio of seawater
Widespread deposition of black shale
Sea level
Rising
Falling
LATEST CRETACEOUS
LATE CRETACEOUS
Gondwanaland landmass breaks apart
LATE JURASSIC

modern types of bivalve and gastropod mollusks populated Late Cretaceous seas along with enormous coiled oysters and other now-extinct sedentary bivalves. Diverse fishes of the modern kind occupied the same waters as a variety of ammonoids, belemnoids, and reptilian sea monsters—none of which have any close living relatives. Dinosaurs, however, continued to rule the land; mammals remained very small by modern standards.

Pelagic Life

The appearance of new groups of single-celled organisms gave marine plankton a thoroughly modern character by the end of Cretaceous time. The primary change among the phytoplankton was the evolutionary expansion of the **diatoms** (see Figure 3-16). Diatoms may have existed during the Jurassic Period, but they did not radiate extensively until mid-Cretaceous time. Together with dinoflagellates and, in warm seas, calcareous nannoplankton, diatoms must have accounted for most of the marine photosynthesis in Cretaceous time, as they do today (p. 112). Today diatoms are the dominant contributors to the siliceous oozes of the deep sea, and their abundant accumulation in deep-sea sediment began during the Cretaceous Period (p. 148).

Higher in the pelagic food web, the modern planktonic foraminifera diversified greatly for the first time. This group has a meager fossil record in Jurassic rocks; not until the upper part of the Lower Cretaceous System is it well enough represented to be of great value in biostratigraphy (Figure 17-1).

Late Cretaceous adaptive radiations of two of the single-celled planktonic groups altered depositional patterns in the pelagic realm: since mid-Cretaceous time, both foraminifera and calcareous nannoplankton have contributed vast quantities of calcareous sediment to oceanic areas (see Figure 5-35), whereas before about 100 million years ago, little or no calcareous ooze was present on the deep seafloor.

Recall that during Late Cretaceous time calcareous nannoplankton were so abundant in warm seas that the small plates that armored their cells accumulated in huge volumes as the fine-grained limestone commonly known as chalk; it appears that the great abundance of calcareous nannoplankton, which secrete calcite, resulted from the low magnesium-calcium ratio of Late Cretaceous seas (see Box 10-1). Cretaceous chalk has, in fact, been widely used for writing on blackboards. The most famous chalk deposits in the world crop out along the southeastern coast of England, where they are known formally as the Chalk (p. 464). Similar chalks formed in shal-

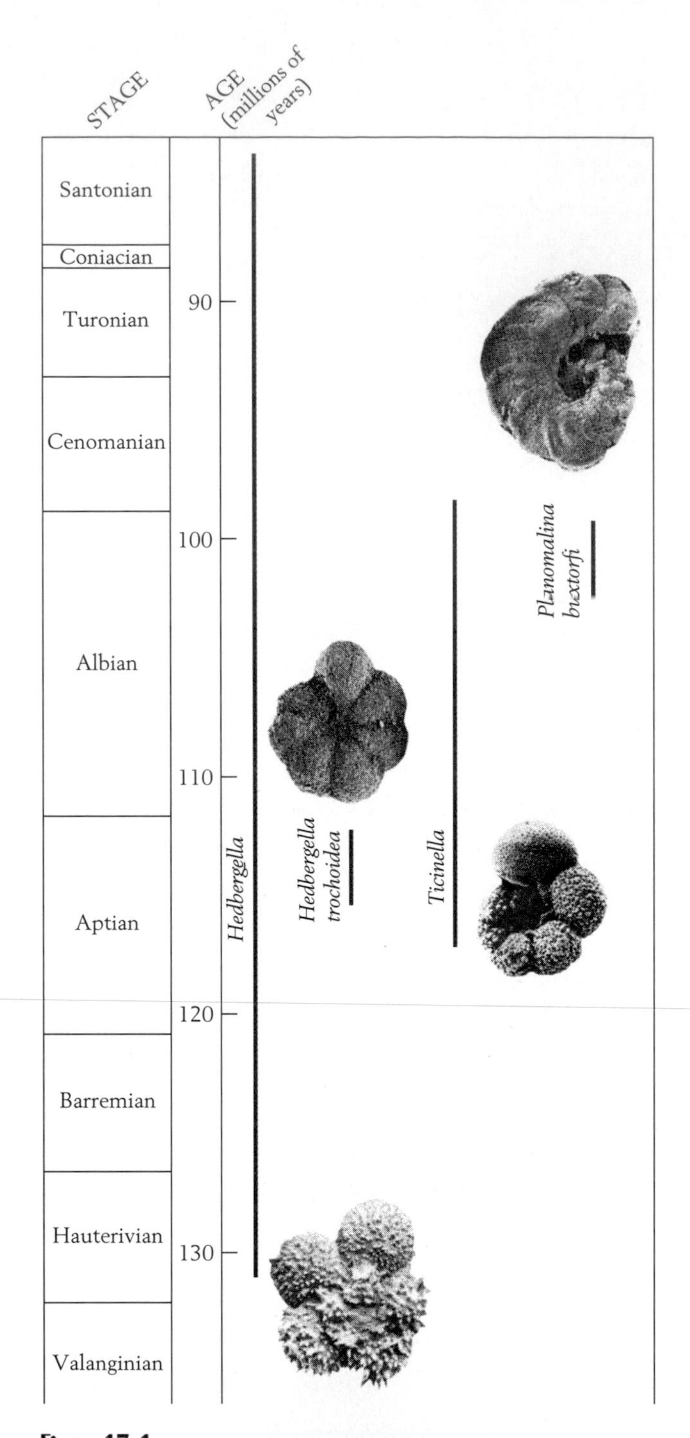

Figure 17-1 **Early planktonic foraminifera (Globigerinacea).** Cretaceous stages are shown at the left. These species average about $\frac{1}{2}$ millimeter ($\frac{1}{50}$ of an inch) in diameter. *(After A. Boersma.)*

Figure 17-2 **Cretaceous invertebrate fossils.** *A.* A coiled oyster, the size of a small grapefruit. *B.* A burrowing bivalve mollusk. *C.* A crab. *D.* A carnivorous snail (gastropod). *E.* An ammonoid.

low seas in Kansas and nearby regions and along the Gulf Coast of the United States.

Still higher in the pelagic food web of Late Cretaceous time, the ammonoids and belemnoids persisted as major swimming carnivores. The ammonoids serve as valuable index fossils for the Cretaceous System, just as they do for the Triassic and Jurassic. Among the Cretaceous ammonoids were many species with straight, cone-shaped shells and others with coiled shells (Figure 17-2).

New on the scene in Cretaceous time were the **teleost fishes,** a subclass that today is the dominant group of marine and freshwater fishes. Teleosts are characterized by such features as symmetrical tails, round scales, specialized fins, and short jaws that are often adapted to take particular kinds of food. By Late Cretaceous time, a wide variety of teleosts already existed, including the largest species known from the fossil record (Figure 17-3). This group also included close relatives of the modern sunfish, carp, and eel, as well as members of the salmon family. Similarly, Cretaceous sharks resembled present-day forms.

Most of the top carnivores of Cretaceous pelagic habitats, however, were not at all modern. Whereas whales of one kind or another have occupied the "top carnivore" adaptive zone during most of the Cenozoic Era, reptiles were the largest marine carnivores until the end of Cretaceous time. Ichthyosaurs and marine crocodiles were rare by this time, but plesiosaurs still thrived, some exceeding 10 meters (35 feet) in length. Plesiosaurs are depicted in Figure 16-8 along with other members of the Late Cretaceous pelagic community of the western interior of the United States. Huge marine lizards known as

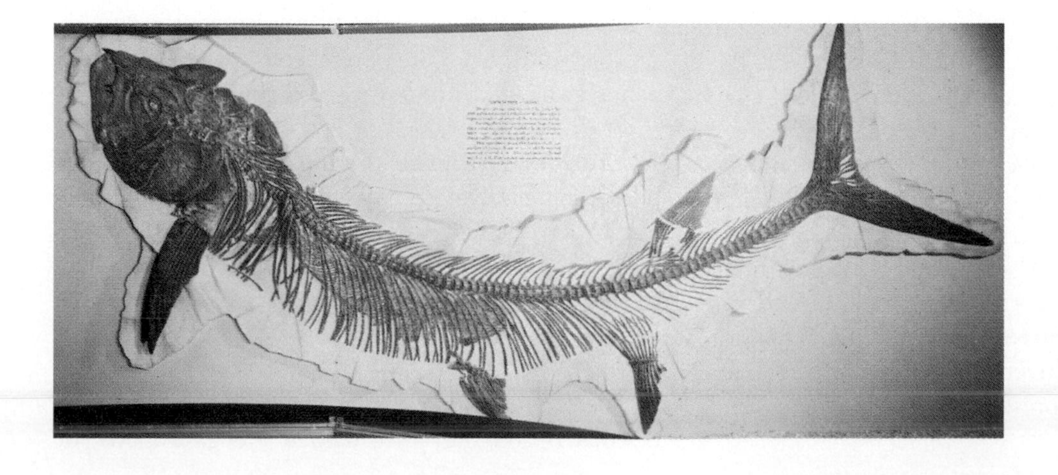

Figure 17-3 ***Xiphactinus,* a Cretaceous fish.** At about 5 meters (16 feet) in length, this is the largest known teleost. A careful look reveals that the animal shown here died with a good-sized fish in its belly.

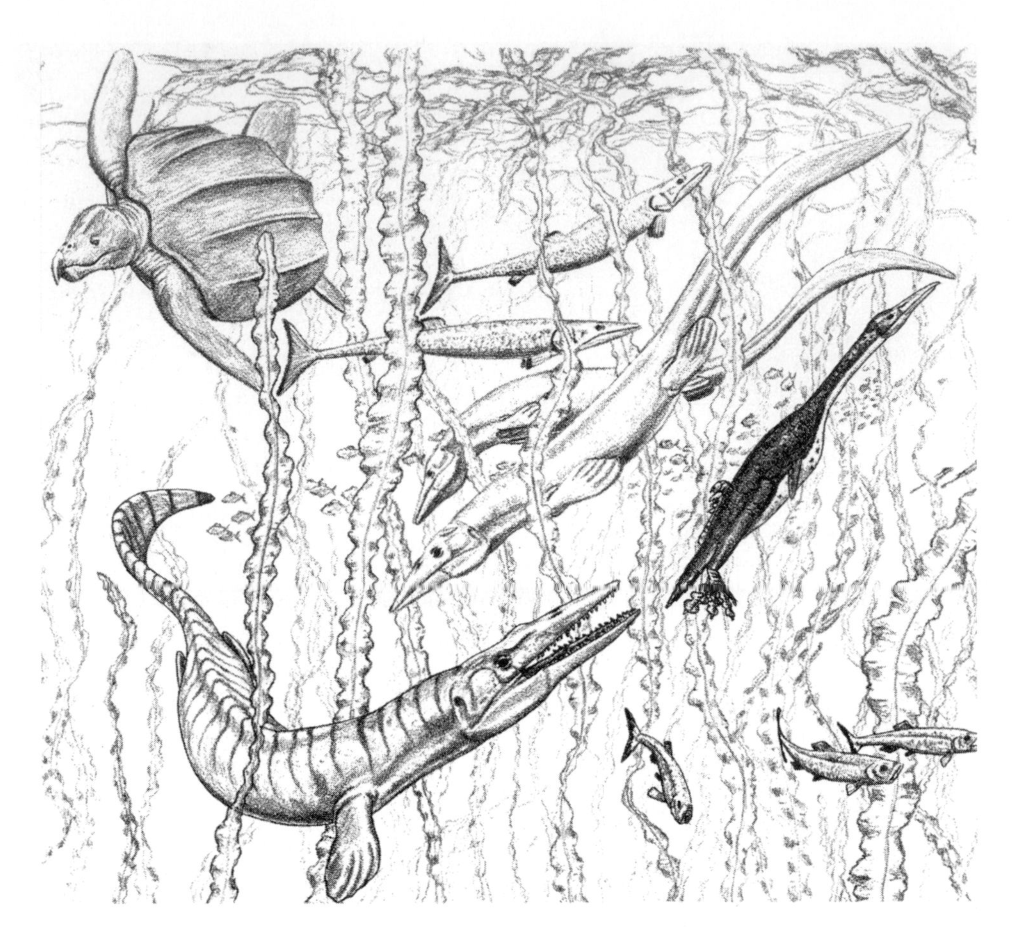

Figure 17-4 Reconstruction of marine life preserved in the Upper Cretaceous Pierre Shale of the western interior of the United States. The animals are shown swimming in a bed of kelp, which are algae of large proportions. The giant turtle at the upper left is *Archelon*, which reached a length of almost 4 meters (13 feet). The striped animal at the lower left is the mosasaur *Clidastes*, and beyond it is a pair of mosasaurs of the genus Platecarpus. *Clidastes* is in pursuit of the diving bird *Hesperornis*. The teleost fishes are *Cimolichthyes* (the pikelike pair near the turtle) and *Enchodus* (the small fishes on the lower right). *(Drawing by Gregory S. Paul.)*

mosasaurs were probably the most formidable marauders of Cretaceous seas; some grew to be longer than 15 meters (45 to 50 feet). Unlike the plesiosaurs, whose forms are those of distance swimmers, mosasaurs probably lurked inconspicuously and ambushed their prey. Figure 17-4 also portrays the flightless diving bird *Hesperornis* and a species of marine turtle that grew to a length of nearly 4 meters (13 feet).

Seafloor Life

Life on the seafloor began to take on a modern appearance during the Cretaceous Period. One noteworthy feature was the decline of the brachiopods, which had suffered greatly in the mass extinction at the end of the Paleozoic Era but had experienced a moderate expansion again early in Mesozoic time. Sea urchins and the hexacorals diversified but underwent no startling adaptive changes. Other major groups produced distinctive new representatives that have survived to the present. Some of them are described below.

Foraminifera A large percentage of the families of bottom-dwelling foraminifera in existence today appeared during the Cretaceous Period, so that this group had a modern aspect (Figure 17-5).

Bryozoans The most abundant modern bryozoans are the cheilostomes, which commonly encrust marine surfaces, including the hulls of boats, in the form of low-growing mats (see Figure 3-31). Cheilostomes originated in Jurassic time but did not enjoy success until the Late Cretaceous, when they expanded to include more than 100 genera.

Burrowing Bivalve Mollusks Early Cretaceous burrowing bivalves resembled those of the Jurassic, but by the end of the period new genera had appeared as well, including many that were rapid burrowers or burrowed deep into the sediment (see Figure 17-2).

Gastropod Mollusks, or Snails During the Cretaceous Period the aptly named Neogastropoda, or

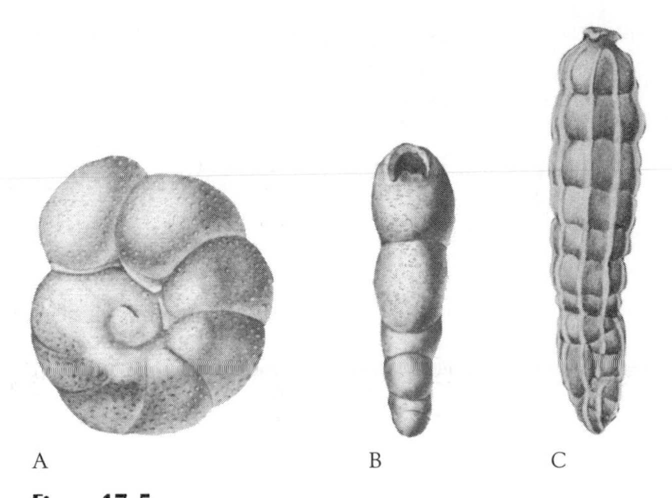

Figure 17-5 **Genera of bottom-dwelling foraminifera that arose during the Cretaceous Period.** *A. Anomalinoides* (×68). *B. Pleurostomella* (×48). C. *Siphogenerinoides* (×46). *(H. Tappan.)*

"new snails," produced many modern families and genera (see Figure 17-2). Unlike most earlier snails, these animals are generally carnivorous, feeding on such prey as worms, bivalves, and other snails. Some live in the sediment, others on the sediment surface. Many modern seashells popular with collectors belong to neogastropod species.

Crabs A more or less modern type of crab had evolved during the Jurassic Period, but a much greater variety of crabs appeared during the Cretaceous Period (see Figure 17-2).

Surface-Dwelling Bivalve Mollusks Among bivalve mollusks living on the surface of the substratum, coiled oysters and other groups that had existed during Jurassic time evolved species of enormous size (see Figure 17-2). Of these large forms the rudists were of special significance because they lived like corals, forming large tropical reefs (Figure 17-6). Rudists grew a cone-shaped lower shell and a lidlike upper shell. These curious animals attached to hard objects (often other rudists) and grew upward, some reaching heights of more than 1 meter (~3 feet).

Before their demise the rudists apparently flourished at the expense of reef-building corals. Shallow-water reefs built in Early Cretaceous time, like those built during the Jurassic Period and in the modern world, were formed primarily by hexacorals. In mid-Cretaceous time, however, rudist bivalves displaced corals to assume the dominant role in tropical reef growth. Rudists, like reef-building corals, apparently grew rapidly by feeding on symbiotic algae that lived and multiplied in their tissues (see Box 4-1 and Figure 10-15). Only after the end of the Cretaceous Period, when rudists, like dinosaurs, became extinct, did corals prevail on reefs once more.

The corals declined at a time when the magnesium-calcium ratio of seawater reached its lowest Phanerozoic level (see Figure 10-18). This chemical condition may have hampered corals in their precipitation of aragonitic skeletons (p. 282). Rudists secreted shells that included both calcite and aragonite. Seawater chemistry probably had little influence on their shell secretion, however, because, as mollusks, they were more advanced animals than corals and secreted their shells from fluids excreted by their bodies– fluids that had a different chemical composition than seawater. Mollusks grew both calcitic and aragonitic shells throughout Phanerozoic time, without apparent influence from seawater.

Figure 17-6 **A reef formed by a population of rudists of the genus *Durania* preserved in the Upper Cretaceous of Egypt.** The cap-shaped upper valves that fitted on the cone-shaped lower valves have been eroded away.

The Rise of Modern Marine Predators

Many of the general changes that occurred in bottom-dwelling marine life during Jurassic and Cretaceous time seem to have been related to the great expansion of modern types of marine predators. Among the new predators were the advanced teleost fishes (see Figure 17-3), modern kinds of crabs, and carnivorous gastropods (see Figure 17-2). Many of these new predators were efficient at penetrating shells: fishes by biting, crabs by crushing or peeling with their claws, and some of the gastropods by drilling holes.

The contrast between Paleozoic and Mesozoic predation on the seafloor is exemplified by the absence during Paleozoic time of large arthropods with crushing claws and by the virtual absence of holes drilled by predators in fossilized Paleozoic brachiopods and bivalve shells.

The decline of brachiopods and stalked crinoids, both of which were moderately well represented in early Mesozoic seas, probably resulted from the diversification of modern predators. The few species of stalked crinoids that survive today live in deep water; in shallow waters, predation by fishes is probably too severe to permit their existence. Similarly, more species of brachiopods today live in temperate seas than in tropical seas, where predation by crabs, fishes, and snails is severe. By the end of the Mesozoic Era, relatively few immobile species of animals lived on the surface of the seafloor in the mode typical of many groups of Paleozoic brachiopods (see Figure 13-13). The ability to swim or to burrow actively appears to have been the best defense against predation, except for species that had defensive spines or unusually heavy protective shells.

Flowering Plants Conquer the Land

The greatest change in terrestrial ecosystems during the Cretaceous Period was the ascendance of the flowering plants (angiosperms), although gymnosperm floras resembling those of Triassic and Jurassic age continued to dominate the land during Cretaceous time. The most conspicuous change during Early Cretaceous time was in the types of gymnosperms that predominated: conifers became the most numerous species of trees, and the Age of Cycads came to a close. The angiosperms that now made their appearance included not only plants with conspicuous flowers but also hardwood trees, such as maples and oaks, and grasses. The key reproductive feature that distinguishes angiosperms from gymnosperms (naked-seed plants) is the enclosure of the seed (p. 76).

The Earliest Floras Fossils of the Atlantic Coastal Plain in Maryland document the early phase of the evolutionary radiation of flowering plants. Here, within a sedimentary interval representing only about 10 million years of mid-Cretaceous time, both fossil leaves and fossil pollen increase in variety and in complexity of form (Figure 17-7). The early leaves have simple, smooth outlines, and their supporting veins branch in irregular patterns. Later leaves include varieties with many marginal lobes and veins that follow more regular geometric patterns. Probably the more regular patterns gave the leaves greater strength to with stand tearing.

Secrets of Success One special feature of the flowering plants is their ability to provide a food supply for their seeds by a process known as double fertilization. One fertilization event produces a seed within the ovary. A second fertilization event, also within the ovary, produces a supply of stored food for the seed, such as the nutritional part of a kernel of corn or a grain of wheat. The rapid manufacture of this food supply allows for the quick release of a well-fortified seed. Because gymnosperms lack this double-fertilization mechanism, it takes much longer for the parent plants to supply their seeds with enough food to enable the progeny to survive on their own. As a result, most gymnosperms have reproductive cycles of 18 months or longer. In contrast, thousands of flowering plants grow from a seed and then release seeds of their own in just a few weeks.

A second reproductive mechanism of flowering plants that has contributed enormously to their success is the flowers' attraction of insects. Insects benefit from the nutritious nectar that the flowers provide, and the flowers benefit because the insects unknowingly carry pollen from one flower to another, fertilizing the plants on which they feed. This attraction is often specialized: a particular kind of insect feeds on a particular kind of plant, providing a unique mechanism for speciation. If a flower of a new shape, color, or scent develops within a small population of plants, the flower may attract a different kind of insect than the one that visited its ancestors. The plants with the new kind of flower will thus be reproductively isolated from their ancestral species; in other words, the new forms will become a

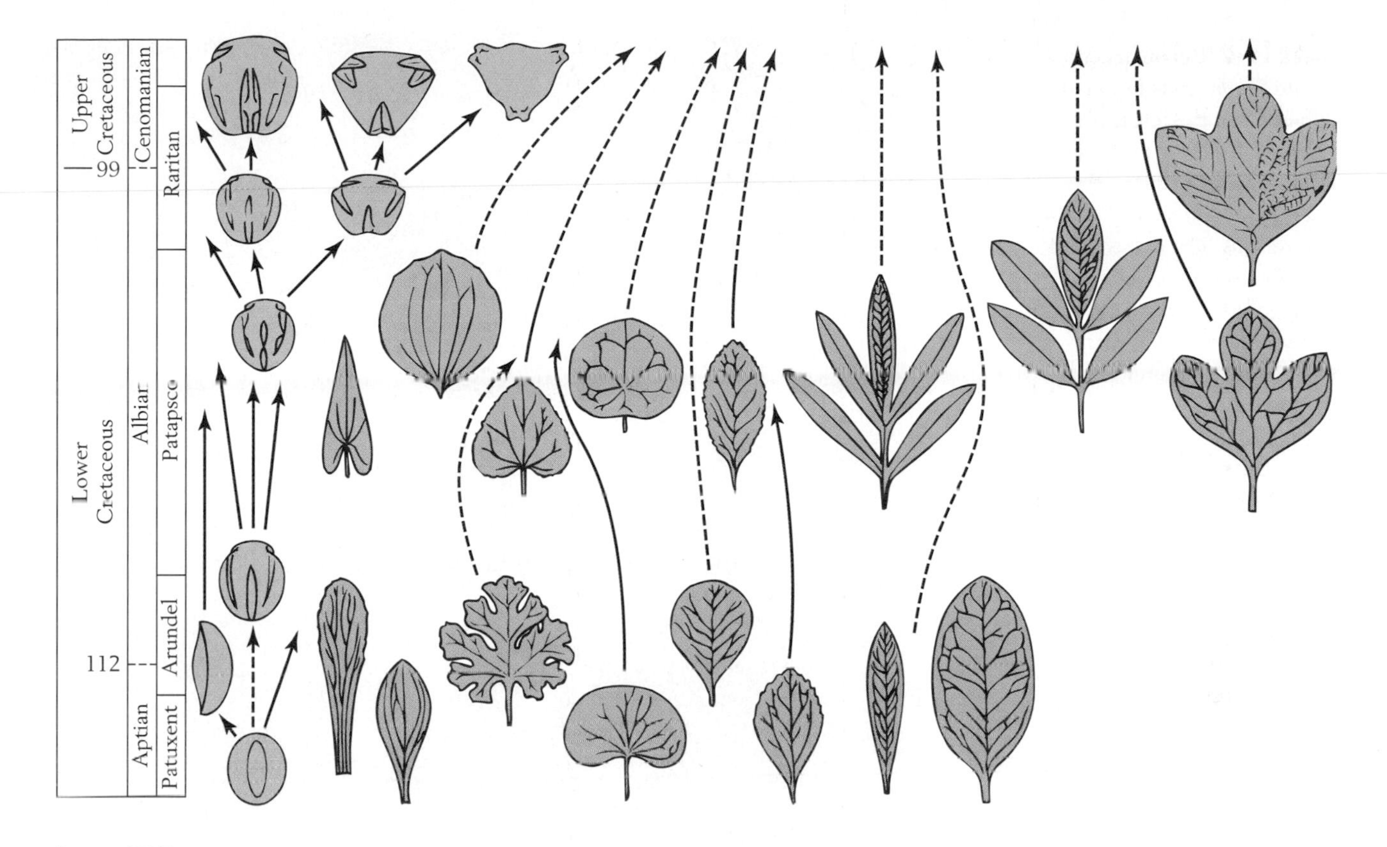

Figure 17-7 The pattern of initial adaptive radiation of flowering plants. These fossil leaves and pollen are found in formations (Patuxent through Raritan) of the Cretaceous Potomac Group of Maryland. Both pollen (left) and leaves exhibit an increase in complexity and variety of form through time. *(After J. A. Doyle and L. J. Hickey, in C. B. Beck [ed.], Origin and Early Evolution of the Angiosperms, Columbia University Press, New York, 1976.)*

new species. In general, new kinds of insects create opportunities for the development of new species of plants (with new kinds of flowers); and similarly, new kinds of plants create feeding opportunities for new species of insects. This reciprocity has apparently accelerated rates of speciation in both flowering plants and insects. High rates of speciation have permitted the frequent development of new adaptations. Thus the symbiotic (mutually beneficial) relationship between flowering plants and insects has played a major role in the great success of both groups since mid-Cretaceous times. The fossil record of flowers, though meager, shows a diversification of these showy organs in Late Cretaceous time– a diversification that clearly reflects the association of particular species of angiosperms with particular kinds of insects. A few other kinds of animals, including hummingbirds, also visit flowers and fertilize angiosperms today.

Flowering plants diversified late in Cretaceous time, but for unknown reasons they were unable to dislodge gymnosperms and ferns from many terrestrial habitats. Plants that were buried catastrophically where they stood by a volcanic eruption in Wyoming provide a snapshot of the composition of a flora near the end of Cretaceous time. This flora was dominated by ferns, gymnosperms, and a single species of low-growing palm (a flowering plant). Many other species of flowering plants were present, but in low abundance. This and other Cretaceous floras indicate that angiosperms were quite diverse in Late Cretaceous time but remained ecologically marginal.

The fossil record indicates that angiosperms evolved in the tropics and spread poleward during the Cretaceous. They flourished primarily in unstable habitats, especially along riverbanks, while gymnosperms and ferns continued to dominate most terrestrial environments. Sycamores were one of the most abundant groups of Cretaceous flowering plants; interestingly, even today sycamores tend to grow in unstable environments along streams.

Figure 17-8 Reconstruction of a Late Cretaceous fauna of Alberta. On the left is the armored herbivorous dinosaur *Edmontonia* in front of the duck-billed herbivore *Kritosaurus*. The duck-billed herbivores to their right belong to the genus *Corythosaurus*. The ferocious carnivore to the right of center is *Tyrannosaurus;* it confronts horned dinosaurs of the genus *Chasmosaurus,* with the large head shields, and *Monoclonius,* with the long horn. Passing overhead in the foreground are pterosaurs (flying reptiles) of the genus *Quetzalcoatlus*. The water birds flying in the distance have feathered wings, in contrast to the naked wings of the pterosaurs. *(Drawing by Gregory S. Paul.)*

Large Dinosaurs and Small Mammals

Owing to a patchy fossil record, Early Cretaceous vertebrate faunas are poorly known, but Late Cretaceous faunas are well known from collecting sites in Wyoming, Montana, Alberta, and Asia.

In the American West, Late Cretaceous dinosaurs formed a community that has been compared with the modern mammal faunas of the African plains. Instead of antelopes, zebras, and wildebeests, there were many species of duck-billed dinosaurs (Figure 17-8). These fast-running herbivores probably traveled in herds and may have trumpeted signals to one another by passing air through complex chambers in their skulls (see Box 16-1). They also tended their young after birth (Figure 17-9). In place of rhinoceroses, there were horned dinosaurs with beaks and teeth for cutting harsh vegetation (see Figure 17-8). Sharing the Late Cretaceous plains with these herbivores

Figure 17-9 **Painting of a maiasaur dinosaur tending her young.**

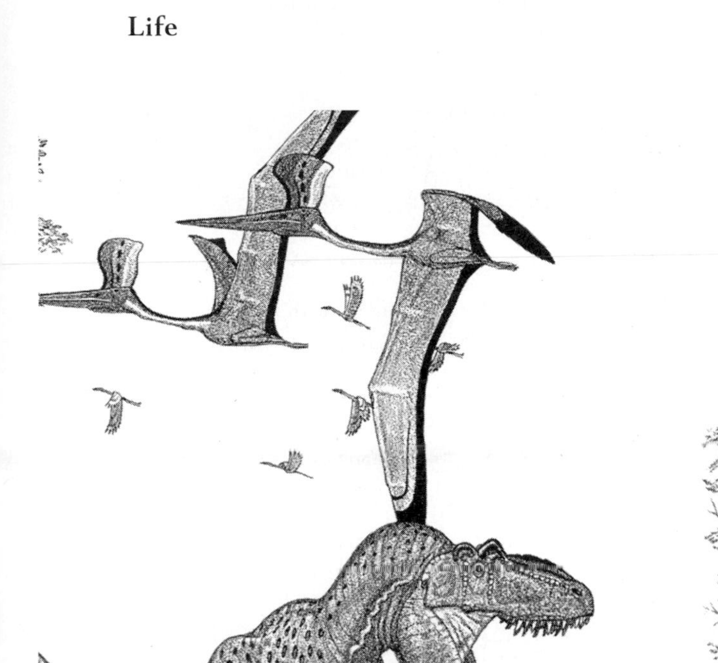

were fearsome predators, including the largest carnivorous land animals of all time, such as *Tyrannosaurus*. Here, too, were terrestrial crocodiles (Figure 17-10) that grew to the remarkable length of 15 meters (45 to 50 feet).

The skies above the plains where the dinosaurs roamed were populated– perhaps sparsely– by the two groups of flying vertebrates that had evolved earlier in the Mesozoic: birds and flying reptiles (see Figures 16-17 and 16-18). Most Cretaceous birds were large wading birds and shorebirds that lived like modern herons and cranes; there were no songbirds of the kind that surround us today. Flying reptiles were among the most spectacular of all Cretaceous animals. While they may have relied heavily on passive transport, soaring on the wind, it appears that many and perhaps all species at times flapped their wings in flight. The largest known species, represented by fossils from the uppermost Cretaceous of Texas,

Figure 17-10 **Skull of a huge terrestrial crocodile, *Phobosuchus,* which probably fed on Late Cretaceous dinosaurs of small and medium size.** The length of the head equaled the height of a large man.

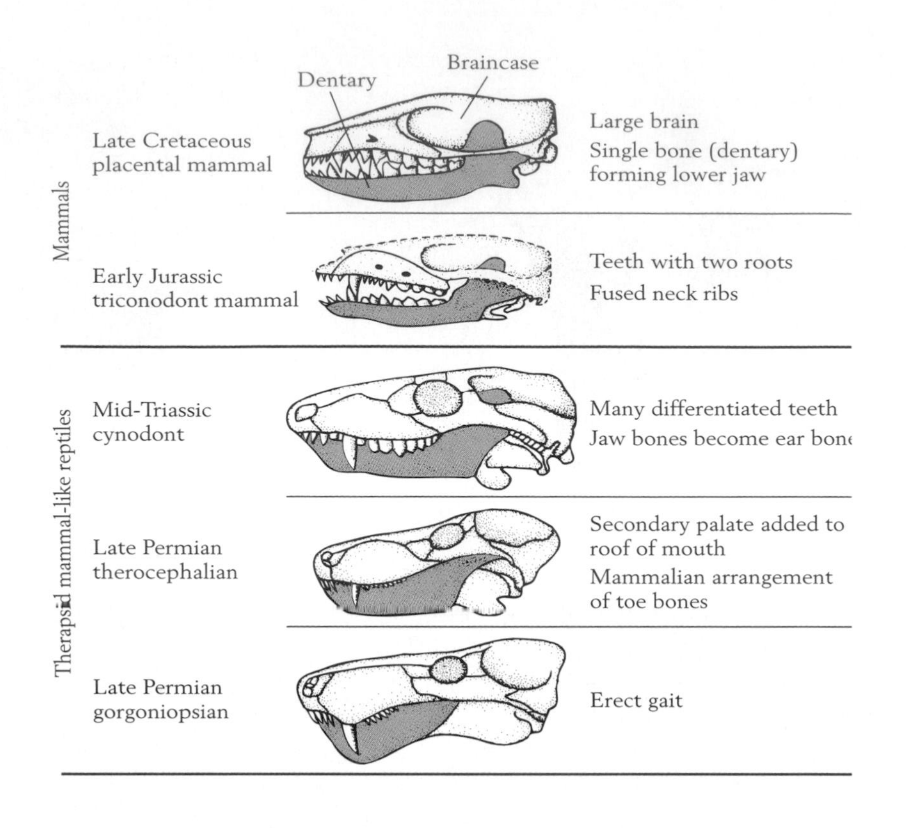

Figure 17-11 Stages in the evolution of mammals from mammal-like reptiles. Among the most important changes were the evolution of more highly differentiated, specialized teeth; enlargement of the brain; modification of jawbones into ear bones; and reduction of the number of bones forming the jaw to one, the dentary (shown in color). *(Adapted from R. E. Sloan.)*

is estimated to have had a wingspan of 11 meters (~35 feet) or more. Members of this species, like modern vultures, may have soared through the sky in search of carrion in the form of dinosaur carcasses.

Throughout the Mesozoic Era, mammals evolved new traits, some of which are shown in Figure 17-11. Nonetheless, as described in Box 17-1, mammals remained quite small until the end of the Cretaceous; only after the dinosaurs disappeared did the mammals diversify markedly to develop the wide variety of adaptations that are so familiar to us today.

The Paleogeography of the Cretaceous World

Because the Cretaceous System has undergone less metamorphism and erosion than older geologic systems, it is represented on modern continents by an extensive record of shallow marine and nonmarine sediments and fossils. Cretaceous sediments and fossils are also widespread in the deep sea, in contrast to the sparse deep-sea records for the Triassic and Jurassic periods. This difference reflects the fact that movements of plates across Earth's surface are rapid enough so that by now a large percentage of deep-sea sediments older than the Cretaceous System have been swallowed up along subduction zones. The relative abundance of Cretaceous sediments in the ocean basins and on the continents helps us to interpret paleogeographic patterns and global events of the period. Additional information is drawn from the Upper Cretaceous fossil record of flowering plants. As we saw in Chapter 4, these organisms are particularly sensitive to climatic conditions.

New Continents and Oceans

Although Pangea had begun to break apart early in the Mesozoic Era, the smaller continents that had formed from the supercontinent remained tightly clustered at the beginning of the Cretaceous Period. The continued fragmentation of Pangea and the dispersion of its daughter continents were among the most important developments in Cretaceous global geography. At the start of the Cretaceous Period, Gondwanaland was still intact. By Late Cretaceous time, however, South America, Africa, and peninsular India had all become discrete entities; of the present-day continents that represent fragments of Gondwanaland, only Antarctica and Australia remained attached to each other (Figure 17-12).

For the Record 17-1

The Meek Did Inherit the Earth

Workers who opened a quarry at Stonesfield, England, in 1822 soon discovered a one-ton dinosaur and a tiny mammal in Jurassic strata. These early fossil discoveries told the most important story about Mesozoic terrestrial vertebrates that could not swim or fly: huge dinosaurs ruled Earth, and no mammal grew larger than a modern domestic cat. The great success of the mammals would come only after the dinosaurs had vanished.

Purgatorius was about the size of a rat.

The oldest known mammals are of Late Triassic age. The bones that formed the joint between the jaw and the skull establish their identity. Different bones form this joint in mammal-like reptiles. With bodies only about 15 centimeters (6 inches) long and pointed snouts, the earliest mammals resembled modern shrews. Their fossil remains reveal a remarkable amount about their mode of life. Their pointed, cutting teeth show that they were carnivorous, and their small size would have restricted them to a diet made up largely of insects. Their mouth structure indicates that they were endothermic: they had a secondary palate, the bony structure that in all mammals separates the nasal air passages from the mouth so that they can breathe while they eat. Endothermy entails such a high metabolic rate that breathing cannot be interrupted for long. Reptiles, in contrast, can suspend breathing temporarily during their meals. Fossil skulls reveal that the brains of the earliest mammals were large for their overall size, and that large regions of the brain were associated with hearing and smell. The fact that these senses are particularly useful after dark suggests that these small creatures were nocturnal, avoiding the much larger dinosaurs, which presumably were active in daylight. Early mammals appear also to have suckled their young, as modern mammals do. Here the pattern of tooth development is the evidence. Lower vertebrates have functional teeth early in life, and many species replace worn or lost teeth more than once. Mammals, in contrast, do not have functional teeth for quite some time, because early in life their only food is their mother's milk. Because mammals do not need teeth until long after birth, they generally have only two sets of teeth, the baby teeth that appear in infancy and the adult teeth that replace them. Finally, the earliest mammals had rear feet that were adapted for grasping; this characteristic points to a life of tree climbing.

Although mammals remained small and relatively inconspicuous until the end of the Cretaceous, they did diversify to a degree. The first herbivorous forms, which evolved during the Jurassic Period, had gnawing teeth like those of modern rodents. By Late Cretaceous time, the two large modern groups of mammals were present. These were the placental forms, which include most modern species, and the marsupials, which carry their young in pouches and are the dominant group in Australia today (p. 85).

Among the small Cretaceous mammals was the genus *Purgatorius*. This animal would have had no special interest for a human observer during its lifetime, but from our modern perspective it had great significance. *Purgatorius* belonged to a group of animals that was ancestral to modern primates, including humans, and some workers even assign this group to the primate order. When a mass extinction at the end of the Cretaceous Period swept away the dinosaurs, many small mammals were fortunate enough to survive. Among them was *Purgatorius*, whose fossil remains are found in very early Cenozoic deposits. With the oppressive dinosaurs gone, the surviving mammals proliferated in a spectacular adaptive radiation, which produced the great diversity of mammals in the modern world. Had the group of ratlike animals to which *Purgatorius* belonged died out with the dinosaurs, we humans would never have evolved.

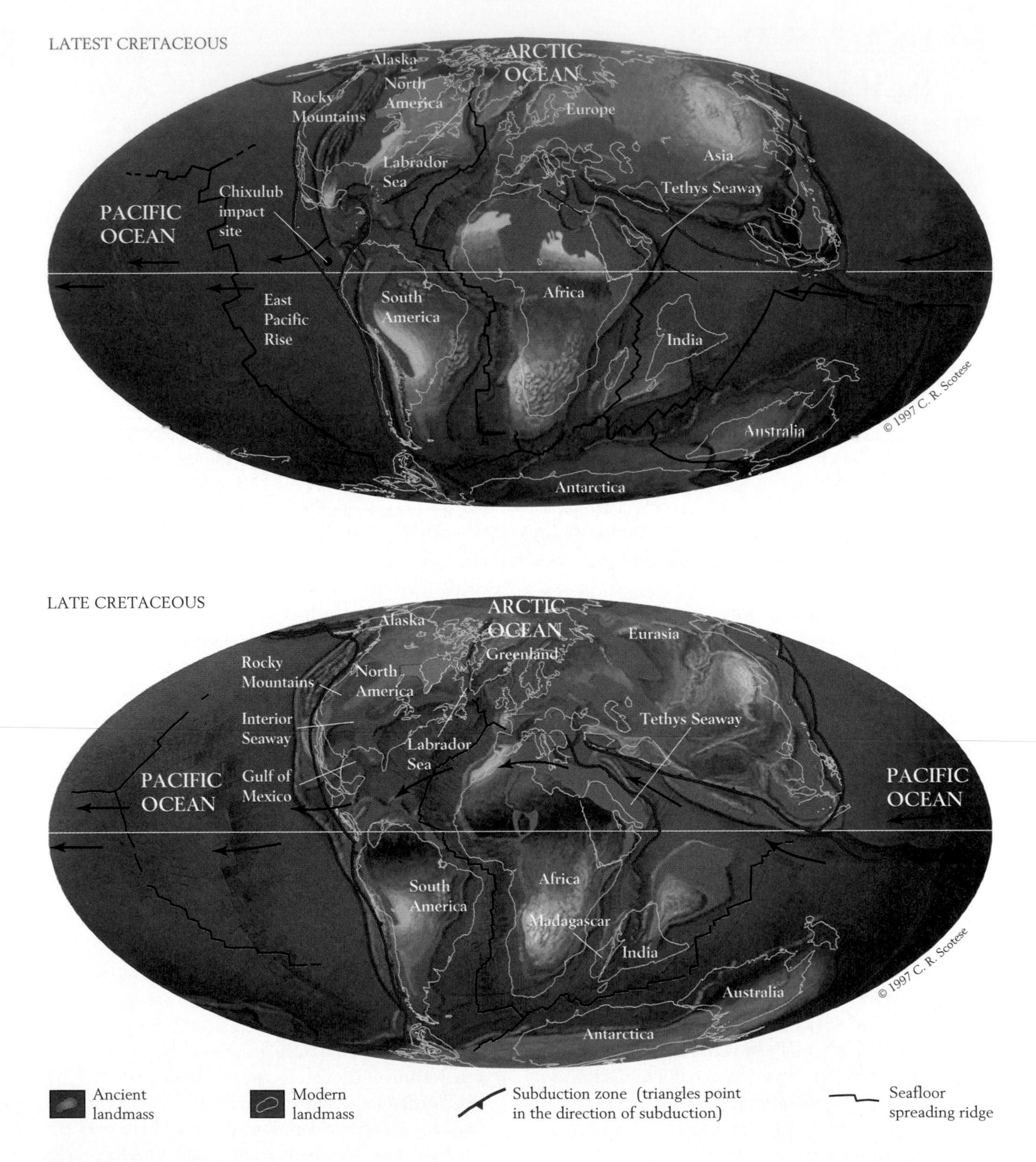

Figure 17-12 Changes in global geography during Late Cretaceous time. Fragments of Pangea moved farther apart. Greenland rifted away from North America, forming the Labrador Sea. Early in Late Cretaceous time, the Interior Seaway of North America extended from the Gulf of Mexico to the Arctic Ocean, and a strong current flowed westward through the Tethys Seaway. In latest Cretaceous time, the Interior Seaway of North America retreated from the Arctic Ocean. Northward movement of Africa and lowering of sea level constricted the flow of water through the Tethys. *(Adapted from paleogeographic maps by C. R. Scotese, PALEOMAP Project, University of Texas at Arlington, 1997.)*

The fragmentation and separation of continents during Cretaceous time caused new oceans to form and narrow oceans to widen. Greenland finally broke away from North America, but remained attached to Scandinavia, with Great Britain wedged in between them (see Figure 17-12). A small, triangular piece of lithosphere that had been part of Gondwanaland remained attached to North America. It forms a portion of the southeastern United States, including Florida (see Figures 15-17 and 17-12).

Also notable were the Early Cretaceous openings of the South Atlantic, the Gulf of Mexico, and the Caribbean Sea. As we have seen, evaporites had formed during the Jurassic Period, when marine waters spilled into the rifts that later widened to form the Gulf of Mexico and the South Atlantic (see Figure 16-20). Early in the Cretaceous Period, these basins remained narrowly connected to the rest of the world's oceans. The evaporites accumulated along the basin margins in the restricted bodies of water, especially in the trade wind belt (Figure 17-13).

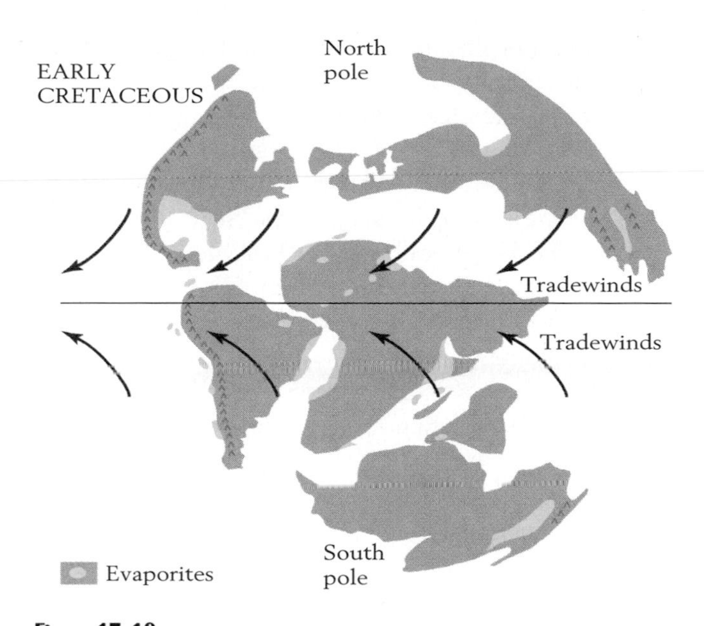

Figure 17-13 **Sites of Early Cretaceous evaporite deposition.** Most evaporites accumulated along margins of restricted shallow seas recently formed by the breakup of Pangea. Evaporites are best developed in the trade wind belt.

Sea Level, Ocean Circulation, and Climates

Throughout Early Cretaceous time sea level rose throughout the world, with only minor interruptions (see p. 470). As a result, sea level stood perhaps as high throughout mid-Cretaceous time as at any other time in the Phanerozoic, and extensive marine deposits blanketed most continents. On the North American craton these deposits constitute the Zuni sequence (see Figure 6-22). The rise in sea level apparently resulted from exceptionally active seafloor spreading: swelling along mid-ocean ridges pushed oceans upward (Figure 17-14), producing a low

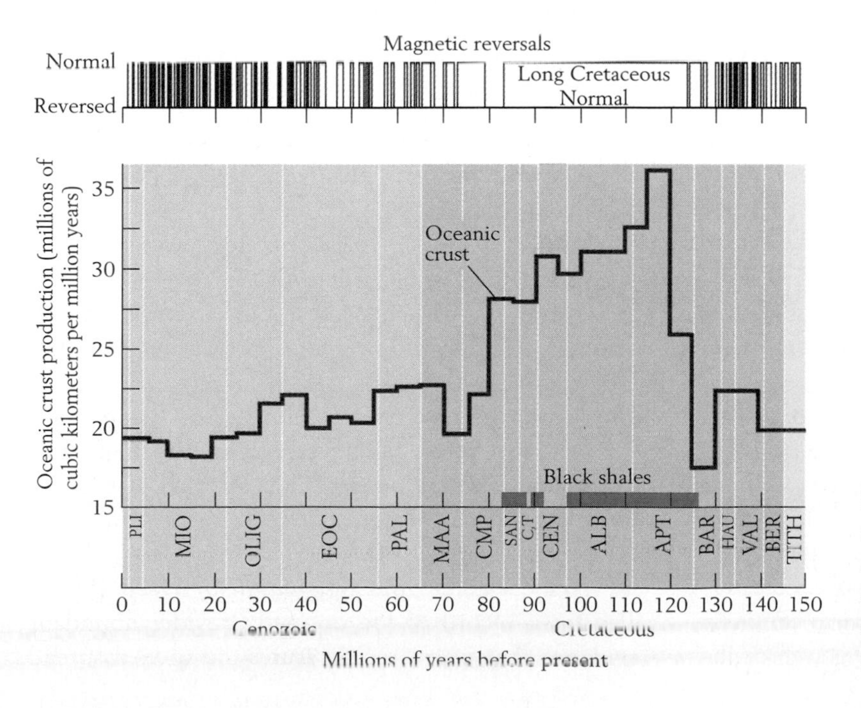

Figure 17-14 **Correspondence in mid-Cretaceous time of a high rate of oceanic crust production with accumulation of abundant black shales and absence of magnetic reversals.** *(After R. L. Larson, Geology 19:963–986, 1991.)*

magnesium-calcium ratio in seawater (p. 468). In particular, broad areas of seafloor originated in the eastern Pacific Ocean during the interval from about 125 million to 80 million years ago. This interval coincided with a remarkably long interval when Earth's magnetic field failed to reverse its polarity (see Figure 17-14). Recall that the magnetic field originates through movements in Earth's core (p. 158). It has therefore been suggested that movement of a large plume of magma upward from near the core-mantle boundary may have altered movements within the core while simultaneously building extensive oceanic crust.

The Tethys A dominant feature of the Cretaceous world was the tropical Tethys Seaway, whose waters trade winds drove westward without obstruction by large landmasses. These waters warmed while flowing across the broad equatorial Pacific. The channel between Eurasia and Africa deflected the Tethys so that it carried heat from the Pacific to a latitude about 40° north of the equator, across continental crust that now forms southern Europe (see Figure 17-12). Tropical rudist reefs flourished throughout the Tethys; oxygen isotope ratios for Tethyan rudists from Greece and Turkey are displayed in Figure 10-15.

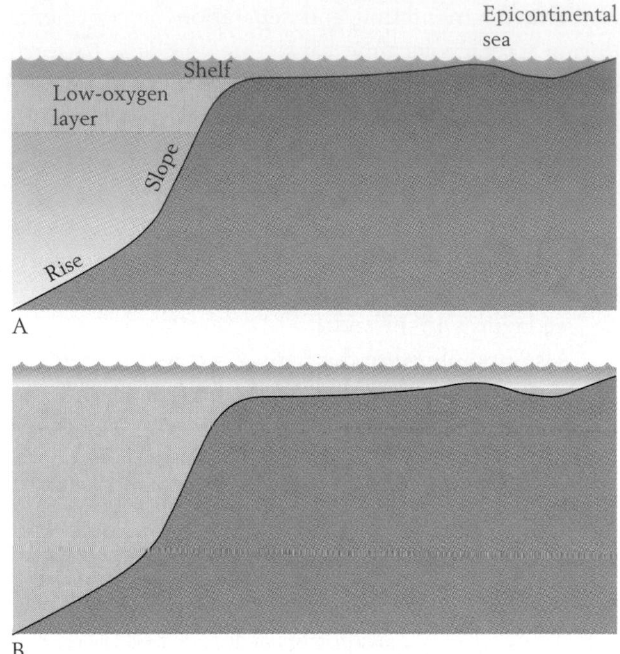

Figure 17-15 Expansions of the low-oxygen layer in the ocean at a time when the deep sea is warm. At times like the present, when cold water at the poles is sinking to the deep sea, it supplies deep waters with oxygen, so that the low-oxygen layer is relatively thin *(A)*. When polar waters are warmer and thus are of low density, they do not sink, so they fail to supply oxygen to the deep sea. Under these conditions, the warm waters below the depth of wave activity are relatively stagnant and the low-oxygen layer thickens, extending even into some epicontinental seas *(B)*. *(Modified from A. G. Fischer and M. A. Arthur, Soc. Econ. Paleont. and Mineral. Spec. Publ. No. 25:19–50, 1977.)*

Widespread Black Shales The middle part of the Cretaceous Period was marked by intervals when marine muds rich in organic matter accumulated on continents (see Figure 10-7). Some of the abundant organic matter within the black shales has been transformed into large volumes of petroleum. The dark muds accumulated because of unusually poor circulation within the ocean and stagnation of much of the water column. As Figure 17-15 indicates, these waters must at times have spilled over from oceanic areas into shallow seas, leading to the epicontinental deposition of black muds.

Stagnation of mid-Cretaceous seas contrasts to the strong mixing of the oceans today. In modern seas, cold waters in polar regions sink to the deep sea and spread along the seafloor toward the equator, carrying with them oxygen from the atmosphere (p. 110). Only a thin zone of the ocean is characterized by a low concentration of oxygen (see Figure 17-15*A*). The light color of the sediments on the present seafloor reflects the presence of oxygen. In mid-Cretaceous time polar seas were apparently too warm for their surface waters to descend and spread oxygen throughout the deep waters of the ocean. As a result, the low-oxygen zone was greatly expanded (see Figure 17-15*B*).

The Tethys may have supplied the waters of the deep sea during mid-Cretaceous time. At every time in Earth's history the densest waters of the ocean sink to form the water mass of the deep sea. During the mid-Cretaceous interval, when polar regions were relatively warm, the densest shallow waters were probably bodies of hypersaline water formed by high rates of evaporation along continental margins in the dry trade wind belt.

Polar Warmth and Stagnation of Deep Waters The fossil record of life on land provides abundant evidence that warm temperatures did indeed spread to high latitudes during mid-Cretaceous time. Fossil leaves of warm-adapted plant species occur in Cretaceous deposits of northern Alaska and Greenland (Figure 17-16). Dinosaurs also lived within about 15° of the Cretaceous south pole. Their fossil remains have been discovered near Melbourne, Australia, along with fossil plants that suggest an average annual temperature of about 10°C (50°F). These dinosaurs had to endure several weeks of darkness in winter, but the warm climate must have provided them with food throughout the year.

Some scientists favor the idea that hypersaline waters warmed polar regions during the Cretaceous. While strongly hypersaline waters descended to the deep sea, less hypersaline waters may have descended to intermediate depths at low latitudes and flowed poleward. Upward mixing of these intermediate waters may have delivered much of the heat that warmed climates at high latitudes.

The presence of warm climates at high latitudes had important consequences for wind-driven circulation in the oceans. At times like the present, when steep temperature gradients extend from the equator to the poles, large regional temperature contrasts produce strong winds. (Cold, dense air masses push under warm air masses, for example.) Upwelling caused by strong winds stirs the ocean (p. 114). During mid-Cretaceous time, when temperature gradients were gentle, winds blowing over the ocean would have been weaker, on average, than they are today. Upwelling would therefore have been weaker as well. Thus the absence of strong wind-driven circulation contributed to the general stagnation of the lower portion of the mid-Cretaceous ocean.

The Global Climate The mid-Cretaceous world has sometimes been described as a greenhouse world, in the sense that the average surface temperature was high as a result of a high concentration of atmospheric carbon dioxide. In fact, there is no evidence that Earth was warmer overall than during Triassic or Jurassic time. Furthermore, some calculations indicate that the atmospheric carbon dioxide was on the decline during the Cretaceous (see Figure 10-13). High rates of plate tectonic activity during the Cretaceous may have resulted in the emission of substantial quantities of carbon dioxide along subduction zones (see Figure 10-11*A*). Other conditions may have offset this flux of carbon dioxide to the atmosphere, however. The accumulation of widespread black muds

A B

Figure 17-16 A fossil leaf of a warm-adapted plant from the Cretaceous System of Greenland *(A)* that resembles a modern breadfruit leaf *(B)*.

extracted large amounts of carbon from the carbon cycle (see Figure 10-5). In addition, continental relief was relatively low during the Cretaceous, so that weathering extracted carbon dioxide from the atmosphere at a relatively low rate (p. 270).

Circulation Changes and Extinction in the Ocean Carbon and oxygen isotope ratios for both planktonic foraminifera and foraminifera that lived on the floor of the deep sea shifted significantly between about 71 million and 70 million years ago, during the final Cretaceous stage, named the Maastrichtian. These isotopic shifts, measured for specimens extracted from deep-sea cores, signal a general change in ocean circulation. A worldwide drop in sea level about this time may have caused this oceanographic change by eliminating shallow seas where saline waters had formed and by causing constriction of the Tethys Seaway between Africa and Eurasia. Additional constriction came from Africa's northward movement toward Europe. In any event, a shift in the composition of terrestrial floras at high latitudes points to polar cooling about this time.

Cooling of tropical seas may have been responsible for an extinction event that decimated the rudist reef fauna early in Maastrichtian time. As a result, the rudists were already an impoverished group when the mass extinction that also killed off the dinosaurs brought about their total demise. This catastrophe brought the Mesozoic Era to an end.

The Terminal Cretaceous Extinction

The Mesozoic Era came to a dramatic end with a mass extinction that was quite sudden on a geologic scale of time. Many forms of life that had played major ecologic roles for tens of millions of years disappeared. The most prominent in the minds of modern humans were the dinosaurs, but many other groups of animals and plants died out as well. Both the gymnosperms and the angiosperms suffered heavy losses. In the ocean, ammonoids disappeared, as did reptilian "sea monsters," including mosasaurs, plesiosaurs, and giant turtles (see Figure 17-4). About 90 percent of all species of calcareous nannoplankton and planktonic foraminifera died out, and on the seafloor many groups of mollusks disappeared, including the remaining groups of reef-building rudists. The extinction of the rudists, like earlier extinctions of the Phanerozoic Eon, exemplified the fragility of reef ecosystems in general (see Box 4-1).

Discoveries during the past few years have established beyond a reasonable doubt that the impact of an asteroid with Earth caused the terminal Cretaceous mass extinction.

Evidence of an Asteroid Impact

In 1981 a team of workers led by the physicist Luis Alvarez and his son Walter, a geologist, discovered an abnormally high concentration of the element iridium precisely at the level of the Cretaceous-Paleogene boundary in the stratigraphic section at Gubbio, Italy. Soon a comparable "iridium anomaly" was found at the same stratigraphic level at many other places in the world, in rocks of both marine and terrestrial origins (Figure 17-17). Since iridium is very rare on Earth but fairly abundant in meteorites, the Alvarez team advanced the hypothesis that a large meteorite struck Earth at the end of the Cretaceous, producing a great explosion that dispersed dust relatively rich in iridium high into the atmosphere. The dust from such an explosion would have spread around the globe and then settled to produce a layer of iridium-rich sediment in nearly all depositional environments. On a geologic scale of time, these events would have been instantaneous. A meteorite of average composition would have had to be about 10 kilometers (6 miles) in diameter to produce the total amount of iridium that forms the worldwide anomaly.

Three additional kinds of evidence favor the idea that the source for the widespread iridium anomaly was extraterrestrial. Each is a type of grain that occurs at the terminal Cretaceous boundary and can form only under very intense heat or pressure of the kind that develops when a large extraterrestrial body collides with Earth:

1 One kind of grain displays groups of parallel welded fractures that formed under enormous pressure (Figure 17-18). Grains of this kind occur at sites where meteorites are known to have formed craters on earth. These "shocked" grains have turned up in many parts of the world at the level of the iridium anomaly.

Figure 17-17 Location of the terminal Cretaceous iridium anomaly in terrestrial deposits near Drumheller, Alberta, Canada. The anomaly is located in the thin clay bed at the lower end of the white pen that is perched on the outcrop. The last dinosaur bones occur just below the iridium anomaly.

2 A second kind of grain is a "microspherule," a nearly spherical grain that resembles window glass in its molecular structure (Figure 17-19). In other words, it is largely uncrystallized: it cooled so rapidly after having been liquefied that its chemical elements failed to assemble into a consistent geometric pattern. Microspherules, like shocked grains, occur where meteorites are known to have struck Earth. They formed when droplets of rock, liquefied by the enormous heat generated during the impact, were thrown into the atmosphere, where they cooled very quickly.

3 Grains of the third type are microscopic diamonds, which have been discovered at the level of the iridium anomaly at several sites in North America. Diamonds form only at extremely high pressures (p. 37).

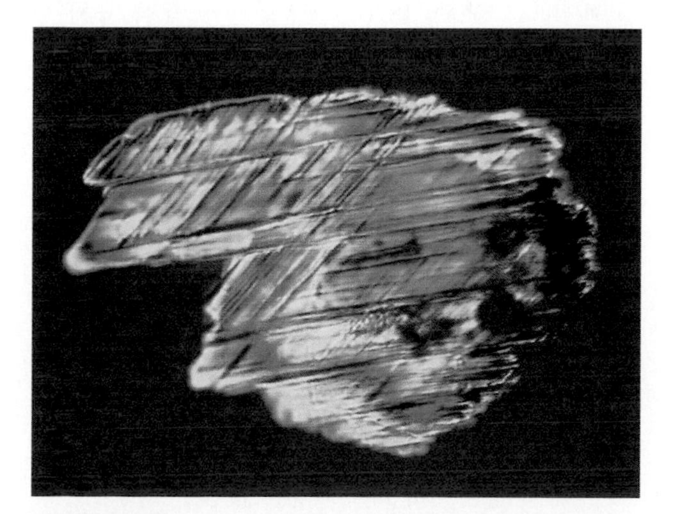

Figure 17-18 Shocked quartz grain from uppermost Cretaceous deposits in Montana, showing sets of planar lamellae.

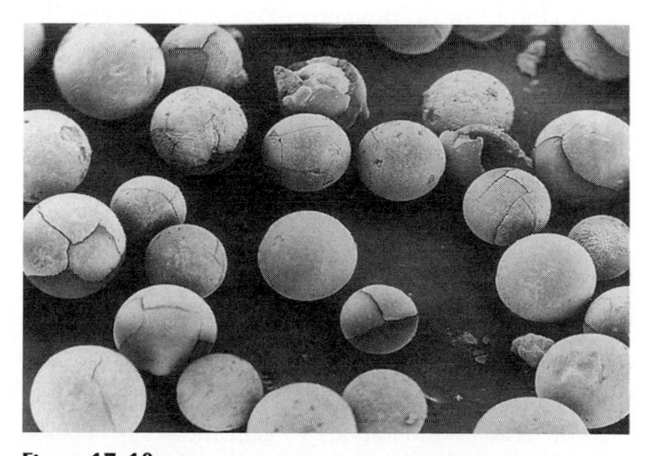

Figure 17-19 Microspherules from a thin clay layer at the terminal Cretaceous boundary in Wyoming. These grains, which have undergone chemical alteration, apparently once were glassy structures that cooled rapidly from droplets of molten rock.

Completing one of the greatest triumphs of modern geology, researchers have actually succeeded in locating the crater that the deadly asteroid created on impact. This so-called Chicxulub crater is a ringlike structure, about 200 kilometers (120 miles) in diameter, in the Gulf of Mexico just offshore from the Yucatan Peninsula of Mexico (Figure 17-20; see also Figure 17-12). The impact caused an explosion that opened a cavity about 100 kilometers (60 miles) in diameter in the center of the ringlike structure. To produce such a large crater, the asteroid had to be about 10 kilometers (6 miles) in diameter. A core obtained by drilling the crater from a ship yielded rocks formed by cooling of magma produced by the heat of impact. Argon-argon radiometric dating of these rocks shows that they formed 65 ± 0.4 million years ago. This time of remelting of the regional rocks is within the narrow range of possible dates established for the upper Cretaceous boundary elsewhere in the world.

The huge Chicxulub crater represents one of the largest impact structures produced by an extraterrestrial body in the last 4 billion years– since the interval in Earth's history when large impacts occurred frequently (see Box 11-1). Given the unusually large size of the Cretaceous impactor, it is no surprise that its effects were so devastating.

The Chicxulub impact structure is asymmetrical, indicating that the asteroid arrived from the southeast and struck at a low angle– at 20° to 30° from Earth's surface. This trajectory should have driven a fiery vapor cloud toward the northwest, creating a corridor of incineration across west-central North America. Two patterns of extinction support this scenario:

First, terrestrial floras of western North America suffered especially heavy extinction; about 75 percent of all species died out. In contrast, plants in Australia and New Zealand were virtually unscathed. Second, the abundance of microspheres at the top of the Cretaceous System decreases away from the Chicxulub site. At localities in Mexico relatively close to the impact structure, the microspherules are concentrated in a layer about 1 meter (3 feet) thick; the equivalent layer in Texas is only about 10 centimeters (4 inches) thick; and still farther away, in New Jersey, the layer is a mere 5 centimeters (2 inches) thick. Much smaller concentrations of microspherules are found in regions of the world even more distant from the Chicxulub site.

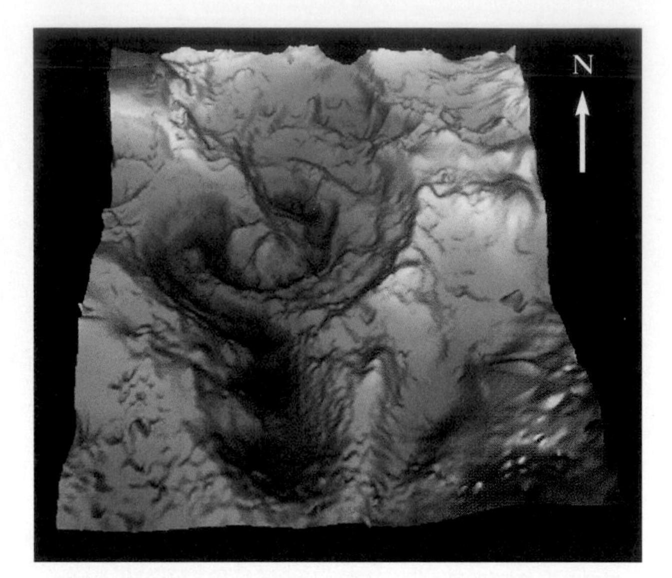

Figure 17-20 The Chicxulub crater, as portrayed by variations in the strength of Earth's gravitational field. The crater's outer ring has a diameter of about 200 kilometers. The crater flares out to the northwest, indicating that the impactor arrived from the southeast with a low-angle trajectory.

Consequences of the Impact

Opinions have varied as to how the impact of an asteroid 10 kilometers in diameter would disturb environments on Earth. A landing in the oceanic realm would have different consequences than one on land or in a shallow sea. Even a bolide that made an oceanic landing would plunge through the water column and penetrate the lithosphere. Here are some consequences that many scientists have predicted for the Chicxulub impact:

1 *Perpetual night.* Dust particles would have blown high into the atmosphere, spread around the world, and screened out nearly all sunlight. Much of the dust would have remained aloft for several months, preventing plants from conducting photosynthesis.

2 *Months of global refrigeration.* The atmospheric dust– and tiny particles of liquid called *aerosols* blown into the atmosphere with it– would have darkened Earth, plunging the entire planet into cold, wintery weather for several months.

3 *Delayed greenhouse warming.* The aerosols would have remained in the atmosphere long after the

dust had settled, and by trapping solar radiation they would have intensified the greenhouse effect (p. 257). In other words, after experiencing a severe cold snap, life might have been subjected to abnormal warmth.

4 *Acid rain.* The asteroid penetrated limestones and sulfate evaporites in forming the Chicxulub crater. Its impact must have released oxides of sulfur dioxide from the evaporites. Chemical reaction of these compounds with water in the atmosphere would have produced sulfurous and sulfuric acid. The result would have been acid rain that might have harmed many forms of life.

5 *Fires.* Wildfires should have raged across the land, triggered by the fiery cloud that burst from the impact site. The trajectory of the impact toward the northwest should have concentrated the most severe fire damage in North America, where, as we have seen, the heaviest extinction of plants actually occurred.

Although the most extreme devastation was to the northwest of the impact site, lethal consequences encompassed the entire globe. Creatures as different as marine ammonoids and terrestrial dinosaurs vanished from both hemispheres. Interestingly, freshwater life in west-central North America—fishes, amphibians, turtles, and crocodiles—suffered relatively little extinction at the end of the Cretaceous. These animals presumably benefited from being surrounded by water, which because of its high heat capacity would have undergone less heating than the air above. In the oceans, virtually all dinoflagellate species survived the terminal Cretaceous crisis. Presumably their ability to form protective cysts under unfavorable circumstances allowed them to survive (p. 71).

Lessons for Humanity

In contemplating the consequences of a large bolide impact, scientists have gained new insights into global disturbances in general. For example, the idea that atmospheric dust would refrigerate the world after an impact led to the conclusion that a large-scale nuclear war would produce a nuclear winter. The dust and soot from explosions and fires caused by nuclear explosions would darken and cool the planet, causing massive death of humans and other forms of life far from areas of nuclear attack.

Fossils and the Timing of Extinction

When it was first suggested that an extraterrestrial impact caused the terminal Cretaceous extinction, some aspects of the fossil record seemed to conflict with this idea. Several fossil groups appeared to have died out over several million years. Few species of dinosaurs, for example, were found in the uppermost few meters of Cretaceous sediment in Montana and nearby regions of Canada, where the fossil record of latest Cretaceous dinosaurs is the best in the world. Thus it appeared that the dinosaurs died out gradually near the end of the Cretaceous Period. The impact hypothesis, however, stimulated paleontologists to scour the dinosaurs' uppermost fossil record more thoroughly than they had done before. The result was the discovery that many species of dinosaurs once thought to have died out long before the end of the Cretaceous may well have survived to the very end: their fossils have now been found within a meter of the iridium anomaly (see Figure 17-17). The lesson here is that the imperfection of the fossil record—or our imperfect knowledge of the record—can fool us.

Environments where the last members of the species lived may not be represented in the sediments that we have available for study. We may also have failed to find some fossils that are actually preserved in these sediments. The new discoveries of species close to the boundary indicate that this second problem previously led to underestimation of the number of extinctions that occurred at the very end of the Cretaceous Period.

The Aftermath

Most of the groups of animals and plants that survived the terminal Cretaceous crisis at reduced diversity expanded again during the Cenozoic Era. For a time, however, life on Earth was impoverished in many ways. Among the survivors, the calcareous nannoplankton exhibit an especially interesting ecologic pattern. Immediately after many species in this group died out or became quite rare at the very end of the Cretaceous, a few species blossomed to great abundance in the ocean. Perhaps these ecologic opportunists (p. 97) were especially tolerant of abnormal conditions. After new species evolved during the Cenozoic Era, the opportunists declined in abundance.

Figure 17-21 **The calcareous nannoplankton species *Braarudisphaera bigelowi* (× 3135 times).** *(From B. U. Haq and A. Boersma [eds.], Introduction to Micropaleontology, Elsevier, New York, 1978.)*

One species, however, survived for more than 150 million years: *Braarudisphaera bigelowi* (Figure 17-21) exists even today but is confined to marginal marine lagoons. The fossil record shows that this form has occasionally spread to the open ocean and undergone population explosions, apparently because unusual conditions have briefly favored it.

A similar pattern of opportunism is evident among plants. A phenomenon called a *fern spike* is present at the level of the anomaly in terrestrial sediments of western North America. Pollen becomes very rare in the sediment close to the anomaly, but then within an interval just a few centimeters thick spores of ferns heavily dominate the assemblage of microfossils that represent terrestrial plants. This pattern suggests that communities of flowering plants died out suddenly and were replaced in the landscape by a heavy growth of ferns. Often when a fire or volcanic eruption wipes out a forest today, ferns readily invade the cleared land. Thus ferns are ecologic opportunists (p. 257). Above the fern spike and iridium anomaly are fossil pollen and leaves of flowering plants that represent a flora quite different from the kind that existed during Late Cretaceous time.

On the land both angiosperms and mammals were beneficiaries of the terminal Cretaceous extinction. Angiosperms quickly rose to dominance over gymnosperms, and in the absence of dinosaurs mammals underwent a spectacular diversification (see Box 17-1).

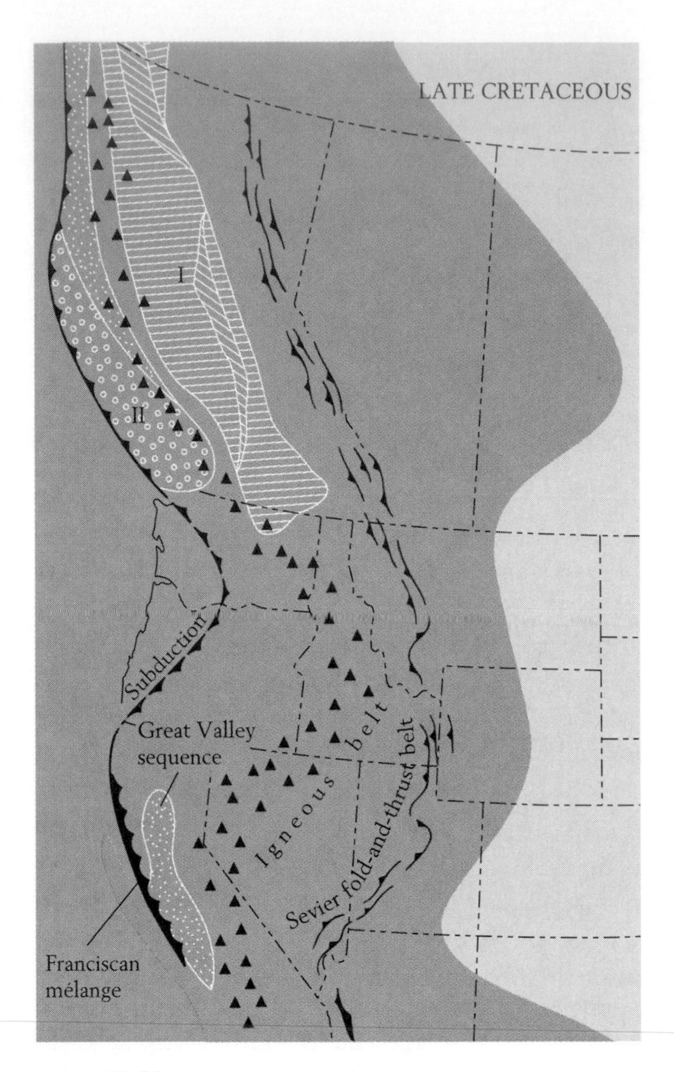

Figure 17-22 **Late Cretaceous geologic features of western North America.** Subduction produced the Franciscan mélange in California. North of California, igneous activity resulting from subduction was located far to the east of the continental margin; this activity, together with the folding and thrusting to the east, represented the latter part of the Sevier orogeny. In Canada, the margin of the continent consisted of two blocks of exotic terrane (I and II) that had been sutured to North America earlier in the Mesozoic Era; each of these blocks consisted of two or more slivers of crust that were welded together to form the block before it was attached to North America.

North America in the Cretaceous World

Mountain building continued in western North America during the Cretaceous Period, and it produced an enormous foreland basin that became flooded by the seaway that extended from the Gulf Coast to the Arctic Ocean. The Gulf Coast itself was fringed by rudist reefs, and a rudist-rimmed carbonate bank also stretched along a large segment of the adjacent Atlantic coast until midway through the Cretaceous Period, when it gave way to the deposition of mud and sand that continues today.

Cordilleran Mountain Building Continues

During Cretaceous time an important change took place in the pattern of igneous activity in western North America. Subduction of the Franciscan complex along the western margin of the continent continued, as did the associated igneous activity (Figure 17-22; see also Figure 16-31). By Late Cretaceous time, however, although volcanic and plutonic activity persisted in the Sierra Nevada region, the northern igneous activity had shifted eastward to Nevada and Idaho (Figure 17-23). This pattern contrasted with that of the Late Jurassic Epoch, when igneous activity in the north had been centered near the coast, in northern California and Oregon (see Figure 16-32). The eastward migration of igneous activity in the northern United States must have resulted from a decrease in the angle of subduction. This change would have resulted from an increased rate of westward movement of the North American plate, which caused faster rollback of the subducted plate (see Figure 9-16). The subducted crust therefore failed to sink deep enough to melt until it had extended far inland (see Figure 17-23). The fold-and-thrust belt in front of the mountainous igneous regions also shifted inland in the northern United States. By Late Cretaceous time, folding and thrusting extended eastward as far as the Idaho-Wyoming border.

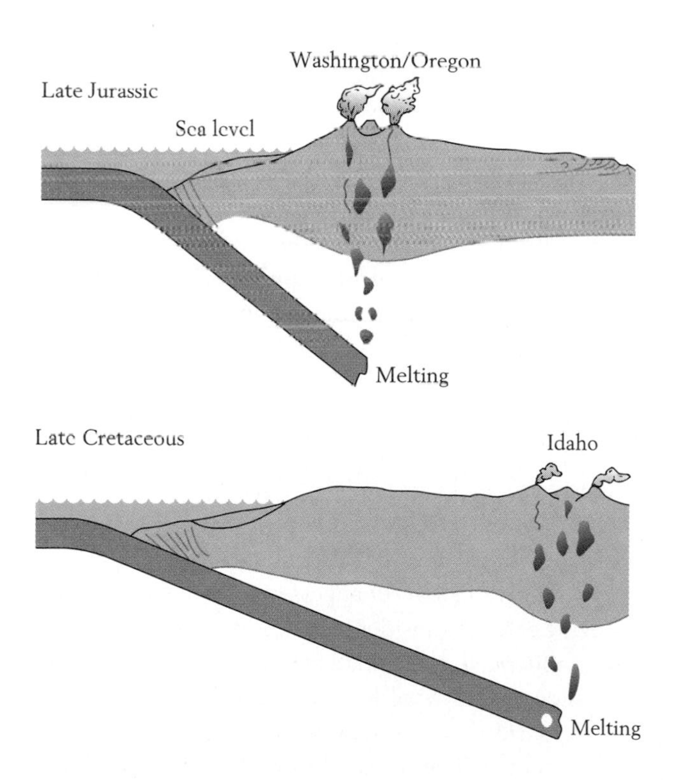

Figure 17-23 A likely explanation for the eastward migration of igneous activity in the Cordillera during Cretaceous time. As suggested here, the subducted plate began to pass downward at a reduced angle, so that it reached the depth of melting only after passing far to the east.

A major episode of igneous activity and eastward folding and thrusting coincided approximately with the Cretaceous Period; although this episode was not entirely divorced from earlier and later tectonic activity, it is separately identified as the *Sevier orogeny.* East of this belt lay a vast foreland basin, which in Late Cretaceous time was occupied by a narrow seaway stretching from the Gulf of Mexico to the Arctic Ocean.

The orogenic belt that occupied western North America during the latter half of Cretaceous time was unusually broad, apparently because of low-angle subduction. In its development of a foreland basin and certain other features, however, the orogenic belt was typical. Like the modern Andes, for example, it was symmetrical (see Figure 9-15): the Sevier folding and faulting cast of the belt of igneous activity mirrored on a larger scale the Franciscan deformation at the continental margin (see Figure 17-22).

The Mesozoic history of western Canada is far more complicated. Recall that during the Jurassic Period, a sizable microcontinent was sutured to this region of North America (see Figure 16-28). This theme of continental accretion continued into the Cretaceous Period, when a small microcontinent was attached along the western margin of the first. This new landmass, like the one accreted during the Jurassic, was a composite of two or more terranes. They had become amalgamated during the Jurassic Period and were attached to North America during Cretaceous time.

The Gulf Coast and the Interior Seaway

Shortly before the end of Early Cretaceous time, during Albian time, Arctic waters spread southward, flooding a large area of western North America with the Mowry Sea (Figure 17-24). This body was named for the Mowry Formation that accumulated within

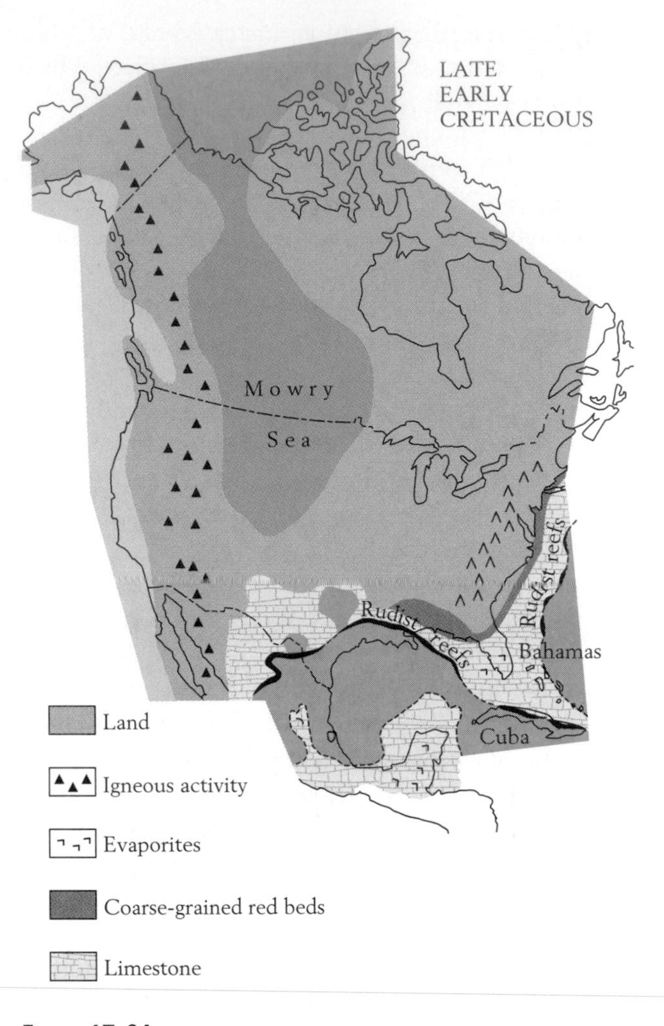

Figure 17-24 Geography of North America late in Early Cretaceous time. The Mowry Sea, where black muds were deposited, spread southward from the Arctic Ocean. A carbonate platform bordered by rudist reefs encircled the Gulf of Mexico, and carbonate deposition extended far to the north along the East Coast. *(After G. D. Williams and C. R. Stelck, Geol. Soc. Canada Spec. Paper No. 13, 1975.)*

it. The Mowry consists mostly of oil shale, a well-laminated shale in which dark layers rich in fish bones and scales alternate with thicker, lighter layers. The Mowry Sea formed as a part of the great mid-Cretaceous marine transgression that resulted in the deposition of black shales on many continents (see Figure 10-7). To the south, the Gulf of Mexico was part of the tropical Tethyan realm, and rudist reefs flourished around its margin.

The Mowry Sea made brief and intermittent contact with the Gulf of Mexico before the end of Early Cretaceous time, but an enduring connection was established at the start of the Late Cretaceous. The result of this contact was the enormous Cretaceous Interior Seaway, which occupied the foreland basin to the east of the Sevier orogenic belt. Until just before the end of the Cretaceous Period, this seaway extended from the Gulf of Mexico to the Arctic Ocean (see Figure 17-12). Most of the sediments deposited here were shed from the Cordilleran mountains that formed to the west. The history of the seaway is especially well understood because of excellent stratigraphic correlations based on abundant fossil ammonoids; in addition, ash falls from volcanic eruptions to the west provided numerous marker horizons, many of which can be dated radiometrically (see Figure 6-14).

Barrier islands bounded much of the seaway. Behind them stood lagoons bordered by broad swamps; on the western margin of the seaway the swamps gave way to alluvial plains, which were succeeded near the mountains by alluvial fans.

The western shoreline of the seaway shifted back and forth, primarily in response to the rate of sediment supply. At all times conglomerate sediments were shed eastward from the neighboring mountains as clastic wedges, but at times of particularly active thrusting or uplift these wedges prograded especially far to the east (Figure 17-25). Because of the great weight of sediments on the western side of the seaway, subsidence was more rapid there than farther east. Along the western margin, in nonmarine environments, Late Cretaceous dinosaurs left a rich fossil record.

The Upper Cretaceous strata of the Interior Seaway represent large depositional cycles, one of which is illustrated in Figure 17-26. Each cycle consists of an interval of transgression followed by an interval of regression. In addition to changing rates of sediment supply, global changes in sea level and the changing rate of subsidence of the seaway floor must have influenced these patterns of transgression and regression. At times of low supply and maximum lateral expansion of the seaway, chalks were laid down in the center. The most famous of these deposits is the Niobrara Chalk, which occupies the middle of a transgressive-regressive cycle. The Niobrara has yielded beautifully preserved fossil vertebrates (see Figures 17-3 and 17-4).

Just before the end of the Cretaceous Period, the seas retreated southward from the Interior Seaway,

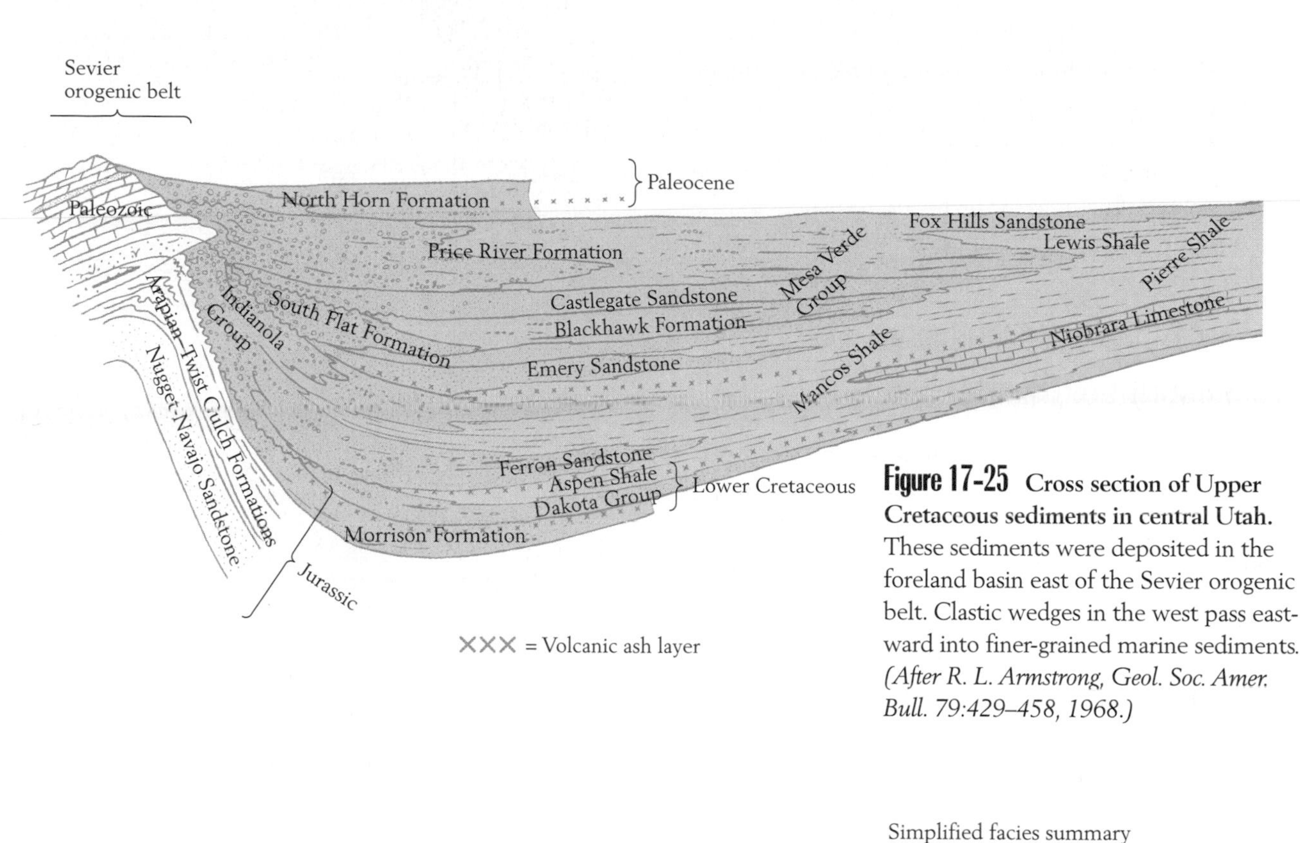

Figure 17-25 Cross section of Upper Cretaceous sediments in central Utah. These sediments were deposited in the foreland basin east of the Sevier orogenic belt. Clastic wedges in the west pass eastward into finer-grained marine sediments. *(After R. L. Armstrong, Geol. Soc. Amer. Bull. 79:429–458, 1968.)*

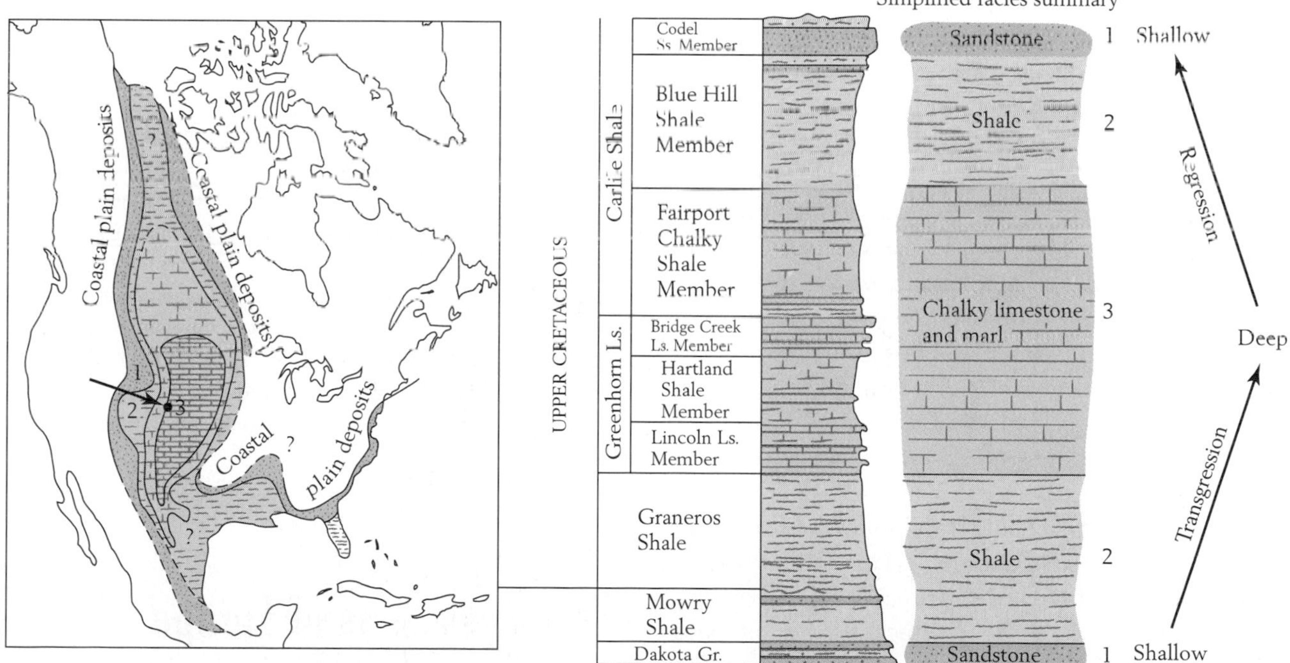

Figure 17-26 The early Late Cretaceous "Greenhorn" depositional cycle of the North American Interior Seaway. The three facies in the simplified facies summary are shown in map view on the left. The stratigraphic section represents the vicinity of eastern Colorado (arrow on map) where the cycle developed by an oscillation of the shoreline (transgression and regression). During the transgression, the area of chalky limestone and marl (limey clay) deposition in the center of the basin (facies 3) expanded; first facies 2 and then facies 3 spread into eastern Colorado. During the regression, the area of deposit of facies 3 contracted, and facies 2 and then facies 1 shifted into eastern Colorado. *(Modified from E. G. Kauffman, Mountain Geologist 6:227–245, 1969.)*

and a new pulse of mountain building began along its western margin. This Laramide orogeny continued well into the Cenozoic Era. Except for a brief and less extensive incursion just after the beginning of the Cenozoic Era, the seas have never returned to the western interior of North America.

The East Coast: Development of the Modern Continental Shelf

Seismic studies reveal a great thickness of sediments beneath the continental shelf in eastern North America (Figure 17-27). These sediments consist of deposits laid down during the early Mesozoic episodes of rifting that formed the modern Atlantic Ocean. At the base are fault basin deposits like those of the Newark Supergroup that are exposed on the continent to the west (see Figure 16-23). Next come large thicknesses of Jurassic carbonates that accumulated in the narrow, young Atlantic Ocean as passive margin deposition commenced (see Figure 17-27). Above the Jurassic carbonates are more carbonates from the Early Cretaceous interval, when, under much warmer climatic conditions than exist today, reef-rimmed carbonate banks bordered the ocean from Florida to New Jersey (see Figure 17-24).

Before the end of Early Cretaceous time, reef growth gave way to deposition of predominantly siliciclastic sediments. This change marked the beginning of the growth of the large clastic wedge that forms the modern continental shelf. The clastic wedge consists largely of sands and muds from the Appalachian region laid down in nonmarine and shallow marine settings. Off the coast of New Jersey (see Figure 17-27) the total thickness of Cretaceous and Cenozoic sediments is approximately 3 kilometers (2 miles). Only to the south, in southern Florida, did carbonate deposition persist to the present. Here about 3 kilometers of sediments, consisting mainly of carbonates, accumulated during the Cretaceous Period, and another 2 kilometers or so were added during the Cenozoic Era.

Sediments of the clastic wedge along the East Coast of North America were apparently supplied by uplands produced by renewed uplift of the Appalachian Mountain belt to the west, after it had been largely leveled during Jurassic time.

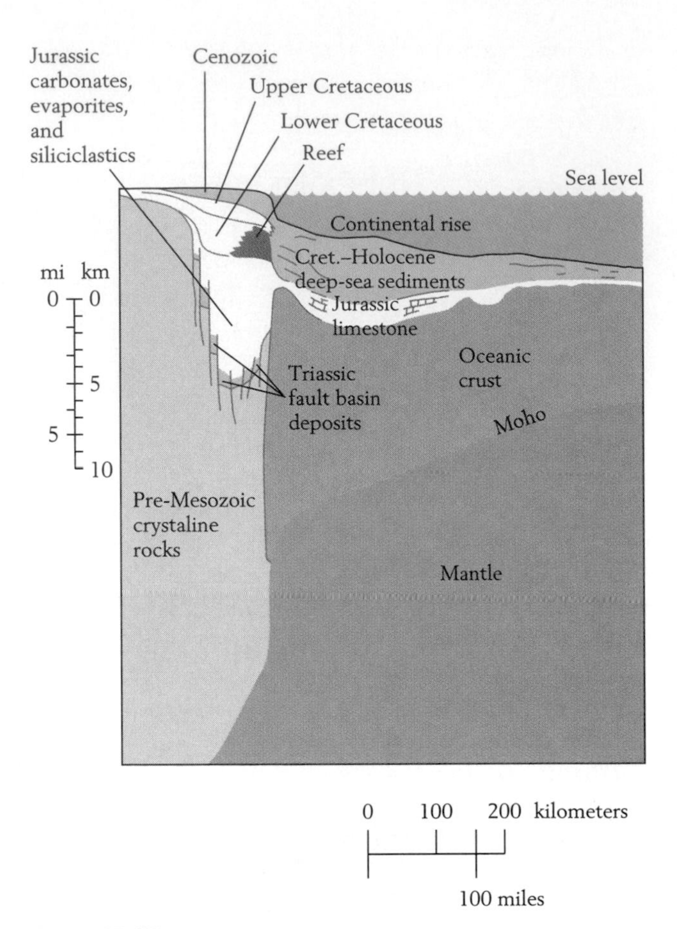

Figure 17-27 Cross section of the continental shelf and deep sea off the coast of New Jersey. Here the opening of the Atlantic is recorded by early Mesozoic sediments deposited within fault-bounded basins. During Early Cretaceous time, a reef-rimmed carbonate platform extended this far north under the influence of Tethyan ocean currents (see Figure 14-35). During Late Cretaceous time, carbonate deposition gave way to siliciclastic deposition, which has predominated to the present day. *(After R. E. Sheridan et al., The Geology of Continental Margins, Springer-Verlag, New York, 1974.)*

The Chalk Seas of Europe

To many geologists the term *chalk* refers specifically to the soft, fine-grained limestones of western Europe, although, as we have seen, chalky rocks are found elsewhere, including the Cretaceous System of North America. The Cretaceous chalks of Europe are spectacularly displayed as the White Cliffs of Dover and coastal cliffs of Denmark (see p. 464 and Box 10-1). They accumulated throughout almost all of Late Cretaceous time, when high seas led to extensive flooding of western Europe.

Except in the newly forming Alps to the south, tectonic activity was largely absent from western Europe during Late Cretaceous time, when the chalk accumulated. Here and there, massifs stood relatively high as islands or, during transgressions, as shallow seafloors (Figure 17-28). Chalk accumulated almost continuously in the basins that surrounded the massifs. The lower portions of the chalk often contain clay, but the remainder is relatively pure calcium carbonate, consisting of minute skeletal debris that is about 75 percent planktonic in origin.

The general presence of oxidizing conditions on the floor of the chalk seas is demonstrated by the widespread occurrence of a fauna of bottom-dwelling species in the chalk—among them bryozoans, ostracods, foraminifera, brachiopods, bivalves, echinoids, and soft-bodied burrowers. Knowing the typical thickness of the European chalk and the total time elapsed during its deposition, we can estimate that it accumulated at a rate perhaps as great as 15 centimeters (6 inches) per 1000 years. Because the very low magnesium-calcium ratio of Late Cretaceous seawater favored the precipitation of calcite, productivity of calcareous nannoplankton was greater in both the chalk seas of Europe and the Interior Seaway of North America than it has ever been anywhere in the ocean since Cretaceous time (see Box 10-1).

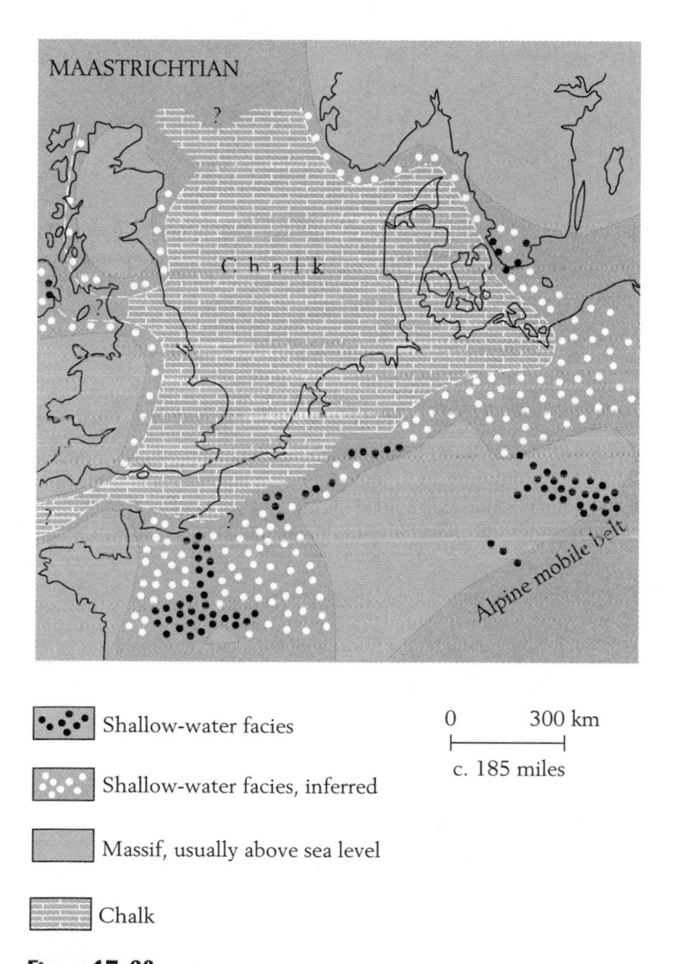

Figure 17-28 Paleogeography of northwestern Europe during Maastrichtian time. Chalk deposition was centered in the North Sea basin. Several stable blocks (massifs) formed islands around which marginal facies developed; these facies consisted mainly of coarse limestones but included siliciclastics as well. *(Modified from E. Hakansson et al., Spec. Publ. Internat. Assoc. Sedimentol. 1:211–233, 1979.)*

Chapter Summary

1 With the evolutionary expansion of the dinoflagellates, diatoms, and calcareous nannoplankton during the Cretaceous Period, the phytoplankton assumed a modern character. Similarly, the diversification of the planktonic foraminifera contributed to the modernization of the zooplankton.

2 Benefiting from a very low magnesium-calcium ratio in seawater, calcareous nannoplankton rained down on the seafloor to produce thick deposits of chalk in western Europe and elsewhere.

3 On the seafloor, predators such as crabs, teleost fishes, and carnivorous snails diversified markedly during Cretaceous time, altering the nature of the marine ecosystem.

4 In mid-Cretaceous time, rudist bivalves temporarily displaced corals as the primary builders of organic reefs.

5 On the land, angiosperms, or flowering plants, diversified, but gymnosperms and ferns remained more abundant in most environments.

6 Gondwanaland broke apart during the Cretaceous Period, forming the South Atlantic and other oceans.

7 The Tethys was a tropical seaway that carried warm waters from the equatorial Pacific Ocean through the Mediterranean region.

8 Latitudinal temperature gradients were gentle during most of Cretaceous time, ocean circulation was sluggish, and much of the ocean was depleted of oxygen. High rates of production of oceanic crust elevated sea level during mid-Cretaceous time, and organic-rich black muds accumulated in epicontinental seas.

9 In western North America, orogenic activity shifted eastward from its Jurassic position. To the east of the orogenic belt, an interior seaway stretched from the Gulf of Mexico to the Arctic Ocean.

10 At the end of the Cretaceous Period, a meteorite landed in the Gulf of Mexico, causing a mass extinction that eliminated the ammonoids, rudists, marine reptiles, and dinosaurs and devastated many other groups of organisms.

Review Questions

1 What accounts for the abundance of chalk in the Cretaceous System?

2 What were the most prominent groups of swimming predators in Cretaceous seas?

3 What general climatic and oceanographic conditions characterized the mid-Cretaceous world?

4 How did climatic and oceanographic conditions change a few million years before the end of the Cretaceous Period? What were the consequences for the reef-building rudists?

5 What conditions may account for the formation of widespread black shales in seas that spread over continental surfaces at certain times during the Cretaceous Period?

6 What modern continents that were once part of Gondwanaland remained attached to each other at the end of the Cretaceous Period?

7 Why did thick siliciclastic deposits accumulate in the western interior of North America during Cretaceous time?

8 How did Greenland become separated from North America?

9 What happened along the passive margin of the eastern United States during the Cretaceous Period?

10 What evidence is there that a meteorite struck Earth at the end of the Cretaceous Period?

11 Major physical and chemical events altered life during and at the very end of the Cretaceous Period. Using the Visual Overview on page 466 and what you have learned in this chapter, describe these events and their biological consequences.

Additional Reading

Alvarez, W., *T. Rex and the Crater of Doom*, Princeton University Press, Princeton, N.J., 1997.

Currie, P. J., and K. Padian, *Encyclopedia of Dinosaurs*, Academic Press, San Diego, 1997.

Novacek, M., *Dinosaurs of the Flaming Cliffs*, Doubleday, New York, 1996.

Vickers-Rich, P., and T. Rich, "Australia's Polar Dinosaurs," *Scientific American*, July 1993.

CHAPTER 18

The Paleogene World

The close of the Cretaceous Period marked a major transition in Earth's history. Scarcely any belemnoids survived, and ammonoids, rudists, and marine reptiles disappeared from the seas. What remained were marine taxa that persist as familiar inhabitants of modern oceans, among them bottom-dwelling mollusks and teleost fishes. On the land, the flowering plants of the Paleogene resembled those of latest Cretaceous time in many ways, but animal life changed dramatically. Taking the place of the dinosaurs were the mammals, which were universally small and inconspicuous at the start of the Paleogene interval but in many ways resembled modern mammals by the period's end.

The most profound geographic change during Paleogene time was a refrigeration of Earth's polar regions, which resulted in a chilling of the deep sea and, later in the Cenozoic Era, in widespread glaciation. Paleogene mountain-building events in western North America foreshadowed Neogene uplifts of such ranges as the Sierra Nevada and the Rocky Mountains. For the most part, the sediments that record these and other Paleogene events are unconsolidated, or soft, although most carbonates are lithified.

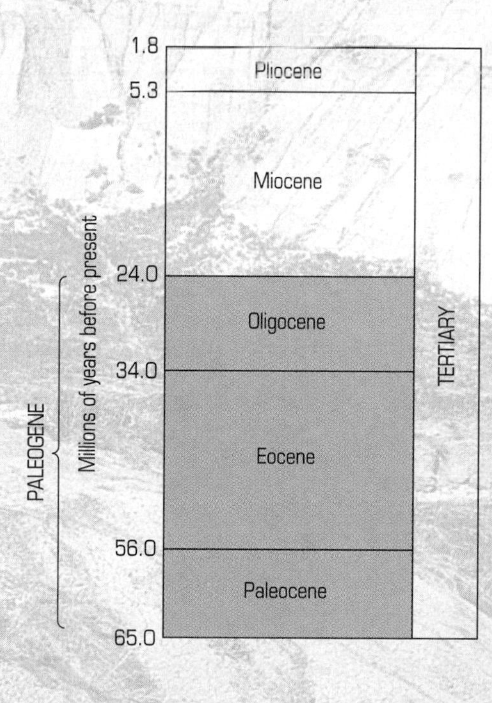

Pinkish terrestrial Eocene sediments rest atop tan marine Cretaceous sediments in Badlands National Park of South Dakota.

Visual Overview

Major Events of the Paleogene

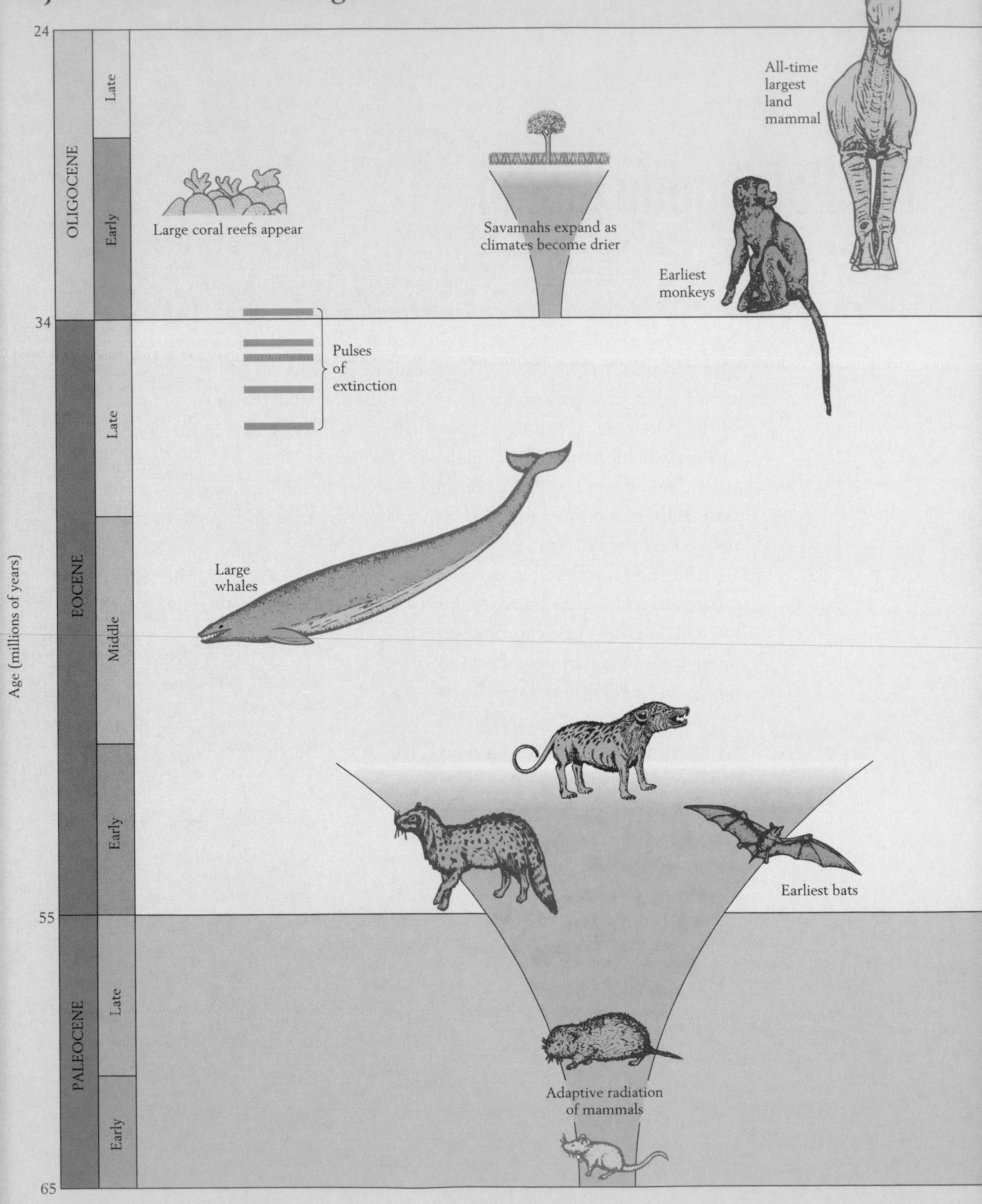

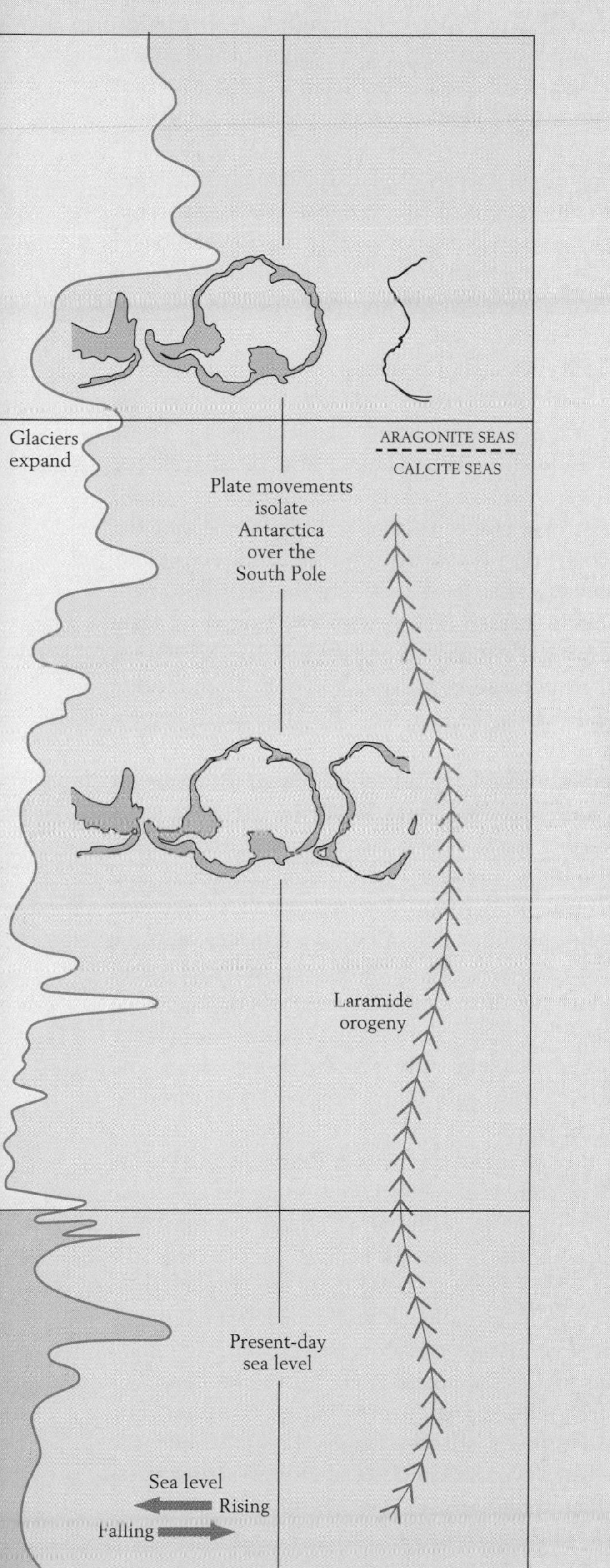

MIDDLE MIOCENE

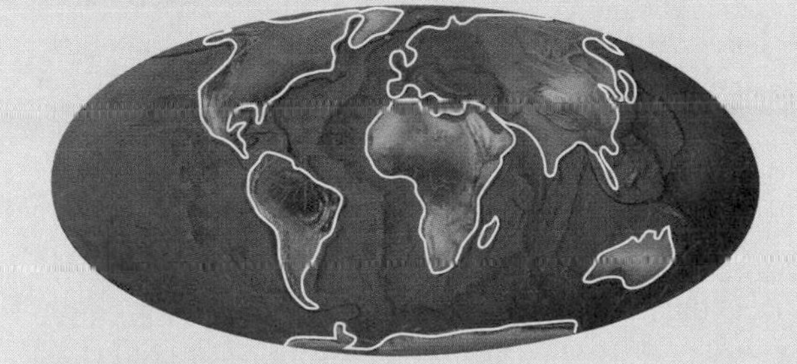

MIDDLE EOCENE

LATE CRETACEOUS

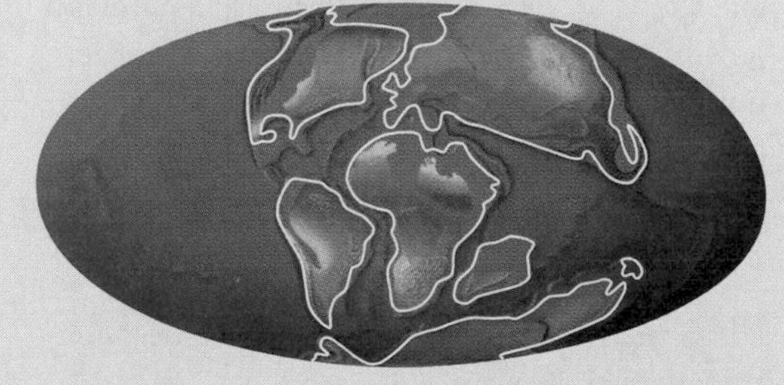

Early in the history of modern geology the marked difference between Mesozoic and Cenozoic biotas was readily apparent. Subdivision of the Cenozoic Era itself is less clear-cut. Today many geologists divide the era into two periods: the Paleogene Period, which includes the Paleocene, Eocene, and Oligocene epochs, and the Neogene Period, which includes the Miocene, Pliocene, Pleistocene, and Holocene epochs. This Paleogene-Neogene classification has become increasingly popular during the past two decades. Traditionally, however, the Cenozoic Era has been divided into two periods of quite different lengths: the Tertiary, which encompasses the interval from the Paleocene through the Pliocene, and the Quaternary, which includes only the Pleistocene and Holocene epochs (an interval of less than 2 million years). The advantage of the Paleogene-Neogene division is that it yields two periods of more similar duration.

The first formally recognized Paleogene epoch was the Eocene, which Charles Lyell established in 1833 on the basis of deposits found in the Paris and London basins (see Figure 6-1). Lyell named and described the Eocene Series in his great book *Principles of Geology,* the work that popularized the uniformitarian view of geology (p. 5). It was not until 1854 that Heinrich Ernst von Beyrich distinguished the Oligocene Series from the Eocene in Germany and Belgium on the basis of fossils, and then, in 1874, W. P. Schimper established the Paleocene Series on the basis of distinctive fossil assemblages of terrestrial plants in the Paris Basin.

Worldwide Events

Paleogene life is so familiar to us that it requires no special introduction; its most interesting features are the major expansions and contractions of certain taxonomic groups in all parts of the globe. Nor do we need a special section on Paleogene paleogeography, because the Paleogene lasted no more than 42 million years—a span in which paleogeographic changes occurred on such a small scale that they are best considered as regional events.

The Evolution of Marine Life

The present marine ecosystem is for the most part populated by groups of animals, plants, and single-celled organisms that survived the extinction at the end of the Mesozoic Era to expand during the Cenozoic. Many benthic foraminifera, sea urchins, cheilostome bryozoans, crabs, snails, bivalves, and teleost fishes survived in sufficiently large numbers to assume prominent ecologic positions in Paleogene seas.

Only three species of planktonic foraminifera appear to have survived the terminal Cretaceous crisis, but they gave rise to a remarkably rapid evolutionary radiation—one that yielded 17 new species, assigned to 8 new genera, within the first 100,000 years of Paleocene time.

Calcareous nannoplankton, which had suffered severe losses at the end of the Cretaceous Period, also rediversified rapidly during the Paleogene. These forms as well as the diatoms and dinoflagellates, which were not so adversely affected, have accounted for most of the ocean's productivity throughout the Cenozoic Era, just as they did in Cretaceous time. Nonetheless, after the Cretaceous the calcareous nannoplankton formed conspicuous chalk deposits again only early in Paleogene time (see Box 10-1). Presumably the sharp rise in the magnesium-calcium ratio of seawater during the Cenozoic Era reduced their productivity (see Figure 10-18)

Interestingly, having been displaced as dominant reef builders by the rudists during the Cretaceous, the corals failed to take advantage of the rudists' demise: they built no massive reefs during Paleocene and Eocene time, only small, scattered ones. The influence of the magnesium-calcium ratio of seawater on these aragonitic forms seems to have been the opposite of the effect on the calcareous nannoplankton, which secrete calcite. Only during the Oligocene, when the magnesium-calcium ratio reached a high level, did Cenozoic corals begin to form massive reefs throughout tropical seas.

Although many elements of Paleogene marine life closely resembled those of Late Cretaceous age, some forms were dramatically new. Perhaps the most distinctive marine organisms of this period were the whales, which evolved during the Eocene Epoch from carnivorous land mammals and quickly achieved success as large marine predators (see Figure 7-15). Joining the whales as replacements for the reptilian "sea monsters," the top carnivores of the Mesozoic Era, were enormous sharks (Figure 18-1). Unlike the whales, however, sharks descended from similar creatures that lived during Cretaceous time.

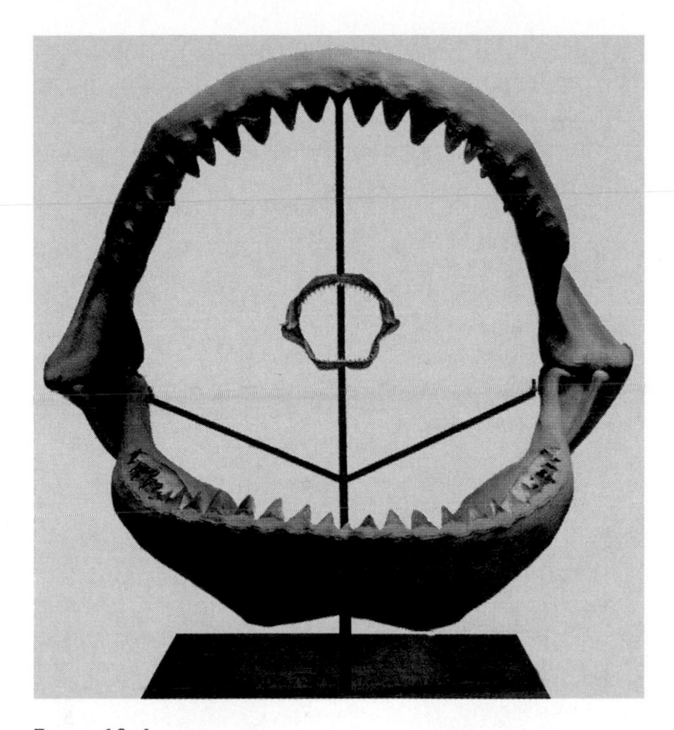

Figure 18-1 Jaws of the enormous fossil shark *Carcharodon* from the Eocene Epoch engulf the jaws of a modern shark. The jaws of the fossil shark were more than 2 meters (6.5 feet) across.

The marine ecosystem expanded during Paleogene time to include new niches along the fringes of the oceans. Sand dollars, for example, which are the only sea urchins able to live along sandy beaches, evolved at this time from biscuit-shaped ancestors (Figure 18-2), and new kinds of bivalve mollusks also invaded exposed sandy coasts. Both of these new groups have successfully inhabited shifting sands by virtue of their ability to burrow quickly into the sand again after they have been dislodged. Other newcomers to the ocean margins were the penguins, a group of swimming birds of Eocene origin, and possibly the pinnipeds, a group that includes walruses, seals, and sea lions. It is widely believed that the pinnipeds evolved before the beginning of the Neogene Period, although this group left no known Paleogene fossil record.

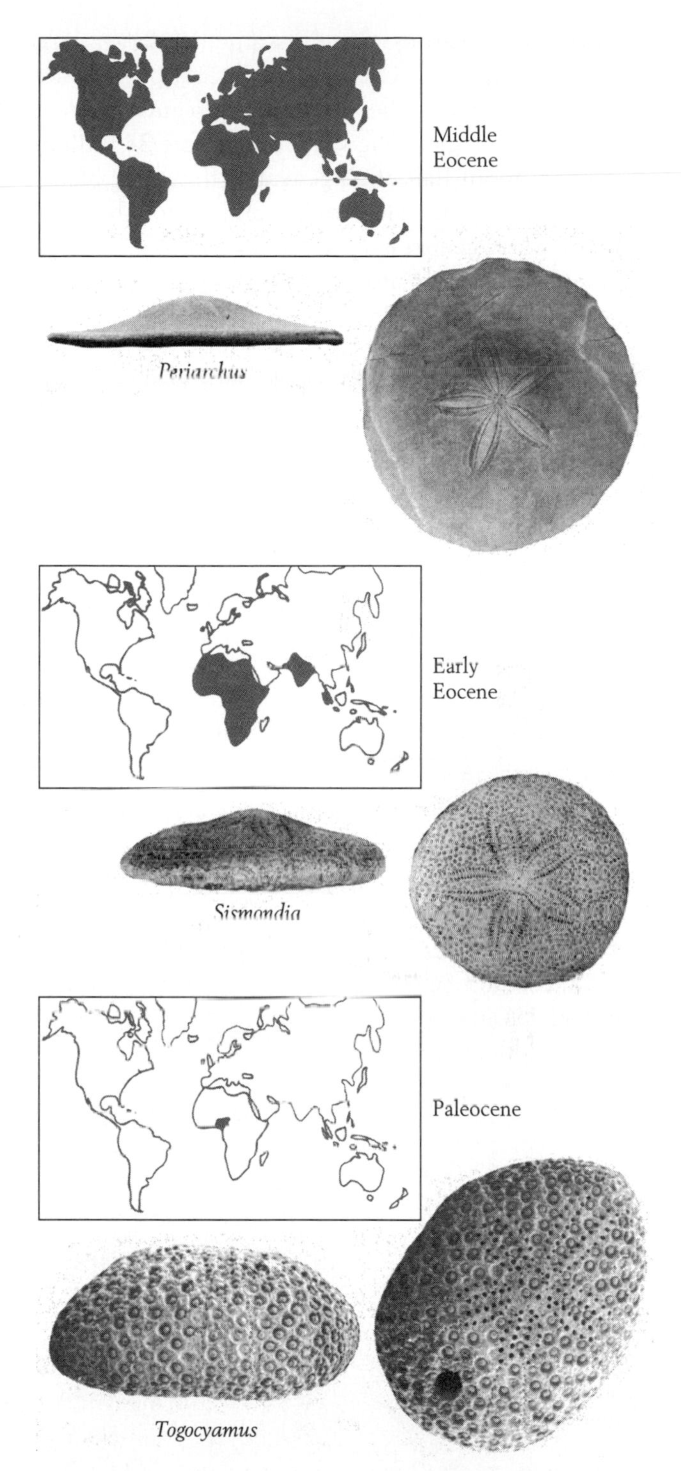

Figure 18-2 Specimens illustrating the evolution of the sand dollars from biscuit-shaped Paleocene ancestors *(Togocyamus)*, which are known only from Africa. True sand dollars *(Periarchus)* were present throughout the world by middle Eocene time. *Sismondia* is an intermediate genus that has been found in Africa and India. These animals were the size of small cookies.

The Evolution of Terrestrial Plants

The story of land plants is quite different. The transition to the Paleogene was apparently not marked by any drastic change in the character of terrestrial floras, but flowering plants assumed a larger role after the terminal Cretaceous extinction. In the process, modern families of flowering plants evolved. By the beginning of Oligocene time, some 34 million years ago, about half of all genera of flowering plants were ones that are alive today, and although many modern plant genera had not yet evolved, forests had taken on a distinctly modern appearance.

One major event in plant evolution that did take place during the Paleogene interval was the origin of the grasses, although these usually low-growing flowering plants did not reach their full ecologic potential until late Oligocene and Miocene times. Early grasses may have been confined to wooded or swampy areas. The mode of growth of early grasses, like that of the modern sedges that form marshlands along continental coastlines, did not allow their leaves to grow continuously and thus to recover from heavy grazing by animals of the sort that inhabit open country in large numbers. It was only an adaptive breakthrough—the origin of the continuous growth process, which forces us to cut our lawns every week or two—that ultimately enabled grasses to invade open country with great success. Once they were able to survive the effects of heavy grazing by animals, grasses quickly spread over vast expanses to form grasslands.

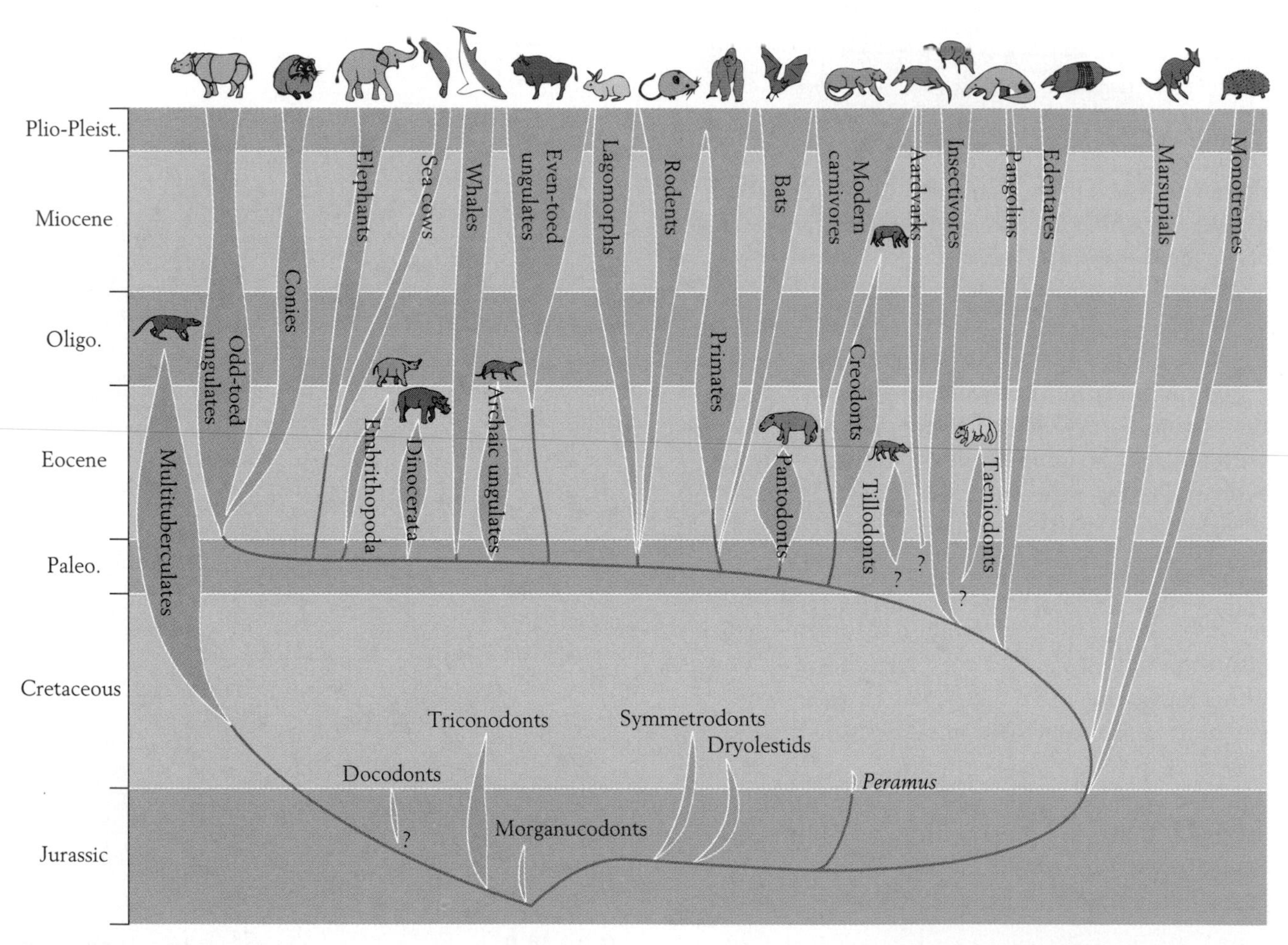

Figure 18-3 The pattern of adaptive radiation of mammals that gave the Cenozoic Era its informal name, the Age of Mammals. It is thought that the two large groups of modern mammals, the placentals and the marsupials, had a common ancestry in the Cretaceous Period. The multituberculates, which failed to survive into Neogene time, apparently evolved separately, as did the Monotremata, primitive survivors of which include the Australian platypus, which lays eggs instead of giving birth to live young. For the placental mammals, each vertical bar represents an order. Note that most of the new Cenozoic orders had already evolved by the beginning of the Eocene Epoch, about 10 million years after the dinosaurs disappeared. *(D. R. Prothero.)*

Figure 18-4 A fossil of an Eocene bat from the Green River Formation of Wyoming.

Early Paleogene Terrestrial and Freshwater Animals

In Chapter 7 and Box 17-1 we saw that the mammals, having inherited the world from the dinosaurs, underwent a remarkably rapid adaptive radiation during the early part of the Cenozoic Era. It was probably through both competition and predation that the dinosaurs had prevented the mammals from undergoing any great evolutionary expansion during Mesozoic time.

Early in the Paleocene Epoch, when they first had the world to themselves, mammals were small creatures, most of which resembled modern rodents; no mammal seems to have been substantially larger than a good-sized dog. Furthermore, most mammal species tended to remain generalized in both feeding and locomotory adaptations; those that dwelled on the ground generally retained a primitive limb structure that caused the heels of the hind feet and the "palms" of the front feet to touch the ground as the animals moved about. Perhaps 12 million years later, however, by the end of early Eocene time, mammals had diversified to the point at which most of their modern orders were in existence (Figure 18-3). Bats already fluttered through the night air (Figure 18-4), for example, and, as we have seen, large whales swam the oceans.

Also included among the Paleocene mammals were groups that had survived from Cretaceous time, such as marsupials, multituberculates, and placental mammals called insectivores. Some experts have assigned other Paleocene mammals to the Primates, the order to which humans belong. Although these small animals were quite different from monkeys, apes, or humans in many respects, by early Eocene time they did climb with grasping hind limbs and forelimbs that foreshadowed our own hands and feet (Figure 18-5). By mid-Paleocene time, true mammalian carnivores—members of the living order Carnivora—had emerged (Figure 18-6). This is the order to which nearly all living carnivorous placental mammals belong. By the end of Paleocene time, the earliest members of the horse family had evolved as well; these animals were no larger than small dogs (Figure 18-7), but by the end of the epoch, larger herbivorous mammals, some the size of cows, had appeared.

The variety of mammals continued to increase in the Eocene Epoch. The number of mammalian

Figure 18-5 Reconstruction of the early primate *Cantius*, a small genus of early Eocene time. This arboreal animal had large toes on its hind feet and nails much like our own, and it apparently jumped from limb to limb. *(After R. T. Bakker.)*

Figure 18-6 Reconstruction of a mid-Paleocene biota of New Mexico. The large trees represent the sycamore genus *Platanus,* which has survived to the present time. A small insectivore *(Deltatherium)* rests on a small branch. On the ground, mesonychid carnivores of the genus *Ancalagon* feed on the small crocodile *Allognathosuchus. Ancalagon* is closely related to the mesonychids that gave rise to the whales. Ferns and sable palms (or fan palms) constitute the undergrowth in the background. *(Drawing by Gregory S. Paul.)*

Figure 18-7 *Hyracotherium* ("eohippus"), the earliest genus of the horse family. This animal, which was present in Late Paleocene and Eocene time, was no larger than a small dog. It had four toes on each front foot and three on each hind foot.

families doubled to nearly 100, approximating that of the world today. In addition, more modern varieties of hoofed herbivores appeared. Most animals of this kind are known as **ungulates,** and they are divided into **odd-toed ungulates** (living horses, tapirs, and rhinos) and **even-toed ungulates,** or cloven-hoofed animals (cattle, antelopes, sheep, goats, pigs, bisons, camels, and their relatives). The odd-toed ungulates expanded before the even-toed group did, but primitive even-toed ungulates were also present early in the Eocene (Figure 18-8); of the modern even-toed types, camels

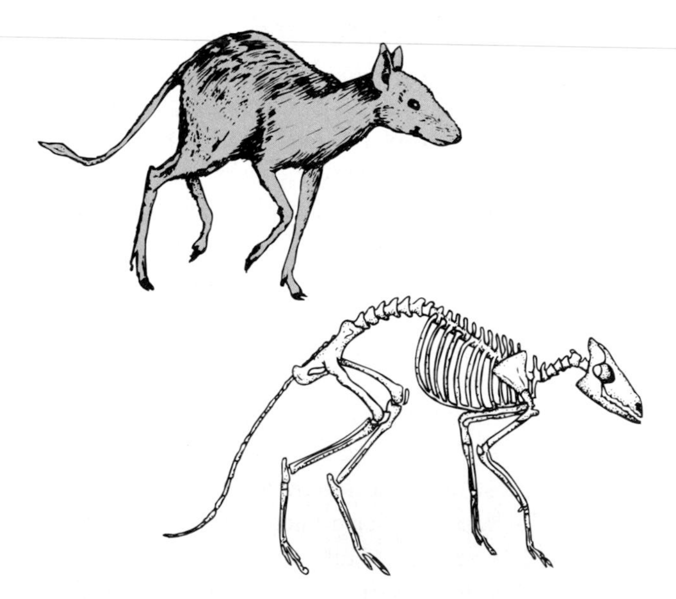

Figure 18-8 *Diacodexis,* an early even-toed ungulate, or cloven-hoofed herbivore. *Hyracotherium* (Figure 18-7), in contrast, was an early odd-toed ungulate. The limb structure of *Diacodexis* shows that it was an unusually adept runner and leaper for early Eocene time. *(After K. D. Rose, Science 216:621–623, 1984.)*

Figure 18-9 Early members of the elephant group. *A. Moeritherium*, which evolved in late Eocene time, was about a meter (3 feet) tall. It probably wallowed in shallow waters and grubbed for roots or other low-growing vegetation. *B. Palaeomastodon*, which was about twice as tall, possessed a rudimentary trunk and tusks.

and relatives of the present-day chevrotain (the Oriental "mouse deer") evolved before the end of the epoch. In addition, the earliest members of the elephant order appeared during early Eocene time; *Moeritherium*, the earliest genus well known from the fossil record, was a bulky animal slightly larger than a pig, with rudimentary tusks and a short snout rather than a fully developed trunk (Figure 18-9*A*). The **rodents,** which had originated in the Paleocene, continued to diversify as well, but they may have attained their success at the expense of the archaic multituberculates, which were also specialized for gnawing seeds and nuts. As the rodents expanded during the Eocene, the multituberculates declined; they finally became extinct early in the Oligocene Epoch.

Among the animals that preyed on these herbivorous mammals were groups that originated in the Paleocene Epoch. They include the superficially doglike mesonychids and the diatrymas, which were huge flightless birds with powerful clawed feet and enormous slicing beaks (Figure 18-10). The diatrymas disappeared toward the end of the Eocene Epoch. At that time the mesonychids were joined by primitive members of three familiar modern carnivore groups: the dog, cat, and weasel families.

The monstrous diatrymas were not the only birds of the Eocene, but flying birds were much less diverse then than they are today. Most species were shore birds that waded in shallow water when they were not in flight (Figure 18-11). Not yet present were many other modern kinds of birds, including the songbirds that are so numerous today.

As for other forms of vertebrate life, reptiles and amphibians were relatively inconspicuous during Paleogene time. The first record of the Ranidae, the largest family of living frogs, is in the Eocene Series, but the fossil record of this group of fragile animals is not good, so we do not know precisely when the Ranidae originated or attained high diversity.

There is no question that the insects took on a modern appearance with the origin of several modern families in Paleogene time. Several Oligocene forms, which have been preserved with remarkable precision in amber, closely resemble living species.

Figure 18-10 Large terrestrial predators that had evolved by early Eocene time. The animals that superficially resemble dogs are giant mesonychids of the genus *Pachyhyaena*, which were the size of small bears. The flightless birds guarding their chicks are members of the genus *Diatryma*, which stood about 2.4 meters (8 feet) tall. *(Drawing by Gregory S. Paul.)*

Figure 18-11 **The long-legged Eocene duck *Presbyornis*.** The large numbers of fossils of this wading bird found in Green River sediments of Wyoming and Utah suggest that it lived in enormous colonies. *Presbyornis* left tracks revealing that webbed feet supported it when it walked in mud. Along with the tracks are lines of probing marks made by the beak while the bird searched for food. Its legs were too long to allow it to swim, but modern ducks have inherited its webbed feet and employ them as paddles to propel them through the water. *(After a drawing by J. P. O'Neill.)*

Mammals of the Oligocene Epoch

Mammals became increasingly modern during the Oligocene Epoch. As many Eocene families died out, many living (Holocene) groups expanded. The horse family had disappeared from Eurasia during the Eocene, but a few horse species survived in North America. Other odd-toed ungulates that enjoyed greater success during the Oligocene Epoch than in previous intervals were the rhinos, which included the largest land mammal of all time (Figure 18-12), and the rhinolike brontotheres (Figure 18-13). Many more large mammals populated the land during the Oligocene Epoch than during the Eocene.

As the Oligocene Epoch progressed, odd-toed ungulates were outnumbered for the first time by even-toed ungulates, including deerlike animals and pigs, which became especially diverse. This trend has continued to the present day, when there are many more species belonging to the deer and antelope families than to the horse and rhino families. Elephants became larger early in the Oligocene and developed rudimentary trunks and tusks (see Figure 18-9*B*).

Among the carnivores, the dog, cat, and weasel families, which had their origins in Eocene time, radiated during the Oligocene Epoch and produced more advanced forms, including large saber-toothed cats

Figure 18-12 ***Paraceratherium*, which, as far we now know, was the largest mammal ever to exist on Earth.** This Oligocene giant from Asia belonged to the rhinoceros family and stood about 5.5 meters (18 feet) at the shoulder. This is the height of the top of the head of a good-sized modern giraffe. *(Drawing by Gregory S. Paul.)*

Figure 18-13 **An Early Oligocene mammal fauna from Nebraska and South Dakota.** The huge animal at the top of the painting is the brontothere *Brontotherium*. Below it is the early tapir *Protapirus*, and to its right is *Subhyracodon*, an early rhinoceros. Below *Subhyracodon*, in succession, are *Merycodon*, a sheeplike animal; *Hyaenodon*, an archaic hyaenalike animal eating a lizard; and the sabertooth cat *Hoplophoneus*. At the upper left is the small rhinoceros *Hyracodon*, and to its right are the large, piglike *Archaeotherium* and the ancestral camel *Proebrotherium*, in front of which is the horned herbivore *Protoceras (Painting by Jay Matternes.)*.

(Figure 18-14), bearlike dogs, and animals that resembled modern wolves.

An especially important aspect of the modernization of mammals during the Oligocene Epoch was the appearance of monkeys and apelike primates. The genus *Aegyptopithecus*, an arboreal animal the size of a cat, had teeth resembling those of an ape but a head and a tail resembling those of a monkey (Figure 18-15). In Neogene time, before the arrival of humans, apes attained considerable diversity in Africa and Eurasia.

Climatic Change and Mass Extinction

Early in Paleogene time, the average global temperature rose higher than at any other time in the

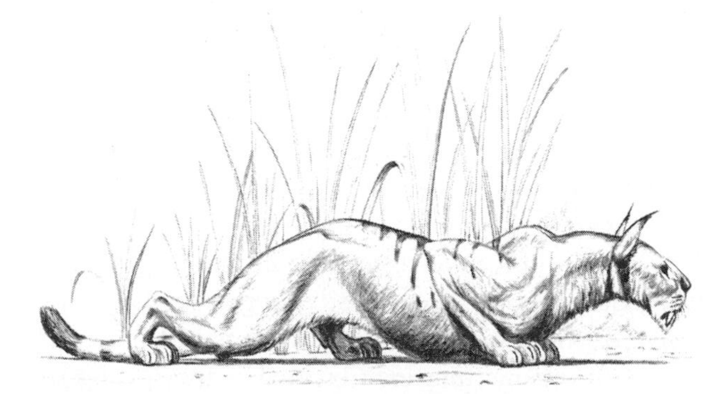

Figure 18-14 **The Oligocene cat *Dinictis*, which was approximately the size of a modern lynx.** This animal possessed advanced adaptations for running and springing upon prey. Its canine teeth were elongated for stabbing, but *Dinictis* was not a member of the true saber-toothed cat group. *(Drawing by Gregory S. Paul.)*

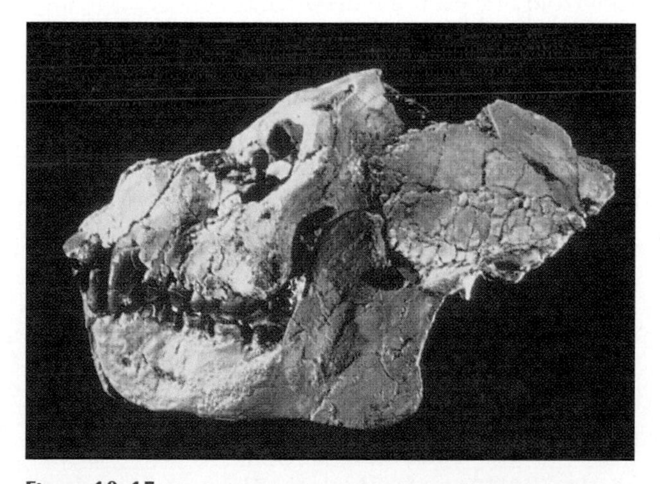

Figure 18-15 **The Oligocene primate *Aegyptopithecus*, whose name reflects its discovery in Egypt.** The skull resembles that of a monkey, but the teeth are apelike. The brain of *Aegyptopithecus* was unusually large for the size of the animal, perhaps reflecting a high level of intelligence for the Oligocene world. *(Drawing by S. F. Kimbrough.)*

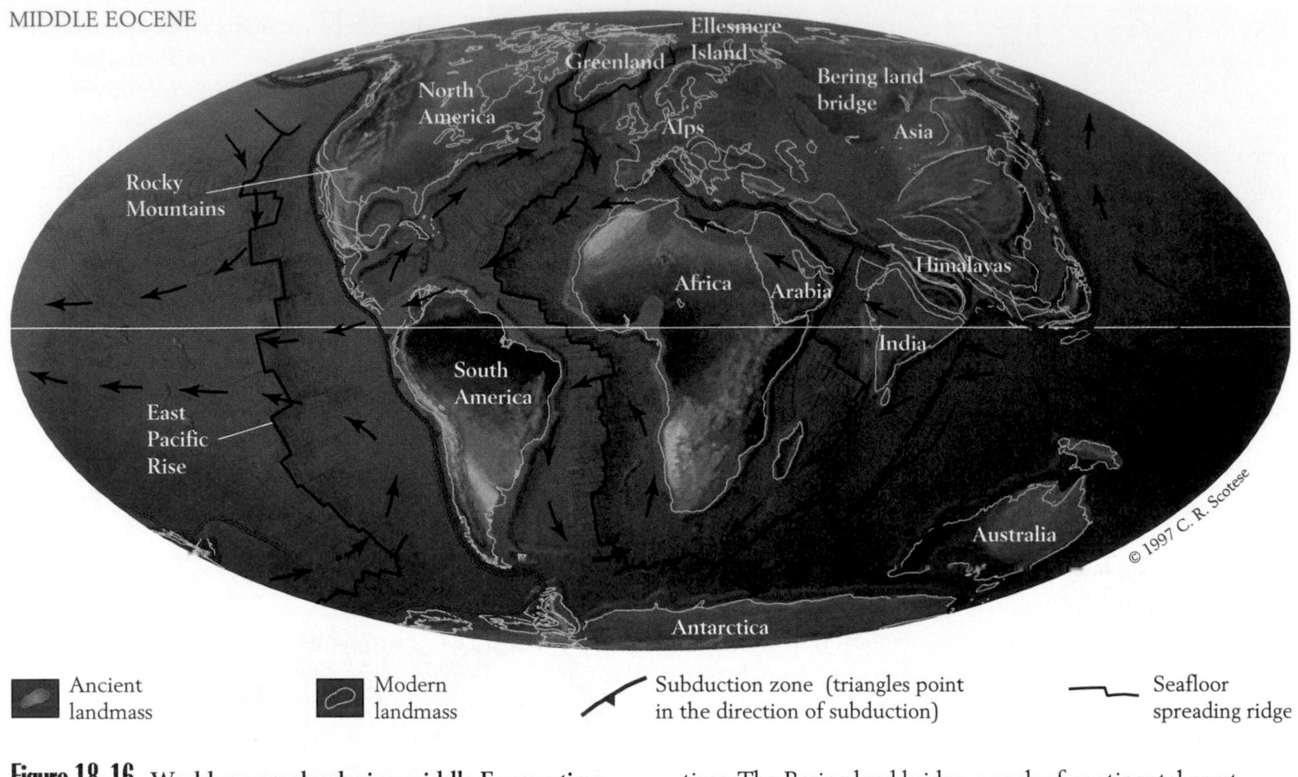

Figure 18-16 **World geography during middle Eocene time. Substantial areas of continental crust were inundated during this interval.** Relatively warm climates prevailed even as far north as Ellesmere Island, where a diverse fauna of terrestrial plants and vertebrate animals flourished in middle Eocene time. The Bering land bridge, a neck of continental crust, stood above sea level throughout the Eocene interval, permitting animals to migrate between North America and Eurasia. *(Adapted from paleogeographic map by C. R. Scotese, PALEOMAP Project, University of Texas at Arlington, 1997.)*

Cenozoic Era; even the deep sea briefly became very warm. During the second half of the Paleogene Period, a mass extinction struck both on the land and in the sea. This catastrophe was not so severe as some earlier ones. It did eliminate many genera and species, but relatively few higher taxa disappeared. Actually, it was not a single event but a series of pulses of extinction, brought about by the onset of cooler conditions in many regions—and also on the land by widespread drying of climates.

Early Eocene Warmth The very warm interval of the Paleogene began during the transition between the Paleocene and the Eocene epochs. Marking the boundary between these two epochs in cores from the deep sea is an abrupt shift of oxygen isotope ratios for foraminifera toward lighter values. This shift occurs for both planktonic species and species that lived on the floor of the deep sea. The total volume of glacial ice on Earth was quite small at that time, so the isotope shift does not reflect melting of glaciers and return to the ocean of waters enriched in oxygen 16 (see Figure 10-16). Instead, the shift signals a temperature change. It indicates that even near Antarctica both surface and deep-sea waters warmed by several degrees to a temperature of about 18°C (~64°F) within less than 3000 years. More than 70 percent of the foraminifera species of the deep sea died out, probably because the warming terminated the downward flow of cool polar waters that had previously brought oxygen to the deep sea. In other words, the foraminifera probably asphyxiated. Low levels of oxygen must have remained, because deep-sea sediments did not turn dark with organic matter. Presumably the species of foraminifera that survived were ones that could tolerate low oxygen conditions.

The global climate remained warm throughout early Eocene time. Isotope ratios for planktonic foraminifera suggest that although tropical seas were only about as warm as they are today, very warm conditions prevailed at high latitudes: the average annual

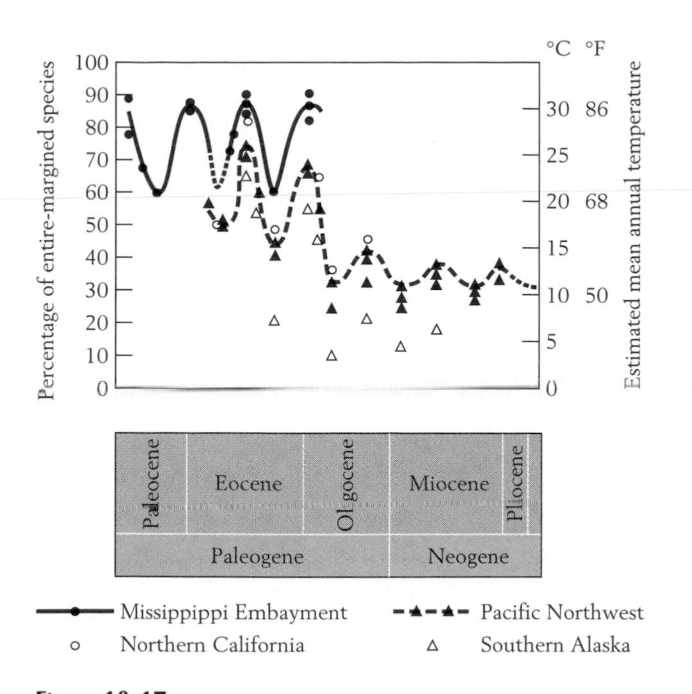

Figure 18-17 Estimated changes in temperature in four areas of North America in the course of Cenozoic time, based on percentages of entire-margined leaves in fossil floras (see also Figure 4-20). The fact that the curves follow parallel paths where they overlap in time suggests that the trends hold for large areas of Earth's surface. *(After J. A. Wolfe, American Scientist 66:694–703, 1978.)*

temperature of the sea surface near Antarctica was about 15°C (almost 60°F).

Fossil floras reveal that in early Eocene time southeastern England, which was positioned close to its present high latitude, was cloaked in a virtually tropical jungle (Box 18-1). Fossil floras on Ellesmere Island, inside the Arctic Circle (Figure 18-16), point to a climate similar to that of southern California today.

Possibly greenhouse warming elevated Earth's average temperature early in the Eocene, but there is no independent evidence of an enhanced greenhouse effect. In addition, the gentle temperature gradient from equator to poles in the early Eocene is puzzling. Researchers debate what caused the spread of so much heat to high latitudes. One factor may have been the poleward flow of warm saline waters at intermediate depths in the oceans, followed by upward mixing that heated the surface waters and atmosphere (see p. 481). Another factor was certainly the presence of dense forests at high latitudes. Forests have a low albedo; that is, they absorb much light, which turns to heat (p. 99). They also insulate Earth's surface, trapping heat beneath their canopy. Additional factors, not yet uncovered, may have played important roles in bringing warmth to high latitudes.

Global Climatic Change Fossil plants attest to dramatic changes in Earth's climate after the early Eocene interval of global warmth. Flowering plants (angiosperms) are commonly viewed as the thermometers of the past 100 million years. As we have seen, their value derives in part from the strong correlation between mean annual temperature and percentage of species within a flora that have leaves with entire margins (see Figure 4-20). Data on the leaf margins of fossil floras of North America reveal that major climatic changes took place throughout the world after Early Eocene time. Three pulses of cooling occurred, each more severe than the one before (Figure 18-17).

The events that took place in the deep sea provide an explanation for the episodes of cooling. Study of deep-sea cores reveals that at this time there was a large increase in the isotope ratio of oxygen for benthic foraminifera of the deep sea (Figure 18-18). Bottom waters cooled at this time, but the large size of the isotope shift also reflects growth of glaciers on Antarctica, with oxygen 16 accumulating

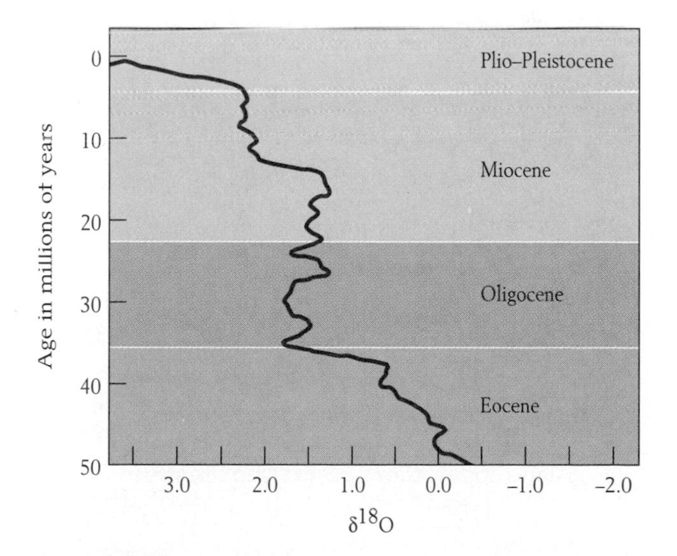

Figure 18-18 Oxygen isotope curve for deep-sea benthic foraminifera. The strong shift toward high values of oxygen 18 near the end of the Eocene signals both cooling of deep-sea waters and buildup of ice sheets. *(After K. S. Miller, R. G. Fairbanks, and G. S. Mountain, Paleoceanography 2:201–19, 1987.)*

For the Record 18-1

Global Warming in the Eocene

Just how warm was the Eocene world? This question is especially pertinent as we contemplate the substantial warming that humans will soon inflict on Earth by adding carbon dioxide to the atmosphere.

The climatic history of England tells us that warm conditions spread to very high latitudes during the Eocene. Today England has a stable but relatively cool climate. It lies farther north than any American state but Alaska. It was similarly positioned during Early Eocene time, yet it was cloaked by a virtually tropical jungle that shared many taxa with the modern equatorial region of Malaysia. Needless to say, England has never been this warm again.

Even more northerly regions were bathed in balmy climates during the Eocene. Fossil palm leaves are known from as far north as southern Alaska. These palms were part of a diverse subtropical flora that spread from Asia around the northern rim of the Pacific to North America as northern climates heated up early in the Cenozoic Era.

An especially interesting Eocene flora comes from Ellesmere Island, which lies well within the Arctic Circle, north of eastern Canada. Even this biota is subtropical, sharing many taxa with Eocene biotas of western North America and Eurasia. During the interval of Eocene warmth, more than 40 percent of the plant species of the Ellesmere flora had leaves with smooth margins—an indication that the average temperature resembled that of southern California today. Also present were large tortoises and alligators, which require warm winter temperatures. On Ellesmere Island today, temperatures hover around −18°C (about 0°F) throughout the year. In late Eocene time, however, the island remained above freezing even in winter.

Temperatures near the equator were similar to those of today. If during the Eocene warming they had risen as much as temperatures at high latitudes, few plants or animals could have tolerated the heat. Instead, the temperature gradients from the equator to the poles were relatively gentle, because an enormous amount of heat was somehow transported toward the poles. Warm climatic zones of the Eocene world were so broad that Earth's average temperature must have been higher than it is today, but we do not know how much higher. Perhaps this condition resulted from greenhouse warming, but if so, we have not yet discovered what caused it.

The greenhouse warming of the future will produce climatic patterns very different from those of the Eocene, because it will develop within a world very different from that of the Eocene. Not only is the equator-to-pole temperature gradient much steeper today, but most regions display much greater seasonal contrasts in temperature. In addition, larger areas at mid-latitudes are perennially or seasonally arid. The result is that a traveler to the Arctic from any point at the equator will encounter at least five distinctive floras (p. 101). The complex climatic and floral patterns of the modern world will amplify the impact of future greenhouse warming on life. The problem is that today many regions lie near boundaries between floras. Warming, and the changes it produces in patterns of precipitation, will transform the floras, and hence the faunas, of these numerous marginal regions. Some species in these regions will suffer from exposure to new competitors or predators, or from confinement to smaller habitats. Many agricultural and recreational areas will be adversely affected as well. Unfortunately, we cannot yet reliably predict the regional patterns of climatic and biotic change that various degrees of greenhouse warming will produce.

Fossil palm frond from the Eocene Green River Formation of Wyoming, far north of where palms live today.

preferentially in the glacial ice (see Figure 10-16). The growth of glacial ice on Antarctica was a prelude to the greater expansion of high-latitude ice sheets near the end of the Cenozoic Era, during what is informally known as the recent Ice Age.

It was during the late Eocene cooling episode that the cold bottom layer of the ocean came into being. Cold, dense polar water began to sink to the deep sea, as it does today (see Figure 4-23). Bottom water temperatures dropped by perhaps 4° to 5°C (7° to 9°F). Fossils in deep-sea cores show that changes in the bottom-dwelling fauna occurred in less than 100,000 years. Thus the cooling was quite rapid.

Mass Extinction Deep-sea cores also provide the most detailed record of the pattern of late Paleogene extinction—a record that consists of microfossils. The record of planktonic foraminifera reveals five successive pulses of change between about 40 million and 31 million years ago (between middle Eocene and early Oligocene time). The final pulse of extinction appears to have coincided with the final episode of cooling in North America, as indicated by changes in terrestrial floras (see Figure 18-17). The species that disappeared in these pulses were mostly spiny forms adapted to warm conditions. The pattern of extinction of calcareous nannoplankton is not yet documented in great detail, but during the final 7 million years of Eocene time the total number of species in the world declined markedly. Pulses of extinction also occurred on the seafloor. Many of the species that died out were adapted primarily to warm conditions.

Many species of mammals disappeared from both North America and Europe at the end of the Eocene Epoch. It seems evident that climatically induced changes in the terrestrial flora played a major role in the mammalian extinctions. Many of these floral changes resulted more from increased aridity than from cooling. Before the climatic change at the end of the Eocene, moist tropical and subtropical forests cloaked much of North America and Eurasia. During the Oligocene Epoch dry woodlands with large, grassy clearings spread across large areas of continents. The shrinking of forests must have created a positive feedback for the general climatic change. Forests not only have a low albedo and trap heat; they also retain moisture. Loss of forest cover during the Eocene-Oligocene transition would have accentuated the cooling and drying of climates, leading to further contraction of forests. The mammal species that died out in North America and Europe included many tree climbers and also many herbivores whose teeth required a diet of soft leaves. Among the latter were the huge brontotheres (see Figure 18-13). Many of the new Oligocene species possessed molar teeth well suited to grinding coarser vegetation.

Regional Events

In examining important regional events of the Paleogene Period, we will travel to the ends of Earth—first to the south pole, where Antarctica became separated from Australia and developed its icy cover, and then toward the north pole, where land areas of North America and Eurasia were more closely connected than they are today. Then we will look at the history of the Cordilleran region of North America, where the Laramide orogeny yielded many structures that remain conspicuous in the Rocky Mountains today. We will also examine the nature of deposition along the Gulf Coast of North America, where large volumes of petroleum are trapped in soft Paleogene sediments.

Antarctica and Global Change

Cooling of Antarctica and expansion of glaciers there during the Eocene Epoch resulted from the movement of plates that had previously formed parts of Gondwanaland. Both South America and Australia rifted away from Antarctica, leaving it isolated in a polar position.

Even before its isolation, Antarctica had been centered over the south pole, but it had remained warm because its shores were bathed in relatively warm waters from lower latitudes (Figure 18-19*A*). Movement of South America and Australia away from Antarctica automatically formed the circumpolar current, which has persisted to the present day (see Figure 4-22). Water from the South Atlantic, Indian, and South Pacific oceans becomes trapped in this current, becoming progressively colder as it cycles around and around Antarctica. Initially, when the passageway around Antarctica was constricted and cooling was relatively weak, glaciers covered only part of Antarctica (Figure 18-19*B*). As South America and Australia moved farther away from Antarctica, the circumpolar current must have strengthened, causing further cooling and glacial expansion. In fact, a sudden mid-Oligocene eustatic decline must have resulted from substantial expansion of Antarctic glaciers (see p. 21).

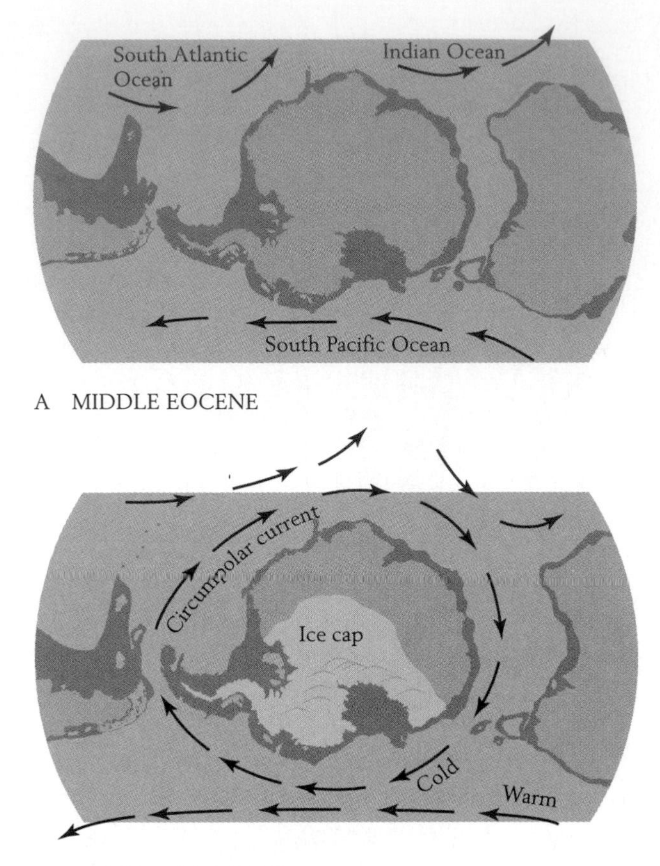

Figure 18-19 Origin of the circumpolar current around Antarctica. *A*. In middle Eocene time warm currents from the southern oceans warmed Antarctica. *B*. By early Oligocene time South America and Australia had moved far enough from Antarctica to allow water from the southern oceans to become trapped in a current around Antarctica, where they became very cold. As a result, glaciers grew on Antarctica. *(From L. A. Lawver, L. M. Gahagan, and M. F. Coffin, Antarctic Research Series, Amer. Geophys. Union 56:7–30, 1992.)*

The Top of the World: Changing Positions of Land and Sea

Major geologic and geographic changes also took place during the Paleogene Period within 30° latitude of the north pole. Many of these changes can be read from the ages of various segments of the deep-sea floor. The Arctic deep-sea basin existed during the Cretaceous Period, but until nearly the end of Cretaceous time it remained separated from the Atlantic, except by way of shallow seas, because North America, Greenland, and Eurasia were still united as a single landmass. Recall that the Mid-Atlantic Rift proceeded northward during Cretaceous time, splitting Greenland from North America (see Figure 17-12). Early in the Paleogene, this rifting shifted to a younger spreading zone between Greenland and Scandinavia, eventually establishing a broad connection between the Arctic Ocean and the Atlantic Ocean (see Figure 18-16).

In contrast, continental crust has continued to separate the Arctic and Pacific basins; Alaska and Siberia have remained connected by a stretch of continental crust despite the fact that this connecting segment now lies submerged beneath the Bering Sea. During much of the Cenozoic Era, this segment, known as the *Bering land bridge*, stood above sea level and served as a land corridor between North America and Eurasia. This corridor remained open throughout Paleogene time, allowing mammals and land plants to migrate between Asia and North America.

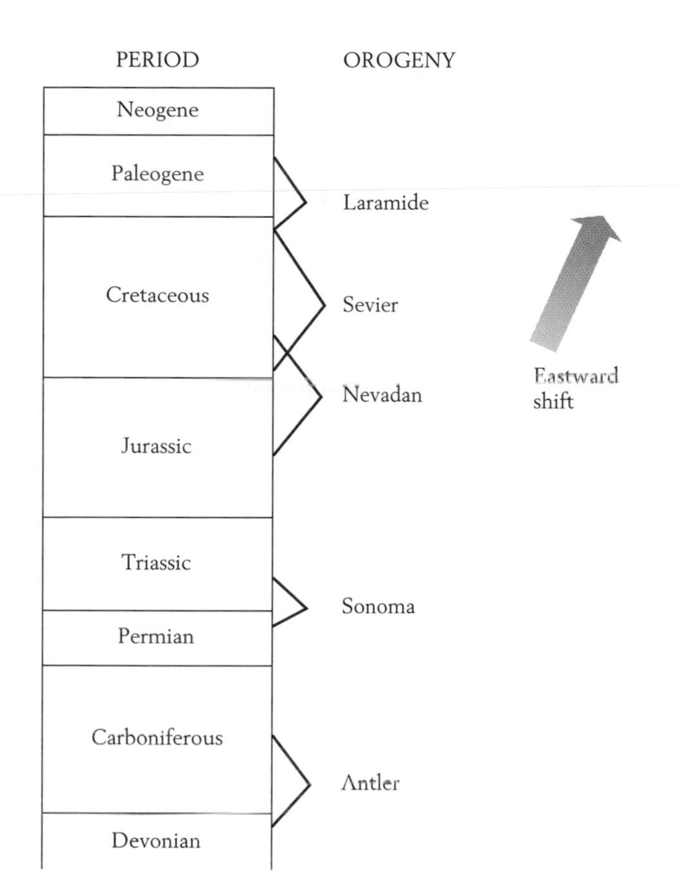

Figure 18-20 Summary of major orogenic events in the eastern Cordilleran region. Between Jurassic and Paleogene time, orogenic activity migrated eastward (see Figure 17-23).

Tectonics of Western North America

Mountain-building activity in the Cordilleran region of North America continued into the Paleogene Period, but with a number of changes. Figure 18-20, which summarizes the orogenic history of the eastern Cordilleran region, shows that the Sevier episode spanned almost the entire Cretaceous Period. In latest Cretaceous time, however, a different style of tectonic activity was initiated in the American West. This activity persisted through the Paleocene and well into Eocene time.

The Laramide Orogeny The episode characterized by the new tectonic style is known as the *Laramide orogeny.* The northern and southern segments of the Laramide were typical of orogenies in general. In the north, extending from the United States into Canada, there remained an active belt of igneous activity and, inland from it, an active fold-and-thrust belt (Figure 18-21). Thrust sheets of enormous proportions are spectacularly exposed in the Canadian Rockies (Figure 18-22). A similar pattern of tectonism persisted both in the southern United States and in Mexico.

The unusual features of the Laramide orogeny were in the central part of the western United States, where a broad area of tectonic quiescence extended from the Great Valley of California to the Colorado Plateau (see Figure 18-21). East of this inactive region, in a strange pattern of tectonism, large blocks of underlying crystalline rock were uplifted in a belt extending from Montana to Mexico. The largest of these blocks were centered in Colorado, where the Ancestral Rocky Mountains had been uplifted more than 200 million years earlier, late in the Paleozoic Era.

The Pattern of Subduction What created the unusual tectonic pattern that characterized the central part of the Cordilleran region during the Laramide orogeny? Note that the Paleogene uplifts were, for

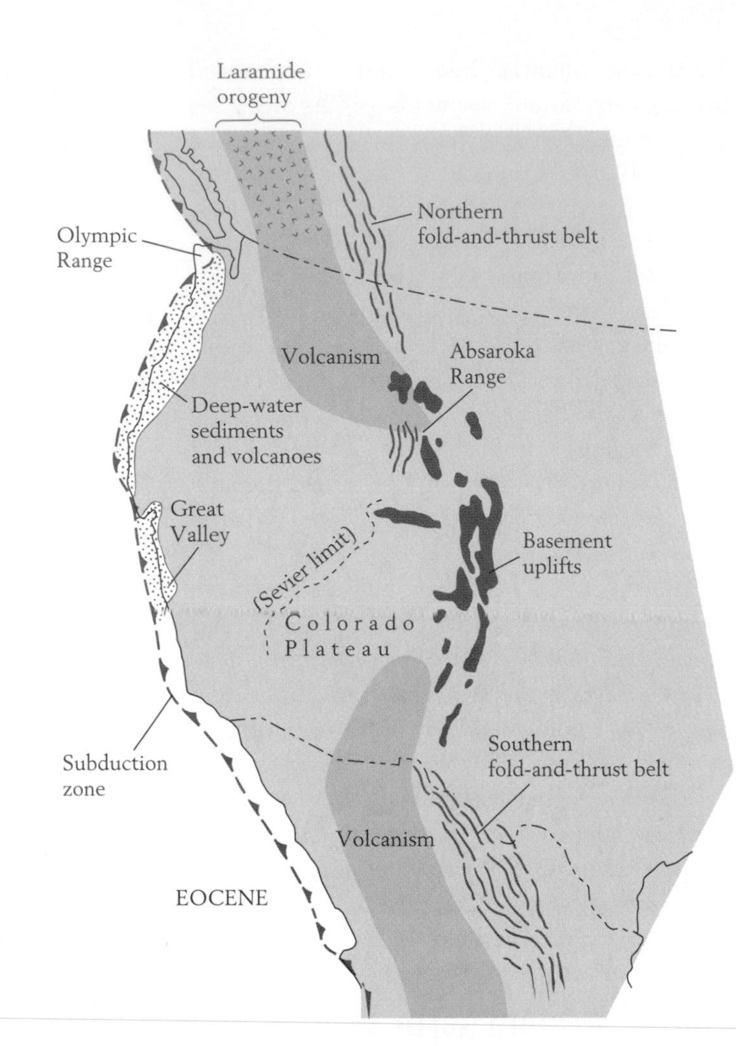

Figure 18-21 Geologic features of western North America during Eocene time. Subduction continued along the west coast. Marine sediments were deposited in the Great Valley of California, and deep-water sediments and volcanics accumulated in a forearc basin to the north, in Washington and Oregon. Farther inland in the north and south, the Laramide orogeny produced a band of volcanism and, still farther inland, a belt of folding and thrusting. In Colorado and adjacent regions, however, the orogeny was expressed as a series of crystalline uplifts that extended far to the east of the Cretaceous Sevier orogenic belt. They may have formed by a slight clockwise rotation of the Colorado Plateau in relation to the continental interior.

Figure 18-22 The Lewis Thrust Fault in the fold-and-thrust belt of the northern Rocky Mountains. In this view the fault is exposed on the side of a mountain in Glacier National Park, Montana. The tree line follows the fault, separating bare Proterozoic rocks from the Cretaceous rocks over which they have been thrust.

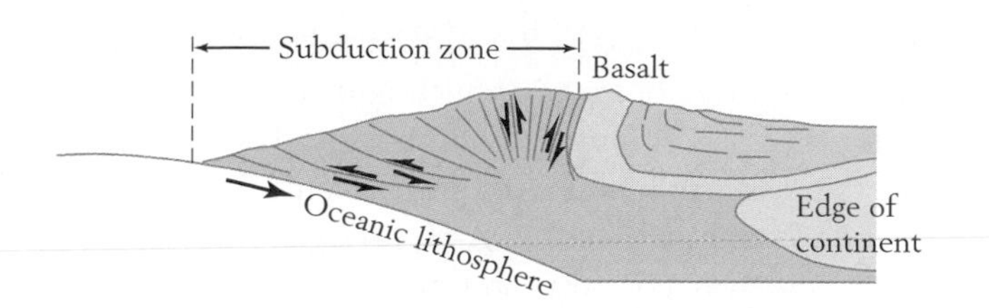

Figure 18-23 Formation of the Olympic Range in an embayment of the Pacific border of the state of Washington. Subduction piled sediments against basaltic rocks. *(After D. E. Kari and G. F. Sharman, Geol. Soc. Amer. Bull. 86:377–389, 1975.)*

the most part, positioned well to the east of Sevier orogenic activity (see Figure 18-21), and recall in addition that an eastward migration of orogenic activity had taken place during the Mesozoic Era, culminating in the Sevier orogeny (p. 487). The widely favored explanation for the earlier eastward shift applies to the Laramide shift as well: a central segment of the subducted plate that passed beneath North America assumed a still lower angle, extending a great distance eastward before becoming deep enough to melt and to create igneous activity within the overlying crust (see Figure 17-23). As in Cretaceous time, the angle of subduction must have resulted from an increase in the westward movement of North America and more rapid rollback of the subducted plate (see Figure 9-16).

Farther west, along the coast, the Great Valley of California continued to receive marine sediment, while northern California and the Sierra Nevada region remained as highlands. A separate basin in Washington and Oregon received deep-water sediments and layers of pillow lava. Here the Olympic Range began to form along a sharp inward bend of the subduction zone (Figure 18-23; and see also Figure 18-21).

The Eastern Uplifts and Basins The eastern belt of uplifts stretches from Montana to New Mexico. Because this is the region in which the central and southern Rocky Mountains developed during Neogene time, Paleogene events that preceded the uplift of this segment warrant special attention. Deformation here began in latest Cretaceous time with the origin of ranges and basins trending from north to south (Figure 18-24); in Utah and Wyoming these structures lay along the eastern margin of the northern fold-and-thrust belt. The uplift farthest to the east formed the Black Hills of South Dakota (see Box 11-1 and Figure 9-23). In Colorado, many ranges were formed by the elevation of large bodies of rock along thrust faults. It has been suggested that these uplifts were produced by a slight clockwise rotation of the Colorado Plateau, which behaved as a rigid crustal block, absorbing some of the convergence along the subduction zone to the west (see Figure 18-21).

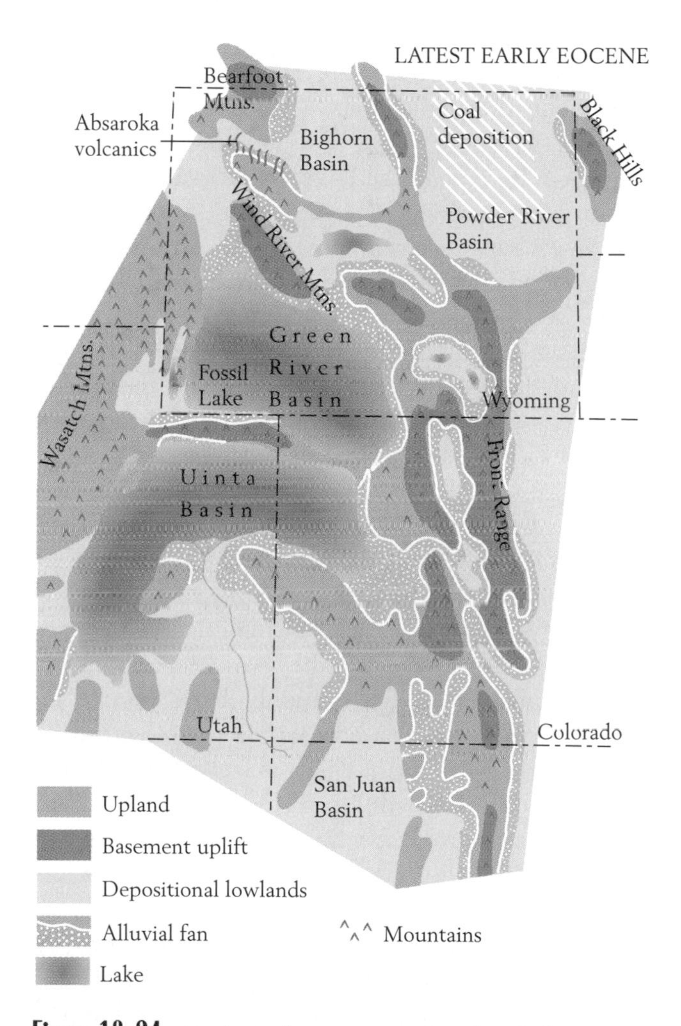

Figure 18-24 Geologic features associated with the Laramide uplifts of Colorado and adjacent regions at the end of early Eocene time. The cores of major uplifts consist of Precambrian crystalline rock. The Green River Formation, which is well known for its oil shales and splendid fossils, was accumulating in the Green River and Uinta basins. To the north, volcanism formed the Absaroka Mountains, where Yellowstone Park is now located. The Black Hills of South Dakota represent the easternmost uplift.

Today, of course, many peaks and ridges of the central and southern Rocky Mountains stand at very high elevations; the Front Range uplift, for example, rises far above the high plains of eastern Colorado (see Figure 15-28). It is inappropriate, however, to evaluate the effects of the Laramide orogeny simply by viewing the Rocky Mountains today, because, as we will see, the high elevations of the modern Rockies reflect post-Laramide uplift. During the Laramide orogeny, erosion nearly kept pace with uplift in most areas of the United States, so that the regional topography remained less rugged than it is today. Basins in front of elevated areas were receiving large volumes of rapidly eroding material.

By the end of early Eocene time the regional north–south pattern had weakened, and individual basins were experiencing independent histories. Most of these basins received alluvial and swamp deposits with abundant fossil mammal remains, and at times some were occupied by lakes.

Later in early Eocene time, lakes came to occupy most of the areas within the basins, and these lakes survived, sometimes at reduced size, throughout much of the Eocene Epoch (see Figure 18-24). The famous Green River deposits accumulated in and around their margins. Plant remains in these sediments, including fossil palms, reveal that climates in this region were much warmer in Eocene time than they are today, even subtropical at times (see Box 18-1). These lake deposits, which are extremely well laminated, have commonly been termed *oil shales,* because algal material within them has broken down to yield vast quantities of petroleum (see Figure 5-3). Unfortunately, this petroleum is disseminated throughout the rock and thus has proved difficult to extract. Nonetheless, the Green River deposits—the largest body of ancient lake sediments known—may eventually serve as a valuable source of fuel. The fine undisturbed lamination of the Green River lake deposits accounts for the remarkable preservation of a host of animal and plant fossils, including delicate creatures such as bats (see Figure 18-4).

The Yellowstone Hot Spot Another interesting group of rocks found in the region of Paleogene basins and uplifts are the volcanics that form the Absaroka Range in western Wyoming and Montana. A large portion of Yellowstone National Park lies within the Absarokas; here the still-active geysers and hot springs serve as evidence that igneous activity has not ceased completely. Recall that Yellowstone seems to represent a hot spot in which igneous activity is localized (p. 227). During Eocene time this area stood at the eastern margin of the volcanic belt of the Pacific Northwest (see Figure 18-21). Volcanism at this time was episodic, and all of the volcanic episodes were catastrophic, destroying entire forests and, we must assume, the animal life within them. Fossilized leaves, needles, cones, and seeds reveal the presence of lowlands with subtropical vegetation. Today the remnants of the Eocene forests can be seen at high elevations in Yellowstone National Park, where trees are preserved upright as stumps buried by lavas, mud flows, and flood-deposited volcanic debris (Figure 18-25). More than 20 successive forests, all of them killed in this way, can be identified (Figure 18-26).

Figure 18-25 Petrified stumps standing upright in the Absaroka volcanics at Specimen Ridge, Yellowstone National Park. Some logs here have diameters of about 1 meter (3 feet).

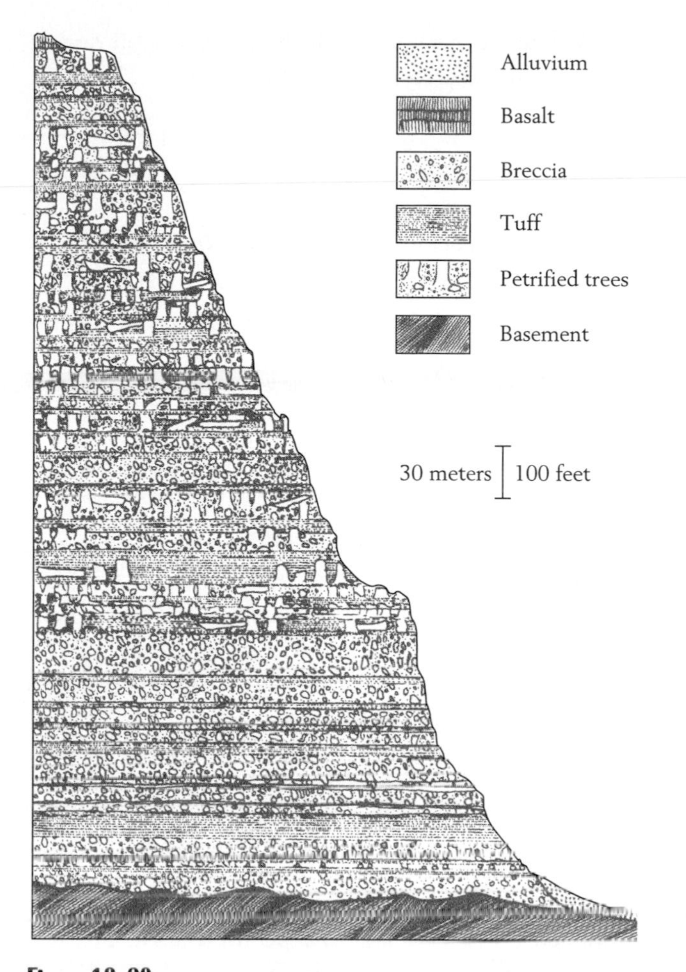

Figure 18-26 Succession of 27 petrified forests now exposed along the Lamar Valley in Yellowstone National Park. Each of these Eocene forests was destroyed by volcanic eruption. *(After E. Dorf, Scientific American, April 1964.)*

As the Eocene Epoch drew to a close, the level of volcanism declined sharply in every part of the northwestern United States except Oregon and Washington. In addition, most of the depositional basins that lay between Montana and New Mexico were filled with sediment by the end of Eocene time. The Laramide orogeny was completed, and the uplands that it had produced were largely leveled; thus, as the Oligocene Epoch dawned, a monotonous erosion surface stretched across western and central North America, interrupted by only a few isolated hills.

The Present-Day Rockies The modern Rocky Mountains are the product of a renewed uplift during Neogene time. In many areas in the Rockies today, a person can look into the distance and view a flat surface formed by the tops of mountains (Figure 18-27). This is what remains of the broad erosional surface that existed at the end of the Eocene Epoch, but this so-called subsummit surface now stands high above the Great Plains as a result of Neogene uplift.

During the Oligocene Epoch, although most of the Laramide uplifts had been leveled, a thin veneer of deposits spread as far east as South Dakota. The Badlands of South Dakota consist of rugged terrain carved mostly from Eocene and Oligocene deposits (p. 494). These deposits have yielded rich faunas of fossil mammals, and changes in the nature of their fossil soils reveal that aridity increased here, as it did in many parts of the world. Forests of late Eocene age gave way to open woodlands and finally, in late Oligocene time, to still drier savannahs.

Figure 18-27 The subsummit surface, formed by the flat-topped Rocky Mountains. This surface, seen here near Derby Peak, Colorado, formed near the end of the Eocene Epoch and has since been uplifted and dissected by erosion.

The Gulf Coast

Unlike the Cordilleran region, the Gulf Coast of North America has remained an area of tectonic quiescence throughout the Cenozoic Era. The sea retreated southward in latest Cretaceous time, draining the Interior Seaway (see Figure 17-12). During the Paleocene, marine waters still occupied the Mississippi Embayment, an inland extension of the Gulf of Mexico where a thick sequence of Eocene marine sediments accumulated (Figure 18-28). In the Oligocene Epoch the seas withdrew to the approximate position of the present shoreline of the Gulf of Mexico and then spread inland, but less far than they had during most of Eocene time. The total thickness of Paleogene sediments near the present coastline of the Gulf of Mexico exceeds 5 kilometers (3 miles), largely because of the enormous quantities of sediment carried to the region by the Mississippi River system. The Cenozoic clastic wedge in this region, though largely buried, has been studied in great detail during the successful search for petroleum there.

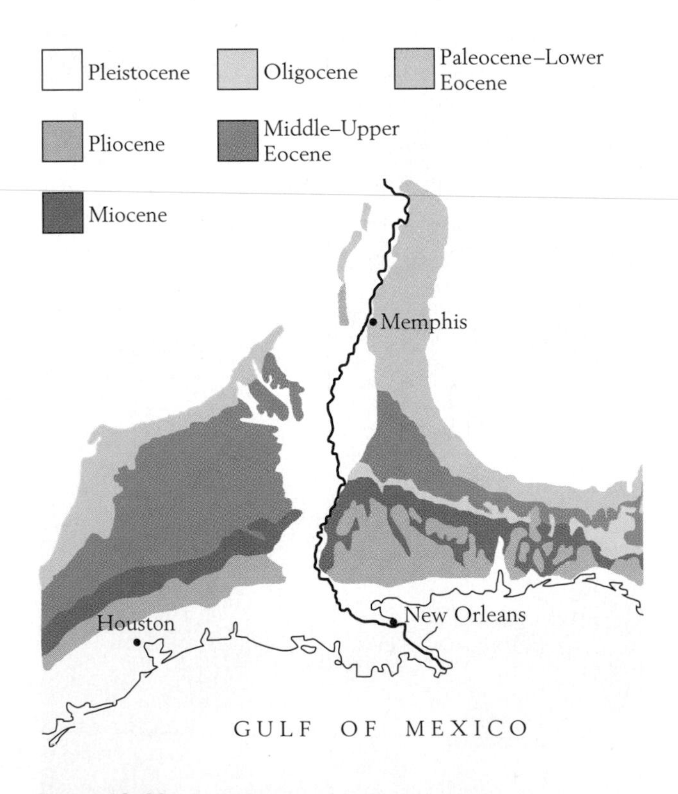

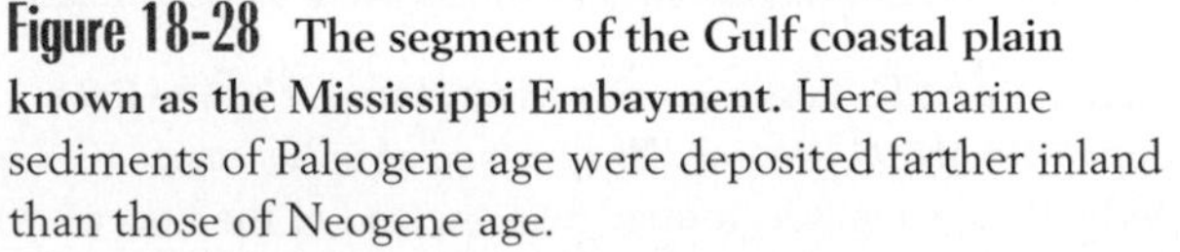

Figure 18-28 The segment of the Gulf coastal plain known as the Mississippi Embayment. Here marine sediments of Paleogene age were deposited farther inland than those of Neogene age.

Chapter Summary

1 Marine life of the Paleogene Period resembled that of the modern world.

2 Plants on land also resembled those of the present time, except that grasses were not widespread until late in Paleogene time.

3 Terrestrial vertebrate animals were more primitive than those of the present. Although most of the mammalian orders alive today existed in Paleogene time, many Paleogene families and genera are now extinct. Most species of birds were wading animals that resembled storks or herons.

4 Fossil floras and oxygen isotope ratios for planktonic foraminifera show that very warm climates extended to high latitudes early in the Eocene Epoch.

5 Late in Eocene time, polar regions cooled, and cold, dense waters began to descend to form the cold layer of the deep sea that has persisted to the present time.

6 Climates cooled in at least some regions during late Eocene time, and cooling and drying persisted into Oligocene time. Partly as a result of these trends, dry woodlands and savannahs replaced dense forests in many parts of the world. Many species became extinct both on the land and in the sea.

7 Both north and south polar regions underwent plate tectonic changes in mid-Paleogene time. In the south, South America and Australia broke away from Antarctica, which was left isolated over the south pole. In the north, Europe and Greenland separated.

8 In North America the Laramide orogeny produced northern and southern fold-and-thrust belts that were separated by a curious zone of uplifts. By the end of the Eocene Epoch, the Laramide orogeny had ended, and the mountains it had produced had been subdued by erosion.

Review Questions

1 What groups of animals that played important roles in Late Cretaceous ecosystems were absent from the Paleocene world?

2 Which group changed more from the beginning of the Paleogene Epoch to the end—marine or land animals? Explain your answer.

3 What seems to have prevented corals from forming massive reefs before Oligocene time?

4 Which animals took the place of Mesozoic reptiles in Paleogene oceans?

5 How does the fossil record of flowering plants reveal climatic change during Paleogene time?

6 What happened to the deep sea at the end of the Paleocene Epoch?

7 How did changes in patterns of ocean circulation bring about glaciation in Antarctica?

8 How did the location of the Laramide orogeny differ from that of the Cretaceous Sevier orogeny? What might explain this change?

9 What did the region of the modern Rocky Mountains look like at the end of the Eocene Epoch?

10 What is the origin of the geologic features that intrigue tourists at Yellowstone National Park?

11 Climatic changes occurred throughout the world during the latter part of the Paleogene Period. Using the Visual Overview on page 496 and what you have learned in this chapter, explain what caused these climatic changes and describe the effects of these changes on life in the sea and on the land.

Additional Reading

Marshall, L. G., "The Terror Birds of South America," *Scientific American*, February 1994.

McGowran, B., "Fifty Million Years Ago," *American Scientist* 78:30–39, 1990.

Prothero, D. R., *The Eocene-Oligocene Transition: Paradise Lost*. Columbia University Press, New York, 1994.

Savage, R. J. G., and M. R. Long, *Mammal Evolution: An Illustrated Guide*, Facts on File, Inc., New York, 1986.

Wolfe, J., "A Paleobotanical Interpretation of Tertiary Climates in the Northern Hemisphere," *American Scientist* 66:694–903, 1978.

CHAPTER 19

The Neogene World

Because it leads up to the present, the Neogene Period holds special interest for us. It was during the Neogene that the modern world took shape—that is, global ecosystems acquired their present configuration and prominent topographic features assumed the forms we are familiar with today.

No mass extinction marked the transition from the Paleogene to the Neogene. During the scant 24 million years of the Neogene, however, life and Earth's physical features have changed significantly. The most far-reaching biotic changes were the spread of grasses and weedy plants and the modernization of vertebrate life. Snakes, songbirds, frogs, rats, and mice expanded dramatically too, and humans evolved from apes. The Rocky Mountains and the less rugged Appalachians took shape during Neogene time, as did the imposing Himalayas. The Mediterranean Sea almost disappeared and then rapidly formed again. The most widespread physical changes on Earth, however, were climatic. Glaciers expanded across large areas of North America and Eurasia late in Neogene time. Although this glacial interval is commonly thought of as corresponding to the Pleistocene Epoch, it actually began during the Pliocene Epoch and presumably will continue long into the future. The spreading of glaciers has been episodic, and today we live during the latest of numerous intervals between pulses of severe glaciation. Today ice caps in the far north remain poised to spread southward, as they have done repeatedly during the past 2 million years or so.

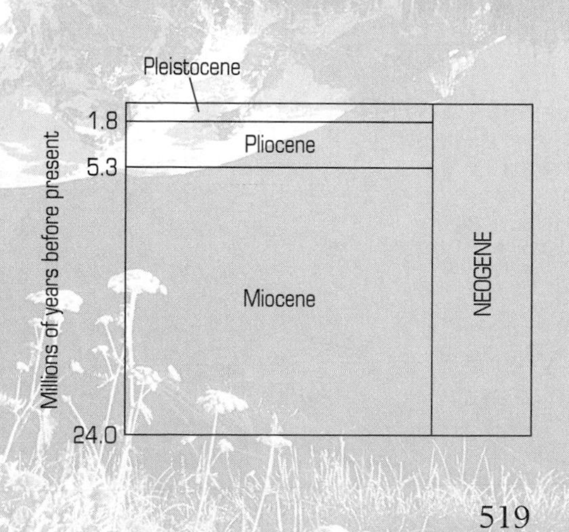

Mt. Rainier, a volcanic peak in the state of Washington that formed during the Neogene Period.

Visual Overview

Major Events of the Neogene

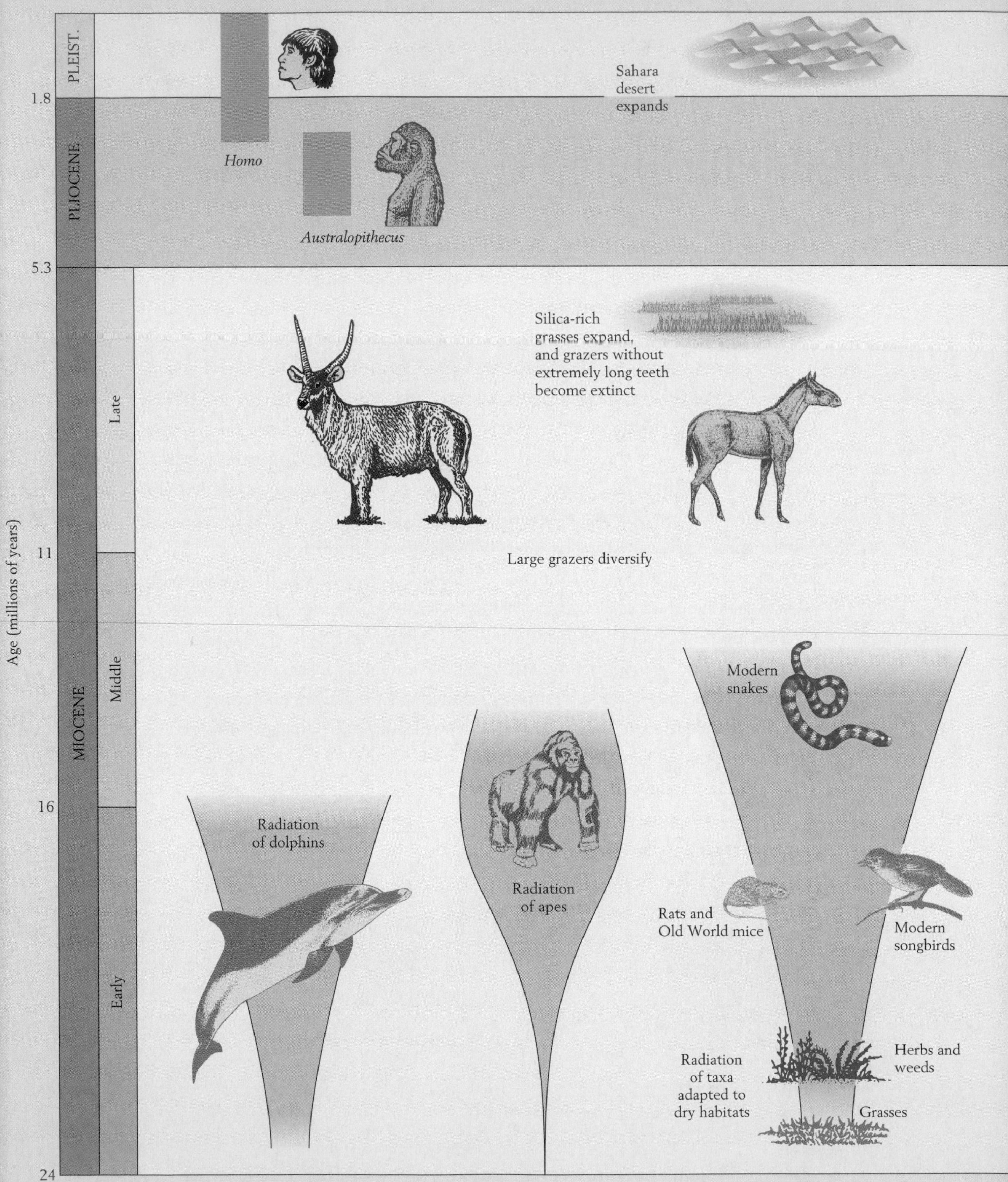

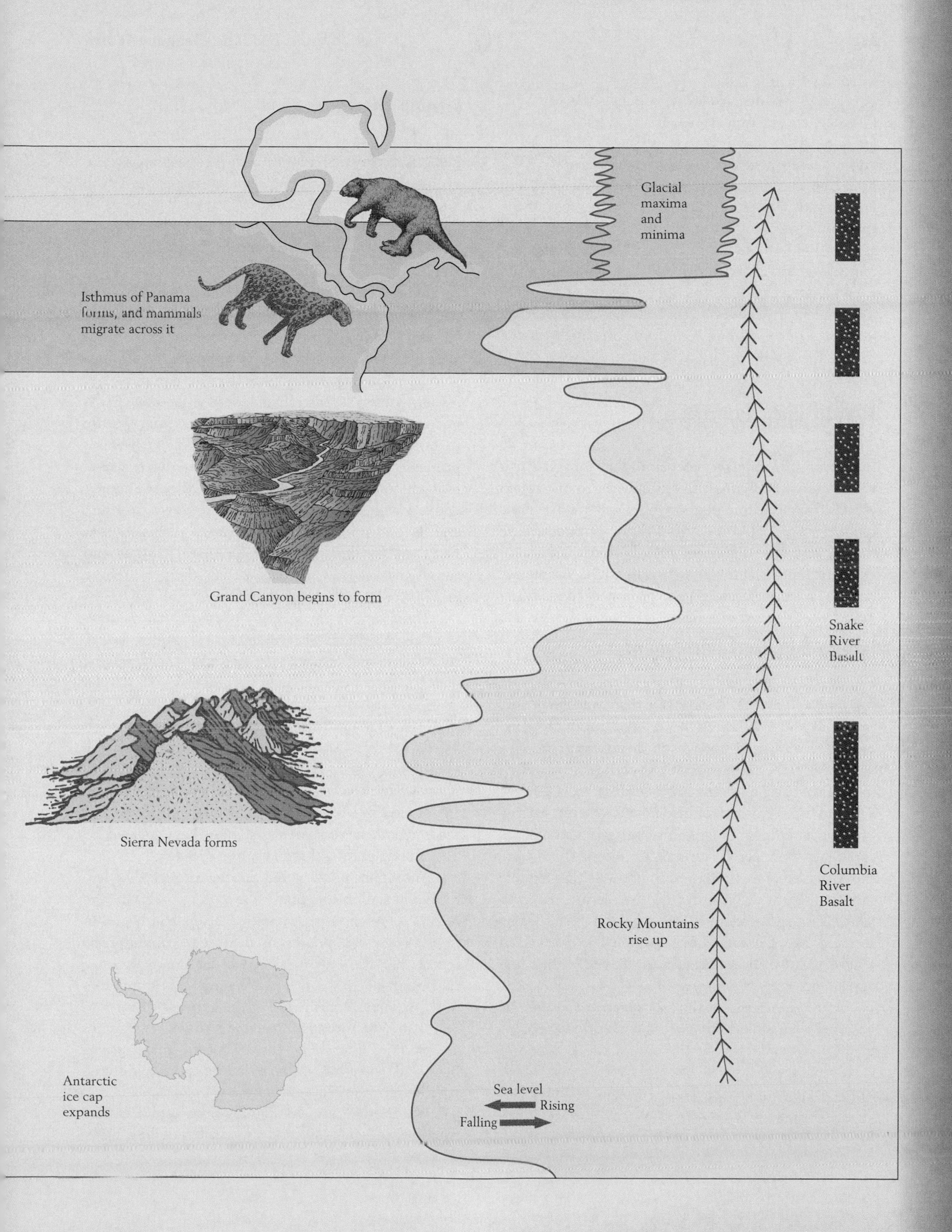
Glacial maxima and minima
Isthmus of Panama forms, and mammals migrate across it
Grand Canyon begins to form
Snake River Basalt
Sierra Nevada forms
Columbia River Basalt
Rocky Mountains rise up
Antarctic ice cap expands
Sea level
Rising
Falling

The first three Neogene epochs—the Miocene, Pliocene, and Pleistocene—were founded by Charles Lyell in 1833 in his *Principles of Geology.* Lyell distinguished the epochs of the Neogene Period on the basis of his observations of marine strata and fossils in France and Italy, noting that about 90 percent of the molluscan species found in Pleistocene strata are still alive in modern oceans, but that Pliocene strata contain fewer surviving species and that Miocene strata contained fewer still. It was not until later in the nineteenth century, however, that glacial deposits were recognized on land and found to correlate with the marine Pleistocene record.

Worldwide Events

In general, the animals and plants that inhabit Earth today are representative of Neogene life, so once again we are dealing with a period whose general life forms require no special introduction. Also, continents have moved very little during the brief Neogene interval of geologic time. On the other hand, shifts in climate have led to major changes in environments and life.

Life in Aquatic Environments

Charles Darwin observed long ago that invertebrate life tends to evolve less rapidly than vertebrate life; thus the short Neogene Period has produced only modest evolutionary changes in invertebrate life. Not surprisingly, the most dramatic evolutionary development in Neogene oceans has been the expansion of a group of vertebrate animals—the whales. During the Miocene Epoch a large number of whale species came into existence (Figure 19-1); among them were the earliest representatives of the groups that include modern sperm whales, which are carnivores with large teeth, and modern baleen whales, which feed by straining zooplankton from seawater. The specialized whales that we know as dolphins also made their first appearance early in Miocene time.

At the other end of the size spectrum for pelagic life, the planktonic foraminifera had suffered greatly in the mass extinction at the end of the Eocene but expanded again early in the Miocene Epoch. These forms serve as index fossils for oceanic sediments of Neogene age. On the seafloor, evolutionary changes from Paleocene time were relatively minor.

Life on the Land

Climatic changes exerted profound influence over Neogene terrestrial biotas, and the geographic and evolutionary modifications of biotas preserved in the fossil record help us reconstruct these changes. As in Late Cretaceous and Paleogene times, flowering plant fossils represent our best gauge of climatic shifts.

Flowering Plants: Climatic Deterioration and an Explosion of Herbs In the world of plants, the Neogene Period might be described as the Age of Herbs. Herbs, or herbaceous plants, are small, nonwoody plants that die back to the ground after releasing their seeds. (Defined in this way, herbs include many more plants than the few that we use to season our food.) The recent success of herbs is primarily the result of the worldwide deterioration of climates during Oligocene and Miocene times, when cooler, drier conditions caused forests to shrink and opened up new environments to plants such as herbs and grasses, which prefer open habitats and can withstand low rainfall. Today there are some 10,000 species of grasses alone.

The Compositae, a family of herbs that includes such seemingly distinct members as daisies, asters, sunflowers, and lettuces, appeared near the beginning of the Neogene Period, only 20 million or 25 million years ago, yet today this family contains some 13,000 species, including the plants that ecologists refer to as weeds (p. 97). As any gardener knows, weeds are exceptionally good invaders of bare ground. They may not compete successfully against other plants to retain the space they invade, but they soon disperse their seeds to other bare areas that have been cleared by fires, floods, or droughts and spring up anew.

The appearance in the Southeast Pacific of ice-rafted coarse sediments shows that early in the Miocene Epoch large Antarctic glaciers had begun to flow to the sea. Cores of deep-sea sediment also reveal that the belt of siliceous diatomaceous ooze that encircled Antarctica (see Figure 5-35) simultaneously expanded northward at the expense of carbonate ooze, which tends to accumulate where climates are warmer. Ever since early Miocene time, assemblages of plankton at high latitudes have differed greatly from those at low latitudes.

Water evaporates less rapidly from cool seas than from warm seas, resulting in transfer of less water

Figure 19-1 **Reconstruction of the marine fauna represented by fossils of the middle Miocene Calvert Formation of Maryland.** Representing the whale family were early baleen whales *(Pelocetus)*, which strained minute zooplankton from the sea (center); long snouted dolphins (*Eurhinodelphis*, lower left); and short-snouted dolphins (*Kentriodon*, upper right). Sharks include the six-gilled shark *Hexanchus* (lower right). *(Drawing by Gregory S. Paul.)*

from the ocean to the land. Thus, Miocene cooling reduced rainfall in many continental regions.

Analyses of leaf margins from terrestrial floras of North America reveal only modest fluctuations during Neogene time, a trend that contrasts sharply with the dramatic drop in temperatures that marked the Eocene-Oligocene interval (see Figure 18-17). A slight cooling trend began early in the Oligocene Epoch, however, and continued until the Ice Age began, late in the Pliocene Epoch. During this interval, climates not only became slightly cooler but also grew drier and more seasonal. As a result, dry grasslands expanded into areas that had once been open woodlands and dense forests.

Modernization of Terrestrial Vertebrates Most of us are so interested in the origins of large mammals that we often ignore the great success of smaller creatures. In fact, the Neogene Period might well be called the Age of Frogs, the Age of Rats and Mice, the Age of Snakes, or the Age of Songbirds, because all four of these groups have undergone tremendous evolutionary radiations over the past few million years (see p. 520).

Many Late Neogene food webs include plants, herbivores, and carnivores that these evolutionary radiations produced. This ecologic relationship indicates that the radiations of plants stimulated those of herbivores, and that these radiations in turn stimulated the diversification of snakes. Many species of rats and mice dig burrows in dry terrain and eat the seeds of grasses and herbs, for example. To a large extent, the success of these small rodents during Neogene time resulted from that of both the grasses and the Compositae and also, more fundamentally, from the drying and cooling of climates that favored these plants. Perhaps we can attribute the success of modern frogs and toads, whose species number about 2000, to their remarkable ability to catch insects through the quick protrusion of their long tongues. In any case, the snakes have obviously flourished largely because of the proliferation of frogs and rodents; few other predators can pursue small rodents down their burrows without digging. Before the start of the Neogene Period, there were few snakes except for members of the primitive boa constrictor group. Today, however, the more advanced snakes of the family Colubridae include about 1400 species, many of which are poisonous.

Also poorly represented before Neogene time were the passerine birds, or songbirds and their relatives, which are highly conspicuous today. These birds have also benefited from the diversification of seed-bearing species of herbs, but, like frogs, they owe some of their success to their ability to capture flying insects. It is probable that many types of flying insects were not heavily preyed upon until groups of passerine birds took to the air.

Of course, groups of large animals also developed their modern characteristics during the Neogene Period. Among the herbivores, the horse and rhinoceros families dwindled after mid-Miocene time in a continuation of the general decline of the odd-toed ungulates. Meanwhile, the even-toed, or cloven-hoofed, ungulates expanded, especially through the adaptive radiation of both the deer family and the family called the Bovidae, which includes cattle, antelopes, sheep, and goats. The giraffe family and the pig family also radiated during the Miocene Epoch, but the number of species in these families has since declined. Similarly, many types of elephants, including those with long trunks, experienced great success during the Miocene and Pliocene intervals but later declined. Today only two elephant species survive: the large-eared African elephant and the smaller, more docile Indian elephant—the species that is commonly trained to perform in circuses.

Carnivorous mammals also assumed their modern character in the course of the Neogene Period; this group included the dog and cat families, both of which had appeared during Paleogene time. The bear and hyena families were other important Miocene additions to the carnivore group.

Many of the Neogene mammal groups expanded successfully because of the spread of grassy woodlands (Figure 19-2). Several herbivore groups, such as antelopes, cattle, and horses, evolved many species

Figure 19-2 **Reconstruction of the so-called *Hipparion* fauna of Asia.** This diverse fauna occupied open woodlands in Asia about 10 million years ago, in late Miocene time. The galloping horse is *Hipparion*. The elephant on the left, with downward-directed tusks, is *Dinotherium*. In the foreground, short-legged hyenas of the genus *Percrocuta* look on from their den. *(Drawing by Gregory S. Paul.)*

that were well adapted for long-distance running over open terrain and that grazed on harsh grasses with the aid of high-crowned molar teeth. Many grasses contain tiny fragments of silica, and only long teeth can tolerate the resulting wear. Also on the increase were the groups of rodents that are adapted for burrowing in prairies and the elephants, which require open habitats simply to move around. As we might expect, the diversification of herbivores in savannahs and woodlands in turn fostered the success of groups of carnivores that were well adapted for attacking herbivores in open country—groups such as hyenas, lions, cheetahs, and long-legged dogs.

What about primates, the group to which we humans belong? In general, primates favor forests over savannahs; in fact, most live in trees. As we have seen, monkeys were present by Oligocene time; the oldest group includes the so-called Old World monkeys, which now live in Africa and Eurasia. Before the end of the Oligocene interval, however, a distinctive group of monkeys reached South America. How they got there remains uncertain. These New World monkeys, which differed from their Old World counterparts in that most possessed prehensile (or grasping) tails, probably had a separate evolutionary origin. In any event, monkeys on both sides of the Atlantic underwent adaptive radiations during Neogene time.

Apes, which evolved in the Old World, flourished for a time but have since declined in number of species. We will discuss apes and apelike animals when we examine the origins of humans, which belong to the same superfamily, the Hominoidea (see Figure 3-7). The most recent phases of human evolution have taken place within the climatic context of the recent Ice Age; therefore, before we discuss the Hominoidea, it is appropriate to examine the major global events of this fascinating interval of geologic time and of the Pliocene Epoch that preceded it.

The Spread of C_4 Grasses and the Extinction of Large Herbivores Between 7 million and 6 million years ago, near the end of the Miocene Epoch, North American mammals suffered the largest extinction event that has struck them since the crisis at the end of the Eocene (p. 509). Species of large herbivores that lacked extremely tall molar teeth formed the largest group of victims. A similar extinction event occurred in Asia. It was not a spread of grasslands that eliminated these species, because they were predominantly forms whose tall molars were adapted to grazing on grasses. It was only grazing species whose molars were extremely tall that survived, however.

What changes in the environment could have caused this kind of selective extinction? As it turns out, the extinction event coincided with a major change in the kind of grasses that formed grasslands. The evidence of this change comes from stable isotopes (carbon 12 and carbon 13). When plant material decays, it releases carbon into the soil. The isotope ratio of this carbon reflects the ratio in the plants that produced it. Similarly, the carbon isotope ratio in herbivores reflects that of the plants they eat. Analysis of ancient grassland soils and the teeth of the herbivores that grazed on those grasslands has revealed that a pronounced shift toward heavier carbon isotope ratios occurred on many continents between 7 million and 6 million years ago (Figure 19-3). This shift reflects the partial replacement of the group of grasses known as C_3 grasses by the group known as C_4 grasses. These two groups of grasses have different physiologies. When the C_4 grasses extract carbon dioxide from the atmosphere, they assimilate a larger fraction of carbon 13 than C_3 grasses do.

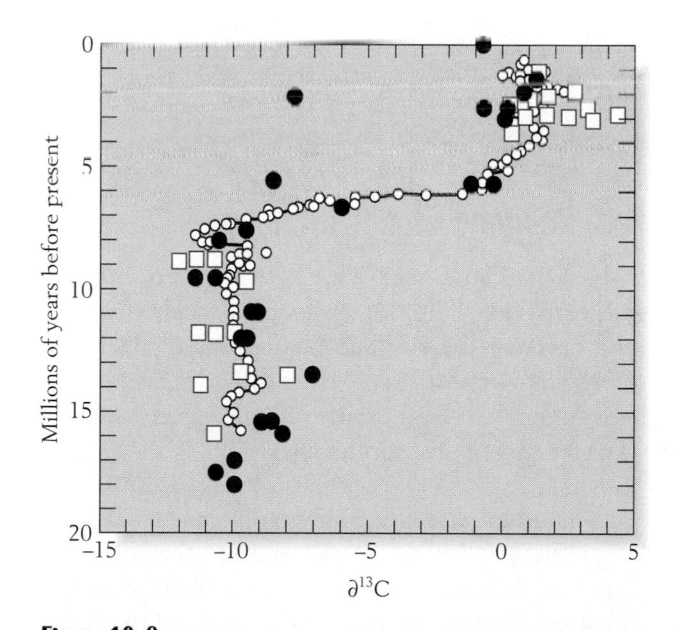

Figure 19-3 **Major shifts in carbon isotopes indicating the spread of C_4 grasses between 7 million and 6 million years ago.** The plotted values are carbon isotope ratios from ancient soils and mammal teeth from Pakistan and North America. *(After T. E. Cerling, Y. Wang, and J. Quade, Nature 361:344–346, 1993.)*

Recall that silica in grasses is what makes them highly abrasive to herbivores. It happens that C_4 grasses have about five times as much silica in their leaves as C_3 grasses. Clearly, the expansion of C_4 grasses near the end of the Miocene caused the extinction of grazing herbivores that lacked extremely long teeth. The teeth of grazing animals wear down during their lifetimes, and these animals end up suffering from malnutrition, which leads to their death. This is why looking a gift horse in the mouth is actually a good idea: a horse with heavily worn teeth is an old animal with a short life expectancy. Apparently when C_4 grasses first spread into many regions of the world near the end of the Miocene, they abraded the teeth of grazers that lacked very long molars and so caused their extinction.

Why C_4 grasses expanded their geographic range so rapidly remains unclear. Because they can prosper at lower atmospheric concentrations of carbon dioxide than C_3 plants can, some researchers favor the idea that a decline in atmospheric carbon dioxide caused their sudden spread, but there is no independent evidence that such a decline occurred. Because C_4 plants also tolerate warm, dry conditions, other researchers favor the idea that a climatic change triggered their expansion, and in fact there is evidence of this kind of climatic change near the end of the Miocene. Finally, there is the possibility that the evolution of new kinds of C_4 grasses triggered the floral change. Unfortunately, the fossil record of grasses is too poor to permit this idea to be tested.

Late Neogene Climatic Change

Early in the Pliocene Epoch, relatively warm climates spread to high latitudes. Partway through the Pliocene, however, the northern hemisphere plunged into the modern Ice Age, and climates in many regions have been cooler and drier ever since, even at times when ice sheets have shrunk back.

Pliocene Equability Stratigraphic unconformities in various parts of the world indicate that near the end of the Miocene Epoch, about 6 million years ago, global sea level fell by perhaps 50 meters (165 feet). This event probably resulted from removal of water from the ocean by the expansion of glaciers. Cooling and drying of climates at this time may have caused the spread of C_4 grasses.

As the Pliocene Epoch got under way, about 5 million years ago, sea level rose again, leaving marine deposits inland of coastlines in such areas as California, eastern North America, and countries bordering the North Sea and the Mediterranean. Fossil faunas and floras also reveal that global climates during this time were more equable than they are today; pollen analyses, for example, indicate that southeastern England was subtropical, or nearly so. Especially in the Northern Hemisphere, however, this warm interval came to a sudden close with the start of the modern Ice Age, about 3.2 million years ago.

Continental Glaciation Figure 19-4 depicts Earth at a time of full glacial expansion, or what is termed a **glacial maximum.** Glacial maxima have alternated with times of glacial recession, such as the present. During these warmer intervals, or **glacial minima,** the Greenland ice cap is the only continental glacier to survive in the northern hemisphere (see Figure 4-13). A wide variety of evidence documents the recent Ice Age, revealing details of the timing and geographic distribution of continental glaciation.

Figure 19-4 Reconstruction of glaciers as they existed during a typical glacial interval of the Pleistocene Epoch. Large continental glaciers were centered in North America, Greenland, and Scandinavia. *(After a drawing by A. Sotiropoulos in J. Imbrie and K. P. Imbrie, Ice Ages, Enslow, Short Hills, N.J., 1979.)*

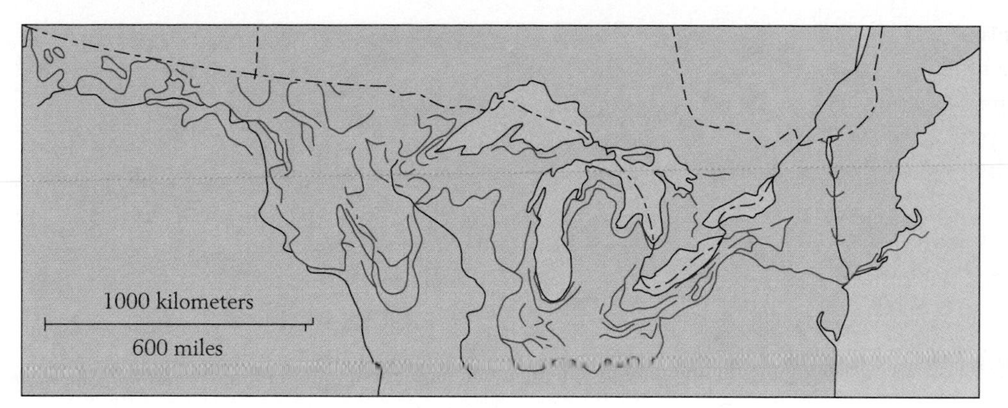

Figure 19-5 **Positions of moraines that mark the southern limits of ice sheets in eastern North America during the most recent glacial maximum, about 20,000 years ago.** *(After D. M. Mickelson et al., in Quaternary Environments of the United States, vol. 1, S. C. Porter [ed.], Longman, London, 1983.)*

1 *Erratic boulders* Large rocks that sit on Earth's surface far from exposures of the bedrock from which they have broken, so-called **erratic boulders** are too large to have been transported by rivers, and it is difficult to imagine that any agent other than continental glaciers might have transported them.

2 *Glacial till* A mixture of boulders, pebbles, sand, and mud that has been plowed up, transported, and then deposited by glaciers, till is difficult to confuse with sediments deposited by other mechanisms, especially where it rests at Earth's surface and forms ridges known as moraines or is associated with outwash deposits (p. 127). Such moraines form much of Cape Cod, Massachusetts, where they extend into the marine realm. Retreating glaciers commonly left terminal moraines behind them, and when these glaciers melted back, they often left shallow basins in which water accumulated behind the moraines. The Great Lakes of North America occupy such basins; they did not exist before the modern Ice Age (Figure 19-5).

3 *Depression of the land* Earth's crust remains depressed in regions that lay beneath large glaciers a few thousand years ago. Hudson Bay, the only epicontinental sea that exists today in North America, occupies such a depressed region in eastern Canada (see Figure 12-1).

4 *Glacial scouring* Glaciers smoothed the sides of mountains that they scraped past. Mount Monadnock, in New Hampshire, stood partially above surrounding ice sheets, as some mountains of Antarctica do today (Figure 19-6). The lower part of Mount Monadnock, which was smoothed by flowing glaciers, stands in sharp contrast to the upper part, which remains rugged. Small glaciers, known as alpine or mountain glaciers, left their marks along valleys within the Rockies and other mountain chains, where they flowed during the recent Ice Age. Here the most spectacular products of glaciers are U-shaped valleys that they sculpted from valleys that were once shaped like a V (see Figure 12-17).

5 *Lowering of sea level* One important effect of each major expansion of ice sheets during the Pleistocene Epoch was a profound lowering of sea level as great quantities of water were locked up on the land. During major glacial expansions, most of the surfaces that now form continental shelves stood above sea level. Rivers cut rapidly downward through the soft sediments of many continental

Figure 19-6 **Dark rock of the Prince Charles Mountains projecting above the surface of the modern Antarctic ice cap.** Many North American mountains were partly buried in ice during the Pleistocene Epoch.

shelves to form valleys that exist today as submarine canyons, having been excavated further by submarine turbidity currents. During some glacial episodes, sea level dropped slightly more than 100 meters (330 feet) below its present position.

Today there remain only two ice caps of the sort that expanded to cover broad areas many times during the Pleistocene Epoch. One of these modern ice caps covers much of Greenland (see Figure 4-13) and the other covers nearly all of Antarctica (see Figure 19-6). Today about three-quarters of the world's fresh water is locked up in glacial ice, and about 90 percent of it belongs to the Antarctic ice cap. It may seem impressive that glaciers now contain about 25 million cubic kilometers of ice, but it has been estimated that the volume of ice was nearly three times as great during glacial advances of the Pleistocene Epoch, with larger ice sheets averaging about 2 kilometers (1.2 miles) in thickness. The total volume of ice has been calculated from the volume of water that was removed from the ocean to lower sea level slightly more than 100 meters. Shelves of ice projected into the sea, and these shelves, together with the icebergs and pack ice that broke loose from them, spread over half the world's oceans.

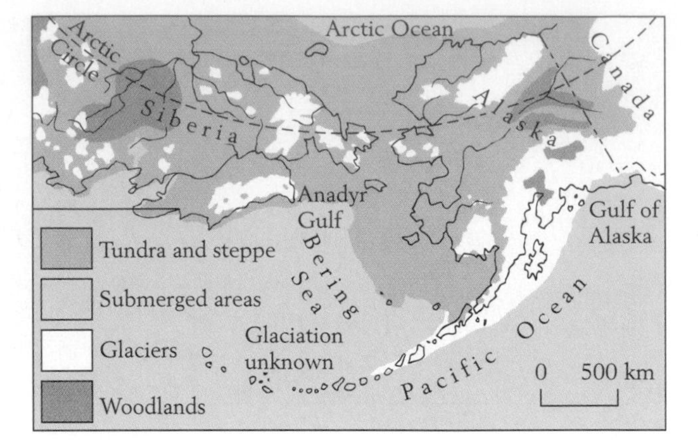

Figure 19-7 The geography of Beringia during the most recent glacial interval (the Wisconsin, or Riss-Wurm, Stage). Though positioned at a high latitude, Beringia was dry and free of large glaciers. *(After D. M. Hopkins, in D. M. Hopkins [ed.], The Bering Land Bridge, Stanford University Press, Stanford, Calif., 1967.)*

6 *Migration of species* As ice sheets have expanded repeatedly and sea level has dropped, the resulting geographic changes have allowed species to migrate to new regions. Regression of the seas during glacial episodes turned the Bering Strait into a land corridor between Asia and North America, and it was by this land bridge that many mammals, including the first humans, entered the New World. Ironically, this region, which was hospitable to terrestrial mammals during the height of glaciation (Figures 19-7 and 19-8), included por-

Figure 19-8 Reconstruction of the mammalian fauna that occupied the steppe in the Alaska portion of Beringia, about 12,000 years ago, during the most recent glacial interval. Of 61 species depicted here, 11 are extinct; among them are the woolly mammoth, the American mastodon, the long-horned bison, a lion, and a saber-toothed cat. *(Painting by Jay H. Matternes.)*

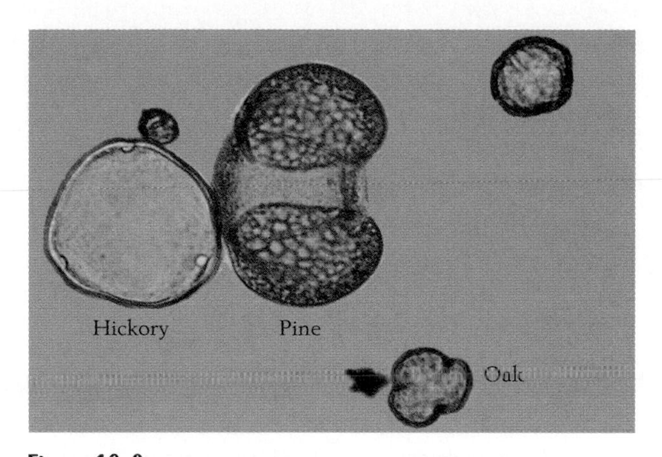

Figure 19-9 Three common forms of pollen found in Pleistocene sediments of eastern North America.

tions of Siberia and Alaska—areas that we now view as inhospitable to most species but that remained unglaciated because prevailing weather patterns brought them little snow.

The alternations of glacial maxima and minima have caused climatic belts and the floras and faunas that occupy them to shift over distances measured in hundreds of kilometers. Fossils of mammals such as the muskrat, which today does not range south of Georgia, reveal that climates in Florida were cool when glaciers pushed southward into the northern United States. Other fossil occurrences, such as those of hippopotamuses in Britain, show that during at least some interglacial intervals, climates were warmer than they are today.

One of the most useful fossil indicators of Pleistocene climates is the pollen of terrestrial plants (Figure 19-9). Pollen assemblages reveal climatic change by indicating the shifting of floras to the north or south. Figure 19-10 shows the southward movement of floras in Europe by about 20° latitude during the most recent glacial interval there.

Studies of the distribution of fossil pollen have revealed, however, that whole floras have not migrated as units during recent climatic changes. Rather, individual species have shifted their ranges independently. The result has been constant reshuffling of species within communities. The discovery of this phenomenon has overturned the long-popular idea that each major plant community on Earth today, such as the deciduous forest of the eastern United States or the Douglas fir forest of the Pacific Northwest, is an ancient entity whose component species have evolved in association with one another. Instead, such communities are transient assemblages that have formed since the last glacial maximum.

The Chronology of Glaciation The Pleistocene Epoch is often thought of as the modern Ice Age, but the Ice Age actually began long before the end of the Pliocene Epoch. The most detailed chronology of the glaciation comes from oxygen isotope ratios of

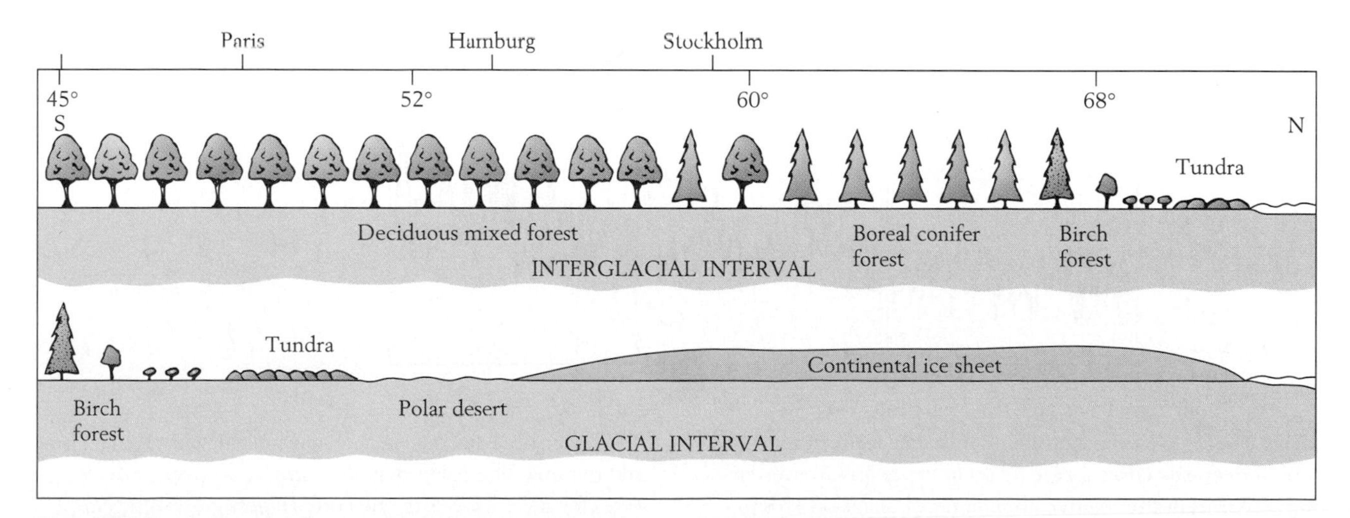

Figure 19-10 North–south migration of vegetation in Europe during the Pleistocene Epoch. During glacial intervals (lower diagram) continental glaciers spread southward to the vicinity of Hamburg, and tundra shifted to the latitude of Paris. *(After T. Van Der Hammen, in K. K. Turekian [ed.], The Late Cenozoic Glacial Ages, Yale University Press, New Haven, Conn., 1971.)*

foraminiferan skeletons preserved in deep-sea sediments. Recall that this ratio shifts toward heavier values during glacial episodes for two reasons. First, a disproportionate amount of the lighter isotope, oxygen 16, accumulates in glaciers, leaving the ocean enriched in oxygen 18. Second, as temperatures decline, foraminifera take up a larger percentage of oxygen 18 from the seawater in which they live.

Slightly before 3 million years ago there were marked increases in the ratio of oxygen 18 to oxygen 16 in the skeletons of foraminifera in many oceanic areas (Figure 19-11). This change resulted from a brief episode of widespread cooling, which terrestrial floras also document. In northwestern Europe, for example, several subtropical species of land plants, including palms, disappeared.

Continental glaciers began to expand slightly thereafter, and large oscillations of oxygen isotope ratios in planktonic foraminifera indicate that these glaciers expanded and contracted every few tens of thousands of years (see Figure 19-11). The isotope ratios indicate that by about 2.5 million years ago the northern hemisphere had moved fully into the Ice Age. In addition, deep-sea deposits of this age in the North Atlantic record the first occurrence of numerous sand grains released by melting icebergs. In other

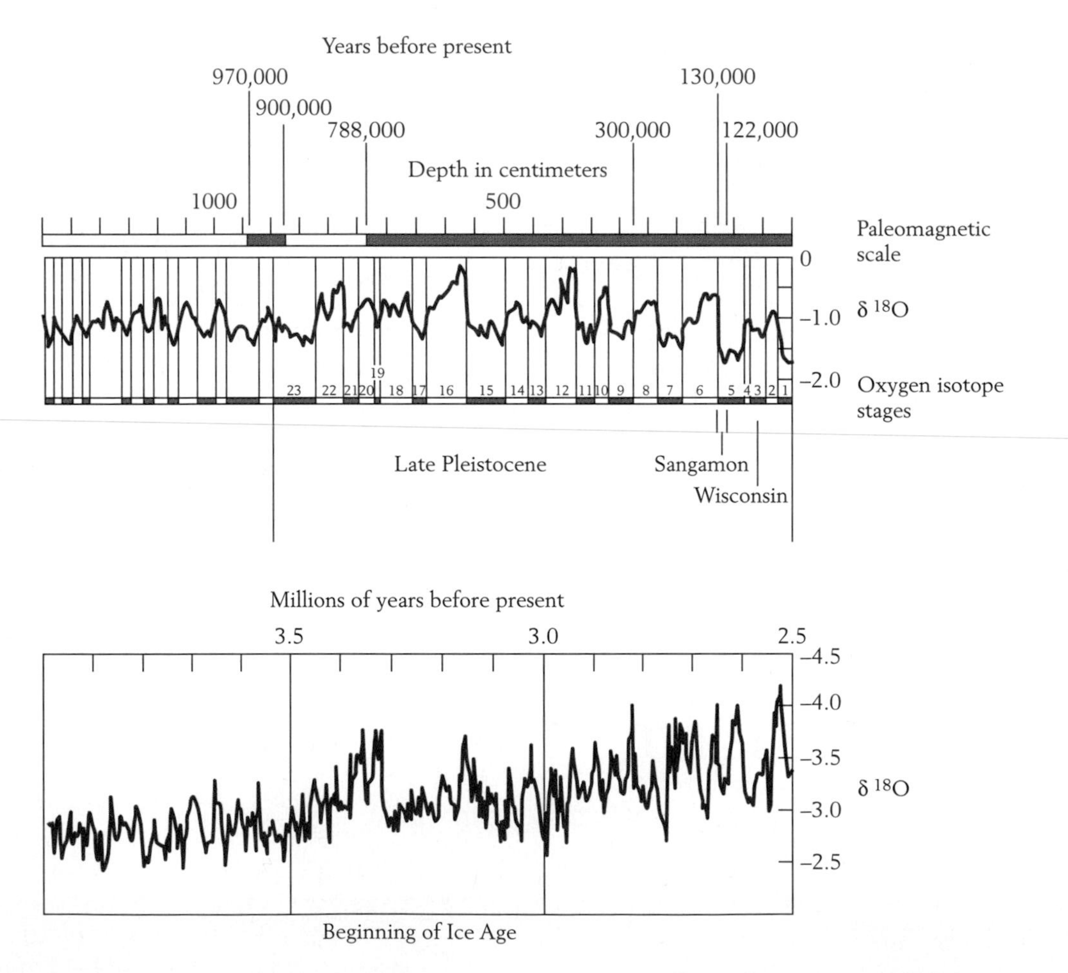

Figure 19-11 Oxygen isotope fluctuations for foraminifera within deep-sea cores. Large peaks in the graphs, representing increases in the relative abundance of oxygen 18, represent glacial maxima. The lower graph shows the initial buildup of ice sheets between 3.5 million and 3.0 million years ago. Below the upper graph are oxygen isotope stages (1–23) for the late Pleistocene, representing glacial maxima and minima. The paleomagnetic time scale (top) provides dates for several levels in the core. *(Upper graph after N. J. Shackelton and N. D. Opdyke, Geol. Soc. Amer. Mem. 145: 449–464, 1976; lower graph after R. Tiedmann, M. Sarnthein, and N. J. Shackleton, Paleoceanography 9:619–638.)*

words, continental glaciers were now flowing to the North Atlantic, releasing sediment-laden blocks of ice. Terrestrial floras underwent dramatic changes as well. About 2.5 million years ago, northwestern Europe lost the last of many subtropical plant taxa that it had shared with Malaysia earlier in Pliocene time.

The Drying of Climates It was not only cooling of climates that affected floras, but also increased aridity. Many regions became drier as climates cooled because cooler sea surfaces released less water to the atmosphere through evaporation. Increased aridity in Africa led to a great expansion of the Sahara Desert. Pollen from deep-sea cores off the west coast of Africa reveals that before about 3 million years ago tropical forest frequently extended to a latitude quite close to the present southern limit of the Sahara. The Sahara itself was at the time a small desert far from the ocean.

Three large glacial centers developed in the northern hemisphere—one in North America, one in Greenland, and one in Scandinavia (see Figure 19-4). Because the northern Atlantic Ocean was adjacent to these three glacial centers, it was more profoundly affected by the Pleistocene glaciers than any of the world's other major oceans except the Arctic, and land areas adjacent to the Atlantic also suffered marked climatic change. When glaciers grew to their maximum extent, pack ice choked large areas of the North Atlantic (Figure 19-12), just as it now occupies bays adjacent to northern Canada. Farther south, along the east coast of North America, glaciers flowed southward to New Jersey, and tundra occupied what is now Washington, D.C.

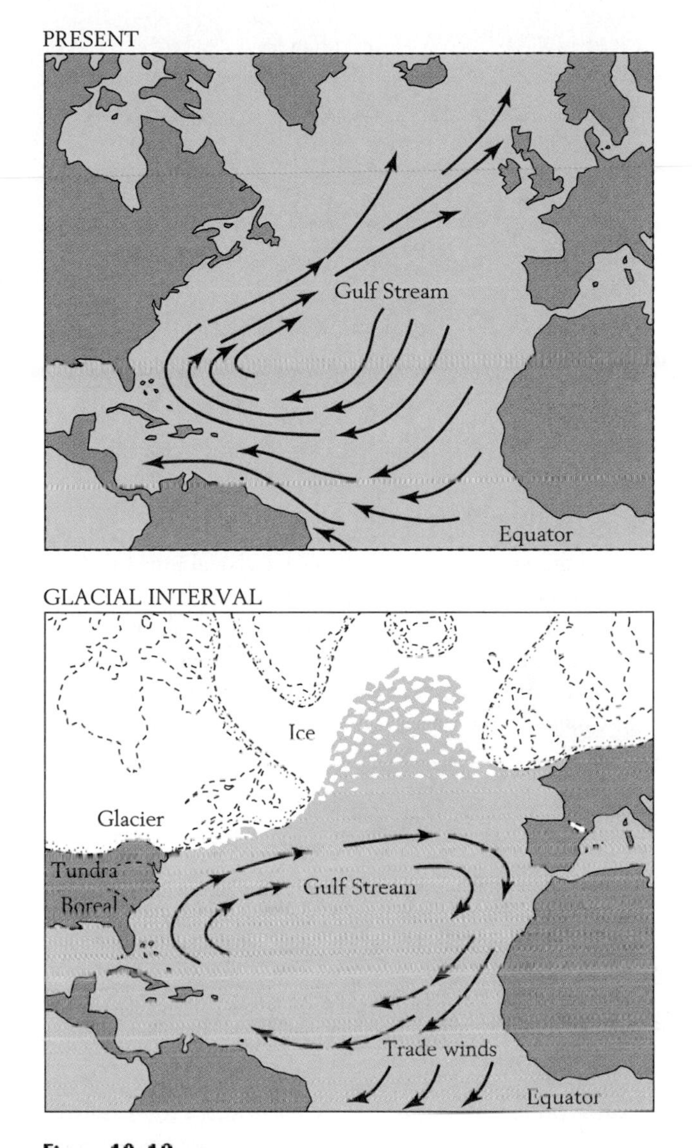

Figure 19-12 **The Atlantic Ocean and neighboring regions today and during the most recent Pleistocene glacial interval.**

The Nature of Glacial Expansions and Contractions Glacial maxima and glacial minima are represented by peaks and valleys in oxygen isotope curves (see Figure 19-11). Today we live during the glacial minimum established when glaciers melted back between about 15,000 and 10,000 years ago. The most recent glacial maximum peaked between about 35,000 and 15,000 years ago, during the Wisconsin Stage. It was during this glacial maximum that sea level dropped to at least 100 meters (330 feet) below its present level. This interval was preceded by the Sangamon glacial minimum, about 125,000 years ago, when sea level stood slightly higher than it does today. Because of its relative recency, the last interval of glacial expansion (the late Wisconsin) has left the best record of glacial deposits. Furthermore, the time of maximum expansion of Wisconsin glaciers, about 35,000 to 15,000 years ago, places their organic deposits, including fossilized wood, well within the range of radiocarbon dating. Detailed stratigraphic studies in the region of the Great Lakes have revealed that the late Wisconsin glacial advance was a complex event, consisting of pulses of glacial expansion separated by partial retreats (see Figure 19-5). In addition, individual lobes of the North American ice sheets did

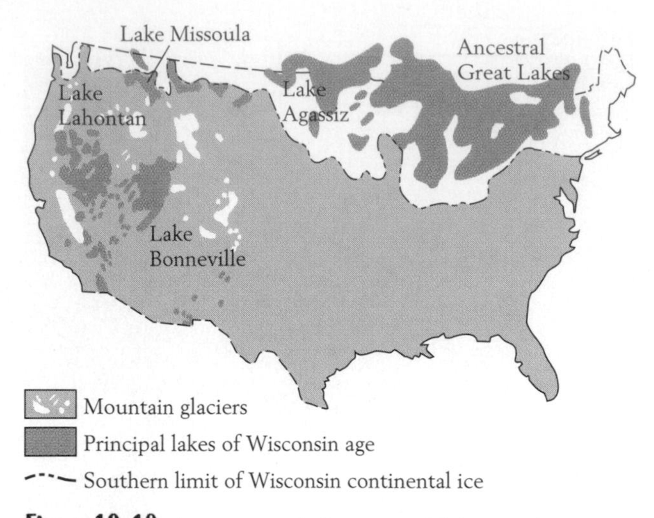

Figure 19-13 Locations of glaciers and lakes in the United States during the most recent glacial interval. Lake Agassiz and the ancestral Great Lakes formed to the south of the continental glacier as it retreated northward near the end of the glacial interval. *(C. B. Hunt, Natural Regions of the United States and Canada, W. H. Freeman and Company, New York, 1974.)*

not always expand and contract at the same rate. Ultimately, however, all the glacial lobes retreated into large basins that became the Great Lakes when the ice sheets melted back into Canada (Figure 19-13).

Conditions during Glacial Maxima Figure 19-14 shows world geography during the most recent glacial maximum. Having expanded when the Ice Age was getting under way, the Sahara grew still larger during glacial maxima because of reduced evaporation from cool seas. Figure 19-15 shows the southern expansion of dune activity in the Sahara Desert during the most recent glacial maximum. Deserts expanded in a similar way on other continents.

Rain forests also shrank at the start of the Ice Age and then even more drastically during glacial maxima. The African rain forest was repeatedly restricted to three small areas. Populations of rain forest species were therefore fragmented, and some, including the gorilla, up to the present day have failed to recolonize the entire rain forest, remaining in separate areas that served as refuges when the rain forest contracted. In their isolation, the two gorilla populations evolved into separate subspecies (see Figure 19-15). Other taxa appear to have evolved many new species as a result of geographic isolation of small populations at times when the rain forest contracted.

During the Wisconsin glacial interval, north-south temperature gradients steepened in the northern hemisphere both in shallow seas and on land. Winter temperatures fell by a few degrees in most tropical areas but plummeted at latitudes north of 30° in the northern hemisphere.

There were exceptions to the general pattern of increased aridity during glacial maxima. Among the exceptions was the Great Basin in the American West, where water accumulated to form numerous lakes in areas that are now arid (see Figure 19-13). Apparently the great mountain of ice to the north deflected winds from the Pacific Ocean, causing them to follow a more southerly course and bring moisture to the Great Basin. The Great Salt Lake in Utah is a remnant of the largest western Ice Age lake, known as Bonneville.

Shockingly Rapid Climatic Shifts Detailed studies of the last glacial maximum, between 80,000 and 15,000 years ago, have revealed astoundingly sudden climatic changes separated by longer cooling trends. Part of the evidence comes from a particular species of planktonic foraminifera that survives today and is known to flourish in very cold water. Additional evidence comes from oxygen isotope ratios in annual layers of ice that can be observed in cores of the Greenland ice cap. A thick layer of ice accumulated every year in summer and a thin layer accumulated in winter, when surface waters of the ocean supplied less moisture for precipitation (Figure 19-16). Oxygen isotopes in the ice reflect temperature changes from year to year in the Greenland region.

As Figure 19-17 shows, the foraminiferan and isotope records are in accord, showing that climatic oscillations were grouped into long-term cooling cycles that lasted an average of 10,000 to 15,000 years. The average temperature declined within a cycle, but each cycle ended with an abrupt warming event, during which the temperature jumped by several degrees Celsius within only about ten years. Interestingly, not long before each pulse of warming, a Heinrich event occurred. A *Heinrich event* is a massive discharge of icebergs that release sedimentary debris to the sea as they melt. Heinrich layers are conspicuous in North Atlantic deep-sea cores (see Figure 19-17). Heinrich events occurred when climates were very cold and

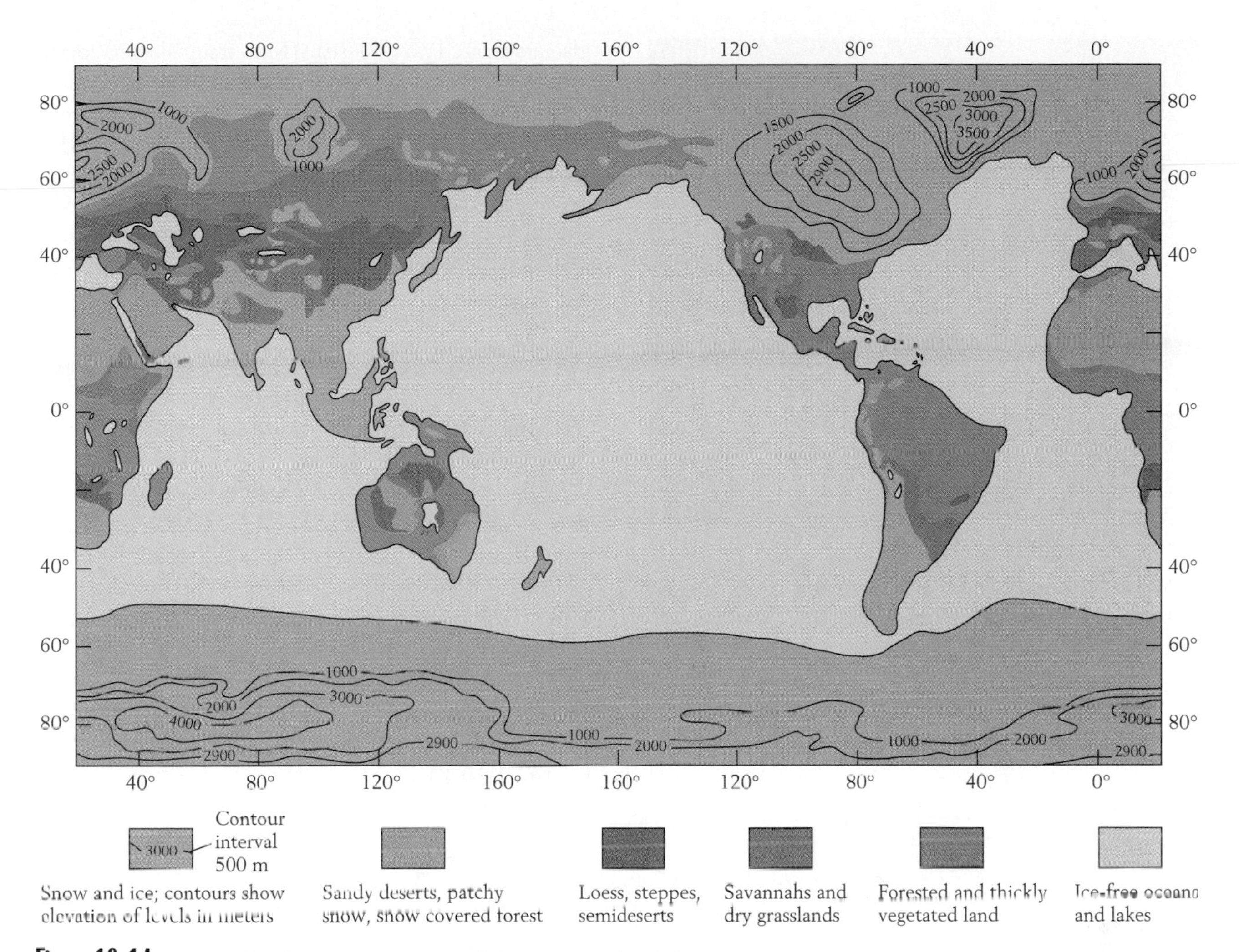

Figure 19-14 Geographic features reconstructed from the height of the most recent glacial interval, about 18,000 years ago, when sea level was about 85 meters (280 feet) below its present level. *(Modified from CLIMAP, Science 191:1131–1144, 1976.)*

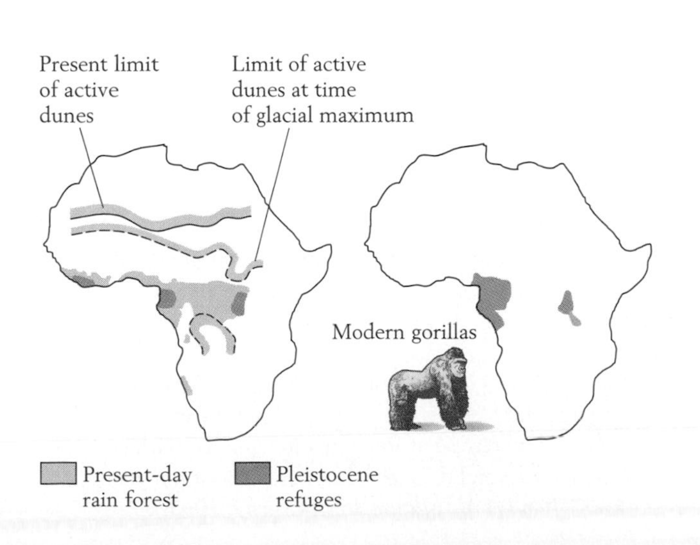

Figure 19-15 Environmental changes in Africa during the most recent glacial maximum. Dune activity shifted toward the equator, marking the southward expansion of the Sahara Desert. The rain forest fragmented and shrank into three small areas. Repeated contractions of this kind have left gorillas divided into two populations that have evolved to become separate subspecies.

Figure 19-16 **Annual bands of ice are visible in this glacier in the Andes of South America.**

glaciers surged to the sea. (Mountain glaciers occasionally surge in the present world, though on a much smaller scale, and sometimes they launch numerous icebergs into the sea; see Figure 4-14.) Why a sudden warming followed is unclear, as is the lengthening of the cooling cycles. Whatever may have caused the abrupt warming, such sudden changes serve warning that the global climate can change dramatically without human influence within the space of a single decade.

The Cause of Glaciation The cause of the late Neogene glaciation in the northern hemisphere is a matter of debate.

One possibility is that changes in ocean circulation brought about the Late Neogene ice age. The key event here would have been the uplift of the Isthmus of Panama (Figure 19-18). The presence of the isthmus today makes the North Atlantic Ocean much

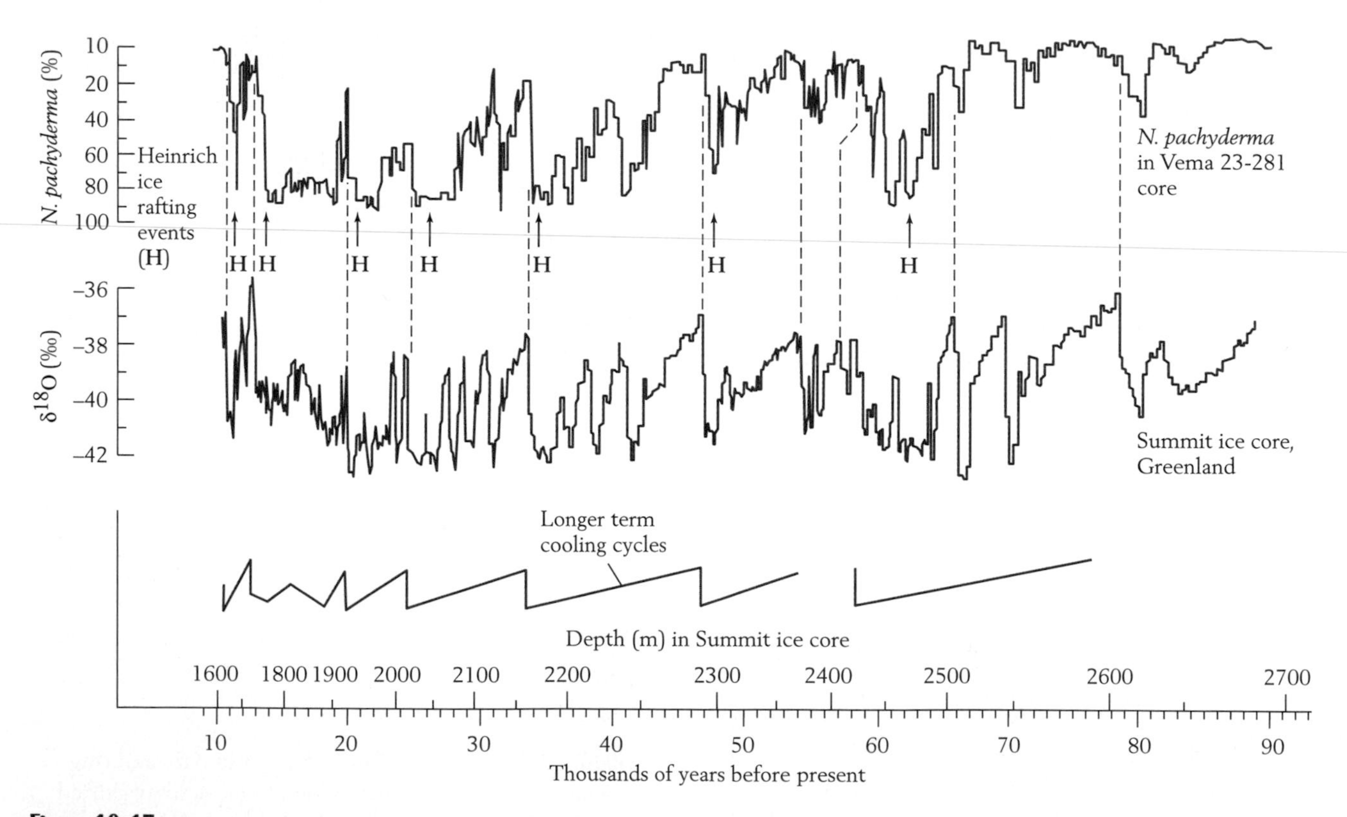

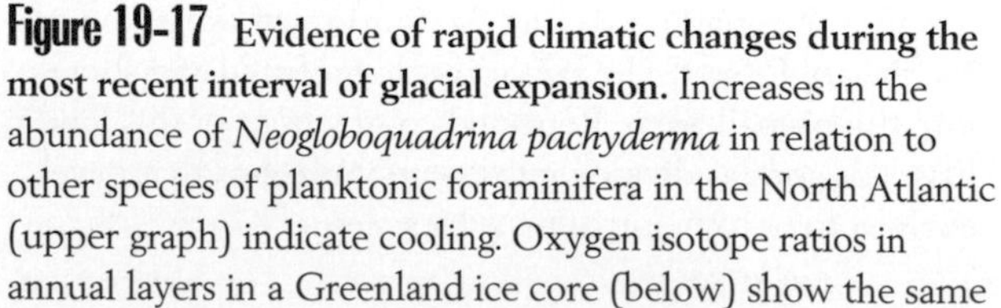

Figure 19-17 **Evidence of rapid climatic changes during the most recent interval of glacial expansion.** Increases in the abundance of *Neogloboquadrina pachyderma* in relation to other species of planktonic foraminifera in the North Atlantic (upper graph) indicate cooling. Oxygen isotope ratios in annual layers in a Greenland ice core (below) show the same pattern. Intervals of net cooling in the north Atlantic, averaging 10,000 to 15,000 years in duration and punctuated by temperature oscillations, ended with sudden warming by several degrees within only about ten years. *(After G. Bond et al., Nature 365:143–147, 1993.)*

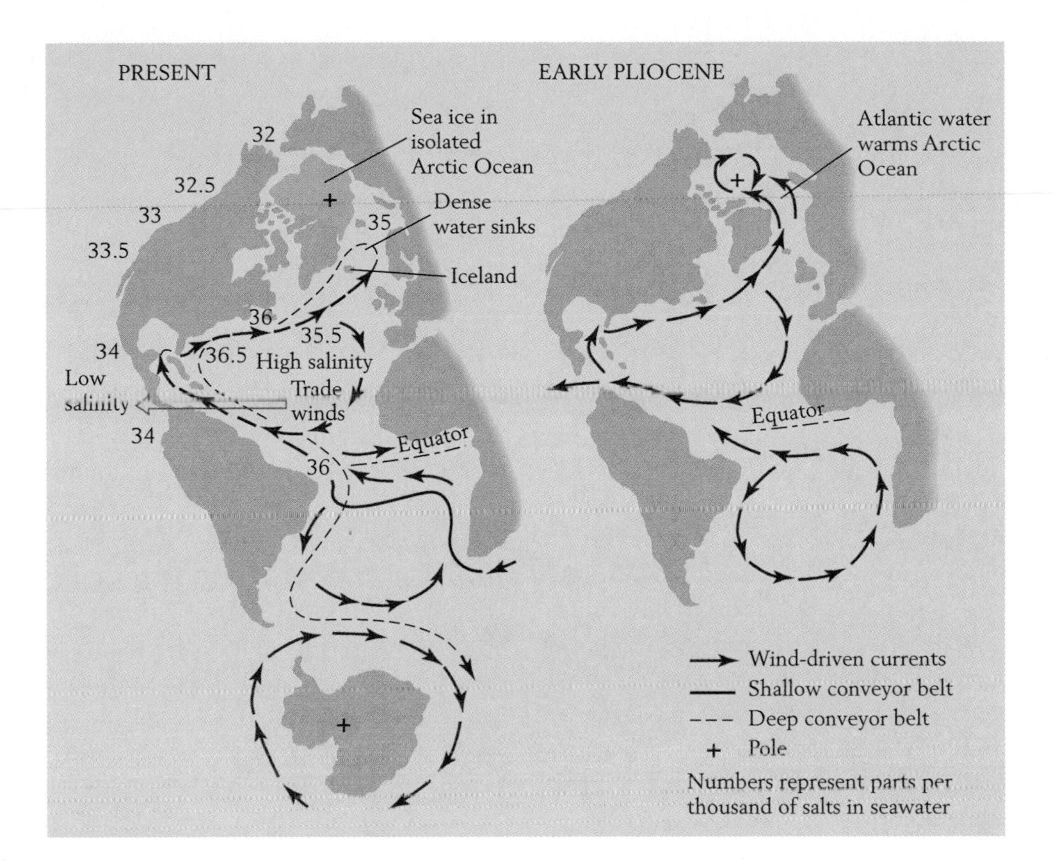

Figure 19-18 The conveyor belt of the present world (left) and the oceanographic pattern that may have existed in the North Atlantic during the early Pliocene (right), before the modern Ice Age of the northern hemisphere began. Today the trade winds carry to the Pacific water that they evaporate from the surface of the Atlantic. (Numbers show salinities for the two oceans.) The dense, saline Atlantic water sinks north of Iceland, driving the conveyor belt. During early Pliocene time, before the Isthmus of Panama was in place, mixing with Pacific waters should have lowered the salinity of the Atlantic. The more buoyant Atlantic waters may then have flowed northward into the Arctic, keeping the polar region warm. The uplift of the Isthmus of Panama may have triggered the Ice Age by elevating the salinity of Atlantic waters and causing them to sink north of Iceland, as they do today; this change would have deprived the Arctic Ocean of heat from the Atlantic. *(After S. M. Stanley, Jour. of Paleontology 69:999–1007, 1995.)*

saltier than the North Pacific. The Atlantic is saltier because the dry trade winds blowing westward from the Sahara Desert evaporate large amounts of water from its surface. Not only does this evaporation increase the salinity of the Atlantic, but it results in lower salinity for the Pacific because much of the evaporated water is transported across Central America and enters the eastern Pacific through rainfall.

The clockwise circulation of the North Atlantic carries the saline water formed in the trade wind belt northward. This dense saline water cools as it moves northward in the Gulf Stream and thus becomes even denser. Finally it sinks just north of Iceland. This sinking is the primary driving force of a huge loop of moving water known as the *oceanic conveyor belt.* The water that sinks north of Iceland flows back to the south and bends eastward into the Pacific, where it surfaces and then returns to the Atlantic, completing the loop of the conveyor belt. Because the waters of the North Pacific are less saline and therefore less dense than those of the North Atlantic, they do not sink to form a conveyor belt.

Because the waters of the Atlantic sink at the brink of the Arctic Ocean and loop back to the south, the Arctic Ocean is deprived of Atlantic warmth. Without inflow of warm Atlantic surface waters, the upper Arctic Ocean assumes the cold temperature that is appropriate for a relatively isolated body of water at a very high latitude. The key role of the isthmus is as a barrier that prevents the saline Atlantic waters from mixing with the waters of the Pacific. Thus it keeps the Atlantic waters dense enough to sink just north of Iceland.

A variety of evidence, including the presence of nearshore mollusks of mid-Pliocene age, indicates that the Isthmus of Panama was emplaced by plate movements between 3.5 million and 3 million years ago, about the time the Ice Age of the northern hemisphere got under way. Before the isthmus formed, Atlantic waters should have flowed freely into the Pacific through the gap between North and South America (see Figure 19-18). Mixing of the waters of the Atlantic and Pacific should have maintained the two oceans at similar levels of salinity: the Pacific should have been more saline and the Atlantic less saline. Being less dense than they are today, North Atlantic waters may have flowed into the Arctic Ocean before cooling sufficiently to became dense enough to sink. By flowing into the Arctic, the Atlantic waters would have kept this ocean warmer than it is today. The uplift of the Isthmus of Panama changed the pattern. If it suddenly caused waters of the North Atlantic to descend north of Iceland by increasing their salinity (and density), then the sudden isolation and cooling of the Arctic Ocean may have cooled the entire Arctic region. This cooling may have brought on the Ice Age.

There is, in fact, evidence that when the isthmus formed, the Atlantic water to its east became more saline. Oxygen isotope ratios for planktonic foraminifera of this region became heavier between 4 million and 3 million years ago, although temperatures underwent little change and continental glaciers had not yet expanded. The increase in oxygen 18 to the east of the isthmus apparently resulted from an increase in the salinity of the waters there, with evaporation preferentially removing oxygen 16, the lighter isotope (p. 277).

An alternative hypothesis is that the Ice Age was brought on by a decline in greenhouse warming, but there is no clear evidence that atmospheric carbon dioxide declined between 4 million and 3 million years ago.

Glacial Cycles Just as interesting as the cause of the modern Ice Age is the source of the glacial oscillations that have characterized the Ice Age since its inception. It is now generally agreed that changes in Earth's relationship with the sun have caused these oscillations. Figure 19-11 indicates that the oscillations were more frequent during the early part of the Ice Age than later on.

Early in the Ice Age, glacial oscillations seem to have corresponded to the so-called tilt cycle of Earth's axis of rotation. The axis is always tilted slightly away from vertical with respect to the plane of Earth's orbit around the sun, but the angle of tilt oscillates through time, with a periodicity of about 41,000 years. At the point in the tilt cycle when the axis is farthest from vertical, the polar regions are aimed most directly toward the sun during the summer and receive a maximum amount of sunlight and solar heating.

Beginning about 800,000 years ago, glacial oscillations became less frequent, shifting to a periodicity of 90,000 to 100,000 years. This new periodicity appears to have corresponded to changes in the shape of Earth's orbit. These changes result from oscillations in the gravitational pull of other planets on Earth, which have a periodicity of about 92,500 years. When the orbit changes so as to bring Earth closer to the sun, it receives more solar heat than it does when it is farther away. It is not known why, about 800,000 years ago, Earth's orbital oscillations came to govern the expansion and contraction of glaciers, overshadowing the tilt cycle.

Regional Events

The history of the western United States in the Neogene Period is highlighted not only by the elevation of imposing mountains that form part of our scenery today—the Cascade Range, the Sierra Nevada, and the Rocky Mountains—but also by climatic changes that resulted from the uplifting of these mountains. Tectonic movements were milder in and around the western Atlantic Ocean, but the passive margin of eastern North America accumulated sediments that contain a rich fossil record.

Development of the American West

The pre-Neogene history of mountain building in the Cordilleran region, described in earlier chapters, is summarized in Figure 18-20. By late Paleogene time, uplifts resulting from the final mountain-building episode of the western interior, the Laramide orogeny, had been largely subdued by erosion, which set the stage for the Neogene events that produced the Rocky Mountains. In the broad region west of the Rockies, the Neogene Period was a time of widespread tectonic and igneous activity, which built most of the mountains standing there today.

Provinces of the American West Lying between the Great Plains and the Pacific Ocean are several distinctive physiographic provinces that have taken shape largely in Neogene time, primarily as a result of uplift and igneous activity (Figure 19-19). Let us briefly review the present characteristics of these provinces before considering how they have come into being.

The lofty, rugged peaks of the Rocky Mountains, some of which stand more than 4.5 kilometers (14,000 feet) above sea level, could only be of geologically recent origin. We have seen that the widespread subsummit surface of the Rockies was all that remained of the Laramide uplifts by the end of the Eocene Epoch, about 34 million years ago (see Figure 18-27). One question we must answer, then, is how the Rocky Mountain region became mountainous during the Neogene Period.

Adjacent to the Rockies in the "four corners" area where Colorado, Utah, New Mexico, and Arizona meet is the oval-shaped Colorado Plateau, much of which stands about 1.5 kilometers (1 mile) above sea level. The Phanerozoic sedimentary units here are not intensively deformed. Some, however, are gently folded in a steplike pattern, and others, especially to the west, are offset by block faults (Figure 19-20). Cutting through the plateau is the spectacular Grand Canyon of the Colorado River (see Figure 1-5), but about 15 million years ago neither the Colorado Plateau nor the Grand Canyon existed. The origin of these features forms another part of our story.

West of the Rockies and the Colorado Plateau, within the belt of Mesozoic orogeny, lies the Basin and Range Province (see Figure 19-19). This is an area of north-south-trending block-fault valleys and

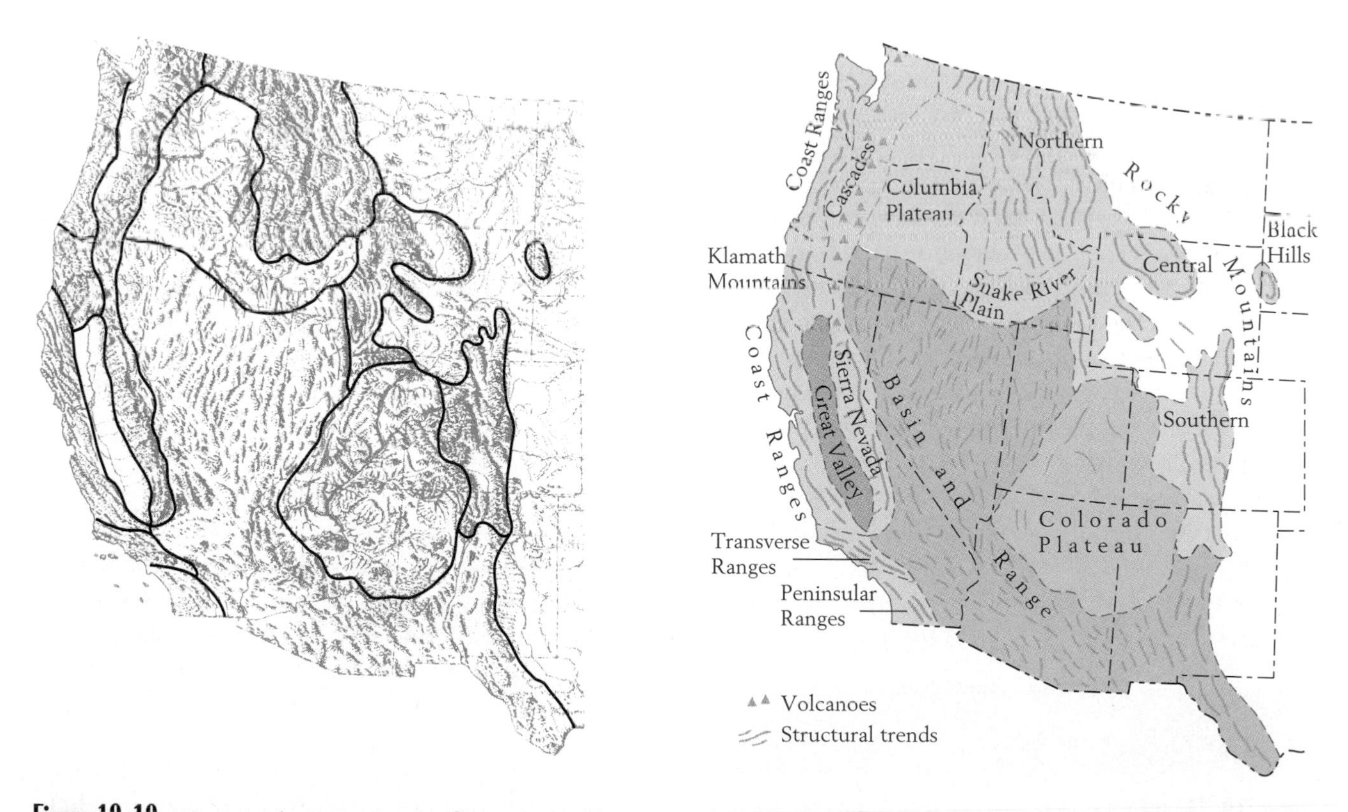

Figure 19-19 Major geologic provinces of western North America. The map at the left shows the relations of the provinces to topographic features. *(Topographic map based on U.S. Geological Survey, National Atlas of the United States of America.)*

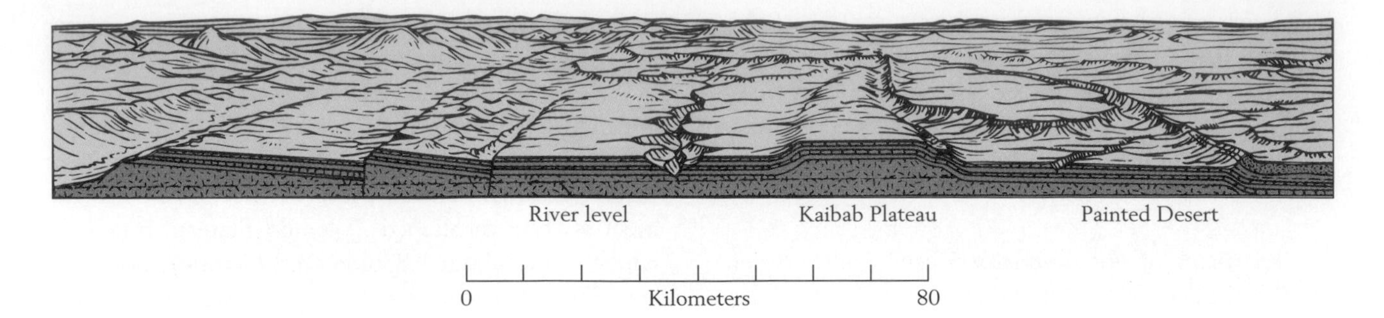

Figure 19-20 The western part of the Colorado Plateau north of the Grand Canyon. This high-standing region is characterized by block faulting (left) and gentle, steplike folds (right). *(After P. B. King, The Evolution of North America, Princeton University Press, Princeton, N.J., 1977.)*

intervening ridges (Figure 19-21)—features of Neogene origin. A large area of this province forms the Great Basin, an arid region of interior drainage (p. 129). Volcanism has been associated with some faulting episodes here. The thickness of Earth's crust in the Basin and Range Province ranges from about 20 to 30 kilometers, in contrast to thicknesses of 35 to 50 kilometers in the Colorado Plateau. The thinning and block faulting in the Basin and Range Province point to extension of the crust by at least 65 percent and perhaps by as much as 100 percent.

Farther north, centered in Oregon, is a broad area covered by volcanic rocks of the Columbia River and Snake River plateaus (see Figure 19-19). Today the climate here is cool and semiarid; only about one-quarter of the plateau area is cloaked in forest and woodland, and sagebrush and drier conditions characterize about half of the terrain. In Oligocene time, however, lavas had not yet blanketed the region, and, as remains of fossil plants reveal, a large forest of redwood trees grew there.

Along the western margin of the Columbia Plateau stand the lofty peaks of the Cascade Range (p. 519). These cone-shaped volcanoes represent the volcanic arc associated with subduction of the Pacific plate along the western margin of the continent. Volcanism began here in Oligocene time and continues to the present.

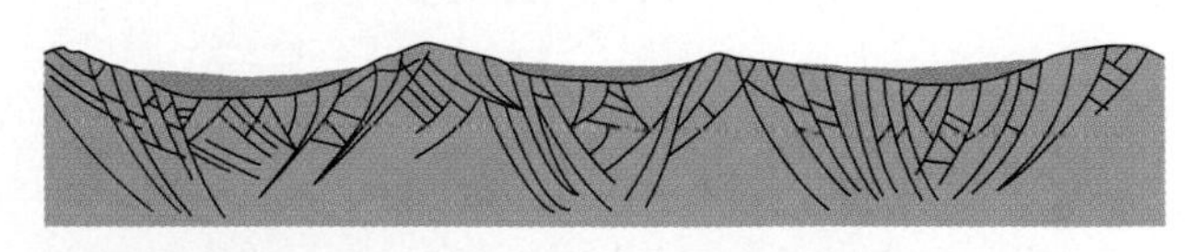

Figure 19-21 The possible pattern of the block faulting in the Basin and Range Province that might have been responsible for lateral extension of the crust.

The Cascade volcanic belt passes southward into the Sierra Nevada, a mountain-sized fault block of granitic rocks. The plutons forming the Sierra Nevada were emplaced in east-central California during Mesozoic time, before igneous activity at this latitude shifted inland. As we will see, however, the present topography of the Sierra Nevada is of Neogene origin. This mountain range is unusual in that, throughout its length of some 600 kilometers (350 miles), it is not breached by a single river. This is why it represented such a formidable obstacle to early pioneers attempting to reach the Pacific (Figure 19-22).

The Sierra Nevada stands between the Basin and Range Province to the east and the Great Valley of California to the west (see Figure 19-19). The Great Valley is an elongate basin containing large volumes of Mesozoic sediment (the Great Valley Sequence; see Figure 16-31) eroded from the plutons of the Sierra Nevada region long before the modern Sierra Nevada formed by block faulting (Figure 19-23). Resting on top of these sediments are Cenozoic deposits, some of which accumulated during marine invasions of the Great Valley and others during times of nonmarine sedimentation.

West of the Great Valley are the California Coast Ranges, which consist of slices of crust that include crystalline rocks representing Mesozoic orogenic activity, Franciscan rocks of deep-water origin (Figure

Figure 19-22 The eastern face of the Sierra Nevada. This is a fault scarp that is partly dissected by youthful valleys. The view is from the Owens Valley, Inyo County, California.

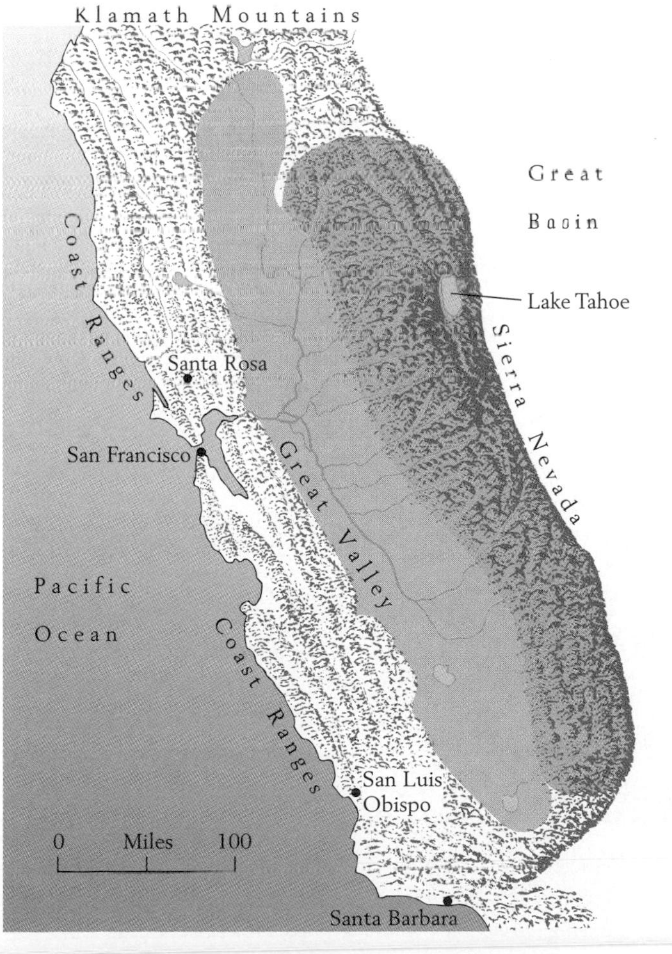

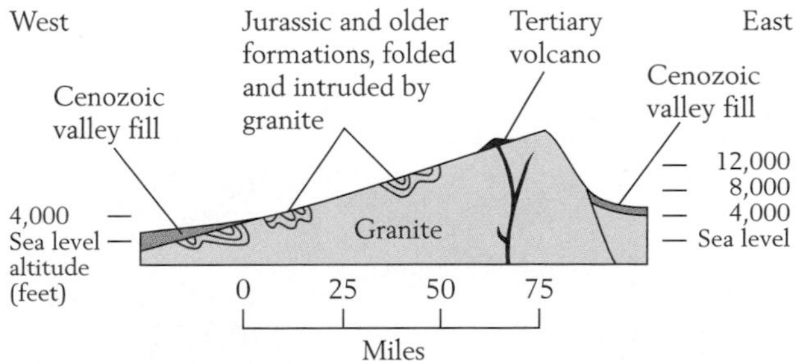

Figure 19-23 The Sierra Nevada fault block of California and Nevada. Sediments of the Great Valley lap onto the gentle western slope of the Sierra Nevada. *(After C. B. Hunt, Natural Regions of the United States and Canada, W. H. Freeman and Company, New York, 1974.)*

19-24), and Tertiary rocks. To the south, the Transverse and Peninsular ranges are formed of similarly faulted and deformed rocks, but these ranges lie inland of the main belt of Franciscan rocks in the region of intensive Mesozoic igneous activity. Striking features of all of these mountainous terrains are the great faults that divide the crust into sliver-shaped blocks. The longest and most famous of these faults is the San Andreas, which extends for about 1600 kilometers (1000 miles). Until the great San Francisco earthquake of 1906, it was not widely recognized that the San Andreas was still active. The earthquake of 1906 was produced by a sudden horizontal movement of up to 5 meters (16 feet) along the fault. Geologic features cut by the San Andreas fault show that its total movement during the past 15 million years has amounted to about 315 kilometers (190 miles). Continued movement at this rate for the next 30 million years or so would bring Los Angeles northward to the latitude of San Francisco, through which the fault passes. As we will see, the faulting and uplifting of the Coastal Ranges of California are probably related not only to the Neogene uplift of the Sierra Nevada but also to the origins of the Basin and Range topography to the east.

The Olympic Mountains of Washington have quite a different history. These relatively low mountains, which lie to the west of the Cascade volcanics, consist of oceanic sediments and volcanics that were deformed primarily during Eocene time in association with subduction along the continental margin (see Figure 18-23).

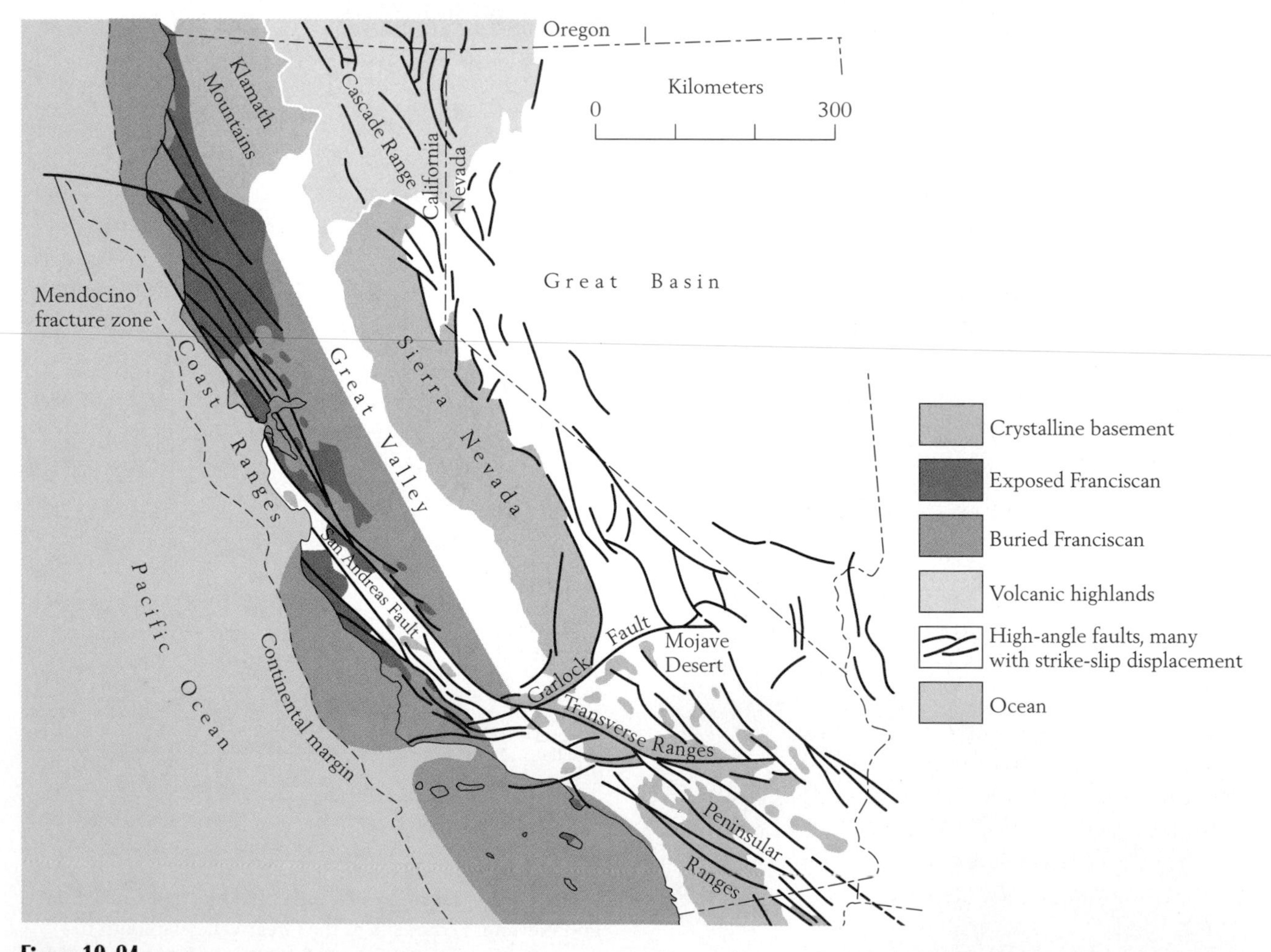

Figure 19-24 Major geologic features of California. The Franciscan terrane was attached to the coast late in the Jurassic Period (see Figures 16-31 and 16-32). The many faults depicted here, including the famous San Andreas, are of Neogene age. *(Modified from P. B. King, The Evolution of North America, Princeton University Press, Princeton, N.J., 1977.)*

Developments in Western North America Geologic features of the far west in Miocene time are shown in Figure 19-25. Subduction continued beneath the continental margin in the northwestern United States, and the resulting volcanic arc produced peaks in the Cascade Range, where volcanism continues today. To the south, in California, the mid-Miocene interval was a time of faulting and mountain building; elements of the modern Coast Ranges and other nearby mountains were raised, and the seas were driven westward. Meanwhile, as in Paleogene time, the Great Valley remained a large embayment, and during Miocene time it received great thicknesses of siliciclastic sediments, most of which were shed from the Sierra Nevada, which rose to at least its present height during Miocene time.

During Miocene time, the Basin and Range Province began forming to the east of the Sierra Nevada. Volcanism began during Paleogene time, and the Basin and Range topography began to form near the beginning of the Miocene. To the north of the Basin and Range Province, great volumes of basalt spread from fissures at the site of a hot spot. Most of the great Columbia Plateau formed by outpourings of lava between about 16 million and 13 million years ago (see Figure 2-13); individual basalt flows of this plateau range in thickness from 30 to 150 meters (100 to 500 feet), and in places the total accumulation reaches about 5 kilometers (3 miles).

Geologists have reconstructed the histories of the Colorado Plateau and Rocky Mountains by studying the time at which rivers have cut through well-dated volcanic rocks. Many of the rivers of these regions existed before uplift began in the Miocene Epoch, and they cut rapidly downward as the land rose, producing deep gorges. A large part of the Grand Canyon, for example, was incised during the rapid elevation of the Colorado Plateau between about 10 million and 8 million years ago. Uplift in the Rockies began slightly earlier, in Early Miocene time, and terrain that now forms the southern Rockies has since risen between 1.5 and 3.0 kilometers (1 to 2 miles).

During the Pliocene and Pleistocene epochs, igneous activity continued in the volcanic provinces of Oregon, Washington, and Idaho (Figure 19-26). Many of the scenic volcanic peaks of the Cascades have formed within the past 2 million years or so (see p. 518). Beginning in late Miocene time and continuing sporadically to the present, the flow of basalt from

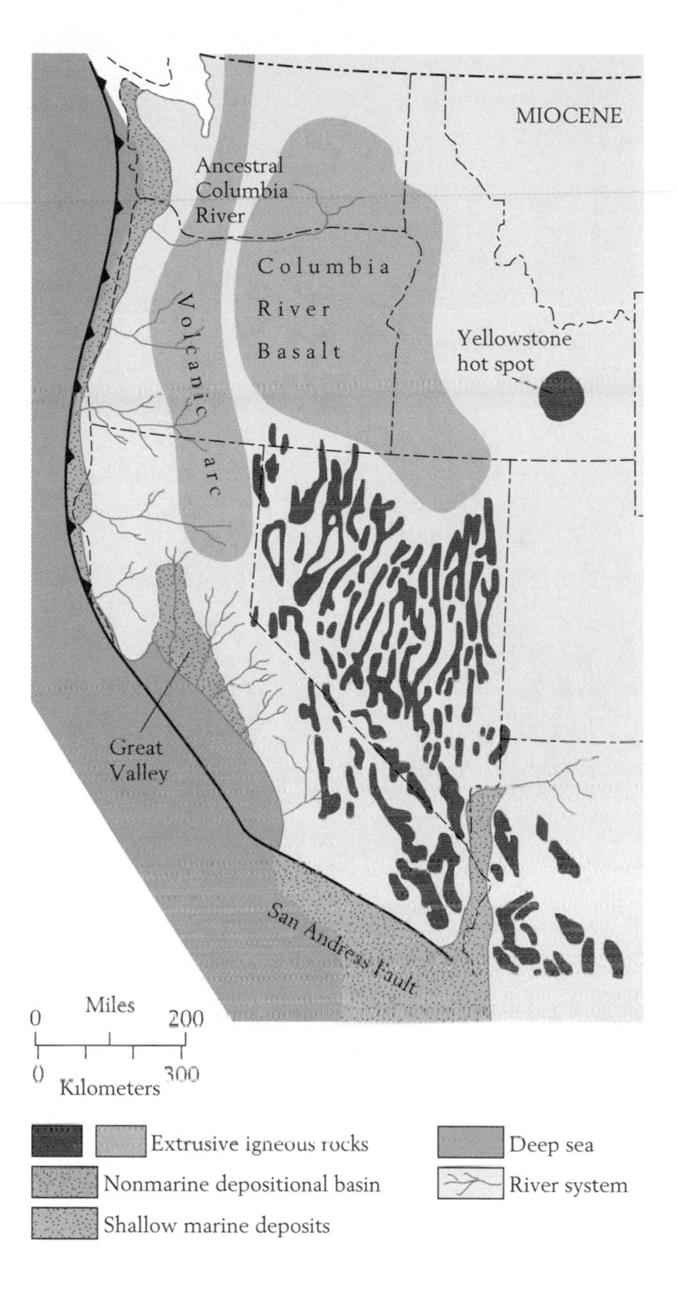

Figure 19-25 Geologic features of western North America in Miocene time. West of the San Andreas fault, coastal southern California lay farther south than it does today. The area currently occupied by the Great Valley of California was, for the most part, a deep-water basin from which a nonmarine depositional basin extended to the north. Volcanoes of the early Cascade Range formed along a volcanic arc inland from the subduction zone along the continental margin. Igneous rocks were extruded along north–south-trending faults in the Great Basin, and farther north the Columbia River Basalts spread over a large area. *(Modified from J. M. Armentrout and M. R. Cole, Soc. Econ. Paleont. and Mineral. Pacific Coast Paleogeog. Symp. 3:297–323, 1979.)*

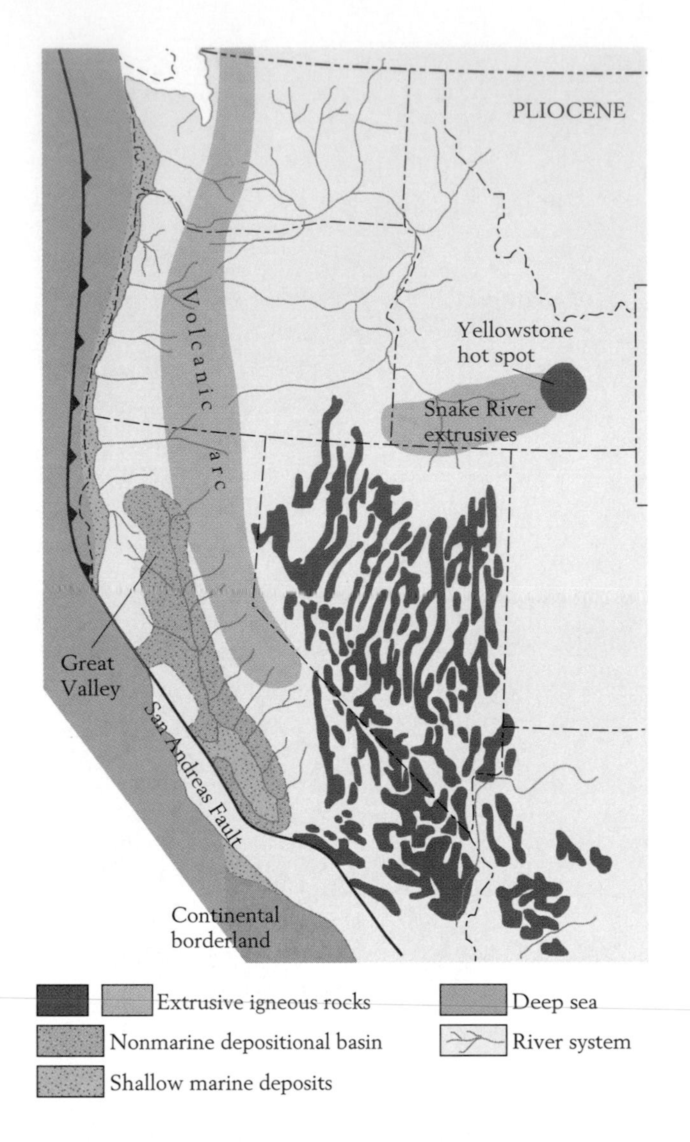

Figure 19-26 Geologic features of western North America in Pliocene time. The Great Valley of California was a shallow basin that received nonmarine sediments except where a shallow sea flooded its southern portion. The volcanic arc continued to form volcanoes of the Cascade Range, and igneous rocks continued to be extruded along faults in the Great Basin. The Snake River extrusives spread over a large area of southern Idaho west of Yellowstone. *(Modified from J. M. Armentrout and M. R. Cole, Soc. Econ. Paleont. and Mineral. Pacific Coast Paleogeog. Symp. 3:297–323, 1979.)*

fissures has produced the Snake River Plain, which amounts to an eastward extension of the Columbia Plateau and has been formed by the same hot spot (see Figures 19-25 and 19-26).

Faulting and deformation continued in California during the Pliocene and Pleistocene epochs. Since the beginning of the Pliocene Epoch, about 5 million years ago, the sliver of coastal California that includes Los Angeles has moved northward on the order of 100 kilometers (60 miles). The Great Valley has, of course, remained a lowland to the present day, but during Pliocene and Pleistocene time it became transformed from a marine basin into a terrestrial one. Early in the Pliocene Epoch, seas flooded the basin from both the north and the south, but as the epoch progressed, uplift associated with movement along the San Andreas fault eliminated the southern connection. Eventually nonmarine deposition prevailed throughout the Great Valley, which is now one of the world's richest agricultural areas as well as a site of large reservoirs of petroleum.

Major climatic changes occurred in western North America in latest Miocene and Pliocene time. The Basin and Range Province, which had been covered by forests throughout most of Miocene time, became carpeted by savannah and eventually turned largely into a desert. Researchers in the past attributed this change toward more arid conditions to a rain-shadow effect of the rising Sierra Nevada to the west. The idea of a strengthened rain shadow has been refuted, however, by new evidence that the Sierra Nevada stood tall during the Miocene. Instead, it appears that increased aridity throughout much of the western United States during late Neogene time has resulted largely from a global trend toward cooler and drier climates.

Sediments derived from the Rocky Mountains spread eastward late in Miocene time, forming the Ogallala Formation. Caliche nodules are abundant in many parts of the Ogallala, indicating the presence of seasonally arid climates (p. 124). The Ogallala is a thin, largely sandy unit, most of which lies buried under the Great Plains from Wyoming to Texas, and it serves as a major source of groundwater. Unfortunately, this is ancient water that is not being renewed as rapidly as it is drawn from Earth. As a result, severe water shortages may one day strike many areas of the central United States.

With the onset of the Ice Age in the northern hemisphere, frigid conditions brought glaciation to mountainous regions of the western United States just as they foster glaciation in Alaskan mountains today (see Figure 4-14). The Sierra Nevada, for example, was heavily glaciated, as were portions of the

Rocky Mountains (see Figure 19-13 and Box 19-1). Today broad U-shaped valleys in both mountain systems testify to the scouring activity of Pleistocene glaciers (see Figure 12-17).

Possible Mechanisms of Uplift and Igneous Activity What has led to the many tectonic and igneous events of Neogene time in the American West? It might seem likely that the secondary uplift of the Colorado Plateau and the Rocky Mountains, which took place long after the Laramide orogeny, resulted from simple isostatic adjustment (see Figure 1-16). Geophysical studies, however, show that these uplifts do not have deep roots that might have caused them to bob up. Instead, for some reason swelling of Earth's mantle below seems to have elevated broad areas of the American west.

The block faulting of the Basin and Range requires a different explanation. Basin and Range events, the spreading of the Columbia River Basalt, and the extensive faulting and folding along the California coast all began in Miocene time and seem to be related in some way to plate tectonic movements along the Pacific coast.

How these movements have produced the Basin and Range Province is a controversial issue. The most popular idea relates to the famous San Andreas fault. The San Andreas is a transform fault (see Figure 9-1) associated with the East Pacific Rise, a large oceanic rift that passes into the Gulf of California and breaks up as it passes inland through thick continental crust (Figure 19-27). Spreading along the rise should have ceased when the rise came into contact with the subduction zone at the western boundary of the North American plate; instead, movement must have been propagated along one or more transform faults such as the San Andreas, passing along the continental margin. Crustal shearing adjacent to a strike-slip fault such as the San Andreas will automatically cause extensional

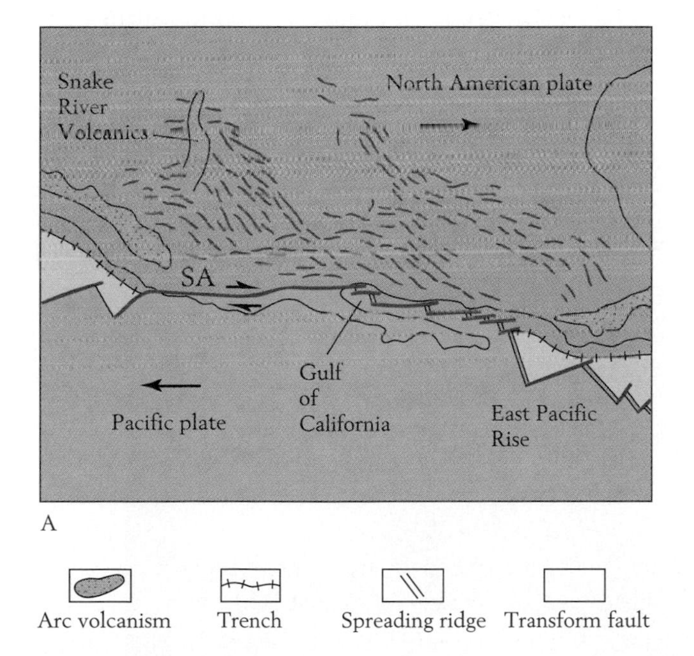

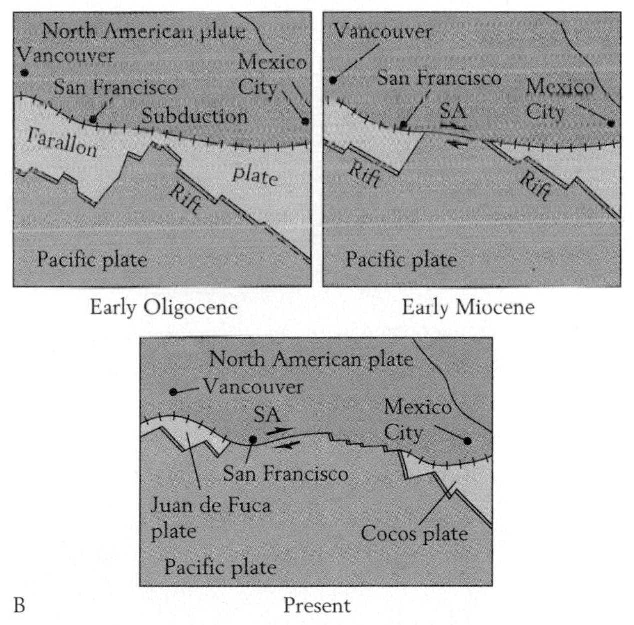

Figure 19-27 Plate tectonic features that may account for the Basin and Range structure in western North America. *A.* At present the spreading ridge known as the East Pacific Rise passes into the Gulf of California. The East Pacific Rise abuts the North American continent and is offset westward along the San Andreas fault (SA). It appears that shearing forces resulting from relative movement of terrain on either side of the San Andreas fault (heavy arrows) pull the crust apart, producing the north–south-trending faults of the Basin and Range Province. *B.* In the second alternative, the rift zone between the Pacific and Farallon plates encountered the thick crust of North America along the subduction zone that bordered the continent. Unable to pass inland, the rift was divided along a strike-slip fault (the San Andreas). *(Based on J. H. Stewart, Geol. Soc. Amer. Mem. 152:1–31, 1975. After Atwater.)*

For the Record 19-1

The Biggest Flood on Record

One of the greatest controversies of modern geology erupted in 1923, when J. Harlan Bretz, a professor at the University of Chicago, advanced what he modestly called the "outrageous hypothesis" that catastrophic floods had swept across a broad region of the northwestern United States as Earth emerged from the last glacial maximum. Far from being outrageous, Bretz's bold idea turned out to be correct.

Bretz based his idea of catastrophic flooding on his studies of the so-called channeled scablands in the eastern part of the state of Washington. The channeled scablands are a landscape of bare rock that has obviously been scoured by water. Its topography includes water-carved channels, some of which are offset by steps that are the sites of ancient waterfalls. The channels of the scablands also display remarkable depositional features, including giant ripples of gravelly sediment that are typically about 5 meters (16 feet) tall and spaced about 100 meters (330 feet) apart. So little soil carpets the scablands that Bretz concluded that they must have formed quite recently, at the time when Wisconsin glaciers were melting back. Radiometric dating has since shown that the scablands were indeed formed between about 20,000 and 11,000 years ago.

When Bretz proposed the catastrophic flooding, critics argued that there was no source for the voluminous floodwaters that his hypothesis required. Soon, however, Bretz and others recognized that the likely source was Lake Missoula, a body of water that had formed in front of the glaciers that covered the Rocky Mountains (see Figure 19-3). The configuration of the lake was well known from well-layered sediments that display annual varves (see Figure 5-6). Lake Missoula was dammed by a lobe of glacial ice. The volume of the lake is estimated to have been about 2000 cubic kilometers (500 cubic miles).

Years after Bretz presented his argument, calculations showed that catastrophic collapse of Lake Missoula's ice dam would have produced currents deep and swift enough to have formed the enormous ripples of the scablands. Unconformities in the rippled sediments and in the sediments of Lake Missoula indicate that as many as 40 catastrophic flows may have

Scabland topography in eastern Washington State, where floodwaters carved a channel 100–150 meters (~330–500 feet) deep in the Columbia River Basalt.

Giant ripples, averaging 8 meters (~27 feet) in height, that were deposited during catastrophic flooding in eastern Washington State.

occurred. According to one proposal, the ice that dammed the lake stretched across a valley, and a flood occurred when the level of the lake rose to a point at which the water pressure at the bottom separated the ice from the rocks on which it rested. The ice dam then collapsed and the waters of the lake burst through, rushing westward to the Pacific Ocean, scouring the landscape, and depositing gravelly sediment in the form of giant ripples. Floating ice then began to pile up again in the constricted valley, forming a new dam, and the sequence of events repeated itself. After about 40 episodes of this kind, the glaciers so far receded that they no longer supplied enough meltwater and floating ice to re-form Lake Missoula. All that remained of the lake was a sequence of well-layered sediments, punctuated by many unconformities.

Today J. Harlan Bretz is widely viewed as a hero for advancing a novel idea and defending it rationally on the basis of sound observations. The floods that he brought to light stand as the largest ever identified on Earth.

faulting similar to that of the Great Basin. This hypothesis is deficient in one regard: it fails to account for the broad elevation of the Basin and Range Province during Neogene time.

We are more certain about the general pattern of tectonism along the Pacific coast. North America encountered the Pacific plate near the beginning of the Miocene Epoch. Movements along the San Andreas and other faults that have formed since that time account for the complex slivering and deformation in the Coast Ranges and neighboring areas (see Figure 19-24).

The Western Atlantic Ocean and Its Environs

Although the margins of the Atlantic Ocean were relatively quiescent during the Neogene Period, they did experience mild vertical tectonic movements—and these movements, together with more profound changes in sea level, had major effects on shoreline positions.

Global sea level has never stood as high during the Neogene Period as it did during much of Cretaceous or Paleogene time. For this reason, Neogene marine sediments along the Atlantic Ocean stand above sea level in only a few low-lying areas. Among the most impressive of the Miocene deposits found here are those of the Chesapeake Group, which form cliffs along the Chesapeake Bay in Maryland. The Chesapeake Group accumulated in the Salisbury Embayment during a worldwide high stand of sea level between about 16 million and 14 million years ago. The Salisbury Embayment is one of several downwarps of the American continental margin (Figure 19-28). Inhabiting the waters of the Salisbury Embayment was a rich fauna that included many large vertebrates, especially whales, dolphins, and sharks (see Figure 19-1). The fact that most of the fossils of baleen whales represent juvenile animals suggests that the embayment was a calving ground. Perhaps sharks were numerous because the young whales were especially vulnerable prey. Land-mammal bones are also found here and there in the Chesapeake Group, indicating that the waters of the embayment were shallow. Pollen from nearby land plants settled in the Salisbury Embayment, leaving a fossil record that shows a warm temperate flora near

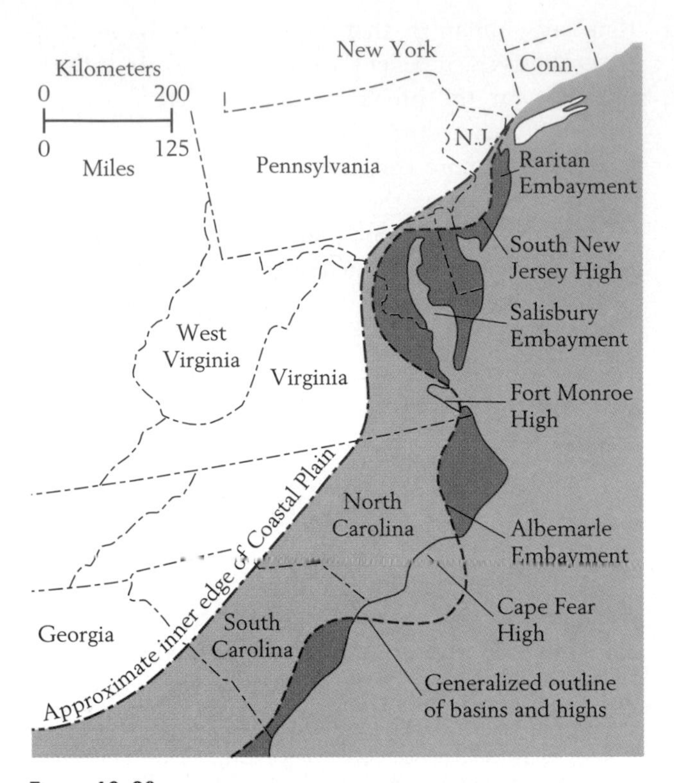

Figure 19-28 Elevated regions and depositional embayments along the mid-Atlantic coast during the Miocene Epoch. *(After J. P. Menard et al., Geol. Soc. Amer. Northeast-Southeast Sections, Field Trip Guidebook 7a, 1976.)*

The Birth of the Caribbean Sea

Although the Caribbean Sea is now an embayment of the Atlantic Ocean, it was once connected to the Pacific (Figure 19-30). During the Cretaceous Period, the floor of the Caribbean, which consists of oceanic rocks, was a small segment of the Pacific plate that was pushing toward the Atlantic, but during the Cenozoic Era, the Caribbean seafloor has lain along the north coast of South America while the Atlantic plate has been subducted beneath it. The Caribbean plate became a discrete entity late in Cenozoic time, when a new subduction zone came to connect the subduction zone bordering North America with the one bordering South America. The Greater Antilles—Cuba, Puerto Rico, Jamaica, and Hispaniola—represent an ancient mountain belt that is actually the southern end of the North American Cordillera. The Lesser Antilles represent an island arc west of the subduction zone, together with islands formed by deformation associated with subduction. The Yucatan Peninsula is a broad carbonate platform that lies to the west of the Caribbean. Offshore from this peninsula is the site of the asteroid impact that caused the mass extinction at the end of the Cretaceous Period (see Figure 17-12). As we have seen, the Bahamas are an ancient carbonate platform positioned farther north (see Figure 5-30).

the base of the Chesapeake Group slowly giving way upward in the sedimentary sequence to a flora adapted to slightly cooler conditions.

The Chesapeake Group and buried deposits of earlier age to the south consist primarily of siliciclastic sediments shed from the Appalachians to the west. Erosional features associated with the Appalachians reveal that these ancient mountains have a complex history. Like the modern Rockies, the existing topographic mountains that we call the Appalachians are the products of secondary isostatic uplift. The Appalachian orogenic belt was largely leveled by erosion long before the end of the Mesozoic Era. In many regions there were three or more additional intervals of uplift and erosion during the Cenozoic Era. The erosion that followed intervals of uplift left ridges of resistant folded rock standing above elongate valleys, but when erosion was especially intense, preexisting rivers cut through ridges as they and their tributaries carved out the valleys (Figure 19-29).

The Great American Interchange of Mammals

Before the Isthmus of Panama formed, slightly before 3 million years ago, a few species of mammals had passed between North and South America early in the Neogene Period—perhaps by swimming or floating on logs—but the terrestrial faunas of the two continents had remained largely separate. South America had been a great island continent and, like Australia, was populated by many marsupial mammals. The marsupials of Australia and South America have common ancestors that populated these continents as well as Antarctica when all three were part of a single Mesozoic landmass; in fact, Eocene mammal faunas resembling those of South America occur in Antarctica. By Pliocene time, however, when the isthmus developed, South American marsupials differed greatly from Australian marsupials, and the South American fauna included several groups of placental mammals

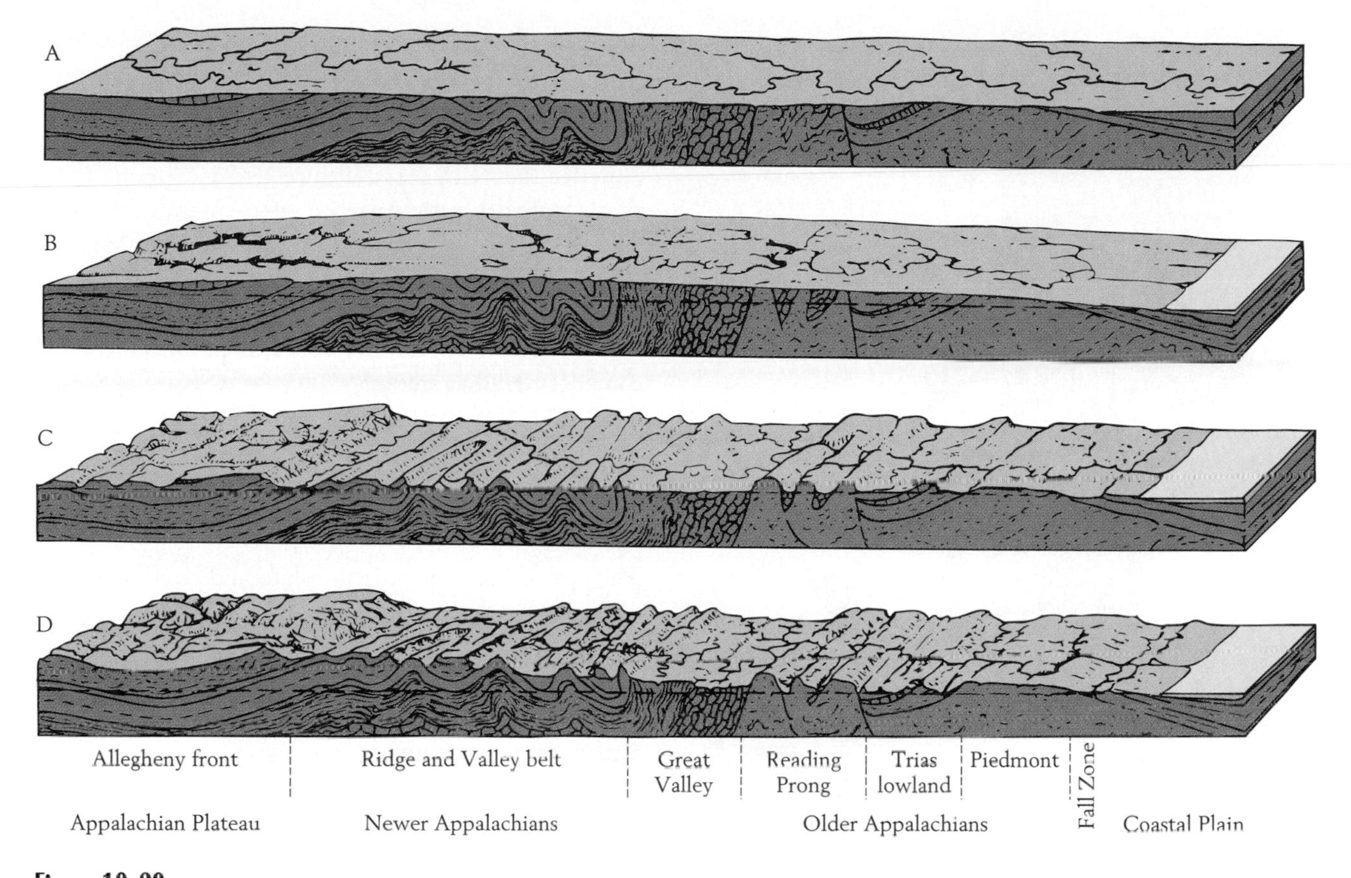

Figure 19-29 **The way episodic uplift has rejuvenated topography in the Valley and Ridge Province of the Appalachians.** *A* depicts the modest amount of relief that characterized some regions early in the Cretaceous Period. Later intervals of uplift and erosion *(B–D)* have produced the modern topography, characterized by ridges of resistant rocks, such as sandstone, and valleys of easily eroded rock, including shale. *(After D. W. Johnson, Stream Sculpture on the Atlantic Slope, Columbia University Press, New York, 1931.)*

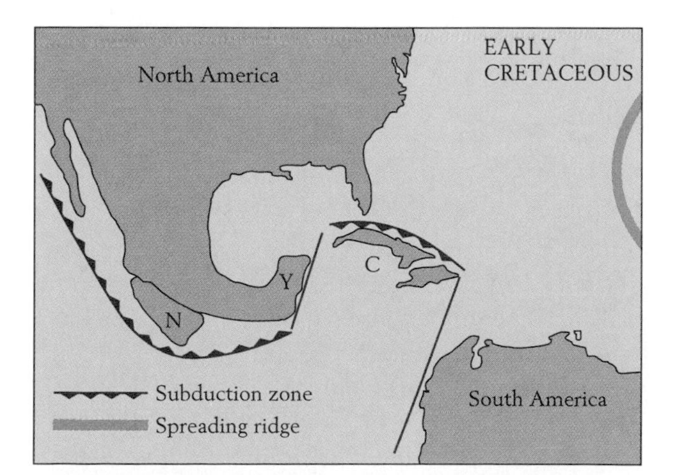

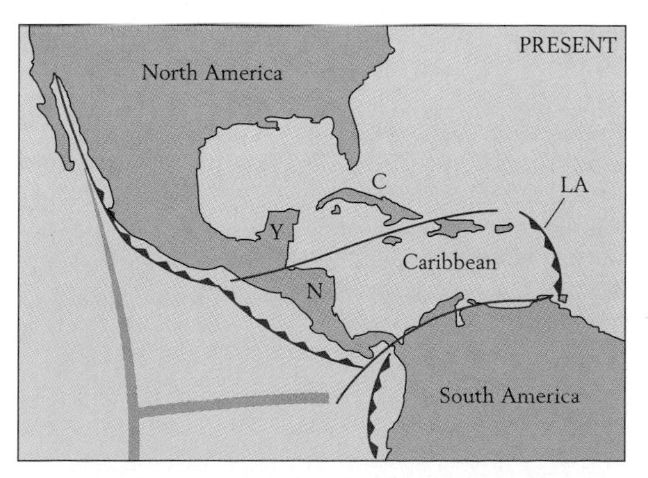

Figure 19-30 **Tectonic development of the Caribbean Sea as a segment of Pacific oceanic crust that has overridden Atlantic crust along a subduction zone that now extends northward from South America east of the Lesser Antilles.** N = Nicaragua; Y = Yucatan; C = Cuba; LA = Lesser Antilles. *(After G. W. Moore and L. Del Castillo, Geol. Soc. Amer. Bull. 85:607–618, 1974.)*

Figure 19-31 **Animals that took part in the great faunal interchange between North and South America when the Isthmus of Panama was elevated, connecting the continents.** The animals shown in North and Central America are immigrants from the south; they include armadillos, sloths, porcupines, and opossums. The animals shown in South America are immigrants from the north; among them are rabbits, elephants, deer, camels, and members of the bear, dog, and cat families. More animals migrated southward than northward. *(Drawing by M. Hill Werner, courtesy of L. G. Marshall.)*

whose ancestors had reached the continent from the north at the beginning of Cenozoic time or even earlier. Among the South American marsupials present when the land bridge formed were members of the opossum family, and among the placentals were sloths and armadillos that dwarf their relatives in the modern world (see Figure 7-5).

More North American species invaded South America than vice versa (Figure 19-31). Among those that reached South America were members of the camel, pig, deer, horse, elephant, tapir, rhino, rat, skunk, squirrel, rabbit, bear, dog, raccoon, and cat families. Migrating in the opposite direction were monkeys, anteaters, armadillos, porcupines, opossums, and some less familiar animals.

The Destruction of the Tethyan Seaway

The collision of the African plate and southern India with Eurasia during the Cenozoic Era destroyed what remained of the Tethyan Seaway. Today only vestiges of the seaway remain in the form of the isolated Mediterranean, Black, Caspian, and Aral seas. As Africa moved northward, movements of small plates in the Mediterranean region uplifted the Alps (see Figure 9-19). Recall that, farther east, the collision of what is now peninsular India with Asia has produced the exceptionally high Himalayan mountains (see Figure 9-20).

A Salinity Crisis in the Mediterranean At the end of the Miocene Epoch the Mediterranean Sea underwent spectacular changes. The first strong hint that geologists had of these changes was the discovery in 1961 of pillar-shaped structures in seismic profiles of the Mediterranean seafloor. These structures looked very much like the salt domes of Jurassic age in the Gulf of Mexico (see Figure 16-21), but if the strange features were indeed salt domes, the salt could have formed only by evaporation of Mediterranean waters. In 1970 the presence of evaporites here was confirmed by drilling that brought up anhydrite in cores of latest Miocene age. The idea that the Mediterranean had somehow turned into a shallow hypersaline basin was confirmed by the discovery of halite (rock salt) near the center of the eastern Mediterranean basin.

Further evidence that the Mediterranean shrank by evaporation at the end of Miocene time was the discovery of deep valleys filled with Pliocene sediments lying beneath the present beds of such rivers as the Rhône in France, the Po in Italy, and the Nile in Egypt. Rivers such as the Rhône and the Nile were already flowing into the Mediterranean earlier in the Miocene Epoch, and when the waters of the sea fell, the rivers cut deep canyons. In attempting to find solid footing for the Aswan Dam, Soviet geologists discovered a canyon buried beneath the present Nile delta and judged it to rival the modern Grand Canyon of Arizona in size. Clearly, at the end of the Miocene Epoch the single narrow connection between the Mediterranean Sea and the Atlantic Ocean nearly closed, probably as a result of a lowering of sea level in the Atlantic. Rates of evaporation similar to those of the Mediterranean region today would dry up an isolated sea as deep as the Mediterranean in a mere thousand years. During the crisis enough water must have flowed into the Mediterranean from the Atlantic to keep it from drying up altogether.

All of this happened between about 6 million years ago, when the eastern passage to the Atlantic closed, and 5 million years ago, when the Mediterranean basin refilled with deep water. Five-million-year-old deep-water microfossils in sediments on top of evaporites attest to the refilling. Apparently the connection with the Atlantic was enlarged again when the natural barrier at Gibraltar was suddenly breached. It has therefore been suggested that the first Atlantic waters must have been carried into the deep basin by a waterfall that would have dwarfed Niagara Falls.

Events to the East Farther east, the Tethyan Seaway was interrupted during Miocene time by the attachment of the Indian Peninsula to the Eurasian plate (p. 249), where molasse that shed southward from the newly forming Himalayas produced a lengthy and relatively complete fossil record for mammals. The famous Siwalik beds of Pakistan and India, for example, provide a nearly continuous record for the interval from 11 million to 1 million years ago (see Figure 9-21). The Siwalik beds document the composition of the rich faunas that occupied the spreading savannahs of late Miocene and Pleistocene age (see Figure 19-2). The great rivers of eastern Asia, which flow from the Himalayas to the sea, also formed in Miocene time during the uplift of the Himalayan region. Because of the high relief and abundant rainfall of the region, the Indus and Ganges of India and the several large rivers of Indochina contribute huge volumes of sediment to the ocean each year.

Human Evolution

In addition to the single species that now constitutes the human family, the superfamily Hominoidea currently consists of just four species of the ape family—the common chimpanzee, the pigmy chimpanzee, the gorilla, and the orangutan—together with six species of the gibbon family (see Figure 3-7). The human family, Hominidae, did not evolve from the modern ape family, Pongidae; instead, the two families have followed independent lines of evolution. Although it is possible that Hominidae and Pongidae evolved independently from a single family of primitive apes, the case is not clear.

Early Apes in Africa and Asia

Although an extensive Plio-Pleistocene fossil record has been uncovered for the Hominidae, very few known fossil remains of any age represent the Pongidae. Furthermore, the fossil record of the superfamily Hominoidea in latest Miocene time (8 million to 5 million years ago) is almost entirely barren. Sediments representing this interval in Africa, where apes and human ancestors evolved, are rare and poorly studied.

Farther back in the Miocene series are fossils of two extinct hominoid families of species that we can loosely call apes. Among these early forms must be ancestors of both modern apes and modern humans, but the evolutionary connections are not yet understood. The oldest fossils of these early apes come from African sediments about 20 million years old. Both of these extinct families first spread from Africa to Eurasia about 15 million to 16 million years ago. This was just a few million years after Africa, having moved thousands of kilometers northward after the breakup of Gondwanaland, finally collided with Eurasia and allowed the exchange of mammals between the two landmasses. Not only hominoids made their way northward to Eurasia during mid-Miocene time; so did many other previously isolated groups of African mammals, including elephants and giraffes (both of which have since become extinct in Eurasia).

The early apes underwent large adaptive radiations in both Africa and Eurasia. Some species were so large that they must have spent most of their time on the ground, but others were probably arboreal (tree-climbing) animals.

Throughout most of Miocene time the Old World was much more heavily populated with apes than Africa is today. By the end of the Miocene, however, only a single genus of primitive apes seems to have survived. This was the aptly named *Gigantopithecus,* a gorilla-sized creature that lived on into the Pleistocene. After the decline of apes, very close to the boundary between the Miocene and Pliocene epochs, there was an evolutionary event of great significance: the emergence of the earliest hominids from some unknown group of apes. These hominids constitute a distinct subfamily within the Hominidae and are informally termed the australopithecines. They are of special interest to modern humans because we are their only living descendants.

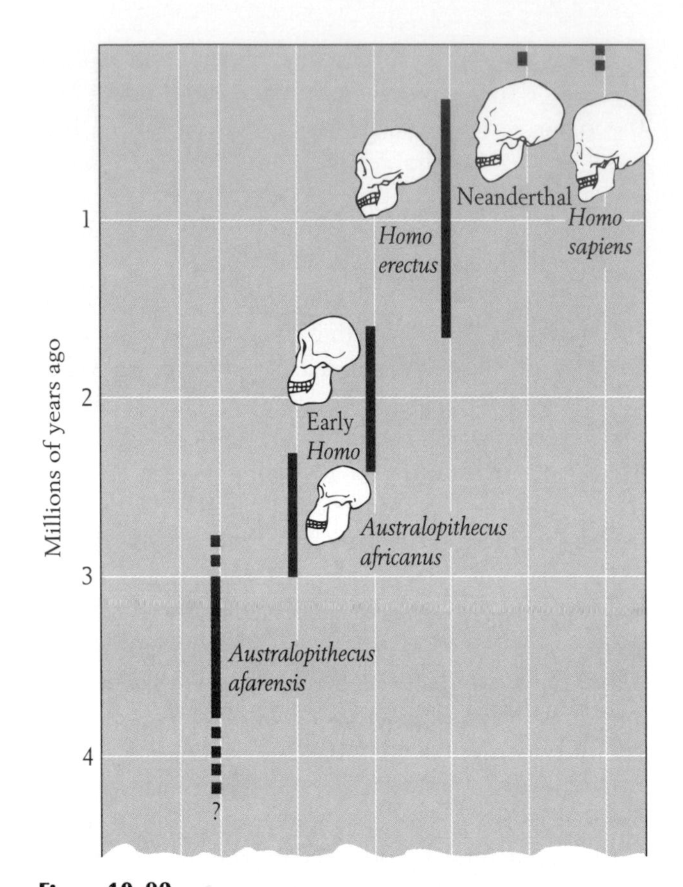

Figure 19-32 **Stratigraphic ranges of species of the Hominidae (human family) as recognized from fossil data.** Early *Homo* includes more than one species.

The Australopithecines

The australopithecines include *Australopithecus* and the closely related genus *Paranthropus.* Members of the two genera weighed roughly the same as chimpanzees (the males were much larger than the females) and their brain was only slightly larger than that of a chimp. It appears that, like chimpanzees, they were active tree climbers. When they were on the ground, however, they walked upright like humans rather than on all fours in the manner of apes. *Australopithecus* was the immediate ancestor of our genus, *Homo* (Figure 19-32).

Although *Australopithecus* did not evolve from modern apes, it was in many respects intermediate in form between apes and humans, resembling one group more in some ways and the other group more in others. Its skull was more like that of an ape than that of a human. In addition to having a relatively small brain, *Australopithecus* had an apelike face, with heavy bony ridges above the eye sockets and a large, projecting jaw (Figure 19-33). In fact, the bony ridges strengthened the skull for the attachment of heavy muscles that operated the large jaw. These features and large molars indicate that *Australopithecus* fed largely on coarse food, probably fruits and seedpods. Apparently it did not fashion stone tools, presumably because it lacked the intelligence to do so.

Australopithecus was much shorter than an average modern human. Females appear to have averaged slightly more than a meter (3.5 feet) in height, and males were about 30 percent taller (4.5 feet). The famous skeleton called Lucy, found in Ethiopia in 1973, was in life a female of the species *Australopithecus afarensis.* Lucy's bones and those of less fully preserved individuals reveal that the australopithecine body, unlike the skull, had some strikingly human features. The pelvis was broad, for support of the body in a vertical posture, unlike the narrow, elongate pelvis of an ape (see Figure 19-33). Tracks beautifully preserved in volcanic ash in the geographic region occupied by *Australopithecus afarensis* provide direct evidence of upright, two-legged walking. In fact, they are remarkably similar to the footprints of modern humans (Figure 19-34).

Although *Australopithecus* walked much the way we do, it displays many features suggesting that it spent a great deal of time in trees (see Figure 19-33). Its long arms, strong wrists, and long, curved fingers and toes were all apelike features that would have

Figure 19-33 Reconstruction of *Australopithecus afarensis* from partial skeletal remains of the individual named Lucy. This skeleton, from Hadaf, Ethiopia, is about 3.2 million years old. Although Lucy walked upright, many of her other anatomical features suggest that she frequently climbed trees.

been useful for climbing. It also had an upward-directed shoulder joint of the kind that improves the climbing ability of apes. Its legs were shorter than ours in relation to its body weight. These legs and the long toes would have been useful for climbing but detrimental to endurance in running on the ground. In addition, the inner ear bones of *Australopithecus* were like those of an ape, not like those that provide humans with the balance required for running on two legs. Probably the fastest gait for *Australopithecus* was a lope or jog. In short, the locomotory adaptations of *Australopithecus* represented an adaptive compromise between the need to move about effectively on the ground and the need to climb trees adeptly. This creature was less agile in the trees than a chimpanzee and less capable on the ground than a modern human.

It is not difficult to understand why *Australopithecus* would have climbed trees frequently. First of all, many of the fruits and seedpods that it ate must have grown on trees. Second, in order to avoid predators, it must have needed to sleep in trees and occasionally to flee into them during the day. A band of *Australopithecus* individuals probably slept in a grove of trees and fed in and around it during the day, staying close enough to the grove to flee into the trees if predators approached. When the local food supply dwindled, they would have been forced to migrate across dangerous open country to a new grove. Today this behavior characterizes both baboons, which are very large monkeys, and chimpanzees. *Australopithecus* would have been as defenseless against lions and hyenas as are these other large primates. Like them, it lacked advanced weapons and was slower on the ground than large four-legged herbivores such as antelopes and zebras.

Australopithecus survived for the better part of 2 million years, from at least 4 million to about 2.3 million years ago (see Figure 19-32). From this genus, sometime between 2.5 million and 2 million years ago, the modern human genus, *Homo*, evolved.

Figure 19-34 Tracks made in volcanic ash at Laetolil, Tanzania, more than 3 million years ago.

The Human Genus Makes Its Appearance The oldest bones thus far assigned to *Homo* are about 2.4 million years old. We do not know which of the two species of *Australopithecus* was its immediate ancestor. It appears that by about 2 million years ago, two or more species of *Homo* were in existence. Because the taxonomy of these early forms is not yet well established, it is convenient to group them all under the informal label "early *Homo*."

Some fossil skulls of early *Homo* reveal a large brain capacity, one of *Homo*'s trademarks. The average volume of an *Australopithecus* skull is about 450 cubic centimeters; the volume of the early *Homo* skull shown in Figure 19-35, in contrast, is about 760 cubic centimeters, and fragments of less well preserved skulls indicate brain capacities well above 800. The skull of early *Homo* exhibits additional features that make it more human in form than the skull of *Australopithecus*. The teeth, for example, are smaller.

Fossil remains of the pelvis and thigh bone of early *Homo* do not differ greatly from those of modern humans. These bones appear to have belonged to large individuals that spent almost all of their time on the ground.

Early *Homo* appears to have put its large brain to good use in the manufacture of stone tools. In fact, some types of fossils included within this group have been assigned to a species named *Homo habilis*, or "handy man," because stone tools have been found in some of the deposits from which such fossils have been collected. These tools include sharp flakes of stone and many-sided "core" stones that are what was left after the flakes were broken away (Figure 19-36). Such tools are referred to as Oldowan, because they were first found at Olduvai Gorge, the site of many fossil hominid discoveries. The oldest known jawbone of *Homo* and also the oldest known Oldowan tools, which come from farther north, in Ethiopia, are about 2.4 million years old.

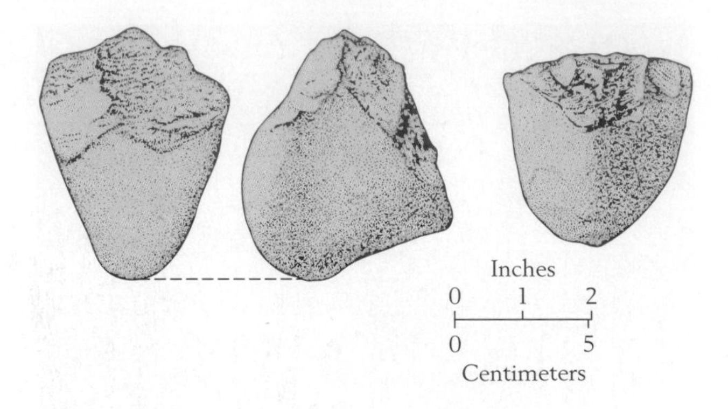

Figure 19-36 Simple stone implements from Olduvai Gorge, Tanzania, representing the Oldowan culture, which existed about 2.5 million years ago. These are "core" implements from which flakes were struck. *(After K. P. Oakley, Man the Toolmaker, British Museum of Natural History, London, 1950.)*

Perhaps australopithecines, like modern chimpanzees, supplemented their largely vegetarian diet by devouring small animals, but meat apparently formed a much larger part of early *Homo*'s diet. Scratched bones of other mammals found in association with Oldowan stones indicate that early *Homo* used its stone tools to sever meat from bones. It is uncertain whether most of this meat was the product of active hunting or of scavenging on carcasses killed by other animals. In any event, the rather abrupt appearance of *Homo* requires an explanation. *Australopithecus* had existed with little evolutionary change for at least 1.5 million years before it gave rise to the human genus. One theory holds that the sudden global climatic change that occurred about 2.5 million years ago led to the origin of *Homo*. This explanation relates to the way *Homo*'s large brain evolved.

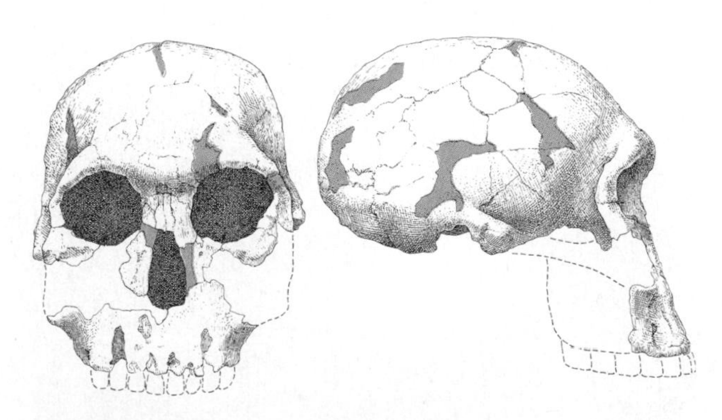

Figure 19-35 Front and side views of the skull of *Homo habilis*. *(From A. Walker and R. E. F. Leakey, "The Hominids of East Turkana." Copyright 1978 by Scientific American, Inc. All rights reserved.)*

Brain Size, Climbing, and the Shrinkage of Forests The size of early *Homo*'s brain can be attributed mainly to a change in the pattern of infant development. The brains of all newborn primates (including monkeys, apes, and humans) account for about

10 percent of their total body weight—a very large proportion. The brains of all these species grow rapidly before birth. In monkeys and apes, however, this high rate of brain growth slows dramatically shortly after birth, so that the brains of adults are only moderately larger than those of newborns. In *Homo*, however, the brain continues to grow rapidly for about a year after birth. This is the main reason why adult humans have such large brains. The size of the human brain increases only moderately after the age of 1, but an average year-old human infant already has a very large brain capacity—more than twice that of an adult chimpanzee.

The extension of the high rate of brain growth through the first year of life was achieved through a general evolutionary delay of maturation. Our permanent teeth do not replace our baby teeth, for example, until we are much older than apes are when they undergo these changes. A key feature of the human pattern of development is that it produces great intelligence at a very early age, so that small children can engage in relatively advanced learning.

We humans develop so slowly that our infants are helpless much longer than the infants of any other species of mammals. The disadvantage of the need to provide years of care to our offspring is more than offset, however, by the enormous advantages that result from the expansion of the brain: humans are much more intelligent than any other species of mammals, and for this reason we are able to cope with a host of environmental problems. We are neither physically powerful nor fleet of foot, yet we have come to dominate Earth.

Nonetheless, it is quite evident why australopithecines did not evolve a large brain for more than 1.5 million years of existence. As we have seen, they were required to spend a significant portion of their lives in trees. A mother who needed both arms free to climb trees could not have carried a helpless infant about with her. In other words, the large human brain could evolve only after human ancestors stopped climbing trees habitually. Why did they finally stop doing so? Recall that climates became cooler and drier over broad regions of the world about 2.5 million years ago, when continental glaciers expanded in the northern hemisphere. Fossil pollen reveals that at this time terrestrial floras in Africa underwent a dramatic change. Forests shrank and grasslands expanded. This is exactly the kind of change that would be expected to force animals such as australopithecines to abandon dependence on tree climbing. Presumably the australopithecines faced an ecological crisis: when they no longer had trees to climb, they lost important sources of food and were suddenly more exposed to such predators as lions and hyenas. Probably many populations of *Australopithecus* died out during this crisis, but apparently at least one population evolved into *Homo*. A large brain was so valuable for avoiding predators and developing advanced hunting techniques that its value overshadowed the problems that resulted from the prolonged helplessness of human infants.

Homo Erectus, Our Recent Ancestor

About 1.6 million years ago or slightly earlier, *Homo erectus* evolved from early *Homo* (see Figure 19-32). This new species seems not to have differed greatly from early *Homo* but it was slightly more similar to modern humans. In addition, *Homo erectus* was the first hominid species to have migrated beyond Africa. Known as *Pithecanthropus* before its similarities to modern humans were fully acknowledged, *Homo erectus* lived not only in Africa and Europe but also in China, where it has been referred to as "Peking man," and in Java, where it has been called "Java man." Dates for fossil skulls of *Homo erectus* encompass a long interval of time, from about 1.6 million years ago to perhaps 300,000 years ago. *Homo erectus* produced magnificent hand axes as part of its distinctive stone culture, known as **Acheulian** (Figure 19-37).

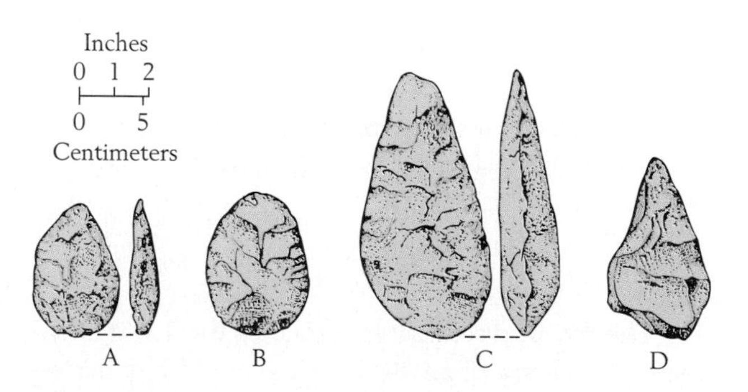

Figure 19-37 Tools of the widespread Acheulian culture, attributed to *Homo erectus*. *A.* A twisted oval tool from Saint-Acheul near Amiens, France. *B.* An oval hand ax from south of Wadi Sidr, Israel. C. A large hand ax from Orgesailie, Kenya. *D.* A hand ax from Hoxne, Suffolk, England. *(After K. P. Oakley, Man the Toolmaker, British Museum of Natural History, London, 1950.)*

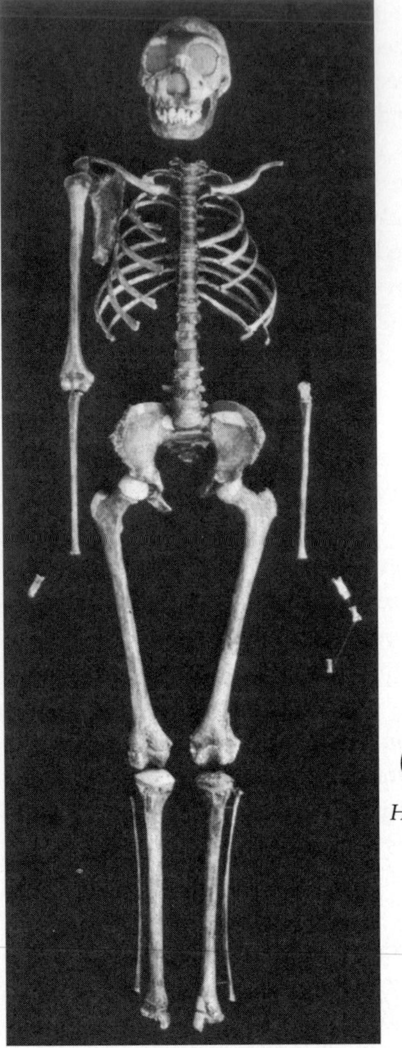

H. erectus H. sapiens

Figure 19-38 Nearly complete skeleton of an 11- or 12-year-old *Homo erectus* boy from strata 1.6 million years old. His height was about 5 feet 3 inches (1.6 meters), and had he lived to adulthood, he would have grown to a height of about 6 feet (1.8 meters). On the right, the neck of this boy's femur is seen to be much longer than that of a femur belonging to a typical modern human, who has a wider pelvis.

The recent discovery in Africa of the 1.6-million-year-old skeleton of an 11- or 12-year-old boy has revealed the remarkable resemblance between *Homo erectus* and modern humans (Figure 19-38). This species resembled us in body size but had a smaller brain. Its cranial capacity averaged about 1000 cubic centimeters, compared to 1400 for modern male humans. The pelvis of *Homo erectus* was narrower than ours, and here we see why this extinct species necessarily had a smaller brain. The size of the pelvis limits the size of the brain in hominids, because a baby's head must pass through its mother's pelvis during birth. The narrow pelvis of *Homo erectus* would not have permitted the birth of a modern baby of average size. Evolution has provided us with a mechanism to develop a brain larger than *Homo erectus*'s: our larger pelvis allows our brain to grow larger before birth, so we have a head start in the growth of the adult brain. Not only was the brain of *Homo erectus* smaller than ours, but it had a different shape, as reflected in a low, sloping forehead and elongate skull form that it inherited from *Australopithecus.* Other primitive features of the skull were prominent brow ridges, a projecting mouth, and a heavy lower jaw.

Neanderthal, Our Cousin

Fossils of the creature known as Neanderthal appear in sediments about 100,000 years old (see Figure 19-32). Although for a time regarded as a variety of our species, Neanderthal is now known to represent a separate species that goes by the name *Homo neanderthalensis.* Researchers recently extracted DNA from a limb bone of a Neanderthal, and a comparison with the DNA of modern humans uncovered so many differences that the two forms clearly do not represent a single species. Furthermore, the number of genetic differences between *Homo sapiens* and *Homo neanderthalensis* suggests that the lineages of these two species branched from some common ancestor more than 500,000 years ago. This finding is not surprising because fossils that resemble Neanderthal and apparently represent populations ancestral to it range back several hundred thousand years before the present.

The record of Neanderthal itself extends from Spain to central Asia, ranging in time up to about 35,000 years ago. Enough of its bones and artifacts have been found in caves to suggest that Neanderthal frequently took shelter in caves. Neanderthal resembled *Homo erectus* in its long, low skull with prominent brow ridges, projecting mouth, and receding chin (Figure 19-39). On the other hand, its brain was quite large—slightly larger, on average, than that of modern humans. Actually, however, this brain was smaller in relation to body size than ours. Neanderthal, though of somewhat shorter stature than a

Figure 19-39 **Burial of a young Neanderthal, whose skeleton shows that the head was cradled on one arm.** Also present are the charred bones of animals that apparently were left as food for the dead individual in an afterlife. *(Painting by Z. Burian under the supervision of Professor J. Augusta, Professor J. Filipa, and Dr. J. Moleho.)*

modern human, was more heavily built, so that a large proportion of its brain functioned simply to operate its massive body.

Neanderthals developed a distinctive stone culture known as *Mousterian,* which is characterized by flakes of stone fashioned into knives and scrapers—far more sophisticated tools than the Acheulian implements of *Homo erectus*—but never attached stone points to spears. Instead, they hunted animals with wooden spears sharpened at one end. Neanderthals obviously had a difficult life under the Ice Age conditions in Europe. Their skeletons display many wounds, including broken bones that healed. They nurtured infirm members of their society, however, as indicated by skeletons of individuals who had survived for many years in a crippled state and must have lived under the care of others. Neanderthals also appear to have had religion. Burial sites reveal that Neanderthals sometimes prepared their dead for a future life by interring them with flint tools and cooked meat (see Figure 19-39). In the Zagros Mountains of Iraq, a Neanderthal man who died after a skull injury was buried in a bed of boughs and flowers that can be identified from the pollen they left beneath the skeleton. It is fortunate for us that Neanderthals buried their dead. In effect, they intentionally produced much of their own fossil record, preserving many complete skeletons to the benefit of modern science.

Both genetic studies of modern humans and a few fossil skulls indicate that *Homo sapiens* evolved in Africa approximately 150,000 years ago. Sophisticated tools made of bone and stone appear in the fossil record of East Africa in sediments nearly 100,000 old. Perhaps our species then required several tens of thousands of years to make its way northward to the icy climates of Europe. The oldest remains of modern humans in Europe date to about 33,000 years ago. It is a striking fact that the Neanderthals then disappeared within about 3000 years after *Homo sapiens* appeared in Europe. Could the timing be a matter of chance? Certainly it is possible that our species' aggressive behavior caused the Neanderthals' extinction. Perhaps *Homo sapiens* was the victor in competition for occupancy of caves or control of territory in general.

In any event, our species brought to Europe far more advanced technology than that of Neanderthal. The culture of *Homo sapiens* that quickly emerged in Europe is known as the *Cro-Magnon culture.* It at first incorporated Mousterian elements but then developed into the more sophisticated *Late Neolithic culture,* in which a variety of specialized tools were invented.

Even more innovative was the artwork of the Cro-Magnon people, which reflected new use of the powers of imagination. Their magnificent cave paintings, primarily of animals, can still be admired in France and Spain (Figure 19-40). Other artifacts show that their artistic efforts included modeling in clay, carving friezes, decorating bones, and fabricating jewelry from teeth and shells. All of this reflected one unique aspect of the history of *Homo sapiens*: rapid

Figure 19-40 A cave painting of the Cro-Magnon people.

cultural evolution. Since early in the history of *Homo sapiens,* its culture has been changing rapidly. In contrast, cultures of earlier species of the human family, once in place, remained almost stagnant. A second unique aspect of our culture that has strengthened through time is an ability to manipulate Earth's environment. As we will see in Chapter 20, this expanding power now threatens our well-being.

Chapter Summary

1 Invertebrate life in the oceans underwent only minor changes during Neogene time, but whales radiated rapidly.

2 On land, grasses and herbs, benefiting from a cooling of climates, diversified and occupied more territory during Miocene time.

3 Terrestrial mammals assumed their modern character during Neogene time, and frogs, rats and mice, snakes, and songbirds underwent major adaptive radiations.

4 Apes radiated during the Miocene Epoch but the number of their species has since dwindled.

5 The modern Ice Age began about 3.2 million years ago with the expansion of continental glaciers in the northern hemisphere. Climates in many regions became cooler and drier, and grassy habitats expanded while forests shrank.

6 The modern Ice Age has been characterized by glacial maxima and minima. During the past 900,000 years, continental glaciers have expanded and contracted 9 times, with expansions lowering sea level by as much as 100 meters. Glaciers are now in a contracted state, but they will probably expand again.

7 During Neogene time the American West has been affected by extensive volcanism, and major uplifts have produced the Sierra Nevada, the Rocky Mountains, and the Colorado Plateau.

8 In the western Atlantic region, the Caribbean Sea, which is bounded on the east by an island arc, developed its modern configuration during Neogene time, and the Isthmus of Panama was uplifted, permitting extensive biotic interchange between North and South America.

9 The Mediterranean Sea became weakly connected to the Atlantic Ocean about 6 million years ago and briefly became hypersaline. Soon thereafter reenlargement of the connection allowed the Mediterranean to fill with normal marine waters again.

10 *Australopithecus,* the oldest known genus of the human family, evolved at least 4 million years ago. It possessed a small body, a relatively small brain, and apelike teeth. *Homo,* the modern human genus, evolved in Africa about 2.5 million years ago, perhaps as a consequence of the shrinkage of forests. About 150,000 years ago, humans of the modern type were present.

Review Questions

1 In what ways did mammals become modernized during the Neogene Period?

2 What factors have influenced the isotopic composition of oxygen in skeletons of marine organisms during the Pleistocene Epoch?

3 What changes in the geographic distribution of land animals did the uplift of the Isthmus of Panama produce?

4 How did the Rocky Mountains develop their present configuration in the course of the Neogene Period?

5 How and when did the Sierra Nevada form?

6 What kinds of volcanic activity occurred in the American West during Neogene time?

7 How did climatic change alter the general distribution of terrestrial vegetation since early in Pliocene time?

8 How did the Appalachian Mountains develop their present configuration in the course of Neogene time?

9 What evidence is there that about 7 million years ago a major change occurred in the kinds of grasses that populate the world? How did this change affect animals?

10 How did members of the genus *Australopithecus* differ from modern humans?

11 Two phases of global climatic change occurred during the Neogene Period. The first one was gradual, the second one sudden. Using the Visual Overview on page 520 and what you have learned in this chapter, review each of these phases of climatic change and describe how each affected life on Earth.

Additional Reading

Delcourt, H., and P. A. Delcourt, *Pleistocene Environments of the British Isles,* Chapman & Hall, London, 1991.

Jones, K. L., and D. H. Keen, *Pleistocene Environments in the British Isles,* Chapman & Hall, London, 1993.

Lowe, J. J., and M. J. C. Walker, *Reconstructing Quaternary Environments,* Longman, Harlow, Great Britain, 1997.

Savage, R. J. G., and M. R. Long, *Mammal Evolution: An Illustrated Guide,* Facts on File, Inc., New York, 1986

Stanley, S. M., *Children of the Ice Age: How a Global Catastrophe Allowed Humans to Evolve,* W. H. Freeman and Company, New York, 1998.

CHAPTER 20

The Holocene

The Holocene, sometimes called the Recent interval, is the time that extends back from the present to the retreat of continental glaciers in the northern hemisphere. The Holocene was originally established as the final epoch of the Cenozoic Era. Whereas the Miocene, Pliocene, and Pleistocene epochs, which preceded the Holocene, are assigned to the Neogene Period, however, the Holocene currently belongs to no geologic period. Technically, then, it cannot be formally recognized as an epoch, which must be a subdivision of a period. On the other hand, the Holocene could itself be elevated to the status of a geologic period, although it would be a very short one. It stands out as a unique geologic interval—the interval during which humans have altered Earth's environment, first by hunting, later by cutting down trees and planting crops, and eventually by building cities, burning fossil fuels, and creating vast networks of communication and transportation. The boundary between the Pleistocene and the Holocene remains loosely defined. We will begin this chapter with events that took place slightly before 12,000 years ago because those events marked the beginning of the end of the Ice Age in the northern hemisphere and were quickly followed by the first major impacts of humans on Earth's ecosystems.

Holocene sediments and fossils are fully within the range of radiocarbon dating, so they can be dated with great precision. In addition, because the vast majority of Holocene species of plants, animals, and protists survive today, paleontologists can use the fossil records of these species to characterize habitats with great accuracy.

Holocene history holds special importance for us today because it reveals how the Earth system has approached its present state.

The upper limit of trees and the lower limit of a glacier in Olympic National Park in the state of Washington.

Visual Overview

Major Events of the Holocene

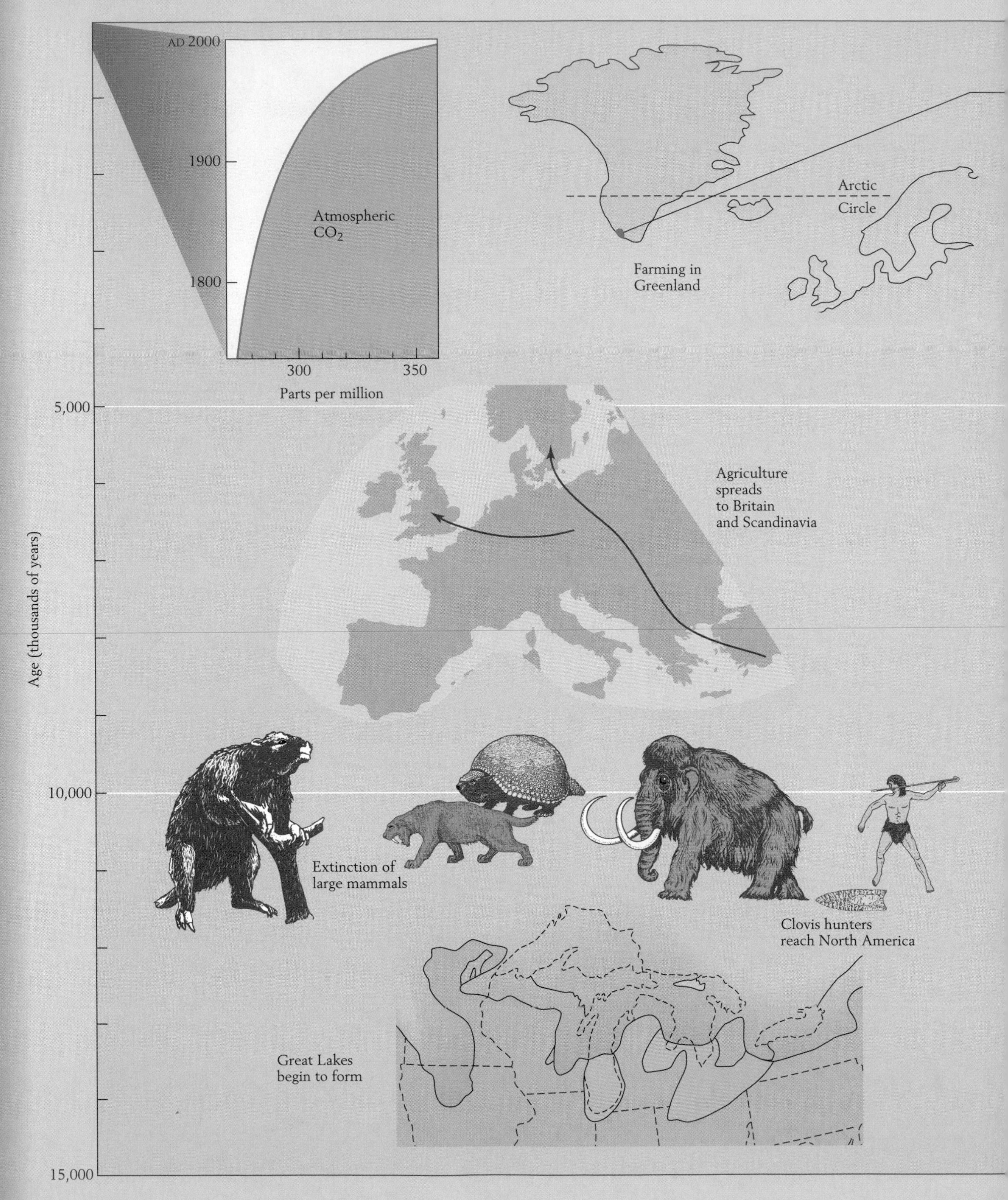

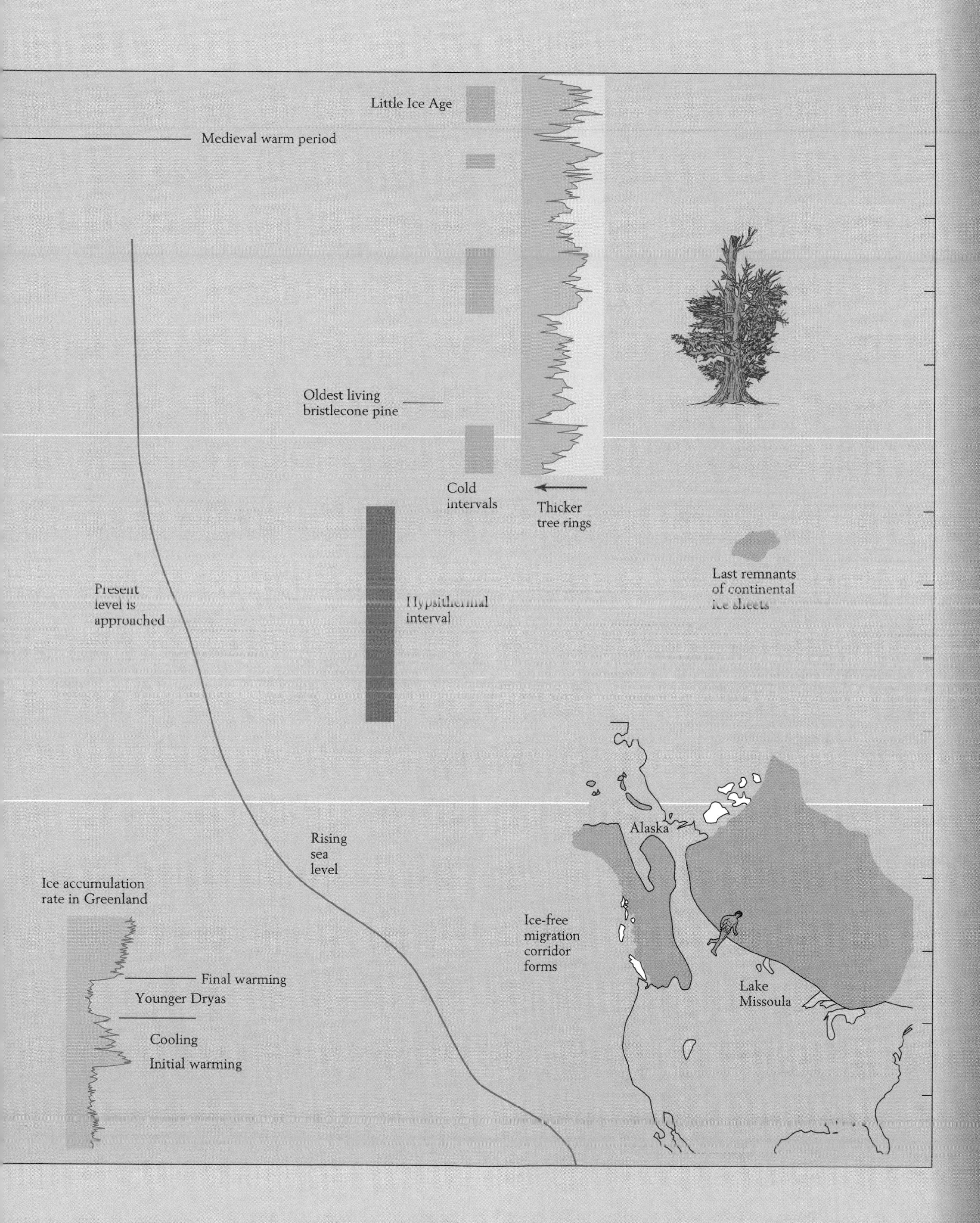

Little Ice Age
Medieval warm period
Oldest living
bristlecone pine
Cold
intervals
Thicker
tree rings
Present
level is
approached
Hypsithermal
interval
Last remnants
of continental
ice sheets
Alaska
Rising
sea
level
Ice accumulation
rate in Greenland
Ice-free
migration
corridor
forms
Lake
Missoula
Final warming
Younger Dryas
Cooling
Initial warming

Future changes in the global environment build on trends of the recent past. The geologic record of the Holocene reveals the speed with which environments can change and the ways in which the changes affect the species that still populate our planet. This record, together with other patterns of change seen in the geologic record of the more distant past, provides lessons that will help us confront environmental changes in the future.

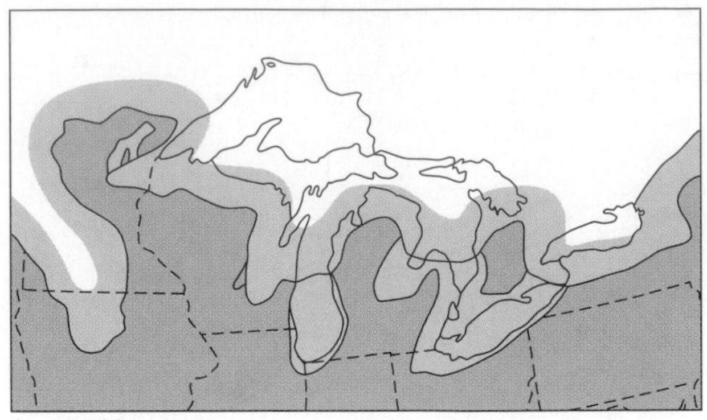

A 13,500 years ago

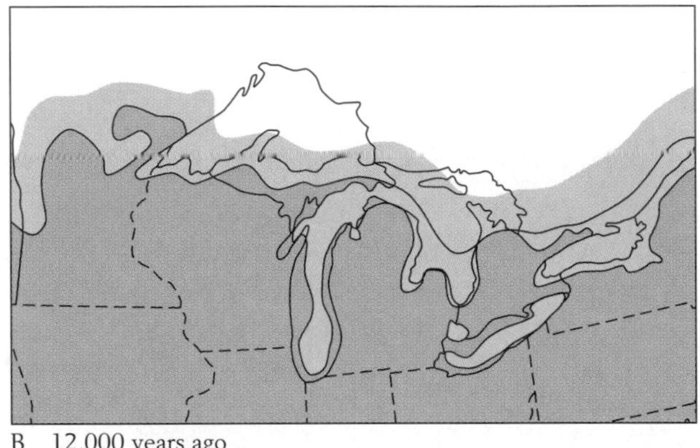

B 12,000 years ago

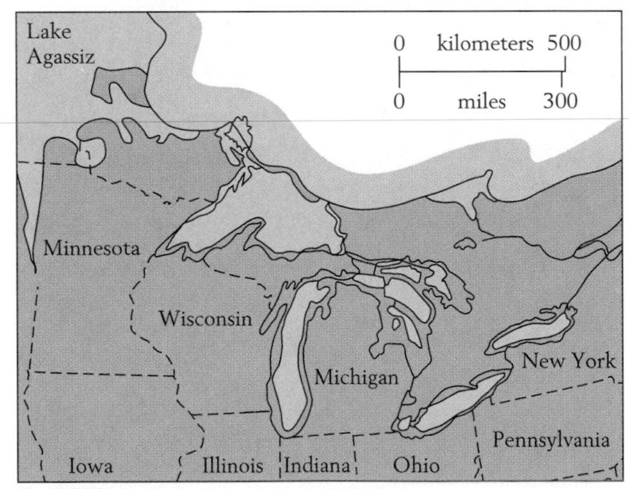

C 9,500 years ago

Figure 20-1 The retreat of glaciers from eastern North America after the last glacial maximum. The Great Lakes occupied depressions left when the glaciers retreated.

The Retreat of Glaciers

Soon after the last glacial maximum, about 20,000 years ago, continental glaciers began to waste away. Fossil insects, dated by the radiocarbon method, show that temperatures in the Rocky Mountain region began to rise about 15,000 years ago; the insect species that provide this information survive today, and their temperature tolerances are well known. The meltwaters of the shrinking continental glaciers flowed to the sea, which therefore began to rise. Moraines reveal that the retreat of glaciers was slow at first but accelerated after about 15,000 years ago (Figure 20-1). As the great North American ice sheet melted back, the waters that drained from its southern border formed lakes that were ancestral to the modern Great Lakes. Farther west, a broad, shallow body of water known as Lake Agassiz formed slightly after 12,000 years ago, constantly shifted its shape, and then disappeared about 4000 or 5000 years later (see Figure 20-1*A*). Mounds of ice a few meters across remained for a time in Minnesota, the Dakotas, and southern Canada. Many of these remnants created depressions that remain today as what are called *prairie potholes:* small ponds and marshes where many species of waterfowl stop during their seasonal migrations (Figure 20-2).

Tundra, which bordered the continental glaciers, shifted northward with them. Surprisingly, insects migrated independently of vegetation. About 13,000 years ago, assemblages of insect species that today inhabit taiga (northern evergreen forest) migrated north to occupy the tundra habitat in southern Canada. The insects that now inhabit taiga migrated into this habitat only after it approached its present distribution.

Farther south, taxa of deciduous trees, such as beech, hickory, and maple, migrated northward as climates warmed (Figure 20-3). As happened throughout the Pleistocene, the various species migrated at different rates, so that forests were continually restructured. The climate to the south of the retreating ice sheets must also have differed from any climate of the present world because fossil pollen reveals that the transitional climate supported an evergreen forest

Figure 20-2 Prairie potholes in southern Canada. These depressions, which harbor ponds and marshes, formed when chunks of ice left by retreating glaciers melted after becoming partly buried in sediment.

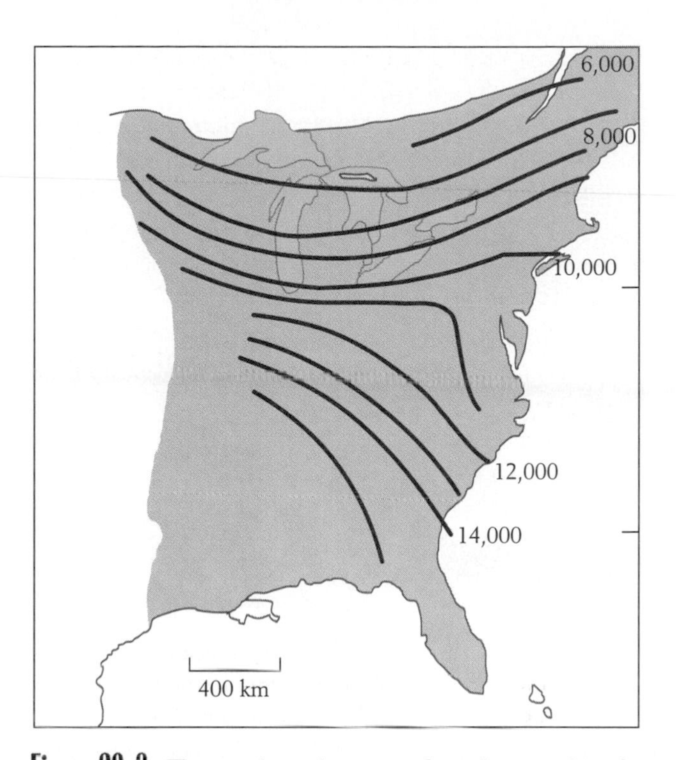

Figure 20-3 The northward retreat of maple trees (*Acer*) at the end of the last glacial maximum, as indicated by pollen occurrences. Lines indicate northern limits for maples at various times. *(After M. B. Davis, in Late Quaternary Environments of the United States, H. E. Wright [ed.], University of Minnesota Press, Minneapolis, 1983.)*

unlike any modern flora. Spruce trees were common, but pine trees, an abundant component of modern northern evergreen forests, were absent. On the other hand, such hardwood trees as oak, elm, and ash abounded. As sea ice melted back in the North Atlantic and the salinity of the seawater increased, downwelling there created the modern pattern of conveyor belt flow (see Figure 19-18).

Abrupt Global Events of the Early Holocene

Earth's emergence from the last glacial maximum was not smooth. The transition to the present world was marked by temporary reversals to cold conditions and, more generally, by abrupt climatic changes. Between 15,000 and 7000 years ago three abrupt rises of sea level probably corresponded to sudden global shifts in climate.

Fossil corals on the island of Barbados in the Caribbean provide evidence of the three Holocene episodes of rapid elevation of sea level. Here scientists have made use of the reef-building "moosehorn" coral, *Acropora palmata*, which always grows close to sea level. Fossil specimens of this species, which can be dated by the uranium-thorium method, indicate the level of the sea in relation to the island for various past intervals. Barbados has been rising tectonically since late Pleistocene time. Colonies of *Acropora palmata* that grew during previous glacial minima, when sea level was close to its present position, now stand many meters above sea level. Their position indicates that Barbados has been rising tectonically at an average rate of more that 30 centimeters per year for the past 125,000 years. Cores obtained by drilling through the limestones of Barbados have yielded fossil *Acropora palmata* from many depths below the surface (Figure 20-4). Radiometric dating by the radiocarbon and uranium-thorium methods (p. 166) has established the ages of these corals. Their depths

Figure 20-4 Holocene rise in sea level, indicated by occurrences of *Acropora palmata*, a reef-building coral that lives only close to the sea surface. Dots in the graph show occurrences of this species dated by the radiocarbon method (black) and uranium-thorium method (white). *(Graph after P. Blanchon and J. Shaw, Geology 23:4–8, 1995.)*

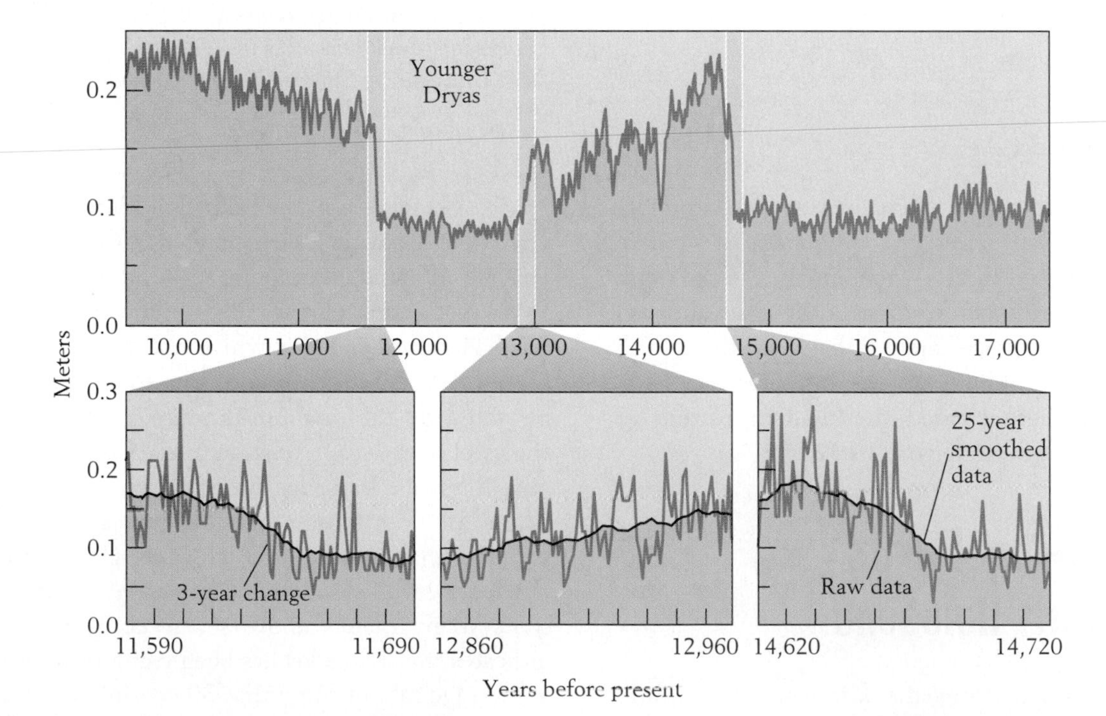

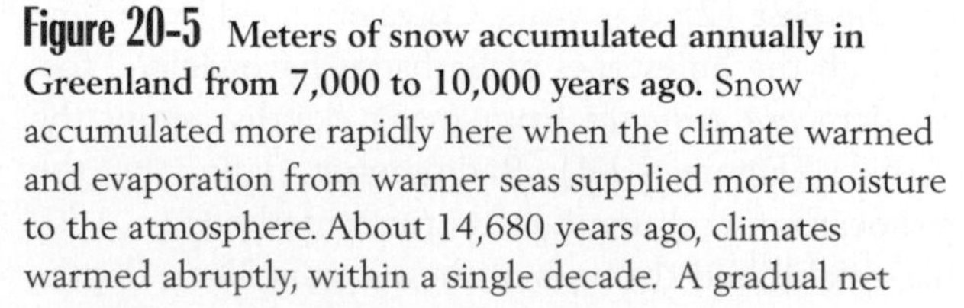

Figure 20-5 Meters of snow accumulated annually in Greenland from 7,000 to 10,000 years ago. Snow accumulated more rapidly here when the climate warmed and evaporation from warmer seas supplied more moisture to the atmosphere. About 14,680 years ago, climates warmed abruptly, within a single decade. A gradual net cooling trend followed, and then an episode of more rapid cooling began about 13,000 years ago and lasted about 200 years, producing the Younger Dryas cold interval. The Younger Dryas ended about 11,600 years ago, in the space of only about three years. *(After R. B. Alley et al., Nature 312:527–529.)*

below the surface, corrected for tectonic uplift of the island, then indicate the position of sea level back to the glacial maxima, about 18,000 years ago. Sea level rose rapidly slightly after 15,000, 12,000, and 8,000 years ago.

Possibly sea level occasionally rose rapidly by a few meters during the general eustatic rise of the past 18,000 years. Certainly there have been abrupt pulses of warming in the far north. Sometimes the Greenland glacier has responded differently than other ice sheets to global warming: when climatic warming has caused ice sheets to shrink in North America and Europe, annual layers of glacial ice in Greenland have thickened because of the increased supply of moisture from the North Atlantic Ocean (p. 532). About 14,680 years ago, in the space of only about ten years, annual layers in the Greenland glacier approximately doubled in thickness (Figure 20-5). The sudden warming that produced the change remains unexplained, but about 500 years later a massive movement of ice into the North Atlantic created Heinrich layers.

For unknown reasons, a long cooling trend began after the sudden warming event. This trend was accentuated slightly before 13,000 years ago, when the northern hemisphere began to shift back toward full glacial conditions. The shift took place over about 200 years, and the cold conditions persisted for more than 1000 years (see Figure 20-5). This cold interval is called the Younger Dryas, after a genus of flowers (*Dryas*) that spread southward with the frigid conditions. Sea ice also spread southward, and planktonic foraminifera from deep-sea cores show that the relative amount of oxygen 18 increased markedly in the North Atlantic, reflecting a reduction in salinity that was perhaps a consequence of an influx of glacial meltwaters. Warm saline water from the south apparently failed to reach the cold North Atlantic. As a result, the downwelling that had driven the conveyor belt shifted southward.

Younger Dryas cooling was global in scale, causing mountain glaciers to expand well below their present limits in New Zealand, for example (Figure 20-6). The Younger Dryas suddenly ended about 11,600 years ago, however. Amazingly, oxygen isotope measurements in ice cores reveal that the climatic shift took place within just three years, most of it during a single year (see Figure 20-5). Heinrich layers also formed, pointing to a massive launching of icebergs, and isotope studies suggest that the climate of Greenland warmed by about 7°C. The abrupt end of the Younger Dryas moved Earth fully into the Holocene interval of glacial retreat.

The final sudden rise in sea level deduced from the Barbados reef rock took place about 7600 years ago. No Heinrich layers mark this event; in fact, too little of the Laurentide ice sheet remained to release icebergs. It seems most likely that a catastrophic loss of ice from the Antarctic ice cap elevated sea level. At this time climates were restructured in ways that

Figure 20-6 A tree-covered terminal moraine in southern New Zealand that formed during the Younger Dryas cooling event.

are not fully understood. Many parts of Africa, for example, became drier. The eustatic rise about 7600 years ago brought global sea level very close to its present position; after this time very little of the Laurentide ice sheet remained to contribute additional meltwater to the ocean.

Global environmental changes of the early Holocene occurred with frightening speed. Knowledge of the instability of the natural system must heighten our concern about the future effects of human activities, including the liberation of greenhouse gases and the destruction of rain forests, which retain moisture in the regions where they grow.

The First Americans

About the time of the Younger Dryas, humans colonized North America on a grand scale. The timing of this migration was no accident. After moving northward to Eurasia, modern humans had spread rapidly across this vast continent. By 30,000 years ago they reached Siberia, where climates were very cold but a weak supply of moisture prevented large glaciers from forming. Soon thereafter humans reached the Bering land bridge, where many large animals roamed (see Figures 19-7 and 19-8), but glaciers blocked easy passage to Alaska. A small number of archeological sites farther south suggest that a few bands of humans made their way to the Americas about 30,000 years ago. Only after glaciers were melting rapidly, however, did a large corridor open through Alaska and western Canada to unglaciated portions of North America. This icefree avenue passed between the glacier that still occupied the northern Rocky Mountains and the retreating Laurentide ice sheet to the east (p. 561). Along it trod the Clovis people, who developed the first widespread culture in the New World and became the ancestors of nearly all Native Americans. They reached North America slightly before 11,000 years ago.

Archeological studies have revealed little about the Clovis people, except that they hunted adeptly in groups to take down elephants and bison with spears. Each of three species of elephants occupied a particular North American habitat when the Clovis hunters arrived (Figure 20-7). The woolly mammoth occupied tundra and grassy terrain close to the great ice sheets in both Eurasia and North America. Frozen carcasses in Siberia and Alaska retain long hair that insulated the woolly mammoth's body. These carcasses also display small ears and a short trunk, features that would have reduced heat loss. Stomach contents are also preserved, revealing a diet of grasses and tundra plants. Populations of the mastodon, a heavily built elephant, were concentrated in eastern forests, where twigs and conifers often formed much of its diet. Great herds of the great southern mammoth roamed prairies of the Midwest and Southwest. This animal stood as tall as 3.4 meters (more than 11 feet) at the shoulder and had strongly curved molars that served well for grinding up harsh prairie grasses.

Clovis hunters used a distinctive spear point (Figure 20-8). It was fluted; that is, a channel was

Figure 20-7 The last three species of elephants to inhabit North America. The southern mammoth (*Mammuthus columbi*) formed great herds on the prairies; the mastodon (*Mammut americanum*) occupied eastern forests, and the woolly mammoth (*Mammuthus primigenius*) lived in cold northern regions.

Figure 20-8 **Clovis hunting.** *A.* A spear point found at Folsom, New Mexico, between the ribs of a giant bison. *B.* A reconstruction of Clovis hunters attacking a giant bison.

chipped into each side opposite the point to accommodate a slot in a spear shaft. The spears of Clovis hunters were cut short so that they could be hurled by a spear thrower—a hand-held device that generated greater velocity than a hunter could produce by throwing a spear he held in his hand. The archeological record offers abundant evidence of Clovis hunters' ability to take down large animals. In the 1920s a cowboy stumbled upon a bison skeleton near Folsom, New Mexico, with a spear point lodged between its ribs. Since then, anthropologists have discovered Clovis sites displaying spear points lodged in mammoth skulls or vertebrae, or otherwise associated with mammoth bones.

Sudden Extinction of Large Mammals

Between about 12,000 and 10,000 years ago, many species of large mammals disappeared from North and South America. The extinction of these creatures, which transformed the terrestrial ecosystem, has been the subject of great controversy. Were the species killed off by Clovis hunters or by widespread climatic change? Before considering the arguments on each side, let us look at the victims. The bones of many of them have been found in the La Brea tar pits in Los Angeles, along with remains of mammals that survived to the present day (Figure 20-9). Here large herbivores became mired in sticky pools of tar that oozed up from sediments below. The trapped creatures attracted numerous carnivores, many of whom also became entrapped and sank into the soft tar.

The most striking feature of the mammal species that disappeared early in the Holocene was their large average body size. Among the large species of North American herbivores to disappear were all three species of American elephants (see Figure 20-7); a beaver that was as large as a black bear; five species of horses; three members of the deer family; two species of wild oxen; three species of musk oxen; the only surviving North American camel, which stood about 20 percent taller than the modern Mideastern camel; and a large bison, whose horns spread to more than 2 meters (~7 feet) (see Figure 20-8); two species of giant armadillos, each of which weighed about a ton; and several species of large ground sloths (see Figure 7-5 and p. 180), one of which weighed about three tons.

Figure 20-9 **Mammals of the La Brea tar pits, now located in the city of Los Angeles.** In the foreground a sabertooth cat defends a zebra carcass against huge dire wolves. Beyond them a lion pounces on a bison calf. Imperial mammoths roam in the distance. On the right are tall camels and in front of them are a giant ground sloth and a scavenging condor. *(Painting by Mark Hallet.)*

Several species of large North American carnivores also died out: the so-called short-faced bear, which stood nearly 2 meters (~6 feet) tall at the shoulder and was probably capable of running down horses and bison; the dire wolf, a fearsome creature that was probably 30 percent heavier than the living wolf species; a lion that was perhaps a subspecies of the modern African lion but was about 50 percent heavier; a species of cheetah; and three species of large sabertooth cats.

It may be that the big carnivores died out because their prey disappeared. This is especially likely to be the case for the sabertooth cats, which probably used their bladelike teeth to pierce the thick hides and fatty tissues of elephants. These teeth were fragile and would have broken easily against the shoulder blade or rib of a horse or deer. The Freisenbahn cave in Texas served as a lair for *Homotherium*, one of the American sabertooth cats. From this single cave paleontologists have recovered the skeletal remains of more than 30 skeletons of these cats—many representing juveniles—along with the bones of about 70 juvenile mammoths. Clearly the sabertooths favored vulnerable young mammoths as prey and dragged them into the cave. Probably all three species of sabertooth cats died out because their elephant prey disappeared.

Interestingly, several species of large birds, including eagles, vultures, and a condor, disappeared from North America along with the large mammals. These forms probably scavenged on carcasses of large herbivores after carnivores were through with them (see Figure 20-9); and when many species of large herbivores died out, the scavenging birds began to starve along with the carnivores.

The Overkill Hypothesis

The question remains: Why did the large herbivores die out? The *overkill hypothesis* focuses on human hunting as the cause. According to this hypothesis, large bands of proficient Clovis hunters invaded regions that few, if any, humans had entered before. The prey they encountered, having never seen humans before, were easy marks, and when a band of hunters exhausted the supply of large prey in one area, they moved on to another. Thus according to the overkill hypotheses, a wave of Clovis hunters swept across North America and into South America, devastating populations of elephants, bison, oxen, horses, camels, and ground sloths.

Researchers who favor the overkill hypothesis note that the Clovis people possessed an unprecedented level of hunting skill and expanded rapidly throughout the New World. These scientists cite several additional patterns of supporting evidence:

1 Hunters favor large animals as prey because a single kill provides abundant meat.

2 Large animals are especially conspicuous to hunters.

3 Extinctions of large animals were especially numerous in North and South America, where humans had recently arrived and animals may not have feared them. Few extinctions occurred in Eurasia and Africa, where large animals had long experience of human hunters.

4 Those species of large North American herbivores that did not become extinct had recently migrated from Eurasia, where they had learned to fear human hunters. These species included familiar species of the present: mountain lion, moose, elk, caribou, deer, musk ox, bighorn sheep, and mountain goat.

Opponents of the overkill hypothesis argue that early humans could not have overwhelmed an entire species of large animals. Besides, the extinctions did not begin in the north and spread southward, as would be expected if a wave of hunters had swept from Alaska to South America. Finally, there is no archeological evidence that Clovis people killed any large mammals other than elephants and bison.

The Climate Hypothesis

Some researchers have proposed that changes in climate and vegetation caused the extinction of large mammals. They point to certain changes as potential agents of extinction:

1 The Younger Dryas interval began and ended suddenly, and temperatures shifted abruptly.

2 Certain habitats that had supported many kinds of large herbivores disappeared at the beginning of the Holocene. In particular, the northern grassy region that supported a variety of nutritious herbs and shrubs (and many species of large mammals; see Figure 19-8) gave way to less nutritious prairies.

Opponents of the idea that climatic change caused the extinctions argue that species that died out had weathered comparable changes when glaciers had waxed and waned throughout the Pleistocene.

A Possible Compromise: Both Hunting and Climate

The overkill and climate scenarios may both be partly correct. Most species of large animals are inherently vulnerable to extinction because their populations are small. Perhaps the extinctions between 12,000 and 10,000 years ago resulted from the combined impact of sudden shifts of vegetation during the Younger Dryas and a simultaneous intensification of human hunting.

Many kinds of large marsupial mammals died out in Australia, among them two species of kangaroos that stood over 2.5 meters (~8 feet tall) and a lionlike marsupial with no present-day counterpart. Unfortunately, the timing of the relevant events in Australia is not yet pinned down. Humans apparently became well established there about 35,000 years ago, and the extinctions occurred soon thereafter, between about 30,000 and 25,000 years ago. On the other hand, Australia was also becoming cooler and dryer when the huge mammals disappeared.

Climatic Fluctuations of the Last 10,000 Years

Even after the Younger Dryas event, climates did not shift along a smooth path toward their present state. An early warm interval has been followed by a net decline in global temperatures to the present day, and along the way there have been several intervals of marked cooling. We began to emerge from the most recent cold spell just a few decades ago and are now in a warming phase.

The Beginnings of Agriculture

One of the most significant transitions in human history began when some population abandoned a mobile life of hunting and gathering and settled down to domesticate plants and animals for food. These activities spread throughout Europe between about 9000 and 6000 years ago.

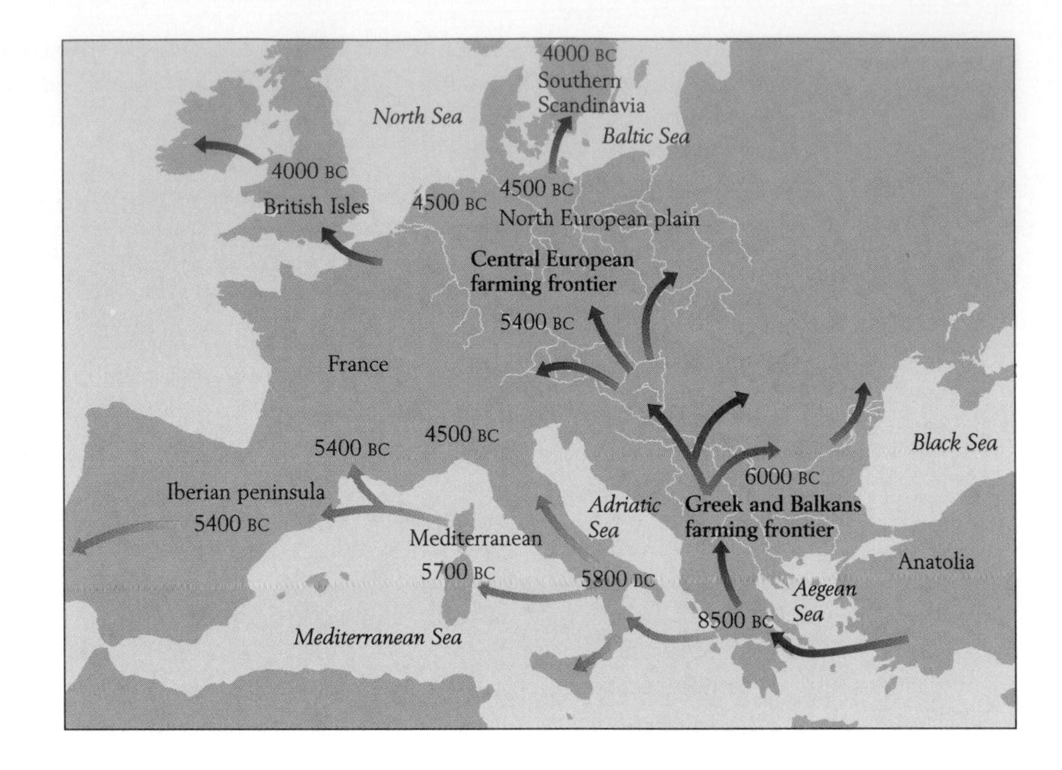

Figure 20-10 Timing of the initial spread of agriculture throughout Europe from Greece. *(After P. Bogucki, American Scientist 84:242–253, 1996.)*

Agriculture first emerged in the Zagros Mountains, near the borders between modern Iraq, Iran, and Turkey. Archeological sites reveal that populations here lived in caves before 12,000 years ago and hunted wild animals, especially sheep and goats. Pollen in lake sediments reveals that climates of the region became moister after 12,000 years ago and trees invaded what had been a dry, grassy habitat. Wild wheat and barley probably migrated into the region about the same time. These events may have coincided with the end of the Younger Dryas cold interval (see Figure 20-5). In any event, grinding tools found in caves indicate that local human populations were producing flour between 10,000 and 9000 years ago. Remains of houses and domestic animals reveal that people were living in the open, rather than in caves, before 9000 years ago; they were raising sheep, goats, pigs, and dogs in Mesopotamia, near the Zagros Mountains.

Farmers reached Greece from Turkey before 8000 years ago by colonizing one island after another in the Aegean Sea (Figure 20-10). From Greece, agriculture spread across Europe by fits and starts, reaching the British Isles and Scandinavia by 6000 years ago. Animal bones preserved at archeological sites reveal that cattle replaced sheep, goats, and pigs as the main large domesticated animals in Europe before 7000 years ago, perhaps reflecting an increase in the use of dairy products.

There is no evidence that regional climatic changes guided the initial expansion of agriculture. On the other hand, agriculture could not have spread throughout Europe until glacial and near-glacial conditions gave way to temperate climates. Certainly the abrupt onset of the hypsithermal (high heat) episode provided favorable conditions for farming.

The Hypsithermal Interval

Continental glaciers all but disappeared between about 9000 and 6000 years ago, and climates became warmer than they have ever been since. During this *hypsithermal interval*, eastern North America warmed enough that dwarf birch shrubs replaced tundra in some areas. Fossil needles of hemlock reveal that during the hypsithermal interval this conifer species was able to invade mountainous elevations of New England where the mean annual temperature today is 2°C too low for hemlock to survive. From this condition and additional evidence from fossil pollen,

A

B

Figure 20-11 Tree rings reveal the climates in which very old pines have grown. *A*. Annual rings in a cross section of the trunk of a Ponderosa pine. *B*. A very old bristlecone pine in the White Mountains of California.

researchers have estimated that mean annual temperatures in North America and Europe during the hypsithermal were about 2°C warmer than today. Temperatures were similarly elevated in the southern hemisphere, as far south as Antarctica.

Because huge masses of ice melt slowly, remnants of continental glaciers persisted into the hypsithermal interval in North America and Europe. Glaciers have existed only in mountainous regions of these large northern continents since the last remnants of continental ice sheets disappeared about 6500 years ago.

Glaciers, Tree Lines, and Tree Rings

Mountain glaciers behave like thermometers, expanding down valleys when the climate cools and retreating when the climate warms. Radiocarbon dating of spruce stumps provides a chronology for the upward and downward movements of tree lines in mountainous regions. Similarly, radiocarbon dating of fossil wood collected from moraines reveals the vertical movements of glaciers since the hypsithermal interval. As it turns out, the major movements of tree lines and glaciers in mountains of Alaska and Europe have generally paralleled one another, indicating that these movements reflect climatic changes throughout the northern hemisphere.

Studies of rings of growth in ancient trees provide an independent record of climatic conditions that generally supports the record of glaciers and tree lines (Figure 20-11*A*). The trunks of trees grow between the bark and preexisting wood. In nontropical areas, which have a distinct warm growing season, tree growth is rapid early in this season, when moisture is abundant; at this time large, thin-walled cells form. Growth slows in the late summer and fall, when moisture is less abundant; at this time small, thick cells form. The result is one wide and one narrow ring of growth each year. When climates become warmer, if they do not also become very dry, the growing season lengthens and growth accelerates; wider rings result. Thus trees provide a record of climatic warming and cooling. The bristlecone pine trees of western North America have exceptionally long lives and therefore provide exceptionally lengthy records of climatic change (Figure 20-11*B*). Rings counted in small cores extracted from the trunk of the oldest known bristlecone, aptly named Methuselah, show it to be about 4600 years old. This tree is probably the oldest living plant or animal on Earth

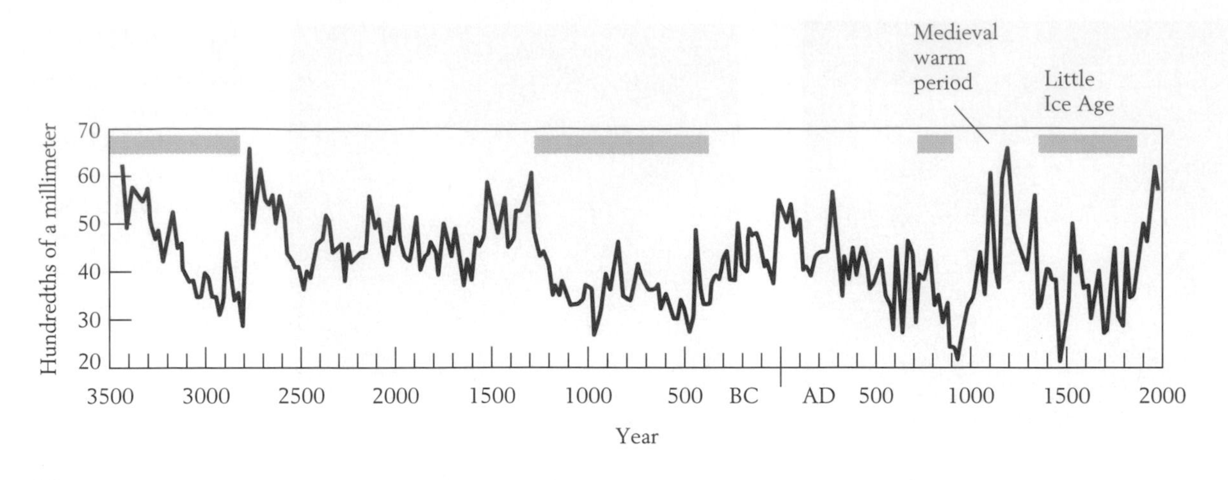

Figure 20-12 Cold intervals of the past 3500 years recorded by widths of tree rings in bristlecone pines near the upper tree line of the White Mountains of California. *(Based on data from V. C. La Marche in H. H. Lamb, Climate, History and the Modern World, Routledge, London, 1995.)*

Rings measured by coring of bristlecone pines from near the tree line in the White Mountains of California have yielded estimates of temperature changes for an interval extending back more than 3400 years. Although regional climatic variations are to be expected, the overall pattern is in remarkable agreement with that derived from elevations of tree lines and glaciers in both Alaska and Europe (Figure 20-12). These three independent kinds of evidence reveal four intervals of marked global cooling during the past 6000 years.

Temperatures Since the Hypsithermal Interval

About 5800 years ago the hypsithermal episode gave way to the first major cold interval since the retreat of continental glaciers (see Figure 20-12). Fossil pollen reveals that the white pine, a species adapted to cold climates, migrated northward in eastern North America at this time. Similarly, the lodgepole pine spread northward in the Pacific Northwest. This cold interval ended about 4900 years ago. A second such interval began about 3300 years ago and ended almost 2400 years ago. A third cold interval extended from about AD 700 to AD 900. After this third pulse of cooling came a warm interval. This so-called *Medieval warm period* permitted the Vikings to flourish in the northern Atlantic region, exploiting other countries by sea between about AD 700 and AD 1200. In AD 985 they established an outpost in western Greenland and raised cattle and herded sheep there, where frigid conditions prevent such activities today. The climate began to cool in the thirteenth century, however, and by about 1500 the Norse colony died out after sea ice isolated Greenland from Europe.

The climatic deterioration that destroyed the Norse colony in Greenland marked the beginning of a long, cold interval that lasted until about 1850. During this so-called *Little Ice Age* glaciers expanded along mountain valleys not only in North America and Europe but also in New Zealand. New England and northern Europe suffered bitter winters and short summers that produced major crop failures. Potatoes replaced wheat on many European farms. Figure 20-13 is a scene painted in February 1575, during the first of many severe winters that descended on Holland over a 200-year period. George Washington and his shivering troops at Valley Forge in 1777 and 1778 were actually fortunate to have experienced a winter that was relatively mild for the times.

A global warming trend occurred during the first half of the nineteenth century. A warm, moist summer in 1846 favored the spread of the potato fungus in Ireland and led to the famine that caused many deaths and the emigration of many Irish people to America. The Little Ice age came to an end about this time, and Earth's average annual temperature remained relatively stable for the rest of the nineteenth century.

Figure 20-13 ***Hunters in the Snow.*** Pieter Brueghel painted this scene in 1575, late in the first severe winter that Holland experienced during the Little Ice Age.

Episodes of Drying

Moisture conditions, like temperatures, have varied greatly in many regions during the Holocene. Fossils of lake-dwelling life, along with pollen blown into lakes, reveal four intervals when African lake levels were low, as a result of some combination of reduced rainfall and increased evaporation.

Pictures taken from satellites reveal that numerous dunes, now stabilized by vegetation, were active near the Colorado-Nebraska border at four times within the past 10,000 years (Figure 20-14). Exactly when the dunes were active is not well established, but the most recent activity was less than 1000 years ago. At that time the climate must have been too dry to support the short prairie grass that grows in the area today.

Analyses of tree rings on the western slope of the Sierra Nevada have indicated that during the past 1000 years California has seldom received as much

Figure 20-14 **Satellite images of ancient dunes that are now stabilized, adjacent to the Platte River near Greeley, Colorado.** These dunes were active earlier in the Holocene, when climates in this region were drier than they are today.

precipitation as it does today. Narrow rings during the growing season indicate dry conditions. Mono Lake, which occupies this region of the Sierra, was lowered in 1940, when the city of Los Angeles began to divert streams that flow into it. The lake's decline exposed ancient stumps that are the remnants of trees that previously grew on dry land when the climate was drier than it is today and the lake level was very low. Radiocarbon dating has shown that some of those trees died about AD 1112 and others about AD 1350. Both groups of trees had lived at the site more than 50 years, while the climate remained very dry.

The recent history of climatic changes in both California and the Great Plains serves as a warning to present-day inhabitants of these regions. California is already suffering from a water shortage and will face severe problems if its climate returns to its more customary Holocene state. Similarly, it is possible that the Great Plains will again experience climates like those that a few centuries ago made them even drier than they were in the 1930s, when drought turned them into the so-called Dust Bowl.

Sea Level

Although the melting of glaciers after the last glacial maximum caused sea level to rise dramatically (see Figure 20-4), the pattern of changes in sea level in relation to the land has varied from place to place. Local shorelines have shifted not only because of the eustatic rise but also because of regional elevation and subsidence of the land. Thus, even at a time of eustatic rise, sea level has declined in relation to the land in some regions: the shoreline has regressed (see Figure 6-21).

Consequences of the Early Holocene Sea-Level Rise

Of course, the initial Holocene rise of sea level by more than 100 meters as continental glaciers melted had profound effects along continental margins. Between about 8500 and 7600 years ago shallow-water reefs became established off Florida and many Caribbean islands at depths of 30 to 15 meters below present sea level. These reefs died abruptly about 7600 years ago, when sea level rose further, flooding broad areas behind them (see Figure 20-4). Settling of suspended sediment from the newly formed shelves behind the reefs may have killed them. In any event, modern reef tracts in Florida and the Caribbean became established when sea level approached its present level, slightly after 7600 years ago.

Rising seas also flooded broad river valleys that had formed when sea level dropped more than 100 meters at times of glacial expansion and rivers flowed across what are now broad continental shelves. Sedimentation has not totally filled these drowned river valleys, so that today they form broad estuaries such as Chesapeake, Delaware, and Mobile bays.

In the west, small, steep rivers that drain the Coast Ranges of California cut downward into the narrow continental shelf to the west during the last glacial maximum. The Holocene eustatic rise has turned these valleys into estuaries by flooding them and causing them to partially fill with sediment. Tectonic subsidence has accentuated the flooding of some of these estuaries, including San Francisco Bay.

In northern regions, rising seas invaded broad valleys that had been widened by glaciers. The resulting fingerlike inland extensions of the sea, known as fjords, are best developed along the coast of Norway.

The sea level curve for Barbados (see Figure 20-4) was produced under the assumption that the island has been undergoing tectonic uplift at a constant rate. To the degree that this rate has varied, the curve is distorted. Attempts to produce eustatic curves in other regions suffer from similar uncertainties. In fact, many experts believe that sea level rose to virtually its present level by about 7000 years ago, when nearly all of the meltwater of continental glaciers had been released. These workers believe that no significant eustatic change has occurred since that time, only regional changes in relative sea level that have resulted from elevation or subsidence of coastal zones.

Coastlines of the Past 7000 Years

The distribution of glaciers has contributed to regional differences in relative sea level change during the last several thousand years. Where large glaciers stood 20,000 years ago, their enormous weight depressed the continental lithosphere, and since their retreat this lithosphere has been rebounding (Figure 20-15). As a result, some coastlines of the North Atlantic Ocean, including those of Scotland and much of

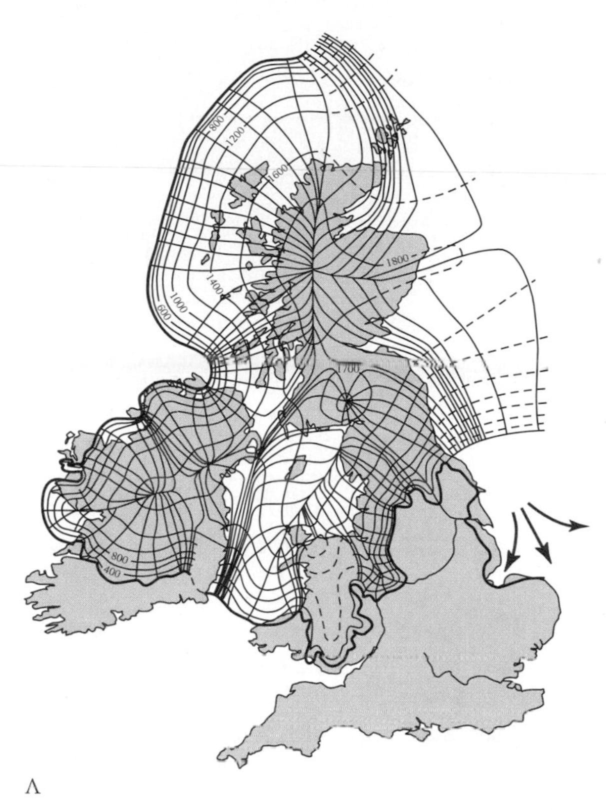

A

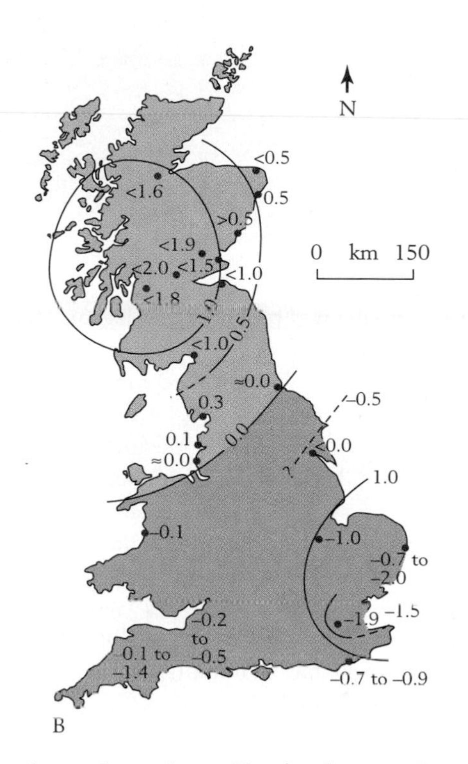

B

Figure 20-15 **Effects of the weight of glaciers on elevations in Great Britain.** *A.* Reconstruction of surface topography of glaciers, in meters above present sea level, during the most recent glacial maximum. *B.* Estimated current rates of crustal movement, in millimeters per year. The northern region, having been depressed by glacial ice, is rebounding; the southern region, having been elevated by the glacier to the north as a peripheral bulge, is sinking. *(A. After G. S. Boulton et al., in British Quaternary Studies: Recent Advances, F. W. Shotton [ed.], Clarendon Press, Oxford, 1971; B. After I. Shennan, Jour. Quaternary Sci. 4:77–89, 1989.)*

Scandinavia, are still rising and the seas are regressing from the land.

Ironically, in other areas, Earth's surface is subsiding because ice sheets have melted away. These are areas that were elevated when large glaciers grew nearby. The elevation occurred because as Earth's surface subsided beneath the central portion of the large ice sheet, material of the lithosphere squeezed out in all directions. This material accumulated around the margins of the ice sheet, where it elevated the crust to form what is called a *peripheral bulge*. (Note that such a bulge is comparable to the one that can form in front of the foreland basin produced by a mountain chain; see Figure 13-26). Peripheral bulges that formed during the last glacial maximum are still subsiding as a result of the disappearance of the ice sheets. One of these bulges is now subsiding in southern Great Britain (see Figure 20-15*B*). Another passes along the northeastern margin of North America; because of its subsidence, the sea is transgressing over the land from Maine to Delaware (see Figure 6-20).

The coast of New England continues to subside as part of the peripheral bulge that formed southeast of the Laurentide glacier (see Figure 6-20). Intertidal marshes that became established there about 7600 years ago have transgressed inland since that time, but the rate of transgression has decreased as the subsidence of the land has slowed.

Forces unrelated to glaciers also cause vertical movements of coastal terrain. For example, glacial tectonic uplift has probably augmented glacial rebound in elevating land in the vicinity of Seattle, Washington. Here glacial sediments deposited in shallow seas during the final retreat of glaciers from Puget Sound now stand about 50 meters above sea level. In contrast, subsidence of the Gulf Coast in the

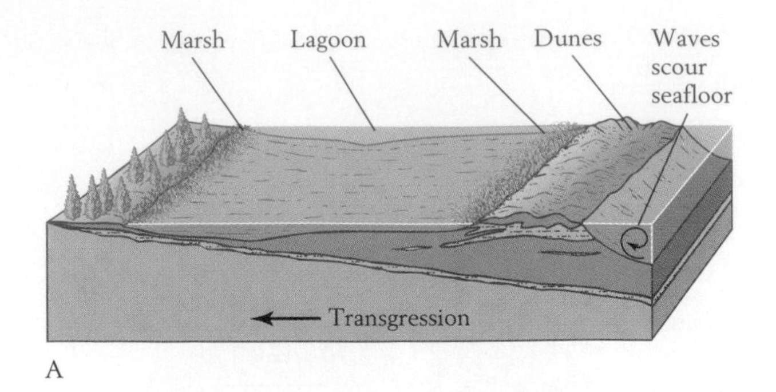

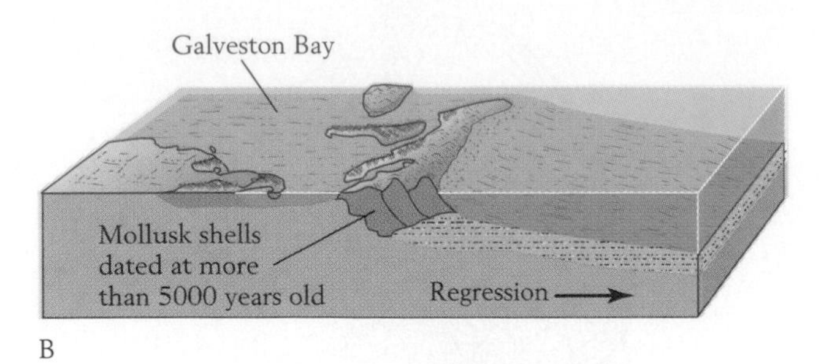

Figure 20-16 Movement of barrier island–lagoon complexes during the Holocene, after sea level approached its present position. *A.* In New Jersey the complex is shifting inland because the land is subsiding, having been elevated as part of the peripheral bulge during the most recent advance of the Laurentide glacier; with little sediment being supplied, the sea is transgressing over the land. *B.* Galveston Island, Texas, is positioned in a region that has not been subsiding and has been receiving an abundant supply of sediment; it has been prograding seaward for more than 5000 years, as indicated by radiocarbon ages of mollusk shells. *(A. After A. G. Fischer, Amer. Assoc. Petroleum Geol. Bull. 45:1656–1666, 1961. B. After H. A. Bernard and R. J. LaBlanch, in Quaternary Geology of the United States, H. E. Wright and D. G. Frey [eds.], Princeton University Press, Princeton, N.J. 1965.)*

region of the Mississippi delta is producing a transgression. This region has been subsiding for millions of years under the weight of the sediment that has been accumulating there (p. 135).

We have seen that humans have accelerated the rate at which sea level is rising in relation to the land in the vicinity of the Mississippi delta (see Box 5-1). First, by extracting water from wells, humans are accelerating the rate at which the coastal plain subsides. Second, by damming tributaries of the Mississippi, they have reduced the influx of river-borne sediment to marginal marine environments. A stronger influx of sediments would at least partly offset the effect of subsidence.

Along the coast of New Jersey, Holocene subsidence of the peripheral bulge that formed south of the Laurentide ice sheet has caused barrier island–lagoon complexes to transgress over older sediments of the Coastal Plain (Figure 20-16*A*). Rivers here deposit too little sand to build a gentle shore face in front of the barrier islands. Instead, storm waves scour the seafloor, creating a steep shore face and destroying the transgressive record of lagoon deposits. In other words, processes of destruction are winning the battle over processes of accumulation.

In contrast to the transgressive pattern along the New Jersey coast, a regressive pattern is evident along Galveston Island in Texas (Figure 20-16*B*). Here radiocarbon dates from mollusk shells obtained from boreholes show that a regression has occurred during the Holocene. A high rate of sand supply has caused Galveston Island to expand by about 4 kilometers (2.5 miles) seaward during the past 5300 years, while relative sea level has changed very little.

The Twentieth and Twenty-First Centuries: The Impact of Humans

It is debated whether humans have yet raised global temperatures significantly by increasing the concentration of so-called greenhouse gases in the atmosphere. There is no question, however, that substantial greenhouse warming will soon result from our continued burning of fossil fuels, which oxidize carbon to produce CO_2.

Human Activities and Greenhouse Warming

Figure 20-17 shows the buildup of CO_2 in the atmosphere since the acceleration of the industrial revolution in the nineteenth century. The data for the interval before 1957 are measurements of CO_2 trapped in Antarctic ice and extracted from cores, and those for the interval since 1957 are measurements of CO_2 taken directly from air samples. At present the rampant destruction of forests, especially in the tropics, may be contributing about half as much CO_2 to the atmosphere as the burning of fossil fuels. Trees are often burned during deforestation, or if they are cut and removed, plant debris that remains is burned. Even the rotting of abandoned debris releases CO_2 through respiration (p. 263).

Only about half of the CO_2 produced by human activities ends up in the atmosphere. Roughly a quarter of it ends up in the oceans, where part of it remains as dissolved CO_2 and part is converted to biomass through photosynthesis by phytoplankton. The remaining quarter of the CO_2 generated by humans is probably turned into plant biomass through the expansion of forests. These forests may actually benefit from the increased atmospheric concentration of CO_2, a raw material for their production of sugars.

Has the increase of atmospheric CO_2 elevated global temperatures significantly during the past century? Figure 20-18 shows that since 1900 the average temperature at Earth's surface has increased by about one-half of 1°C. Figure 20-18 also displays the results of the computer model devised to estimate the temperature effects of carbon dioxide emissions resulting from human activities. An international commission, the Intergovernmental Panel on Climate Change, concluded that the model on which Figure 20-18 was based may offer meaningful predictions for the future. Predicted future temperatures shown in Figure 20-18 are based on the premise that civilization will exercise no controls over future CO_2. The low and high estimations depict the range of uncertainty in the model. Note that the low estimation of the model for the period from 1860 to 1990 approximately matches the actual pattern of global temperature change. This finding suggests that the model exaggerates global warming from CO_2 emissions. The reason for this discrepancy is unclear, but even the low estimation predicts warming of about 2°C between 1980 and 2100.

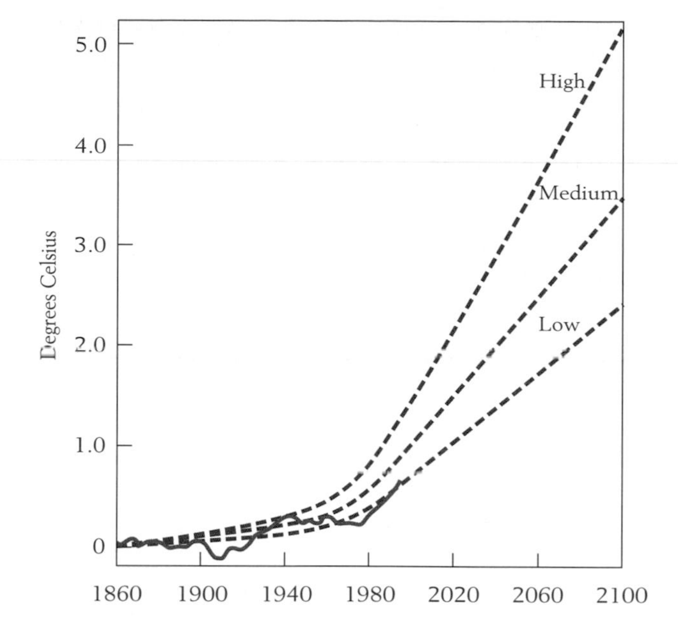

Figure 20-18 Increase in global average temperature predicted by the Intergovernmental Panel on Climate Change in 1992. The predicted change (dashed lines) is based on the assumption that humans will make little effort to reduce carbon dioxide emissions. The model employed to make the prediction also provided estimates of past increases in global temperature from humans' enhancement of greenhouse warming. The "medium" line represents the group's best estimate, and the "high" and "low" lines show the estimated range of uncertainty. The solid line indicates the actual increase that occurred over about 130 years. Note that the actual net increase nearly matches the low prediction.

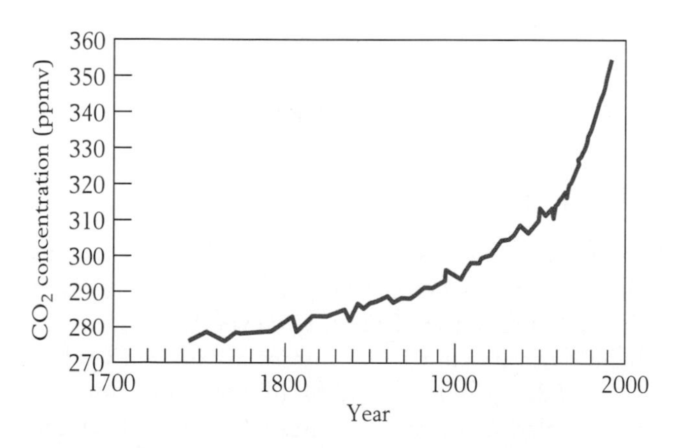

Figure 20-17 The increase in concentration of carbon dioxide in the atmosphere since 1700. *(After R. T. Watson et al., in Climate Change, the IPCC Scientific Assessment, J. T. Houghton, G. J. Jenkins, and J. J. Ephraums [eds.], CUP, 1990.)*

Methane from Tundra

Carbon dioxide in not the only greenhouse gas in Earth's atmosphere. In fact, methane has about 7.5 times more effect than carbon dioxide on greenhouse warming. The low concentration of methane in the present atmosphere makes it an unimportant greenhouse gas, but its role could increase in the future. Huge quantities of methane are frozen in soils deep beneath the tundra at high northern latitudes.

Most of this methane was produced by bacteria in marshy environments during Arctic summers. If future global warming releases enough methane gas from the tundra, this gas may add significantly to the greenhouse warming caused by carbon dioxide.

Consequences of Future Climate Change

For people now living in regions with long, cold winters, the prospect of a warmer climate may seem welcome. Unfortunately, such a simplistic view ignores many likely consequences of future global warming.

Migration The Pleistocene record shows that when climates change, communities of plants and animals are reshuffled (p. 529). Thus global warming will yield new communities of plants and animals. Some populations will suffer from a diminished supply of resources or from exposure to new competitors, predators, or diseases. Even simple dislocations of plant and animal populations will be problematic. Some people may be pleased if optimal conditions for certain crops shift to their region, but what of the farmers in other regions who can no longer depend on those crops for their livelihood?

Temperature changes alone will damage some ecosystems. Although coral reefs require warm seas, they cannot tolerate extreme warmth (see Box 4-1). If tropical waters heat up by 2° or 3°C, the warmest regions of the Pacific will become inhospitable to reef-building corals. Reefs may expand to other regions that were previously too cool to support them, but they will require centuries to become well established.

Water Supply in Terrestrial Environments On the land, global warming will create severe problems of water supply for both natural and human communities. Scientists have offered widely varying predictions of the geographic patterns of increased and decreased wetness that will result from global warming. These predictions are based in part on evidence of patterns of wetness inferred for the hypsithermal interval of the early Holocene (p. 572). Measures of wetness (and dryness) take into account not only the amount of precipitation but also rate of water loss through evaporation and through transpiration by plants (p. 21). Despite many uncertainties, we can make three generalizations about wetness and global warming:

1. Many areas that derive most of their precipitation from nearby oceans will become wetter, because warm seas supply more water through evaporation than cold seas. Thus the West Coast and Gulf Coast of North America may become wetter.

2. Monsoonal rains will increase because seasonal warming of the land, which draws them inland, will intensify (p. 107). As a result, southern India will experience even heavier torrential rains during summer months than it does today.

3. Many areas in the middle of large continents will become drier, primarily because higher temperatures will cause rates of evaporation to increase. Thus farmers in the American Midwest may face hard times.

Increased evaporation will make some semiarid regions even drier. *Desertification* of this kind is accelerated by a positive feedback: when plants begin to disappear, erosion increases and moisture is lost through reduction of plant cover and soil. Thus the system becomes even drier and still more vegetation disappears. Figure 20-19 identifies the semiarid regions of the present world that will be susceptible to desertification. One such region is southern Africa, which has been exceptionally warm since 1980 and has experienced two major droughts in the 1990s. This association suggests that dry conditions will accompany global warming in southern Africa. Increased aridity would place human populations there in jeopardy because they already suffer from a shortage of water.

Direct Effects of Increased CO_2 on Plants

Future increases in atmospheric CO_2 will actually "fertilize" plants because they use this compound to manufacture sugars. Recall, however, that C_3 plants benefit more than C_4 plants from increased CO_2 levels (p. 526). Among domestic plants, wheat and rice are C_3 plants, whereas corn is a C_4 plant. Wheat and rice may therefore be expected to prosper as atmospheric CO_2 rises, but the situation is more complex than this. As C_3 plants, wheat and rice will not tolerate as well as corn the hot, dry conditions that will

Figure 20-19 **Dry regions of the modern world.** The dry grasslands are in danger of future desertification.

result from global warming in many areas. Obviously, predictions of the biotic effects of greenhouse warming are fraught with uncertainties.

Rising Sea Level

The amount of warming depicted in Figure 20-18 would cause a eustatic rise of about 50 centimeters (~20 inches) by the year 2100. This figure compares to an estimated rise of 4–10 centimeters during the past 100 years. As we have seen, the relative rise in sea level would vary from place to place (p. 173). Melting of mountain glaciers would contribute a significant portion of the increase in volume, but an even larger amount would result from the expansion of the ocean as it heats up slightly. Surprisingly, partial melting of ice caps may contribute little to the eustatic rise. The Greenland ice cap may shrink only slightly. A predicted increase in snowfall for the south polar region may actually expand the Antarctic ice cap; on the other hand, increased ice flow to the ocean may diminish the Antarctic ice cap. Furthermore, a collapse of Antarctic ice shelves could produce a sudden, catastrophic sea-level rise (Box 20-1). Fossil marine diatoms beneath the Ross ice shelf reveal that this ice shelf collapsed sometime during late Pleistocene time, and the place where this shelf now rests was covered by a shallow sea.

Even a 1-meter rise of sea level would endanger the homes of about a billion people, as well as one-third of the world's crop-growing areas. Rising seas during storms would pose the greatest threat. Some areas would be adversely affected by a eustatic rise of only 0.5 meters. Some low coastal cities, including the architecturally rich Italian city of Venice, would be destroyed (Figure 20-20). The country of Bangladesh, about half of which is already within 5 meters (~16 feet) of sea level, would experience even more frequent disastrous floods than it has suffered in recent years. Many of the world's wetlands would be

For the Record 20-1

A Future Meltdown in Antarctica?

The ice cap that covers 99 percent of Antarctica contains 90 percent of the world's ice. If all of this frozen water were to melt, global sea level would rise by more than 60 meters (197 feet). Were future global warming to melt away even a small fraction of the Antarctic ice, many cities and towns would be flooded.

The Antarctic ice cap, which is in places more than 4 kilometers (~2.5 miles) thick, is divided into the east and west Antarctic ice sheets. Each of these sheets is dome-shaped, and ice flows outward to the sea from the elevated center along ice streams at rates as high as a few kilometers per year. The flowing ice forms shelves that float in bays along nearly half the Antarctic coastline.

The ice shelves are frozen to the rocky sides of the bays in which they reside, and their undersurfaces are also attached to bedrock. Because the ice shelves are attached to the land, they resist the flow of the ice streams that feed them. Ice nonetheless flows through the central regions of the ice shelves, and icebergs sporadically break from them and float away. Thus Antarctica is continually losing glacial ice. On the other side of the ledger, it gains ice through the accumulation of snow. Antarctica amounts to a frigid desert, however, receiving an average of only about one meter (~3 feet) of snow per year. One giant iceberg can carry away the equivalent of more than a year's precipitation for the entire continent! The question is: What is happening and what will happen in the next few decades to rates of addition and subtraction of Antarctic ice?

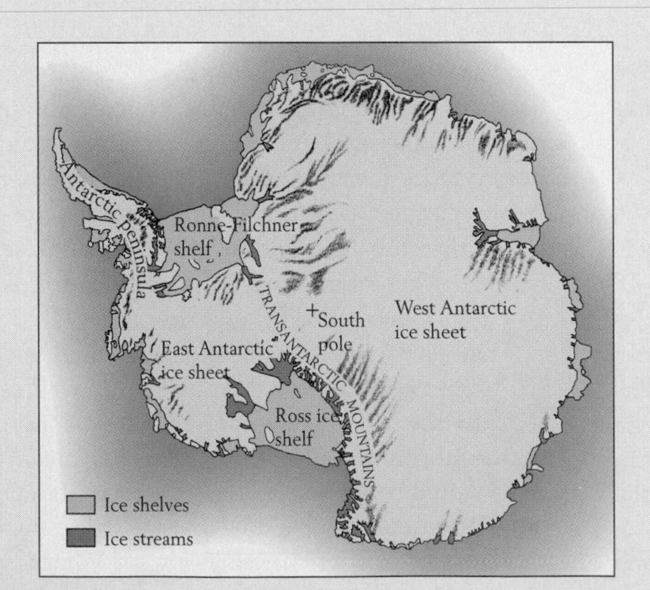

During the past 50 years, five small ice shelves have retreated dramatically along the Antarctic peninsula, which extends toward South America. One of these, the Larsen ice shelf, abruptly shrank by about 4200 square kilometers (1500 square miles) in January 1995 through a massive discharge of icebergs. Antarctic ice shelves survive only in areas where the average annual temperature is below −5° Celsius (−23° Fahrenheit). Increased temperatures have clearly caused the disintegration of the small Antarctic ice shelves. Since 1957, the average temperature in Antarctica has risen by nearly one-quarter of a degree Celsius per decade. During this time, the areas where the ice shelves have disintegrated, being relatively far from the south pole, have become too warm to sustain ice shelves.

Warming not only melts the surfaces of ice shelves, but also weakens them and increases the rate at which they release icebergs. Might future warming from an enhanced greenhouse effect greatly diminish the volume of Antarctic glacial ice and elevate sea level? The loss of ice shelves alone would have little effect on sea level because floating ice is, in effect, part of

An ice shelf and icebergs in Antarctica.

the ocean even before it melts. The question is: Might warming diminish the size of the Antarctic ice sheets, releasing stored water to the ocean? The East Antarctic ice sheet, being small and centered close to the ocean, appears to be especially vulnerable. The Ronne-Filcher and Ross ice shelves hold back the East Antarctic ice sheet. They are each the size of France and lie relatively close to the south pole, so that a very large amount of warming would be required to destabilize them. Furthermore, while global warming will increase the rate at which ice shelves shed icebergs, it may compensate by increasing the supply of snowfall to glacial ice. Warmer temperatures will increase the rate of evaporation from the sea surface adjacent to Antarctica. The resulting increase in atmospheric water vapor will result in more snow.

The truth is, we do not know whether global warming will shrink the Antarctic ice cap by accelerating its loss of ice to the ocean or whether global warming will instead expand the ice cap by producing a large increase in regional snowfall. Our uncertainty about the future effects of Antarctic ice on sea level is more than a little unsettling.

Figure 20-20 **Flooding in Venice, Italy, during a very high tide in 1990.**

lost because obstructions that humans have emplaced, including landfills and bulkheads, would prevent coastal marshes from migrating inland with shorelines (Figure 20-21).

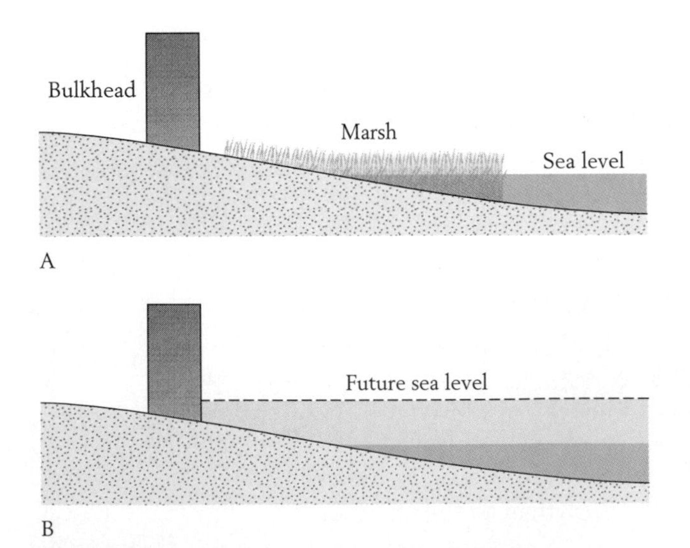

Figure 20-21 **A bulkhead built slightly landward of a coastal marsh at a time when sea level is rising.** After sea level rises from its initial position (*A*), the presence of the bulkhead eliminates the marginal marine environment where the marsh must grow (*B*).

Chapter Summary

1 Continental glaciers began to retreat after 15,000 years ago, during Earth's emergence from the last glacial maximum.

2 Plant species migrated northward as glaciers retreated, but at varying rates, so that plant communities were continually restructured.

3 Abrupt global climatic changes, many of them temporary, occurred during Earth's transition to the glacial minimum condition of the present day.

4 Humans may have reached North America as early as 30,000 years ago, but large populations of advanced hunters did not arrive until slightly before 11,000 years ago.

5 Numerous species of large mammals disappeared from North America between 12,000 and 11,000 years ago. Clovis hunters may have contributed to their extinction, but so may climatic changes associated with the Younger Dryas interval.

6 Agriculture spread rapidly throughout Europe when climates warmed after the end of the Younger Dryas cold interval.

7 Continental glaciers disappeared between 9000 and 6000 years ago, and climates soon became warmer than they have been since that time.

8 Trees generally grow more rapidly when the climate warms, lengthening the growing season. Widths of annual rings in the trunks of very old trees therefore provide a record of climatic changes for the last 3400 years.

9 The Medieval warm period was followed by the Little Ice Age, which began after 1300 and lasted until about 1850.

10 At many times in the past 10,000 years, climates in many regions have been drier than they are today.

11 Global sea level approached its present position about 7000 years ago. Since that time, changes in the positions of coastlines have resulted largely from regional patterns of elevation and subsidence of the land.

12 Humans' burning of fossil fuels will warm climates in the future. An important consequence of this global warming will be increased aridity in inland regions where rates of evaporation increase.

13 Elevation of sea level as a result of human-induced global warming may pose severe problems for coastal populations.

Review Questions

1 How did the Great Lakes form?

2 How does fossil pollen indicate that modern plant communities are temporary associations of species?

3 How do fossil corals record Holocene sea level changes?

4 What kinds of large mammals disappeared from North America between 12,000 and 11,000 years ago?

5 What evidence favors the idea that human hunters contributed to the extinction of the large mammals?

6 What evidence favors the idea that climatic changes contributed to the extinction of the large mammals?

7 How does glacial ice in Greenland that is thousands of years old provide a record of climatic change?

8 Why is it not surprising that agriculture spread from the Middle East to Europe, rather than in the opposite direction?

9 What evidence is there that at times during the last 10,000 years climates in many areas have been drier than they are today?

10 What conditions have affected relative sea level in particular regions during the past 7000 years, since global sea level approached its present position?

11 Global environmental change is of great concern in the world today. Using the Visual Overview on page 560 and what you have learned in this chapter, review the causes and effects of changes in global climate and sea level that have occurred during the past 14,000 years or that may occur during the next few centuries.

Additional Reading

Dawson, A. G., *Ice Age Earth: Late Quaternary Geology and Climate,* Routledge, London, 1992.

Goudie, A., *Environmental Change,* Clarendon Press, Oxford, 1992.

Houghton, J., *Global Warming: A Complete Briefing,* Lion Publishing, Oxford, 1994.

Lamb, H. H., *Climate, History and the Modern World,* Routledge, London, 1995.

Pielou, E. C., *After the Ice Age: The Return of Life to Glaciated North America,* University of Chicago Press, Chicago, 1991.

Wright, H. E., et al., *Global Climates since the Last Glacial Maximum,* University of Minnesota Press, Minneapolis, 1993.

APPENDIX

Stratigraphic Stages

In many parts of the world the geologic record has been divided into stages. As discussed in Chapter 6, stages are time-stratigraphic units. For the most part, the stages recognized in Europe have become the standard stages with which stages defined elsewhere are correlated. Correlations remain imperfect, however, as do estimates of the absolute ages of stage boundaries. This appendix is a reference for students who encounter unfamiliar stage names in their studies. Figure 1 lists major Paleozoic and Mezozoic stages that were first defined in Europe and shows how a number of North American stages are currently believed to correlate with them. Figure 2 presents the same kind of information for Cenozoic stages, showing how European stages are thought to relate to American stages that are based on biostratigraphic zones for fossil land mammals.

Era	AGE (Million years)	SYSTEM		SERIES	STAGE (European)	STAGE (North American)
CENOZOIC			QUATERNARY	PLEISTOCENE	(See Figure A2)	
		NEOGENE	TERTIARY	PLIOCENE		
				MIOCENE		
	24	PALEO-GENE		OLIGOCENE		
				EOCENE		
				PALEOCENE		
MESOZOIC	65	CRETACEOUS		UPPER	Maastrichtian	
					Campanian	
					Santonian	
					Coniacian	
					Turonian	
					Cenomanian	
				LOWER	Albian	
					Aptian	
					Barremian	
					Hauterivian	
					Valanginian	
					Berriasian	
	142	JURASSIC		UPPER	Tithonian	
					Kimmeridgian	
					Oxfordian	
				MIDDLE	Callovian	
					Bathonian	
					Bajocian	
					Aalenian	
				LOWER (LIAS)	Toarcian	
					Pliensbachian	
					Sinemurian	
					Hettangian	
	206	TRIASSIC		UPPER	Rhaetian	
					Norian	
					Carnian	
				MIDDLE	Ladinian	
					Anisian	
				LOWER	Scythian	
	251					

ERA	AGE (Million years)	SYSTEM		SERIES	STAGE (European)	STAGE (North American)	(SERIES)
PALEOZOIC		PERMIAN		UPPER	Tatarian	Ochoan	
					Ufimian/Kazanian	Guadalupian	
				LOWER	Kungurian, Artinskian	Leonardian	
					Sakmarian, Asselian	Wolfcampian	
	290	CARBONIFEROUS	PENNSYLVANIAN	UPPER	Stephanian	Virgilian	
					Westphalian	Missourian, Desmoinesian, Atokan, Morrowan	
	323		MISSISSIPPIAN	LOWER	Namurian	Springerian, Chesterian	
					Visean	Meramecian, Osagean	
					Tournaisian	Kinderhookian	
	354	DEVONIAN		UPPER	Famennian	Chautauquan	
					Frasnian	Senecan	
				MIDDLE	Givetian, Eifelian	Erian	
				LOWER	Emsian, Siegenian, Gedinnian	Ulsterian	
	417	SILURIAN		UPPER	Ludlovian	Cayugan	
				LOWER	Wenlockian	Niagaran	
					Llandoverian	Medinan	
	443	ORDOVICIAN		UPPER	Ashgillian	(shaded)	Cincinnatian / Mohawkian
					Caradocian	(shaded)	
				LOWER	Llandeilian	Chazyan	Champlanian
					Llanvirnian	Whiterockian	
					Arenigian, Tremadocian		Canadian
	495	CAMBRIAN		UPPER	Dolgellian, Maentwrogian	Trempealeauan, Franconian, Dresbachian	Croixan
				MIDDLE	Menevian, Solvan		Albertan
				LOWER	Lenian, Aldabanian, Tommotian, Nemakitian-Daldynian		Waucoban
	544						

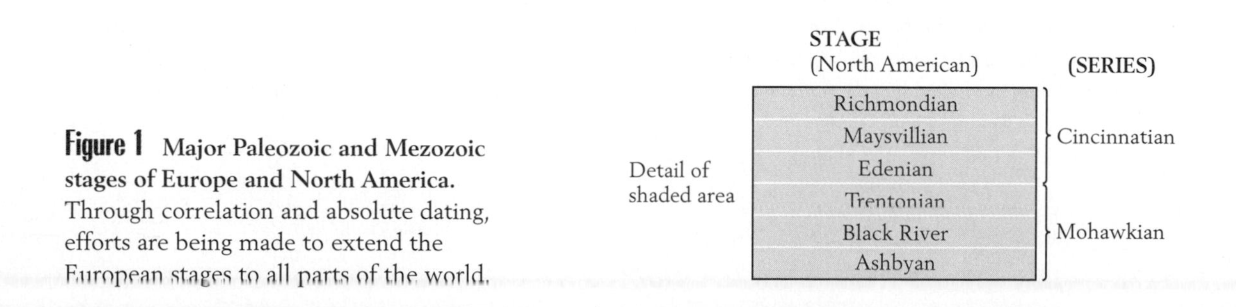

Figure 1 **Major Paleozoic and Mezozoic stages of Europe and North America.** Through correlation and absolute dating, efforts are being made to extend the European stages to all parts of the world.

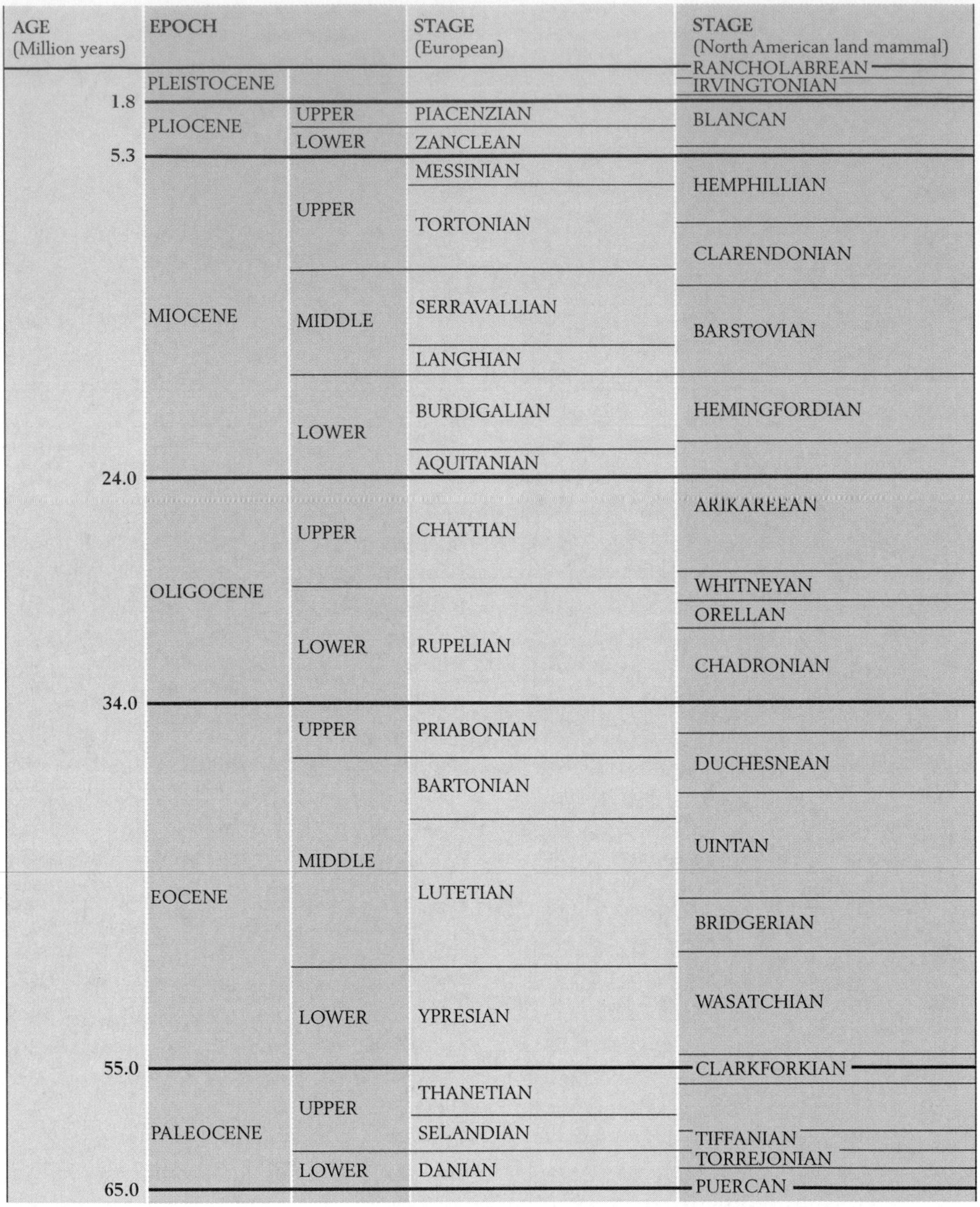

AGE (Million years)	EPOCH		STAGE (European)	STAGE (North American land mammal)
	PLEISTOCENE			RANCHOLABREAN; IRVINGTONIAN
1.8	PLIOCENE	UPPER	PIACENZIAN	BLANCAN
		LOWER	ZANCLEAN	
5.3	MIOCENE	UPPER	MESSINIAN	HEMPHILLIAN
			TORTONIAN	CLARENDONIAN
		MIDDLE	SERRAVALLIAN	BARSTOVIAN
			LANGHIAN	
		LOWER	BURDIGALIAN	HEMINGFORDIAN
			AQUITANIAN	
24.0	OLIGOCENE	UPPER	CHATTIAN	ARIKAREEAN
		LOWER	RUPELIAN	WHITNEYAN; ORELLAN; CHADRONIAN
34.0	EOCENE	UPPER	PRIABONIAN	
		MIDDLE	BARTONIAN	DUCHESNEAN
			LUTETIAN	UINTAN; BRIDGERIAN
		LOWER	YPRESIAN	WASATCHIAN
55.0	PALEOCENE	UPPER	THANETIAN	CLARKFORKIAN
			SELANDIAN	TIFFANIAN
		LOWER	DANIAN	TORREJONIAN
65.0				PUERCAN

Figure 2 Major Cenozoic stages of Europe and North America. The North American stages, which are currently only crudely correlated with the European stages, are based largely on fossil occurrences of land mammals in the Midwest and West. The most recent epoch, not listed in this diagram, is known as the Holocene or Recent. This brief epoch began at the end of the Pleistocene, roughly 12,000 years ago.

Glossary

Terms used in these definitions that are also defined in this glossary are in many instances *italicized* for the readers convenience.

Abyssal plain The broad expanse of seafloor lying between about 3 and 6 kilometers (~ 2 to 4 miles) below sea level.

Acanthodians Small, elongate marine and freshwater fishes of middle Paleozoic age with jaws, numerous paired fins supported by sharp spines, and scales rather than bony plates.

Accretionary wedge A body of *rocks* that have accumulated above an oceanic *plate* undergoing *subduction*. Slices of *mélange* pile up along *thrust faults* to form the wedge.

Acheulian culture The tool culture of *Homo erectus*, a species of the human family that lived during the Pleistocene Epoch.

Acritarchs An extinct group of apparently *eukaryotic phytoplankton* whose earliest representatives are in Proterozoic *rocks*.

Active lobe (of a delta) The site on a *delta* where functioning *distributary channels* cause the delta to grow seaward.

Active margin The border of a continent along which *subduction* occurs, producing igneous activity and deformation.

Actualism The interpretation of ancient *rocks* by applying the results of analyses of modern-day geologic processes in accordance with the principle of *uniformitarianism*.

Adaptation A feature of an organism that serves one or more functions useful to the organism.

Adaptive breakthrough An evolutionary innovation that affords a group of organisms a special ecologic opportunity and often leads to the *evolutionary radiation* of that group.

Age, geologic The division of geologic time smaller than an *epoch*.

Albedo The percentage of solar radiation reflected from Earth's surface. That percentage is higher for ice than for land or water, and usually higher for land than for water.

Alluvial fan A low, cone-shaped structure that forms where an abrupt reduction in slope—for example, the transition from a highland area to a broad valley—causes a stream to slow down.

Amniote egg The type of egg laid by *reptiles* and *birds*, having a nutritious yolk and a hard outer shell to protect the embryo from the dry environment. The amniote egg is named for the amnion, the sac that contains the embryo.

Amphibians Vertebrate animals that hatch and spend their juvenile period in the water, then usually metamorphose into air-breathing, land-dwelling adults, but return to the water to lay their eggs. Among modern amphibians are frogs, toads, salamanders, and their relatives.

Angiosperm A plant that produces flowers or flowerlike structures and that produces covered seeds.

Angular unconformity An *unconformity* separating horizontal *strata*, above, from older strata that had been tilted and eroded.

Anhydrite The *mineral* that consists of calcium sulfate ($CaSO_4$), or the *rock* composed of that mineral.

Animalia A kingdom of multicellular *eukaryotes* whose members digest food outside their cells and then absorb the products.

Anomalocarid One of a group of Cambrian *arthropods* that propelled itself through the water with flaps positioned along its body and impaled prey on daggerlike spines along its frontal appendages.

Anticline A fold that is concave in a downward direction—that is, the vertex is the highest point.

Apparent polar wander A hypothetical migration of Earth's magnetic pole that would account for the changing orientation, with age, of *paleomagnetism* in *rocks* at a particular fixed location. Most geologists believe that it is the continents that have moved, not the magnetic pole—that polar wander is indeed only apparent, not real.

Aragonite A form of calcium carbonate that precipitates from watery solutions in nature and is secreted by some organisms to form a skeleton. Aragonite is an important *mineral* in many *limestones*.

Aragonite needles Slender crystals of the *mineral aragonite* that constitute most *carbonate muds* in the modern ocean. Some of the needles form by direct precipitation from *seawater* and some by the collapse of the skeletons of organisms, principally algae.

Archaeobacteria Old bacteria; a group of *prokaryotes* that forms one of the six kingdoms of organisms.

Archaeocyathids Primary frame builders of Early Cambrian *reefs*, probably *suspension-feeding* sponges that pumped water through holes in their vase-shaped and bowl-shaped skeletons.

Arkose A *rock* consisting primarily of *sand*-sized particles of feldspar. Most arkose accumulates close to the source area of the feldspar, because feldspar weathers quickly to *clay* and seldom travels far.

Arthropod Any of a variety of *invertebrates* (*insects*, spiders, *crustaceans*) with a segmented body and an external skeleton.

Asthenosphere The *ultramafic* layer of Earth lying below the *lithosphere*. The asthenosphere is marked by low seismic velocities, an indication that it is partly molten.

Atmosphere The envelope of gases that surrounds Earth.

Atoll A circular or horseshoe-shaped organic *reef* growing on a submerged volcano.

Axial plane An imaginary plane that cuts through a fold, dividing it as symmetrically as possible.

Axis of a fold The line of intersection between the *axial plane* of a fold and the *beds* of folded *rock*.

Backswamp A broad vegetated area that lies adjacent to a *meandering river* and becomes covered with water when the river overflows its banks.

Banded iron formation Complex *rocks* that consist of oxides, sulfides, or carbonates of iron interlayered with thin beds of *chert*. Most rocks of this type are older than about 2 billion years.

Barrier island An elongate island composed of *sand* heaped up by ocean waves that lies approximately parallel to the shoreline.

Barrier island–lagoon complex The set of marginal marine environments that consists of a *barrier island*, the *lagoon* behind it, and (usually) *tidal flats, marshes*, and sandy beaches.

Barrier reef An elongate organic *reef* that parallels a coastline and is large enough to dissipate ocean waves, leaving a quiet-water *lagoon* on its landward side.

Basalt A fine-grained, *extrusive, mafic igneous rock;* the dominant rock of oceanic *crust*.

Bed A distinct layer (*stratum*) of *sediment* thicker than 1 centimeter.

Bedding The arrangement of a *sedimentary rock* into discrete layers (*strata*) thicker than 1 centimeter (*beds*).

Benthic (or **benthonic**) **life** *See* **Benthos**.

Benthos The bottom-dwelling life of an ocean or freshwater environment.

Big bang The enormous explosion that created the expanding universe.

Biogenic sediment *Sediment* consisting of *mineral* grains that were once parts of organisms.

Biogeography The study of the distribution and abundance of organisms on a broad geographic scale.

Biostratigraphic unit A body of *rock*, such as a *zone*, defined on the basis of its *fossil* content and having approximately time-parallel upper and lower boundaries.

Biota The *flora* and *fauna* of an *ecosystem*.

Biozone *See* **Zone, biostratigraphic**.

Birds A class of *endothermic vertebrates* that evolved from *dinosaurs* and are characterized by feather-covered bodies and by forelimbs that have evolved into wings.

Bivalves A group of aquatic *mollusks* with shells divided into two halves (valves). Among this group are clams, mussels, oysters, and scallops.

Boulder A piece of *gravel* larger than 256 millimeters (~ 10 inches).

Boundary stratotype A *stratigraphic unit* where the boundary between two *time-rock units* is formally recognized.

Brachiopods A group of marine *mollusks* with shells divided into two halves and a frilly loop-shaped structure that pumps water and sieves food particles from it.

Brackish water Water whose *salinity* is lower than that of normal *seawater* and higher than that of *freshwater*, ranging from 30 to 0.5 parts salt per 1000 parts water.

Braided stream A stream that has many intertwining channels separated by bars of coarse *sediment*. Braided streams develop where sediment is supplied to the stream system at a very high rate—on an *alluvial fan*, for example, or in front of a melting *glacier*.

Breccia A *rock* that contains large amounts of *gravel* that is angular.

Bryozoans A group of aquatic *invertebrates* that reproduce by budding and form colonies.

Burial metamorphism *Metamorphism* produced when *rocks* are buried so deep that they are exposed to temperatures and pressures high enough to change their chemical composition.

Calcareous nannoplankton Small, nearly spherical unicellular algae that secrete minute, overlapping, shieldlike plates of calcium carbonate that serve as armor against attackers.

Calcite A form of calcium carbonate that precipitates from watery solutions in nature and is secreted by some organisms to form a skeleton. Calcite is an important *mineral* in many *limestones*.

Calcrete *See* **Caliche.**

Caliche Nodular calcium carbonate that accumulates in the layer of *soil* below the *topsoil* in warm climates that are relatively dry.

Carbonate mineral A *mineral* in which the basic building block is a carbon atom linked to three oxygen atoms. *Calcite, aragonite,* and *dolomite* are the most abundant carbonate minerals found in *sediments* and *sedimentary rocks.*

Carbonate mud An accumulation of *aragonite needles* formed when calcareous skeletons collapse or when aragonite precipitates directly from shallow tropical seas; a major component of *sediments* that harden to form *limestone*.

Carbonate platform A marine structure that is composed largely of calcium carbonate and that stands above the neighboring seafloor on at least one of its sides.

Carbonate rock A *sedimentary rock* that consists primarily of *carbonate minerals*. The dominant mineral is nearly always either *calcite*, in which case the rock is *limestone*, or dolomite, in which case the rock is *dolomite*.

Carbonate sediment Unconsolidated *sediment* that consists primarily of *carbonate minerals,* usually *aragonite* or *calcite*.

Carbonization The process by which a residue of carbon remains on the surface of an *impression* of an organism after liquids and gases have escaped during *fossilization*.

Carnivore An animal that feeds on other animals or animal-like organisms; a *consumer* that feeds on other *consumers*.

Catastrophism The outmoded doctrine that sudden, violent, and widespread events caused by supernatural forces formed most of the *rocks* that are visible at Earth's surface.

Cementation The *lithification* of *sediment* by the precipitation of *minerals* from watery solutions that percolate through the pores between grains of sediment.

Cephalopods A group of marine *mollusks* (including squids, octopuses, chambered nautiluses) that pursue prey by jet propulsion, squirting water out through a small opening in the body.

Chemical sediment *Sediment* created by precipitation of one or more inorganic materials from natural waters, sometimes as a result of evaporation.

Chemosynthesis The breakdown of simple chemical compounds within a cell for the production of energy.

Chert (flint) An impure *rock*, often gray, that consists primarily of extremely small quartz crystals precipitated from watery solutions.

Chlorite A green, micalike *mineral* that occurs in *metamorphic rocks,* primarily in *schist;* it is a prominent constituent of Archean *greenstone belts*.

Chloroplast A body within a plant cell or plantlike cell that serves as the site of *photosynthesis* within the cell. Chloroplasts are apparently evolutionary descendants of *cyanophytes* that became trapped in other single-celled organisms.

Chordate Any of a group of animals characterized by a notochord, a flexible rodlike structure that may develop into a vertebral column. *Vertebrates* are chordates.

Chromosome One of several elongate bodies in which *DNA* is concentrated within the nucleus of a cell.

Chronostratigraphic unit *See* **Time-rock unit.**

Circumpolar current The circular flow of water around Antarctica, resulting from the juncture of the *west-wind drifts* of the Atlantic, Pacific, and Indian oceans.

Clade A cluster of *species* that have a single origin.

Clast A fragment of rock produced by destructive forces. Clasts are sometimes referred to as *detritus* or detrital material.

Clastic wedge A wedge-shaped body of *molasse*.

Clay *Siliciclastic sedimentary* particles that are smaller than $\frac{1}{256}$ millimeter; also a member of the clay *mineral* family, which includes *silicates* that resemble micas.

Cnidarian A member of an aquatic *species* whose body consists of two layers of tissue separated by a jellylike inner layer, and which captures food with tentacles armed with stinging cells. Examples are jellyfishes, sea pens, and modern *corals*. Cnidarians reproduce both sexually and asexually, by budding.

Coal *Rock* formed by the low-grade *metamorphism* of stratified plant debris. It burns readily because organic carbon compounds account for more than 50 percent of its composition.

Cobble A piece of *gravel* measuring between 8 and 256 millimeters (~ 10 inches).

Coelom The body cavity that houses the internal organs of most animals.

Coelomates Animals more advanced that very simple worms, with a body cavity (a *coelom*) that houses internal organs.

Community, ecologic Populations of several *species* living together in a *habitat*.

Compaction (of sediment) The process in which grains of *sediment* are squeezed together beneath the weight of overlying sediment.

Competition, ecologic The condition in which two *species* vie for an environmental resource, such as food or space, that is in limited supply.

Components, principle of The principle that a body of *rock* is younger than any other body of rock from which any of its components is derived.

Conglomerate A *rock* that contains large amounts of *gravel* that is rounded.

Conifers Plants (pines, spruces, redwoods, and their relatives) whose seeds are exposed on cones rather than covered, like the seeds of flowering plants.

Consumer, ecologic An animal or animal-like organism that obtains nutrition by consuming the organic material of other forms of life.

Contact metamorphism Local *metamorphism* caused by *igneous intrusion* that bakes nearby *rocks*.

Continental drift The movement of continents with respect to one another over Earth's surface.

Continental accretion The marginal growth of a continent along a *subduction zone* by mountain building or by addition of a *microplate*.

Continental margin *See* **Continental shelf.**

Continental rise A gently sloping region along the base of the *continental slope*. The continental rise is formed of *sediment* transported down the slope, often by *turbidity currents*.

Continental shelf An extension of a continental landmass beneath the sea; also called a continental margin.

Continental slope The sloping submarine portion of a continent, extending from the *continental margin* to the *continental rise* or the *abyssal plain*.

Convection Rotational flow of a fluid resulting from an imbalance in densities. Convection often occurs because the fluid below is heated and becomes less dense than the fluid above or because the fluid above is cooled and becomes more dense than the fluid below.

Convective cell One of a number of rotational units believed to operate within Earth's *mantle* as a result of *convection*.

Cope's rule The generalization that body size tends to increase during the *evolution* of a group of animals.

Corals *Cnidarians* that secrete skeletons of calcium carbonate. Some form large *colonies* of interconnected individuals by budding.

Cordaites A group of Late Carboniferous *gymnosperms*, trees that often reached 30 meters (100 feet) in height and formed large woodlands resembling modern pine forests.

Core (of Earth) The central part of Earth below a depth of 2900 kilometers. It is composed largely of iron and is molten on the outside, with a solid central region.

Coring The process of inserting a tube into *sediments* or *rocks* and then extracting the tube along with a core, or plug, of material for study.

Coriolis effect The tendency of a current of air or water flowing over Earth's surface to bend to the right in the northern hemisphere and to the left in the southern hemisphere.

Correlation The procedure of demonstrating correspondence between geographically separated parts of a *stratigraphic unit*.

Craton The portion of a continent that has not undergone *tectonic* deformation since Precambrian or early Paleozoic time.

Crinoid Any of a group of *echinoderms* that sieve food from the water with featherlike arms and pass it to the centrally positioned mouth with tube feet. Some species swim by waving their arms; others attach to the seafloor by a long, flexible stalk.

Crocodiles Marine *reptiles* that evolved in Triassic time as terrestrial animals but became adapted to aquatic environments.

Cross-bedding (cross-stratification) A *sedimentary structure* in which groups of *strata* lie at angles to the horizontal.

Cross-stratification *See* **Cross-bedding**.

Crust The outermost layer of the *lithosphere*, consisting of *felsic* and *mafic rocks* less dense than the rocks of the *mantle* below.

Crustacean An *arthropod* with a head formed of five fused segments, behind which are a thorax and an abdomen formed of additional segments. Among the crustaceans are lobsters, shrimps, and crabs.

Crystalline rocks *Igneous* or *metamorphic rocks*.

Cyanobacteria (cyanophytes) Photosynthetic *prokaryotes* that originated in Archean time and that form *stromatolites*.

Cyanophytes. *See* **Cyanobacteria.**

Cycadeoids A group of *gymnosperms* that were prominent in the Mesozoic Era.

Cycads A diverse group of *gymnosperms* whose few modern descendants superficially resemble palm trees.

Cyclothems *Sedimentary cycles* that include *coal beds*. Most cyclothems are of Late Carboniferous (Pennsylvanian) age.

Declination, magnetic The angle that a compass needle makes with the line running to the geographic north pole, reflecting the fact that the magnetic pole (to which the compass needle points) does not coincide with the geographic pole.

Deep-focus earthquake An earthquake produced by movements along or within a subducted slab of *lithosphere* more than 300 kilometers (~ 190 miles) below Earth's surface.

Deep-sea floor The *continental slope* and *abyssal plain*.

Degassing The loss of gases by Earth early in its history, when it became liquefied.

Delta A depositional body of *sand*, *silt*, and *clay* formed when a river discharges into a body of standing water so that its current dissipates and drops its load of *sediments*. This structure takes its name from the Greek letter Δ, which it resembles in shape.

Delta, river-dominated A *delta* that projects far out into the ocean because its construction from river-borne *sediment* prevails over the destructive forces of the sea.

Delta front The submarine slope of a *delta* extending downward from the *delta plain*. The delta front is usually the site of accumulation of *silt* and *clay*.

Delta plain The upper surface of a *delta*, characterized by *distributary channels* and their *natural levees* and intervening swamps.

Deposit feeders Seafloor dwellers that extract organic matter from *sediment*.

Desert A terrestrial environment that receives less than about 25 centimeters (10 inches) of rain per year and consequently supports only a few kinds of plants.

Diagenesis The set of processes, including solution, that alter *sediments* at low temperatures after burial.

Diatoms Unicellular algae that secrete skeletons of opal in two parts, which fit together like the top and bottom of a petri dish.

Dike A sheetlike or tabular body of *intrusive igneous rock* that cuts upward through sedimentary layers or *crystalline rocks*.

Dinoflagellates Unicellular algae that employ two whiplike structures (flagella) for limited locomotion, but are transported chiefly by movements of the water in which they drift.

Dinosaur Any of a group of extinct terrestrial *vertebrates* that evolved from *reptiles* and were confined to the Mesozoic Era. Dinosaurs are defined by their distinctive pelvic structure.

Dip The angle that a tilted *bed* or *fault* forms with the horizontal.

Disconformity An *unconformity* above *rocks* that underwent *erosion* before the *beds* above the unconformity were deposited. The *strata* above and below a disconformity are horizontal.

Distributary channels Channels on a *delta plain* that radiate out from the mainland, carrying river water to the ocean in several directions.

Diversity, biotic The variety of *species* that live together in a *community* or belong to a single *taxonomic* group.

DNA Deoxyribonucleic acid, the double helix molecule that carries chemically coded genetic information and is passed from generation to generation.

Dollo's law The rule that any substantial evolutionary change is virtually irreversible because genetic changes are not likely to be reversed in an order exactly opposite to the order in which they originally developed.

Dolomite A *mineral* that consists of calcium magnesium carbonate, with calcium and magnesium present in nearly equal proportions, or the *sedimentary rock* that consists largely of this mineral.

Dropstone A stone dropped to the bottom of a lake or ocean from a melting body of ice afloat on the surface.

Easterlies Winds that form near Earth's poles, where cold, dense air descends and flows toward the west under the influence of the *Coriolis effect*.

Echinoderm A group of marine *invertebrates* characterized by fivefold radial symmetry, radial rows of tube feet that terminate in suction cups, and an internal skeleton or skeletal elements formed of *calcite* plates.

Ecology The study of the factors that govern the distribution and abundance of organisms in natural environments.

Ecosystem An environment together with the group of organisms that live within it.

Ectothermic Cold-blooded: characterized by a body temperature that is not internally regulated but controlled by the environment.

Endothermic Warm-blooded: characterized by a body temperature that is internally regulated.

Eocrinoids Evolutionary ancestors of *crinoids* (sea lilies), abundant in Cambrian seas, that formed simple *communities*.

Eon The largest unit of geologic time. There are three eons: the Archean, Proterozoic, and Phanerozoic.

Epicontinental sea A shallow sea formed when ocean waters flood an area of a continent far from the *continental margin*.

Epoch, geologic A division of geologic time shorter than a *period*.

Equatorial countercurrent The eastward-flowing global ocean current that carries the water that has been piled up by the *equatorial current*.

Equatorial current The global ocean current pushed westward along the equator by the *trade winds*.

Era, geologic A division of geologic time shorter than an *eon* but including two or more *periods*.

Erathem A *time-rock unit* consisting of all the *rocks* that represent a geologic *era*.

Erosion The group of processes that loosen *rock* and transport the resulting products.

Erratic boulder A boulder that a *glacier* has transported from its place of origin to a distant location.

Eubacteria True bacteria; a group of *prokaryotes* that forms one of the six kingdoms of organisms.

Eukaryote An organism whose cells are characterized by a nucleus with *chromosomes*, *mitochondria*, and other complex internal structures. All organisms except bacteria and *cyanophytes* are of this type.

Eukaryotic cell A cell that has a nucleus with *chromosomes*, *mitochondria*, and other complex internal structures. This is the kind of cell that forms higher organisms (all organisms but bacteria and *cyanophytes*).

Eustatic event A change in sea level throughout the world.

Even-toed ungulates A group of cloven-hoofed herbivores that includes cattle, antelopes, sheep, goats, pigs, bisons, camels, and their relatives.

Evaporite A *mineral* or *rock* formed by precipitation of crystals from evaporating water.

Event stratigraphy The use of geologic records of sudden events, such as the relocation of shorelines or the deposition of volcanic ash, to correlate rocks of widely separated regions.

Evergreen coniferous forest A high-latitude forest, often adjacent to *tundra* and always dominated by *conifers*, such as spruces, pines, and firs.

Evolution The process by which particular forms of life give rise to other forms by way of genetic changes.

Evolutionary convergence The evolution of similar features in two or more different biological groups, or *taxa*.

Evolutionary radiation The rapid origin of many new *species* or higher *taxa* from a single ancestral group.

Exotic terrane A block of *lithosphere* that has been sutured to a much larger continent.

Exposure *See* **Outcrop.**

Exterior drainage A drainage pattern in which lakes and rivers carry runoff from a region beyond the borders of that region.

Extinction The total disappearance of a *species* or higher *taxon*.
Extrusive (volcanic) igneous rock An *igneous rock* that has been erupted onto Earth's surface.

Facies The set of characteristics of a *rock* that represents a particular local environment.
Failed rift A *rift* that projects inland from a *continental margin* but that fails to divide the continent into two separate landmasses.
Fault A surface along which *rocks* have broken and moved.
Fauna The animals and *protozoans* of an *ecosystem*.
Feedback A consequence of some change that retards or accelerates the change.
Felsic rock A silicon-rich *igneous rock* that contains only a small percentage of iron and magnesium. Granite is the most abundant example. Felsic rocks dominate the *crust* of continents.
Fern A seedless *vascular plant* that reproduces by means of *spores*.
Fissile Having the property of fissility, or a tendency to break along *bedding* surfaces; a property of some *sedimentary rocks*, especially *shales*.
Fission-track dating The dating of a *rock* according to the number of fission tracks produced by the decay of uranium 238. In the process of decaying, uranium 238 atoms eject subatomic particles that leave microscopic tracks in the surrounding rock.
Fissure A crack in a body of *rock*, often filled with *minerals* or *intrusive igneous rock*.
Flint *See* **Chert.**
Flood basalt *Extrusive rocks* of *mafic* composition that have flowed widely over Earth's surface.
Flora The plants and plantlike *protists* of an *ecosystem*.
Flux The rate at which a reservoir gains or loses its contents, and thus expands or contracts.
Flysch *Shales* and *turbidites* that accumulate in deep water within a *foreland basin* bordering an active mountain system.
Focus A point within Earth from which *seismic waves* emanate as *rocks* move against other rocks along a *fault*.
Fold-and-thrust belt The tectonic zone of a mountain chain characterized by folds and *thrust faults* and positioned adjacent to the *metamorphic belt* and farther away than the metamorphic belt from the igneous core of the mountain chain.
Folding Tectonic bending of *rocks* into *anticlines* and *synclines* or other contorted configurations.
Foliation The alignment of platy *minerals* in *metamorphic rocks*, caused by the high pressure applied during *metamorphism*.
Food chain The sequence of nutritional steps in an *ecosystem*, with *producers* at the bottom and *consumers* at the top.
Food web The nutritional structure of an *ecosystem* in which more than one *species* occupies each level. Thus there are usually several *producer* species and several *consumer* species in a food web.
Foraminifera Marine protozoans that form a chambered skeleton by secreting *calcite* or cementing grains of *sand* together. Long filaments of their protoplasm extend through pores in the skeleton and interconnect to form a sticky net in which they catch food.
Foreland basin An elongate depositional basin that lies between an igneous arc and the associated *accretionary wedge*.
Formation The fundamental *rock unit*. A body of rock characterized by a particular set of *lithologic* features and given a formal name.
Fossil The remains or tangible traces of an ancient organism preserved in *sediment* or *rock*.
Fossil fuel Condensed and altered organic matter that can be burned to supply energy for human use. Examples are *coal*, petroleum, and natural gas.
Fossilization The group of processes by which *fossils* form.
Fossil succession The vertical ordering of *fossil taxa* in the geologic record, reflecting the operation of *evolution* and *extinction*.
Fresh water Natural water that contains less than 0.5 percent salt by weight.
Fringing reef An elongate organic *reef* that fringes a coastline and has no *lagoon* on its landward side.
Fungi A kingdom of unicellular and multicellular *eukaryotes* whose members absorb food materials into their cells and digest them there.
Fusulinids A group of large *foraminifera* that lived on Paleozoic shallow seafloors.

Gabbro A *mafic igneous rock;* the coarse-grained, *intrusive* equivalent of *basalt*.
Gamete A sex cell (egg or sperm) that carries half the normal complement of *chromosomes* and combines with another sex cell to produce a new individual possessing the normal complement.
Gastropods Snails, the largest and most varied class of *mollusks*.
Gene A unit of inheritance consisting of a segment of *DNA* that performs a particular function.
Gene pool The sum total of the genetic components of a population.
Geochronologic unit *See* **Time unit.**
Geologic system The *rocks* that represent a particular period of geologic time.
Glacial maximum The point of an ice cap's maximum advance.
Glacial minimum The point of an ice cap's farthest retreat.
Glacier A large mass of ice that creeps over Earth's surface.
Gneiss A high-grade *metamorphic rock* whose intergrown crystals resemble those of *igneous rock*, being granular rather than platy, but whose *minerals* tend to be segregated into wavy layers.
Graben A valley bounded by *normal faults* along which a central block has slipped downward.
Grade, metamorphic A classification system based on the level of temperature and pressure responsible for *metamorphism*. *Metamorphic rocks* may be of high grade, medium grade, or low grade.
Graded bed A *sedimentary structure* in which grain size decreases from the bottom to the top.
Gradualistic model The theory that most evolutionary change takes place in small steps within well-established *species*.
Granule A small piece of *gravel* (between 2 and 4 millimeters).
Grassland *See* **Savannah.**
Gravel *Sediment* larger than *sand* (larger than 2 millimeters).
Gravity spreading The lateral spreading of a mountain chain when it becomes so tall that the *rocks* within it deform under their own weight.
Graywacke A dark-gray *siliciclastic rock* consisting of *sand*- and *silt*-sized grains that include dark rock fragments, and substantial amounts of *clay*.
Greenhouse effect The warming of Earth's surface and lower *atmosphere* caused by the accumulation of carbon dioxide and other gases in the atmosphere, which act in the same manner as the glass in a greenhouse, allowing solar radiation to pass to Earth's surface and then preventing much of the resulting heat from escaping from the lower atmosphere.

Greenstone belt A podlike body of *rock* characteristic of Archean *terranes*. It consists of volcanic rocks and associated *sediments* that have commonly been metamorphosed so that they have a greenish color.
Groundwater Water in the vast *reservoir* formed by the upper 4 kilometers of the *lithosphere*.
Group A *rock unit* of a rank higher than *formation*.
Guide fossil *See* **Index fossil**.
Guyot A flat-topped volcanic seamount in the deep sea. It appears that guyots form in shallow water when wave action truncates the upper part of a volcano and that they are transported to deeper water by lateral *plate* movements.
Gymnosperms Plants whose seeds are lodged in exposed positions on cones or on other reproductive organs.
Gypsum A *mineral* that consists of calcium sulfate with water molecules attached ($CaSO_4 \cdot H_2O$), or the *rock* that consists primarily of that mineral.
Gyre A large-scale circular flow of winds or ocean currents.

Habitat Setting on or close to Earth's surface that is inhabited by life.
Half-life The time required for a particular radioactive *isotope* to decay to half its original amount. That time is consistent for any isotope, regardless of the amount of the isotope present at the outset.
Halite The *mineral* that consists of sodium chloride (NaCl), popularly known as rock salt, or the *rock* that consists primarily of that mineral.
Herbivore An animal that feeds on plants or plantlike organisms; a *consumer* that feeds on *producers*.
Hexacorals A group of *corals* that includes colonial reef builders and solitary species and that flourishes in present-day seas.
Homology The presence in two different groups of animals or plants of organs that have the same ancestral origin but serve different functions.
Hot spot A small area of heating and igneous activity in Earth's *crust* where a *thermal plume* rises from the *mantle*. The Hawaiian Islands represent a hot spot.
Humus Organic matter in *soils*, formed largely by the decay of leaves, woody tissues, and other plant materials.
Hydrothermal metamorphism Local *metamorphism* caused by the percolation of hot, watery fluids through *rocks*, as happens along a *mid-ocean ridge*, where seawater circulates through the hot, newly formed *lithosphere*.
Hypersaline water Water that is higher in *salinity* that normal seawater (contains more than 40 parts salt per 1000 parts water).
Hypsometric curve A graph that displays the proportions of Earth's surface that lie at various altitudes above and various depths below sea level.

Ichthyosaurs Fishlike *reptiles* that resembled modern dolphins, which are *mammals*. Ichthyosaurs bore live young.
Igneous rocks *Rocks* formed by the cooling of molten material.
Impression A *fossil* that consists of the flattened imprint of a soft or semihard organism, such as an *insect* or a leaf.
Index fossil (guide fossil) A *species* or genus of *fossils* that provides for especially precise *correlation*. An ideal index fossil is easily distinguished from other *taxa*, is geographically widespread, is common in many kinds of *sedimentary rocks*, and is restricted to a narrow stratigraphic interval.
Insects A group of *arthropods* that breathe air through a system of tubes, have bodies divided into head, thorax, and abdomen, and usually have two pairs of wings.
Interior drainage A drainage pattern in which lakes and rivers fail to carry runoff from a region beyond the borders of the drainage area. The rainfall is so light that streams and rivers are temporary, drying up at intervals.
Intertidal zone The belt that is alternately exposed and flooded as the *tide* ebbs and flows along a coast.
Intertropical convergence zone The tropical zone where the northern and southern trade winds converge. Because of the tilt of Earth's axis, this zone shifts seasonally with the location of maximum solar heating, from a few degrees north of the equator during the northern summer to a few degrees south of the equator during the southern summer.
Intrusion (pluton) A body of coarse-grained *igneous rock* that formed within Earth and displaced or melted its way into preexisting rock.
Intrusive igneous rock A *rock* formed by the cooling of *magma* within Earth.
Intrusive relationships, principle of The principle that an *intrusive igneous rock* is always younger than the rock that it invades.
Invertebrate An animal that lacks a backbone.
Iron meteorite A *meteorite* of which iron is the primary component.
Island arc A curved chain of islands produced by volcanism at a site where *magma* rises through the *lithosphere* from a *subducted plate*.
Isotope One of two or more varieties of an element that differ in number of neutrons within the atomic nucleus.
Isotope stratigraphy The dating of rock strata by measurement of the isotopic composition of mineral grains and fossils.
Istostasy The mechanism whereby areas of Earth's *crust* rise or subside to keep the crust in gravitational equilibrium as it floats on the *mantle*. Thus a mountain is balanced by a root of crustal material.

Key bed A sedimentary *bed*, such as a bed of volcanic ash, that is of nearly the same age everywhere and thus is useful for *correlation*.

Lagoon A ponded body of water along a marine coastline, usually landward of a *barrier island* or organic *reef*.
Late Neolithic culture The tool culture of early modern humans.
Laterite A *soil* rich in oxides of aluminum, iron, or both of these elements. Iron gives laterite a rusty red color.
Lava Molten *rock* (*magma*) that has reached Earth's surface.
Life habit The mode of life of an organism, or the way it functions within its *ecologic niche* — how it obtains nutrients or food, reproduces, and stations itself or moves about within its environment.
Limestone A *sedimentary rock*, either *biogenic* or chemical in origin, consisting primarily of calcium carbonate.
Limiting factor, ecologic An environmental condition, such as temperature, that restricts the distribution of a *species* in nature.
Lithification The consolidation of loose *sediment* by *compaction*, precipitation of *mineral* cement, or a combination of those processes to form a *sedimentary rock*.
Lithologic correlation A *correlation* between *stratigraphic units* based on *rock* type.
Lithology The physical and chemical characteristics of *rock*.
Lithosphere Earth's outer rigid shell, situated above the *asthenosphere* and consisting of the *crust* and upper *mantle*. The lithosphere is divided into *plates*.
Lithostratigraphic unit *See* **Rock unit.**
Lobe-finned fishes A group of largely

freshwater fishes with lungs and paired fins whose bones are attached to their bodies by a single shaft. They declined after the Devonian Period but are the ancestors of all terrestrial vertebrates.

Longshore current An ocean current that flows along a coast, often sweeping *sand* in a direction parallel to the coastline.

Lungfishes A group of *lobe-finned fishes* with lungs, which allow them to gulp air when they are trapped in stagnant pools during the dry season. They were abundant in the Devonian Period but only three genera survive today.

Mafic rock A dark, dense *igneous rock* that is relatively poor in silicon and rich in iron and magnesium. *Basalt,* the characteristic igneous rock of oceanic *crust,* is an example.

Magma Naturally occurring molten *rock* found within Earth.

Magma ocean Molten silicates that rose to Earth's surface when denser material sank to the center to form a predominantly iron *core* and a *mantle* of denser silicates.

Magnetic field, Earth's The field of magnetism that results from motions of Earth's iron-rich outer *core;* those motions cause Earth to behave like a giant bar magnet, with a north and south pole. A reversal in polarity provides for accurate *correlation* throughout the world.

Magnetic stratigraphy The use of magnetic properties in *rocks* to establish *correlation* of the rocks.

Mammal Any of a class of *endothermic vertebrates* characterized by body hair, legs positioned fully under the body, and sweat glands, some of which are modified to secrete milk to nourish their young, which they bear live.

Mantle The zone of Earth's interior between the *core* and the *crust,* ranging from depths of approximately 40 to 2900 kilometers. It is composed of dense *ultramafic* silicates and divided into concentric layers.

Marble A homogeneous, granular *metamorphic rock* that consists of *calcite, dolomite,* or a mixture of the two and forms by the *metamorphism* of sedimentary carbonates.

Maria (singular, **mare**) The large craters on the surface of the moon.

Marker bed *See* **Key bed.**

Marsh, intertidal A *habitat* along the seashore that is dominated by low-growing plants and is alternately flooded by the *tide* and exposed to the air. The remains of the plants usually accumulate to form *peat.*

Marsupial mammals *Mammals* that carry their immature offspring in a pouch.

Mass extinction An episode of large-scale *extinction* in which large numbers of *species* disappear in a few million years or less.

Meandering river A river that winds back and forth like a ribbon, depositing *sediment* on the inside of each curve and eroding sediment on the outside.

Mediterranean climate A climate characterized by dry summers and wet winters, often found along coasts lying about 40° from the equator. Much of California and much of the Mediterranean region of Europe have this kind of climate.

Mélange A chaotic, deformed mixture of *rocks,* such as often forms where *subduction* occurs along a *deep-sea trench.*

Meltwater The water that issues from the front of a melting *glacier.*

Member A *rock unit* of a rank lower than *formation.*

Metamorphic belt The metamorphic zone parallel to the long axis of a mountain chain and near the igneous core of the mountain chain. This is a zone of *regional metamorphism.*

Metamorphic rock A *rock* formed by *metamorphism.*

Metamorphism The alteration of *rocks* within Earth under conditions of temperature and pressure high enough to change their chemical composition.

Meteorite An extraterrestrial object that has crashed to Earth's surface after being captured by Earth's gravitational field.

Microplate A small lithospheric *plate,* usually of predominantly *felsic* composition.

Mid-ocean ridge A ridge on the ocean floor where oceanic *crust* forms and from which it moves laterally in each direction.

Mineral A naturally occurring inorganic solid element or compound with a particular chemical composition or range of compositions and a characteristic internal structure.

Mitochondrion (plural, **mitochondria**) A body within a *eukaryotic cell* in which complex compounds are broken down by oxidation to yield energy and, as a by-product, carbon dioxide. The mitochondrion is apparently an evolutionary descendant of a small bacterium that became trapped within a larger one.

Moho *See* **Mohorovičić discontinuity.**

Mohorovičić discontinuity (Moho) The boundary between the *crust* and *mantle,* marked by a rapid increase in the velocity of *seismic waves.*

Molasse Nonmarine and shallow marine sediments—representing such environments as *alluvial fans,* river systems, and *barrier island–lagoon complexes*—that accumulate in front of a mountain system after heavy sedimentation from the mountains has driven deep marine waters from the *foreland basin* there.

Mold A *fossil* that consists of a three-dimensional imprint of an organism or part of an organism.

Molecule The smallest unit of a substance, composed of one or more atoms, that retains all the properties of the substance.

Mollusks The group of *invertebrates* that includes snails, clams, and octopuses. Most mollusks secrete skeletons of calcium carbonate.

Monoplacophorans The most primitive mollusks, with cap-shaped shells and a broad foot.

Monsoon Strong onshore or offshore winds near the margin of a continent, caused by the difference in temperature between land and water.

Moraine, glacial A ridge of *till* plowed up in front of a *glacier.*

Mososaurs Marine lizards of the Cretaceous Period that sometimes reached 15 meters (45 to 50 feet) in length.

Mousterian culture The tool culture of Neanderthal, a late Pleistocene humanoid creature.

Mud An aggregate consisting of *silt-* or *clay-*sized *siliciclastic sedimentary* particles or a combination of the two.

Mudcracks *Sedimentary structures* that form, often in hexagonal patterns, as fine-grained, *clay-*rich sediments dry out and shrink.

Mudstone *Rocks* formed largely of *mud.*

Mutation A chemical change in a genetic feature. Such changes provide much of the variability on which *natural selection* operates.

Natural levee A gentle ridge bordering a *meandering river* or a *distributary channel* of a *delta* and composed of *sand* and *silt* deposited by the river or distributary channel when it overflows its banks.

Natural selection The process recognized by Charles Darwin as the primary mechanism of *evolution.* The selection process, which operates on heritable variability, results from differences among individuals in longevity and in rate of production of offspring.

Nekton Fishes and other marine animals that move through the water primarily by swimming.

Niche, ecologic The ecologic position of a species in its environment, including its requirements for certain kinds of food and physical and chemical conditions and its interactions with other species.

Nonconformity An *unconformity* separating *bedded rocks*, above, from *crystalline rocks*, below.

Normal fault A *fault* whose *dip* is steeper than 45° and along which the *rocks* above have moved downward in relation to the rocks below.

Nothosaurs Marine *reptiles* of the Triassic Period, that had paddlelike limbs resembling those of modern seals.

Oceanic realm The portion of the ocean that lies above the *deep-sea floor*.

Odd-toed ungulates A group of herbivorous animals that includes horses, tapirs, and rhinoceroses.

Onychophorans Animals intermediate in form between *segmented worms* and *arthropods*.

Oolite A *sediment* consisting of nearly spherical grains (ooliths) that grow by accumulating *aragonite needles* as they roll about in shallow water, where the seafloor is agitated by strong water movements.

Ooze, deep-sea Fine-grained *sediment* in the deep sea that consists of calcareous or siliceous skeletons of dead *planktonic* organisms.

Ophiolite A segment of seafloor that is elevated so as to rest on continental *crust*. An ophiolite usually includes *turbidites*, black *shales*, *cherts*, and *pillow basalts* along with *ultramafic rocks* from the *mantle*.

Opportunistic species A *species* that specializes in the rapid invasion of newly vacated *habitats*, where there is little *competition* from other species.

Original horizontality, principle of The principle enunciated in the seventeenth century that all *strata* are horizontal when they form. (A more accurate statement would be that almost all strata are initially more nearly horizontal than vertical.)

Original lateral continuity, principle of The principle that similar *strata* found on opposite sides of a valley or some other erosional feature were originally connected.

Orogenesis The process of mountain building.

Orogenic stabilization The compression and *metamorphism* of *sediments* that have accumulated along a *continental shelf*, which thickens the *crust* and hardens sediments and *sedimentary rocks*.

Ostracoderms A group of small fishes of early and middle Paleozoic time with paired eyes, bony armor, and small mouths, but no jaws.

Outcrop A portion of a body of *rock* that is visible at Earth's surface. (Some geologists restrict this term to rocks laid bare by natural processes and apply the term *exposure* to artificially exposed areas of rock.)

Outwash, glacial Well-stratified glacial *sediment* deposited by a stream of *meltwater* issuing from a melting *glacier*.

Overturned fold A fold in which at least one limb has been rotated more than 90° from its original position.

Oxide mineral A *mineral* consisting of a compound formed by combination of oxygen with one or more positive ions.

Paleomagnetism The magnetism of a *rock*, developed from Earth's *magnetic field* when the rock formed.

Parasite An organism that derives its nutrition from other organisms without killing them.

Particulate inheritance The presence of hereditary factors called *genes* that retain their identity while being passed on from parent to offspring.

Passive margin A continental margin that is not affected by *rifting*, *subduction*, *transform faulting*, or other large-scale *tectonic* processes, but instead forms a shelf that accumulates *sediments*.

Patch reef A small mound or *reef* growing in the *lagoon* behind a *barrier reef*.

Peat Plant debris that is not buried deeply enough to have metamorphosed into *coal*. It accumulates in water than contains little oxygen, and therefore few bacteria that cause decay.

Pebble A piece of *gravel* measuring between 4 and 8 millimeters.

Pelagic life Oceanic life that exists above the seafloor.

Pelagic sediment Fine-grained *sediment* that settles through the oceanic water column to the deep sea. Some of this sediment is *biogenic*.

Pelycosaurs Fin-backed *reptiles* and their relatives, which were related to therapsids, the ancestors of *mammals*.

Period, geologic The most commonly used unit of geologic time, representing a subdivision of an *era*.

Permineralization The mode of *fossilization* in which spaces within part of an organism (such as bony or woody tissue) become filled with *mineral* material.

Photic zone The upper layer of the ocean, where enough light penetrates the water to permit *photosynthesis*.

Photosynthesis The process by which plants and single-celled plantlike organisms employ the compound chlorophyll to convert carbon dioxide and water from their environment into energy-rich sugar, which fuels essential chemical reactions.

Phylogeny A segment of the tree of life that includes two or more evolutionary branches.

Phytoplankton *Plankton*, or floating aquatic life, that is photosynthetic. Most phytoplankton *species* are single-celled algae.

Pillow basalt *Basalt* with a hummocky surface formed by rapid cooling of *lava* beneath water.

Placental mammals Mammals whose prenatal offspring are nourished by a placenta, an internal organ that unites the fetus to the mothers uterus.

Placoderms Large, heavily armored, jawed fishes of the Paleozoic.

Placodonts Blunt-toothed *reptiles* of early Mesozoic seas, with broad, armored bodies that gave them the appearance of enormous turtles.

Plankton Organisms that float in the ocean or in lake waters.

Plantae A kingdom of multicellular *eukaryotes* whose members produce their own food by means of *photosynthesis*.

Plate A segment of the *lithosphere* that moves independently over Earth's interior.

Plate tectonics The movements and interactions of lithospheric *plates*.

Playa lake A temporary lake in a region of *interior drainage*. When such a lake dries out, *evaporite* deposits form.

Plesiosaurs Aquatic *reptiles* that evolved from the *nothosaurs* in mid-Triassic time and in Cretaceous time attained the proportions of modern predatory whales, reaching some 12 meters (40 feet) in length.

Plunging fold A fold whose *axis* plunges (lies at an angle to the horizontal) so that the *beds* of the fold have a curved *outcrop* pattern if they are truncated by *erosion*.

Pluton *See* **Intrusion.**

Point bar An accretionary body of *sand* on the inside of a bend of a *meandering river*.

Polarity time-rock unit A *time-rock unit* in which the polarity of Earth's *mag-*

netic field was either the same as it is today (a so-called normal interval) or the opposite of what it is today (a reversed interval).

Population A group of individuals that live in the same area and interbreed.

Precambrian shield The Precambrian portion of a *craton* that is exposed at Earth's surface.

Predation The eating of an animal by one of another species.

Prodelta The gently seaward-sloping bottom-set area of a *delta front* where *clay* accumulates in deep water.

Producer, ecologic A plant or plantlike organism that manufactures its own food.

Prograde To grow seaward by the accumulation of *sediment* or *sedimentary rocks. Deltas* often prograde, as do organic *reefs*. Progradation produces *regression*, or seaward migration, of the shoreline.

Prokaryote An organism whose cells contain no nucleus or certain other internal structures characteristic of the cells of higher organisms.

Protista A kingdom of *eukaryotes* whose members do not fit within any of the other three eukaryotic kingdoms. These simple, mostly unicellular *species* include the groups that were ancestral to plants, fungi, and animals.

Protozoan A single-celled animal-like *eukaryote*.

Pseudoextinction The disappearance of a *species*, not by dying out but by evolving to the point at which it is recognized as a different species.

Punctuation model The theory that most evolutionary change occurs rapidly, through *speciation*.

Quartzite A *metamorphic rock* formed by the *metamorphism* of quartz *sandstone* and consisting of almost pure quartz.

Radioactive decay The spontaneous breakdown of certain kinds of atomic nuclei into one or more nuclei of different elements, with a release of energy and subatomic particles.

Radiocarbon dating *Radiometric dating* by means of carbon 14, a radioactive *isotope* with a *half-life* so short that its decay can be used to date materials younger than about 70,000 years.

Radiolarians Marine protozoans that capture food with threadlike extensions of protoplasm that radiate from skeletons of opal.

Radiometric dating Measurement of the amount of naturally occurring radioactive *isotopes* in *rocks* in relation to their daughter isotopes (products of *radioactive decay*) to ascertain the ages of the rocks.

Rain forest, tropical A jungle that develops in an equatorial region where heavy, regular rainfall results from the cooling of air that has ascended after being warmed and picking up moisture near Earth's surface.

Rain shadow A region that is on the downwind side of a mountain and receives little rain because the winds rise as they pass over the mountain, cooling and dropping most of their moisture before they reach the other side.

Ray-finned fishes A group of jawed fishes that dominated Mesozoic and Cenozoic seas and are widely represented today, with thin bones that radiate from the body and support the fins. Among modern ray-finned fishes are trout, bass, herring, and tuna.

Red beds *Sediments* of any grain size that are reddish, usually because of the presence of iron oxide cement.

Redshift An increase in the length of light waves as they travel through space.

Reef, organic A solid but porous *limestone* structure standing above the surrounding seafloor and constructed by living organisms, some of which contribute skeletal material to the reef framework.

Reef flat The flat upper surface of a *reef*, usually standing close to sea level (often in the *intertidal zone*).

Regional metamorphism The metamorphism of *rocks* over an area whose dimensions are measured in hundreds of kilometers. The phenomenon is usually associated with mountain building.

Regional strike In deformed terrain, the prevailing orientation of fold *axes* or of the lines of *outcrop* of tilted *beds*.

Regression A seaward migration of a marine shoreline and of nearby environments.

Relict distribution The localized occurrence of a taxonomic group after it has died out throughout most of the geographic area that it previously occupied.

Remobilization *Regional metamorphism* and deformation that affect a segment of *crust* previously altered by similar processes.

Reptile Any of a class of air-breathing, *ectothermic vertebrates* that evolved from *amphibians* through the development of eggs with protective shells. Among them are turtles, lizards, snakes, and *crocodiles*.

Reservoir A body of one or more chemical entities that occupies a particular space.

Rift A juncture between two *plates* where *lithosphere* forms and the plates diverge.

Ripples Small dunelike structures formed on the surface of *sediment* by moving water or wind.

Rock An aggregate of interlocking or bonded grains, most of which are composed of a single *mineral*.

Rock cycle The endless pathway along which *rocks* of various kinds change into rocks of other kinds.

Rock unit A body of *rock* that is formally recognized as a *formation*, *member*, *group*, or *supergroup*.

Salinity The saltiness of natural water. The salinity of normal seawater is 35 parts salt per 1000 parts water.

Sand *Sediment* ranging in diameter from $\frac{1}{16}$ to 2 millimeters.

Sand dune A hill of *sand* that has been piled up by the wind. The sand within a dune is characterized by *trough cross-stratification*.

Sandstone A *siliciclastic sedimentary rock* consisting primarily of *sand*—usually sand that is predominantly quartz.

Sauropods Lizard hipped *herbivores* of the Jurassic Period that moved about on all fours; the largest of the *dinosaurs*.

Savannah A broad grassland, which typically forms where there is enough rainfall to sustain grass but not enough to sustain the trees that form woodlands or forests.

Scavenger An organism that feeds on other organisms after they have died of causes other than the scavengers predation.

Schist A *metamorphic rock* of low to medium *grade* that consists largely of platy grains of *minerals*, often including mica; because of its strong *foliation*, schist tends to break along parallel surfaces.

Sea urchin Any of a group of *echinoderms* that have a rigid external skeleton to which numerous spines attach by ball-and-socket joints.

Sediment Material deposited on Earth's surface by water, ice, or air.

Sedimentary cycle A composite sedimentary unit that is repeated many times in succession within a given region. The unit includes two or more characteristic *beds* or groups of beds arranged in a characteristic vertical sequence that often reflects *Walther's law*.

Sedimentary rock A *rock* formed by the consolidation of loose *sediment* or by precipitation from a watery solution.

Sedimentary structure A distinctive arrangement of grains in a *sedimentary rock*.

Seed A reproductive structure of a plant—produced by the union of *gametes* and then released from the plant—that has the potential to grow into a new plant. The seed is actually a juvenile stage of the *spore*-bearing generation of the plant.

Seed ferns *Ferns* of the Carboniferous Period that reproduced by means of *seeds* rather than *spores*. They varied widely in size, from small bushy plants to large treelike ones.

Segmented worm An advanced worm with a series of segments, each with its own fluid-filled *coelom*, which serves as a primitive skeleton under the pressure of muscular contractions. Many segmented worms are marine, but others, including earthworms, are terrestrial.

Seismic stratigraphy The study of *sedimentary rocks* by means of seismic reflections generated when artificially produced *seismic waves* bounce off physical discontinuities within buried *sediments*.

Seismic wave A large vibration that travels through Earth as a consequence of a natural earthquake or an artificial disturbance, such as a nuclear explosion.

Sequence A large body of marine *sediment* deposited on a continent when the ocean rose in relation to the level of the continental surface and then receded again.

Series A *time-rock unit* consisting all the *rocks* that represent a geologic *epoch*.

Sexual recombination The mixing of *chromosomes* from generation to generation, which continually creates new genetic combinations and hence new kinds of individuals on which *natural selection* can operate.

Shale A *fissile sedimentary rock* consisting primarily of *clay*.

Shark Any of a group of cartilaginous fishes that were well represented in early Mesozoic seas.

Shelf break The edge of a *continental shelf*, where it meets the *continental slope*.

Silicates The *mineral* group that includes the most abundant minerals in Earth's *crust* and *mantle*. The basic building block of silicates is a tetrahedral structure consisting of four oxygen atoms surrounding a silicon atom.

Siliciclastic rock *Sedimentary rock* composed of *clasts* of *silicate minerals*.

Siliciclastic sediment Detrital *sediment* consisting of *silicate minerals*. This is the most abundant kind of sediment of Earth.

Sill A sheetlike or tabular body of *intrusive igneous rock* that has been injected between sedimentary layers.

Silt *Siliciclastic sedimentary* particles with diameters between $\frac{1}{256}$ and $\frac{1}{16}$ millimeters.

Slab An area of *lithosphere* that is subducted.

Slate A fine-grained *metamorphic rock* of very low *metamorphic grade* that is *fissile*, like the *sedimentary rock shale*, but whose fissility results from alignment of platy materials by deformational pressures rather than by depositional orientation of particles.

Soil Loose *sediment* that accumulates in contact with the *atmosphere*.

Solar nebula A dense cloud of cosmic dust, the remains of a *supernova*.

Speciation The origin of a new *species* from two or more individuals of a preexisting species.

Species A group of individuals that interbreed or have the potential to interbreed in nature and that do not breed with other interbreeding groups.

Sphenopsids *Spore* plants characterized by branches that radiate from discrete nodes along the vertical stem and by horizontal underground stems that bear roots.

Spore A reproductive structure, not produced from *gametes*, that is released from a plant and has the potential to grow into a new plant.

Spreading zone A zone along which new *lithosphere* forms as *mafic magma* of relatively low density rises from the *ultramafic asthenosphere* and cools.

Stage The *time-rock unit* that ranks below a *series* and consists of all the *rocks* that represent a geologic *age*.

Starfish Any of a group of flexible *echinoderms* that grasp their prey with their tube feet.

Stony meteorite A *meteorite* of rocky composition.

Stony-iron meteorite A *meteorite* consisting of a mixture of rocky and metallic material.

Stratification The arrangement of *sedimentary rocks* in discrete layers (or *strata*).

Stratigraphic section A local *outcrop* or series of adjacent outcrops that displays a vertical sequence of *strata*.

Stratigraphic unit A *stratum* or group of adjacent strata distinguished by some physical, chemical, or paleontological property or the unit of time that is based on the age of such strata.

Stratigraphy The study of stratified *rocks*, especially their geometric relations, compositions, origins, and age relations.

Stratum (plural, **strata**) A distinct layer of *sediment*.

Strike The compass direction that lies at right angles to the *dip* of a tilted *bed* or *fault;* that is, the compass direction of a horizontal line lying in the plane of a tilted bed or fault.

Strike-slip fault A high-angle *fault* along which the *rocks* on one side move horizontally in relation to rocks on the other side with a shearing motion.

Stromatolite An organically produced sedimentary structure that consists of alternating layers of organic-rich and organic-poor *sediment*. The organic-rich layers have usually been formed by sticky threadlike algae, which have trapped the sediment of organic-poor layers.

Stromatoporoids A group of sponges that secreted massive skeletons of *calcite* and were important *reef* builders beginning in Ordovician time.

Subduction Descent of a *slab* of *lithosphere* into the *asthenosphere* along a *deep-sea trench*.

Subduction zone A region where *subduction* of the *lithosphere* occurs.

Substratum The surface—*sediment*, *rock*, or another organism—on which or within which a *benthic* aquatic organism lives.

Subtidal zone The belt positioned seaward of the *intertidal zone*.

Sulfate mineral A *mineral* in which the basic building block is a sulfur atom linked to four oxygen atoms. Most sulfate minerals are highly soluble in water and form by the evaporation of natural waters.

Supergroup A *rock unit* of a rank higher than *group*.

Supernova An exploding star that casts off matter of low density.

Superposition, principle of The principle that in an undisturbed sequence of *strata*, the oldest lie at the bottom and the progressively younger strata are successively higher.

Supratidal zone The belt along a coast just landward of the *intertidal zone* and flooded only occasionally, during storms or unusually high *tides*.

Suspension feeder A member of an aquatic species that strains small particles of food from the water in which it lives.

Suture The juncture between two continents along a *subduction zone*.

Suturing The unification of two continents along a *subduction zone*.

Syncline A fold that is concave in an upward direction—that is, the vertex is the lowest point.

System A *time-rock unit* consisting of all the *rocks* representing a geologic *period*.

Tabulate corals An extinct group of *corals* that secreted colonial skeletons of *calcite* and were important *reef* builders of Ordovician, Silurian, and Devonian time.

Tabulate-strome reef A reef formed by *tabulate corals* and *stromatoporoid* sponges during Ordovician, Silurian, or Devonian time.

Talus, reef The pile of rubble sloping seaward from the living surface of a *reef*.

Taxon (plural, **taxa**) **(taxonomic group)** A formally named group of related organisms of any rank, such as phylum, class, family, or *species*.

Taxonomic group *See* Taxon.

Taxonomy The study of the composition and relationships of *taxa* of organisms.

Tectonics The study of *rock* deformation entailing large-scale features such as mountains.

Teleost fishes Marine and *freshwater* fishes characterized by symmetrical tails, round scales, specialized fins, and short jaws. Most living fish species belong to this group.

Temperate forest A forest dominated by deciduous trees (trees that lose their leaves in winter). This kind of forest typically grows under slightly warmer climatic conditions than an *evergreen coniferous forest*.

Terrane A geologically distinctive region of Earth's *crust* that has behaved as a coherent crustal block.

Thecodonts Triassic ancestors of the *dinosaurs*, with four legs positioned beneath their bodies rather than sprawling to the side, like those of mammal-like *reptiles*, and adapted for speedy two-legged running.

Therapsids An extinct group of animals that occupied an intermediate evolutionary position between *reptiles* and *mammals*.

Thermal plume A column of *magma* rising from the *mantle* through the *lithosphere*.

Thrust fault A low-angle *reverse fault*. During many mountain-building episodes, large slices of *rock* (*thrust sheets*) move hundreds of kilometers over rigid unrelated rocks.

Thrust sheet A large, tabular segment of the *crust* that moves along a *thrust fault* during mountain building.

Tidal flat A surface where *mud* or *sand* accumulates in the *intertidal zone*.

Tide A major movement of the ocean that results primarily from the gravitational attraction of the moon. Tides ebb and flow in particular regions as Earth rotates beneath a bulge of water created by the pull of the moon.

Tillite Lithified *till*.

Time unit The interval during which a *time-rock unit* formed.

Time-rock unit A *stratigraphic unit* that includes all the strata in the world that were deposited during a particular interval of time.

Top carnivore The carnivorous animal positioned at the top of a *food chain* or *food web*.

Topsoil The upper zone of many *soils*, consisting primarily of *sand* and *clay* mixed with *humus*.

Trace fossil A track, trail, or burrow left in the geologic record by a moving animal.

Trade winds Winds that in each hemisphere blow diagonally westward toward the equator between about 20° and 30° latitude of the equator. They result from a zone of high pressure than forms where air that has risen near the equator builds up.

Transform fault A *strike-slip fault* along which two segments of *lithosphere* move in relation to each other. Many transform faults offset *mid-ocean ridges*.

Transgression A landward migration of a marine shoreline and of nearby environments.

Transpiration The emission of water by plants into the *atmosphere*, primarily through their leaves.

Trilobite Any of an extinct group of marine *arthropods* with a body divided into three lobes.

Triple junction A point where three lithospheric *plates* meet.

Tropical climate A climate in which the average annual temperature is in the range of 18 to 20°C (64 to 68°F) or higher. Most tropical climates lie within about 30° of the equator.

Trough cross-stratification *Cross-stratification* of *sediments* in which one set of *beds* is truncated by *erosion* in such a way that the next set of beds laid down accumulates on a curved surface.

Tuff A *rock* deposited as a *sediment* but consisting of volcanic particles.

Tundra A terrestrial environment where air temperatures rise above freezing during the summer but a layer of soil beneath the surface remains frozen. Tundras are characterized by low-growing plants that require little moisture.

Turbidite A *graded bed*, with poorly sorted coarse sediment at the base and *mud* at the top, formed when a *turbidity current* slowed down and spread out.

Turbidity current A flow of dense, *sediment*-charged water that moves down a slope under the influence of gravity.

Ultramafic rock A very dense *rock* that is even poorer in silicon and richer in iron and magnesium than a *mafic rock*. Ultramafic rocks characterize Earth's *mantle*.

Unconformity A surface between a group of sedimentary *strata* and the *rocks* beneath them, representing an interval of time during which *erosion* occurred rather than deposition.

Ungulates Hoofed herbivorous mammals, which may be either *odd-toed ungulates* or *even-toed ungulates*.

Uniformitarianism The principle that there are inviolable laws of nature that have not changed in the course of time.

Upwelling Ascent of cold water from the deep sea to the *photic zone*, usually providing nutrients for rich growth of *phytoplankton*. Upwelling is most common where ocean currents drag surface waters away from *continental margins*.

Varves Alternating layers of coarse and fine *sediments* that accumulate in a lake in front of a *glacier*. A coarse layer forms each summer, when streams of *meltwater* carry *sand* and *silt* to the lake. A fine layer forms each winter, when the surface of the lake is covered by ice, so that only *clay* and organic matter settle to the bottom.

Vascular plant A plant that has vessels in its stem to transport water and nutrients, and to distribute the food it manufactures from those raw materials.

Vent An opening in Earth's *crust* through which *lava* emerges at the surface.

Vertebrate An animal that possesses a backbone.

Vestigial organ An organ that serves no apparent function but was functional in ancestors of the organism that now possesses it.

Volcanic igneous rock *See* **Extrusive igneous rock.**

Walther's law The principle that when depositional environments migrate laterally, *sediments* of one environment come to lie on top of sediments of an adjacent environment.

Weathering The various aspects of *erosion* that take place before transport. Some weathering is physical and some is chemical.

Westerlies Winds that flow toward the northeast in each hemisphere between about 30° and 60° from the equator. These winds result from the same high-pressure zone that produces the *trade winds,* which flow in the opposite direction.

West-wind drift The eastward-flowing ocean current near the north or south pole that is created by the major ocean *gyres* and the reinforcing *westerly* winds.

Zone, biostratigraphic A body of *rocks* whose upper and lower boundaries are based on the ranges of one or more *taxa*—usually *species*—in the stratigraphic record.

Zooplankton *Plankton,* or floating aquatic life, that is animal-like in mode of nutrition (feeds on other organisms).

Credits

iii: David Parker/Photo Researchers

CHAPTER 1 0: Chapter opener, Al Merrill/Breck P. Kent. **3**: Figure 1-1A, Raymond Siever. Figure 1-1B, Reg Morrison/Auscape International. **6**: Figure 1-2, Breck P. Kent. Figure 1-3, Martin G. Miller/Earth Lens. **7**: Figure 1-4, Anne E. Hubbard/Photo Researchers. Figure 1-5, Tom Bean. **8**: Figure 1-6, Dan Guravich/Photo Researchers. **11**: Figure 1-10, Peter Kresan. **12**: Figure 1-12, courtesy Smithsonian Institution, photo by Chip Clark. **24**: Figure 1-22, S. Stanley. **25**: Figure 1-23, USGS, photo by David W. Houseknecht, Reston, Virginia.

CHAPTER 2 **28**: Lawrence Hardie. **35**: Figure 2-6B, Breck P. Kent. Figure 2-6D, Chip Clark. **36**: Figure 2-7A, E. R. Degginger/Photo Researchers. Figure 2-7B, Ian G. Macintyre, National Museum of Natural History, Smithsonian Institution. Figure 2-8, courtesy Smithsonian Institution, photo by Chip Clark. **39**: Figure 2-9, amphibole, Breck P. Kent; mica, Chip Clark; clay, W. D. Keller; quartz and pink feldspar, Chip Clark. **40**: Figure 2-10, Chip Clark. **42**: Figure 2-12, Peter Kresan. Figure 2-13, Calvin Larsen/Photo Researchers. **43**: Figure 2-14, S. Stanley. **44**: Figure 2-15, conglomerate, Glenn M. Oliver/Visuals Unlimited; sandstone and siltstone, Breck P. Kent; claystone and shale, D. Cavagnaro/Visuals Unlimited. **45**: Figure 2-18, top, E. R. Degginger/Photo Researchers; bottom, Rex R. Elliott. **46**: Figure 2-19, Peter Kresan. **47**: Figure 2-20, Breck P. Kent. Figure 2-21, bottom, E. F. McBride. **48**: For the Record 2-1, Michael Collier. **49**: For the Record 2-1, A. J. Copley/Visuals Unlimited. **50**: Figure 2-22A, S. Stanley. Figure 2-22B, A. J. Copley/Visuals Unlimited. **51**: Figure 2-23, S. Stanley. **52**: Figure 2-24C, Peter Kresan. **53**: Figure 2-25A, Breck P. Kent. Figure 2-25B, S. Stanley. Figure 2-25C, Visuals Unlimited. Figure 2-26A, Tom Bean. Figure 2-26B, Joyce Photographics/Photo Researchers.

CHAPTER 3 **56**: Opener, Douglas Faulkner/Photo Researchers. **60**: Figure 3-1, left, Chip Clark; right, courtesy Smithsonian Institution, photo by Chip Clark. **61**: Figure 3-2, Nature Museum, Senkenberg, Germany. Figure 3-3, Ken Lucas/Visuals Unlimited. **62**: Figure 3-4, Martin Land/Science Photo Library/Photo Researchers. Figure 3-5, Dinosaur State Park, Rockey Hill, Connecticut. **71**: Figure 3-13, Phil Farnes/Photo Researchers. Figure 3-14, Sinclair Stamers/Science Photo Library/Photo Researchers. **72**: Figure 3-15, right, Michael Abbey/Photo Researchers; left, M. Walker/NHPA; center, Eric Grave/Photo Researchers. Figure 3-16A, Michael Hoban, California Academy of Science. Figure 3-16B, C. L. Stein. Figure 3-16C, Mitch Covington, Florida Geological Survey. **73**: Figure 3-17A, Breck P. Kent. Figure 3-17B, R. J. Goldstein/Visuals Unlimited. Figure 3-18, Manfred Kage/Peter Arnold. **74**: Figure 3-19, Eric Grave/Photo Researchers. **75**: Figure 3-21, William Ormerod/Visuals Unlimited. **77**: Figure 3-24A, Gerry Ellis/ENP Images. **79**: Figure 3-27, courtesy Smithsonian Institution, photo by Chip Clark. **80**: Figure 3-28A, Michael Fodgen/Bruce Coleman. Figure 3-28B, courtesy Smithsonian Institution. **82**: Figure 3-30A, A. J. Copley/Visuals Unlimited. Figure 3-30B, James R. McCullagh/Visuals Unlimited. **83**: Figure 3-31, Ed Robinson/Tom Stack & Associates. Figure 3-32, Brian Parker/Tom Stack & Associates. Figure 3-33, David L. Meyer, University of Cincinnati. **84**: Figure 3-34B, Heather Angel. Figure 3-35, Tom McHugh/Photo Researchers. **85**: Figure 3-36, left, John Green/Visuals Unlimited; right, S. R. Maglione/Photo Researchers. **86**: For the Record 3-1, A. J. Copley/Visuals Unlimited. **87**: For the Record 3-1, Dr. J. K. Ingham, Hunterian Museum & Art Gallery, Glasgow.

CHAPTER 4 **90**: Robert W. Hernandez/Photo Researchers. **103**: Figure 4-11, Andre Bartschi. Figure 4-12, Peter Kresan. **104**: Figure 4-14, Tom Bean. **105**: Figure 4-15, Michael Francis/Wildlife Collection. **108**: Figure 4-19A, Gerald Cubitt. **112**: Figure 4-25, Bates Littlehales. **116**: For the Record 4-1, Roger Steene/ENP Images.

CHAPTER 5 **120**: Paul M. (Mitch) Harris with permission of Chevron Petroleum Technology Company. **125**: Figure5-1, top, Lawrence Hardie, Johns Hopkins University. Figure 5-2A, Carnegie Museum of Natural History. **126**: Figure 5-3, S. Stanley. **127**: Figure 5-4, Peter Kresan. Figure 5-5, Martin G. Miller/Earth Lens. **128**: Figure 5-6A, P. F. Karrow, Ontario Department of Mines. Figure 5-6B, F. J. Pettijohn. Figure 5-7, D. A. Lindsey, U.S. Geological Survey. **129**: Figure 5-8, Peter Kresan. Figure 5-9, Michael Collier. **130**: Figure 5-10, Peter Kresan. Figure 5-11, Tom Bean. **131**: Figure 5-12C, Tom Bean. **132**: Figure 5-14, Tom Bean. **133**: Figure 5-15, Peter Kresan. Figure 5-16B, F. J. Pettijohn and P. E. Potter. **137**: For the Record 5-1, Donald Davis, Louisiana Geological Society. **141**: Figure 5-25, P. E. Playford, Geological Survey of Western Australia. **142**: Figure 5-26, Ralph and Daphne Keller/NHPA. **143**: Figure 5-28, Truchet/Tony Stone Worldwide. **145**: Figure 5-32, Paul F. Hoffman. **146**: Figure 5-33A, B, Earle F. McBride. **147**: Figure 5-35B, Heather Angel/Biofotos.

CHAPTER 6 **150**: David M. Dennis/Tom Stack & Associates. **166**: Figure 6-9, Rosen and Miller, Rensselear. **170**: Figure 6-13A, Breck P. Kent. **171**: Figure 6-14, Erle Kauffmann, University of Colorado. Figure 6-15, R. Y. Anderson et al., *Geol. Soc. Amer. Bull.* 83:59–86, 1972.

CHAPTER 7 **180**: Field Museum of Natural History, Chicago, Neg. GEO8460c. **184**: Figure 7-2, Michael Viard, Auscape International. **185**: Figure 7-3, painting by John Collier, National Portrait Gallery, London. Figure 7-4, painting by John Gould, from C. Darwin, *Zoology of the Voyage of H.M.S. Beagle*, Smith and Elder, London, 1838–1843. **186**: Figure 7-5A, Jerry Ellis/The Wildlife Collection. Figure 7-5B, Field Museum of Natural History, Chicago, Neg. CK20T; painting by C. R. Knight. **190**: Figure 7-8, M. Grumbach and A. Morishima. **197**: For the Record 7-1, Jim Brandenburg/Minden Pictures. **202**: Figure 7-17B, Stephen Dalton/Photo Researchers. Figure 7-17C, Michael Durham/ENP Images. **203**: Figure 7-18, A. Hallam.

CHAPTER 8 **206**: Danilo G. Donadoni/Bruce Coleman. **211**: Figure 8-3, Jim Frazier/Mantis Wildlife Films Pty Ltd. Figure 8-4, painting by Heinrich C. Berann; Bruce C. Heezen and Marie Tharp, World Ocean Floor, 1977, © Marie Tharp. **215**: Figure 8-10, © 1999 Mark Hallett Illustrations, Salem, Oregon. **224**: For the Record 8-1, Kall Muller/Woodfin Camp & Associates. **225**: Figure 8-23, John Wakabayashi, Hayward, California.

CHAPTER 9 **230**: Opener, Jock Montgomery/Bruce Coleman. **234**: Figure 9-1, Wallace/USGS. **238**: Figure 9-5, Richard W. Allmendinger, Cornell University. **247**: For the Record 9-1, Paul Karl Link, Idaho State University.

CHAPTER 10 **256**: From David Middleton, *Ancient Forests* © 1992, published by Chronicle Books, San Francisco. **270**: Figure 10-12, Farrell Grehan/Photo Researchers. **275**: Figure 10-15, T. Steuber, Institut für Geologie und Mineralogie, Der Universität Erlangen, Nürnberg;

from T. Steuber, *Geology* 24:315–318, 1996. **280**: For the Record 10-1, Anthony Ekdale, University of Utah.

CHAPTER 11 **284**: R.S. Hildebrand, Geological Survey of Canada. **289**: Figure 11-3, NASA. **290**: Figure 11-4, R. A. Oriti. Figure 11-5, Anglo-Australian Telescope Board. **291**: Figure 11-7, J. Hester and P. Scowen (AZ State Univ.), NASA, Hubble Space Telescope. **293**: Figure 11-10, Jet Propulsion Laboratory, California Institute of Technology. **295**: Figure 11-12, painting by Alfred T. Kamajian, *Scientific American*, July 1994, cover. **296**: Figure 11-13, Lick Observatory. **299**: For the Record 11-1, painting by William K. Hartmann. **300**: For the Record 11-1, Johnson Space Center, NASA. **302**: Figure 11-18, CSIRO Division of Exploration Geoscience, Australia. **303**: Figure 11-19, Paul F. Hoffman. Figure 11-20, Preston Cloud and David Pierce. **306**: Figure 11-24, J. William Schopf, University of California, Los Angeles. **307**: Figure 11-25, Gary Byerly, Louisiana State University. **310**: Figure 11-28, Woods Hole Oceanographic Institute.

CHAPTER 12 **314**: Cornelius Klein, The University of New Mexico. **320**: Figure 12-3, Paul F. Hoffman, Geological Survey of Canada. **322**: Figure 12-6, Bruce Runnegar, University of California. **323**: Figure 12-7A, Andrew H. Knoll. Figure 12-7B, Nicholas Butterfield. **324**: Figure 12-8, Bruce Simonson, Oberlin College. **326**: Figure 12-9, Charles Byers, University of Wisconsin, Madison. Figure 12-10A, B, N. L. Banks. **327**: Figure 12-11A–D, M. F. Glaessner. Figure 12-11E, Ben Waggoner, University of Central Arkansas. **328**: Figure 12-12, S. Bengtson & Yue Zhao, 1991, Predatorial borings in Late Precambrian mineralized exoskeletons, *Science* 257:367–369. **332**: Figure 12-17, Martin G. Miller/Earth Lens. **333**: For the Record 12-1, Paul Horestead. **335**: Figure 12-19, N. K. Huber/U.S. Geological Survey.

CHAPTER 13 **340**: Paul Copper, Laurentian University, Sudbury, Ontario. **346**: Figure 13-4, T. P. Crimes. Figure 13-5, Chip Clark. **347**: Figure 13-7, Val Gunther, Brigham City, Utah. **349**: Figure 13-10, U.S. Geological Survey. **350**: Figure 13-11A–C, courtesy Smithsonian Institution. **352**: Figure 13-12, courtesy Smithsonian Institution, photo by Chip Clark. **365**: Figure 13-28, H. E. Cook and M. W. Taylor, U.S. Geological Survey, from Cook and Mullins, 1983, *AAPG Memoir 33*, p. 586, Fig. 80.

CHAPTER 14 **368**: John Forbes/PEP. **374**: Figure 14-2, S. Stanley. **375**: Figure 14-3, Carnegie Museum of Natural History. Figure 14-5, Chip Clark. **377**: Figure 14-9, Chip Clark. **380**: Figure 14-12, bottom right, Swedish Museum of Natural History. **382**: Figure 14-14, Francis M. Haeber, Smithsonian Institution. Figure 14-15C, C. D. Klees/PEP. **383**: Figure 14-16A, after C. R. Beck, *Biol. Rev.* 45:379–400, 1970. Figure 14-17, Chase Studio, Inc. **391**: Figure 14-25, The British Institute of Geological Sciences. **394**: Figure 14-28, S. Stanley.

CHAPTER 15 **398**: Field Museum of Natural History, Chicago, Neg. GEO75400c. **403**: Figure 15-1B, courtesy Smithsonian Institution. Figure 15-2, Chase Studios, Inc. **404**: Figure 15-3, S. Stanley. Figure 15-4, courtesy Smithsonian Institution, photo by Chip Clark. Figure 15-5, external photograph courtesy Smithsonian Institution; cross section photograph, R. C. Douglass, U.S. Geological Survey. **405**: Figure 15-6, Chip Clark. Figure 15-7A, B, photographs of leaf scars, Field Museum of Natural History, Chicago, Negs. 75444 and 75410. Figure 15-7C, The British Institute of Geological Sciences. **406**: For the Record 15-1, Bates Littlehales. **408**: Figure 15-8, Theodore Clutter/Photo Researchers. Figure 15-9A, James L. Amos/Photo Researchers. Figure 15-9C, Jeff Foott/Survival Anglia. **409**: Figure 15-11A, Museum für Naturkunde, Humboldt-University, Berlin. Figure 15-11B, Phil A. Dotson/Photo Researchers. **412**: Figure 15-15, Field Museum of Natural History, Chicago, Neg. GEO85820. **416**: Figure 15-18, after a photograph by J. M. Schopf. **417**: Figure 15-20, David M. Dennis/Tom Stack & Associates. **425**: Figure 15-28, Ann Duncan/Tom Stack & Associates. **428**: Figure 15-32, photo, National Park Service.

CHAPTER 16 **434**: Tom Bean. **439**: Figure 16-2, George Stanley. **440**: Figure 16-4A, Field Museum of Natural History, Chicago, Neg. GEO80826. Figure 16-4B, Chip Clark. **441**: Figure 16-5A, C, Museum Hauff, Holzmaden, Germany. Figure 16-6A, A. Dan Varner. **442**: Figure 16-7, Museum für Geologie und Paläontologie, Tübingen, Germany. Figure 16-8, Museum für Geologie und Paläontologie, Tübingen, Germany. **443**: Figure 16-9, Staatliches Museum für Naturkunde. **444**: Figure 16-10A, A. Dave Watts/NHPA. Figure 16-10B, Martin Land/Science Photo Library/Photo Researchers. **448**: Figure 16-16, Chip Clark. Figure 16-17, Franz Hoeck, Bayerische Staatssammlung fur Palaontologie und Geologie Munchen. **449**: Figure 16-18A, J. H. Ostrom, Peabody Museum of Natural History, Yale. Figure 16-18B, Walther-Arndt-Fonds, Fordererkreis der naturwissenschaftlichen Museen Berlins e.V. **450**: For the Record 16-1, Royal Tyrell Museum of Paleontology/Alberta Culture and Multiculturalism. **451**: For the Record 16-1, top, Museum of the Rockies, Montana State University, Bozeman; bottom left and right, Mick Ellison, Department of Vertebrate Paleontology, American Museum of Natural History. **457**: Figure 16-25, Breck P. Kent. **458**: Figure 16-26, Tom Bean. **460**: Figure 16-30, Peter Kresan. **461**: Figure 16-31B, Martin G. Miller/Earth Lens. **462**: Figure 16-33, Dinosaur Nature Association.

CHAPTER 17 **464**: Daivd Woodfall/NHPA. **469**: Figure 17-2, courtesy Smithsonian Institution, photo by Chip Clark. Figure 17-3, courtesy Smithsonian Institution. **471**: Figure 17-6, T. Steuber, Institüt fur Geologie und Mineralogie, Der Universität Erlangen, Nürnberg. **474**: Figure 17-9, Museum of the Rockies, Montana State University, Bozeman. **475**: Figure 17-10, British Museum of Natural History. **481**: Figure 17-16A, left, Swedish Museum of Natural History, photo by Yvonne Arremo, Stockholm. Figure17-16B, right, Gerry Ellis/Wildlife Collection. **483**: Figure 17-17, Francois Gohier, San Diego. Figure 17-18, Glen A. Izett, U.S. Geological Survey. Figure 17-19, Bruce F. Bohor, U.S. Geological Survey. **484**: Figure 17-20, V. L. Sharpton, Lunar Planetary Institute, Houston.

CHAPTER 18 **494**: Tom Bean. **499**: Figure 18-1, Field Museum of Natural History, Chicago. Figure 18-2, photos, P. M. Kier, Smithsonian Institution. **501**: Figure18-4, Senckenberg Museum. **502**: Figure 18-7, Field Museum of Natural History. **504**: Figure 18-11, right, S. Stanley. **505**: Figure 18-13, courtesy Smithsonian Institution, painting by Jay H. Matternes. Figure 18-15, top, drawing by S. F. Kimbrough. Figure 18-15, bottom, photograph by E. Simons. **509**: For the Record 18-1, Chip Clark. **512**: Figure 18-22, Martin G. Miller/Visuals Unlimited. **514**: Figure 18-25, Joe Case/Visuals Unlimited. **515**: Figure 18-27, Michael Collier.

CHAPTER 19 **518**: Mark J. Bilak/Visuals Unlimited. **527**: Figure 19-6, D. Parer and E. Parer-Cook/Auscape International. **528**: Figure 19-8, courtesy Smithsonian Institution, painting by Jay H. Matternes. **529**: Figure 19-9, Paige Newby, Brown University. **534**: Figure 19-16, Lonnie G. Thompson, Byrd Polar Research Institute, Ohio State University. **539**: Figure 19-22, Greg Vaughn/Tom Stack & Associates. **544**: For the Record 19-1, Victor R. Baker, The University of Arizona, Tucson. **545**: For the Record 19-1, Victor R. Baker, The University of Arizona, Tucson. **551**: Figure 19-33, Denver Museum of Natural History, photo by Rick Wicker. Figure 19-34, J. Reader/National Geographic Society. **554**: Figure 19-38, David L. Brill, Fairburn, Georgia. **556**: Figure 19-40, Ferrero/Labat/Auscape International.

CHAPTER 20 **558**: Breck P. Kent. **563**: Figure 20-2, Bates Littlehales. **564**: Figure 20-4, Marty Snyderman/Visuals Unlimited. **565**: Figure 20-6, George Denton, University of Maine. **567**: Figure 20-8A, B, Denver Museum of Natural History. **568**: Figure 20-9, © 1999 Mark Hallett Illustrations, Salem, Oregon. **571**: Figure 20-11A, Tom Bean. Figure 20-11B, Anna E. Zuckerman/Tom Stack & Associates. **573**: Figure 20-13, Bruegel, Pieter the Elder, from the series of paintings of "The Seasons," oil on oakwood, 1565. Erich Lessing/Art Resource. Figure 20-14, image courtesy of Roberta H. Yuhas, CSES/CIRES, Univ. of Colorado at Boulder, copyright 1997, all rights reserved. **581**: For the Record 20-1, Tom Brakefield/The Stock Market. **582**: Figure 20-20, Luigi Tazzari/Gamma Liaison.

Index

Page numbers in *italics* indicate illustrations.

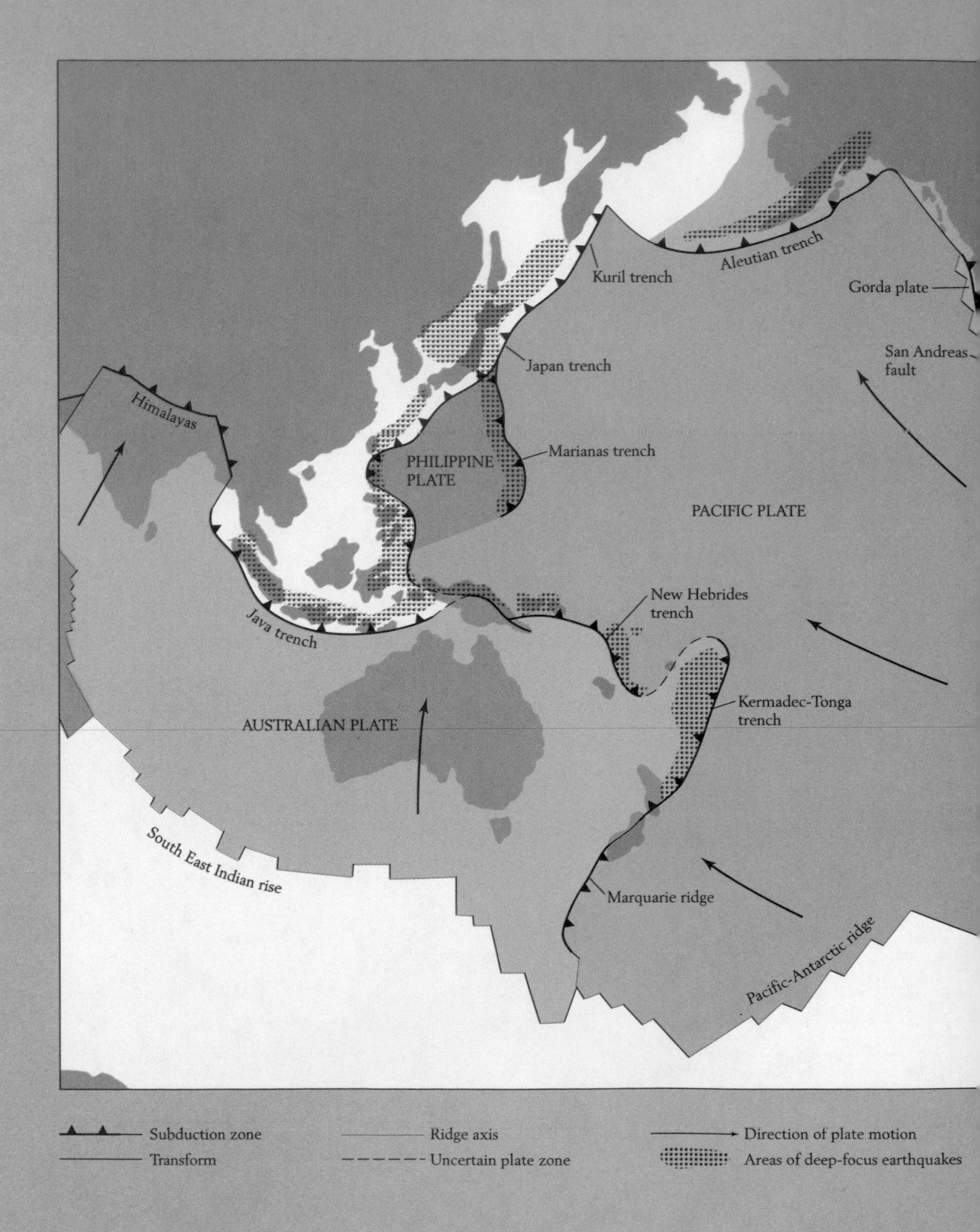
Aleutian trench
Kuril trench
Gorda plate
San Andreas fault
Japan trench
Himalayas
PHILIPPINE PLATE
Marianas trench
PACIFIC PLATE
New Hebrides trench
Java trench
AUSTRALIAN PLATE
Kermadec-Tonga trench
South East Indian rise
Marquarie ridge
Pacific-Antarctic ridge
Subduction zone
Transform
Ridge axis
Uncertain plate zone
Direction of plate motion
Areas of deep-focus earthquakes

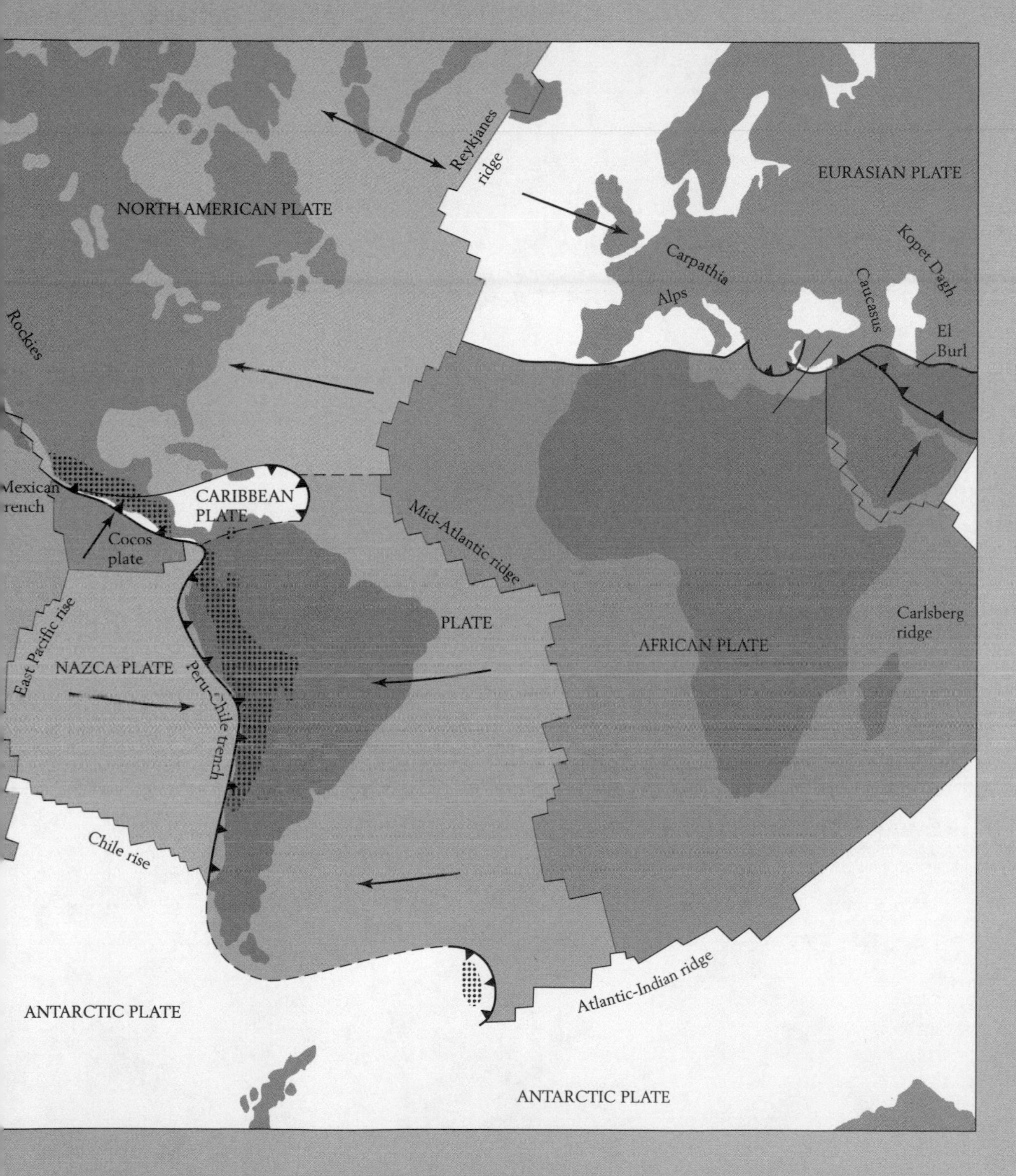

After "Plate Tectonics" by J. F. Dewey.